Advancing Sustainable Agriculture Through Arbuscular Mycorrhizal Fungi

Plant Biology, Sustainability, and Climate Change

Advancing Sustainable Agriculture Through Arbuscular Mycorrhizal Fungi

Edited by

Allah Ditta
Associate Professor Department of Environmental Sciences, Shaheed Benazir Bhutto University Sheringal, Upper Dir, Khyber Pakhtunkhwa, Pakistan; Adjunct Senior Research Fellow School of Biological Sciences, The University of Western Australia, Perth, WA, Australia

Sajid Mehmood
Senior Associate Professor, School of Tropical Agriculture and Forestry, Hainan University, Haikou, Hainan, China

Muhammad Imtiaz
Senior Scientist, Soil and Environmental Biotechnology Division, National Institute for Biotechnology and Genetic Engineering (NIBGE), Faisalabad, Punjab, Pakistan

ELSEVIER

ACADEMIC PRESS
An imprint of Elsevier
elsevier.com/books-and-journals

Academic Press is an imprint of Elsevier

125 London Wall, London EC2Y 5AS, United Kingdom
50 Hampshire Street, 5th Floor, Cambridge, MA 02139, United States

ISBN: 978-0-443-45130-0

For Information on all Academic Press publications visit our website at https://www.elsevier.com/books-and-journals

Publisher: Mara Conner
Acquisitions Editor: Andre Wolff
Editorial Project Manager: Ritika Dagar
Production Project Manager: Rachel E. G. Michael
Cover Designer: Raman Kumar

Typeset by Aptara, New Delhi, India

Contents

11. Arbuscular mycorrhizal fungi as plant growth enhancers: Mechanisms and applications

Ghulam Murtaza, Muhammad Usman, Zeeshan Ahmed, Abdul Razzaq, Rashid Iqbal and Javed Iqbal

12. Arbuscular mycorrhizal fungi and drought tolerance in plants

Anoo Gunasekera, Dinoo Gunasekera and Disna Ratnasekera

13. Arbuscular mycorrhizal fungi and salinity stress tolerance in plants

Saba Mazhar, Huma Dawood, Maryam Shahzadi, Sajjad Hyder and Bekri Xhemali

14. Arbuscular mycorrhizal fungi and temperature stress tolerance in plants

Sana Ashraf, Meshal Saeed, Muhammad Kashif Irshad, Sajid Rashid Ahmad, Anam Tariq, Amna Maheen, Eshal Zainab, Sobia Ashraf and Umer Farooq

15. Arbuscular mycorrhizal fungi and biotic stress tolerance in plants

Sohaib Mehmood, Muhammad Rashid, Hamida Akbar and Natasha Amjad

19. Arbuscular mycorrhizal fungi as biofertilizers for sustainable agriculture

Anoo Gunasekera, Dinoo Gunasekera, W.J.A.P.M. Jayasooriya and Disna Ratnasekera

20. Enhancing crop production through arbuscular mycorrhizal fungi

Muhammad Shehzad, Muhammad Mazhar Iqbal, Aqeel Ahmad, Khalid Mehmood, Nosheen Zahid, Mehdi Maqbool, Hafiz Muhammad Rashad Javeed, Khuram Mubeen, Muhammad Abdul Qayyum and Tayyaba Naz

21. Arbuscular mycorrhizal fungi and organic farming

Khalid Mehmood, Muhammad Shehzad, Muhammad Mazhar Iqbal, Majid Mahmood Tahir, Hafiz Muhammad Rashad Javeed, Ansa Tamkeen, Omer Farooq and Tayyaba Naz

22. Arbuscular mycorrhizal fungi and urban agriculture

Sana Ashraf, Meshal Saeed, Muhammad Kashif Irshad, Sajid Rashid Ahmad, Sobia Ashraf, Umer Farooq, Maimoona Shahbaz, Noor Fatima and Humra Imran

23. Arbuscular mycorrhizal fungi and rehabilitation in degraded lands: Supporting plant growth potential in a degraded environment

Muhammad Usama, Muniba Farhad, Muhammad Iqbal, Shahbaz Ali Khan, Hina Rizvi, Hafiz Syed Tanzeem-ul-Haq, Hadia Ahsan, Hafiz Muhammad Tauqeer Khan, Alisha Amdad and Zartasha

Contributors

Banzeer Ahsan Abbasi, Department of Botany, Rawalpindi Women University, Rawalpindi, Pakistan

Affan Afzal, Department of Microbiology, Faculty of Veterinary and Animal Sciences, The Islamia University of Bahawalpur, Bahawalpur, Pakistan

Khoula Afzaal, Department of Botany, Government College Women University, Sialkot, Pakistan

Aqeel Ahmad, Department of Agronomy, Faculty of Agriculture, University of Poonch, Rawalakot, Pakistan

Nabeel Ahmad, Soil and Environmental Biotechnology Division, National Institute for Biotechnology and Genetic Engineering (NIBGE), Faisalabad, Pakistan

Sajid Rashid Ahmad, College of Earth and Environmental Sciences, University of the Punjab, Lahore, Pakistan

Shakeel Ahmad, Department of Soil and Environmental Sciences, Muhammad Nawaz Shareef University of Agriculture, Multan, Pakistan

Wazir Ahmed, Department of Soil and Environmental Sciences, Muhammad Nawaz Shareef University of Agriculture, Multan, Pakistan

Zeeshan Ahmed, College of Life Science, Shenyang Normal University, Shenyang, Liaoning, China

Hadia Ahsan, Department of Environmental Sciences, Government College University Faisalabad, Faisalabad, Pakistan

Hamida Akbar, Institute of Botany, Bahauddin Zakariya University, Multan, Pakistan

Kamran Akbar, Department of Botany, Bacha Khan University Charsadda, Charsadda, Pakistan

Naveed Akhtar, Department of Environmental Sciences, COMSATS University Islamabad, Vehari Campus, Vehari, Pakistan

Muhammad Sohail Akram, Department of Botany, Faculty of Life Sciences, Government College University Faisalabad, Faisalabad, Pakistan

Sana Akram, Department of Botany, Faculty of Life Sciences, Government College University Faisalabad, Faisalabad, Pakistan

Aamir Ali, Department of Botany, University of Sargodha, Sargodha, Pakistan

Ayesha Ali, Department of Soil and Environmental Sciences, Muhammad Nawaz Shareef University of Agriculture, Multan, Pakistan

Muhammad Ali, Institute of Chemistry, University of Sargodha, Sargodha, Pakistan

Alisha Amdad, Department of Environmental Sciences, Government College University Faisalabad, Faisalabad, Pakistan

Natasha Amjad, University of Management and Technology Lahore Sub Campus Sialkot, Sialkot, Pakistan

Muhammad Arshad, Department of Entomology, University of Sargodha, Sargodha, Pakistan

Parmal Arshad, Institute of Chemistry, University of Sargodha, Sargodha, Pakistan

Alyan Ashraf, Pakistan Environmental Protection Agency, Ministry of Climate Change & Environmental Coordination, Islamabad, Pakistan

Sana Ashraf, College of Earth and Environmental Sciences, University of the Punjab, Quaid-e-Azam Campus, Lahore, Pakistan

Sobia Ashraf, Department of Environmental Sciences, Government College University, Faisalabad, Pakistan

Muhammad Talha Aslam, Department of Agronomy, University of Agriculture Faisalabad, Faisalabad, Pakistan

Nasir Assad, Institute of Chemistry, University of Sargodha, Sargodha, Pakistan

Ayesha Noor Awan, Department of Botany, Government College University, Lahore, Pakistan

Saba Babar, Microelement Research Center, College of Resources and Environment, Huazhong Agricultural University, Wuhan, Hubei, China

Muhammad Amjad Bashir, Department of Plant Protection, Ghazi University, Dera Ghazi Khan, Pakistan

Saqib Bashir, Department of Soil and Environmental Sciences, Ghazi University, Dera Ghazi Khan, Pakistan

Tanzeela Bashir, Department of Environmental Sciences, COMSATS University Islamabad, Vehari Campus, Vehari, Pakistan

Aima Iram Batool, Department of Zoology, University of Sargodha, Sargodha, Pakistan

Munaza Batool, Department of Soil and Environmental Sciences, Ghazi University, Dera Ghazi Khan, Pakistan

Samah Ait Benichou, Laboratory of Health, Environment and Biotechnologies, Ain Chock's Sciences Faculty, Hassan II University of Casablanca, Casablanca, Morocco

Saboor Muarij Bunny, Pakistan Environmental Protection Agency, Ministry of Climate Change & Environmental Coordination, Islamabad, Pakistan

Muhammad Umer Chattha, Department of Agronomy, University of Agriculture Faisalabad, Faisalabad, Pakistan

Hassan Javed Chaudhary, Department of Plant Sciences, Faculty of Biological Sciences, Quaid-i-Azam University, Islamabad, Pakistan

Usman Khalid Chaudhry, Pakistan Environmental Protection Agency, Ministry of Climate Change & Environmental Coordination, Islamabad, Pakistan

Mounia Cherki, Laboratory of Health, Environment and Biotechnologies, Ain Chock's Sciences Faculty, Hassan II University of Casablanca, Casablanca, Morocco

Vignesh Chinnaiya, Department of Biotechnology, Rathnavel Subramaniam (RVS) College of Arts and Science, Sulur, Coimbatore, Tamil Nadu, India

Huma Dawood, Department of Botany, Faculty of Natural Sciences, Government College Women University Sialkot, Pakistan

Somayeh Dehghani, Soil and Water Research Department, Agricultural Research, Education and Extension Organization (AREEO), Bushehr Agricultural and Natural Resources Research and Education Center, Bushehr, Iran

Allah Ditta, Department of Environmental Sciences, Shaheed Benazir Bhutto University Sheringal, Sheringal, Upper Dir, Khyber Pakhtunkhwa, Pakistan; School of Biological Sciences, The University of Western Australia, Perth, WA, Australia

Hossam S. El-Beltagi, Agricultural Biotechnology Department, College of Agriculture and Food Sciences, King Faisal University, Al Hofuf, Saudi Arabia

Karima El Bouqdaoui, Laboratory of Health, Environment and Biotechnologies, Ain Chock's Sciences Faculty, Hassan II University of Casablanca, Casablanca, Morocco; Laboratory of Care, Health and Sustainable Development, Higher Institute of Nursing Professions and Health Techniques, Casablanca, Morocco

Zeinab El-Desouki, Microelement Research Center, College of Resources and Environment, Huazhong Agricultural University, Wuhan, Hubei, China; Department of Soil Sciences, Faculty of Agriculture, Ain Shams University, Cairo, Egypt

Loubna El Fels, Laboratory of Water Sciences, Microbial Biotechnologies, and Natural Resources Sustainability (AQUABIOTECH), Unit of Microbial Biotechnologies, Agrosciences, and Environment (BIOMAGE)-CNRST Labeled Research Unit N°4, Faculty of Sciences Semlalia, University Cadi Ayyad, Marrakech, Morocco

Ayman EL Sabagh, Department of Field Crops, Faculty of Agriculture, Siirt University, Siirt, Turkey

Muniba Farhad, Department of Chemistry, Government College University Faisalabad, Faisalabad, Pakistan

Omer Farooq, Department of Agronomy, Institute of Agronomy, Bahauddin Zakariya University, Multan, Pakistan

Umer Farooq, Department of Food Engineering, University of Agriculture Faisalabad, Faisalabad, Pakistan

Noor Fatima, College of Earth and Environmental Sciences, University of the Punjab, Quaid-e-Azam Campus, Lahore, Pakistan

Muhammad Fayyaz ur Rehman, Institute of Chemistry, University of Sargodha, Sargodha, Pakistan

Saira Ghafoor, Key Laboratory of Crop Physiology, Ecology, and Genetic Breeding, Ministry of Education/College of Agronomy, Jiangxi Agricultural University, Nanchang, China

Dong-Xing Guan, State Key Laboratory of Soil Pollution Control and Safety, College of Environmental and Resource Sciences, Zhejiang University, Hangzhou, China

Hansa Gul, Department of Zoology, University of Mianwali, Mianwali, Pakistan

Sanober Gul, Department of Plant Breeding and Genetics, Ghazi University, Dera Ghazi Khan, Pakistan

Anoo Gunasekera, Department of Biomedical Sciences, Faculty of Health Sciences, University of Ruhuna, Matara, Sri Lanka

Dinoo Gunasekera, Department of Information and Communication Technology, Faculty of Technology, University of Ruhuna, Matara, Sri Lanka

Muhammad Zulqurnain Haider, Department of Botany, Faculty of Life Sciences, Government College University, Faisalabad, Pakistan

Waseem Hassan, Department of Soil and Environmental Sciences, Muhammad Nawaz Shareef University of Agriculture, Multan, Pakistan

Misbha Hina, Department of Environmental Sciences, COMSATS University Islamabad, Vehari Camous, Vehari, Pakistan

Muhammad Baqir Hussain, Department of Soil and Environmental Sciences, Muhammad Nawaz Shareef University of Agriculture, Multan, Pakistan

Sajjad Hyder, Department of Botany, Faculty of Natural Sciences, Government College Women University Sialkot, Pakistan

Raina Ijaz, Department of Horticulture, University of Poonch, Rawalakot, Pakistan

Shumaila Ijaz, Department of Plant Sciences, Faculty of Biological Sciences, Quaid-i-Azam University Islamabad, Islamabad, Pakistan; School of Biomedical Engineering, Shenzhen University Medical School, Shenzhen University, Shenzhen, China

Humra Imran, College of Earth and Environmental Sciences, University of the Punjab, Quaid-e-Azam Campus, Lahore, Pakistan

Muhammad Imran, Department of Soil and Environmental Sciences, Muhammad Nawaz Shareef University of Agriculture, Multan, Pakistan

Madiha Imtiaz, Department of Botany, Bacha Khan University Charsadda, Charsadda, Pakistan

Muhammad Imtiaz, Soil and Environmental Biotechnology Division, National Institute for Biotechnology and Genetic Engineering (NIBGE), Faisalabad, Punjab, Pakistan

Naila Inayat, Department of Botany, Shaheed Benazir Bhutto Woman University Peshawar, Pakistan

Javed Iqbal, Departemnt of Botany, Bacha Khan University, Charsadda, Pakistan

Muhammad Iqbal, Department of Environmental Sciences, Government College University Faisalabad, Faisalabad, Pakistan

Muhammad Mazhar Iqbal, Department of Soil and Environmental Sciences, College of Agriculture, University of Sargodha, Sargodha, Pakistan

Rashid Iqbal, Department of Agronomy, Faculty of Agriculture and Environment, The Islamia University of Bahawalpur, Bahawalpur, Pakistan; Department of Life Sciences, Western Caspian University, Baku, Azerbaijan

Muhammad Kashif Irshad, Department of Environmental Sciences, Government College University, Faisalabad, Pakistan; Department of Environmental and Energy Engineering, Yonsei University, Wonju, South Korea

Mahboobeh Jalali, Department of Soil Sciences, Faculty of Agriculture, Lorestan University, Khorramabad, Iran

Muhammad Usman Jamshaid, Department of Soil and Environmental Sciences, Muhammad Nawaz Shareef University of Agriculture, Multan, Pakistan; School of Agriculture and Biology, Shanghai Jiao Tong University, Shanghai, China

Muhammad Tariq Javed, Department of Botany, Faculty of Life Sciences, Government College University, Faisalabad, Pakistan

Hafiz Muhammad Rashad Javeed, Department of Environmental Sciences, COMSATS University Islamabad, Vehari Campus, Vehari, Pakistan

W.J.A.P.M. Jayasooriya, Faculty of Graduate Studies, University of Ruhuna, Matara, Sri Lanka

Parimala Gnana Soundari Arockiam Jeyasundar, Department of Biotechnology, Rathnavel Subramaniam (RVS) College of Arts and Science, Sulur, Coimbatore, Tamil Nadu, India

Cuncang Jiang, Microelement Research Center, College of Resources and Environment, Huazhong Agricultural University, Wuhan, Hubei, China

Rizwana Kausar, Soil and Water Testing Laboratory for Research, Sargodha, Pakistan

Kamal Khalkhal, Soil and Water Research Department, Agricultural Research, Education and Extension Organization (AREEO), Kermanshah Agricultural and Natural Resources Research and Education Center, Kermanshah, Iran

Hafiz Muhammad Tauqeer Khan, Department of Environmental Sciences, University of Gujrat, Gujrat, Pakistan

Imran Khan, Department of Agronomy, University of Agriculture Faisalabad, Faisalabad, Pakistan

Muhammad Aamir Khan, Obstetrics and gynaecology Department, Ayub Teaching Hospital, Abbottabad, Khyber Pakhtunkhwa, Pakistan

Muhammad Nauman Khan, Department of Botany, University of Chakwal, Chakwal, Punjab, Pakistan; Department of Botany, Islamia College Peshawar, Peshawar, Pakistan; Biology Laboratory, University Public School, University of Peshawar, Peshawar, Pakistan

Nazish Huma Khan, Department of Environmental Sciences, University of Swabi, Khyber Pakhtunkhwa, Pakistan

Shahbaz Ali Khan, National Compliance Centre, Ministry of Commerce, Islamabad, Pakistan

Amna Maheen, College of Earth and Environmental Sciences, University of the Punjab, Lahore, Pakistan

Imran Manzoor, Department of Environmental Sciences, COMSATS University Islamabad, Vehari Camous, Vehari, Pakistan

Mehdi Maqbool, Department of Horticulture, Faculty of Agriculture, University of Poonch, Rawalakot, Pakistan

Rizwan Maqbool, Department of Agronomy, University of Agriculture Faisalabad, Faisalabad, Pakistan

Saba Mazhar, Department of Botany, Faculty of Natural Sciences, Government College Women University Sialkot, Pakistan

Khalid Mehmood, Department of Agronomy, Faculty of Agriculture, University of Poonch, Rawalakot, Pakistan

Sohaib Mehmood, Institute of Botany, Bahauddin Zakariya University, Multan, Pakistan

Saima Parveen Memon, Key Laboratory of Cultivated Land Quality Monitoring and Evaluation, Ministry of Agriculture and Rural Affairs, College of Environmental Science and Engineering, Yangzhou University, Yangzhou, China

Khuram Mubeen, Department of Agronomy, Muhammad Nawaz Shareef University of Agriculture, Multan, Pakistan

Qasim Muhammad, Microelement Research Center, College of Resources and Environment, Huazhong Agricultural University, Wuhan, Hubei, China

Memoona Mukhtar, Institute of Chemistry, University of Sargodha, Sargodha, Pakistan

Ghulam Murtaza, School of Agriculture, Yunnan University, Kunming, Yunnan , China

Muhammad Mustaqeem, Institute of Chemistry, University of Sargodha, Sargodha, Pakistan

Mohammad Nafees, Department of Environmental Sciences, University of Peshawar, Peshawar, Khyber Pakhtunkhwa, Pakistan

Naima Huma Naveed, Department of Botany, University of Sargodha, Sargodha, Pakistan

Nida Naz, Department of Environmental Sciences, Abdul Wali Khan University, Mardan, Khyber Pakhtunkhwa, Pakistan

Tayyaba Naz, Institute of Soil and Environmental Sciences, University of Agriculture Faisalabad, Faisalabad, Pakistan

Salma Noureen, Department of Botany, Bacha Khan University Charsadda, Charsadda, Pakistan

Elmostapha Outamamat, Laboratory of Water Sciences, Microbial Biotechnologies, and Natural Resources Sustainability (AQUABIOTECH), Unit of Microbial Biotechnologies, Agrosciences, and Environment (BIOMAGE)-CNRST Labeled Research Unit N°4, Faculty of Sciences Semlalia, University Cadi Ayyad, Marrakech, Morocco

Ayyappa Das Madhubala Parameswaran, Department of Biotechnology, Rathnavel Subramaniam (RVS) College of Arts and Science, Coimbatore, Tamil Nadu, India

Subha Bharathi Pitchaimuthu, Department of Biotechnology, Rathnavel Subramaniam (RVS) College of Arts and Science, Coimbatore, Tamil Nadu, India

Rafi Qamar, Department of Agronomy, College of Agriculture, University of Sargodha, Sargodha, Pakistan

Muhammad Abdul Qayyum, Department of Soil and Environmental Sciences, Ghazi University, Dera Ghazi Khan, Pakistan

Li-Xue Qiu, State Key Laboratory of Soil Pollution Control and Safety, College of Environmental and Resource Sciences, Zhejiang University, Hangzhou, China

Anas Raklami, Laboratory of Water Sciences, Microbial Biotechnologies, and Natural Resources Sustainability (AQUABIOTECH), Unit of Microbial Biotechnologies, Agrosciences, and Environment (BIOMAGE)-CNRST Labeled Research Unit N°4, Faculty of Sciences Semlalia, University Cadi Ayyad, Marrakech, Morocco

Muhammad Rashid, University of Management and Technology Lahore Sub Campus Sialkot, Sialkot, Pakistan

Disna Ratnasekera, Department of Agricultural Biology, Faculty of Agriculture, University of Ruhuna, Matara, Sri Lanka

Abdul Razzaq, Faculty of Management and Economics, Kunming University of Science and Technology, Kunming, China

Hasan Riaz, Department of Plant Pathology, Muhammad Nawaz Shareef University of Agriculture, Multan, Pakistan

Hina Rizvi, Department of Environmental Sciences, Government College University Faisalabad, Faisalabad, Pakistan

Saliha Rukhsar, Department of Botany, Government College University, Lahore, Pakistan

Meshal Saeed, College of Earth and Environmental Sciences, University of the Punjab, Lahore, Pakistan

Samra Saeed, Department of Botany, Government College University, Lahore, Pakistan

Saleem Sajjad, Department of Soil and Environmental Sciences, College of Agriculture, University of Sargodha, Sargodha, Pakistan

Muneeba Saleem, Research Center for Plant Functional Genes and Tissue Culture Technology, Jiangxi Agricultural University, Nanchang, China

Suleman Haider Shah, Department of Soil and Environmental Sciences, Muhammad Nawaz Shareef University of Agriculture, Multan, Pakistan; Department of Agriculture (Extension & Adaptive Research), Government of Punjab, Murree, Punjab, Pakistan

Maimoona Shahbaz, College of Earth and Environmental Sciences, University of the Punjab, Quaid-e-Azam Campus, Lahore, Pakistan

Muhammad Shahid, Department of Bioinformatics and Biotechnology, Government College University Faisalabad, Faisalabad, Pakistan

Maryam Shahzadi, Department of Botany, Faculty of Natural Sciences, Government College Women University Sialkot, Pakistan

Minahil Shehzad, Institute of Chemistry, University of Sargodha, Sargodha, Pakistan

Muhammad Shehzad, Department of Agronomy, Faculty of Agriculture, University of Poonch, Rawalakot, Pakistan

Umar Sheraz, Department of Soil and Environmental Sciences, College of Agriculture, University of Sargodha, Sargodha, Pakistan

Majid Mahmood Tahir, Department of Soil and Environmental Sciences, Faculty of Agriculture, University of Poonch, Rawalakot, Pakistan

Ansa Tamkeen, Department of Agronomy, Faculty of Agriculture, University of Poonch, Rawalakot, Pakistan

Kashif Tanwir, Department of Botany, Faculty of Life Sciences, Government College University, Faisalabad, Pakistan

Hafiz Syed Tanzeem-ul-Haq, Department of Environmental Sciences, Government College University Faisalabad, Faisalabad, Pakistan

Anam Tariq, College of Earth and Environmental Sciences, University of the Punjab, Lahore, Pakistan

Hafiz Muhammad Tauqeer, Department of Environmental Sciences, University of Gujrat, Gujrat, Pakistan

Kayalvizhi Thirumalaisamy, Department of Biotechnology, Rathnavel Subramaniam (RVS) College of Arts and Science, Coimbatore, Tamil Nadu, India

Minisha Udhayakumar, Department of Biotechnology, Rathnavel Subramaniam (RVS) College of Arts and Science, Coimbatore, Tamil Nadu, India

Shanmugapriya Udhayakumar, Department of Biotechnology, Rathnavel Subramaniam (RVS) College of Arts and Science, Coimbatore, Tamil Nadu, India; School of Biotechnology and Genetic Engineering, Bharathiar University, Coimbatore, Tamil Nadu, India

Barkat Ullah, Department of Botany, University of Chakwal, Chakwal, Punjab, Pakistan; Biology Laboratory, University Public School, University of Peshawar, Peshawar, Pakistan

Muhammad Irfan Ullah, Department of Entomology, University of Sargodha, Sargodha, Pakistan

Zakir Ullah, Department of Plant Sciences, Faculty of Biological Sciences, Quaid-i-Azam University Islamabad, Islamabad, Pakistan

Zahid Ullah, Department of Botany, Bacha Khan University Charsadda, Charsadda, Pakistan

Rehan Ullah, Institute of Chemistry, University of Sargodha, Sargodha, Pakistan

Amjad Ur Rahman, Department of Conservation, WWF Pakistan, Peshawar, Pakistan

Muhammad Usama, Department of Environmental Sciences, Government College University Faisalabad, Faisalabad, Pakistan

Keyvan Valizadeh-Rad, Department of Soil Sciences, Faculty of Agriculture, Lorestan University, Khorramabad, Iran

Muhammad Wakeel, Department of Environmental Sciences, COMSATS University Islamabad, Vehari Camous, Vehari, Pakistan

Jiyuan Wang, Microelement Research Center, College of Resources and Environment, Huazhong Agricultural University, Wuhan, Hubei, China

Xiangling Wang, Microelement Research Center, College of Resources and Environment, Huazhong Agricultural University, Wuhan, Hubei, China

Bekri Xhemali, Faculty of Agribusiness, University Haxhi Zeka, Pejë, Kosovo; Laboratory of Plant Protection, Kosovo Institute of Agriculture, Pejë, Kosovo

Xiaoyang Xia, Microelement Research Center, College of Resources and Environment, Huazhong Agricultural University, Wuhan, Hubei, China

Sana Yaqoob, Institute of Soil & Environmental Sciences, University of Agriculture, Faisalabad, Pakistan

Tabassum Yaseen, Department of Botany, Bacha Khan University, Charsadda, Pakistan

Muhammad Younas, Department of Crop Breeding and Genetics, Agriculture College, Yangzhou University, Yangzhou, Jiangsu, China

Nousheen Yousaf, Department of Botany, Government College University, Lahore, Pakistan

Bakhtawar Zafar, Department of Plant Sciences, Quaid-I-Azam University, Islamabad, Pakistan

Nosheen Zahid, Department of Horticulture, Faculty of Agriculture, University of Poonch, Rawalakot, Pakistan

Muskan Zahra, Institute of Chemistry, University of Sargodha, Sargodha, Pakistan

Eshal Zainab, College of Earth and Environmental Sciences, University of the Punjab, Lahore, Pakistan

Ghazala Zamani, Department of Biotechnology, Bacha Khan University Charsadda, Pakistan

Farishta Zarshan, Department of Botany, Bacha Khan University Charsadda, Charsadda, Pakistan

Zartasha, Department of Environmental Sciences, Government College University Faisalabad, Faisalabad, Pakistan

Muhammad Zia Ul Haq, Department of Agronomy, University of Agriculture Faisalabad, Faisalabad, Pakistan

About the Editors

Dr. Allah Ditta has been working as an associate professor (tenured) in the Department of Environmental Sciences, Shaheed Benazir Bhutto University (SBBU) Sheringal, Upper Dir, Khyber Pakhtunkhwa, Pakistan, since December 5, 2021. Earlier, he worked as an assistant professor at the Department of Environmental Sciences, SBBU Sheringal, Upper Dir, Khyber Pakhtunkhwa, Pakistan (December 5, 2016, to December 5, 2021) and as an assistant professor at the Department of Environmental Sciences, PMAS, Arid Agriculture University, Rawalpindi (PMAS-UAAR), Pakistan (March 10, 2015, to March 9, 2016). During his tenure, he has supervised twelve MPhil (two at PMAS-UAAR and ten at SBBU) and eight students' research works. He has been among top 2% of scientists in the world, ranking by Stanford University, since 2021 and among the 5% of scientists in the world (SciRank Global Index 2025). Recently, he has served as a chief organizer of two international conferences in 2024 and 2025 at Shaheed Benazir Bhutto University (SBBU) Sheringal, Upper Dir, Khyber Pakhtunkhwa, Pakistan. He has also participated in various national and international conferences and presented 29 abstracts as oral presentations. Based on his research experience, he was selected as a senior adjunct research fellow in the School of Biological Sciences, The University of Western Australia, Australia, from July 17, 2018, to July 16, 2024. Recently, he has been promoted to senior adjunct research fellow in the School of Biological Sciences, The University of Western Australia, Australia, since July 17, 2024. Regarding his qualities, he is very hard-working and committed. Moreover, he is flexible in getting advanced knowledge, which could be required through training. He is enthusiastic and conducive to learning the latest advanced techniques and technologies. Dr. Ditta has a strong publication profile. He has edited three books with internationally well-recognized publishers such as Elsevier and CRC (Taylor and Francis), authored 191 research (141), reviews (23), and book chapters (27) with an accumulative impact factor of 397.20. My Google Scholar citations are 6715, with an h-index of 43 and a hi-index of 110. His research work has encompassed soil nutrients using chemical, organic, and biofertilizers; plant growth under abiotic stress; C-sequestration; nanonutrition of crop plants; bioremediation; public health; and health risk assessment of different contaminants. He has published in various well-reputed journals such as Renewable and Sustainable Energy Reviews, Science of the Total Environment, Chemosphere, Plant Physiology and Biochemistry, Environmental Science and Pollution Research, Environmental Pollution, Journal of Soils and Sediments, and Journal of Soil Science and Plant Nutrition. He has also edited special issues as a guest editor in MDPI-Sustainability. Moreover, he has a strong collaboration with different researchers around the world.

Dr. Allah Ditta
Associate Professor
Department of Environmental Sciences
Shaheed Benazir Bhutto University (SBBU) Sheringal
Upper Dir, Khyber Pakhtunkhwa, Pakistan

Adjunct Senior Research Fellow
School of Biological Sciences
The University of Western Australia
Perth, WA, Australia

Dr. Muhammad Imtiaz is a senior scientist at Soil and Environmental Biotechnology Division, National Institute for Biotechnology and Genetic Engineering (NIBGE), and his research is focusing on sustainable approaches for soil conservation and health improvement for the betterment of crop production under abiotic stresses using microbes.

From 2005 to 2011, he obtained a bachelor's degree and a master's degree in soil science from University of Sargodha, Sargodha, Punjab, Pakistan. In 2015 he received a doctoral degree (Plant Nutrition) from the Huazhong Agricultural University, Wuhan, Hubei, China. After the completion of his doctoral degree, China Scholarship Council (CSC) awarded him "Distinguished Scholar" award on the basis of his academic achievements and research publications. He also won a project funded by National Natural Science Foundation China (NSFC) from 2015 to 2017 during his postdoctoral fellowship (Soil Ecology) at Huazhong Agricultural University, Wuhan, China.

In 2017 Dr. Imtiaz was selected as "associate professor" under the program of "Young Talent Hunt Program" at School of Environmental Science and Engineering, Guangzhou University, Guangzhou, China.

Dr. Imtiaz has produced about 98 research publications in various well-reputed scientific impact factor journals with a total impact factor of 337 and 6542 citations. He has participated in national and international conferences and seminars and presented his research achievements.

Dr. Muhammad Imtiaz
Senior Scientist
Soil and Environmental Biotechnology Division,
National Institute for Biotechnology
and Genetic Engineering (NIBGE),
Faisalabad, Punjab, Pakistan

Dr. Sajid Mehmood Born on May 9, 1991, in Pakistan, Dr. Sajid Mehmood is an internationally recognized soil scientist whose work bridges environmental chemistry, sustainable agriculture, and green remediation technologies. His academic foundation was built at Pir Mehr Ali Shah Arid Agriculture University, where he earned both his B.Sc. (Hons.) in Agriculture and M.Sc. (Hons.) in Soil Science. His early research on biochar and soil microbial dynamics reflected a commitment to innovative soil amendments and ecological sustainability.

Seeking advanced scientific training, he pursued his Ph.D. at Huazhong Agricultural University (2015–18), specializing in soil properties and heavy-metal environmental behavior. His doctoral work laid the foundation for his long-term focus on contaminated soil remediation and sustainable nutrient management.

Dr. Mehmood has authored more than 70 peer-reviewed publications, accumulating over 5300 citations, with 37 papers as corresponding or leading author. His research, published in leading Q1 journals such as Science of the Total Environment, Environmental Pollution, and Journal of Hazardous Materials, focuses on soil and water remediation using biochar, nanoparticles, and nature-based materials to immobilize toxic metals and enhance ecosystem resilience.

Following his Ph.D., he continued his academic career in China, serving as research associate under NSFC-supported projects and later as a postdoctoral research fellow at Guangzhou University and Hainan University. In July 2024, he was appointed associate professor at Hainan University, where he leads research on tropical agriculture, biodiversity, and climate change mitigation while mentoring undergraduate and graduate students.

Beyond research, Dr. Mehmood has contributed as a book editor and chapter author with international publishers, including Elsevier and CRC Press, shaping discourse on soil remediation, nanotechnology in agriculture, and heavy-metal pollution. An invited international speaker at major scientific conferences, he actively fosters global collaborations across Asia and beyond.

Dr. Sajid Mehmood, Ph.D.
Senior Associate Professor
School of Tropical Agriculture and Forestry
Hainan University, Haikou, Hainan, China

Chapter 1

Arbuscular mycorrhizal fungi: Understanding the basis of symbiosis

Kamal Khalkhal[a], Somayeh Dehghani[b], Keyvan Valizadeh-Rad[c], Mahboobeh Jalali[c] and Allah Ditta[d,e]

[a] *Soil and Water Research Department, Agricultural Research, Education and Extension Organization (AREEO), Kermanshah Agricultural and Natural Resources Research and Education Center, Kermanshah, Iran,* [b] *Soil and Water Research Department, Agricultural Research, Education and Extension Organization (AREEO), Bushehr Agricultural and Natural Resources Research and Education Center, Bushehr, Iran,* [c] *Department of Soil Sciences, Faculty of Agriculture, Lorestan University, Khorramabad, Iran,* [d] *Department of Environmental Sciences, Shaheed Benazir Bhutto University Sheringal, Sheringal, Pakistan,* [e] *School of Biological Sciences, The University of Western Australia, Perth, Australia*

1.1 Introduction: The arbuscular mycorrhizal symbiosis

Arbuscular mycorrhizal fungi (AMF) are a group of soil-borne microorganisms belonging to the phylum Glomeromycota that form a mutualistic symbiotic association with the roots of the vast majority of terrestrial plants, including angiosperms, gymnosperms, pteridophytes, and some bryophytes (Schüßler et al., 2001). Taxonomically, they are classified within the subkingdom Mucoromycota (Diagne et al., 2020; Tedersoo et al., 2018). This ancient symbiosis, dating back over 400 million years, is characterized by the formation of unique, tree-like structures within the cortical cells of plant roots called arbuscules. These arbuscules are the primary sites of nutrient exchange between the fungus and the host plant, where carbon compounds from the plant are traded for mineral nutrients, particularly phosphorus and nitrogen, acquired by the extraradical fungal hyphae from the soil (Gou et al., 2024; Smith & Read, 2008). Fossil evidence suggests that this mutualism appeared 400–460 million years ago and played an important role in the adaptation of plants to terrestrial ecosystems (Árevalo et al., 2025; Parniske, 2008).

1.1.1 Key importance: Why is this relationship vital?

The AMF symbiosis is a cornerstone of terrestrial ecosystem functioning. Its importance extends far beyond the individual plant, influencing community dynamics and ecosystem processes. Plant–microbial symbiosis is one of the earliest forms of interaction that remains critical to plant health, survival, and resistance (Wasudeo et al., 2026). These fungi play a crucial role in plant nutrition and growth under stressed conditions and enhance several essential ecosystem processes (Diagne et al., 2020; Siddiqui & Pichtel, 2008). AMF are fundamental because they

- **Enhance plant nutrition and growth:** By extending a network of fine hyphae into the soil, AMF dramatically increase the root surface area, enabling plants to access water and nutrients, especially immobile ones like phosphorus, from beyond the root depletion zone (Allen, 2011; Posta & Duc, 2020). This is particularly crucial in soils with low to moderate fertility (Valverde-Barrantes et al., 2018). AMF not only absorb phosphorus and microelements more efficiently but also modulate diverse pathways of plant fitness, making plants resilient to unfavorable environmental conditions (Nikalje et al., 2025; Wasudeo et al., 2026).
- **Increase plant resistance to abiotic and biotic stresses:** AMF colonized plants exhibit enhanced tolerance to a wide spectrum of environmental stresses, including drought, salinity, extreme temperatures, heavy metal toxicity, and

Advancing Sustainable Agriculture Through Arbuscular Mycorrhizal Fungi. DOI: https://doi.org/10.1016/B978-0-443-45130-0.00007-2

pathogen attack (Diagne et al., 2020; Lenoir et al., 2016). This protective role contributes to plant survival and ecosystem stability under changing environmental conditions. AMF help plants to cope with these stresses, and a plant colonized by AMF will use more adaptation mechanisms compared to a plant without mycorrhizae (Diagne et al., 2020). Under climate change-driven resource limitation, plant–microbe partnerships enhance stress tolerance through defined physiological and molecular mechanisms, including improved water-use efficiency, ion homeostasis, antioxidant activation, and stress–responsive hormonal signaling (Sharma et al., 2025; Wasudeo et al., 2026).

- **Drive ecosystem processes:** AMF regulate nutrient cycles (Gou et al., 2024), improve soil structure by producing the glomalin-related soil protein (GRSP) that binds soil particles into stable aggregates (Lehmann et al., 2014), and influence plant community composition and productivity (Van Der Heijden et al., 1998). They are a key component of the "wood wide web," facilitating communication and resource transfer between different plants (Gorzelak et al., 2015). Furthermore, AMF contribute significantly to carbon sequestration in terrestrial ecosystems. Their external hyphae deposit carbon deep in the soil that interacts with minerals, and the production of stable glomalin pools enhances long-term carbon storage (Árevalo et al., 2025; Rillig et al., 2001).
- **Support ecosystem restoration:** Because of their multiple benefits, AMF are recognized as a sustainable and effective biological tool for accelerating the recovery of degraded ecosystems, improving plant establishment, diversity, and overall ecosystem functionality (Asmelash et al., 2016; Neuenkamp et al., 2019). The incorporation of AMF in restoration projects is an effective strategy for improving soil health and favoring reforestation (Árevalo et al., 2025; Smith & Read, 2008).

1.1.2 A historical perspective: Discovery and early identification

The study of mycorrhizas began in the late 19th century. The German forest pathologist A.B. Frank is credited with coining the term "mycorrhiza" (from Greek: *mukes* = fungus, *rhiza* = root) in 1885 to describe the mutualistic association between fungi and tree roots. However, the specific type now known as arbuscular mycorrhiza was first observed even earlier. In the 1840s researchers such as Nägeli and Reissek noted the presence of fungal structures within plant roots, which were later termed "vesicles" and "spores." It was Janse in 1897 who first correctly interpreted the intracellular structures as a beneficial fungal association. The term "arbuscular" was adopted later to distinguish this ancient and widespread group from other mycorrhizal types, and the development of modern molecular techniques in the late 20th century led to the establishment of the separate phylum Glomeromycota (Schüßler et al., 2001). Paleobotanical and phylogenetic data have since demonstrated that plants and AMF have coevolved for more than 400 million years, highlighting the profound antiquity of this partnership (Winagraski et al., 2019). Based on fossil records and molecular data, this symbiosis dates back to the first appearance of land plants, about 400–450 million years ago (Diagne et al., 2020; Smith & Read, 2008). Early research on AMF focused mainly on their role in nutrient acquisition by plants, but the focus gradually expanded to explore their diversity and ecological roles, potential applications in agriculture and forestry, and the impact of environmental changes on this symbiosis (Árevalo et al., 2025; Clark & Zeto, 2000; Koide & Mosse, 2004).

1.1.3 Aim of the chapter

This chapter aims to provide a comprehensive overview of the arbuscular mycorrhizal symbiosis. Readers will journey from the fundamental biology of the fungal partners and the intricate molecular dialogue that establishes the symbiosis to the core nutritional exchange and the holistic benefits for plant health. We will explore the ecological and evolutionary context of this relationship, the modern tools used to study it, and its applied perspectives in sustainable agriculture and ecosystem restoration. Finally, we will look ahead to unresolved mysteries and future frontiers in AMF research.

1.2 The players: Identity and biology of arbuscular mycorrhizal fungi

1.2.1 Taxonomic placement: The phylum Glomeromycota

AMF are not a diverse group within the fungal kingdom but constitute a distinct monophyletic phylum, the Glomeromycota. This classification, established by Schüßler et al. (2001), separated them from other fungi based on molecular phylogenetic evidence. Within this phylum, they are further classified into orders, families, and genera (e.g., *Glomus, Rhizophagus, Funneliformis, Diversispora, Acaulospora*, and *Scutellospora*). It is estimated that there are approximately 250–300 described morphospecies, though molecular techniques suggest the actual diversity is much higher. Taxonomically, AMF are classified within the subkingdom Mucoromycota and the phylum Glomeromycota, which includes three classes:

Glomeromycetes, Archaeosporomycetes, and Paraglomeromycetes (Diagne et al., 2020; Tedersoo et al., 2018). AMF belong to 11 families and 25 genera (Diagne et al., 2020; Schüßler et al., 2001; Spatafora et al., 2016). Tropical forests harbor approximately 75% of all identified *Glomeromycotina* species (Árevalo et al., 2025; Marinho et al., 2018).

1.2.2 Key structures: Hyphae, arbuscules, vesicles, and spores

Several distinct structures characterize the AMF life cycle and function:

- **Hyphae:** The vegetative body of the fungus consists of thread-like, coenocytic (aseptate) hyphae. These can be divided into two functional types:
- **Extraradical hyphae:** These grow out from the root into the surrounding soil, forming an extensive network that can explore up to 5–15 cm beyond the root depletion zone. Their very thin diameter (2–7 μm) allows them to penetrate soil pores inaccessible to roots, scavenging for water and nutrients (Allen, 2011). AMF extend the functional root system through extraradical hyphal networks that enhance phosphorus, zinc, and water acquisition (Ahmed et al., 2025; Wasudeo et al., 2026). It is calculated that the rate of water transport from external hyphae to the root ranges from 0.1 to 0.76 μL H_2O h^{-1} per hyphal infection point (Allen, 1982; Diagne et al., 2020; Faber et al., 1991; Ruth et al., 2011).
- **Intraradical hyphae:** These grow between and within the cortical cells of the host root, serving as conduits for nutrient transport and giving rise to the symbiotic structures.
- **Arbuscules:** These are the defining and functionally most important structures. They are highly branched, tree-like structures formed within the lumen of root cortical cells by repeated dichotomous branching of the intraradical hyphae. The arbuscule is surrounded by the plant-derived peri-arbuscular membrane (PAM), creating a large interfacial surface area that is the primary site of bidirectional nutrient exchange (carbon for minerals). They are ephemeral structures, collapsing after a few days to a couple of weeks.
- **Vesicles:** These are thin-walled, lipid-filled, globular structures formed by some AMF families (e.g., Glomeraceae) within the roots or in the soil. They are thought to function as storage organs for the fungus, containing lipids and other carbon-rich compounds derived from the plant.
- **Spores:** These are thick-walled, asexual resting structures (chlamydospores or azygospores) produced terminally on hyphae, either in the soil or within the root. They are crucial for the survival of the fungus in the absence of a host and for dispersal. Spore morphology (size, color, wall structure) has been the traditional basis for species identification. Spore density and species richness are often used as indicators of soil health and restoration success (MacColl et al., 2024; Rodríguez-León et al., 2021).

1.2.3 Defining characteristic: Obligate biotrophy

A key feature that sets AMF apart from many other fungi is that they are obligate biotrophs. This means they are completely dependent on a living host plant to complete their life cycle and cannot be cultured in the absence of the host. Glomeromycota are obligate symbionts that rely on the carbon substrates provided by their host plants (up to 20% of plant-fixed carbon) to survive (Diagne et al., 2020; Johns, 2014; Siddiqui & Pichtel, 2008). Analysis of the AMF genome reveals why: they lack the genes encoding for key enzymes involved in the breakdown of complex organic polymers, such as cellulases, pectinases, and lignin-degrading peroxidases, which would be necessary for a saprotrophic lifestyle (Chen et al., 2018). Although the ancestors of Glomeromycota and contemporary AMF lack a strong degradation machinery (Chen et al., 2018; Tedersoo et al., 2018), certain ascomycete and basidiomycete ectomycorrhiza (EcM) and ericoid mycorrhiza (ErM) symbionts can produce enzymes to release N from organic polymers (Tedersoo et al., 2018). In exchange for the carbon they receive (up to 20% of the plant's photosynthetic output), they have a full complement of genes for nutrient uptake and transport, including high-affinity phosphate transporters, ammonium transporters (AMTs), and nitrate transporters (NTs) (Chen et al., 2018).

1.3 The symbiotic dialogue: From first contact to establishment

The establishment of the AMF symbiosis is a finely tuned process involving a continuous molecular dialogue between the two partners.

1.3.1 Stage 1: Plant signaling (secretion of strigolactones)

The dialogue begins with the host plant. Under phosphate and nitrogen-limiting conditions, plant roots release a group of signaling molecules into the rhizosphere known as strigolactones. These hormones act as a "cry for help," signaling to the AMF the presence of a potential host. Strigolactones stimulate the germination of AMF spores, induce hyphal branching and metabolism, and increase the fungus's metabolic activity and mitochondrial density, thereby preparing it for contact with the root (Mostofa et al., 2018; Ruiz-Lozano et al., 2015).

1.3.2 Stage 2: Fungal response (secretion of Myc factors)

In response to strigolactones the AMF releases its own set of signaling molecules, known as "Myc factors." These are a mixture of lipochitooligosaccharides (LCOs) and short-chain chitin oligomers. The plant root perceives these Myc factors through specific receptors, triggering a signal transduction pathway that leads to the activation of the plant's symbiotic program. This includes the induction of the "common symbiosis signaling pathway (CSSP)" (also shared with rhizobial symbiosis in legumes), leading to calcium spiking in the root epidermal nuclei and the expression of genes required for symbiosis. Recent studies have demonstrated that the CSSP is conserved among diverse plant species and plays an important role in plant symbiotic interactions (Wang et al., 2022; Wasudeo et al., 2026). LCO signals from AMF (Myc factors) activate the CSSP through perception by LysM-domain receptor-like kinases (Fukuda et al., 2024; Wasudeo et al., 2026).

1.3.3 Stage 3: Penetration and colonization (formation of the hyphopodium)

Once the fungus reaches the root surface, it forms a swollen, flattened, and adherent structure called a hyphopodium (or appressorium) on the surface of an epidermal cell. This structure is not for puncturing the cell but rather for facilitating entry. From the hyphopodium a penetration peg develops, and the fungus enters the root. It then grows between the cortical cells (intercellularly) toward the inner cortex, guided by prepenetration apparatuses built by the plant cell. Signal recognition initiates SYMRK-associated signaling and nuclear calcium oscillations, which are decoded by the calcium- and calmodulin-dependent protein kinase CCaMK in coordination with CYCLOPS, NIN, and RAM1 to regulate infection thread formation, cortical cell division, and arbuscule development (Wasudeo et al., 2026; Zhang et al., 2024).

1.3.4 Stage 4: Arbuscule differentiation

Upon reaching the inner cortex the fungus receives further signals from the plant that trigger a profound morphogenetic change. The intraradical hypha penetrates the cell wall of a cortical cell (but not the plasma membrane) and begins to undergo repeated dichotomous branching to form the arbuscule. Simultaneously the plant cell undergoes massive reorganization: the vacuole fragments, the nucleus moves to a central position, and the plasma membrane invaginates and proliferates enormously to surround the developing arbuscular branches. This new membrane, the PAM, is the functional interface of the symbiosis. Nuclear pore complex components, including NUP85 and NUP133, are required for calcium spiking and nuclear signal integration, underscoring the subcellular control of CSSP (Genre & Russo, 2016; Wasudeo et al., 2026).

1.4 The core exchange: A mutualistic marketplace

The central function of the AMF symbiosis is the reciprocal exchange of resources, a tightly regulated "marketplace" at the arbuscular interface.

1.4.1 The carbon flow: From plant to fungus (lipids and sugars)

As the obligate biotroph the AMF is entirely dependent on its host for its carbon supply. It receives up to 20% of the plant's photosynthetically fixed carbon. Initially, it was thought that the fungus received primarily sugars (e.g., glucose). However, genomic and lipid profiling studies have revealed that lipids are a major form of carbon transferred from the plant to the fungus. The plant synthesizes fatty acids and then transfers them across the PAM to the arbuscule, where the fungus incorporates them into its own extensive membranes and stores them in vesicles and spores. The plant also provides sugars, which the fungus uses for energy. In several Ectomycorrhizal fungi (ECM) model systems, plant-derived sucrose is transferred to the apoplast and converted to glucose and fructose by means of fungi-encoded invertase, whereas in AMF, invertase is not known, and therefore C transfer may be more under plant control (Casieri et al., 2013; Tedersoo et al., 2018).

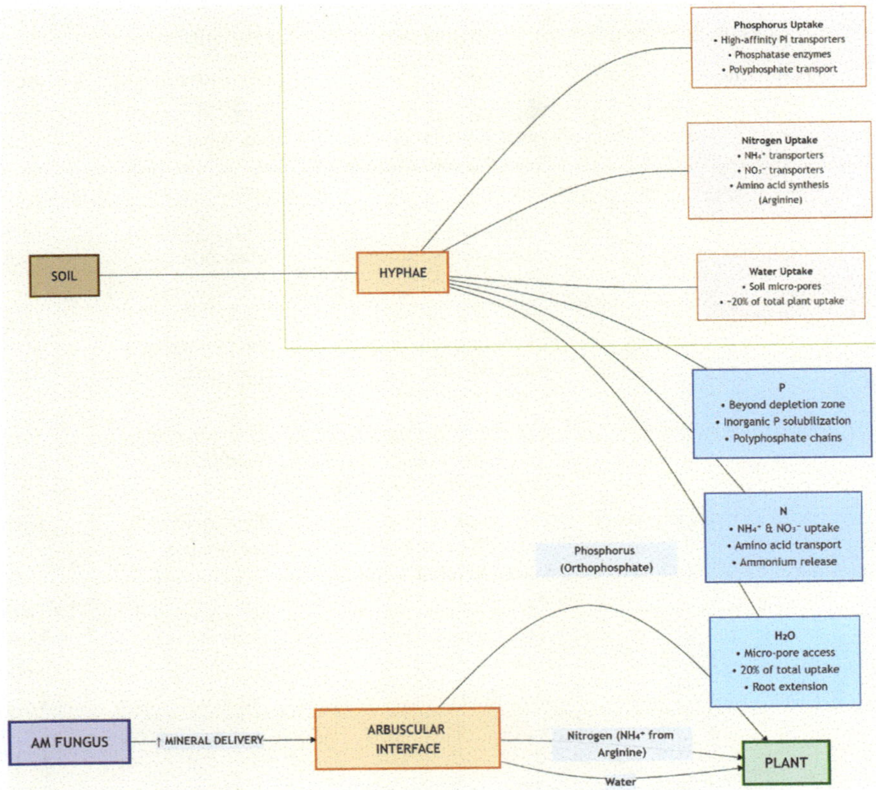

FIGURE 1.1 Mineral flow from AM fungus to plant. The extraradical hyphal network acquires phosphorus (P), nitrogen (N), and water from the soil and delivers them to the plant across the arbuscular interface. Phosphorus is taken up via high-affinity Pi transporters and phosphatase enzymes, transported as polyphosphate, and released as orthophosphate. Nitrogen is absorbed as NH_4^+ and NO_3^-, incorporated into amino acids (e.g., arginine), and released as NH_4^+ at the arbuscule. Water uptake through soil micropores accounts for approximately 20% of total plant water uptake.

1.4.2 The mineral flow: From fungus to plant (phosphorus, nitrogen, and water)

In return for carbon the AMF delivers mineral nutrients and water acquired from the soil by its extraradical hyphae. Fig. 1.1 provides a schematic overview of this three-way mineral transfer, highlighting the key transporters and uptake mechanisms involved.

- **Phosphorus (P):** This is the most well-documented benefit. The fine hyphae can access soil P far beyond the P-depletion zone that quickly forms around roots. AMF possess high-affinity phosphate transporters and secrete phosphatase enzymes to solubilize inorganic P (Casieri et al., 2013; Nehls & Plassard, 2018). P is taken up by the hyphae, incorporated into polyphosphate chains for transport, and then transferred to the plant across the arbuscule interface as orthophosphate.
- **Nitrogen (N):** AMF hyphae can take up both ammonium (NH_4^+) and nitrate (NO_3^-) from the soil. Although they have a limited capacity to release N from complex organic polymers themselves, they can access N from organic matter decomposed by other soil microorganisms. N is incorporated into amino acids (like arginine) within the hyphae, transported, and then broken down to release ammonium for the plant at the arbuscule (Chen et al., 2018). Both fungi of different functional guilds and plants possess multiple NH_4^+ and NO_3^- and amino acid transporters to acquire simple N compounds (Casieri et al., 2013; Nehls & Plassard, 2018; Tedersoo et al., 2018).
- **Water:** The extensive hyphal network acts as an extension of the root system, facilitating water uptake from soil micropores. Studies estimate that approximately 20% of a plant's total water uptake can be delivered by AMF hyphae (Allen, 1982; Faber et al., 1991; Ruth et al., 2011).

1.4.3 The exchange hub: The arbuscule-per-arbuscular membrane interface

The arbuscule, surrounded by the PAM, is the dedicated exchange hub. This interface is packed with specialized transporter proteins. On the plant side (PAM), there are transporters for exporting carbon (e.g., sugar and lipid transporters) and

TABLE 1.1 Key genes and transporters induced during arbuscular mycorrhizal symbiosis.

Gene/transporter	Full name	Function	Expression site	Key references
PT	Phosphate transporter	High-affinity inorganic phosphate (Pi) uptake from the peri-arbuscular space	Peri-arbuscular membrane	Casieri et al. (2013), Chen et al. (2018)
AMT	Ammonium transporter	Ammonium (NH_4^+) transport from fungus to plant	Peri-arbuscular membrane	Chen et al. (2018), Diagne et al. (2020)
AQ	Aquaporin	Water transport across membranes regulates root hydraulic conductivity	Plasma membrane of root cells	Diagne et al. (2020), Maherali and Klironomos (2007)
STR/STR2	ABC transporters	Lipid transfer from plant to fungus (required for arbuscule development)	Peri-arbuscular membrane	Smith and Read (2008), Gorzelak et al. (2015)
NFR1/NFR5	LysM receptor kinases	Perception of Myc factors (LCO signals) from AM fungi	Epidermal cell membrane	Fukuda et al. (2024)
SYMRK	Symbiosis receptor kinase	Signal transduction in the common symbiosis pathway	Epidermal and cortical cells	Genre and Russo (2016), Wang et al. (2022)
CCaMK	Calcium/calmodulin-dependent protein kinase	Decoding of nuclear calcium oscillations	Nucleus	Genre and Russo (2016), Parniske (2008)
CYCLOPS/IPD3	Transcription factor	Phosphorylation target of CCaMK activates symbiosis-related genes	Nucleus	Genre and Russo (2016)
RAM1	GRAS domain transcription factor	Regulates arbuscule branching and differentiation	Cortical cells	Ho-Plagaro and Garcia-Garrido (2022), Genre and Russo (2016)
NIN	Nodule inception protein	Transcription factor regulating symbiotic responses	Nucleus of cortical cells	Delaux and Gutjahr (2024), Fu et al. (2022)

LCO, lipochitooligosaccharide.

importing nutrients (e.g., plant phosphate and AMTs induced specifically in arbusculated cells). On the fungal side (arbuscule membrane), there are transporters for taking up carbon and releasing mineral nutrients.

1.4.4 Molecular control of exchange

The exchange is not passive but is tightly regulated by both partners to ensure fair trade. The plant can "reward" more beneficial fungal partners with more carbon and "punish" fewer cooperative ones by restricting carbon flow. This control is exerted at the genetic level, with the expression of specific phosphate and AMTs in the PAM being closely linked to the carbon status of the root cell. The fungal genome also contains a battery of genes for transporters and interacting proteins that are upregulated during symbiosis to facilitate and control the nutrient fluxes (Casieri et al., 2013; Chen et al., 2018). It is well-known that AMF symbiosis specifically induces the expression of transporters such as the plant aquaporin genes, Pi transporters (PT), AMT, NT, sulfur (S) transporter, Zn transporter, carbon transporter, and protein transporter (Allen & Shachar-Hill, 2008; Burleigh & Bechmann, 2002; Diagne et al., 2020; Giovannetti et al., 2014; Nguyen et al., 2019; Sieh et al., 2012). Reciprocal rewarding and sanctioning mechanisms for nutrient-to-carbon exchange have evolved in AM systems (Tedersoo et al., 2018).

Table 1.1 summarizes the key genes and transporters that are specifically induced or upregulated during the AMF symbiosis, highlighting their functions and sites of expression.

1.5 Benefits beyond nutrition: Holistic plant health

The advantages of the AMF symbiosis extend far beyond improved mineral nutrition, contributing to enhanced overall plant health and resilience.

1.5.1 Resistance to drought and salinity stress

AMF colonization significantly enhances plant tolerance to both drought and salinity, two major abiotic stresses limiting global crop production (Diagne et al., 2020; Posta & Duc, 2020). AMF communities change in composition in response to abiotic stress, with stresses reducing AMF diversity and altering community composition, resulting in an AMF community with a higher proportion of species that are more tolerant of that specific abiotic stress (Diagne et al., 2020; Millar & Bennett, 2016). The mechanisms involved are multifaceted:

- **Improved water and nutrient uptake:** The extensive extraradical hyphal network accesses water from soil pores unavailable to roots and enhances the uptake of nutrients like P, whose deficiency is exacerbated under dry/saline conditions (Bernardo et al., 2019).
- **Osmotic adjustment:** AMF-colonized plants accumulate higher concentrations of compatible solutes (osmolytes) such as proline, sugars, and glycine betaine, which help maintain cell turgor and protect cellular structures (Posta & Duc, 2020; Yamato et al., 2008).
- **Ion homeostasis:** Under salinity, AMF help maintain a high K^+/Na^+ ratio in plant tissues by selectively promoting K^+ uptake while restricting the uptake and transport of toxic Na^+ and Cl^- ions, acting as an "ionic filter" (Evelin et al., 2009; Li et al., 2020; Santander et al., 2019).
- **Hormonal regulation:** AMF modulate the levels of stress-related hormones such as abscisic acid and jasmonic acid (JA), which regulate stomatal closure, root hydraulic conductivity, and the expression of stress–responsive genes (De Ollas & Dodd, 2016; Ouledali et al., 2019).
- **Antioxidant activity:** AMF enhance the activity of antioxidant enzymes (e.g., catalase, superoxide dismutase, and peroxidase), which scavenge reactive oxygen species and reduce oxidative damage caused by stress (Latef & Chaoxing, 2011; Talaat & Shawky, 2014; Yizhu et al., 2020). AMF also enhance stress tolerance through modulation of host physiological and molecular responses by promoting the accumulation of osmoprotectants and enhancing antioxidant defense systems (Ullah et al., 2025; Wasudeo et al., 2026).

1.5.2 Protection against pathogens (biotic stress)

AMF can prime the plant's immune system, a phenomenon known as mycorrhiza-induced resistance (MIR). MIR provides broad-spectrum protection against a range of soil-borne pathogens, including fungi, nematodes, and some bacteria (Nguvo & Gao, 2019; Pozo & Azcón-Aguilar, 2007). Symbiotic microorganisms enhance biotic stress resistance primarily via induced systemic resistance (ISR), a JA- and ethylene-dependent immune priming mechanism, distinct from salicylic acid-mediated systemic acquired resistance (Verma et al., 2022; Wasudeo et al., 2026). Well-characterized ISR inducers promote transcriptional reprogramming of defense pathways, increasing the accumulation of pathogenesis-related proteins, antimicrobial metabolites, and structural barriers (Santoyo et al., 2024; Wasudeo et al., 2026).

- **Direct and indirect mechanisms:** Protection is achieved through a combination of mechanisms: (1) improved nutrition and vigor, making the plant less susceptible to infection; (2) competition for resources and colonization sites in the roots; (3) anatomical changes, such as lignification of cell walls; and (4) priming of defense responses, allowing for a faster and stronger activation of defense genes upon pathogen attack. For instance, AMF colonization has been shown to reduce the severity of *Fusarium* wilt in tomato and soybean (Pozo et al., 2010) and decrease galling caused by root-knot nematodes (*Meloidogyne* spp.) (Pellegrino & Bedini, 2014). Colonization by AMF has a protective effect, and studies have proven that AMF reduces the damage caused by various plant pathogens (Diagne et al., 2020; Nguvo & Gao, 2019; Vos et al., 2012). In parallel, symbionts directly suppress pathogens through antimicrobial production, iron sequestration via siderophores, competitive niche exclusion, and, in some cases, mycoparasitism (Aloo et al., 2023; Mengistu, 2020; Wasudeo et al., 2026).

1.5.3 Improving soil structure and aggregate stability

AMF play a critical role as soil "ecosystem engineers." Their extraradical hyphae, which can be extensive (reaching tens of meters per gram of soil), enmesh soil particles, physically binding them together into macroaggregates. Furthermore, they

TABLE 1.2 Functional roles of arbuscular mycorrhizal fungi (AMF) in enhancing plant stress tolerance and improving soil ecosystem processes.

Functional category	AMF mechanisms	Plant/ecosystem outcomes	Key references
Drought tolerance	Increased water uptake via extraradical hyphae; osmotic adjustment through accumulation of proline and sugars; enhanced antioxidant activity	Improved water-use efficiency, reduced oxidative damage, and higher survival under drought	Diagne et al. (2020), Bernardo et al. (2019), Posta and Duc (2020)
Salinity tolerance	Regulation of ion homeostasis (higher K^+/Na^+ ratio); modulation of ABA and JA signaling	Reduced Na^+ toxicity, improved stomatal regulation, and enhanced growth under salinity	Evelin et al. (2009), Santander et al. (2019), Ouledali et al. (2019)
Biotic stress resistance	Mycorrhiza-induced resistance; ISR activation; competition for infection sites	Reduced pathogen infection, decreased nematode galling, enhanced systemic immunity	Pozo and Azcón-Aguilar (2007), Vos et al. (2012), Nguvo and Gao (2019)
Soil structure improvement	Production of glomalin-related soil proteins; hyphal enmeshment of soil particles	Increased aggregate stability, improved porosity, enhanced carbon sequestration	Lehmann et al. (2014), Rillig et al. (2001)

ABA, abscisic acid; *ISR*, induced systemic resistance; *JA*, jasmonic acid.

produce a hydrophobic, glycoproteinaceous substance called GRSP, which acts as a "glue," stabilizing these aggregates (Lehmann et al., 2014; Rillig & Mummey, 2006). AMF increase soil aggregation by adhering to minerals and organic matter, thus improving soil porosity, aeration, and water retention (Kennedy et al., 2012; Wasudeo et al., 2026). Improved soil aggregation enhances aeration, water infiltration and retention, and resistance to erosion, creating a better environment for root growth and other soil organisms (Nazir et al., 2023; Thind et al., 2022). Soil proteins closely related to AMF-derived glomalin (GRSPs) are extremely stable and contribute to carbon accumulation (Rillig et al., 2001; Wasudeo et al., 2026). The multifunctional contributions of AMF to plant stress tolerance and soil ecosystem processes are summarized in Table 1.2.

1.6 Ecological and evolutionary context

1.6.1 Role in the early evolution of land plants

Fossil evidence and molecular clock analyses place the origin of the AMF symbiosis at over 460 million years ago, coinciding with the colonization of land by the first plants. Early land plants, such as the early Devonian fossil *Aglaophyton major*, lacked complex root systems and were poorly equipped to acquire water and nutrients from the primitive terrestrial substrate. It is widely hypothesized that a symbiotic association with Glomeromycotan fungi was crucial for this transition. The fungi would have provided essential mineral nutrients and water from the soil in exchange for photosynthetically derived carbon from the plants, effectively functioning as a root system for this early flora. This coevolutionary event was so successful that the partnership has been maintained in over 70% of extant vascular plants, underscoring its fundamental role in plant evolution (Winagraski et al., 2019). Based on fossil records and molecular data, this symbiosis dates back to the first appearance of land plants, about 400–450 million years ago (Diagne et al., 2020; Parniske, 2008; Smith & Read, 2008). Mycorrhizal associations are classified into four principal mycorrhizal types with remarkable differences in anatomy, evolutionary history, and functioning (Smith & Read, 2008; Tedersoo et al., 2018). Arbuscular mycorrhiza (AM) evolved in early land plants, associating with members of Glomeromycota and Endogonomycetes to improve carbon-to-nutrient exchange via intracellular structures termed "arbuscules" (Orchard et al., 2017; Strullu-Derrien et al., 2018; Tedersoo et al., 2018).

Table 1.3 provides a comparative overview of the four principal mycorrhizal types, including their fungal partners, host plants, key structures, main benefits, and evolutionary origins.

1.6.2 The underground network: "Wood Wide Web" and plant communication

The extraradical mycelium of AMF is not confined to a single plant root. A single fungus can simultaneously colonize the roots of two or more different plant individuals, often of different species. This creates a common mycorrhizal network

TABLE 1.3 Principal mycorrhizal types and their key characteristics.

Mycorrhizal type	Fungal partners	Host plants	Key structures	Main benefits	Evolutionary origin
Arbuscular mycorrhiza (AM)	Glomeromycota	~80% of vascular plants (cereals, legumes, vegetables, and grasses)	Arbuscules, vesicles, intraradical, and extraradical hyphae	Phosphorus and nitrogen uptake, water transport, stress tolerance	~400–460 million years ago
Ectomycorrhiza	Basidiomycota, Ascomycota	~2% of vascular plants (oak, birch, pine, and eucalypt)	Fungal mantle, Hartig net, rhizomorphs	Nitrogen acquisition from organic matter, physical protection, and pathogen defense	~200 million years ago (multiple origins)
Ericoid mycorrhiza	Ascomycota (Helotiales)	*Ericaceae* (heather, Rhododendron, and blueberry)	Hyphal coils in epidermal cells	Nitrogen mobilization from organic compounds in acidic soils	~100 million years ago
Orchid mycorrhiza	Basidiomycota (mostly)	Orchidaceae	Pelotons (digestible coils)	Carbon supply for seed germination, mineral nutrition	~80–100 million years ago

(CMN), often referred to as the "Wood Wide Web." These networks are not just passive conduits but are thought to have significant ecological functions (Gorzelak et al., 2015). Recent studies have further emphasized the ecological importance of CMNs, particularly in mediating the transport of bioactive compounds such as allelochemicals, as well as facilitating the movement of phytohormones and signaling molecules (Árevalo et al., 2025; Figueiredo et al., 2021; Watson-Guido & Rivera-Mendez, 2024). These networks promote plant growth, increase nutrient uptake, and activate defense mechanisms against pathogens and environmental stressors (Árevalo et al., 2025; Marinho et al., 2018), thereby playing a central role in chemical communication and in shaping plant community structure (Achatz et al., 2014; Árevalo et al., 2025; Barto et al., 2011).

- **Resource redistribution:** CMNs can facilitate the transfer of carbon, nitrogen, phosphorus, and water between connected plants. For example, carbon can move from a mature, shaded tree to a young, light-limited seedling, improving its survival.
- **Chemical signaling:** Plants connected by a CMN can warn each other of impending danger. For instance, if one plant is attacked by an herbivore, it can send a chemical signal (e.g., defensive compounds) through the network to neighboring plants, which then prime their own defenses (Gorzelak et al., 2015).
- **Influencing plant community structure:** By differentially connecting and benefiting certain plant species over others, CMNs can influence plant community composition, diversity, and succession dynamics (Van Der Heijden et al., 1998). These networks, also called CMNs, can integrate multiple plant species and multiple fungal species that interact, provide feedback, and adapt, which comprise a complex adaptive social network (Diagne et al., 2020; Gorzelak et al., 2015). However, the same networks that support ecological recovery can also be exploited by invasive plant species, which often manipulate AMF to gain competitive advantages in disturbed environments (Árevalo et al., 2025; Bunn et al., 2015; Reinhart & Callaway, 2006).

1.7 Methodological toolkit: Observing the unseen

Studying the hidden world of AMF requires a diverse set of traditional and modern tools.

1.7.1 Traditional methods: Microscopy and spore isolation

- **Spore isolation and morphological identification:** For over a century the primary method for studying AMF communities was isolating their large, robust spores from soil via wet sieving and decanting. Spores are then counted and identified to the genus or species level based on morphological characteristics such as size, color, wall structure, and hyphal attachment (e.g., Schüßler et al., 2001). This method provides information on spore density and the "potential" community but is time-consuming, requires expert taxonomic knowledge, and misses nonsporulating or rare taxa. It is important to note that relying solely on spore morphology may underestimate true AMF diversity compared to molecular techniques (Winagraski et al., 2019).

- **Root colonization assessment:** The level of fungal colonization in cleared and stained roots is a standard measure of the symbiotic activity. Roots are cleared with **Potassium hydroxide (KOH)**, acidified, and stained with dyes such as trypan blue or ink and vinegar. The percentage of root length containing hyphae, arbuscules, and vesicles is then quantified using a grid line intersect method.

1.7.2 Molecular methods: DNA metabarcoding and sequencing

The advent of molecular tools has revolutionized our ability to identify and study AMF communities. Recent developments in high-throughput sequencing technologies (e.g., massively parallel sequencing) have enhanced our ability to explore AMF diversity, phylogeny, community composition, and functional traits (Árevalo et al., 2025; Tedersoo et al., 2010). These tools have also revealed how different types of disturbance can alter AMF community structure, thereby informing targeted restoration strategies (Árevalo et al., 2025; Lekberg et al., 2013; Pagano & Gupta, 2016).

- **DNA metabarcoding:** This is the most common high-throughput approach. DNA is extracted directly from environmental samples (roots or soil). A specific genetic marker region (a "barcode"), typically part of the ribosomal RNA genes (e.g., the ITS region or a fragment of the 18S rDNA), is amplified by Polymerase Chain Reaction (PCR) using AMF-specific primers. The resulting mix of PCR products (amplicons) is then sequenced using next-generation sequencing platforms. The thousands to millions of sequences obtained are clustered into operational taxonomic units based on sequence similarity, providing a comprehensive profile of the AMF community composition and diversity. Omics technologies have made a remarkable impact on plant–microbial symbiosis studies through extensive knowledge of key genes, gene expression dynamics, and key metabolites (Dubey et al., 2025; Fitzpatrick et al., 2020; Wasudeo et al., 2026). Through the use of comparative transcriptomics, ChIP-seq, single-cell RNA-seq, and integrative proteomics, researchers can discern the symbiotic gene expression at a resolution that was previously unattainable (Serrano et al., 2024; Wasudeo et al., 2026).
- **Whole genome sequencing:** Advances in sequencing technology have enabled the sequencing of several AMF genomes (e.g., *Rhizophagus irregularis*). This has provided profound insights into their biology, revealing the genetic basis of their obligate biotrophy, their transport capabilities, and their metabolic pathways. High-throughput omics platforms have led to the identification of essential genes, regulatory pathways, and molecular signals involved in nodulation, arbuscular mycorrhizal (AM) formation, and stress-adaptive symbioses (Dubey et al., 2025; Ho-Plagaro & Garcia-Garrido, 2022; Wasudeo et al., 2026).

1.7.3 Tracking techniques: Stable isotopes

Stable isotope probing and labeling are powerful techniques for tracking nutrient flows within the symbiosis.

- **Isotope labeling:** Plants can be grown in an atmosphere enriched with $^{13}CO_2$, a stable isotope of carbon. By subsequently measuring the appearance of ^{13}C in the AMF hyphae and spores, researchers can quantify the amount of carbon transferred from the plant to the fungus (Olsson et al., 2002). Similarly, applying ^{15}N- or ^{33}P-labeled nutrients to the soil in a hyphal compartment accessible only by the fungus allows for direct measurement of nutrient uptake and transfer to the plant (Smith & Smith, 2011). This has confirmed the direct mycorrhizal pathway for nutrient uptake. Stable isotope analyses are based on heavy isotope discrimination during photosynthesis and nutrient transport (Hobbie & Högberg, 2012; Tedersoo et al., 2018). Across biomes, relative ^{15}N enrichment is highest in nonmycorrhizal plants, followed by AM, EcM, and ErM plants because of differential access to soil organic N pools (Craine et al., 2009; Tedersoo et al., 2018).

1.8 Applied perspectives: Harnessing the symbiosis

Given its multiple benefits, there is immense interest in harnessing the AMF symbiosis for applied purposes.

1.8.1 Sustainable agriculture: Reducing fertilizer inputs

The use of AMF as biofertilizers is a cornerstone of sustainable agriculture. By enhancing the uptake of phosphorus and other nutrients, AMF can reduce the need for synthetic fertilizers, which are energy-intensive to produce and can cause environmental pollution (Berruti et al., 2016; Kuyper & Jansa, 2023). The application of beneficial microbes as biofertilizers, biopesticides, and biostimulants supports sustainable agriculture by activating symbiosis-associated signaling pathways to enhance nutrient acquisition, stress resilience, and pathogen suppression (Alzate Zuluaga et al., 2024; Wasudeo et al., 2026). Metaanalyses and multilocation field trials report crop yield increases of 8%–25%, improvements in nutrient-use efficiency of 10%–30%, and reductions in synthetic fertilizer inputs of 20%–40% (Dissanayaka et al., 2025; Hanif et al., 2024; Wasudeo et al., 2026). Inoculating crops with effective AMF strains has been shown to improve growth, yield, and nutritional quality (e.g., increased Zn, Se, and Fe) in a variety of crops, including tomato, sorghum, wheat, and watermelon (Bona et al., 2016; Golubkina et al., 2020; Kaya et al., 2003; Nakmee et al., 2016). Furthermore, their role in enhancing stress tolerance (drought and salinity) makes them a valuable tool for maintaining productivity under challenging environmental conditions. Rhizobia fix atmospheric nitrogen in legumes through symbiotic nitrogen fixation, whereas mycorrhizal fungi expand the effective root absorptive surface, thereby improving phosphorus uptake and mobilization of micronutrients such as zinc and iron (Ahmed et al., 2025; Diagne et al., 2020; Wasudeo et al., 2026).

1.8.2 Restoration of degraded ecosystems

AMF are increasingly recognized as a key tool for restoring degraded lands, such as postmining sites, eroded areas, and agricultural lands taken out of production (Asmelash et al., 2016). Restoration success is often linked to the recovery of the soil microbial community, and AMF play a pivotal role by

- **Enhancing plant establishment and survival:** AMF improve the growth, nutrient uptake, and stress tolerance of native plants used in restoration efforts, increasing seedling survival and accelerating vegetation cover (Smith et al., 1998; Zhang et al., 2012). Three years of field inoculation experiments in a central Asian desert demonstrated that inoculation with AMF significantly improved the biomass, density, and cover of ephemeral plants, accelerating the regeneration process of degraded spring grassland (Zhang et al., 2012). The incorporation of AMF in restoration projects is an effective strategy for improving soil health and favoring reforestation (Árevalo et al., 2025; Smith & Read, 2008).
- **Increasing plant diversity:** A diverse AMF community can support a more diverse plant community by differentially benefiting various plant species, which is a primary goal of ecological restoration (Koziol & Bever, 2017; Neuenkamp et al., 2019). Inoculation with AMF has been shown to alter plant community composition, increase the proportion of mycorrhizal plants, and enhance plant diversity and species richness in degraded ecosystems (Zhang et al., 2012). Plants, plant communities, and ecosystems rely heavily on the diversity of AMF (Árevalo et al., 2025; Asmelash et al., 2016; Schultz et al., 2001). AMF play a vital role in ecosystem recovery by facilitating plant establishment, enhancing soil health, and supporting successional processes (Árevalo et al., 2025; Kardol et al., 2007; Koziol et al., 2018; Mangan & Adler, 2002).
- **Rebuilding soil structure:** AMF hyphae and glomalin production help stabilize soil aggregates, reduce erosion, and improve water infiltration, which are critical first steps in soil rehabilitation (Lehmann et al., 2014). Studies have shown that restoration practices can significantly increase AMF spore density, species richness, and diversity, with these fungal community parameters serving as indicators of restoration success (MacColl et al., 2024; Rodríguez-León et al., 2021).
- **Mechanisms of action in restoration:** AMF contribute to ecosystem recovery by facilitating plant establishment, enhancing soil health, and supporting successional processes. Incorporating native AMF species into restoration strategies has been shown to accelerate outcomes and improve ecosystem quality (Árevalo et al., 2025). The resilience of AMF communities to disturbance is a key factor; they can exhibit resistance (no change), resilience (recovery over time), or functional redundancy (different species performing similar roles) (Amalia et al., 2021). Fig. 1.2 illustrates the potential responses of arbuscular mycorrhizal fungal communities to disturbance, including resistance, resilience, functional redundancy, and complete community shift (adapted from Amalia et al., 2021). Disturbances such as tillage or fertilization can alter community composition, often favoring disturbance-tolerant "ruderal" species with rapid colonization abilities (Amalia et al., 2021). Furthermore, AMF are not host-specific but are selected by environmental conditions, with some species being ubiquitous across biomes (e.g., *Acaulospora scrobiculata* and *Claroideoglomus etunicatum*), whereas others are rare (Winagraski et al., 2019). This knowledge is crucial for selecting effective inoculants for restoration. The positive effects of AMF on plant growth and reestablishment have been demonstrated in various contexts, including roadside prairie restoration (Smith et al., 1998), abandoned agricultural fields in semiarid regions (Richter & Stutz, 2002), and mine waste sites (Ryszka & Turnau, 2007). Despite advances in our understanding of AMF ecology, several knowledge gaps still hinder the effectiveness of restoration campaigns—particularly in tropical ecosystems (Árevalo et al., 2025; Koziol et al., 2018; Tonietto & Larkin, 2018). Key uncertainties remain regarding the specificity of AMF–plant interactions, the influence of environmental factors on AMF dynamics, optimal inoculation

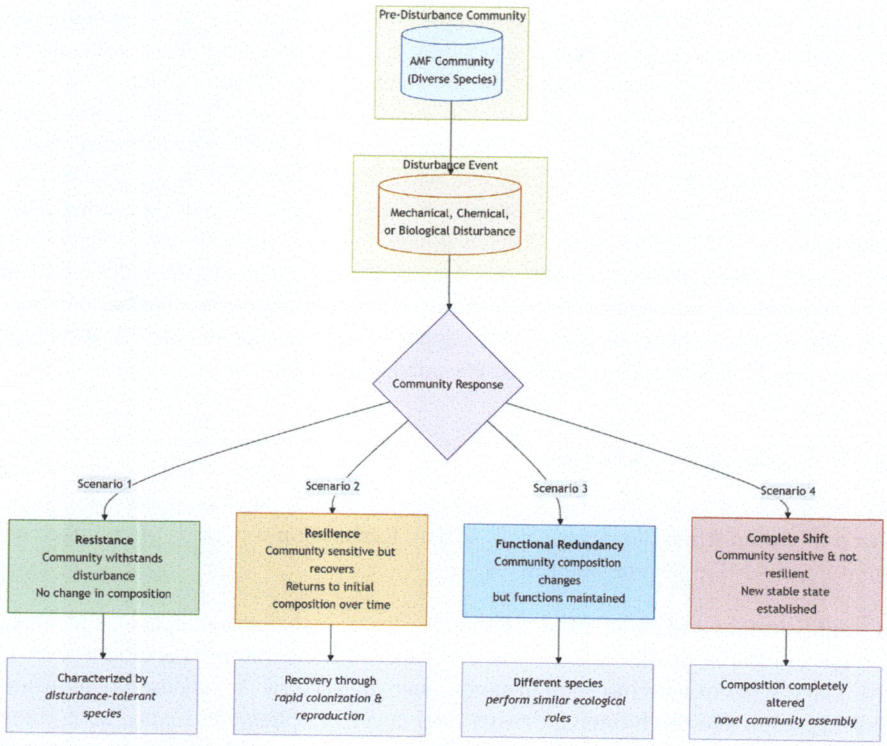

FIGURE 1.2 Potential impacts of disturbance on AMF community structure and function. The diagram illustrates four possible outcomes based on community sensitivity and recovery capacity: (1) resistance—the community withstands disturbance with no significant change in composition; (2) resilience—the community temporarily changes but recovers to its original composition over time; (3) functional redundancy—community composition changes but ecological functions are maintained by new species; (4) complete shift—the community is sensitive, does not recover, and transitions to a new stable state with altered composition and function. *AMF*, Arbuscular mycorrhizal fungi.

strategies, and the long-term outcomes of inoculation success (Árevalo et al., 2025; Ohsowski et al., 2012; Pagano & Gupta, 2016).

1.8.3 Challenges in inoculum production and application

Despite the great potential, there are significant challenges to the widespread and effective use of AMF inoculants:

- **Obligate biotrophy:** Because AMF cannot be cultured without a host, inoculum production is expensive and slow, and requires large amounts of space and resources (e.g., in pot cultures with trap plants). This contrasts sharply with the cheap, industrial-scale fermentation used for other bioinoculants.
- **Inoculum quality and stability:** Producing high-quality, contaminant-free, and viable inoculum with a known concentration of infective propagules (spores, hyphae, and colonized root fragments) is a major hurdle.
- **Efficacy under field conditions:** A strain that works wonders in a controlled greenhouse pot may not establish or perform as well in a complex field environment with competition from native AMF communities and variable soil conditions. The "helper" role of native versus allochthonous strains is a key area of research (Ortiz et al., 2015). Yield gains are most consistent in low-input and stress-prone agroecosystems, where microbial-mediated immune priming and stress mitigation confer measurable advantages, whereas high-input systems often exhibit reduced or variable responses (Ge & Wang, 2025; Sharma et al., 2023; Wasudeo et al., 2026).
- **Specificity and compatibility:** The effectiveness of an AMF isolate can depend on the plant–fungus combination. What works for sorghum may not be optimal for wheat, requiring a careful selection process (Kim et al., 2017). The extent to which a host plant benefits varies with the AMF species used (Diagne et al., 2020; Kim et al., 2017). Identifying ubiquitous AMF species that are adapted to a range of environmental conditions, such as *A. scrobiculata, Acaulospora foveata, C. etunicatum*, and *Glomus macrocarpum*, offers a promising strategy for developing broadly effective inoculants for forest restoration (Winagraski et al., 2019). The application of this knowledge in real environments and

according to biogeographical zones becomes essential to promote their industrial production on a large scale and increase their impact (Diagne et al., 2020).

1.9 Unresolved mysteries and future frontiers

Despite over a century of research, many fundamental questions about the AMF symbiosis remain unanswered.

1.9.1 Unanswered questions in the regulation of exchange

How exactly is the "fair trade" of carbon for nutrients enforced at the molecular level? Although the "biological market" theory is compelling, the precise signaling and regulatory mechanisms that allow a plant to detect a "cheater" fungus that takes carbon without providing adequate nutrients, and vice versa, are still being uncovered. The complex interplay of transporter regulation, sugar and lipid partitioning, and potential sanctions remains a vibrant area of research (Wang, 2017). The degree to which CSSP components can be safely engineered without disrupting immunity or developmental plasticity needs to be elucidated (Ho-Plagaro & Garcia-Garrido, 2022; Wasudeo et al., 2026).

1.9.2 Functional diversity within the arbuscular mycorrhizal fungi community

We know that different AMF species can have different effects on plant growth and stress tolerance. For instance, some are better at alleviating cold stress (Hajiboland et al., 2019), whereas others are more effective against salinity (Wang et al., 2018). Macro- and micronutrient uptake could depend partly not only on the fungal partner but also on the host plant (Diagne et al., 2020; Trouvelot et al., 2015). However, our understanding of the full extent of this functional diversity is limited. AMF communities show varied responses to disturbance; some species are sensitive and decline, whereas others are tolerant and may even increase in abundance (Amalia et al., 2021). Applying life-history strategy frameworks (competitor, stress-tolerant, and ruderal) to AMF can help predict their responses to environmental changes and their roles in succession (Amalia et al., 2021). Both AM and EcM fungi display great within-group differences in the extent of mycelium and hence in potential maintenance cost to the host (Agerer, 2001; Maherali & Klironomos, 2007; Tedersoo et al., 2018). Which fungal genes and traits confer specific benefits? How does the functional composition of the AMF community, not just its taxonomic diversity, shape ecosystem function? Connecting phylogeny and function is a critical goal (Maherali & Klironomos, 2007). Comparative genomics analysis demonstrates that guilds of mycorrhizal fungi serve mostly similar functions in mineral nutrient uptake and stress amelioration of host plants but display relatively greater differences in CAZymes and oxidative enzymes to mobilize and take up macronutrients from organic material (Tedersoo et al., 2018).

1.9.3 Potentials for engineering the symbiosis

Could we "engineer" or manage the AMF symbiosis to a greater extent? This could involve the following:

- **Breeding crops for optimal mycorrhizal responsiveness:** Modern crop breeding under high-fertilizer conditions may have inadvertently selected against traits that promote beneficial mycorrhizal associations. Reintroducing these traits could enhance the efficiency of the symbiosis in low-input agriculture.
- **Developing synthetic communities:** Instead of using single strains, could we design and apply synthetic communities of compatible AMF species and other beneficial soil microbes (e.g., plant growth-promoting bacteria) that work synergistically to provide a wider range of benefits (Plant growth-promoting rhizobacteria (PGPR)–AMF interactions as noted in Yadav et al., 2023). Several studies have reported that the responses of plants to colonization by AMF vary depending on inoculum composition, and a combination of mycorrhizal fungi is often more effective than a monospecific inoculum (Diagne et al., 2020; Gogoi, 2011; Ibijbijen et al., 1996; Kim et al., 2017; Nakmee et al., 2016). In modern agriculture the application of synthetic microbial consortia (SMC) is an emerging technology for promoting plant growth, protection against abiotic stress, nutrient cycling, and disease suppression (Cirillo et al., 2023; Dubey et al., 2025; Wasudeo et al., 2026). The principles of SMC design bring together features such as metabolic complementarity, microbial resiliency, host compatibility, and ecological viability (Dubey et al., 2025; Wasudeo et al., 2026).
- **Managing CMNs in natural and agricultural systems:** How can we manage land-use practices (e.g., tillage, crop rotation, and cover cropping) to preserve and promote beneficial CMNs, thereby enhancing ecosystem resilience and resource use efficiency (Van Der Heijden et al., 1998). A key step is to move beyond controlled experiments and test the efficacy of selected, often ubiquitous, AMF strains in real-world, large-scale restoration, and agricultural settings

across different biogeographical zones (Diagne et al., 2020; Winagraski et al., 2019). The application of this knowledge in real environments and according to biogeographical zones becomes essential to promote their industrial production on a large scale and increase their impact (Diagne et al., 2020). Advances in molecular biotechnologies have provided a mechanistic understanding of the plant–microbe symbiotic interactions, highlighting significant opportunities for genetic engineering (Ku et al., 2024; Wasudeo et al., 2026). The cutting-edge gene-editing techniques like CRISPR/Cas9 have further led to the manipulation of genes associated with plant–microbe symbiosis, such as NFR1, DMI1, SYM, CCaMK, and RAM1, thereby improving recognition, intracellular colonization, and signaling (Fu et al., 2022; Ho-Plagaro & Garcia-Garrido, 2022; Wasudeo et al., 2026).

1.10 Chapter summary and conclusion

1.10.1 Summary

This chapter has provided a comprehensive overview of the ancient and essential symbiosis between AMF and the majority of terrestrial plants. We began by defining symbiosis and establishing its fundamental importance for plant health, ecosystem function, and the history of life on land. We explored the biology of the fungal partners—the Glomeromycota—their characteristic structures, and their unique lifestyle as obligate biotrophs (Schüßler et al., 2001). AMF play a crucial role in plant nutrition and growth in stressed conditions and enhance many essential ecosystem processes (Diagne et al., 2020).

The establishment of this intimate relationship is a finely choreographed molecular dialogue involving plant strigolactones and fungal Myc factors, culminating in the formation of the arbuscule, the dedicated hub for nutrient exchange (Ruiz-Lozano et al., 2015). At this interface a core marketplace operates: the plant supplies up to 20% of its photosynthetically fixed carbon, including lipids and sugars, in exchange for essential mineral nutrients—primarily phosphorus and nitrogen—and water, acquired by the extensive extraradical hyphal network (Smith & Read, 2008). Central to both rhizobial and mycorrhizal associations is the CSSP, a conserved, modular network that integrates microbial signal perception with host transcriptional reprogramming (Delaux & Gutjahr, 2024; Wasudeo et al., 2026).

The benefits of this partnership extend beyond simple nutrition, bestowing upon the plant a state of enhanced health and resilience against a spectrum of abiotic stresses (drought, salinity, and heavy metals) and biotic stresses (pathogens and nematodes) through mechanisms involving hormonal regulation, osmotic adjustment, and immune priming (Diagne et al., 2020; Pozo & Azcón-Aguilar, 2007). At the ecosystem level, AMF act as keystone organisms, shaping plant community composition, improving soil structure via glomalin production, and forming underground CMNs—the "wood wide web"—that facilitates interplant communication and resource sharing (Gorzelak et al., 2015; Lehmann et al., 2014; Van Der Heijden et al., 1998). They are also key players in carbon sequestration (Árevalo et al., 2025).

Modern science has developed a powerful toolkit, from traditional spore morphology to cutting-edge DNA metabarcoding and stable isotope tracing, to unravel the secrets of this hidden world. Recent developments in high-throughput sequencing technologies have enhanced our ability to explore AMF diversity, phylogeny, community composition, and functional traits (Árevalo et al., 2025; Tedersoo et al., 2010). This knowledge is now being translated into applied strategies, harnessing the symbiosis for sustainable agriculture as a natural biofertilizer and for the ecological restoration of degraded lands, where AMF are pivotal in rebuilding plant communities and soil health (Asmelash et al., 2016; Berruti et al., 2016; MacColl et al., 2024). Field experiments have demonstrated that AMF inoculation can significantly accelerate the restoration process of degraded grasslands by improving plant biomass, density, cover, and diversity (Zhang et al., 2012). Understanding how AMF communities respond to disturbance—whether through resistance, resilience, or functional redundancy—is critical for predicting restoration outcomes (Amalia et al., 2021). The identification of ubiquitous AMF species offers a promising path for developing effective and broadly applicable inoculants (Winagraski et al., 2019). Bibliometric analysis reveals that scientific research on AMF in ecosystem restoration has grown annually, with key topics including phytoremediation, heavy metals, bioremediation, and mycorrhiza (Árevalo et al., 2025; Saeed et al., 2025).

1.10.2 Recapitulation of basic principles

The AMF symbiosis is fundamentally governed by a few core principles:

1. **Reciprocity:** It is a mutually beneficial exchange based on carbon for nutrients. Reciprocal rewarding and sanctioning mechanisms for nutrient-to-carbon exchange have evolved in AM systems (Tedersoo et al., 2018).
2. **Obligacy:** The fungus is completely dependent on the plant host.
3. **Universality:** It is widespread, associated with the vast majority of land plants.

4. **Functionality:** It provides a suite of benefits, from improved nutrition to stress tolerance and soil building.
5. **Specificity and context-dependency:** The outcomes are context-dependent, varying with plant and fungal genotypes and, critically, with environmental conditions and disturbance regimes (Amalia et al., 2021; Winagraski et al., 2019). The extent to which a host plant benefits varies with the AMF species used (Diagne et al., 2020).

1.10.3 Emphasis on the integrative importance of this topic

The study of AMF is not an isolated subdiscipline of mycology or plant biology. It is an inherently integrative science that sits at the nexus of plant physiology, soil science, ecology, evolution, genomics, and agronomy. The partnership between plants and microbiota serves as a natural and sustainable way for increasing crop production, thus indirectly helping global food security in an ecofriendly manner (Ortiz & Sansinenea, 2022; Ullah et al., 2025; Wasudeo et al., 2026). Understanding this symbiosis is crucial for addressing some of the most pressing challenges of the 21st century, including food security, sustainable agriculture, and the restoration of degraded ecosystems in the face of global climate change. The partnerships between plants and microbes could lead to exciting transformations for agriculture that are both sustainable and resilient to climate challenges (Ge & Wang, 2025; Wasudeo et al., 2026). As we continue to uncover the mysteries of this ancient partnership, we open new frontiers for harnessing its power to build more resilient and sustainable ecosystems.

References

Achatz, M., Morris, E. K., Müller, F., Hilker, M., & Rillig, M. C. (2014). Soil hypha-mediated movement of allelochemicals: Arbuscular mycorrhizae extend the bioactive zone of juglone. *Functional Ecology, 28*, 1020–1029.

Agerer, R. (2001). Exploration types of ectomycorrhizae. *Mycorrhiza, 11*, 107–114.

Ahmed, N., Li, J., Li, Y., Deng, L., Deng, L., Chachar, M., & Umrani, J. H. (2025). Symbiotic synergy: How arbuscular mycorrhizal fungi enhance nutrient uptake, stress tolerance, and soil health through molecular mechanisms and hormonal regulation. *IMA Fungus, 16*, e144989.

Allen, J. W., & Shachar-Hill, Y. (2008). Sulfur transfer through an arbuscular mycorrhiza. *Plant Physiology, 149*, 549–560.

Allen, M. F. (1982). Influence of vesicular-arbuscular mycorrhizae on water movement through *Bouteloua gracilis* (HBK) Lag ex Steud. *New Phytologist, 91*(2), 191–196.

Allen, M. F. (2011). Linking water and nutrients through the vadose zone: A fungal interface between the soil and plant systems. *Journal of Arid Land, 3*, 155–163.

Aloo, B. N., Desseuault-Rompré, J., Tripathi, V., Nyongesa, B. O., & Were, B. A. (2023). Signaling and crosstalk of rhizobacteria and plant hormones that mediate abiotic stress tolerance in plants. *Frontiers in Microbiology, 14*, 1171104.

Alzate Zuluaga, M. Y., Fattorini, R., Cesco, S., & Pii, Y. (2024). Plant-microbe interactions in the rhizosphere for smarter and more sustainable crop fertilization: The case of PGPR-based biofertilizers. *Frontiers in Microbiology, 15*, 1440978.

Amalia, A. V., Dewi, N. R., Heiryanti, A. P., Daeni, F., & Atunnisa, R. (2021). A review of arbuscular mycorrhizal fungal communities in response to disturbance. *Journal of Physics: Conference Series, 1968*(1), 012001.

Arévalo, Y., Avila-Salem, M. E., Loján, P., Urgiles-Gómez, N., Pucha-Cofrep, D., Aguirre, N., & Benavidez-Silva, C. (2025). Arbuscular mycorrhizal fungi in the ecological restoration of tropical forests: A bibliometric review. *Forests, 16*(8), 1266.

Asmelash, F., Bekele, T., & Birhane, E. (2016). The potential role of arbuscular mycorrhizal fungi in the restoration of degraded lands. *Frontiers in Microbiology, 7*, 1095.

Barto, E. K., Hilker, M., Müller, F., Mohney, B. K., Weidenhamer, J. D., & Rillig, M. C. (2011). The fungal fast lane: Common mycorrhizal networks extend bioactive zones of allelochemicals in soils. *PLoS ONE, 6*, e27195.

Bernardo, L., Carletti, P., Badeck, F., Rizza, F., Morcia, C., Ghizzoni, R., & Lucini, L. (2019). Metabolomic responses triggered by arbuscular mycorrhiza enhance tolerance to water stress in wheat cultivars. *Plant Physiology and Biochemistry, 137*, 203–212.

Berruti, A., Lumini, E., Balestrini, R., & Bianciotto, V. (2016). Arbuscular mycorrhizal fungi as natural biofertilizers: Let's benefit from past successes. *Frontiers in Microbiology, 6*, 1559.

Bona, E., Cantamessa, S., Massa, N., Manassero, P., Marsano, F., Copetta, A., & Berta, G. (2016). Arbuscular mycorrhizal fungi and plant growth-promoting pseudomonads improve yield, quality, and nutritional value of tomato: A field study. *Mycorrhiza, 27*, 1–11.

Bunn, R. A., Ramsey, P. W., & Lekberg, Y. (2015). Do native and invasive plants differ in their interactions with arbuscular mycorrhizal fungi? A meta-analysis. *Journal of Ecology, 103*, 1547–1556.

Burleigh, S. H., & Bechmann, I. E. (2002). Plant nutrient transporter regulation in arbuscular mycorrhizas. *Plant and Soil, 244*, 247–251.

Casieri, L., Lahmidi, N. A., Doidy, J., Veneault-Fourrey, C., Migeon, A., Bonneau, L., & Brun, A. (2013). Biotrophic transportome in mutualistic plant-fungal interactions. *Mycorrhiza, 23*, 597–625.

Chen, W., Koide, R. T., & Eissenstat, D. M. (2018). Root morphology and mycorrhizal type strongly influence root production in nutrient hot spots of mixed forests. *Journal of Ecology, 106*, 148–156.

Cirillo, V., Romano, I., Woo, S. L., Di Stasio, E., Lombardi, N., Comite, E., & Maggio, A. (2023). Inoculation with a microbial consortium increases soil microbial diversity and improves agronomic traits of tomato under water and nitrogen deficiency. *Frontiers in Plant Science, 14*, 1304627.

Clark, R., & Zeto, S. (2000). Mineral acquisition by arbuscular mycorrhizal plants. *Journal of Plant Nutrition, 23*, 867–902.

Craine, J. M., Elmore, A. J., Aidar, M. P., Bustamante, M., Dawson, T. E., Hobbie, E. A., & Wright, I. J. (2009). Global patterns of foliar nitrogen isotopes and their relationships with climate, mycorrhizal fungi, foliar nutrient concentrations, and nitrogen availability. *New Phytologist, 183*, 980–992.

De Ollas, C., & Dodd, I. C. (2016). Physiological impacts of ABA-JA interactions under water-limitation. *Plant Molecular Biology, 91*, 641–650.

Delaux, P. M., & Gutjahr, C. (2024). Evolution of small molecule-mediated regulation of arbuscular mycorrhiza symbiosis. *Philosophical Transactions of the Royal Society B: Biological Sciences, 379*(1914), 20230369.

Diagne, N., Ngom, M., Djighaly, P. I., Fall, D., Hocher, V., & Svistoonoff, S. (2020). Roles of arbuscular mycorrhizal fungi on plant growth and performance: Importance in biotic and abiotic stress regulation. *Diversity, 12*(10), 370.

Dissanayaka, N. S., Udumann, S. S., Nuwarapaksha, T. D., & Atapattu, A. J. (2025). Microbial partnerships in agriculture: Boosting crop health and productivity. *Circular Agricultural Systems, 5*, e013.

Dubey, A., Malla, M. A., Vimal, S. R., Kumar, A., Prasad, S. M., & Khan, M. L. (2025). Plant-microbiome engineering: Synergistic microbial partners for crop health and sustainability. *Plant Growth Regulation, 105*, 1825–1839.

Evelin, H., Kapoor, R., & Giri, B. (2009). Arbuscular mycorrhizal fungi in alleviation of salt stress: A review. *Annals of Botany, 104*, 1263–1280.

Faber, B. A., Zasoski, R. J., Munns, D. N., & Shackel, K. (1991). A method for measuring hyphal nutrient and water uptake in mycorrhizal plants. *Canadian Journal of Botany, 69*, 87–94.

Figueiredo, A., Boy, J., & Guggenberger, G. (2021). Common mycorrhizae network: A review of the theories and mechanisms behind underground interactions. *Frontiers in Fungal Biology, 2*, 735299.

Fitzpatrick, C. R., Salas-Gonzalez, I., Conway, J. M., Finkel, O. M., Gilbert, S., Russ, D., & Dangl, J. L. (2020). The plant microbiome: From ecology to reductionism and beyond. *Annual Review of Microbiology, 74*, 81–100.

Fu, M., Sun, J., Li, X., Guan, Y., & Xie, F. (2022). Asymmetric redundancy of soybean Nodule Inception (NIN) genes in root nodule symbiosis. *Plant Physiology, 188*, 477–489.

Fukuda, H., Mamiya, R., Akamatsu, A., & Takeda, N. (2024). Two LysM receptor-like kinases regulate arbuscular mycorrhiza through distinct signaling pathways in *Lotus japonicus*. *New Phytologist, 243*, 519–525.

Ge, A. H., & Wang, E. (2025). Exploring the plant microbiome: A pathway to climate-smart crops. *Cell, 188*, 1469–1485.

Genre, A., & Russo, G. (2016). Does a common pathway transduce symbiotic signals in plant-microbe interactions? *Frontiers in Plant Science, 7*, 96.

Giovannetti, M., Tolosano, M., Volpe, V., Kopriva, S., & Bonfante, P. (2014). Identification and functional characterization of a sulfate transporter induced by both sulfur starvation and mycorrhiza formation in *Lotus japonicus*. *New Phytologist, 204*, 609–619.

Gogoi, P. (2011). Differential effect of some arbuscular mycorrhizal fungi on the growth of *Piper longum* L. (Piperaceae). *Indian Journal of Science and Technology, 4*, 119–125.

Golubkina, N. A., Krivenkov, L., Sekara, A., Vasileva, V., Tallarita, A., & Caruso, G. (2020). Prospects of arbuscular mycorrhizal fungi utilization in the production of *Allium* plants. *Plants, 9*, 279.

Gorzelak, M. A., Asay, A. K., Pickles, B., & Simard, S. W. (2015). Inter-plant communication through mycorrhizal networks mediates complex adaptive behaviour in plant communities. *AoB Plants, 7*, plv044.

Gou, X., Hu, Y., Ni, H., Wang, X., Qiu, L., Chang, X., & Wei, X. (2024). Arbuscular mycorrhizal fungi alleviate erosional soil nitrogen loss by regulating nitrogen cycling genes and enzymes in experimental agro-ecosystems. *Science of the Total Environment, 906*, 167425.

Hajiboland, R., Joudmand, A., Aliasgharzad, N., Tolrá, R., & Poschenrieder, C. (2019). Arbuscular mycorrhizal fungi alleviate low-temperature stress and increase freezing resistance as a substitute for acclimation treatment in barley. *Crop & Pasture Science, 70*(3), 218–233.

Hanif, M. S., Tayyab, M., Baillo, E. H., Islam, M. M., Islam, W., & Li, X. (2024). Plant microbiome technology for sustainable agriculture. *Frontiers in Microbiology, 15*, 1500260.

Hobbie, E. A., & Högberg, P. (2012). Nitrogen isotopes link mycorrhizal fungi and plants to nitrogen dynamics. *New Phytologist, 196*, 367–382.

Ho-Plagaro, T., & Garcia-Garrido, J. M. (2022). Molecular regulation of arbuscular mycorrhizal symbiosis. *International Journal of Molecular Sciences, 23*, 5960.

Ibijbijen, J., Urquiaga, S., Ismaili, M., Alves, B. J. R., & Boddey, R. M. (1996). Effect of arbuscular mycorrhizal fungi on growth, mineral nutrition, and nitrogen fixation of three varieties of common beans (*Phaseolus vulgaris*). *New Phytologist, 134*, 353–360.

Johns, C. D. (2014). *Agricultural application of mycorrhizal fungi to increase crop yields, promote soil health, and combat climate change*. Future Directions International.

Kardol, P., Cornips, N., Van Kempen, M., Bakx-Schotman, J., & Van Der Putten, W. (2007). Microbe-mediated plant-soil feedback causes historical contingency effects in plant community assembly. *Ecological Monographs, 77*, 147–162.

Kaya, C., Higgs, D., Kirnak, H., & Taş, I. (2003). Mycorrhizal colonisation improves fruit yield and water use efficiency in watermelon (*Citrullus lanatus* Thunb.) grown under well-watered and water-stressed conditions. *Plant and Soil, 253*, 287–292.

Kennedy, P. G., Smith, D. P., Horton, T. R., & Molina, R. J. (2012). *Arbutus menziesii* (Ericaceae) facilitates regeneration dynamics in mixed evergreen forests by promoting mycorrhizal fungal diversity and host connectivity. *American Journal of Botany, 99*, 1691–1701.

Kim, S. J., Eo, J.-K., Lee, E.-H., Park, H., & Eom, A.-H. (2017). Effects of arbuscular mycorrhizal fungi and soil conditions on crop plant growth. *Mycobiology, 45*, 20–24.

Koide, R. T., & Mosse, B. (2004). A history of research on arbuscular mycorrhiza. *Mycorrhiza, 14*, 145–163.

Koziol, L., & Bever, J. D. (2017). The missing link in grassland restoration: Arbuscular mycorrhizal fungi inoculation increases plant diversity and accelerates succession. *Journal of Applied Ecology, 54*(5), 1301–1309.

Koziol, L., Schultz, P. A., House, G. L., Bauer, J. T., Middleton, E. L., & Bever, J. D. (2018). The plant microbiome and native plant restoration: The example of native mycorrhizal fungi. *Bioscience, 68*, 996–1006.

Ku, Y. S., Liao, Y. J., Chiou, S. P., Lam, H. M., & Chan, C. (2024). From trade-off to synergy: Microbial insights into enhancing plant growth and immunity. *Plant Biotechnology Journal, 22*, 2461–2471.

Kuyper, T. W., & Jansa, J. (2023). Arbuscular mycorrhiza: Advances and retreats in our understanding of the ecological functioning of the mother of all root symbioses. *Plant and Soil, 489*, 41–88.

Latef, A. A. H. A., & Chaoxing, H. (2011). Effect of arbuscular mycorrhizal fungi on growth, mineral nutrition, antioxidant enzymes activity, and fruit yield of tomato grown under salinity stress. *Scientia Horticulturae, 127*, 228–233.

Lehmann, A., Veresoglou, S. D., Leifheit, E. F., & Rillig, M. C. (2014). Arbuscular mycorrhizal influence on zinc nutrition in crop plants—A meta-analysis. *Soil Biology and Biochemistry, 69*, 123–131.

Lekberg, Y., Gibbons, S. M., Rosendahl, S., & Ramsey, P. W. (2013). Severe plant invasions can increase mycorrhizal fungal abundance and diversity. *ISME Journal, 7*, 1424–1433.

Lenoir, I., Fontaine, J., & Sahraoui, A. L.-H. (2016). Arbuscular mycorrhizal fungal responses to abiotic stresses: A review. *Phytochemistry, 123*, 4–15.

Li, Z., Wu, N., Meng, S., Wu, F., & Liu, T. (2020). Arbuscular mycorrhizal fungi (AMF) enhance the tolerance of *Euonymus maackii* Rupr. at a moderate level of salinity. *PLoS ONE, 15*, e0231497.

MacColl, K. A., Tosi, M., Chagnon, P. L., MacDougall, A. S., Dunfield, K. E., & Maherali, H. (2024). Prairie restoration promotes the abundance and diversity of mutualistic arbuscular mycorrhizal fungi. *Ecological Applications, 34*(5), e2981.

Maherali, H., & Klironomos, J. N. (2007). Influence of phylogeny on fungal community assembly and ecosystem functioning. *Science, 316*, 1746–1748.

Mangan, S. A., & Adler, G. H. (2002). Seasonal dispersal of arbuscular mycorrhizal fungi by spiny rats in a Neotropical forest. *Oecologia, 131*, 587–597.

Marinho, F., Da Silva, I., Oehl, F., & Maia, L. (2018). Checklist of arbuscular mycorrhizal fungi in tropical forests. *Sydowia, 70*, 107–127.

Mengistu, A. A. (2020). Endophytes: Colonization, behaviour, and their role in defense mechanisms. *International Journal of Microbiology, 2020*, 6927219.

Millar, N. S., & Bennett, A. E. (2016). Stressed out symbionts: Hypotheses for the influence of abiotic stress on arbuscular mycorrhizal fungi. *Oecologia, 182*, 625–641.

Mostofa, M. G., Li, W., Nguyen, K. H., Fujita, M., & Tran, L.-S. P. (2018). Strigolactones in plant adaptation to abiotic stresses: An emerging avenue of plant research. *Plant, Cell & Environment, 41*, 2227–2243.

Nakmee, P. S., Techapinyawat, S., & Ngamprasit, S. (2016). Comparative potentials of native arbuscular mycorrhizal fungi to improve nutrient uptake and biomass of Sorghum bicolor Linn. *Agriculture and Natural Resources, 50*, 173–178.

Nazir, A., Sarfraz, W., Ditta, A., Khalid, N., Farid, M., Shafiq, M., Firdaus-e-Bareen, Rizvi, Z. F., & Naeem, N. (2023). Synergistic impact of two autochthonous saprobic fungi (*A. niger* and *T. pseudokoningii*) on the growth, ionic contents, and metals uptake in *Brassica juncea* L. and *Vigna radiata* L. under tannery solid waste contaminated soil. *International Journal of Phytoremediation, 25*(11), 1488–1500.

Nehls, U., & Plassard, C. (2018). Nitrogen and phosphate metabolism in ectomycorrhizas. *New Phytologist, 220*(4), 1047–1058.

Neuenkamp, L., Prober, S. M., Price, J. N., Zobel, M., & Standish, R. J. (2019). The benefits of mycorrhizal inoculation to ecological restoration depend on plant functional type, restoration context, and time. *Fungal Ecology, 40*, 140–149.

Nguvo, K. J., & Gao, X. (2019). Weapons hidden underneath: Bio-control agents and their potentials to activate plant-induced systemic resistance in controlling crop *Fusarium* diseases. *Journal of Plant Disease Protection, 126*(3), 177–190.

Nguyen, T. D., Cavagnaro, T. R., & Watts-Williams, S. J. (2019). The effects of soil phosphorus and zinc availability on plant responses to mycorrhizal fungi: A physiological and molecular assessment. *Scientific Reports, 9*, 14880.

Nikalje, G. C., Yadav, K., Vashisht, M., Srivastava, S., Shelake, R. M., Kadam, U. S., & Suprasanna, P. (2025). Synergistic activity of endophytic phytobiome enhances stress tolerance and secondary metabolite production in plants. *Plant Stress, 16*, 101140.

Ohsowski, B. M., Klironomos, J. N., Dunfield, K. E., & Hart, M. M. (2012). The potential of soil amendments for restoring severely disturbed grasslands. *Applied Soil Ecology, 60*, 77–83.

Olsson, P. A., Jakobsen, I., & Wallander, H. (2002). Foraging and resource allocation strategies of mycorrhizal fungi in a patchy environment. In M. G. A. van der Heijden, & I. R. Sanders (Eds.), *Mycorrhizal ecology* (pp. 93–115). Springer.

Orchard, S., Standish, R. J., Dickie, I. A., Renton, M., Walker, C., Moot, D., & Ryan, M. H. (2017). Fine root endophytes under scrutiny: A review of the literature on arbuscule-producing fungi recently suggested to belong to the Mucoromycotina. *Mycorrhiza, 27*, 619–638.

Ortiz, C., & Sansinenea, E. (2022). The role of beneficial microorganisms in soil quality and plant health. *Sustainability, 14*, 5358.

Ortiz, N., Armada, E., Duque, E., Roldan, A., & Azcón, R. (2015). Contribution of arbuscular mycorrhizal fungi and/or bacteria to enhancing plant drought tolerance under natural soil conditions: Effectiveness of autochthonous or allochthonous strains. *Journal of Plant Physiology, 174*, 87–96.

Ouledali, S., Ennajeh, M., Ferrandino, A., Khemira, H., Schubert, A., & Secchi, F. (2019). Influence of arbuscular mycorrhizal fungi inoculation on the control of stomata functioning by abscisic acid (ABA) in drought-stressed olive plants. *South African Journal of Botany, 121*, 152–158.

Pagano, M. C., & Gupta, V. K. (2016). Overview of the recent advances in mycorrhizal fungi. In M. C. Pagano (Ed.), *Recent advances on mycorrhizal fungi* (pp. 8–17). Springer.

Parniske, M. (2008). Arbuscular mycorrhiza: The mother of plant root endosymbioses. *Nature Reviews Microbiology, 6*, 763–775.

Pellegrino, E., & Bedini, S. (2014). Enhancing ecosystem services in sustainable agriculture: Biofertilization and biofortification of chickpea (*Cicer arietinum* L.) by arbuscular mycorrhizal fungi. *Soil Biology and Biochemistry, 68*, 429–439.

Posta, K., & Duc, N. H. (2020). Benefits of arbuscular mycorrhizal fungi application to crop production under water scarcity. *Drought detection and solutions*. IntechOpen.

Pozo, M. J., & Azcón-Aguilar, C. (2007). Unraveling mycorrhiza-induced resistance. *Current Opinion in Plant Biology, 10*(4), 393–398.

Pozo, M. J., López-Ráez, J. A., Azcón-Aguilar, C., & García-Garrido, J. M. (2010). Phytohormones as integrators of environmental signals in the regulation of mycorrhizal symbioses. *New Phytologist, 185*(1), 68–79.

Reinhart, K., & Callaway, R. (2006). Soil biota and invasive plants. *New Phytologist, 170*, 445–457.

Richter, B. S., & Stutz, J. C. (2002). Mycorrhizal inoculation of big sacaton: Implications for grassland restoration of abandoned agricultural fields. *Restoration Ecology, 10*, 607–616.

Rillig, M. C., Wright, S. F., Nichols, K. A., Schmidt, W. F., & Torn, M. S. (2001). Large contribution of arbuscular mycorrhizal fungi to soil carbon pools in tropical forest soils. *Plant and Soil, 233*, 167–177.

Rodríguez-León, C. H., Peña-Venegas, C. P., Sterling, A., Muñoz-Ramirez, H., & Virguez-Díaz, Y. R. (2021). Changes in soil-borne communities of arbuscular mycorrhizal fungi during natural regrowth of abandoned cattle pastures are indicative of ecosystem restoration. *Agronomy, 11*(12), 2468.

Ruiz-Lozano, J. M., Aroca, R., Zamarreño, Á. M., Molina, S., Andreo-Jimenez, B., Porcel, R., & López-Ráez, J. A. (2015). Arbuscular mycorrhizal symbiosis induces strigolactone biosynthesis under drought and improves drought tolerance in lettuce and tomato. *Plant, Cell & Environment, 39*, 441–452.

Ruth, B., Khalvati, M., & Schmidhalter, U. (2011). Quantification of mycorrhizal water uptake via high-resolution online water content sensors. *Plant and Soil, 342*, 459–468.

Ryszka, P., & Turnau, K. (2007). Arbuscular mycorrhiza of introduced and native grasses colonizing zinc wastes: Implications for restoration practices. *Plant and Soil, 298*, 219–229.

Saeed, H. S., Munir, M., Samra, S., Farooq, A., Mahmood, H., Ahmad, I., Akram, M. I., Iqbal, R., Elsadek, M. F., Almutairi, S. M., Ditta, A., & Daraz, U. (2025). Screening of biosurfactants producing fungi from infected *Citrus sinensis* (L.). *Polish Journal of Environmental Studies, 34*(5), 5261–5270.

Santander, C., Sanhueza, M., Olave, J., Borie, F., Valentine, A., & Cornejo, P. (2019). Arbuscular mycorrhizal colonization promotes the tolerance to salt stress in lettuce plants through an efficient modification of ionic balance. *Journal of Soil Science and Plant Nutrition, 19*, 321–331.

Santoyo, G., Orozco-Mosqueda, M. D. C., Afridi, M. S., Mitra, D., Valencia-Cantero, E., & Macías-Rodríguez, L. (2024). *Trichoderma* and *Bacillus* multifunctional allies for plant growth and health in saline soils: Recent advances and future challenges. *Frontiers in Microbiology, 15*, 1423980.

Schultz, P. A., Miller, R. M., Jastrow, J. D., Rivetta, C. V., & Bever, J. D. (2001). Evidence of a mycorrhizal mechanism for the adaptation of *Andropogon gerardii* (Poaceae) to high- and low-nutrient prairies. *American Journal of Botany, 88*, 1650–1656.

Schüßler, A., Schwarzott, D., & Walker, C. (2001). A new fungal phylum, the Glomeromycota: Phylogeny and evolution. *Mycological Research, 105*, 1413–1421.

Serrano, K., Tedeschi, F., Andersen, S. U., & Scheller, H. V. (2024). Unraveling plant-microbe symbioses using single-cell and spatial transcriptomics. *Trends in Plant Science, 29*, 1356–1367.

Sharma, A., Das, N., Pandey, P., & Shukla, P. (2025). Plant-microbiome responses under drought stress and their metabolite-mediated interactions towards enhanced crop resilience. *Current Plant Biology, 3*, 100513.

Sharma, S. B., Raverkar, K. P., Wani, S. P., Bagyaraj, D. J., Kannapalli, A., Kandula, D. R. W., & Sharma, R. (2023). Role of the plant-microbiome partnership in environmentally harmonious 21st century agriculture. *Microorganisms, 13*, 2839.

Siddiqui, Z. A., & Pichtel, J. (2008). Mycorrhizae: An overview. In Z. A. Siddiqui, M. S. Akhtar, & K. Futai (Eds.), *Mycorrhizae: Sustainable agriculture and forestry* (pp. 1–35). Springer.

Sieh, D., Watanabe, M., Devers, E. A., Brueckner, F., Hoefgen, R., & Krajinski, F. (2012). The arbuscular mycorrhizal symbiosis influences sulfur starvation responses of *Medicago truncatula*. *New Phytologist, 197*, 606–616.

Smith, F. A., & Smith, S. E. (2011). Roles of arbuscular mycorrhizas in plant nutrition and growth: New paradigms from cellular to ecosystem scales. *Annual Review of Plant Biology, 62*, 227–250.

Smith, M. R., Charvat, I., & Jacobson, R. L. (1998). Arbuscular mycorrhizae promote the establishment of prairie species in a tallgrass prairie restoration. *Canadian Journal of Botany, 76*, 1947–1954.

Smith, S. E., & Read, D. J. (2008). *Mycorrhizal symbiosis* (3rd ed.). Elsevier Academic Press.

Spatafora, J. W., Chang, Y., Benny, G. L., Lazarus, K., Smith, M. E., Berbee, M. L., & Grigoriev, I. V. (2016). A phylum-level phylogenetic classification of zygomycete fungi based on genome-scale data. *Mycologia, 108*(5), 1028–1046.

Strullu-Derrien, C., Selosse, M. A., Kenrick, P., & Martin, F. M. (2018). The origin and evolution of mycorrhizal symbioses: From palaeomycology to phylogenomics. *New Phytologist, 220*, 1012–1030.

Talaat, N. B., & Shawky, B. T. (2014). Protective effects of arbuscular mycorrhizal fungi on wheat (*Triticum aestivum* L.) plants exposed to salinity. *Environmental and Experimental Botany, 98*, 20–31.

Tedersoo, L., Nilsson, R. H., Abarenkov, K., Jairus, T., Sadam, A., Saar, I., & Kõljalg, U. (2010). 454 Pyrosequencing and Sanger sequencing of tropical mycorrhizal fungi provide similar results but reveal substantial methodological biases. *New Phytologist, 188*, 291–301.

Tedersoo, L., Sánchez-Ramírez, S., Kõljalg, U., Bahram, M., Döring, M., Schigel, D., & Abarenkov, K. (2018). High-level classification of the Fungi and a tool for evolutionary ecological analyses. *Fungal Diversity, 90*(1), 135–159.

Thind, S., Chaudhary, M. S., Ditta, A., Hussain, I., Parveen, A., Ullah, N., Mahmood, Q., Al-ashkar, I., & Sabagh, A. E. L. (2022). Impact of mycorrhizal fungi from different rhizospheric soils on fungal colonization, growth, and chlorophyll contents of *Cenchrus ciliaris*. *Agronomy, 12*(11), 2644.

Tonietto, R. K., & Larkin, D. J. (2018). Habitat restoration benefits wild bees: A meta-analysis. *Journal of Applied Ecology, 55*, 582–590.

Trouvelot, S., Bonneau, L., Redecker, D., Van Tuinen, D., Adrian, M., & Wipf, D. (2015). Arbuscular mycorrhiza symbiosis in viticulture: A review. *Agronomy for Sustainable Development, 35*, 1449–1467.

Ullah, F., Ali, S., Siraj, M., Akhtar, M. S., & Zaman, W. (2025). Plant microbiomes alleviate abiotic stress-associated damage in crops and enhance climate-resilient agriculture. *Plants, 14*, 1890.

Valverde-Barrantes, O. J., Smemo, K. A., Feinstein, L. M., Kershner, M. W., & Blackwood, C. B. (2018). Patterns in spatial distribution and root trait syndromes for ecto- and arbuscular mycorrhizal temperate trees in a mixed broadleaf forest. *Oecologia, 186*, 731–741.

Van Der Heijden, M. G. A., Klironomos, J. N., Ursic, M., Moutoglis, P., Streitwolf-Engel, R., Boller, T., & Sanders, I. R. (1998). Mycorrhizal fungal diversity determines plant biodiversity, ecosystem variability, and productivity. *Nature, 396*, 69–72.

Verma, A., Shameem, N., Jatav, H. S., Sathyanarayana, E., Parray, J. A., Pocazi, P., & Sayyed, R. Z. (2022). Fungal endophytes to combat biotic and abiotic stresses for climate-smart and sustainable agriculture. *Frontiers in Plant Science, 13*, 953836.

Vos, C., Tesfahun, A., Panis, B., De Waele, D., & Elsen, A. (2012). Arbuscular mycorrhizal fungi induce systemic resistance in tomato against the sedentary nematode *Meloidogyne incognita* and the migratory nematode *Pratylenchus penetrans*. *Applied Soil Ecology, 61*, 1–6.

Wang, D., Dong, W., Murray, J., & Wang, E. (2022). Innovation and appropriation in mycorrhizal and rhizobial symbioses. *Plant Cell, 34*, 1573–1599.

Wang, F. (2017). Occurrence of arbuscular mycorrhizal fungi in mining-impacted sites and their contribution to ecological restoration: Mechanisms and applications. *Critical Reviews in Environmental Science and Technology, 47*(20), 1901–1957.

Wang, Y.-H., Wang, M., Li, Y., Wu, A., & Huang, J. (2018). Effects of arbuscular mycorrhizal fungi on growth and nitrogen uptake of *Chrysanthemum morifolium* under salt stress. *PLoS ONE, 13*, e0196408.

Wasudeo, N. G., Kumar, G. H., Kumar, M. A., & Suprasanna, P. (2026). Plant-microbial symbiosis: Molecular insights and applications in sustainable agriculture. *Current Plant Biology*, 100587.

Watson-Guido, W., & Rivera-Mendez, W. (2024). Comunicación en las asociaciones simbióticas: Mecanismos entre hongos micorrízicos arbusculares, plantas y organismos edáficos. *Agronomía Mesoamericana, 35*, 1–16.

Winagraski, E., Kaschuk, G., Monteiro, P. H. R., Auer, C. G., & Higa, A. R. (2019). Diversity of arbuscular mycorrhizal fungi in forest ecosystems of Brazil: A review. *CERNE, 25*(1), 25–35.

Yadav, V. K., Kumar, D., Jha, R. K., Bairwa, R. K., Singh, J. P., Kumar, A., & Jayaswal, D. (2023). Mycorrhizae set the stage for plants to produce a higher production of biomolecules and stress-related metabolites: A sustainable alternative to agrochemicals to enhance the quality and yield of beetroot (*Beta vulgaris* L.). *Frontiers in Microbiology, 14*, 1196101.

Yamato, M., Ikeda, S., & Iwase, K. (2008). Community of arbuscular mycorrhizal fungi in a coastal vegetation on Okinawa Island and effect of the isolated fungi on the growth of sorghum under salt-treated conditions. *Mycorrhiza, 18*, 241–249.

Yizhu, L., Imtiaz, M., Ditta, A., Rizwan, M. S., Ashraf, M., Mehmood, S., Aziz, O., Mubeen, F., Ali, M., Elahi, N. N., Ijaz, R., S, Lelel, Shuang, C., & Tu, S. (2020). Response of growth, antioxidant enzymes, and root exudates production towards as stress in *Pteris vittata* and *Astragalus sinicus* colonized by arbuscular mycorrhizal fungi. *Environmental Science and Pollution Research, 27*, 2340–2352.

Zhang, J., Sun, J., Chiu, C. H., Landry, D., Li, K., Wen, J., & Feng, F. (2024). A receptor required for chitin perception facilitates arbuscular mycorrhizal associations and distinguishes root symbiosis from immunity. *Current Biology, 34*(8), 1705–1717.

Zhang, T., Sun, Y., Shi, Z., & Feng, G. (2012). Arbuscular mycorrhizal fungi can accelerate the restoration of degraded spring grassland in Central Asia. *Rangeland Ecology & Management, 65*(4), 426–432.

Chapter 2

Taxonomy, ecology, and evolution of arbuscular mycorrhizal fungi: A key component of soil microbiome

Muhammad Nauman Khan [a,b], Hansa Gul [c], Barkat Ullah [a], Tabassum Yaseen [d], Nasir Assad [e] and Affan Afzal [f]

[a] *Department of Botany, University of Chakwal, Chakwal, Punjab, Pakistan,* [b] *Department of Botany, Islamia College Peshawar, Peshawar, Pakistan,* [c] *Department of Zoology, University of Mianwali, Mianwali, Pakistan,* [d] *Department of Botany, Bacha Khan University, Charsadda, Pakistan,* [e] *Institute of Chemistry, University of Sargodha, Sargodha, Pakistan,* [f] *Department of Microbiology, Faculty of Veterinary and Animal Sciences, The Islamia University of Bahawalpur, Bahawalpur, Pakistan*

2.1 Introduction

Root exudates are used by plants to stabilize the imbalance in the mineral nutrient transporter in the concept of detecting the nutrient available and subsequently the direct provision of nutrients to the demand (Yizhu et al., 2020). The fungal community discovered on the roots influences how the root exudates will be based on their abundance and composition. Root exudation is there at the onset, which controls the communication channel between the plant roots and the associated soil microbes, such as the arbuscular mycorrhizal fungi (AMF). Moreover, the root exudates have the task of being a signal used to trigger the formation of AMF symbiosis (Koshila Ravi & Muthukumar, 2024). AMF (Subphylum: Glomeromycotina) are an old yet species-poor system of root symbionts. The current ability to identify AMF taxa relies mainly on sequencing technology, because many AMF species cannot be cultured. The unified approaches to unraveling AMF taxa and sampling sample material are available and highly applied, which means that the mechanisms and patterns of global diversity distribution can be learned (Ma et al., 2023). Lots of species of AMF have been evaluated throughout the globe; thus the most utilized species as a model are the following: *Funneliformis mosseae* (formerly recognized as *Glomus mosseae*), *Gigaspora rosea, Gigaspora margarita, Gigaspora gigantea*, and *Rhizophagus irregularis* (previously known as *Glomus intraradices* and *Glomus irregulare*). AMF are obligate mutualists rather than parasites, relying entirely on a compatible host plant to fulfill their developmental cycle (Fall et al., 2022).

AMF forms mutual relationships with plants in which both organisms gain benefits. AMF increase plant uptake of water and other essential nutrients, especially phosphorus (P); meanwhile, plants allocate about 10.20% of photosynthetically fixed carbon to the fungi. These endophytic fungi are found in almost 85% of the plant families and play a valuable role in the sustenance of the ecosystem. A lot of stresses, such as biotic and abiotic stressors, have an adverse effect on plant growth and productivity (Khaliq et al., 2022). Nevertheless, the AMF-plant association enhances the flexibility and efficiency of the two partners, making the mutualism concept strong. AMF enhance the absorption of soil nutrients, particularly phosphorus (P), by plants, while the plants, in return, provide carbohydrates to the fungi. Even though in some cases the interchange of the resources might occur asymmetrically, AMF in most cases improves the mineral obtaining, with nutrient competitions being relieved by these among plants (Montesinos-Navarro et al., 2012). Carbon is taken through photosynthesis and shaped into products, which are now obtained by these fungi, and in return, the fungi deposit essential nutrients into the host. Slow and organic fertilizers usually result in enhanced AMF activity within the rhizosphere; thus a variety of synthetic fertilizers have

Advancing Sustainable Agriculture Through Arbuscular Mycorrhizal Fungi. DOI: https://doi.org/10.1016/B978-0-443-45130-0.00038-2

been observed to hinder their establishment and activity. AMF are very important in the phosphorus uptake, thus enhancing the P levels in the plants. Moreover, AMF promote the translocation of inorganic phosphate and other low-mobility nutrients into the roots of plants. Therefore the essence of AMF is the provision of colonized roots with phosphorus (Seleiman & Hardan, 2021).

AMF are symbionts used to alleviate different abiotic stresses experienced by the host plant, such as drought, salinity, exposure to extreme temperatures, deficiency of nutrients, and heavy metal (HM) toxicity (Saeed et al., 2025). The HM-induced stress is relieved by AMF by their secretion of compounds such as glycoproteins (particularly glomalin), globulins, and organic acids in the rhizosphere, thus helping in the chelation of HM. Such secretions lower the bioavailability of HMs in the soil, hence minimizing their absorption by plants (Li et al., 2023). Studies show that AMF are abundant in ultramafic soils and have the capability of building functional symbiosis with almost all plant species within ultramafic ecosystems (Nazir et al., 2023; Thind et al., 2022). The colonization of AMF has also occurred in Ni-hyperaccumulating plants and even the sedges, which have been traditionally regarded as nonmycorrhizal, although on a rare basis in New Caledonia in the ultramafic environments. These properties of fungal communities have adapted to extreme edaphic conditions, preconditioning the rise of metal tolerance, especially to nickel and the development of novel AMF taxa (Amir et al., 2023). It is the expansion of industrial agriculture worldwide that has been blamed for destroying the native ecosystems and the loss of biodiversity. Nevertheless, its effect on microbial diversity and functionality, including AMF, has not been well studied (Albornoz et al., 2022). Agronomically, AMF are everywhere and form part of the basic microbial machinery in all ecosystems, agricultural fields, forests, grasslands, and deserts, fulfilling key roles in the ecosystem. These are improvements in the plant growth, cycling of the nutrients, and stability of the soil structure via an extensive network of symbiosis. The interdependence of AMF and host plants has an important impact on species richness as well as community structure and spatial distribution. Thus, in terms of sustainable agricultural development, AMF and their host plant partners can constitute the key leveraging units toward attaining a better community composition and building a higher native inoculum potential within the local agricultural systems (Liu et al., 2024).

2.2 Soil microbiome context

In the One Health framework, soil is one of the major components that maintains numerous and rich communities of organisms. The holistic knowledge of the diversity and functions of microbes is key in the realization of healthy soil systems to support sustainable farming operations (Zhang et al., 2024). Soil microbiomes are the key to agroecosystem functionality because they mediate soil fertility, crop yield, and plant resilience. The soil structure, especially the formation of aggregates and pore connectivity, shows complex relations with how the microbiome is constructed, thus regulating water, oxygen, and nutrient movement within the soil matrix (Hartmann & Six, 2023). Although mycorrhizal growth response is a standardized statistic used to evaluate the success of AMF inoculation on crop yields, inoculation efficacy remains inconsistent because knowledge is lacking about the success of the inoculation of introduced populations of AMF (Lutz et al., 2023). Acting as symbionts the AMF enter into mutually beneficial relationships with most land plant species. The absorption capacity of roots becomes significantly greater because of the enlargement of the functional zone of absorption with the help of AMF, and the fungi receive the necessary carbohydrates with the help of the host plants to complete their life cycle. Induced complex networks of ecology occur because AMF continually interact with a wide variety of microbial taxa, which include endobacteria, plant growth-promoting rhizobacteria, mycorrhiza helper bacteria, plant parasitic nematodes, saprophytic fungi, and other rhizosphere and hyphosphere organisms (Kumar et al., 2024). Most plant systems also comprise a hyphosphere, which has been defined as the soil area around the extraradical hyphae of AMF. These dedicated interfaces enable the following processes to occur, which have critical roles: better mineral nutrition and water uptake, better biotic and abiotic stress tolerance, faster mineral weathering, formation and stabilization of soil macroaggregates, fixation of carbon, and interaction with complex communities of soil microbes (Wang et al., 2022).

The last decades have observed the growing scientific interest in the hyphosphere of the AMF, which are nonexclusive symbionts of plants. Because AM fungal networks substantially pervade soil matrices (82–111 m cm^{-3} prairie ecosystems and 52–81 m cm^{-3} ungrazed pasture systems) and destabilize 4%–20% of the overall plant-synthesized carbon, such a research focus is essential to investigate the role of AM fungus functioning, especially because of their reduced ability to utilize organic nutrients and, thus, rely on bacterial collaboration in hyphosphere ecosystems (Jin et al., 2024). It is assumed that hyphosphere microbial associations enable AMF to access organic nutrients by decomposing complex organic materials that are not readily broken down by the AMF. On the other hand, AMF fronts all energy-rich microenvironments augmented with hyphal exudates that enhance microbial growth and motility at hyphosphere regions (Faghihinia et al., 2023). Mycorrhizospheres are colonized by various microorganisms such as bacteria, fungi, protozoa, nematodes, and arthropods, among others (Priyadharsini et al., 2017). It has been reported by several researchers that AMF improves the

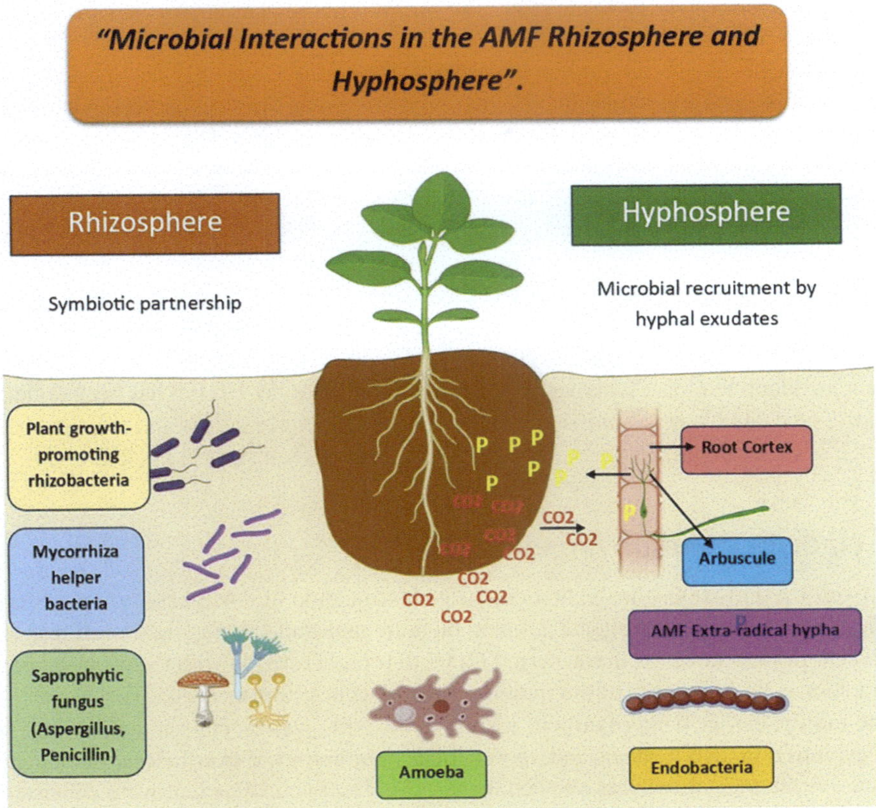

FIGURE 2.1 The diagram illustrates the role of arbuscular mycorrhizal fungi (AMF) in enhancing plant growth through improved nutrient. It highlights key interactions at the root interface, including arbuscule formation and symbiotic exchange between plant and fungus.

development of plants. The advantages of these useful microorganisms may be achieved by using individual applications or consortia of these microbes to maximize the benefits of the plant. In particular the coinoculation of AMF and plant growth-promoting rhizobacteria has proved to have positive effects on different species of crops (Desai et al., 2016) (Fig. 2.1).

2.3 Historical morphology-based taxonomy

AMF (formerly classified in division Zygomycetes, order Glomales) are involved in significant mutualistic relationships with most species of terrestrial plants. Genotypic traits of AMF have been used as the basis of their taxonomic group classification into seven groups with five families. It is, however, unclear whether existing taxonomic schemes—mainly founded on morphological characters—are effective predictors of ecological functioning. New data have indicated the existence of differences in the ecologies of AMF taxa, apparent both in ecosystem-level functionality and on a smaller scale in physiology and morphology (Hart & Reader, 2002). Recent phylogenetic verification has determined AMF as a monophyletic group of the phylum Glomeromycota. The taxonomic and systematic history of these obligate biotrophic organisms has four developmental phases. The first discovery period (1845–1974) was marked mainly by the description and explanation of sporocarp-forming species as well as proposals of a preliminary classification of these fungi. Morphological backbones to identify species and continually classify them were built during the next period of alpha taxonomy (1975–89) and gave rise to numerous species descriptions of previously unknown species but also required codification of nomenclature to subcellular parts of spores (Stürmer, 2012). The species diversity and population dynamics of AMF are made variable by geographic location and rhizosphere conditions. In addition, AMF have heterogeneous morphological and physiological properties, so taxonomic classification is of extreme importance (Susilawati et al., 2022). According to a survey of Indonesian agricultural soils recently done by the authors, there were three AMF isolates identified as MV 5_, MV 17, and MV 18 using morphological and molecular methods. In the morphology analysis, it was observed that sporiferous saccule necks in MV 5 produced spores, as well as the bulbous suspensors of MV 17 and the hyphae that subtended it, MV 18. These *Acaulospora longula* (MV 5), *G. margarita* (MV 17), and *Glomus etunicatum* (MV 18) could be identified by nested-PCR amplification of SSU rRNA fragments. The correspondence of molecular and morphological approaches proved their complementary nature in the AMF taxonomy (Rini et al., 2021).

Classic classification of AMF by spore morphology is multifaceted and frequently incongruent, and in sensitive spore morphology, it requires professional interpretation. Current progress prefers molecular methods, particularly Polymerase Chain Reaction (PCR), of ribosomal-DNA areas (such as Small Subunit (SSU), Large Subunit (LSU), and ITS-Internal Transcribed Spacer) because of the conserved and distinct sequence. These identifiers give greater resolution with regard to differentiating AMF species. Nevertheless, individually, none of the markers can have sufficient discriminatory potential. Thus the merging of several regions of the genome is optimal to adequately identify the taxonomy of organisms (Sedaghati et al., 2021).

Morphological plasticity in spore morphology of AMF, such as size, shape, and color, is under environmental variation, as well as the propagation techniques. As an illustration, *Rhizophagus intraradices* and *R. irregularis* are closely like identical species (so it is difficult to differentiate them) unless rDNA is used to carry out the identification. Analysis of SSU–ITS–LSU regions can help correct misidentifications; for example, *G. intraradices* DAOM 197198 has been reidentified through DNA analysis as *G. irregulare*. One can observe similar fluctuations in *Rhizophagus clarus* and *Scutellospora pellucida*, which accentuates the presence of heritable phenotypic diversity in AMF spores (Kokkoris et al., 2024).

2.4 Advent of cladistics and phylogenetics

The cladistics period from 1990 to 2000 saw the first cladistic classification of AMF based on phenotypic characters only. At the end of this period, genetic characters played a role in defining taxa and elucidating evolutionary relationships within the group. The most recent phylogenetic synthesis period (2001 to present) started with the proposal of a new classification based on genetic characters using sequences of the multicopy rRNA genes (Stürmer, 2012).

The Gigasporaceae family, earlier with Gigaspora and Scutellospora, was recently redefined into several families and genera following the evolution of rRNA genes and spore shield morphology. Nevertheless, phylogenetics of 25S rRNA, beta-tubulin genes, and morphological features showed disparities. The two clades that were consistently supported were *Gigaspora* and *Racocetra*. The majority of proposed taxa were not well supported and, as such, were phylogenetically ambiguous, so far as there was a homoplastic nature of characters and insufficient species sampling. As a result the original ranking of the *Gigasporaceae* was reinstated, comprising *Gigaspora, Racocetra*, and *Scutellospora* (Morton & Msiska, 2010).

Glomeromycota and Endogonomycetes include relatively few described species, and many species and higher taxonomic groups, particularly arbuscular mycorrhizal fungi (AMF), still remain undescribed. A synthesized morphological and molecular (DNA) classification scheme has been suggested to increase taxonomic understanding and compatibility with phylogenetic variables. The study used environmental DNA and long-read sequencing as type material to describe 15 novel species and their genera (e.g., *Pseudoentrophosporakesseensis, Hoforsarebekkae*, and *Kungsaengenashadiae*) as well as the novel genus *Parvocarpum*. It also added 19 new families and 17 orders (e.g., *Hoforsales, Langduoales*, and *Unemaeeales*) and recombined 6 groups of species, based on strong phylogenetic support (Tedersoo et al., 2024b). The polyphyletic nature of the genus *Glomus* has been repeatedly determined through molecular phylogeny studies that used sequences of ribosomal DNA (rDNA). Taxa of AMF should include *Glomus*-type spore producers that are scattered among six distinct evolutionary lineages in the phylum Glomeromycota. Such taxonomic imperfection lends support to the inefficiencies of morphologically based systems of classification in the correct representation of phylogenetic relationships. Therefore molecular solutions have proved necessary in determining the veritable evolutionary relationships between these fungal symbionts (Ouhaddou et al., 2025).

2.5 Consensus frameworks

It has been estimated that nearly 350 AMF species have been officially described as per the Botanical Nomenclature Code (Yurkov et al., 2023). Based on molecular and morphological data, species in the AMF show only low endemism and are taxonomically assigned to four orders (*Diversisporales, Glomerales, Archaeosporales*, and *Paraglomerales*), 11–16 families, and 39–49 genera, depending on the classification model used (Stürmer & Kemmelmeier, 2021).

The most elaborate AMF biogeographical study was written based on synthesized works and included both molecular and morphological identification of species analysis, revealing that Diversisporaceae, Glomeraceae, and Paraglomeraceae families are dominant in boreal regions; Glomeraceae and Paraglomeraceae families prevail in temperate zones; and Acaulosporaceae, Gigasporaceae, and Ambisporaceae families dominate in tropical areas (Delavaux et al., 2025).

Modern taxonomic names are constantly updated using phylogenetic approaches and data contained in new databases and data sets, as well as taxonomic alterations in phylogeny-based hierarchical schemes (Tedersoo et al., 2024a). Digital archives such as MaarjAM include curated phylogenetic reconstructions and metadata of thousands of AMF sequences representing a wide variety of diverse global environments. The platforms allow metaanalytical methods, which allow an investigator to study the biogeographical distribution of AMF as well as host relationships (Morman, 2021).

2.6 Evolutionary origins and genomics

2.6.1 Deep time and plant colonization

In 1885, Frank described them as mycorrhiza (Greek: fungus root). It is a symbiotic interaction of a plant and a fungus, which starts on the roots and moves into the rhizosphere. Mycorrhizae are divided into three groups: ectomycorrhiza and endomycorrhiza, as well as ectoendomycorrhiza (a blend of both). The most common occurs in the arbuscular mycorrhizal (AM) association (Kumar et al., 2022).

The evidence suggests that the ancestors of arbuscular mycorrhizal fungi (AMF) date back to at least the Devonian period, approximately 400 million years ago, based on their associations with early terrestrial vascular plants preserved in the Rhynie chert fossil deposits. (Stürmer & Kemmelmeier, 2021). The broad-scale surveys worldwide, combined with molecular and cytological studies, have defined that *Mucoromycotina* symbionts, together with Glomeromycotina, are widely distributed in both complex and simple thalloid liverworts and in the whole hornwort family, with physiological investigations indicating mycorrhizal-like aspects of liverwort symbioses and significant functional distinctions in *Mucoromycotina* and Glomeromycotina symbioses. It has been demonstrated that the liverworts and the symbiotic organisms of *Mucoromycotina* are seen to have a better role to play in the nitrogen metabolism of plants, which is yet to be explained in all the other plant groups, one of which is called angiosperms (Pressel et al., 2021). Both palaeobotanical and molecular evidence point to the emergence of an association between terrestrial plants and a glomonoid fungus over 460 million years ago, 300–400 million years before the development of root nodules by terrestrial plants that occur through symbiosis with diazotrophic bacteria. Existing studies concerning the evolutionary history of AMF and mycorrhizal association with plants mainly dwell on genetic and molecular information. It is the fossil record that has played a big role in informing our knowledge of the pattern of mycorrhizal symbiosis distribution in ancestral ecosystems (Medina-García & San José de las Lajas, 2022).

During the initial phase of plant colonization of land, the root system evolved into a diverse range of multicellular and often complex organs, primarily specialized for anchorage and nutrient uptake. Events of interaction between fungi and early trees are, however, unknown. Regrettably, we are not able to observe fungal linkages because of the kind or level of preservation. These important aspects may begin to become evident based on newly found fossils in Eurasia, which we are currently investigating. Having been found right in the habitat they would have inhabited, virtually complete original anatomy of the whole tree base remains as casts. In response to the question of whether caloric restriction suppresses testosterone levels (Kumar et al., 2022).

The onset of the earliest terrestrial plant-dominated ecosystems at the beginning of the Paleozoic exerted an immense impact on the pattern of development of soils, the shape of the sedimentary system, and also a shift in the world's biogeochemical cycling of processes. This was partly due to the development in plants of sophisticated below-ground (root-like) anchorage systems, which, by increasing and expanding the surface area of plant–mineral contact, enhanced nutrient uptake, accelerated rock weathering, and contributed to the stabilization of surface sediments. However, very little is understood about the mechanics of these microscale processes because there is a relative scarcity of in situ plant material preserved in sedimentary rocks/palaeosols, testimonies to the effect of the said interactions. Some of the same features of the very earliest terrestrial plants are seen in some of the more recent plants (e.g., liverworts, mosses, and lycophytes), namely, uni- or multicellular rhizoid-like anchorage systems, or simple roots, and the possibility of forming below-ground networks of association by prostrate axes and close association with fungi, hence being good analogues (Pressel, 2025). The formation of life occurred in the water. That was followed by the supremacy of the dry land; it is terrestrialization. This was a necessity, because the occupation of any potential space was one of the characteristics of life. The plants had to go before the animals and the further evolution of the process of territorialization in the coevolution course. Such movement of life outside the waters to the land could have occurred in a number of diverse stages. This most probably happened in two waves of terrestrialization (Lüttge, 2020).

The proxy-reconstructed dramatic decrease in CO_2 in the atmosphere in the Devonian (416.0359.2 Ma) and the gradual CO_2 decrease beginning in the Cretaceous (145.565.5 Ma), respectively, have been attributed to the expansion of deeply rooted trees and the emergence of angiosperms. However, such a paradigm ignores the coevolution of roots with major categories of their symbiotic fungal partners that have prevailed over terrestrial ecosystems throughout Earth's history. The change to land was followed by the colonization of land by plants and the proliferation of AMF, but the Cenozoic (c.65.5 Ma) saw the increase of ectomycorrhizal fungi, which are coupled with the roots of both gymnosperms and angiosperms in the check (Taylor et al., 2009).

Fungal–plant relationships started with the conquest of land in terms of colonization, where fossils of the AMF have been discovered in the roots of the earliest terrestrial plants in the Rhynie chert. These fungi helped in the absorption of nutrients, especially phosphorus and nitrogen, in nutrient-poor soils, helping the early plants to flourish. The formation of AMF by the bryophytes and other nonvascular plants is considered to help them in the reduction of desiccation stress and nutrient availability. The relationship of mutualistic nutrient exchange of carbon supply also established the background of the evolution and diversification of vascular plants so that they could adapt to diverse landforms. These fossils showed AMF colonizing the roots of early vascular plants, including Aglaophyton major, and these showed the significance of fungi in creating the background in which plants could occupy land (Hajji-Hedfi et al., 2025).

2.6.2 Multigenomic lifestyle

There are also specificities of the AMF life history that are relevant to the AMF genetics and its evolutionary implications. The asexual spores and hyphae are coenocytic, and so they are multinucleate. This varies extremely in different species, with an upper limit of about 50,000 nuclei in spores in species of *Gigaspora* to about 800–2000 in different species of *Glomus* and *Scutellospora*. In this way, AMF have the potential to transmit numerous nuclei to future generations through spores and the formation of new symbionts by building new symbioses through colonization of new roots and hemodynamic hyphae (Sanders, 2002).

The pregenome period and AMF genetics have been conceptually characterized as non-Mendelian and, more specifically, as a property of variations of the relative abundance or loss through segregation of distinct nuclei (or certain markers) in an individual. It was hypothesized that this genetic background would change over time either through the conduct of over nuclear transfers, bottlenecks, or drifts among individual spore progeny and/or among AMF individuals, which will eventually produce different results in plant health, expression of plant symbiosis-specific genes, and AMF phenotypes. It was strikingly different from what was known about the genetics of multinucleate fungi: as it turned out, fungal geneticists were to express growing dissatisfaction with this interpretation (Oliveira et al., 2024).

AMF can distribute the tons of nuclei into the next generation with the help of the multinucleated spores. Research evidence on *G. irregulare* indicates that spores develop between 30 and 60 days and, over the period, develop a greater number of nuclei because of nuclear migration and mitosis. Remarkably, both origins are found in the nuclei of spores; thus there is no stage of a single nucleus, and this is a possible indication that AMF does not undergo a genetic bottleneck that is typical of other eukaryotes (Marleau et al., 2011). AMF produce multinucleated spores and aseptate hyphae, and thus, whether their nuclei are genetically distinct (heterokaryosis) or homogenous (homokaryosis). Heterokaryosis used to be supported by molecular studies, but the more recent whole-genome support questions this, and the actual nature of AMF nuclear organization remains unknown (Ropars et al., 2016).

AMF reproduce asexually and form numerous spores, which enable them to multiply rapidly even under various environmental stresses (Ebbisa, 2022). AMF forms numerous spores that multiply faster even under situations of different pressures (Ebbisa, 2022). The AMF have also been determined to reproduce asexually to develop large multinucleated spores, which develop into a new mycelium. During the period of spore formation, there is a massive migration of the nuclei into the new structures, and it is assumed that the same undergoes multiplication with the help of mitosis as well. The nuclear relocation procedure has experienced the insult of being called a stochastic process and is able to bring about a degree of value in the quantity of nuclei that are moved into the many spores of an AMF. However, such variation did not have a significant impact on their germination potential. New findings flowing in as a result of the DNA sequencing procedure have been indicating the asexuality of AMF. On the one hand the genetic recombination remnants were observed in some clones of AMF with the same origin (one field), and this meant that there could have been a cryptic sexual reproduction (Joaquim, 2023). This kind of genomic pattern is considered because it helped adapt to various environmental conditions and other plants as the host and prefers the ecological plasticity of these plants.

2.6.3 Modern genomic insights

Possible sexuality was indicated by both the finding of a possible mating-type locus (an MAT locus) in *R. irregularis* and the recombination already proposed between nuclei of a dikaryon *R. irregularis* isolate. No excessive evolution of MAT-locus diversity is expected unless the MAT locus is likely to be put into the circulation of sexual reproduction quite often. Moreover,

MAT-locus evolution in sexual species does not coevolve with the evolution of random genome-wide loci (Aparicio Chacón et al., 2024).

In the six sequenced AMFs the study determined 72 conserved components of four nutrient-sensing pathways (cAMP/PKA-Cyclic Adenosine Monophosphate / Protein Kinase A pathway, SNF1-Sucrose Non-Fermenting 1 pathway, TOR-Target of Rapamycin pathway, and PHO-Phosphate signaling pathway). Similar data on gene expression support the hypothesis that these pathways contribute to reactions of AMF to nutrient conditions and maintain the important biological mechanisms of nutrient movement, lipid utilization, and resistance to stress (Zhou et al., 2021). The application of AMF greatly facilitated the growth of plants and supported disease resistance through changes in enzyme activities, increased production of lignin, and flavonoids. The *R. intraradices* enters into stable symbiosis with *Lycium barbarum* roots and improves the biomass of plants, their chlorophyll content, and resistance to the disease. It increases the activity of enzymes of the phenylpropanoids, pathogenesis-related proteins, and defense signaling compounds and inhibits the occurrence and severity of root rot (Li et al., 2024).

2.7 Ecology of arbuscular mycorrhizal fungi

Arbuscular mycorrhiza (AM) is the oldest and most common form of plant–fungi mutualistic symbiosis (evolution began more than 400 million years ago); AM is a mutualistic relation between plants and soil fungi (Brito et al., 2021). Arbuscular mycorrhizal fungi (AMF) form intracellular connections with root cortical cells and develop an extensive extraradical mycelium (ERM) in the soil. This ERM is vital in increasing the ecological footprint of the host plant through the increase that they get in terms of soil nutrients and water. AMF associations do not exist in an isolated setting or environment; they are world-distributed and determine plant productivity, biodiversity, and nutrient cycling in terrestrial ecosystems. The ecological importance of AMF is further promulgated by the fact that they assist in the aggregation of soil, stabilization of organic matter, and provision of support to the microbial diversity (Chowdhary & Songachan, 2025; Morton & Msiska, 2010).

The fresh molecular technologies, especially high-throughput sequencing and metagenomics, have contributed a lot to the knowledge of AMF ecology, which turns out to be much more diverse than initially supposed through morphological means. Moreover, metabolomic studies currently allow us to understand the ways in which AMF mediate the flux of metabolites between the soil and the plant, especially under the conditions of low nutrients or drought (Holátko et al., 2021; Zhang et al., 2018). Much of this research, however, is constrained to basic laboratory conditions. The potential adaptation of AMF to the agrosystematic settings and the effect of agricultural activities upon the native community composition and the activity of AMF are still underrepresented (Brito et al., 2012; Ciccolini et al., 2016).

AMF have been considered quite minimally in the management of agriculture, in part because of the absence to date of consistently improved yields in major crops such as wheat, despite their known ability to increase plant nutrition and plant stress tolerance (Ryan & Graham, 2018). However, as AMF also improve micronutrient content, pathogen defense, and soil stability, their indirect contributions to agroecosystem health and sustainability are increasingly being recognized (Garland et al., 2021; Wang et al., 2021).

2.7.1 Functional roles

The main advantage that AMF benefits host plants is the increased nutrient uptake, especially that of phosphorus as well as other slow-moving ions, because of the ERM that amplifies the volume of soil reachable by plant roots by a large (Jin et al., 2024) margin. It also promotes water transport to the fungal network, which improves the propagation of plants in the face of drought. AMF also increases the vigor and growth of the plants during disease attack by biotic (e.g., soil-borne diseases or pathogens) and abiotic (e.g., HM toxicity) conditions (Brito et al., 2019; Magkourilou et al., 2024).

These fungi additionally control the inner control of the host plant through their expression of genes and regulation of plant hormone patterning (Cameron et al., 2013). AMF significantly contribute to soil multifunctionality improvement through their ability to raise enzyme activity, retention capabilities of nutrients, and stabilization of organic matter, especially through the synthesis of glomalin-related soil proteins that lead to soil aggregation and carbon sequestration (Holátko et al., 2021; Zhou et al., 2020).

Indirectly, AMF enhance plant health by enhancing the soil structure by enmeshing soil particles (Rillig & Mummey, 2006) and providing microhabitats to beneficial microbes, including rhizobacteria, which can stimulate uptake of nutrients and resistance to pathogens. Such complex functions portray the possibilities of AMF as prominent biological agents in sustainable farming and healing of ecosystems (Chowdhary & Songachan, 2025; Wagg et al., 2019).

TABLE 2.1 Summary of studies on the persistence of introduced AMF inoculants in field conditions.

Introduced AMF strain(s)	Host plant(s) and location	Key findings on persistence	Molecular tool(s) used	References
Funneliformis mosseae (BEG12 and AZ225C)	Alfalfa, Italy	Both strains persisted, with AZ225C being more abundant than BEG12	SSU, ITS, LSU rRNA sequencing	Pellegrino et al. (2022)
Rhizophagus irregularis	Multiple crops, Canada	Establishment is highly site-specific, not linked to crop management	Metabarcoding	Kokkoris et al. (2019)
R. irregularis (IR27)	*Ziziphus mauritiana*, Senegal	Inoculant persisted with sustained growth benefits	Not specified	Thioye et al. (2021)
F. mosseae (BEG167)	*Plantago lanceolata*, United Kingdom	Persistence noted, but not correlated with early plant gains	PCR with specific primers	Farmer et al. (2007)
Commercial inoculant (*Funneliformis, Septoglomus*)	Tomato, Italy	Inoculant failed to establish	Metabarcoding	Berruti et al. (2017)
Commercial inoculant	*Stimulation*	Likely driven by coformulated microbes	Metabarcoding	Berruti et al. (2017)
F. mosseae	*Suppression*	Native *R. irregularis* and *Rhizophagus intraradices* were dominant	PCR-(T)-RFLP	Pellegrino et al. (2012)
Not specified	*Conditional*	Environmental factors modulate inoculation effects	Not specified	Symanczik et al. (2015)

2.7.2 Functional diversity and biodiversity

Any new techniques in metabolomics and metagenomics have shown a larger level of diversity in functional and taxonomic diversity of AMF than had been realized before. The ability of AMF species to colonize to a great extent, build ERM, and provide stress endurance to the host plants differs extremely (Lax et al., 2011). As a result a greater AMF species diversity positively contributes to the variety and efficacy of the benefits of the symbiosis provided to the plants (Ellouze et al., 2014). The similarity in functional complementarity of AMF species increases the adaptive potential of changed environmental conditions or the growing plants.

By comparison, native AMF strains in the soil can be better than the commercial inoculants, as they have adapted to the local conditions. Nonetheless the commercial inocula have produced mixed results in the field; thus their application in agriculture is contentious (Elliott et al., 2021). Numerous commercial goods are expensive and usually possess minimal and common AMF species with low viability and infectivity (Chen et al., 2018). Although some production systems such as greenhouses, nurseries, or rehabilitation of the worn-out land might be helpful upon inoculation, subjecting it to it at the field level is not reliable.

It is seen that with intact ERM, propagules are mostly the most infective type that enables faster and stronger colonizing, which is crucial when crops experience early-season stressors (Goss et al., 2017). Early colonization, as facilitated by ERM, gives a benefit in resistance against both abiotic and biotic stresses. These points to the fact that native AMF communities should be promoted by land management practices that are sustainable as opposed to the use of exogenous inocula (Brito et al., 2021). The persistence and establishment of introduced AMF strains are highly variable and context-dependent, often failing to establish or persisting at low abundance compared to native communities, as summarized in several key studies (Table 2.1). Although inoculation may be beneficial in some production systems such as greenhouses, nurseries, or degraded land rehabilitation, its application at the field scale is not always reliable.

2.7.3 Ecological interactions and network dynamics

AMF form complex ecological networks with plants and other soil organisms. One of their most remarkable ecological functions is the development of common mycorrhizal networks (CMNs), which allow multiple plants to share nutrients and biochemical signals (Jakobsen & Hammer, 2015; Luo et al., 2023). This network-based interaction enables nutrient transfer between plants and enhances defense priming, contributing to greater stability and resilience in plant communities. Such interactions are particularly critical in resource-limited or degraded ecosystems (Magkourilou et al., 2024).

Host preference in AMF associations is well-established; specific plant species tend to associate with particular AMF genotypes, shaping distinct fungal assemblages even when grown in the same soil (Wipf et al., 2019). The identity and physiological traits of the host plant influence the composition of its mycorrhizal partners, whereas environmental factors such as climate also play a role (Enebe & Erasmus, 2023). For example, *Hepatica nobilis* maintained similar AMF consortia across locations with comparable climates, supporting the influence of both host and environmental filtering. Furthermore, AMF communities within a plant can change during its life cycle, ranging from only a few to over 30 operational taxonomic units.

The use of exotic AMF inoculants poses ecological risks, including the displacement of native fungal species and the alteration of native plant–fungal interactions. The inoculation of AMF in natural systems is a risky practice that can disrupt ecological balance. Consequently, reliance on native AMF communities, which are adapted, diverse, and effective, is a more sustainable strategy (Brito et al., 2021). Managing agricultural systems to support and preserve native AMF diversity—through practices such as reduced tillage, crop rotation, and selection of compatible plant species—offers a promising path toward sustainable production and ecosystem resilience.

2.8 Applications and knowledge gaps

The AMF symbiosis improves plant response to abiotic stressors by increasing the level of antioxidants and stabilizing the biosynthesis of lipids, as well as by promoting the uptake of low-mobilizable trace elements (Al-Arjani et al., 2020). These are K, Ca, Mg, Mn, Co, Cu, Zn, and Fe, which are considered important elements that plants use in metabolism. The AMF creates expansions of the production root structure with extra external hyphae, which enable it to reach more soil and better nourishment absorption. Another significant way AMF enhances the bioavailability of these valuable components is through biochemical methods that include solubilizing and mobilizing nutrients (Seutra Kaba et al., 2021).

Plant productivity and the stability of the environment are crippled by abiotic stresses. AMF lead to plant adaptation by modulating gene expression and hormonal networks that mediate growth, development, and metabolism. In conditions of drought the AMF reduce oxidative stress, stimulating the accumulation of osmolytes and increasing the absorption of water via extraradical mycelium, increasing photosynthetic capacity and root biomass (Saharan et al., 2022). They also induce stress-responsive genes and synthesize chaperone proteins that maintain cellular activities in conditions of water shortage. The plants inoculated with AMF perform physiologically better (higher level of chlorophyll, both in volume and number of roots, and antioxidant activity) even during drought (Begum et al., 2019; Robdrup et al., 2024).

AMF improve the structure of roots and attenuate the oxidative damage because of enhanced antioxidant enzyme activity and accumulation of osmolytes under waterlogging. The colonization of AMF increases the level of proline and sugars, enhances the content of chlorophyll, and induces the expression of stress-related genes, leading to enhanced survival of plants under hypoxic conditions (Chowdhary & Songachan, 2025).

During salinity stress, AMF can balance the ions and regulate their uptake (Na^+ and K^+) and promote the production of chlorophyll, the efficiency of water use, and antioxidant protection. There is enhanced photosynthesis, as AMF prevent PSII damage and enhance PSI reversion in the salty medium. Research studies in crops such as *Antirrhinum majus, Ocimum basilicum*, and *Allium sativum* illustrate that AMF-inoculated plants exhibit an increase in biomass, nutrient quantity, and photosynthetic accomplishments in the presence of salinity stress (Antunes et al., 2025; Shah et al., 2021; Wahab et al., 2023).

The agricultural promise of AMF is the capacity to decrease the amount of chemical fertilizers used and increase the probability of high yields, nutrient uptake, and stress tolerance. Inoculant specific to the crop/soil types using AMF may boost the tolerance to drought conditions and salinized soils and ease the absorption of trace elements in already deficient soils. The strategy facilitates the concept of sustainable agriculture with its efficiency in the use of water and enhancement of soil structure.

Another major role of AMF is in the area of ecological restoration. Planting of ecoconsortia with AMF can speed up the process of replacing vegetation cover and the restoration of degraded and low-lying lands. Their role in the aggregation of soils, stabilization of organic matter, and the balancing of microbial communities renders them useful in reclaiming soil fertility and soil ecosystem functioning. Inoculation with AMF contributes to the establishment of the native plants and their resilience, making them an important mechanism in the restoration of the habitats that have been eroded, polluted, or turned into intensive agricultural lands (Antunes et al., 2025). As illustrated the extraradical mycelium dramatically expands the plant's access to soil resources, which is the primary mechanism behind the observed growth benefits (Fig. 2.2).

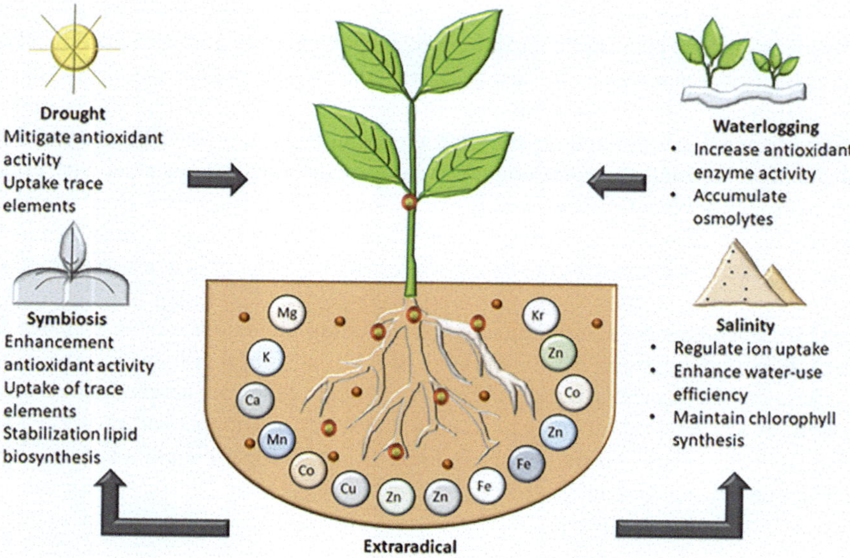

FIGURE 2.2 Role of arbuscular mycorrhizal fungi (AMF) in plants during stressful conditions.

Although AMF have experienced great breakthroughs, there are a number of important gaps that characterize the future of research. The first requirement is the complete characterization of AMF-plant genetic compatibility, especially given the new evidence demonstrating extreme intraspecific genetic diversity within AMF. To give an example, most of them are available in the homokaryotic or dikaryotic forms, where the latter possess 2 characteristic nuclear genotypes with different genomic structures and epigenetic fingerprints. It is shown that these nuclear combinations seem to have a great impact on symbiotic responses of the plants with the potato cultivars, which displayed negligible mutualistic performance in the dikaryotic strains when compared to homokaryotic ones. These results can highlight the significance of considering host fungus compatibility below the species level to perhaps genotype pairs to networking and the necessity to experimentally address the role of a diversity of AMF nuclear architecture in affecting plant reactions in an environmentally inclined manner in a wide range of hosts (Antunes et al., 2025).

High-resolution molecular taxonomy is also important for elucidating the AMF biodiversity and activity. The importance of low-resolution morphological and genetic markers has been shown by the fact that both homokaryotic and dikaryotic strains within a single species can coexist (Serghi et al., 2021). Such a cryptic diversity requires high-throughput genomic tools to be unraveled and to be laid onto ecologically relevant functional traits. Genome size here becomes a compelling but underutilized trait. Although genome size has had little attention in fungi, it has received substantial consideration in plants, including associations with climatic adaptation, flowering phenology, and water preference and with a set of morphological and physiological traits (Bhadra et al., 2023). On AMF the reported data are scarce, yet the variations are statistically remarkable (116 Mb in *R. clarus* and 740 Mb in *Gigasporaceae* species) (Antunes et al., 2025).

The other urgent frontier is to know more about AMF ecology of phosphorus-poor or climatically fluctuating environments. In this case, fungal ecological strategies can be investigated by using genetic factors like copy number of rDNA. The hypothesis is the low rRNA operon copy number selection has been associated with low growth rates and inability to adapt to resource-restricted environments, whereas high values are positively linked to fast rates of growth that require nutrient-rich conditions (Roller et al., 2016). Fungi demonstrate great diversity in copy numbers of rDNA; namely, they vary between 14 and 1400+ across taxa (Lofgren et al., 2019). However, *R. irregularis* in AMF has been analyzed so far, which surprisingly shows a low number of only 11 copies (Maeda et al., 2018). This insignificant amount is similar to those that have been considered conventional traits of symbiotic and asexual organisms, and thus rDNA copy number could be a good predictor of ecological specialization, responsiveness to nutrients, or efficiency of colonization across different AMF species (Antunes et al., 2025).

And lastly, though the role of hyphal networks in nutrient foraging and exchange is understood well, the role of the hyphal networks in mediation of host-to-host signaling is speculative (Fig. 2.3). The key question that the future research should address rigorously is whether observed interactions between hosts are mirrors of actual biochemical signaling or rather are mere artifacts of physical fungal connectivity. In this respect, genomic features, including guanine and cytosine (G)

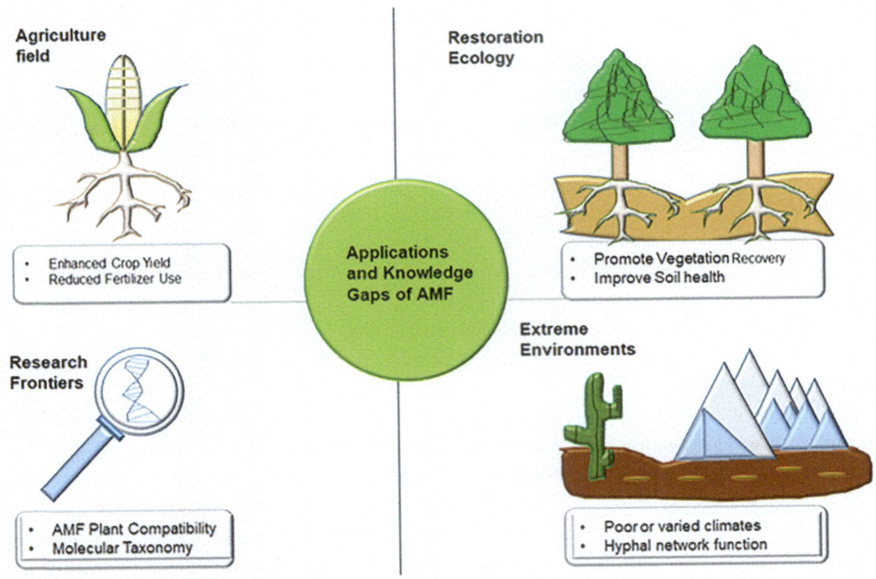

FIGURE 2.3 Major uses and knowledge gaps of arbuscular mycorrhizal fungi (AMF) in four fields: agriculture, restoration ecology, research frontiers, and extreme environments. All four quadrants identify the unique opportunities and research gaps that are necessary to tap into the potential of AMF in ensuring sustainable ecosystem management.

content in AMF DNA, 25% and 36% (Malar et al., 2022), can reveal information about their wider ecological and genomic approaches. Other fungal systems have associated variation in G + C content with niche specialization, pathogenicity, and genome stability. As rapid progress is being made in single-cell genomics and the sequencing of spores, it is becoming more possible to collect and analyze such genomic traits upon direct sampling within the environment (Antunes et al., 2025).

2.9 Conclusion

The AMF are ancient representatives of the phylum Glomeromycota, which play a role in the structure and functioning of terrestrial habitats. Their cooperation with plants, which lasted more than 400 million years, has been one of the keys to maintaining and evolving plants and their ecosystem. The chapter has discussed the evolutionary background, taxonomy, ecological functions, and feasible usage of AMF with a focus on the significance of AMF as key-stone organisms of the soil microbiome. Taxonomically, AMF still are a complicated lot. Molecular biology and phylogenetics have shown that there has been more diversity in cryptic species, which is problematic in terms of traditional classifications based on morphology. A multigene approach, which emphasizes accurate species identification and a better understanding of evolutionary relationships, is indispensable. This information is essential to establishing a relationship between AMF diversity and ecosystem functions and optimal utilization in farming and restoration. At the ecological level, AMF increases the uptake of nutrients by plants, particularly phosphorus, and the resistance to environmental stress, including drought, salinity, and pollution by HMs. They build the soil structure and water retention and create a hyphosphere, a living environment around and between roots via their hyphal networks and CMNs. The most promising application of AMF is in sustainable agriculture, where the reliance on chemical fertilizers is eased and the obligation to the reclamation of degraded soils is enhanced. Nonetheless, there are two issues that point to the necessity of site-specific, ecologically minded approaches: the erratic success of commercial inoculants and the ecological costs of naturalizing foreign strains. In the future, genetics ought to be given emphasis too on the genetic programs of AMF as far as implications of genome size variation and nuclear structure (homokaryons and dikaryons) are concerned. Molecular tools of high resolution will play a crucial role in determining native AMF communities and associating them with functional traits. Finally, it could be advantageous and longer term to increase the diversity of native AMF by using knowledgeable land management, as opposed to using commercial strains. The step toward a better perception of AMF is vital to the development of food security worldwide and ecosystem stability.

References

Al-Arjani, A. F., Hashem, A., & Abd Allah, E. F. (2020). Arbuscular mycorrhizal fungi modulates dynamics tolerance expression to mitigate drought stress in *Ephedra foliata* Boiss. *Saudi Journal of Biological Sciences, 27*(1), 380–394. https://doi.org/10.1016/j.sjbs.2019.10.008.

Albornoz, F. E., Ryan, M. H., Bending, G. D., Hilton, S., Dickie, I. A., Gleeson, D. B., & Standish, R. J. (2022). Agricultural land-use favours Mucoromycotinian, but not Glomeromycotinian, arbuscular mycorrhizal fungi across ten biomes. *New Phytologist, 233*(3), 1369–1382.

Amir, H., Cavaloc, Y., Crossay, T., Bourles, A., Gensous, S., Lagrange, A., Burtet-Sarramegna, V., & Guentas, L. (2023). Importance and roles of arbuscular mycorrhizal fungi in New Caledonian ultramafic soils. *Botany Letters, 170*(3), 449–458.

Antunes, P. M., Stürmer, S. L., Bever, J. D., Chagnon, P.-L., Chaudhary, V. B., Deveautour, C., Fahey, C., Kokkoris, V., Lekberg, Y., Powell, J. R., Aguilar-Trigueros, C. A., & Zhang, H. (2025). Enhancing consistency in arbuscular mycorrhizal trait-based research to improve predictions of function. *Mycorrhiza, 35*(2), 14. https://doi.org/10.1007/s00572-025-01187-7.

Aparicio Chacón, M. V., Hernández Luelmo, S., Devlieghere, V., Robichez, L., Leroy, T., Stuer, N., De Keyser, A., Ceulemans, E., Goossens, A., & Goormachtig, S. (2024). Exploring the potential role of four *Rhizophagus irregularis* nuclear effectors: Opportunities and technical limitations. *Frontiers in Plant Science, 15*, 1384496.

Begum, N., Ahanger, M. A., Su, Y., Lei, Y., Mustafa, N. S. A., Ahmad, P., & Zhang, L. (2019). Improved drought tolerance by AMF inoculation in maize (Zea mays) involves physiological and biochemical implications. *Plants (Basel), 8*(12), 579.

Berruti, A., Lumini, E., & Bianciotto, V. (2017). AMF components from a microbial inoculum fail to colonize roots and lack soil persistence in an arable maize field. *Symbiosis, 72*(1), 73–80.

Bhadra, S., Leitch, I. J., & Onstein, R. E. (2023). From genome size to trait evolution during angiosperm radiation. *Trends in Genetics, 39*(10), 728–735.

Brito, I., Carvalho, M., & Goss, M. J. (2021). Managing the functional diversity of arbuscular mycorrhizal fungi for the sustainable intensification of crop production. *Plants, People, Planet, 3*(5), 491–505.

Brito, I., Goss, M., Alho, L., Brígido, C., van Tuinen, D., Félix, M., & Carvalho, M. (2019). Agronomic management of AMF functional diversity to overcome biotic and abiotic stresses—The role of plant sequence and intact extraradical mycelium. *Fungal Ecology, 40*, 72–81.

Brito, I., Goss, M. J., de Carvalho, M., Chatagnier, O., & Van Tuinen, D. (2012). Impact of tillage system on arbuscular mycorrhiza fungal communities in the soil under Mediterranean conditions. *Soil and Tillage Research, 121*, 63–67.

Cameron, D. D., Neal, A. L., van Wees, S. C., & Ton, J. (2013). Mycorrhiza-induced resistance: More than the sum of its parts? *Trends in Plant Science, 18*(10), 539–545.

Chen, M., Arato, M., Borghi, L., Nouri, E., & Reinhardt, D. (2018). Beneficial services of arbuscular mycorrhizal fungi–from ecology to application. *Frontiers in Plant Science, 9*, 1270.

Chowdhary, N. A., & Songachan, L. S. (2025). Boosting sustainable agriculture by arbuscular mycorrhizal fungi under abiotic stress condition. *Plant Stress, 17*, 100945. https://doi.org/10.1016/j.stress.2025.100945.

Ciccolini, V., Ercoli, L., Davison, J., Vasar, M., Öpik, M., & Pellegrino, E. (2016). Land-use intensity and host plant simultaneously shape the composition of arbuscular mycorrhizal fungal communities in a Mediterranean drained peatland. *FEMS Microbiology Ecology, 92*(12), fiw186.

Delavaux, C. S., Aellen, A., Stürmer, S. L., Primieri, S., Schütte, U. M., Drown, D. M., Ramos, R. J., Crowther, T. W., & Bever, J. D. (2025). Uncovering diversity within the Glomeromycota: Novel clades, family distributions, and land use sensitivity. *Ecology and Evolution, 15*(1), e70597.

Desai, S., Kumar, G. P., Amalraj, L. D., Bagyaraj, D., & Ashwin, R. (2016). Exploiting PGPR and AMF biodiversity for plant health management. Microbial inoculants in sustainable agricultural productivity*: Vol. 1: Research* perspectives, 145–160.

Ebbisa, A. (2022). Arbuscular mycorrhizal fungi (AMF) in optimizing nutrient bioavailability and reducing agrochemicals for maintaining sustainable agroecosystems. *Arbuscular mycorrhizal fungi in agriculture-new insights*. IntechOpen.

Elliott, A. J., Daniell, T. J., Cameron, D. D., & Field, K. J. (2021). A commercial arbuscular mycorrhizal inoculum increases root colonization across wheat cultivars but does not increase assimilation of mycorrhiza-acquired nutrients. *Plants, People, Planet, 3*(5), 588–599.

Ellouze, W., Esmaeili Taheri, A., Bainard, L. D., Yang, C., Bazghaleh, N., Navarro-Borrell, A., Hanson, K., & Hamel, C. (2014). Soil fungal resources in annual cropping systems and their potential for management. *BioMed Research International, 2014*(1), 531824.

Enebe, M. C., & Erasmus, M. (2023). Symbiosis—A perspective on the effects of host traits and environmental parameters in arbuscular mycorrhizal fungal richness, colonization and ecological functions. *Agriculture, 13*(10), 1899.

Faghihinia, M., Jansa, J., Halverson, L. J., & Staddon, P. L. (2023). Hyphosphere microbiome of arbuscular mycorrhizal fungi: A realm of unknowns. *Biology and fertility of soils, 59*(1), 17–34.

Fall, A. F., Nakabonge, G., Ssekandi, J., Founoune-Mboup, H., Apori, S. O., Ndiaye, A., Badji, A., & Ngom, K. (2022). Roles of arbuscular mycorrhizal fungi on soil fertility: Contribution in the improvement of physical, chemical, and biological properties of the soil. *Frontiers in Fungal Biology, 3*, 723892.

Farmer, M.-J., Li, X., Feng, G., Zhao, B., Chatagnier, O., Gianinazzi, S., Gianinazzi-Pearson, V., & van Tuinen, D. (2007). Molecular monitoring of field-inoculated AMF to evaluate persistence in sweet potato crops in China. *Applied Soil Ecology, 35*(3), 599–609.

Garland, G., Banerjee, S., Edlinger, A., Miranda Oliveira, E., Herzog, C., Wittwer, R., Philippot, L., Maestre, F. T., & van Der Heijden, M. G. A. (2021). A closer look at the functions behind ecosystem multifunctionality: A review. *Journal of Ecology, 109*(2), 600–613.

Goss, M. J., Carvalho, M., & Brito, I. (2017). *Functional diversity of mycorrhiza and sustainable agriculture: Management to overcome biotic and abiotic stresses*. Academic Press.

Hajji-Hedfi, L., Wannassi, T., & El-Maradny, Y. A. (2025). Fungal endophytes: Insight into evolution, classification, and ecological functions in plants. *Microbial Biosystems, 10*(1), 51–67.

Hart, M. M., & Reader, R. J. (2002). Taxonomic basis for variation in the colonization strategy of arbuscular mycorrhizal fungi. *New Phytologist, 153*(2), 335–344.

Hartmann, M., & Six, J. (2023). Soil structure and microbiome functions in agroecosystems. *Nature Reviews Earth & Environment, 4*(1), 4–18.

Holátko, J., Brtnický, M., Kučerík, J., Kotianová, M., Elbl, J., Kintl, A., Kynický, J., Benada, O., Datta, R., & Jansa, J. (2021). Glomalin–truths, myths, and the future of this elusive soil glycoprotein. *Soil Biology and Biochemistry, 153*, 108116.

Jakobsen, I., & Hammer, E. C. (2015). Nutrient dynamics in arbuscular mycorrhizal networks. In *Mycorrhizal networks* (pp. 91–131). Springer.

Jin, Z., Jiang, F., Wang, L., Declerck, S., Feng, G., & Zhang, L. (2024). Arbuscular mycorrhizal fungi and *Streptomyces*: Brothers in arms to shape the structure and function of the hyphosphere microbiome in the early stage of interaction. *Microbiome, 12*(1), 83.

Joaquim, C. C. (2023). *Quantitative genetic variation in symbiotic fungi and its effect on fungal gene expression*. University of Lausanne.

Khaliq, A., Perveen, S., Alamer, K. H., Haq, M. Z. U., Rafique, Z., Alsudays, I. M., Althobaiti, A. T., Saleh, M. A., Hussain, S., & Attia, H. (2022). Arbuscular mycorrhizal fungi symbiosis to enhance plant–soil interaction. *Sustainability, 14*(13), 7840.

Kokkoris, V., Banchini, C., Paré, L., Abdellatif, L., Séguin, S., Hubbard, K., Findlay, W., Dalpé, Y., Dettman, J., Corradi, N., & Corradi, N. (2024). *Rhizophagus irregularis*, the model fungus in arbuscular mycorrhiza research, forms dimorphic spores. *New Phytologist, 242*(4), 1771–1784.

Kokkoris, V., Li, Y., Hamel, C., Hanson, K., & Hart, M. (2019). Site specificity in establishment of a commercial arbuscular mycorrhizal fungal inoculant. *Science of the Total Environment, 660*, 1135–1143.

Koshila Ravi, R., & Muthukumar, T. (2024). Root exudates and their importance in arbuscular mycorrhizal symbiosis and nutrients navigation from inaccessible soil: An efficient mediator of mineral acquisition in nutrient deprived soil. In *Mycorrhizal symbiosis and agroecosystem restoration* (pp. 101–123). Springer.

Kumar, D., Kour, S., Ali, M., Sharma, R., Parkirti, Vikram, Changotra, H., Manhas, R. K., & Ohri, P. (2024). Interactions between arbuscular mycorrhizal fungi and other microorganisms in the rhizosphere and hyphosphere. In G. J. Ahammed, & R. Hajiboland (Eds.), *Arbuscular mycorrhizal fungi and higher plants: Fundamentals and applications* (pp. 37–66). Springer.

Kumar, R., Bhardwaj, A. K., Chandra, K., & Singh, A. K. (2022). Mycorrhizae: An historical journey of plant association. *Chhattisgarh Journal of Science & Technology, 19*, 437–447.

Lax, P., Becerra, A. G., Soteras, F., Cabello, M., & Doucet, M. E. (2011). Effect of the arbuscular mycorrhizal fungus *Glomus intraradices* on the false root-knot nematode *Nacobbus aberrans* in tomato plants. *Biology and Fertility of Soils, 47*(5), 591–597.

Li, H., Zhang, L., Wu, B., Li, Y., Wang, H., Teng, H., Wei, D., Yuan, Z., & Yuan, Z. (2023). Physiological and proteomic analyses reveal the important role of arbuscular mycorrhizal fungi on enhancing photosynthesis in wheat under cadmium stress. *Ecotoxicology and Environmental Safety, 261*, 115105.

Li, N., Chen, W., Wang, B., Zhang, C., Wang, Y., Li, R., Yan, Y., & He, J. (2024). Arbuscular mycorrhizal fungi improve the disease resistance of *Lycium barbarum* to root rot by activating phenylpropane metabolism. *Frontiers in Plant Science, 15*, 1459651.

Liu, R., Li, M., Guo, S., & Chen, Y. (2024). The role of AMF community composition, diversity, and distribution in sustainable agroecosystems. In *Arbuscular mycorrhizal fungi in sustainable agriculture: Inoculum production and application* (pp. 281–317). Springer.

Lofgren, L. A., Uehling, J. K., Branco, S., Bruns, T. D., Martin, F., & Kennedy, P. G. (2019). Genome-based estimates of fungal rDNA copy number variation across phylogenetic scales and ecological lifestyles. *Molecular Ecology, 28*(4), 721–730.

Luo, X., Liu, Y., Li, S., & He, X. (2023). Interplant carbon and nitrogen transfers mediated by common arbuscular mycorrhizal networks: Beneficial pathways for system functionality. *Frontiers in Plant Science, 14*, 1169310.

Lüttge, U. (2020). Terrestrialization: The conquest of dry land by plants. In *Progress in botany: 83* (pp. 65–89). Springer.

Lutz, S., Bodenhausen, N., Hess, J., Valzano-Held, A., Waelchli, J., Deslandes-Hérold, G., Schlaeppi, K., & van der Heijden, M. G. A. (2023). Soil microbiome indicators can predict crop growth response to large-scale inoculation with arbuscular mycorrhizal fungi. *Nature Microbiology, 8*(12), 2277–2289.

Ma, X., Xu, X., Geng, Q., Luo, Y., Ju, C., Li, Q., & Zhou, Y. (2023). Global arbuscular mycorrhizal fungal diversity and abundance decreases with soil available phosphorus. *Global Ecology and Biogeography, 32*(8), 1423–1434.

Maeda, T., Kobayashi, Y., Kameoka, H., Okuma, N., Takeda, N., Yamaguchi, K., Bino, T., Shigenobu, S., & Kawaguchi, M. (2018). Evidence of non-tandemly repeated rDNAs and their intragenomic heterogeneity in *Rhizophagus irregularis*. *Communications Biology, 1*(1), 87.

Magkourilou, E., Bell, C., Daniell, T., & Field, K. J. (2024). The functionality of arbuscular mycorrhizal networks across scales of experimental complexity and ecological relevance. *Functional Ecology, 39*(6), 1384–1399.

Malar, C., Wang, Y., Stajich, J., Kokkoris, V., Villeneuve-Laroche, M., Yildirir, G., & Corradi, N. (2022). Early branching arbuscular mycorrhizal fungus *Paraglomus occultum* carries a small and repeat-poor genome compared to relatives in the Glomeromycotina. *Microbial Genomics, 8*, 000810.

Marleau, J., Dalpé, Y., St-Arnaud, M., & Hijri, M. (2011). Spore development and nuclear inheritance in arbuscular mycorrhizal fungi. *BMC Evolutionary Biology, 11*(1), 51.

Medina-García, L. R., & San José de las Lajas, M. (2022). Arbuscular mycorrhizal fungi and their role in agroecosystems. *Cultivos Tropicales, 43*(1), e14.

Montesinos-Navarro, A., Segarra-Moragues, J. G., Valiente-Banuet, A., & Verdú, M. (2012). The network structure of plant–arbuscular mycorrhizal fungi. *New Phytologist, 194*(2), 536–547.

Morman, K. E. (2021). *Impacts of soil origin and fire on arbuscular mycorrhizal fungi associated with hawkweeds and tall tussock (Unpublished MSc thesis)*. Auckland, New Zealand: Auckland University of Technology.

Morton, J. B., & Msiska, Z. (2010). Phylogenies from genetic and morphological characters do not support a revision of Gigasporaceae (Glomeromycota) into four families and five genera. *Mycorrhiza, 20*(7), 483–496.

Nazir, A., Sarfraz, W., Allah, D., Khalid, N., Farid, M., Shafiq, M., Firdaus-E-Bareen, Rizvi, Z. F., & Naeem, N. (2023). Synergistic impact of two autochthonous saprobic fungi (*A. niger* and *T. pseudokoningii*) on the growth, ionic contents, and metals uptake in *Brassica juncea* L. and *Vigna radiata* L. under tannery solid waste contaminated soil. *International Journal of Phytoremediation, 25*(11), 1488–1500. https://doi.org/10.1080/15226514.2023.2166457.

Oliveira, J., Yildirir, G., & Corradi, N. (2024). From chaos comes order: Genetics and genome biology of arbuscular mycorrhizal fungi. *Annual Review of Microbiology, 78*, 147–168.

Ouhaddou, R., Anli, M., Ben-Laouane, R., Boutasknit, A., Baslam, M., & Meddich, A. (2025). The importance of the *Glomus* genus as a potential candidate for sustainable agriculture under arid environments: A review. *International Journal of Plant Biology, 16*(1), 32.

Pellegrino, E., Nuti, M., & Ercoli, L. (2022). Multiple arbuscular mycorrhizal fungal consortia enhance yield and fatty acids of Medicago sativa: A two-year field study on agronomic traits and tracing of fungal persistence. *Frontiers in Plant Science, 13*, 814401.

Pellegrino, E., Turrini, A., Gamper, H. A., Cafà, G., Bonari, E., Young, J. P. W., & Giovannetti, M. (2012). Establishment, persistence and effectiveness of arbuscular mycorrhizal fungal inoculants in the field revealed using molecular genetic tracing and measurement of yield components. *New Phytologist, 194*(3), 810–822.

Pressel, S. (2025). Terrestrial surface stabilisation by modern analogues of the earliest land plants: A multi-dimensional imaging study. *Geobiology, 21*(4), 454–473.

Pressel, S., Bidartondo, M. I., Field, K. J., & Duckett, J. G. (2021). Advances in understanding of mycorrhizal-like associations in bryophytes. *Bryophyte Diversity and Evolution, 43*(1), 284–306.

Priyadharsini, P., Rojamala, K., Ravi, R. K., Muthuraja, R., Nagaraj, K., & Muthukumar, T. (2017). Mycorrhizosphere: The extended rhizosphere and its significance. In *Plant-microbe interaction: An approach to sustainable agriculture* (pp. 97–124). Springer.

Rillig, M. C., & Mummey, D. L. (2006). Mycorrhizas and soil structure. *New Phytologist, 171*(1), 41–53.

Rini, M. V., Yelli, F., Tambunan, D. L., & Damayanti, I. (2021). Morphological and molecular identifications of three native arbuscular mycorrhizal fungi isolated from the rhizosphere of *Elaeis guineensis* and *Jatropha curcas* in Indonesia. *Biodiversitas Journal of Biological Diversity, 22*(11), 4940–4947.

Robdrup, M., Hubbard, M., Gorim, L. Y., & Gorzelak, M. A. (2024). Arbuscular mycorrhizal fungi under intercrop, regenerative, and conventional agriculture systems. In *Arbuscular mycorrhizal fungi and higher plants: Fundamentals and applications* (pp. 287–318). Springer.

Roller, B. R., Stoddard, S. F., & Schmidt, T. M. (2016). Exploiting rRNA operon copy number to investigate bacterial reproductive strategies. *Nature Microbiology, 1*(11), 1–7.

Ropars, J., Toro, K. S., Noel, J., Pelin, A., Charron, P., Farinelli, L., Marton, T., Krüger, M., Fuchs, J., Brachmann, A., & Brachmann, A. (2016). Evidence for the sexual origin of heterokaryosis in arbuscular mycorrhizal fungi. *Nature Microbiology, 1*(6), 1–9.

Ryan, M. H., & Graham, J. H. (2018). Little evidence that farmers should consider abundance or diversity of arbuscular mycorrhizal fungi when managing crops. *New Phytologist, 220*(4), 1092–1107.

Saharan, B. S., Brar, B., Duhan, J. S., Kumar, R., Marwaha, S., Rajput, V. D., & Minkina, T. (2022). Molecular and physiological mechanisms to mitigate abiotic stress conditions in plants. *Life (Basel), 12*(10). https://doi.org/10.3390/life12101634.

Sanders, I. R. (2002). Ecology and evolution of multigenomic arbuscular mycorrhizal fungi. *The American Naturalist, 160*(S4), S128–S141.

Sania Saeed, H., Munir, M., Samra, S., Farooq, A., Mahmood, H., Ahmad, I., Irfan Akram, M., Iqbal, R., Elsadek, M. F., Almutairi, S. M., Ditta, A., & Daraz, U. (2025). Screening of biosurfactants producing fungi from infected *Citrus sinensis* (L.). *Polish Journal of Environmental Studies, 34*(5), 5261–5270. https://doi.org/10.15244/pjoes/190588.

Sedaghati, E., Yazdanpanah, M., & Nadi, M. (2021). A review of taxonomic studies of arbuscular mycorrhizal fungi in Iran. *Mycologia Iranica, 8*(2), 15–30.

Seleiman, M. F., & Hardan, A. N. (2021). Importance of mycorrhizae in crop productivity. In *Mitigating environmental stresses for agricultural sustainability in Egypt* (pp. 471–484). Springer.

Serghi, E. U., Kokkoris, V., Cornell, C., Dettman, J., Stefani, F., & Corradi, N. (2021). Homo- and dikaryons of the arbuscular mycorrhizal fungus *Rhizophagus irregularis* differ in life history strategy. *Frontiers in Plant Science, 12*, 715377.

Seutra Kaba, J., Abunyewa, A. A., Kugbe, J., Kwashie, G. K., Owusu Ansah, E., & Andoh, H. (2021). Arbuscular mycorrhizal fungi and potassium fertilizer as plant biostimulants and alternative research for enhancing plants adaptation to drought stress: Opportunities for enhancing drought tolerance in cocoa (Theobroma cacao L.). *Sustainable Environment, 7*(1), 1963927.

Shah, W., Ullah, S., Ali, S., Idrees, M., Khan, M. N., Ali, K., Khan, A., Ali, M., & Younas, F. (2021). Effect of exogenous alpha-tocopherol on physio-biochemical attributes and agronomic performance of lentil (*Lens culinaris* Medik.) under drought stress. *PLoS One, 16*(8), e0248200.

Stürmer, S. L. (2012). A history of the taxonomy and systematics of arbuscular mycorrhizal fungi belonging to the phylum Glomeromycota. *Mycorrhiza, 22*(4), 247–258.

Stürmer, S. L., & Kemmelmeier, K. (2021). The Glomeromycota in the neotropics. *Frontiers in Microbiology, 11*, 553679.

Susilawati, E., Chan, S. R. O. S., Achmad, B. S., & Maulina, F. (2022). Exploration and morphology identification of spores asbuscular mycorrhizal fungi from horticultural plantation. *Journal of Applied Agricultural Science and Technology, 6*, 20–30.

Symanczik, S., Courty, P.-E., Boller, T., Wiemken, A., & Al-Yahya'ei, M. N. (2015). Impact of water regimes on an experimental community of four desert arbuscular mycorrhizal fungal (AMF) species, as affected by the introduction of a non-native AMF species. *Mycorrhiza, 25*(8), 639–647.

Taylor, L., Leake, J., Quirk, J., Hardy, K., Banwart, S., & Beerling, D. (2009). Biological weathering and the long-term carbon cycle: Integrating mycorrhizal evolution and function into the current paradigm. *Geobiology, 7*(2), 171–191.

Tedersoo, L., Hosseyni Moghaddam, M. S., Mikryukov, V., Hakimzadeh, A., Bahram, M., Nilsson, R. H., Yatsiuk, I., Geisen, S., Schwelm, A., Piwosz, K., Prous, M., Sildever, S., Chmolowska, D., Rueckert, S., Skaloud, P., Laas, P., Tines, M., Jung, J.-H., Choi, J. H., … Piwosz, K. (2024a). EUKARYOME: The rRNA gene reference database for identification of all eukaryotes. *Database, 2024*, baae043.

Tedersoo, L., Magurno, F., Alkahtani, S., & Mikryukov, V. (2024b). Phylogenetic classification of arbuscular mycorrhizal fungi: New species and higher-ranking taxa in Glomeromycota and Mucoromycota (class Endogonomycetes). *MycoKeys, 107*, 273.

Thind, S., Chaudhary, M. S., Ditta, A., Hussain, I., Parveen, A., Ullah, N., Mahmood, Q., Al-ashkar, I., & El-Sabagh, A. (2022). Impact of mycorrhizal fungi from different rhizospheric soils on fungal colonization, growth, and chlorophyll contents of *Cenchrus ciliaris*. *Agronomy, 12*(11). https://doi.org/10.3390/agronomy12112644.

Thioye, B., Sanguin, H., Kane, A., Ndiaye, C., Fall, D., Sanogo, D., Duponnois, R., de Faria, S. M., Sylla, S. N., & Bâ, A. (2021). Mycorrhizal inoculation increases fruit production without disturbance of native arbuscular mycorrhizal community in jujube tree orchards (Senegal). *Symbiosis, 83*(3), 361–372.

Wagg, C., Schlaeppi, K., Banerjee, S., Kuramae, E. E., & van der Heijden, M. G. A. (2019). Fungal-bacterial diversity and microbiome complexity predict ecosystem functioning. *Nature Communications, 10*(1), 4841.

Wahab, A., Muhammad, M., Munir, A., Abdi, G., Zaman, W., Ayaz, A., Khizar, C., & Reddy, S. P. P. (2023). Role of arbuscular mycorrhizal fungi in regulating growth, enhancing productivity, and potentially influencing ecosystems under abiotic and biotic stresses. *Plants (Basel), 12*(17), 3102.

Wang, F., Zhang, L., Zhou, J., Rengel, Z., George, T. S., & Feng, G. (2022). Exploring the secrets of hyphosphere of arbuscular mycorrhizal fungi: Processes and ecological functions. *Plant and Soil, 481*(1), 1–22.

Wang, J., Wang, X., Liu, G., Zhang, C., & Wang, G. (2021). Bacterial richness is negatively related to potential soil multifunctionality in a degraded alpine meadow. *Ecological Indicators, 121*, 106996.

Wipf, D., Krajinski, F., van Tuinen, D., Recorbet, G., & Courty, P. E. (2019). Trading on the arbuscular mycorrhiza market: From arbuscules to common mycorrhizal networks. *New Phytologist, 223*(3), 1127–1142.

Yizhu, L., Imtiaz, M., Ditta, A., Shahid Rizwan, M., Ashraf, M., Mehmood, S., Aziz, O., Mubeen, F., Ali, M., Noor Elahi, N., Ijaz, R., Lele, S., Shuang, C., & Tu, S. (2020). Response of growth, antioxidant enzymes and root exudates production towards As stress in *Pteris vittata* and in *Astragalus sinicus* colonized by arbuscular mycorrhizal fungi. *Environmental Science and Pollution Research, 27*(2), 2340–2352. https://doi.org/10.1007/s11356-019-06785-5.

Yurkov, A. P., Kryukov, A. A., Gorbunova, A. O., Kudriashova, T. R., Kovalchuk, A. I., Gorenkova, A. I., Bogdanova, E. M., Laktionov, Y. V., Zhurbenko, P. M., Mikhaylova, Y. V., Puzanskiy, R. K., Bagrova, T. N., Yakhin, O. I., Rodionov, A. V., & Shishova, M. F. (2023). Diversity of arbuscular mycorrhizal fungi in distinct ecosystems of the North Caucasus, a temperate biodiversity hotspot. *Journal of Fungi, 10*(1), 11.

Zhang, J., Ruotong, Z., Xia, L., & Zhang, J. (2024). Potential of arbuscular mycorrhizal fungi for soil health: A review. *Pedosphere, 34*(2), 279–288.

Zhang, Z., Xiao, J., Yuan, Y., Zhao, C., Liu, Q., & Yin, H. (2018). Mycelium- and root-derived C inputs differ in their impacts on soil organic C pools and decomposition in forests. *Soil Biology and Biochemistry, 123*, 257–265.

Zhou, J., Zang, H., Loeppmann, S., Gube, M., Kuzyakov, Y., & Pausch, J. (2020). Arbuscular mycorrhiza enhances rhizodeposition and reduces the rhizosphere priming effect on the decomposition of soil organic matter. *Soil Biology and Biochemistry, 140*, 107641.

Zhou, X., Li, J., Tang, N., Xie, H., Fan, X., Chen, H., Tang, M., & Xie, X. (2021). Genome-wide analysis of nutrient signaling pathways conserved in arbuscular mycorrhizal fungi. *Microorganisms, 9*(8), 1557.

Chapter 3

Arbuscular mycorrhizal fungi: Plant nutrition and health for sustainable agriculture

Minisha Udhayakumar[a], Shanmugapriya Udhayakumar[a,b], Vignesh Chinnaiya[c], Subha Bharathi Pitchaimuthu[a], Kayalvizhi Thirumalaisamy[a], Ayyappa Das Madhubala Parameswaran[a] and Parimala Gnana Soundari Arockiam Jeyasundar[a]

[a] *Department of Biotechnology, Rathnavel Subramaniam (RVS) College of Arts and Science, Coimbatore, Tamil Nadu, India,* [b] *School of Biotechnology and Genetic Engineering, Bharathiar University, Coimbatore, Tamil Nadu, India,* [c] *Department of Biotechnology, Rathnavel Subramaniam (RVS) College of Arts and Science, Sulur, Coimbatore, Tamil Nadu, India*

3.1 Introduction

Global agriculture faces unprecedented challenges in meeting the food security demands of an exponentially growing population while simultaneously addressing the environmental degradation caused by intensive farming practices (Thind et al., 2022). Modern agricultural systems have increasingly relied on synthetic chemical inputs, including nitrogen, phosphorus, and potassium (NPK) fertilizers, pesticides, and fungicides, to sustain crop productivity, yet these practices have resulted in severe environmental contamination, soil degradation, water pollution, and threats to human and animal health (Gianinazzi-Pearson, 1996; Nazir et al., 2023; Smith & Read, 2010). The profuse use of phosphate fertilizers and chemicals poses significant pollution problems and health hazards, with approximately 75% of applied phosphorus fertilizer not being utilized by crops but instead being converted to forms unavailable to plants, representing both an economic loss and an environmental burden (Mosse et al., 1973). However, there is an urgent imperative to develop and implement sustainable agricultural practices that reduce dependency on synthetic chemical inputs while maintaining or enhancing crop productivity and soil health (Grant et al., 2005).

Arbuscular mycorrhizal fungi (AMF), obligate biotrophic soil organisms belonging to the phylum Glomeromycota, represent a natural and biological solution to these multifaceted agricultural challenges (Yizhu et al., 2020). These fungi, comprising more than 200 identified species, have evolved alongside land plants for over 450 million years, establishing mutualistic symbiotic relationships with the roots of approximately 80% of terrestrial plant species (Brachmann, 2006). The symbiotic association between AMF and plants is characterized by the formation of specialized fungal structures called arbuscules, finely branched hyphal structures that penetrate plant root cells and facilitate bidirectional nutrient exchange, with the fungus receiving photosynthetically fixed carbon from the plant, whereas the plant obtains essential mineral nutrients, particularly phosphorus, nitrogen, zinc, and copper (Garg & Chandel, 2011). Beyond their primary role in nutrient acquisition, AMF provide plants with enhanced tolerance to various biotic and abiotic stresses, including pathogenic infections, drought, salinity, heavy metal contamination, and extreme temperatures, while simultaneously improving soil structure and ecosystem function through the production of glomalin-related soil proteins (GRSP) and the regulation of carbon and nutrient cycling (Chen et al., 2018; Ruiz-Lozano, 2003).

Recent advances in AMF research and agricultural applications have demonstrated remarkable commercial and environmental potential. A few studies have shown that AMF inoculation increases barley yields by 17.27% and exhibits notable benefits in wheat and maize under saline and stressful soil conditions, while simultaneously reducing synthetic fertilizer requirements by up to 50% without compromising crop performance (Alotaibi et al., 2024) Nanjundappa et al., 2019). The combination of AMF with plant growth–promoting rhizobacteria (PGPR), such as *Bacillus* and *Pseudomonas*,

Advancing Sustainable Agriculture Through Arbuscular Mycorrhizal Fungi. DOI: https://doi.org/10.1016/B978-0-443-45130-0.00036-9

creates synergistic microbial consortia that further enhance nutrient mobilization and stress tolerance, representing an emerging frontier in sustainable agriculture technology (Saeed et al., 2025; Zhang et al., 2024). However, the successful exploitation of AMF potential in commercial agriculture is not straightforward and depends critically on appropriate agronomic management practices, species–environment matching, and a comprehensive understanding of the factors that influence AMF colonization, functionality, and persistence in agricultural soils (Bethlenfalvay & Schüepp, 1994). Although substantial progress has been made in characterizing AMF biology and documenting their benefits, significant research gaps persist regarding glomalin dynamics in soil, long-term effects of AMF on carbon and nitrogen cycling, tolerance mechanisms to heavy metal stress, and the maintenance of mutualistic functionality under intensive agricultural management (Nadeem et al., 2014).

This chapter synthesizes current knowledge on AMFs, their roles in sustainable crop production, soil health enhancement, and ecosystem services, while identifying critical research gaps and proposing future research directions essential for translating AMF science into practical, scalable agricultural solutions. By examining the multifaceted contributions of AMF to plant nutrient acquisition, stress tolerance, soil structure stabilization, and disease suppression, as well as the agronomic and environmental factors influencing their effectiveness, this work provides a comprehensive foundation for understanding and optimizing the integration of AMF-based technologies into next-generation regenerative and productive farming systems.

3.2 Current status of agricultural soil and crop production

AMF has become a major resource of agricultural research and commercial development, such as soil health enhancement, crop production, and yield performance (Wahab et al., 2023). It improves plant nutrient uptake efficiency by extending fungal hyphae to the reach of plant roots. This increased exploration enables plants to reach phosphorus and other nutrients stored in soil that would otherwise be unavailable (Kafle et al., 2019). AMF inoculation can increase nutrient availability; synthetic NPK fertilizer requirements can be lowered by up to 50% without losing crop performance (Fall et al., 2023). This combination benefits maintaining yields while reducing fertilizer inputs, making AMF both commercially and environmentally appealing. One of AMF's most significant contributions to long-term soil health is the creation of GRSP. These glycoproteins act as biological "glue," holding soil particles together in stable macroaggregates. This better aggregation increases soil porosity, water infiltration, aeration, and root penetration (Bronick & Lal, 2005). More crucially, AMF functions as a "rhizospheric carbon engineer," directing plant photosynthates into the soil and encouraging the development of long-lasting organic carbon pools. This carbon sequestration capacity is especially beneficial in arid and semiarid areas, where soil carbon is often sparse. According to studies, AMF inoculation increases microbial biomass carbon and promotes long-term carbon storage, helping to mitigate climate change (Shukla et al., 2025).

Recent studies have shown that AMF consistently and significantly improve a variety of crops and soil types. Field tests reveal that AMF inoculation increases barley yields by 17.27%, whereas wheat and maize exhibit notable benefits in saline soils and stressful conditions (Wahab et al., 2023). AMF improves salinity resilience in marginal soils, alleviates phosphorus scarcity in low-fertility tropical soils, and increases drought tolerance through extended hyphal networks and physiological changes, all while concurrently inducing systemic disease resistance and lowering nematode pressure (Tian et al., 2021). The most significant recent invention combines AMF with plant growth–promoting bacteria (PGPR), such as *Bacillus* and *Pseudomonas*, resulting in synergistic microbial consortia that cut synthetic fertilizer requirements by 50% while preserving crop performance. However, success depends on strategic implementation. AMF is sensitive to intense tillage, synthetic fertilizers, and fungicides, necessitating suitable agronomic techniques, such as conservation tillage, crop rotation, and organic amendments (Chaudhary et al., 2025). Species–environment matching is crucial, as effectiveness is affected by soil pH, temperature, moisture, microbiota, and inoculation time. The greatest benefits occur in low-to-moderate agrochemical systems, though AMF improves sustainability even in traditional agriculture. AMF reduces input costs while boosting soil carbon sequestration, reducing agricultural pollution, and increasing biodiversity, making it a key component of next-generation productive and regenerative farming systems (Khangura et al., 2023).

3.3 Role of plant nutrients for sustainable crop production

The AMF play a key role in sustainable crop production and limit plant growth by enhancing nutrient acquisition (Clark et al., 2020). AMF is encouraging uptake of plant nutrients, especially P uptake. The benefits of AMF symbiosis are particularly pronounced in agricultural systems characterized by low phosphorus availability, where the fungal symbiont's extensive extraradical mycelium significantly enhances early phosphorus uptake, improving crop yield potential without the necessity for starter phosphorus fertilizer application (Grant et al., 2005). The relative dependency of crop plants on AMF for nutrient

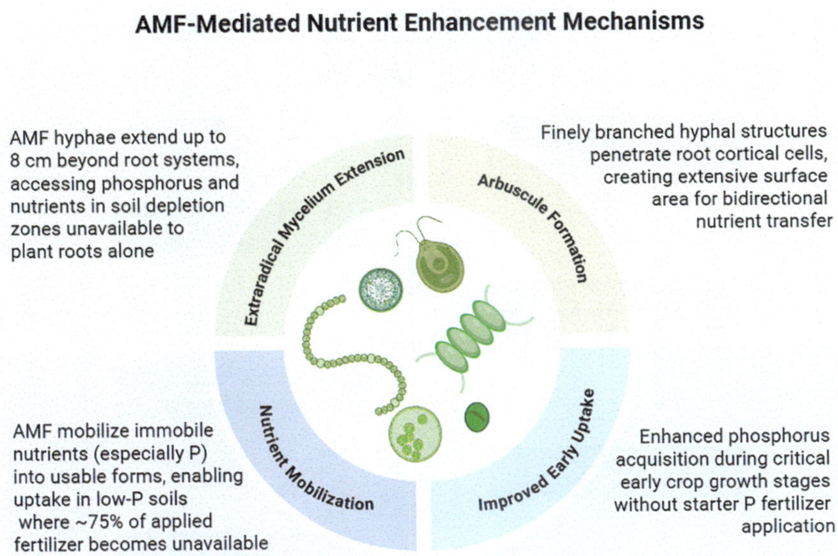

FIGURE 3.1 AMF-mediated nutrient enhancement mechanisms. This figure displays the nutrient enhancement mechanisms in plat through AMF. *AMF*, Arbuscular mycorrhizal fungi.

acquisition varies according to root morphological factors, such as surface area, root hair abundance and length, growth rate, and responses to soil conditions; crops, including corn (*Zea mays*) and flax (*Linum usitatissimum*), are highly dependent on AMF to meet their early phosphorus requirements, whereas legumes, beans, potatoes, barley, wheat, and oats also benefit significantly from mycorrhizal symbiosis (Smith & Read, 2010). Tropical agricultural systems particularly benefit from AMF symbiosis because of the prevalence of phosphorus-deficient soils in these regions, and research indicates that approximately 75% of applied phosphorus fertilizer is not utilized by crops but instead is converted to forms unavailable to plants, highlighting the importance of mycorrhizal-mediated nutrient mobilization in sustainable production (Mosse et al., 1973). By promoting efficient nutrient cycling through AMF associations, agricultural systems can achieve improved crop productivity while simultaneously reducing environmental pollution and health hazards associated with the excessive application of phosphate fertilizers and chemical inputs (Fig. 3.1).

By harnessing the natural abilities of soil through AMF symbiosis, it is possible to reduce the need for costly and environmentally harmful chemical inputs, including water and fuel inputs, while simultaneously reducing input overheads and increasing production efficiency (Ebbisa, 2022). The benefits of AMF symbiosis are particularly pronounced in agricultural systems characterized by low phosphorus availability, where the fungal symbiont's extensive extraradical mycelium significantly enhances nutrient mobilization into usable forms, improving crop yield potential without the necessity for starter phosphorus fertilizer application (Wahab et al., 2023). The relative dependency of crop plants on AMF for nutrient acquisition varies according to root morphological factors, such as surface area, root hair abundance and length, growth rate, and responses to soil conditions. AMF meet their advanced phosphorus requirements, whereas legumes, beans, and other crops also benefit significantly from mycorrhizal symbiosis (Chandrasekaran, 2022). Tropical agricultural systems particularly benefit from AMF symbiosis because of the prevalence of phosphorus-deficient soils in these regions, and research indicates that approximately 75% of applied phosphorus fertilizer is not utilized by crops but instead is converted to forms unavailable to plants, highlighting the importance of mycorrhizal-mediated nutrient mobilization (Jatana et al., 2021). By promoting efficient nutrient cycling and establishing symbiotic relationships that enable plants to access more essential nutrients, including phosphorus and nitrogen, from the soil, AMF associations can efficiently boost agricultural productivity while reducing chemical inputs, such as insecticides and fertilizers, thereby maintaining soil ecosystem health and ensuring long-term sustainability (Ebbisa, 2022).

3.4 Overview of arbuscular mycorrhizal fungi

AMF are obligate biotrophic soil organisms belonging to the phylum Glomeromycota (Glomeromycotina within Mucoromycota), comprising more than 200 identified species that are characterized by the presence of arbuscules, specialized,

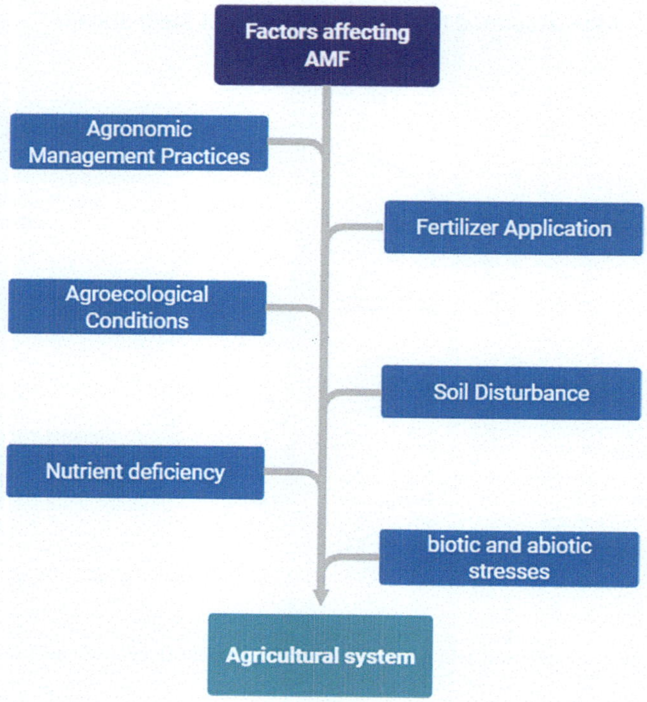

FIGURE 3.2 Factors affecting AMF in agriculture ecosystem. This chart shows the different factors that influence AMF in agriculture. *AMF*, Arbuscular mycorrhizal fungi.

finely branched hyphal structures responsible for nutrient exchange between the fungal symbiont and its plant host (Brachmann, 2006). These fungi establish a mutualistic symbiosis with the roots of the majority of land plants, colonizing the root cortex by penetrating the root system and forming intracellular arbuscules that penetrate the plant cell membrane, providing an extensive surface area for bidirectional nutrient transfer (Gianinazzi-Pearson, 1996; Smith & Read, 2010). The arbuscules are fine hyphal structures that function as the primary site for nutrient exchange between partners, with the fungus obtaining carbon compounds in the form of carbohydrates or lipids from the host plant, whereas the plant receives improved mineral nutrition, particularly phosphorus, nitrogen, zinc, and copper (Akhtar & Panwar, 2011).

The life cycle of AMF includes distinct developmental phases that are critical to understanding their symbiotic relationships with plants. The process begins with the asymbiotic stage initiated by spore germination in soil, followed by the presymbiotic stage, where germ tubes elongate in response to plant-derived signal molecules, such as strigolactones and flavonoids (Lanfranco et al., 2018). Once the germinated spore hyphae make contact with compatible plant roots, either directly through root hairs or via hyphopodia, the symbiotic phase commences, during which intraradical mycelium colonizes the root cortex and forms arbuscules and other specialized structures, including vesicles and hyphal coils (Parihar et al., 2020). Simultaneously the extraradical mycelium extends beyond the roots, functioning as an extension of the root system by forming an extensive hyphal network that can penetrate up to 8 cm into the surrounding soil, thereby increasing the plant's capacity to forage for nutrients and water beyond the depletion zone (Giovannetti et al., 2006). AMF possess unique genomic characteristics, including heterokaryotic nuclei within coenocytic hyphae and the capacity for anastomosis between related strains, which contribute to significant genetic diversity and adaptability within populations despite their ancient evolutionary origins dating back over 450 million years (Barea et al., 2002; Chen et al., 2018).

3.5 Role of arbuscular mycorrhizal fungi in agricultural ecosystems

AMF play a fundamental role in sustainable agricultural systems through their biofertilizing action, which represents the most important benefit of the AMF–plant symbiosis in agricultural contexts (Gianinazzi-Pearson, 1996). The extensive extraradical mycelium significantly improves plant uptake of essential nutrients that are typically present at limiting concentrations in field soils, particularly the macronutrients phosphorus and nitrogen, as well as micronutrients, such as zinc and copper (Bolan, 1991). Beyond nutrient acquisition, mycorrhizal associations discuss substantial nonnutritional benefits to plants, including enhanced tolerance to various biotic and abiotic stresses (Fig. 3.2), such as pathogenic infections,

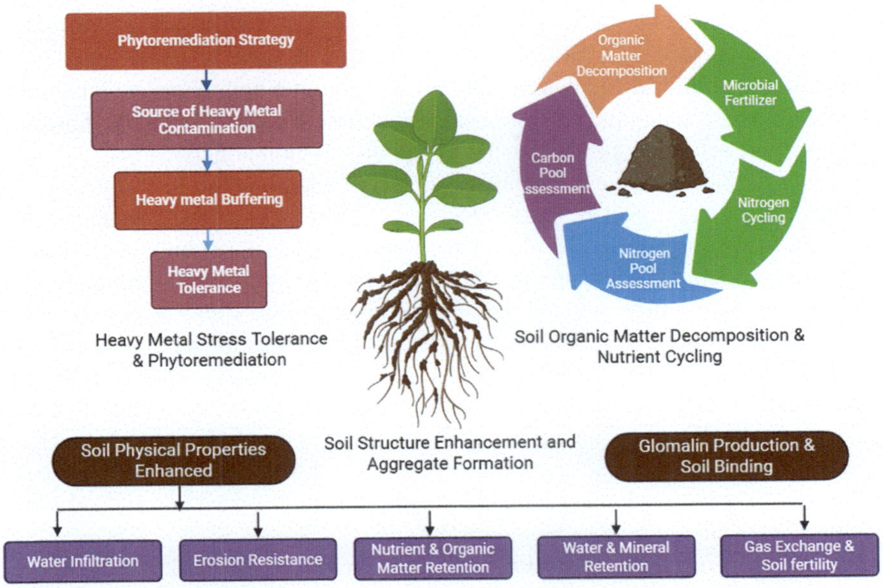

FIGURE 3.3 Improvement of soil health and structure. This diagram displays the multiple mechanisms of how the AMFs influence the soil health and structures. *AMF*, Arbuscular mycorrhizal fungi.

drought conditions, salinity, heavy metal contamination, and extreme temperatures (Chen et al., 2018; Ruiz-Lozano, 2003). The effectiveness of mycorrhizal fungi applications in agricultural systems will depend on numerous factors, as these fungal species are affected by different agronomic management practices, uses of fertilizers, and various agroecological conditions, making knowledge of various agriculture systems essential for optimizing their use as plant protection agents (Akhtar & Panwar, 2011).

Beyond their direct effects on plant nutrition and stress tolerance, AMF substantially contribute to the maintenance and enhancement of soil structure and ecosystem function within agricultural systems. The extraradical hyphae physically enmesh and stabilize soil particles into water-stable aggregates, whereas the glycoprotein glomalin, secreted by functioning hyphae and potentially comprising up to 1.5% of soil dry weight, plays a critical role in aggregate stability and water retention (Nichols, 2003). Furthermore, AMF regulate carbon flow and nutrient cycling within the soil microbial community through effects on root exudation, thereby influencing the composition and activity of rhizosphere bacterial populations and functioning as determinants of the structure and function of plant communities (Zaman et al., 2025). The integration of AMF into agricultural management practices represents a strategic approach to developing sustainable crop production systems with reduced chemical inputs, as modern intensive agricultural practices including synthetic fertilization, continuous monoculture, and tillage significantly impact AMF colonization and functionality, underscoring the importance of understanding agricultural effects on these beneficial fungal communities in developing management regimes for optimized agroecosystem health and productivity (Ellouze et al., 2014).

3.6 Improvement of soil health and quality

AMFs play an important role in improving soil structure and maintaining health and their quality (Fig. 3.3). Mycelia occur in large quantities within soils (de Novais et al., 2019), and their hyphae contribute to the formation of stable soil aggregates. These fungi function as prolonged soil-binding agents through the production of a glycoprotein known as glomalin, which is produced by the extraradical mycelium (Singh et al., 2022). Glomalin is a hydrophobic and thermally stable compound that is against too high soil temperatures. Its hydrophobic nature increases the resistance of soil aggregates to water, and its production is greatest in senescent mycelia. Although glomalin is steadily biodegradable by soil bacteria and fungi, it plays a crucial role in stabilizing soil aggregates (Hu et al., 2023). It is characterized as a natural "glue"; glomalin binds soil microaggregates (<250 μm in diameter) into stable macroaggregates (Lehmann et al., 2020). These macroaggregates upgrade soil physical properties by increasing water infiltration, decreasing surface runoff and soil erosion, reducing nutrient and organic matter losses, enhancing gas exchange, and increasing the retention of water and minerals, particularly potassium, thereby contributing to upgraded crop productivity (Parihar et al., 2020). In addition, the AMF mycelial network undergoes continuous renewal, and even after mycelial death, its structural contribution continues until decomposition occurs

(Gianinazzi-Pearson, 1996). Dead mycelia further contribute to soil organic matter and serve as physical binding agents, taking part in aggregate formation. Moreover, these processes decrease the risk of soil compaction and increase overall soil fertility (Clark et al., 2020). Thus AMFs enhance soil structure through both chemical and biophysical mechanisms, including soil particle enmeshment and alignment. However, gaps persist in understanding the longevity of glomalin in soil and the impacts of anthropogenic activities, such as bush burning, on soil glomalin dynamics.

AMF are extensive plant symbionts and are frequently regarded as "microbial fertilizers" because of their advantageous effects on crop growth and productivity (Sui et al., 2019; Zhang et al., 2024). Enhancing evidence indicates that mycorrhizal fungi play a significant role in soil organic matter decomposition and nutrient cycling, particularly nitrogen, which are vital processes for achieving agricultural sustainability. Although AMF are generally considered to have positive effects on host plant performance and soil quality, detailed agricultural practices are frequently associated with a reduction in mycorrhizal functioning (Thirkell et al., 2017). Conditions, such as high nutrient inputs from chemical fertilizers, excessive soil moisture, and soil disturbance, can reduce the mutualistic role of mycorrhizas, rendering them neutral symbionts or even shifting them toward parasitic behavior (Trejo et al., 2016). These responses mess up the effective application of AMF inoculation in agricultural systems. However, further research is needed to clarify how plant productivity and soil organic matter decomposition respond to environmental stress in the presence of AMF inoculation. In particular, understanding the effects of AMF on soil carbon and nitrogen pools is essential for assessing their contribution to soil quality and the prolonged sustainability of agriculture, especially in semiarid regions (Dai et al., 2022).

AMFs also play a vital role in increasing plant tolerance to environmental stress, particularly in the revegetation of naturally or industrially metal-contaminated sites (Udhayakumar et al., 2024). AMF are often found in metal-rich soils and contribute significantly to phytoremediation strategies (Krishnamoorthy et al., 2019). In addition, phosphate fertilizers, one of the major sources of heavy metal inputs, can contaminate soils and adversely affect agricultural systems (Nziguheba & Smolders, 2008). AMF not only increase phosphorus intake and mobilization but also exert buffering effects on cadmium intake, thereby decreasing its toxicity to plants (Rivera-Becerril et al., 2002). However, high concentrations of heavy metals have been noted to suppress, delay, or even eliminate AMF populations (Dhalaria et al., 2020). Negative relationships between zinc levels in sludge-amended soils and AMF colonization have also been reported (Sayın et al., 2019). Conversely, other studies have shown that the application of metal-containing sludge does not necessarily impair AMF development. These contrasting findings suggest that variations among studies may be attributed to differences in AMF ecotypes, which show varying degrees of tolerance to heavy metals (Singh et al., 2022).

AMF play a vital role in plant disease control through a variety of well-reported mechanisms. AMF inoculation has been shown to decrease the pathogenic potential of huge soil-borne pathogens while, at the same time, increasing plant tolerance to pathogen attack (Ravnskov et al., 2020). Both direct and indirect mechanisms contribute to the biomanagement of plant diseases by AMF. Direct effects include competition for nutrients and space, which limits pathogen propagation and establishment. Indirectly, AMF enhance host plant tolerance by stimulating defense responses, inducing systemic resistance, and altering root exudation patterns, thereby reshaping the rhizosphere environment in ways that are unfavorable to pathogens. These mechanisms do not act in isolation; rather, an integrated combination of processes operates collectively to suppress plant pathogens (Mathur & Jajoo, 2020).

3.7 Plant nutrient acquisition via arbuscular mycorrhizal fungi networks

Enhanced mineral nutrition is considered the primary advantage of AMF symbiosis, particularly in boosted phosphorus and nitrogen absorption (Fig. 3.4). These two essential macronutrients are needed in large quantities by plants, yet most plants repeatedly face low levels of available N and P in natural environments (Elser et al., 2007). Plants capable of forming AMF symbiosis are regarded as possessing a specialized "mycorrhizal phosphate uptake" pathway, along with the standard phosphate uptake by root epidermal cells, which typically provides a lot of their phosphate (Tarafdar, 2022). Nutrient exchange is a main function in the relationship between AMF and their host plants. The extraradical mycelium of the fungi absorbs mineral nutrients from the soil and discharges some of these nutrients at the fungus–plant interface, whereas the plant in revoke gives carbon to the fungus (Battini et al., 2017). Nutrient exchange in the relationship between plant roots and AMF emerges through two fungal structures, the intraradical mycelium (IM), which grows within the root cortex, and the extraradical mycelium (EM), which extends into the soil. Within the root cortex the IM forms arbuscules inside cortical cells, and it is primarily over the close interface between these arbuscules and the plant cell membrane that nutrient uptake occurs (Harrison et al., 2002). The extraradical mycelium (EM) forms short-lived, branched absorbing structures, as described by Williams et al. (2017), which continue to increase in number under high phosphorus concentrations in the neighboring medium.

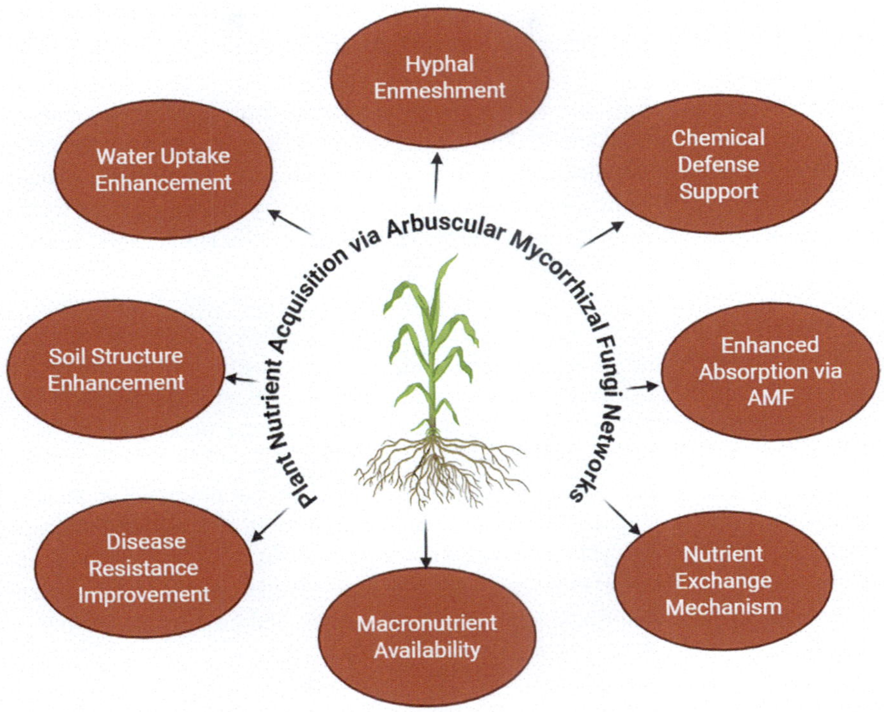

FIGURE 3.4 Plant nutrient acquisition via AMF network. The image shows the ways of plant nutrient acquisition by AMF. *AMF*, Arbuscular mycorrhizal fungi.

AMF can also apply nonnutritional effects on plants, some of which may be just as crucial as their role in nutrient exchange in upgrading plant fitness and influencing community structure and ecosystem dynamics (Smith & Read, 2010). In addition to interchange nutrients for carbon, AMF offer various nonnutritional benefits to plants, such as increasing water uptake, enhancing soil structure, improving disease resistance, supporting chemical defenses, and helping allelopathic interactions. The extraradical mycelial networks not only transport nutrients but also contribute to greater water absorption (Fricker et al., 2017) and assist in the distribution of allelochemicals (Lehmann et al., 2020). AMFs also provide GRSP, which promote soil aggregation, helping to reduce erosion and improve water stability (Rillig et al., 2019). Moreover, AMF hyphae aid in stabilizing soil macroaggregates through a process known as hyphal "enmeshment," where soil particles are physically bound by fungal hyphae (Rillig, 2004). Beyond structural benefits, AMF can upgrade plant defense mechanisms through nonnutritional pathways. They can prime plants to activate defense-related genes, leading to mycorrhizal-induced resistance and an enhanced defensive state triggered by AMF colonization (Azcón-Aguilar & Barea, 1992).

3.8 Current research gap and future research directions

Despite the substantial advances in understanding AMF and their applications in sustainable agriculture, several significant research gaps persist that limit the optimization and widespread adoption of AMF-based technologies. A critical gap exists in understanding the longevity and dynamics of glomalin in soil, particularly regarding how anthropogenic activities, such as bush burning and other land management practices, affect the persistence and efficacy of GRSP in stabilizing soil aggregates. Furthermore a comprehensive understanding of how plant productivity and soil organic matter decomposition respond to environmental stress in the presence of AMF inoculation remains incomplete, especially concerning the distinct effects of AMF on soil carbon and nitrogen pools, knowledge which is essential for assessing their contribution to soil quality and the prolonged sustainability of agriculture, particularly in semiarid regions. In addition, although recent studies have demonstrated that AMF success depends on strategic implementation and species–environment matching, detailed information regarding the precise mechanisms of how different soil pH, temperature, moisture regimes, and microbiota compositions interact with specific AMF species to determine colonization success and functional efficacy remains limited. Research on the variable responses of different AMF ecotypes to heavy metal stress is inconclusive, with contrasting findings regarding the tolerance levels of various ecotypes to metal contamination, suggesting that more systematic characterization of AMF genetic diversity and metal tolerance mechanisms is urgently needed. Finally, further research is required to clarify the circumstances under which high nutrient inputs from chemical fertilizers, excessive soil moisture, and soil disturbance transform AMF from mutualistic symbionts into neutral or parasitic organisms and how to maintain their beneficial functions under intensive agricultural management.

Future research should prioritize the molecular and mechanistic characterization of AMF–plant–microbe interactions, particularly focusing on how combined inoculants of AMF with plant growth–promoting bacteria (PGPR) synergistically enhance nutrient acquisition and stress tolerance across diverse crop species and soil types. Longitudinal field studies conducted across different agroecological zones should be designed to quantify the long-term effects of AMF inoculation on soil carbon sequestration, nitrogen cycling dynamics, and the persistence of glomalin in soils under various management practices and climatic conditions. Research should also investigate the genetic basis of AMF tolerance to heavy metals and the mechanisms underlying their phytoremediation potential in metal-contaminated sites, with particular attention to identifying ecotypes with enhanced tolerance and bioaccumulation capabilities. Future studies must develop predictive models that integrate soil and environmental parameters (pH, temperature, moisture, texture, and microbiota composition) with AMF species-specific characteristics to enable precision matching of fungal inoculants to specific agricultural systems and ensure consistent field performance. In addition, research should focus on understanding how to maintain AMF mutualistic functionality under intensive agricultural conditions, including the development of agronomic practices that minimize negative impacts of high fertilizer inputs, soil disturbance, and chemical fungicide applications while optimizing AMF colonization and function. Finally, field validation studies of recent laboratory findings demonstrating the synergistic effects of AMF–PGPR consortia on reducing synthetic fertilizer requirements by 50% while maintaining crop yields should be scaled across diverse crops, climates, and soil types to establish practical guidelines for farmers and develop standardized commercial inoculant formulations suitable for next-generation regenerative and productive farming systems.

3.9 Conclusion

AMFs represent a pivotal biological resource for addressing the interconnected challenges of sustainable crop production, soil health deterioration, and environmental contamination in modern agriculture. Their multifaceted contributions, encompassing enhanced plant nutrient acquisition, improved stress tolerance, soil structure stabilization through glomalin production, carbon sequestration, disease suppression, and phytoremediation, collectively position AMF as fundamental components of regenerative and productive agricultural systems. The evidence from recent field studies demonstrates that AMF inoculation can increase crop yields (barley by 17.27% and wheat and maize in saline conditions) while reducing synthetic fertilizer requirements by up to 50%, making them both economically and environmentally advantageous. However, the successful exploitation of AMF potential is contingent upon understanding and implementing appropriate agronomic practices, including conservation tillage, crop rotation, and organic amendments, as well as achieving precise species–environment matching tailored to specific soil conditions and agricultural systems. The formation of synergistic microbial consortia combining AMF with plant growth–promoting bacteria represents an emerging frontier that promises enhanced sustainability outcomes while maintaining crop performance. Despite substantial progress in characterizing AMF biology and their ecological roles, critical research gaps persist regarding glomalin dynamics, long-term carbon and nitrogen cycling effects, heavy metal tolerance mechanisms, and the maintenance of mutualistic functionality under intensive management. Addressing these gaps through coordinated research initiatives and field validation studies is essential for transitioning from laboratory-based findings to practical, farmer friendly applications that can be scaled globally. The integration of AMF-based technologies into mainstream agricultural practice offers a pathway toward reducing agricultural pollution, enhancing soil ecosystem services, increasing biodiversity, and ensuring the long-term sustainability of agricultural production systems while mitigating the impacts of climate change and supporting food security for growing global populations.

References

Akhtar, M., & Panwar, J. (2011). Arbuscular mycorrhizal fungi and opportunistic fungi: Efficient root symbionts for the management of plant parasitic nematodes. *Advanced Science, Engineering and Medicine, 3*, 165–175.

Alotaibi, M.M., Aljuaid, A., Alharbi, M.M., Qumsani, A.T., Alzuaibr, F.M., Alsubeie, M.S., ... & Awad-Allah, M.M. (2024). The effects of bio-fertilizer by arbuscular mycorrhizal fungi and phosphate solubilizing bacteria on the growth and productivity of barley under deficit of water irrigation conditions. Agronomy, 14(9), 1973.

Azcón-Aguilar, C., & Barea, J. M. (1992). *Interactions between Mycorrhizal Fungi and Other Rhizosphere Microorganisms*. Mycorrhizal functioning: an integrative plant-fungal process (pp. 163–198). New York: Chapman and Hall.

Barea, J. M., Gryndler, M., Lemanceau, P., Schüepp, H., & Azcón, R. (2002). The rhizosphere of mycorrhizal plants. In *Mycorrhizal technology in agriculture: From genes to bioproducts* (pp. 1–18). Springer.

Battini, F., Grønlund, M., Agnolucci, M., Giovannetti, M., & Jakobsen, I. (2017). Facilitation of Phosphorus Uptake in Maize Plants by Mycorrhizosphere Bacteria. *Sci. Rep., 7*(1), 4686.

Bethlenfalvay, G. J., & Schüepp, H. (1994). Arbuscular mycorrhizas and agrosystem stability. In *Impact of arbuscular mycorrhizas on sustainable agriculture and natural ecosystems* (pp. 117–131). Springer.

Bolan, N. S. (1991). A critical review on the role of mycorrhizal fungi in the uptake of phosphorus by plants. *Plant and Soil, 134*, 189–207.

Brachmann, A. (2006). Plant-fungal symbiosis en gros and en detail. *New Phytologist, 171*, 242–246.

Bronick, C. J., & Lal, R. (2005). Soil structure and management: A review. *Geoderma, 124*, 3–22.

Chandrasekaran, M. (2022). Arbuscular mycorrhizal fungi mediated enhanced biomass, root morphological traits and nutrient uptake under drought stress: A meta-analysis. *Journal of Fungi, 8*, 660.

Chaudhary, A., Poudyal, S., & Kaundal, A. (2025). Role of arbuscular mycorrhizal fungi in maintaining sustainable agroecosystems. *Applied Microbiology, 5*, 6.

Chen, M., Arato, M., Borghi, L., Nouri, E., & Reinhardt, D. (2018). Beneficial services of arbuscular mycorrhizal fungi–from ecology to application. *Frontiers in Plant Science, 9*, 1270.

Clark, A. L., Jahn, C. E., & Norton, A. P. (2020). Initiating plant herbivory response increases impact of fungal pathogens on a clonal thistle. *Biological Control, 143*, 104207.

Dai, F., Rong, Z., Wu, Q., Abd-Allah, E. F., Liu, C., & Liu, S. (2022). Mycorrhiza Improves Plant Growth and Photosynthetic Characteristics of Tea Plants in Response to Drought Stress. *Biocell, 46*(5), 1339–1346.

de Novais, C. B., Giovannetti, M., de Faria, S. M., & Sbrana, C. (2019). Two herbicides, two fungicides and spore-associated bacteria affect *Funneliformis mosseae* extraradical mycelium structural traits and viability. *Mycorrhiza, 29*, 341–349.

Dhalaria, R., Kumar, D., Kumar, H., Nepovimova, E., Kuča, K., Torequl Islam, M., & Verma, R. (2020). Arbuscular mycorrhizal fungi as potential agents in ameliorating heavy metal stress in plants. *Agronomy, 10*(6), 815.

Ebbisa, A. (2022). Arbuscular mycorrhizal fungi (AMF) in optimizing nutrient bioavailability and reducing agrochemicals for maintaining sustainable agroecosystems. *Arbuscular mycorrhizal fungi in agriculture-new insights*. IntechOpen.

Ellouze, W., Esmaeili Taheri, A., Bainard, L. D., Yang, C., Bazghaleh, N., Navarro-Borrell, A., Hanson, K., & Hamel, C. (2014). Soil fungal resources in annual cropping systems and their potential for management. *BioMed Research International, 2014*, 531824.

Elser, J. J., Bracken, M. E. S., Cleland, E. E., Gruner, D. S., Harpole, W. S., Hillebrand, H., ... Smith, J. E. (2007). Global Analysis of Nitrogen and Phosphorus Limitation of Primary Producers in Freshwater, Marine and Terrestrial Ecosystems. *Ecol. Lett., 10*(12), 1135–1142.

Fall, A. F., Nakabonge, G., Ssekandi, J., Founoune-Mboup, H., Badji, A., Ndiaye, A., Ndiaye, M., Kyakuwa, P., Anyoni, O. G., & Kabaseke, C. (2023). Combined effects of indigenous arbuscular mycorrhizal fungi (AMF) and NPK fertilizer on growth and yields of maize and soil nutrient availability. *Sustainability, 15*, 2243.

Fricker, M. D., Heaton, L. L. M., Jones, N. S., & Boddy, L. (2017). The Mycelium as a Network. *The fungal kingdom*, 335–367.

Garg, N., & Chandel, S. (2011). Arbuscular mycorrhizal networks: Process and functions. In *Sustainable agriculture volume 2* (pp. 907–930). Springer.

Gianinazzi-Pearson, V. (1996). Plant cell responses to arbuscular mycorrhizal fungi: Getting to the roots of the symbiosis. *Plant Cell, 8*, 1871.

Giovannetti, M., Avio, L., Fortuna, P., Pellegrino, E., Sbrana, C., & Strani, P. (2006). At the root of the wood wide web: Self recognition and nonself incompatibility in mycorrhizal networks. *Plant Signaling & Behavior, 1*, 1–5.

Grant, C., Bittman, S., Montreal, M., Plenchette, C., & Morel, C. (2005). Soil and fertilizer phosphorus: Effects on plant P supply and mycorrhizal development. *Canadian Journal of Plant Science, 85*, 3–14.

Harrison, M. J., Dewbre, G. R., & Liu, J. (2002). A Phosphate Transporter from Medicago Truncatula Involved in the Acquisition of Phosphate Released by Arbuscular Mycorrhizal Fungi. *Plant Cell, 14*(10), 2413–2429.

Hu, X., Yang, P., Chai, C., Liu, J., Sun, H., Wu, Y., Zhang, M., Zhang, M., Liu, X., & Yu, H. (2023). Structural and mechanistic insights into fungal β-1, 3-glucan synthase FKS1. *Nature, 616*, 190–198.

Jatana, B. S., Kitchens, C., Ray, C., Gerard, P., & Tharayil, N. (2021). Dual inoculation with arbuscular mycorrhizal fungi and phosphorus solubilizing fungi synergistically enhances the mobilization and plant uptake of phosphorus from meat and bone meal. *Frontiers in Soil Science, 1*, 757839.

Kafle, A., Cope, K. R., Raths, R., Krishna Yakha, J., Subramanian, S., Bücking, H., & Garcia, K. (2019). Harnessing soil microbes to improve plant phosphate efficiency in cropping systems. *Agronomy, 9*, 127.

Khangura, R., Ferris, D., Wagg, C., & Bowyer, J. (2023). Regenerative agriculture—A literature review on the practices and mechanisms used to improve soil health. *Sustainability, 15*, 2338.

Krishnamoorthy, R., Venkatramanan, V., Senthilkumar, M., Anandham, R., Kumutha, K., & Sa, T. (2019). Management of Heavy Metal Polluted Soils: Perspective of Arbuscular Mycorrhizal Fungi. In Sustainable green technologies for environmental management (pp. 67–85). Springer.

Lanfranco, L., Fiorilli, V., Venice, F., & Bonfante, P. (2018). Strigolactones cross the kingdoms: Plants, fungi, and bacteria in the arbuscular mycorrhizal symbiosis. *Journal of Experimental Botany, 69*, 2175–2188.

Lehmann, A., Zheng, W., Ryo, M., Soutschek, K., Roy, J., Rongstock, R., Maaß, S., & Rillig, M. C. (2020). Fungal traits important for soil aggregation. *Frontiers in Microbiology, 10*, 2904.

Mathur, S., & Jajoo, A. (2020). Arbuscular Mycorrhizal Fungi Protects Maize Plants from High Temperature Stress by Regulating Photosystem II Heterogeneity. *Ind. Crops Prod., 143*, 111934.

Mosse, B., Hayman, D. S., & Arnold, D. J. (1973). Plant growth responses to vesicular-arbuscular mycorrhiza V. Phosphate uptake by three plant species from P-deficient soils labelled with 32P. *New Phytologist, 72*, 809–815.

Nadeem, S. M., Ahmad, M., Zahir, Z. A., Javaid, A., & Ashraf, M. (2014). The role of mycorrhizae and plant growth promoting rhizobacteria (PGPR) in improving crop productivity under stressful environments. *Biotechnology Advances, 32*, 429–448.

Nanjundappa, A., Bagyaraj, D. J., Saxena, A. K., Kumar, M., & Chakdar, H. (2019). Interaction between arbuscular mycorrhizal fungi and Bacillus spp. in soil enhancing growth of crop plants. *Fungal biology and biotechnology, 6*(1), 23.

Nazir, A., Sarfraz, W., Allah, D., Khalid, N., Farid, M., Shafiq, M., Firdaus-E-Bareen, Rizvi, Z. F., & Naeem, N. (2023). Synergistic impact of two autochthonous saprobic fungi (*A. niger* and *T. pseudokoningii*) on the growth, ionic contents, and metals uptake in *Brassica juncea* L. and *Vigna radiata* L. under tannery solid waste contaminated soil. *International Journal of Phytoremediation, 25*(11), 1488–1500. doi:10.1080/15226514.2023.2166457.

Nichols, K. A. (2003). *Characterization of glomalin, a glycoprotein produced by arbuscular mycorrhizal fungi*. University of Maryland.

Nziguheba, G., & Smolders, E. (2008). Inputs of Trace Elements in Agricultural Soils via Phosphate Fertilizers in European Countries. *Science of the total environment, 390*(1), 53–57.

Parihar, M., Rakshit, A., Meena, V. S., Gupta, V. K., Rana, K., Choudhary, M., Tiwari, G., Mishra, P. K., Pattanayak, A., & Bisht, J. K. (2020). The potential of arbuscular mycorrhizal fungi in C cycling: A review. *Archives of Microbiology, 202*, 1581–1596.

Ravnskov, S., Cabral, C., & Larsen, J. (2020). Mycorrhiza Induced Tolerance in Cucumis Sativus against Root Rot Caused by Pythium Ultimum Depends on Fungal Species in the Arbuscular Mycorrhizal Symbiosis. *Biological Control, 141*, 104133.

Rillig, M. C. (2004). Arbuscular Mycorrhizae, Glomalin, and Soil Aggregation. *Can. J. Soil Sci., 84*(4), 355–363.

Rillig, M. C., Aguilar-Trigueros, C. A., Camenzind, T., Cavagnaro, T. R., Degrune, F., Hohmann, P., … Yang, G. (2019). Why farmers should manage the arbuscular mycorrhizal symbiosis. *The New Phytologist, 222*(3), 1171–1175.

Rivera-Becerril, F., Calantzis, C., Turnau, K., Caussanel, J., Belimov, A. A., Gianinazzi, S., … Gianinazzi-Pearson, V. (2002). Cadmium Accumulation and Buffering of Cadmium-induced Stress by Arbuscular Mycorrhiza in Three Pisum Sativum L. *Genotypes. J. Exp. Bot., 53*(371), 1177–1185.

Ruiz-Lozano, J. M. (2003). Arbuscular mycorrhizal symbiosis and alleviation of osmotic stress. New perspectives for molecular studies. *Mycorrhiza, 13*, 309–317.

Saeed, H. S., Munir, M., Samra, S., Farooq, A., Mahmood, H., Ahmad, I., Irfan Akram, M., Iqbal, R., Elsadek, M. F., Almutairi, S. M., Ditta, A., & Daraz, U. (2025). Screening of biosurfactants producing fungi from infected *Citrus sinensis* (L.). *Polish Journal of Environmental Studies, 34*(5), 5261–5270. doi:10.15244/pjoes/190588.

Sayın, F. E., Khalvati, M. A., & Erdinçler, A. (2019). Effects of Sewage Sludge Application and Arbuscular Mycorrhizal Fungi (G. Mosseae and G. Intraradices) Interactions on the Heavy Metal Phytoremediation in Chrome Mine Tailings. *Front. Plant Sci., 112*, 217–224.

Shukla, S., Didwania, N., & Choudhary, R. (2025). Arbuscular mycorrhizal fungi (AMF): A pathway to sustainable soil health, carbon sequestration, and greenhouse gas mitigation. *Journal of the Saudi Society of Agricultural Sciences, 24*, 22.

Singh, P., Mugunthan, J., Patel, R., & Mukherjee, S. (2022). Mycorrhizal product glomalin: A proficient agent of nutrient sequestration and soil fertility restoration under jeopardized agroecosystem. In *Microbes and microbial biotechnology for green remediation* (pp. 781–794). Elsevier.

Smith, S. E., & Read, D. J. (2010). *Mycorrhizal symbiosis*. Academic Press.

Sui, X., Liu, Y., Yang, L., Li, M., Zhang, S., Zhang, T., Wang, J., Cui, X., & Ni, H. (2019). Response of soil AMF diversity to nitrogen deposition in a Calamagrostis augustifolia wetland of Sanjiang plain, China. *International Journal of Agriculture and Biology, 22*, 1565–1572.

Tarafdar, J. C. (2022). Impact of Nanofertilizer on Arid Crops Grown Under Rainfed Conditions. *Current Agriculture, 39*(2).

Thind, S., Chaudhary, M. S., Ditta, A., Hussain, I., Parveen, A., Ullah, N., Mahmood, Q., Al-Ashkar, I., & El-Sabagh, A. (2022). Impact of mycorrhizal fungi from different rhizospheric soils on fungal colonization, growth, and chlorophyll contents of *Cenchrus ciliaris*. *Agronomy, 12*(11), 2644. doi:10.3390/agronomy12112644.

Thirkell, T. J., Charters, M. D., Elliott, A. J., Sait, S. M., & Field, K. J. (2017). Are mycorrhizal fungi our sustainable saviours? Considerations for achieving food security. *Journal of Ecology, 105*, 921–929.

Tian, J., Ge, F., Zhang, D., Deng, S., & Liu, X. (2021). Roles of phosphate solubilizing microorganisms from managing soil phosphorus deficiency to mediating biogeochemical P cycle. *Biology (Basel), 10*, 158.

Trejo, D., Barois, I., & Sangabriel-Conde, W. (2016). Disturbance and land use effect on functional diversity of the arbuscular mycorrhizal fungi. *Agroforestry Systems, 90*, 265–279.

Udhayakumar, M., Varadharajan, M., Pachu, A. V., Krishna, T. P. A., Ceasar, S. A., Jeyasundar, P. G. S. A., & Das Madhubala Parameswaran, A. (2024). Exploring the Role of Beneficial Soil Microbiomes Associated with Mangroves of Kerala in India. *Biologia (Bratisl), No.*, 0123456789. https://doi.org/10.1007/s11756-024-01821-3.

Wahab, A., Muhammad, M., Munir, A., Abdi, G., Zaman, W., Ayaz, A., Khizar, C., & Reddy, S. P. P. (2023). Role of arbuscular mycorrhizal fungi in regulating growth, enhancing productivity, and potentially influencing ecosystems under abiotic and biotic stresses. *Plants, 12*, 3102.

Yizhu, L., Imtiaz, M., Ditta, A., Rizwan, M. S., Ashraf, M., Mehmood, S., Aziz, O., Mubeen, F., Ali, M., Elahi, N. N., Ijaz, R., Lele, S., Shuang, C., & Tu, S. (2020). Response of growth, antioxidant enzymes and root exudates production towards as stress in *Pteris vittata* and in *Astragalus sinicus* colonized by arbuscular mycorrhizal fungi. *Environmental Science and Pollution Research, 27*(2), 2340–2352. https://link.springer.com/journal/11356. doi:10.1007/s11356-019-06785-5.

Zaman, F., Ali, A., Khattak, W. A., Khan, H., Anwar, Z., Rahimi, M., & Chen, F.-S. (2025). Diversity of arbuscular mycorrhizal fungi (AMF) and its and role in nutrient recycling and availability in subtropical forest ecosystem of China: A review. *Eurasian Soil Science, 58*, 1–17.

Zhang, X., Wang, Z., Lu, Y., Wei, J., Qi, S., Wu, B., & Cheng, S. (2024). Sustainable remediation of soil and water utilizing arbuscular mycorrhizal fungi: A review. *Microorganisms, 12*, 1255.

Chapter 4

Arbuscular mycorrhizal fungi: Improving soil fertility through enhancement of nutrient availability

Muniba Farhad[a], Hadia Ahsan[b], Muhammad Usama[b], Hafiz Syed Tanzeem-ul-Haq[b], Muhammad Iqbal[b] and Hafiz Muhammad Tauqeer[c]

[a] *Department of Chemistry, Government College University Faisalabad, Faisalabad, Pakistan,* [b] *Department of Environmental Sciences, Government College University Faisalabad, Faisalabad, Pakistan,* [c] *Department of Environmental Sciences, University of Gujrat, Gujrat, Pakistan*

4.1 Introduction

Soil fertility is a fundamental component for sustainable agricultural production, as it determines the availability of essential nutrients required for optimal plant growth and development (Chaudhry et al., 2024). However, the excessive use of chemical fertilizers leads to soil nutrient imbalances, soil degradation, and declining crop productivity (Herath et al., 2024). Alternatively, biological approaches that enhance soil fertility while maintaining ecological stability are receiving increasing attention, among which arbuscular mycorrhizal fungi (AMF) play a pivotal role (Agnihotri et al., 2022).

AMF are soil-borne, ubiquitous, obligate biotrophic microbes that establish symbiotic associations with up to 90% of terrestrial plant species across diverse ecosystems (Ahammed & Hajiboland, 2024). They belong to the phylum Glomeromycota and form specialized structures, known as arbuscules, within root cortical cells, which serve as active sites for nutrient exchange among AMF and host plants (Kumar et al., 2024; Shi et al., 2023). Through this symbiotic relationship, AMF improve the acquisition of water and mineral nutrients from the soil. AMF's extensive extraradical hyphal networks also enhance nutrient bioavailability in soil and plant nutrient uptake (Kakouridis et al., 2022).

AMF significantly contribute to soil fertility by increasing the availability of macronutrients, including phosphorus (P), nitrogen (N), potassium (K), calcium (Ca), magnesium (Mg), and sulfur (S), crucial for plant metabolic processes (Herath et al., 2024). AMF facilitate N mobilization by enhancing inorganic N availability in the soil and promoting beneficial microbial interactions, thereby improving N uptake and utilization efficiency in plants (Fall et al., 2022; Xing et al., 2025). Likewise, AMF contribute to P acquisition by solubilizing insoluble P forms and transporting phosphate to host plants through specialized phosphate transporters (Qi et al., 2022; Wipf et al., 2019). In addition to N and P, AMF substantially increase K content in roots and shoots by regulating K transporter genes in plants (Ma et al., 2024). Calcium uptake is also enhanced by AMF colonization, owing to the extensive extraradical hyphal network and the regulation of transporter genes, which collectively promote improved nutrient absorption, plant growth, and development (Fu et al., 2023; Nazir et al., 2023; Saeed et al., 2025; Thind et al., 2022; Yizhu et al., 2020). AMF induce modifications in root architecture and overall plant health, facilitating improved soil exploration and nutrient acquisition and supporting plant growth (Zhou et al., 2025). Magnesium acquisition through AMF transport mechanisms enables plants to access this essential element, which is important for chlorophyll synthesis and stress tolerance (Ahmed et al., 2023). Furthermore, AMF enhance S availability and uptake by stimulating sulfur-mobilizing microorganisms and upregulating sulfate transporter genes in plant roots (Narayan et al., 2023; Wu et al., 2024).

Advancing Sustainable Agriculture Through Arbuscular Mycorrhizal Fungi. DOI: https://doi.org/10.1016/B978-0-443-45130-0.00035-7

Beyond macronutrients, AMF play an essential role in increasing the bioavailability of micronutrients such as iron (Fe), zinc (Zn), copper (Cu), and manganese (Mn), which are needed for plant growth and development (Ahmed et al., 2025). They enhance micronutrient uptake through different mechanisms such as chelation, organic acid secretion, siderophore production, and the regulation of nutrient transporter genes, thereby improving plant nutritional status and stress resilience (Moreno Jiménez et al., 2024; Watts-Williams et al., 2023). AMF also indirectly support soil fertility through the secretion of a compound known as glomalin (Kalamulla & Yapa, 2024; Li et al., 2022; Matos et al., 2021). In addition, AMF enhance enzymatic activities involved in nutrient cycling and organic matter decomposition (Ahmed et al., 2025).

The integration of AMF into soil management practices offers a sustainable way to boost crop productivity while maintaining long-term soil health (Koshila Ravi et al., 2024). Practices like minimum tillage help preserve soil structure and mycorrhizal mycelial networks, supporting nutrient uptake and soil aggregation (Koshila Ravi et al., 2024). Likewise, crop rotation with such crops known to support mycorrhizal symbiosis and the addition of cover crops encourage AMF colonization, increase spore density, and enhance soil carbon (C) dynamics, thereby collectively improving nutrient bioavailability (de Oliveira et al., 2024; Paranavithana et al., 2021). This chapter provides insight into the contribution of AMF to improving soil fertility by enhancing nutrient bioavailability, regulating soil microbial processes, improving soil physicochemical properties, and supporting sustainable plant nutrition.

4.2 Arbuscular mycorrhizal fungi

4.2.1 Classification and symbiosis with plants

AMF are widespread soil-dwelling fungi that live with plant roots as obligate symbiotic partners, known to establish symbiosis with up to 90% of plants grown on different soils and environmental situations (Ahammed & Hajiboland, 2024). The term "AMF" highlights the unique symbiosis and its associated functions between the host plant rhizosphere and mycorrhizal fungi. Arbuscular mycorrhiza is derived from two words, where "arbuscules" stands for a special hyphal component within the root inner cortex responsible for nutrient and/or water exchange between AMF and the host plant root. Mycorrhiza further represents two words: "myco," meaning fungus, and "rhiza," meaning root (Shi et al., 2023). According to biological classification, AMF is included in the phylum Glomeromycota and subkingdom Mucoromycota, consisting of three classes (Archaeosporomycetes, Glomeromycetes, and Paraglomeromycetes) (Kumar et al., 2024), including 25 genera, 11 families, and more than 250 species (Husna et al., 2022).

Generally, during AMF-plant symbiosis, AMF assist host plant roots in absorbing nutrients and water from soil beyond their reach through an extended hyphal network, ultimately reducing the host plant's reliance on irrigation and synthetic fertilization (Kakouridis et al., 2022). By facilitating active nutrient translocation, AMF support plant growth and resilience to various stresses, such as nutrient deficiency, salinity, and heavy metal stress (Shi et al., 2023). AMF provide nutrients to colonized plants through direct and indirect pathways. In the direct pathway, nutrient acquisition takes place directly by root hairs and the epidermis of root cells. However, in the mycorrhizal pathway, AMF hyphae facilitate nutrient transport to the root cells from the rhizosphere soil (Ahammed & Hajiboland, 2024).

4.3 Roles of arbuscular mycorrhizal fungi for enhancing nutrient bioavailability in soil and their accessibility by plants

Soil fertility is a key factor in resilient agriculture, primarily related to the availability of essential nutrients in the soil (Chaudhry et al., 2024). For development, growth, and sustaining several physiological functions, nutrients are needed by the plants. Nutrients are categorized as macronutrients such as N, P, K, Ca, Mg, and S, needed in excess over the micronutrients [Fe, Mn, Zn, Cu, boron (B), molybdenum (Mo), and chlorine (Cl)] for plants. All of these nutrients are needed for photosynthesis, protein synthesis, respiration, and nucleic acid and other molecules synthesis (Herath et al., 2024; Sparks & Benfey, 2017).

4.3.1 Macronutrients

4.3.1.1 Carbon

As the most crucial nutrient, C sustains each aspect of life via its mobility between biotic and abiotic entities in the environment (Ahammed & Hajiboland, 2024). One of AMF's leading roles in improving soil health is the acquisition and accumulation of various organic C sorts as rhizodeposits (Herath et al., 2024). AMF receive nearly 10%–20% of photosynthetic C from the plants (Ahammed & Hajiboland, 2024). In an increased CO_2 condition, AMF can enhance plant development because of active nutrient absorption, acquisition of photosynthetic C from host plants, and improved

C availability to roots, thereby enhancing the photosynthetic rate owing to the augmented C-sink strength in mycorrhizal plants (Shi et al., 2021). Another result indicated AMF role in increasing plant growth, nutrient accessibility, and uptake in soil. Moreover, 28% more organic C was also noticed in comparison to soil without AMF (Akter et al., 2024). Diverse soil microbial associations and AMF play an essential role in C cycling through different mechanisms such as immobilization, mineralization, respiration, and decomposition (Parihar et al., 2020; Philippot et al., 2013). The addition of organic residues by means of intra- and extracellular mycelium, glomalin, and other hyphal derivatives, AMF increase the soil organic C pool. The photosynthetic products produced by shoots are shifted to roots, which are consumed by AMF and resultantly are released into the soil in the form of amino acids and extracellular polymers, ultimately raising soil C reserve (Aliasgharzad & Malekzadeh, 2024).

4.3.1.2 Nitrogen

Nitrogen, a highly crucial nutrient, serves its role in the growth and development of plants (Shrivastav et al., 2020). Nitrogen exists in the soil in mineral (nitrites, nitrates, and ammonium ions) and organic forms. Plants prefer nitrate (NO_3^-) uptake over the ammonium (NH_4^+) form of N. AMF support N uptake by the plants via mobilizing soil inorganic form of N (NH_4^+) (Fall et al., 2022). Interestingly, nearly 33% of the root protein N is because of the contribution by AMF (Ahammed & Hajiboland, 2024). The high reproduction rate of proteobacteria, which is the dominant AMF bacterial population (Gram-negative bacteria), is responsible for the conversion of organic N into inorganic N (NH_4^+ and NO_3^-) for boosting the fertility of soil, compared to the control (Jiang et al., 2021; Xing et al., 2025). Similarly, applying two AMF species (*Funneliformis mosseae* and *Rhizophagus intraradices*) in the soil increased 35.9%–63.3% NH_4^+ and 5.2%–17.8 % inorganic N contents in white mulberry, than noninoculated ones (Xing et al., 2025). AMF promoted N uptake in cotton plants, as AMF colonization increased root NH_4^+-N and NO_3^--N, respectively, in N-free treatments (Wang et al., 2025).

A symbiotic pathway for NO_3 uptake is supported by specialized transporters after consuming energy. These transporters belong to the Nitrate and Peptide Transporter Families (NPF) (NRT1/PTR family), NRT2, and NRT3 families (Wipf et al., 2019). AMF access N sources from the external root zone and supply them to the plants via the common mycorrhizal networks, thereby improving plant nitrogen-use efficiency (NUE) (Luo et al., 2023). By upregulating the ZmAMT3;1 gene, AMF increase NH_4^+ translocation, thereby enhancing plant growth under low N conditions (Hui et al., 2022). Moreover, AMF decline loss of N via boosting the N-fixing genes or rising N_2O-consuming denitrifiers, supporting plant N uptake (Yangyang et al., 2024).

4.3.1.3 Phosphorus

Phosphorus is a vital nutrient for the growth and development of plants but is commonly present in the soil as insoluble and nonavailable forms for the plants (Narayanan & Vyshnavi, 2025). It is well known that ~90% of plants' P is supplied by AMF (Ahammed & Hajiboland, 2024). AMF provide P to plants by solubilizing and mobilizing it in soil (Qi et al., 2022). AMF species such as *R. intraradices* have the potential to support active P uptake during its insoluble state (Qi et al., 2022). Elevated P accumulation was observed in roots, shoots, and fruits by 87%, 4%, and 98%, respectively, compared with non-AMF treatments (Akter et al., 2024). Similarly, in cotton plants, the application of AMF (*Rhizophagus irregularis CD1*) improved P uptake by 13.65%–43.27%, compared with the control (Marwanto et al., 2024).

AMF dissolve P, which is inaccessible to the plants, by releasing organic acids (low-molecular weight) from extraradical hyphae (Andrino et al., 2021). Moreover, AMF employ a variety of other mechanisms to convert inorganic P into its soluble state, such as exchange reactions, chelation, acidification, and the release of metabolites, hydrogen ion (H^+), and organic acids (Fall et al., 2022). In low P soils, AMF increase plant growth through hormonal induction and P solubilization, thereby boosting P acquisition by the plants (Zou et al., 2023). In addition, AMF provide P to plants from surrounding soil by phosphate transporter gene, such as (PHT1), present on extraradical hyphae in root cortical cells (Wipf et al., 2019). Moreover, by upregulating P transporter genes, including SlPT3, AMF increase plants' P acquisition capability (Rui et al., 2023).

4.3.1.4 Potassium

Potassium plays a leading role in plant metabolism and development, as it is a vital nutrient widely available in soil. Considering its abundance, nearly 95% of soil K is bound to various soil minerals, whereas only 2%–3% is present in a free, soluble form available to plants (Olaniyan et al., 2022). The efficacy and productivity of natural ecosystems are affected by K deficiency (Muthuraja & Muthukumar, 2021). The application of AMF increased K acquisition (by 99.5 ± 25.8 mg/kg) in pulses when compared to non-AMF treatment (Noreen et al., 2023). Similarly, improved K contents in roots and shoots by 19.7% and 22.4%, respectively, were noticed with AMF inoculation in watermelon, in contrast to the treatment where AMF were not present (Ma et al., 2024). Another study indicated a 2% higher P content in Unnayan tomatoes treated with AMF (*F. mosseae*) over control (Alam et al., 2023). Likewise, AMF inoculation, compared with non-AMF treatment, increased P acquisition by 1.97, 1.93, and 1.45 folds in okra fruit, root, and shoot, respectively (Akter et al., 2024). The ectomycorrhizal fungus *Pisolithus microcarpus*–treated seedlings increased P channels and transporters and mobilized P from clay minerals with ectomycorrhizal exudates (Sardans et al., 2023). AMF increase K acquisition by regulating various K transporter genes (LbHAK) (Zhang et al., 2024). The expression of nutrient transporter channel genes like (*preClKC2*) by AMF was also responsible for high K concentration in watermelon roots (Ma et al., 2024).

4.3.1.5 Calcium

Calcium ions (Ca^{2+}) serve as essential secondary messengers in signal transduction pathways, responding to various environmental stimuli, including cold, drought, salinity, and oxidative stress. These factorsnegatively influence plant growth and development (Jiamin et al., 2025). Plants can also experience Ca deficiency because of low Ca mobility in the soil, even in Ca-rich soil (Fu et al., 2023). Plants experiencing a Ca^{2+} shortage can suffer from meristem mortality, stunted root and shoot development, and leaf curling. Moreover, high Ca^{2+}-consuming plant species may face elevated reactive oxygen species levels in Ca-deficient conditions, leading to oxidative stress (Ligaba-Osena et al., 2022). AMF inoculation improved plant growth by supporting Ca uptake in aerial parts and enabling Ca absorption by the roots from soil (Fu et al., 2023). The total Ca^{2+} ions increased (1964 mg 100 g^{-1}) in papaya leaves after 6 months of AMF addition when compared to non-AMF plants (668 mg/100g) (Maffo et al., 2025). Similarly, AMF species (*F. mosseae, Rhizophagus clarus, and R. intraradices*) significantly enhanced Ca content in tomato by 13% compared with the control (Alam et al., 2023). AMF supported nutrient cycling by promoting the release of root exudates like organic acids from the roots, which lowered the pH of the rhizosphere soil and stimulated rhizosphere microbes (Fu et al., 2023; Zhang et al., 2015). Moreover, the deposition of Ca oxalate crystals on the rhizomorph surface, after it has been solubilized from apatite, promoted Ca uptake in the plants with the presence of ectomycorrhizal fungi (Sardans et al., 2023). AMF promote Ca acquisition by the plants through a unique mechanism, including upregulation of genes related to the expression of Ca^{2+} signal, channels, and transporters (Fu et al., 2023).

4.3.1.6 Magnesium

Magnesium is among the most common elements found in the Earth's crust, ranking eighth in abundance. Plants typically require about 1.5–3.5 g/kg of Mg for proper development (Sanz et al., 2022; Xie et al., 2021). Magnesium is important for plant growth and productivity because it plays a key role in biochemical and developmental processes, such as protein synthesis, partitioning of carbohydrates, and photosynthesis (Xie et al., 2021). The intensive use of K^+ and NH_4^+ fertilizers reduces Mg acquisition by the plants, thereby inducing Mg deficiency. In acidic soils, plant growth faces challenges as H^+, Al^{3+}, and Mn^{2+} ions are abundant in such soils, whereas leaching of basic cations like Mg causes deficiency (Zhang et al., 2021).

Previous research has indicated the leading potential of AMF in mobilizing Mg in soil and its acquisition by the host plants (Jarosz et al., 2021; Sardans et al., 2023). AMF inoculation significantly increased the tomato plant's potential to uptake Mg in comparison to nonmycorrhizal plants (Liu et al., 2023). Interestingly the rice, Mexican hat, and Sorghum plants experienced 24.5%, 20.4%, 10.6% increase in Mg^{2+} contents, respectively, in leaf tissues with AMF (Zhou et al., 2025). Similarly the nutritional worth and plant growth of tomatoes were found to be increased with enhanced Mg concentration in response to the inoculation with five AMF species (*R. intraradices, R. clarus, Acaulospora morrowiae, Pyrodictium occultum, and F. mosseae*) compared to the control (Alam et al., 2023). The AMF increased Mg acquisition by improving root architecture and overall plant health, which led to extensive soil exploration for elevated uptake of nutrients (Zhou et al., 2025). AMF improve Mg acquisition by the plants via a transport mechanism, allowing them to utilize this vital nutrient to endure stress and chlorophyll synthesis (Ahmed et al., 2023).

4.3.1.7 Sulfur

Sulfur serves as a vital element for living beings. It is a component of several important molecules, such as key protein-forming amino acids, including methionine and cysteine, vitamins like biotin and thiamine, phytochelatins, coenzyme A, chlorophyll, and S-adenosyl-methionine (Li et al., 2020; Nakai & Maruyama-Nakashita, 2020). Moreover, it is essential for plant development, growth, and physiological processes, and it is also a key part of plant secondary metabolites (Narayan et al., 2023). AMF were reported to promote S concentrations in the root, shoot, and fruit of *Abelmoschus esculentus* L. plant by 159%, 183%, and 87%, respectively, compared to noninoculated plants (Akter et al., 2024). Better soybean growth and S absorption were stimulated by AMF inoculation because of increased expression of *soxY, yedZ*, and *aspA* (by 236.04%, 436.64%, and 143.23%, respectively), including elevated content of available S from 12.88 to 17.55 mg/kg and arylsulfatase activity in soil [39.17–49.82 μg/(mL•h)] (Wu et al., 2024). Similarly the application of AMF promoted S supply by increasing root colonization and the abundance of cultivable sulfonate-mobilizing bacteria in *Lolium perenne* plant (Gahan et al., 2015).

The increase in bioavailability of S by AMF is because of enhancements in the activities of sulfur-solubilizing bacteria in the host plant rhizosphere by AMF, which boosts S uptake (Nadeem et al., 2023). AMF colonization upregulates sulfate transporter genes (MtSULTR 1.2 and MtSULTR 1.1) expression in the roots of the host plant, which increases the plant's S absorption rate. Furthermore, through extraradical mycelia, AMF support direct acquisition and transport of S to the host plant (Ma et al., 2022; Wu et al., 2018). Under S-deficient conditions, plants rapidly absorb SO_4^{2-}, resulting in the development of a sulfate-depletion zone near the roots. In response, the extraradical hyphae of AMF extend beyond this depleted region and access sulfate from nearby area of soil, thereby contributing to S supply in low S environments (Narayan et al., 2023).

4.3.2 Micronutrients

Micronutrients, such as Zn, Cu, Fe, and Mn, are involved in various biological functions, such as reproduction, enzymatic metabolism, and the induction of tolerance to pathogens in plants (Moreno Jiménez et al., 2024). Plants experience reduced productivity and yield because of limited availability of micronutrients, which weaken the root system and reduce chlorophyll content, photosynthesis, and plant growth (Vijayakumar et al., 2022). Plants employ various mechanisms, such as creating symbiosis with AMF and releasing exudates in the rhizosphere, which increase the roots' capacity to take up soil nutrients (Sardans et al., 2023). This AMF-plant symbiosis promotes the acquisition of essential micronutrients to support the development and growth of plants (Ahmed et al., 2025). The application of AMF (*R. intraradices, F. mosseae*, and *R. clarus*) increased micronutrient concentrations over control in tomato plants (Alam et al., 2023). Similarly, in watermelon, AMF inoculation improved Cu and Zn concentrations in roots, Fe content in shoots, and Mn concentrations in both roots and shoots compared with noninoculated plants (Ma et al., 2024). AMF (*Paraglomus laccatum* and *R. irregularis*) stimulated Cu uptake and plant stress tolerance in poplars through species-specific effects (Soltangheisi et al., 2024). In *Rhus chinensis*, the accumulation of Mn in roots was enhanced by *F. mosseae* (Pan et al., 2023).

Through extended extraradical hyphae, AMF support the translocation of essential micronutrients Fe, Zn, Mn, and Cu from soil to the host plant by obtaining C from it as true biotrophs (Fu et al., 2023). AMF directly upregulate nutrient transporter genes for micronutrient absorption, ultimately inducing stress tolerance, improving growth, and overall plant health (Ahmed et al., 2025). The enhanced Zn solubility in the rhizosphere of the host plants is associated with the release of organic acids by AMF. Moreover, AMF's role in stimulating Zn acquisition by upregulating Zn transporter genes such as BoZIP1 and MtZIP14 is also a vital mechanism (Arifuzzaman et al., 2024; Watts-Williams et al., 2023). In addition, Cu acquisition even in Cu-contaminated soils is regulated by AMF via immobilization of excess Cu (Ahmed et al., 2025). AMF also release compounds similar to siderophores, organic acids, and phenols that sequester metals (Fe and Zn) in the soil in a chelate-like manner, increasing their solubility, thus facilitating their access to fungi and plants (Moreno Jiménez et al., 2024).

4.4 Arbuscular mycorrhizal fungi–released organic compounds in enhancing soil nutrient bioavailability

4.4.1 Glomalin

Glomalin is a unique glycoprotein that is ubiquitous, hydrophobic, insoluble, and recalcitrant, produced by AMF spores and hyphae and later deposited in soil after their decomposition (Li et al., 2022). A positive correlation exists between glomalin and indicators of soil fertility (soil N, P, and C), as glomalin contains 3%–5% N, 3%–4% P, 1%–9% Fe, 30%–40% C, and 1%–2% nutrients such as Ca, Mg, Zn, Cu, S, and K (Singh et al., 2022). A rise in the content of glomalin in the soil also represents higher AMF abundance in soil, which, in turn, provides nutrients (N and P) and water to the plants through their extensive hyphal network (Chandrasekaran et al., 2019; Rosier et al., 2008). Glomalin enhances the capacity of soil microbes to release growth hormones, facilitating nutrient cycling and promoting better plant growth and development (Farhad et al., 2024).

4.4.2 Soil enzymes

Soil enzymes are essential for nutrient cycling, including N, P, and K, and affect their bioavailability to plants (Murtaza et al., 2025). AMF promote the activities of soil enzymes that are responsible for the decomposition of organic matter and nutrient cycling. For instance, AMF produce phosphatases, which transform organic P into a bioavailable form by mineralization, a vital mechanism of P availability in P-deficient soils (Deng et al., 2024; Jin et al., 2024). Symbiosis between AMF and *Artemisia annua* enhanced soil acid phosphatase activity, promoted the synthesis of soil organic P compounds, facilitated their transformation, and increased the efficiency of P assimilation in plants. Moreover, by boosting sucrose activity in soil, AMF support soil maturation and fertility (Ma et al., 2021). Microbial catabolic functions are enhanced by AMF-released glomalin, which, in turn, enhances the capabilities of soil microbes to secrete soil enzymes such as peroxidase, fluorescein diacetate, catalase, and phosphatase (Matos et al., 2021).

AMF promote β-glucosidase activity in soil, which is responsible for the conversion of cellulose into glucose, which serves as a source for soil microbes to contribute to the soil organic C pool (Ahmed et al., 2017). Nitrogen metabolism is improved by AMF by stimulating soil urease and nitrate reductase, which are responsible for improving N acquisition and minimizing N losses (Gou et al., 2024; Yin et al., 2009). In addition, AMF increase antioxidant enzymes i.e., superoxide dismutase (SOD) and catalase (CAT) activities, which help protect plants and soil microbes from oxidative damage, thus supporting plants under stress (Ahmed et al., 2025; Begum et al., 2019). With all of these enzymatic mechanisms, AMF contribute to regulating nutrient cycling and soil organic matter accumulation, ultimately improving soil health (Ahmed et al., 2025).

4.4.3 Sidrophores

Siderophores are produced by fungi as tiny molecules for Fe chelation. Siderophores can tightly bind bioavailable Fe, forming an Fe-siderophore complex that attaches to specific surface receptors on cells, enabling the transportation of Fe in the cell (Kalamulla & Yapa, 2024). In addition to plants, AMF adopt specific mechanisms to chelate insoluble Fe by unleashing siderophores to meet their Fe requirements (Upadhayay et al., 2019). Coapplied AMF (*Acaulospora, Glomus*, and *Scutellospora*) and plant growth-promoting rhizobacteria (PGPR) in *Sorghum bicolor* and *Pennisetum glaucum* plants resulted in increased absorption of Fe because of siderophores (Kalamulla & Yapa, 2024). Similarly, AMF secreted siderophores in the rhizosphere of alfalfa, which increased Fe bioavailability and solubility in the rhizosphere during Fe-deficient conditions, ultimately raising the uptake of Fe by the plants (Rahman et al., 2020). Interestingly the unidentified involvement of mycorrhizal siderophores is presumed, as Fe absorption rates are higher in plants in the presence of AMF (Ibrahim et al., 2024). In addition, AMF mycelium contains PGPR that produce siderophores, which facilitate plant growth and soil health (Fall et al., 2022).

4.5 Arbuscular mycorrhizal fungi–mediated improved soil physicochemical properties for better soil nutrient bioavailability

By altering the rhizospheric physicochemical environment, the AMF positively influence microbial activity, thereby promoting nutrient availability in soil (Fall et al., 2022). AMF-derived glomalin has been reported to improve soil porosity and aggregate stability while reducing bulk density, thereby positively improving the nutritional status of soil (Singh et al., 2022). AMF generally proliferate on soil particles and in pores, which influence the soil physicochemical characteristics because of the release of several biochemical compounds, thus supporting the redistribution of particles (Singh et al., 2022). In addition, AMF enhance physical and chemical attributes of soil by adopting different mechanisms such as (1) elevated root colonization with extensive hyphal network for physical gripping of soil particles and (2) production of glomalin for soil particle aggregation, lowering bulk density and organic C sequestration and enhancing soil structure, porosity, and available water content, ultimately promoting bioavailability of nutrients and their acquisition (Farhad et al., 2024).

4.6 Strategies to improve arbuscular mycorrhizal fungi abundance, proliferation, and activities in soil

Nowadays, environmentally sustainable agriculture has become a global imperative. Excessive agrochemical usage raises the price for crop production, overuses groundwater, increases soil salinity, pollutes the environment, and deteriorates the food quality (Aulakh et al., 2022). Conventional agricultural practices disrupt the development of the extraradical hyphal network, thereby imposing prolonged periods during which AMF lack access to host plant roots and are consequently deprived of C. Rehabilitation of AMF diversity and population density can be achieved either by adopting different management strategies that favor AMF proliferation or by the addition of AMF inoculum in the soil (Elliott et al., 2021).

4.6.1 Tillage

Tillage disrupts soil layers, microbial communities, soil density, and moisture and negatively impacts AMF mycelial networks, ultimately reducing the activities of soil mycorrhizal communities and crop performance (de Oliveira et al., 2024). In comparison to conventional tillage, AMF are positively influenced by reduced or no tillage (Lu et al., 2018). The initial study of AMF highlighted functionality regarding crop tillage. Whereas recent neotropical agrosystems analysis highlights that conventional systems negatively influence the functions of AMF (de la Cruz-Ortiz et al., 2020). Hence as reported by various recent studies, an increase in soil C, water-stable aggregates, and soil aggregation can be achieved through minimum tillage (Koshila Ravi et al., 2024; Wang et al., 2018). Similarly, under varied tillage treatments, AMF parameters were improved in soil and maize roots under reduced or no tillage. In contrast, decreased AMF spore density was reported in intensive or deep tillage (Gu et al., 2020). Furthermore the composition of AMF communities in the subsoil was affected by deep tillage in contrast to no tillage (Bowles et al., 2017). So, reduced or no-tillage is recommended for a favorable AMF proliferation zone in the rhizosphere (Koshila Ravi et al., 2024).

4.6.2 Crop management

In annual monoculture copping practice, the functional dynamics and composition of rhizosphere microbial communities can be positively influenced by selecting various plant species and adopting crop rotation (de Oliveira et al., 2024). Commonly, AMF create symbiosis with the majority of crops, so crop rotation never disturbs AMF–plant symbiosis, as most of the crops exhibit symbiotic relations with AMF (Benami et al., 2020). A comparative analysis revealed that AMF root colonization, spore number (large-sized), microbial biomass carbon, water-soluble C, and total organic C were increased in rice–soybean crop rotation (Paranavithana et al., 2021). Reduced soil erosion, improved nutrient-use efficiency, and reduced nutrient loss by minimizing their leaching are the salient features of cover crops (de Oliveira et al., 2024). In an intercropping experiment, AMF perpetuation and proliferation in terms of root colonization, hyphal length, soil nutrients, and overall richness of fungal community were observed in contrast to the traditional method (Hontoria et al., 2019). Moreover, the presence of AMF below tillage depths in crops, with deep roots and cover crops, also enhances AMF beneficial traits (Sosa-Hernández et al., 2019). Hence the soil productivity and stability, and resource utilization efficiency of AMF communities can be improved by adopting multiple cropping technologies (Koshila Ravi et al., 2024).

4.6.3 Organic amendments and arbuscular mycorrhizal fungi inocula

Applying vermicompost and manures originated from farmyards and green plants raises the nutrients status (N, P, K, Ca, Mg, Fe, Mo, Zn, and Cu) of soil and plant yield (Aulakh et al., 2022). AMF symbiosis is greatly enhanced in an organic farming system, as narrated by previous research (Liu et al., 2020a, 2020b). The application of *Funneliformis geosporum, Claroideoglomus etunicatum,* and *F. mosseae* with poultry compost increased spore density and root colonization by AMF in maize plants (Abdullahi et al., 2015). Likewise, soybean plants showed enhanced AMF extraradical hyphal density in soil, followed by improved fungal colonization and spore density because of adding compost in soil (Yang et al., 2018). Furthermore, biochar application increased soil glomalin content, microbial biomass carbon, and AMF root colonization in peas (Farhad et al., 2024) and mycorrhizal intensity and frequency in *Zea mays* (Minkosse et al., 2023). Arbuscules in the root system were increased by 29.72%, mycelium by 11.79%, vesicles by 41.80%, and the number of AMF spores in soil by 5.34%, respectively, because of conditioning soil with biochar (Hashem et al., 2019). The addition of AMF inoculant resulted in an increased P uptake and AMF root colonization (up to 30%) in all wheat cultivars (Elliott et al., 2021). Because of various beneficial services, AMF inoculants are gaining importance for ensuring resilient agriculture. For efficient AMF inoculum production, the following factors are considered critical: inoculum potential, colonization ability, and efficiency (Agnihotri et al., 2022).

4.6.4 Pesticides, fertilizers, and wastewater

Biocides, such as pesticides, fungicides, insecticides, and herbicides, are used to achieve agricultural growth and productivity. However, their drastic effects on AMF are also crucial (Koshila Ravi et al., 2024). The harmful impacts of these fungicides, as evidenced by various research, resulted in diminished P acquisition and hyphal transportation from soil to plants (de Novais et al., 2019). The application of nonsystemic and systemic fungicides for managing plant diseases has an adverse impact on AMF root colonization, hyphal development, germination of spores, and acquisition of nutrients (Buysens et al., 2015; Channabasava et al., 2015). These fungicides can decline crop development as they act as a barrier for nutrient uptake linked with AMF (Kuila & Ghosh, 2022). For example, azoxystrobin, a systemic fungicide, reduced AMF root colonization when applied to the soil (Vuyyuru et al., 2018).

Inorganic fertilizers are applied to soil to achieve agricultural productivity. However, long-term reliance on these fertilizers adversely affects soil microbes, including AMF, causing nutrient leaching, and leads to soil deterioration (Koshila Ravi et al., 2024). Likewise, AMF root colonization was notably decreased because of intensive P fertilization, which led to high cadmium (Cd) deposition in tissues of the wheat plant (Yazici et al., 2021). Furthermore, under variable irrigation status, increased N fertilizer application resulted in reduced AMF spore density, community composition, extraradical mycelial density, and intraradical colonization rate in wheat roots (Babalola et al., 2022). Alarmingly, in developing countries, wastewater is commonly used to meet agricultural needs (Khalid et al., 2018). Unfortunately, this practice of applying wastewater to the soil may result in the accumulation of heavy metals [Cd, Zn, nickel (Ni), lead (Pb), arsenic (As), and mercury (Hg)] in it (Farhad et al., 2024). Reportedly, decreased water-holding capacity (WHC), hydraulic conductivity, infiltration rate, and poor soil structure are linked with irrigating soil with wastewater (Sikhau et al., 2023). Moreover, unusual changes in soil cation exchange capacity, pH levels, and content of organic matter are also noted, which decrease microbial activities of soil (Guadie et al., 2021).

4.7 Conclusion

The beneficial services of AMF in sustainable agricultural systems are evident from studies depicting their role in improving soil fertility, notably in nutrient-deficient soils or soils with low nutrient bioavailability. Through symbiotic associations with plant roots, AMF support nutrient cycling and improve the plant nutrient assimilation efficiency. Overall, the integration of AMF into soil management practices supports long-term soil fertility and promotes sustainable agriculture. It is recommended to introduce nutrient-specific AMF formulations to address the deficiency of targeted nutrients. Moreover, AMF inocula with diverse AMF species, which have the potential to mobilize different nutrients, could also be a viable option to overcome plant deficiencies in multiple nutrients. High siderophore, glomalin, and enzyme-secreting AMF species should be screened and deployed in agriculture to improve soil and plant nutrition. Such strategies can ultimately help local farmers rely on natural resources to improve crop yields.

Abbreviations

AMF	arbuscular mycorrhizal fungi
As	arsenic
B	boron
Cd	cadmium
Ca	calcium
C	carbon
CEC	cation exchange capacity
Cl	chlorine
Cu	copper
Fe	iron
Mg	magnesium
Mn	manganese
Hg	mercury
Mo	molybdenum
Ni	nickel
N	nitrogen
NUE	nitrogen-use efficiency
P	phosphorus
PGPR	plant growth-promoting bacteria
K	potassium
S	sulfur
WHC	water holding capacity
Zn	zinc

References

Abdullahi, R., Lihan, S., & Edward, R. (2015). Effect of arbuscularmycorrhizal fungi and poultry manure on growth and nutrient uptake by maize under field conditions. *International Journal of Agriculture Innovations and Research, 4*(1), 158–163.

Agnihotri, R., Sharma, M. P., Bucking, H., Dames, J. F., & Bagyaraj, D. J. (2022). Methods for assessing the quality of AM fungal bio-fertilizer: Retrospect and future directions. *World Journal of Microbiology and Biotechnology, 38*(6), 97.

Ahammed, G. J., & Hajiboland, R. (2024). Introduction to arbuscular mycorrhizal fungi and higher plant symbiosis: Characteristic features, functions, and applications. In *Arbuscular mycorrhizal fungi and higher plants: Fundamentals and applications* (pp. 1–17). Springer Nature Singapore.

Ahmed, A., Aslam, M., & Ashraf, M. (2017). Microbial β-glucosidases: Screening, characterization, cloning, and applications. *Journal of Applied & Environmental Microbiology, 5*(2), 57–73.

Ahmed, N., Li, J., Li, Y., Deng, L., Deng, L., Chachar, M., & Tu, P. (2025). Symbiotic synergy: How arbuscular mycorrhizal fungi enhance nutrient uptake, stress tolerance, and soil health through molecular mechanisms and hormonal regulation. *IMA Fungus, 16*, e144989.

Ahmed, N., Zhang, B., Bozdar, B., Chachar, S., Rai, M., Li, J., & Tu, P. (2023). The power of magnesium: Unlocking the potential for increased yield, quality, and stress tolerance of horticultural crops. *Frontiers in Plant Science, 14*, 1285512.

Akter, S., Kamruzzaman, M., Sarder, M. P., Amin, M. S., Joardar, J. C., Islam, M. S., & Halder, M. (2024). Mycorrhizal fungi increase plant nutrient uptake, aggregate stability, and microbial biomass in the clay soil. *Symbiosis, 93*(2), 163–176.

Alam, M. Z., Choudhury, T. R., & Mridha, M. A. U. (2023). Arbuscular mycorrhizal fungi enhance biomass growth, mineral content, and antioxidant activity in tomato plants under drought stress. *Journal of Food Quality, 2023*(1), 2581608.

Aliasgharzad, N., & Malekzadeh, E. (2024). Glomalin and carbon sequestration in terrestrial ecosystems. In *Arbuscular mycorrhizal fungi and higher plants: Fundamentals and applications* (pp. 239–258). Springer Nature Singapore.

Andrino, A., Guggenberger, G., Kernchen, S., Mikutta, R., Sauheitl, L., & Boy, J. (2021). Production of organic acids by arbuscular mycorrhizal fungi and their contribution to the mobilization of phosphorus bound to iron oxides. *Frontiers in Plant Science, 12*, 661842.

Arifuzzaman, M., Bagchi, R., Hasan, M. R., Rahman, M. A., Rahman, M. M., Rahman, M. A., & Kabir, A. H. (2024). Arbuscular mycorrhizal fungi induce Zn uptake and antioxidant efficiency in broccoli (*Brassica oleracea L.*) exposed to Zn deficiency. *Plant Stress, 11*, 100415.

Aulakh, C. S., Sharma, S., Thakur, M., & Kaur, P. (2022). A review of the influences of organic farming on soil quality, crop productivity, and produce quality. *Journal of Plant Nutrition, 45*(12), 1884–1905.

Babalola, B. J., Li, J., Willing, C. E., Zheng, Y., Wang, Y. L., Gan, H. Y., & Guo, L. D. (2022). Nitrogen fertilisation disrupts the temporal dynamics of arbuscular mycorrhizal fungal hyphae but not spore density and community composition in a wheat field. *New Phytologist, 234*(6), 2057–2072.

Begum, N., Qin, C., Ahanger, M. A., Raza, S., Khan, M. I., Ashraf, M., & Zhang, L. (2019). Role of arbuscular mycorrhizal fungi in plant growth regulation: Implications in abiotic stress tolerance. *Frontiers in Plant Science, 10*, 1068.

Benami, M., Isack, Y., Grotsky, D., Levy, D., & Kofman, Y. (2020). The economic potential of arbuscular mycorrhizal fungi in agriculture. In *Grand challenges in fungal biotechnology* (pp. 239–279). Springer International Publishing.

Bowles, T. M., Jackson, L. E., Loeher, M., & Cavagnaro, T. R. (2017). Ecological intensification and arbuscular mycorrhizas: A meta-analysis of tillage and cover crop effects. *Journal of Applied Ecology, 54*(6), 1785–1793.

Buysens, C., Dupré de Boulois, H., & Declerck, S. (2015). Do fungicides used to control *Rhizoctonia solani* impact the non-target arbuscular mycorrhizal fungus *Rhizophagus irregularis*? *Mycorrhiza, 25*(4), 277–288.

Chandrasekaran, M., Chanratana, M., Kim, K., Seshadri, S., & Sa, T. (2019). Impact of arbuscular mycorrhizal fungi on photosynthesis, water status, and gas exchange of plants under salt stress: A meta-analysis. *Frontiers in Plant Science, 10*, 457.

Channabasava, A., Lakshman, H. C., & Jorquera, M. A. (2015). Effect of fungicides on association of arbuscular mycorrhiza fungus *Rhizophagus fasciculatus* and growth of proso millet (*Panicum miliaceum* L.). *Journal of Soil Science and Plant Nutrition, 15*(1), 35–45.

Chaudhry, H., Vasava, H. B., Chen, S., Saurette, D., Beri, A., Gillespie, A., & Biswas, A. (2024). Evaluating the soil quality index using three methods to assess soil fertility. *Sensors, 24*(3), 864.

de la Cruz-Ortiz, Á. V., Álvarez-Lopeztello, J., Robles, C., & Hernández-Cuevas, L. V. (2020). Tillage intensity reduces the arbuscular mycorrhizal fungi attributes associated with *Solanum lycopersicum*, in the Tehuantepec Isthmus (Oaxaca), Mexico. *Applied Soil Ecology, 149*, 103519.

Deng, L., Tu, P., Ahmed, N., Zhang, G., Cen, Y., Huang, B., & Yuan, H. (2024). Biochar-based phosphate fertilizer improves phosphorus bioavailability, microbial functioning, and citrus seedling growth. *Scientia Horticulturae, 338*, 113699.

de Novais, C. B., Avio, L., Giovannetti, M., de Faria, S. M., Siqueira, J. O., & Sbrana, C. (2019). Interconnectedness, length, and viability of arbuscular mycorrhizal mycelium as affected by selected herbicides and fungicides. *Applied Soil Ecology, 143*, 144–152.

de Oliveira, I. F., Campolino, M. L., de Oliveira, R. G., de Paula Lana, U. G., Gomes, E. A., & de Sousa, S. M. (2024). Agronomic practices for optimizing the AMF abundance and diversity for sustainable food production. In *Arbuscular mycorrhizal fungi in sustainable agriculture: Nutrient and crop management* (pp. 55–76). Springer Nature Singapore.

Elliott, A. J., Daniell, T. J., Cameron, D. D., & Field, K. J. (2021). A commercial arbuscular mycorrhizal inoculum increases root colonization across wheat cultivars but does not increase assimilation of mycorrhiza-acquired nutrients. *Plants, People,Planet, 3*(5), 588–599.

Fall, A. F., Nakabonge, G., Ssekandi, J., Founoune-Mboup, H., Apori, S. O., Ndiaye, A., & Ngom, K. (2022). Roles of arbuscular mycorrhizal fungi on soil fertility: Contribution to the improvement of physical, chemical, and biological properties of the soil. *Frontiers in Fungal Biology, 3*, 723892.

Farhad, M., Noor, M., Yasin, M. Z., Nizamani, M. H., Turan, V., & Iqbal, M. (2024). Interactive suitability of rice stubble biochar and arbuscular mycorrhizal fungi for improving wastewater-polluted soil health and reducing heavy metals in peas. *Sustainability, 16*(2), 634.

Fu, W., Yan, M., Zhao, L., Zeng, X., Cai, B., Qu, S., & Wang, S. (2023). Inoculation with arbuscular mycorrhizal fungi increases calcium uptake in *Malus robusta*. *Scientia Horticulturae, 321*, 112295.

Gahan, J., & Schmalenberger, A. (2015). Arbuscular mycorrhizal hyphae in grassland select for a diverse and abundant hyphospheric bacterial community involved in sulfonate desulfurization. *Applied Soil Ecology, 89*, 113–121.

Gou, X., Hu, Y., Ni, H., Wang, X., Qiu, L., Chang, X., & Wei, X. (2024). Arbuscular mycorrhizal fungi alleviate erosional soil nitrogen loss by regulating nitrogen cycling genes and enzymes in experimental agro-ecosystems. *Science of the Total Environment, 906*, 167425.

Gu, S., Wu, S., Guan, Y., Zhai, C., Zhang, Z., Bello, A., & Yang, W. (2020). The arbuscular mycorrhizal fungal community was affected by tillage practices rather than residue management in the black soil of northeast China. *Soil and Tillage Research, 198*, 104552.

Guadie, A., Yesigat, A., Gatew, S., Worku, A., Liu, W., Ajibade, F. O., & Wang, A. (2021). Evaluating the health risks of heavy metals from vegetables grown on soil irrigated with untreated and treated wastewater in Arba Minch, Ethiopia. *Science of the Total Environment, 761*, 143302.

Hashem, A., Kumar, A., Al-Dbass, A. M., Alqarawi, A. A., Al-Arjani, A. B. F., Singh, G., & Abd_Allah, E. F. (2019). Arbuscular mycorrhizal fungi and biochar improve drought tolerance in chickpea. *Saudi Journal of Biological Sciences, 26*(3), 614–624.

Herath, B. M., Kalamulla, K. W. Y. R., Mayadunna, T. A. N., Perera, M. A. E. G., Jayamanna, I. K., & Yapa, P. N. (2024). Applications of arbuscular mycorrhizal fungi for sustainable agricultural systems. In *Arbuscular mycorrhizal fungi and higher plants: Fundamentals and applications* (pp. 319–340). Springer Nature Singapore.

Hontoria, C., García-González, I., Quemada, M., Roldán, A., & Alguacil, M. M. (2019). The cover crop determines the AMF community composition in soil and in the roots of maize after a ten-year continuous crop rotation. *Science of the Total Environment, 660*, 913–922.

Hui, J., An, X., Li, Z., Neuhäuser, B., Ludewig, U., Wu, X., & Yuan, L. (2022). The mycorrhiza-specific ammonium transporter ZmAMT3; 1 mediates mycorrhiza-dependent nitrogen uptake in maize roots. *The Plant Cell, 34*(10), 4066–4087.

Husna, H., Tuheteru, F. D., Albasri, A., Arif, A., Basrudin, B., Nurdin, W. R., & Safitri, I. (2022). Diversity of arbuscular mycorrhizal fungi of *Kalappia celebica*: An endemic and endangered plant species in Sulawesi, Indonesia. *Biodiversitas Journal of Biological Diversity, 23*(10), 5290–5297.

Ibrahim, D. S., Riad, S. N., Abo-Elyousr, K. A., Nashwa, S. M., Khalil Bagy, H. M., Abdelrazek, S., & Abdellatif, A. A. (2024). Unraveling the mysteries of mycorrhiza-plant interactions: Mechanisms of protection and ecological factors influencing symbioses. In *Mycorrhizal symbiosis and agroecosystem restoration* (pp. 197–226). Springer Nature Singapore.

Jarosz, Z., Michalojc, Z., Pitura, K., Dzida, K., & Koter, M. (2021). Influence of fertilization and mycorrhizae on the nutritional status of *Rhododendron* (Rhododendron hybridum) in a nursery. *Agriculture, 11*, 538.

Jiamin, S., Kexin, D., Yu, Z., Zhang, Y., Ni, J., & Wu, Y. (2025). Aquorin bioluminescence-based Ca^{2+} imaging reveals differential calcium signaling responses to abiotic stresses in physcomitrella patens. *Plants, 14*(8), 1178.

Jiang, F., Zhang, L., Zhou, J., George, T. S., & Feng, G. (2021). Arbuscular mycorrhizal fungi enhance mineralisation of organic phosphorus by carrying bacteria along their extraradical hyphae. *New Phytologist, 230*(1), 304–315.

Jin, Z., Wang, G., George, T. S., & Zhang, L. (2024). Potential role of sugars in the hyphosphere of arbuscular mycorrhizal fungi to enhance organic phosphorus mobilization. *Journal of Fungi, 10*(3), 226.

Kakouridis, A., Hagen, J. A., Kan, M. P., Mambelli, S., Feldman, L. J., Herman, D. J., & Firestone, M. K. (2022). Routes to roots: Direct evidence of water transport by arbuscular mycorrhizal fungi to host plants. *New Phytologist, 236*(1), 210–221.

Kalamulla, R., & Yapa, N. (2024). Co-inoculation of AMF and other microbial biofertilizers for better nutrient acquisition from the soil system. In M. Parihar, A. Rakshit, A. Adholeya, & Y. Chen (Eds.), *Arbuscular mycorrhizal fungi in sustainable agriculture: Nutrient and crop management* (pp. 99–111). Springer.

Khalid, S., Shahid, M., Natasha, Bibi, I. , Sarwar, T. , Shah, A. , H., & Niazi, N. K (2018). A review of environmental contamination and health risk assessment of wastewater use for crop irrigation with a focus on low and high-income countries. *International Journal of Environmental Research and Public Health, 15*(5), 895.

Koshila Ravi, R., Balachandar, M., & Muthukumar, T (2024). Multifaceted role of arbuscular mycorrhizal fungi in crop growth promotion: An overview. *Arbuscular mycorrhizal fungi in sustainable agriculture: Nutrient and crop management*, 1–54.

Kuila, D., & Ghosh, S. (2022). Aspects, problems, and utilization of arbuscular mycorrhizal (AM) application as bio-fertilizer in sustainable agriculture. *Current Research in Microbial Sciences, 3*, 100107.

Kumar, D., Kour, S., Ali, M., Sharma, R., Parkirti, Vikram, & Ohri, P (2024). Interactions between arbuscular mycorrhizal fungi and other microorganisms in the rhizosphere and hyphosphere. In G. J. Ahammed, & R. Hajiboland (Eds.), *Arbuscular mycorrhizal fungi and higher plants: Fundamentals and applications* (pp. 37–66). Springer Nature Singapore.

Li, Q., Gao, Y., & Yang, A. (2020). Sulfur homeostasis in plants. *International Journal of Molecular Sciences, 21*(23), 8926.

Li, Y., Xu, J., Hu, J., Zhang, T., Wu, X., & Yang, Y. (2022). Arbuscular mycorrhizal fungi and glomalin play a crucial role in soil aggregate stability in Pb-contaminated soil. *International Journal of Environmental Research and Public Health, 19*(9), 5029.

Ligaba-Osena, A., Salehin, M., Numan, M., Wang, X., Choi, S. C., Jima, D., & Guo, W. (2022). Genome-wide transcriptome analysis of the orphan crop tef (*Eragrostis tef* (Zucc.) Trotter) under long-term low calcium stress. *Scientific Reports, 12*(1), 19552.

Liu, A., Ku, Y. S., Contador, C. A., & Lam, H. M. (2020a). The impacts of domestication and agricultural practices on legume nutrient acquisition through symbiosis with rhizobia and arbuscular mycorrhizal fungi. *Frontiers in Genetics, 11*, 583954.

Liu, J., Fang, L., Pei, W., Li, F., & Zhao, J. (2023). Effects of magnesium application on the arbuscular mycorrhizal symbiosis in tomato. *Symbiosis, 89*(1), 73–82.

Liu, Z., Wang, H., Xu, W., & Wang, Z. (2020b). Isolation and evaluation of the plant growth-promoting rhizobacterium *Bacillus methylotrophicus* (DD-1) for growth enhancement of rice seedlings. *Archives of Microbiology, 202*(8), 2169–2179.

Lu, X., Lu, X., & Liao, Y. (2018). Effect of tillage treatment on the diversity of soil arbuscular mycorrhizal fungi and soil aggregate-associated carbon content. *Frontiers in Microbiology, 9*, 2986.

Luo, X., Liu, Y., Li, S., & He, X. (2023). Interplant carbon and nitrogen transfers mediated by common arbuscular mycorrhizal networks: Beneficial pathways for system functionality. *Frontiers in Plant Science, 14*, 1169310.

Ma, J., Ma, Y., Wei, Z., Wu, J., Sun, C., Yang, J., & Huang, J. (2021). Effects of arbuscular mycorrhizal fungi symbiosis on microbial diversity and enzyme activities in the rhizosphere soil of *Artemisia annua*. *Soil Science Society of America Journal, 85*(3), 703–716.

Ma, J., Zhao, Q., Zaman, S., Anwar, A., & Li, S. (2024). The transcriptomic analysis revealed the molecular mechanism of arbuscular mycorrhizal fungi (AMF) inoculation in watermelon. *Scientia Horticulturae, 332*, 113184.

Ma, Q., Chadwick, D. R., Wu, L., & Jones, D. L. (2022). Arbuscular mycorrhiza fungi colonisation stimulates uptake of inorganic nitrogen and sulphur but reduces utilisation of organic forms in tomato. *Soil Biology and Biochemistry, 172*, 108719.

Maffo, A. F., Ngonkeu, E. L. M., Temegne, C. N., Chaintreuil, C., Ntsomboh-Ntsefong, G., & Youmbi, E. (2025). Effect of arbuscular mycorrhizal fungi combined with exogenous calcium on papaya (*Carica papaya* L.) plant growth. *Asian Journal of Research in Crop Science, 10*(3), 116–133.

Marwanto, M., Bustamam, H., Handajaningsih, M., & Anggraini, S. (2024). Co-application of arbuscular mycorrhizae via seed coating and phosphorus fertilizer for enhancing growth, yield, and nutrient uptake in ultisols for maize. *TERRA Journal of Land Restoration, 7*(1), 8–13.

Matos, P. S., Silva, C. F. D., Damian, J. M., Cerri, C. E. P., Pereira, M. G., & Zonta, E. (2021). Beneficial services of glomalin and arbuscular mycorrhizal fungi in degraded soils in Brazil. *Scientia Agricola, 79*(5), e20210064.

Minkosse, C., Langenfeld, A., Azzaz, A. A., Jeguirim, M., El-Bassi, L., Akrout, H., & Nassr, N. (2023). Short-term effects of olive-mill-wastes-derived biochars amendment and arbuscular mycorrhizal fungi inoculation on growth of maize (*Zea mays*) and mycorrhizal colonization. *Comptes Rendus Chimie, 26*(S1), 183–198.

Moreno Jiménez, E., Ferrol, N., Corradi, N., Peñalosa, J. M., & Rillig, M. C. (2024). The potential of arbuscular mycorrhizal fungi to enhance metallic micronutrient uptake and mitigate food contamination in agriculture: Prospects and challenges. *New Phytologist, 242*(4), 1441–1447.

Murtaza, G., Ahmed, Z., Iqbal, R., & Deng, G. (2025). Biochar from agricultural waste as a strategic resource for promotion of crop growth and nutrient cycling of soil under drought and salinity stress conditions: A comprehensive review with context of climate change. *Journal of Plant Nutrition, 48*(11), 1832–1883.

Muthuraja, R., & Muthukumar, T. (2021). Isolation and characterization of potassium-solubilizing *Aspergillus* species isolated from saxum habitats and their effect on maize growth in different soil types. *Geomicrobiology Journal, 38*(8), 672–685.

Nadeem, S. M., Hanif, A., Khan, M. Y., Waqas, M. R., Ahmad, Z., Ashraf, M. R., & Naveed, M. (2023). Elemental sulphur with sulphur-oxidizing bacteria enhances phosphorus availability and improves growth and yield of wheat in calcareous soil. *Archives of Agronomy and Soil Science, 69*(9), 1494–1502.

Nakai, Y., & Maruyama-Nakashita, A. (2020). Biosynthesis of sulfur-containing small biomolecules in plants. *International Journal of Molecular Sciences, 21*(10), 3470.

Narayan, O. P., Kumar, P., Yadav, B., Dua, M., & Johri, A. K. (2023). Sulfur nutrition and its role in plant growth and development. *Plant Signaling & Behavior, 18*(1), 2030082.

Narayanan, N., & Vyshnavi, K. (2025). Sustainable agriculture through phosphate-solubilizing microorganisms: A pathway to eco-friendly farming. In *Advancing global food security with agriculture 4.0 and 5.0* (pp. 257–288). IGI Global Scientific Publishing.

Nazir, A, Sarfraz, W, Ditta, A, Khalid, N, Farid, M, Shafiq, M, Firdaus-e-Bareen, ZF Rizvi, & Naeem, N (2023). Synergistic impact of two autochthonous saprobic fungi (*A. niger* and *T. pseudokoningii*) on the growth, ionic contents, and metals uptake in *Brassica juncea* L. and *Vigna radiata* L. under tannery solid waste contaminated soil. *International Journal of Phytoremediation, 25*(11), 1488–1500.

Noreen, S., Yaseen, T., Iqbal, J., Abbasi, B. A., Farouk Elsadek, M., Eldin, S. M., & Ali, I. (2023). Morphological and molecular characterizations of arbuscular mycorrhizal fungi and their influence on soil physicochemical properties and plant nutrition. *ACS Omega, 8*(36), 32468–32482.

Olaniyan, F. T., Alori, E. T., Adekiya, A. O., Ayorinde, B. B., Daramola, F. Y., Osemwegie, O. O., & Babalola, O. O. (2022). The use of soil microbial potassium solubilizers in potassium nutrient availability in soil and its dynamics. *Annals of Microbiology, 72*(1), 45.

Pan, G., Wang, W., Li, X., Pan, D., & Liu, W. (2023). Revealing the effects and mechanisms of arbuscular mycorrhizal fungi on manganese uptake and detoxification in *Rhus chinensis*. *Chemosphere, 339*, 139768.

Paranavithana, T. M., Marasinghe, S., Perera, G. A. D., & Ratnayake, R. R. (2021). Effects of crop rotation on enhanced occurrence of arbuscular mycorrhizal fungi and soil carbon stocks of lowland paddy fields in seasonally dry tropics. *Paddy and Water Environment, 19*(1), 217–226.

Parihar, M., Rakshit, A., Meena, V. S., Gupta, V. K., Rana, K., Choudhary, M., & Jatav, H. S. (2020). The potential of arbuscular mycorrhizal fungi in C cycling: A review. *Archives of Microbiology, 202*(7), 1581–1596.

Philippot, L., Raaijmakers, J. M., Lemanceau, P., & Van Der Putten, W. H. (2013). Going back to the roots: The microbial ecology of the rhizosphere. *Nature Reviews Microbiology, 11*(11), 789–799.

Qi, S., Wang, J., Wan, L., Dai, Z., da Silva Matos, D. M., Du, D., & Moles, A. T. (2022). Arbuscular mycorrhizal fungi contribute to phosphorus uptake and allocation strategies of *Solidago* canadensis in a phosphorus-deficient environment. *Frontiers in Plant Science, 13*, 831654.

Rahman, M. A., Parvin, M., Das, U., Ela, E. J., Lee, S. H., Lee, K. W., & Kabir, A. H. (2020). Arbuscular mycorrhizal symbiosis mitigates iron (Fe)-deficiency retardation in alfalfa (*Medicago sativa* L.) through the enhancement of Fe accumulation and sulfur-assisted antioxidant defense. *International Journal of Molecular Sciences, 21*(6), 2219.

Rosier, C. L., Piotrowski, J. S., Hoye, A. T., & Rillig, M. C. (2008). Intraradical protein and glomalin as a tool for quantifying arbuscular mycorrhizal root colonization. *Pedobiologia, 52*(1), 41–50.

Rui, W., Ma, J., Wei, N., Zhu, X., & Li, Z. (2023). Genome-wide analysis of the PHT gene family and its response to mycorrhizal symbiosis in tomatoes under phosphate starvation conditions. *International Journal of Molecular Sciences, 24*(12), 10246.

Saeed, HS, Munir, M, Samra, S, Farooq, A, Mahmood, H, Ahmad, I, Akram, MI, Iqbal, R, Elsadek, MF, Almutairi, SM, Ditta, A, & Daraz, U (2025). Screening of biosurfactants producing fungi from infected *Citrus sinensis* (L.). *Polish Journal of Environmental Studies, 34*(5), 5261–5270.

Sanz, J., Tomasa, O., Jiménez-Franco, A., & Sidki-Rius, N. (2022). Magnesium (Mg)[Z= 12]. In *Elements and mineral resources* (pp. 127–130). Springer Nature Switzerland.

Sardans, J., Lambers, H., Preece, C., Alrefaei, A. F., & Penuelas, J. (2023). Role of mycorrhizas and root exudates in plant uptake of soil nutrients (calcium, iron, magnesium, and potassium): Has the puzzle been completely solved? *The Plant Journal, 114*(6), 1227–1242.

Shi, J., Wang, X., & Wang, E. (2023). Mycorrhizal symbiosis in plant growth and stress adaptation: From genes to ecosystems. *Annual Review of Plant Biology, 74*(1), 569–607.

Shi, S., Luo, X., Dong, X., Qiu, Y., Xu, C., & He, X. (2021). Arbuscular mycorrhization enhances nitrogen, phosphorus and potassium accumulation in *Vicia faba* by modulating soil nutrient balance under elevated CO_2. *Journal of Fungi, 7*(5), 361.

Shrivastav, P., Prasad, M., Singh, T. B., Yadav, A., Goyal, D., Ali, A., & Dantu, P. K. (2020). Role of nutrients in plant growth and development. In M. Naeem, A. Ansari, & S. Gill (Eds.), *Contaminants in agriculture: Sources, impacts and management* (pp. 43–59). Springer International Publishing.

Sikhau, T., Tanga, M., Adetunji, A., Howell, C., Mulidzi, R., & Lewu, F. (2023). Sustainable winery wastewater management for improving soil quality, environmental health, and crop yield. In M. Fanadzo, N. Dunjana, H. A. Mupambwa, & E. Dube (Eds.), *Towards sustainable food production in Africa: Best management practices and technologies* (pp. 153–170).

Singh, A. K., Zhu, X., Chen, C., Wu, J., Yang, B., Zakari, S., & Liu, W. (2022). The role of glomalin in the mitigation of multiple soil degradation problems. *Critical Reviews in Environmental Science and Technology, 52*(9), 1604–1638.

Soltangheisi, A., Hales-Henao, A., Pena, R., & Tibbett, M. (2024). Species-specific effects of mycorrhizal symbiosis on *Populus trichocarpa* after a lethal dose of copper. *Ecotoxicology and Environmental Safety, 272*, 116112.

Sosa-Hernández, M. A., Leifheit, E. F., Ingraffia, R., & Rillig, M. C. (2019). Subsoil arbuscular mycorrhizal fungi for sustainability and climate-smart agriculture: A solution right under our feet? *Frontiers in Microbiology, 10*, 744.

Sparks, E. E., & Benfey, P. N. (2017). The contribution of root systems to plant nutrient acquisition. In *Plant macronutrient use efficiency* (pp. 83–92). Academic Press.

Thind, S., Chaudhary, M. S., Ditta, A., Hussain, I., Parveen, A., Ullah, N., Mahmood, Q., Al-ashkar, I., & Sabagh, A. E. (2022). Impact of mycorrhizal fungi from different rhizospheric soils on fungal colonization, growth, and chlorophyll contents of *Cenchrus ciliaris*. *Agronomy, 12*(11), 2644.

Upadhayay, V. K., Singh, J., Khan, A., Lohani, S., & Singh, A. V. (2019). Mycorrhizal mediated micronutrients transportation in food-based plants: A biofortification strategy. *Mycorrhizosphere and pedogenesis*, 1–24.

Vijayakumar, S., Gobinath, R., Gompa, S. J., Surekha, K., Kumar, R. M., & Sundaram, R. M. (2022). Management of micronutrient deficiency and toxicity in rice. *Indian Farming, 72*(10), 11–14.

Vuyyuru, M., Sandhu, H. S., McCray, J. M., & Raid, R. N. (2018). Effects of soil-applied fungicides on sugarcane root and shoot growth, rhizosphere microbial communities, and nutrient uptake. *Agronomy, 8*(10), 223.

Wang, H., Wang, Y., Cheng, X., He, Y., Shen, Z., Zhang, W., & Pu, X. (2025). Arbuscular mycorrhizal fungi colonization facilitates nitrogen uptake in cotton under nitrogen-reducing conditions. *BMC Plant Biology, 25*(1), 1129.

Wang, Y., Zhang, Y., Zhou, S., & Wang, Z. (2018). Meta-analysis of the no-tillage effect on wheat and maize water use efficiency in China. *Science of the Total Environment, 635*, 1372–1382.

Watts-Williams, S. J., Wege, S., Ramesh, S. A., Berkowitz, O., Xu, B., Gilliham, M., & Tyerman, S. D. (2023). The function of the *Medicago truncatula* ZIP transporter MtZIP14 is linked to arbuscular mycorrhizal fungal colonization. *Plant, Cell & Environment, 46*(5), 1691–1704.

Wipf, D., Krajinski, F., van Tuinen, D., Recorbet, G., & Courty, P. E. (2019). Trading on the arbuscular mycorrhiza market: From arbuscules to common mycorrhizal networks. *New Phytologist, 223*(3), 1127–1142.

Wu, S., Hu, Y., Zhang, X., Sun, Y., Wu, Z., Li, T., & Chen, B. (2018). Chromium detoxification in arbuscular mycorrhizal symbiosis mediated by sulfur uptake and metabolism. *Environmental and Experimental Botany, 147*, 43–52.

Wu, Y., Sun, Z., Liu, R., Wang, L., & Cai, B. (2024). Enhancing sulfur absorption in soybean rhizosphere through arbuscular mycorrhizal fungi inoculation: Implications for soil health and crop growth. *Journal of Cleaner Production, 463*, 142759.

Xie, K., Cakmak, I., Wang, S., Zhang, F., & Guo, S. (2021). Synergistic and antagonistic interactions between potassium and magnesium in higher plants. *The Crop Journal, 9*(2), 249–256.

Xing, D., Zhang, H., Cheng, H., Zhang, F., Shi, Y., Peng, S., & Twagirayezu, G. (2025). Arbuscular mycorrhizal fungi decrease soil nitrate by accelerating mulberry uptake and dissimilatory nitrate reduction to ammonium. *Environmental Technology & Innovation, 38*, 104139.

Yang, W., Gu, S., Xin, Y., Bello, A., Sun, W., & Xu, X. (2018). Compost addition enhanced hyphal growth and sporulation of arbuscular mycorrhizal fungi without affecting their community composition in the soil. *Frontiers in Microbiology, 9*, 169.

Yangyang, J. I. A., Van Der Heijden, M. G., Valzano-Held, A. Y., Jocher, M., & Walder, F. (2024). Mycorrhizal fungi mitigate nitrogen losses of an experimental grassland by facilitating plant uptake and soil microbial immobilization. *Pedosphere, 34*(2), 399–410.

Yazici, M. A., Asif, M., Tutus, Y., Ortas, I., Ozturk, L., Lambers, H., & Cakmak, I. (2021). Reduced root mycorrhizal colonization as affected by phosphorus fertilization is responsible for high cadmium accumulation in wheat. *Plant and Soil, 468*(1), 19–35.

Yin, D, Hong, S, Wen-qian, L, et al. (2009). Effects of AMF on key enzymes of nitrogen as simulated in maize under different ammonium to nitrate ratios. *Journal of Plant Nutrition and Fertilizers, 15*(6), 1380–1385.

Yizhu, L., Imtiaz, M., Ditta, A., Rizwan, M. S., Ashraf, M., Mehmood, S., Aziz, O., Mubeen, F., Ali, M., Elahi, N. N., Ijaz, R., Lelel, S, Shuang, C., & Tu, S. (2020). Response of growth, antioxidant enzymes, and root exudates production towards As stress in *Pteris vittata* and *Astragalus sinicus* colonized by arbuscular mycorrhizal fungi. *Environmental Science and Pollution Research, 27*, 2340–2352.

Zhang, Q., Tang, D., Yang, X., Geng, S., He, Y., Chen, Y., & Ruan, J. (2021). Plant availability of magnesium in typical tea plantation soils. *Frontiers in Plant Science, 12*, 641501.

Zhang, Y., Han, X., Ren, W., Zhang, H., & Tang, M. (2024). Arbuscular mycorrhizal fungi improve *Lycium barbarum* potassium uptake by activating the expression of LbHAK. *Plants (Basel), 13*, 1244.

Zhang, Y., Yao, Q., Li, J., Wang, Y., Liu, X., Hu, Y., & Chen, J. (2015). Contributions of an arbuscular mycorrhizal fungus to growth and physiology of loquat (*Eriobotrya japonica*) plants subjected to drought stress. *Mycological Progress, 14*(10), 84.

Zhou, R., Raza, A., Song, J., Janiad, S., Li, Q., Huang, M., & Hassan, M. A. (2025). Growth-promoting effects of arbuscular mycorrhizal fungus *Funneliformis mosseae* in rice, sesame, sorghum, Egyptian pea, and Mexican hat plant. *Frontiers in Microbiology, 16*, 1549006.

Zou, Y. N., Xu, Y. J., Liu, R. C., Huang, G. M., Kuča, K., Srivastava, A. K., & Wu, Q. S. (2023). Two different strategies of *Diversispora* spurca-inoculated walnut seedlings to improve leaf P acquisition at low and moderate P levels. *Frontiers in Plant Science, 14*, 1140467.

Chapter 5

Arbuscular mycorrhizal fungi and major nutrients cycling

Saleem Sajjad[a], Umar Sheraz[a], Rizwana Kausar[b], Raina Ijaz[c], Muhammad Imtiaz[d], Allah Ditta[e,f] and Nabeel Ahmad[d]

[a] *Department of Soil and Environmental Sciences, College of Agriculture, University of Sargodha, Sargodha, Pakistan,* [b] *Soil and Water Testing Laboratory for Research, Sargodha, Pakistan,* [c] *Department of Horticulture, University of Poonch, Rawalakot, Pakistan,* [d] *Soil and Environmental Biotechnology Division, National Institute for Biotechnology and Genetic Engineering (NIBGE), Faisalabad, Pakistan,* [e] *Department of Environmental Sciences, Shaheed Benazir Bhutto University Sheringal, Sheringal, Pakistan,* [f] *School of Biological Sciences, The University of Western Australia, Perth, Australia*

5.1 Introduction

5.1.1 Overview of soil–plant–microbe interactions

Human society is based on agriculture, which depends on highly volatile and complex relationships between soil and plants (Hurni et al., 2015). These processes have a significant impact on nutrient supply, soil structure, plant growth, and overall crop health (Khan et al., 2023). A comprehensive knowledge of these complex processes is key to sustainable crop production that will help achieve maximum yields without compromising soil health and environmental stability (Shah & Wu, 2019). Over the past few decades, advances in soil science have emphasized the importance of microbial communities, nutrient cycling, and root–soil interactions in achieving sustainable agriculture (Das et al., 2022). These aspects are covered in this chapter and how they relate to the long-term sustainability of agriculture. Soil is a finite, nonrenewable natural resource that is critical in supporting human life (Bhattacharya et al., 2022). The increase in agricultural activities in the world because of the increasing demand for food in the last several decades has also led to large-scale soil degradation (Kopittke et al., 2019). The degradation has resulted in a decrease in soil fertility, a decrease in water retention capacity, a loss of organic carbon, a decrease in biodiversity, and an instability of nutrient cycling (Lal, 2015).

Plants, soil constituents, and microorganisms interact in a complex manner and significantly affect soil health and productivity (Kumar & Verma, 2018). Microorganisms such as bacteria, fungi, and other microscopic life forms in soil have many interactions with each other as well as with plant roots (Xing et al., 2025). These interactions help maintain a wide variety of biological functions that are essential for ensuring ecological balance and sustainability in the soil environment . Interactions between plants and microbes can be both positive and negative for plant growth (Schirawski & Perlin, 2018). Microbial activity that is detrimental to plant growth is considered antagonistic or detrimental (Wang et al., 2022). In contrast, microbial associations that enhance plant survival, enhance nutrient uptake, and increase crop yield are considered to produce synergy or benefit (Das et al., 2022). To ensure and enhance soil health a harmonious and balanced relationship between soil microorganisms and plant systems is essential (Ortiz & Sansinenea, 2022).

Because of the growing awareness of the negative effects of environmental degradation and the need to maintain soil quality, the importance of soil microorganisms in agribusiness and the functioning of ecosystems in general has become increasingly evident (Timmis & Ramos, 2021). This has led to increased awareness that has promoted the use of organic amendments such as compost, green manure, and animal waste as alternatives to synthetic chemical fertilizers (Bremaghani, 2024). Not only do organisms reduce soil degradation, but they also increase the diversity of microorganisms, improve soil structure, and improve long-term agricultural productivity (Fig. 5.1).

Advancing Sustainable Agriculture Through Arbuscular Mycorrhizal Fungi. DOI: https://doi.org/10.1016/B978-0-443-45130-0.00019-9

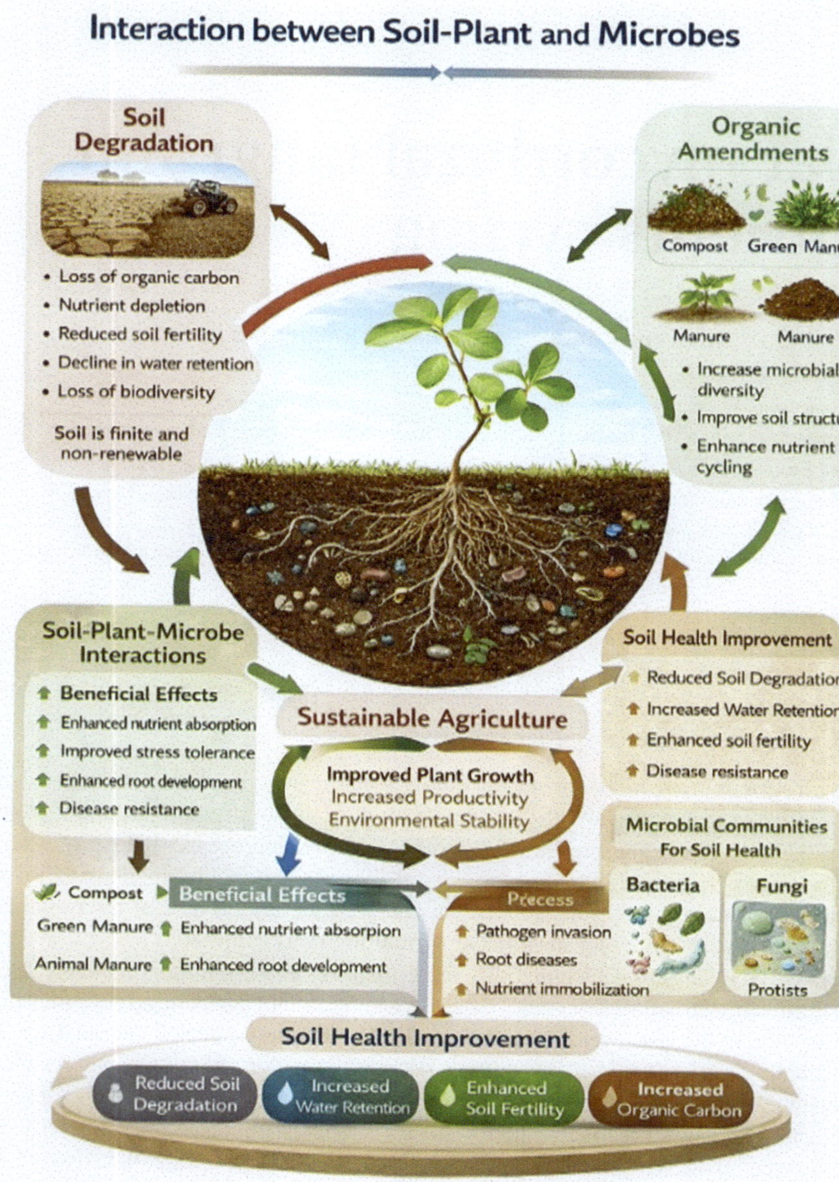

FIGURE 5.1 Interaction between soil–plant and microbes.

5.1.2 Importance of nutrient cycling in agroecosystems

Nutrient cycles can be seen as the relationship between an agricultural system and the rest of the environment. Such cycles can be confined to an agricultural ditch, stream, or watershed. Nutrient cycles are linked to climate change, water and air pollution, and geological emissions from agricultural activities at regional and global scales (Imran et al., 2025). The nutrient cycle has been dramatically altered over the past 50 years because of the high rate of synthetic fertilizer use. High levels of fertilizer use have multiple environmental impacts, many of which come with social and economic costs in terms of reduced ecosystem services (Srivastav et al., 2021). These problems arise in part from the fact that most agricultural production models, which are based on high fertilizer use, aim to produce arable crops in short periods of time (Toor et al., 2024). However, in the context of their overall ecological matrix, or agroecosystem, it is clear that agricultural nutrient cycles can be enhanced through farm management to also provide other ecosystem services, including water quality and flow control, soil fertility, carbon storage, climate regulation, and biodiversity.

The arbuscular mycorrhizal (AM) fungi (AMFs) play a significant role in plant bodies to uptake nutrients, more than 60% of terrestrial plants. Because of their widespread presence, AMF can make up as much as 50% of the soil's microbial biomass (Uddin & Kebreab, 2020). AMF is a very old symbiotic relationship (Etienne et al., 2018). AMF is the most common soil bacterium that forms symbiotic relationships with over 80% of plants, except for a few plant families that do not demonstrate any interaction, including Amaranthaceae, Brassicaceae, Cruciferae, Chenopodiaceae, Caryophyllaceae, Juncaceae, Cyperaceae, and Polygonaceae. AMFs are obligatory symbionts that require a host plant to complete their life cycle; they are not parasites. AMFs are fungi that grow in the soil and can greatly increase plant uptake of nutrients (Shukla et al., 2025). The properties of AMF's mycelium are responsible for the increase in host plant nutrient absorption. AMF has been shown to enhance a number of soil properties, including soil aggregation, soil nutrient availability, water retention, microbial activity, nitrogen, carbon, and phosphorus cycling, and soil acidity correction, in addition to having an effect on plant growth and productivity (Begum et al., 2019). By boosting the uptake of water and nutrients such as potassium (K), phosphorus (P), and nitrogen (N), they increase crop productivity. These mycorrhizal hyphae have a larger surface area than nonmycorrhizal roots and absorb nutrients osmotrophically (Begum et al., 2020). Furthermore, AMFs play a role in their host plant through means other than nutrition by improving its defenses against a variety of biotic (pathogens and herbivores) and abiotic (drought, salt, and heavy metals) challenges (Bell et al., 2024). As "ecosystem engineers" of plant communities, AMFs are crucial to the survival and fitness of plants. The purpose of this chapter is to synthesize the current understanding of AMF symbiosis, specifically the advantageous impacts on soil and the major role in macronutrients within plants.

5.2 Biology and ecology of arbuscular mycorrhizal fungi

5.2.1 Taxonomy and classification of arbuscular mycorrhizal fungi (phylum Glomeromycota)

Until the end of the last millennium the classification of the *Glomeromycota* was largely morphologically motivated. Fungi of the genus *Glomeromycota* are known to form arbuscular mycorrhizae, except for one genus. They were identified using the shape of their spores, their spore formation, and the structure of their spore wall (Gerdemann & Trappe 1974). However, molecular phylogenetic tools were quickly incorporated into taxonomic analyses and became the driving force behind the creation of a new classification in the near future. Arbuscular mycorrhiza–forming fungi were classified into three families (Acaulosporaceae, Gigasporaceae, and Glomeraceae) and six genera (*Acaulospora, Entrophospora, Gigaspora, Glomus, Sclerocystis*, and *Scutellospora*) belonging to the order *Zygomycota* without the aid of molecular aspects. This classification was made using spore morphology and spore formation morphology (*Oculosporidiosis, Enterospore, Gigaspora, Glomoid, Radial-glomiid*, and *Scutellospora*). There were differences in the structure of the spore walls that were used at the species level.

We recognize 3 classes (Archaeosporomycetes, Glomeromycetes, and Paraglomeromycetes), 5 orders (*Archaeosporales, Diversisporales, Gigasporales, Glomerales*, and *Paraglomerales*), 49 families, 29 genera, and approximately 230 crores (30 crores). Until recently, it was not known whether *Glomeroid* and *Gigaspora* species could be further divided into different morphological groups according to the major phylogenetic clades obtained as a result of molecular studies. Not all mycologists accepted the first revision based on the homologous morphology and phylogenetic features of the sporogenous cell formation (*Gigaspora* and *Scutellospora*) of glomeromycetes (Oehl et al., 2008). Subsequent work based on a larger database (Goto et al., 2010), however, confirmed that *Scutellospora* is monophyletic between the revised genus *Scutellospora* and the newly described *Racocetra, Cetraspora, Dentiscutata*, and *Orbispora*.

5.2.2 Morphology and life cycle

AMFs have been around for about 600 million years on Earth and are likely to have evolved as a result of ancient symbiosis between fungi and plants that can no longer exist independently and whose life cycle must be completed only in the presence of plants. Approximately 80% of terrestrial plant species form mycorrhizal symbioses with AMF (Wang & Qiu, 2006), and the most common form of mycorrhizal symbiosis, which is devoid of host plant specificity, is AM symbiosis. However, as reported by Husband et al. (2002), the name arose because of the specific structure of arbuscules (Lat. arbuscula = small tree, shrub) found in the cortical cells of most plant roots. These structures are thought to diagnose AM symbiosis with storage vesicles, and these fungi are classified under the phylum Glomeromycota (Schussler et al., 2001). AMFs are obligate symbionts, dependent on host plants and colonizing plant roots with spores, hyphae, or infected root fragments (Kelinomos & Hart, 2002). The fungal mycelium of AMF invades plants by expanding the root system, which enables it to increase the use of water and minerals in the soil. Plants colonized by AMFs are less susceptible to environmental stresses, including drought, cold, and pollution (Juniper & Abbott, 2006), and are more resistant to attacks by bacterial and fungal pathogens (Selvaraj & Chellappan, 2006). Mycorrhizal symbiosis maintains and enhances plant growth, drastically reduces the use of synthetic fertilizers, improves soil condition, elevates soil macrofloral diversity, and reduces soil erosion.

The life cycle of AMF begins with the germination of the fungus in the soil when the environment is favorable, and the host plant is not involved. Fungal colonies grow several centimeters in length and develop characteristic growth structures (Giovannetti et al., 1994). This stage of symbiosis becomes presymbiosis, which is characterized by extensive offshoring of hyphae because of the presence of the host plant (Giovannetti et al., 1993). This is a crucial stage of the AMF life cycle, relying on the chemotactic abilities of the AMF to allow the hyphae to spread to the roots of the host plant, making it an important mechanism for the establishment of root space, appressorium development, and symbiosis (Sbrana & Giovanne, 2005). The symbiosis phase begins with the interaction of fungal hyphae with plant roots, and with their help, the fungus penetrates the cortex (Giovannetti et al., 1993) to form morphologically distinct specialized structures—inter- and intracellular hyphae, vesicles, and arbuscules. Arbuscules mark the site of bidirectional nutrient exchange between plants and fungi (Requena et al., 2007) and play a key role in AM symbiosis. Hyphae penetrate the roots, outside the soil, to form extraradical mycelium, and the life cycle is completed with the formation of new asexual spores in the extraradical mycelium (Requena & Breuninger, 2004). AMF promote plant growth and reproduction under such symbiosis by increasing nutrient (P and N) and soil water uptake through extra- and intraradical mycelium (Mishra et al., 2018). On the other hand the plant has carbon in the form of saccharides, which are the result of photosynthesis and are transferred to the fungi through active or passive processes through intraradical fungal structures. After AMF invade plants, they persist along the root system and can be transferred to other soils.

5.3 Mechanisms of nutrient uptake and transfer

5.3.1 Properties and benefits of macro- and micronutrients

Macronutrients play a crucial role in the survival of plants and human beings as they perform various metabolic functions (Yuan et al., 2023). Nutrients come in two categories: the major macronutrients—N, P, and K; and secondary macronutrients—Ca, Mg, and S. These play a significant role in plant metabolism by protecting the plants against various stresses and improving the crop yield, performance, and quality (Sedaghati et al., 2023). More recently, researchers have been looking at the different roles these nutrients play a significant role in plant life and their potential in the future. Plant macronutrients have been described to have vital functions in normal development and growth of plants since their primary macromolecular role is as structural components as well as redox-sensitive reagents (Finucane et al., 2020). Nitrogen assists in improving seed and fruit yield and improves leaves, good-quality forage crops, and rapid vegetation of the plants. Whereas phosphorus has been known to be an essential part of the cell membrane of plant where it exists as phosphate and serves a significant purpose because phosphorus is a part of adenosine triphosphate, ribonucleic acid, and deoxyribonucleic acid (Riaz et al., 2021). Potassium is highly mobile and is an important macronutrient element of a plant that is significantly located in every younger portion of the plant body. Calcium (Ca) is involved in cell wall structure and the membrane. It helps in the growth of leaves and roots. Magnesium (Mg) is the most important part of chlorophyll that is crucial in photosynthesis. Mg also helps to activate the enzyme. Proteins and amino acids depend on sulfur (S) as a macronutrient, in addition to its role in the synthesis of chlorophyll (Motamedi et al., 2022). Furthermore, these elements in plants specifically fulfill their nutritional role in a balanced proportion, which is essential for the growth and development of different species (Ngosong et al., 2022). To improve nutrients availability for plant uptake, it is very crucial to support nutrient availability through fertilizers or the alteration of the soil environment, which significantly affects the nutrient availability in addition to regulating the destructive plant diseases in an integrated pest management system. Plants diseases seem to be fatal in the management system of crops as nutrient-based beneficial plants seem to die on a big-business basis.

5.3.2 Arbuscular mycorrhizal fungi and phosphorus cycling

Phosphorus is a nutrient that is crucial for plant growth and other processes in cells, such as nutrient transport, sugar breakdown, and transmission of genetic information. P can be found in the soil in inorganic soluble, inorganic insoluble, and organic forms (Gosling et al., 2013). AMFs transfer inorganic P compounds to host plants via arbuscules (Gao et al., 2001). Under P-deficient conditions, it has been established that AMFs increase the uptake of more P in plants. AMFs have positive effects on P uptake and translocation, and their effect improves plant growth under different environmental conditions. AM fungal hyphae are capable of delivering plant P over distances of up to 100 mm (up to 80%) (Jansa et al., 2005). This demonstrates the importance of AM fungal hyphae in the area of long-distance P transport to

TABLE 5.1 Role of AMF in phosphorus uptake and translocation in different plants under various conditions.

Plant	Type of AM fungi	Function	Condition	References
Wheat	*Glomus intraradices*	Enhanced total P uptake	Under the P application conditions	Li et al. (2006)
Maize	*Funneliformis mosseae*	Increase P uptake	Under drought stress conditions	Zhang et al. (2025)
Barley	*Mixed-Glomus species*	Enhance plant P acquisition	Under salinity stress	Masrahi et al. (2023)
Rice	*G. intraradices*	Increase P transport	Under arsenate stress	Chen et al. (2013)
Cotton	*Rhizophagus clarus*	Increase P content in the plant	Under field conditions	Cely et al. (2016)

AMF, arbuscular mycorrhizal fungi.

plant roots (Plassard et al., 2019). Fungal hyphae provide AM P and also help in its uptake via the MP pathway and transport P over long distances to plant roots (Smith et al., 2011). Their distribution and growth rate affect the level of P transport through hyphae (Adeyemi et al., 2021). However, the efficiency of P transport via the MP pathway in AMFs varies between species and also with plant type. For example, P acquisition by spores of *Glomus mosseae* is more efficient than that of *Glomus fasciculatum* (Wahid et al., 2020). *G. mosseae* was better at P uptake in wheat than *Glomus monosporum*. Both AMFs, *Glomus manihotis* and *Glomus occultum*, were highly active in promoting P uptake in cassava (*Manihot esculenta*) roots (Straker et al., 2010). Furthermore, AMFs, which are related to bacteria that accumulate on their spores, have been observed to enhance P uptake through a mycorrhizal pathway (Etesami et al., 2021). The effectiveness of fungal spores may vary depending on hyphal structure and growth, spore species and their colonization, soil compatibility with native plants, fungal metabolism, and enzyme composition (Mensah et al., 2015). AM species, such as *Rhizophagus irregularis* and *Glomus intraradices*, are particularly productive in low-P soils. In contrast, others can better cope with high-P environments. But the efficiency of P uptake and transport in the MP pathway is species-specific and depends on the interaction of the final species potential, enzymatic activity, and host plant compatibility. AMFs enhance P uptake at different stages of plant development and enhance physical and biochemical activities. AMFs, together with P-solubilizing bacteria, such as *Enterobacter agglomerans*, increase P uptake and growth and abiotic resistance (Gosling et al., 2013). Under P-deficient conditions, AM symbiosis provides plants with additional P and enhances the production of phytohormones in stress responses (Bhantana et al., 2021). An example is *Gigaspora margarita*, which was observed to increase phenolic compounds in onions (*Allium cepa*) under low P conditions (Table 5.1). Similarly, in tomatoes (*Lycopersicon esculentum*), *Glomus etunicatum* increased P uptake and plant tolerance in sterile soil (Balliu et al., 2015). Furthermore, AMFs increase P, antioxidant activity, and plant adaptation to salt and water stress. Nevertheless, field experiments on AMFs are needed to further understand the effects of AMFs on crops that require high P concentrations.

5.3.3 Role of arbuscular mycorrhizal fungi in nitrogen cycling

Plants require average levels of nitrogen for optimal growth and development (Chen et al., 2020). An ideal N uptake can promote photosynthesis, leaf area, growth, and yield. However, N can be removed either from the deep tiller or from the root zone at a distance from the plant root. Rhizosphere N uptake and plant root transport can occur via hyphae of AMFs (Tian et al., 2010). Recent studies have shown that AMFs can contribute significantly to N uptake and concentration in different plants. AM hyphae can provide 30%–40% of the total N in AM-inoculated plants. Some studies have shown that up to 25% of the measured variation in plant N of field-grown plants can be attributed to variation in AM fungal associations (Ngwene et al., 2010). AM-inoculated plants show improved growth with N treatments. For example, the use of *G. mosseae* allows Chinese celery (*Apium graveolens*) to absorb more 15 N from the soil than uninoculated plants (Rouydel et al., 2021). Similarly, onion (*A. cepa*) was enhanced to take up 15 N by *G. fasciculatum* as indicated by increased N levels in roots and shoots (Zhu et al., 2016). In alfalfa (*Medicago sativa*), *G. mosseae* increased N uptake, leaf N concentration, and overall growth. But research shows that although N is crucial for hyphal growth, the N provided by hyphae is not sufficient to meet plant needs. AMFs also coexist with N-fixing bacteria that enhance nodulation, especially in fruits. *G. mosseae* is an example that greatly improved nodulation in soybean (*Glycine max*), resulting in significant plant growth, yield, and productivity in AM soybean plants versus non-AM plants (Liu et al., 2017). These results suggest that nitrogen uptake in plants, especially legumes, can be greatly improved by the use of AMFs.

In addition to direct effects on N uptake, AMFs enhance N use efficiency and nutrient uptake under stress conditions (Table 5.2). As an example, *G. fasciculatum* enhanced N uptake in lettuce (*Lactuca sativa*) under water-stressed environments (Ghonjalipour-Goshki et al., 2021). Similarly, *G. intraradices* is another AM fungus that promoted N uptake

TABLE 5.2 Role of AM fungi in enhancing nitrogen uptake in various plants under different conditions.

Plant	Type of AM fungi	Function	Condition	References
Wheat	*Rhizophagus irregularis*	Improved N uptake and N use efficiency	Under elevated CO_2	Zhu et al. (2016)
Maize	*Funneliformis mosseae*	Increased total N uptake	Under high water conditions	Wu et al. (2021)
Rice	*Glomus species*	Increased N accumulation and distribution in vegetative tissue	Under the C and N application	Zhang et al. (2017)
Cotton	*Rhizophagus clarus*	Increase N content in the plant	Under field conditions	Cely et al. (2016)

AM, arbuscular mycorrhizal.

in maize (*Zea mays*) during drought stress. Growth-promoting effects on N uptake have been noted in wheat (*Triticum aestivum*) through AM symbiosis, which led to increased biomass growth and improved utilization of P, N, and Zinc (Zn) during field studie (Khan et al., 2022). In addition, *Funneliformis mosseae* has been reported to increase N uptake in wheat in saline soils. *Glomus claroideum* is another AM fungus that increased wheat biomass and chlorophyll content by increasing N availability during drought (Abdulfatah & Israr, 2012). Inoculation with *Rhizophagus intraradices* increased N levels in the roots and shoots of *Leymus chinensis* and therefore promoted growth with or without nitrogen fertilizer application. When using *Glomus luteum*, maize produced higher glomeruli with high N fertilization, which further increased the N content in the plant (Eddiwal et al., 2014). These examples illustrate the importance of AMFs in improving nitrogen uptake, plant growth, and survival under diverse environmental conditions. The effect of N application on AM fungal colonization and spore development, according to recent studies, is highly dependent on soil conditions. For example, the AM fungal genera *Gigaspora* and *Acaulospora* are severely affected by poor soil conditions. Cotton (2018) explains that where N availability is low, AM fungal composition will shift to *Glomeraceae*, as opposed to *Gigasporaceae*. Interestingly the class Mucoromycotina, previously viewed as Glomeromycotina, was given to be unaffected by high N conditions (Fellbaum et al., 2012). However, the response of most AM fungal groups and classes to N fertilization has not been well studied. The addition of ammonium (NH4 þ) to soil through high N fertilization causes mitochondrial expansion and cytoplasmic bursting of plant roots (Eddiwal Saidi et al., 2014). Ammonium, which AMFs can readily absorb, can cause the fungus to take up large amounts of carbohydrates, which can result in undesirable growth and development of the host plant (Bennett & Groten, 2022).

Furthermore, too much N can damage the abnormal hyphae of the fungus, leading to the destruction of the symbiotic interaction between the fungi and the host plant (Ghorbanpour et al., 2018). In contrast, moderate N application has been shown to promote colonization of AMFs, which are known to directly promote plant growth and productivity. According to cost–benefit models (Verzeaux et al., 2017), symbiotic demand between plants and AMFs may be reduced if plants can obtain adequate N. Changes in belowground nutrient dynamics allow plants to allocate energy more efficiently, reducing the photosynthetic products accessible to mycorrhizae. According to the functional equilibrium model the higher the density of AM fungal spores, the higher the colonization rates. In nutrient-rich soils, plants allocate more energy to aboveground growth, thus promoting the growth of their stems and leaves (Aguilar-Trigueros et al., 2019). However, in soils with high N, the relative distribution between both arbuscules and extraradical hyphae may likely be reduced. In addition, N accumulation has been mentioned to suppress the Gadgil effect (where mycorrhizae suppress saprotrophic activity), which could potentially enhance saprotrophic decomposition. However, these results are in contrast to data on reduced decomposition in the presence of N (Liang et al., 2022). The effect of moderate N accumulation on EcMF activity and C storage is still not well understood, and, therefore, it is difficult to identify the net effect of N-mediated changes in mycorrhizal types on C storage along the soil profile.

5.3.4 Arbuscular mycorrhizal fungi and potassium cycling

Such additional and complementary strategies can be offered by potassium (K) fertilizers and native AMFs that can be used in combination with others currently under development. Notably, K has been identified to improve the drought resistance of various annual crops in the presence of water stress (Moradi, 2016). K has been documented to play a role in plant water stress resistance because of its ability to stabilize the osmotic potential of cells and to regulate stomatal activity, properties that enhance drought resistance in plants. K plays a protective role by maintaining high pH in the stroma and preventing photooxidative injury to chloroplasts (Wang et al., 2013). In addition, K controls the activation of many enzyme systems that control photosynthesis, water use efficiency and movement, nitrogen uptake, and protein synthesis in plants.

Romheld and Kirkby (2010) also pointed out that adequate potassium supply has a significant effect on root growth and, consequently, root surface area, with a relative increase in water uptake in the presence of water deficit. They are known to enter symbionts with the roots of host plant species in exchange for carbon in relation to AMF (Farhaoui et al., 2025). The role played by AMF in enhancing nutrient and water uptake, increasing nutrient use efficiency, and achieving higher biomass and yield is well documented in the literature.

5.4 Arbuscular mycorrhizal fungi: A natural solution for soil restoration and sustainable farming

AMFs are among the most important beneficial soil microorganisms that may coexist with plants in a mutualistic symbiotic relationship (Diagne et al., 2020; Nazir et al., 2023; Saeed et al., 2025; Yizhu et al., 2020). AMFs are obligatory symbionts that can form a mutualistic symbiotic relationship with most (more than 80%) vascular plants, including bryophytes, hepatics, pteridophytes, gymnosperms, and angiosperms in terrestrial ecosystems (Boyno et al., 2024). The carbohydrates and the lipids that support the survival of the photosynthetically generated carbohydrates by the host plant are provided to AMF so that 10%–40% carbohydrates can be assimilated by fungi (Garcia et al., 2020). AMF, in turn, enhances the uptake of the vital mineral nutrients (especially phosphorus and nitrogen) in exchange water supply (Li et al., 2022). This is done through the development of a branching mycelial network in the soil capable of increasing the surface area of its host roots by 40-fold (Zhou et al., 2020). These fungi are fundamentally other sources of benefits that include carbon recycling, enhancement of soil aggregation, plant growth hormone production, increased photosynthetic efficiency of the host, and aid to the plant to withstand abiotic (such as drought, salinity, and heavy metal contamination) and biotic stresses (pathogens and insects) by use of several mechanisms.

5.4.1 Macronutrient dynamics in soil: Foundations for sustainable agriculture

The relation between soil health and plant nutrition is key to sustainable agriculture and to improving long-term productivity and ecological sustainability. Soils that are in good health and have balanced physical, chemical, and biological properties offer a solid foundation for plant growth and efficient nutrient recycling (Wang et al., 2022). Ecofriendly practices of crop rotation, cover cropping, reduced tillage, and organic amendments promote soil energy and microbial diversity, thereby promoting soil structure and nutrient availability (Garg and Singh, 2018). The use of balanced fertilizers and health-related techniques such as precision agriculture provides efficient nutrient management, meaning crops grow with essential nutrients without harming the environment (Xiao et al., 2022). Stimulating soil biodiversity, limiting soil erosion and soil composition, and practicing water conservation techniques further enhance the sustainability and productivity of arable land. Finally, it is important to protect soil health and maximize plant nutrition for future food. Productivity, maintaining environmental quality, and providing food security amidst global challenges such as population growth and climate change (Wahid et al., 2020). Sustainable agriculture has the potential to achieve environmental balance, economic stability, and social justice by incorporating soil and nutrient management recommendations.

There are several ways in which agricultural activities greatly affect the environment and the state of natural resources. There are limitations and challenges associated with macronutrient management in sustainable agriculture, such as environmental impacts related to nutrient runoff, soil degradation, and the need to shift toward green farming practices (Marín et al., 2016). This is further complicated by the demand to reconcile food production with environmental sustainability and the difficulties of implementing new farming systems.

5.4.2 Mechanisms of arbuscular mycorrhizal fungi in soil health and sustainable agriculture

A sustainable agroecosystem is a complex and dynamic system that integrates ecological processes with an agricultural landscape to make the best possible use of resources, to increase and maintain biodiversity, and to provide long-term productivity in situations in which economic viability and environmental stewardship are also assured (Johan et al., 2021). Among other microorganisms the mycorrhizal fungi are quite relevant since they have a symbiotic relationship with the host plants (Fig. 5.2).

The mycorrhizal fungi have been found to protect the host plants from other environmental changes, such as metal toxicity, root pathogens, salinity, acidity, and temperature (Sarkodee et al., 2020). The versatile advantages given by the AMFs onto their host plants play a great role in the resilience and sustainability of agro-systems (Mitra et al., 2023). Healthier soils can also play a significant role in the reduction of carbon dioxide emissions, and AMF is a key organism helping in mediating the benefits of the healthier soils (Anandan et al., 2021). Mycorrhizal fungi enhance the structure of the soil by

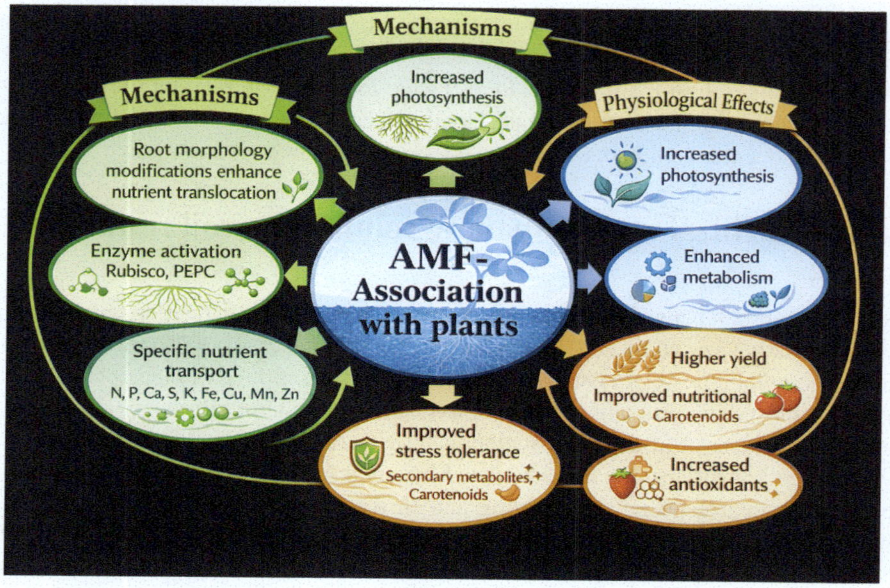

FIGURE 5.2 Mechanisms of AMF association with plants. *AMF*, Arbuscular mycorrhizal fungi.

enhancing the aggregation of soil particles that reduce soil erosion, improve the water retention of the soil, and increase the porosity of the soil (Narwal et al., 2021). The soil organic matter compound that is produced as a result of AMF activity can be used as a carbon reservoir, thus sequestering CO_2. Besides, AMF assists in the soil cycling of nutrients. It makes more of the important nutrients like phosphorus available to grow plants (Baki et al., 2021). This causes the ability of such plants to sequester more carbon over the long run and hence requires less use of synthetic fertilizers, which is correlated to increased release of GHGs.

The symbiotic benefits that AMFs provide to successive crops are eroded because heavy tillage breaks up the delicate hyphal networks on which the fungi rely to sustain plant health and nutrient acquisition (Chandrasekaran et al., 2019). Disruption of the network structures of these communities when tillage splits them apart hinders the operation of mycorrhizae and affects the overarching soil ecosystem, including the diminution of soil organic matter and interrupting key nutrient cycling processes (Dutta et al., 2024). All these are disadvantageous to soil health and development, which affects the capacity of the soil to promote a healthy plant and ecological balance in the system. Because of the extensive use of chemical fertilizers and pesticides in conventional farming systems, there are harmful effects of these activities on the number and activities of AMFs (Aguilar et al., 2023). Moreover, soil microbial composition can be disrupted because of the use of pesticides and herbicides; the beneficial AMFs are at risk in communities. According to Helander et al. (2018) the practice of herbicidal management reduces colonization of AMFs in perennial grass species, meaning that the composition and productivity of plant communities are also altered. Munyanziza et al. (1997) have also looked at the consequences of transforming natural systems to agricultural systems on the AMFs. They hypothesized that intensive farming practices are deleterious to the AMFs. Therefore it is desirable to teach low-input sustainable farming methods to growers to increase the AMFs' population in soil. It is based on maintaining a sustainable agroecosystem.

5.5 Integrated effect of macronutrients and arbuscular mycorrhizal fungi on soil health and sustainable agriculture

5.5.1 Synergistic effects of macronutrients and arbuscular mycorrhizal fungi

Mycorrhiza is the symbiotic relationship between the root systems of plants and the fungi in the soil. AMFs are the most widely occurring symbiosis between a plant and the various forms of mycorrhizae (Quan et al., 2023). The fungus absorbs nutrients in the soil with its unusual mycelium and transfers them to the soil. They live within the host root within the intraradical mycelium and trade nutrients with the host's carbon in a special plant–fungal interaction (Sarker et al., 2021). In AMF symbiosis, host plants grow better and become more nutritious and less dependent on fertilizers. The attachment of mycorrhizae to the roots of plants enables exploring larger soil volume, hence increasing the uptake of immobile mineral elements such as phosphorus (P), water absorption, water use efficiency, and resistance to pathogens in soil

(Fayaz & Zahedi, 2021). In addition to phosphorus, the excretion of many other immobile mineral elements, such as copper, zinc, and, in certain circumstances, potassium, calcium, magnesium, sulfur, and iron, is increased (Carella et al., 2018). Therefore it is evident that AMF has a synergistic impact on the availability of macronutrients in soil.

5.5.2 Impacts on soil management practices

Soil properties and microbial diversity are greatly influenced by soil management practices. The application of mineral fertilizers, chemical pesticides, and herbicides is critical to keeping up crop nutrition, and they replenish the soil nutrients pool removed by or extracted by the crop, weed control, and pest management (Liu et al., 2020). Nevertheless, high- and long-term exposure to inorganic fertilizers reduces plant dependence on AMF and, consequently, mycorrhizal diversity and abundance. Soil phosphorus is the most important factor to control plant mycorrhizal symbiosis (Mbodj et al., 2018). On the other hand the low-dose application of fertilizer allows the plant to form a mycorrhizal symbiosis. Particularly, as observed by Liu et al. (2016) in nutrient-poor soil, application of P enhances the mycorrhizal-related advantages relayed to the plant. Moreover, the use of organic fertilizers is associated with both positive and negative effects on AMF diversity. Zhu et al. (2016) state that organic matter enhances the structure of AMF communities in the maize rhizosphere. Similarly the effects of inorganic applications of pesticides and herbicides have both positive and negative effects on AMF. Such effects are mediated by the release of active substances taken up by plants via root or hyphae out of the rhizosphere (Bhantana et al., 2021). Besides the chemical application, heavy tillage has a significant effect on the community composition of AMF, decreasing mycelium extension, rate of colonization, and diversity structure. This is because tillage causes a significant influence on soil permeability, structure, and microbial food substrates, which ultimately affect the activity of soil microbiota and their habitat type characteristics (Tan et al., 2015). There are a few studies that have found that conservation or zero tillage enhances AMF diversity and abundance, leading to better plant growth (Pattnaik et al., 2021). These conflicting findings show that the AMF effects of various chemical applications and tillage practices still require more investigations to determine the influence of various soil management regimes on soil AMF.

5.5.3 Improvements in plant growth, yield, and quality

Because of its enormous impact on plant systems, the use of AMFs has emerged as an interesting strategy for environmentally responsible agriculture. AMFs can retain approximately 90% of the P, 60% of the N, and 20% of the carbon in the plant. According to Khan et al. (2021), AMFs enhance the N, P, calcium (Ca), and other nutrients such as sulfur (S), K, iron (Fe), copper (Cu), manganese (Mn), and zinc (Zn) levels in different plants. AMFs may act as parasites, deriving carbon but contributing very little or no P to the host plant. AMFs could activate physiological and biochemical processes induced in reactions of cereal plants at different growth stages, enabling them to grow fast and support them against stress (Hussain et al., 2021). An increased photosynthetic rate may contribute to increased metabolism of host plants, which could increase their performance. Conversely, AMFs support enzyme activity and enhance the ratio of chlorophyll and photosynthetic rate in cereal plants (Khalloufi et al., 2017). Moreover, under AMF inoculation, there is an increase in various enzymes such as phosphoenolpyruvate carboxylase (PEPC). Rubisco may be optimized, which would increase the photosynthesis rate.

Beneficial microorganisms in the rhizosphere not only can raise the nutrient level of crops as explained above, but they can also increase the yield quality of crops (Bijalwan et al., 2021). For example, AMF-colonized strawberries had a higher concentration of secondary metabolites that led to increased antioxidant properties. Bona et al. (2017) also noted positive outcomes of AMF treatment on the quality of tomatoes. Another way AMF can help is by improving the production of beneficial phytochemicals in food crops and making them susceptible to healthy food production chains (Chu et al., 2016). There is a need to examine the role/function of AMF spores in many plants by engaging in additional research studies. Moreover, molecular processes and genetic peculiarities of each species ought to be studied to better understand the differences between different functions that different species of AMFs possess (Nogales et al., 2020). AMF have the potential to improve the nutritional value of crops, which is influenced by the production and synthesis of carotenoids and some volatile compounds (Fig. 5.2).

5.6 Environmental benefits and potential applications

A sustainable agroecosystem is a complex dynamic system that combines ecological processes in a given agricultural landscape to maximize the use of resources, increase biodiversity, and ensure long term productivity, economic sustainability, and environmental stewardship (Pedranzani et al., 2016). Among these microorganisms, mycorrhizal fungi are especially important because of their relationship as a symbiont with a host plant. It has been shown that mycorrhizal fungus offers protection to host plants against many environmental stressors, which include metal toxicity, root pathogens, salinity, drought, acidity, and temperature variations (Aalipour et al., 2020). In addition, AMFs augment the nutrient absorption of plants, particularly nitrogen (N), phosphorus (P), and potassium (K), and stimulate general plant growth and development (Chen et al., 2020). The complex gains enjoyed by AMFs on their host plants act in favor of resilience and agroecosystem sustainability. The research indicates that employing the symbiotic relationship between AMFs and plants can produce sustainable agriculture and better plant health management practices.

5.7 Case studies and experimental evidence

5.7.1 Field trials and practical applications

AM symbiosis is common in agriculture. It is believed to ameliorate plant mineral nutrition, to enhance water stress tolerance, and to contribute to a better soil aggregate formation, which is important for soil structure and stability against erosion (Li et al., 2023). These are major determinants of effective low-input farming. Therefore the establishment and operation of the AM symbiosis seem to be significant in sustainable agriculture (Dhiman et al., 2022). Different management practices, including the type and the extent of management of arable land, influence AM symbiosis of fertilization, crop protection, rotation, fallow period, and soil cultivation (Dhiman et al., 2022). Whereas the individual effects on the formation of mycorrhizae are usually studied, it is possible that interactions of these factors are also significant. The influence of AM symbiosis can be especially relevant to agroecosystems where the use of synthetic mineral fertilizers is substituted with organic ones, at a moderate rate, as in the case of biological farming (Jabborova et al., 2021).

Replicated field experiments of AM symbiosis comparing biological and conventional agricultural practices till now have only been used in short-term observations. According to Werner et al. (1990) a higher degree of AM root colonization was observed in a plot trial with strawberries in response to a switch to biological management in only the second year postconversion. Similarly, field experiments on rotation with potatoes and wheat showed that switching from traditional, high-input farming to a low-input farming system caused a three- to fourfold increase in AM root colonization 4 years after the change.

In a standard greenhouse project, AM colonization of plant roots increased by 2–10 times. Soil samples taken in the eighth and ninth years of two low-input farming systems were compared with soil samples in the conventional high-input farming system, although the results from plants grown in corresponding field plots were supportive (Zai et al., 2021). A greater variety of crop rotation in the low-input systems, which involved cover crops between the cash crops, is thought to have played a critical role in the enhanced AM infection potential of such soils.

5.8 Limitation and concluding remarks

Nowadays, pesticides and fertilizers are actively used in farming. High-yielding crops are gaining more popularity because of their susceptibility to diseases compared to their wild counterparts. Overuse of fertilizers, pesticides, tillage, and crop rotation using nonmycorrhizal crops can all lead to decreased AM activity, diversity, and association (Porcel et al., 2015). Agricultural soils exhibit distinct and less diversified AM flora relative to natural soil. High P concentration in the soil can inhibit mycorrhizal symbiosis in plants. Using high-P fertilizer can enable plants to absorb phosphorus without losing carbohydrates (Ghorchiani et al., 2018). AMFs in exchange for phosphate require carbon from plants. Both sides need to cooperate with AM. The plant selects the most feasible and compatible strains by exchanging resources. Both pot and field trials have shown that colonization is reduced using nitrogen-based fertilizers (Kuila & Ghosh, 2022). Under low or moderate colonization the AMFs stimulate plant growth and root development. An increased level of application of nitrogen fertilizer can inhibit AM colonization in plants. On the other hand, high concentrations of potassium interfere with AM transmission, reduce root exudation, and enhance soluble carbohydrates at the cortex (Farooq et al., 2022). Agricultural productivity is not affected by using agrochemicals, which may have a negative impact on the environment, soil, and human health. Fungicides and pesticides are used to control leaf, seed, and soil diseases (Hitayezu et al., 2016). Conversely a high proportion of them adversely affect phosphatase activity efficiency, extraradical hyphal growth, sporulation, colonization, and AM spore germination. Fungicides that restrain nutrient transport, especially in dry soils, can restrain plant development that needs AMF (Khatri et al., 2019). There is a possibility that soil disturbances and tillage can cause disturbance of the hyphal network, and AM spores would be crushed, resulting in decreased root colonization. Combining root fragments of colonized hyphal networks and networks alters the capability of AMFs to eliminate the soil (Choden & Ghaley, 2021). The current favorable conditions of AMF species are affected by the migration of soil layers. The traditional tillage systems interfere with the extraradical mycelium net during the early colonization process, which contributes to water and nutrient uptake through AM activity and glomalin-related soil aggregate formation, as well as resisting pathogens (Ismail et al., 2018). Long spells of barren land and crop rotation, in particular that of nonmycorrhizal crops, harm propagule density, activity, and AMF population (Ryschawy et al., 2019). The propagule density and quantity of AMF are also influenced by the fallow period and nonhost species, which influence the subsequent crop production (Hundal et al., 2023). Waterlogged soil in paddy fields suppresses AM activity, as it is incapable of forming in saturated soils. In this regard the benefits of conventional farming cause the destruction of the symbiosis and efficacy of AM.

5.9 Conclusion

AMF has gained a lot of attention because of its capacity to support soil organisms to uptake nutrients. Recent studies indicate that plants inoculated with AMF may have a tolerance to various environmental signals and drought, including salinity, nutrient stress, cold stress, alkali stress, and high and low temperatures, contributing to crop and vegetable production. It is essential to encourage AMF usage for sustainable global agriculture. Agricultural development using the AMF can have immense potential in reducing the number of pesticides and other synthetic fertilizers that are used, thus making biohealthy farming processes. The AMF-enhanced growth and productivity of crop plants will enable the plants to meet the growing consumption demands of the world. AMF is an environment-friendly application of this technology that works aggressively and needs serious support.

References

Aalipour, H., Nikbakht, A., Etemadi, N., Rejali, F., & Soleimani, M. (2020). Biochemical response and interactions between arbuscular mycorrhizal fungi and plant growth-promoting rhizobacteria during establishment and stimulating growth of Arizona cypress (*Cupressus arizonica* G.) under drought stress. *Scientia Horticulturae, 261*, 108923.

Abdel-Fattah, G. M., & Asrar, A. W. A. (2012). Arbuscular mycorrhizal fungal application to improve growth and tolerance of wheat (*Triticum aestivum* L.) plants grown in saline soil. *Acta Physiologiae Plantarum, 34*(1), 267–277.

Adeyemi, N. O., Atayese, M. O., Sakariyawo, O. S., Azeez, J. O., Olubode, A. A., Ridwan, M., & Ibrahim, I. (2021). Influence of different arbuscular mycorrhizal fungi isolates in enhancing growth, phosphorus uptake, and grain yield of soybean in a phosphorus-deficient soil under field conditions. *Communications in Soil Science and Plant Analysis, 52*(10), 1171–1183.

Aguilar-Paredes, A., Valdés, G., Araneda, N., Valdebenito, E., Hansen, F., & Nuti, M. (2023). Microbial community in the composting process and its positive impact on the soil biota in sustainable agriculture. *Agronomy, 13*(2), 542.

Aguilar-Trigueros, C. A., Hempel, S., Powell, J. R., Cornwell, W. K., & Rillig, M. C. (2019). Bridging reproductive and microbial ecology: A case study in arbuscular mycorrhizal fungi. *The ISME Journal, 13*(4), 873–884.

Anandan, A., Pradhan, S. K., Panda, S., Dash, S. K., Panneerselvam, P., Meher, J., & Patra, B. C. (2021). Aerobic dry direct-seeded rice: A system of rice cultivation for water-shortfall irrigated and lowland areas. *Proceedings of the NRRI Research Bulletin*, (33), 32.

Baki, M. Z. I., Suzuki, K., Takahashi, K., Chowdhury, S. A., Asiloglu, R., & Harada, N. (2021). Molecular genetic characterization of arbuscular mycorrhizal fungi associated with upland rice in Bangladesh. *Rhizosphere, 18*, 100357.

Balliu, A., Sallaku, G., & Rewald, B. (2015). AMF inoculation enhances growth and improves the nutrient uptake rates of transplanted, salt-stressed tomato seedlings. *Sustainability, 7*(12), 15967–15981.

Begum, N., Ahanger, M. A., & Zhang, L. (2020). AMF inoculation and phosphorus supplementation alleviate drought-induced growth and photosynthetic decline in *Nicotiana tabacum* by up-regulating antioxidant metabolism and osmolyte accumulation. *Environmental and Experimental Botany, 176*, 104088.

Begum, N., Qin, C., Ahanger, M. A., Raza, S., Khan, M. I., Ashraf, M., & Zhang, L. (2019). Role of arbuscular mycorrhizal fungi in plant growth regulation: implications in abiotic stress tolerance. *Frontiers in Plant Science, 10*, 1068.

Bell, C. A., Magkourilou, E., Ault, J. R., Urwin, P. E., & Field, K. J. (2024). Phytophagy impacts the quality and quantity of plant carbon resources acquired by mutualistic arbuscular mycorrhizal fungi. *Nature Communications, 15*(1), 801.

Bennett, A. E., & Groten, K. (2022). The costs and benefits of plant–arbuscular mycorrhizal fungal interactions. *Annual Review of Plant Biology, 73*(1), 649–672.

Bhantana, P., Rana, M. S., Sun, X. C., Moussa, M. G., Saleem, M. H., Syaifudin, M., & Hu, C. X. (2021). Arbuscular mycorrhizal fungi and their major role in plant growth, zinc nutrition, phosphorus regulation, and phytoremediation. *Symbiosis, 84*(1), 19–37.

Bhattacharyya, S. S., Ros, G. H., Furtak, K., Iqbal, H. M., & Parra-Saldívar, R. (2022). Soil carbon sequestration – An interplay between soil microbial community and soil organic matter dynamics. *Science of the Total Environment, 815*, 152928.

Bijalwan, P., Jeddi, K., Saini, I., Sharma, M., Kaushik, P., & Hessini, K. (2021). Mitigation of saline conditions in watermelon with mycorrhiza and silicon application. *Saudi Journal of Biological Sciences, 28*(7), 3678–3684.

Bona, E., Cantamessa, S., Massa, N., Manassero, P., Marsano, F., Copetta, A., & Berta, G. (2017). Arbuscular mycorrhizal fungi and plant growth–promoting pseudomonads improve yield, quality, and nutritional value of tomato: A field study. *Mycorrhiza, 27*(1), 1–11.

Boyno, G., Yerli, C., Çakmakci, T., Sahin, U., & Demir, S. (2024). The effect of arbuscular mycorrhizal fungi on carbon dioxide (CO_2) emission from turfgrass soil under different irrigation intervals. *Journal of Water and Climate Change, 15*(2), 541–553.

Bremaghani, A. (2024). Utilization of organic waste in compost fertilizer production: Implications for sustainable agriculture and nutrient management. *Law and Economics, 18*(2), 86–98.
Carella, P., Gogleva, A., Tomaselli, M., Alfs, C., & Schornack, S. (2018). Phytophthora palmivora establishes tissue-specific intracellular infection structures in the earliest divergent land plant lineage. *Proceedings of the National Academy of Sciences, 115*(16), E3846–E3855.
Cely, M. V., De Oliveira, A. G., De Freitas, V. F., De Luca, M. B., Barazetti, A. R., Dos Santos, I. M., & Andrade, G. (2016). Inoculant of arbuscular mycorrhizal fungi (*Rhizophagus clarus*) increase yield of soybean and cotton under field conditions. *Frontiers in Microbiology, 7*, 720.
Chandrasekaran, M., Chanratana, M., Kim, K., Seshadri, S., & Sa, T. (2019). Impact of arbuscular mycorrhizal fungi on photosynthesis, water status, and gas exchange of plants under salt stress – A meta-analysis. *Frontiers in Plant Science, 10*, 457.
Chen, H., Jia, Y., Xu, H., Wang, Y., Zhou, Y., Huang, Z., & Guo, J. (2020). Ammonium nutrition inhibits plant growth and nitrogen uptake in citrus seedlings. *Scientia Horticulturae, 272*, 109526.
Chen, W., Meng, P., Feng, H., & Wang, C. (2020). Effects of arbuscular mycorrhizal fungi on growth and physiological performance of *Catalpa bungei* CA Mey. under drought stress. *Forests, 11*(10), 1117.
Chen, X. W., Wu, F. Y., Li, H., Chan, W. F., Wu, C., Wu, S. C., & Wong, M. H. (2013). Phosphate transporters expression in rice (*Oryza sativa* L.) associated with arbuscular mycorrhizal fungi (AMF) colonization under different levels of arsenate stress. *Environmental and Experimental Botany, 87*, 92–99.
Choden, T., & Ghaley, B. B. (2021). A portfolio of effective water and soil conservation practices for arable production systems in Europe and North Africa. *Sustainability, 13*(5), 2726.
Chu, X. T., Fu, J. J., Sun, Y. F., Xu, Y. M., Miao, Y. J., Xu, Y. F., & Hu, T. M. (2016). Effect of arbuscular mycorrhizal fungi inoculation on cold stress-induced oxidative damage in leaves of *Elymus nutans* Griseb. *South African Journal of Botany, 104*, 21–29.
Cotton, T. A. (2018). Arbuscular mycorrhizal fungal communities and global change: An uncertain future. *FEMS Microbiology Ecology, 94*(11), fiy179.
Das, P. P., Singh, K. R., Nagpure, G., Mansoori, A., Singh, R. P., Ghazi, I. A., & Singh, J. (2022). Plant-soil-microbes: A tripartite interaction for nutrient acquisition and better plant growth for sustainable agricultural practices. *Environmental Research, 214*, 113821.
Dhiman, M., Sharma, L., Kaushik, P., Singh, A., & Sharma, M. M. (2022). Mycorrhiza: An eco-friendly bio-tool for better survival of plants in nature. *Sustainability, 14*(16), 10220.
Diagne, N., Ngom, M., Djighaly, P. I., Fall, D., Hocher, V., & Svistoonoff, S. (2020). Roles of arbuscular mycorrhizal fungi on plant growth and performance: Importance in biotic and abiotic stress regulation. *Diversity, 12*(10), 370.
Dutta, B., Goswami, A., Medhi, B. K., & Deka, K. (2024). Soil health and plant nutrition for sustainable agriculture. In *Role of agronomy in cultivating a sustainable food future* (pp. 19–28). Cornous Publications LLP.
Eddiwal, E., Saidi, A., Lenin, I., Husin, E. F., & Rasyidin, A. (2014). Potential selection of arbuscular mycorrhizal fungi (AMF) indigenous ultisols through the production of glomalin. *Journal of Tropical Soils, 19*(3), 181–189.
Etesami, H., Jeong, B. R., & Glick, B. R. (2021). Contribution of arbuscular mycorrhizal fungi, phosphate–solubilizing bacteria, and silicon to P uptake by plants. *Frontiers in Plant Science, 12*, 699618.
Etienne, P., Diquelou, S., Prudent, M., Salon, C., Maillard, A., & Ourry, A. (2018). Macro and micronutrient storage in plants and their remobilization when facing scarcity: The case of drought. *Agriculture, 8*(1), 14.
Farhaoui, A., Taoussi, M., Laasli, S. E., Legrifi, I., El Mazouni, N., Meddich, A., & Lahlali, R. (2025). Arbuscular mycorrhizal fungi and their role in plant disease control: A state-of-the-art. *The Microbe, 8*, 100438.
Farooq, M. S., Uzair, M., Raza, A., Habib, M., Xu, Y., Yousuf, M., & Ramzan Khan, M. (2022). Uncovering the research gaps to alleviate the negative impacts of climate change on food security: A review. *Frontiers in Plant Science, 13*, 927535.
Fayaz, F., & Zahedi, M. (2021). Beneficial effects of arbuscular mycorrhizal fungi on wheat (*Triticum aestivum* L.) nutritional status and tolerance indices under soil salinity stress. *Journal of Plant Nutrition, 45*(2), 185–201.
Fellbaum, C. R., Gachomo, E. W., Beesetty, Y., Choudhari, S., Strahan, G. D., Pfeffer, P. E., & Bücking, H. (2012). Carbon availability triggers fungal nitrogen uptake and transport in arbuscular mycorrhizal symbiosis. *Proceedings of the National Academy of Sciences, 109*(7), 2666–2671.
Finucane, M. L., Acosta, J., Wicker, A., & Whipkey, K. (2020). Short-term solutions to a long-term challenge: Rethinking disaster recovery planning to reduce vulnerabilities and inequities. *International Journal of Environmental Research and Public Health, 17*(2), 482.
Gao, L. L., Delp, G., & Smith, S. E. (2001). Colonization patterns in a mycorrhiza-defective mutant tomato vary with different arbuscular-mycorrhizal fungi. *New Phytologist, 151*(2), 477–491.
Garcia de Leon, D., Vahter, T., Zobel, M., Koppel, M., Edesi, L., Davison, J., & Öpik, M. (2020). Different wheat cultivars exhibit variable responses to inoculation with arbuscular mycorrhizal fungi from organic and conventional farms. *PLoS One, 15*(5), e0233878.
Garg, N., & Singh, S. (2018). Arbuscular mycorrhiza, *Rhizophagus irregularis*, and silicon modulate growth, proline biosynthesis, and yield in *Cajanus cajan* L. Millsp. (pigeonpea) genotypes under cadmium and zinc stress. *Journal of Plant Growth Regulation, 37*(1), 46–63.
Gerdemann, J. W., & Trappe, J. M. (1974). The Endogonaceae in the Pacific Northwest. *Mycologia Memoirs, 5*, 76.
Ghonjalipour Goshki, M., Abdollahi, F., & Sadeghi Lari, A. (2021). Effect of mycorrhiza fertilizer on physiological traits and economic yield of lettuce (*Lactuca sativa* L.) under water stress conditions. *Journal of Vegetables Sciences, 5*(9), 157–173.
Ghorbanpour, M., Omidvari, M., Abbaszadeh-Dahaji, P., Omidvar, R., & Kariman, K. (2018). Mechanisms underlying the protective effects of beneficial fungi against plant diseases. *Biological Control, 117*, 147–157.
Ghorchiani, M., Etesami, H., & Alikhani, H. A. (2018). Improvement of growth and yield of maize under water stress by co-inoculating an arbuscular mycorrhizal fungus and a plant growth-promoting rhizobacterium together with phosphate fertilizers. *Agriculture, Ecosystems & Environment, 258*, 59–70.

Giovannetti, M., Ayio, L., Sbrana, C., & Silvia Citernesi, A. (1993). Factors affecting appressorium development in the vesicular–arbuscular mycorrhizal fungus *Glomus mosseae* (Nicol. & Gerd.) Gerd. & Trappe. *New Phytologist, 123*(1), 115–122.

Giovannetti, M., Sbrana, C., & Logi, C. (1994). Early processes involved in host recognition by arbuscular mycorrhizal fungi. *New Phytologist, 127*(4), 703–709.

Gosling, P., Mead, A., Proctor, M., Hammond, J. P., & Bending, G. D. (2013). Contrasting arbuscular mycorrhizal communities colonizing different host plants show a similar response to a soil phosphorus concentration gradient. *New Phytologist, 198*(2), 546–556.

Goto, B. T., Alves da Silva, G., Maia, L. C., & Oehl, F. (2010). Dentiscutata colliculosa, a new species in the Glomeromycetes from Northeastern Brazil with colliculate spore ornamentation. *Nova Hedwigia, 90*(3–4), 383–393.

Helander, M., Saloniemi, I., Omacini, M., Druille, M., Salminen, J. P., & Saikkonen, K. (2018). Glyphosate decreases mycorrhizal colonization and affects plant-soil feedback. *Science of the Total Environment, 642*, 285–291.

Hitayezu, P., Zegeye, E. W., & Ortmann, G. F. (2016). Farm-level crop diversification in the Midlands region of KwaZulu-Natal, South Africa: Patterns, microeconomic drivers, and policy implications. *Agroecology and Sustainable Food Systems, 40*(6), 553–582.

Hundal, G. S., Laux, C. M., Buckmaster, D., Sutton, M. J., & Langemeier, M. (2023). Exploring barriers to the adoption of Internet of Things-based precision agriculture practices. *Agriculture, 13*(1), 163.

Hurni, H., Giger, M., Liniger, H., Studer, R. M., Messerli, P., Portner, B., & Breu, T. (2015). Soils, agriculture and food security: The interplay between ecosystem functioning and human well-being. *Current Opinion in Environmental Sustainability, 15*, 25–34.

Husband, R., Herre, E. A., Turner, S. L., Gallery, R., & Young, J. P. W. (2002). Molecular diversity of arbuscular mycorrhizal fungi and patterns of host association over time and space in a tropical forest. *Molecular Ecology, 11*(12), 2669–2678.

Hussain, T., Li, J., Feng, X., Asrar, H., Gul, B., & Liu, X. (2021). Salinity induced alterations in photosynthetic and oxidative regulation are ameliorated as a function of salt secretion. *Journal of Plant Research, 134*(4), 779–796.

Imran, M., Tahir, M., Murtaza, S., Nadeem, M., Ashraf, M. I., Waris, A. A., & Ayub, M. A. (2025). Influence of climate change on soil nutrient cycling: Consequences for ecosystem sustainability and agricultural production. *Journal of Health and Climate Change, 4*(1). https://jhcc.supp.journalrmc.com/index.php/public/article/view/17.

Ismail, H. M., Al-Sadek, A. F., Ashmawy, A. A., Khalil, M. A. I., & Elsayed, M. (2018). Computational model to improve dairy animal feeding under a mixed farming system (crops/livestock) as a case study. *Arab Universities Journal of Agricultural Sciences, 26*(Special issue (2A)), 1053–1064.

Jabborova, D., Annapurna, K., Paul, S., Kumar, S., Saad, H. A., Desouky, S., & Elkelish, A. (2021). Beneficial features of biochar and arbuscular mycorrhiza for improving spinach plant growth, root morphological traits, physiological properties, and soil enzymatic activities. *Journal of Fungi, 7*(7), 571.

Jansa, J., Mozafar, A., & Frossard, E. (2005). Phosphorus acquisition strategies within the arbuscular mycorrhizal fungal community of a single field site. *Plant and Soil, 276*(1), 163–176.

Johan, P. D., Ahmed, O. H., Omar, L., & Hasbullah, N. A. (2021). Phosphorus transformation in soils following co-application of charcoal and wood ash. *Agronomy, 11*(10), 2010.

Juniper, S., & Abbott, L. K. (2006). Soil salinity delays germination and limits the growth of hyphae from propagules of arbuscular mycorrhizal fungi. *Mycorrhiza, 16*(5), 371–379.

Khalloufi, M., Martínez-Andújar, C., Lachaâl, M., Karray-Bouraoui, N., Pérez-Alfocea, F., & Albacete, A. (2017). The interaction between foliar GA3 application and arbuscular mycorrhizal fungi inoculation improves growth in salinized tomato (*Solanum lycopersicum* L.) plants by modifying the hormonal balance. *Journal of Plant Physiology, 214*, 134–144.

Khan, D., Kabiraj, A., Biswas, R., Roy, R. K., & Bandopadhyay, R. (2023). Plant-microbe community dynamics in rhizosphere: Reviewing the grassroots ecology towards sustainable agriculture. *The Indian Journal of Agricultural Sciences, 93*, 247–257.

Khan, Y., Shah, S., & Hui, T. (2022). The roles of arbuscular mycorrhizal fungi in influencing plant nutrients, photosynthesis, and metabolites of cereal crops: A review. *Agronomy, 12*(9), 2191. https://doi.org/10.3390/agronomy12092191.

Khan, Y., Yang, X., Zhang, X., Yaseen, T., Shi, L., & Zhang, T. (2021). Arbuscular mycorrhizal fungi promote plant growth of *Leymus chinensis* (Trin.) Tzvelev by increasing the metabolomics activity under nitrogen addition. *Grassland Science, 67*(2), 128–138.

Khatri-Chhetri, A., Pant, A., Aggarwal, P. K., Vasireddy, V. V., & Yadav, A. (2019). Stakeholders' prioritization of climate-smart agriculture interventions: Evaluation of a framework. *Agricultural Systems, 174*, 23–31.

Klironomos, J. N., & Hart, M. M. (2002). Colonization of roots by arbuscular mycorrhizal fungi using different sources of inoculum. *Mycorrhiza, 12*(4), 181–184.

Kopittke, P. M., Menzies, N. W., Wang, P., McKenna, B. A., & Lombi, E. (2019). Soil and the intensification of agriculture for global food security. *Environment International, 132*, 105078.

Kuila, D., & Ghosh, S. (2022). Aspects, problems, and utilization of arbuscular mycorrhizal (AM) application as bio-fertilizer in sustainable agriculture. *Current Research in Microbial Sciences, 3*, 100107.

Kumar, A., & Verma, J. P. (2018). The role of microbes in improving crop productivity and soil health. In *Ecological wisdom-inspired restoration engineering* (pp. 249–265). Springer.

Lal, R. (2015). Restoring soil quality to mitigate soil degradation. *Sustainability, 7*(5), 5875–5895.

Li, G., Tang, X., Hou, Q., Li, T., Xie, H., Lu, Z., & Wen, X. (2023). Response of soil organic carbon fractions to legume incorporation into cropping system and the factors affecting it: A global meta-analysis. *Agriculture, Ecosystems & Environment, 342*, 108231.

Li, H., Smith, S. E., Holloway, R. E., Zhu, Y., & Smith, F. A. (2006). Arbuscular mycorrhizal fungi contribute to phosphorus uptake by wheat grown in a phosphorus-fixing soil even in the absence of positive growth responses. *New Phytologist, 172*(3), 536–543.

Li, J., Meng, B., Yang, X., Cui, N., Zhao, T., Chai, H., & Sun, W. (2022). Suppression of AMF accelerates N_2O emission by altering soil bacterial community and gene abundance under varied precipitation conditions in a semiarid grassland. *Frontiers in Microbiology, 13*, 961969.

Liang, S., Wang, W., Zeng, X., Wu, R., & Chen, W. (2022). Enhanced suppression of saprotrophs by ectomycorrhizal fungi under high levels of nitrogen fertilization. *Frontiers in Ecology and Evolution, 10*, 974449.

Liu, J., Chen, J., Xie, K., Tian, Y., Yan, A., Liu, J., & Xu, G. (2020). A mycorrhiza-specific H+-ATPase is essential for arbuscule development and symbiotic phosphate and nitrogen uptake. *Plant, Cell & Environment, 43*(4), 1069–1083.

Liu, M., Sun, J., Li, Y., & Xiao, Y. (2017). Nitrogen fertilizer enhances the growth and nutrient uptake of *Medicago sativa* inoculated with *Glomus tortuosum* grown in Cd-contaminated acidic soil. *Chemosphere, 167*, 204–211.

Liu, W., Zhang, Y., Jiang, S., Deng, Y., Christie, P., Murray, P. J., & Zhang, J. (2016). Arbuscular mycorrhizal fungi in soil and roots respond differently to phosphorus inputs in an intensively managed calcareous agricultural soil. *Scientific Reports, 6*(1), 24902.

Marín, C., Aguilera, P., Cornejo, P., Godoy, R., Oehl, F., Palfner, G., & Boy, J. (2016). Arbuscular mycorrhizal assemblages along contrasting Andean forests of Southern Chile. *Journal of Soil Science and Plant Nutrition, 16*(4), 916–929.

Masrahi, A. S., Alasmari, A., Shahin, M. G., Qumsani, A. T., Oraby, H. F., & Awad-Allah, M. M. (2023). Role of arbuscular mycorrhizal fungi and phosphate-solubilizing bacteria in improving yield, yield components, and nutrient uptake of barley under saline soil. *Agriculture, 13*(3), 537.

Mbodj, D., Effa-Effa, B., Kane, A., Manneh, B., Gantet, P., Laplaze, L., & Grondin, A. (2018). Arbuscular mycorrhizal symbiosis in rice: establishment, environmental control, and impact on plant growth and resistance to abiotic stresses. *Rhizosphere, 8*, 12–26.

Mensah, J. A., Koch, A. M., Antunes, P. M., Kiers, E. T., Hart, M., & Bücking, H. (2015). High functional diversity within species of arbuscular mycorrhizal fungi is associated with differences in phosphate and nitrogen uptake and fungal phosphate metabolism. *Mycorrhiza, 25*(7), 533–546.

Mishra, V., Ellouze, W., & Howard, R. J. (2018). Utility of arbuscular mycorrhizal fungi for improved production and disease mitigation in organic and hydroponic greenhouse crops. *Journal of Horticultural Sciences, 5*(03), 1000237.

Mitra, D., Nayeri, F. D., Sansinenea, E., Ortiz, A., Bhatta, B. B., Adeyemi, N. O., & Panneerselvam, P. (2023). Unraveling arbuscular mycorrhizal fungi interaction in rice for plant growth development and enhancing phosphorus use efficiency through recent development of regulatory genes. *Journal of Plant Nutrition, 46*(13), 3184–3220.

Moradi, P. (2016). Key plant products and common mechanisms utilized by plants in water deficit stress responses. *Botanical Sciences, 94*(4), 657–671.

Motamedi, M., Zahedi, M., Karimmojeni, H., Motamedi, H., & Mastinu, A. (2022). Effect of rhizosphere bacteria on antioxidant enzymes and some biochemical characteristics of *Medicago sativa* L. subjected to herbicide stress. *Acta Physiologiae Plantarum, 44*(8), 84.

Munyanziza, E., Kehri, H. K., & Bagyaraj, D. J. (1997). Agricultural intensification, soil biodiversity and agro-ecosystem function in the tropics: The role of mycorrhiza in crops and trees. *Applied soil ecology, 6*(1), 77–85.

Narwal, E., Annapurna, K., Choudhary, J., Dhakar, R., & Singh, Y. (2021). Bioprospecting aerobic rice (*Oryza sativa*) and mycorrhizal interaction for nutrient uptake and plant growth. *Indian Journal of Agriculture Science, 91*, 1236–1277.

Nazir, A, Sarfraz, W, Ditta, A, Khalid, N, Farid, M, Shafiq, M, Firdaus-e-Bareen, ZF Rizvi, & Naeem, N (2023). Synergistic impact of two autochthonous saprobic fungi (*A. niger* and *T. pseudokoningii*) on the growth, ionic contents, and metals uptake in *Brassica juncea* L. and *Vigna radiata* L. under tannery solid waste contaminated soil. *International Journal of Phytoremediation, 25*(11), 1488–1500.

Ngosong, C., Tatah, B. N., Olougou, M. N. E., Suh, C., Nkongho, R. N., Ngone, M. A., & Ruppel, S. (2022). Inoculating plant growth-promoting bacteria and arbuscular mycorrhiza fungi modulates rhizosphere acid phosphatase and nodulation activities and enhances the productivity of soybean (*Glycine max*). *Frontiers in Plant Science, 13*, 934339.

Ngwene, B., George, E., Claussen, W., & Neumann, E. (2010). Phosphorus uptake by cowpea plants from sparingly available or soluble sources as affected by nitrogen form and arbuscular-mycorrhiza-fungal inoculation. *Journal of Plant Nutrition and Soil Science, 173*(3), 353–359.

Nogales, A., Ribeiro, H., Nogales-Bueno, J., Hansen, L. D., Gonçalves, E. F., Coito, J. L., & Cardoso, H. (2020). Response of mycorrhizal'Touriga Nacional 'variety grapevines to high temperatures measured by calorespirometry and near-infrared spectroscopy. *Plants, 9*(11), 1499.

Oehl, F., de Souza, F. A., & Sieverding, E. (2008). Revision of *Scutellospora* and description of five new genera and three new families in the arbuscular mycorrhiza-forming *Glomeromycetes*. *Mycotaxon, 106*(1), 311–360.

Ortiz, A., & Sansinenea, E. (2022). The role of beneficial microorganisms in soil quality and plant health. *Sustainability, 14*(9), 5358.

Pattnaik, S., Mohapatra, B., & Gupta, A. (2021). Plant growth-promoting microbe-mediated uptake of essential nutrients (Fe, P, K) for crop stress management: Microbe–soil–plant continuum. *Frontiers in Agronomy, 3*, 689972.

Pedranzani, H., Rodríguez-Rivera, M., Gutiérrez, M., Porcel, R., Hause, B., & Ruiz-Lozano, J. M. (2016). Arbuscular mycorrhizal symbiosis regulates the physiology and performance of *Digitaria eriantha* plants subjected to abiotic stresses by modulating antioxidant and jasmonate levels. *Mycorrhiza, 26*(2), 141–152.

Plassard, C., Becquer, A., & Garcia, K. (2019). Phosphorus transport in mycorrhiza: How far are we? *Trends in Plant Science, 24*(9), 794–801.

Porcel, R., Redondo-Gómez, S., Mateos-Naranjo, E., Aroca, R., Garcia, R., & Ruiz-Lozano, J. M. (2015). Arbuscular mycorrhizal symbiosis ameliorates the optimum quantum yield of photosystem II and reduces non-photochemical quenching in rice plants subjected to salt stress. *Journal of Plant Physiology, 185*, 75–83.

Quan, L., Duan, K., Wei, Z., Li, W., Chen, Y., Duan, W., & Xia, Y. (2023). Beneficial effects of arbuscular mycorrhizae on Cu detoxification in *Mimosa pudica* L. grown in Cu-polluted soils. *Environmental Science and Pollution Research, 30*(10), 25755–25763.

Requena, N., & Breuninger, M. (2004). The old arbuscular mycorrhizal symbiosis in the light of the molecular era. In *Progress in botany: Genetics, physiology, systematics, ecology* (pp. 323–356). Springer.

Requena, N., Serrano, E., Ocón, A., & Breuninger, M. (2007). Plant signals and fungal perception during arbuscular mycorrhiza establishment. *Phytochemistry, 68*(1), 33–40.

Riaz, M., Kamran, M., Fang, Y., Wang, Q., Cao, H., Yang, G., & Wang, X. (2021). Arbuscular mycorrhizal fungi-induced mitigation of heavy metal phytotoxicity in metal-contaminated soils: A critical review. *Journal of Hazardous Materials, 402*, 123919.

Römheld, V., & Kirkby, E. A. (2010). Research on potassium in agriculture: Needs and prospects. *Plant and soil, 335*(1), 155–180.

Rouydel, Z., Barin, M., Rasouli-Sadaghiani, M. H., Khezri, M., Vetukuri, R. R., & Kushwaha, S. (2021). Harnessing the potential of symbiotic endophytic fungi and plant growth-promoting rhizobacteria to enhance soil quality in saline soils. *Processes, 9*(10), 1810.

Ryschawy, J., Moraine, M., Péquignot, M., & Martin, G. (2019). Trade-offs among individual and collective performances related to crop–livestock integration among farms: A case study in southwestern France. *Organic Agriculture, 9*(4), 399–416.

Saeed, HS, Munir, M, Samra, S, Farooq, A, Mahmood, H, Ahmad, I, Akram, MI, Iqbal, R, Elsadek, MF, Almutairi, SM, Ditta, A, & Daraz, U (2025). Screening of biosurfactants producing fungi from infected *Citrus sinensis* (L.). *Polish Journal of Environmental Studies, 34*(5), 5261–5270.

Sarker, M. M. R., Ali, M. A., Islam, M. S., Huda, M. S., Pun, I., Podder, R., & Hossain, A. (2021). Use of different eco-friendly management approaches for controlling root-knot nematode (*Meloidogyne incognita* L.) in tomato (*Solanum lycopersicum* L.). *Thai Journal of Agricultural Science, 54*(2), 89–103.

Sarkodee-Addo, E., Yasuda, M., Lee, C. G., Kanasugi, M., Fujii, Y., Ansong Omari, R., & Okazaki, S. (2020). Arbuscular mycorrhizal fungi associated with rice (*Oryza sativa* L.) in Ghana: effect of regional locations and soil factors on diversity and community assembly. *Agronomy, 10*(4), 559.

Sbrana, C., & Giovannetti, M. (2005). Chemotropism in the arbuscular mycorrhizal fungus *Glomus mosseae*. *Mycorrhiza, 15*(7), 539–545.

Schirawski, J., & Perlin, M. H. (2018). Plant–microbe interaction 2017—The good, the bad, and the diverse. *International Journal of Molecular Sciences, 19*(5), 1374.

Schüßler, A., Schwarzott, D., & Walker, C (2001). A new fungal phylum, the *Glomeromycota*: Phylogeny and evolution. *Mycological research, 105*(12), 1413–1421.

Sedaghati, E., Yazdanpanah, M., & Nadi, M. (2021). A review of taxonomic studies of arbuscular mycorrhizal fungi in Iran. *Mycologia Iranica, 8*(2), 15–30.

Selvaraj, T., & Chellappan, P. (2006). Arbuscular mycorrhizae: A diverse personality. *Journal of Central European Agriculture, 7*(2), 349–358.

Shah, F., & Wu, W. (2019). Soil and crop management strategies to ensure higher crop productivity within sustainable environments. *Sustainability, 11*(5), 1485.

Shukla, S., Didwania, N., & Choudhary, R. (2025). Arbuscular mycorrhizal fungi (AMF): A pathway to sustainable soil health, carbon sequestration, and greenhouse gas mitigation. *Journal of the Saudi Society of Agricultural Sciences, 24*(4), 22.

Smith, S. E., Jakobsen, I., Grønlund, M., & Smith, F. A. (2011). Roles of arbuscular mycorrhizas in plant phosphorus nutrition: Interactions between pathways of phosphorus uptake in arbuscular mycorrhizal roots have important implications for understanding and manipulating plant phosphorus acquisition. *Plant physiology, 156*(3), 1050–1057.

Srivastav, A. L., Dhyani, R., Ranjan, M., Madhav, S., & Sillanpää, M. (2021). Climate-resilient strategies for sustainable management of water resources and agriculture. *Environmental Science and Pollution Research, 28*(31), 41576–41595.

Straker, C. J., Hilditch, A. J., & Rey, M. E. C. (2010). Arbuscular mycorrhizal fungi associated with cassava (*Manihot esculenta* Crantz) in South Africa. *South African Journal of Botany, 76*(1), 102–111.

Tan, Z., Liu, S., Sohl, T. L., Wu, Y., & Young, C. J. (2015). Ecosystem carbon stocks and sequestration potential of federal lands across the conterminous United States. *Proceedings of the National Academy of Sciences, 112*(41), 12723–12728.

Tian, C., Kasiborski, B., Koul, R., Lammers, P. J., Bücking, H., & Shachar-Hill, Y. (2010). Regulation of the nitrogen transfer pathway in the arbuscular mycorrhizal symbiosis: Gene characterization and the coordination of expression with nitrogen flux. *Plant Physiology, 153*(3), 1175–1187.

Timmis, K., & Ramos, J. L. (2021). The soil crisis: The need to treat as a global health problem and the pivotal role of microbes in prophylaxis and therapy. *Microbial Biotechnology, 14*(3), 769–797.

Toor, M. D., Basit, A., Okorie, B., Nath, D., Din, M. M. U., Kumar Verma, P., & Mohamed, H. I (2024). Earthworms as catalysts for climate-resilient agriculture: Enhancing food security and water management in the face of climate change. *Water, Air, & Soil Pollution, 235*(12), 779.

Uddin, M. E., & Kebreab, E. (2020). Impact of food and climate change on pastoral industries. *Frontiers in Sustainable Food Systems, 4*, 543403.

Verzeaux, J., Hirel, B., Dubois, F., Lea, P. J., & Tétu, T. (2017). Agricultural practices to improve nitrogen use efficiency through the use of arbuscular mycorrhizae: Basic and agronomic aspects. *Plant Science, 264*, 48–56.

Wahid, F., Fahad, S., Danish, S., Adnan, M., Yue, Z., Saud, S., & Datta, R. (2020). Sustainable management with mycorrhizae and phosphate-solubilizing bacteria for enhanced phosphorus uptake in calcareous soils. *Agriculture, 10*(8), 334.

Wang, B. Q. I. U., & Qiu, Y. L. (2006). Phylogenetic distribution and evolution of mycorrhizas in land plants. *Mycorrhiza, 16*(5), 299–363.

Wang, F., Zhang, L., Zhou, J., Rengel, Z., George, T. S., & Feng, G. (2022). Exploring the secrets of the hyphosphere of arbuscular mycorrhizal fungi: Processes and ecological functions. *Plant and Soil, 481*(1), 1–22.

Wang, M., Zheng, Q., Shen, Q., & Guo, S. (2013). The critical role of potassium in plant stress response. *International Journal of Molecular Sciences, 14*(4), 7370–7390.

Wang, Y., Liu, H., Shen, Z., Miao, Y., Wang, J., Jiang, X., & Li, R. (2022). Richness and antagonistic effects co-affect plant growth promotion by synthetic microbial consortia. *Applied Soil Ecology, 170*, 104300.

Werner, M. R., Kluson, R. A., & Gliessman, S. R. (1990). Colonization of strawberry roots by VA mycorrhizal fungi in agroecosystems under conventional and transitional organic management. *Biological Agriculture & Horticulture, 7*(2), 139–151.

Wu, Y., Chen, C., Li, J., & Wang, G. (2021). Effects of arbuscular mycorrhizal fungi on maize nitrogen uptake strategy under different soil water conditions. *Plant and Soil, 464*(1), 441–452.

Xiao, L., Lai, S., Chen, M., Long, X., Fu, X., & Yang, H. (2022). Effects of grass cultivation on soil arbuscular mycorrhizal fungi community in a tangerine orchard. *Rhizosphere*: 24.

Xing, Y., Wang, X., & Mustafa, A. (2025). Exploring the link between soil health and crop productivity. *Ecotoxicology and Environmental Safety, 289*, 117703.

Yizhu, L., Imtiaz, M., Ditta, A., Rizwan, M. S., Ashraf, M., Mehmood, S., Aziz, O., Mubeen, F., Ali, M., Elahi, N. N., Ijaz, R., Lelel, S, Shuang, C., & Tu, S. (2020). Response of growth, antioxidant enzymes, and root exudates production towards as stress in *Pteris vittata* and *Astragalus sinicus* colonized by arbuscular mycorrhizal fungi. *Environmental Science and Pollution Research, 27*, 2340–2352.

Yuan, Y., Liu, Q., Zheng, H., Li, M., Liu, Y., Wang, X., & Xing, B. (2023). Biochar as a sustainable tool for improving the health of salt-affected soils. *Soil & Environmental Health, 1*(3), 100033.

Zai, X. M., Fan, J. J., Hao, Z. P., Liu, X. M., & Zhang, W. X. (2021). Effect of co-inoculation with arbuscular mycorrhizal fungi and phosphate-solubilizing fungi on nutrient uptake and photosynthesis of beach palm under salt stress environment. *Scientific Reports, 11*(1), 5761.

Zhang, B., Li, X., Bao, J., Tian, Z., Zhang, F., & Zhang, M. (2025). A combined strategy using *funneliformis mosseae* and phosphorus addition for enhancing oat drought tolerance. *Agronomy, 15*(9), 2033.

Zhang, X., Wang, L., Ma, F., Yang, J., & Su, M. (2017). Effects of arbuscular mycorrhizal fungi inoculation on carbon and nitrogen distribution and grain yield and nutritional quality in rice (*Oryza sativa* L.). *Journal of the Science of Food and Agriculture, 97*(9), 2919–2925.

Zhou, J., Zang, H., Loeppmann, S., Gube, M., Kuzyakov, Y., & Pausch, J. (2020). Arbuscular mycorrhiza enhances rhizodeposition and reduces the rhizosphere priming effect on the decomposition of soil organic matter. *Soil Biology and Biochemistry, 140*, 107641.

Zhu, C., Ling, N., Guo, J., Wang, M., Guo, S., & Shen, Q. (2016). Impacts of fertilization regimes on arbuscular mycorrhizal fungal (AMF) community composition were correlated with organic matter composition in maize rhizosphere soil. *Frontiers in Microbiology, 7*, 1840.

Zhu, X., Song, F., Liu, S., & Liu, F. (2016). Arbuscular mycorrhiza improve growth, nitrogen uptake, and nitrogen use efficiency in wheat grown under elevated CO_2. *Mycorrhiza, 26*(2), 133–140.

Chapter 6

Arbuscular mycorrhizal fungi in soil microbiome and ecosystem functioning

Minisha Udhayakumar, Parimala Gnana Soundari Arockiam Jeyasundar and Ayyappa Das Madhubala Parameswaran

Department of Biotechnology, Rathnavel Subramaniam (RVS) College of Arts and Science, Sulur, Coimbatore, Tamil Nadu, India

6.1 Introduction

Soils are a nonrenewable resource that serves as an essential life-supporting component of the biosphere. Soil health is an essential aspect of "One Health," which strives to sustainably balance and optimize the health of people, animals, and ecosystems (Bansal & Mukerji, 1994). About 80% of terrestrial plant species have close symbiotic relationships with arbuscular mycorrhizal (AM) fungi (AMFs), a crucial functional group of soil bacteria that belongs to the phylum Glomeromycota (Smith & Read, 2010; Wijayawardene et al., 2022). Mycorrhizal fungi colonize host plants in these interactions, facilitating their access to water and nutrients that are located much beyond their roots. The fungus provides its host plants with photosynthetic lipids and carbohydrates in exchange. In addition to acting as a link between plants, soils, and people, this carbon nutrient nexus represents the foundation of the soil–plant–microbiome interaction (Saia & Jansa, 2022). In addition to their well-known impacts on plant nutrition, mycorrhizas have a significant role in controlling vital soil processes and soil health, both of which are critical for sustainable agriculture (Barrios, 2007; Jeffries et al., 2003; Rillig, 2004).

AMFs promote plant development and resistance to environmental stressors. AMFs create symbiotic interactions with plant roots, and the period of colonization is crucial for nutrient intake and plant growth (Barceló et al., 2020). Understanding the elements influencing AMF-driven plant growth promotion, such as experimental circumstances, plant type, and root–shoot ratio, is vital for increasing plant productivity and reducing reliance on chemical fertilizers. Mycorrhizal associations can provide various advantages to host plants, including boosting nutritional status and assisting the crop in overcoming metal stress, particularly in well-established colonization (Lokhande, 2025). Among the several types of AMF inoculum, intact extra radicular mycelium (ERM) resulted in earlier and faster root colonization, leading to resistance to metal stress (Dhalaria et al., 2020). The possible roles of AMFs have been investigated, and AM fungal species and isolates have been sequenced (Adachi et al., 2025; Morin et al., 2019; Tisserant et al., 2013; Venice et al., 2021). AMFs contain fewer genes encoding enzymes that break down carbohydrates than saprophytic and ectomycorrhizal fungi (Miyauchi et al., 2020; Tisserant et al., 2013). Phytases, which hydrolyze phytate, a significant component of soil organic phosphorus, are not encoded by genes found in AMF (Liu et al., 2024). These imply that soil organic nutrients, particularly organic P, are not well mobilized by AMFs. The turnover of soil organic nutrients may be significantly aided by the strong capacity of soil bacteria colonizing in the hyphosphere to produce a variety of enzymes that break down soil organic nitrogen and phosphorus (Falkowski & Matysiak, 2010; Zhang et al., 2024). Thus the AM fungal genome may be functionally supplemented by the hyphosphere microbiome.

AMF are well-known for their beneficial effects on plant development and health. AMFs are obligate biotrophs that form mutualistic symbioses with the roots of most terrestrial plants, including the most important food crops, including grains, legumes, potatoes, fruit trees, vegetables, and medicinal herbs (Bhantana et al., 2021). They obtain carbon from host plants in exchange for facilitating the uptake and transfer of mineral nutrients from the soil, such as phosphorus, nitrogen,

Advancing Sustainable Agriculture Through Arbuscular Mycorrhizal Fungi. DOI: https://doi.org/10.1016/B978-0-443-45130-0.00033-3

sulfur, potassium, calcium, copper, and zinc, via an extensive network of hyphae spreading from colonized roots into the surrounding environment (Yizhu et al., 2020). AMF not only improves mineral nutrition, but it also increases plant resistance to biotic and abiotic stressors and provides a variety of ecosystem services in both natural and agricultural situations, ranging from biogeochemical cycle completion to biological soil fertility management (Thind et al., 2022). In addition to these multifunctional roles, AMF alter plant secondary metabolism by stimulating the biosynthesis of phytochemical compounds such as polyphenols and carotenoids, resulting in the production of safe and high-quality foods that can benefit human health (Nazir et al., 2023; Sbrana et al., 2014).

The potential of soil bacteria to influence AMF spore germination and hyphal development was investigated as early as 1959 (Mosse, 1959). Their functional importance in promoting mycorrhizal activity was validated by other studies Azcón-Aguilar and Barea (1992) and Frey-Klett et al. (2007) suggested calling these bacteria "mycorrhiza helpers." Although *Klebsiella pneumoniae* and *Trichoderma* spp. boosted the growth of germlings in *Glomus deserticola* and *Glomus mosseae* (Calvet et al., 1995; Will & Sylvia, 1990), numerous actinobacteria isolated from the soil, specifically species from the genera *Streptomyces* and *Corynebacterium*, as well as species of *Pseudomonas*, increased the germination of *Glomus versiforme, G. mosseae*, and Gigaspora *margarita* spores (Carpenter-Boggs et al., 1995; Mayo et al., 1986). Spore germination, germling development, and AMF root colonization were all improved by bacteria isolated from the mycorrhizosphere, either from mycorrhizal roots or from AMF spores and hyphae (Mayo et al., 1986; Sbrana et al., 2014; Xavier & Germida, 2003). The capacity of certain bacteria to break down insoluble biopolymers, such as chitin and chitosan, which are the two primary components of AMF spore walls, to promote germination, is one of the potential mechanistic reasons for the phenomena. The frequent isolation of chitinolytic bacteria from *Glomus macrocarpum, G. mosseae, and Rhizophagus intraradices* spores lends credence to this theory (Battini et al., 2017; Filippi et al., 1998).

The formation of ERM, the delicate absorbing network of hyphae that surrounds the roots, may be enhanced by mycorrhizosphere and sporosphere bacteria, which may also function as "mycorrhiza helpers." According to in vitro research, *Paenibacillus rhizosphaerae, Rhizobium etli*, and other strains of *Azospirillum* and *Pseudomonas* enhanced the ERM length of *Glomus intraradices* and *Rhizophagus irregularis* (Bidondo et al., 2011; Ordoñez et al., 2016). As a result, in vivo studies verified significant ERM development, which was enhanced by *Streptomyces* species in *Glomus caledonium, G. intraradices,* and *R. irregularis*, as well as *Pseudomonas fluorescens, Burkholderia cepacia,* and *Sinorhizobium meliloti* (Battini et al., 2017; Ravnskov et al., 1999). In addition, three actinobacteria species, *Paenibacillus favisporus, S. meliloti,* and *Fictibacillus barbaricus,* were among the bacteria that were isolated from *G. mosseae* and *R. irregularis* spores and were found to produce indole acetic acid (Battini et al., 2017; Bidondo et al., 2011). It is still necessary to look into the mechanical explanation of ERM development promotion.

6.2 Overview of soil microbiome

The rhizosphere is the zone around roots where root exudates play an important role in regulating communication among soil microbiomes. Soil microbiomes are mostly composed of bacteria and fungi, and they play important roles in maintaining soil health and fertility (Fig. 6.1). Bacterial and fungal components of the root microbiome have been shown to have commensal, pathogenic, or beneficial connections with both their hosts and one another. Soil microbiomes play important roles in biogeochemical cycles, including nitrogen, carbon, sulfur, and phosphorus. In the nitrogen cycle, *Rhizobium* sp. are known to fix nitrogen in legumes in return for fixed carbon and sugar. *Azotobacter* is a free-living bacterium capable of nitrogen fixation (Faghihinia et al., 2023).

AMF promotes sulfur absorption in soil (Wu et al., 2009). In alkaline soil, P precipitates are easily accessible (Tian et al., 2021), and phosphate-solubilizing microorganisms (PSMs) convert insoluble organic and inorganic phosphorus compounds into easily digested forms (Ditta et al., 2015, 2018a, 2018b; Siddique et al., 2023). PSM contains several well-known strains from the genera *Rhizobium, Pseudomonas*, and *Bacillus*, as well as *Aspergillus* and *Penicillium*, AMF, and actinomycetes (Tian et al., 2021). Other metabolic activities that the soil microbiome participates in include antibiotic synthesis for pathogen defense, biomass breakdown, biodegradation, soil structure preservation, nutrient absorption, and stress tolerance (Saeed et al., 2025). A few possible bacteria and fungi have been investigated to aid in the growth of agricultural plants such as tomato, wheat, and rice.

Fungal communities in soil play a significant role in P solubilization and N_2 absorption by hosts (Sun et al., 2022). Mycorrhizal fungi play a vital role in nutrient intake because the symbiotic connection improves water and nutrient uptake efficiency in olive trees. It also improves resistance to biotic and abiotic stress (Porras-Soriano et al., 2009). The presence of AMF colonizing plant roots favors various crops under *Verticillium assault*. AMF can improve micronutrient absorption and tolerance to a wide range of abiotic stressors (Wahab et al., 2023). The majority of AMF come from the subphylum

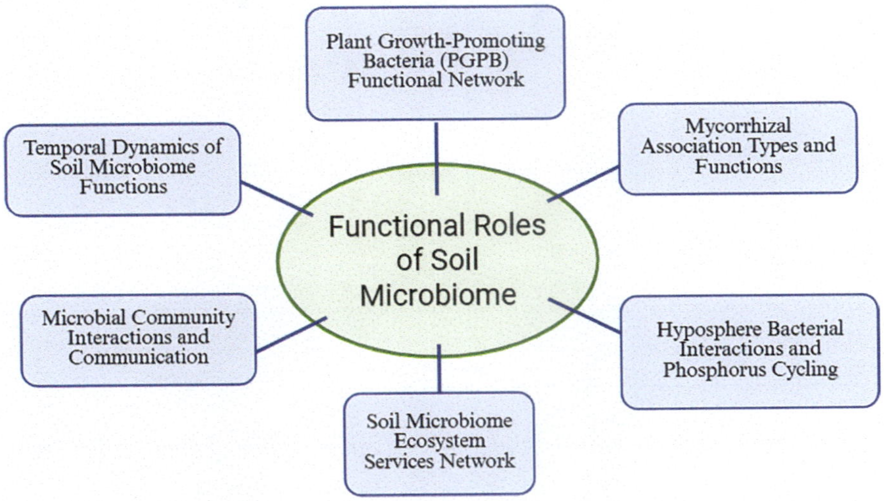

FIGURE 6.1 Functional networks and ecological roles of the soil microbiome. The figure emphasizes the networked and dynamic nature of soil microbial functions in regulating soil fertility and ecosystem processes.

Glomeromycotina and the phylum Mucoromycota (Tedersoo et al., 2024). They take up photosynthetic products and lipids as obligatory biotrophs to complete their life cycle. AMF-mediated development protects plants from fungal diseases while also allowing them to absorb water and nutrients from the surrounding soil (Wahab et al., 2023). In reality, AMF are crucial endosymbionts that help plants grow and operate properly. Vesicular arbuscular mycorrhizal (VAM) fungi, on the other hand, form symbiotic relationships in the roots of host plants and may be used to improve the rate of phosphate absorption from the soil and raise phosphorus levels, both of which are necessary for plant growth (Dar & Reshi, 2017; Gianinazzi-Pearson & Gianinazzi, 1983; Lalitha et al., 2017). The VAM fungus generates and releases organic molecules (siderophores) that improve P desorption in the P pool of labile soil. VAM can employ organic acids to dissolve insoluble and low-soluble P sources, which are part of the soil's mineral crystalline structure (Rakshit & Bhadoria, 2010).

6.3 Functional role of soil microbiome

Plant growth–promoting bacteria, particularly those inhabiting the rhizosphere [plant growth–promoting rhizobacteria (PGPR)], represent a diverse group, including genera such as *Azotobacter, Bacillus, Rhizobium*, and *Pseudomonas*. These bacteria enhance plant development through multiple mechanisms, including production of phytohormones like auxins that stimulate root elongation and lateral root formation, gibberellins that regulate plant growth, and cytokinins that influence cell division and differentiation (Carpenter-Boggs et al., 1995). Beyond hormonal regulation, several bacterial species offer remarkable environmental benefits (Fig. 6.2). Organisms, such as *Bacillus cereus, Bacillus subtilis*, and *Pseudomonas aeruginosa*, demonstrate bioremediation capabilities through enzyme production (such as amylase and cellulase) and heavy metal sequestration, addressing industrial contamination challenges. Other species, including members of genera such as *Acinetobacter, Enterobacter,* and *Serratia*, produce 1-AminoCyclopropane-1-Carboxylate-deaminase, an enzyme that regulates ethylene levels in plants, thereby alleviating stress-induced growth inhibition during challenging environmental conditions (Mayo et al., 1986). Endophytic bacteria represent another critical group that colonizes internal plant tissues without causing disease. These include Proteobacteria (*Pantoea, Pseudomonas, Klebsiella*), *Firmicutes* (various *Bacillus* species), and Actinobacteria (*Curtobacterium, Microbacterium, Frankia*). These endophytes deliver multiple benefits, including phytohormone production, nitrogen fixation, phosphorus solubilization, and even conferring tolerance to trace metals, effectively functioning as internal plant growth promoters (Battini et al., 2016; Ditta & Khalid, 2016; Khan et al., 2025).

Fungal symbionts further expand the microbial partnership network through various mycorrhizal associations. Ectomycorrhizal fungi (including *Amanita, Boletus*, and *Ramaria*) primarily associate with woody plants, forming sheaths around root surfaces and contributing to soil aggregation, carbon cycling, and nutrient acquisition. AMF, represented by genera such as *Glomus, Funneliformis*, and *Gigaspora*, form the most widespread plant–fungal symbiosis, creating extensive hyphal networks that dramatically increase the plants' effective root surface area for nutrient absorption, particularly

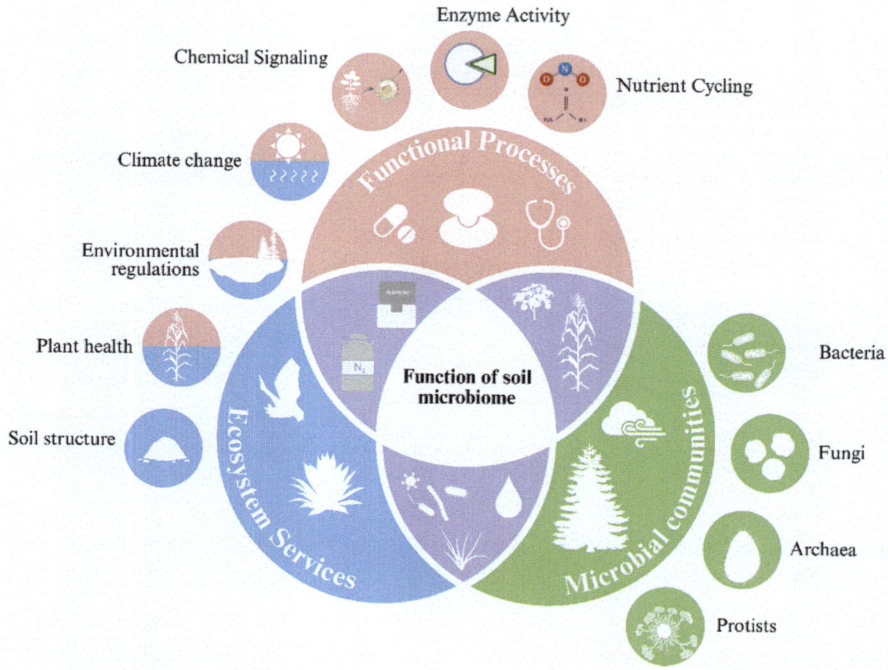

FIGURE 6.2 Integrated functions of the soil microbiome linking processes, communities, and ecosystem services. The figure represents the integrated functions of soil microbiomes through the interaction of functional processes, microbial communities, and ecosystem services.

phosphorus (Battini et al., 2017). More specialized mycorrhizal relationships include ericoid mycorrhizae (associated with plants in the Ericaceae family) and orchid mycorrhizae, which are essential for seedling establishment and survival in these plant groups. Beyond mycorrhizae, plant growth–promoting fungi such as *Aspergillus* and *Penicillium* assist in phosphate mobilization, whereas endophytic fungi such as *Colletotrichum tropicale* enhance plant immunity. Antagonistic fungi, including *Trichoderma* and *Gliocladium species*, produce antibiotics, phytotoxins, and lytic enzymes that suppress pathogenic microorganisms, functioning as biological control agents. Collectively, these diverse microbial communities form a biological foundation for plant health, nutrient cycling, and ecosystem function, representing nature's toolkit for sustainable agriculture and ecosystem management (Mayo et al., 1986).

In recent years the roles of hyphosphere bacteria have drawn a lot of attention because of their importance in promoting organic P acquisition for the mycorrhizal pathway (Zhang et al., 2024). First, prior research demonstrated that AM fungus extraradical hyphae could absorb P from an organic P patch and enhance plant P nutrition when they developed into the patch (Peng et al., 2024; Tarafdar & Marschner, 1994). Additional research revealed that the presence of phosphate-solubilizing bacteria (PSB) and AM fungus greatly boosted plant P content. This finding showed that the hyphosphere interaction between PSB and AMFs improved soil organic P utilization for plants (Zhang et al., 2024). Second, histochemical staining revealed phosphatase activity, which is generated by bacteria and compensates for the functional flaws in AM fungal P mineralization (Joner & Johansen, 2000). According to recent research, AMFs can attract bacteria that carry phosphatase genes and alter the bacterial population to boost the activity of alkaline phosphatase on their hyphal surface. In addition, recent research has shown that AM fungal hyphae function as a "highway" for bacterial migration to soil organic P patches, improving soil organic P utilization (Jiang et al., 2021; Keyes et al., 2022). Furthermore the hyphosphere microbiome that various AM fungal species recruit may serve varied purposes in the mobilization of soil organic P for AMFs (Zhou et al., 2020). Two isolates of *R. irregularis* used distinct tactics to increase maize P absorption at the medium plant-accessible P level, according to our most recent study: To investigate soil inorganic P, for instance, *R. irregularis* MUCL 41833 produces a lot of hyphae, but *R. irregularis* MUCL 43194 recruits a higher number of *Betaproteobacteria* to mobilize organic P (Zhang et al., 2016). The solubilization of P, a crucial mineral nutrient that is insufficiently available to plants in the majority of agricultural soils, because of its immobilization and precipitation with other soil minerals, calcium in alkaline soils, and iron and aluminum in acidic soils, represents another crucial function performed by AMF-associated microbiota. *Streptomyces* and *Leifsonia* species, as well as *Streptomyces* spp., *Bacillus pumilus, Lisinobacillus fusiformis,* and *S. meliloti* (Battini et al., 2017), were identified as the P-mobilizing bacteria that were recovered from AMF spores of *G. mosseae* and *R. intraradices*. According to Battini et al. (2017), these bacteria provide a sustainable method of mobilizing the soil P pool and facilitating mycorrhizal plant P absorption.

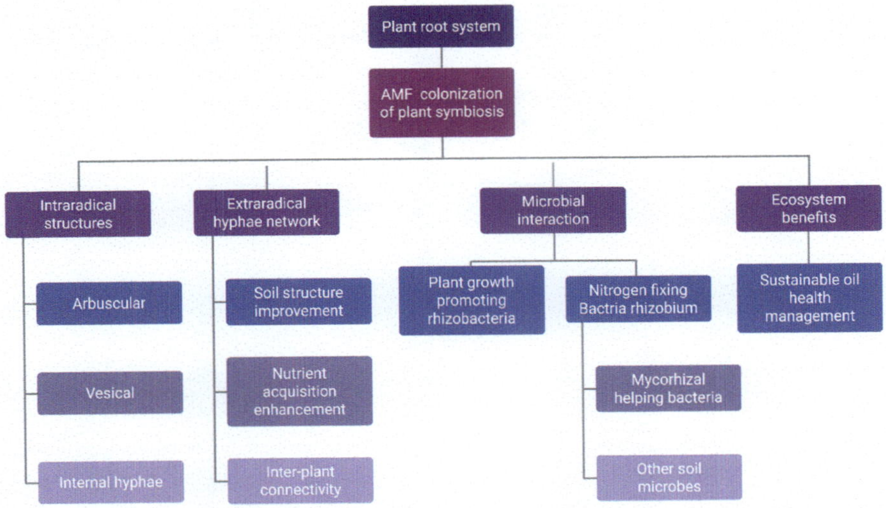

FIGURE 6.3 Functional framework of arbuscular mycorrhizal fungi in plant–microbe–soil interactions. Diagram depicting the colonization of plant root systems by arbuscular mycorrhizal fungi and their associated functional pathways. *AMF*, Arbuscular mycorrhizal fungi.

6.4 Role of arbuscular mycorrhizal fungi in ecosystem functioning

6.4.1 Soil health management

The use of microorganisms such as AMF has become more significant in biological remediation procedures in recent years. The majority of agricultural, horticultural, and hardwood crop species are among the 80%–85% of plants that form symbioses (Fig. 6.3) with the roots of AMFs, which are common rhizosphere microflora and members of the Glomeromycota (Filippi et al., 1998; Prasad et al., 2017). By establishing mutually beneficial relationships in which plants receive nutrients such as P and N in exchange for photosynthates necessary for the growth of mycorrhizal fungi, AMF play a significant role in ecosystem functioning, such as nutrient cycling, carbon sequestration, and plant growth promotion (Kaiser et al., 2015). Because of their extramatrical hyphae, mycorrhizal roots increase the root surface area, which allows the plant to absorb water and nutrients, particularly P from the soil, more effectively than nonmycorrhizal plants (Nadeem et al., 2014). Marschner and Dell (1994) claim that the AM fungus can provide around 80% of the phosphorus that the mycorrhizal plants absorb overall. AM improves the absorption of additional nutrients, including N, Mg, Cu, K, and Zn, in addition to increasing P acquisition, especially in soils where these elements are present in less accessible forms (Clark & Zeto, 2000; Smith & Read, 2010). As mycorrhizal hyphae spread from the roots of one infected plant to the roots of other nearby plants, it has been observed that assimilations and nutrients are transported from one plant to another through AM hyphal connections (Heap & Newman, 1980). Furthermore, because of their mutually beneficial and synergistic interactions, they are thought to be extremely advantageous to legume plant species. The symbiotic interaction between legume, mycorrhiza, and *Rhizobium* improves rhizobia's N_2-fixing activity and plant development. Mycorrhizal fungi contribute to soil aggregation by producing exudates such as citric, malic, and oxalic acids, as well as the glycoprotein glomalin (Gonzalez-Chavez et al., 2004).

AMF interacts with a variety of species in the rhizosphere. Azcón-Aguilar and Barea (1992) and Rillig (2004) found that AM interactions can be favorable, neutral, or negative depending on the rhizosphere component. Mycorrhizal roots' rhizosphere, also known as mycorrhizosphere, varies from nonmycorrhizal roots in both quantity and quality (Bansal & Mukerji, 1994). Bacteria, divided into saprophytes and symbionts, interact with AM fungus in various ways, which can be beneficial, neutral, or deleterious (Johansson et al., 2004). Mycorrhiza helper bacteria, often present in temperate and tropical habitats (Frey-Klett et al., 2007), promote AM fungal root colonization, mycelia development, and spore germination (Gryndler, 2000). The interaction between AMFs and nodulating nitrogen fixers has attracted significant study because of the high P requirement involved in nitrogen fixation. Under low-fertility circumstances, the two symbionts work together to increase N and P content in plants. This is more effective than inoculating them separately. AM fungal inoculation, even at low water potential, increases nodulation and N_2 fixation by the bacterial symbiont (Goicoechea et al., 1998), neutralizing the effects of salinity. Evelin et al. (2009) found that early interaction of seedlings with AM fungus can reduce host stress. AM fungal inoculation has been shown to reduce premature nodule senescence in soybean during drought circumstances (Porcel et al., 2003). AMFs and PGPR have been shown to interact synergistically (Muthukumar & Udaiyan, 2006). However, the benefits of engagement may differ.

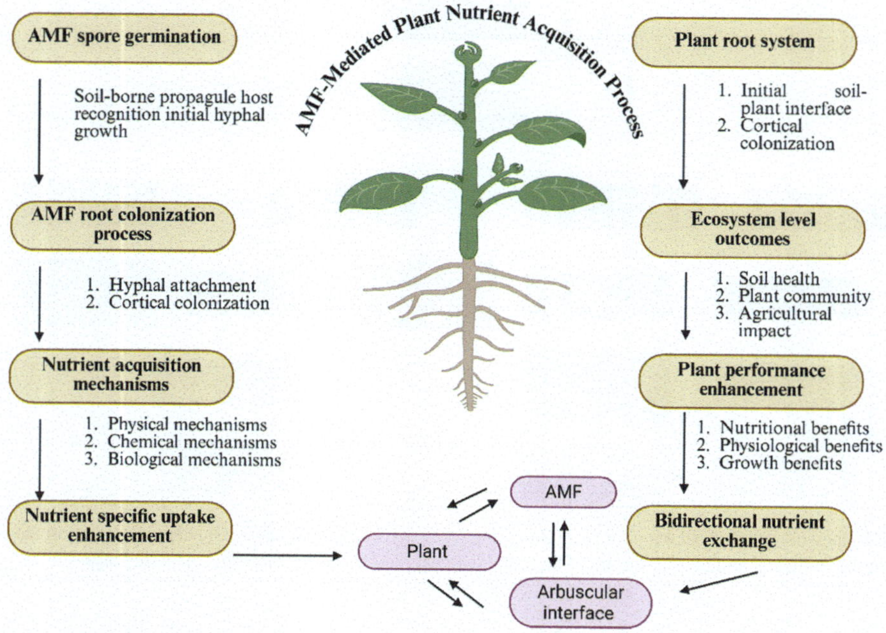

FIGURE 6.4 AMF-mediated nutrient acquisition in plants. This figure demonstrates the AMF colonization and nutrient acquisition in plant systems. *AMF*, Arbuscular mycorrhizal fungi.

Chandanie et al. (2006) found that combining *Trichoderma* with an AM fungus enhanced plant development. When *Penicillium* was coinoculated with the same AM fungus, there was no noticeable influence on plant growth.

6.4.2 Plant nutrient management

AMF promote nutrient absorption by increasing plant surface area. The absorbing hyphae's small width enhances nitrogen uptake from soil solutions. Plant roots extract minerals from the soil solution, resulting in nutrient depletion zones surrounding them (Fig. 6.4). Poorly mobile ions, like phosphate, form a small depletion zone at the root. AMF's hyphae can expand outside the depletion zone and absorb phosphate from the soil (Li et al., 1991). AMFs promote nutrient absorption, including N, K, and micronutrients, which are often restricted in the soil for different reasons. AM fungus efficiently absorbs nutrients from soil because of their tiny hyphae width and extended lifetime compared to root hairs. The diffusion gradient of nutrients is inversely proportional to the radius of the absorbing unit. Therefore soil solutions should be less depleted at the surface of narrow absorbing units like hyphae. Narrow hyphae can develop into microscopic soil pores inaccessible to roots and root hairs (O'Keefe & Sylvia, 1991). Crops having well-developed root systems, such as wheat, barley, and oats (*Avena sativa*), are less susceptible to AM colonization (Ryan & Graham, 2002).

Mycorrhizal associations play a crucial role in enhancing plant nutrient uptake through various physiological and morphological modifications in root systems (Fig. 6.4). Proper nutrition contributes to host endurance against different abiotic and biotic stresses, with plant species benefiting significantly from mycorrhizal association because of greater efficiency in nutrient and water uptake (Daft & Nicolson, 1969; Gerdemann, 1968; Harley, 1959). Mycorrhizal roots demonstrate increased nutrient absorption capacity through several mechanisms, including increased effective absorption surface area because of proliferated hyphae; exploration of greater soil volume; longevity of absorbing roots; enhanced utilization of low-available nutrients; and better retention of soluble nutrients (Clark & Zeto, 2000). For poorly mobile ions such as phosphate, zinc, and copper, a sharp and narrow depletion zone develops close to the root, which hyphae can readily bridge by growing into soil with an adequate supply of these nutrients (Bagheri et al., 2012; Fattah, 2013; Tavasolee & Aliasgharzad, 2009).

The phosphorus acquisition mechanism involves physiochemical release of inorganic and organic phosphorus by organic acids, including citric, ascorbic, mallic, and oxalic acid, through the action of low-molecular-weight organic anions such as oxalate (Jalali et al., 2022). These compounds can replace phosphorus sorbed at metal-hydroxide surfaces through ligand-exchange reactions, dissolve metal-oxide surfaces that sorbed phosphorus, and dissolve complex metals in solution to prevent precipitation of metal phosphates. In addition, mycorrhizal fungi release inorganic phosphorus through mineralization of organic matter via phosphatase-mediated hydrolysis of organic phosphate ester bonds, with mycorrhizal associations intensifying phosphatase activity in the rhizosphere. Ericoid and ectomycorrhizal fungi demonstrate special roles in nitrogen mineralization, possessing a wide range of hydrolytic and oxidative enzymes capable of depolymerizing organic nitrogen from plant litter that typically has a high C:N ratio and is rich in lignin and tannins. Furthermore, mycorrhizal associations enhance *Rhizobium* nodulation and atmospheric N-fixation capacity of leguminous plants, creating synergistic effects for better nutrient uptake while increasing production of plant growth hormones such as cytokinins and gibberellins responsible for increased plant growth.

6.4.3 Plant defense and disease management

Mycorrhizal fungi serve as effective biocontrol agents against various plant pathogens through multiple defense mechanisms (Weng et al., 2022). As a biological control method to control a variety of phytopathogenic fungi, AMF has been extensively employed, with studies showing that mycorrhizal cotton plants were more resistant to infection with the pathogen *Thielaviopsis basicola* than those with sterile roots (Weng et al., 2022). The relationship between mycorrhizal infection levels and pathogen development demonstrates an inverse correlation, with later research revealing that the level of mycorrhizal infection and *T. basicola* chlamydospore output were inversely related (Soukayna, 2012). Combined inoculation strategies prove particularly effective, with AMF and *Trichoderma* combinations showing superior results compared to single inoculations. Studies demonstrate that combined inoculation increased aboveground biomass of *Solanum lycoperdicum* by 11.6%–69.7% compared with single inoculation of AMF mixed inoculum or *Trichoderma harzianum*, with the combination showing obvious synergistic effects on plant disease control. Tomato incidence rates decreased from 70.0% with *Fusarium oxysporum* inoculation to 20.0% after inoculation with *G. mosseae* and *Acaulospora laevis* and were further reduced to only 10.0% when one AMF was combined with *Trichoderma viride*, with complete disease inhibition achieved through combined inoculation of two AMF and *Trichoderma* species (Soukayna, 2012).

The combination of AMF with other beneficial microorganisms also demonstrates significant biocontrol potential. AMF and *Pseudomonas* combinations show obvious synergistic effects on improving plant disease resistance, with studies showing that while single inoculation of *Glaphyropoma spinosum, Gibbula albida,* and *P. fluorescens* reduced *Phaseolus vulgaris* incidence rates by 50.5%–52.8%, combined inoculation achieved 68.9%–69.2% reduction (Almuslimawi et al., 2024). Similarly, AMF and *Bacillus* combinations demonstrate excellent control effects on plant root diseases, with mixed inoculation of *G. mosseae* and *B. subtilis* reducing tomato *fusarium* root rot severity by 85.0%–93.4% while improving plant nutrients, including nitrogen, phosphorus, potassium, calcium, magnesium, iron, and zinc (Cai et al., 2021). The mechanisms underlying these biocontrol effects include advancement of crop development, decrease in colonization sites accessible to aggressors, alteration in root morphology, increase in damage compensation, changes in root exudate and rhizosphere microbiome composition, and stimulation of plant immune structures (Giri & Singh, 2017).

6.4.4 Plant growth and development

Mycorrhizal associations significantly enhance plant growth and development through improved ecological fitness and stress tolerance mechanisms (Shi et al., 2023). Mycorrhizal plants demonstrate superior competitiveness and environmental stress tolerance compared to nonmycorrhizal plants, with this characteristic being utilized for rehabilitating degraded lands, developing pasture lands, and successful afforestation in challenging environments, including coal mines, industrially polluted areas, and other wasteland-category soils. It is well established that tree seedlings without mycorrhizae fail to grow in nutrient-poor soils but establish successfully when inoculated with suitable mycorrhizal fungi (Harley, 1959). The enhanced growth response varies based on soil nutrient status, with inoculation of ectomycorrhizae in apple seedlings showing enhanced uptake of nutrients, including N, P, K, Fe, Mg, Zn, and Cu, where the symbiotic association response was higher in nutrient-poor orchard soil compared to nutrient-rich basin soil. Mycorrhizal roots of seedlings, including fir (*Abies pindrow*), spruce (*Picea smithiana*), and chilgoza pine (*Pinus gerardiana*), showed greater stress tolerance than nonmycorrhizal seedlings. Different mycorrhizal fungi species exhibit varying degrees of drought tolerance under moisture stress conditions, with many possessing specific traits regarding tolerance to soil temperature extremes, pH, moisture, low fertility, salinity, and toxicants, providing host plants with ecological competitive advantages that facilitate increased survival, growth, nutrition, and yield under stress conditions (Al-Karaki, 2000; Trappe, 2010).

The growth enhancement mechanisms extend beyond nutrient acquisition to include soil structure improvement, with extramatrical mycelium of AMF being important in binding sand grains and stabilizing dunes and sandy soils. The development of soil aggregates approximately 2 mm in diameter depends on binding by mycorrhizal hyphae, contributing to improved soil physical properties that support plant growth (Lehmann et al., 2017). Agricultural plants have been reported to utilize nitrates as nitrogen sources compared to ectomycorrhizal plants, which preferentially use ammonium, with this ability being particularly important in soils with low pH and low nitrification rates, where the large surface area provided by hyphae and hyphal strands enables sufficient uptake of rather immobile ammonium (Bledsoe & Rygiewicz, 1986). The comprehensive growth promotion effects result from the integration of improved nutrient acquisition, enhanced stress tolerance, soil structure modification, and hormonal regulation that collectively optimize plant development under various environmental conditions.

6.5 Opportunities and future prospects

The future of AMF in soil microbiome and ecosystem functioning presents unprecedented opportunities for sustainable agriculture and environmental restoration. As our understanding of AMF evolves from viewing them as "ubiquitous, generalist symbionts" to recognizing them as specialized organisms with comprehensive impacts on community dynamics, new avenues for their application continue to emerge. The shift toward precision agriculture offers significant opportunities for developing targeted AMF inoculants that are specifically selected based on their functional traits and genetic characteristics, using advanced molecular tools to match specific AMF taxa with particular crops and environmental conditions (Trappe, 2010). The development of consortium-based inoculants represents a promising frontier, as coinoculation strategies utilizing AMF from different families have demonstrated superior performance compared to mono-inoculations, particularly under multiple stress conditions. This functional complementarity between species provides greater resilience and adaptability to varying environmental challenges, offering enhanced genetic variability that enables AMF communities to better adapt to prevailing conditions. The integration of spore-associated bacteria with AMF presents another innovative opportunity, combining phosphate solubilization and nitrogen fixation capabilities to create multifunctional biofertilizer systems that maximize agricultural benefits (Al-Karaki, 2000).

Climate change adaptation strategies offer substantial prospects for AMF applications, particularly in developing drought and salinity-resistant crop systems (Devi et al., 2023). Given that drought has caused massive yield losses over the past 35 years, understanding AMF-mediated mechanisms that improve crop productivity under extreme environmental conditions becomes crucial for future food security. The unique ability of different AMF communities to adapt to specific stress conditions—whether water stress or high salinity—provides opportunities for developing specialized inoculants tailored to climate-specific challenges. The integration of next-generation sequencing and other molecular technologies opens new possibilities for quality control, inoculant tracking, and understanding complex soil microbiome interactions. These tools enable researchers to screen for beneficial microbial candidates with specific functional traits and gain insights into how AMF interact with native flora, facilitating the development of more effective and predictable biofertilizer products. The potential for incorporating AMF responsiveness as a selection criterion in plant breeding programs offers another avenue for enhancing crop-inoculant compatibility and maximizing agricultural outcomes (Battini et al., 2016).

Ecosystem restoration presents significant opportunities for AMF applications beyond agriculture, particularly in remediating degraded habitats where well-established AMF communities and their complex interactions have been disrupted. The multifaceted nature of AMF in soil remediation, including their ability to remediate heavy metal contamination through specialized physiological and molecular mechanisms, positions them as valuable tools for environmental restoration projects. Their role in maintaining soil structure, enhancing plant community composition and productivity, and influencing greenhouse gas emissions highlights their potential contribution to ecosystem-level sustainability (O'Keefe & Sylvia, 1991). The democratization of AMF technology through farmer-produced inoculants using native soils represents an important opportunity for making this technology more accessible and affordable, particularly for small-scale farmers in tropical regions where phosphorus-deficient soils would benefit most from AMF applications. This approach, combined with the development of standardized quality control protocols and regulatory frameworks, could facilitate wider adoption while maintaining effectiveness and safety standards. The establishment of technical support systems and increased government funding for research and development could accelerate the transition from laboratory success to field-scale implementation, ultimately realizing the full potential of AMF as a cornerstone of sustainable agricultural and ecosystem management practices (Saia & Jansa, 2022).

6.6 Conclusion

AMFs represent a cornerstone of sustainable ecosystem functioning, serving as critical mediators in the complex interactions between plants, soil microbiomes, and environmental processes. The comprehensive evidence presented demonstrates that AMF function far beyond simple plant–fungal partnerships, orchestrating intricate networks of beneficial microorganisms that collectively enhance plant health, nutrient cycling, and ecosystem resilience. The symbiotic relationships formed between AMF and approximately 80% of terrestrial plant species (Wijayawardene et al., 2022) establish these fungi as fundamental components of the soil–plant–microbiome continuum. The multifaceted roles of AMF in nutrient management, particularly their collaboration with hyphosphere bacteria for organic phosphorus mobilization and their enhancement of nitrogen fixation in leguminous plants, underscore their importance in developing sustainable agricultural systems. The documented synergistic effects of AMF combinations with beneficial bacteria such as *Trichoderma, Pseudomonas*, and *Bacillus* species provide compelling evidence for their potential in integrated disease management strategies, offering biological alternatives to chemical interventions. These findings are particularly significant in the context of reducing agricultural dependency on synthetic fertilizers and pesticides while maintaining or improving crop productivity. The growth enhancement mechanisms mediated by AMF, extending from improved nutrient acquisition to enhanced stress tolerance and soil structure modification, position these fungi as essential tools for addressing contemporary agricultural challenges, including climate change adaptation, soil degradation, and food security. The ability of mycorrhizal associations to facilitate plant establishment in challenging environments, from nutrient-poor soils to contaminated lands, demonstrates their potential for ecological restoration and sustainable land management practices. Moving forward the integration of AMF-based biotechnologies with modern agricultural practices represents a promising pathway toward achieving the dual objectives of agricultural productivity and environmental sustainability. The complex interactions within the mycorrhizosphere, involving diverse bacterial communities that complement AMF functionality, highlight the importance of adopting holistic approaches to soil microbiome management. As our understanding of these intricate microbial networks continues to evolve, AMF and their associated microorganisms will undoubtedly play increasingly important roles in developing resilient, sustainable agricultural and ecological systems that can meet the challenges of the 21st century while preserving soil health and ecosystem integrity for future generations.

References

Adachi, K., Kobayashi, T., Suzuki, T., & Kosugi, T. (2025). Sweet potatoes' relationship with arbuscular mycorrhizal fungi and endophytic diazotrophic bacteria: A co-inoculation experiment. *Soil Microorganisms, 79*, 95–107.

Al-Karaki, G. N. (2000). Growth of mycorrhizal tomato and mineral acquisition under salt stress. *Mycorrhiza, 10*, 51–54.

Almuslimawi, A. A. A., Kuchár, B., Navas, S. E. A., Turóczi, G., & Posta, K. (2024). The effect of combined application of biocontrol microorganisms and arbuscular mycorrhizal fungi on plant growth and yield of tomato (*Solanum lycopersicum* L.). *Agriculture, 14*, 768.

Azcón-Aguilar, C., & Barea, J. M. (1992). Interactions between mycorrhizal fungi and other rhizosphere microorganisms. In *Mycorrhizal functioning: An integrative plant-fungal process* (pp. 163–198). Chapman and Hall.

Bagheri, H., El-Soda, M., van Oorschot, I., Hanhart, C., Bonnema, G., Jansen-van den Bosch, T., Mank, R., Keurentjes, J. J. B., Meng, L., & Wu, J. (2012). Genetic analysis of morphological traits in a new, versatile, rapid-cycling *Brassica rapa* recombinant inbred line population. *Frontiers in Plant Science, 3*, 183.

Bansal, M., & Mukerji, K. G. (1994). Positive correlation between VAM-induced changes in root exudation and mycorrhizosphere mycoflora. *Mycorrhiza, 5*, 39–44.

Barceló, M., van Bodegom, P. M., Tedersoo, L., den Haan, N., Veen, G. F., Ostonen, I., Trimbos, K., & Soudzilovskaia, N. A. (2020). The abundance of arbuscular mycorrhiza in soils is linked to the total length of roots colonized at ecosystem level. *PLoS One, 15*, e0237256.

Barrios, E. (2007). Soil biota, ecosystem services and land productivity. *Ecological Economics, 64*, 269–285.

Battini, F., Grønlund, M., Agnolucci, M., Giovannetti, M., & Jakobsen, I. (2017). Facilitation of phosphorus uptake in maize plants by mycorrhizosphere bacteria. *Scientific Reports, 7*, 4686.

Battini, F., Turrini, A., Sgherri, C., Malorgio, F., & Quartacci, M. F. (2016). Dual inoculation with AMF and associated bacteria improves nutraceutical value of sweet basil grown under commercial conditions. *Agrochimica, 60*(2), 81–99.

Bhantana, P., Rana, M. S., Sun, X. C., Moussa, M. G., Saleem, M. H., Syaifudin, M., Shah, A., Poudel, A., Pun, A. B., Bhat, M. A., & Mandal, D. L. (2021). Arbuscular mycorrhizal fungi and its major role in plant growth, zinc nutrition, phosphorous regulation and phytoremediation. *Symbiosis, 84*(1), 19–37.

Bidondo, L. F., Silvani, V., Colombo, R., Pérgola, M., Bompadre, J., & Godeas, A. (2011). Pre-symbiotic and symbiotic interactions between *Glomus intraradices* and two *Paenibacillus* species isolated from AM propagules. In vitro and in vivo assays with soybean (AG043RG) as plant host. *Soil Biology and Biochemistry, 43*, 1866–1872.

Bledsoe, C. S., & Rygiewicz, P. T. (1986). Ectomycorrhizas affect ionic balance during ammonium uptake by Douglas-fir roots. *New Phytologist, 102*, 271–283.

Cai, X., Zhao, H., Liang, C., Li, M., & Liu, R. (2021). Effects and mechanisms of symbiotic microbial combination agents to control tomato *fusarium* crown and root rot disease. *Frontiers in Microbiology, 12*, 629793.

Calvet, C., Pinochet, J., Camprubí, A., & Fernández, C. (1995). Increased tolerance to the root-lesion nematode *Pratylenchus vulnus* in mycorrhizal micropropagated BA-29 quince rootstock. *Mycorrhiza, 5*, 253–258. doi:10.1007/BF00204958.

Carpenter-Boggs, L., Loynachan, T. E., & Stahl, P. D. (1995). Spore germination of *Gigaspora margarita* stimulated by volatiles of soil-isolated actinomycetes. *Soil Biology and Biochemistry, 27*, 1445–1451.

Chandanie, W. A., Kubota, M., & Hyakumachi, M. (2006). Interactions between plant growth promoting fungi and arbuscular mycorrhizal fungus *Glomus mosseae* and induction of systemic resistance to anthracnose disease in cucumber. *Plant and Soil, 286*, 209–217.

Clark, R. B., & Zeto, S. K. (2000). Mineral acquisition by arbuscular mycorrhizal plants. *Journal of Plant Nutrition, 23*, 867–902.

Daft, M. J., & Nicolson, T. H. (1969). Effect of endogone mycorrhiza on plant growth: II. Influence of soluble phosphate on endophyte and host in maize. *New Phytologist, 68*, 945–952.

Dar, M. H., & Reshi, Z. A. (2017). Vesicular arbuscular mycorrhizal (VAM) fungi-as a major biocontrol agent in modern sustainable agriculture system. *Russian Agricultural Sciences, 43*, 138–143.

Devi, P., Dey, S. R., Saini, L., Kumar, P., Panigrahi, S., & Dwivedi, P. (2023). Toward sustainable agriculture: Strategies involving phytoprotectants against reactive oxygen species. In *Reactive oxygen species: Prospects in plant metabolism* (pp. 229–247). Springer Nature Singapore.

Dhalaria, R., Kumar, D., Kumar, H., Nepovimova, E., Kuča, K., Torequl Islam, M., & Verma, R. (2020). Arbuscular mycorrhizal fungi as potential agents in ameliorating heavy metal stress in plants. *Agronomy, 10*, 815.

Ditta, A., Arshad, M., Zahir, Z. A., & Jamil, A. (2015). Comparative efficacy of rock phosphate enriched organic fertilizer vs. mineral phosphatic fertilizer for nodulation, growth and yield of lentil. *International Journal of Agriculture and Biology, 17*(3), 589–595. http://www.fspublishers.org/published_papers/5595_.pdf. doi:10.17957/IJAB/17.3.14.954.

Ditta, A., Imtiaz, M., Mehmood, S., Rizwan, M. S., Mubeen, F., Aziz, O., Qian, Z., Ijaz, R., & Tu, S. (2018a). Rock phosphate-enriched organic fertilizer with phosphate-solubilizing microorganisms improves nodulation, growth, and yield of legumes. *Communications in Soil Science and Plant Analysis, 49*(21), 2715–2725. http://www.tandfonline.com/loi/lcss20. doi:10.1080/00103624.2018.1538374.

Ditta, A., & Khalid, A. (2016). Organic fertilizers- From basic concepts to applied outcomes. In M. Larramendy, & S. Soloneski (Eds.), *Bio-organo-phos: A sustainable approach for managing phosphorus deficiency in agricultural soils*. IntechOpen. doi:10.5772/62473.

Ditta, A., Muhammad, J., Imtiaz, M., Mehmood, S., Qian, Z., & Tu, S. (2018b). Application of rock phosphate enriched composts increases nodulation, growth and yield of chickpea. *International Journal of Recycling of Organic Waste in Agriculture, 7*(1), 33–40. www.springer.com/environment/pollution+and+remediation/journal/40093. doi:10.1007/s40093-017-0187-1.

Evelin, H., Kapoor, R., & Giri, B. (2009). Arbuscular mycorrhizal fungi in alleviation of salt stress: A review. *Annals of Botany, 104*, 1263–1280.

Faghihinia, M., Jansa, J., Halverson, L. J., & Staddon, P. L. (2023). Hyphosphere microbiome of arbuscular mycorrhizal fungi: A realm of unknowns. *Biology and Fertility of Soils, 59*, 17–34.

Falkowski, G., & Matysiak, B. (2010). The use of arbuscular mycorrhizal fungi in container production of selected ornamental conifers under organic-mineral fertilization level. *Journal of Fruit and Ornamental Plant Research, 18*(2), 335–348.

Fattah, O. A. (2013). Effect of mycorrhiza and phosphorus on micronutrients uptake by soybean plant grown in acid soil. *International Journal of Agronomy and Plant Production, 4*(3), 429–437.

Filippi, C., Bagnoli, G., Citernesi, A. S., & Giovannetti, M. (1998). Ultrastructural spatial distribution of bacteria associated with sporocarps of *Glomus mosseae*. *Symbiosis, 24*, 1–12.

Frey-Klett, P., Garbaye, J., & Tarkka, M. (2007). The mycorrhiza helper bacteria revisited. *New Phytologist, 176*, 22–36.

Gerdemann, J. W. (1968). Vesicular-arbuscular mycorrhiza and plant growth. *Annual Review of Phytopathology, 6*, 397–418.

Gianinazzi-Pearson, V., & Gianinazzi, S. (1983). The physiology of vesicular-arbuscular mycorrhizal roots. *Plant and Soil, 71*, 197–209.

Giri, S., & Singh, A. K. (2017). Ecological and human health risk assessment of agricultural soils based on heavy metals in mining areas of Singhbhum copper belt, India. *Human and Ecological Risk Assessment: An International Journal, 23*, 1008–1027.

Goicoechea, N., Szalai, G., Antolín, M. C., Sánchez-Díaz, M., & Paldi, E. (1998). Influence of arbuscular mycorrhizae and *Rhizobium* on free polyamines and proline levels in water-stressed alfalfa. *Journal of Plant Physiology, 153*, 706–711.

Gonzalez-Chavez, M. C., Carrillo-Gonzalez, R., Wright, S. F., & Nichols, K. A. (2004). The role of glomalin, a protein produced by arbuscular mycorrhizal fungi, in sequestering potentially toxic elements. *Environmental Pollution, 130*, 317–323.

Gryndler, M. (2000). Interactions of arbuscular mycorrhizal fungi with other soil organisms. In *Arbuscular mycorrhizas: Physiology and function* (pp. 239–262). Springer.

Harley, J. L. (1959). *The biology of mycorrhiza* (p. 233). Leonard Hill.

Heap, A. J., & Newman, E. I. (1980). The influence of vesicular-arbuscular mycorrhizas on phosphorus transfer between plants. *New Phytologist, 85*, 173–179.

Jalali, M., & Jalali, M. (2022). Effect of low-molecular-weight organic acids on the release of phosphorus from amended calcareous soils: Experimental and modeling. *Journal of Soil Science and Plant Nutrition, 22*(4), 4179–4193.

Jeffries, P., Gianinazzi, S., Perotto, S., Turnau, K., & Barea, J.-M. (2003). The contribution of arbuscular mycorrhizal fungi in sustainable maintenance of plant health and soil fertility. *Biology and Fertility of Soils, 37*, 1–16.

Jiang, S., An, X., Shao, Y., Kang, Y., Chen, T., Mei, X., Dong, C., Xu, Y., & Shen, Q. (2021). Responses of arbuscular mycorrhizal fungi occurrence to organic fertilizer: A meta-analysis of field studies. *Plant and Soil, 469*, 89–105.

Johansson, J. F., Paul, L. R., & Finlay, R. D. (2004). Microbial interactions in the mycorrhizosphere and their significance for sustainable agriculture. *FEMS Microbiology Ecology, 48*, 1–13.

Joner, E. J., & Johansen, A. (2000). Phosphatase activity of external hyphae of two arbuscular mycorrhizal fungi. *Mycological Research News, 104*, 81–86.

Kaiser, C., Kilburn, M. R., Clode, P. L., Fuchslueger, L., Koranda, M., Cliff, J. B., Solaiman, Z. M., & Murphy, D. V. (2015). Exploring the transfer of recent plant photosynthates to soil microbes: Mycorrhizal pathway vs direct root exudation. *New Phytologist, 205*, 1537–1551.

Keyes, S., Van Veelen, A., McKay Fletcher, D., Scotson, C., Koebernick, N., Petroselli, C., Williams, K., Ruiz, S., Cooper, L., & Mayon, R. (2022). Multimodal correlative imaging and modelling of phosphorus uptake from soil by hyphae of mycorrhizal fungi. *New Phytologist, 234*, 688–703.

Khan, K. S., Naveed, M., Ditta, A., Alam, T., Mumtaz, M. Z., Abbas, A., Manzoni, S., & Li, L. (2025). Soil C/P stoichiometry mediated microbially driven phosphorus mobilization: New insight in soil P management. *Journal of Soil Science and Plant Nutrition, 25*(3), 7691–7709. https://www.springer.com/journal/42729. doi:10.1007/s42729-025-02624-4.

Lalitha, M., Anil Kumar, K. S., Dharumarajan, S., Balakrishnan, N., Srinivasan, R., Nair, K. M., Hegde, R., & Singh, S. K. (2017). Role of vesicular-arbuscular mycorrhizae in mobilization of soil phosphorus. In Agriculturally important microbes for sustainable agricultur*e: Volume I: Plant-soil-microbe nexus* (pp. 317–331). Springer.

Lehmann, A., Leifheit, E. F., & Rillig, M. C. (2017). Mycorrhizas and soil aggregation. In *Mycorrhizal mediation of soil* (pp. 241–262). Elsevier.

Li, X.-L., George, E., & Marschner, H. (1991). Extension of the phosphorus depletion zone in VA-mycorrhizal white clover in a calcareous soil. *Plant and Soil, 136*, 41–48.

Liu, C.-Y., Hao, Y., Wu, X.-L., Dai, F.-J., Abd-Allah, E. F., Wu, Q.-S., & Liu, S.-R. (2024). Arbuscular mycorrhizal fungi improve drought tolerance of tea plants via modulating root architecture and hormones. *Plant Growth Regulation, 102*, 13–22.

Lokhande, M. S. (2025). *Mycorrhiza in plant growth and soil health*. AG Publishing House (AGPH Books).

Marschner, H., & Dell, B. (1994). Nutrient uptake in mycorrhizal symbiosis. *Plant and Soil, 159*, 89–102.

Mayo, K., Davis, R. E., & Motta, J. (1986). Stimulation of germination of spores of *Glomus versiforme* by spore-associated bacteria. *Mycologia, 78*, 426–431.

Miyauchi, S., Kiss, E., Kuo, A., Drula, E., Kohler, A., Sánchez-García, M., Morin, E., Andreopoulos, B., Barry, K. W., & Bonito, G. (2020). Large-scale genome sequencing of mycorrhizal fungi provides insights into the early evolution of symbiotic traits. *Nature Communications, 11*, 5125.

Morin, E., Miyauchi, S., San Clemente, H., Chen, E. C. H., Pelin, A., de La Providencia, I., Ndikumana, S., Beaudet, D., Hainaut, M., & Drula, E. (2019). Comparative genomics of *Rhizophagus irregularis, R. cerebriforme, R. diaphanus* and *Gigaspora rosea* highlights specific genetic features in *Glomeromycotina*. *New Phytologist, 222*, 1584–1598.

Mosse, B. (1959). The regular germination of resting spores and some observations on the growth requirements of an *Endogone* sp. causing vesicular-arbuscular mycorrhiza. *Transactions of the British Mycological Society, 42*(3), 273–286.

Muthukumar, T., & Udaiyan, K. (2006). Growth of nursery-grown bamboo inoculated with arbuscular mycorrhizal fungi and plant growth promoting rhizobacteria in two tropical soil typeswith and without fertilizer application. *New Forest, 31*, 469–485.

Nadeem, S. M., Ahmad, M., Zahir, Z. A., Javaid, A., & Ashraf, M. (2014). The role of mycorrhizae and plant growth promoting rhizobacteria (PGPR) in improving crop productivity under stressful environments. *Biotechnology Advances, 32*, 429–448.

Nazir, A., Sarfraz, W., Allah, D., Khalid, N., Farid, M., Shafiq, M., Firdaus-E-Bareen, Rizvi, Z. F., & Naeem, N. (2023). Synergistic impact of two autochthonous saprobic fungi (*A. niger* and *T. pseudokoningii*) on the growth, ionic contents, and metals uptake in *Brassica juncea* L. and *Vigna radiata* L. under tannery solid waste contaminated soil. *International Journal of Phytoremediation, 25*(11), 1488–1500. doi:10.1080/15226514.2023.2166457.

Ordoñez, Y. M., Fernandez, B. R., Lara, L. S., Rodriguez, A., Uribe-Velez, D., & Sanders, I. R. (2016). Bacteria with phosphate solubilizing capacity alter mycorrhizal fungal growth both inside and outside the root and in the presence of native microbial communities. *PLoS One, 11*, e0154438.

Peng, Z., Zulfiqar, T., Yang, H., Wang, M., & Zhang, F. (2024). Effect of arbuscular mycorrhizal fungi (AMF) on photosynthetic characteristics of cotton seedlings under saline-alkali stress. *Scientific Reports, 14*, 8633.

Porcel, R., Barea, J. M., & Ruiz-Lozano, J. M. (2003). Antioxidant activities in mycorrhizal soybean plants under drought stress and their possible relationship to the process of nodule senescence. *New Phytologist, 157*, 135–143.

Porras-Soriano, A., Soriano-Martín, M. L., Porras-Piedra, A., & Azcón, R. (2009). Arbuscular mycorrhizal fungi increased growth, nutrient uptake and tolerance to salinity in olive trees under nursery conditions. *Journal of Plant Physiology, 166*, 1350–1359.

Prasad, R., Bhola, D., Akdi, K., Cruz, C., KVSS, S., Tuteja, N., & Varma, A. (2017). Introduction to mycorrhiza: Historical development. In *Mycorrhiza-function, diversity, State of the Art* (pp. 1–7). Springer.

Rakshit, A., & Bhadoria, P. S. (2010). Role of VAM on growth and phosphorus nutrition of maize with low soluble phosphate fertilization. *Acta Agrobotanica, 59*, 111–118.

Ravnskov, S., Larsen, J., Olsson, P. A., & Jakobsen, I. (1999). Effects of various organic compounds on growth and phosphorus uptake of an arbuscular mycorrhizal fungus. *New Phytologist, 141*, 517–524.

Rillig, M. C. (2004). Arbuscular mycorrhizae, glomalin, and soil aggregation. *Canadian Journal of Soil Science, 84*, 355–363.

Ryan, M. H., & Graham, J. H. (2002). Is there a role for arbuscular mycorrhizal fungi in production agriculture? *Plant and Soil, 244*, 263–271.

Saeed, H. S., Munir, M., Samra, S., Farooq, A., Mahmood, H., Ahmad, I., Akram, M. I., Iqbal, R., Elsadek, M. F., Almutairi, S. M., Ditta, A., & Daraz, U. (2025). Screening of biosurfactants producing fungi from infected *Citrus sinensis* (L.). *Polish Journal of Environmental Studies, 34*(5), 5261–5270. https://www.pjoes.com/pdf-190588-115894?filename=Screening%20of.pdf. doi:10.15244/pjoes/190588.

Saia, S., & Jansa, J. (2022). Editorial: Arbuscular mycorrhizal fungi: The bridge between plants, soils, and humans. *Frontiers in Plant Science, 13*, 875958.

Sbrana, C., Avio, L., & Giovannetti, M. (2014). Beneficial mycorrhizal symbionts affecting the production of health-promoting phytochemicals. *Electrophoresis, 35*, 1535–1546.

Shi, J., Wang, X., & Wang, E. (2023). Mycorrhizal symbiosis in plant growth and stress adaptation: From genes to ecosystems. *Annual Review of Plant Biology, 74*(1), 569–607.

Siddique, M. BA., Khalid, A., Ditta, A., Mahmood, S., Alataway, A., Dewidar, A. Z., & Mattar, M. A. (2023). Climate change variables modify microbial community structure and soil enzymes involved in nitrogen and phosphorus metabolism. *Rhizosphere, 28*, 100793. doi:10.1016/j.rhisph.2023.100793.

Smith, S. E., & Read, D. J. (2010). *Mycorrhizal symbiosis*. Academic Press.

O'Keefe, D. M., & Sylvia, D. M. (1991). *Mechanisms of the vesicular-arbuscular mycorrhizal plant-growth response. Handbook of applied mycology. Soil and plants, 1*, 35–53.

Soukayna, H. (2012). *Mycorrhiza-Induced Resistance Against Thielaviopsis* basicola *in the ornamental crop petunia hybrida*.

Sun, R., Zhang, W., Liu, Y., Yun, W., Luo, B., Chai, R., Zhang, C., Xiang, X., & Su, X. (2022). Changes in phosphorus mobilization and community assembly of bacterial and fungal communities in rice rhizosphere under phosphate deficiency. *Frontiers in Microbiology, 13*, 953340.

Tarafdar, J. C., & Marschner, H. (1994). Phosphatase activity in the rhizosphere and hyphosphere of VA mycorrhizal wheat supplied with inorganic and organic phosphorus. *Soil Biology and Biochemistry, 26*, 387–395.

Tavasolee, A. R., & Aliasgharzad, N. (2009). Effect of arbuscular mycorrhizal fungi on nutrient uptake and Onion yield in a saline soil at field conditions. *Water and Soil Science, 19*, 145–158.

Tedersoo, L., Magurno, F., Alkahtani, S., & Mikryukov, V. (2024). Phylogenetic classification of arbuscular mycorrhizal fungi: New species and higher-ranking taxa in *Glomeromycota* and *Mucoromycota* (class Endogonomycetes). *MycoKeys, 107*, 273.

Thind, S., Chaudhary, M. S, Ditta, A., Hussain, I., Parveen, A., Ullah, N., Mahmood, Q., Al-Ashkar, I., & El-Sabagh, A. (2022). Impact of mycorrhizal fungi from different rhizospheric soils on fungal colonization, growth, and chlorophyll contents of *Cenchrus ciliaris*. *Agronomy, 12*(11), 2644. doi:10.3390/agronomy12112644.

Tian, J., Ge, F., Zhang, D., Deng, S., & Liu, X. (2021). Roles of phosphate solubilizing microorganisms from managing soil phosphorus deficiency to mediating biogeochemical P cycle. *Biology (Basel), 10*, 158.

Tisserant, E., Malbreil, M., Kuo, A., Kohler, A., Symeonidi, A., Balestrini, R., Charron, P., Duensing, N., Frei dit Frey, N., & Gianinazzi-Pearson, V. (2013). Genome of an arbuscular mycorrhizal fungus provides insight into the oldest plant symbiosis. *Proceedings of the National Academy of Sciences, 110*, 20117–20122.

Trappe, J. M. (2010). James Wessell Gerdemann, 1921–2008. *Mycologia, 102*, 1518–1522.

Venice, F., Chialva, M., Domingo, G., Novero, M., Carpentieri, A., Salvioli di Fossalunga, A., Ghignone, S., Amoresano, A., Vannini, C., & Lanfranco, L. (2021). Symbiotic responses of *Lotus japonicus* to two isogenic lines of a mycorrhizal fungus differing in the presence/absence of an endobacterium. *The Plant Journal, 108*, 1547–1564.

Wahab, A., Muhammad, M., Munir, A., Abdi, G., Zaman, W., Ayaz, A., Khizar, C., & Reddy, S. P. P. (2023). Role of arbuscular mycorrhizal fungi in regulating growth, enhancing productivity, and potentially influencing ecosystems under abiotic and biotic stresses. *Plants, 12*, 3102.

Weng, W., Yan, J., Zhou, M., Yao, X., Gao, A., Ma, C., Cheng, J., & Ruan, J. (2022). Roles of arbuscular mycorrhizal fungi as a biocontrol agent in the control of plant diseases. *Microorganisms, 10*, 1266.

Wijayawardene, N. N., Mešić, A., Pošta, A., Tkalčec, Z., & Thines, M. (2022). Outline of fungi and fungus-like taxa–2021. *Mycosphere, 13*, 53–453.

Will, M. E., & Sylvia, D. M. (1990). Interaction of rhizosphere bacteria, fertilizer, and vesicular-arbuscular mycorrhizal fungi with sea oats. *Applied and Environmental Microbiology, 56*, 2073–2079.

Wu, D., Hugenholtz, P., Mavromatis, K., Pukall, R., Dalin, E., Ivanova, N. N., Kunin, V., Goodwin, L., Wu, M., & Tindall, B. J. (2009). A phylogeny-driven genomic encyclopaedia of bacteria and *Archaea*. *Nature, 462*, 1056–1060.

Xavier, L. J. C., & Germida, J. J. (2003). Bacteria associated with *Glomus clarum* spores influence mycorrhizal activity. *Soil Biology and Biochemistry, 35*, 471–478.

Yizhu, L., Imtiaz, M., Ditta, A., Rizwan, M. S., Ashraf, M., Mehmood, S., Aziz, O., Mubeen, F., Ali, M., Elahi, N. N., Ijaz, R., Lele, S., Shuang, C., & Tu, S. (2020). Response of growth, antioxidant enzymes and root exudates production towards as stress in *Pteris vittata* and in *Astragalus sinicus* colonized by arbuscular mycorrhizal fungi. *Environmental Science and Pollution Research, 27*(2), 2340–2352. https://link.springer.com/journal/11356. doi:10.1007/s11356-019-06785-5.

Zhang, J., Ruotong, Z., Xia, L. I., & Zhang, J. (2024). Potential of arbuscular mycorrhizal fungi for soil health: A review. *Pedosphere, 34*, 279–288.

Zhang, J., Wang, F., Che, R., Wang, P., Liu, H., Ji, B., & Cui, X. (2016). Precipitation shapes communities of arbuscular mycorrhizal fungi in Tibetan alpine steppe. *Scientific Reports, 6*, 23488.

Zhou, J., Zang, H., Loeppmann, S., Gube, M., Kuzyakov, Y., & Pausch, J. (2020). Arbuscular mycorrhiza enhances rhizodeposition and reduces the rhizosphere priming effect on the decomposition of soil organic matter. *Soil Biology and Biochemistry, 140*, 107641.

Chapter 7

Arbuscular mycorrhizal fungi and carbon sequestration

Nida Naz[a], Nazish Huma Khan[b], Mohammad Nafees[c], Muhammad Aamir Khan[d] and Amjad Ur Rahman[e]

[a] *Department of Environmental Sciences, Abdul Wali Khan University, Mardan, Khyber Pakhtunkhwa, Pakistan,* [b] *Department of Environmental Sciences, University of Swabi, Khyber Pakhtunkhwa, Pakistan,* [c] *Department of Environmental Sciences, University of Peshawar, Peshawar, Khyber Pakhtunkhwa, Pakistan,* [d] *Obstetrics and gynaecology Department, Ayub Teaching Hospital, Abbottabad, Khyber Pakhtunkhwa, Pakistan,* [e] *Department of Conservation, WWF Pakistan, Peshawar, Pakistan*

7.1 Introduction

Arbuscular mycorrhiza (AM) is a notable adaptation that emerged some 480 million years ago and is widespread among terrestrial plants in diverse ecological environments. Some plants have developed into obligatory mycoheterotrophs, which are totally dependent on AM fungi for sustenance, but the majority are facultative symbionts that benefit from AM fungi. On the other hand, some taxa, such as Chenopodiaceae and Brassicaceae, have evolved other feeding methods and lost their capacity to interact with AM fungi, becoming symbiotic (Chen et al., 2018).

A key component of soil health, soil organic carbon (SOC) affects water storage, food production, and climate change mitigation. It is crucial for nutrient pools that improve soil fertility and stability and make up 50%–58% of soil organic matter (SOM). By decreasing nutrient availability and food resources for soil organisms, the depletion of SOC reduces soil diversity, endangering plant development and food security. By generating glomalin-related soil proteins (GRSPs), arbuscular mycorrhizal fungus (AMF) contributes to the improvement of SOM (Shukla et al., 2025), as shown in Fig. 7.1.

7.2 Direct role of AMF in carbon sequestration

7.2.1 Contribution of AMF in the essential carbon reservoir and primary productivity

In terrestrial ecosystems the pedosphere is an essential carbon reservoir that influences climate change and the global carbon cycle. Net primary productivity (NPP) and the distribution of photosynthetic carbon in biomass have an impact on the store of SOC. By developing symbiotic associations with more than 80% of terrestrial plant species (Jia et al. 2023), arbuscular mycorrhizal fungi (AMF) improve primary productivity and carbon cycling. Each year, AMF transfers about 5 Pg of carbon to the soil. As a soil supplement that improves agricultural output and removes contaminated soils, AMF also has a major impact on the root microbiome. In addition, AMF inoculation may improve plant vitality and efficiently capture carbon, especially in stressful situations (Wu et al. 2024; Tao & Liu, 2024).

7.2.2 Carbon allocation from plants to AMF and reciprocal nutrient exchange

Plant partners in arbuscular mycorrhizal (AM) symbiosis exchange carbon (C) for vital nutrients, including nitrogen (N) and phosphorus (P), in a reciprocal reward system that benefits both parties. About 5%–30% of the products of AMF plants' photosynthesis are given to the fungi, which then supply soil nutrients such as N and P (Hawkins et al., 2023). As AMF depend on hosts but are not necessary, host parasitism may result if fungi are unable to provide advantages. By improving the uptake of inorganic N and P and releasing nutrients from organic matter through "priming effects," including interactions

Advancing Sustainable Agriculture Through Arbuscular Mycorrhizal Fungi. DOI: https://doi.org/10.1016/B978-0-443-45130-0.00026-6

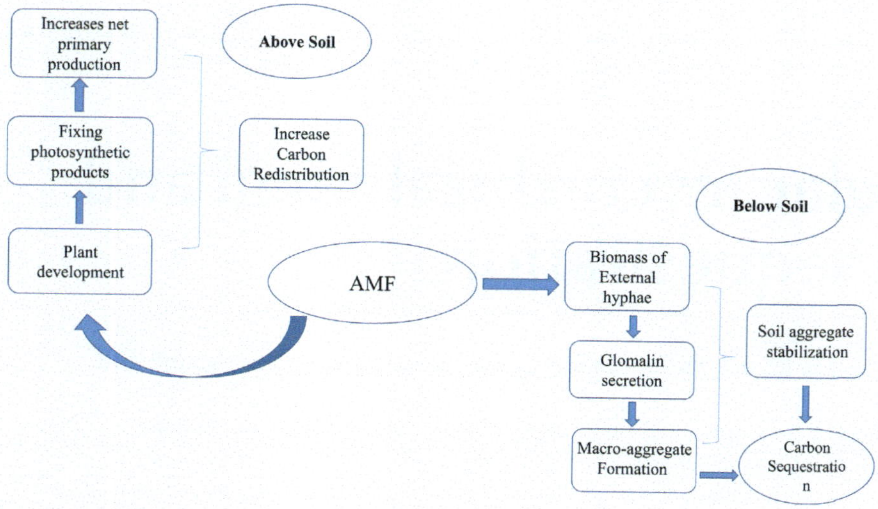

FIGURE 7.1 Contribution of AMF toward the environment.

TABLE 7.1 Mechanism of carbon flow from plant to AMF to soil.

Source of carbon	Transfer form	Mechanism	Soil fate	SOC
Fungal biomass	Necromass	Hyphal turnover	Carbon associated with minerals	Long-term stabilization
Glomalin	GRSP	Soil aggregation	Aggregate protection	Physical protection of SOC
Photosynthate	Lipids	Direct transfer of lipids	Hyphal growth	Structural input of carbon
Photosynthate	Sugars	Arbuscular exchange	Fungal biomass	Short-term carbon pool

Wahab et al. (2023).
SOC, soil organic carbon; *GRSP*, glomalin-related soil protein.

with the hyposphere bacteria, AMF helps reduce carbon loss. In addition, by encouraging unstable carbon deposits and attracting symbiotic bacteria, AMF can quicken the turnover of SOC (Rozmos et al., 2022).

According to the research the colonization of arbuscular mycorrhizal fungus (AMF) in plant roots considerably uses plant-fixed carbon to establish a carbon sink requirement. Before evaluating plants' capability for sequestering carbon, it is crucial to comprehend the translocation of carbon from plants to AMF (Ali et al., 2023; Sarfraz et al., 2019). Through chemical cues (strigolactones) that promote spore germination and hyphal development toward plant roots, AMF root colonization establishes a symbiotic relationship. By penetrating root cells the hyphae create arbuscules encased in a periarbuscular membrane, facilitating the flow of carbon and nutrients. According to recent research, both lipids and sugars from host plants are transported to AMF (Parihar, Rakshit, Rana, Tiwari, & Jatav, 2020). Glomalin is not a specific gene product; rather, it is a soil fraction distinguished by its extractability and immunoreactive qualities. The origin and functions of glomalin and GRSPs are unclear, according to recent research. With half-lives ranging from a few years to decades, they are important for soil quality and act as a steady carbon sink. By encouraging soil particle aggregation and aiding in carbon storage, GRSPs add to the total amount of organic soil carbon and prevent the destruction of other carbon forms. Arbuscular mycorrhizal (AM) fungi's hyphal network also promotes root formation and plant growth, further preventing soil erosion (Chen et al., 2018). The given Table 7.1 shows the mechanism of carbon flow from the plant to AMF to soil.

7.2.3 Stabilization mechanism of carbon transfer

AMF contribute significantly to increased carbon sequestration because they move carbon from the highly respiring zones close to the roots to soil aggregates (Shukla et al., 2025). Increased carbon dioxide in the environment causes plants to transmit more carbon to arbuscular mycorrhizal fungus, which further encourages the expansion and advancement of their sort (Buckeridge et al. 2020). To improve nutrition and water absorption, these fungi spread hyphal networks into the soil and colonize plant root systems, forming structures such as vesicles and arbuscules. They highlight their evolutionary relevance

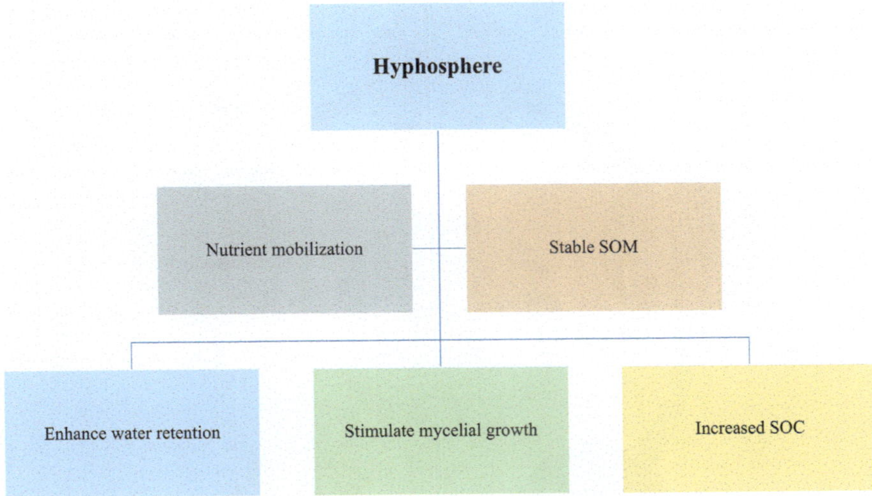

FIGURE 7.2 Functions of the hyposphere inside the soil. *SOC*, Soil organic carbon; *SOM*, soil organic matter.

and contribution to ecosystem health by facilitating the uptake of difficult-to-reach ions, especially inorganic phosphate, in exchange for carbon derived from photosynthesis (Shukla et al., 2025).

7.2.4 Processes of the hyphosphere and soil carbon stabilization

The hyphosphere, the area of soil surrounding fungal hyphae, is crucial for soil carbon (C) cycling. Microorganisms in the hyphosphere enhance nutrient mobilization, aiding arbuscular mycorrhizal fungi (AMF) in nutrient acquisition and stimulating mycelial growth. AMF extraradical hyphal networks constitute a significant portion of soil microbial biomass, as shown in Fig. 7.2. Their rapid turnover directly supports SOC formation and cycling, whereas their chitin-based cell walls resist decomposition and accumulate in soil (Wang et al. 2022). In addition, AMF promote soil C sequestration by enhancing soil aggregation through the binding of soil particles, forming stable aggregates that physically protect SOC. However, AMF inoculation can also lead to increased decomposition of fresh residues by influencing soil microbial communities (Tao & Liu, 2024).

7.2.5 Mineral interaction and AMF mediation of carbon stabilization

The main way that arbuscular mycorrhizal fungi (AMF) stabilize rhizo-C is by interacting with minerals and hyphal aggregates, which improve soil stability. AMF can change the metabolic characteristics and root structure of host plants or directly affect soil aggregates through their extraradical hyphae (ERH). To preserve SOM the *Glomus* genus helps move monosaccharides from plant roots to soil aggregates. By means of adsorption procedures, minerals such as goethite further improve aggregate formation and stability. AMF also contributes to the mineralization of SOM by using SOM's nitrogen to satisfy their nutrient needs in response to rhizo-C increases and nutrient deficits. This nitrogen fixation encourages a favorable rhizosphere priming effect, especially when nutrients are scarce. Nevertheless, little is known about the complex processes by which AMF affect microbial populations and SOM mineralization (Jeewani et al., 2021).

7.2.6 Contribution of AMF biomass and necromass in soil organic carbon stabilization

Arbuscular mycorrhizal fungi (AMF) use both living and dead fungal biomass to affect the dynamics of SOC. By physically entangling soil particles and producing aggregates that restrict decomposers' access to organic molecules, their ERH improve the stability and storage of SOM. Stable organo-mineral compounds are produced when SOM is bound to minerals by weathering processes facilitated by ERH (Basiru & Hijri, 2024). In addition, ERH facilitates SOM redistribution, which leads to the accumulation of microbial residue. ERH networks, which account for 15%–30% of soil microbial biomass, are crucial to soil microbial necromass, which makes up more than 50% of soil organic C (Basiru & Hijri, 2024; Wu et al., 2023). AMF necromass offers low-molecular-weight molecules for carbon inputs despite its quick mineralization. Furthermore, by

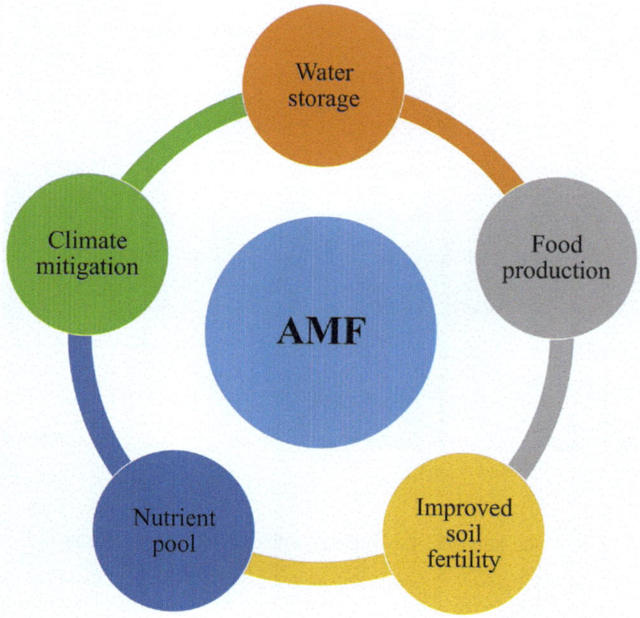

FIGURE 7.3 Working of AMF above and below the soil.

creating mineral-associated organic matter, fungal residues with high aromatic aliphatic C ratios may improve long-term C sequestration (Basiru & Hijri, 2024; Liu et al., 2023).

7.2.7 Exchange of nutrients and resource acquisition

Arbuscular mycorrhizal fungi (AMF) play a vital role in nutrient uptake by plants, particularly in the absorption of phosphorus and nitrogen. The amount of soil that may be accessed for nutrient uptake is greatly increased by AMF hyphae, which extend beyond the root systems. Phosphatase activity from AMF breaks down organic phosphorus, increasing the amount of phosphorus available to plants. AMF hyphae can directly absorb nitrogen from organic sources without being negatively impacted by synthetic nitrification inhibitors, according to research showing that AMF facilitates nitrogen uptake in mycorrhizal plants independent of nitrification inhibitors (Shukla et al., 2025).

7.3 Indirect role of AMF in carbon sequestration

7.3.1 Enhanced plant productivity

By encouraging plant nutrient uptake through mycorrhizal networks the arbuscular mycorrhizal fungi (AMF) improve the ecosystem's capacity to store carbon. This results in healthier plant development and greater photosynthesis, which increases CO_2 capture. When AMF colonize plant roots, they form vast mycelial networks that are effective in long-term carbon storage because they are resistant to breakdown and may preserve organic carbon (Fig. 7.3). AMF further supports soil carbon retention by aiding in soil aggregation and the retention of stable organic matter (Deluz et al. 2020; Liu et al. 2023; Shukla et al., 2025).

7.3.2 Trade-offs between soil organic carbon and plant biomass

Trade-offs between SOC and plant biomass or nutrient gains have been observed with higher plant biomass or nutrient acquisition under AM symbiosis, leading to SOC loss and CO_2 reemission (Terrer et al., 2021). For instance a metaanalysis of global forest datasets by Yang et al. (2024) revealed that, in warmer conditions, plant biomass increased by 48.3%, but SOC composition decreased by 11.5%. When vital mineral nutrients such as P and N are running low, as well as when CO_2 and global warming are present, there is a significant chance of a trade-off (Jeewani et al., 2021).

7.3.3 Enhanced plant productivity and photosynthesis carbon input

By improving plant carbon sinks, arbuscular mycorrhizal fungi (AMF) can reduce carbon loss from the breakdown of SOM and boost the capacity of plants to fix CO_2 (Laza et al., 2023). It has been demonstrated that AMF inoculation enhances the net photosynthetic efficiency of a variety of plant species, encouraging increased soil carbon storage. According to recent calculations, over 36% of the yearly CO_2 emissions from fossil fuels are caused by carbon entering fungal mycelium (Hawkins et al., 2023). Under increased CO_2 and warmth, mycorrhizal plants' higher photosynthetic rate is anticipated to boost their growth and productivity, which will assist fungal biomass and the sequestration of SOC. The results validate theories indicating a trade-off between plant benefits and AMF-influenced SOC storage, with possible counterbalances from increased fungal symbiont carbon deposition, particularly under nutrient-limited circumstances (Basiru & Hijri, 2024), as shown in Table 7.1.

7.4 . AMF in climate smart and sustainable agriculture

7.4.1 Inoculation technologies and scaling challenges

Different soil types and crop reactions make it difficult to use arbuscular mycorrhizal fungi (AMF) in agriculture. Pelletized AMF has shown promise in cereal crops, and recent research aims to increase AMF administration through better inoculation techniques and bioformulations. Scaling up AMF inoculant production is essential to meeting the increasing demand for sustainable practices. This calls for improving culture methods for increased output and viability, encouraging stakeholder collaboration, and obtaining funding for expansions and training programs (Shi et al. 2023; Shukla et al., 2025; Zhao et al. 2022).

7.4.2 Genetic improvement of AMF

It is anticipated that soil microbiome sequencing and arbuscular mycorrhizal fungal (AMF) genetic engineering will improve AMF performance. Furthermore, biotechnological methods to increase the stress tolerance of AMF strains have been established (Shukla et al., 2025; Smith et al., 2024). Arbuscular mycorrhizal fungi (AMF) are classified as "microbial-based plant biostimulants" after an investigation of the toxicity, ecotoxicology, environmental persistence, and plant residues associated with AM inoculants (Ghorui et al. 2024; Shukla et al., 2025). Compliance, standardization, education, certification, monitoring, and cost challenges must be resolved to fully reap the benefits of AM inoculants. To maximize the effects of AMF on agricultural practices, which can result in better soil health, higher crop yields, and a more sustainable agricultural sector, enhanced monitoring is essential (Shukla et al., 2025).

7.4.3 Bioremediation and management of heavy metals

The potential of AM fungi for bioremediation, assisting plants in lowering soil contamination from organic chemicals and heavy metals, has been investigated in recent decades. By collecting and sequestering harmful metal ions, protecting the host plant, or promoting the delivery of metals such as copper and zinc, which can cause heavy metal accumulation in the host, AM fungi can improve plant growth in polluted soils. These techniques, which call for fungi that can withstand heavy metals and tolerate host plants, enable either minimal crop contamination or the phytoextraction of heavy metals (Chen et al., 2018). Few companies provide AM fungal products, and whereas numerous laboratory studies have examined the bioremediation potential of AM fungi, there are few field studies on their large-scale applicability. Reduced AM colonization because of high pollution levels and the predominance of tiny, non-AM host plants among heavy metal-accumulating species are obstacles (Chen et al., 2018).

7.5 Environmental and plant factors influencing AMF-mediated carbon sequestration

7.5.1 Responses of mycorrhizal fungi in plants

Arbuscular mycorrhizal fungi (AMF) and plants have symbiotic connections that are impacted by plant characteristics such as photosynthetic pathways and root shape. Tap-root plants are more dependent on AMF than fibrous-rooted ones, yet some root traits may not affect how AMF inoculation affects growth. Although C_4 plants are more efficient at photosynthetic processes than C_3 plants, it is still unknown how AMF inoculation affects these plant species (Duan et al. 2023). Because

TABLE 7.2 AMF mitigating environmental stresses.

Temperature stress	Heavy metals stress	Salinity stress	Drought stress
Enhance the efficiency of photosynthesis	Heavy metal chelation	Glomalin production	Increase photosynthetic efficiency
Metabolism of carbohydrates	Regulatory genes are involved in heavy metal stress	Accumulation of dry matter	Enhance tolerance to oxidative stress
Water relation and secondary metabolism	Phytoextraction and Phytostabilization	Osmotic adjustment	Osmotic adjustment and enhanced compatible osmolyte
Improves water and nutrient uptake	Prevent nutrient deficiency, increase growth and biomass	Uptake of nutrients	Improve the assimilation of nitrogen and phosphorus
Plasma membrane permeability	The fungal cell wall absorbs microflora and pH	Production of growth hormone and antioxidants	Changes in the level of phytohormones

mycorrhizal results are determined by interactions between plants, microbes, and soil, soil conditions are also very important (Tao & Liu, 2024).

Crop plants and arbuscular mycorrhizal (AM) fungi have a crucial interaction that increases crops' resistance to biotic and abiotic environmental challenges. Abiotic stressors are exacerbated by elements such as climate change and traditional farming methods, which have a detrimental impact on crop productivity and quality. By promoting biochemical regulation processes such as osmotic control and the synthesis of proline, glutathione, and soluble sugars, AM fungi help plants adapt to a variety of stresses, such as pollutants, salinity, drought, and extreme temperatures (Chaudhary et al., 2025; Nazir et al., 2023; Saeed et al., 2025; Thind et al., 2022; Yizhu et al., 2020), as shown in Table 7.2.

7.5.2 Effects of nutrients and pH on AMF

Plants invest less in AMF in nutrient-rich soils, which is inversely correlated with AMF colonization; in contrast, plants become more dependent on AMF in nutrient-deficient soils, which may cause the relationship to change from mutualism to parasitism. Although results on the relationship between pH and AMF abundance are mixed, soil pH affects nutrient availability and interacts with AMF (Jerbi et al. 2021; Tao & Liu, 2024).

7.5.3 Functional specialization, host identity, and plant AMF compatibility

AM fungi have a range of effects on their plant hosts; some combinations are advantageous, whereas others are neutral or detrimental. Host identity affects AM fungal proliferation, suggesting functional specialization in relationships. According to a study, mycorrhizal growth response can vary from −50% to +50%, and growth depression can result from a variety of combinations. Phylogeny and mutualistic potential do not correlate, indicating separate adaptation mechanisms. Coadaptation is demonstrated by the superior performance of combinations from the same location. More plant community diversity is supported by a variety of AM fungi in soil than by a small number, indicating that each plant species may require certain AM fungal partners even when host specificity is minimal in lab settings. Plant production and biodiversity are impacted by this functional specialization (Chen et al., 2018).

7.5.4 Impacts of climate change on the community of AMF and carbon cycling

According to recent research, over the past 30 years, arbuscular mycorrhizal (AM) dominance has been promoted in the eastern United States by human activities, including nitrogen deposition and fire control in conjunction with climate change (Jo, Fei, Oswalt, Domke, & Phillips, 2019). According to (Steidinger et al., 2019), ectomycorrhizal (ECM) trees may fall by up to 10% by 2070 as a result of climate change, particularly in the boreal-temperate transition zones. Although it is yet unknown when climate change may affect ecosystems, current research points to a loss of important ECM hosts in the future (Reich et al., 2015). Furthermore, compared to AM fungi, ECM fungi exhibit a higher increase in mycorrhizal biomass in high CO_2 environments (Dong, Wang, Sun, Yang, & Xu, 2018). Although the impact of climate change on forest symbiosis has been acknowledged, little research has been done on how these effects might feed back into the nitrogen and carbon cycles. The differences in NPP between models utilizing current maps and those that consider future ECM distributions (Braghiere et al., 2021; Steidinger et al., 2019). Large parts of South America, particularly savanna regions, are expected to show the largest negative feedback effects on NPP because of climate change, with boreal forests following. In contrast, tropical forests and regions in China may see benefits from changes in plant symbiotic relationships. However, these findings should be approached with caution because of limitations in the forest plot training data utilized by (Steidinger et al., 2019). Machine learning predictions suggest an increase in ECM fungi in tropical areas, possibly influenced by elevated CO_2 conditions (Braghiere et al., 2021).

7.6 Conclusion

Terrestrial ecosystems depend on arbuscular mycorrhizal fungi (AMF), which establish symbiotic connections with the majority of plants to exchange resources that are vital for soil health and productivity. Through structures known as arbuscules, plants boost their development and nutrient exchange capacity by allocating 5%–30% of their fixed carbon to AMF. This connection facilitates increased nutrient absorption and supports plant growth by improving the acquisition of nitrogen and phosphorus. AMF also promotes the durability of soil aggregates and nitrogen cycling by stimulating microbial populations. Many variables, such as plant characteristics and environmental circumstances, affect how well carbon and nutrients are exchanged. Overall, AMF are essential for maintaining carbon sequestration and ecosystem production, underscoring their importance in sustainable farming methods.

Recommendations

The following recommendations are for AMF:

1. High-quality arbuscular mycorrhizal fungus (AMF) inoculant development must be prioritized to improve agricultural production and sustainability.
2. Choosing fungal strains that are tolerant of environmental stressors and compatible with native plants is part of this.
3. To minimize the need for synthetic fertilizers, AMF management should be in line with soil fertility practices, incorporating techniques such as cover crops and reduced tillage.
4. Enhancing AMF functionality through plant diversification promotes nutrient cycling. AMF can facilitate bioremediation in contaminated soils.
5. To fully comprehend AMF mechanisms in carbon transport and interactions with soil bacteria, more research is required.
6. To maximize AMF advantages in agriculture, capacity training, stakeholder participation, and ongoing monitoring of AMF dynamics in the face of climate change are essential.

List of abbreviations

AM	arbuscular mycorrhiza
AMF	arbuscular mycorrhizal fungi
SOC	soil organic carbon
SOM	soil organic matter
ERH	extraradical hyphae
GRSPs	glomalin-related soil proteins
NPP	net primary production
ECM	ectomycorrhizal
RPE	rhizosphere priming effect
C	carbon
N	nitrogen
P	phosphorus

References

Ali, M, Javeed, H. M. R., Tariq, M., Khan, A. A., Qamar, R., Nawaz, F., Masood, N., Ditta, A., Abbas, T., Zamir, M. S. I., Sabagh, A. E, Shahzad, M., Naeem, R., & Mubeen, M (2023). Use of biochar for biological carbon sequestration. In W. N. Jatoi, M. Mubeen, M. Z. Hashmi, S. Ali, S. Fahad, & K. Mahmood (Eds.), *Climate change impacts on agriculture* (pp. 393–411). Springer.

Basiru, S., & Hijri, M. (2024). Trade-off between soil organic carbon sequestration and plant nutrient uptake in arbuscular mycorrhizal symbiosis. *Fungal Biology Reviews, 49*, 100381.

Braghiere, R. K., Fisher, J. B., Fisher, R. A., Shi, M., Steidinger, B. S., Sulman, B. N., ... Crowther, T. W. (2021). Mycorrhizal distributions impact global patterns of carbon and nutrient cycling. *Geophysical Research Letters, 48*(19) e2021GL094514.

Buckeridge, K. M., Mason, K. E., McNamara, N. P., Ostle, N., Puissant, J., Goodall, T., & Whitaker, J. (2020). Environmental and microbial controls on microbial necromass recycling, an important precursor for soil carbon stabilization. *Communications Earth & Environment, 1*(1), 36.

Chaudhary, A., Poudyal, S., & Kaundal, A. (2025). Role of arbuscular mycorrhizal fungi in maintaining sustainable agroecosystems. *Applied Microbiology, 5*(1), 6.

Chen, D., Liu, X., Bian, R., Cheng, K., Zhang, X., Zheng, J., ... Li, L. (2018). Effects of biochar on availability and plant uptake of heavy metals–A meta-analysis. *Journal of Environmental Management, 222*, 76–85.

Deluz, C., Nussbaum, M., Sauzet, O., Gondret, K., & Boivin, P. (2020). Evaluation of the potential for soil organic carbon content monitoring with farmers. *Frontiers in Environmental Science, 8*, 113.

Dong, Y., Wang, Z., Sun, H., Yang, W., & Xu, H. (2018). The response patterns of arbuscular mycorrhizal and ectomycorrhizal symbionts under elevated CO2: a meta-analysis. *Frontiers in microbiology, 9*, 1248.

Duan, D., Feng, X., Wu, N., Tian, Z., Dong, X., Liu, H., & Chen, T. (2023). Drought eliminates the difference in root trait plasticity and mycorrhizal responsiveness of two semiarid grassland species with contrasting root systems. *International Journal of Molecular Sciences, 24*(12), 10262.

Ghorui, M., Chowdhury, S., Balu, P., & Burla, S. (2024). Arbuscular mycorrhizal inoculants and their regulatory landscape. *Heliyon, 10*(9), e30359.

Hawkins, H. J., Cargill, R. I., Van Nuland, M. E., Hagen, S. C., Field, K. J., Sheldrake, M., & Kiers, E. T. (2023). Mycorrhizal mycelium as a global carbon pool. *Current Biology, 33*(11), R560–R573.

Jeewani, P. H., Luo, Y., Yu, G., Fu, Y., He, X., Van Zwieten, L., & Xu, J. (2021). Arbuscular mycorrhizal fungi and goethite promote carbon sequestration via hyphal-aggregate mineral interactions. *Soil Biology and Biochemistry, 162*, 108417.

Jerbi, M., Labidi, S., Bahri, B. A., Laruelle, F., Tisserant, B., Jeddi, F. B., & Sahraoui, A. L. H. (2021). Soil properties and climate affect arbuscular mycorrhizal fungi and soil microbial communities in Mediterranean rainfed cereal cropping systems. *Pedobiologia, 87*, 150748.

Jia, Y. Y., Qin, W. H., Zhang, T., & Feng, G. (2023). Progress on mechanisms underlying arbuscular mycorrhizal fungi maintaining desert ecosystem stability under climate change. *Science Bulletin, 68*, 3172–3184.

Jo, I., Fei, S., Oswalt, C. M., Domke, G. M., & Phillips, R. P. (2019). Shifts in dominant tree mycorrhizal associations in response to anthropogenic impacts. *Science advances, 5*(4), eaav6358.

Laza, H. E., Acosta-Martinez, V., Cano, A., Baker, J., Mahan, J., Gitz, D., & Payton, P. (2023). Elevated [CO_2] enhances soil respiration and AMF abundance in a semiarid peanut agroecosystem. *Agriculture, Ecosystems & Environment, 355*, 108592.

Liu, L., Gunina, A., Zhang, F., Cui, Z., & Tian, J. (2023). Fungal necromass increases soil aggregation and organic matter chemical stability under improved cropland management and natural restoration. *Science of the Total Environment, 858*, 159953.

Liu, Z., Yu, Z., Song, B., Li, Y., Fang, J., Guo, Y., & Adams, J. M. (2023). Elevated CO_2 and temperature increase arbuscular mycorrhizal fungal diversity, but decrease root colonization, in maize and wheat. *Science of the Total Environment, 873*, 162321.

Nazir, A., Sarfraz, W., Allah, A., Khalid, N., Farid, M., Shafiq, M., Firdaus-E-Bareen, Rizvi, Z. F., & Naeem, N. (2023). Synergistic impact of two autochthonous saprobic fungi (*A. niger* and *T. pseudokoningii*) on the growth, ionic contents, and metals uptake in *Brassica juncea* L. and *Vigna radiata* L. under tannery solid waste contaminated soil. *International Journal of Phytoremediation, 25*(11), 1488–1500.

Parihar, M., Rakshit, A., Rana, K., Tiwari, G., & Jatav, S. S. (2020). Arbuscular mycorrhizal fungi mediated salt tolerance by regulating antioxidant enzyme system, photosynthetic pathways and ionic equilibrium in pea (Pisum sativum L.). *Biologia Futura, 71*(3), 289–300.

Reich, P. B., Sendall, K. M., Rice, K., Rich, R. L., Stefanski, A., Hobbie, S. E., & Montgomery, R. A. (2015). Geographic range predicts photosynthetic and growth response to warming in co-occurring tree species. *Nature Climate Change, 5*(2), 148–152.

Rozmoš, M., Bukovská, P., Hršelová, H., Kotianová, M., Dudáš, M., Gančarčíková, K., & Jansa, J. (2022). Organic nitrogen utilisation by an arbuscular mycorrhizal fungus is mediated by specific soil bacteria and a protist. *The ISME Journal, 16*(3), 676–685.

Saeed, H. S., Munir, M., Samra, S., Farooq, A., Mahmood, H., Ahmad, I., Akram, M. I., Iqbal, R., Elsadek, M. F., Almutairi, S. M., Ditta, A., & Daraz, U. (2025). Screening of biosurfactants producing fungi from infected *Citrus sinensis* (L.). *Polish Journal of Environmental Studies, 34*(5), 5261–5270.

Sarfraz, R., Hussain, A., Sabir, A., Fekih, I. B., Ditta, A., & Xing, S. (2019). Role of biochar and plant growth-promoting rhizobacteria to enhance soil carbon sequestration—A review. *Environmental Monitoring and Assessment, 191*, 251.

Shi, J., Wang, X., & Wang, E. (2023). Mycorrhizal symbiosis in plant growth and stress adaptation: From genes to ecosystems. *Annual Review of Plant Biology, 74*(1), 569–607.

Shukla, S., Didwania, N., & Choudhary, R. (2025). Arbuscular mycorrhizal fungi (AMF): A pathway to sustainable soil health, carbon sequestration, and greenhouse gas mitigation. *Journal of the Saudi Society of Agricultural Sciences, 24*(4), 22.

Smith, P., Poch, R. M., Lobb, D. A., Bhattacharyya, R., Alloush, G., Eudoxie, G. D., & Hallett, P. (2024). Status of the world's soils. *Annual Review of Environment and Resources, 49*(1), 73–104.

Steidinger, B. S., Crowther, T. W., Liang, J., Van Nuland, M. E., Werner, G. D., Reich, P. B., ... Hérault, B. (2019). Climatic controls of decomposition drive the global biogeography of forest-tree symbioses. *Nature, 569*(7756), 404–408.

Tao, J., & Liu, X. (2024). Does arbuscular mycorrhizal fungi inoculation influence soil carbon sequestration? *Biology and Fertility of Soils, 60*(2), 213–225.

Terrer, C., Phillips, R. P., Hungate, B. A., Rosende, J., Pett-Ridge, J., Craig, M. E., & Jackson, R. B. (2021). A trade-off between plant and soil carbon storage under elevated CO_2. *Nature, 591*(7851), 599–603.

Thind, S., Chaudhary, M. S., Ditta, A., Hussain, I., Parveen, A., Ullah, N., Mahmood, Q., Al-ashkar, I., & Sabagh, A. E. L. (2022). Impact of mycorrhizal fungi from different rhizospheric soils on fungal colonization, growth, and chlorophyll contents of *Cenchrus ciliaris*. *Agronomy, 12*(11), 2644.

Wahab, A., Muhammad, M., Munir, A., Abdi, G., Zaman, W., Ayaz, A., & Reddy, S. P. P. (2023). Role of arbuscular mycorrhizal fungi in regulating growth, enhancing productivity, and potentially influencing ecosystems under abiotic and biotic stresses. *Plants, 12*(17), 3102.

Wang, F., Zhang, L., Zhou, J., Rengel, Z., George, T. S., & Feng, G. (2022). Exploring the secrets of the hyphosphere of arbuscular mycorrhizal fungi: Processes and ecological functions. *Plant and Soil, 481*(1), 1–22.

Wu, H., Wan, S., Ruan, C., Wan, W., Han, M., Chen, G., & Wang, G. (2023). Soil microbial necromass: The state-of-the-art, knowledge gaps, and future perspectives. *European Journal of Soil Biology, 115*, 103472.

Wu, S., Fu, W., Rillig, M. C., Chen, B., Zhu, Y. G., & Huang, L. (2024). Soil organic matter dynamics mediated by arbuscular mycorrhizal fungi–an updated conceptual framework. *New Phytologist, 242*(4), 1417–1425.

Yang, K., Zhang, Q., Zhu, J., Wang, Q., Gao, T., & Wang, G. G. (2024). Mycorrhizal type regulates trade-offs between plant and soil carbon in forests. *Nature Climate Change, 14*(1), 91–97.

Yizhu, L., Imtiaz, M., Ditta, A., Rizwan, M. S., Ashraf, M., Mehmood, S., Aziz, O., Mubeen, F., Ali, M., Elahi, N. N., Ijaz, R., Lelel, S., Shuang, C., & Tu, S. (2020). Response of growth, antioxidant enzymes, and root exudates production towards as stress in *Pteris vittata* and *Astragalus sinicus* colonized by arbuscular mycorrhizal fungi. *Environmental Science and Pollution Research, 27*, 2340–2352.

Zhao, Y., Cartabia, A., Lalaymia, I., & Declerck, S. (2022). Arbuscular mycorrhizal fungi and production of secondary metabolites in medicinal plants. *Mycorrhiza, 32*(3), 221–256.

Chapter 8

Role of arbuscular mycorrhizal fungi in improving soil structure and water holding capacity

Qasim Muhammad[a], Saba Babar[a], Saima Parveen Memon[b], Muhammad Younas[c], Zeinab El-Desouki[a,d], Xiaoyang Xia[a], Jiyuan Wang[a], Xiangling Wang[a], Rashid Iqbal[e,f] and Cuncang Jiang[a]

[a] *Microelement Research Center, College of Resources and Environment, Huazhong Agricultural University, Wuhan, Hubei, China,* [b] *Key Laboratory of Cultivated Land Quality Monitoring and Evaluation, Ministry of Agriculture and Rural Affairs, College of Environmental Science and Engineering, Yangzhou University, Yangzhou, China,* [c] *Department of Crop Breeding and Genetics, Agriculture College, Yangzhou University, Yangzhou, Jiangsu, China,* [d] *Department of Soil Sciences, Faculty of Agriculture, Ain Shams University, Cairo, Egypt,* [e] *Department of Agronomy, Faculty of Agriculture and Environment, The Islamia University of Bahawalpur, Bahawalpur, Pakistan,* [f] *Department of Life Sciences, Western Caspian University, Baku, Azerbaijan*

8.1 Introduction

Arbuscular mycorrhizal fungi (AMF) form mutualistic relations with plants via the development of intracellular structures within root cortical cells called "arbuscules," enabling them to absorb and exchange essential nutrients, especially phosphorus (P), nitrogen (N), potassium (K), etc. (Ahmed et al., 2025; Lammel et al., 2025). They are among the most prominent microbes and are commonly used as biostimulants in sustainable agriculture (Jatuwong et al., 2024). They are frequently employed as biofertilizers in agroecosystems because of their provision of multiple benefits, including nitrogen (N) fixation, soil carbon (C) cycling, soil erosion control, pollutant remediation, biodiversity support, plant water regulation, and increased carbon sequestration (Ebbisa, 2023; Saeed et al., 2025). Their symbiotic relationship is critical for terrestrial ecosystems, as AMF improve soil structure, form hyphal networks to stabilize aggregates, and enhance nutrient and water movement (Leifheit et al., 2021). Moreover, AMF affect the composition of bacterial and fungal communities in soil and also strengthens plant resilience to a range of abiotic stresses (Bhupenchandra et al., 2024; Nazir et al., 2023; Thind et al., 2022). For instance, AMF enhances root hydraulic conductivity under drought, regulates osmotic balance under saline conditions, and detoxifies heavy metals under contaminated soil conditions via sequestration and chelation (Gong et al., 2023). Furthermore, AM fungus affects plant metabolism and stress response at the molecular level as symbiosis with AMF initiates transcriptional programming in plants by activating different genes associated with nutrient transport, stress adaptation, and hormonal signaling (Gao et al., 2023). All these changes capacitate the plants to counter abiotic stresses, reflecting the multifaceted benefits of using AMF in agricultural systems (Ahmed et al., 2025). AMF symbiosis functions and establishment are coordinated by intricate molecular signaling pathways that govern fungal colonization and nutrient exchange. The interaction starts upon the release of strigolactones from plant roots, which induces fungal spore germination as well as hyphal branching (Slimani et al., 2024). The significance of AMF and strigolactones in plant-microbe interactions lies in their potential to positively affect plant growth, development, and overall health (Boyno & Demir, 2022; Mitra et al., 2021). Mechanically, AMF have been reported to affect the production as well as release of strigolactones from plant roots (Yoneyama & Brewer, 2021), which serve as attractants for beneficial microorganisms, and in return, colonization and establishment of AMF within plant roots are modulated by strigolactones (Boyno et al., 2023). The significance of this scientific interaction becomes evident on account of its potential to improve plant nutrient utilization, crop productivity, and plants' resilience in the face of different environmental challenges (Badri et al., 2009; Mitra et al., 2021). The AMF spore

Advancing Sustainable Agriculture Through Arbuscular Mycorrhizal Fungi. DOI: https://doi.org/10.1016/B978-0-443-45130-0.00001-1

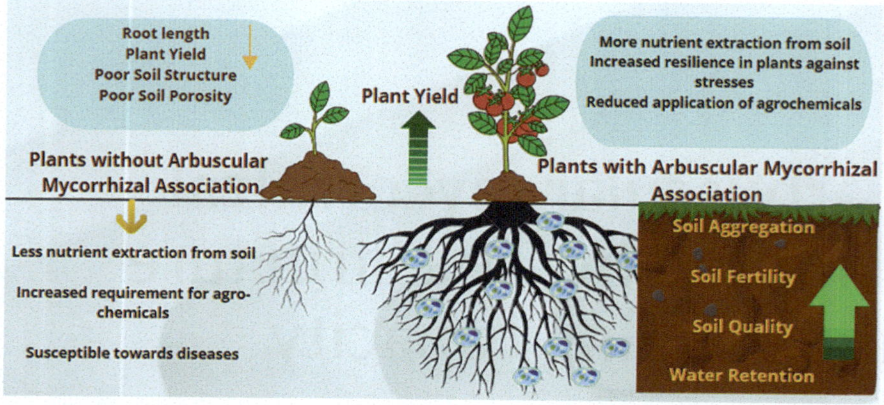

FIGURE 8.1 Effect of AMF inoculation on soil health and plant growth. *AMF*, Arbuscular mycorrhizal fungi.

dynamics exist at higher densities in rhizosphere soil, and they can solubilize nutrients from soil particles and organic matter (OM), increasing their availability for plant uptake (Wen et al., 2022). Moreover, AMF release certain enzymes to break down complex organic compounds and release essential nutrients for plants (Qin et al., 2022).

As a group of ubiquitous fungi, AMF may offer imperative roles in regulating soil multifunctionality, including primary productivity, water regulation and purification, nutrient cycling, habitat for biodiversity, pest and disease control, pollutant degradation and detoxification, and C-regulation and climate control, through a series of pathways particularly contributing to soil and plant health (Wang, Rengel, Cheng, & Shen, 2024). The hyphal network formed by AMF enhances soil structure, nutrient cycling, and water retention, contributing to prolonged sustainability (Herath et al., 2024). This makes them invaluable for sustainable agriculture as tools for reducing fertilizer application in agriculture or specific crop improvement (Kumar et al., 2025). Besides nutrient cycling, AMF play substantial roles in improving soil health, particularly in nutrient-poor environments, where they can increase plant growth and stress resilience (Enebe & Erasmus, 2023; Hu et al., 2024) (Fig. 8.1). Furthermore, AMF-mediated improved soil structure enhances the water retention capacity of soil and aeration, assisting the plants to thrive in the adverse impacts of different environmental stressors such as drought and salinity (Mwampashi et al., 2024). Hence AMF is an integral component of a sustainable agricultural system aimed at improving soil fertility, enhancing crop yields, and favoring environmental sustainability (Djotan et al., 2024; Neuenkamp et al., 2019).

Research aimed at highlighting the potential of indigenous AMF species as effective biofertilizers has demonstrated their efficacy in enhancing soil structure and improving crop yields across diverse soil conditions (Kalamulla et al., 2022). Indigenous AMF are better suited to local environments, and this adaptability contributes to healthier plants and improved nutrient cycling (Vieira et al., 2020). Recent molecular techniques, such as functional genomics, have identified specific AMF strains that improve soil structure and abiotic stress resilience (Lahlali et al., 2021). The inoculation of plants with these beneficial AMF strains can result in the development of a more resilient agricultural system capable of withstanding harsh environmental conditions such as degraded soils and water deficit conditions. Furthermore, ecological insights into AMF also play an inevitable role in ecosystem restoration and management (Yurkov et al., 2023). The current chapter highlights the beneficial aspects of AMF in terms of improving soil structure and its water retention capacity and draws a link between improved soil structure and plant growth under AMF–plant symbiosis.

8.1.1 Soil structure under arbuscular mycorrhizal fungi inoculation

Soil structure reflects the arrangement of soil particles into stable aggregates, which determine its properties and suitability for agricultural purposes. Soil health is governed by soil quality and, therefore, affects human health by increasing crop yields (Gupta, 2020). Unsustainable soil management practices such as bush burning, overgrazing, tillage practices, and continuous crop cultivation have dramatically increased the depletion rate of soil fertility (Shahzad et al., 2024). The most significant threats to soil functions on a global scale are soil erosion, excessive use of agricultural inputs, soil organic C (SOC), and nutritional imbalance (Fall et al., 2022). In this regard, the application of AMF has proved to be an ideal and ecofriendly approach for improving soil structure and its fertility status, as reported by multiple researchers (Aminzadeh et al., 2025; Souza et al., 2025). AMF application exerts a profound beneficial effect on soil structure as fungal mycelia prevail in soil in massive numbers, which have the capacity to form stable soil aggregates (de Novais et al., 2019). AMF forms soil

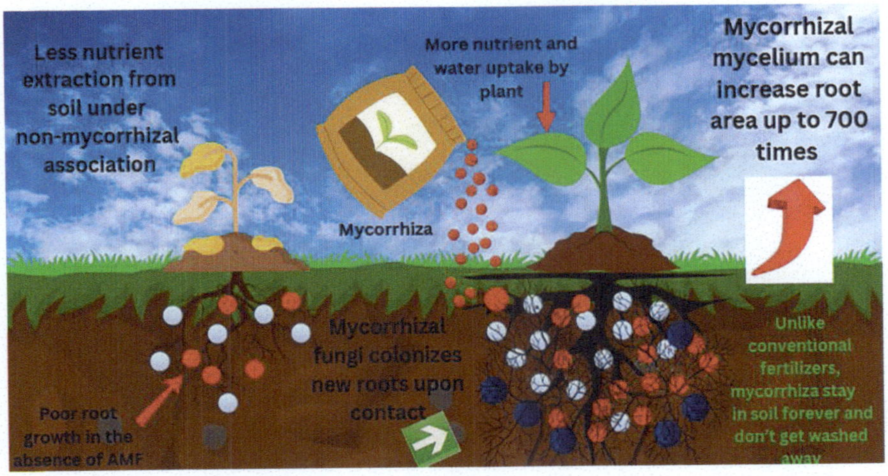

FIGURE 8.2 Effect of AMF inoculation on soil quality. *AMF*, Arbuscular mycorrhizal fungi.

microaggregates via the secretion of hydrophobic glycoproteins and glomalin, which are water-stable and serve as long-term binding agents (Fall et al., 2022). The glomalin is deposited by AMF between outer hyphal walls and adjacent soil particles to form micro- as well as macroaggregates, hence forming the backbone for soil aggregation (Jeffries et al., 2003; Li et al., 2022). The hydrophobicity of glomalin makes it suitable for conferring resistance to microaggregates against water, and the production of this substance reaches its maximum in senescent mycelia (Singh et al., 2022). Furthermore, glomalin is prone to slow degradation by bacteria and fungi in soil, and its main function is to stabilize soil aggregates (Hu et al., 2022). These stable aggregates result in better soil infiltrations; decrease surface runoff; regulate soil erosion; minimize nutrient and OM losses; increase gas exchange; and ensure better retention of nutrients, that is, potassium (K) and water, and thereby enhance crop productivity (Demenois et al., 2018; Parihar et al., 2020). Resultantly an increase in the water retention potential, soil volume, air flow, and OM level is enhanced (Kalamulla et al., 2022). This is particularly feasible and important under drought conditions, as AMF's hyphal networks increase the soil water-holding capacity, permitting the plants to maintain hydration and assisting them in thriving under water-limited conditions (Ahmed et al., 2025). Moreover, in agricultural lands, minimal soil disturbance and no-tillage operations are required to take advantage of the maintenance provided by these aggregates (Fall et al., 2022). The fungal mycelial network continuously renews itself, and dead mycelia also protect soil structure until degradation, as they contribute to stocks of OM and physical binder associated with soil aggregation (Hamel & Plenchette, 2017). All these phenomena reduce the risk of soil compaction and enhance soil fertility (Norton, 2021). Hence it can be stated that AMF enhance soil structure via biophysical or chemical mechanisms such as alignment and enmeshment (Lehmann et al., 2020). Fig. 8.2 illustrates the role of AMF in improving soil quality and compares it with a nonmycorrhizal fungus.

The AMF exhibit both positive and negative ecological rhizosphere interactions and contact positively with soil microorganisms, that is, phosphate (P) solubilizing bacteria, nitrogen (N) fixing bacteria, as well as plant growth–promoting rhizobacteria (Etesami, Jeong, & Glick, 2021; Li et al., 2023). At the same time, AMF interacts negatively with multiple soil-borne pathogens and pests to suppress them (Nanjundappa et al., 2019). All these processes are directly and indirectly involved in improving the soil structure. In addition, by ensuring soil structure and avoiding soil erosion, AMF help in maintaining the integrity of the soil profile, which is essential for supporting plant growth and minimizing the degradation of agricultural lands (Agnihotri et al., 2022). Moreover, AMF contribute to C-sequestration via production of organic acids and glomalin, which stabilize soil aggregates, protecting them against erosion (Fig. 8.3).

Li et al. (2024) performed a 7-year-old field experiment where AMF inoculation was carried out under half and full film mulching. Long-term application of AMF enhanced the SOC contents from 7.9 to 8.4 g kg^{-1} along with the mean weight diameter of aggregates (1.21–1.35), causing aggregate stability. AMFs are a crucial fungal group for C sequestration because of hyphal-mediated photosynthetic C transport, production of glomalin-related protein, and indirect function of increasing plant growth to fix more carbon dioxide (CO_2) (Morris et al., 2019). Increased soil C-storage as a result of AMF activities underscores their role in mitigating climate change (Yang et al., 2023). This increased plant–AMF symbiotic interaction is very crucial toward increasing plant C input and sustaining SOC retention in the dryland ecosystem (Ren et al., 2022). In a study conducted by Bi et al. (2025) a study employed 3D CT-scanning to examine the effect of AMF inoculation along with mixed planting on soil pore structure and its infiltration capacity in Northern China. They observed a significant

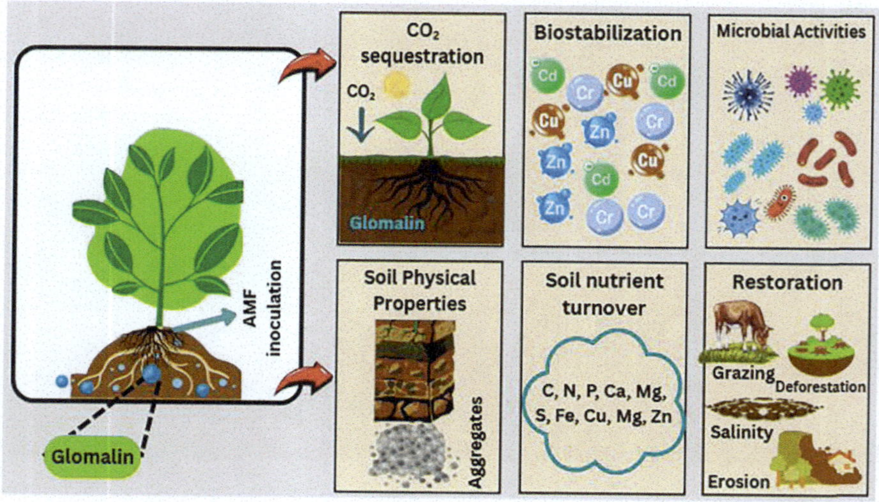

FIGURE 8.3 Functions of mycorrhizal-induced glomalin. *AMF*, Arbuscular mycorrhizal fungi.

improvement in these two attributes, as they yielded the most uniform distribution of connected pores within the soil core and displayed the highest model complexity in correlation analyses. AMF alter the physical and chemical properties of soil via the secretion of different enzymes, organic acids, and other substances, creating a conducive environment for other microorganisms (Fan et al., 2024). Multiple studies have highlighted the beneficial effects of AMF inoculation in improving the soil structure and related attributes (Al-Maliki & Ebreesum, 2020; Bao et al., 2022; Dierks et al., 2021; Pauwels et al., 2023; Yizhu et al., 2020).

In a study conducted by Qiang et al. (2023), soil samples from secondary succession, grasslands, shrubs, secondary, and primary forests were taken to draw a comparison regarding the contribution of AMF to aggregate structure. They observed that proportions of large-sized macroaggregates were enhanced by 10%, whereas small macroaggregates were reduced by 9% during vegetation succession, increasing the mean weight diameter by 17%. Furthermore the Simpson index reflecting the AMF diversity was found to be positively correlated with soil structure. They concluded that biological aggregation mechanisms in the studied region followed ecosystem succession, with AMF colonies being the critical factors influencing the soil structure. Similarly, Al-Maliki and Ebreesum (2020) strived to determine the interaction between AMF and algal biomass and their interaction effects on the C-mineralization, bacterial biomass C, proportion of large macroaggregates (>2000 μm), small macroaggregates (250–2000 μm), and microaggregates (53–250 μm) in maize. Their results showed that a combination of AMF and algae was efficient in provoking C-decomposition and aggregate formation in soil by lowering soil pH and increasing bacterial biomass C. Zheng et al. (2023) used a quantified imaging technique to evaluate the effect of AMF and fine roots on soil macropores properties such as connectivity, tortuosity, fractal dimension, and microporosity along a 55-year-old vegetation restoration in the Loess Plateau. They observed a significant enhancement in all these properties with vegetation succession. The effect of the AMF community on macropore features was indirectly mediated via root chemical properties. Their study provided useful insights into the association between AMF, plant roots, and soil structure (macropores) in a restored ecosystem. The AMF contribute indirectly to ensuring C sequestration by favoring soil aggregation by promoting soil aggregation. Glomalin acts as a long-term binding agent, depositing on the hyphal wall and adjacent soil particles, leading to soil aggregation (Morris et al., 2019). Long-term (7 years) and short-term experiments increased the labile C sequestration (Ren et al., 2021; Wilson et al., 2009). In the same way, AMF-mediated enhancement in the soil structure via the release of different substances, soil C sequestration, and improvement in the soil fertility status have also been reported (Mishra et al., 2024).

8.2 Mechanism of action

The AMF can interfere with the soil aggregation process at various levels, such as plant roots and plant communities, along with the fungal mycelium-mediated effects (Table 8.1). They have been reported to not only affect the macroaggregate formation but also the microaggregates, which are thought to be formed within the microaggregates. Hence AMF-induced improvement in the macroaggregate formation would indirectly influence the microaggregate formation. The AMF can influence the composition of plant communities via their well-established potential, that is, via the provision of several

TABLE 8.1 Effect of AMF inoculation on soil structural improvement.

Experiment	Plant	AMF strains	Effect on soil structure	References
Pot experiment	*Zea mays* L.	*Glomus mosseae* L.	A consistent increase in the aggregate formation was observed under AMF inoculation, where proportions of large macroaggregates (>2000 μm) were increased by 31.5%, macroaggregates (2000–250 μm) were increased by 40%, and microaggregates (250–53 μm) were increased by 34%, respectively	Al-Maliki and Ebreesum (2020)
Pot experiment	*Nitraria tangutorum*	*G. mosseae* L.	Significant increase in the percentage of water-stable aggregates in the >5 mm range was observed upon coinoculation of AMF and PGPR, which were also found to be significant in terms of increasing the mean weight diameter (45.67%), geometric mean diameter (57.3%), and ratio of macroaggregates (25.26%) as compared to the control	Pan et al. (2022)
Glasshouse experiment	*Urochloa brizantha*	*Acaulospora colombiana, Acaulospora longula,* and *Acaulospora morrowiae*	Mean geometric diameter of dry aggregates and wet aggregates were significantly increased upon AMF inoculation, where *A. colombiana* (59%) yielded the most significant outcomes, followed by *A. longula* (33%), as compared to the control treatment	Barbosa et al. (2019)
Field experiment	*Taxodium "zhongshan-shan"*	*Funneliformis mosseae*	Soil capillary porosity upon addition of AMF was increased by 3.17% at soil depth of 0–20 cm. In addition the fractal dimensions of soil aggregates was significantly reduced by 19.06% at the same soil depth and by 13.71% in the 40–60 cm layer	Zheng et al. (2023)
Field experiment	*Z. mays/Glycine max* intercropping	*Glomus_f_Glomeraceae, Paraglomus,* and *unclassified-c-Glomeromycetes*	Intercropping enhanced the AMF population and diversity in the studied area, which, in turn, increased the stability of soil aggregates and favored the formation of larger sized aggregates	Zhang et al. (2022)
Pot experiment	*Bidens parviflora*	*F. mosseae*	AMF inoculation significantly enhanced the percentage of soil particles >2 mm, 2–1 mm, and 1–0.25 mm	Li et al. (2022)
Pot experiment	*Oryza sativa* L.	–	More soil aggregate stability was observed in AMF-inoculated plants, at 5 and 3.2%, respectively, in sterilized and nonsterilized soils, and was found to positively correlate with spore formation and AMF-induced total glomalin production	Syamsiyah et al. (2018)
Field experiment	*Citrus sinensis*	*Diversispora spurca* and *Diversispora versiformis*	AMF inoculation enhanced the distribution of water-stable aggregates, mainly in the size range between 2 and 4 mm, and at the same time, enhanced the production of glomalin-related soil proteins, which were found to be positively correlated with aggregate stability	Cheng et al. (2022)
Greenhouse experiment	*Pennisetum glaucum*	*Rhizophagus aggregatus* and *F. mosseae*	Positive correlation between AMF inoculation and soil aggregation was observed regardless of the temperature treatment, as evident by the increased ratio of root-adhering soil to root biomass (MAS/RB)	Ndeko et al. (2022)

(continued on next page)

TABLE 8.1 Effect of AMF inoculation on soil structural improvement—cont'd

Experiment	Plant	AMF strains	Effect on soil structure	References
Pot experiment	*F. mosseae, Glomus versiforme*, and *Rhizophagus intraradices*	*Toona sinensis* and *Delavaya toxocarpa*	The percentage of water-stable aggregates (>0.25 mm) in the rhizosphere soil of inoculated seedlings was significantly enhanced by 15.6–20.5. Furthermore an increase in the quantity of aggregates having a diameter > 2 mm, concomitant with a decline in the aggregates having a size between 0.25 and 0.5 mm, was observed under AMF inoculation	Zhang et al. (2019)
Field experiment	*Elymus nutans, Leymus secalinus, Roegneria thoroldiana*, and *Poa crymophila*	*Glomeromycetes*	Extraradical mycelium was observed to be positively correlated with soil aggregates > 4 mm and negatively correlated with aggregates 0.25–0.038 mm	Qin et al. (2020)
Growth chamber experiment	*Astragalus adsurgens*	*R. intraradices (BEG141)* and *F. mosseae (BEG12)*	Significant increment in the water-stable aggregates with a simultaneous decline in the macroaggregate disruption and microaggregate dispersion rate, which were attributed to hyphal length and glomalin-related soil protein	Ji et al. (2024)
Field experiment	Six-stage vegetation succession from grasslands and shrublands to primary forest	–	The proportion of large macroaggregates was enhanced by 10%, and at the same time, the proportion of small microaggregates reduced by 9%, thereby increasing the MWD by 17%. In addition, the Simpson index for AMF exhibited a positive linear relationship with soil structure	Qiang et al. (2023)
Pot experiment	*Poncirus trifoliata*	*G. mosseae, G. versiforme*, and *G. diaphanum*	AMF inoculation enhanced the fractions of water-stable aggregates (>2, 1–2, and >0.25 mm) but reduced the fraction of water-stable aggregates (0.25–0.5 mm). Positive correlation between hyphal length densities and water-stable aggregates explained this increase	Wu and Zou (2009)
Pot experiment	*P. trifoliata*	*D. versiformis*	Significantly higher percentage of water-stable aggregates in 0.25–0.50, 0.50–1.00, and 1–2 mm with a simultaneous reduction in the proportion of 2–4 mm aggregates was observed. Furthermore, high aggregate stability in terms of MWD under saline and nonsaline conditions was also pragmatic	Zhang et al. (2018)
Pot experiment	*Olea europaea*	*Rhizophagus irregularis, Septoglomus deserticola*, and *Gigaspora gigantea*	Only inoculation of *R. irregularis* increased the percentage of water-stable aggregates, with a corresponding increase in hyphal length by 81% to the control	Kohler et al. (2017)

AMF, arbuscular mycorrhizal fungi; *MWD*, mean weight diameter; *PGPR*, plant growth-promoting rhizobacteria.

benefits to their hosts (Klironomos et al., 2000). Alterations in the plant community composition can exert their effects on the soil structure. AMF diversity has also shown significant contributions in controlling the productivity of plant communities owing to their effects on community composition (Van der Heijden et al., 1998). Similarly, plant roots have also been reportedly involved in affecting soil structure mainly via root penetration, alteration in soil water regime, root decomposition, rhizodeposition, and root entanglement of soil particles (Six et al., 2004), and mycorrhizae can influence each of these processes. Mycorrhizal infection can lead to morphological changes in roots, and roots in turn can create shear stress that may reach up to 2 MPa (Goss et al., 1993), resulting in localized soil compression and reorientation of clay particles around root surfaces. Ultimately, the spatial constraints on microaggregate formation are eliminated (Six et al., 2004). Moreover, soil water content also affects the soil structure, as decreased water contents enhance contact points between soil organic matter and primary particles, causing enhanced soil strength and cohesion (Horn et al., 1994). The binding between root exudates and clay particles is improved via localized soil drying in close proximity to plant roots (Reid & Goss, 1982), which facilitates microaggregate formation. Mycorrhizal fungi not only affect overall plant growth but also exhibit different water relations from their nonmycorrhizal counterparts (Augé, 2004). For instance, higher stomatal conductance and transpiration can take place in the presence of mycorrhizae, exploring more water, causing more extreme dry/wet cycles, and leading to stronger impacts on soil aggregation (Duan et al., 1996). In addition, more carbon (C) fixation during photosynthesis may increase C inputs to the soil, causing more organic content in the soil and improved soil structure (Rillig & Mummey, 2006). Rhizodeposition, the release of exudates from living roots, can be significantly impacted by mycorrhizae, as they can affect C metabolism of plants. These root exudates supply much of the C known to stimulate aggregate formation; that is, root mucilage can assist in particle sticking, causing short-term stabilization of soil aggregates. Similarly, C addition to soil can stimulate the microbial activities and give rise to the rhizosphere, which, in turn, can contribute significantly toward aggregate formation. Furthermore, root decomposition also adds C to the soil, causing soil aggregation. Mycorrhizal colonization can interfere with the root chemistry and thereby change the below-ground litter quality (Langley & Hungate, 2003), which can affect the rate and nature of decomposition products.

8.1.2 Soil water retention and arbuscular mycorrhizal fungi

AMF affect plant water relations, and there exists sufficient evidence from numerous studies regarding the role of AMF in improving the soil water contents (SWC) and conferring drought tolerance in crops (Pauwels et al., 2023). However, drought tolerance is a relative term, and therefore mycorrhizal symbiosis can be better illustrated as an emergent property for enhanced water scavenging from dried soils (Table 8.2). AMF assist the plant in absorbing water and nutrients, and such a symbiotic relationship dates back 400 million years (Khaliq et al., 2022; Selosse et al., 2016). Soil water retention reflects the relationship between soil water potential and SWC and, in association with the hydraulic conductivity of soil, sets limits to plant water extraction from soil (Carminati et al., 2020; Kakouridis et al., 2022). Both these attributes rely on soil texture, structure, and wetting traits of the soil's solid phase. According to the United States Department of Agriculture (USDA), soil texture indicates the relative proportions of sand, silt, and clay in soils (Irshad et al., 2024). Soil texture restrains soil water retention, whereas hydraulic conductivity, as particle size, gives rise to pore size (Cui et al., 2024). Soil hydraulic conductivity is affected when AMF grow in soil pores that were previously devoid of it. The AMF inoculation significantly improves soil structure, and this structural improvement not only increases the soil permeability but also increases its water retention abilities and, at the same time, creates conducive environments for root growth, resulting in their increased penetration and expansion (Gong et al., 2024). Furthermore, AMF-mediated restructuring of soil particles and biochemical stabilizing of soil aggregates via the release of root exudates increase the soil permeability as well as water retention capacity (Du et al., 2024). In addition, the formation of large-sized macropores under the influence of AMF also improves the water infiltration capacity of soil (Fayuan and Zed, 2024). Improvement in the root morphology under AMF inoculation establishes a structural foundation for pore formation in compact soils, and in arid to semiarid regions, this development of an intricate 3D root network stabilizes the soil matrix and improves soil water retention (Hu et al., 2024). Increased water use efficiency and improved soil water retention under AMF inoculation also assist the plants in thriving in the water-deficient conditions (Bokhari et al., 2023; Chen et al., 2023). Extension of hyphae into soil horizons helps the plant roots to explore water from deeper soil layers (Jabborova et al., 2022). Glomalin produced by AMF activities enhances the soil aggregate formation, and resultantly, water retention is improved, which is particularly vital during osmotic stress, as it permits the soil to conserve moisture for a longer period of time, ensuring the provision of a water reservoir for plant uptake (Fig. 8.4). This increased water-holding capacity aids in maintaining soil moisture levels and ensuring sustained and prolonged water supply to plants, lessening the effects of osmotic stress (Alotaibi et al., 2023).

Soil pore structure is a crucial factor determining water retention, solute transport, and gas exchange (Fatichi et al., 2020). A well-structured soil pore system ensures increased soil moisture availability, nutrient retention, and root proliferation,

TABLE 8.2 Effect of AMF inoculation on soil water retention capacity.

Experiment	Plant	AMF strains	Effect on soil water-holding capacity	References
Pot experiment	*Solanum lycopersicum*	*Rhizoglomus irregulare*	AMF inoculation reduced saturated water contents and total porosity but enhanced the air-filled porosity. Plant available water contents were positively correlated with soil protein contents, which in turn were highest in pots containing strong hyphal colonization	Bitterlich et al. (2018)
Glasshouse experiment	*Zea mays* L.	*Rhizophagus irregularis*	Strong influence of AMF inoculation on soil available water contents was observed only under low soil water contents and was found to be increased by 7.7-fold in sandy soil as compared to water retention in loam soil	Pauwels et al. (2023)
Greenhouse experiment	*Oryza sativa* L.	*Glomus claroideum*, *Glomus fasciculatum*, *Funneliformis mosseae*, *Glomus etunicatum*, and *Aucolospora rogusa*	Reduction in the soil bulk density by 38% along with a simultaneous increment in the soil porosity by 61% under the combined application of AMF and N fertilizer was pragmatic. Increased porosity under AMF inoculation was attributed to enhanced soil water retention	Mulyadi and Jiang (2023)
Field experiment	*Amorpha fruticosa*	*F. mosseae*	AMF inoculation, along with mixed planting, significantly enhanced the soil porosity and root development. Furthermore, most uniform distribution of connected pores within soil cores, along with increased fractal dimensions and permeability to ensure increased water retention	Bi et al. (2025)
Greenhouse experiment	*Z. mays* L.	*Rhizophagus intraradices*	Increased water uptake by different plant parts under AMF-inoculated treatment was observed, which was correlated with AMF-mediated alterations in the soil water retention properties	Zhao et al. (2015)
Glasshouse experiment	*Solanum tuberosum*	*Glomus* sp.	An increase in the total pore percentage by 0.38% and soil water availability by 12% was observed upon AMF inoculation as compared to the control	Nurbaity et al. (2013)
Macrocosm study	*Plantago lanceolata*	*Glomus geosporum*, *Glomus mosseae*, and *Glomus intraradices*	Increased soil porosity by 12.89% as compared to the control treatment enhanced the overall water retention, which, in turn enhanced the water sorptivity by soil aggregates	Martin et al. (2012)
Field experiment	Different tree and shrub species	*Gigaspora margarita*, *Acaulospora scrobiculata*, and *G. etunicatum*	Balanced distribution of macro- (25%) and micropores (24%) was observed in the presence of AMF, and the percentage contribution of microporosity toward total porosity was 36.6, which enhanced the soil water retention capacity	Kimura and Scotti (2016)
Growth chamber study	*S. lycopersicum*	*R. irregularis*	AMF colonization regulated leaf water potential in tomato plants consistent with increased transpiration rates. This increase was correlated with AMF's potential to enhance effective root radius, increasing the soil–plant hydraulic conductivity and decreasing the water flux at the root–soil interface	Abdalla and Ahmed (2021)
Greenhouse experiment	*S. lycopersicum*	*R. irregularis*	Increased root hydraulic properties assisted the plant roots in drawing water from larger soil volumes, and at the same time, increased expression of aquaporins facilitated the water uptake	David et al. (2024)

AMF, arbuscular mycorrhizal fungi.

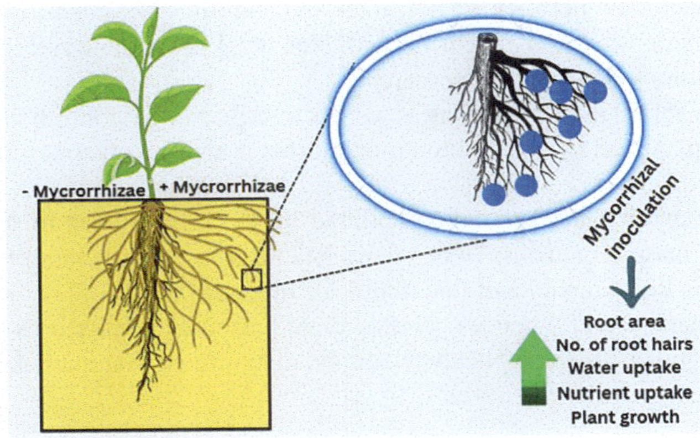

FIGURE 8.4 Mycorrhizal inoculation and soil water-holding capacity.

ultimately causing prolonged stability of the ecosystem (Phalempin et al., 2022). AMF establishes a mutualistic relationship with the roots of higher terrestrial plants and thereby results in enhanced water and nutrient uptake through their specialized mycorrhizal structures (Keyes et al., 2022). An extensive mycelium of AMF penetrates micropores and widens water transport pathways, improves soil permeability, reduces water loss, and enhances soil water retention capacity (Tang et al., 2024; Yao et al., 2024). In a study conducted by Bi et al. (2025) on the effect of AMF inoculation on root morphology of *Amorpha fruticosa*, soil pore structure, and soil infiltration rate were checked. They reported that AMF inoculation significantly enhanced soil porosity and plant root development. Furthermore, AMF in association with mixed planting resulted in the most uniform number of connected pores within soil cores and also maximized fractal dimensions as well as permeability with a concomitant reduction in tortuosity and increment in connectivity. They concluded that the application of AMF in synergy with mixed planting practices significantly optimized soil pore architecture and increased the infiltration dynamics, predominantly via their stimulatory effects on the plant root system. The relationship between AMF and plants is also important in lieu of improving plant growth and drought resilience under water deficit conditions (Hashem et al., 2025; Wu & Zou, 2009). Glomalin produced by AMF serves as super glue to increase soil aggregation its structure, as well as water-holding capacity (Behrooz et al., 2019; Gupta, 2020). Chen et al. (2020) performed a pot experiment to check the effect of AMF species *Rhizophagus intraradices* inoculation in improving the drought tolerance in the ornamental plant *Catalpa bungei*. They observed a significant enhancement in the overall growth and yield attributes of the plant under drought stress by increasing the root mass and root/shoot ratio as well as root morphology, which helped the plants to take up essential nutrients, improve their physiological matrices, and ultimately increase their yield under drought stress conditions. Similarly, Du et al. (2024) observed a 5.77%–9.66% increment in the SWC in wheat, which was grown in the presence of AMF *Funneliformis mosseae* inoculation under drought stress, ultimately enhancing the grain yield of wheat by 28.5% under water deficit conditions. However, they did not find any significant variation in these attributes under well-watered conditions.

8.3 Future prospects

The mycorrhizae seem to be one of the most important microorganisms to take into account, as they are involved in soil structural stability and mineral and nutrient acquisition as well as protection from different abiotic and biotic constraints. Regardless of the well-established importance of mycorrhizae in improving soil health and fertility, our understanding of the key underlying mechanisms is limited. Regarding the role of mycorrhizae in improving soil structure and water retention capacity, the current chapter summarizes the beneficial roles played by mycorrhizae in improving overall soil health and highlights various mechanisms involved in fungi-mediated soil health improvements. Future studies should take into account the following aspects.

- The role of glomalin in improving soil aggregation and structure under various climate and soil types should be considered to expedite its use in resolving soil degradation issues that are reported to be worsened by prevailing climate scenarios.
- Similarly, assessing the life span and accumulation of glomalin in soil under different climate, management, and land-use conditions is also of crucial significance.

- It is well-reported that the interaction between mycorrhizae and soil microorganisms is very important with respect to soil fertility. However, the interaction of mycorrhizae with beneficial nematodes and their impacts on soil health and structure under stressed conditions needs further research.
- Furthermore the role of mycorrhizae in the synthesis as well as transport of phytohormones such as auxins, gibberellins, and cytokinins from plant to plant and toward soil microorganisms is also poorly understood, and hence, its exploration is of paramount importance.
- Advancements in our understanding of the functions offered by the mycorrhizae in soil are partially limited by the obligate biotrophic nature of these organisms. Hence there is a need to conduct more field experiments regarding the impact of plant biodiversity on the diversity and functioning of mycorrhizae.
- Regarding the agricultural management practices, studies should also be conducted to investigate the impact of these soil practices, such as tillage and chemical applications, on the mycorrhizal functioning and performance in soil.

8.4 Conclusion

AMF appears to be one of the most important soil organisms owing to its symbiotic relationship with roots of higher plants. They are involved in plant mineral nutrition and water absorption and, at the same time, improve soil structure, aggregate stability, soil water retention, soil porosity, and hydraulic conductivity. All these processes are directly and indirectly involved in supporting plant growth. Through their complex hyphal system, AMF exude water-stable hydrophobic glycoproteins and glomalin, which serve as long-term binding agents and promote soil aggregation. Furthermore, by increasing the number of pores and decreasing their tortuosity, AMF help to enhance the soil water retention capacity and assist the plant to survive even under water-stressed conditions. In addition, AMF-mediated root growth helps the plants to explore farther areas for nutrients and moisture and, in turn, helps them to grow under diverse soil conditions. In current document, we highlighted the beneficial role of AMF in improving soil structure, aggregation, and soil moisture-holding capacity and emphasized their potential to support plant growth via improving these soil attributes. Future research should focus on the molecular mechanisms behind interactions of glomalin and soil particles, which can provide insights into their adhesive properties and their associated contribution toward soil stability. Furthermore, the indirect role of glomalin and other glycoproteins in improving the soil porosity via increased aggregation needs to be further exploited, and the role of soil texture should also be identified in this regard.

References

Abdalla, M., & Ahmed, M. A. (2021). Arbuscular mycorrhiza symbiosis enhances water status and soil-plant hydraulic conductance under drought. *Frontiers in Plant Science, 12*, 722954.

Agnihotri, P., Maitra, M., & Mitra, A. (2022). Isolation, characterization and identification of an as (V)-resistant plant growth promoting rhizobacterium associated with the rhizosphere of *Azolla microphylla*. *Journal of Microbiology, Biotechnology and Food Sciences, 12*(2), e4728.

Ahmed, N., Li, J., Li, Y., Deng, L., Deng, L., Chachar, M., Chachar, Z., Chachar, S., Hayat, F., Raza, A., Umrani, J. H., Gong, L., & Tu, P. (2025). Symbiotic synergy: How arbuscular mycorrhizal fungi enhance nutrient uptake, stress tolerance, and soil health through molecular mechanisms and hormonal regulation. *IMA Fungus, 16*, e144989.

Al-Maliki, S., & Ebreesum, H. (2020). Changes in soil carbon mineralization, soil microbes, roots density and soil structure following the application of the arbuscular mycorrhizal fungi and green algae in the arid saline soil. *Rhizosphere, 14*, 100203. doi:10.1016/j.rhisph.2020.100203.

Alotaibi, M. O., Ikram, M., Alotaibi, N. M., Hussain, G. S., Ghoneim, A. M., Younis, U., Naz, N., & Danish, S. (2023). Examining the role of AMF-biochar in the regulation of spinach growth attributes, nutrients concentrations, and antioxidant enzymes in mitigating drought stress. *Plant Stress, 10*. www.journals.elsevier.com/plant-stress. doi:10.1016/j.stress.2023.100205.

Aminzadeh, A., Dorostkar, V., & Asghari, H. R. (2025). Soil structural stability improvement using arbuscular mycorrhizal fungi and biochar in water repellent and non-water repellent soil. *Soil Use and Management, 41*(1), e70024.

Augé, R. M. (2004). Arbuscular mycorrhizae and soil/plant water relations. *Canadian Journal of Soil Science, 84*(4), 373–381. http://pubs.aic.ca/toc/cjss/current. doi:10.4141/S04-002.

Badri, D. V., Weir, T. L., Van der Lelie, D., & Vivanco, J. M. (2009). Rhizosphere chemical dialogues: Plant-microbe interactions. *Current Opinion in Biotechnology, 20*(6), 642–650. doi:10.1016/j.copbio.2009.09.014.

Bao, X., Zou, J., Zhang, B., Wu, L., Yang, T., & Huang, Q. (2022). Arbuscular mycorrhizal fungi and microbes interaction in rice mycorrhizosphere. *Agronomy, 12*(6), 1277. doi:10.3390/agronomy12061277.

Barbosa, M. V., Pedroso, D. D. F., Pinto, F. A., Santos, J. V. D., & Carneiro, M. A. C. (2019). Arbuscular mycorrhizal fungi and Urochloa brizantha: Symbiosis and spore multiplication. *Pesquisa Agropecuária Tropical, 49*, e54530.

Behrooz, A., Vahdati, K., Rejali, F., Lotfi, M., Sarikhani, S., & Leslie, C. (2019). Arbuscular mycorrhiza and plant growth-promoting bacteria alleviate drought stress in walnut. *HortScience, 54*(6), 1087–1092. doi:10.21273/hortsci13961-19.

Bhupenchandra, I., Chongtham, S. K., Devi, A. G., Dutta, P., Sahoo, M. R., Mohanty, S., Kumar, S., Choudhary, A. K., Devi, E. L., Sinyorita, S., Devi, S. H., Mahanta, M., Kumari, A., Devi, H. L., Josmee, R. K., Pusparani, A., Pathaw, N., Gupta, S., Meena, M., … Swapnil, P. (2024). Unlocking the potential of arbuscular mycorrhizal fungi: Exploring role in plant growth promotion, nutrient uptake mechanisms, biotic stress alleviation, and sustaining agricultural production systems. *Journal of Plant Growth Regulation*. https://www.springer.com/journal/344. doi:10.1007/s00344-024-11467-9.

Bi, Y., Tian, L., Du, X., & Yin, K. (2025). Optimizing soil pore structure in mined land: Integrating arbuscular mycorrhizal fungi and mixed planting for ecological restoration. *Plant and Soil*. doi:10.1007/s11104-025-07666-2.

Bitterlich, M., Franken, P., & Graefe, J. (2018). Arbuscular mycorrhiza improves substrate hydraulic conductivity in the plant available moisture range under root growth exclusion. *Frontiers in Plant Science, 9*, 301.

Bokhari, S. S., Farhat, H., Ali, S. A., Urooj, F., Rahman, A., Ara, J., Irfan, M., & Ehteshamul-Haque, S. (2023). Role of mycorrhizospheric fluorescent *Pseudomonas* in suppressing the root rot disease, enhancement of vesicular arbuscular mycorrhizal (VAM) population and phosphorus uptake in sunflower. *Pakistan Journal of Botany, 55*(2), 779–790. http://www.pakbs.org/pjbot/. doi:10.30848/PJB2023-2(36).

Boyno, G., Danesh, Y. R., Demir, S., Teniz, N., Mulet, J. M., & Porcel, R. (2023). The complex interplay between arbuscular mycorrhizal fungi and strigolactone: Mechanisms, sinergies, applications and future directions. *International Journal of Molecular Sciences, 24*(23), 16774. doi:10.3390/ijms242316774.

Boyno, G., & Demir, S. (2022). Plant-mycorrhiza communication and mycorrhizae in inter-plant communication. *Symbiosis, 86*(2), 155–168. doi:10.1007/s13199-022-00837-0.

Carminati, A., Ahmed, M. A., Zarebanadkouki, M., Cai, G., Lovric, G., & Javaux, M. (2020). Stomatal closure prevents the drop in soil water potential around roots. *New Phytologist, 226*(6), 1541–1543. http://onlinelibrary.wiley.com/journal/10.1111/(ISSN)1469-8137. doi:10.1111/nph.16451.

Chen, L., Wang, F., Zhang, Z., Chao, H., He, H., Hu, W., Zeng, Y., Duan, C., Liu, J., & Fang, L. (2023). Influences of arbuscular mycorrhizal fungi on crop growth and potentially toxic element accumulation in contaminated soils: A meta-analysis. *Critical Reviews in Environmental Science and Technology, 53*(20), 1795–1816. doi:10.1080/10643389.2023.2183700.

Chen, W., Meng, P., Feng, H., & Wang, C. (2020). Effects of arbuscular mycorrhizal fungi on growth and physiological performance of *Catalpa bungei* C.A.Mey. under drought stress. *Forests, 11*(10), 1117. doi:10.3390/f11101117.

Cheng, X. F., Xie, M. M., Li, Y., Liu, B. Y., Liu, C. Y., Wu, Q. S., & Kuča, K. (2022). Effects of field inoculation with arbuscular mycorrhizal fungi and endophytic fungi on fruit quality and soil properties of Newhall navel orange. *Applied Soil Ecology, 170*, 104308.

Cui, L. X., Cheng, Q., So, P. S., Tang, C. S., Tian, B. G., & Li, C. Y. (2024). Relationship between root characteristics and saturated hydraulic conductivity in a grassed clayey soil. *Journal of Hydrology, 645*. https://www.sciencedirect.com/science/journal/00221694. doi:10.1016/j.jhydrol.2024.132231.

Demenois, J., Carriconde, F., Bonaventure, P., Maeght, J. L., Stokes, A., & Rey, F. (2018). Impact of plant root functional traits and associated mycorrhizas on the aggregate stability of a tropical Ferralsol. *Geoderma, 312*, 6–16. www.elsevier.com/inca/publications/store/5/0/3/3/3/2. doi:10.1016/j.geoderma.2017.09.033.

David, P., Jana, R., Radka, S., Jan, J., & Michael, B. (2024). Soil compaction reversed the effect of arbuscular mycorrhizal fungi on soil hydraulic properties. *Mycorrhiza, 34*(4), 361–368.

de Novais, C. B., de Oliveira, J. R., Siqueira, J. O., de Faria, S. M., da Silva, E. M. R., Aquino, A. M., & Saggin Júnior, O. J. (2019). Trophic relationships between the earthworm *Pontoscolex corethrurus* and three tropical arbuscular mycorrhizal fungal species. *Applied Soil Ecology, 135*, 9–15. www.elsevier.com/inca/publications/store/5/2/4/5/1/8/index.htt. doi:10.1016/j.apsoil.2018.11.001.

Dierks, J., Blaser-Hart, W. J., Gamper, H. A., Nyoka, I. B., Barrios, E., & Six, J. (2021). Trees enhance abundance of arbuscular mycorrhizal fungi, soil structure, and nutrient retention in low-input maize cropping systems. *Agriculture, Ecosystems and Environment, 318*. www.elsevier.com/inca/publications/store/5/0/3/2/9/8. doi:10.1016/j.agee.2021.107487.

Djotan, A. K. G., Matsushita, N., & Fukuda, K. (2024). Year-round dynamics of arbuscular mycorrhizal fungi communities in the roots and surrounding soils of *Cryptomeria japonica*. *Mycorrhiza, 34*(1), 119–130.

Du, Y., Yang, H., Tang, J., Hu, L., & Chen, X. (2024). Arbuscular mycorrhizal fungi affect root morphology locally but not systemically through altering nutrient- and phytohormone-related gene expressions at low soil P level. *Plant and Soil*. doi:10.1007/s11104-024-07150-3.

Duan, H. X., Luo, C. L., Zhou, R., Zhao, L., Zhu, S. G., Chen, Y., & Xiong, Y. C. (2024). AM fungus promotes wheat grain filling via improving rhizospheric water & nutrient availability under drought and low density. *Applied Soil Ecology, 193*, 105159.

Duan, X., Neuman, D. S., Reiber, J. M., Green, C. D., Saxton, A. M., & Augé, R. M. (1996). Mycorrhizal influence on hydraulic and hormonal factors implicated in the control of stomatal conductance during drought. *Journal of Experimental Botany, 47*(303), 1541–1550. http://jxb.oxfordjournals.org/. doi:10.1093/jxb/47.10.1541.

Ebbisa, A. (2023). *Arbuscular mycorrhizal fungi (AMF) in optimizing nutrient bioavailability and reducing agrochemicals for maintaining sustainable agroecosystems*. IntechOpen. doi:10.5772/intechopen.106995.

Enebe, M. C., & Erasmus, M. (2023). Symbiosis—A perspective on the effects of host traits and environmental parameters in arbuscular mycorrhizal fungal richness, colonization and ecological functions. *Agriculture, 13*(10), 1899.

Etesami, H., Jeong, B. R., & Glick, B. R. (2021). Contribution of arbuscular mycorrhizal fungi, phosphate–solubilizing bacteria, and silicon to P uptake by plant. *Frontiers in Plant Science, 12*, 699618.

Fall, A. F., Nakabonge, G., Ssekandi, J., Founoune-Mboup, H., Apori, S. O., Ndiaye, A., Badji, A., & Ngom, K. (2022). Roles of arbuscular mycorrhizal fungi on soil fertility: Contribution in the improvement of physical, chemical, and biological properties of the soil. *Frontiers in Fungal Biology, 3*. www.frontiersin.org/journals/fungal-biology. doi:10.3389/ffunb.2022.723892.

Fan, L., Zhang, P., Cao, F., Liu, X., Ji, M., & Xie, M. (2024). Effects of AMF on maize yield and soil microbial community in sandy and saline soils. *Plants, 13*(15), 2056. doi:10.3390/plants13152056.

Fatichi, S., Or, D., Walko, R., Vereecken, H., Young, M. H., Ghezzehei, T. A., Hengl, T., Kollet, S., Agam, N., & Avissar, R. (2020). Soil structure is an important omission in Earth system models. *Nature Communications, 11*(1). http://www.nature.com/ncomms/index.html. doi:10.1038/s41467-020-14411-z.

Fayuan, W., & Zed, R. (2024). Disentangling the contributions of arbuscular mycorrhizal fungi to soil multifunctionality. *Pedosphere, 34*(2), 269–278. doi:10.1016/j.pedsph.2023.12.015.

Gao, Y., An, T., Kuang, Q., Wu, Y., Liu, S., Liang, L., Yu, M., Macrae, A., & Chen, Y. (2023). The role of arbuscular mycorrhizal fungi in the alleviation of cadmium stress in cereals: A multilevel meta-analysis. *Science of the Total Environment, 902*, 166091. doi:10.1016/j.scitotenv.2023.166091.

Gong, M., Bai, N., Wang, P., Su, J., Chang, Q., & Zhang, Q. (2023). Co-inoculation with arbuscular mycorrhizal fungi and dark septate endophytes under drought stress: Synergistic or competitive effects on maize growth, photosynthesis, root hydraulic properties and aquaporins? *Plants, 12*(14), 2596. doi:10.3390/plants12142596.

Gong, Y., & Qian, Y. (2024). Deriving an unsaturated hydraulic conductivity function based on ergun equation. *IOP Conference Series: Earth and Environmental Science, 1334*(1), 012006.

Goss, M. J., Miller, M. H., Bailey, L. D., & Grant, C. A. (1993). Root growth and distribution in relation to nutrient availability and uptake. *European Journal of Agronomy, 2*(2), 57–67. doi:10.1016/s1161-0301(14)80135-4.

Gupta, M. M. (2020). *Arbuscular mycorrhizal fungi: The potential soil health indicators* (pp. 183–195). Springer Science and Business Media LLC. doi:10.1007/978-3-030-44364-1_11.

Hamel, C., & Plenchette, C. (2017). Implications of past, current, and future agricultural practices for mycorrhiza-mediated nutrient flux. In *Mycorrhizal mediation of soil: Fertility, structure, and carbon storage* (pp. 175–186). Elsevier BV. doi:10.1016/b978-0-12-804312-7.00010-3.

Hashem, A., Almutairi, K. F., Alshaikh, N. A., Kumar, A., Wu, Q. S., Ibramimova, U., & Abd-Allah, E. F. (2025). Role of arbuscular mycorrhizal fungi in plant growth promotion and biotic stress management. In *Management of mycorrhizal symbiosis for mycoremediation and phytostabilization* (pp. 145–155). Elsevier. https://www.sciencedirect.com/book/9780443248634. doi:10.1016/B978-0-443-24863-4.00008-X.

Herath, B. M., Kalamulla, K. W. Y. R., Mayadunna, T. A. N., Perera, M. A. E. G., Jayamanna, I. K., & Yapa, P. N. (2024). Applications of arbuscular mycorrhizal fungi for sustainable agricultural systems. *Arbuscular mycorrhizal fungi and higher plants: Fundamentals and applications*, 319–340. https://link.springer.com/book/9789819982202. doi:10.1007/978-981-99-8220-2_14.

Horn, R., Taubner, H., Wuttke, M., & Baumgartl, T. (1994). Soil physical properties related to soil structure. *Soil and Tillage Research, 30*(2–4), 187–216. doi:10.1016/0167-1987(94)90005-1.

Hu, J., Li, S., Zhang, Y., Du, D., & Zhu, X. (2024). Potential regulatory effects of arbuscular mycorrhizal fungi on lipid metabolism of maize in response to low-temperature stress. *Journal of Agricultural and Food Chemistry*. doi:10.1021/acs.jafc.4c06908.

Hu, Y., Pandey, A. K., Wu, X., Fang, P., & Xu, P. (2022). The role of arbuscular mycorrhiza fungi in drought tolerance in legume crops: A review. *Legume Research—An International Journal, 45*. doi:10.18805/lrf-660.

Jabborova, D., Annapurna, K., Azimov, A., Tyagi, S., Pengani, K. R., Sharma, P., Vikram, K. V., Poczai, P., Nasif, O., Ansari, M. J., & Sayyed, R. Z. (2022). Co-inoculation of biochar and arbuscular mycorrhizae for growth promotion and nutrient fortification in soybean under drought conditions. *Frontiers in Plant Science, 13*. https://www.frontiersin.org/journals/plant-science. doi:10.3389/fpls.2022.947547.

Jatuwong, K., Aiduang, W., Kiatsiriroat, T., Kamopas, W., & Lumyong, S. (2024). Effects of biochar and arbuscular mycorrhizal fungi on soil health in Chinese Kale (*Brassica oleracea* var. *alboglabra* L.) Cultivation. *Microbiology Research, 15*(1), 404–421. doi:10.3390/microbiolres15010027.

Jeffries, P., Gianinazzi, S., Perotto, S., Turnau, K., & Barea, J. M. (2003). The contribution of arbuscular mycorrhizal fungi in sustainable maintenance of plant health and soil fertility. *Biology and Fertility of Soils, 37*(1), 1–16. link.springer.de/link/service/journals/00374/index.htm. doi:10.1007/s00374-002-0546-5.

Ji, L., Chen, X., Huang, C., & Tan, W. (2024). Arbuscular mycorrhizal hyphal networks and glomalin-related soil protein jointly promote soil aggregation and alter aggregate hierarchy in Calcaric Regosol. *Geoderma, 452*, 117096.

Kakouridis, A., Hagen, J. A., Kan, M. P., Mambelli, S., Feldman, L. J., Herman, D. J., Weber, P. K., Pett-Ridge, J., & Firestone, M. K. (2022). Routes to roots: Direct evidence of water transport by arbuscular mycorrhizal fungi to host plants. *New Phytologist, 236*(1), 210–221. http://onlinelibrary.wiley.com/journal/10.1111/(ISSN)1469-8137. doi:10.1111/nph.18281.

Kalamulla, R., Karunarathna, S. C., Tibpromma, S., Galappaththi, M. C. A., Suwannarach, N., Stephenson, S. L., Asad, S., Salem, Z. S., & Yapa, N. (2022). Arbuscular mycorrhizal fungi in sustainable agriculture. *Sustainability, 14*(19), 12250.

Keyes, S., van Veelen, A., McKay Fletcher, D., Scotson, C., Koebernick, N., Petroselli, C., Williams, K., Ruiz, S., Cooper, L., Mayon, R., Duncan, S., Dumont, M., Jakobsen, I., Oldroyd, G., Tkacz, A., Poole, P., Mosselmans, F., Borca, C., Huthwelker, T., … Roose, T. (2022). Multimodal correlative imaging and modelling of phosphorus uptake from soil by hyphae of mycorrhizal fungi. *New Phytologist, 234*(2), 688–703. doi:10.1111/nph.17980.

Khaliq, A., Perveen, S., Alamer, K. H., Ul Haq, M. Z., Rafique, Z., Alsudays, I. M., Althobaiti, A. T., Saleh, M. A., Hussain, S., & Attia, H. (2022). Arbuscular mycorrhizal fungi symbiosis to enhance plant–soil interaction. *Sustainability, 14*(13), 7840. doi:10.3390/su14137840.

Kimura, A. C., & Scotti, M. R. (2016). Soil aggregation and arbuscular mycorrhizal fungi as indicators of slope rehabilitation in the São Francisco River Basin (Brazil). *Soil & Water Research, 11*(2).

Klironomos, J. N., McCune, J., Hart, M., & Neville, J. (2000). The influence of arbuscular mycorrhizae on the relationship between plant diversity and productivity. *Ecology Letters, 3*(2), 137–141. doi:10.1046/j.1461-0248.2000.00131.x.

Kohler, J., Roldán, A., Campoy, M., & Caravaca, F. (2017). Unraveling the role of hyphal networks from arbuscular mycorrhizal fungi in aggregate stabilization of semiarid soils with different textures and carbonate contents. *Plant and Soil, 410*(1), 273–281.

Kumar, V. M. V. P., Chandrapati, A., Mandlik, A., Rao, Y. D., Tripathi, P., & Maurya, S. (2025). A comprehensive review on the concept and dynamics of earth's mycobiome: AMF networks and processes. *International Journal of Research in Agronomy, 8*(2), 246–252.

Lahlali, R., Ibrahim, D. S. S., Belabess, Z., Roni, M. Z. K., Radouane, N., Vicente, C. S. L., Menendez, E., Mokrini, F., Barka, E. A., de Melo e Mota, M. G., & Peng, G. (2021). High-throughput molecular technologies for unraveling the mystery of soil microbial community: Challenges and future prospects. *Heliyon, 7*(10). http://www.journals.elsevier.com/heliyon/. doi:10.1016/j.heliyon.2021.e08142.

Lammel, D. R., Kim, S. W., Rong, L., Chen, H., Ingraffia, R., & Rillig, M. C. (2025). Effects of microplastic types and shapes on the community structure of arbuscular mycorrhizal fungi in different soil types. *Environmental Science and Pollution Research, 32*(20), 12504–12512. https://www.springer.com/journal/11356. doi:10.1007/s11356-025-36408-1.

Langley, J. A., & Hungate, B. A. (2003). Mycorrhizal controls on belowground litter quality. *Ecology, 84*(9), 2302–2312. http://esajournals.onlinelibrary.wiley.com/hub/journal/10.1002/%28ISSN%291939-9170/issues/. doi:10.1890/02-0282.

Lehmann, A., Zheng, W., Ryo, M., Soutschek, K., Roy, J., Rongstock, R., Maaß, S., & Rillig, M. C. (2020). Fungal traits important for soil aggregation. *Frontiers in Microbiology, 10*. https://www.frontiersin.org/journals/microbiology#. doi:10.3389/fmicb.2019.02904.

Leifheit, E. F., Lehmann, A., & Rillig, M. C. (2021). Potential effects of microplastic on arbuscular mycorrhizal fungi. *Frontiers in Plant Science, 12*, 626709.

Li, J., Wang, Q., Wang, Y., Zhang, Q., Luo, J., Jiang, X., Liu, W., & Zhao, X. (2022). Effects of arbuscular mycorrhizal fungi on improvement of degraded landscape soil in an ionized rare earth mining area, subtropical China. *Soil Science Society of America Journal, 86*(2), 275–292. doi:10.1002/saj2.20363.

Li, N., Wen, J., Wu, R., Hu, D., Zhang, L., Zhang, W., & Zhang, M. (2023). Dual effects of plant growth-promoting rhizobacteria (PGPR) on the Moso bamboo-soil system: Plant growth promotion and microbial community stability. *Industrial Crops and Products, 203*, 117151.

Li, Y., Li, Y., Wang, Y., Yang, Y., Qi, M., Su, T., Li, R., Liu, D., Gao, Y., Qi, Y., & Qiu, L. (2024). Flg22-facilitated PGPR colonization in root tips and control of root rot. *Molecular Plant Pathology, 25*(11). http://onlinelibrary.wiley.com/journal/10.1111/(ISSN)1364-3703. doi:10.1111/mpp.70026.

Martin, S. L., Mooney, S. J., Dickinson, M. J., & West, H. M. (2012). The effects of simultaneous root colonisation by three Glomus species on soil pore characteristics. *Soil Biology and Biochemistry, 49*, 167–173.

Mishra, A. K., Bhowmick, M. K., Das, R., Pidikiti, P., Maurya, P. K., Sharma, S., Peramaiyan, P., & Singh, S. (2024). Potentials and prospects of AMF for soil carbon sequestration and nutrient cycling in rice-based cropping system. In *Arbuscular mycorrhizal fungi in sustainable agriculture: Nutrient and crop management* (pp. 113–130). Springer Nature. https://link.springer.com/book/10.1007/978-981-97-0300-5. doi:10.1007/978-981-97-0300-5_5.

Mitra, D., Djebaili, R., Pellegrini, M., Mahakur, B., Sarker, A., Chaudhary, P., & Mohapatra, P. K. D. (2021). Arbuscular mycorrhizal symbiosis: Plant growth improvement and induction of resistance under stressful conditions. *Journal of Plant Nutrition, 44*(13), 1993–2028.

Morris, E. K, Morris, D. J. P., Vogt, S., Gleber, S.-C., Bigalke, M., Wilcke, W., & Rillig, M. C (2019). Visualizing the dynamics of soil aggregation as affected by arbuscular mycorrhizal fungi. *The ISME Journal, 13*(7), 1639–1646. doi:10.1038/s41396-019-0369-0.

Mulyadi, & Jiang, L. (2023). Combined application of arbuscular mycorrhizal fungi (AMF) and nitrogen fertilizer alters the physicochemical soil properties, nitrogen uptake, and rice yield in a polybag experiment. *Agriculture, 13*(7), 1364.

Mwampashi, L. L., Magubika, A. J., Ringo, J. F., Theonest, D. J., Tryphone, G. M., Chilagane, L. A., & Nassary, E. K. (2024). Exploring agro-ecological significance, knowledge gaps, and research priorities in arbuscular mycorrhizal fungi. *Frontiers in Microbiology, 15*. https://www.frontiersin.org/journals/microbiology#. doi:10.3389/fmicb.2024.1491861.

Nanjundappa, A., Bagyaraj, D. J., Saxena, A. K., Kumar, M., & Chakdar, H. (2019). Interaction between arbuscular mycorrhizal fungi and *Bacillus* spp. in soil enhancing growth of crop plants. *Fungal Biology and Biotechnology, 6*(1). fungalbiolbiotech.biomedcentral.com/. doi:10.1186/s40694-019-0086-5.

Nazir, A., Sarfraz, W., Allah, D., Khalid, N., Farid, M., Shafiq, M., Firdaus-E-Bareen, Rizvi, Z. F, & Naeem, N. (2023). Synergistic impact of two autochthonous saprobic fungi (*A. niger* and *T. pseudokoningii*) on the growth, ionic contents, and metals uptake in *Brassica juncea* L. and *Vigna radiata* L. under tannery solid waste contaminated soil. *International Journal of Phytoremediation, 25*(11), 1488–1500. doi:10.1080/15226514.2023.2166457.

Ndeko, A. B., Founoune-Mboup, H., Kane, A., & Cournac, L. (2022). Arbuscular mycorrhizal fungi alleviate the negative effect of temperature stress in millet lines with contrasting soil aggregation potential. *Gesunde Pflanzen, 74*(1), 53–67.

Neuenkamp, L., Prober, S. M., Price, J. N., Zobel, M., & Standish, R. J. (2019). Benefits of mycorrhizal inoculation to ecological restoration depend on plant functional type, restoration context and time. *Fungal Ecology, 40*, 140–149. http://www.elsevier.com. doi:10.1016/j.funeco.2018.05.004.

Norton, J. (2021). Rangeland soil health: Interactions with vegetative recovery. Presented by Extension and the Animal Science Departments of South Dakota. *State University*, 67–78.

Nurbaity, A., Hidayat, C., Hudaya, D., & Sauman, J. (2013). Mycorrhizal fungi and organic matter affect some physical properties of andisols. *Soil Water Journal, 2*(2), 639–644.

Pan, J., Huang, C., Peng, F., Wang, T., Liao, J., Ma, S., … Xue, X. (2022). Synergistic combination of arbuscular mycorrhizal fungi and plant growth-promoting rhizobacteria modulates morpho-physiological characteristics and soil structure in Nitraria tangutorum bobr. under saline soil conditions. *Research in Cold and Arid Regions, 14*(6), 393–402.

Parihar, M., Rakshit, A., Meena, V. S., Gupta, V. K., Rana, K., Choudhary, M., Tiwari, G., Mishra, P. K., Pattanayak, A., Bisht, J. K., Jatav, S. S., Khati, P., & Jatav, H. S. (2020). The potential of arbuscular mycorrhizal fungi in C cycling: A review. *Archives of Microbiology, 202*(7), 1581–1596. link.springer.de/link/service/journals/00203/index.htm. doi:10.1007/s00203-020-01915-x.

Pauwels, R., Graefe, J., & Bitterlich, M. (2023). An arbuscular mycorrhizal fungus alters soil water retention and hydraulic conductivity in a soil texture specific way. *Mycorrhiza, 33*(3), 165–179.

Phalempin, M., Landl, M., Wu, G.-M., Schnepf, A., Vetterlein, D., & Schlüter, S. (2022). Maize root-induced biopores do not influence root growth of subsequently grown maize plants in well aerated, fertilized and repacked soil columns. *Soil and Tillage Research, 221*, 105398. doi:10.1016/j.still.2022.105398.

Qiang, W., Gunina, A., Kuzyakov, Y., He, L., Zhang, Y., Liu, B., & Pang, X. (2023). Contributions of mycorrhizal fungi to soil aggregate formation during subalpine forest succession. *CATENA, 221*, 106800. doi:10.1016/j.catena.2022.106800.

Qiang, W., Gunina, A., Kuzyakov, Y., He, L., Zhang, Y., Liu, B., & Pang, X. (2023). Contributions of mycorrhizal fungi to soil aggregate formation during subalpine forest succession. *Catena, 221*, 106800.

Qin, M., Li, L., Miranda, J. P., Tang, Y., Song, B., Oosthuizen, M. K., & Wei, W. (2022). Experimental duration determines the effect of arbuscular mycorrhizal fungi on plant biomass in pot experiments: A meta-analysis. *Frontiers in Plant Science, 13*. https://www.frontiersin.org/journals/plant-science. doi:10.3389/fpls.2022.1024874.

Qin, M., Zhang, Q., Pan, J., Jiang, S., Liu, Y., Bahadur, A., … Feng, H. (2020). Effect of arbuscular mycorrhizal fungi on soil enzyme activity is coupled with increased plant biomass. *European Journal of Soil Science, 71*(1), 84–92.

Reid, J. B., & Goss, M. J. (1982). Interactions between soil drying due to plant water use and decreases in aggregate stability caused by maize roots. *Journal of Soil Science, 33*(1), 47–53. doi:10.1111/j.1365-2389.1982.tb01746.x.

Ren, A. T., Mickan, B. S., Li, J. Y., Zhou, R., Zhang, X. C., Ma, M. S., Wesly, K., & Xiong, Y. C. (2021). Soil labile organic carbon sequestration is tightly correlated with the abundance and diversity of arbuscular mycorrhizal fungi in semiarid maize fields. *Land Degradation and Development, 32*(3), 1224–1236. http://onlinelibrary.wiley.com/journal/10.1002/(ISSN)1099-145X. doi:10.1002/ldr.3773.

Ren, C., Liu, K., Dou, P., Shao, X., Zhang, D., Wang, K., Liu, X., Li, J., & Wang, K. (2022). Soil nutrients drive microbial changes to alter surface soil aggregate stability in typical grasslands. *Journal of Soil Science and Plant Nutrition, 22*(4), 4943–4959. doi:10.1007/s42729-022-00972-z.

Rillig, M. C., & Mummey, D. L. (2006). Mycorrhizas and soil structure. *New Phytologist, 171*(1), 41–53. doi:10.1111/j.1469-8137.2006.01750.x.

Saeed, H. S., Munir, M., Samra, S., Farooq, A., Mahmood, H., Ahmad, I., Akram, M. I., Iqbal, R., Elsadek, M. F., Almutairi, S. M., Ditta, A., & Daraz, U. (2025). Screening of biosurfactants producing fungi from infected *Citrus sinensis* (L.). *Polish Journal of Environmental Studies, 34*(5), 5261–5270. https://www.pjoes.com/pdf-190588-115894?filename=Screening%20of.pdf. doi:10.15244/pjoes/190588.

Selosse, M. A., Bocayuva, M. F., Kasuya, M. C. M., & Courty, P. E. (2016). Mixotrophy in mycorrhizal plants: Extracting carbon from mycorrhizal networks. *Molecular mycorrhizal symbiosis*, 451–471. http://onlinelibrary.wiley.com/book/10.1002/9781118951446. doi:10.1002/9781118951446.ch25.

Shahzad, L., Waheed, A., Sharif, F., & Ali, M. (2024). *Soil fertility and soil biodiversity health under different agroforestry systems* (pp. 3–14). Springer Science and Business Media LLC. doi:10.1007/978-981-99-7282-1_1.

Singh, A. K., Zhu, X., Chen, C., Wu, J., Yang, B., Zakari, S., & Liu, W. (2022). The role of glomalin in mitigation of multiple soil degradation problems. *Critical Reviews in Environmental Science and Technology, 52*(9), 1604–1638.

Six, J., Bossuyt, H., Degryze, S., & Denef, K. (2004). A history of research on the link between (micro)aggregates, soil biota, and soil organic matter dynamics. *Soil and Tillage Research, 79*(1), 7–31. doi:10.1016/j.still.2004.03.008.

Slimani, A., Oufdou, K., & Meddich, A. (2024). Combining intercropping and co-inoculation of AMF and PGPR mitigate salinity in barley and alfalfa by improving photosynthetic capacity, nutrient acquisition, and redox potential. *Plant Biosystems-An International Journal Dealing With all Aspects of Plant Biology, 158*(5), 1115–1124.

Souza, T., Lima dos Santos, J. B., & Batista, D. S. (2025). An assessment of plant growth and physiological responses in annual crops grown in P deficient soils inoculated with indigenous arbuscular mycorrhizal fungi. *Brazilian Journal of Microbiology, 56*(2), 1241–1251.

Syamsiyah, J., Herawati, A., & Mujiyo (2018). The potential of arbuscular mycorrhizal fungi application on aggregrate stability in alfisol soil. *IOP conference series: Earth and environmental science, 142*, 012045.

Tang, L., Lu, Z., Zhao, T., Li, Z., Qiu, P., Yu, Y., Jin, L., Li, Y., & Jia, H. (2024). Exploring strength deterioration mechanism of soil-rock mixture based on pore structure characteristics under freeze–thaw cycles. *Cold Regions Science and Technology, 217*, 104040. doi:10.1016/j.coldregions.2023.104040.

Thind, S., Chaudhary, M. S, Ditta, A., Hussain, I., Parveen, A., Ullah, N., Mahmood, Q., Al-Ashkar, I., & El-Sabagh, A. (2022). Impact of mycorrhizal fungi from different rhizospheric soils on fungal colonization, growth, and chlorophyll contents of *Cenchrus ciliaris*. *Agronomy, 12*(11), 2644. doi:10.3390/agronomy12112644.

Van der Heijden, M. G. A., Boller, T., Wiemken, A., & Sanders, I. R. (1998). Different arbuscular mycorrhizal fungal species are potential determinants of plant community structure. *Ecology, 79*(6), 2082–2091. http://esajournals.onlinelibrary.wiley.com/hub/journal/10.1002/%28ISSN%291939-9170/issues/. doi:10.1890/0012-9658(1998)079[2082:DAMFSA]2.0.CO;2.

Vieira, L. C, Silva, D. K. A., Escobar, I. E. C., Silva, J. M., Moura, I. A., Oehl, F., & da Silva, G. A. (2020). Changes in an arbuscular mycorrhizal fungi community along an environmental gradient. *Plants, 9*(1), 52. doi:10.3390/plants9010052.

Wang, L., Rengel, Z., Cheng, L., & Shen, J. (2024). Coupling phosphate type and placement promotes maize growth and phosphorus uptake by altering root properties and rhizosphere processes. *Field Crops Research, 306*, 109225.

Wen, Z., Chen, Y., Liu, Z., & Meng, J. (2022). Biochar and arbuscular mycorrhizal fungi stimulate rice root growth strategy and soil nutrient availability. *European Journal of Soil Biology, 113*, 103448. doi:10.1016/j.ejsobi.2022.103448.

Wilson, G. W. T., Rice, C. W., Rillig, M. C., Springer, A., & Hartnett, D. C. (2009). Soil aggregation and carbon sequestration are tightly correlated with the abundance of arbuscular mycorrhizal fungi: Results from long-term field experiments. *Ecology Letters, 12*(5), 452–461. doi:10.1111/j.1461-0248.2009.01303.x.

Wu, Q. S., & Zou, Y. N. (2009). Mycorrhiza has a direct effect on reactive oxygen metabolism of drought-stressed citrus. *Plant, Soil and Environment, 55*(10), 436–442.

Yang, Y., Zhang, J., He, J., Shen, Y., & Yang, X. (2023). Glomalin-related soil proteins respond negatively to fertilization and fungicide application in China's arid grassland. *European Journal of Soil Biology, 119*, 103557. doi:10.1016/j.ejsobi.2023.103557.

Yao, N. N., Liu, A., Wang, X. J., Deng, H. H., Fan, S. Q., Zhang, H. J., Liao, H. X., Peng, S. L., & Chen, B. M. (2024). Influence of arbuscular mycorrhizal fungi on root foraging and competitive ability depends on soil phosphorus distribution: Evidence from two pairs of invasive and native plants. *Applied Soil Ecology, 201*. https://www.sciencedirect.com/science/journal/09291393. doi:10.1016/j.apsoil.2024.105507.

Yizhu, L., Imtiaz, M., Ditta, A., Rizwan, M. S., Ashraf, M., Mehmood, S., Aziz, O., Mubeen, F., Ali, M., Elahi, N. N., Ijaz, R., Lele, S., Shuang, C., & Tu, S. (2020). Response of growth, antioxidant enzymes and root exudates production towards as stress in *Pteris vittata* and in *Astragalus sinicus* colonized by arbuscular mycorrhizal fungi. *Environmental Science and Pollution Research, 27*(2), 2340–2352. https://link.springer.com/journal/11356. doi:10.1007/s11356-019-06785-5.

Yoneyama, K., & Brewer, P. B. (2021). Strigolactones, how are they synthesized to regulate plant growth and development? *Current Opinion in Plant Biology, 63*, 102072. doi:10.1016/j.pbi.2021.102072.

Yurkov, A. P., Kryukov, A. A., Gorbunova, A. O., Kudriashova, T. R., Kovalchuk, A. I., Gorenkova, A. I., & Shishova, M. F. (2023). Diversity of arbuscular mycorrhizal fungi in distinct ecosystems of the North Caucasus, a temperate biodiversity hotspot. *Journal of Fungi, 10*(1), 11.

Zhang, L., Shi, N., Fan, J., Wang, F., George, T. S., & Feng, G. (2018). Arbuscular mycorrhizal fungi stimulate organic phosphate mobilization associated with changing bacterial community structure under field conditions. *Environmental Microbiology, 20*(7), 2639–2651.

Zhang, Y., Yang, X., Luo, Z. X., Lai, J. L., Li, C., & Luo, X. G. (2022). Effects of polystyrene nanoplastics (PSNPs) on the physiology and molecular metabolism of corn (Zea mays L.) seedlings. *Science of the Total Environment, 806*, 150895.

Zhang, Z., Mallik, A., Zhang, J., Huang, Y., & Zhou, L. (2019). Effects of arbuscular mycorrhizal fungi on inoculated seedling growth and rhizosphere soil aggregates. *Soil and Tillage Research, 194*, 104340.

Zhao, R., Guo, W., Bi, N., Guo, J., Wang, L., Zhao, J., & Zhang, J. (2015). Arbuscular mycorrhizal fungi affect the growth, nutrient uptake and water status of maize (Zea mays L.) grown in two types of coal mine spoils under drought stress. *Applied Soil Ecology, 88*, 41–49.

Zheng, X., Liu, Q., Chen, X., Cao, M., Wu, F., Li, W., Zhang, L., Liu, S., & Jiang, J. (2023). Arbuscular mycorrhizal fungi decrease soil ammonium availability and nitrous oxide emissions under nitrogen input. *Agricultural and Forest Meteorology, 333*. www.elsevier.com/inca/publications/store/5/0/3/2/9/5. doi:10.1016/j.agrformet.2023.109385.

Chapter 9

Role of arbuscular mycorrhizal fungi in the remediation of soil pollutants

Suleman Haider Shah [a,b], Muhammad Baqir Hussain [a] and Allah Ditta [c,d]

[a] *Department of Soil and Environmental Sciences, Muhammad Nawaz Shareef University of Agriculture, Multan, Pakistan,*
[b] *Department of Agriculture (Extension & Adaptive Research), Government of Punjab, Murree, Punjab, Pakistan,*
[c] *Department of Environmental Sciences, Shaheed Benazir Bhutto University Sheringal, Sheringal, Khyber Pakhtunkhwa, Pakistan,* [d] *School of Biological Sciences, The University of Western Australia, Perth, WA, Australia*

9.1 Introduction

The intensification of anthropogenic activities, including industrialization, mining, and high-input agriculture, has resulted in the accumulation of hazardous pollutants in soil ecosystems. Soil contamination is no longer an environmental issue; it is an operational constraint for agriculture by affecting an estimated 33% of global land area that is moderately to highly degraded, with chemical pollution among the major soil threats (Food and Agriculture Organization of the United Nations, & Intergovernmental Technical Panel on Soils, 2015; Food and Agriculture Organization of the United Nations, & United Nations Environment Program, 2021). The primary contaminants include potentially toxic elements (PTEs) such as cadmium (Cd), lead (Pb), and arsenic (As), as well as organic xenobiotics such as polycyclic aromatic hydrocarbons (PAHs) and pesticide residues. PTEs accumulate in soils because of mining, smelting activities, wastewater irrigation, industrial fallout, and phosphate-fertilizer application. Organic pollutants from PAHs, petroleum products, and pesticides are brought into the soil by human activities, for example, spills, combustion, chemical-intensive agriculture, and industries. These soil pollutants deteriorate soil health, disturb microbial community structure, lower yield, and cause bioaccumulation in the food chain that exposes human health to severe health-related risks.

Traditional methods of remediation, such as soil excavation, chemical leaching, thermal desorption, and vitrification, showed efficient results on soil remediation, but these methods are very costly and are not practically feasible to remove contaminants from large areas (Nazir et al., 2023; Saeed et al., 2025; Thind et al., 2022; Yizhu et al., 2020). Nonadaptation of these methods on a large scale shifted the focus of research on phytoremediation techniques in which plants are grown in contaminated soils to extract, stabilize, and degrade pollutants into non/less harmful forms (Akhtar et al., 2023; Dean et al., 2022; Jan et al., 2021; Ullah et al., 2023). The plants used in phytoremediation sometimes suffer from toxicity symptoms that decrease their remediation efficiency. This drawback of using plants for soil remediation shifted the research focus to using microbes in remediation techniques. Arbuscular mycorrhizal fungi (AMF) were found to be the best option used in phytoremediation of heavy metals and organic pollutants in contaminated soils.

Mycoremediation uses a living process layer to improve crop growth and soil aggregation and reduce bioavailability and mobility of pollutants. AMF can make symbiotic associations with approximately 80% of all land plants (Smith & Read, 2008), and they play an important role in plant nutrient uptake (Chapter 1) and carbon sequestration (Chapter 7). The ability of AMF to improve plants' buffering capacity against pollutant toxicity enables them to play a pivotal role in sustainable remediation strategies.

The constraints in implementing field-level remediation strategy can play a significant role because the results cannot be consistent if the inoculum quality is substandard, the soil is highly contaminated, PTEs are high, or AMF strains and host plants mismatch. AMF can also increase uptake of some elements under specific conditions, so monitoring for risk assessment is necessary. Here, we will discuss the mechanisms of AMF-mediated remediation techniques, differentiating

Advancing Sustainable Agriculture Through Arbuscular Mycorrhizal Fungi. DOI: https://doi.org/10.1016/B978-0-443-45130-0.00011-4

the core mechanisms involving the inorganic metals' immobilization and the organic pollutants' dissipation. We will also evaluate their potential in reclaiming contaminated soils for sustainable agriculture.

9.2 Pollutant fate and bioavailability

The selection of an appropriate remediation strategy depends on the nature of pollutant spread and its stay in the soil. Total concentration of any pollutant in soil does not correctly reflect its phytotoxicity or its ecological risk. The bioavailable fraction is the correct portion of pollutant concentration that reaches to plant body and then is transferred to the human food chain (Dai et al., 2018).

Soil pollutants are divided into two major classes based on their chemical nature, physical structure, and recalcitrance. First class includes inorganic pollutants that include heavy metals such as cadmium (Cd), lead (Pb), mercury (Hg), and arsenic (As), and the second class are organic pollutants that include PAHs and pesticide residues. Inorganic pollutants cannot be converted into harmless products, so the focus remains on biostabilization through AMF that facilitate conversion of mobile forms into immobile inert forms within microbial cell walls or vacuoles (Kushwaha et al., 2018). Boorboori and Zhang (2022) emphasize that the high surface-to-volume ratio of extraradical hyphae provides extensive ligands for the complexation of these inorganic ions, effectively removing them from the soil pore water. Second, organic xenobiotics include petroleum hydrocarbons, PAHs, and legacy/modern pesticide residues, and these are subject to microbial catabolism. AMF enhances this process by acting as "metabolic facilitators." By exuding primary metabolites, AMF stimulate the growth of heterotrophic bacteria capable of cleaving complex rings. The degradation of phenanthrene in the mycorrhizosphere is significantly higher than in bulk soil, as the fungal partner provides the energetic "priming" necessary for bacterial mineralization of these recalcitrant structures into CO_2 and H_2O (Rajtor et al., 2016;Yu et al., 2011).

A dynamic equilibrium always exists between the solid (adsorbed) and the liquid phase (dissolved) of the bioavailable pollutant fraction that is shifted through the rhizosphere modification by AMF. Heavy metals' mobility is associated with soil pH, and a decrease in pH results in an increase in protonation on soil exchange sites that leads to metal cations' release into the soil solution. AMF modifies rhizosphere pH by secretion of organic acids and alteration of CO_2 partial pressure that facilitates uptake of some nutrients such as P and Zn and precipitates toxic metals such as Al and Pb. This local modification of soil pH by Glomeromycota supports them to survive in contaminated mine tailings under acidic conditions (Meier et al., 2012).

The cation exchange process at the soil surface because of organic matter and clay minerals serves as a sorption sink for pollutants because high organic matter soils have the capacity to reduce pollutant toxicity by binding them on exchange sites, slowly releasing them, which results in "legacy effect" (Nuruzzaman et al., 2025). AMF-produced glomalin-related soil proteins (GRSPs) are stable binding agents that support heavy metals' fixation in stable soil aggregates and resultantly reduce the bioavailable fraction of heavy metals and ultimately reduce the leaching of these pollutants to groundwater (Chen et al., 2022).

In wetland agricultural systems, such as paddy fields, the oxidation–reduction (redox) status governs the mobility and toxicity of metalloids, such as arsenic (As) and chromium (Cr). For example, under anaerobic conditions, As(V) can be reduced to As(III), which is typically more mobile and toxic. AMF colonization can influence the rhizosphere redox microenvironment (e.g., via effects on root physiology and associated microbial activity) and promote iron plaque formation on root surfaces, which can adsorb As and reduce its entry into plant tissues (Li et al., 2013).

In the agricultural system, remediation success is increasingly defined by associated risks rather than complete removal of contaminants. The priority is to keep PTEs concentrations in edible tissues (grain, fruit, and tubers) below maximum permissible levels while restoring key soil functions (e.g., aggregation and microbial activity) that limit dust, runoff, and leaching pathways (Food and Agriculture Organization of the United Nations, & United Nations Environment Program, 2021; Li et al., 2024). The following subsection outlines practical exposure pathways, monitoring metrics, and decision thresholds for food-safety-aligned AMF-assisted remediation.

9.3 Food safety, exposure pathways, and risks

A successful soil remediation strategy does not target complete reductions of pollutant concentrations in soil, but it focuses on the risks linked with food safety and the capacity to implement control measures to mitigate its exposure. Total pollutant concentration is calculated from strong-acid digestion, and it does not correctly correlate with plant uptake and its ability for human exposure. Therefore any remediation strategy should focus on the bioavailable/phytoavailable fractions of pollutants and their potential concentrations that could translocate into plant edible parts (Lindsay & Norvell, 1978; Manouchehri et al., 2006). This differentiation between total pollutant concentration and bioavailable fraction is highly important for PTEs that are translocated and immobilized in the plant body after mineralization (Boorboori & Zhang, 2022; Meier et al., 2012).

TABLE 9.1 Risk monitoring metrics for AMF-assisted remediation in agricultural systems.

Risk tiers	Primary objective	Success criteria	Minimum monitoring metrics	Interpretation for AMF-assisted management
Tier 1—soil risk proxy	Reduce exposure potential by limiting mobility and bioavailability	Declining bioavailable/phytoavailable pool and stable mobility controls	DTPA- or EDTA-extractable metals; soil pH and OM; redox proxy in flooded systems	If the extractable pool does not decline, reassess the host-AMF fit
Tier 2—plant safety	Prevent contaminant entry into edible tissues	Edible tissues remain below limits; TF/BCF reduced	Edible tissue concentration (grain/tuber/leaf); TF (shoot/root); BCF (plant/soil)	If yield rises but TF/BCF increases, shift to phytostabilization targets, adjust crop choice, or restrict edible cropping
Tier 3—stability and ecosystem recovery	Ensure risk reduction is durable and agronomically sustainable	Structural recovery and reduced off-site transport risk	Aggregate stability proxy; infiltration/erosion indicators; colonization verification (optional)	If structure does not improve or disturbance reverses gains, implement reduced tillage and extend the monitoring window

AMF, arbuscular mycorrhizal fungi.

A pollutant risk prevention program is also designed, starting with the mapping of different pollutant exposure pathways like contaminated edible plant part intake as food, inhaling contaminated dust, and consuming underground water contaminated via pollutant leaching and runoff. This program involves the application of AMF to reduce the bioavailable fraction of pollutants in soil water, translocation of pollutants from root to shoot, and removal of pollutants out of the contaminated site through soil erosion and leaching via soil structure improvements (Boorboori & Zhang, 2022; Rillig & Mummey, 2006). The magnitude and direction of AMF effects depend on the pollutant's nature, host type, and the soil. These effects are always verified through monitoring of the whole remediation program instead of the outcomes from the AMF colonization alone (Meier et al., 2012).

A risk prevention program is based on three tiers (Table 9.1). (1) First tier measures the changes in active bioavailable fractions via DTPA/EDTA extraction under key soil properties such as pH, organic matter content, and soil redox potentials that control metal pollutant mobility (Lindsay & Norvell, 1978; Manouchehri et al., 2006); (2) it measures pollutant concentration in edible plant parts and calculates translocation and bioconcentration factor of pollutants that represent the measure of plant vigor to resist pollutant entry into edible plant parts (Boorboori & Zhang, 2022; Meier et al., 2012); and (3) it evaluates whether risk reduction is durable, using indicators of aggregation and transport risk and, where cocontaminants exist, functional biodegradation potential in the mycorrhizosphere (Kohlmeier et al., 2005; Leifheit et al., 2014).

To operationalize these endpoints at the field scale, risk management can be paired with a zoning workflow (e.g., low, moderate, and high risk) derived from high-resolution site assessment and tiered risk management principles (Environment Agency, 2025). Under such governance, AMF inoculation is treated as a managed intervention with defined decision thresholds: if bioavailable pools are not declining or if edible tissues approach regulatory limits, the system should escalate to additional stabilization measures (e.g., AMF-biochar stacking, crop switching, or restricted edible cropping) rather than assuming remediation success based on colonization or yield recovery alone (Food and Agriculture Organization of the United Nations, & United Nations Environment Programme, 2021; U.S. Environmental Protection Agency, 2000).

Note. Bioavailable fraction is defined by the extractant and protocol (e.g., DTPA or EDTA), so method consistency is essential for time series interpretation (Lindsay & Norvell, 1978; Manouchehri et al., 2006). Tiered risk management and decision thresholds should follow structured, risk-based management (Environment Agency, 2025; U.S. Environmental Protection Agency, 2000). Aggregation outcomes are supported by AMF-soil structure literature (Leifheit et al., 2014), and microbial dispersal along hyphae supports indirect biodegradation facilitation (Kohlmeier et al., 2005).

9.4 Arbuscular mycorrhizal fungi biology and diversity in contaminated soils

Within these boundary conditions, being nonpassive soil inhabitants, Glomeromycota function as active metabolic filters under extreme edaphic stress. Fungi within the phylum Glomeromycota exhibit high plasticity. Under high edaphic stress (e.g., heavy metal toxicity or salinity), AMF utilize several "avoidance" and "tolerance" strategies: In highly toxic

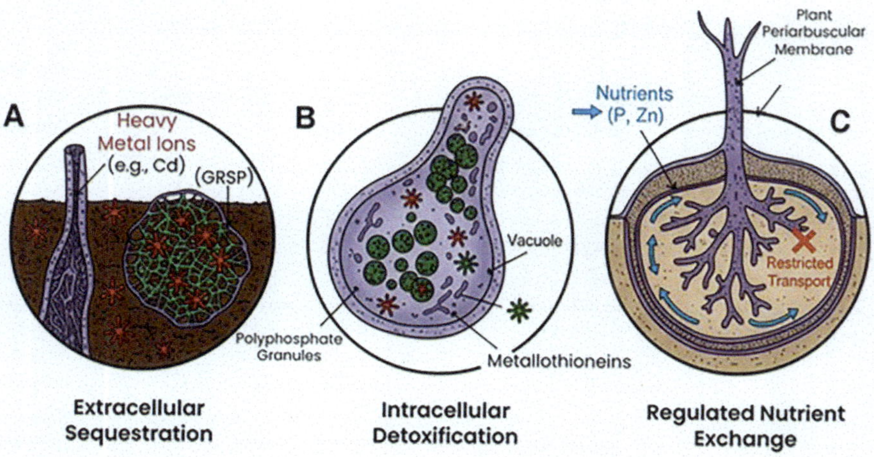

FIGURE 9.1 Multitiered defense mechanisms of AMF against potentially toxic elements (PTEs). (A) Physical and chemical immobilization in the soil matrix, (B) cellular sequestration and detoxification, and (C) selective transport at the host–fungus interface. *AMF*, Arbuscular mycorrhizal fungi; *GRSP*, glomalin-related soil protein.

"hotspots," AMF may remain as thick-walled spores, waiting for environmental conditions to improve or for a suitable host plant to signal colonization. Many stress-tolerant AMF produce melanin in their hyphal walls, which provides physical protection and additional binding sites for toxic metal cations.

The selective pressure of edaphic stress fundamentally reshapes AMF community composition, favoring specialized taxa that can maintain symbiotic functionality under extreme conditions. In arid and chemically stressed desert ecosystems the identification of native AMF from perennial grass roots has revealed a specialized "core" microbiome that persists despite high salinity and nutrient deficiency (Madouh et al., 2025). These findings underscore the importance of evolutionary filtering, where specific fungal lineages develop localized niche adaptations that are essential for restoring functionality to degraded agricultural or arid lands.

A critical issue in commercial mycoremediation is the use of exotic or lab-reared strains. Native AMF isolated from contaminated sites (e.g., mine tailings or industrial-fringe farms) often exhibit significantly higher remediation efficiency than commercial strains (Chenchouni et al., 2020; Wu et al., 2021). As the indigenous strains of AMF work as natural primers to support nutrient uptake and sequester pollutants, their use in any remediation strategy is critical to make it effective and scalable. These indigenous strains show a high success rate of symbiosis with host plants, as they are already adapted to the local ecosystem and plant phenology (Silva-Castro et al., 2023).

9.5 Mechanisms of arbuscular mycorrhizal fungi–assisted heavy metals immobilization

The main issue in managing heavy metals such as Cd, Pb, and As in agricultural soils is their persistent nature because metals cannot be degraded (Kushwaha et al., 2018). AMF facilitate metal remediation through phytostabilization by reducing their bioavailability and mobility within the soil–plant system (Li et al., 2020). The interface of the AMF hyphae and the soil solution is the first line of defense against the metal pollutants, as the AMF hyphae extend more than the host plant root zone to serve as a physical and chemical filter by holding pollutants in their cell walls rich in chitin, chitosan, and melanin to stop them from entering the plant body via roots. The functional groups such as carboxyl, hydroxyl, and amino groups on these biopolymers have a high affinity for divalent and trivalent metal cations and keep them held (Hildebrandt et al., 2007). Boorboori and Zhang (2022) showed that a large fraction of soil Pb and Zn is immobilized via surface adsorption onto the AMF hyphal body, preventing them from entering the host plant.

One of the major mechanisms for soil remediation is the secretion of GRSP by the AMF hyphae (Fig. 9.1). GRSP is a recalcitrant, hydrophobic glycoprotein, and it can permanently immobilize heavy metal pollutants in soil microaggregates (Dhalaria et al., 2020). In addition to its role in heavy metal immobilization, GRSP serves as a major component of soil health by increasing stability and improving water-holding capacity of the soil that ultimately supports plant growth in contaminated agro-ecosystems (Irving et al., 2021). AMF bind the metal ions into stable soil aggregates to reduce the risk of heavy metal pollutant leaching into groundwater and their uptake by plants (Cornejo et al., 2008).

When pollutant metal ions are adsorbed at the interface of AMF hyphae and soil, their movement is controlled by AMF-assisted intracellular mechanisms that maintain low cytosolic concentrations of pollutant ions (Dhalaria et al., 2020). The AMF cell vacuole is the primary site to store the excess pollutant ions and make their complexes with polyphosphate granules that prevent them from interfering with vital cytoplasmic metabolic processes (Ezawa & Saito, 2018). AMF also form low molecular weight and cysteine-rich metallothioneins to bind Cd^{2+} and Cu^{2+} and neutralize them before they reach the host plants (Hildebrandt et al., 2007).

In a symbiotic state the fungus acts as a gatekeeper by facilitating the transport of essential nutrients like phosphorus (P) to the plant and restricting the flux of toxic metals. This is quantified by the translocation factor, which is typically lower in mycorrhizal plants compared to nonmycorrhizal controls. Khan et al. (2023) observed that AMF colonization in berseem crops significantly reduced Cd translocation to the shoots, ensuring that the edible biomass remained within safe limits for livestock consumption.

9.6 Mechanisms of arbuscular mycorrhizal fungi–assisted degradation of organic xenobiotics

Organic pollutant remediation in agricultural soils is associated with processes within the rhizosphere. Various enzymatic oxidation and ring-cleavage pathways transform organic pollutants such as PAHs, anilines, and phenols during remediation. AMF increase organic pollutant dissipation by modifying the rhizosphere into a mycorrhizosphere that has high microbial activity and improved access to hydrophobic substrates (Rajtor et al., 2016). Extraradical hyphae can provide physical conduits (fungal highways) that facilitate bacterial dispersal through soil pores and increase contact with pollutant hotspots (Kohlmeier et al., 2005).

Mycorrhizal colonization can enrich bacterial taxa and functional genes associated with hydrocarbon and pesticide degradation (e.g., alkB) in the hyphosphere and rhizosphere. Hyphae also provide a continuous, liquid-coated surface that supports bacterial movement across air-filled pores, increasing encounter rates between degraders and hydrophobic compounds that are otherwise diffusion-limited in micropores (Kohlmeier et al., 2005).

AMF can further enhance the overall metabolic capacity of the rhizosphere. For example, Huang et al. (2009) reported high atrazine dissipation in AMF-colonized maize, accompanied by higher dehydrogenase and phosphatase activities. AMF can also modify plant water relations in soil and transpiration from the plant body that may affect the plant-controlled transfer of some soluble organic compounds within the soil–plant system (Augé, 2001). Along with biodegradation processes, AMF improve long-term soil aggregation and reduce soil erosion, reducing runoff risks and redistributing residual contaminants across the field soil and water (Meier et al., 2012; Nuruzzaman et al., 2025).

AMF increase the structural integrity of soil via physical attachment of soil particles with hyphae and the secretion of many sticky substances. AMF hyphae develop a three-dimensional framework that binds unstable micro soil aggregates and converts them into stable macro aggregates (Leifheit et al., 2014; Rillig, 2004; Rillig & Mummey, 2006). GRSP and water-stable aggregation play a key role in increasing the effect of this AMF framework to bind the pollutants within stable soil aggregates (Chen et al., 2022; Cornejo et al., 2008; Irving et al., 2021). This stable soil aggregation improves the soil structure and restores soil aeration and water-holding capacity that results in good plant growth in contaminated soils (Meier et al., 2012).

9.7 Molecular and genetic basis of mycoremediation

AMF regulates pollutant transport, chelation, and redox balance in contaminated soils through coordinated stress–responsive pathways during their association with host plants (Hildebrandt et al., 2007; Meier et al., 2012). These functions of AMF in their association with host plants support better selection of efficient AMF-host plant combinations in contaminated soils, as many edaphic stresses reduce AMF colonization and plant performance (Dhalaria et al., 2020).

The specialized transporters mediate the movement of toxic metal ions through the AMF membranes. Gonzalez-Guerrero et al. (2005) identified the GintZnT1 gene in *Rhizophagus irregularis*, which encodes a zinc transporter critical for Zn homeostasis. Under elevated Zn, upregulation of GintZnT1 contributes to intracellular sequestration and homeostatic control, thereby protecting the fungal cytoplasm. Reviews have further described high-affinity transporters and ABC-type systems that help regulate internal concentrations of both essential and potentially toxic micronutrients in AMF (Hildebrandt et al., 2007).

Beyond membrane transport, metal exposure activates binding and sequestration mechanisms that lower free ionic activity in cells. In plants, AMF frequently enhance phytochelatins and metallothionein-based complexation, while in AMF, metal-binding proteins and polyphosphate-associated vacuolar storage contribute to detoxification (Dhalaria et al., 2020; Hildebrandt et al., 2007).

Transcriptomic and physiological studies indicate that AMF colonization can reprogram host stress responses under PTE exposure. A consistent outcome is improved antioxidant capacity, including higher activities and/or expression of superoxide dismutase, catalase, and ascorbate peroxidase, which is associated with reduced lipid peroxidation (MDA) in metal-stressed plants (Dhalaria et al., 2020; Li et al., 2020; Zhao et al., 2024).

AMF can also restrict apoplastic pollutant movement and slow down their translocation to the plant body through their effects on the structural barriers in the host plant root system, such as lignification and cell wall modifications (Dhalaria et al., 2020; Meier et al., 2012). The AMF-mediated changes in rhizosphere redox balance and better plant root-related function can improve plant growth during the soil remediation process (Li et al., 2020; Zhao et al., 2024).

9.8 Synergistic mycoremediation approaches

AMF do not show consistent performance in highly contaminated agricultural soils. Therefore many recent reviews suggest synergistic strategies by integrating AMF with amendments and microbial inoculants for remediation in cocontaminated soils and high-toxicity hotspots (Fang et al., 2024; Meier et al., 2012; Zhuo et al., 2020). Using biochar as a coamendment for soil remediation serves dual benefits; it chemically absorbs the pollutants, and its porous structure supports the growth of AMF by providing a microhabitat for AMF survival and physical protection to the hyphae from soil predators and environmental hazards. These positive effects enhance AMF and host plants' symbiosis in contaminated soils (Fang et al., 2024; Huang et al., 2023; Jaafar, 2014; Zhuo et al., 2020).

Synergistic interaction between AMF and plant growth-promoting rhizobacteria (PGPR) improves plant growth as AMF increase phosphorus uptake and PGPRs produce phytohormones (Etesami et al., 2021). This nutritional and physiological support by synergistic interactions between AMF and PGPR increases plant growth and rhizosphere activity and improves phytoextraction and phytostabilization efficiency in contaminated agricultural soils (Hnini et al., 2024). Some PGPR can support AMF root colonization and hyphal development that ultimately improve soil remediation (Khan et al., 2023). Many field and lab studies showed that indigenous or soil-adapted AMF consortia can perform better than commercial inoculants in colonization efficiency and plant growth-promoting activities under specific edaphic conditions in contaminated soils. This supports the selection of locally adapted inocula (Chenchouni et al., 2020; Silva-Castro et al., 2023).

Synergistic use of phosphate-solubilizing bacteria (PSB) in AMF-mediated remediation significantly increases the effectiveness of the remediation, as this symbiosis among AMF, host plant, and PSB consortia develops a strong buffer against the edaphic stress in contaminated soils. Recent findings by Rahayu and Al-Najar (2024) showed that synergistic use of AMF with PSB consortia increases plant nutrient uptake and reduces heavy metal pollutant translocation in contaminated soils. This synergistic application of AMF and PSB solubilizes the fixed soil phosphorus that AMF transport to the host plant and improves its overall growth and capacity to resist or store heavy metal ions in roots.

The use of engineered nanoparticles (NPs) to clean contaminated soil is an emerging soil remediation technology. Some NPs are toxic to fungi, but their application at low concentrations supports pollutant transformation or enhanced sorption and potentially supports phytoremediation. Iron-based NPs promote the transformation of some pesticides by driving redox reactions. However, microbial-mediated processes govern longer term stabilization and biodegradation of transformed pollutant products (Kalia & Singh, 2020; Tian et al., 2019).

9.9 Applications of mycoremediation

Mycoremediation is designed as a complete risk-managed program instead of a single treatment with clear objectives of reduced pollutant uptake by plants, reduced leaching risk, and improved soil health (Environment Agency, 2025; U.S. Environmental Protection Agency, 2000). The assessment of contaminated soil depends on the difference between various land areas and the pathways of their exposure to the pollutants. The best approach before designing any remediation strategy is to arrange contaminated areas into different management zones based on risk level and land utilization objectives: low risk (green), moderate risk (amber), and high risk (red) areas. The specific remediation strategies are devised and implemented considering the specific requirements of different zones (Environment Agency, 2025).

As a poor relationship exists between pollutant total concentrations and their plant uptake and environmental risk, the focus always remains on the bioavailable fraction. Under specific soil pH, organic matter content, and soil texture, the DTPA extractable fraction of heavy metals represents the plant available pool, whereas the EDTA extractable fraction represents the chelatable pool (Lindsay & Norvell, 1978; Manouchehri et al., 2006).

TABLE 9.2 Inoculum QA/QC checklist and reproducibility safeguards for AMF-assisted remediation.

Deployment stage	QA/QC objective	Minimum acceptance criteria	Verification method	Common failure mode	Corrective measures
Inoculum procurement	Ensure the inoculum is fit for purpose and clean	Declared AMF taxa/consortium; viable propagule density; clean carrier; traceable source	Supplier certificate of analysis plus internal spot checks; chain-of-custody documentation	Low propagule viability; contaminated carrier	Reject batch; source alternative; require viability/propagule assay
Host-AMF compatibility (pilot)	Confirm colonization and plant response in the representative soil	Colonization established; no severe growth penalty; early indication of reduced TF/BCF	Greenhouse microcosm in site soil; early tissue screening	Host mismatch; suppression because of high phosphorus or toxicity	Adjust host species; switch to site-adapted AMF; add stabilization amendment
Field inoculation	Achieve consistent delivery and establishment	Correct dose and placement in the root zone; adequate moisture for establishment	Field protocol checklist; plot-level records	Uneven inoculation; drought stress	Reinoculate hotspots; improve moisture management; adjust delivery method
Early colonization verification	Confirm symbiosis before key growth stages	Site-defined colonization threshold met	Root staining and quantification; early plant nutrient status	Phosphorus oversupply suppresses colonization; tillage disruption	Reduce phosphorus; band phosphorus; implement reduced-/no-till; repeat verification
Contaminant transfer safeguard	Ensure biomass gains do not increase food-chain risk	TF/BCF not increasing; edible tissues trending below limits	Tissue assays; TF/BCF calculation	Increased translocation under specific soil-element conditions	Switch to phytostabilization crop; restrict edible harvest; escalate to AMF-biochar stack
Seasonal stability	Maintain functional hyphal network and soil structure benefits	Stable aggregation indicators; reduced erosion/runoff risk	Aggregate stability proxy; infiltration/erosion observations	Disturbance reverses structure gains	Extend monitoring; reinforce conservation practices; adjust rotation and cover crops

AMF, arbuscular mycorrhizal fungi.

Cultural management practices in the field should always be adjusted to support AMF colonization and hyphal network growth. High P application in soil decreases plant dependence on symbiosis with AMF, which suppresses AMF growth and colonization; therefore P applications are always strictly monitored to not exceed plant essential requirements (Smith & Read, 2008). Excessive tillage can shear hyphae; reduced-till or no-till practices help maintain the fungal network and limit the mobilization of contaminated particles. For PTEs, select plant–AMF combinations that favor phytostabilization (high root retention and low shoot translocation) and consider stacked stabilization in hotspots (Boorboori & Zhang, 2022; Fang et al., 2024; Zhuo et al., 2020).

9.10 Inoculum QA/QC and field reproducibility

Field performance is the primary constraint on scaling AMF-assisted remediation from controlled experiments to agricultural deployment. In practice, inconsistent outcomes are frequently attributable to inoculum quality, colonization failure, and management-induced disruption of hyphal networks. Accordingly, AMF deployment should be treated as a quality-controlled biological input with explicit acceptance criteria and verification steps (Meier et al., 2012; Rillig & Mummey, 2006).

A reproducible program begins with inoculum specification (Table 9.2). At minimum, inoculum should be characterized by propagule density (e.g., spores and infective propagules per unit mass or per seedling), carrier type, viability window, and absence of unintended contaminants (e.g., weed seeds and pathogenic fungi). Field success is often improved when inocula are locally adapted (indigenous or site-primed) rather than generic commercial strains, provided they are produced under controlled conditions and matched to the target host and soil chemistry (Meier et al., 2012). Where indigenous AMF are used, sourcing should follow a documented chain of custody and be supported by a greenhouse pilot to confirm compatibility under representative soil conditions.

Second, remediation success requires verification of colonization and network establishment. Colonization should be quantified in roots (e.g., percent root length colonized) and interpreted alongside functional indicators (plant nutrient status, growth response, and contaminant transfer behavior). Colonization alone is not sufficient because AMF effects on uptake and translocation can be element- and context-dependent (Boorboori & Zhang, 2022; Meier et al., 2012). Early monitoring will help to take corrective measures before the edible tissues are developed.

Third, agronomic management must be aligned with AMF biology. Excess phosphorus inputs can reduce host dependence on AMF and suppress colonization; thus phosphorus should be managed to sustain productivity while maintaining symbiotic demand (Smith & Read, 2008). Intensive tillage disrupts extraradical hyphae and can reduce the biofilter footprint; reduced- or no-till practices are recommended where feasible (Rillig & Mummey, 2006). In cocontaminated systems, amendment stacking (e.g., biochar) can moderate acute toxicity and improve establishment, but inoculum-amendment interactions should be pilot-tested to avoid carrier incompatibility.

Finally, reproducibility improves when inoculation is embedded in a stage-gated deployment workflow: baseline site characterization, inoculum selection and specification, greenhouse compatibility screening, field inoculation, early colonization verification, and adaptive management. This workflow reduces failure risk, supports auditability, and enables transparent decision-making when endpoints are not met (Environment Agency, 2025; U.S. Environmental Protection Agency, 2000).

Note. Inoculum performance and colonization outcomes are influenced by soil chemistry and management (e.g., phosphorus inputs and tillage) and should be treated as stage-based deliverables rather than assumed benefits (Rillig & Mummey, 2006; Smith & Read, 2008). Tiered risk governance supports documented workflows with decision thresholds and corrective actions (Environment Agency, 2025; U.S. Environmental Protection Agency, 2000).

9.11 Conclusion and future perspectives

Evidence synthesized in this chapter suggests that AMF can function as a biologically mediated filter, reducing exposure to PTEs under a risk-gated remediation objective. At the soil interface, extraradical hyphae and GRSP contribute to metal binding and aggregate stabilization, which can lower dissolved metal activity and plant-available pools in many systems (Irving et al., 2021). Within fungal tissues, metal transport regulation and compartmentalization (e.g., vacuolar sequestration) help maintain low cytosolic metal concentrations and can reduce root-to-shoot transfer in the host plant, depending on element, soil chemistry, and host–fungus pairing (Zhao et al., 2024).

For organic xenobiotics, AMF effects are typically indirect and mediated through the mycorrhizosphere. Hyphal networks can increase microbial habitat and dispersal and stimulate degradative consortia, which have been associated with faster dissipation of PAHs and certain pesticides relative to nonmycorrhizal controls (Rajtor et al., 2016). In practice the remediation outcome is best viewed as an emergent property of the plant–AMF–microbiome system rather than a single-organism process, and effect sizes remain context-dependent.

From an agricultural perspective, AMF deployment can support food-safety-aligned management by sustaining plant nutrition and biomass while lowering contaminant transfer to edible tissues in suitable crop–soil combinations, consistent with risk-based standards used in food systems (Gío-Trujillo & Alvarado-López, 2025; Tan et al., 2023). This is particularly relevant where cultivation continues on marginal or legacy-contaminated land because of land scarcity and historical pollution burdens (Food and Agriculture Organization of the United Nations, & United Nations Environment Program, 2021; Li et al., 2024). Evidence from recent studies and metaanalyses suggests that combining AMF with complementary interventions (e.g., biochar amendments and PGPR) can improve establishment and remediation reliability, especially under cocontamination or high-stress conditions (Huang et al., 2023; Rahayu & Al-Najar, 2024).

The effectiveness of any mycoremediation strategy in the field depends on various factors, including inoculum quality, suitability to the habitat, and management of risks. Indigenous and sited-adapted AMF inocula are efficient as compared to commercially grown ones, but the results are much associated with the type of host plants and edaphic condition of contaminated soils (Khan et al., 2023; Madouh et al., 2025). Advancements in the identification of host recognition and metal handling processes because of high-tech functional genomics are improving microbial strain selection and inoculum formulation. To better align with successful remediation and minimum negative impacts on the ecosystem, regulatory and monitoring systems are more focused on pollutant bioavailability and bioaccessibility than total concentration in soil.

AMF remediation strategies combined with pollutant risk reduction play a key role in the climate-smart and pollutant-resilient agriculture that results in improvement in the structure and biological function of soil. These strategies are more efficient in rehabilitating the ecosystem services as compared to traditional engineering-intensive options.

References

Akhtar, M. S., Aslam, S., Ditta, A., Albalawi, B. F. A., Oki, Y., & Nakashima, Y. (2023). Interspecific variability in growth characteristics and phytoremediation of Cu by free-floating *Azolla* macrophytes. *Sustainability, 15*(1), 497. https://doi.org/10.3390/su15010497.

Augé, R. M. (2001). Water relations, drought, and vesicular-arbuscular mycorrhizal symbiosis. *Mycorrhiza, 11*(1), 3–42. https://doi.org/10.1007/s005720100097.

Boorboori, M. R., & Zhang, H. Y. (2022). Arbuscular mycorrhizal fungi are an influential factor in improving the phytoremediation of arsenic, cadmium, lead, and chromium. *Journal of Fungi, 8*(2), 176. https://doi.org/10.3390/jof8020176.

Chen, H., Xiong, J., Fang, L., Han, F., Zhao, X., Fan, Q., & Tan, W. (2022). Sequestration of heavy metals in soil aggregates induced by glomalin-related soil protein: A five-year phytoremediation field study. *Journal of Hazardous Materials, 437*, 129445. https://doi.org/10.1016/j.jhazmat.2022.129445.

Chenchouni, H., Mekahlia, M. N., & Beddiar, A. (2020). Effect of inoculation with native and commercial arbuscular mycorrhizal fungi on growth and mycorrhizal colonization of olive (*Olea europaea* L.). *Scientia Horticulturae, 261*, 108969. https://doi.org/10.1016/j.scienta.2019.108969.

Cornejo, P., Meier, S., Borie, G., Rillig, M. C., & Borie, F. (2008). Glomalin-related soil protein in a Mediterranean ecosystem affected by a copper smelter and its contribution to Cu and Zn sequestration. *Science of the Total Environment, 406*(1–2), 154–160. https://doi.org/10.1016/j.scitotenv.2008.07.045.

Dai, Y., Nasir, M., Zhang, Y., Gao, J., Lv, Y., & Lv, J. (2018). Comparison of DGT with traditional extraction methods for assessing arsenic bioavailability to *Brassica chinensis* in different soils. *Chemosphere, 191*, 183–189. https://doi.org/10.1016/j.chemosphere.2017.10.035.

Dean, S., Akhtar, M. S., Ditta, A., Valipour, M., & Aslam, S. (2022). Microcosm study on the potential of aquatic macrophytes for phytoremediation of phosphorus-induced eutrophication. *Sustainability, 14*(24), 16415. https://doi.org/10.3390/su142416415.

Dhalaria, R., Kumar, D., Kumar, H., Nepovimova, E., Kuča, K., Islam, M. T., & Verma, R. (2020). Arbuscular mycorrhizal fungi as potential agents in ameliorating heavy metal stress in plants. *Agronomy, 10*(6), 815. https://doi.org/10.3390/agronomy10060815.

Environment Agency. (2025). *Land contamination risk management (LCRM)*. GOV.UK https://www.gov.uk/government/publications/land-contamination-risk-management-lcrm.

Etesami, H., Jeong, B. R., & Glick, B. R. (2021). Contribution of arbuscular mycorrhizal fungi, phosphate–solubilizing bacteria, and silicon to P uptake by plant. *Frontiers in Plant Science, 12*, 699618. https://doi.org/10.3389/fpls.2021.699618.

Ezawa, T., & Saito, K. (2018). How do arbuscular mycorrhizal fungi handle phosphate? New insight into fine-tuning of phosphate metabolism. *New Phytologist, 220*(4), 1116–1121. https://doi.org/10.1111/nph.15187.

Fang, X., Lee, X., Twagirayezu, G., Cheng, H., Lu, H., Huang, S., Deng, L., & Ji, B. (2024). A critical review of the effectiveness of biochar coupled with arbuscular mycorrhizal fungi in soil cadmium immobilization. *Journal of Fungi, 10*(3), 182. https://doi.org/10.3390/jof10030182.

Food and Agriculture Organization of the United Nations, & Intergovernmental Technical Panel on Soils. (2015). *Status of the world's soil resources: Main report*. FAO https://openknowledge.fao.org/items/f16010ce-1874-4108-bd03-a6a592e2e53a.

Food and Agriculture Organization of the United Nations, & United Nations Environment Programme. (2021). *Global assessment of soil pollution: Summary for policymakers*. FAO https://doi.org/10.4060/cb4827en.

Gío-Trujillo, J. A., & Alvarado-López, C. J. (2025). *Enhancing nutritional and sensory quality of plant-based foods through arbuscular mycorrhizal fungi*. IntechOpen https://doi.org/10.5772/intechopen.1007162.

González-Guerrero, M., Azcón-Aguilar, C., Mooney, M., Valderas, A., MacDiarmid, C. W., Eide, D. J., & Ferrol, N. (2005). Characterization of a *Glomus intraradices* gene encoding a putative Zn transporter of the cation diffusion facilitator family. *Fungal Genetics and Biology, 42*(2), 130–140. https://doi.org/10.1016/j.fgb.2004.10.007.

Hildebrandt, U., Regvar, M., & Bothe, H. (2007). Arbuscular mycorrhiza and heavy metal tolerance. *Phytochemistry, 68*(1), 139–146. https://doi.org/10.1016/j.phytochem.2006.09.023.

Hnini, M., Rabeh, K., & Oubohssaine, M. (2024). Interactions between beneficial soil microorganisms (PGPR and AMF) and host plants for environmental restoration: A systematic review. *Plant Stress, 11*, 100391. https://doi.org/10.1016/j.stress.2024.100391.

Huang, H., Zhang, S., Wu, N., Luo, L., & Christie, P. (2009). Influence of *Glomus etunicatum/Zea mays* mycorrhiza on atrazine degradation, soil phosphatase and dehydrogenase activities, and soil microbial community structure. *Soil Biology and Biochemistry, 41*(4), 726–734. https://doi.org/10.1016/j.soilbio.2009.01.009.

Huang, K., Li, M., Li, R., Rasul, F., Shahzad, S., Wu, C., Shao, J., Huang, G., Li, R., Almari, S., Hashem, M., & Aamer, M. (2023). Soil acidification and salinity: The importance of biochar application to agricultural soils. *Frontiers in Plant Science, 14*, 1206820. https://doi.org/10.3389/fpls.2023.1206820.

Irving, T. B., Alptekin, B., Kleven, B., & Ané, J. M. (2021). A critical review of 25 years of glomalin research: A better mechanical understanding and robust quantification techniques are required. *New Phytologist, 232*(4), 1572–1581. https://doi.org/10.1111/nph.17713.

Jaafar, N. M. (2014). Biochar as a habitat for arbuscular mycorrhizal fungi. In Z. Solaiman, L. Abbott, & A. Varma (Eds.). In *Mycorrhizal fungi: Use in sustainable agriculture and land restoration. Soil biology: 41* (pp. 273–283). Springer.

Jan, A. U., Hadi, F., Shah, A., Ditta, A., Nawaz, M. A., & Tariq, M. (2021). Plant growth regulators and EDTA improve phytoremediation potential and antioxidant response of *Dysphania ambrosioides* (L.) Mosyakin & Clemants in a Cd-spiked soil. *Environmental Science and Pollution Research, 28*, 43417–43430. https://doi.org/10.1007/s11356-021-13772-2.

Kalia, A., & Singh, S. (2020). Myco-decontamination of azo dyes: Nano-augmentation technologies. *3 Biotech, 10*, 384. https://doi.org/10.1007/s13205-020-02378-z.

Khan, S. N., Nafees, M., & Imtiaz, M. (2023). Phytoremediation potential of arbuscular mycorrhizal fungi and berseem in soil irrigated with industrial wastewater. *Journal of Hunan University Natural Sciences, 50*(4). https://doi.org/10.55463/issn.1674-2974.50.4.6.

Kohlmeier, S., Smits, T. H., Ford, R. M., Keel, C., Harms, H., & Wick, L. Y. (2005). Taking the fungal highway: Mobilization of pollutant-degrading bacteria by fungi. *Environmental Science & Technology, 39*(12), 4640–4646. https://doi.org/10.1021/es047979z.

Kushwaha, A., Hans, N., Kumar, S., & Rani, R. (2018). A critical review on speciation, mobilization, and toxicity of lead in soil-microbe-plant system and bioremediation strategies. *Ecotoxicology and Environmental Safety, 147*, 1035–1045. https://doi.org/10.1016/j.ecoenv.2017.09.049.

Leifheit, E. F., Veresoglou, S. D., Lehmann, A., Morris, E. K., & Rillig, M. C. (2014). Multiple factors influence the role of arbuscular mycorrhizal fungi in soil aggregation—a meta-analysis. *Plant and Soil, 374*(1), 523–537. https://doi.org/10.1007/s11104-013-1899-2.

Li, H., Chen, X. W., Wu, L., Luo, N., Huang, W. X., Mo, C. H., Li, Y. W., Xiang, L., Zhao, H. M., & Cai, Q. Y. (2020). Effects of arbuscular mycorrhizal fungi on redox homeostasis of rice under Cd stress. *Plant and Soil, 455*(1), 121–138. https://doi.org/10.1007/s11104-020-04678-y.

Li, H., Man, Y. B., Ye, Z. H., Wu, C., Wu, S. C., & Wong, M. H. (2013). Do arbuscular mycorrhizal fungi affect arsenic accumulation and speciation in rice with different radial oxygen loss? *Journal of Hazardous Materials, 262*, 1098–1104. https://doi.org/10.1016/j.jhazmat.2012.05.044.

Li, H., Wang, T., Du, H., Guo, P., Wang, S., & Ma, M. (2024). Research progress in the joint remediation of plants–microbes–soil for heavy metal-contaminated soil in mining areas: A review. *Sustainability, 16*(19), 8464. https://doi.org/10.3390/su16198464.

Lindsay, W. L., & Norvell, W. A. (1978). Development of a DTPA soil test for zinc, iron, manganese, and copper. *Soil Science Society of America Journal, 42*(3), 421–428. https://doi.org/10.2136/sssaj1978.03615995004200030009x.

Madouh, T., Suleiman, M., Quoreshi, A., & Davidson, M. (2025). Diversity of arbuscular mycorrhiza fungi in the arid desert ecosystems of Kuwait: Detection and identification from perennial native grass roots. *Diversity, 17*(2), 130. https://doi.org/10.3390/d17020130.

Manouchehri, N., Besançon, S., & Bermond, A. (2006). Major and trace metal extraction from soil by EDTA: Equilibrium and kinetic studies. *Analytica Chimica Acta, 559*(1), 105–112. https://doi.org/10.1016/j.aca.2005.11.050.

Meier, S., Borie, F., Bolan, N., & Cornejo, P. (2012). Phytoremediation of metal-polluted soils by arbuscular mycorrhizal fungi. *Critical Reviews in Environmental Science and Technology, 42*(7), 741–775. https://doi.org/10.1080/10643389.2010.528518.

Nazir, A., Sarfraz, W., Ditta, A., Khalid, N., Farid, M., Shafiq, M., Firdaus-E-Bareen, Rizvi, Z. F., & Naeem, N. (2023). Synergistic impact of two autochthonous saprobic fungi (*A. niger* and *T. pseudokoningii*) on the growth, ionic contents, and metals uptake in *Brassica juncea* L. and *Vigna radiata* L. under tannery solid waste contaminated soil. *International Journal of Phytoremediation, 25*(11), 1488–1500.

Nuruzzaman, M., Bahar, M. M., & Naidu, R. (2025). Diffuse soil pollution from agriculture: Impacts and remediation. *Science of the Total Environment, 962*, 178398. https://doi.org/10.1016/j.scitotenv.2025.178398.

Rahayu, Y. S., & Al-Najar, H. (2024). Role of mycorrhizal fungi and phosphate-solubilizing bacteria in plant uptake of essential and metal elements in heavy metal-contaminated soil. *Revista Chapingo. Serie Horticultura, 30*(2), 27–42. https://doi.org/10.5154/r.rchsh.2024.02.001.

Rajtor, M., & Piotrowska-Seget, Z. (2016). Prospects for arbuscular mycorrhizal fungi (AMF) to assist in phytoremediation of soil hydrocarbon contaminants. *Chemosphere, 162*, 105–116. https://doi.org/10.1016/j.chemosphere.2016.07.071.

Rillig, M. C. (2004). Arbuscular mycorrhizae, glomalin, and soil aggregation. *Canadian Journal of Soil Science, 84*(4), 355–363. https://doi.org/10.4141/S04-003.

Rillig, M. C., & Mummey, D. L. (2006). Mycorrhizas and soil structure. *New Phytologist, 171*(1), 41–53. https://doi.org/10.1111/j.1469-8137.2006.01750.x.

Saeed, H. S., Munir, M., Samra, S., Farooq, A., Mahmood, H., Ahmad, I., Akram, M. I., Iqbal, R., Elsadek, M. F., Almutairi, S. M., Ditta, A., & Daraz, U. (2025). Screening of biosurfactants producing fungi from infected *Citrus sinensis* (L.). *Polish Journal of Environmental Studies, 34*(5), 5261–5270.

Silva-Castro, G. A., Cano, C., Moreno-Morillas, S., Bago, A., & García-Romera, I. (2023). Inoculation of indigenous arbuscular mycorrhizal fungi as a strategy for the recovery of long-term heavy metal-contaminated soils in a mine-spill area. *Journal of Fungi, 9*(1), 56. https://doi.org/10.3390/jof9010056.

Smith, S. E., & Read, D. J. (2008). *Mycorrhizal symbiosis* (3rd ed.). Academic Press.

Tan, Q., Guo, Q., Wei, R., Zhu, G., Du, C., & Hu, H. (2023). Influence of arbuscular mycorrhizal fungi on bioaccumulation and bioavailability of As and Cd: A meta-analysis. *Environmental Pollution, 316*, 120619. https://doi.org/10.1016/j.envpol.2022.120619.

Thind, S., Chaudhary, M. S., Ditta, A., Hussain, I., Parveen, A., Ullah, N., Mahmood, Q., Al-ashkar, I., & Sabagh, A. E. L. (2022). Impact of mycorrhizal fungi from different rhizospheric soils on fungal colonization, growth, and chlorophyll contents of *Cenchrus ciliaris*. *Agronomy, 12*(11), 2644.

Tian, H., Ghorbanpour, M., & Kariman, K. (2019). Are nanoparticles a threat to mycorrhizal and rhizobial symbioses? A critical review. *Frontiers in Microbiology, 10*, 1660. https://doi.org/10.3389/fmicb.2019.01660.

U.S. Environmental Protection Agency. (2000). *Introduction to phytoremediation (EPA/600/R-99/107)*. National Risk Management Research Laboratory, Office of Research and Development https://www.epa.gov/sites/default/files/2015-04/documents/phytoremprimer.pdf.

Ullah, S., Liu, Q., Wang, S., Jan, A. U., Sharif, H. M. A., Ditta, A., Wang, G., & Cheng, H. (2023). Sources, impacts, factors affecting Cr uptake in plants, and mechanisms behind phytoremediation of Cr-contaminated soils. *Science of the Total Environment, 899*, 165726. https://doi.org/10.1016/j.scitotenv.2023.165726.

Wu, Y. H., Wang, H., Liu, M., Li, B., Chen, X., Ma, Y. T., & Zhang, X. (2021). Effects of native arbuscular mycorrhizae isolated on root biomass and secondary metabolites of *Salvia miltiorrhiza* Bge. *Frontiers in Plant Science, 12*, 617892. https://doi.org/10.3389/fpls.2021.617892.

Yizhu, L., Imtiaz, M., Ditta, A., Rizwan, M. S., Ashraf, M., Mehmood, S., Aziz, O., Mubeen, F., Ali, M., Elahi, N. N., Ijaz, R., Lelel, S., Shuang, C., & Tu, S. (2020). Response of growth, antioxidant enzymes, and root exudates production towards as stress in *Pteris vittata* and *Astragalus sinicus* colonized by arbuscular mycorrhizal fungi. *Environmental Science and Pollution Research, 27*, 2340–2352.

Yu, X.-Z., Wu, S.-C., Wu, F.-Y., & Wong, M.-H. (2011). Enhanced dissipation of PAHs from soil using mycorrhizal ryegrass and PAH-degrading bacteria. *Journal of Hazardous Materials, 186*(2–3), 1206–1217. https://doi.org/10.1016/j.jhazmat.2010.11.116.

Zhao, S., Yan, L., Kamran, M., Liu, S., & Riaz, M. (2024). Arbuscular mycorrhizal fungi-assisted phytoremediation: A promising strategy for cadmium-contaminated soils. *Plants, 13*(23), 3289. https://doi.org/10.3390/plants13233289.

Zhuo, F., Zhang, X.-F., Lei, L.-L., Yan, T.-X., Lu, R.-R., Hu, Z.-H., & Jing, Y.-X. (2020). The effect of arbuscular mycorrhizal fungi and biochar on the growth and Cd/Pb accumulation in *Zea mays*. *International Journal of Phytoremediation, 22*(10), 1009–1018. https://doi.org/10.1080/15226514.2020.1725867.

Chapter 10

Impact of arbuscular mycorrhizal fungi on soil microbial communities

Bakhtawar Zafar[a] **and Khoula Afzaal**[b]

[a] *Department of Plant Sciences, Quaid-I-Azam University, Islamabad, Pakistan,* [b] *Department of Botany, Government College Women University, Sialkot, Pakistan*

10.1 Introduction

The soil presents a highly complex and biological ecosystem, which harbors unimaginable microorganisms (Nannipieri, 2020). Microbes are important in the formation of soil aggregates and soil stability that provide the soil with fertility and productivity. The surface soil is never deficient in the native microbial population (Gupta et al., 2016).

Soil microbiota (i.e., bacteria, fungi, and archaea) form the largest population count and biomass of soil organisms (GSBI, 2020). Microorganisms in the soil are also involved in the important processes of ecosystems, including the carbon and nitrogen cycles (Mercado-Blanco et al., 2018). The recycling and loss of nitrogen are controlled by ammonia oxidation, denitrification, and ammonification to convert nitrogen between organic, inorganic, and gaseous forms. Microbial communities have a crucial role in the decomposition of several substances, such as organic pollutants, including pesticides (Barra Caracciolo & Grenni, 2022), and in stimulating plant growth and disease management (Saleem et al., 2019). The key regulators of the basic processes involved in the ecosystem are the diversity and biomass of soil microbial communities, which assist in crop production (Saleem et al., 2019). Indeed the essential conditions for plant growth and, accordingly, crop production include a good quality of soil, that is, the presence of multiple and rich microbial communities and their activity (Nazir et al., 2023; Paz-Ferreiro & Fu, 2016; Saeed et al., 2025; Thind et al., 2022; Yizhu et al., 2020). Specifically the presence of microbial biomass, activity, and diversity in the soil depicts the fertility of the soil and the ability of an ecosystem to remain productive. This is the reason why they are employed as markers of soil health and quality. The populations of microbes differ with the various abiotic factors, including the soil type, the existence/absence of plants, and the climate (Dinca et al., 2022).

Mycorrhiza is a symbiotic relationship between the roots and biotrophic mycorrhizal fungi in which the fungi form a dense hyphal network around or inside roots. This network enhances effective root surface area, which enhances water and essential nutrient uptake to the soil, especially phosphorus and micronutrients (Kumar et al., 2022). Arbuscular mycorrhiza (AM) is one of the diverse groups of AM. These are microscopic and multifid structures of the cortical cells of roots (Basu et al., 2018). The influence of AM fungi in this symbiotic association has massive potential, which has been well-recorded in terms of plant species to enhance agricultural ecosystems in terms of resilience, productivity, and sustainability (Lahrach et al., 2024). Overall, 80% of vascular plants contain these AM fungi, and they may undertake various functions. Mycorrhizae increase the levels of nutrient uptake by plants, thus mediating the growth of plants and enhancing soil properties and conditions, which affect the overall functioning of the ecosystem (Zaman et al., 2025).

The success of AM fungi is ecological because they have the capacity to obtain nutrients with extraradical foraging hyphae that inhabit the soil, alleviating abiotic stress, decreasing pathogen attack, and increasing hyphal growth. Relationships between AM fungi and soil microbiota can be negative as well as positive, which contribute significantly to this success (Fig. 10.1). These soil microbes are closely or loosely interacting with arbuscular mycorrhizal fungi (AMF) mycelia (Cruz & Ishii, 2011; Tarkka et al., 2018). The members of the Glomeromycotina were sequenced, and the genome lineage disclosed that the lineage does not possess some important genes. These include carbon assimilation, such as lignocellulose

Advancing Sustainable Agriculture Through Arbuscular Mycorrhizal Fungi. DOI: https://doi.org/10.1016/B978-0-443-45130-0.00025-4

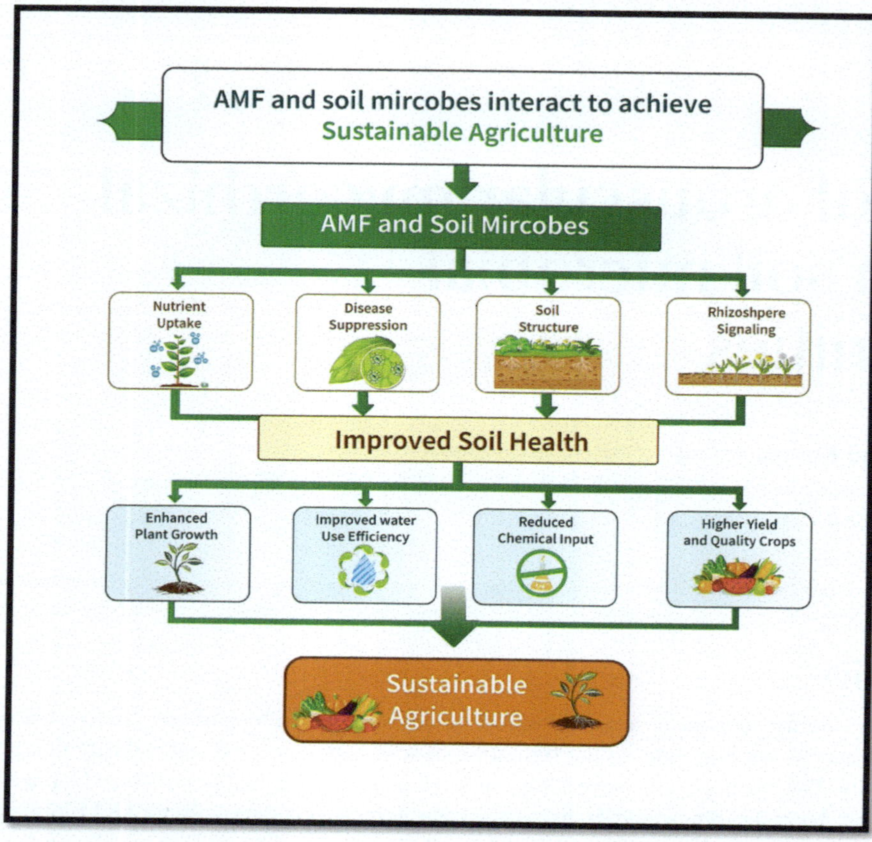

FIGURE 10.1 Synergic functional roles of AM fungi and soil microbes. *AM*, Arbuscular mycorrhiza; *AMF*, arbuscular mycorrhizal fungi.

breakdown; production of fatty acids and secondary metabolites; and thiamin biosynthesis genes (Lahrach et al., 2024). In addition, AMF cannot secrete invertase or sucrose transporters and are unable to mineralize insoluble nutrients (Lee et al., 2013). AM fungi mitigate this constraint by forming an obligate symbiosis with vegetation and mobilizing or colonizing useful soil microbiota.

The chapter is devoted to the multifunctionality of AMF in the health of vegetation and economically sustainable agriculture. It describes how AM fungi can increase the defense response of plants to biotic and abiotic stresses, such as drought, heavy metals, and pathogens. Other benefits of using AM fungi, as discussed in the chapter, refer to their ecological and synergistic relationships with other valuable microorganisms and the environmental benefits of their application, which include enhanced soil health or decreased dependence on chemical inputs.

10.2 Overview of arbuscular mycorrhizal fungi

The molecular evidence and fossil record indicate that AM symbiosis is evolutionarily old, initially emerging in the Devonian period, about 353–462 million years ago. Following that, their presence has been observed in a variety of geographical locations, and they have acquired new relationships with various plant species (Alrajhi et al., 2024). All angiospermic families contain AMF except for a few, including Betulaceae, Urticaceae, Commelinaceae, Cyperaceae, and Polygonaceae. Some significant genera include *Gigaspora, Glomus, Entrophospora, Acaulospora, Scutellospora*, and *Sclerocystis* (Aggarwal et al., 2011).

AMF develop special microscopic structures in the root cortex that protrude into the surrounding soil and carry out direct symbiotic activity. AMF are characterized by the existence of arbuscules, which are typical, much-branched hyphal structures developed between the cell wall and the plasma membrane of the plant (Souza, 2015). The main places of interaction of nutrients between the fungus and the host plant are the arbuscules. Conversely, AMF produces oval-shaped vesicles, which are spherical in shape, and which act as storage organs and are used for the storage of lipids and glycogen. They grow between or inside cortical cells and are observed in all major fungal families, including Acaulosporaceae and Glomeraceae (Souza, 2015).

10.2.1 Classification and taxonomy of arbuscular mycorrhizal fungi

The AMF were previously grouped by the order Glomales of the Zygomycota and then Glomeromycota. All of them are now united under the subphylum Glomeromycotina. Previously the fungal species were referred to as endomycorrhiza and then vesicular-AMs and finally AMF (Alrajhi et al., 2024).

Commonly the taxonomy of AM fungi has been determined by the morphological features of the spores, such as color, size, and cell wall layers. The spores of AM fungi possess simple structures and have few morphological characteristics. Moreover, spore samples obtained from the agricultural soil are not in their natural state, and various species of AM fungi do not develop spores in laboratory-controlled conditions. Such issues regarding the morphological characterizations of AM fungi allow the species identity to be confirmed only by taxonomists and have hampered the taxonomy and ecological studies of AMF (Lee et al., 2013). Today, rapid growth in genetic research and the development of molecular markers improved our understanding of AMF identity and diversity and their role in ecosystem functioning and promotion of plant growth. Molecular markers can be classified into non-Polymerase chain reaction (non-PCR)based and PCR-based approaches. They both have their merits and demerits. Hence the identification of AMF requires different markers, and AMF identified using two or more genetic markers is more efficiently identified than a single marker (Krishnamoorthy et al., 2017).

10.2.2 Plant-arbuscular mycorrhizal fungi interaction

AMF are mandatory mutualistic entities, and they rely on the roots of plants to complete their whole life cycle (Piliarova et al., 2019). The symbiotic plants maintain a relationship with AMF whereby they supply the essential sources of carbon and mineral nutrients that induce spore formation. Similarly, species of fungi help plants through their developed hyphal system in absorbing available water from the soil. Therefore it is not surprising that AM fungal species are highly abundant in the plants located in xeric and desert environments (Ma et al., 2022). Nevertheless, assisting the plant with water or supplying the fungi with C sources is a simplification of the structure–function relationship in terms of host association (Luo et al., 2023).

AMF play the primary role of increasing nutrient and water supply to the host plants. The nutrient that AMF has provided the most is phosphorus (P), whose extraradical hyphae reach out to the soil areas that have already depleted phosphorus, thereby enabling them to absorb phosphorus effectively within the soil (Qi et al., 2022). Moreover, AMF have properties to inhibit the destruction of nematode species and facilitate the general growth of the plant. As a result, mycorrhizal interaction between AMF and plants becomes multifunctional as an important indicator of soil quality, sustainability of the ecosystem, and agricultural resilience (Alrajhi et al., 2024).

10.3 Soil microbial diversity

10.3.1 Soil bacterial phyla

The investigations of the composition of the soil bacterial community using molecular tools have been extensively employed (Table 10.1). Such studies have revealed that even though bacteria have been grouped into over 100 phyla, they are not always present in large numbers in soils, with less than 10 phyla being popular in soils (Aislabie et al., 2013). These dominant phyla may occur in relative abundance across different soils or samples, which is a manifestation of the differences in the soil properties, environmental factors, and ecological niches. The members of the phyla α-, β-, γ-, and δ-Proteobacteria, Acidobacteria, and Actinobacteria are considered to be the most common and abundant, whereas the representatives of the Verrucomicrobia, Bacteroidetes, Firmicutes, Chloroflexi, Planctomycetes, and Gemmatimonadetes phyla are usually rare in the soil environment (Kim et al., 2021).

Proteobacteria are a metabolically and functionally diverse group of organisms and include several subphyla that carry out a broad ecological repertoire. These include the α-, β-, γ-, and δ-Proteobacteria subphyla, which are the most reported in soils. These bacteria play vital roles in soil that include nutrient cycling, decomposition, fixation of nitrogen, and plant and other soil microorganism interactions (Aislabie et al., 2013).

10.3.2 Soil fungal phyla

The fungal communities in soil are taxonomically diverse and play essential roles in soil processes, nutrient cycling, and plant health (Table 10.2). The prevailing fungal phyla in the soil are Ascomycota, Basidiomycota, Glomeromycota, Mortierellomycota, Mucoromycota, and Chytridiomycota, each having specific ecological functions (Tedersoo et al., 2014; Treseder & Lennon, 2015). Ascomycota and Basidiomycota are the most abundant groups of soil-dwelling fungi. They

TABLE 10.1 Taxonomy, examples, and ecological functions of dominant soil bacterial diversity,

Bacterial phylum	Subphylum/class	Representative members	Functions
Proteobacteria	α-Proteobacteria β-Proteobacteria δ-Proteobacteria	*Rhizobium, Bradyrhizobium* *Burkholderia, Nitrosomonas* *Pseudomonas, Azotobacter*	Nitrogen fixation, phosphate solubilization, siderophore production, organic matter decomposition (Fierer, 2017; Philippot et al., 2013)
Actinobacteria	Actinobacteria	*Micromonospora, Streptomyces*	Organic matter decomposition, nutrient cycling (Barka et al., 2016)
Firmicutes	Bacilli	*Bacillus, Paenibacillus*	Decompose biomass, promote plant growth (Huang et al., 2023)
Acidobacteria	Acidobacteria	*Acidobacterium, Solibacter*	Decompose organic matter, cycle nutrients (Kalam et al., 2020)
Planctomyces	Planctomycetia	*Planctomyces*	Nitrogen cycle involving anaerobic ammonium oxidation (Fuerst et al., 2011)
Bacteroidetes	Bacteroidia	*Flavobacterium*	Nutrient cycling degrades complex polysaccharides (Wexler, 2007)
Verrucomicrobia	Verrucomicrobiae	*Opitutus*	Polysaccharide degradation, carbon cycling (Bergmann et al., 2011)
Cyanobacteria	Cyanophyceae	*Anabaena, Nostoc*	Oxygenic photosynthesis, nitrogen fixation (Whitton et al., 2012; Zehr, 2011)

TABLE 10.2 Taxonomy, examples, and ecological functions of dominant soil fungal diversity.

Fungal phylum	Subphylum/Class	Representative examples	Functions
Ascomycota	Sordariomycetes Eurotiomycetes Dothideomycetes	*Tricoderma, Fusarium* *Penicillium, Aspergillus* *Cladosporium, Alternaria*	Organic matter decomposition (Maharachchikumbura et al., 2016) Organic matter decomposition, secondary metabolite production (Abdel-Azeem et al., 2019) Soil organic matter degradation (Wang et al., 2023a)
Basidiomycota	Agaricomycetes	*Pleurotus, Agaricus*	Lignin and plant polymer decomposition, nutrient cycling (Manici et al., 2024)
Glomeromycota	Glomeromycetes	*Rhizophagus, Funneliformis*	Form AM symbioses and enhance plant nutrients and water uptake (Helgason et al., 2009)
Mucoromycota	Mucoromycetes	*Rhizopus, Mucor*	Plant symbiosis, nutrient transfer (Rimington et al., 2020)
Mortierellomycota	Mortierellomycetes	*Mortierella*	Organic matter decomposition, fatty acid production (Li et al., 2020)
Chytridiomycota	Chytridiomycetes	*Rhizophlyctis*	Zoospore-mediated nutrient cycling (Gleason et al., 2017)

AM, arbuscular mycorrhiza.

are principal drivers in degrading plant litter, lignin, and cellulose, hence modulating carbon cycling and soil organic matter dynamics (Tedersoo et al., 2014). AMF, or Glomeromycota, are mutualistic and form associations dominantly with the roots of land plants and contribute to the acquisition of phosphorus, uptake of micronutrients, and soil aggregation through the production of glomalin (Chaudhary et al., 2025). Fast-growing saprotrophs, Mortierellomycota and Mucoromycota, facilitate the initial stages of organic matter breakdown and nutrient release, especially under nutrient-enriched conditions (Treseder & Lennon, 2015). Chytridiomycota are relatively rare in terrestrial ecosystems, but they contribute to the degradation of complex polymers such as chitin and cellulose, particularly in moist or waterlogged soils

(Tedersoo et al., 2014). These fungal phyla coexist with plants, bacterial species, and AMF to ensure the stability of the ecosystem and agricultural productivity.

10.3.3 Soil archaeal phyla

Archaea are also not as common, although microscopic, and play crucial parts in the biogeochemical cycles, contributing to the nutrient transformation process and overall ecosystem performance. The archaeal groups that are the most common in soils are Euryarchaeota and Nanoarchaeota, whereas in terrestrial environments, the most common are Thaumarchaeota and Crenarchaeota (Offre et al., 2013; Xiong et al., 2018). The Thaumarchaeota phylum has a significant role in ammonia oxidation, and they are also important in catalyzing nitrification in acidic and nutrient-limiting soil where nitrifiers are not as widespread as bacteria (Prosser & Nicol, 2012). The cause of greenhouse gas emissions is anaerobic methane production by methanogenic archaea on waterlogged or flooded soils (Le Mer & Roger, 2001). Although Crenarchaeota have been known to only establish themselves in harsh environments, they are very prolific in the problem of soil, nitrogen cycles, and the transformation of organic matter (Offre et al., 2013). Nanoarchaeota are less common and may be symbionts or parasites of other archaeal groups and may influence microbial interactions and networks, thus affecting the soil ecosystem (Xiong et al., 2018). Their complex interaction with the community of soil bacteria and fungi affects nutrient turnover and soil fertility as well as stability of terrestrial ecosystems.

10.4 Influence of arbuscular mycorrhiza fungi on the diversity and function of soil microbes

The AMF contribute to the increased microbial diversity and richness of the soil habitat by facilitating colonization of rhizosphere microbes along their mycelial networks. It is known that the chemical properties of soil may greatly affect the structure and activity of the bacterial population (Huang et al., 2019). Nevertheless, these chemical properties can be significantly changed by root exudates that are released into the hyphae area of the rhizosphere. As a result, plant–root interactions with AMF can alter microbial communities within a specified root zone, as well as those of other ecologically related environments, and thus alter the overall soil microbial structure (Wang et al., 2024).

Moreover, AMF also uses glomalin, which is a glycoprotein that contributes significantly to soil aggregation, structural stability, and water retention (Rillig, 2004). AMF releases glomalin, which stabilizes the particles of soil in the outer hyphae and on the walls of the spores with organic matter (Atakan & Özkaya, 2021). When fungal biomass metabolizes with time and replenishes the soil, the glomalin is located within the soil matrix and is a substantial part of the soil organic matter (Magurno et al., 2019). Carbon sequestration, nutrient cycle, soil aggregation, maintenance of biodiversity, immobilization of heavy metals, binding of organic pollutants, and recovery of degraded ecosystems are some of the ecosystem activities that have been carried out by glomalin. It also supplies nutrients to both plants and microorganisms in soil. It has been established that glomalin can be used to increase the stability of soil macroaggregates (Ji et al., 2019). The higher levels of glomalin relate to better physical characteristics of the soil, such as improved porosity, large water capacity, soil moisture, and lower bulk density. Moreover, glomalin depicts the improvement of nutrient content, the enhancement of microbes, and the general enhancement of soil fertility (Pal & Pandey, 2014).

The colonization of mycorrhizae triggers the activation of the defense-related genes, and the response helps to increase the resistance of plants against bacterial and fungal pathogens (Jung et al., 2012). AM fungi enhance the development of helpful rhizosphere-related microbes through altering the exudation pattern of roots and the synthesis of the antimicrobial efficacy of secondary metabolites. AM fungi improve the resistance to biotic stress in plants by boosting their immune system, encouraging the competence of plants, and boosting their physiological functioning (Rodriguez-Caballero et al., 2017). Rhizobacteria also go an extra mile to mobilize unavailable nutrients, fix nitrogen in the atmosphere, produce growth-promoting phytohormones, suppress soil-borne diseases, and enhance plant resistance to stress (Singh et al., 2020). It is widely known that the plant growth-promoting rhizobacteria (PGPR) to solubilize unavailable phosphorus through various mechanisms, such as acidification, chelation, ion exchange, phosphatases, and organic acid secretion, which are the processes that are important in the mobilization of nutrients in soils (Maldonado et al., 2020). AM fungi and PGPR, when used together, form a synergistic relationship that increases nutrient availability and improves the biological and physical properties of the soil, which results in healthier and more productive ecosystems (Kumar et al., 2025). The interactions enhance plant disease resistance through complicated signaling that is connected to induced systemic resistance (ISR). AMF used in combination with ISR processes promotes the defense responses in plants (Hussain et al., 2024; Yu et al., 2022). AMF control the ISR in plants by triggering the activation of plant defense system , and making it possible to respond to the attack of pathogens quite quickly and effectively. (Badrbani et al., 2024). In addition, AMF balance biochemical defense systems result in the production of defensive molecules in plants (Tao et al., 2016). The coinoculation of rhizobacteria and AM fungi was demonstrated to affect the productivity of plants, their cycling of nutrients, the sequestration of carbon, the microbial diversity, and the functional capacity of the microbial community (Boyno et al., 2025).

The mycorrhizal fungi have an extensive influence on the microbial population of soil. Plant rhizosphere and associated microorganisms' communication involve the mechanism by which the diversity and structural composition of the microbial population in the rhizosphere are manipulated. The colonization by mycorrhizal fungi also changes the structure of the soil microorganisms because it provides the bacteria and other fungi with additional carbon and other nutrients (Miransari, 2011). These will enhance the circulation of nutrients, biodegradation of organic products, and the general ecosystem processes. In addition, AMF were also observed to increase soil biodiversity by modulating the microbial community structure, colonizing potentially useful microorganisms, and improving mycorrhizal networks (Powell & Rillig, 2018).

The growth and functioning of AMF are supported by the presence of microorganisms in the soils and their interaction pattern. It is also demonstrated that AMF can transport nutrients between plants and the surrounding microorganisms through the established symbiotic associations, which can be experimentally verified with a stable isotope tracing technique (Duan et al., 2024). This type of nutrient exchange augments nutrient cycling in the soil ecosystems and improves the quality of the soil because it leads to an increase in the activity of microbes. It is also mentioned that AMF increases the bioaccumulation of organic matter and the overall activity of the biological environment in the soils, which improves the health and performance of the soil (Zhou et al., 2020).

10.5 Interactions between arbuscular mycorrhizal fungi and the other microbial communities

The importance of microorganisms in soil is that they promote tolerability of the plants to the stresses in the environment as well as offer long-term sustainable solutions (Demir et al., 2022). One of such microorganisms is the PGPR, which has an intimate relationship with the plant root systems and whose presence improves nutrient uptake efficacy through a number of biological processes. These beneficial microbes can also have an effect of increasing efficiency of water use as they reduce dependency of plants on water in unfavorable conditions, such as abiotic stress, that is, drought (Ullah et al., 2019). AMF also contribute to this role by improving the bioavailability of key nutrients, especially phosphorus, which is a key macronutrient essential in the growth of plants, energy transfer, and general metabolic processes. But generally, it does not have easy access to the soil. AMF enhances phosphorus solubility and transfer, thereby boosting its uptake in plants (Bhantana et al., 2021). Furthermore, AMF form hyphae networks, which are so extensive as to exceed the root zone, thus increasing the efficiency of water and nutrient circulation in the circuitry of the soil (Kuyper et al., 2021). On the same note the rhizobacteria increase the accessibility of the nutrients and collaborate in symbiosis with the plants to promote the plant growth and nutrient uptake (Pii et al., 2015). AMF also augment the consumption of potassium by expanding the useful root surface area, enabling crops to access the nonavailable potassium stores (Shin, 2014). It is confirmed that AMF, along with the presence of potassium, might influence the morphology of roots and increase the capacity to take up nutrients (Yuan et al., 2023).

Recent evidence shows that AMF has the direct ability to regulate potassium uptake by stimulating root potassium transporters; for example, the LbHAK gene was discovered in the *Lycium barbarum* (Zhang et al., 2024). The uptake of potassium is listed among the valuable processes contributing to the improved tolerance of plants to the abiotic stressors (Zhang et al., 2024). AMF, when exposed to saline, can regulate sodium and chloride ion channel expression, and this further describes how it mitigates the effects of abiotic stressors to plants (Wang et al., 2023b). This has enhanced the application of microbial inoculants, which include AM fungi, PGPR, species of *Trichoderma*, or their combined use to enhance soil fertility, plant growth, and ecosystem maintenance in the field of sustainable agriculture (Basiru & Hijri, 2022). The relationships between AM fungi and rhizosphere bacteria are intricate and multidimensional in terms of nutrient exchange, signaling pathways, mutual development, and activity of both sides, which jointly improve plant health and soil ecosystem activities, greatly useful to the growth and well-being of plants (Wahab et al., 2023). Interactions of such sorts can inhibit pathogens in several ways (Perez-de-Luque et al., 2017). Plant-related rhizobacteria express protective peptides, including antimicrobial peptides, siderophores, and organic acids, and assist virulence reduce the activity of pathogens within the rhizosphere. Some rhizobacteria induce systemic plant defense responses, including salicylic acid-mediated responses, to enhance the overall plant health and development (Zhu et al., 2022).

It is a series of procedures that AMF uses to safeguard the root of plants against external attacks. Mycorrhizal networks strengthen the root defense mechanisms and make them less vulnerable to pathogenic organisms (Boyno & Demir, 2022). Besides, phytohormones are also synthesized because of AMF–rhizobacteria interaction, and they

TABLE 10.3 Synergic effect of AM fungi and PGPR on different plant species.

AM fungus	PGPR	Interaction effects
Grammostola rosea	*Pseudomonas putida*	Plant growth promoted, PGPR density reduced
Glomus fasciculatum	*Azospirillum brasilense*	The growth of the plant improved, and the population of PGPR in the rhizosphere increased
Glomus intraradices	*Pseudomonas alcaligenes* *Bacillus pumilus*	Dry mass of shoots increased, AMF root colonization improved
G. intraradices	*Bacillus thuringiensis*	Root growth improved, AMF root colonization and *arbuscule* richness also increased
G. intraradices	*Bacillus subtilis*	Plant growth improved, reducing root biomass
G. fasciculatum	*Pseudomonas* sp.	Root colonization of AMF improved, and the biomass of the plant increased
G. intraradices	PGA is the combination of various *Bacillus* species	In the field, the amount of nitrogen improved, and the capacity of the P ion also increased

Adopted from Gopal et al. (2012); Saldajeno and Hyakumachi (2011)
AM, arbuscular mycorrhiza; *AMF*, arbuscular mycorrhizal fungi; *PGA*, plant growth activator; *PGPR*, plant growth-promoting rhizobacteria.

directly enhance vegetative growth (Kumar et al., 2015). The rhizobacteria can also secrete plant growth regulators, including auxins, cytokinins, ethylene, and gibberellins, which enhance the growth of the plant (shoot and root) and make the plant vigorous (Vejan et al., 2016). The AMF–rhizobacteria may be synergistic to boost the manufacture and action of these hormones in enhancing the growth performance under both favorable and unfavorable environments (Sbrana et al., 2014). Moreover, the AM fungi mycelial networks contribute to the efficient movement of the phytohormones in the soil–root network and the improvement of the physiological response of the plants (Sun et al., 2024).

AMF also includes mycelial networks that are associated with improved soil structure by holding the soil particles together and forming aggregates that increase the aeration and stability of soil (Fall et al., 2022). Such hyphal networks also play a role in water conservation by lowering water evaporation and enhancing water retention for plants during droughts. The second benefit of Rhizobacteria is that it enhances the strength of soil aggregates and the integrity of soil (Naseem et al., 2018). These interactions can help plants to either survive and grow in drought and other abiotic stressors; the AMF hyphae and their spores are central to the morphogenesis of plants. PGPR has also been depicted to induce germination and growth of AM fungi spores and mycelia, which also contribute to further improvement of the performance of the plants (Gopal et al., 2012). Previous studies have demonstrated that crops (e.g., basil and lettuce) react to inoculation with PGPR synthesizing flavonoids (Boyno et al., 2025). These flavonoids are a signaling molecule in mycorrhizal settlement and have a considerable positive association with AMF root colonization, which implies that these flavonoids are a significant process through which AMF root settlement is regulated utilizing PGPR (Zeng et al., 2025).

Such facilitative interactions enable AMF to be an increasingly successful colonizer of plant roots and also to stimulate plant growth under harsh conditions (Table 10.3). Nevertheless, certain doubts concerning the use of AMF and PGPR simultaneously exist. Although many studies have reported a higher AMF colonization and enhanced growth in plants with coinoculation, other studies have observed that there is no notable difference in AMF colonization when there are specific conditions or treatments (Begum et al., 2022). These variations might be attributed to the variations in the plant species, microbial species, the environmental conditions, and microbial performance under various conditions of stress. Regardless of these variations, both the synergistic use of the AMF and the PGPR have a lot of potential, particularly in improving drought stress tolerance. A study has shown that AMF and PGPR interaction in drought conditions induces growth and photosynthetic rate in plants because they activate the antioxidant defense system and increase the metabolism of nutrients, such as those observed in tobacco plants (Begum et al., 2019). Similarly the AMF coexists with spore-associated bacteria, and they regulate genes involved in ion transport and K^+/Na^+ homeostasis to enhance AM fungi symbiosis and increase salinity stress tolerance in maize (Selvakumar et al., 2018). Nevertheless, the mechanisms associated with AM fungi and PGPR interactions in the pathogen resistance are poorly characterized. Although some fixed amount of research studies indicate that the expression of the systemic defense by the presence of the PGPR increases the resistance of plants to invasion by the pathogen, it is not always connected to better colonization of AMF, which is why further research is needed (Savastano & Bais, 2024).

FIGURE 10.2 Benefits of the synergistic effect of AMF and soil microbes on agriculture and the ecosystem. *AMF*, Arbuscular mycorrhizal fungi.

10.6 Application of arbuscular mycorrhiza fungi and multiple microbial agents in agriculture

Plant pathogens that are transmitted by soil are mainly fungi and bacteria and should be managed using effective long-term management strategies. They can be eradicated culturally, chemically, biologically, and physically. Nevertheless, the ongoing application of synthetic pesticides in the continuous cropping farm has also been found to be convoluted and mostly inefficient, which has attracted more significance to the use of friendly and sustainable technologies in managing diseases (Panth et al., 2020).

In the earlier methods the use of one microbial agent was the primary method of biological control. The recent works, however, suggest that there are more efficient and more reliable effects when multiple microorganisms are used together in the prevention of diseases and promoting plant growth (Boyno et al., 2025). The types of symbiotic microorganisms that inhabit plant roots are AM fungi, PGPR, endophytic fungi, and species of *Trichoderma*. These useful microbes can prevent the infection of pathogens and, at the same time, increase the growth of the plants. Individual biological control agents can suppress some of the pathogens, even though their effect is usually enhanced when they are used in conjunction. A study revealed that by using AM fungi and *Bacillus* spp. in the field together, it was possible to reduce the recommended amount of NPK fertilizer by 50% without affecting the growth of plants, their yield, or the nutrient status (Nanjundappa et al., 2019). In the same line, coinoculation with *Bacillus* spp. and *Funneliformis mosseae* or with *Bacillus* spp. and *Glomus versiforme* induces resistance against bacterial wilt disease and results in significantly increased root length, causing potato growth (Shu-Peng et al., 2015).

The plant–bacteria, fungus–bacteria, and plant–fungus systems represent natural examples of mutualistic symbiotic systems, which act in an integrated biological unit (Fig. 10.2). The collective symbiotic interactions usually work better than single organisms do when implemented. AMF tests in the field have also proved that the combination of AMF and *Trichoderma longibrachiatum* can be used with success to reduce soil-borne diseases and improve the quality of crops (Yang et al., 2022). In the second field study, Wang et al. (2019) stated that *Trichoderma harzianum* in conjunction with *Bacillus cereus* offered better management of tomato root-knot nematodes. Moreover, *T. harzianum* has antagonistic behavior against *Rhizoctonia solani*, and *Bacillus subtilis* neutralizes the same pathogen in other ways. These complementary inhibitory effects were further crossed to give positive inhibition of black scurf potato disease, which was caused by *Alternaria solani* (Boyno et al., 2022).

10.7 Ecological importance of arbuscular mycorrhiza fungi

AMF have several benefits regarding agriculture and ecosystem management because of the special features of these organisms. AMF also has promoted the development of plants by establishing a symbiotic relationship with plant roots and improving the resistance of the plants to adverse environmental factors and enhancing their survival in adverse environments (Begum et al., 2019). The result of this symbiotic relationship is a massive network of hyphae that increases the interface of the soil and root that triggers the enhanced absorption of important macro- and micronutrients (Khaliq et al., 2022). In addition, AMF also improves environmental sustainability by reducing the effects of drought, salinity, and heavy metal contamination with increased phosphorus and nutrient solubility (Khan et al., 2023).

Besides the increased growth of plants and the alleviation of stress, AMF also have a crucial role in improving the soil's quality and environmental sustainability. Another important contribution of the AMF is the production of glomalin, a glycoprotein that enhances the aggregate stability of the soil, antierosion resistance, and the soil structure (Zhang, Dong, & Shangguan, 2023). The above features render AMF a handy tool in making sure that the agricultural practices incur a smaller carbon footprint, besides minimizing climate change challenges (Rillig, 2004). Being able to coexist in symbiosis with plant roots, AMF allows the absorption of other nutrients, such as phosphorus, which are normally limited in soil, and reduces the necessity to use chemical fertilizers (Wahab et al., 2023). It is a natural gain of nutrients that lowers the production cost and lowers the pollution of soil and water (Bhantana et al., 2021). In addition, AMF develops a biological shield near plant roots that prevents pathogens, thus avoiding the use of pesticides (Harrier & Watson, 2004). As a result, AMF helps to decrease the level of environmental degradation and helps to switch to the environmentally oriented agricultural systems (Gosling et al., 2006). Overall the diminishing of the number of chemical inputs related to AMF utilization is one of the main approaches to sustainable agriculture. Moreover, AMF increases biodiversity in the ecosystem by engaging with the microbiome that is associated with roots and facilitating the growth of useful microorganisms that enhance the health of plants and keep the soil ecologically balanced (Zhang et al., 2024). All these benefits emphasize AMF as significant players in better crop growth, a decrease in production costs, and environmental protection. Their wide usage provides a strategic solution to increasing crop productivity and the sustainability of the ecosystem. Thus the popularization of AM fungi-based technologies is a significant direction toward the attainment of agriculture and ecological sustainability (de Oliveira et al., 2024).

AMF are obligatory symbionts that are incapable of growing in pure culture and developing their life cycle in the absence of a host plant. This is a biological limit that renders the mass production of AMF inoculum time-consuming and technologically demanding. Inoculum is usually prepared using soil taken in root zone of AMF-colonized plants, which contains infected root fragments, hyphal structure, and spores. Moreover, in the preparation of crude inoculum, soil-borne spores can be applied as inoculants. The crude inoculum production can be achieved through the cultivation of an established AMF isolate in the presence of a host trap plant, followed by massive colonization in an inert growth medium optimally suited to propagate AMF (Boyno et al., 2025). AMF inoculants are priced based on formulation and quality, yet high propagule density inoculum products are typically used when large quantities of crop inoculants are required in large-scale inoculation. Moreover, another source of inoculum is colonized root fragments of known AMF hosts, which can be used instead (Ghorui et al., 2025).

In addition, not every crop species forms an equally efficient symbiosis with AM fungi, and some of the modern cultivars with high yields have a diminished responsiveness because of their dependency on high fertilizer levels. The environmental factors affecting the colonization efficiency and functional activity of AMF comprise environmental factors such as soil moisture, temperature, and seasonal variation. Practical issues concerning the mass production of inocula, their formulation, storage, and use in the field may further reduce the uniformity and consistency of AMF-based biofertilizers. Hence, although AMF are a prospective and ecofriendly solution to sustainable agriculture, their implementation needs to be fully adjusted to local agricultural systems to ensure that the benefits are maximized for the plant and the possible limitations or constraints are minimized (Igiehon & Babalola, 2017).

10.8 Advances in microbiome research and omics technologies

The studies of microbiomes have formed a central cornerstone to the study of the complicated interactions between the plant root systems and the surrounding microorganisms. The rhizosphere is a very dynamic zone of the soil that is directly affected by plant roots and that has a variety of microorganisms, such as bacteria, fungi, and archaea (Bulgarelli et al., 2013).

The advancement of omics-based methodologies, including metagenomics, metaproteomics, and metatranscriptomics, has significantly increased the information regarding microbial communities as well as their functional capabilities in various ecosystems (Qi et al., 2023).

These high-end measures allow the study of microbial interactions in detail in terms of competition, cooperation, and nutrient exchange using gene expression dynamics and metabolic networks in microbial consortia. With such approaches, our perceptions of microbiomes in various environments have transformed because it has revealed certain significant roles played by ecological and physiological processes that control the dynamics of microbial communities. As an example the rhizosphere microbiome of crop plants can be studied to design beneficial bacterial groups (e.g., nitrogen-fixing rhizobia) and the genes involved in nitrogen fixation, which will help understand how the microbial communities can help the plants to acquire nutrients (Swarupa et al., 2016).

Tian et al. (2015) made a comparative analysis of rhizospheric microbiomes of healthy tomato plants and nematode-infected tomato plants. Their metagenomic evaluation revealed that microbial communities inside root galls were enhanced in genes working in the degradation of plant polysaccharides, carbohydrate and protein metabolism, and nitrogen fixation. Moreover, comparative and metagenomic studies of plant-associated microbiomes have provided new information on the phylogenetic distribution of these communities and the complex functional interactions between host plants and these communities (Meena et al., 2017).

The basic microbial community that surrounds the roots of the plants is important in facilitating the development and health of the plants. Characterization of core microbiomes in the endosphere, rhizosphere, and phyllosphere has generally been carried out using taxonomic markers. Network-based analyses provide a broader view into the taxonomic composition and functional capacity of the microbial communities more than the perceptions based on the research conducted on the individual taxa. The combination of taxonomic data with functional gene information has aided in clear visualization of the roles of groups of microbes in specific processes, including nitrogen fixation and carbon sequestration, as well as in the degradation of pollutants. These interactions of networks underline the collaborative character of the microbial communities in the functioning of the ecosystems. The variations in the correlation between microbial and plant diversity can be explained by various factors, and interactions between the plants and the microbes differ across different groups of microbes. The site-level examinations have also demonstrated that the population of nitrogen-fixing bacteria is more strongly connected with plant diversity as compared to other microorganisms (Liu et al., 2020).

Wu et al. (2017) found that the distribution of bacterial diversity in arable soils is highly dependent on soil pH, whereby higher levels of diversity are usually found in neutral soils than in acidic soils. They also showed that the structure of microbial communities in agricultural systems is determined by their acid-base balance, the stability of the sorption complex, the level and the quality of organic matter, and the availability of nutrients. Moreover, the abiotic factors responded differently to prokaryotic and fungal communities in orchards and vineyards. The microbiomes of vineyards were more resistant to herbicide application compared to the orchard soils. Herbicides in orchard systems have also changed the abundance of *Chthoniobacterales, Flavobacterium*, and *Monographella* significantly, yet possibly pathogenic fungi, including *Nectria* and *Thelonectria,* have also grown. The findings are very informative in enhancing better practice in soil management to ensure the maintenance of healthy microbiomes in agriculture.

10.9 Conclusion

Molecular research has demonstrated that although a few microbial groups (phyla) comprise a majority of the soil communities, they are important in the regulation of soil functionality and ecosystem stability. AMF enhances nutrient uptake, soil structure, and plant stability, resulting in soil fertility, ecosystem stability, and agricultural sustainability. Their impact on the environment is manifested through the increased microbial activity and soil organic matter, as well as plant resistance to environmental stress. The recent advancements in the technologies of omics have created the possibility to examine AMF-related microbial networks in detail and learn how they contribute to the process of nutrient cycling and carbon sequestration and how they sustain healthy and sustainable soil ecology. The results indicate how dire the situation with AMF is to promote productive and sustainable agroecosystems.

References

Abdel-Azeem, A. M., Abdel-Azeem, M. A., Abdul-Hadi, S. Y., & Darwish, A. G. (2019). *Aspergillus*: Biodiversity, ecological significances, and industrial applications. In *Recent advancements in white biotechnology through fungi: Volume 1: Diversity and enzymes perspectives* (pp. 121–179). Springer International Publishing.

Aggarwal, A., Kadian, N., Tanwar, A., Yadav, A., & Gupta, K. K. (2011). Role of arbuscular mycorrhizal fungi (AMF) in global sustainable development. *Journal of Applied and Natural Science, 3*(2), 340.

Aislabie, J., Deslippe, J. R., & Dymond, J. (2013). Soil microbes and their contribution to soil services. In *Ecosystem services in New Zealand–conditions and trends: 1* (pp. 143–161). Manaaki Whenua Press.

Alrajhi, K., Bibi, S., & Abu-Dieyeh, M. (2024). Diversity, distribution, and applications of arbuscular mycorrhizal fungi in the Arabian Peninsula. *Saudi Journal of Biological Sciences, 31*(2), 103911.

Atakan, A., & Özkaya, H. Ö. (2021). Arbuscular mycorrhizal fungi and glomalin. *Turkish Journal of Agriculture-Food Science and Technology, 9*(12), 2371–2375.

Badrbani, A. H., Amini, J., Sharifi, R., & Karimi, K. (2024). Arbuscular mycorrhizal fungi induce resistance in tomato plants against *Fusarium* wilt through transferring an underground warning signal. *Physiological and Molecular Plant Pathology, 133*, 102380.

Barka, E. A., Vatsa, P., Sanchez, L., Gaveau-Vaillant, N., Jacquard, C., Klenk, H. P., & van Wezel, G. P. (2016). Taxonomy, physiology, and natural products of Actinobacteria. *Microbiology and Molecular Biology Reviews, 80*(1), 1–43.

Barra Caracciolo, A., & Grenni, P. (2022). Bioremediation of soil ecosystems from triazine herbicides. In *Pesticides in soils: Occurrence, fate, control and remediation* (pp. 353–377). Springer International Publishing.

Basiru, S., & Hijri, M. (2022). The potential applications of commercial arbuscular mycorrhizal fungal inoculants and their ecological consequences. *Microorganisms, 10*(10), 1897.

Basu, S., Rabara, R. C., & Negi, S. (2018). AMF: The prospect for sustainable agriculture. *Physiological and Molecular Plant Pathology, 102*, 36–45.

Begum, N., Qin, C., Ahanger, M. A., Raza, S., Khan, M. I., Ashraf, M., & Zhang, L. (2019). Role of arbuscular mycorrhizal fungi in plant growth regulation: Implications in abiotic stress tolerance. *Frontiers in Plant Science, 10*, 1068.

Begum, N., Wang, L., Ahmad, H., Akhtar, K., Roy, R., Khan, M. I., & Zhao, T. (2022). Co-inoculation of arbuscular mycorrhizal fungi and the plant growth-promoting rhizobacteria improves growth and photosynthesis in tobacco under drought stress by up-regulating antioxidant and mineral nutrition metabolism. *Microbial Ecology, 83*, 971–988.

Bergmann, G. T., Bates, S. T., Eilers, K. G., Lauber, C. L., Caporaso, J. G., Walters, W. A., & Fierer, N. (2011). The under-recognized dominance of Verrucomicrobia in soil bacterial communities. *Soil Biology and Biochemistry, 43*(7), 1450–1455.

Bhantana, P., Rana, M. S., Sun, X.-c., Moussa, M. G., Saleem, M. H., Syaifudin, M., Shah, A., Poudel, A. P., Pun, A. B., Alam, M., Mandal, D. L., Shah, S., Dong, Z., Tan, Q., & Hu, C.-X. (2021). Arbuscular mycorrhizal fungi and their major role in plant growth, zinc nutrition, phosphorus regulation, and phytoremediation. *Symbiosis, 84*, 19–37.

Boyno, G., & Demir, S. (2022). Plant-mycorrhiza communication and mycorrhizae in inter-plant communication. *Symbiosis, 86*, 155–168.

Boyno, G., Demir, S., & Danesh, Y. R. (2022). Effects of some biological agents on the growth and biochemical parameters of tomato plants infected with *Alternaria solani* (Ellis & Martin) Sorauer. *European Journal of Plant Pathology, 162*(1), 19–29.

Boyno, G., Rezaee Danesh, Y., Çevik, R., Teniz, N., Demir, S., Demirer Durak, E., & Mulet, J. M. (2025). Synergistic benefits of AMF: Development of a sustainable plant defense system. *Frontiers in Microbiology, 16*, 1551956.

Bulgarelli, D., Schlaeppi, K., Spaepen, S., Van Themaat, E. V. L., & Schulze-Lefert, P. (2013). Structure and functions of the bacterial microbiota of plants. *Annual Review of Plant Biology, 64*(1), 807–838.

Chaudhary, A., Poudyal, S., & Kaundal, A. (2025). Role of arbuscular mycorrhizal fungi in maintaining sustainable agroecosystems. *Applied Microbiology, 5*(1), 6.

Cruz, A. F., & Ishii, T. (2011). Arbuscular mycorrhizal fungal spores host bacteria that affect nutrient biodynamics and biocontrol of soil-borne plant pathogens. *Biology Open, 1*(1), 52–57.

Demir, S., Danesh, Y. R., Boyno, G., & Najafi, S. (2022). Arbuscular mycorrhizal fungi in biotic and abiotic stress conditions: Function and management in horticulture. In *Sustainable horticulture* (pp. 157–183). Elsevier.

de Oliveira, I. F., Campolino, M. L., de Oliveira, R. G., de Paula Lana, U. G., Gomes, E. A., & de Sousa, S. M. (2024). Agronomic practices for optimizing the AMF abundance and diversity for sustainable food production. In M. Parihar, A. Rakshit, A. Adholeya, & Y. Chen (Eds.), *Arbuscular mycorrhizal fungi in sustainable agriculture: Nutrient and crop management* (pp. 55–76). Springer Nature Singapore.

Dincă, L. C., Grenni, P., Onet, C., & Onet, A. (2022). Fertilization and soil microbial community: A review. *Applied Sciences, 12*(3), 1198.

Duan, S., Feng, G., Limpens, E., Bonfante, P., Xie, X., & Zhang, L. (2024). Cross-kingdom nutrient exchange in the plant–arbuscular mycorrhizal fungus–bacterium continuum. *Nature Reviews Microbiology, 22*, 773–790.

Fall, A. F., Nakabonge, G., Ssekandi, J., Founoune-Mboup, H., Apori, S. O., Ndiaye, A., & Ngom, K. (2022). Roles of arbuscular mycorrhizal fungi on soil fertility: Contribution to the improvement of physical, chemical, and biological properties of the soil. *Frontiers in Fungal Biology, 3*, 723892.

Fierer, N. (2017). Embracing the unknown: Disentangling the complexities of the soil microbiome. *Nature Reviews Microbiology, 15*(10), 579–590.

Fuerst, J. A., & Sagulenko, E. (2011). Beyond the bacterium: Planctomycetes challenge our concepts of microbial structure and function. *Nature Reviews Microbiology, 9*(6), 403–413.

Ghorui, M., Chowdhury, S., & Burla, S. (2025). Recent advances in the commercial formulation of arbuscular mycorrhizal inoculants. *Frontiers in Industrial Microbiology, 3*, 1553472.

Gleason, F. H., Scholz, B., Jephcott, T. G., van Ogtrop, F. F., Henderson, L., Lilje, O., & Macarthur, D. J. (2017). Key ecological roles for zoosporic true fungi in aquatic habitats. *Microbiology Spectrum, 5*(2), 10–1128.

Gopal, S., Chandrasekaran, M., Shagol, C., Kim, K. Y., & Sa, T. M. (2012). Spore-associated bacteria (SAB) of arbuscular mycorrhizal fungi (AMF) and plant growth-promoting rhizobacteria (PGPR) increase nutrient uptake and plant growth under stress conditions. *Korean Journal of Soil Science and Fertilizer, 45*(4), 582–592.

Gosling, P., Hodge, A., Goodlass, G., & Bending, G. D. (2006). Arbuscular mycorrhizal fungi and organic farming. *Agriculture, Ecosystems & Environment, 113*, 17–35.

GSBI, S. (2020). State of knowledge of soil biodiversity-status, challenges and potentialities. Summary for policy makers.

Gupta, A., Gupta, R., & Singh, R. L. (2016). Microbes and environment. In *Principles and applications of environmental biotechnology for a sustainable future* (pp. 43–84). Springer Singapore.

Harrier, L. A., & Watson, C. A. (2004). The potential role of arbuscular mycorrhizal (AM) fungi in the bioprotection of plants against soil-borne pathogens in organic and/or other sustainable farming systems. *Pest Management Science, 60*, 149–157.

Helgason, T., & Fitter, A. H. (2009). Natural selection and the evolutionary ecology of the arbuscular mycorrhizal fungi (Phylum Glomeromycota). *Journal of Experimental Botany, 60*(9), 2465–2480.

Huang, J., Gao, K., Yang, L., & Lu, Y. (2023). Successional action of Bacteroidota and Firmicutes in decomposing straw polymers in a paddy soil. *Environmental Microbiome, 18*(1), 76.

Huang, Q., Wang, J., Wang, C., & Wang, Q. (2019). The 19 years of inorganic fertilization increased bacterial diversity and altered bacterial community composition and potential functions in a paddy soil. *Applied Soil Ecology, 144*, 60–67.

Hussain, H. A., Uddin, S., Hussain, S., Rahman, A., Khalid, S., Hussain, W., et al. (2024). Crosstalk between phytohormones and environmental signals in the regulation of mycorrhizal symbiosis. In *Essential minerals in plant-soil systems* (pp. 197–214). Elsevier.

Igiehon, N. O., & Babalola, O. O. (2017). Biofertilizers and sustainable agriculture: Exploring arbuscular mycorrhizal fungi. *Applied Microbiology and Biotechnology, 101*, 4871–4881.

Ji, L., Tan, W., & Chen, X. (2019). Arbuscular mycorrhizal mycelial networks and glomalin-related soil protein increase soil aggregation in Calcaric Regosol under well-watered and drought stress conditions. *Soil Tillage Res, 185*, 1–8.

Jung, S. C., Martinez-Medina, A., Lopez-Raez, J. A., & Pozo, M. J. (2012). Mycorrhiza-induced resistance and priming of plant defenses. *Journal of Chemical Ecology, 38*, 651–664.

Kalam, S., Basu, A., Ahmad, I., Sayyed, R. Z., El-Enshasy, H. A., Dailin, D. J., & Suriani, N. L. (2020). Recent understanding of soil acidobacteria and their ecological significance: A critical review. *Frontiers in Microbiology, 11*, 580024.

Khaliq, A., Perveen, S., Alamer, K. H., Zia Ul Haq, M., Rafique, Z., Alsudays, I. M., Althobaiti, A. T., Saleh, M. A., Hussain, S., & Attia, H (2022). Arbuscular mycorrhizal fungi symbiosis to enhance plant–soil interaction. *Sustainability, 14*(13), 7840.

Khan, F., Siddique, A. B., Shabala, S., Zhou, M., & Zhao, C. (2023). Phosphorus plays key roles in regulating plants' physiological responses to abiotic stresses. *Plants, 12*, 2861.

Kim, H. S., Lee, S. H., Jo, H. Y., Finneran, K. T., & Kwon, M. J. (2021). Diversity and composition of soil Acidobacteria and Proteobacteria communities as a bacterial indicator of past land-use change from forest to farmland. *Science of the Total Environment, 797*, 148944.

Krishnamoorthy, R., Premalatha, N., Karthik, M., Anandham, R., Senthilkumar, M., Gopal, N. O., & Sa, T. (2017). Molecular markers for the identification and diversity analysis of arbuscular mycorrhizal fungi (AMF). In *Molecular markers in mycology: Diagnostics and marker developments* (pp. 177–199). Springer International Publishing.

Kumar, M., Kaur, A., Pachouri, C. U., & Singh, J. (2015). Growth-promoting characteristics of rhizobacteria and AM fungi for biomass amelioration of *Zea mays*. *Archives of Biological Sciences, 67*(3), 877–887.

Kumar, S., Chaudhary, T., Diksha, S. S., Chaudhary, D., & Kumar, R. (2022). Mycorrhizal fungi: An eco-friendly input for the sustenance of soil fertility and plant health. In *Microbes for humanity and its applications* (pp. 21–74). Daya Publishing House.

Kumar Yadava, A., Assouguem, A., Mehta, C. M., & Lahlali, R. (2025). Exploring synergistic interactions between arbuscular mycorrhizal fungi (AMF) and beneficial microbes in plant health and productivity. *CABI Reviews, 20*(1), 0047.

Kuyper, T. W., Wang, X., & Muchane, M. N. (2021). The interplay between roots and arbuscular mycorrhizal fungi influencing water and nutrient acquisition and use efficiency. In Z. Rengel, & I. Djalovic (Eds.), *The root systems in sustainable agricultural intensification* (pp. 193–220). Wiley.

Lahrach, Z., Legeay, J., Ahmed, B., & Hijri, M. (2024). The composition of the arbuscular mycorrhizal fungal bacteriome is species dependent. *Environmental Microbiome, 19*(1), 77.

Lee, E. H., Eo, J. K., Ka, K. H., & Eom, A. H. (2013). Diversity of arbuscular mycorrhizal fungi and their roles in ecosystems. *Mycobiology, 41*(3), 121–125.

Le Mer, J., & Roger, P (2001). Production, oxidation, emission, and consumption of methane by soils: A review. *European Journal of Soil Biology, 37*(1), 25–50.

Li, F., Zhang, S., Wang, Y., Li, Y., Li, P., Chen, L., & Han, Y. (2020). Rare fungus, *Mortierella capitata*, promotes crop growth by stimulating primary metabolisms related genes and reshaping the rhizosphere bacterial community. *Soil Biology and Biochemistry, 151*, 108017.

Liu, H., Wang, H., Nie, Z., Tao, Z., Peng, H., Shi, H., & Liu, H. (2024). Combined application of arbuscular mycorrhizal fungi and selenium fertilizer increased wheat biomass under cadmium stress and shaped rhizosphere soil microbial communities. *BMC Plant Biology, 24*(1), 359.

Luo, X., Liu, Y., Li, S., & He, X. (2023). Interplant carbon and nitrogen transfers mediated by common arbuscular mycorrhizal networks: Beneficial pathways for system functionality. *Frontiers in Plant Science, 14*, 1169310.

Ma, J., Wang, W., Yang, J., Qin, S., Yang, Y., Sun, C., & Huang, J. (2022). Mycorrhizal symbiosis promotes the nutrient content accumulation and affects the root exudates in maize. *BMC Plant Biology, 22*(1), 64.

Magurno, F., Malicka, M., Posta, K., Wozniak, G., Lumini, E., & Piotrowska-Seget, Z. (2019). Glomalin gene as a molecular marker for functional diversity of arbuscular mycorrhizal fungi in soil. *Biology and Fertility of Soils, 55*(4), 411–417.

Maharachchikumbura, S. S., Hyde, K. D., Jones, E. G., McKenzie, E. H. C., Bhat, J. D., Dayarathne, M. C., & Wijayawardene, N. N. (2016). Families of sordariomycetes. *Fungal Diversity, 79*(1), 1–317.

Maldonado, S., Rodríguez, A., Ávila, B., Morales, P., González, M. P., Araya Angel, J. P. A., Olalde, V., Bravo, J., Jana, C., Sierra, C., & Stoll, A. (2020). Enhanced crop productivity and sustainability by using native phosphate-solubilizing rhizobacteria in the agriculture of arid zones. *Frontiers in Sustainable Food Systems, 4*, 607355.

Manici, L. M., Caputo, F., De Sabata, D., & Fornasier, F. (2024). The enzyme patterns of Ascomycota and Basidiomycota fungi reveal their different functions in soil. *Applied Soil Ecology, 196*, 105323.

Meena, K. K., Sorty, A. M., Bitla, U. M., Choudhary, K., Gupta, P., Pareek, A., & Minhas, P. S. (2017). Abiotic stress responses and microbe-mediated mitigation in plants: The omics strategies. *Frontiers in Plant Science, 8*, 172.

Mercado-Blanco, J., Abrantes, I., Barra Caracciolo, A., Bevivino, A., Ciancio, A., Grenni, P., & Proença, D. N. (2018). Belowground microbiota and the health of tree crops. *Frontiers in Microbiology, 9*, 1006.

Miransari, M. (2011). Interactions between arbuscular mycorrhizal fungi and soil bacteria. *Applied Microbiology and Biotechnology, 89*, 917–930.

Nanjundappa, A., Bagyaraj, D. J., Saxena, A. K., Kumar, M., & Chakdar, H. (2019). Interaction between arbuscular mycorrhizal fungi and *Bacillus* spp. in soil enhancing growth of crop plants. *Fungal Biology and Biotechnology, 6*, 23.

Nannipieri, P. (2020). Soil is still an unknown biological system. *Applied Sciences, 10*(11), 3717.

Naseem, H., Ahsan, M., Shahid, M. A., & Khan, N. (2018). Exopolysaccharides-producing rhizobacteria and their role in plant growth and drought tolerance. *Journal of Basic Microbiology, 58*, 1009–1022.

Nazir, A., Sarfraz, W., Ditta, A., Khalid, N., Farid, M., Shafiq, M., Firdaus-E-Bareen, Rizvi, Z. F., & Naeem, N. (2023). Synergistic impact of two autochthonous saprobic fungi (*A. niger* and *T. pseudokoningii*) on the growth, ionic contents, and metals uptake in *Brassica juncea* L. and *Vigna radiata* L. under tannery solid waste contaminated soil. *International Journal of Phytoremediation, 25*(11), 1488–1500.

Offre, P., Spang, A., & Schleper, C. (2013). Archaea in biogeochemical cycles. *Annual Review of Microbiology, 67*, 437–457.

Pal, A., & Pandey, S. (2014). Role of glomalin in improving soil fertility. *International Journal of Plant and Soil Science, 3*, 112–129.

Panth, M., Hassler, S. C., & Baysal-Gurel, F. (2020). Methods for management of soilborne diseases in crop production. *Agriculture, 10*, 16.

Paz-Ferreiro, J., & Fu, S. (2016). Biological indices for soil quality evaluation: Perspectives and limitations. *Land Degradation & Development, 27*, 14–25.

Pérez-de-Luque, A., Tille, S., Johnson, I., Pascual-Pardo, D., Ton, J., & Cameron, D. D. (2017). The interactive effects of arbuscular mycorrhiza and plant growth-promoting rhizobacteria synergistically enhance host plant defences against pathogens. *Scientific Reports, 7*, 16409.

Philippot, L., Raaijmakers, J. M., Lemanceau, P., & Van Der Putten, W. H. (2013). Going back to the roots: the microbial ecology of the rhizosphere. *Nature Reviews Microbiology, 11*(11), 789–799.

Pii, Y., Mimmo, T., Tomasi, N., Terzano, R., Cesco, S., & Crecchio, C. (2015). Microbial interactions in the rhizosphere: beneficial influences of plant growth-promoting rhizobacteria on the nutrient acquisition process. A review. *Biology and Fertility of Soils, 51*, 403–415.

Piliarová, M., Ondreičková, K., Hudcovicová, M., Mihálik, D., & Kraic, J. (2019). Arbuscular mycorrhizal fungi–their life and function in the ecosystem. *Agriculture, 65*(1), 3–15.

Powell, J. R., & Rillig, M. C. (2018). Biodiversity of arbuscular mycorrhizal fungi and ecosystem function. *New Phytologist, 220*, 1059–1075.

Prosser, J. I., & Nicol, G. W. (2012). Archaeal and bacterial ammonia-oxidisers in soil: The quest for niche specialisation and differentiation. *Trends in Microbiology, 20*(11), 523–531.

Qi, S., Wang, J., Wan, L., Dai, Z., da Silva Matos, D. M., Du, D., & Moles, A. T. (2022). Arbuscular mycorrhizal fungi contribute to phosphorus uptake and allocation strategies of *Solidago canadensis* in a phosphorus-deficient environment. *Frontiers in Plant Science, 13*, 831654.

Qi, S., Wang, J., Zhang, Y., Naz, M., Afzal, M. R., Du, D., & Dai, Z. (2023). Omics approaches in invasion biology: understanding mechanisms and impacts on ecological health. *Plants, 12*(9), 1860.

Rillig, M. C. (2004). Arbuscular mycorrhizae, glomalin, and soil aggregation. *Canadian Journal of Soil Science, 84*, 355–363.

Rimington, W. R., Duckett, J. G., Field, K. J., Bidartondo, M. I., & Pressel, S. (2020). The distribution and evolution of fungal symbioses in ancient lineages of land plants. *Mycorrhiza, 30*(1), 23–49.

Rodríguez-Caballero, G., Caravaca, F., Fernández-González, A. J., Alguacil, M. M., Fernández-López, M., & Roldán, A. (2017). Arbuscular mycorrhizal fungi inoculation mediated changes in rhizosphere bacterial community structure while promoting revegetation in a semiarid ecosystem. *Science of the Total Environment, 584–585*, 838–848.

Saeed, H. S., Munir, M., Samra, S., Farooq, A., Mahmood, H., Ahmad, I., Akram, M. I., Iqbal, R., Elsadek, M. F., Almutairi, S. M., Ditta, A., & Daraz, U. (2025). Screening of biosurfactants producing fungi from infected *Citrus sinensis* (L.). *Polish Journal of Environmental Studies, 34*(5), 5261–5270.

Saldajeno, M. G. B., & Hyakumachi, M. (2011). Arbuscular mycorrhizal interactions with rhizobacteria or saprotrophic fungi and their implications to biological control of plant diseases. In *Mycorrhizal fungi: Soil, agriculture and environmental implications* (pp. 187–212). Nova Science Publishers.

Saleem, M., Hu, J., & Jousset, A. (2019). More than the sum of its parts: Microbiome biodiversity as a driver of plant growth and soil health. *Annual Review of Ecology, Evolution, and Systematics, 50*(1), 145–168.

Savastano, N., & Bais, H. (2024). Synergism or antagonism: Do arbuscular mycorrhizal fungi and plant growth-promoting rhizobacteria work together to benefit plants? *International Journal of Plant Biology, 15*, 944–958.

Sbrana, C., Avio, L., & Giovannetti, M. (2014). Beneficial mycorrhizal symbionts affect the production of health-promoting phytochemicals. *Electrophoresis, 35*, 1535–1546.

Selvakumar, G., Shagol, C. C., Kim, K., Han, S., & Sa, T. (2018). Spore-associated bacteria regulate maize root K+/Na+ ion homeostasis to promote salinity tolerance during arbuscular mycorrhizal symbiosis. *BMC Plant Biology, 18*, 109.

Shin, R. (2014). Strategies for improving potassium use efficiency in plants. *Molecules and Cells, 37*, 575–584. doi:10.14348/molcells.2014.0141.

Shu-Peng, T. A. N., Wen-xian, S. U. N., & Run-jin, L. I. U (2015). Combination of *Glomus* spp. and *Bacillus* sp. M3-4 promotes plant resistance to bacterial wilt in potato. *Acta Phytopathologica Sinica, 45*, 661–669.

Singh, V., Sharma, S., Kunal, S. K. , G., Choudhary, R., Singh, R., Adholeya, A., & Singh, B. (2020). Synergistic use of plant growth—Promoting rhizobacteria, arbuscular mycorrhizal fungi, and spectral properties for improving nutrient use efficiencies in wheat (*Triticum aestivum* L.). *Commun. Soil Sci. Plant Anal, 51*, 14–27.

Souza, T. (2015). *Handbook of arbuscular mycorrhizal fungi*: Vol. 1030. Cham: Springer.

Sun, D., Shang, X., Cao, H., Lee, S. J., Wang, L., Gan, Y., & Feng, S. (2024). Physio-biochemical mechanisms of arbuscular mycorrhizal fungi enhancing plant resistance to Abiotic Stress. *Agriculture, 14*(12), 2361.

Swarupa, V., Pavitra, K., Shivashankara, K. S., & Ravishankar, K. V. (2016). Omics-driven approaches in plant–microbe interaction. In *Microbial inoculants in sustainable agricultural productivity: Vol. 1: Research perspectives* (pp. 61–84). Springer India.

Tao, L., Ahmad, A., de Roode, J. C., & Hunter, M. D. (2016). Arbuscular mycorrhizal fungi affect plant tolerance and chemical defences to herbivory through different mechanisms. *Journal of Ecology, 104*(2), 561–571.

Tarkka, M. T., Drigo, B., & Deveau, A. (2018). Mycorrhizal microbiomes. *Mycorrhiza, 28*(5), 403–409.

Tedersoo, L., Bahram, M., Põlme, S., Kõljalg, U., Yorou, N. S., Wijesundera, R., & Abarenkov, K. (2014). Global diversity and geography of soil fungi. *Science, 346*(6213), 1256688.

Thind, S, Chaudhary, M. S., Ditta, A., Hussain, I., Parveen, A., Ullah, N., Mahmood, Q., Al-ashkar, I., & Sabagh, A. E. L. (2022). Impact of mycorrhizal fungi from different rhizospheric soils on fungal colonization, growth, and chlorophyll contents of *Cenchrus ciliaris*. *Agronomy, 12*(11), 2644.

Tian, B. Y., Cao, Y., & Zhang, K. Q. (2015). Metagenomic insights into communities, functions of endophytes, and their association with infection by root-knot nematode, *Meloidogyne incognita*, in tomato roots. *Scientific Reports, 5*(1), 17087.

Treseder, K. K., & Lennon, J. T. (2015). Fungal traits that drive ecosystem dynamics on land. *Microbiology and Molecular Biology Reviews, 79*(2), 243–262.

Ullah, H., Santiago-Arenas, R., Ferdous, Z., Attia, A., & Datta, A. (2019). Improving water use efficiency, nitrogen use efficiency, and radiation use efficiency in field crops under drought stress: A review. *Advances in Agronomy, 156*, 109–157.

Vejan, P., Abdullah, R., Khadiran, T., Ismail, S., & Nasrulhaq Boyce, A. (2016). Role of plant growth-promoting rhizobacteria in agricultural sustainability—A review. *Molecules, 21*(5), 573.

Wahab, A., Muhammad, M., Munir, A., Abdi, G., Zaman, W., Ayaz, A., & Reddy, S. P. P. (2023). Role of arbuscular mycorrhizal fungi in regulating growth, enhancing productivity, and potentially influencing ecosystems under abiotic and biotic stresses. *Plants, 12*(17), 3102.

Wang, B., Xiao, Q., Geng, X., Lin, K., Li, Z., Li, Y., & Li, X. (2024). Arbuscular mycorrhizal fungi alter rhizosphere bacterial diversity, network stability, and the function of lettuce in barren soil. *Scientia Horticulturae, 323*, 112533.

Wang, Y., Dong, F., Chen, H., Xu, T., & Tang, M. (2023b). Effects of arbuscular mycorrhizal fungus on sodium and chloride ion channels of *Casuarina glauca* under salt stress. *International Journal of Molecular Sciences, 24*, 3680.

Wang, Y., Hao, J., Guo, T., Zhao, L., Chai, B., & Jia, T. (2023a). Fungal community characteristics and driving factors in *Bothriochloa ischaemum* litter in a copper mining area. *Fungal Biology, 127*(10–11), 1426–1438.

Wang, Z., Li, Y., Zhuang, L., Yu, Y., Liu, J., Zhang, L., Gao, Z., Wu, Y., Gao, W., Din, G.-c., & Wang, Q. (2019). A rhizosphere-derived consortium of *Bacillus subtilis* and *Trichoderma harzianum* suppresses common scab of potato and increases yield. *Computational and Structural Biotechnology Journal, 17*, 645–653.

Wexler, H. M. (2007). Bacteroides: The good, the bad, and the nitty-gritty. *Clinical Microbiology Reviews, 20*(4), 593–621.

Whitton, B. A. (Ed.). (2012). *Ecology of cyanobacteria II: Their diversity in space and time*. Springer Science & Business Media.

Wu, Y., Zeng, J., Zhu, Q., Zhang, Z., & Lin, X. (2017). pH is the primary determinant of the bacterial community structure in agricultural soils impacted by polycyclic aromatic hydrocarbon pollution. *Scientific Reports, 7*(1), 40093.

Xiong, W., Jousset, A., Guo, S., Karlsson, I., Zhao, Q., Wu, H., & Geisen, S. (2018). Soil protist communities form a dynamic hub in the soil microbiome. *The ISME Journal, 12*(2), 634–638.

Yang, R., Qin, Z., Wang, J., Zhang, X., Xu, S., Zhao, W., & Huang, Z. (2022). The interactions between arbuscular mycorrhizal fungi and *Trichoderma longibrachiatum* enhance maize growth and modulate root metabolome under increasing soil salinity. *Microorganisms, 10*, 1042.

Yizhu, L., Imtiaz, M., Ditta, A., Rizwan, M. S., Ashraf, M., Mehmood, S., Aziz, O., Mubeen, F., Ali, M., Elahi, N. N., Ijaz, R., Lelel, S, Shuang, C., & Tu, S. (2020). Response of growth, antioxidant enzymes, and root exudates production towards As stress in *Pteris vittata* and *Astragalus sinicus* colonized by arbuscular mycorrhizal fungi. *Environmental Science and Pollution Research, 27*, 2340–2352.

Yu, Y., Gui, Y., Li, Z., Jiang, C., Guo, J., & Niu, D. (2022). Induced systemic resistance for improving plant immunity by beneficial microbes. *Plants, 11*(3), 386.

Yuan, J., Shi, K., Zhou, X., Wang, L., Xu, C., Zhang, H., et al. (2023). Interactive impact of potassium and arbuscular mycorrhizal fungi on the root morphology and nutrient uptake of sweet potato (*Ipomoea batatas* L.). *Frontiers in Microbiology, 13*, 1075957.

Zaman, F., Ali, A., Khattak, W. A., Khan, H., Anwar, Z., Rahimi, M., & Chen, F. S. (2025). Diversity of arbuscular mycorrhizal fungi (AMF) and their role in nutrient recycling and availability in the subtropical forest ecosystem of China: A review. *Eurasian Soil Science, 58*(11), 145.

Zehr, J. P. (2011). Nitrogen fixation by marine cyanobacteria. *Trends in Microbiology, 19*(4), 162–173.

Zeng, W., Xiang, D., Li, X., Gao, Q., Chen, Y., Wang, K., et al. (2025). Effects of combined inoculation of arbuscular mycorrhizal fungi and plant growth-promoting rhizosphere bacteria on seedling growth and rhizosphere microecology. *Frontiers in Microbiology, 15*, 1475485.

Zhang, Y., Dong, L., & Shangguan, Z. (2023). Appropriate N addition improves soil aggregate stability through AMF and glomalin-related soil proteins in a semiarid agroecosystem. *Land Degradation & Development, 34*(3), 710–722.

Zhang, Y., Han, X., Ren, W., Zhang, H., & Tang, M. (2024). Arbuscular mycorrhizal fungi improve *Lycium barbarum* potassium uptake by activating the expression of LbHAK. *Plants, 13*, 1244.

Zhou, J., Zang, H., Loeppmann, S., Gube, M., Kuzyakov, Y., & Pausch, J. (2020). Arbuscular mycorrhiza enhances rhizodeposition and reduces the rhizosphere priming effect on the decomposition of soil organic matter. *Soil Biology and Biochemistry, 140*, 107641.

Zhu, L., Huang, J., Lu, X., & Zhou, C. (2022). Development of plant systemic resistance by beneficial rhizobacteria: Recognition, initiation, elicitation and regulation. *Frontiers in Plant Science, 13*, 952397.

Chapter 11

Arbuscular mycorrhizal fungi as plant growth enhancers: Mechanisms and applications

Ghulam Murtaza[a], Muhammad Usman Jamshaid[b], Zeeshan Ahmed[c], Abdul Razzaq[d], Rashid Iqbal[e,f] and Javed Iqbal[g]

[a] *School of Agriculture, Yunnan University, Kunming, Yunnan , China,* [b] *School of Agriculture and Biology, Shanghai Jiao Tong University, Shanghai, China,* [c] *College of Life Science, Shenyang Normal University, Shenyang, Liaoning, China,* [d] *Faculty of Management and Economics, Kunming University of Science and Technology, Kunming, China,* [e] *Department of Agronomy, Faculty of Agriculture and Environment, The Islamia University of Bahawalpur, Bahawalpur, Pakistan,* [f] *Department of Life Sciences, Western Caspian University, Baku, Azerbaijan,* [g] *Departemnt of Botany, Bacha Khan University, Charsadda, Pakistan*

11.1 Introduction

AMF are soil microbes that establish a symbiotic association with 80% of vascular plant species and 90% of agricultural plants, encompassing the majority of crops. They exhibit a pervasive distribution across global ecosystems, predominantly characterized by the worldwide distribution of identified plant hosts (Murtaza et al., 2025). AMFs are included in the phylum Glomeromycota, which encompasses three classes: Paraglomeromycetes, Archaeosporomycetes, and Glomeromycetes. AMF comprises 25 genera, 11 families, and around 250 species. Glomeromycota constitute obligate symbioses that depend on C substrates supplied by their host plants, which can account for up to 20% of plant-fixed C, for their survival (Delgado et al., 2021). The fungi enhance the availability of nutrients and water, including nitrogen and phosphate, to the host plant via intraradical and extraradical hyphae. This symbiosis originated with the initial emergence of terrestrial plants, around 400 million years ago, as evidenced by fossil records (Dueñas et al., 2022). AMF symbiosis is perhaps the most prevalent advantageous relationship between plants and microbes. Numerous studies have indicated their vital involvement in plant growth and nutrition under stress situations, as well as their enhancement of various critical ecosystem functions (Ge et al., 2025).

In recent times the swift advancement of biological mycorrhizal theories and methodologies has stimulated comprehensive research on the control of AMF in augmenting the absorption of advantageous elements by crops (Priyadarshani et al., 2025). Numerous variables that induce mycorrhizae have been found from both molecular and biological perspectives. A systematic and complete analysis of these factors is still absent (Banerjee et al., 2025). Additional investigation and integration of this developing body of information are required to enhance comprehension of the diverse factors affecting mycorrhizal relationships in crops. Recent research has explored the complex mechanisms by which AMF facilitate nutrient uptake in crops, revealing prospective strategies for optimizing agricultural operations and promoting plant development (Ghorui et al., 2025). This chapter synthesizes domestic and international studies on the regulation of arbuscular mycorrhizal fungi (AMFs) in relation to plant mineral uptake. This summary addresses the production and separation of macro and trace elements in AMF-influenced plants, as well as the use of AMF to mitigate stresses in plants, while also highlighting existing challenges and inadequacies. The objective is to establish a scientific foundation for comprehending the absorption mechanisms of beneficial components in AMF crops and the farming utilization of biological fertilizers.

This chapter aims to summarize knowledge regarding arbuscular mycorrhizal connections, specifically beneficial impacts on soil and host plants. The function of AMFs in plant nutrition, development, and resilience to stressors is examined. The interactions between AMF and other soil microbes are ultimately analyzed (Fig. 11.1).

Advancing Sustainable Agriculture Through Arbuscular Mycorrhizal Fungi. DOI: https://doi.org/10.1016/B978-0-443-45130-0.00028-X

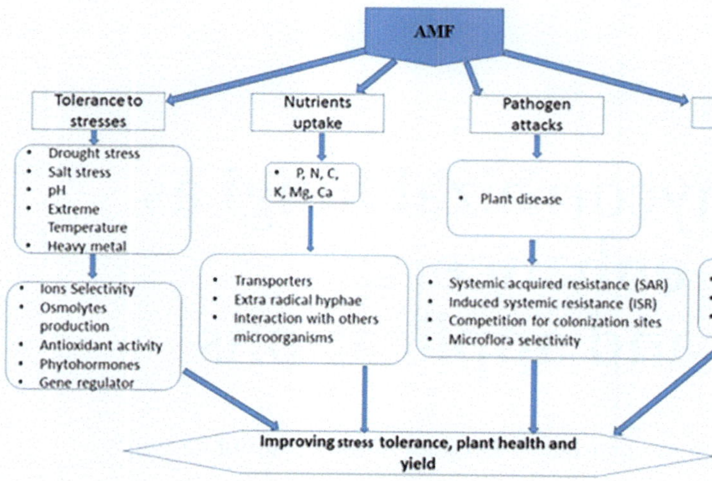

FIGURE 11.1 AMF role on plant growth. *AMF*, Arbuscular mycorrhizal fungi. (*From (Stewart et al., 2025).*)

11.2 Arbuscular mycorrhizal fungi background

AMF are soil-dwelling fungi that can markedly enhance plant nutrient absorption and bolster resilience to various abiotic stressors (Sharma et al., 2024). The predominant species of AMF are classified within the subphylum Glomeromycotina of the phylum Mucoromycota (Stewart et al., 2025). This subphylum comprises four orders of AMF: Glomerales, Archaeosporales, Paraglomerales, and Diversisporales, together with 25 taxa (Aguilar et al., 2025). They are obligate biotrophs that consume plant photosynthetic products (Bago, Pfeffer, & Shachar-Hill, 2000) and lipids to complete their life cycle (Ganugi et al., 2019). AMF-mediated growth enhancement occurs not only through augmentation of water and mineral nutrient absorption from surrounding soil but also by protecting against fungal diseases (Castaño et al., 2025). Consequently, AMFs are essential endosymbionts that significantly contribute to plant productivity and ecosystem functionality. They are crucial for sustainable agricultural enhancement (Figueiredo de Oliveira et al., 2025).

11.2.1 Attributes of arbuscular mycorrhizal fungi symbiosis

The symbiotic relationship between AMF and plants was documented 400 million years ago. These linkages are formed through a series of biological processes that result in various beneficial effects in both natural ecosystems and agricultural biotas (Ramana et al., 2025). The symbiotic connection of AMF exemplifies mutualism, capable of regulating plant growth. The mycelial network of fungi stretches beneath plant roots and facilitates the absorption of otherwise inaccessible nutrients. The fungal mycelium colonizes the roots of several plant species, resulting in a common mycorrhizal network. This common mycorrhizal network is regarded as a fundamental element of terrestrial ecosystems, exerting considerable influence on various plant communities, especially invasive species, and facilitating fungal-based transport of phosphate and nitrogen to plants (Akbar et al., 2025). In addition, communal nutrients transfer from fungi to plants, along with other associated effects, likely explaining why AMF enhance plant tolerance to biotic and abiotic stressors (Hashem et al., 2025). They possess the capacity to enhance soil properties, hence promoting plant growth under optimal and adverse conditions. AMF colonization enhances plant tolerance to stressors by inducing various alterations in their morpho-physiological characteristics (Chiasson et al., 2024; Liu et al., 2020). AMFs are regarded as natural growth regulators for most terrestrial plants. AMFs serve as bioinoculants, and researchers advocate for their application as significant biofertilizers in sustainable agricultural productivity (Sonbol et al., 2025). Moreover, AMF-inoculated soil exhibits more consistent masses and much greater extraradical hyphal mycelium compared to non-AMF-treated soils (Dagher et al., 2025). Moreover, the inoculation of AMF stimulates dry matter buildup and increases water moisture absorption, hence augmenting plant resilience to stressors such as drought and salinity. The utilization of AMF for plant growth across diverse biological habitats can significantly enhance organic cultivation for growth enhancement and yield optimization (Fig. 11.2).

11.2.2 Role of arbuscular mycorrhizal fungi on plant growth and nutrition

By augmenting the root absorption area, AMFs enhance the overall absorption surface of infected plants, so facilitating improved access to nutrients, especially those with low mobility in ionic forms or that exist in minimal concentrations in soil solution (Nohesara et al., 2025). Moreover, AMF accounts for roughly 20% of total plant water absorption, underscoring the

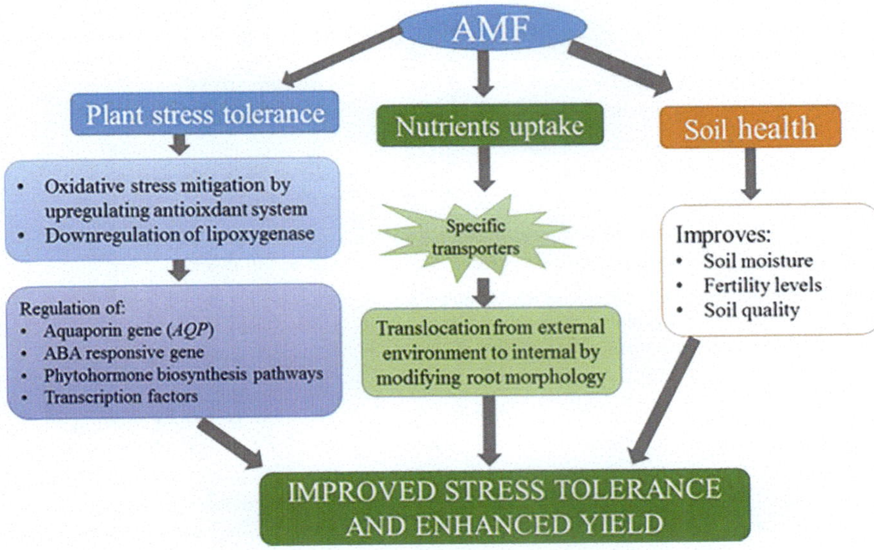

FIGURE 11.2 A diagrammatic illustration of mycorrhizal roles in regulating many ecological processes and promoting plant development under abiotic stress conditions. *AMF*, Arbuscular mycorrhizal fungi. (*From (Begum et al., 2019).*)

significance of this symbiosis in the hydration state of host plants (Thind et al., 2022). AMF markedly enhanced development and metabolism, including carbohydrate levels in root and shoot systems of treated plants relative to control treatment (Madouh & Davidson, 2024). AMFs play a significant role in biological fortification (Jaiswal et al., 2022). AMF inoculation may influence selenium absorption from soil and antioxidant chemical levels in vegetable crops, including green asparagus. Investigations conducted by Upadhayay et al. (2019) showed an augmentation of selenium concentration in wheat grain via inoculation. Research by Zhang et al. (2022) indicates that AMF alter the content and placement of nutrients in crop grains. AMF inoculation enhances the nutritional composition of grains in terms of protein, iron, and zinc (Gautam et al., 2025).

Under specific climatic conditions, Jing et al. (2025) determined that AMF symbiosis favorably influenced the zinc levels in diverse crop plant tissues. AMF can significantly enhance the zinc nutrition of grains; nevertheless, AMF's role in zinc uptake is contingent upon the functional suitability between the AMF strain and cultured cereal species (Kaur et al., 2025). AMF symbiosis is recognized for specifically inducing various transporter expression, including crop aquaporin genes, phosphate transporters, ammonium transporters, nitrate transporters, sulfur transporters, zinc transporters, protein, and carbon transporters (Saboor et al., 2021). In addition, it was shown that fungi establish networks known as mycorrhizal networks that enhance nutrient transport between plants via the spread of fungal mycelium (Liu et al., 2024). Commonly referred to as common mycorrhizal structures, these mycorrhizal networks can combine many plant species and multiple fungus species that communicate, offer feedback, and adapt, forming a complex, adaptable social network (Quan et al., 2021). The findings presented in Hassan et al. (2025) validate the function of AMF in facilitating biological interactions between adjacent plants.

11.2.3 Arbuscular mycorrhizal fungi's role in crop drought resilience

Drought is a significant stressor that can substantially diminish plant output (Murtaza et al., 2025). Water limitations induce stomatal closure, leading to a diminished inflow of carbon dioxide, which then reduces photosynthesis and carbon allocation, ultimately resulting in decreased agricultural yield. Research has shown AMF enhances plant performance under drought stress (Li et al., 2025). Table 11.1 illustrates AMF impacts on the drought resistance of crops. Mycorrhizal plants manage water scarcity via drought mitigation and tolerance to drought (Nader et al., 2024). A drought mitigation method is facilitated by indirect AMF benefits and improved water absorption, whereas tolerant to drought includes a blend of direct AMF advantages that strengthen the plant's inherent capacity to withstand stress (Yang et al., 2025).

The enhancement of plant health by AMF during drought mitigation may be attributed to augmented surface area for water absorption afforded by AMF hyphae, enhanced access to diminutive soil pores, or improved apoplastic water movement. Multiple investigations have connected the mechanisms of stress mitigation (Basyal, 2024). Research by Aili et al. (2023) indicates that in field-cultivated plants, the colonization of roots by the AMF facilitates optimal growth under drought stress by enhancing nutrient levels and WUE (WUE). Additional mechanisms contribute to plant resistance to

TABLE 11.1 AMF role in assisting plants to manage drought stress.

AMF-mediated mechanism	Physiological/biochemical response in the host plant	Drought stress alleviation outcome	Representative evidence
Enhanced root system architecture	Increased root length, surface area, lateral roots, and fine root proliferation	Improved soil exploration and water uptake under limited moisture	Murtaza et al. (2025)
Improved water uptake via extraradical hyphae	The hyphal network accesses micropores inaccessible to roots	Maintains plant water status and delays wilting	Li et al. (2025)
Regulation of aquaporin expression	Upregulation of plasma membrane intrinsic protein (PIP) and tonoplast intrinsic protein (TIP) aquaporins in roots and leaves	Increased root hydraulic conductivity and water transport efficiency	Nader et al. (2024)
Osmotic adjustment	Accumulation of proline, soluble sugars, and glycine betaine	Maintains cell turgor and metabolic activity under drought	Yang et al. (2025)
Enhanced nutrient uptake (P, K, Mg, Zn)	Improved adenosine triphosphate(APT) synthesis, enzyme activation, and stomatal regulation	Sustains photosynthesis and growth under water deficit	Liu et al. (2024)
Improved photosynthetic efficiency	Higher chlorophyll content, Fv/Fm, and CO_2 assimilation rate	Reduced drought-induced photoinhibition	Quan et al. (2021)
Modulation of stomatal conductance	Optimized stomatal opening via hormonal and hydraulic signals	Reduced transpiration losses while maintaining carbon fixation	Jaiswal et al. (2022)
Hormonal regulation (ABA, indole acetic acid (IAA), cytokinins)	Balanced abscisic acid (ABA) accumulation and enhanced root-to-shoot signaling	Improved drought perception and adaptive responses	Gautam et al. (2025)
Antioxidant defense enhancement	Increased superoxide dismutase (SOD), catalase (CAT), POD (peroxidase), and ascorbate peroxidase (APX) activities and nonenzymatic antioxidants	Reduced ROS accumulation and membrane damage	Nohesara et al. (2025)

AMF, arbuscular mycorrhizal fungi; *ROS*, reactive oxygen species.

drought. Under these systems, osmotic adjustment enables plants to sustain their transpiration and physiological functions by accumulating suitable solutes, including sugars, proline, organic acids, and glycine betaine. Drought stress, under saline conditions, stimulates the generation of reactive oxygen species (ROS) (Chauhan et al., 2024).

11.2.4 Arbuscular mycorrhizal fungi's role in crop salt stress resilience

AMFs are recognized to occur spontaneously in saline conditions (Iwuala et al., 2025). Their role in enhancing the growth of a variety of plants under saline circumstances is well documented (Table 11.2). This primarily pertains to an amalgamation of physiological and biochemical and dietary influences (Boorboori & Lackóová, 2025). The mechanisms contributing to salinity tolerance in AMF-inoculated plants include enhanced nutrient uptake and water absorption, osmoregulator accumulation, maintenance of ionic homeostasis, and reduced uptake of sodium and chloride (Pooja et al., 2025). Furthermore, it has been shown that AMF colonization enhances stomatal conductivity and mitigates oxidative harm in crops subjected to salt (Kakabouki et al., 2023). Under salt stress, AMF inoculation markedly decreased oxidative harm in wheat. They additionally documented elevated gas exchange ability, stomatal conductivity, and proline, amino acids, and sugars in plants associated with AMF (Jiang et al., 2025). Maize plants inoculated individually with three indigenous AMF exhibited enhanced yields of biomass and elevated shoot proline and potassium levels, in contrast to mycorrhizal-free plants (Khoza et al., 2025).

AMF has been shown to mitigate the adverse impacts of salt on *Acacia* species growth. This mostly pertains to enhanced nutrient uptake, increased total chlorophyll, glucose, and proline levels, raised K/Na ratios in tissues, and alterations in

TABLE 11.2 Role of AMF in improving plant tolerance to salt stress.

AMF-mediated mechanism	Physiological/biochemical response in the host plant	Salt stress mitigation outcome	Supporting evidence
Enhanced root system architecture	Increased root length, branching, and root surface area	Improved soil exploration and water acquisition under saline conditions	Paramaputra et al. (2025)
Improved water uptake and hydraulic conductivity	Maintenance of higher relative water content and leaf water potential	Alleviation of osmotic stress and delayed salt-induced dehydration	Hamzehzadeh et al. (2025)
Regulation of ion homeostasis	Reduced Na^+ uptake and enhanced K^+ and Ca^{2+} acquisition	Lower ionic toxicity and improved metabolic stability	Hui et al. (2025)
Na^+ sequestration and compartmentalization	Increased Na^+ retention in roots and vacuoles	Protection of photosynthetic tissues from Na^+ toxicity	Khoza et al. (2025)
Enhanced nutrient uptake	Improved ATP production, enzyme activity, and chlorophyll synthesis	Sustained growth and photosynthesis under salinity	Kakabouki et al. (2023)
Osmotic adjustment	Increased accumulation of proline, soluble sugars, and polyols	Maintenance of cell turgor and osmotic balance	Jiang et al. (2025)
Improvement in photosynthetic performance	Higher chlorophyll content, net photosynthetic rate, and Fv/Fm	Reduced salt-induced photoinhibition	Kakabouki et al. (2023)
Antioxidant defense enhancement	Increased activities of SOD, CAT, POD, APX, and glutathione reductase (GR)	Reduced ROS accumulation and oxidative damage	Pooja et al. (2025)
Maintenance of membrane stability	Lower electrolyte leakage and malondialdehyde content	Protection of cellular membranes under salinity stress	Iwuala et al. (2025)

AMF, arbuscular mycorrhizal fungi; *ROS*, reactive oxygen species.

root shape in mycorrhizal plants than nonmycorrhizal plants. The outcomes were more pronounced when salt-stressed plants were inoculated with AMF (Hui et al., 2025). It is important to note that in research examining several strains, the degree of AMF reaction on plant development and root colonization differed according to fungal species and salt levels (Hamzehzadeh et al., 2025). The authors discovered AMF have different impacts on plants under salinity, with *Diversispora versiformis* demonstrating greater activity than *Funneliformis mosseae*. Concerning ionic homeostasis, AMF enhance ionic equilibrium by the filtering capabilities of their structures in soil and roots, which inhibit the ingress of harmful Na^+ ions (Paramaputra et al., 2025).

11.2.5 Arbuscular mycorrhizal fungi role in crop heavy metals stress resilience

Contaminated sites with heavy metals have AMFs that have been tailored to soil contamination by toxic metals (Chiasson et al., 2024; Nazir et al., 2023). Multiple studies indicate that over 80% of plants surveyed in mining areas are colonized by AMF, with a significant variety of AMF types and diversity present in different mining-affected locations (Khatoon et al., 2024; Saeed et al., 2025). The authors reviewed studies indicating that AMFs demonstrate substantial positive impacts, including higher plant survival, increased development and sustenance, and enhanced soil structure. Numerous investigations indicated that mycorrhizae can function as a stress-mitigating agent in soils contaminated with heavy metals, aiding plant survival under such adverse conditions (Table 11.3). AMF can remediate heavy metals by hyphal "metal binding," hence diminishing the bioavailability of elements such as copper, lead, cobalt, cadmium, and zinc (Shukla et al., 2025). The mitigation of heavy metal toxicity by AMF is contingent upon the fungal partner, plant development conditions, and specific toxic metal (Mazumder et al., 2025). Inoculating with AMF yielded optimal outcomes for seed germination percentage, seedling viability, and fresh and dry biomass. In two distinct heavy metal-contaminated soils, root colonization of plants by *Glomus* isolates diminished heavy metal levels in shoots and roots, while enhancing the

TABLE 11.3 The role of AMF in alleviating heavy metal stress in plants.

AMF-mediated mechanism	Physiological/biochemical response in the host plant	Heavy metal stress alleviation outcome	Supporting evidence
Immobilization of heavy metals in the rhizosphere	Binding of metals to fungal cell walls (chitin, glomalin) and soil particles	Reduced metal bioavailability and uptake by plants	Gu et al. (2025)
Metal sequestration within fungal structures	Accumulation of metals in hyphae, vesicles, and arbuscules	Protection of host plant tissues from metal toxicity	Sahu and Verma (2025)
Regulation of metal uptake and translocation	Reduced root-to-shoot translocation of Cd, Pb, As, and Cr	Lower metal accumulation in aerial parts	Wen et al. (2025)
Improved nutrient acquisition (P, Ca, Mg, Fe, Zn)	Enhanced mineral nutrition counteracting metal-induced deficiencies	Maintenance of growth and metabolic functions	Mazumder et al. (2025)
Enhanced antioxidant defense system	Increased activities of SOD, CAT, APX, POD, and GR	Reduced oxidative stress and ROS-mediated damage	Shukla et al. (2025)
Maintenance of membrane integrity	Reduced malondialdehyde content and electrolyte leakage	Preservation of cellular membrane stability	Chiasson et al. (2024)
Osmotic and metabolic adjustment	Accumulation of proline, soluble sugars, and organic acids	Stabilization of proteins and membranes under metal stress	Khatoon et al. (2024)
Regulation of metal transporter genes	Downregulation of metal influx transporters and activation of vacuolar sequestration (e.g., HMA heavy metal-transporting ATPases (ATP), ATP-binding cassette (ABC) transporters)	Controlled metal detoxification and compartmentalization	Sahu and Verma (2025)
Hormonal modulation (IAA, ABA, cytokinins)	Improved root growth and stress signaling	Enhanced tolerance and recovery under metal stress	Wen et al. (2025)
Improvement of photosynthetic efficiency	Increased chlorophyll content and photosynthetic rate	Reduced metal-induced photosynthetic inhibition	Mazumder et al. (2025)
Rhizosphere modification and soil aggregation	Glomalin production improves soil structure and metal stabilization	Long-term reduction of metal mobility	Khatoon et al. (2024)
Enhanced plant growth and biomass production	Improved shoot and root biomass despite metal exposure	Increased phytostabilization potential	Sahu and Verma (2025)

AMF, arbuscular mycorrhizal fungi; *ROS*, reactive oxygen species.

levels of important nutrients such as potassium, phosphorus, and magnesium in the roots (Wen et al., 2025). AMF exhibited that several metal transporters are essential for metal regulation. Recently, many zinc transporters have been discovered in AMF, including *Rhizophagus irregularis* (Sahu & Verma, 2025). Numerous candidate genes encoding for iron, copper, and zinc transporters have also been discovered (Gu et al., 2025). These transporters may have a role in heavy metal resistance in crops treated with AMF.

11.2.6 Arbuscular mycorrhizal fungi and integrated abiotic stresses

AMF is well recognized for its potential to mitigate different stressors, including drought, salt, temperature fluctuations, nutritional deficiencies, and heavy metal toxicity. Exposure to crops to a combination of salinity and drought results in increased formation of ROS, which can be detrimental to plants (Diagne et al., 2020). The detoxification of ROS is facilitated by enzymes such as superoxide dismutase, catalase, peroxidase, and glutathione reductase (Singh et al., 2022). Scarce studies reporting exist in the literature illustrating the role of AMF in alleviating the combined effects of many stresses. AMF symbiosis safeguards plants from diverse abiotic stresses through mechanisms including enhanced photosynthetic efficiency, increased mineral nutrient absorption and storage, osmoprotectant accumulation, upregulation of antioxidant enzyme activity, and alterations in the rhizosphere ecosystem (Jajoo & Mathur, 2021). Numerous studies indicate enhanced nutritional status of AMF crops under osmotic stress circumstances because of salinity and water stress (Abdel Latef & Miransari, 2014). Similarities in resistance mechanisms may arise in response to AMF-driven combined stress responses. It is suggested that AMF-induced modifications in phytohormone profiles, mineral absorption and assimilation, secondary metabolites and compatible osmolytes buildup, and enhancement of the antioxidant system may represent common mechanisms activated under various stress conditions. Nonetheless, certain mechanisms such as compartmentalization and sequestration of harmful ions, synthesis of phytochelatins, and protein expression can be distinct and demonstrate considerable variation depending on the type of stress and the species of AMF involved. Modifications in root traits, such as hydraulic conductivities, can significantly enhance resistance to osmotic stress (Bhupenchandra et al., 2024). Mitra et al. (2021) demonstrated that AMF safeguarded castor bean against salt stress by modifying gas exchange characteristics and concentrations of certain metabolites. The aforementioned features of AMF may enhance the nutraceutical quality of crops and hold significant agronomic relevance for the production and maintenance of various prospective crops. Nevertheless, comprehensive research is necessary to elucidate the function of AMF in mitigating the impacts of coupled stressors.

11.2.7 Mechanisms of action of arbuscular mycorrhizal fungi

The inoculation of AMF altered biochemical processes of lemongrass, resulting in enhanced growth and increased suppression of *Fusarium* root rot in greenhouse conditions. Furthermore, essential oils derived from mycorrhizal lemongrass significantly inhibited the development of *Fusarium solani* Fs4, more so than those from noninoculated plants (Eke et al., 2020). AMF can establish a symbiotic relationship with a majority of vegetable plants, encompassing primary crops from various families, including Apiaceae, Amaryllidaceae, Cucurbitaceae, Asteraceae, Fabaceae, and Solanaceae (Baum et al., 2015). The primary advantageous effects of AMF include enhanced absorption of mineral elements, including phosphorus, calcium, zinc, copper, iron, and sulfur, as well as improved resilience to various forms of stress (Siddiky et al., 2012). One investigation indicated that α-diversity of AMF and diazotroph communities, nitrogen-related enzymes, and sediment nutrient components are primary factors influencing the biological nitrogen fixation process (Yu et al., 2021). The concurrent application of chitosan nanoparticles and AMF enhanced concentrations of tocopherols, polyphenols, and carotenoids in the roots, thereby augmenting antioxidant capacity by 33%. Consequently, chitosan nanoparticles may serve as efficacious biofertilizers to enhance plant growth and vitality, as demonstrated by the enhancement of health-promoting constituents in wheat (Saleh et al., 2023). With AMF treatment, phosphorus levels in the leaves of plants were elevated under both watering conditions. In addition, even under restricted watering, the concentrations of calcium, potassium, zinc, magnesium, and manganese reached levels comparable to those of untreated plants (Leventis et al., 2021). AMF enhanced nutrient absorption, especially phosphorus and other less mobile elements, mostly by expanding the absorption surface of hyphal exploration, and functioned as agents for soil restoration and biocontrol in polluted soils (Tekaya et al., 2021). AMFs are used in symbiosis with *Miscanthus* plants for phytoremediation and for cultivation in zinc-deficient soils. AMF enhanced plant development in the presence of harmful trace element stress. The role of AMF in modulating metal transport is contingent upon concentration, and the transport direction of cadmium and zinc in plants influenced by AMF varies with trace elements. AMF enhanced the absorption of zinc by growth dispersion and cadmium via biological enrichment, with the interaction among cadmium and zinc influenced by AMF and other factors (Sarkar et al., 2017).

Mycorrhizal fungi exhibit varied responses to metal pollution. *F. mosseae* enhances dry yields and photosynthesis rates in pepper under copper stress, whereas *Glomus versiforme* improves phosphorus nutrition and antioxidant capacity in lotus under cadmium stress. *Rhizophagus intraradices* accumulates chromium in the roots and reduces its translocation to *Brachiaria mutica* shoots under chromium stress. *R. irregularis* may immobilize copper and lead in soil for willow cultivation. In addition, *R. intraradices* regulates arsenic transporters in root epidermal cells, thereby decreasing arsenic uptake in pea, and *Claroideoglomus* species reduce cadmium buildup in wheat (Baghaie et al., 2019). *F. mosseae* is a notable mycorrhizal fungus among numerous AMF species because of its symbiotic relationship with roots of several plant species, potentially enhancing nutrient exchange via root extension via mycelial outer membrane and promoting root development (Sbrana et al., 2014). Significant instances of mycorrhizal fungi alleviating soil nutrient deficiencies include *R. irregularis* enhancing phosphorus and zinc availability in *Medicago truncatula* through the upregulation of MtZIP5 and MtPT4 genes; *R. irregularis* increasing root absorption area and phosphorus accessibility in maize (Nguyen et al., 2019); *R. irregularis* altering the ZIP transporter action in barley and enhancing grain zinc bioavailability; *F. mosseae* facilitating manganese, magnesium, zinc, iron, calcium, sulfur, nitrogen, phosphorus, and potassium uptake in cucumber; and *R. irregularis* promoting MtZIP6 gene expression and expanding root absorption area in *M. truncatula* (Watts-Williams & Cavagnaro, 2018). Numerous physical and biochemical mechanisms have been identified as significant contributors to the enhancement of stress resilience in bean plants through AMF inoculation against *Rhizoctonia solani* root rot disease, including enhanced plant growth, improved nutritional uptake, cytoplasmic granulation, and increased cell wall thickening (Abdel-Fttah et al., 2011). AMF enhanced cadmium buildup in wheat roots while markedly reducing cadmium levels in grains and shoots. Furthermore, improving the levels of stomatal conductivity, photosynthesis rates, chlorophyll, and carbohydrate buildup under cadmium stress was enhanced by AMF symbiosis (Li et al., 2023).

11.3 Concussion and prospective considerations

AMF significantly enhance plant response to biotic and abiotic stimuli and mitigate the impact of these stresses on plants. Their function in enhancing plant growth and yield, disease resistance, and tolerance to stress offers an ecofriendly alternative to diminishing reliance on harmful pesticides and synthetic fertilizers. Nevertheless, further investigation is required to validate the laboratory and greenhouse findings in real-world conditions. The use of this knowledge in practical settings and in accordance with biogeographical zones is crucial for enhancing industrial production on a wide scale and maximizing its influence to secure sufficient food for all individuals globally, both now and in the future. To promote the utilization of AMF as an ecofriendly approach, researchers and both private and governmental sectors must enhance their production, especially in developing countries where AMF inocula are neither accessible nor economical.

Future studies must aim to enhance AMF inoculants for various agroecosystems by identifying host-specific strains and assessing environmental tolerance. Transcriptomics can elucidate tripartite interactions among AMF, plants, and diseases. Field trials under diverse climatic and soil conditions will assess efficacy, whereas precision agriculture techniques can incorporate AMF for targeted application. The examination of synergistic effects resulting from interactions with different biocontrol agents by combinatorial testing will improve disease control strategies. Longitudinal research on soil fertility and carbon sequestration via glomalin production is essential. Ultimately, the enhancement of AMF manufacturing techniques will facilitate the commercial feasibility of sustainable agriculture.

References

Abdel Latef, A. A. H., & Miransari, M. (2014). The role of arbuscular mycorrhizal fungi in alleviation of salt stress. In *Use of microbes for the alleviation of soil stresses: Volume 2: Alleviation of soil stress by PGPR and mycorrhizal fungi* (pp. 23–38). Springer New York.

Abdel-Fttah, G. M., El-Haddad, S. A., Hafez, E. E., & Rashad, Y. M. (2011). Induction of defense responses in common bean plants by arbuscular mycorrhizal fungi. *Microbiological Research, 166*, 268–281.

Aguilar, C. H., Pires, D., Cortaga, C., Peja, R., Jr, Cruz, M. A., Langres, J., & Balendres, M. A. (2025). Harnessing legume productivity in tropical farming systems by addressing challenges posed by legume diseases. *Nitrogen, 6*(3), 65.

Aili, Y., Chen, X., Gao, W., Wang, H., Dawuti, M., & Ma, X. (2023). Response of *Alhagi sparsifolia* seedlings to AMF inoculation and nitrogen addition under drought stress. *Atmosphere, 14*(3), 446.

Akbar, K., Yaseen, T., Abbasi, B. A., Iqbal, J., Ullah, Z., Ijaz, S., & Iqbal, R. (2025). Role and genetic diversity of arbuscular mycorrhizal fungi in improving the productivity and enhancing nutrient absorption in chickpea (*Cicer arietinum* L.) genotypes. *Genetic Resources and Crop Evolution, 72*(2), 2459–2473.

Baghaie, A. H., Aghili, F., & Jafarinia, R. (2019). Soil-indigenous arbuscular mycorrhizal fungi and zeolite addition to soil synergistically increase grain yield and reduce cadmium uptake of bread wheat (through improved nitrogen and phosphorus nutrition and immobilization of Cd in roots). *Environmental Science and Pollution Research International, 26*, 30794–30807.

Bago, Berta, Pfeffer, P. E., & Shachar-Hill, Y. (2000). Carbon metabolism and transport in arbuscular mycorrhizas. *Plant Physiology, 124*(3), 949–958.

Banerjee, S., Mandal, J., Sarkar, D., Datta, R., & Bhattacharyya, P. (2025). A review and meta-analysis of the efficacy of arbuscular mycorrhizal fungi in remediating toxic metals in mine-affected soils. *Frontiers in Environmental Science, 12*, 1532169.

Basyal, B. (2024). Plant-arbuscular mycorrhizal fungi association under drought stress. In *Arbuscular mycorrhizal fungi in sustainable agriculture: Nutrient and crop management* (pp. 211–221). Springer Nature Singapore.

Baum, C., El-Tohamy, W., & Gruda, N. (2015). Increasing the productivity and product quality of vegetable crops using arbuscular mycorrhizal fungi: A review. *Scientia Horticulturae, 187*, 131–141.

Begum, N., Qin, C., Ahanger, M. A., Raza, S., Khan, M. I., Ashraf, M., & Zhang, L. (2019). Role of arbuscular mycorrhizal fungi in plant growth regulation: Implications in abiotic stress tolerance. *Frontiers in Plant Science, 10*, 1068.

Bhupenchandra, I., Chongtham, S. K., Devi, A. G., Dutta, P., Sahoo, M. R., Mohanty, S., & Swapnil, P. (2024). Unlocking the potential of arbuscular mycorrhizal fungi: Exploring role in plant growth promotion, nutrient uptake mechanisms, biotic stress alleviation, and sustaining agricultural production systems. *Journal of Plant Growth Regulation, 44*, 1–39.

Boorboori, M. R., & Lackóová, L. (2025). Arbuscular mycorrhizal fungi and salinity stress mitigation in plants. *Frontiers in Plant Science, 15*, 1504970.

Castaño, C., Glimskär, A., Hallin, S., Maaroufi, N. I., Skånes, H., Taylor, A., & Clemmensen, K. E. (2025). Fungal and arbuscular mycorrhizal communities are unique to old grasslands in a Swedish agricultural landscape. *Applied Soil Ecology, 213*, 106233.

Chauhan, P. K., Upadhyay, S. K., Rajput, V. D., Dwivedi, P., Minkina, T., & Wong, M. H. (2024). Fostering plant growth performance under drought stress using rhizospheric microbes, their gene editing, and biochar. *Environmental Geochemistry and Health, 46*(2), 41.

Chiasson, M. A., Gumiere, T., & Gumiere, S. J. (2024). The abiotic stress management in agroecosystems through AMF technology. In *Arbuscular mycorrhizal fungi in sustainable agriculture: Nutrient and crop management* (pp. 193–210). Springer Nature Singapore.

Dagher, D., Taskos, D., Mourouzidou, S., & Monokrousos, N. (2025). Microbial-enhanced abiotic stress tolerance in grapevines: Molecular mechanisms and synergistic effects of arbuscular mycorrhizal fungi, plant growth-promoting rhizobacteria, and endophytes. *Horticulturae, 11*(6), 592.

Delgado, E. F., Valdez, A. T., Covarrubias, S. A., Tosi, S., & Nicola, L. (2021). Soil fungal diversity of the Aguarongo Andean forest (Ecuador). *Biology, 10*(12), 1289.

Diagne, N., Ngom, M., Djighaly, P. I., Fall, D., Hocher, V., & Svistoonoff, S. (2020). Roles of arbuscular mycorrhizal fungi on plant growth and performance: Importance in biotic and abiotic stress regulation. *Diversity, 12*(10), 370.

Dueñas, J. F., Hempel, S., Homeier, J., Suárez, J. P., Rillig, M. C., & Camenzind, T. (2022). Root-associated fungal lineages of a tropical montane forest show contrasting sensitivities to the long-term addition of nitrogen and phosphorus. *Environmental Microbiology Reports, 14*(5), 775–784.

Eke, P., Adamou, S., Fokom, R., Nya, V. D., Fokou, P. V. T., Wakam, L. N., Nwaga, D., & Boyom, F. F. (2020). Arbuscular mycorrhizal fungi alter antifungal potential of lemongrass essential oil against *Fusarium solani*, causing root rot in common bean (*Phaseolus vulgaris* L.). *Heliyon, 6*, e05737.

Figueiredo de Oliveira, I., Ferreira Simeone, M. L., Gomes de Paula Lana, U., de Carvalho Guimarães, C., & Morais de Sousa Tinôco, S. (2025). Enhancing sorghum growth: Influence of arbuscular mycorrhizal fungi and sorgoleone. *Microorganisms, 13*(2), 423.

Ganugi, P., Masoni, A., Pietramellara, G., & Benedettelli, S. (2019). A review of studies from the last twenty years on plant–arbuscular mycorrhizal fungi associations and their uses for wheat crops. *Agronomy, 9*(12), 840.

Gautam, A., Sharma, V., Chauhan, C., Thakur, A., Pandey, M. K., Priscilla, K., & Kumar, R. (2025). Fortifying crops with micronutrients for sustainable global nutritional security. *Agronomy Journal, 117*(3), e70086.

Ge, S., Chen, Y., Wang, Z., Li, Z., Shen, C., Zhang, T., & Wang, J. (2025). Deep insights into the diversified cropping and their impact on arbuscular mycorrhizal fungi: A global meta-analysis. *Agriculture, Ecosystems & Environment, 383*, 109537.

Ghorui, M., Chowdhury, S., Balu, P., Das, K., & Sunar, K. (2025). Boosting plant immunity: The functional role and mechanism of arbuscular mycorrhizal fungi in resistance. In *Plant microbiome and biological control: Emerging trends and applications* (pp. 195–219). Springer Nature Switzerland.

Gu, Y., Wang, H., Yang, Y., Chen, H., Chen, C., & Cheng, W. (2025). Metabonomics reveals the mechanism of stress resistance in *Vetiveria zizanioides* inoculated with AMF under copper stress. *Scientific Reports, 15*(1), 6005.

Hamzehzadeh, H., Abbaspour, H., Safipour Afshar, A., & Hamdi, S. M. M. (2025). AMF-mediated salinity adaptation in pistachio plants: Photosynthetic efficiency and ionic balance. *Biologia, 80*(1), 1–17.

Hashem, A., Almutairi, K. F., Alshaikh, N. A., Kumar, A., Wu, Q. S., Ibramimova, U., & Abd-Allah, E. F. (2025). Role of arbuscular mycorrhizal fungi in plant growth promotion and biotic stress management. In *Management of mycorrhizal symbiosis for mycoremediation and phytostabilization* (pp. 145–155). Elsevier.

Hassan, A., Azid, A., Hamid, F. S., Pariatamby, A., & Ossai, I. C. (2025). Inoculation of arbuscular mycorrhizal fungi improved the phytoremediation ability of *Jatropha multifida* L. in metal/metalloid contaminated landfill soil. *International Journal of Environmental Science and Technology, 22*, 1–26.

Hui, Q., Li, S., Chen, Z., Zhang, R., Hou, R., Efrose, R., … Rennenberg, H. (2025). Inoculation with salt-tolerant rhizobia and AMF regulates K^+/Na^+ homeostasis in *Robinia pseudoacacia* L. Alleviating salinity. *Plant, Cell & Environment, 49*, 4116–4133.

Iwuala, E., Abiodun, I., Odjegba, V., Popoola, K., Adekunle, A., Ayenigba, E., & Alam, A. (2025). Salt stress tolerance enhanced by symbiotic mycorrhizal fungi in two mungbean landraces. *Journal of Crop Health, 77*(3), 88.

Jaiswal, D. K., Krishna, R., Chouhan, G. K., de Araujo Pereira, A. P., Ade, A. B., Prakash, S., & Verma, J. P. (2022). Bio-fortification of minerals in crops: Current scenario and prospects for sustainable agriculture and human health. *Plant Growth Regulation, 98*(1), 5–22.

Jajoo, A., & Mathur, S. (2021). Role of arbuscular mycorrhizal fungi as an underground savior for protecting plants from abiotic stresses. *Physiology and Molecular Biology of Plants, 27*(11), 2589–2603.

Jiang, D., Yan, Y., Li, J., Zhang, C., Huangfu, S., Sun, Y., & Tian, L. (2025). Arbuscular mycorrhizal fungi play more important roles in saline–sodic soil than in black soil of the paddy field in Northeast China. *Agriculture, 15*(9), 951.

Jing, Y. A. N. G., Yuanzhe, M. A., Yunmei, W. U., Yifan, L. I. U., Huini, D. U., & Fuyong, W. U. (2025). Influence of arbuscular mycorrhizal fungi and zinc oxide nanoparticles on cadmium uptake and accumulation in winter wheat growing in naturally Cd-contaminated soil. *Pedosphere, 36*, 669–682.

Kakabouki, I., Stavropoulos, P., Roussis, I., Mavroeidis, A., & Bilalis, D. (2023). Contribution of arbuscular mycorrhizal fungi (AMF) in improving the growth and yield performances of flax (*Linum usitatissimum* L.) to salinity stress. *Agronomy, 13*(9), 2416.

Kaur, T., Tejoprakash, N., & Reddy, M. S. (2025). Arbuscular mycorrhizal fungi ameliorate selenium stress and increase antioxidant potential of *Zea mays* in seleniferous soil. *Biological Trace Element Research, 203*(6), 3392–3411.

Khatoon, Z., Orozco-Mosqueda, M., & Santoyo, G. (2024). Microbial contributions to heavy metal phytoremediation in agricultural soils: A review. *Microorganisms, 12*, 1945.

Khoza, T., Masenya, A., Khanyile, N., & Thosago, S. (2025). Alleviating plant density and salinity stress in *Moringa oleifera* using arbuscular mycorrhizal fungi: A review. *Journal of Fungi, 11*(4), 328.

Leventis, G., Tsiknia, M., Feka, M., Ladikou, E. V., Papadakis, I. E., Chatzipavlidis, I., Papadopoulou, K., & Ehaliotis, C. (2021). Arbuscular mycorrhizal fungi enhance growth of tomato under normal and drought conditions, via different water regulation mechanisms. *Rhizosphere, 19*, 100394.

Li, H., Zhang, L., Wu, B., Li, Y., Wang, H., Teng, H., Wei, D., Yuan, Z., & Yuan, Z. (2023). Physiological and proteomic analyses reveal the important role of arbuscular mycorrhizal fungi on enhancing photosynthesis in wheat under cadmium stress. *Ecotoxicology and Environmental Safety, 261*, 115105.

Li, J., Zhou, L., Chen, G., Yao, M., Liu, Z., Li, X., & Chen, X. (2025). Arbuscular mycorrhizal fungi enhance drought resistance and alter microbial communities in maize rhizosphere soil. *Environmental Technology & Innovation, 37*, 103947.

Liu, Y. Z., Imtiaz, M., Ditta, A., Rizwan, M. S., Ashraf, M., Mehmood, S., Aziz, O., Mubeen, F., Ali, M., Elahi, N. N., Ijaz, R., Lelel, S., Shuang, C., & Tu, S. (2020). Response of growth, antioxidant enzymes, and root exudates production towards as stress in *Pteris vittata* and *Astragalus sinicus* colonized by arbuscular mycorrhizal fungi. *Environmental Science and Pollution Research, 27*, 2340–2352.

Liu, H., Wang, H., Nie, Z., Tao, Z., Peng, H., Shi, H., & Liu, H. (2024). Combined application of arbuscular mycorrhizal fungi and selenium fertilizer increased wheat biomass under cadmium stress and shaped rhizosphere soil microbial communities. *BMC Plant Biology, 24*(1), 359.

Madouh, T. A., & Davidson, M. K. (2024). Management of environmental pollution: Hyperaccumulator plants, arbuscular mycorrhizal fungi (AMF), and biochar in heavy metal remediation. In *Ecosystem management: Climate change and sustainability* (pp. 115–169). Wiley.

Mazumder, S., Bhattacharya, D., Lahiri, D., & Nag, M. (2025). Rhizobacteria and arbuscular mycorrhizal fungi (AMF) community in growth management and mitigating stress in millets: A plant-soil microbe symbiotic relationship. *Current Microbiology, 82*(6), 242.

Mitra, D., Djebaili, R., Pellegrini, M., Mahakur, B., Sarker, A., Chaudhary, P., & Mohapatra, P. K. D. (2021). Arbuscular mycorrhizal symbiosis: Plant growth improvement and induction of resistance under stressful conditions. *Journal of Plant Nutrition, 44*(13), 1993–2028.

Murtaza, G., Usman, M., Alwutayd, K. M., & Iqbal, R. (2025). Arbuscular mycorrhizal symbiosis drives rhizosphere-regulated drought tolerance in maize. *Rhizosphere, 37*, 101250.

Nader, A. A., Hauka, F. I., Afify, A. H., & El-Sawah, A. M. (2024). Drought-tolerant bacteria and arbuscular mycorrhizal fungi mitigate the detrimental effects of drought stress induced by withholding irrigation at critical growth stages of soybean (*Glycine max*, L.). *Microorganisms, 12*(6), 1123.

Nazir, A., Sarfraz, W., Ditta, A., Khalid, N., Farid, M., Shafiq, M., Bareen, F. E., Rizvi, Z. F., & Naeem, N. (2023). Synergistic impact of two autochthonous saprobic fungi (*A. niger* and *T. pseudokoningii*) on the growth, ionic contents, and metals uptake in *Brassica juncea* L. and *Vigna radiata* L. under tannery solid waste contaminated soil. *International Journal of Phytoremediation, 25*(11), 1488–1500.

Nguyen, T. D., Cavagnaro, T. R., & Watts-Williams, D. J. (2019). The effects of soil phosphorus and zinc availability on plant responses to mycorrhizal fungi: A physiological and molecular assessment. *Scientific Reports, 9*, 14880.

Nohesara, M., Malekzadeh, E., Motlagh, M. B., & Tatari, A. (2025). Effect of nanocellulose-assisted green-synthesized iron nanoparticles and conventional sources of Fe on pot marigold plants symbiotically with arbuscular mycorrhizal fungus (*Funneliformis mosseae*). *BMC Plant Biology, 25*(1), 721.

Paramaputra, M. H., Khoirunnisa, N. S., Wijaya, L., Kanti, A., Sudiana, I. M., & Napitupulu, T. P. (2025). Co-application of biochar and arbuscular mycorrhizal fungi (AMF) for mitigating salinity stress in mangrove seedlings. *Communications in Soil Science and Plant Analysis, 56*(11), 1654–1669.

Pooja, P., Tallapragada, S., Lamba, A., & Punia, S. (2025). Role played by arbuscular mycorrhizal fungi in amelioration of salinity stress: A review. *Plant and Soil, 511*(1), 27–42.

Priyadarshani, T. D. C., Weerasinghe, P. A., Law, S., & Zakeel, M. C. M. (2025). Hidden allies: Exploring the cutting-edge insights into the role of mycorrhizal fungi in plant health. In *Fungal endophytes volume II: Applications in agroecosystems and plant protection* (pp. 551–585). Springer Nature Singapore.

Quan, L., Zhang, J., Wei, Q., Wang, Y., Qin, C., Hu, F., & Xia, Y. (2021). Promotion of zinc tolerance, acquisition, and translocation of phosphorus in *Mimosa pudica* L. mediated by arbuscular mycorrhizal fungi. *Bulletin of Environmental Contamination and Toxicology, 106*(3), 507–515.

Ramana, J. V., Tylianakis, J. M., Allen, W. J., Ridgway, H. J., Waller, L. P., Orwin, K. H., & Dickie, I. A. (2025). Invasive plants decrease arbuscular mycorrhizal fungal diversity and promote generalist fungal partners. *New Phytologist, 247*(5), 2381–2389.

Saboor, A., Ali, M. A., Husain, S., Tahir, M. S., Irfan, M., Bilal, M., & Glick, B. R. (2021). Regulation of phosphorus and zinc uptake in relation to arbuscular mycorrhizal fungi for better maize growth. *Agronomy, 11*(11), 2322.

Saeed, H. S., Munir, M., Samra, S., Farooq, A., Mahmood, H., Ahmad, I., Akram, M. I., Iqbal, R., Elsadek, M. F., Almutairi, S. M., Ditta, A., & Daraz, U. (2025). Screening of biosurfactants producing fungi from infected *Citrus sinensis* (L.). *Polish Journal of Environmental Studies, 34*(5), 5261–5270.

Sahu, L. K., & Verma, N. (2025). Role of fungi in sustainable agriculture and stress mitigation. In *Mycological inventions for sustainable agriculture and food production* (pp. 1–24). IGI Global Scientific Publishing.

Saleh, A. M., El-Soud, W. M. A., Alotaibi, M. O., Beemster, G. T. S., Mohammed, A. E., & AbdElgawad, H. (2023). Chitosan nanoparticles support the impact of arbuscular mycorrhizae fungi on growth and sugar metabolism of wheat crop. *International Journal of Biological Macromolecules, 235*, 123806.

Sarkar, A., Asaeda, T., Wang, Q., Kaneko, Y., & Rashid, M. H. (2017). Response of *Miscanthus sacchariflorus* to zinc stress mediated by arbuscular mycorrhizal fungi. *Flora, 234*, 60–68.

Sbrana, C., Avio, L., & Giovannetti, M. (2014). Beneficial mycorrhizal symbionts affecting the production of health-promoting phytochemicals. *Electrophoresis, 35*, 1535–1546.

Sharma, P., Sharma, M. M. M., Kapoor, D., Loyal, A., & Husen, A. (2024). Beneficial and adverse effects of silver nanoparticles on rhizosphere biology. In *Plant response to silver nanoparticles: Plant growth, development, production, and protection* (pp. 43–57). Springer Nature Singapore.

Shukla, S., Didwania, N., & Choudhary, R. (2025). Arbuscular mycorrhizal fungi (AMF): A pathway to sustainable soil health, carbon sequestration, and greenhouse gas mitigation. *Journal of the Saudi Society of Agricultural Sciences, 24*(4), 22.

Siddiky, M. R., Schaller, J., Caruso, T., & Rillig, M. C. (2012). Arbuscular mycorrhizal fungi and collembola non-additively increase soil aggregation. *Soil Biology and Biochemistry, 47*, 93–99.

Singh, M., Bisht, S., Singh, S., & Sharma, J. G. (2022). Implications of abiotic stress tolerance in arbuscular mycorrhiza colonized plants: Importance in plant growth and regulation. *Journal of Applied Biology & Biotechnology, 10*(6), 1–11.

Sonbol, H., Korany, S. M., Nhs, M., Abdi, I., Maridueña-Zavala, M. G., Alsherif, E. A., & Elsheikh, S. Y. S. (2025). Exploring the benefits of AMF colonization for improving wheat growth, physiology, and metabolism, and antimicrobial activity under biotic stress from aphid infection. *BMC Plant Biology, 25*(1), 198.

Stewart, J. D., Corrales, A., Canteiro, C., Qin, C., Gupta, M. M., Otgonsuren, B., & Manley, B. F. (2025). Advancing knowledge on the biogeography of arbuscular mycorrhizal fungi to support sustainable development goal 15: Life on land. *FEMS Microbiology Letters, 372*, fnaf055.

Tekaya, M., Dahmen, S., Mansour, M. B., Ferhout, H., Chehab, H., Hammami, M., Attia, F., & Mechri, B. (2021). Foliar application of fertilizers and biostimulants has a strong impact on the olive (*Olea europaea*) rhizosphere microbial community profile and the abundance of arbuscular mycorrhizal fungi. *Rhizosphere, 19*, 100402.

Thind, S., Chaudhary, M. S., Ditta, A., Hussain, I., Parveen, A., Ullah, N., Mahmood, Q., Al-ashkar, I., & Sabagh, A. E. L. (2022). Impact of mycorrhizal fungi from different rhizospheric soils on fungal colonization, growth, and chlorophyll contents of *Cenchrus ciliaris*. *Agronomy, 12*(11), 2644.

Upadhayay, V. K., Singh, J., Khan, A., Lohani, S., & Singh, A. V. (2019). Mycorrhizal-mediated micronutrients transportation in food-based plants: A biofortification strategy. In *Mycorrhizosphere and pedogenesis* (pp. 1–24). Springer Singapore.

Watts-Williams, S. J., & Cavagnaro, T. R. (2018). Arbuscular mycorrhizal fungi increase grain zinc concentration and modify the expression of root ZIP transporter genes in a modern barley (*Hordeum vulgare*) cultivar. *Plant Science, 274*, 163–170.

Wen, Y., Wu, R., Xu, T., Cao, R., Song, G., Qi, D., & Song, F. (2025). AMF and biochar reshape the bacterial network in the rhizosphere soil of *Ricinus communis* under chromium (Cr) stress and improve soil quality. *Journal of Hazardous Materials, 492*, 138122.

Yang, X., Wang, Z., Li, J., Struik, P. C., Jiang, S., Jin, K., & Mu, H. (2025). How do arbuscular mycorrhizal fungi enhance the drought resistance of *Leymus chinensis*? *BMC Plant Biology, 25*(1), 1–14.

Yu, Z., Zhao, X., Su, L., Yan, K., Li, B., He, Y., & Zhan, F. (2021). Effect of an arbuscular mycorrhizal fungus on maize growth and cadmium migration in a sand column. *Ecotoxicology and Environmental Safety, 225*, 112782.

Zhang, Z., Li, B., Liu, Y., He, L., Pang, T., Chen, Z., & Wei, Y. (2022). Arbuscular mycorrhizal fungal inoculation increases organic selenium accumulation in soybean (*Glycine max* (Linn.) Merr.) growing in selenite-spiked soils. *Toxics, 10*(10), 565.

Chapter 12

Arbuscular mycorrhizal fungi and drought tolerance in plants

Anoo Gunasekera [a], Dinoo Gunasekera [b] and Disna Ratnasekera [c]
[a] Department of Biomedical Sciences, Faculty of Health Sciences, University of Ruhuna, Matara, Sri Lanka, [b] Department of Information and Communication Technology, Faculty of Technology, University of Ruhuna, Matara, Sri Lanka, [c] Department of Agricultural Biology, Faculty of Agriculture, University of Ruhuna, Matara, Sri Lanka

12.1 Introduction

By 2050 the global population is projected to exceed 9.7 billion, with over 65% relying primarily on agriculture for their survival and expected to rise to 90% in developing nations (Castañeda et al., 2016). Consequently, both national economies and global food security will heavily depend on agricultural output. However, drought-induced water scarcity poses a major threat, significantly reducing crop yields and jeopardizing food production. Drought stress is a pervasive environmental constraint that transcends geographical boundaries, often emerging without clear warning and significantly impairing plant biomass production, quality, and energy metabolism. As a major abiotic stressor driven by temperature fluctuations, high light intensity, and insufficient rainfall, it exerts cumulative and multifaceted effects on plants (Saleem et al., 2024, 2025). These impacts manifest across morphological, physiological, biochemical, and molecular levels, ultimately reducing photosynthetic efficiency (Seleiman et al., 2021). Drought not only impacts food supply but also depletes groundwater reserves, degrades water quality, and accelerates soil erosion (Wasaya et al., 2024). To meet the food demands of an increasing population amid water scarcity, it is imperative to come up with strategies that provide resilience in crops against drought so that crops can grow sustainably and yields can increase. As part of water scarcity, plants have intricate resistance and water adaptation mechanisms comprising physiological and biochemical responses that vary with species.

Plants use multiple mechanisms to cope with drought stress, including increasing diffusive resistance, reinforcing root systems for greater water uptake, or reducing transpiration losses. The key adaptive mechanism in plants allows them to maintain high water status and survive under limited water availability. Further evidence suggests that soil microbial communities would aid in drought resistance in plants (Feng et al., 2023; Sarkar et al., 2022a, 2022b). The role of fungi mechanistically toward enhancing drought tolerance in plants is still poorly understood. The higher level of complexity and functional diversity of the soil fungal community brings significant difficulty to discovering their importance for strengthening plant resilience under water-stress conditions, despite this being of prime importance.

In particular, arbuscular mycorrhizal (AM) fungi (AMF) enhance plants' ability to withstand and adapt to environmental stresses, mainly drought (Abdalla et al., 2023; Cosme, 2023; Das & Sarkar, 2024). AMF can establish a mutualistic association with nearly 80% of terrestrial plants to facilitate water uptake and physiological adaptation under drought stress by, respectively, hydrating tissues and making them metabolically more efficient (Das & Sarkar, 2024). Plant–fungal symbiosis enhances vital plant traits such as growth, water retention, mineral nutrition, and abiotic stress resistance. The positive impacts of AM symbiosis come through the combined action of physical, nutritional, physiological, and cellular factors. The drought resistance of plants occurs through a wide spectrum of AMF-mediated mechanisms, such as the activation of antioxidant defense systems, phytohormone modulation, adjustment of osmotic potential, accumulation of proline, fungal hyphae-mediated water uptake, morphological alterations, and preservation of photosynthetic efficiency. A thorough knowledge of these mechanisms is essential to enhance crop production under increased drought conditions. Understanding and harnessing AMF-mediated drought tolerance mechanisms can enhance crop resistance, improve water-use efficiency, and support sustainable agricultural productivity under drought.

Advancing Sustainable Agriculture Through Arbuscular Mycorrhizal Fungi. DOI: https://doi.org/10.1016/B978-0-443-45130-0.00027-8

12.2 Drought and its effects on crop plants

12.2.1 Global impact on agriculture and food security

The global effects of droughts, with their high frequency and severity as the direct consequence of climate change, pose an increasingly serious threat to the ecosystem and socioeconomic stability, with agriculture occupying the top of the list of vulnerable sectors (Shehzad et al., 2023). According to the reports of the Food and Agriculture Organization, drought has a hand in about 20% of crop yield losses worldwide, with more severe impacts on rainfed agricultural systems (FAO, 2023). Inadequate precipitation during the development stage of a crop can either dramatically reduce the ability of the crop to produce or, in many cases, may lead to outright harvesting losses. Drought stress causes a cascade of changes in plant morphological, physiological, biochemical, and gene expressions, increasing the interruption of growth and developmental processes that consequently lead to major losses in agricultural productivity (Shelake et al., 2022). The extent of the damage caused is influenced by the severity and duration of the stress and the period of development of the crop under consideration (Chaouachi et al., 2023).

According to future climate projections, droughts will be complex, extreme, and spatially unpredictable, further increasing the global food security challenges (Anderson et al., 2019; Yao et al., 2018). Therefore intensified drought regimes under climate change now stand as one of the greatest challenges for sustainable crop production and water resource planning. Consequently, quantifying drought sensitivity in key agricultural zones and projecting future vulnerabilities is essential for developing adaptive strategies to safeguard worldwide food supplies (Pequeno et al., 2021).

12.3 Drought impacts on plant growth

Drought is defined as a physiological condition in which a plant's water loss through transpiration exceeds its water uptake via the roots, leading to reduced turgor pressure, impaired metabolic functions, and potential growth inhibition. The prolonged and severe water scarcity results in the death of plants, leading to total food losses. Common causes of drought include insufficient rainfall, declining groundwater levels, high salinity (physiological drought), or soil particles holding onto water, reducing its availability to plants (Haghpanah et al., 2024). Under water stress conditions, plants coordinated morpho-anatomical, physiological, and biochemical adaptations to mitigate water loss and maintain cellular hydration. Drought conditions impair plant performance across various developmental stages, disrupting critical physiological processes (Kapoor et al., 2020). Water deficit significantly impairs seed germination and early seedling establishment. For example, key crops such as maize (*Zea mays*), sorghum (*Sorghum bicolor*), and wheat (*Triticum aestivum*) showed malfunctions in germination (Kapoor et al., 2020; Queiroz et al., 2019). The duration and intensity of drought stress determine its impact on plant water status and subsequent metabolic dysfunction during growth (Sharma et al., 2020).

During early vegetative growth a shortage of water generates visible symptoms such as wilting of leaves, retarded plant height, bud formation, and flower formation. Furthermore a drought in soil moisture impedes nutrient uptake, and this further diminishes stem elongation.

To counter drought stress, plants tend to build their root system more to extract water from deeper soil layers. Roots sense water availability and respond by changing their architecture to increase length, lateral root formation, and overall spread of roots (Zulfiqar et al., 2020). These changes help maintain crucial activities regarding water absorption, nutrient uptake, and microbial symbiosis in the rhizosphere. However, an extended deficit of water decreases root growth but increases the root-to-shoot ratio, a way of plants prioritizing root allocation to gain critical resources at the expense of development of shoot foliage.

As the primary site for photosynthesis and a key biomass production facility, water stress substantially reduces leaf number in species like *Andrographis paniculata* (Bhargavi et al., 2017). Being a determinant of photosynthetic efficiency the shrinkage of leaves under drought conditions restricts carbon assimilation and ultimately diminishes crop yield.

Studies underlined that water deficit leads to a reduction in leaf area, which is a physiological adaptation balancing root water uptake with plant water status (Kapoor et al., 2020; Srivastava & Srivastava, 2014). Thus a reduction in leaf area acts as a mechanism of drought avoidance by reducing water loss through transpiration. A few obvious factors restricting leaf expansion are, among others, the inhibition of cell division and impaired turgor pressure. Whereas a reduction in soil moisture yields low water content in leaves, thereby decreasing guard cell turgor and prompting stomatal closure (Deka et al., 2018). This otherwise further restricts gas exchange and increases photosynthetic decline. Meanwhile, drought also initiates early leaf senescence, greatly limiting growth.

12.3.1 Effects of drought on plant physiological and biochemical processes

Drought stress predominantly acts to disrupt almost all kinds of physiological processes in plants, with photosynthesis being the most vulnerable of these (Ahmad et al., 2023; Inal et al., 2023; Raza et al., 2023; Vurayai et al., 2011). Inhibition of photosynthesis severely compromises crop yield. Water deficit conditions induce the overgeneration of reactive oxygen species (ROS), initiating oxidative stress as one of the pivotal plant responses. The oxidative burst results in lipid peroxidation, which primarily damages the cellular membranes and subsequently triggers a chain reaction of disintegration of physiological and biochemical processes (Wang et al., 2019). This metabolic dysregulation eventually acts against plant growth and, therefore, productivity. Drought severely affects cellular homeostasis, including osmotic imbalance, cell dehydration, and loss of turgor pressure (Haghpanah et al., 2024). Such conditions initiate plasma membrane and endosomal disruption, whereas metabolic dysfunction subsequently arises because of inhibited synthesis of metabolites and depletion of Adenosine Triphosphate (ATP). (Oguz et al., 2022). Photosynthetic capacity is reduced because of the closure of stomata and photochemical damage that is sustained, together with oxidative stress produced because of the elevated number of ROS.

Destabilization of cellular membranes is the key issue, with water deficit resulting in a diminished hydrophobic force that precipitates protein denaturation and phase transition from lamellar to inverted hexagonal lipid structures. This loss of membrane integrity is a leading factor in water-deficit-induced cell death during extreme drought conditions. To counteract these deleterious effects, stress-related proteins in plants get activated to restore osmotic balance, stabilize macromolecules, maintain membrane functionality, and thus improve cell resilience (Turner, 2018). Plants combat water deficit stress by utilizing antioxidant systems and osmotic regulation. The antioxidant system consists of enzymes and other substances that help eliminate the toxic effects of ROS under drought stress. Meanwhile, plants accumulate compatible solutes such as proline, glycine betaine, sugars (glucose, fructose), polyols (mannitol, inositol), and some amino acids (valine and isoleucine) to maintain osmotic balance within the cells. These osmo-protectants enhance water potential, thereby preserving molecular integrity and mitigating oxidative damage induced by stress-related free radicals (Hasanuzzaman et al., 2020; Rao et al., 2016).

Moreover, drought stress decreases the gap between endosomal and plasma membranes, promoting lipid exchange and changes in permeability. This cascade triggers ion imbalance, ROS accumulation, and oxidative damage to biomolecules, culminating in cell death (Dossena & Marino, 2021). Mitochondria, with their lipid-rich membranes, are particularly susceptible to ROS-mediated injury. Oxidative stress impairs enzymes, disrupts organelles (mitochondria, chloroplasts), and degrades membranes. Dehydration alters lipid composition, reducing fluidity and stability. Lipid peroxidation, driven by free radicals and ROS, further exacerbates membrane disintegration, compromising cellular integrity under water deficit (Choudhury et al., 2017).

Decreased leaf area, low-stomatal conductance, and the subsequent osmotic stress impair the photosynthetic apparatus, significantly limiting photosynthetic efficiency. The reduction in photosynthetic activity is primarily driven by diminished CO_2 assimilation, which has been demonstrated to impair Rubisco function, suppress nitrate reductase and sucrose phosphate synthase activity, and diminish the capacity for ribulose-1,5-bisphosphate (RuBP) regeneration.

Drought inhibits photosynthesis, though the mechanisms differ between photosynthetic types. In C_3 species, stomatal closure initially limits CO_2 uptake, whereas C_4 plants face metabolic constraints on carbon assimilation (Ripley et al., 2010). For C3 plants, mild drought primarily causes stomatal closure, whereas severe stress reduces CO_2 availability in chloroplasts, elevating photorespiration, where Rubisco oxygenates RuBP instead of carboxylating (Pandey et al., 2023). In addition, drought disrupts thylakoid membrane organization, decreasing grana stacking and reducing photosynthetic efficiency (Pandey et al., 2023).

Drought stress induces significant impairments in both photosystem I and photosystem II (PSII) across different developmental stages. The operational quantum efficiency of PSII (ΦPSII) shows its particular sensitivity to water deficit and thus serves as a reliable physiological indicator for early drought detection (Hu et al., 2023). These photosynthetic activities are emphasized by drought suppression over the enzymes, especially Rubisco, whose lesser activity directly limits the carbon assimilation potential. The ability of C_4 species to adapt during drought is genotype-dependent; drought-tolerant genotypes maintain higher rates of photosynthesis (A) and stomatal conductance (gs) than do susceptible ones under the same level of stress (Saha et al., 2016). This varying response is suggestive of the prominence of genetic diversity in photosynthetic adaptation to drought. Conservation of PSII function, enzyme activity, and stomatal conductance becomes majorly important for drought tolerance in C_4 species (Ripley et al., 2010). The combined effects of stomatal regulation to metabolic and structural changes collectively diminish photosynthetic capacity under water deficit conditions.

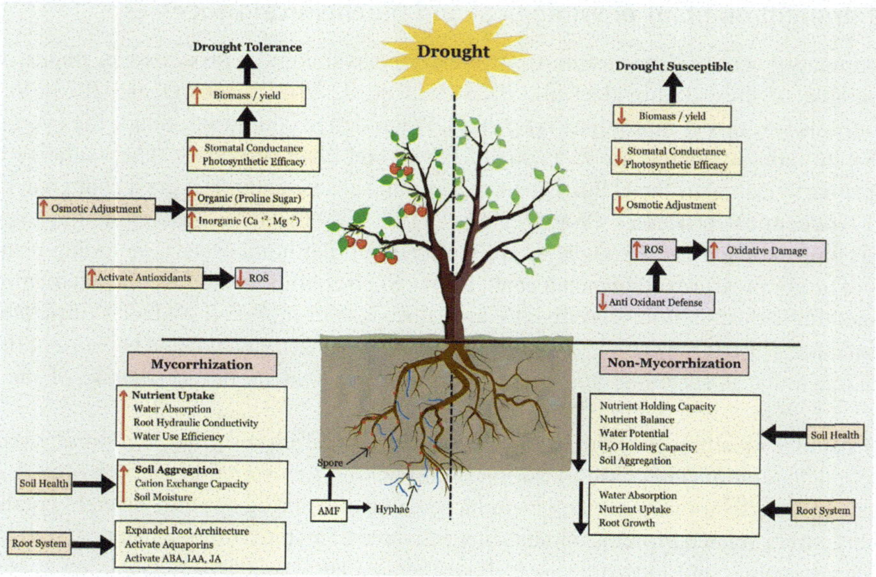

FIGURE 12.1 AMF inoculation and noninoculation: Effects and consequences of drought on plant physiological functions and soil health. *ABA*, Abscisic acid; *AMF*, arbuscular mycorrhizal fungi; *IAA*, indole-3-acetic acid; *JA*, jasmonic acid; *ROS*, reactive oxygen species.

12.4 Arbuscular mycorrhizal fungi and symbiotic associations

AMF belong to *Glomeromycota*, which include various species under seven types that remain uncharacterized (Tedersoo et al., 2024). Among them, AMF are commonly available, obligate root symbionts that establish mutualistic associations with nearly 72% of terrestrial plant species and are characterized by bidirectional nutrient exchange (Genre et al., 2020; Nazir et al., 2023; Thind et al., 2022; Yizhu et al., 2020). AMFs enhance host plant acquisition of mineral nutrients and stress resilience, while deriving fixed carbon in the form of photosynthates (Fig. 12.1).

The structural components of AMs primarily consist of hyphae, arbuscules, and spores, with vesicles and auxiliary cells present in select species. The fungal mycelium differentiates into two distinct compartments: intraradical (within root cortex) and extraradical (soil-borne) hyphae. The extraradical mycelium forms extensive soil networks, extending beyond the root zone to access and transport water and nutrients from soil microsites unreachable to plant roots (Xie et al., 2022). The extraradical mycelium penetrates host root epidermal cells and colonizes cortical tissues, where it undergoes repeated branching to form intricate arbuscular structures. These arbuscules represent the functional core of the AM symbiosis, serving as the primary site for bidirectional nutrient exchange between the fungal partner and host plant. AM symbioses critically influence ecosystem dynamics by optimizing plant physiological processes, including hydromineral uptake, stress response modulation, and phytohormone balance, with demonstrable impacts on primary productivity through enhanced biomass accumulation (Zhang et al., 2024).

AMF also significantly enhance stress tolerance in most vascular plants (Xie et al., 2022), alleviating abiotic stresses through improved water and nutrient use efficiency. Mycelia-synthesized glycoproteins contribute to the formation of a favorable rhizosphere environment, promoting host plant growth by facilitating the absorption and translocation of water and essential mineral nutrients from the surrounding soil (Hu et al., 2017).

AMF establish a direct soil-root interface, enhancing plant nutrient uptake, moisture absorption, and photosynthetic efficiency, thereby mitigating abiotic stresses such as drought, salinity, and nutrient deficit (Begum et al., 2022). Moreover, they facilitate metal complexation, while their extraradical hyphae immobilize metal ions via chelation and filtration mechanisms (Huang et al., 2021a, 2021b). As integral contributors to terrestrial ecosystem functioning, AMF are gaining prominence for their potential applications in sustainable agricultural systems.

Thus, AMF play a critical role in enhancing plant establishment and productivity in nutrient-deficient soils while alleviating abiotic stress (Kakouridis et al., 2022). With 332 formally described species and many more yet unidentified, AMF exhibit significant functional diversity in their symbiotic relationships with host plants (Koch et al., 2017; Wijayawardene et al., 2020). Further, ecosystem functions, including soil aggregation, water balance, and nutrient cycling, were upregulated by AMF–host interactions (Qiu et al., 2022).

Despite the ecological significance of AM fungi, their global biogeographical patterns are not well described. The existing information is primarily concentrated on regions that have been intensively researched, such as Europe, North America, and China, whereas tropical ecosystems and large regions of Asia have received less attention (Větrovský et al., 2023). The*GlobalAMFungi* is a novel database providing the most extensive global atlas of AM fungal distributions, which has the value for informing ecological knowledge of AM fungal taxa (Větrovský et al., 2023). The overall effects and consequences of drought and AMF-mediated alleviation of drought's negative effects are summarized in Fig. 12.1.

12.4.1 Environmental determinants of arbuscular mycorrhizal fungi symbiosis

Environmental factors such as temperature, soil pH, moisture, and nutrient availability critically regulate the establishment and functioning of the symbiosis between AMF and plants (Albornoz et al., 2022). These parameters govern colonization rates and the efficiency of nutrient exchange between AMF and the host plants (Rui et al., 2022). Understanding these interactions is of direct agricultural importance for selecting temperature ranges conducive to symbiosis enhancement and manipulating soil pH for improved nutrient uptake by the crops (Rui et al., 2022).

Seasonal variations in biotic and abiotic factors control AMF communities. For instance, studies show that the spore abundance and diversity of AMF in roots of *Alsophila spinulosa* were greater in the rainy season, whereas root colonization rates were at a maximum in the dry season (Han et al., 2023). A similar seasonal pattern was observed in *Juglans mandshurica*, where AMF diversity in rhizosphere soil was highest in July than in either May or September, with total nitrate nitrogen and water-soluble phenol being the most crucial factors in driving these seasonal changes (Ji et al., 2021). AMF hyphal density, spore density, and root colonization rates were greater in summer than in winter, modifying the community structure of AMF both in soil and in roots because of seasonal changes under water deficit stress (Han et al., 2023).

The host plant species highly influences the sporulation pattern and the community structure of AMF. Various types of experiments across ecosystems show variations depending on the host species of the AMF associations. In Inner Mongolian grasslands and semiarid Mediterranean prairies, different plant species exhibited different levels of mycorrhizal colonization (Han et al., 2023). Medicinal plant studies in Chittagong, Bangladesh, have also found that AMF colonization rates could be influenced by several environmental factors such as climatic conditions, soil characteristics, host plant relationships, and diversity of local species (Halder et al., 2015).

Seasonal changes greatly affect soil microbial communities through fluctuations in temperature, moisture, and nutrient availability. As plants grow and shed litter in different seasons, the deposited litter and organic material modify nutrient dynamics in the soil, creating strong feedback into soil microbes (Ji et al., 2021).

Seasonal shifts essentially affect the diversity and functional gene composition of both bacterial and fungal communities, indicating soil environmental heterogeneity with spatiotemporal bases of microbial dynamics (Bell et al., 2021). These results emphasize that AMF–host interactions and other ecological factors preserve plant diversity in natural systems through species-based colonization. Insertion of knowledge about natural AMF changes in undisturbed environments may greatly benefit bioremediation techniques. Glomalin-related soil proteins, secreted by AMF hyphae, are also known to be important in soil organic carbon storage; thus AMF seasonal dynamics again come into play when considering ecosystem functioning (Kumar et al., 2018). Understanding environmental modulation of AMF interactions enables the prediction of symbiotic responses to climate change and anthropogenic disturbances.

Throughout the AMF-based nutrient exchange mechanisms, assessing threshold values for critical environmental parameters, developing soil amendment strategies in the field, and the climate change resilience of AMF–plant partnerships are areas of attention. This research dimension stands to serve a dual purpose of optimizing the agricultural productivity through targeted management of symbiosis while furthering ecological applications of these essential belowground interactions in the ever-fading environmental scenarios.

12.5 Arbuscular mycorrhizal fungi–mediated drought tolerance mechanisms

Water deficit limits plant growth, development, and biomass accumulation. It stressfully affects the processes in plant metabolism through different chains of events. These may include reduced photosynthesis by limited CO_2 intake through closure of stomata, altered carbon partitioning patterns, an increase in ROS generation, and oxidative stress. Moreover, it limits phosphate bioavailability and nutrient acquisition, thus compromising crop quality.

AM symbiosis renders plants with better adaptation against drought via multifaceted mechanisms that include improvement in soil structure, water and nutrient uptake, and use efficiency; osmotic adjustment (OA); antioxidant defenses against ROS; phytohormone modulation; and activation of stress-responsive genes (Bahadur et al., 2019). The AMF association in effect enhances phenotypic plasticity and thus helps the host plant in adapting better to limited water conditions. The present research shows that AMF drought tolerance is conferred through integrated networks of mechanisms, including physical root architecture and soil modifications, nutrient acquisition and translocation, physiological mechanisms, including adjustments in water relations, and cellular regulation for stress-responsive pathways. Whole enzymatic pathways and plant metabolites involved in drought resistance show the important position of AMF in drought tolerance in plants.

An increase in stomatal conductance, transpiration, and shoot water content was observed in mycorrhizal plants as compared to nonmycorrhizal plants. The AMFs probably superimpose the ability of roots to extract water from dry soil, thereby reducing the onset of stress in leaves and keeping the stomatal conductance and shoot hydration at relatively high levels for a given soil water potential (Chun et al., 2018). Even though the mechanisms remain unclear, AMF symbiosis somehow dictates water retention and uptake in soil, thereby laying down the strategy of inoculation to be followed for AM and non-AM plants.

12.5.1 Arbuscular mycorrhizal fungi–modified root architecture

Root system architecture (RSA) refers to the spatial arrangements and distributions of roots in the soil and plays a very critical role for a plant to adapt to nutrient and water uptake from imperfect soil environments (Chatzistathis et al., 2013; Ma et al., 2018). AMF modifies the fine root system structure according to the host and environmental conditions (Yang et al., 2023). AMFs influence the RSA of the host plants considerably, with the variation in morphology being dependent on the genotype of the plant and fungus, as well as on environmental factors such as water and nutrient stress. Plants show increased root length and surface area under AMF colonization in nutrient-poor conditions, such as in hedgerow plants, winged oil trees, sorghum, soil sedum, and strawberry species (Wu et al., 2016; Yuan et al., 2023). Enhancement in root traits, such as length and surface area, and root hair density increases nitrogen uptake, thus altering the plant's carbon-to-nitrogen balance. Improved root length and surface area extend the capacity to explore soils, thereby intercepting nutrients, a trait strongly associated with high N uptake and accumulation in plants. On the contrary, drought stress severely disables such functions of the RSA in trifoliate orange seedlings; *Glomus mosseae* inoculation, however, overrides this constraint, acting to augment the root absorption zones, both active and total, across diverse soil moisture regimes compared with the nonmycorrhizal control (Wu et al., 2013). Similarly, dual inoculation with AMF and *Frankia* synergistically promotes root branching in *Alnus glutinosa* (Orfanoudakis et al., 2010).

However, the effects are strain-specific; for instance, *Funneliformis mosseae* inoculation did not significantly alter total root length or tip formation in mulberry cuttings (Yan et al., 2020), whereas *Rhizophagus intraradices* reduced root branching and projected area in red beans (Qu et al., 2021). These findings underscore the potential of AMF to enhance plant resilience in nutrient-deficient soils through targeted root architectural modifications in an AMF strain-specific manner.

12.5.2 Arbuscular mycorrhizal fungi–enhanced water uptake: Hyphal network extension and regulation of root cells' aquaporins

AMF colonization expands root architecture, increasing water absorption capacity while enhancing water use efficiency and root hydraulic conductivity. Under arid conditions, AMF hyphae significantly contribute to water uptake, facilitating apoplastic flow within roots (Bárzana et al., 2012). The hyphal narrow diameter enables access to soil micropores beyond root hair reach. Potassium (K^+) plays a pivotal role in hyphal water transport, enhancing root hydraulic conductivity in AMF-colonized plants regardless of moisture status (El-Mesbahi et al., 2012).

RiMsn2 transcription factor in *Rhizophagus irregularis* is a critical gene for arbuscule development and the host plant drought-tolerance mechanisms. This transcription factor interacts with RiHog1 to coregulate the expression of STRE-controlled genes, constituting a crucial signaling module (RiHog1–RiMsn2–STREs) that controls fungal drought responses (Fan et al., 2023). RiSte11, RiPbs2, and RiHog1 in the Hog1-MAPK cascade regulate the expression of drought-resistance-related genes (RiAQPs, RiTPSs, RiNTH1, and Ri14-3-3); gene silencing inhibits arbuscule formation and drought resistance in the host (Wang et al., 2023). Furthermore the 14-3-3-like proteins (Ri14-3-3, Fm201, and RiBMH2) participate in arbuscule development and drought stress responses (Sun et al., 2018), highlighting their dual functional importance in symbiosis and stress.

Mycorrhizal colonization regulates many AQP subfamilies [NOD26-like intrinsic proteins (NIPs); plasma membrane intrinsic proteins (PIPs); tonoplast intrinsic proteins] in host plants, influencing their functions in the transport of water and other essential molecules (CO_2, nutrients, and gases) (Bárzana et al., 2014). This culminates in increased root hydraulic conductivity, which benefits plant water status and drought tolerance (Chitarra et al., 2016). For water conductance, PIPs are of major significance, as they mediate water transport at the cellular level. Symbiosis with AMF regulates the expression of AQPs in the host plant, with dynamics dependent on watering regimes and drought severity. However, some

AQPs could act as multifunctional transporters, allowing the passage of both water and other physiologically important solutes for better functioning of the host (Bahadur et al., 2019). AMF *R. irregularis* inoculation of *Malus hupehensis* seedlings improved drought tolerance, related to the increased mRNA levels of root PIP1-3, PIP1-4, and fungal Rir-AQP1/2 (GintAQPF1/2) genes (Huang et al., 2020). Moreover, drought-induced upregulation of GintAQPF1/2 in AMF hyphae and colonized roots, overexpression of D-myo-inositol-3-phosphate synthase (IPS), and 14-3-3-like protein GF14-mediated drought tolerance in maize were reported (Wang et al., 2024). Complementary research in tomato revealed AMF-induced overexpression of *LeNIP3;1*, encoding NIPs, which correlated with enhanced root hydraulic conductivity (Chitarra et al., 2016).

The expression of these aquaporin genes is dynamically modulated by both soil moisture status and drought intensity (Bárzana et al., 2014; Chitarra et al., 2016). AMF colonization significantly modulates AQP gene expression in host plants, enhancing drought adaptation. In *Robinia pseudoacacia*, AMFs upregulate *RpPIP2;1* expression (Bahadur et al., 2019), potentially facilitating targeted water transport to critical tissues during drought. Contrastingly, lettuce roots exhibit downregulation of *GintAQP1* under water deficit despite AMF colonization, and AMF-colonized maize maintains elevated sap flow rate and root hydraulic conductance during short-term drought, correlating with sustained or increased expression of multiple PIP genes (*ZmPIP1;1, ZmPIP1;2, ZmPIP1;3, ZmPIP1;4, ZmPIP1;6, ZmPIP2;2*, and *ZmPIP2;4*) (Aroca et al., 2007; Bárzana et al., 2014). These findings reveal species-specific AQP regulation patterns under AMF symbiosis, key PIP isoforms maintaining water transport during drought, and the complex interplay between AMF and host aquaporin networks in drought responses.

Under persistent drought, reduced soil water availability decreases hydraulic conductance in AMF-colonized plants, accompanied by downregulation of key *ZmPIP* genes (1;1, 1;3, 1;4, 2;2, and 2;4) to minimize water loss (Porcel et al., 2006). However, AMF aquaporins actively facilitate water deficit tolerance through dual mechanisms, facilitating water transport via extraradical mycelium and periarbuscular membranes and maintaining hydraulic conductivity during stress periods. Fungal aquaporins explain the elevated mycelial-hydraulic conductance during short-term drought and sustained root-hydraulic conductance during prolonged water deficit (Bárzana et al., 2014), demonstrating their critical role in AMF-mediated drought adaptation through coordinated host–fungal water regulation systems.

12.5.3 Arbuscular mycorrhizal fungi–regulated osmotic adjustment

OA serves as a critical physiological adaptation that enhances plant drought tolerance through osmoregulation and osmo-protection. Osmoregulation maintains cellular turgor by accumulating compatible low molecular weight solutes to lower osmotic potential, whereas osmo-protection reinforces stress resilience through antioxidant defense enhancement and ion homeostasis (Barnawal et al., 2019). Key osmo-protectants, including organic compounds (proline, trehalose, glycine betaine, sugars, and polyamines) and inorganic ions (K^+, Ca^{2+}, and Mg^{2+}), collectively reduce osmotic potential while stabilizing cellular integrity by preserving membrane structure, protein function, and cytosolic pH (Wu et al., 2013). Notably, proline serves dual roles as both an osmolyte and a reservoir of carbon/nitrogen for poststress recovery (Barnawal et al., 2019).

Drought disrupts OA by impairing osmoprotectant synthesis, particularly proline, because of reduced nitric acid uptake and translocation (Iqbal et al., 2017). AMFs play a pivotal role by enhancing host plants' osmotic regulation under water deficit by significantly amplifying OA capacity in host plants. For example, in *Macadamia tetraphylla* L., AMF colonization enhanced drought resilience through increased accumulation of osmo-protectants such as proline, free amino acids, and soluble sugars (Yooyongwech et al., 2013). Further, mycorrhizal inoculation improved drought tolerance through more efficient osmotic regulation has been reported in citrus species (Abbaspour et al., 2012). AMF symbiosis improves nutrient acquisition (nitrogen) and upregulates proline biosynthesis, thereby restoring osmotic balance and mitigating drought-induced metabolic constraints (Etesami & Maheshwari, 2018).

Proline accumulation, triggered by AMF, is an adaptation allowing plants to preserve cellular homeostasis and metabolic operations during drought (Chitarra et al., 2016). Proline, a compatible solute, assists enormously in OA by preserving turgor pressure in cells and protecting macromolecular structures (Bahadur et al., 2019). This ability to protect acts in conjunction with proline molecular chaperone activity, which stabilizes proteins, membranes, and other subcellular components under dehydration stress (Yooyongwech et al., 2013). According to research, AMF colonization enhances proline accumulation in tomato plants (*Solanum lycopersicum* L.) under water deficit (Chitarra et al., 2016). This multifunctionality of proline, in osmotic balance, antioxidant activity, and structural stabilization, highlights its essential role in drought tolerance mechanisms in plants. AMF colonization-proline accumulation coordination opens the door for the possibility of mycorrhizal associations improving crop tolerance to water stress by improving the OA capacity.

Proline accumulation varies significantly in response to osmotic stress, with AMF symbiosis playing a key regulatory role. The enhanced proline accumulation suggests AMF colonization improves plant drought resilience through more efficient osmoregulation. Similar patterns have been observed across diverse species, including *Antirrhinum majus, Erythrina variegata, Cyclobalanopsis glauca*, and *Ocimum gratissimum,* underscoring the widespread role of AMF in drought stress adaptation (Chun et al., 2018).

12.5.4 Arbuscular mycorrhizal fungi–regulated plant growth regulators to develop drought tolerance

Signaling molecules such as abscisic acid (ABA), indole-3-acetic acid (IAA), strigolactones (SLs), and jasmonic acid (JA)/jasmonates regulate numerous physiological processes during plant development, including those involved in establishing and maintaining mycorrhizal fungus symbioses. Mycorrhizal plants develop drought tolerance through the variation of hormonal regulations, especially ABA signaling pathways. ABA is useful in arbuscule development in AMF association and improves drought resistance (Bahadur et al., 2019; Martıń-Rodrıǵuez et al., 2016). Various reports showed that AMF can regulate the levels of ABA in roots of the host plant under water deficiency conditions, improving the plant's ability to withstand stress (Aroca et al., 2008; López-Ráez, 2016). For instance, under heavy stress conditions, mycorrhizal plants show increased concentrations of ABA, a hormone crucial in priming the physiology toward resistance against drought stress (Calvo-Polanco et al., 2013). Further, ABA is required for AMF symbiosis establishment by regulating the formation and function of arbuscules (López-Ráez, 2016).

These findings underscore AMF's pivotal role in modulating plant stress responses by optimizing plant growth regulatory mechanisms, offering critical insights for developing sustainable strategies to improve crop performance in water-limited environments.

In addition, molecular analyses revealed an upregulation of ABA-related genes, such as D-myo-inositol-3-phosphate synthase and 14-3-3-like protein GF14, suggesting their role in symbiotic drought resistance (Bahadur et al., 2019). These outcomes emphasize the synergistic communication between AMF and host plants, where fungal-mediated ABA modulation enhances drought tolerance through both hormonal and genetic mechanisms (Table 12.1).

Elevated IAA was observed in mycorrhizal colonization compared to noninoculated controls, demonstrating IAA's regulatory role in promoting root system development. AMFs enhance root system development, particularly lateral root growth, through IAA modulation. Upregulation of IAA and ABA levels in jujube roots was observed by mycorrhizal colonization (Ma et al., 2022). SLs, the derivatives of carotenoids crucial for AMF symbiosis, decline under drought in both AMF-colonized and noncolonized plants (Mitra et al., 2021). Upon SL perception, AMFs activate oxidative metabolism, stimulating hyphal proliferation to enhance host root contact and symbiosis establishment (Bahadur et al., 2019). Notably, increased SL production and enhanced mycorrhizal colonization and drought tolerance were found in overexpressed *MdIAA24* in apple roots (Huang et al., 2020). Particularly, *R. irregularis* upregulates SL biosynthesis in drought-stressed lettuce and tomato, with transcriptional activation of the *SlCCD7* gene (critical for SL production) in tomato roots (Ruiz-Lozano et al., 2016). Both SLs and ABA, derived from apocarotenoids, exhibit coordinated accumulation under stress, functioning as synergistic stress-response hormones in mycorrhizal plants (Aroca et al., 2013; López-Ráez, 2016). This ABA–SL crosstalk, amplified by AMF colonization, modulates physiological adaptations to water deficit. The AMF–SL interaction thus forms a critical drought mitigation mechanism, optimizing plant hydraulic regulation and stress resilience.

JA and its derivatives are critical regulators of AM symbiosis (López-Ráez, 2016). Inoculation with AMF showed elevated endogenous levels of JA, its precursor 12-oxophytodienoic acid (OPDA), and hydroxylated derivatives (11-/12-OH-JA) in *Digitaria eriantha* under drought and salinity stress (López-Ráez, 2016). Studies indicated that combined AMF symbiosis and exogenous methyl jasmonate (MeJA) application alleviates drought-induced damage to root water uptake efficiency in common beans (Sánchez-Romera et al., 2016), suggesting jasmonate signaling as a key mediator of AMF-conferred stress tolerance. These results shed light on the importance of AMF as modulators of the plant stress response through balancing the plant growth regulatory mechanisms, which open up extremely important opportunities to devise sustainable strategies for enhancing crop performance under water-limited environments.

12.5.5 Arbuscular mycorrhizal fungi–enhanced antioxidant defense to alleviate drought stress

Water deficit stress stimulates severe oxidative damage in plants, characterized by excessive accumulation of ROS, including superoxide radicals (O_2^-), hydrogen peroxide (H_2O_2), and hydroxyl radicals (OH•).

The overproduction of ROS triggers oxidative bursts that disrupt cell membranes and biomolecules, leading to cell death (Li et al., 2022). ROS dynamics are therefore influenced dually by AMF; during the early infection process, colonization

TABLE 12.1 Applications of AMF symbiosis for drought mitigation in agriculture.

Crop species	Mechanism	References
Maize (*Zea mays*)	Aquaporin gene GintAQPF1 and GintAQPF2 expressions were enhanced. Improved water uptake	Li et al. (2013)
	IPS and 14-3GF activate 14-3-3 proteins and aquaporins, enhancing drought tolerance in plants	Li et al. (2016)
	The RiHog1-RiMsn2-STREs module regulates drought-responsive genes in AMF, enhancing host plant drought resistance	Fan et al. (2023)
	AM-colonized roots actively converted putrescine to GABA via diamine oxidase activity	Hu et al. (2020)
Citrus (*Citrus reticulata* Blanco)	AM fungal hyphae exhibited increased water uptake rates during drought conditions	Zhang et al. (2018a, 2018b)
Lactuca sativa	Regulating ABA concentrations improves drought tolerance	Aroca et al. (2008)
Wheat (*Triticum aestivum* L.)	Boosting nitrogen fixation and auxin synthesis mitigates drought stress	Arzanesh et al. (2011)
	Elevating chlorophyll, antioxidant enzymes (ascorbic acid), and NPK content enhanced plant physiological responses	Rani et al. (2016)
	Improvement in growth characteristics and photosynthetic pigments	Pal and Pandey (2016)
Pistachio (*Pistacia vera*)	Enhanced PO_4^{2-} and Zn^{2+} uptake for osmotic adjustment and improved nutrient acquisition under drought	Bagheri et al. (2012)
Tobacco (*Nicotiana tabacum* L.)	Enhanced secondary metabolism, osmolyte accumulation, and antioxidant activity to improve drought resistance	Begum et al. (2022)
Cucumber (*Cucumis sativus* L.)	Enhanced expression of RuBisCO, FBPase, F6P, Ru5PK, and associated genes, along with better growth, nutrient absorption, and photosynthetic efficiency	Chen et al. (2017)
Tomato (*Solanum lycopersicum* L.)	The expression levels of ABA and associated genes were modulated, leading to improved tolerance to water stress	Chitarra et al. (2016)
	The association AMF stimulated strigolactone production, enhancing drought resistance	Ruiz-Lozano et al. (2016)
	Enhanced NADP–ME expression and IAA levels in continuous cropping soils, boosting plant growth and root development	Wang et al. (2021)
	Significantly increased plant height, stomatal conductance, water use efficiency, and biomass while decreasing reactive oxygen species and abscisic acid levels	El Maaloum et al. (2020), Chitarra et al. (2016)
Lycium barbarum	Preserved photochemical efficiency and elevated Rir-AQP2 expression, enhancing water stress tolerance	Hu et al. (2017)
Apple (*Malus domestica*)	Modulation of MAPK pathway genes enhanced drought tolerance	Huang et al. (2020)
	MdIAA24 gene modulated strigolactone production and mycorrhizal colonization, resulting in enhanced drought tolerance	Huang et al. (2020)
Lettuce (*L. sativa* L.)	AMF-stimulated strigo-lactone production, enhancing drought resistance	Ruiz-Lozano et al. (2016)
Medicago truncatula	Upregulated Fm201, Ri14-3-3, and RiBMH2 expression enhance tolerance to both drought and salinity stress	Sun et al. (2018)
Astragalus sinicus	The suppression of HOG1-MAPK cascade genes showed reduced expression of RiAQPs, RiTPSs, RiNTH1, and Ri14-3-3, correlating with diminished drought stress tolerance	Wang S. et al. (2023)
Trifoliate orange (*Poncirus trifoliata*)	Elevated fructose and glucose concentrations, reduced sucrose content, and proline accumulation	Wu et al. (2017)
Saccharum arundinaceum	Elevated leaf concentrations of antioxidant enzymes, phenolic compounds, and chlorophyll, accompanied by increased biomass production	Mirshad and Puthur (2016)
Ipomoea batatas	Osmoprotectants regulate osmotic potential	Yooyongwech et al. (2016)

(continued on next page)

TABLE 12.1 Applications of AMF symbiosis for drought mitigation in agriculture—cont'd

Crop species	Mechanism	References
Lycopersicon esculentum, Capsicum annuum	Enhanced biomass, root and shoot elongation, and photosynthetic pigment content, while reducing proline levels	Tallapragada et al. (2016), Padmavathi et al. (2016)
Digitaria eriantha	Elevated shoot dry weight and stomatal conductance while increasing lipid peroxidation and ROS accumulation in both shoot and root tissues	Pedranzani et al. (2016)
Triticum durum	Significantly increased grain biomass, micronutrient concentration, and gliadin content	Goicoechea et al. (2016)
Ephedra foliate Boiss	Elevated antioxidant enzyme activity, proline, glucose, soluble protein, and phytohormone concentrations while enhancing nitrogen metabolism	Al-Bandari et al. (2020)
Arizona cypress, Cupressus arizonica	Plant growth improved while hydrogen peroxide and malondialdehyde concentrations decreased	Aalipour et al. (2020)
Ceratonia siliqua	Significantly enhanced plant growth, nutrient content, stomatal conductance, PSII efficiency, and hydration status	Boutasknit et al. (2020)
Catalpa bungei C.A.Mey.	Root architecture, hydration, biomass, photosynthetic pigments, and most plant hormones (excluding ABA) exhibited significant enhancement	Chauhan et al. (2022)
Cinnamomum migao	The increase in antioxidant enzyme activity and osmo-protectant accumulation resulted in lower malondialdehyde content	Liao et al. (2021), Xiaofeng et al. (2021)
Sesamum indicum L.	Enhanced oil and seed production, soluble protein levels, leaf phosphorus content, and photosynthetic pigment concentrations	Gholinezhad and Darvishzadeh (2021)
Grapevine (*Vitis vinifera* L. cv. Ecolly)	Enhanced photosynthesis, stomatal opening, and transpiration while reducing internal CO_2 levels	Ye et al. (2022)
V. vinifera L. cv. Merlot	Enhanced grapevine vegetative growth, hydration, and photosynthetic efficiency	Torres et al. (2021a)

From (Wang et al. (2024), Wahab et al. (2023)).
ABA, abscisic acid; *AM*, arbuscular mycorrhizal; *AMF*, arbuscular mycorrhizal fungi; *IAA*, indole-3-acetic acid; *PSII*, photosystem II; *ROS*, reactive oxygen species.

requires controlled ROS generation, especially localized production of H_2O_2 in cortical cells (Zou et al., 2021), and after symbiosis is established, mycorrhizae enhance the plant's antioxidant ability, thereby reducing oxidative stresses induced by drought. This balanced ROS regulation underscores AMF's protective role against drought-related cellular damage (Table 12.1).

However, enzymatic antioxidants, including superoxide dismutase (SOD), catalase (CAT), and carotenoids, rapidly scavenge ROS, equilibrating ROS homeostasis in the cellular environment (Zou et al., 2021). Maintaining ROS equilibrium is vital for cellular signaling, stress responses, and plant adaptation to adverse conditions (Li et al., 2022). AMF enhance drought resilience via dual mechanisms: hyphal-mediated water uptake, improving host hydration and reducing ROS accumulation, and symbiotic upregulation of antioxidant systems to counteract oxidative stress (Bahadur et al., 2019).

Heat shock transcription factors (Hsfs) are pivotal regulators of stress-responsive gene expression. Specific Hsf members, including SPL7, HsfA1b, HsfA4a, and HsfA8, have demonstrated modulating ROS homeostasis under drought stress (Hoang et al., 2019). AMFs upregulate heat shock transcription factors (Hsfs) and antioxidant systems to counteract drought-induced oxidative stress. In walnut (*Juglans regia*), AMF (*Diversispora spurca*) elevates JrHsf03/22/24 expression, enhancing drought tolerance (Ma et al., 2022). Further, *D. spurca* inoculation enhanced ascorbate peroxidase (APX), CAT, and SOD activities over noninoculated plants (Ma et al., 2022). Respiratory burst oxidase homologs (Rbohs), which are NADPH oxidases generating ROS, mediate drought responses and other stress adaptations (Chapman et al., 2019). In *Bombax ceiba*, AMF colonization downregulated *Rbohs* under drought but slightly upregulated them in well-watered conditions (Ma et al., 2022). These factors function as ROS sensors, dynamically regulating oxidative bursts during stress responses.

The ascorbate-glutathione cycle is vital for scavenging H_2O_2, with APX using ascorbate to break down H_2O_2 (Bárzana et al., 2015). In drought conditions, AMF-colonized citrus plants are found to have higher activities of CAT, GR, G-POD, and APX, highlighting the AMF's advantage in antioxidant capacity (Wu et al., 2013). Its presence also upregulates enzymatic antioxidants and ascorbate/glutathione biosynthesis genes (Bahadur et al., 2019), thereby helping AMF to fine-tune oxidative stress responses. This entails multiphase transcriptional regulation of antioxidant systems during a mycorrhizal association, thereby presenting a challenge for researchers to find concrete evidence of fungal–plant interactions with ROS scavenging pathways.

12.5.6 Arbuscular mycorrhizal fungi–mediated stomatal conductance and photosynthetic efficacy during drought

Water stress typically causes a decrease in stomatal conductance (gs), but AMF colonization protects against it. Metaanalysis of over 400 studies concludes that under normal watering conditions, AMF intercalation increases gs by an average of 24%, having an even stronger enhancement effect in drought conditions (Table 12.1). Although elevated gs increases transpiration, the concomitant rise in CO_2 assimilation improves carbon gain. This supports root growth in moist soil zones and sustains metabolism during drought-induced stomatal closure. AMF also boosts intrinsic water-use efficiency (iWUE) through enhanced mineral nutrition and reduced oxidative damage, collectively improving photosynthetic efficacy even under water-limited conditions (Geisler et al., 2023).

AMF inoculation with *R. intraradices* in *S. lycopersicum* revealed that higher stomatal numbers occurred under drought, over nonmycorrhizal controls, or *F. mosseae*–colonized plants (Chitarra et al., 2016). This structural adaptation enhances CO_2 assimilation during water stress. Mechanistically, AMF symbiosis activates *LeEPFL9* (a positive regulator of stomatal development), and its transcript levels correlate strongly with elevated stomatal density in *R. intraradices*–associated plants. Concurrently the negative regulators EPF1/EPF2 exhibit AMF-dependent expression, suggesting fine-tuned control of stomatal patterning. These findings demonstrate three key points: that AMF species differentially regulate stomatal development (*R. intraradices* > *F. mosseae*), *LeEPFL9* is a central mediator of AMF-induced stomatal proliferation, and symbiotic fungi optimize plant gas exchange by transcriptional reprogramming of stomatal development pathways under stress (Chitarra et al., 2016). Such insights highlight AMF's potential to enhance drought resilience through targeted modifications of stomatal traits.

The AMF–*Oryza sativa* symbiosis boosts photosynthetic efficiency by >40% under drought stress (Ruiz-Sánchez et al., 2010). Colonized plants exhibit enhanced PSII functionality, higher transpiration rates, and reduced photorespiration, collectively minimizing ROS accumulation (Das & Sarkar, 2024). These adaptations preserve photosynthetic machinery integrity, enabling sustained CO_2 fixation during and after drought. The dual effect of improved PSII efficiency and regulated transpiration likely synergizes to promote growth under water deficit, demonstrating AMF's role in optimizing both light-dependent reactions and carbon assimilation under stress conditions.

12.5.7 Arbuscular mycorrhizal fungi–PGPR inoculation in alleviating drought stress

Coinoculation with AMF and Plant Growth-Promoting Rhizobacteria (PGPR) enhances plant drought tolerance more effectively than single or no inoculation (Anli et al., 2020; Begum et al., 2022). PGPR promotes AMF growth, amplifying drought resilience. Studies showed inoculation of *Bacillus* sp. can improve *F. mosseae* and *R. irregularis* colonization in lettuce (*Lactuca sativa*), mitigating drought effects (Vivas et al., 2003). Coinoculation boosts nutrient and water absorption, photosynthesis, and antioxidant activity while reducing oxidative stress (Anli et al., 2020; Begum et al., 2022). In tobacco (*Nicotiana tabacum*) and date palm, *Glomus versiforme* (AMF) and *Bacillus methylotrophicus* (PGPR) synergistically increased N, P, and K absorption and antioxidant capacity, elevating biomass and height (Begum et al., 2022).

Current commercial AMF inoculants are limited to family Glomeraceae strains, occasionally formulated as multispecies consortia (Basiru & Hijri, 2022). However, such mixtures only improve plant performance if constituent strains provide complementary benefits. Despite evidence supporting AMF–PGPR synergies, few commercial products combine these microbes. Broader adoption of bioinoculants faces challenges, including short shelf life, inconsistent efficacy, and insufficient knowledge of their impacts on native soil microbiomes, hindering their commercial applications in a wider range (Sagar et al., 2021).

12.6 Challenges and limitations of the arbuscular mycorrhizal fungi application

Agricultural practices like excessive fertilizer use (especially phosphorus/nitrogen), pesticides, tillage, and nonmycorrhizal crop rotations reduce AMF diversity, colonization, and activity. These inputs disrupt AM associations, leading to poorer AMF populations in farmed soils versus natural soils.

Particularly fertilization of phosphorus (P) and N has a great impact on AMF colonization. Excessive P fertilization inhibits mycorrhizal symbiosis, which enables plants to obtain sufficient phosphorus independently, reducing carbohydrate allocation to AM fungi (García-Caparr´os et al., 2021). Consequently, high soil P decreases arbuscule formation and active P transfer to plants, impairing AM colonization. Similarly, supraoptimal potassium reduces root exudates, causing carbohydrate accumulation in root cortices that disrupts AM signaling. Nitrogen fertilizers exhibit dose-dependent effects on AM symbiosis; low-to-medium levels enhance colonization, sporulation, and plant growth, whereas excessive application suppresses fungal associations (Getman-Pickering et al., 2021; Lin et al., 2020). Field observations reveal that compromised colonization of onion roots in conventional agriculture shows only mycelial presence, whereas okra (*Abelmoschus esculentus*) in compost-amended systems develops vesicles and spores despite low colonization intensity (Kuila & Ghosh, 2022). Moreover, systemic and nonsystemic pesticides and fungicides are detrimental to AM growth, particularly on spore germination, colonization, sporulation, and mycelial growth.

Tillage exerts dual effects on AMF. The damaging impacts include: physical destruction of spores and hyphal networks, which reduces root colonization; disruption of extraradical mycelium, impairing AMF-mediated phosphorus transfer (especially critical in early growth stages); and alteration of soil stratification that eliminates established AMF microhabitats. Conversely, tillage can benefit soil structure and nutrient availability through organic matter incorporation and aeration. Conservation tillage significantly enhances AMF colonization compared to conventional methods by promoting greater AMF diversity and richness (Higo et al., 2020). These improvements contribute to key soil functions, including aggregation, carbon sequestration, nutrient cycling, and water-use efficiency, ultimately boosting crop yields. However, conventional tillage physically disrupts extraradical hyphal networks during critical early colonization stages, impairing AMF-mediated nutrient/water uptake, glomalin-related soil stabilization, and pathogen protection (Patanita et al., 2020). Although beneficial in the short-term, extended no-till practices may paradoxically reduce AMF propagules because of increased soil compaction and decreased microbial carbon-use efficiency (Schluter et al., 2018), with these effects being highly dependent on local soil conditions, climate, and management duration.

The key hurdles that face AMF implementation in agriculture are **host specificity,** requiring careful strain-plant matching for effective colonization; **drought response variability,** where benefits depend on stress severity and plant–AMF interactions; and **field constraints,** including high inoculum production costs, poor survival in degraded soils, competition from native microbes, and farmer awareness gaps.

12.7 Future research directions

To maximize AMF benefits in agriculture, priority research areas should focus on (1) **genetic enhancement** of AMF strains to boost drought resilience through improved hyphal growth, nutrient transport, and stress-protective compound production; (2) **scalable field implementation** via large-scale trials to identify optimal strain-crop-region combinations while engaging farmers for practical adoption; and (3) **climate-smart integration** by combining AMF inoculants with soil-health practices (conservation tillage, crop rotations) to enhance water/nutrient efficiency and carbon sequestration. Critical knowledge gaps requiring investigation include AMF–plant–microbiome interactions under stress, molecular mechanisms of drought tolerance (hormonal/metabolic pathways), and cost-effective inoculum production for global scalability.

12.8 Conclusion and recommendations

AMF function as vital bioagents that extend far beyond conventional biofertilizer roles, enhancing drought tolerance through improved water/nutrient uptake while synergizing with PGPRs in stressed ecosystems. For optimal implementation, region-specific AMF strains should be matched to crops and local soils, with particular emphasis on native strains in resource-limited regions where laboratory-developed inoculants may be impractical. Successful integration requires careful management of agrochemicals (type, dose, and timing) to avoid disrupting AMF symbiosis while maintaining yields, a balance critical for organic and climate-smart farming systems. Broad-spectrum AMF strains show promise; maximum benefits emerge when combining adapted microbial consortia with complementary practices like conservation tillage and crop rotation. Though a few years of establishment period are typically required, the long-term advantages include enhanced drought resilience, reduced input costs, and improved soil health. This approach offers particular promise for developing economies, where AMF-mediated systems can simultaneously address food security and ecological sustainability with lower technical barriers than conventional intensification.

References

Aalipour, H., Nikbakht, A., Etemadi, N., Rejali, F., & Soleimani, M. (2020). Biochemical response and interactions between arbuscular mycorrhizal fungi and plant growth promoting rhizobacteria during establishment and stimulating growth of *Arizona cypress* (*Cupressus arizonica* G.) under drought stress. *Scientia Horticulturae, 261*, 108923. doi:10.1016/j.scienta.2019.108923.

Abbaspour, H., Saeidi-Sar, S., Afshari, H., & Abdel-Wahhab, M. A. (2012). Tolerance of Mycorrhiza infected pistachio (*Pistacia vera* L.) seedling to drought stress under glasshouse conditions. *Journal of Plant Physiology, 169*(7), 704–709. doi:10.1016/j.jplph.2012.01.014.

Abdalla, M., Bitterlich, M., Jansa, J., Püschel, D., & Ahmed, M. A. (2023). The role of arbuscular mycorrhizal symbiosis in improving plant water status under drought. *Journal of Experimental Botany, 74*(16), 4808–4824. http://jxb.oxfordjournals.org/. doi:10.1093/jxb/erad249.

Ahmad, M., Waraich, E. A., Shahid, H., Ahmad, Z., Zulfiqar, U., Mahmood, N., Al-Ashkar, I., Ditta, A., & Sabagh, A. E. (2023). Exogenously applied potassium enhanced morpho-physiological growth and drought tolerance of wheat by alleviating osmotic imbalance and oxidative damage. *Polish Journal of Environmental Studies, 32*(5), 4447–4459. http://www.pjoes.com/pdf-166352-94155?filename=Exogenously%20Applied.pdf. doi:10.15244/pjoes/166352.

Al-Arjani, A. F., Hashem, A., & Abd_Allah, E. F. (2020). Arbuscular mycorrhizal fungi modulates dynamics tolerance expression to mitigate drought stress in Ephedra foliata Boiss. *Saudi Journal of Biological Sciences, 27*(1), 380–394. https://doi.org/10.1016/j.sjbs.2019.10.008.

Albornoz, F. E., Prober, S. M., Ryan, M. H., & Standish, R. J. (2022). Ecological interactions among microbial functional guilds in the plant-soil system and implications for ecosystem function. *Plant and Soil, 476*(1), 301–313.

Anderson, W. B., Seager, R., Baethgen, W., Cane, M., & You, L. (2019). Synchronous crop failures and climate-forced production variability. *Science Advances, 5*(7), eaaw1976.

Anli, M., Baslam, M., Tahiri, A., Raklami, A., Symanczik, S., Boutasknit, A., Ait-El-Mokhtar, M., Ben-Laouane, R., Toubali, S., Ait Rahou, Y., Ait Chitt, M., Oufdou, K., Mitsui, T., Hafidi, M., & Meddich, A. (2020). Biofertilizers as strategies to improve photosynthetic apparatus, growth, and drought stress tolerance in the date palm. *Frontiers in Plant Science, 11*. doi:10.3389/fpls.2020.516818.

Aroca, R., Del Mar Alguacil, M., Vernieri, P., & Ruiz-Lozano, J. M. (2008). Plant responses to drought stress and exogenous ABA application are modulated differently by mycorrhization in tomato and an ABA-deficient mutant (sitiens). *Microbial Ecology, 56*(4), 704–719. doi:10.1007/s00248-008-9390-y.

Aroca, R., Porcel, R., & Ruiz-Lozano, J. M. (2007). How does arbuscular mycorrhizal symbiosis regulate root hydraulic properties and plasma membrane aquaporins in *Phaseolus vulgaris* under drought, cold or salinity stresses? *New Phytologist, 173*(4), 808–816. doi:10.1111/j.1469-8137.2006.01961.x.

Aroca, R., Ruiz-Lozano, J. M., Zamarreño, a. M., Paz, J. A., García-Mina, J. M., Pozo, M. J., & López-Ráez, J. A. (2013). Arbuscular mycorrhizal symbiosis influences strigolactone production under salinity and alleviates salt stress in lettuce plants. *Journal of Plant Physiology, 170*(1), 47–55. doi:10.1016/j.jplph.2012.08.020.

Aroca, R., Vernieri, P., & Ruiz-Lozano, J. M. (2008). Mycorrhizal and non-mycorrhizal *Lactuca sativa* plants exhibit contrasting responses to exogenous ABA during drought stress and recovery. *Journal of Experimental Botany, 59*(8), 2029–2041. doi:10.1093/jxb/ern057.

Arzanesh, M. H., Alikhani, H. A., Khavazi, K., Rahimian, H. A., & Miransari, M. (2011). Wheat (*Triticum aestivum* L.) growth enhancement by *Azospirillum* sp. under drought stress. *World Journal of Microbiology and Biotechnology, 27*(2), 197–205. doi:10.1007/s11274-010-0444-1.

Bagheri, V., Shamshiri, M. H., Shirani, H., & Roosta, H. R. (2012). Nutrient uptake and distribution in mycorrhizal pistachio seedlings under drought stress. *Journal of Agricultural Science and Technology, 14*, 1591–1604. http://jast.journals.modares.ac.ir/?_action=showPDF&article=623&_ob=32be44cc17fe845c874823d4011075c8&fileName=full_text.pdf.

Bahadur, A., Batool, A., Nasir, F., Jiang, S., Mingsen, Q., Zhang, Q. i, Pan, J., Liu, Y., & Feng, H. (2019). Mechanistic insights into arbuscular mycorrhizal fungi-mediated drought stress tolerance in plants. *International Journal of Molecular Sciences, 20*(17), 4199. doi:10.3390/ijms20174199.

Barnawal, D., Singh, R., & Singh, R. P. (2019). Role of plant growth promoting rhizobacteria in drought tolerance: Regulating growth hormones and osmolytes. In A. K. Singh, A. Kumar, & P. K. Singh (Eds.), *PGPR amelioration in sustainable agriculture: Food security and environmental management* (pp. 107–128). Elsevier BV. doi:10.1016/b978-0-12-815879-1.00006-9.

Bárzana, G., Aroca, R., Bienert, G. P., Chaumont, F., & Ruiz-Lozano, J. M. (2014). New insights into the regulation of aquaporins by the arbuscular mycorrhizal symbiosis in maize plants under drought stress and possible implications for plant performance. *Molecular Plant-Microbe Interactions, 27*(4), 349–363. http://apsjournals.apsnet.org/doi/pdf/10.1094/MPMI-09-13-0268-R. doi:10.1094/MPMI-09-13-0268-R. Spain.

Bárzana, G., Aroca, R., Paz, J. A., Chaumont, F., Martinez-Ballesta, M. C., Carvajal, M., & Ruiz-Lozano, J. M. (2012). Arbuscular mycorrhizal symbiosis increases relative apoplastic water flow in roots of the host plant under both well-watered and drought stress conditions. *Annals of Botany, 109*(5), 1009–1017. doi:10.1093/aob/mcs007.

Bárzana, G., Aroca, R., & Ruiz-Lozano, J. M. (2015). Localized and non-localized effects of arbuscular mycorrhizal symbiosis on accumulation of osmolytes and aquaporins and on antioxidant systems in maize plants subjected to total or partial root drying. *Plant Cell and Environment, 38*(8), 1613–1627. http://onlinelibrary.wiley.com/journal/10.1111/(ISSN)1365-3040. doi:10.1111/pce.12507. Spain.

Basiru, S., & Hijri, M. (2022). The potential applications of commercial arbuscular mycorrhizal fungal inoculants and their ecological consequences. *Microorganisms, 10*(10), 1897. doi:10.3390/microorganisms10101897.

Begum, N., Wang, L., Ahmad, H., Akhtar, K., Roy, R., Khan, M. I., & Zhao, T. (2022). Co-inoculation of arbuscular mycorrhizal fungi and the plant growth-promoting rhizobacteria improve growth and photosynthesis in tobacco under drought stress by up-regulating antioxidant and mineral nutrition metabolism. *Microbial Ecology, 83*(4), 971–988. Link.springer.de/link/service/journals/00248/index.htm. doi:10.1007/s00248-021-01815-7.

Bell, C. A., Magkourilou, E., Urwin, P. E., & Field, K. J. (2021). The influence of competing root symbionts on below-ground plant resource allocation. *Ecology and Evolution, 11*(7), 2997–3003. http://onlinelibrary.wiley.com/journal/10.1002/(ISSN)2045-7758. doi:10.1002/ece3.7292.

Bhargavi, B. (2017). Influence of water stress on morphological and physiological changes in *Andrographis paniculata*. *International Journal of Pure & Applied Bioscience, 5*(6), 1550–1556. doi:10.18782/2320-7051.5932.

Boutasknit, A., Baslam, M., Ait-El-Mokhtar, M., Anli, M., Ben-Laouane, R., Douira, A., El Modafar, C., Mitsui, T., Wahbi, S., & Meddich, A. (2020). Arbuscular mycorrhizal fungi mediate drought tolerance and recovery in two contrasting carob (*Ceratonia siliqua* L.) ecotypes by regulating stomatal, water relations, and (in)organic adjustments. *Plants, 9*(1), 80. doi:10.3390/plants9010080.

Calvo-Polanco, M., Sánchez-Romera, B., & Aroca, R. (2013). *Arbuscular mycorrhizal fungi and the tolerance of plants to drought and salinity* (pp. 271–288). Springer Science and Business Media LLC. doi:10.1007/978-3-642-39317-4_14.

Castaneda, R., Doan, D., Newhouse, D. L., Nguyen, M., Uematsu, H., & Azevedo, J. P. (2016). Who are the poor in the developing world? *World Bank policy research working paper, 7844*. World Bank Group.

Chaouachi, L., Marín-Sanz, M., Kthiri, Z., Boukef, S., Harbaoui, K., Barro, F., Karmous, C., & Salter, W. (2023). The opportunity of using durum wheat landraces to tolerate drought stress: Screening morpho-physiological components. *AoB Plants, 15*(3). doi:10.1093/aobpla/plad022.

Chapman, J. M., Muhlemann, J. K., Gayomba, S. R., & Muday, G. K. (2019). RBOH-dependent ROS synthesis and ROS scavenging by plant specialized metabolites to modulate plant development and stress responses. *Chemical Research in Toxicology, 32*(3), 370–396. http://pubs.acs.org/journal/crtoec. doi:10.1021/acs.chemrestox.9b00028.

Chatzistathis, T., Orfanoudakis, M., Alifragis, D., & Therios, I. (2013). Colonization of Greek olive cultivars' root system by arbuscular mycorrhiza fungus: Root morphology, growth, and mineral nutrition of olive plants. *Scientia Agricola, 70*(3), 185–194. doi:10.1590/s0103-90162013000300007.

Chauhan, S., Mahawar, S., Jain, D., Udpadhay, S. K., Mohanty, S. R., Singh, A., Maharjan, E., & Yadav, A. N. (2022). Boosting sustainable agriculture by arbuscular mycorrhiza under stress condition: Mechanism and future prospective. *BioMed Research International, 2022*(1). doi:10.1155/2022/5275449.

Chen, J., Zhang, H., Zhang, X., & Tang, M. (2017). Arbuscular mycorrhizal symbiosis alleviates salt stress in black locust through improved photosynthesis, water status, and K^+/Na^+ homeostasis. *Frontiers in Plant Science, 8*, 1739. doi:10.3389/fpls.2017.01739.

Chitarra, W., Pagliarani, C., Maserti, B., Lumini, E., Siciliano, I., Cascone, P., Schubert, A., Gambino, G., Balestrini, R., & Guerrieri, E. (2016). Insights on the impact of arbuscular mycorrhizal symbiosis on tomato tolerance to water stress. *Plant Physiology, 171*(2), 1009–1023. http://www.plantphysiol.org/content/171/2/1009.full.pdf. doi:10.1104/pp.16.00307.

Choudhury, F. K., Rivero, R. M., Blumwald, E., & Mittler, R. (2017). Reactive oxygen species, abiotic stress, and stress combination. *The Plant Journal, 90*(5), 856–867.

Chun, S. C., Paramasivan, M., & Chandrasekaran, M. (2018). Proline accumulation influenced by osmotic stress in arbuscular mycorrhizal symbiotic plants. *Frontiers in Microbiology, 9*. https://www.frontiersin.org/journals/microbiology#. doi:10.3389/fmicb.2018.02525.

Cosme, M. (2023). Mycorrhizas drive the evolution of plant adaptation to drought. *Communications Biology, 6*(1). doi:10.1038/s42003-023-04722-4.

Das, S., & Sarkar, S. (2024). Arbuscular mycorrhizal fungal contribution towards plant resilience to drought conditions. *Frontiers in Fungal Biology, 5*. doi:10.3389/ffunb.2024.1355999.

Deka, D., Singh, A. K., & Singh, A. K. (2018). Effect of drought stress on crop plants with special reference to drought avoidance and tolerance mechanisms: A review. *International Journal of Current Microbiology and Applied Sciences, 7*(09), 2703–2721. doi:10.20546/ijcmas.2018.709.336.

Dossena, S., & Marino, A. (2021). Cellular oxidative stress. *Antioxidants, 10*(3), 399. doi:10.3390/antiox10030399.

El Maaloum, S., Elabed, A., Alaoui-Talibi, Z. E., Meddich, A., Filali-Maltouf, A., Douira, A., Ibnsouda-Koraichi, S., Amir, S., & El Modafar, C. (2020). Effect of arbuscular mycorrhizal fungi and phosphate-solubilizing bacteria consortia associated with phospho-compost on phosphorus solubilization and growth of tomato seedlings (*Solanum lycopersicum* L.). *Communications in Soil Science and Plant Analysis, 51*(5), 622–634. http://www.tandfonline.com/loi/lcss20. doi:10.1080/00103624.2020.1729376.

El-Mesbahi, M. N., Azcón, R., Ruiz-Lozano, J. M., & Aroca, R. (2012). Plant potassium content modifies the effects of arbuscular mycorrhizal symbiosis on root hydraulic properties in maize plants. *Mycorrhiza, 22*(7), 555–564. doi:10.1007/s00572-012-0433-3.

Etesami, H., & Maheshwari, D. K. (2018). Use of plant growth promoting rhizobacteria (PGPRs) with multiple plant growth promoting traits in stress agriculture: Action mechanisms and future prospects. *Ecotoxicology and Environmental Safety, 156*, 225–246. http://www.elsevier.com/inca/publications/store/6/2/2/8/1/9/index.htt. doi:10.1016/j.ecoenv.2018.03.013.

Fan, X., Xie, H., Huang, X., Zhang, S., Nie, Y., Chen, H., Xie, X., & Tang, M. (2023). A module centered on the transcription factor Msn2 from arbuscular mycorrhizal fungus *Rhizophagus irregularis* regulates drought stress tolerance in the host plant. *New Phytologist, 240*(4), 1497–1518. doi:10.1111/nph.19077.

FAO. (2023). *The impact of disasters on agriculture and food security. Avoiding and reducing losses through investment in resilience.*

Feng, Q., Cao, S., Liao, S., Wassie, M., Sun, X., Chen, L., & Xie, Y. (2023). Fusarium equiseti-inoculation altered rhizosphere soil microbial community, potentially driving perennial ryegrass growth and salt tolerance. *Science of the Total Environment, 871*, 162153. doi:10.1016/j.scitotenv.2023.162153.

García-Caparros, P., Lao, M. T., Preciado-Rangel, P., & Sánchez, E. (2021). Phosphorus and carbohydrate metabolism in green bean plants subjected to increasing phosphorus concentration in the nutrient solution. *Agronomy, 11*(2), 245. doi:10.3390/agronomy11020245.

Geisler, M., Buerki, S., & Serpe, M. D. (2023). Arbuscular mycorrhizae alter photosynthetic responses to drought in seedlings of *Artemisia tridentata. Plants, 12*(16), 2990. doi:10.3390/plants12162990.

Genre, A., Lanfranco, L., Perotto, S., & Bonfante, P. (2020). Unique and common traits in mycorrhizal symbioses. *Nature Reviews Microbiology, 18*(11), 649–660. doi:10.1038/s41579-020-0402-3.

Getman-Pickering, Z. L., Stack, G. M., & Thaler, J. S. (2021). Fertilizer quantity and type alter mycorrhizae-conferred growth and resistance to herbivores. *Journal of Applied Ecology, 58*(5), 931–940. http://onlinelibrary.wiley.com/journal/10.1111/(ISSN)1365-2664. doi:10.1111/1365-2664.13833.

Gholinezhad, E., & Darvishzadeh, R. (2021). Influence of arbuscular mycorrhiza fungi and drought stress on fatty acids profile of sesame (*Sesamum indicum* L.). *Field Crops Research, 262*, 108035. doi:10.1016/j.fcr.2020.108035.

Goicoechea, N., Bettoni, M. M., Fuertes-Mendizábal, T., González-Murua, C., & Aranjuelo, I. (2016). Durum wheat quality traits affected by mycorrhizal inoculation, water availability and atmospheric CO_2 concentration. *Crop and Pasture Science, 67*(2), 147–155. http://www.publish.csiro.au/nid/40.htm?nid=43&aid=34. doi:10.1071/CP15212.

Haghpanah, M., Hashemipetroudi, S., Arzani, A., & Araniti, F. (2024). Drought tolerance in plants: Physiological and molecular responses. *Plants, 13*(21), 2962. doi:10.3390/plants13212962.

Halder, M., Karim, R., Islam, S., & Akhter, S. (2015). Seasonal variation of arbuscular mycorrhiza fungi colonization with some medicinal plant species of Chittagong BCSIR forest. *Plant Science Today, 2*(3), 87–92. doi:10.14719/pst.2015.2.3.121.

Han, S., Wang, X., Cheng, Y., Wu, G., Dong, X., He, X., & Zhao, G. (2023). Multidimensional analysis reveals environmental factors that affect community dynamics of arbuscular mycorrhizal fungi in poplar roots. *Frontiers in Plant Science, 13*. doi:10.3389/fpls.2022.1068527.

Hasanuzzaman, M. (Ed.). (2020). *Plant ecophysiology and adaptation under climate change: Mechanisms and perspectives II: Mechanisms of adaptation and stress amelioration*. Springer Nature.

Higo, M., Tatewaki, Y., Iida, K., Yokota, K., & Isobe, K. (2020). Amplicon sequencing analysis of arbuscular mycorrhizal fungal communities colonizing maize roots in different cover cropping and tillage systems. *Scientific Reports, 10*(1). doi:10.1038/s41598-020-58942-3.

Hoang, T. V., Vo, K. T. X., Rahman, M. M., Choi, S. H., & Jeon, J. S. (2019). Heat stress transcription factor OsSPL7 plays a critical role in reactive oxygen species balance and stress responses in rice. *Plant Science, 289*. http://www.journals.elsevier.com/plant-science/. doi:10.1016/j.plantsci.2019.110273.

Hu, C., Elias, E., Nawrocki, W. J., & Croce, R. (2023). Drought affects both photosystems in Arabidopsis thaliana. *New Phytologist, 240*(2), 663–675. http://onlinelibrary.wiley.com/journal/10.1111/(ISSN)1469-8137. doi:10.1111/nph.19171.

Hu, W., Zhang, H., Chen, H., & Tang, M. (2017). Arbuscular mycorrhizas influence *Lycium barbarum* tolerance of water stress in a hot environment. *Mycorrhiza, 27*(5), 451–463. doi:10.1007/s00572-017-0765-0.

Hu, Y., Xie, W., & Chen, B. (2020). Arbuscular mycorrhiza improved drought tolerance of maize seedlings by altering photosystem II efficiency and the levels of key metabolites. *Chemical and Biological Technologies in Agriculture, 7*(1). Chembioagro.springeropen.com/. doi:10.1186/s40538-020-00186-4.

Huang, D., Ma, M., Wang, Q., Zhang, M., Jing, G., Li, C., & Ma, F. (2020). Arbuscular mycorrhizal fungi enhanced drought resistance in apple by regulating genes in the MAPK pathway. *Plant Physiology and Biochemistry, 149*, 245–255. doi:10.1016/j.plaphy.2020.02.020.

Huang, D., Wang, Q., Jing, G., Ma, M., Li, C., & Ma, F. (2021a). Overexpression of MdIAA24 improves apple drought resistance by positively regulating strigolactone biosynthesis and mycorrhization. *Tree Physiology, 41*(1), 134–146. http://treephys.oxfordjournals.org/. doi:10.1093/treephys/tpaa109.

Huang, D., Wang, Q., Zou, Y., Ma, M., Jing, G., Ma, F., & Li, C. (2021b). Silencing MdGH3-2/12 in apple reduces cadmium resistance via the regulation of AM colonization. *Chemosphere, 269*, 129407.

Inal, B., Mirzapour, M., Tufekci, E. D., Rustemoglu, M., Kaba, A., Albalawi, M. A., Alalawy, A. I., Sakran, M., Alqurashi, M., & Ditta, A. (2023). Drought-induced miRNA expression correlated with heavy metal, phenolic acid, and protein and nitrogen levels in five chickpea genotypes. *ACS Omega, 8*(39), 35746–35754. Pubs.acs.org/journal/acsodf. doi:10.1021/acsomega.3c03003.

Iqbal, N., Khan, N. A., Ferrante, A., Trivellini, A., Francini, A., & Khan, M. I. R. (2017). Ethylene's role in plant growth, development, and senescence: Interaction with other phytohormones. *Frontiers in Plant Science, 8*, 475.

Ji, L., Yang, Y., Yang, N., Khan, A., & Yang, L. (2021). Seasonal variation of diversity and co-occurrence patterns of arbuscular mycorrhizal fungal communities in mixed broadleaf-conifer forests. *Applied Soil Ecology, 158*, 103782. doi:10.1016/j.apsoil.2020.103782.

Kakouridis, A., Hagen, J. A., Kan, M. P., Mambelli, S., Feldman, L. J., Herman, D. J., Weber, P. K., Pett-Ridge, J., & Firestone, M. K. (2022). Routes to roots: Direct evidence of water transport by arbuscular mycorrhizal fungi to host plants. *New Phytologist, 236*(1), 210–221. http://onlinelibrary.wiley.com/journal/10.1111/(ISSN)1469-8137. doi:10.1111/nph.18281.

Kapoor, D., Bhardwaj, S., Landi, M., Sharma, A., Ramakrishnan, M., & Sharma, A. (2020). The impact of drought in plant metabolism: How to exploit tolerance mechanisms to increase crop production. *Applied Sciences, 10*(16), 5692. doi:10.3390/app10165692.

Koch, A. M., Antunes, P. M., Maherali, H., Hart, M. M., & Klironomos, J. N. (2017). Evolutionary asymmetry in the arbuscular mycorrhizal symbiosis: Conservatism in fungal morphology does not predict host plant growth. *New Phytologist, 214*(3), 1330–1337.

Kuila, D., & Ghosh, S. (2022). Aspects, problems and utilization of arbuscular mycorrhizal (AM) application as bio-fertilizer in sustainable agriculture. *Current Research in Microbial Sciences, 3*, 100107. doi:10.1016/j.crmicr.2022.100107.

Kumar, S., Singh, A. K., & Ghosh, P. (2018). Distribution of soil organic carbon and glomalin related soil protein in reclaimed coal mine-land chronosequence under tropical condition. *Science of the Total Environment, 625*, 1341–1350. Www.elsevier.com/locate/scitotenv. doi:10.1016/j.scitotenv.2018.01.061.

Li, T., Hu, Y. J., Hao, Z. P., Li, H., Wang, Y. S., & Chen, B. D. (2013). First cloning and characterization of two functional aquaporin genes from an arbuscular mycorrhizal fungus *Glomus intraradices*. *New Phytologist, 197*(2), 617–630. doi:10.1111/nph.12011.

Li, T., Sun, Y., Ruan, Y., Xu, L., Hu, Y., Hao, Z., Zhang, X., Li, H., Wang, Y., Yang, L., & Chen, B. (2016). Potential role of D-myo-inositol-3-phosphate synthase and 14-3-3 genes in the crosstalk between *Zea mays* and *Rhizophagus intraradices* under drought stress. *Mycorrhiza, 26*(8), 879–893. doi:10.1007/s00572-016-0723-2.

Li, Z., Zhang, Y., Liu, C., Gao, Y., Han, L., & Chu, H. (2022). Arbuscular mycorrhizal fungi contribute to reactive oxygen species homeostasis of *Bombax ceiba* L. under drought stress. *Frontiers in Microbiology, 13*. doi:10.3389/fmicb.2022.991781.

Liao, X., Chen, J., Guan, R., Liu, J., & Sun, Q. (2021). Two arbuscular mycorrhizal fungi alleviate drought stress and improve plant growth in *Cinnamomum migao* seedlings. *Mycobiology, 49*(4), 396–405.

Lin, C., Wang, Y., Liu, M., Li, Q., Xiao, W., & Song, X. (2020). Effects of nitrogen deposition and phosphorus addition on arbuscular mycorrhizal fungi of Chinese fir (*Cunninghamia lanceolata*). *Scientific Reports, 10*(1). doi:10.1038/s41598-020-69213-6.

López-Ráez, J. A. (2016). How drought and salinity affect arbuscular mycorrhizal symbiosis and strigolactone biosynthesis? *Planta, 243*(6), 1375–1385. Link.springer.de/link/service/journals/00425/index.htm. doi:10.1007/s00425-015-2435-9.

Ma, Z., Guo, D., Xu, X., Lu, M., Bardgett, R. D., Eissenstat, D. M., McCormack, M. L., & Hedin, L. O. (2018). Evolutionary history resolves global organization of root functional traits. *Nature, 555*(7694), 94–97. http://www.nature.com/nature/index.html. doi:10.1038/nature25783.

Ma, Z., Zhao, X., He, A., Cao, Y., Han, Q., Lu, Y., Yong, J. W. H., & Huang, J. (2022). Mycorrhizal symbiosis reprograms ion fluxes and fatty acid metabolism in wild jujube during salt stress. *Plant Physiology, 189*(4), 2481–2499. doi:10.1093/plphys/kiac239.

Martín-Rodríguez, J. A., Huertas, R., Ho-Plágaro, T., Ocampo, J. A., Turečková, V., Tarkowská, D., & García-Garrido, J. M. (2016). Gibberellin–abscisic acid balances during arbuscular mycorrhiza formation in tomato. *Frontiers in Plant Science, 7*, 1273. doi:10.3389/fpls.2016.01273.

Mirshad, P. P., & Puthur, J. T. (2016). Arbuscular mycorrhizal association enhances drought tolerance potential of promising bioenergy grass (*Saccharum arundinaceum* retz.). *Environmental Monitoring and Assessment, 188*(7). Www.wkap.nl/journalhome.htm/0167-6369. doi:10.1007/s10661-016-5428-7.

Mitra, D., Rad, K. V., Chaudhary, P., Ruparelia, J., Sagarika, M. S., Boutaj, H., Mohapatra, P. K. D., & Panneerselvam, P. (2021). Involvement of strigolactone hormone in root development, influence and interaction with mycorrhizal fungi in plant: Mini-review. *Current Research in Microbial Sciences, 2*. Www.journals.elsevier.com/current-research-in-microbial-sciences. doi:10.1016/j.crmicr.2021.100026.

Nazir, A., Sarfraz, W., Allah, D., Khalid, N., Farid, M., Shafiq, M., Firdaus-E-Bareen, Rizvi, Z. F., & Naeem, N. (2023). Synergistic impact of two autochthonous saprobic fungi (*A. niger* and *T. pseudokoningii*) on the growth, ionic contents, and metals uptake in *Brassica juncea* L. and *Vigna radiata* L. under tannery solid waste contaminated soil. *International Journal of Phytoremediation, 25*(11), 1488–1500. doi:10.1080/15226514.2023.2166457.

Oguz, M. C., Aycan, M., Oguz, E., Poyraz, I., & Yildiz, M. (2022). Drought stress tolerance in plants: Interplay of molecular, biochemical and physiological responses in important development stages. *Physiologia, 2*(4), 180–197. doi:10.3390/physiologia2040015.

Orfanoudakis, M., Wheeler, C. T., & Hooker, J. E. (2010). Both the arbuscular mycorrhizal fungus *Gigaspora rosea* and *Frankia* increase root system branching and reduce root hair frequency in *Alnus glutinosa*. *Mycorrhiza, 20*(2), 117–126. doi:10.1007/s00572-009-0271-0.

Padmavathi Tallapragada, P. T., Rashmi Dikshit, R. D., & Swetha Seshagiri, S. S. (2016). Influence of *Rhizophagus* spp. and *Burkholderia seminalis* on the growth of tomato (*Lycopersicon esculatum*) and bell pepper (*Capsicum annuum*) under drought stress. *Communications in Soil Science and Plant Analysis, 47*, 1975–1984.

Pal, A., & Pandey, S. (2016). Role of arbuscular mycorrhizal fungi on plant growth and reclamation of barren soil with wheat (*Triticum aestivum* L.) crop. *International Journal of Soil Science, 12*(1), 25–31. doi:10.3923/ijss.2017.25.31.

Pandey, J., Devadasu, E., Saini, D., Dhokne, K., Marriboina, S., Raghavendra, A. S., & Subramanyam, R. (2023). Reversible changes in structure and function of photosynthetic apparatus of pea (*Pisum sativum*) leaves under drought stress. *Plant Journal, 113*(1), 60–74. http://onlinelibrary.wiley.com/journal/10.1111/(ISSN)1365-313X. doi:10.1111/tpj.16034.

Patanita, M., Campos, M. D., Félix, M. D. R., Carvalho, M., & Brito, I. (2020). Effect of tillage system and cover crop on maize mycorrhization and presence of *Magnaporthiopsis maydis*. *Biology, 9*(3). https://www.mdpi.com/2079-7737/9/3/46/pdf. doi:10.3390/biology9030046.

Pedranzani, H., Rodríguez-Rivera, M., Gutiérrez, M., Porcel, R., Hause, B., & Ruiz-Lozano, J. M. (2016). Arbuscular mycorrhizal symbiosis regulates physiology and performance of *Digitaria eriantha* plants subjected to abiotic stresses by modulating antioxidant and jasmonate levels. *Mycorrhiza, 26*(2), 141–152. doi:10.1007/s00572-015-0653-4.

Pequeno, D. N. L., Hernández-Ochoa, I. M., Reynolds, M., Sonder, K., Moleromilan, A., Robertson, R. D., Lopes, M. S., Xiong, W., Kropff, M., & Asseng, S. (2021). Climate impact and adaptation to heat and drought stress of regional and global wheat production. *Environmental Research Letters, 16*(5). https://iopscience.iop.org/article/10.1088/1748-9326/abd970. doi:10.1088/1748-9326/abd970.

Porcel, R., Aroca, R., Azcón, R., & Ruiz-Lozano, J. M. (2006). PIP aquaporin gene expression in arbuscular mycorrhizal *Glycine max* and *Lactuca sativa* plants in relation to drought stress tolerance. *Plant Molecular Biology, 60*(3), 389–404. doi:10.1007/s11103-005-4210-y.

Qiu, Q., Bender, S. F., Mgelwa, A. S., & Hu, Y. (2022). Arbuscular mycorrhizal fungi mitigate soil nitrogen and phosphorus losses: A meta-analysis. *Science of the Total Environment, 807*. Www.elsevier.com/locate/scitotenv. doi:10.1016/j.scitotenv.2021.150857.

Qu, M. H., Yu, Y. C., Wang, J., Xue, L., Wang, Z. F., & Li, S. (2021). Effects of arbuscular mycorrhizal fungi on biomass distribution and root architecture characters of *Zenia insignis* seedlings in karst soil. *Chinese Journal of Ecology, 40*(3), 766–776.

Queiroz, M. S., Oliveira, C. E. S., Steiner, F., Zuffo, A. M., Zoz, T., Vendruscolo, E. P., Silva, M. V., Mello, B. F. F. R., Cabral, R. C., & Menis, F. T. (2019). Drought stresses on seed germination and early growth of maize and sorghum. *Journal of Agricultural Science, 11*(2), 310. doi:10.5539/jas.v11n2p310.

Rani, B. (2016). *Effect of Arbuscular Mycorrhiza* fungi on biochemical parameters in wheat *(Triticum Aestivum* L.) under drought conditions. *Biochemistry*. CCSHAU.

Rao, D. E., & Chaitanya, K. V. (2016). Photosynthesis and antioxidative defense mechanisms in deciphering drought stress tolerance of crop plants. *Biologia Plantarum, 60*(2), 201–218.

Raza, M. A. S., Ibrahim, M. A., Ditta, A., Iqbal, R., Aslam, M. U., Muhammad, F., Ali, S., Çiğ, F., Ali, B., Muhammad Ikram, R., Muzamil, M. N., Rahman, M. H. u., Alwahibi, M. S., & Elshikh, M. S. (2023). Exploring the recuperative potential of brassinosteroids and nano-biochar on growth, physiology, and yield of wheat under drought stress. *Scientific Reports, 13*(1). https://www.nature.com/srep/. doi:10.1038/s41598-023-42007-2.

Ripley, B., Frole, K., & Gilbert, M. (2010). Differences in drought sensitivities and photosynthetic limitations between co-occurring C3 and C4 (NADP-ME) Panicoid grasses. *Annals of Botany, 105*(3), 493–503. doi:10.1093/aob/mcp307.

Rui, W., Mao, Z., & Li, Z. (2022). The roles of phosphorus and nitrogen nutrient transporters in the arbuscular mycorrhizal symbiosis. *International Journal of Molecular Sciences, 23*(19), 11027.

Ruiz-Lozano, J. M., Aroca, R., Zamarreño, Á. M., Molina, S., Andreo-Jiménez, B., Porcel, R., García-Mina, J. M., Ruyter-Spira, C., & López-Ráez, J. A. (2016). Arbuscular mycorrhizal symbiosis induces strigolactone biosynthesis under drought and improves drought tolerance in lettuce and tomato. *Plant Cell and Environment, 39*(2), 441–452. http://www.blackwellpublishing.com/journals/PCE. doi:10.1111/pce.12631.

Ruiz-Sánchez, M., Aroca, R., Muñoz, Y., Polón, R., & Ruiz-Lozano, J. M. (2010). The arbuscular mycorrhizal symbiosis enhances the photosynthetic efficiency and the antioxidative response of rice plants subjected to drought stress. *Journal of Plant Physiology, 167*(11), 862–869. doi:10.1016/j.jplph.2010.01.018.

Sagar, A., Rathore, P., Ramteke, P. W., Ramakrishna, W., Reddy, M. S., & Pecoraro, L. (2021). Plant growth promoting rhizobacteria, arbuscular mycorrhizal fungi and their synergistic interactions to counteract the negative effects of saline soil on agriculture: Key macromolecules and mechanisms. *Microorganisms, 9*(7). https://www.mdpi.com/2076-2607/9/7/1491/pdf. doi:10.3390/microorganisms9071491.

Saha, P., Sade, N., Arzani, A., Rubio Wilhelmi, M. d. M., Coe, K. M., Li, B., & Blumwald, E. (2016). Effects of abiotic stress on physiological plasticity and water use of *Setaria viridis* (L.). *Plant Science, 251*, 128–138. http://www.journals.elsevier.com/plant-science/. doi:10.1016/j.plantsci.2016.06.011.

Saleem, A., Raza, M. A. S., Khan, I. H., Tahir, M. A., Iqbal, R., Aslam, M. U., Ejaz, Z., Ditta, A., Almunqedhi, B. M., & Al Farraj, D. A. (2024). Amelioration of drought stress in wheat by using plant growth-promoting rhizobacteria and biogas slurry. *Global Nest Journal, 26*(8). https://journal.gnest.org/system/files/2024-10/gnest_05890_final.pdf. doi:10.30955/gnj.005890.

Saleem, A., Raza, M. A. S., Tahir, M. A., Iqbal, R., Aslam, M. U., Toleikiene, M., Khan, M. S., Alwahibi, M. S., Elshikh, M. S., & Ditta, A. (2025). Impact of biogas slurry on physiological and antioxidant mechanisms of wheat under drought stress. *Polish Journal of Environmental Studies, 34*(2), 1721–1731. https://www.pjoes.com/pdf-187777-119877?filename=Impact%20of%20Biogas%20Slurry.pdf. doi:10.15244/pjoes/187777.

Sánchez-Romera, B., Ruiz-Lozano, J. M., Zamarreño, Á. M., García-Mina, J. M., & Aroca, R. (2016). Arbuscular mycorrhizal symbiosis and methyl jasmonate avoid the inhibition of root hydraulic conductivity caused by drought. *Mycorrhiza, 26*(2), 111–122. Link.springer.de/link/service/journals/00572/index.htm. doi:10.1007/s00572-015-0650-7.

Sarkar, S., Kamke, A., Ward, K., Hartung, E., Ran, Q., Feehan, B., & Lee, S. T. (2022a). Pseudomonas cultivated from *Andropogon gerardii* rhizosphere shows functional potential for promoting plant host growth and drought resilience. *BMC Genomics, 23*(1), 784.

Sarkar, S., Kamke, A., Ward, K., Rudick, A. K., Baer, S. G., Ran, Q., & Lee, S. T. (2022b). Bacterial but not fungal rhizosphere community composition differs among perennial grass ecotypes under abiotic environmental stress. *Microbiology Spectrum, 10*(3), e0239121.

Schlüter, S., Großmann, C., Diel, J., Wu, G. M., Tischer, S., Deubel, A., & Rücknagel, J. (2018). Long-term effects of conventional and reduced tillage on soil structure, soil ecology, and soil hydraulic properties. *Geoderma, 332*, 10–19.

Seleiman, M. F., Al-Suhaibani, N., Ali, N., Akmal, M., Alotaibi, M., Refay, Y., Dindaroglu, T., Abdul-Wajid, H. H., & Battaglia, M. L. (2021). Drought stress impacts on plants and different approaches to alleviate its adverse effects. *Plants, 10*(2), 1–25. https://www.mdpi.com/2223-7747/10/2/259/pdf. doi:10.3390/plants10020259.

Sharma, A., Wang, J., Xu, D., Tao, S., Chong, S., Yan, D., Li, Z., Yuan, H., & Zheng, B. (2020). Melatonin regulates the functional components of photosynthesis, antioxidant system, gene expression, and metabolic pathways to induce drought resistance in grafted *Carya cathayensis* plants. *Science of the Total Environment, 713*, 136675. doi:10.1016/j.scitotenv.2020.136675.

Shehzad, M. A., Hussain, I., Akhtar, G., Ahmad, K. S., Nawaz, F., Faried, H. N., & Mehmood, A. (2023). Insights into physiological and metabolic modulations instigated by exogenous sodium nitroprusside and spermidine reveals drought tolerance in *Helianthus annuus* L. *Plant Physiology and Biochemistry, 202*. http://www.journals.elsevier.com/plant-physiology-and-biochemistry/. doi:10.1016/j.plaphy.2023.107935.

Shelake, R. M., Kadam, U. S., Kumar, R., Pramanik, D., Singh, A. K., & Kim, J. Y. (2022). Engineering drought and salinity tolerance traits in crops through CRISPR-mediated genome editing: Targets, tools, challenges, and perspectives. *Plant Communications, 3*(6). Www.cell.com/plant-communications. doi:10.1016/j.xplc.2022.100417.

Srivastava, S., & Srivastava, M. (2014). Morphological Changes and Antioxidant Activity of *Stevia rebaudiana* under water stress. *American Journal of Plant Sciences, 05*(22), 3417–3422. doi:10.4236/ajps.2014.522357.

Sun, Z., Song, J., Xin, X., Xie, X., & Zhao, B. (2018). Arbuscular mycorrhizal fungal 14-3-3 proteins are involved in arbuscule formation and responses to abiotic stresses during AM symbiosis. *Frontiers in Microbiology, 9*. https://www.frontiersin.org/articles/10.3389/fmicb.2018.00091/full. doi:10.3389/fmicb.2018.00091.

Tallapragada, P., Dikshit, R., & Seshagiri, S. (2016). Influence of *Rhizophagus* spp. and *Burkholderia seminalis* on the growth of tomato (*Lycopersicon esculatum*) and bell pepper (*Capsicum annuum*) under drought stress. *Communications in Soil Science and Plant Analysis, 47*(17), 1975–1984.

Tedersoo, L., Magurno, F., Alkahtani, S., & Mikryukov, V. (2024). Phylogenetic classification of arbuscular mycorrhizal fungi: New species and higher-ranking taxa in *Glomeromycota* and *Mucoromycota* (class Endogonomycetes). *MycoKeys, 107*, 273–325.

Thind, S., Chaudhary, M. S., Ditta, A., Hussain, I., Parveen, A., Ullah, N., Mahmood, Q., Al-Ashkar, I., & El-Sabagh, A. (2022). Impact of mycorrhizal fungi from different rhizospheric soils on fungal colonization, growth, and chlorophyll contents of *Cenchrus ciliaris*. *Agronomy, 12*(11), 2644. doi:10.3390/agronomy12112644.

Torres, N., Yu, R., & Kurtural, S. K. (2021). Arbuscular mycorrhizal fungi inoculation and applied water amounts modulate the response of young grapevines to mild water stress in a hyper-arid season. *Frontiers in Plant Science, 11*, 622209. doi:10.3389/fpls.2020.622209.

Turner, N. C. (2018). Turgor maintenance by osmotic adjustment: 40 years of progress. *Journal of Experimental Botany, 69*(13), 3223–3233. http://jxb.oxfordjournals.org/. doi:10.1093/jxb/ery181.

Větrovský, T., Kolaříková, Z., Lepinay, C., Awokunle Hollá, S., Davison, J., Fleyberková, A., Gromyko, A., Jelínková, B., Kolařík, M., Krüger, M., Lejsková, R., Michalčíková, L., Michalová, T., Moora, M., Moravcová, A., Moulíková, Š., Odriozola, I., Öpik, M., Pappová, M., … Kohout, P. (2023). GlobalAMFungi: A global database of arbuscular mycorrhizal fungal occurrences from high-throughput sequencing metabarcoding studies. *New Phytologist, 240*(5), 2151–2163. http://onlinelibrary.wiley.com/journal/10.1111/(ISSN)1469-8137. doi:10.1111/nph.19283.

Vivas, A., Marulanda, A., Ruiz-Lozano, J. M., Barea, J. M., & Azcón, R. (2003). Influence of a *Bacillus* sp. on physiological activities of two arbuscular mycorrhizal fungi and on plant responses to PEG-induced drought stress. *Mycorrhiza, 13*(5), 249–256. doi:10.1007/s00572-003-0223-z.

Vurayai, R., Emongor, V., & Moseki, B. (2011). Effect of water stress imposed at different growth and development stages on morphological traits and yield of bambara groundnuts (*Vigna subterranea* L. Verdc). *American Journal of Plant Physiology, 6*(1), 17–27. http://scialert.net/qredirect.php?doi=ajpp.2011.17.27&linkid=pdf. doi:10.3923/ajpp.2011.17.27. Botswana .

Wahab, A., Muhammad, M., Munir, A., Abdi, G., Zaman, W., Ayaz, A., ... Reddy, S. P. P. (2023). Role of arbuscular mycorrhizal fungi in regulating growth, enhancing productivity, and potentially influencing ecosystems under abiotic and biotic stresses. *Plants, 12*(17), 3102. https://doi.org/10.3390/plants12173102.

Wang, Q., Liu, M., Wang, Z., Li, J., Liu, K. e, & Huang, D. (2024). The role of arbuscular mycorrhizal symbiosis in plant abiotic stress. *Frontiers in Microbiology, 14*. doi:10.3389/fmicb.2023.1323881.

Wang, S., Xie, X., Che, X., Lai, W., Ren, Y., Fan, X., & Chen, H. (2023). Host-and virus-induced gene silencing of HOG1-MAPK cascade genes in *Rhizophagus irregularis* inhibits arbuscule development and reduces resistance of plants to drought stress. *Plant Biotechnology Journal, 21*(4), 866–883. doi:10.1111/pbi.14006.

Wang, X., Liu, H., Yu, F., Hu, B., Jia, Y., Sha, H., & Zhao, H. (2019). Differential activity of the antioxidant defence system and alterations in the accumulation of osmolyte and reactive oxygen species under drought stress and recovery in rice (*Oryza sativa* L.) tillering. *Scientific Reports, 9*(1). doi:10.1038/s41598-019-44958-x.

Wang, Y., Zhang, W., Liu, W., Ahammed, G. J., Wen, W., Guo, S., Shu, S., & Sun, J. (2021). Auxin is involved in arbuscular mycorrhizal fungi-promoted tomato growth and NADP-malic enzymes expression in continuous cropping substrates. *BMC Plant Biology, 21*(1). http://www.biomedcentral.com/bmcplantbiol/. doi:10.1186/s12870-020-02817-2.

Wang, Y., Zou, Y. N., Shu, B., & Wu, Q. S. (2023). Deciphering molecular mechanisms regarding enhanced drought tolerance in plants by arbuscular mycorrhizal fungi. *Scientia Horticulturae, 308*. Www.elsevier.com/inca/publications/store/5/0/3/3/1/6. doi:10.1016/j.scienta.2022.111591.

Wasaya, A., Yaqoob, S., Ditta, A., Yasir, T. A., Sarwar, N., Javaid, M. M., Al-Ashkar, I., Sabagh, A. E. L., & Farooq, M. U. (2024). Exogenous application of β-aminobutyric acid improved water relations, membrane stability index, and achene yield in sunflower hybrids under terminal drought stress. *Polish Journal of Environmental Studies, 33*(5), 5367–5479. https://www.pjoes.com/pdf-177182-111658?filename=Exogenous%20Application%20of.pdf. doi:10.15244/pjoes/177182.

Wijayawardene, N. N., Hyde, K. D., Al-Ani, L. K. T., Tedersoo, L., Haelewaters, D., Rajeshkumar, K. C., Zhao, R. L., Aptroot, A., Leontyev, D. V., Saxena, R. K., Tokarev, Y. S., Dai, D. Q., Letcher, P. M., Stephenson, S. L., Ertz, D., Lumbsch, H. T., Kukwa, M., Issi, I. V., Madrid, H., ... Thines, M. (2020). Outline of Fungi and fungus-like taxa. *Mycosphere, 11*(1), 1060–1456. http://www.mycosphere.org/. doi:10.5943/mycosphere/11/1/8.

Wu, H. H., Zou, Y. N., Rahman, M. M., Ni, Q. D., & Wu, Q. S. (2017). Mycorrhizas alter sucrose and proline metabolism in trifoliate orange exposed to drought stress. *Scientific Reports, 7*(1), 42389.

Wu, Q. S., Srivastava, A. K., & Zou, Y. N. (2013). AMF-induced tolerance to drought stress in citrus: A review. *Scientia Horticulturae, 164*, 77–87. doi:10.1016/j.scienta.2013.09.010.

Wu Qiang Sheng, W. Q., Cao MingQin, C. M., Zou YingNing, Z. Y., Wu Chu, W. C., & He XinHua, H. X. (2016). Mycorrhizal colonization represents functional equilibrium on root morphology and carbon distribution of trifoliate orange grown in a split-root system. *Scientia Horticulturae, 199*, 95–102.

Xiaofeng, L., Jingzhong, C., Ruiting, G., Jiming, L., Qinwen, S., & S. (2021). Two arbuscular mycorrhizal fungi alleviate drought stress and improve plant growth in *Cinnamomum migao*, seedlings. *Mycobiology, 49*(4), 396–405.

Xie, K., Ren, Y., Chen, A., Yang, C., Zheng, Q., Chen, J., & Xu, G. (2022). Plant nitrogen nutrition: The roles of arbuscular mycorrhizal fungi. *Journal of Plant Physiology, 269*, 153591.

Xie, X., Lai, W., Che, X., Wang, S., Ren, Y., Hu, W., & Tang, M. (2022). A SPX domain-containing phosphate transporter from *Rhizophagus irregularis* handles phosphate homeostasis at the symbiotic interface of arbuscular mycorrhizas. *New Phytologist, 234*(2), 650–671. doi:10.1111/nph.17973.

Yan, M., Bu, C., Huang, G., Liu, G., Dong, T., & Xu, X. (2020). Positive effects of arbuscular mycorrhizal fungi (AMF) on the aboveground part of *Morus alba*. *Zhiwu Shengli Xuebao/Plant Physiology Journal, 56*(12), 2647–2654. http://www.plant-physiology.com. doi:10.13592/j.cnki.ppj.2020.0408.

Yang, X., Shen, K., Xia, T., He, Y., Guo, Y., Wu, B., Han, X. u, Yan, J., & Jiao, M. (2023). Invasive and native plants differentially respond to exogenous phosphorus addition in root growth and nutrition regulated by arbuscular mycorrhizal fungi. *Plants, 12*(11), 2195. doi:10.3390/plants12112195.

Yao, Y. B., Yang, J. H., Xiao, G. J., Lei, J., Niu, H. Y., & Zhang, X. Y. (2018). Research advances in the impacts of climate warming on rainfed agriculture and agroecology in Northwest China. *Chinese Journal of Ecology, 37*(7), 2170–2179. http://www.cje.net.cn/EN/article/downloadArticleFile.do?attachType=PDF&id=9002. doi:10.13292/j.1000-4890.201807.011.

Ye, Q., Wang, H., & Li, H. (2022). Arbuscular mycorrhizal fungi improve growth, photosynthetic activity, and chlorophyll fluorescence of *Vitis vinifera* L. cv. ecolly under drought stress. *Agronomy, 12*(7), 1563. doi:10.3390/agronomy12071563.

Yizhu, L., Imtiaz, M., Ditta, A., Rizwan, M. S., Ashraf, M., Mehmood, S., Aziz, O., Mubeen, F., Ali, M., Elahi, N. N., Ijaz, R., Lele, S., Shuang, C., & Tu, S. (2020). Response of growth, antioxidant enzymes and root exudates production towards As stress in *Pteris vittata* and in *Astragalus sinicus* colonized by arbuscular mycorrhizal fungi. *Environmental Science and Pollution Research, 27*(2), 2340–2352. https://link.springer.com/journal/11356. doi:10.1007/s11356-019-06785-5.

Yooyongwech, S., Phaukinsang, N., Cha-um, S., & Supaibulwatana, K. (2013). Arbuscular mycorrhiza improved growth performance in *Macadamia tetraphylla* L. grown under water deficit stress involves soluble sugar and proline accumulation. *Plant Growth Regulation, 69*(3), 285–293. doi:10.1007/s10725-012-9771-6.

Yooyongwech, S., Samphumphuang, T., Tisarum, R., Theerawitaya, C., & Cha-um, S. (2016). Arbuscular mycorrhizal fungi (AMF) improved water deficit tolerance in two different sweet potato genotypes involves osmotic adjustments via soluble sugar and free proline. *Scientia Horticulturae, 198*, 107–117. doi:10.1016/j.scienta.2015.11.002.

Yuan, J., Shi, K., Zhou, X., Wang, L., Xu, C., Zhang, H., Zhu, G., Si, C., Wang, J., & Zhang, Y. (2023). Interactive impact of potassium and arbuscular mycorrhizal fungi on the root morphology and nutrient uptake of sweet potato (*Ipomoea batatas* L.). *Frontiers in Microbiology, 13*. doi:10.3389/fmicb.2022.1075957.

Zhang, F., He, J. D., Ni, Q. D., Wu, Q. S., & Zou, Y. N. (2018a). Enhancement of drought tolerance in trifoliate orange by mycorrhiza: Changes in root sucrose and proline metabolisms. *Notulae Botanicae Horti Agrobotanici Cluj-Napoca, 46*(1), 270–276. http://www.notulaebotanicae.ro/index.php/nbha/issue/archive. doi:10.15835/nbha46110983.

Zhang, F., Zou, Y. N., & Wu, Q. S. (2018b). Quantitative estimation of water uptake by mycorrhizal extraradical hyphae in citrus under drought stress. *Scientia Horticulturae, 229*, 132–136. doi:10.1016/j.scienta.2017.10.038.

Zhang, Y., Feng, H., Druzhinina, I. S., Xie, X., Wang, E., Martin, F., & Yuan, Z. (2024). Phosphorus/nitrogen sensing and signaling in diverse root–fungus symbioses. *Trends in Microbiology, 32*(2), 200–215. https://www.sciencedirect.com/science/journal/0966842X. doi:10.1016/j.tim.2023.08.005.

Zou, Y.-N., Wu, Q.-S., Kuča, K., & Hu, Y. (2021). Unravelling the role of arbuscular mycorrhizal fungi in mitigating the oxidative burst of plants under drought stress. *Plant Biology, 23*(S1), 50–57. doi:10.1111/plb.13161.

Zulfiqar, F., Younis, A., Riaz, A., Mansoor, F., Hameed, M., Akram, N. A., & Abideen, Z. (2020). Morpho-anatomical adaptations of two *Tagetes erecta* L. Cultivars with contrasting response to drought stress. *Pakistan Journal of Botany, 52*(3), 801–810. http://www.pakbs.org/pjbot/papers/1585436460.pdf. doi:10.30848/PJB2020-3(35).

Chapter 13

Arbuscular mycorrhizal fungi and salinity stress tolerance in plants

Saba Mazhar[a], Huma Dawood[a], Maryam Shahzadi[a], Sajjad Hyder[a] and Bekri Xhemali[b,c]

[a] *Department of Botany, Faculty of Natural Sciences, Government College Women University Sialkot, Pakistan,* [b] *Faculty of Agribusiness, University Haxhi Zeka, Pejë, Kosovo,* [c] *Laboratory of Plant Protection, Kosovo Institute of Agriculture, Pejë, Kosovo*

13.1 Introduction

Numerous environmental stresses adversely affect plant growth and metabolism, which ultimately reduce crop productivity (Hashem et al., 2018). Salinity is one of the most important abiotic stresses in the world, which creates serious ecological (such as the loss of biological diversity, the devastation of pastures and forests, desertification, and soil erosion) and agricultural issues, especially in arid and semiarid areas (Jia et al., 2019; Naz et al., 2025; Shahzad et al., 2019). Salt marches are spreading throughout the world because of unsustainable farming practices (such as overuse of chemicals and watering with saline water), climate change, and global warming, which increase soil salinity by increasing transpiration and evaporation and decreasing rainfall (Guo et al., 2020). Salinity-affected areas are increasing by 15–20 million hectares annually, resulting in a 20% reduction in crop productivity (Ahmed et al., 2024; Pan et al., 2020; Shaddam et al., 2024). By 2050 it is anticipated that half of the arable land may be adversely affected by salinization because of climate change (Diao et al., 2021; Janah et al., 2023; Sootahar et al., 2024). Iran, Iraq, Pakistan, China, Thailand, Egypt, Australia, the United States, India, and Argentina are among the nations where soil salinity has had a greater impact on agricultural productivity than other regions of the world (Abeer et al., 2014).

Based on salt sensitivity, plants have two main categories: halophytes and glycophytes (Pan et al., 2020). Glycophytes are sensitive to salt, whereas halophytes are salt-tolerant plants that can survive in soil and water with high saline concentrations (200 mM NaCl or more) (Ditta, 2013; Wang et al., 2019). Halophytes have efficient defense mechanisms against salt stress, whereas glycophytes, which include common crop and forage species in contemporary agriculture, have weak salt tolerance mechanisms (Pan et al., 2020). However, both plants are vulnerable to salinity at their early growth stages (Himabindu et al., 2016). By creating osmotic and ionic stress, altering water relations, and resulting in physiological drought or osmotic stress, salt stress affects plant physiology at both whole-plant and cellular levels (Pirzada et al., 2024). Furthermore, in salinized soils, the buildup of excess ions such as HCO_3^-, Na^+, and Cl^- disrupts enzymatic activity and improves the production of reactive oxygen species (ROS), which eventually hinders plant growth and survival (Chang et al., 2023; Zhang et al., 2022).

One of the major global challenges is to ensure food security, particularly in the context of decreasing agricultural land availability, which requires effective and sustainable approaches to reduce soil salinity (Atta et al., 2023; Mahajan & Pal, 2023). Different management strategies have been used to mitigate salinity stress, such as plant breeding, improving irrigation systems, adding of solutes and growth regulators, growing salt-resistant crops, and providing plant growth-promoting microorganisms (PGPM) (Ali et al., 2024; Diagne et al., 2020; Naveed et al., 2021; Niamat et al., 2019). Among the available strategies, plant inoculation with beneficial soil microorganisms has emerged as a promising approach, as it improves plant tolerance to salt stress while simultaneously enhancing yield (Janah et al., 2023). Arbuscular mycorrhizal fungi (AMF) appear to be a more ecofriendly and cost-effective option among the PGPM (Prasad, 2022; Verma et al., 2023). AMF, which is naturally present in soil, can significantly improve plant nutrient uptake and provide resilience to a range of abiotic stresses (Saeed et al., 2025; Sun et al., 2018). AMF may establish symbiotic relationships with about 90% of plant species, including flowering plants, ferns, and bryophytes. Plant roots are more accessible to a greater region of soil when

Advancing Sustainable Agriculture Through Arbuscular Mycorrhizal Fungi. DOI: https://doi.org/10.1016/B978-0-443-45130-0.00041-2

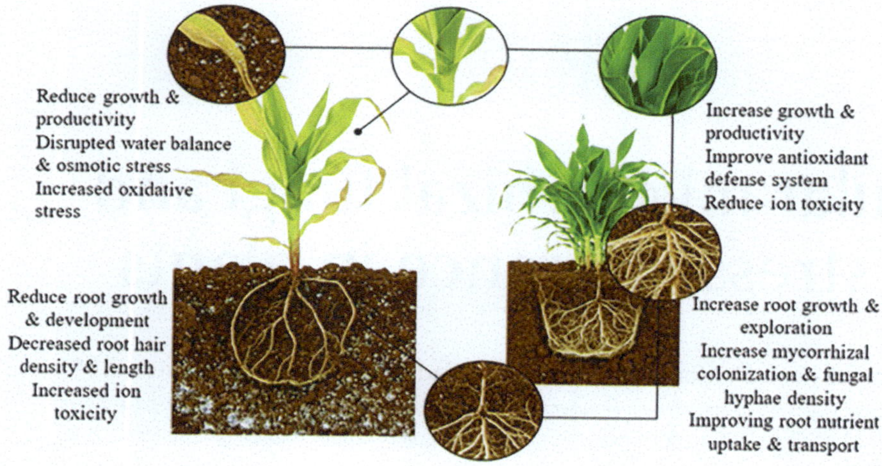

FIGURE 13.1 Comparative response to salinity stress in nonmycorrhizal and mycorrhizal plants.

the AMF forms a hyphal meshwork with them (Nanda et al., 2021). AMF improves the nutritional condition of plants by increasing the availability and movement of various nutrients, particularly immobile P (Khalediyan et al., 2021).

13.2 Arbuscular mycorrhizal fungi: A brief overview

As soil-dwelling fungi, AMF are among the most significant microorganisms (Ahmed et al., 2025; Zhang et al., 2023). By acting as efficient biofertilizers that improve soil health and plant resistance to abiotic stress, AMF play a crucial role in advancing sustainable agricultural systems (Amir & Crossay, 2024; Nie et al., 2024). These symbiotic fungi establish a mutualistic relationship with the roots of most terrestrial plants (Sun et al., 2024). In exchange for the host plant providing carbon to the fungi (Kaya, 2024; Khan et al., 2022), AMF facilitate the uptake of vital nutrients from outside the rhizosphere through a vast network of extraradical hyphae (Ebbisa, 2022). AMF are well known for their capacity to improve host plants' stress tolerance mechanisms to mitigate the negative impacts of abiotic stresses, especially salinity (Mohammadi et al., 2019). Through the enhancement of root colonization, root length density, and root hair formation, AMF markedly improve root system architecture and promote deeper soil exploration for improved water and nutrient acquisition (Pal et al., 2024). This improved root–soil interface enables more effective water acquisition and nutrient cycling, both of which are essential for the best possible crop growth under suboptimal environmental circumstances (Bhupenchandra et al., 2024).

Furthermore, AMF helps reduce stress by promoting the production and accumulation of stress-related proteins and osmo-protectants, which protect plant cells from oxidative damage caused by ROS under saline environments (Zong et al., 2023). In addition, AMF strengthens plant–soil relations and promotes rhizosphere health by secreting extracellular enzymes that mediate nutrient mineralization, organic matter decomposition, and secondary metabolite production (Scavo et al., 2019). AMF play a crucial role in reducing the environmental impact of conventional agriculture because of their ability to reduce dependence on synthetic fertilizers (de Souza & Santos, 2018). Collectively, AMF play a fundamental role in enhancing stress resilience, increasing crop productivity, and advancing the sustainability of agroecosystems (Islam et al., 2026). Fig. 13.1 illustrates the contrasting responses of nonmycorrhizal and mycorrhizal plants when exposed to salinity stress.

13.3 Mechanisms of arbuscular mycorrhizal fungi–mediated salinity stress tolerance

13.3.1 Nutrient uptake and transport

Macronutrients and micronutrients are essential for the growth and development of all plants. Plant roots fail to absorb these vital nutrients under stress conditions. The AMF hyphae facilitate the plant for selective absorption and uptake of nutrients, developing a positive influence on AM plants (Nazir et al., 2023; Thind et al., 2022). The fungal mycelium acts as an extension in the root zone to extract these essential nutrients for plant growth (Etesami et al., 2021). Plants and AMF form a symbiotic relationship as AMF provide inorganic nutrients to plants while getting carbohydrates and lipids from plants because vesicle and spore formation is essential for both partners (Yizhu et al., 2020). Genes that are involved in lipid metabolism are mutated in plants, resulting in reduced vesicle formation, which supports the role of lipid storage vesicles in transferring lipids from the host to AMF (Malar et al., 2022). This process helps AMF in spore germination and early growth after detachment from the host, because plants provide lipid supply through genes such as *KAS III* and *RAM2*, which play an important role in lipid transfer (Lanfranco & Bonfante, 2023). In return, AMF transports vital elements such as phosphorus and nitrogen from the soil to the plants through transporters present in periarbuscular membrane. For the uptake of different nutrients such as ammonia and phosphorus, specialized transporters are required. These transporters serve as markers for cells with functional arbuscules, which are the primary sites for nutrient exchange (Calabrese, 2016; Perotto & Balestrini, 2024). During AMF colonization in *Brassica napus and Capsicum annuum,* genes such as *AMT3;1* and *CaAMT2;1* are activated for ammonium uptake (Fang et al., 2023). Overall, plant nutrient uptake and stress resilience are enhanced by the formation of lipid vesicles and spores, along with the regulation of genes during nutrient exchange, which are crucial for the survival and growth of both partners (Ahmed et al., 2025).

13.3.2 Osmotic adjustment and ion homeostasis

AMF accumulates glycine betaine, proline, and soluble sugars within plant tissues, which helps to maintain the potential of intracellular water. Proline, which acts as an osmo-protectant, membrane stabilizer, and antioxidant, helps prevent protein denaturation. Meanwhile, glycine betaine contributes to membrane stabilization and enhances enzymatic activity, thereby increasing plant tolerance to stress (Evelin et al., 2019). Enhanced proline biosynthesis has been observed in AMF-inoculated plants because of the expression of the **P5CS** gene, which is responsible for increased activity of glutamate dehydrogenase. The accumulation of glycine betaine increases when plants face abiotic stresses, as it acts as an osmolyte in plants. AMF inoculation further enhances its concentration compared to other ammonium compounds (Park et al., 2024). Saline soil negatively affects plant metabolism because of the loss of turgor pressure in plants. AMF regulates ion accumulation in the vacuole to maintain osmotic pressure, which increases water retention within cells and helps avoid water loss (Santander et al., 2017). Ion toxicity in the cytoplasm is reduced by storing Na^+ and Cl^- ions in the vacuoles of plant cells. AMF not only increases Na^+/H^+ antiporters in plant meristematic cells but also promotes the accumulation of K^+ to avoid sodium toxicity. This enhances the performance of enzymatic activities that maintain metabolic functions (Wu et al., 2022). AMF also regulates structural components such as pectin, cellulose, and proteins, which strengthen the cell wall and promote plant growth under stress conditions (Mendoza et al., 2022). AMF prevents water loss in plants, maintains ion homeostasis, and regulates plant metabolism, thereby contributing to plant growth, productivity, and resistance against abiotic stresses (Huang et al., 2023).

13.3.3 Antioxidant defense and reactive oxygen species scavenging

ROS such as superoxide radical (O_2^-), hydrogen peroxide (H_2O_2), hydroxyl radical (OH^-), and singlet oxygen (O^-) are produced in plants, creating oxidative stress when plants are exposed to saline soil. O_2^- and H_2O_2 are the most reactive species and cause oxidative damage to biomolecules, leading to lipid peroxidation, protein denaturation, and DNA mutation (Porcel et al., 2012). Although plants possess different strategies to mitigate salt stress at the cellular level, these ROS disturb the structure and function of plant cells (Alkharabsheh et al., 2021). The enzymes that help eradicate these species are known as ROS-scavenging enzymes. To overcome oxidative stress, plants mitigate both enzymatic and nonenzymatic antioxidant defense mechanisms.

Enzymes such as superoxide dismutase (SOD), peroxidase (POX), catalase (CAT), glutathione reductase (GR), and ascorbate peroxidase (APX) function in the enzymatic defense mechanism, whereas ascorbic acid, glutathione, carotenoids, and alpha-tocopherols function in the nonenzymatic defense mechanism to remove harmful ROS species in plants (Azeem et al., 2023). The ROS-scavenging mechanism is enhanced by AMF, as AMF increases the production and efficiency of these antioxidant molecules within plant tissues (Yadav et al., 2021). Detoxification of ROS is initiated when AMF is inoculated in plant roots, leading to the production of nonenzymatic antioxidant molecules such as glutathione and ascorbate. These molecules contribute to the synthesis of antioxidant enzymes such as SOD, CAT, GR, and POX, which initiate the ROS-scavenging mechanism, whereas molecules such as ascorbic acid, tocopherols, and phenolic compounds, which are water-soluble, help reduce ROS production (Gill & Tuteja, 2010). Several studies have demonstrated increased levels of antioxidant enzymes in pearl millet, date palm, alfalfa, and cucumber. AMF enhances the activities of SOD, CAT, POD, and APX in the leaves of plants exposed to salt stress (Pooja et al., 2025a).

13.3.4 Hormonal regulation and signaling pathways

Salinity causes various changes in plants under stress conditions, affecting the growth, quality, and yield of plant products. Plants naturally produce growth regulators, such as auxins, gibberellins, cytokinins (CTKs), salicylic acid (SA), jasmonic acid (JA), brassinosteroids (BRs), polyamines (PAs), strigolactones (SLs), and nitric oxide. These hormones not only

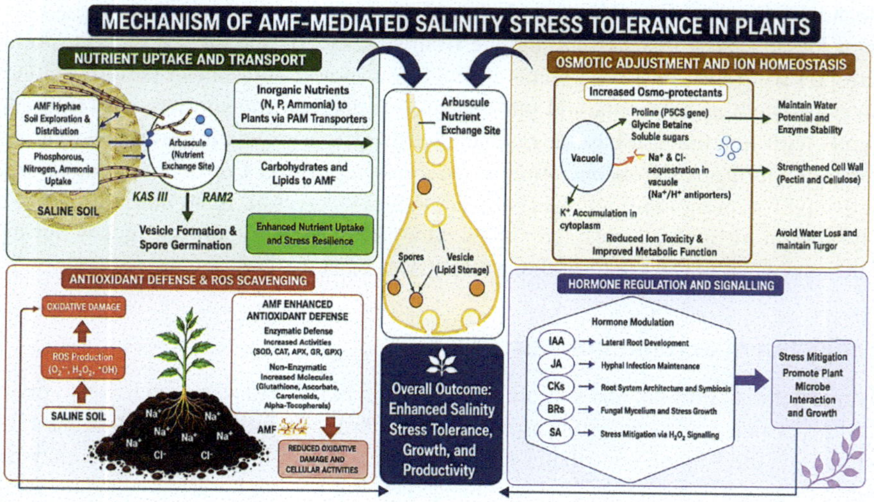

FIGURE 13.2 Mechanisms of AMF-mediated salinity stress tolerance in plants. *AMF*, Arbuscular mycorrhizal fungi.

mitigate stress conditions but also help in developing plant–microbe interactions (Sabagh et al., 2022). AMF inoculation increases levels of indole-3-acetic acid (IAA) in the roots of soybeans, which promotes the development and proliferation of lateral roots (Meixner et al., 2005). ABA and IAA levels are reduced under salt stress in jujube roots when inoculated with AMF, which indicates a complex interaction between AMF symbiosis, hormonal regulation, and stress responses (Ma et al., 2022a).

Metabolomics studies have shown increased concentrations of SA, ABA, and SLs in arbuscular mycorrhizal plants through H_2O_2 signaling, which helps to mitigate the negative effects of salinity stress (Ren et al., 2018). JA is an essential signaling phytohormone involved in stress response mechanisms and plant growth and is also responsible for maintaining hyphal infections of AMF in root zones (Hewedy et al., 2023). AMF plants significantly accumulate different types of CTKs, such as isopentenyladenine, iP riboside, and transzeatin, which work together with SLs and ABA to maintain root system architecture, thereby promoting the mycorrhizal symbiosis process under salt stress conditions (Ma et al., 2022b). AMF also enhances the accumulation of BRs under salinity stress, supporting the formation of fungal mycelium and regulating plant growth and development under both biotic and abiotic stresses (Zhang et al., 2024). The overall mechanisms of AMF-mediated salinity stress tolerance in plants are illustrated in Fig. 13.2.

13.4 Arbuscular mycorrhizal fungi and plant responses to salinity stress

13.4.1 Plant growth and biomass production

The role of AMF in plants under salinity stress has shown that this symbiotic relationship between plants and AMF increases nutrient uptake, regulates phytohormones, enhances the photosynthetic rate, osmoregulation, and ion homeostasis. It promotes the accumulation of sugars, proteins, organic acids, and key metabolic intermediates, while reducing oxidative stress markers such as MDA and H_2O_2, thereby improving plant energy to alleviate salt stress, which depends on the involved AMF species (Wu et al., 2007). The formation of arbuscules and extensive hyphal colonization not only increases water-use efficiency but also strengthens antioxidant defense mechanisms, thereby increasing plant resistance to salinity stress. The growth of AMF-inoculated plants, such as tomato, has been much better than that of noninoculated plants under salinity stress. This improvement is because of AMF symbiosis, which results in higher biomass production and better nutrient uptake. Mycorrhization increases the health of the host plant, resulting in enhanced biomass production (Parvin et al., 2020).

Increased dry root and shoot weight and overall biomass production have been observed in maize, chickpea, fenugreek, and alfalfa plants compared to control plants under saline treatments. Enhanced salt resistance has been reported in seedlings of *Acacia nilotica* with AMF inoculation and improved yield traits, such as seed and stem fiber production, which have been observed in flax plants, demonstrating increased resistance to salinity stress. Moreover, sustainable approaches involving AMF inoculation integrated with nanoparticles have shown increased quality and biomass production in gerbera, sage, and palm beach cultivars under different salinity levels (Pooja et al., 2025a).

13.4.2 Photosynthesis and water relations

Photosynthesis is affected by salt stress in many ways. Salt stress decreases the ability of CO_2 assimilation, reduces stomatal conductance of plant leaves, causes degradation of proteins in the PSII reaction center of photosystem II, and decreases the content of photosynthetic pigments (Evelin et al., 2019). AMF regulates phytohormones such as CTKs, which increase the photosynthetic capacity of the host plant by preventing damage to the thylakoid membrane. The most bioactive CTK is zeatin riboside. AMF releases CTK-like substances that enhance chloroplast development, resulting in higher chlorophyll content in AMF-inoculated plants under salt stress (Cao et al., 2025). AMF inoculation partially prevents leaf aging, leading to improved metabolite accumulation. It also protects against the destruction of plant cell walls, plasma membranes, and chloroplast structures, resulting in higher concentrations of osmolytes such as sugars, betaine, proline, glycine, and PAs. This protection helps maintain CO_2-fixing enzymes and protein complexes involved in photosystem II (Shtark et al., 2019).

AMF inoculation also reduces Na^+ toxicity by inhibiting its transfer to the aerial parts of plants and promotes the activity of the chlorophyll synthase enzyme, thereby enhancing chlorophyll production. A study conducted by Liang et al. (2021) on narrowleaf coneflowers inoculated with *Rhizophagus irregularis* under salt treatments showed chlorophyll content increased by two to three times compared to control treatments. Excessive salt in the root zone also blocks water transfer from roots to aerial plant organs, creating physiological drought conditions. AMF-inoculated plants maintain higher leaf turgidity with lower leaf osmotic potential compared to non-AMF plants under salt stress (Chen et al., 2017).

AMF alters root morphology, facilitating the accumulation of compatible solutes to adjust osmotic potential and balance water transfer in host plants. To regulate the slow movement of water molecules that follows the water potential gradient, there is a certain group of proteins known as "aquaporins." The expression of aquaporin genes is upregulated to allow a higher rate of transcellular water flow in AMF-inoculated plants; however, each aquaporin gene responds differently depending on plant species and tissue type. Overall the transcription of these genes strongly depends on salinity level and gene type, which influence their ability to transport water and other solutes to host plants (Evelin et al., 2019).

13.4.3 Ion uptake and distribution

Plants take up more sodium and potassium ions when the concentration of sodium salts increases in the soil. Although potassium ions are vital for metabolic functioning, a high amount of sodium ions reduces photosynthetic rates and leads to ROS accumulation (Table 13.1). Plants have different strategies to avoid the entry of excessive Na^+ ions into roots and their allocation within leaves under stress conditions (Karaki & Ghazi, 2006). Salt-resistant plants overexpress Na^+/H^+ antiporters that transport sodium toxicity into the vacuole or apoplast. Higher amounts of NaCl disturb nutrient availability and ionic balance, resulting in altered ratios of $Na^+:K^+$, $Na^+:Ca^{2+}$, and $Ca^{2+}:Mg^{2+}$, which disrupt the physiological functioning of plants (Abdel-Fattah & Asrar, 2012).

AMF enhances the sufficient supply of P, N, K, Zn, and Cu through fungal hyphae to mitigate these effects, while regulating organic acids and PAs that help maintain ionic balance and nutrient uptake (Sheng et al., 2011). AMF also enhances phosphorus availability by maintaining phosphorus concentration in the soil, expressing high-affinity phosphate transporters, and facilitating efficient phosphate translocation, which stabilizes membranes and decreases ion toxicity (Evelin et al., 2012). AM plants restrict the transport of sodium ions to shoots to maintain higher $K^+:Na^+$ ratios by sequestering sodium ions in vacuoles and regulating NHX, salt overly sensitive (SOS1), and HKT sodium transporters, which protect photosynthetic tissues. In recent years, maintaining a high cytosolic $K^+:Na^+$ ratio has been increasingly recognized as a key determinant of salinity tolerance. Accordingly, plants have developed multiple mechanisms to restrict Na^+ entry at the root level, promote its sequestration within vacuoles, and regulate its transport and allocation to aerial tissues (Porcel et al., 2012). AMF-mediated symbiosis enhances nitrogen acquisition under saline conditions by stabilizing cellular membranes, stimulating nitrate reductase activity, and increasing the expression of nitrate and ammonium transporter genes (Felicia et al., 2017). In addition, AMF contributes to improved Ca^{2+} and Mg^{2+} uptake and supports the maintenance of optimal $Ca^{2+}:Na^+$ and $Ca^{2+}:Mg^{2+}$ ratios during salt stress, although the precise mechanisms involved are not yet fully understood (Evelin et al., 2012).

13.4.4 Gene expression and molecular responses

Stress signals are continuously monitored by plant cells, and these signals are converted into intracellular responses that enable survival under salinity stress. Calcium ions present in the cytosol act as secondary messengers to activate downstream signaling pathways (Wang et al., 2015). These Ca^{2+} ions function through Ca^{2+} sensor systems, which play a vital role in maintaining cellular ion homeostasis and regulating ion transport via calcineurin B-like proteins (CBLs) and their interacting

TABLE 13.1 Effect of AMF on the absorption of different nutrients by plants under salinity stress.

AMF species name	Plant name	Effect on nutrient uptake	References
Glomus sp.	*Zea mays*	Enhanced maize nutrient absorption by increasing the K^+/Na^+ ratio, indicating better K uptake and reduced Na accumulation	Yang et al. (2022)
Glomus sp.2, *Gigaspora albida*, *Gymnopilus decipiens*	*Eucalyptus camaldulensis*	Enhanced absorption of N, P, and K in eucalyptus while reducing Na accumulation, leading to a much higher K/Na ratio than in noninoculated plants	Klinsukon et al. (2021)
Rhizophagus irregularis	*Cicer arietinum* L.	Improved chickpea nutrient absorption by enhancing uptake of P and K, maintaining better K^+/Na^+ balance compared with noninoculated plants	Azizoglu et al. (2025)
Funneliformis mosseae, *Rhizophagus intraradices*, *Diversispora epigaea*	*Glycine max* (L.) Merr.	Enhanced soybean nutrient absorption by increasing the accumulation of nitrogen (N), phosphorus (P), and potassium (K) in roots, stems, and leaves	Peng et al. (2025)
Laroideoglomus etunicatum, *F. mosseae*, *Gigaspora margarita*, *Scutellospora calospora*, *Claroideoglomus etunicatum*, *F. mosseae*, *G. margarita*	*Lasiurus scindicus* Henr.	Enhanced plant nutrient absorption, notably N, P, and K, under NaCl stress by increasing root surface area and uptake capacity.	Malik et al. (2022)
Glomus intraradices	*Colocasia esculenta* L.	Enhanced nutrient absorption by improving K^+ uptake and maintaining a higher K^+/Na^+ ratio while reducing Na^+ and Cl^- accumulation	Bernal et al. (2022)
Rhizophagus clarus	Cowpea bean	Increased P and K uptake, while reducing Mg, Ca, and S. Mycorrhizal plants maintained higher K at 75–100 mM NaCl and higher P at ≥50 mM NaCl	Tavares et al. (2024)
Glomus viscosum H.T. Nicolson strain A6	*Medicago sativa* L. var.	Mycorrhizal plants absorbed more K and less Na and Cl under salinity; Ca and Mg were unaffected	Campanelli et al. (2013)
Glomus spp.	*Triticum aestivum*	Enhanced uptake of K, Zn, Fe, Cu, and Mn while reducing Na. Improved nutrient balance and ion homeostasis occurred in both roots and shoots	Huang et al. (2023)

AMF, arbuscular mycorrhizal fungi.

kinases (CIPKs). At the same time, ABA-dependent pathways and MAP kinase cascades promote the accumulation of osmolytes, which upregulate the expression of stress–responsive genes and drive metabolic adjustments (Roychoudhury & Banerjee, 2017). ROS and changes in turgor pressure also activate MAPKs, which act as signal amplifiers to regulate ROS scavenging mechanisms, transcription factors, and protein repair systems, thereby restoring cellular equilibrium (de Zelicourt et al., 2016).

The SOS signaling pathway is activated when the concentration of sodium ions increases in plant cells. In this pathway, calcium-bound CBL proteins activate CIPK kinases, which stimulate the expression of the Na^+/H^+ antiporter SOS1. This process promotes the extrusion of sodium ions through NHX (Na^+/H^+ exchangers) and proton pumps. Overall, these signaling pathways, ion transport processes, osmolyte synthesis, and detoxification mechanisms are interconnected in plants and enable them to develop resilience against salt stress and maintain cellular homeostasis (Kim et al., 2013). Recent

TABLE 13.2 AMF species and their effects on plant salinity stress tolerance.

AMF species	Host plant used	Salt stress applied	Impact on the plant	References
Glomus clarium and *Claroideoglomus etunicatum*	*Diospyros lotus*	0, 50, 100, and 150 mM NaCl	Improved growth, nutrient content, and plant physiology	Incesu et al. (2025)
Glomus sp.	*Zea mays* L.	0, 25, 50, and 75 mM NaCl	Enhanced salinity tolerance, overall growth, and chlorophyll content	Shehata et al. (2025)
Rhizophagus fasciculatus	*Cicer arietinum* L.	0, 20, 30, and 40 mM NaCl	Increased antioxidant enzyme activity, photosynthesis, and relative water content	Pooja et al. (2025a)
Rhizophagus clarus	*Vigna unguiculata*	0, 25, 50, 75, and 100 mM NaCl	Enhanced antioxidant activity, higher nutrient content	Tavares et al. (2024)
Glomus mosseae and *Glomus etunicatum*	*Medicago sativa* L.	0, 50, 100, 150, and 200 mM NaCl	Improved salt tolerance, growth, and increased antioxidant enzyme activity	Xu et al. (2024)
Rhizophagus irregularis	*Casuarina glauca*	0 and 600 mM NaCl	Improved salt tolerance and nutrient content	Wang et al. (2023)
Glomus sp.	*Triticum aestivum* L.	200 mM NaCl	Increased antioxidant enzyme activity	Huang et al. (2023)
Funneliformis mosseae and *Corymbiglomus tortuosum*	*Gleditsia sinensis*	0, 50, 100, and 150 mM NaCl	Increased root growth and improved Nitrogen content	Ma et al. (2022c)
Glomus mosseae	*Gerbera jamesonii*	0, 20, and 40 mM NaCl	Increased chlorophyll content, growth, and plant yield	Othman et al. (2022)
R. irregularis and *F. mosseae*	*Arachis hypogaea*	200 mM NaCl	Improved growth parameters, photosynthesis rate, osmolyte production, and relative water content	Qin et al. (2021)
F. mosseae and *C. etunicatum*	*Dracocephalum moldavica*	50 and 150 mM NaCl	Increased physiological parameters and antioxidant production	Alizadeh et al. (2021)
Mixture of *F. mosseae*, *Acaulospora laevis*, and *Gigaspora margarita*	*Oryza sativa*	0, 75, and 120 mM NaCl	Increased chlorophyll content, shoot and root dry biomass, grain yield, and increased nutrient content	Parvin et al. (2020)

AMF, arbuscular mycorrhizal fungi.

studies have shown that genes associated with water-channel proteins (aquaporins), Na^+/H^+ antiporters, Δ^1-pyrroline-5-carboxylate synthetase (LsP5CS), late embryogenesis abundant protein (LsLea), and ABA biosynthesis (Lsnced) regulate stress resistance in AM-inoculated plants. Changes in gene expression under salinity stress promote stress signaling and the production of proteins involved in downregulating stress-induced responses (Saxena et al., 2022).

13.5 Factors influencing arbuscular mycorrhizal fungi–mediated salinity stress tolerance

13.5.1 Arbuscular mycorrhizal fungi species and isolate specificity

AMF has functional diversity, but its effectiveness in mitigating salinity stress highly depends on the mycorrhizal species, different isolates, and plant cultivars (Talaat & Shawky, 2014). Various AMF species and their different isolates can overcome salinity stress by colonizing host roots, developing hyphal networks, and regulating ion networks in different plants (Table 13.2). Previous studies have reported that AMF species such as *Claroideoglomus claroideum, Funneliformis mosseae, Rhizophagus intraradices,* and *Claroideoglomus etunicatum* are particular in alleviating salinity stress by improving the physiological and biochemical parameters of different plants (Boorboori & Lackoowa, 2025).

AMF isolates are very important in determining the role of stress response in different plants. The habitats of AMF isolates from where AMF species are collected have high influence on their performance. The AMF originating from semiarid regions shows higher tolerance (Sene et al., 2012). Similarly, AMF isolates that are isolated from highly saline habitats are comparatively better than the AMF originating from nonsaline habitats (Malik et al., 2022). Host fungus compatibility in AMF-mediated salinity stress tolerance highlights the importance of isolate-specific behavior. Therefore the selection of suitable AMF species and isolates is very important in enhancing the yield and growth of different plants under salinity stress by maximizing the effect of *M. symbiosis*.

13.5.2 Plant species and genotype

The effect of AMF on salinity stress is highly influenced by plant species and genotype (Kuper-Psenisnick & Bennett, 2025). Salt stress is decreased in some cultivars because of AMF inoculation, but in some species, it can increase stress (Royle et al., 2026). Every plant exhibits different root morphology, thus having different nutrient requirements and different abilities to resist salinity stress, which directly influence the AMF root colonization that results in differential plant growth and yield under salinity stress. Previous studies have reported that different crops, for example, *Oryza sativa, Triticum aestivum, Cicer arietinum,* and *Hordeum vulgare,* have increased growth, physiological, and biochemical parameters inoculating AMF under salinity stress (Pooja et al., 2025b).

In contrast to interspecific differences, different genotypes within the same species show different responses under salinity stress with AMF colonization. Different cultivars of the same species in different plants display contrasting levels of mycorrhizal association, ionic balance, and nutrient uptake with AMF colonization under salinity stress. Genotype–AMF interaction highly influenced the degree of growth and yield improvement in different cultivars of the same plant species (e.g., tomato) under salinity stress (Felfoldi et al., 2022). These observations led us to highlight the importance of genetic diversity when colonizing plants with AMF under salinity stress.

13.5.3 Soil properties and environmental conditions

Salinity stress can be mitigated through colonization with AMF, but many factors affect its efficiency. These factors include soil salinity and environmental conditions. Soil salinity expressed in the form of electrical conductivity is an important factor that determines the efficacy of AMF under salinity stress, as increased salinity affects the plant performance by decreasing fungal growth and root colonization (Evelin et al., 2019; Klinsukon et al., 2021). Moderate salinity offers effective root colonization, whereas severe salinity does not induce effective root colonization, thus limiting the beneficial effects of AMF on host plants under salinity stress (Kakabouki et al., 2023). Soil nutrients, particularly under lower phosphorus content, AMF colonization improves and transfers phosphorus to plants, enhancing the growth of plants under saline soil, and higher phosphorus content suppresses the activity of AMF, hence reducing its ability to alleviate salinity stress (Begum et al., 2019).

In addition, soil texture, total organic matter, nitrogen availability, and moisture content of soil affect hyphal growth in AMF colonization, which suppresses the impact of AMF under salinity stress (Mirzaei & Moradi, 2017). Environmental factors such as soil pH, soil chemical factors, climatic conditions such as humidity in air, temperature, altitude, and other stresses further modulate AMF–plant interaction. Field conditions also trigger many kinds of stresses, which negatively impact AMF root colonization under salinity stress (Zhang et al., 2025). Therefore it is very pivotal to consider soil properties and environmental conditions to mediate AMF colonization under salinity stress.

13.6 Recent advances in arbuscular mycorrhizal fungi–based salinity stress management and future directions

13.6.1 Use of arbuscular mycorrhizal fungi as biofertilizers

AMF are now considered biofertilizers because of their ability to enhance soil nutrient content, overall crop production, improve soil health, and mediate uptake of phosphorus, nitrogen, potassium, and other nutrients that are limited in saline soil (Begum et al., 2019). Salinity reduces nutrient uptake and water content of plants, but AMF as biofertilizers enhance nutrient uptake, regulate ion homeostasis, and enhance water uptake and photosynthetic activity. AMF as biofertilizers enhance the resistance of plants toward many biotic and abiotic stresses (Evelin et al., 2019; Madawala, 2021). Under salinity stress, application of AMF as a biofertilizer has shown to improve growth and yield of various plant species such as grains, cereals, and other crops (O et al., 2017). Different application strategies such as soil inoculation and seed coating are used for effective mycorrhizal association. Moreover, inoculation of AMF consortia or AMF and plant growth-promoting rhizobacteria has synergistic effects in enhancing plant growth, antioxidant enzymes, and nutrient uptake under salinity stress (Ditta & Ullah, 2023; Hamid et al., 2021; Hidri et al., 2025). Hence AMF have potential as biofertilizers, but their efficiency depends on factors such as host plant species, soil properties, and many other environmental factors, which directly influence their effectiveness. Therefore it is necessary to develop high-quality, site-specific AMF inoculants and a compatible host–fungus combination, which are mandatory to maximize the performance of AMF biofertilizers in agriculture and agroecosystems (Begum et al., 2019).

13.6.2 Genetic engineering of plants for improved salinity tolerance

Genetic engineering has emerged as a promising technique to enhance salinity stress tolerance of plants by altering or inserting specific genes involved in ion homeostasis, stress signaling pathways, osmotic balance, and antioxidant defense systems. Transgenic methods aiming Na^+/H^+ antiporters (HNX1 and SOS1), regulatory elements (DREB2A and TRF1), genome editing methods (ZFNs, CRISPR–Cas9, and TALENs), RNA interference technologies, and other genes regulating osmo-protectants have shown salt resilience in many crops and improved cellular ions in many plant species (Wani et al., 2020).

Furthermore, genetic engineering of antioxidant enzymes modulates the ROS scavenging activity that enhances the salt salinity stress of plants (Lim et al., 2016). Recent technology advances, particularly CRISPR–Cas9, are used to precisely modify the stress–responsive genes for stress resilience in the plant that enhances the development of salt-resilient crop varieties (Wang et al., 2022). Although genetic engineering offers a promising tool to produce salt-resilient crops, its integration with AMF provides a more sustainable and ecofriendly approach to produce more salt-tolerant crops in a saline environment.

13.6.3 Arbuscular mycorrhizal fungi–based strategies for sustainable agriculture

AMF colonization is a biological approach to reduce salinity stress in plants through increasing nutrient uptake, improving soil quality, and enhancing abiotic stress tolerance, which includes salinity stress tolerance in plants. AMF forms a symbiotic relationship with plant roots and enhances the nutrient uptake, for example, of nitrogen, phosphorus, potassium, and other micronutrients, by reducing the dependence on chemical fertilizers to fulfill the nutrient requirements of plants, thereby minimizing the environmental pollution (Begum et al., 2019). AMF-based strategies offer sustainable agriculture through the production of glomalin and vast hyphal networks, which promote water retention, soil aggregates, and microbial diversity in soil (Boyno et al., 2025). In addition to enhancing nutrient uptake, AMF also improve ion homeostasis, osmotic balance, and an antioxidant defense system of plants to make them salt resilient (Begum et al., 2019). These changes improve the physical and chemical health of the soil, which is necessary for saline soil, in which nutrient availability and water uptake are strictly inhibited.

Furthermore, AMF can be integrated with other sustainable practices such as organic amendments, coinoculation with growth-promoting rhizobacteria, crop rotation, and reduced tillage practices to enhance their effectiveness in the agroecosystem. Such integrated AMF-based strategies can produce an ecofriendly, economical, and sustainable approach to improve crop productivity and soil fertility for a sustainable environment.

References

Abdel-Fattah, G. M., & Asrar, A. W. A. (2012). Arbuscular mycorrhizal fungal application to improve growth and tolerance of wheat (*Triticum aestivum* L.) plants grown in saline soil. *Acta Physiologiae Plantarum, 34*(1), 267–277.

Abeer, H., Abd_Allah, E., Alqarawi, A., El-Didamony, G., Alwhibi, M., Egamberdieva, D., & Ahmad, P. (2014). Alleviation of the adverse impact of salinity on faba bean (*Vicia faba* L.) by arbuscular mycorrhizal fungi. *Pakistan Journal of Botany, 46*(6), 2003–2013.

Ahmed, A. M., Wais, A. H., Ditta, A., Islam, M. R., Chowdhury, M. K., Hosen, M. M., Islam, H. N., Soufan, W., Sabagh, A. E., & Islam, M. S. (2024). Seed germination and early seedling growth of sorghum (*Sorghum bicolor* L. Moench) genotypes under salinity stress. *Polish Journal of Environmental Studies, 33*(3), 3019–3032. https://doi.org/10.15244/pjoes/177180.

Ahmed, N., Li, J., Li, Y., Deng, L., Deng, L., Chachar, M., & Tu, P. (2025). Symbiotic synergy: How Arbuscular Mycorrhizal Fungi enhance nutrient uptake, stress tolerance, and soil health through molecular mechanisms and hormonal regulation. *IMA Fungus, 16*, e144989.

Ali, M. Y., Manj, I. A., Ahmed, I., Iqbal, R., Ditta, A., Kumarasamy, V., Eldin, S. M., Ali, I., Alwahibi, M. S., & Elshikh, M. S. (2024). Characterization of salt-tolerant cultivars of date palm based on morphological and biochemical responses under salinity stress. *Polish Journal of Environmental Studies, 33*(4), 4019–4029. https://doi.org/10.15244/pjoes/177179.

Alizadeh, S., Gharagoz, S. F., Pourakbar, L., Moghaddam, S. S., & Jamalomidi, M. (2021). Arbuscular mycorrhizal fungi alleviate salinity stress and alter phenolic compounds of Moldavian balm. *Rhizosphere, 19*, 100417.

Al-Karaki, G. N. (2006). Nursery inoculation of tomato with arbuscular mycorrhizal fungi and subsequent performance under irrigation with saline water. *Scientia Horticulturae, 109*(1), 1–7.

Alkharabsheh, H. M., Seleiman, M. F., Hewedy, O. A., Battaglia, M. L., Jalal, R. S., Alhammad, B. A., & Al-Doss, A. (2021). Field crop responses and management strategies to mitigate soil salinity in modern agriculture: A review. *Agronomy, 11*(11), 2299.

Amir, H., & Crossay, T. (2024). Functional and practical importance of AMF-mixed inoculants for plant development. In *Arbuscular mycorrhizal fungi in sustainable agriculture: Inoculum production and application* (pp. 319–331). Springer Nature Singapore.

Atta, K., Mondal, S., Gorai, S., Singh, A. P., Kumari, A., Ghosh, T., & Jespersen, D. (2023). Impacts of salinity stress on crop plants: Improving salt tolerance through genetic and molecular dissection. *Frontiers in Plant Science, 14*, 1241736.

Azeem, M., Pirjan, K., Qasim, M., Mahmood, A., Javed, T., Muhammad, H., & Rahimi, M. (2023). Salinity stress improves antioxidant potential by modulating physio-biochemical responses in *Moringa oleifera* Lam. *Scientific Reports, 13*(1), 2895.

Azizoglu, U., Argentel-Martínez, L., Peñuelas-Rubio, O., Meriño-Hernández, Y., Rodríguez-Yon, Y., Dell Amico-Rodriguez, J. M., González-Aguilera, J., & Shin, J. H. (2025). Arbuscular mycorrhizal fungus *Rhizophagus irregularis* enhances antioxidant system mechanisms in chickpea plants under saline conditions. *Plant Growth Regulation, 105*, 883–894.

Baltazar-Bernal, O., Spinoso-Castillo, J. L., Mancilla-Álvarez, E., & Bello-Bello, J. J. (2022). Arbuscular mycorrhizal fungi induce tolerance to salinity stress in Taro plantlets (*Colocasia esculenta* L. Schott) during acclimatization. *Plants, 11*(13), 1780.

Begum, N., Qin, C., Ahanger, M. A., Raza, S., Khan, M. I., Ashraf, M., & Zhang, L. (2019). Role of arbuscular mycorrhizal fungi in plant growth regulation: Implications in abiotic stress tolerance. *Frontiers in Plant Science, 10*, 1068.

Bhupenchandra, I., Chongtham, S. K., Devi, A. G., Dutta, P., Sahoo, M. R., Mohanty, S., … Devi, S. H. (2024). Unlocking the potential of arbuscular mycorrhizal fungi: Exploring role in plant growth promotion, nutrient uptake mechanisms, biotic stress alleviation, and sustaining agricultural production systems. *Journal of Plant Growth Regulation, 44*(12), 6802–6840.

Boorboori, M. R., & Lackóová, L. (2025). Arbuscular mycorrhizal fungi and salinity stress mitigation in plants. *Frontiers in Plant Science, 15*, 1504970.

Boyno, G., Rezaee Danesh, Y., Çevik, R., Teniz, N., Demir, S., Demirer Durak, E., & Mulet, J. M. (2025). Synergistic benefits of AMF: Development of a sustainable plant defense system. *Frontiers in Microbiology, 16*, 1551956.

Calabrese, S. (2016). *Nutrient transport in the arbuscular Mycorrhizal symbiosis: The regulation of nutrient transporters in Rhizophagus irregularis and its host plants Populus trichocarpa and Sorghum bicolor.* https://doi.org/10.5451/UNIBAS-006627849.

Campanelli, A., Ruta, C., De Mastro, G., & Morone-Fortunato, I. (2013). The role of arbuscular mycorrhizal fungi in alleviating salt stress in *Medicago sativa* L. var. icon. *Symbiosis, 59*(2), 65–76.

Cao, Z., He, C., Li, J., Yang, K., Zhang, Y., Fan, X., & Chang, W. (2025). Arbuscular mycorrhizal fungi enhance plant salt tolerance to salt stress using endogenous phytohormones. *Archives of Microbiology, 207*(12), 1–12.

Chang, W., Zhang, Y., Ping, Y., Li, K., Qi, D. D., & Song, F. Q. (2023). Label-free quantitative proteomics of arbuscular mycorrhizal *Elaeagnus angustifolia* seedlings provides insights into salt-stress tolerance mechanisms. *Frontiers in Plant Science, 13*, 1098260.

Chen, J., Zhang, H., Zhang, X., & Tang, M. (2017). Arbuscular mycorrhizal symbiosis alleviates salt stress in black locust through improved photosynthesis, water status, and K+/Na+ homeostasis. *Frontiers in Plant Science, 8*, 282416.

de Souza, T. A. F., & Santos, D. (2018). Effects of using different host plants and long-term fertilization systems on population sizes of infective arbuscular mycorrhizal fungi. *Symbiosis, 76*(2), 139–149.

De Zélicourt, A., Colcombet, J., & Hirt, H. (2016). The role of MAPK modules and ABA during abiotic stress signaling. *Trends in Plant Science, 21*(8), 677–685.

Diagne, N., Ndour, M., Djighaly, P. I., Ngom, D., Ngom, M. C. N., Ndong, G., & Cherif-Silini, H. (2020). Effect of plant growth-promoting rhizobacteria (PGPR) and arbuscular mycorrhizal fungi (AMF) on salt stress tolerance of *Casuarina obesa* (Miq.). *Frontiers in Sustainable Food Systems, 4*, 601004.

Diao, F., Dang, Z., Cui, X., Xu, J., Jia, B., Ding, S., & Guo, W. (2021). Transcriptomic analysis revealed distinctive modulations of arbuscular mycorrhizal fungi inoculation in halophyte *Suaeda salsa* under moderate salt conditions. *Environmental and Experimental Botany, 183*, 104337.

Ditta, A. (2013). Salt tolerance in cereals: Molecular mechanisms and applications. In G. R. Rout, & A. B. Das (Eds.), *Molecular stress physiology of plants* (pp. 133–154). Springer. (ISBN 978-81-322-0807-5). https://doi.org/10.1007/978-81-322-0807-5_5 .

Ditta, A., & Ullah, N. (2023). Perspectives on using plant growth-promoting rhizobacteria under salinity stress for sustainable crop production. In M. Ghorbanpour, & M. A. Shahid (Eds.), *Plant stress mitigators: Types, techniques and functions* (pp. 231–247). Woodhead Publishing Ltd., Elsevier UK. (ISBN 978-0-323-89871-3). https://doi.org/10.1016/B978-0-323-89871-3.00024-0 .

Ebbisa, A. (2022). Arbuscular mycorrhizal fungi (AMF) in optimizing nutrient bioavailability and reducing agrochemicals for maintaining sustainable agroecosystems. *Arbuscular mycorrhizal fungi in agriculture-new insights*. IntechOpen.

Ebrahim Shehata, E. A., Abdelraouf, E., Kishar, H. F. M., & Salim, H. (2025). Alleviating salinity stress on different genotypes of maize (*Zea mays* L.) using mycorrhizae fungi. *Journal of Agricultural and Environmental Sciences, 24*(1), 112–141.

Etesami, H., Jeong, B. R., & Glick, B. R. (2021). Contribution of arbuscular mycorrhizal fungi, phosphate–solubilizing bacteria, and silicon to P uptake by plants. *Frontiers in Plant Science, 12*, 699618.

Evelin, H., Devi, T. S., Gupta, S., & Kapoor, R. (2019). Mitigation of salinity stress in plants by arbuscular *Mycorrhizal symbiosis*: Current understanding and new challenges. *Frontiers in Plant Science, 10*, 470.

Evelin, H., Giri, B., & Kapoor, R. (2012). Contribution of *Glomus intraradices* inoculation to nutrient acquisition and mitigation of ionic imbalance in NaCl-stressed *Trigonella foenum-graecum*. *Mycorrhiza, 22*(3), 203–217.

Fang, L., Wang, M., Chen, X., Zhao, J., Wang, J., & Liu, J. (2023). Analysis of the AMT gene family in chili pepper and the effects of arbuscular mycorrhizal colonization on the expression patterns of CaAMT2 genes. *BMC Genomics, 24*(1), 158.

Felföldi, Z., Vidican, R., Stoian, V., Roman, I. A., Sestras, A. F., Rusu, T., & Sestras, R. E. (2022). Arbuscular mycorrhizal fungi and fertilization influence yield, growth, and root colonization of different tomato genotypes. *Plants, 11*(13), 1743.

Fileccia, V., Ruisi, P., Ingraffia, R., Giambalvo, D., Frenda, A. S., & Martinelli, F. (2017). Arbuscular mycorrhizal symbiosis mitigates the negative effects of salinity on durum wheat. *PloS One, 12*(9), e0184158.

Gill, S. S., & Tuteja, N. (2010). Reactive oxygen species and antioxidant machinery in abiotic stress tolerance in crop plants. *Plant Physiology and Biochemistry, 48*(12), 909–930.

Guo, J., Du, M., Lu, C., & Wang, B. (2020). NaCl improves reproduction by enhancing starch accumulation in the ovules of the euhalophyte *Suaeda salsa*. *BMC Plant Biology, 20*(1), 262.

Hamid, S., Ahmad, I., Akhtar, M. J., Iqbal, M. N., Shakir, M., Tahir, M., Rasool, A., Sattar, A., Khalid, M., Ditta, A., & Zhu, B. (2021). *Bacillus subtilis* Y16 and biogas slurry enhanced potassium to sodium ratio and physiology of sunflower (*Helianthus annuus* L.) to mitigate salt stress. *Environmental Science and Pollution Research, 28*, 38637–38647. https://doi.org/10.1007/s11356-021-13419-2.

Hashem, A., Alqarawi, A. A., Radhakrishnan, R., Al-Arjani, A. B. F., Aldehaish, H. A., Egamberdieva, D., & Abd_Allah, E. F. (2018). Arbuscular mycorrhizal fungi regulate the oxidative system, hormones, and ionic equilibrium to trigger salt stress tolerance in *Cucumis sativus* L. *Saudi Journal of Biological Sciences, 25*(6), 1102–1114.

Hewedy, O. A., Elsheery, N. I., Karkour, A. M., Elhamouly, N., Arafa, R. A., Mahmoud, G. A. E., & Brestic, M. (2023). Jasmonic acid regulates plant development and orchestrates stress response during tough times. *Environmental and Experimental Botany, 208*, 105260.

Hidri, R., Metoui-Ben Mahmoud, O., Debez, A., Zorrig, W., Abdelly, C., Zamarreño, A. M., & Aroca, R. (2025). Dual PGPR-AMF inoculation offsets salinity stress impact on the fodder halophyte *Sulla carnosa* by concomitantly modulating plant ABA content and leaf antioxidant response. *Journal of Plant Growth Regulation, 44*(12), 6967–6985.

Himabindu, Y., Chakradhar, T., Reddy, M. C., Kanygin, A., Redding, K. E., & Chandrasekhar, T. (2016). Salt-tolerant genes from halophytes are potential key players of salt tolerance in glycophytes. *Environmental and Experimental Botany, 124*, 39–63.

Huang, S., Gill, S., Ramzan, M., Ahmad, M. Z., Danish, S., Huang, P., & Alharbi, S. A. (2023). Uncovering the impact of AM fungi on wheat nutrient uptake, ion homeostasis, oxidative stress, and antioxidant defense under salinity stress. *Scientific Reports, 13*(1), 8249.

Igiehon, N. O., & Babalola, O. O. (2017). Biofertilizers and sustainable agriculture: Exploring arbuscular mycorrhizal fungi. *Applied Microbiology and Biotechnology, 101*(12), 4871–4881.

Incesu, M., Cimen, B., Yilmaz, B., Yesiloglu, T., & Ortas, I. (2025). Influences of arbuscular mycorrhizal fungi on *Diospyros lotus* seedlings under salinity stress. *Frontiers in Plant Science, 16*, 1595144.

Islam, M. S., Rahman, M. A., Sultana, M. S., Islam, M. M., Halder, M., & Das, D. (2026). Enhancement of Physio-Biochemical Attributes and Seed Yield of Sunflower Through Integrated Application of Silicon and Mycorrhizal Fungi Under Salt Stress. *Silicon*, 1–13.

Janah, I., Meddich, A., Elhasnaoui, A., Khayat, S., Anli, M., Boutasknit, A., & Loutfi, K. (2023). Arbuscular mycorrhizal fungi mitigate salt stress toxicity in *Stevia rebaudiana* Bertoni through the modulation of physiological and biochemical responses. *Journal of Soil Science and Plant Nutrition, 23*(1), 152–162.

Jia, T., Wang, J., Chang, W., Fan, X., Sui, X., & Song, F. (2019). Proteomics analysis of *E. angustifolia* seedlings inoculated with arbuscular mycorrhizal fungi under salt stress. *International Journal of Molecular Sciences, 20*(3), 788.

Kakabouki, I., Stavropoulos, P., Roussis, I., Mavroeidis, A., & Bilalis, D. (2023). Contribution of arbuscular mycorrhizal fungi (AMF) in improving the growth and yield performances of flax (*Linum usitatissimum* L.) to salinity stress. *Agronomy, 13*(9), 2416.

Kaya, C. (2024). Microbial modulation of hormone signaling, proteomic dynamics, and metabolomics in plant drought adaptation. *Food and Energy Security, 13*(1), e513.

Khalediyan, N., Weisany, W., & Schenk, P. M. (2021). Arbuscular mycorrhizae and rhizobacteria improve growth, nutritional status, and essential oil production in *Ocimum basilicum* and *Satureja hortensis*. *Industrial Crops and Products, 160*, 113163.

Khan, Y., Shah, S., & Tian, H. (2022). The roles of arbuscular mycorrhizal fungi in influencing plant nutrients, photosynthesis, and metabolites of cereal crops—A review. *Agronomy, 12*(9), 2191.

Kim, W.-Y., Ali, Z., Park, H. J., Park, S. J., Cha, J. Y., Perez-Hormaeche, J., Quintero, F. J., Shin, G., Kim, M. R., Qiang, Z., Ning, L., Park, H. C., Lee, S. Y., Bressan, R. A., Pardo, J. M., Bohnert, H. J., & Yun, D.-J. (2013). Release of SOS2 kinase from sequestration with GIGANTEA determines salt tolerance in Arabidopsis. *Nature Communications, 4*, 1352.

Klinsukon, C., Lumyong, S., Kuyper, T. W., & Boonlue, S. (2021). Colonization by arbuscular mycorrhizal fungi improves salinity tolerance of eucalyptus (*Eucalyptus camaldulensis*) seedlings. *Scientific Reports, 11*(1), 4362.

Kuper-Psenicnik, A., & Bennett, J. A. (2025). Intraspecies variation in mycorrhizal response of *Medicago sativa* to *Rhizophagus irregularis* under abiotic stress. *Mycorrhiza, 35*(1), 3.

Lanfranco, L., & Bonfante, P. (2023). Lessons from arbuscular mycorrhizal fungal genomes. *Current Opinion in Microbiology, 75*, 102357.

Liang, B. B., Wang, W. J., Fan, X. X., Kurakov, A. V., Liu, Y. F., Song, F. Q., & Chang, W. (2021). Arbuscular mycorrhizal fungi can ameliorate salt stress in *Elaeagnus angustifolia* by improving leaf photosynthetic function and ultrastructure. *Plant Biology, 23*, 232–241.

Lim, M. Y., Jeong, B. R., Jung, M., & Harn, C. H. (2016). Transgenic tomato plants expressing the strawberry d-galacturonic acid reductase gene display enhanced tolerance to abiotic stresses. *Plant Biotechnology Reports, 10*(2), 105–116.

Ma, J., Xie, Y., Yang, Y., Jing, C., You, X., Yang, J., & Li, Y. (2022a). AMF colonization affects the allelopathic effects of *Zea mays* L. root exudates and community structure of rhizosphere bacteria. *Frontiers in Plant Science, 13*, 1050104.

Ma, S., Zhu, L., Wang, J., Liu, X., Jia, Z., Li, C., & Zhang, J. (2022b). Arbuscular mycorrhizal fungi promote *Gleditsia sinensis* Lam. root growth under salt stress by regulating nutrient uptake and physiology. *Forests, 13*(5), 688.

Ma, Z., Zhao, X., He, A., Cao, Y., Han, Q., Lu, Y., & Huang, J. (2022c). Mycorrhizal symbiosis reprograms ion fluxes and fatty acid metabolism in wild jujube during salt stress. *Plant Physiology, 189*(4), 2481–2499.

Madawala, H.M.S.P. (2021). Arbuscular mycorrhizal fungi as biofertilizers: Current trends, challenges, and prospects. Biofertilizers (pp. 83–93). Advances in Bio-Inoculants. Elsevier. https://doi.org/10.1016/B978-0-12-821667-5.00029-4.

Mahajan, M., & Pal, P. K. (2023). Drought and salinity stress in medicinal and aromatic plants: Physiological response, adaptive mechanism, management/amelioration strategies, and an opportunity for production of bioactive compounds. *Advances in Agronomy, 182*, 221–273.

Malar, C. M., Wang, Y., Stajich, J. E., Kokkoris, V., Villeneuve-Laroche, M., Yildirir, G., & Corradi, N. (2022). Early branching arbuscular mycorrhizal fungus *Paraglomus occultum* carries a small and repeat-poor genome compared to relatives in the Glomeromycotina. *Microbial Genomics, 8*(4), 000810.

Malik, J. A., AlQarawi, A. A., Dar, B. A., Hashem, A., Alshahrani, T. S., AlZain, M. N., & Abd_Allah, E. F. (2022). Arbuscular mycorrhizal fungi isolated from highly saline "sabkha habitat" soil alleviated the NaCl-induced stress and improved *Lasiurus scindicus* Henr. growth. *Agriculture, 12*(3), 337.

Meixner, C., Ludwig-Müller, J., Miersch, O., Gresshoff, P., Staehelin, C., & Vierheilig, H. (2005). Lack of mycorrhizal autoregulation and phytohormonal changes in the supernodulating soybean mutant nts1007. *Planta, 222*(4), 709–715.

Mendoza-Soto, A. B., Rodríguez-Corral, A. Z., Bojórquez-López, A., Cervantes-Rojo, M., Castro-Martínez, C., & Lopez-Meyer, M. (2022). Arbuscular mycorrhizal symbiosis leads to differential regulation of genes and miRNAs associated with the cell wall in tomato leaves. *Biology, 11*(6), 854.

Mirzaei, J., & Moradi, M. (2017). Biodiversity of arbuscular mycorrhizal fungi in Amygdalus scoparia Spach plantations and a natural stand. *Journal of Forestry Research, 28*(6), 1209–1217.

Mohammadi, M., Modarres-Sanavy, S. A. M., Pirdashti, H., Zand, B., & Tahmasebi-Sarvestani, Z. (2019). Arbuscular mycorrhizae alleviate water deficit stress and improve antioxidant response, more than nitrogen-fixing bacteria or chemical fertilizers, in the evening primrose. *Rhizosphere, 9*, 76–89.

Nanda, B., Nandy, S., Mukherjee, A., Pandey, D. K., & Dey, A. (2021). Neoteric trends in medicinal plant-AMF association and elicited accumulation of phytochemicals. In *Recent trends in mycological research: Volume 2: Environmental and industrial perspective* (pp. 359–389). Springer International Publishing.

Naveed, M., Ditta, A., Ahmad, M., Mustafa, A., Ahmad, Z., Conde-Cid, M., Tahir, S., Shah, S. A. A., Abrar, M. M., & Fahad, S. (2021). Processed animal manure improves morpho-physiological and biochemical characteristics of *Brassica napus* L. under nickel and salinity stress. *Environmental Science and Pollution Research, 28*, 45629–45645. https://doi.org/10.1007/s11356-021-14004-3.

Naz, T., Iqbal, M. M., Ullah, I., Mubeen, M. A., & Ditta, A. (2025). Nano-silicon application enhances salt stress tolerance in *Lens culinaris* L.: A comparative study with bulk silicon. *Journal of Plant Nutrition, 48*(19), 3377–3393. https://doi.org/10.1080/01904167.2025.2513404.

Nazir, A., Sarfraz, W., Ditta, A., Khalid, N., Farid, M., Shafiq, M., Firdaus-e-Bareen, Rizvi, Z. F., & Naeem, N. (2023). Synergistic impact of two autochthonous saprobic fungi (*A. niger* and *T. pseudokoningii*) on the growth, ionic contents, and metals uptake in *Brassica juncea* L. and *Vigna radiata* L. under tannery solid waste contaminated soil. *International Journal of Phytoremediation, 25*(11), 1488–1500. https://doi.org/10.1080/15226514.2023.2166457.

Niamat, B., Naveed, M., Ahmad, Z., Yaseen, M., Ditta, A., Mustafa, A., Rafique, M., Bibi, R., & Minggang, X. (2019). Calcium-enriched animal manure alleviates the adverse effects of salt stress on the growth, physiology, and nutrient homeostasis of *Zea mays* L. *Plants, 8*(11), 480. https://doi.org/10.3390/plants8110480.

Nie, W., He, Q., Guo, H., Zhang, W., Ma, L., Li, J., & Wen, D. (2024). Arbuscular mycorrhizal fungi: Boosting crop resilience to environmental stresses. *Microorganisms, 12*(12), 2448.

Othman, Y. A., Tahat, M., Alananbeh, K. M., & Al-Ajlouni, M. (2022). Arbuscular mycorrhizal fungi inoculation improves flower yield and postharvest quality components of *Gerbera* grown under different salinity levels. *Agriculture, 12*(7), 978.

Pal, S. C., Hossain, M. B., Mallick, D., Bushra, F., Abdullah, S. R., Dash, P. K., & Das, D. (2024). Combined use of seaweed extract and arbuscular mycorrhizal fungi for alleviating salt stress in bell pepper (*Capsicum annuum* L.). *Scientia Horticulturae, 325*, 112597.

Pan, J., Peng, F., Tedeschi, A., Xue, X., Wang, T., Liao, J., & Huang, C. (2020). Do halophytes and glycophytes differ in their interactions with arbuscular mycorrhizal fungi under salt stress? A meta-analysis. *Botanical Studies, 61*(1), 13.

Park, H., Kim, K., Walitang, D. I., Sayyed, R., & Sa, T. (2024). Shifts in soil bacterial community composition of jujube orchard influenced by organic fertilizer amendment. *Journal of Microbiology and Biotechnology, 34*(12), 2539.

Parvin, S., Van Geel, M., Yeasmin, T., Verbruggen, E., & Honnay, O. (2020). Effects of single and multiple species inocula of arbuscular mycorrhizal fungi on the salinity tolerance of a Bangladeshi rice (*Oryza sativa* L.) cultivar. *Mycorrhiza, 30*, 431–444.

Peng, Z. C., Xing, Y. X., Ma, Y. H., Li, S. H., Jia, Y. X., Yang, H. C., & Zhang, F. H. (2025). Arbuscular mycorrhizal fungi enhance soybean phosphorus uptake and soil fertility under saline-alkaline stress. *Scientific Reports, 15*(1), 31792.

Perotto, S., & Balestrini, R. (2024). At the core of the endomycorrhizal symbioses: Intracellular fungal structures in orchid and arbuscular mycorrhiza. *New Phytologist, 242*(4), 1408–1416.

Pirzada, A. M., Anwar, T., Qureshi, W. A., Qureshi, H., Siddiqi, E. H., Zaman, W., & Soufan, W. (2024). Salinity stress mitigation in wheat through synergistic application of ascorbic acid, nanoparticles, and *Salvadora oleoides* extract. *Scientific Reports, 14*(1), 30687.

Pooja, P., Devi, S., Tallapragada, S., Ahlawat, Y. K., Sharma, N., Kasnia, P., & Elhindi, K. M. (2025a). Impact of arbuscular mycorrhizal symbiosis on photosynthetic, antioxidant enzyme, and water flux parameters in salt-stressed chickpea (*Cicer arietinum*) plants. *Agronomy, 15*(1), 247.

Pooja, P., Tallapragada, S., Lamba, A., & Punia, S. (2025b). Role played by arbuscular mycorrhizal fungi in amelioration of salinity stress: A review. *Plant and Soil, 511*(1), 27–42.

Porcel, R., Aroca, R., & Ruiz-Lozano, J. M. (2012). Salinity stress alleviation using arbuscular mycorrhizal fungi. A review. *Agronomy for Sustainable Development, 32*(1), 181–200.

Prasad, K. (2022). Potential impression of arbuscular mycorrhizal fungi on agricultural growth, productivity, and environment toward global sustainable development for green technology. In *Applied mycology: Entrepreneurship with fungi* (pp. 111–136). Springer International Publishing.

Qin, W., Yan, H., Zou, B., Guo, R., Ci, D., Tang, Z., & Si, T. (2021). Arbuscular mycorrhizal fungi alleviate salinity stress in peanut: Evidence from pot-grown and field experiments. *Food and Energy Security, 10*(4), e314.

Ren, C. G., Kong, C. C., & Xie, Z. H. (2018). Role of abscisic acid in strigolactone-induced salt stress tolerance in arbuscular mycorrhizal *Sesbania cannabina* seedlings. *BMC Plant Biology, 18*(1), 74.

Roychoudhury, A., & Banerjee, A. (2017). Abscisic acid signaling and involvement of mitogen-activated protein kinases and calcium-dependent protein kinases during plant abiotic stress. *Mechanism of Plant Hormone Signaling Under Stress, 1*, 197–241.

Royle, J. W. L., Farzadfar, S., Wasan, J. P., & Bennett, J. (2026). Genotype-dependent microbial community restructuring under salinity stress outweighs the direct benefits of native microbe inoculation in *Hordeum vulgare*. *Genome*. Canadian Science Publishing.

Sabagh, A. E. L., Islam, M. S., Hossain, A., Iqbal, M. A., Mubeen, M., Waleed, M., & Abdelhamid, M. T. (2022). Phytohormones as growth regulators during abiotic stress tolerance in plants. *Frontiers in Agronomy, 4*, 765068.

Saeed, H. S., Munir, M., Samra, S., Farooq, A., Mahmood, H., Ahmad, I., Akram, M. I., Iqbal, R., Elsadek, M. F., Almutairi, S. M., Ditta, A., & Daraz, U. (2025). Screening of biosurfactants producing fungi from infected *Citrus sinensis* (L.). *Polish Journal of Environmental Studies, 34*(5), 5261–5270. https://doi.org/10.15244/pjoes/190588.

Santander, C., Aroca, R., Ruiz-Lozano, J. M., Olave, J., Cartes, P., Borie, F., & Cornejo, P. (2017). Arbuscular mycorrhiza effects on plant performance under osmotic stress. *Mycorrhiza, 27*(7), 639–657.

Saxena, B., Sharma, K., Kapoor, R., Wu, Q. S., & Giri, B. (2022). Insights into the molecular aspects of salt stress tolerance in mycorrhizal plants. *World Journal of Microbiology and Biotechnology, 38*(12), 253.

Scavo, A., Abbate, C., & Mauromicale, G. (2019). Plant allelochemicals: Agronomic, nutritional, and ecological relevance in the soil system. *Plant and Soil, 442*(1), 23–48.

Sene, G., Thiao, M., Manga, A., Kane, A., Samba-Mbaye, R., Mbaye, M. S., & Sylla, S. N. (2012). Arbuscular mycorrhizal soil infectivity and spores distribution across plantations of tropical, subtropical, and exotic tree species: A case study from the forest reserve of Bandia, Senegal. *African Journal of Ecology, 50*(2), 218–232.

Shaddam, M. O., Islam, M. R., Ditta, A., Ismaan, H. N., Iqbal, M. A., Al-Ashkar, I., Sabagh, A. E., & Islam, M. S. (2024). Genotypic divergences of important mungbean varieties in response to salt stress at germination and early seedling stage. *Polish Journal of Environmental Studies, 33*(5), 5857–5868. https://doi.org/10.15244/pjoes/183567. (Impact Factor = 1.3).

Shahzad, H., Ullah, S., Iqbal, M., Bilal, H. M., Shah, G. M., Ahmad, S., Zakir, A, Ditta, A., Farooqi, M. A., & Ahmad, I. (2019). Salinity types and level-based effects on the growth, physiology, and nutrient contents of maize (*Zea mays*). *Italian Journal of Agronomy, 14*, 199–207. https://doi.org/10.4081/ija.2019.1326.

Sheng, M., Tang, M., Zhang, F., & Huang, Y. (2011). Influence of arbuscular mycorrhiza on organic solutes in maize leaves under salt stress. *Mycorrhiza, 21*(5), 423–430.

Shtark, O. Y., Puzanskiy, R. K., Avdeeva, G. S., Yurkov, A. P., Smolikova, G. N., Yemelyanov, V. V., & Shishova, M. F. (2019). Metabolic alterations in pea leaves during arbuscular mycorrhiza development. *PeerJ, 7*, e7495.

Sootahar, R. K., Sootahar, M. K., Lin, M., Rais, N., Jamro, G. M., Meher-Ul-Nissa, R., Iqbal, R., Ditta, A., Eldin, S. M., Ali, I., Alwahibi, M. S., Elshikh, M. S., & Kumarasamy, V. (2024). In vitro early vegetative growth of tomato (Solanum lycopersicum L.) cultivars under salt stress. *Polish Journal of Environmental Studies, 33*(5), 5879–5885. https://doi.org/10.15244/pjoes/183451.

Sun, D., Shang, X., Cao, H., Lee, S. J., Wang, L., Gan, Y., & Feng, S. (2024). Physio-biochemical mechanisms of arbuscular mycorrhizal fungi enhancing plant resistance to abiotic stress. *Agriculture, 14*(12), 2361.

Sun, Z., Song, J., Xin, X. A., Xie, X., & Zhao, B. (2018). Arbuscular mycorrhizal fungal 14-3-3 proteins are involved in arbuscule formation and responses to abiotic stresses during AM symbiosis. *Frontiers in Microbiology, 9*, 91.

Talaat, N. B., & Shawky, B. T. (2014). Protective effects of arbuscular mycorrhizal fungi on wheat (*Triticum aestivum* L.) plants exposed to salinity. *Environmental and Experimental Botany, 98*, 20–31. https://doi.org/10.1016/j.envexpbot.2013.10.005.

Tavares, D. S., Sant'Anna-Santos, B. F., & Gomes, M. P. (2024). Unleashing the power of fungi: Utilizing the arbuscular mycorrhizal fungi *Rhizophagus clarus* to mitigate salinity stress and boost *Cowpea bean* productivity for food security. *Stresses, 4*(2), 393–410.

Thind, S., Chaudhary, M. S., Ditta, A., Hussain, I., Parveen, A., Ullah, N., Mahmood, Q., Al-ashkar, I., & Sabagh, A. E. L. (2022). Impact of mycorrhizal fungi from different rhizospheric soils on fungal colonization, growth, and chlorophyll contents of *Cenchrus ciliaris*. *Agronomy, 12*(11), 2644. https://doi.org/10.3390/agronomy12112644.

Verma, A. K., Bhojiya, A. A., Upadhyay, S. K., Yadav, V., Singhal, K., & Abbas, K. (2023). Promising role of fungal symbiosis for eco-friendly green technology for environmental health. In *Nano-biofortification for human and environmental health* (pp. 237–266). Springer International Publishing.

Wang, F., Sun, Y., & Shi, Z. (2019). Arbuscular mycorrhiza enhances biomass production and salt tolerance of sweet sorghum. *Microorganisms, 7*(9), 289.

Wang, J. P., Munyampundu, J. P., Xu, Y. P., & Cai, X. Z. (2015). Phylogeny of plant calcium and calmodulin-dependent protein kinases (CCaMKs) and functional analyses of tomato CCaMK in disease resistance. *Frontiers in Plant Science, 6*, 1075.

Wang, Y., Dong, F., Chen, H., Xu, T., & Tang, M. (2023). Effects of arbuscular mycorrhizal fungus on sodium and chloride ion channels of Casuarina glauca under salt stress. *International Journal of Molecular Sciences, 24*(4), 3680.

Wang, Y., Zafar, N., Ali, Q., Manghwar, H., Wang, G., Yu, L., & Jin, S. (2022). CRISPR/Cas genome editing technologies for plant improvement against biotic and abiotic stresses: Advances, limitations, and future perspectives. *Cells, 11*(23), 3928.

Wani, S. H., Kumar, V., Khare, T., Guddimalli, R., Parveda, M., Solymosi, K., & Kavi Kishor, P. B. (2020). Engineering salinity tolerance in plants: Progress and prospects. *Planta, 251*(4), 76.

Wu, N., Li, Z., Wu, F., & Zhen, L. (2022). Sex-specific photosynthetic capacity and Na+ homeostasis in Populus euphratica exposed to NaCl stress and AMF inoculation. *Frontiers in Plant Science, 13*, 1066954.

Wu, Q. S., Zou, Y. N., Xia, R. X., & Wang, M. Y. (2007). Five *Glomus* species affect the water relations of *Citrus* tangerine during drought stress. *Botanical Studies, 48*(2), 147–154.

Xu, W., Liu, Q., Wang, B., Zhang, N., Qiu, R., Yuan, Y., … Cui, G. (2024). Arbuscular mycorrhizal fungi communities and promoting the growth of alfalfa in saline ecosystems of northern China. *Frontiers in Plant Science, 15*, 1438771.

Yadav, V. K., Jha, R. K., Kaushik, P., Altalayan, F. H., Al Balawi, T., & Alam, P. (2021). Traversing arbuscular mycorrhizal fungi and *Pseudomonas fluorescens* for carrot production under salinity. *Saudi Journal of Biological Sciences, 28*(8), 4217–4223.

Yang, R., Qin, Z., Wang, J., Zhang, X., Xu, S., Zhao, W., & Huang, Z. (2022). The interactions between arbuscular mycorrhizal fungi and *Trichoderma longibrachiatum* enhance maize growth and modulate root metabolome under increasing soil salinity. *Microorganisms, 10*(5), 1042.

Yizhu, L., Imtiaz, M., Ditta, A., Rizwan, M. S., Ashraf, M., Mehmood, S., Aziz, O., Mubeen, F., Ali, M., Elahi, N. N., Ijaz, R., Lelel, S., Shuang, C., & Tu, S. (2020). Response of growth, antioxidant enzymes, and root exudates production towards as stress in *Pteris vittata* and *Astragalus sinicus* colonized by arbuscular mycorrhizal fungi. *Environmental Science and Pollution Research, 27*, 2340–2352. https://doi.org/10.1007/s11356-019-06785-5.

Zhang, D. J., Tong, C. L., Wang, Q. S., & Bie, S. (2024). Mycorrhizas affect physiological performance, antioxidant system, photosynthesis, endogenous hormones, and water content in cotton under salt stress. *Plants, 13*(6), 805.

Zhang, M., Shi, Z., Lu, S., & Wang, F. (2023). AMF inoculation alleviates molybdenum toxicity to maize by protecting leaf performance. *Journal of Fungi, 9*(4), 479.

Zhang, Q., Ma, S., Wang, R., Li, L., Zhang, Q., Ju, M., & Gu, P. (2025). Diversity and drivers of arbuscular mycorrhizal fungi in the rhizosphere soil of wine grape in the eastern foot of Helan Mountain in Ningxia, China. *Frontiers in Microbiology, 16*, 1700411.

Zhang, Y., Hou, K., Qian, H., Gao, Y., Fang, Y., Xiao, S., & Ren, W. (2022). Characterization of soil salinization and its driving factors in a typical irrigation area of Northwest China. *Science of the Total Environment, 837*, 155808.

Zong, J., Zhang, Z., Huang, P., & Yang, Y. (2023). Arbuscular mycorrhizal fungi alleviate salt stress in *Xanthoceras sorbifolium* through improved osmotic tolerance, antioxidant activity, and photosynthesis. *Frontiers in Microbiology, 14*, 1138771.

Chapter 14

Arbuscular mycorrhizal fungi and temperature stress tolerance in plants

Sana Ashraf[a], Meshal Saeed[a], Muhammad Kashif Irshad[b,c], Sajid Rashid Ahmad[a], Anam Tariq[a], Amna Maheen[a], Eshal Zainab[a], Sobia Ashraf[b] and Umer Farooq[d]

[a] *College of Earth and Environmental Sciences, University of the Punjab, Lahore, Pakistan,* [b] *Department of Environmental Sciences, Government College University, Faisalabad, Pakistan,* [c] *Department of Environmental and Energy Engineering, Yonsei University, Wonju, South Korea,* [d] *Department of Food Engineering, University of Agriculture Faisalabad, Faisalabad, Pakistan*

14.1 Introduction

The relationship between plants and microbes is a complex and ongoing process since the evolution of plants on Earth. This interaction has resulted in a community of host and nonhost species over millions of years, forming a special ecological unit known as a "holobiont." Plants in both the natural and agricultural settings are frequently exposed to spams by both useful and harmful microbes, primarily bacteria and fungi (Wille et al., 2019). It can be useful to understand the relationships between plants and microbes to reduce root disease resistance breeding in grain legumes. The interactions have positive effects on plant health, growth, defense mechanisms, and ecosystem functioning, which have a significant impact on agricultural productivity and resilience (Tharanath et al., 2024). Among such relationships, positive interactions with microbes colonizing roots are critical to sustainable farming operations (Singh et al., 2020).

The mechanisms established through the beneficial relationships are nutrient transfer in the form of mycorrhizal fungi and rhizobia that settle on roots, which provide plants with essential mineral nutrients and fixed nitrogen. They also directly enhance growth by stimulating phytohormones, resistance against pathogenic microorganisms, and alleviating stress (Liu et al., 2020; Nazir et al., 2023; Saeed et al., 2025; Thind et al., 2022). On the other hand, detrimental interactions are characterized by the development of aggressive microbes into saprophytic ones that lead to the destruction of the plants. Therefore it is important to learn how plants interact with microbes to identify the beneficial and adverse impacts of microbes on plants (Dolatabadian, 2021).

It is suggested that the colonization of land plants is impossible without mycorrhizal associations (Brundrett et al., 2018). Mycorrhiza (meaning fungus-root) is a term that was originally described in the late 19th century by A. B. Frank to describe the type of symbiotic found in trees (Strullu-Derrien et al., 2018b). It has been demonstrated that this symbiosis has deep evolutionary origins, which makes it one of the oldest mutualisms on the planet (Corradi et al., 2012). This is strongly supported by the fact that the existence of fungal structures highly comparable to the modern arbuscular mycorrhizal fungi (AMFs) has been discovered in fossil rhizomes and stems of the early terrestrial vegetation in the Rhynie chert (lower Devonian period, approximately 400–410 million years ago) (Strullu-Derrien et al., 2018b). This finding suggests that AMF symbiosis was a crucial factor that allowed early nonvascular plants to thrive in nutrient-poor and rocky soils (Brundrett et al., 2018).

The genetic and structural conservation of these symbiotic processes among different plant lineages, from basic forms to flowering plants, highlights their fundamental importance (Corradi et al., 2012). Molecular studies show that plants use a largely conserved genetic framework called the common symbiosis signaling pathway (CSSP) to recognize and establish symbiosis with both AMF and nitrogen-fixing rhizobia. This implies that these important advantageous root interactions have a common and ancient evolutionary origin (Strullu-Derrien et al., 2018a). Plants form symbiotic relationships with

Advancing Sustainable Agriculture Through Arbuscular Mycorrhizal Fungi. DOI: https://doi.org/10.1016/B978-0-443-45130-0.00012-6

mutualistic fungi, such as AMF and bacteria, for the exchange of vital nutrients and better adaptation (Wang et al., 2021). AMF help modulate the rhizosphere microbiota to support the symbiosis with legumes.

The symbiotic nitrogen fixation (SNF) associated with legumes provides the main source of biologically fixed nitrogen for agricultural systems. This process involves a group of bacteria typically referred to as rhizobia, characterized by host specificity. Differences in symbiosis between rhizobial strains and legume genotypes influence infection, root nodule development, and nitrogen fixation efficiency. The interaction between rhizobia and legumes is mediated by a lipochitin oligosaccharide produced by rhizobia, known as the "Nod factor" (Drevon et al., 2015).

The most important source of biologically fixed nitrogen to agricultural systems is the SNF linked to legumes. This is done by a group of bacteria, which are commonly known as rhizobia and are host-specific. Symbiosis varied among rhizobial strains, and the genotype of legumes affects infection, the development of root nodules, and the efficiency of nitrogen fixation. The rhizobia–legumes interaction depends on a lipochitin oligosaccharide secreted by the rhizobia, the so-called Nod factor (Drevon et al., 2015)

Phylum Glomeromycota is a group that is defined by its dependence on host plants as fixed sourcesof carbon required during its life cycle and reproduction (Jha et al., 2022). AMF establishes a positive relationship with plants, with both sides benefiting. Fungi increase the plant uptake of water and nutrients, especially phosphorus, and plants deliver approximately 10%–20% of their photosynthates to the fungi (Lugo et al., 2023). AMFs are usually linked to approximately 85% of plant families, and they have a major rolein the sustainability of the ecosystem. Plant growth and productivity are adversely affected by different biotic and abiotic stresses. AMFs improve tolerance of plants to stressors such as drought, salinity, high temperatures, and heavy metals (HMs) (Khaliq et al., 2022). Some bacterial strains are also of significance because of their metabolism and functional peculiarities. These free-living plant growth–promoting rhizobacteria (PGPR) enhance plant growth by colonizing the root and have a long history of study regarding the beneficial influences of these microorganisms on plant growth. PGPR benefits growth by several mechanisms, which include producing substances that are vital to plants and aiding in the uptake of nutrients, as well as resisting diseases.

The fossil record of arbuscular mycorrhizas (AMs) in the lower Devonian Rhynie chert is vital in the explanation of the evolution of mycorrhizal systems. Morphological, molecular, and genetic data are included in research on the evolutionary history of AMFs (Harper et al., 2020). The existence of a good fossil record has greatly enhanced our knowledge of these organisms in the ancient ecosystems (Harper et al., 2020). It is specifically the early Devonian Rhynie chert of Aberdeenshire, Scotland, which was formed in an ancient paleo-ecosystem that was fossilized with silica-rich water of hot springs (Harper et al., 2020), that is particularly interesting. It holds the superbly preserved fungi of glomeromycotan arbuscular fungi in ancient land plants (Taylor et al., 2005), which are valuable fossil evidence of the early evolutionary history of the mycorrhizal system. The existence of AM relationships, which are found in most terrestrial plants, dates to the early Devonian, or more than 400 million years ago, when terrestrial plants started to colonize the land. The research of mycorrhizal relationships is marked by several milestones; one of them was the description of the host epidermis and pellicular wall structure by Hartig. The two concepts were coined by Frank, who introduced the terms "symbiotismus" and "mycorrhiza," and the different morphologies of root and fungal structures were defined.

Agroecosystems are affected by the changing global climate, and hence, the global sustainable development goals are difficult to attain. They are the most widespread phylum of symbionts, with more than 80% of the plant family being affected, probably because of global climatic change. These fungi help plants to take up minerals, thereby supporting their growth and safeguarding them against biotic and abiotic stress. Drought is a significant menace to food security and crop production in the world (Tang et al., 2022). Dryland soil occupies nearly 40% of the total land area in the world because the soil is characterized by low levels of rainfall, intense levels of sun radiation, and changes in temperature.

AMFs assist plants in the ability to endure different conditions of climate-induced stress (Abdela et al., 2021). AMFs are linked to the roots of 70%–90% of plant species in the soil (de Vries et al., 2020), which are beneficial to ecosystems because they mediate water and nutrient cycles (Guo et al., 2019). The processes play a significant role in the biogeochemical cycles of carbon (C), nitrogen (N), and phosphorus (P) (van der Heijden et al., 2015). At low phosphorus concentrations of soil, roots and AMF carry more carbon and phosphorus with each other (Alguacil et al., 2010), leading to 5%–10% of photosynthates of host plants being transported to AMF interactions (Kaur & Suseela, 2020). Under drought stress the AMFs can effectively absorb and deliver nutrients to their hosts through the fine threads (hyphae) (Smith et al., 2011). AMFs also increase the tolerance of their hosts to both biotic and abiotic stresses, particularly when under dry conditions (Heflish et al., 2021). AMF roots also extend the root systems of their host plants, which access the soil water, nutrients, and carbon (Frew et al., 2022). AMF enhances the growth of plants by enhancing nutrient absorption, especially P, and adversity to abiotic stresses such as drought, salinity, and toxic HMs. This has been made possible by the arbuscular mycorrhizal interface through which fungi and plants associate mutually to exchange nutrients, signaling compounds, and protective chemicals

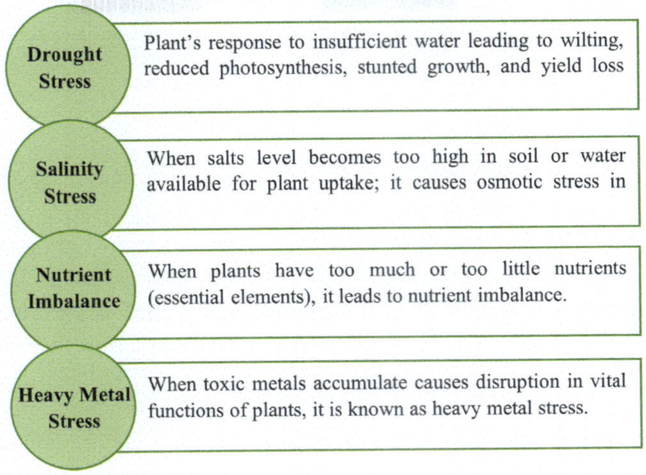

FIGURE 14.1 Influence of abiotic stressors on plant growth.

(Mwampashi et al., 2024). The responses boost plant performance, photosynthetic efficiency, and production of biomass under stress conditions.

The above functions of AMF would render them the crucial natural variables in counteracting the effects of global climate change and environmental degradation (Das & Sarkar, 2024). Their main and most well-known purpose is the enhancement of the phosphorus uptake since the vast system of hyphae enlarges the surface of the roots, enabling the plants to gain access to the phosphorus pools, which otherwise would be inaccessible (Tang et al., 2022). This role is becoming more significant because environmental pressures could contribute to the tightening of resources (Tobin et al., 2022).

AMF improves the soil structure, leading to improved aeration, water infiltration, and reducing the risk of erosion, directly responding to the problematic issues of degraded soils and drought-prone regions (Mwampashi et al., 2024). Moreover, AMF boosts the level of tolerance of plants against various abiotic stresses (Das & Sarkar, 2024). They enhance water use efficiency and protect plants against drought stress through a rise in hydraulic conductivity and drought adaptations and antioxidant mediation (Tang et al., 2022). Moreover, they increase salinity and temperature stress resistance by affecting physiological and genetic reactions of the host (Jian et al., 2024).

14.2 Arbuscular mycorrhizal fungi and abiotic stressors

Abiotic stressors are numerous, such as drought, salinity, temperature, HM, or toxin contamination, as well as nutrient deficiency and imbalance, as demonstrated in Fig. 14.1. All of these factors lead to the stunted growth of plants. AMF is significant in relieving these stressors and upgrading growth and production in plants through symbiotic association.

14.2.1 Drought stress

Inadequate supply of water to roots reduces the rate of transpiration and results in oxidative stress (Saleem et al., 2024, 2025; Wasaya et al., 2024). Drought stress affects the growth of plants adversely because it changes the activity of enzymes, their uptake of ions, and nutrients (Ahmad et al., 2023; Raza et al., 2023). Despite this, AMF has been demonstrated to alleviate drought stress in numerous crops such as wheat, barley, maize, soybean, strawberries, and onions considerably. Much of the soil penetrated by roots and the extraradical hyphae of fungi could be the primary reason why plants become drought-resistant (Begum et al., 2019).

14.2.2 Salinity stress

The salinization of the soil as an abiotic stressor poses a significant threat to the food security of the global system. It is a fact that salinity stress suppresses plant growth in terms of vegetative development and net assimilation rate, which reduces the yield productivity (Ahmed et al., 2024; Ali et al., 2024; Ditta & Ullah, 2023; Naveed et al., 2021). In addition, it promotes excessive generation of reactive oxygen species (ROS). There is research on potential options to improve the productivity of crops in salt-affected soils. One of the potential strategies is the wise adoption of AMF to reduce the adverse impacts of salinity on plants. In several studies, AMF effectiveness in enhancing plant development and productivity during salinity stress has been reported (Begum et al., 2019).

14.2.3 Heavy metal stress

The long-term effects of HMs on plants include their occurrence in soil after penetration (Ahmad et al., 2025; Ditta et al., 2024). The importance of AMF in the ecosystem is immense since it has the potential to significantly enhance the efficiency of the ecosystem by providing fungal structures, such as arbuscules, and an exchange of minerals and inorganic compounds essential in the growth of plants to render them extremely powerful. They also reduce the level of HMs; that is, they serve as a biological filter. They contribute to the active growth of the plant in stressful conditions by enhancing water uptake, photosynthetic rate, and nutrition. HMs normally get immobilized in fungal hemiparasitic hyphae that do not harm the plants (Dhalaria et al., 2020).

14.2.4 Nutrient imbalance

Soils with low fertility, which do not contain nutrients, limit agricultural production, as crops will not be provided with the appropriate environment to grow. Plants depend on their root systems to absorb the water and other nutrients they need to survive in the natural environment and increase their yield and nutritional content in agriculture. Thus one of the strategies to increase the efficiency of mineral resources use in agriculture is the application of microorganisms that contribute to plant growth and form symbiotic relations with plant roots. Mycorrhizas are a symbiosis between soil fungi and the roots of vascular plants, which serve the plant in several ways, such as translocation and absorption of nutrients, especially of phosphate and nitrogen. The positive outcomes of mycorrhizal symbiosis are contingent on the interaction of the plant fungus and the environmental factors, including the supply of phosphorus and carbon to the symbiont (Bertolazi et al., 2018).

14.3 Arbuscular mycorrhizal fungi and biotic stress tolerance

Induced systemic resistance and priming mechanisms: AMF's systemic enhancement of resistance against aboveground and belowground pathogens and insect herbivores is known as "mycorrhiza-induced systemic resistance" (MISR).

14.3.1 Priming mechanisms

In this the priming process is the most well-known of the several mechanisms that have been shown to underlie MISR. Specifically the plant hormones jasmonic acid (JA), salicylic acid (SA), and ethylene (ET) enable mycorrhizal plants to react more quickly and powerfully when confronted by later-arriving herbivores and diseases (Qu et al., 2021).

14.3.2 Defense signaling pathway

Furthermore, many tiny chemicals, including ET, SA, and JA, can regulate defense responses by coordinating the expression of defense-related genes. According to many studies the rapid activation of the JA-dependent signaling pathway (a key regulator in plant defense against necrotrophs and chewing insects) is what makes induced systemic resistance successful. However, on the other hand, plant defense-related reactions to (hemi)biotrophic pathogens are regulated by the SA-dependent pathway. Thus the SA and JA pathways are hostile to one another (Wang et al., 2022).

14.3.3 Synergy with beneficial microbes against pathogens

AMFs work along with other beneficial microorganisms, such as plant growth-promoting bacteria and fungi, to promote plant growth and root development. Mycorrhizal helper bacteria assist these relationships by increasing fungal colonization, nutrient exchange, and hyphal development, as well as regulating plant hormone levels and phosphate intake. For example, coinoculation of AMF with the bacterial strain SMF006 increased the number of beneficial bacterial genera (*Sphingomonas* and *Taibaiella*) and fungal genera (*Trichoderma, Penicillium,* and *Talaromyces*), all of which are known for their antipathogen action (Rotoni et al., 2025).

TABLE 14.1 AMF symbiotic relationship with host plant under low temperature stress.

Mycorrhizal species	Host	Reaction after inoculation	Sources
Glomus etunicatum	*Zea mays* L.	Plant fresh shoot and dry root weight were enhanced by 26.85% and 95.45%, respectively	Zhu et al. (2010b)
Glomus versiforme	*Tectona grandis* L.F.	SOD enzyme activity enhanced by 10.6%	Zhu et al. (2010b)
Funneliformis mosseae	*Cucumis sativus* L.	The H^+-ATPase activity was increased. The ratio of root-shoot, dry mass of root, and fresh mass was enhanced by 26.06%, 31.03%, and 51.33%	Chen et al. (2013), Liu et al. (2014)
G. etunicatum	*Citrus reticulata* Blanco	Gs, Pn, and Tr were enhanced by 46.94%, 22.59%, and 52.08%	Xian et al. (2010)
Clariodeoglomus etunicatum	*Solanum melongena* L.	Dry root and shoot weight were enhanced by 53.85% and 93.62%	Yano et al. (2003)
C. etunicatum	*Camellia sinensis* (L.) O. Kuntze	Caffeine content enhanced by 89.8%	Lei et al. (2024)

AMF, arbuscular mycorrhizal fungi; SOD, superoxide dismutase.

14.3.4 Molecular and physiological mechanisms

More than 80% of terrestrial plants have symbiotic relationships with AMFs. In these interactions, AMF secretes lipochitooligosaccharides and short-chain chitin oligomers to activate the CSSP, which causes calcium spiking, kinase cascades, and transcriptional regulation for arbuscule formation, whereas plants release strigolactones to promote fungal spore germination. Physiologically the AMF colonization impacts the production of secondary metabolites, nutrient distribution, and root architecture, which enhances plant growth and adaptation to biotic stress factors. Reportedly, AMF increases the amount of nutrition consumed, as stated by omics research (transcriptomic, proteomic, and metabolomic), and promotes the concentration of defense-related chemicals, which enhance growth and immunity. These molecular and physiological processes are influenced by environmental factors such as the phosphate content of soil, availability of light, and the developmental stage of plants and influence colonization, transporter expression, and symbiotic efficiency (Kaur & Suseela, 2020).

14.4 Role of arbuscular mycorrhizal fungi on plant growth during temperature stress

14.4.1 Arbuscular mycorrhizal fungi–mediated plant growth during temperature stress

Generally the optimum temperature for higher plant growth is between 20°C and 30°C. If the temperature is below 0°C, plants will not be able to grow. The plant growth rate increases between 0°C and 20°C. If the temperature is above 30°C, the plant growth rate declines or even ends. This shows that plant growth and development can be severely impacted by the factor of temperature stress (Ditta, 2018; Gavito et al., 2005). It has been noted that the growth of fruits, seeds, flowers, roots, leaves, stems, and other organs is impacted by the factor of temperature stress. After AMF inoculation, various plant organs show resilience to the effects of temperature stress.

As a result, AMF inoculation can somewhat guarantee the regular growth and development of the organs of the host plant, as illustrated in Table 14.1. Effects of temperature stress are overgrowth or softening of the stem, declined root growth, and leaf wilting. For instance, in buckwheat, high-temperature stress causes enhanced membrane lipid peroxidation, significant root inhibition, and deteriorated membrane integrity (Tian et al., 2008). High-temperature stress shows negative effects on flowering period, internode length, and plant height in chrysanthemums (Khattak & Pearson, 2006). Joe et al. (1998) reported that at the temperature of 30°C, leaves of young *Chrysanthemum* become fragile and narrow. Effects of low-temperature stress are a declined growth rate, reduced surface area, length, and root dry weight (Luo et al., 2014). In cassava, low temperature softens and bends stems (An et al., 2013).

Low-temperature stress in tomato causes wilting and yellowing (Bloom et al., 2004). AMF-inoculated tomato plants have greater leaf areas, wider crowns, enhanced roots and leaves, and larger leaf relative water content (Abdel Latef et al., 2011; Li et al., 2008), as depicted in Table 14.2. In a condition when low-temperature stress is applied, as compared to the control group, the plants inoculated with AMF show greater surface area and length of root (Zhu et al., 2010b). Under low-temperature stress, AMF inoculation increased the fresh and dry bulk of cucumber roots and shoots (Chen et al., 2013).

TABLE 14.2 AMF symbiotic relationship with host plant under high-temperature stress.

Mycorrhizal species	Host	Reaction after inoculation	Sources
Funneliformis Geosporum, *Funneliformis mosseae*, and *Rhizophagus intraradices*	*Triticum aestivum* L.	Chlorophyll contents were enhanced by 26%	Paradis et al. (1995)
Glomus versiforme and *F. mosseae*	*Lilium brownii* var. *viridulum* Baker	The soluble protein and proline content were enhanced by 15% and 14%	Xing et al. (2018)
Glomus etunicatum; F. geosporum	*Zea mays* L.	Tr, Gs, and Pn were enhanced by 11.91%, 37.65%, and 31.47%; Ci declined by 92.01%; CAT and root SOD activities were enhanced by 79% and 30% Pn	Mathur et al. (2018), Zhu et al. (2010b)
F. mosseae	*Lactuca sativa* var. *ramosa* Hort.	Tr and Pn were enhanced by 5.9% and 7.7%	Yan et al. (2021)
Rhizophagus irregularis, F. mosseae, F. geosporum	*Glycine max* (L.) *Merr.*	Tr and Pn were enhanced by 27% and 32%	Smith et al. (2008)

AMF, arbuscular mycorrhizal fungi; *CAT*, catalase; *SOD*, superoxide dismutase.

Furthermore, compared to plants that were not inoculated, green ash plants that received an AMF inoculation had noticeably longer leaves (Andersen et al., 1987). According to research, AMF is efficient in helping the host plant take up nutrients more efficiently. AMF aids in the accumulation of carbohydrates in eggplants under low-temperature stress (Pasbani et al., 2020).

14.5 Arbuscular mycorrhizal fungi–mediated physiological and biochemical response to temperature stress

14.5.1 Arbuscular mycorrhizal fungi regulation of plant antioxidant systems during temperature stress

During the temperature stress, instability between synthesis and removal of ROS occurs, which causes oxidation of the plant cell membrane, that is, membrane lipids and membrane proteins. All this leads to cell damage (Hossain et al., 2015). The antioxidant system of plants is stimulated to mitigate this. Some enzymes function to combat the deteriorating impacts of factor temperature stress, such as ascorbate peroxidase (APX), catalase (CAT), peroxidase (POD), and superoxide dismutase (SOD) (Buttery et al., 2024). It was found by Zou et al. (2017) that in plant cells, to prevent damage to the cell, all three enzymes, such as POD, CAT, and SOD, work together. The enzyme starts exhibiting a rapid response and greater susceptibility to damage. It was stated that during the high-temperature stress, the activities of POD and SOD were enhanced (Jiang et al., 2019). He confirmed it through the recorded physiological variables of wild lilies. These enzymes reduce the deteriorating effect of high temperature by increasing the ability of the plant to eliminate harmful substances. By this process, they enhance the heat resistance of the plant.

The effects of long-term exposure to low-temperature stress are crystallization in plants. It causes a malfunction in the cell membrane that causes lipid peroxidation. Enzyme activity is affected by temperature. At the start, enzyme activity gets escalated at a lower temperature range, and after that it declines (Zhang et al., 2020; Baek et al., 2003). The antioxidant stress sustains some activity levels to eliminate extra ROS. It maintains standard physiological activities during a plant experiencing temperature stress (Zhu et al., 2010b). The influence of temperature stress on the plant oxidation system is illustrated in Fig. 14.2.

It was found that during high-temperature stress, AM maize plants show increased CAT and SOD activities as compared to non-AM plants. In the activities of enzymes POD and APX in AMF-inoculated lavender, enhanced improvement was observed by Xing et al. during low-temperature stress. It was stated by Abdel Latef et al. (2011) that when low-temperature stress is applied, the antioxidant enzyme activity increases in AM tomato more than in non-AM tomato. During low-temperature stress, in teak seedlings, SOD and POD activity increased by AM colonization (Zhou et al., 2012).

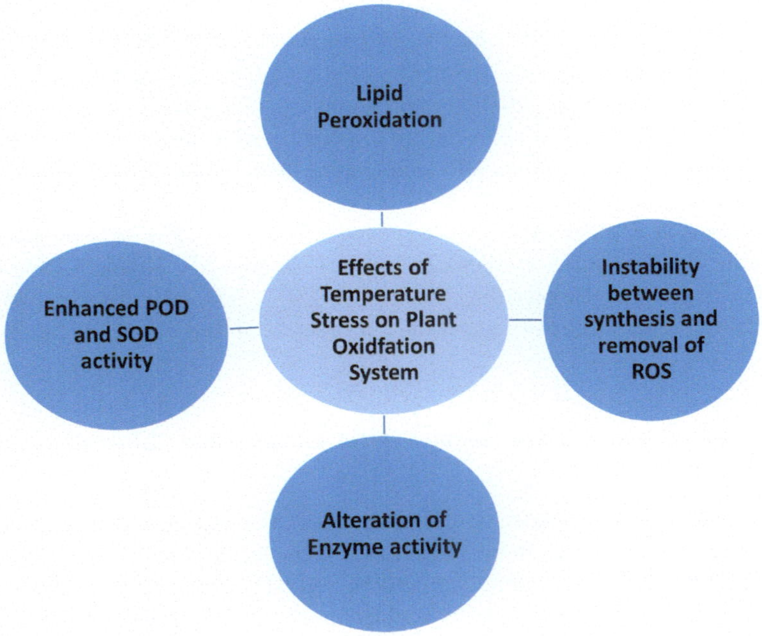

FIGURE 14.2 Influence of temperature stress on plant oxidation system. *POD*, Peroxidase; *ROS*, reactive oxygen species; *SOD*, superoxide dismutase.

14.5.2 Arbuscular mycorrhizal fungi regulation of plant osmotic balance during temperature stress

When the temperature stress is applied to plants, their osmotic control is impacted as oxidation adversely impacts the cell membrane. To preserve intracellular stability, plants store many osmoregulatory compounds, such as proline, soluble sugar, and soluble protein (El-Yazied et al., 2016). It was illustrated by Hong et al. (2023) that *Paeonia ostii's* soluble sugar and proline levels rose during high-temperature stress. It has been demonstrated by Zhang et al. (2011) that rhodendrons increase soluble protein levels under high-temperature stress. Under low-temperature stress, ice crystals may grow in a plant's water. In plants, cold resistance is improved by the help of osmoregulatory compounds that lower the freezing point and stabilize osmotic pressure (Li et al., 2013). For example, when low-temperature stress is applied, alfalfa escalates the concentration of proline and soluble sugar (Haagenson et al., 2003). Similarly, free proline is increased by tobacco, which increases cold tolerance (Bornman et al., 2010). It was discovered that if low-temperature stress is applied to tomatoes, their proline content increases (Chae et al., 2021). In many studies, it has been demonstrated that plants' capability to survive cold is enhanced by increasing soluble sugar, soluble protein, and proline. When high-temperature stress is applied in AMF-inoculated lilies, the concentration of soluble protein and proline increases (Xing et al., 2018). It has been indicated that in mycorrhizal plants during low-temperature stress, soluble sugar and root proline increase in comparison to the control group (Zhu et al., 2010a). It has been stated in a study on sorghum that AMF can decrease damage caused by temperature stress by improving root vitality, raising proline and soluble sugar levels (Ali et al., 2009). Moreover, during low-temperature stress after AMF inoculation, soluble sugar concentration increases (Zhu et al., 2015).

14.6 Arbuscular mycorrhizal fungi regulations of plant photosynthesis during temperature stress

14.6.1 Regulation of plant photosynthetic characteristics during temperature stress

Photosynthesis is very important for plants for their general health, as it helps to metabolize energy and material. The stress resistance in plants is greatly impacted by photosynthesis. The factor of temperature stress influences photosynthesis in two main ways. The first way is that stomatal variables cause stomatal closure that limits CO_2 absorption by minimizing transpiration rate and stomatal conductance. The second way is that nonstomatal substances directly inhibit the photosynthetic mesophyll cells, and they interfere with the photosynthetic process and raise intercellular CO_2 concentration. The temperature stress decreases transpiration rate, stomatal conductance, and net photosynthetic rate while increasing concentration.

According to research, in Nerium oleander, during high-temperature stress, stomatal variables are the main factor influencing Pn (Berry et al., 2003). The reason for wheat's and corn's declined photosynthetic efficiency is that nonstomatal variables were considered main causes (Xu et al., 1995; Khatib et al., 1999). It was demonstrated by research on grapes (Hendrickson et al., 2010), maize (Zhu et al., 2010a), and cucumbers (Chugunova et al., 1980) that the primary cause of decreased photosynthesis is nonstomatal variables during low-temperature stress. At low temperatures, photosynthetic enzymes and mesophyll cells are less active. When low-temperature stress is applied, the concentration of Pn increases in AMF-inoculated maize (Mathur et al., 2018).

When low-temperature stress is applied, AMFs raise photosynthetic rate (Pn), transpiration rate (Tr), and stomatal conductance (Gs) while reducing concentration in maize leaves (Xiang et al., 2010). AMFs, during low temperature, positively impact the Pn, Gs, and Tr of citrus seedlings (Wu et al., 2010). The cucumber seedlings inoculated with AMF indicated better seedling quality, resilience to low temperature, less stomatal restriction, and exhibited greater Pn (Ma et al., 2018).

14.6.2 Arbuscular mycorrhizal fungi–mediated regulation of plant chlorophyll during temperature stress

In plants, photosynthesis takes place in chloroplasts, which have higher chlorophyll content. The chlorophyll content directly influences the photosynthetic process of a plant (Buttery et al., 1977). The factor of temperature stress causes the chloroplasts in plants to become unstable; it also stops the synthesis of chlorophyll. The factor of high temperatures badly impacts the winter wheat cell membranes; it can rupture the integrity of the chloroplast. It has major effects on photosynthesis, lowers the concentration of chlorophyll, and causes expansion and disintegration (Ristic et al., 2007). Similarly, low-temperature stress diminishes the synthesis of chlorophyll. For instance, coffee (de Oliveira et al., 2009) and betel nuts (Kumar et al., 2006) have much less chlorophyll during low-temperature stress in comparison to what they do at normal temperature. According to studies, AMF can guarantee the preservation of the stability of plant chloroplasts and proper chlorophyll (Ma et al., 2020; Zhu et al., 2010b).

AMF-inoculated maize had more chlorophyll than uninoculated plants when high-temperature stress was applied, according to Mathur et al. (2018). Similarly, wheat inoculated with AMF (Paradis et al., 2010) and tomatoes had higher chlorophyll levels than uninoculated plants during low-temperature stress (Caradonia et al., 2019). Consequently, in plants, when temperature stress is applied, AMF can increase the amount of chlorophyll.

In plants, temperature stress triggers protein synthesis and gene transcription. Plants detect the high-temperature signal when the outside temperature rises and produce more heat shock proteins (HSPs). Plants are protected against stress by HSPs, which are broadly dispersed in the cytoplasm, endoplasmic reticulum, mitochondria, and chloroplasts. HSPs participate in the synthesis and renaturation of new proteins in cells under high-temperature stress (Reddy et al., 2010). HSP expression is triggered by high-temperature stress; this process makes plants more resilient to high temperatures (Sailaja et al., 2015; Scharf et al., 2012). When the outside temperature drops, plants experience low-temperature stress, which modifies several genes and initiates the production of new proteins. Low temperatures have been discovered to quickly stimulate CBF1-3, which causes the production of genes regulated by C-repeat Binding Factors (CBF), and it increases the tolerance of plants to frost (Zha et al., 2020). By raising gene expression levels, AMF can improve plants' resistance to temperature stress. According to Tian et al. (2023), AMF can increase the expression of CsPIPs and CsHsp70s at elevated temperatures (Xiao et al., 2023). Moreover, AMF can control the activity of aquaporins and the expression of associated aquaporin genes during temperature stress, which increases the transport of water and decreases the effects of temperature stress in the host plant (Zheng et al., 2011). In a prior study, it was found that in AMF-inoculated cucumbers, the transcription level of some genes increased, which improves the absorption of phosphorus during temperature stress (Buttery et al., 2024; Ma et al., 2015). In addition, it was found that AMF-inoculated rice during cold stress had increased expression levels of various genes, which led to an increase in trehalose concentration and improved cold tolerance in rice (Liu et al., 2014).

14.7 Conclusion

The factor of temperature stress causes a major limitation to plant growth, productivity, and physiological stability in both cold and hot extremes. This chapter highlights the process by which AMFs mitigate the negative effects when temperature stress is applied and preserve functional performance in thermally demanding conditions. AMF enhances plant resilience through a symbiotic relationship with plant roots by coordinating physiological balance, photosynthetic efficiency, growth regulation, and molecular defense responses. AMF are important for sustaining plant growth during temperature stress for the reason they enhance the nutrient uptake in plants, especially phosphorus, the ability to absorb water, and the development of the root system. When the temperature stress interferes with the plant's normal metabolic activity, AMF allows the plant to grow by enhancing the efficiency of resource use. AMF has a major impact on plant stress responses at biochemical and physiological levels by controlling osmotic and oxidative systems. AMF enhances antioxidant defense mechanisms by reducing temperature-induced oxidative damage to macromolecules and cellular membranes and lowering the buildup of ROS. AMF simultaneously helps in osmotic adjustment by regulating ion homeostasis and compatible solute accumulation. AMF symbiosis is essential for preserving photosynthetic function, and photosynthesis is especially vulnerable to temperature stress. To clarify the mechanisms underlying AMF-mediated thermotolerance and to facilitate the successful implementation of AMF in climate-resilient and sustainable agricultural systems, future studies combining physiological, biochemical, and multiomics approaches will be crucial.

List of abbreviations

CSSP	Common symbiosis signaling pathway
SNF	Symbiotic nitrogen fixation
PGPR	Plant growth–promoting rhizobacteria
C	Carbon
N	Nitrogen
P	Phosphorus
WUE	Water use efficiency
HMs	Heavy metals
MISR	Mycorrhiza-induced systemic resistance
JA	Jasmonic acid
SA	Salicylic acid
ET	Ethylene
PGPB	Plant growth–promoting bacteria
MHB	Mycorrhizal helper bacteria
LCOs	Lipochitooligosaccharides
SLs	Strigolactones
ROS	Reactive oxygen species
APX	Ascorbate peroxidase
CAT	Catalase
POD	Peroxidase
SOD	Superoxide dismutase
Pn	Phosphorus mononitride
HSPs	Heat shock proteins
Aps	Activity of aquaporins

References

Ahmad, S., Zulfiqar, T., Yang, H., & Farooq, M. (2025). Strigolactones mitigate drought stress in maize by improving chloroplast protection stomatal function and antioxidant defense mechanisms. *Annals of Agricultural Sciences*, 70(1), 100389.

Abdela, A. A., Barea, J. M., & Aroca, R. (2021). Arbuscular mycorrhizal fungi and plant tolerance to drought stress. *Frontiers in Plant Science, 12*, 648829. https://doi.org/10.3389/fpls.2021.648829.

Abdel Latef, A. A. H., & He, C. (2011). Arbuscular mycorrhizal influence on growth, photosynthetic pigments, osmotic adjustment, and oxidative stress in tomato plants subjected to low temperature stress. *Acta Physiologiae Plantarum, 33*, 1217–1225.

Ahmad, M., Waraich, E. A., Shahid, H., Ahmad, Z., Zulfiqar, U., Mahmood, N., Al-Ashkar, I., Ditta, A., & Sabagh, A. E. (2023). Exogenously applied potassium enhanced morpho-physiological growth and drought tolerance of wheat by alleviating osmotic imbalance and oxidative damage. *Polish Journal of Environmental Studies, 32*(5), 4447–4459. http://www.pjoes.com/pdf-166352-94155?filename=Exogenously%20Applied.pdf. doi:10.15244/pjoes/166352.

Ahmed, A. M., Wais, A. H., Ditta, A., Islam, M. R., Chowdhury, M. K., Pramanik, M. H., Ismaan, H. N., Soufan, W., El Sabagh, A., & Islam, M. S. (2024). Seed germination and early seedling growth of Sorghum (*Sorghum bicolor* L. Moench) genotypes under salinity stress. *Polish Journal of Environmental Studies, 33*(3), 3019–3032. https://www.pjoes.com/pdf-177180-108001?filename=Seed%20Germination%20and.pdf. doi:10.15244/pjoes/177180.

Ali, M. Y., Ashraf, I., Ahmed, I., Iqbal, R., Ditta, A., Kumarasamy, V., Eldin, S. M., Ali, I., Alwahibi, M. S., & Elshikh, M. S. (2024). Characterization of salt-tolerant cultivars of date palm based on morphological and biochemical responses under salinity stress. *Polish Journal of Environmental Studies, 33*(4), 4019–4029. https://www.pjoes.com/pdf-177179-110388?filename=Characterization%20of.pdf. doi:10.15244/pjoes/177179.

Andersen, C. P., Sucoff, E. I., & Dixon, R. K. (1987). The influence of low soil temperature on the growth of vesicular–arbuscular mycorrhizal *Fraxinus pennsylvanica*. *Canadian Journal of Forest Research, 17*, 951–956.

Alguacil, M. M., Lozano, Z., Campoy, M. J., & Roldán, A. (2010). Phosphorus fertilisation management modifies the biodiversity of AM fungi in a tropical savanna forage system. *Soil Biology and Biochemistry, 42*(7), 1114–1122.

Ali, Q., Ashraf, M., & Anwar, F. (2009). Physio-biochemical attributes of arbuscular mycorrhizal sorghum under drought and temperature stress. *Acta Physiologiae Plantarum, 31*(4), 715–723.

An, D., Yang, J., & Zhang, P. (2013). Transcriptome profiling of low temperature-treated cassava apical shoots showed dynamic responses of tropical plant to cold stress. *BMC Genomics, 13*, 64.

Baek, K. H., Skinner, D. Z., & Stougaard, R. N. (2003). Differential expression of antioxidant enzymes under chilling stress in maize seedlings. *Plant Science, 165*(5), 1033–1040.

Berry, J. A., & Bjorkman, O. (1980). Photosynthetic response and adaptation to temperature in higher plants. *Annual Review of Plant Physiology, 31*, 491–543. https://doi.org/10.1146/annurev.pp.31.060180.002423.

Begum, N., Qin, C., Ahanger, M. A., Raza, S., Khan, M. I., Ashraf, M., Ahmed, N., & Zhang, L. (2019). Role of arbuscular mycorrhizal fungi in plant growth regulation: Implications in abiotic stress tolerance. *Frontiers in Plant Science, 10*. https://www.frontiersin.org/journals/plant-science. doi:10.3389/fpls.2019.01068.

Bertolazi, A. A., da Silva Folli-Pereira, M., Caione, G., Passamani, L. Z., Colodete, C. M., de Souza, S. B., Ramos, A. C., Rasool, N., de Freitas Seben Júnior, G., & Schoninger, E. L. (2018). Linking plant nutritional status to plant-AMF interactions. In D. Egamberdieva, & P. Ahmad (Eds.). In Plant microbiome: Stress response. *Microorganisms for sustainability: 5* (pp. 351–384). Springer. https://www.springer.com/series/14379. doi:10.1007/978-981-10-5514-0_16.

Bloom, A. J., Zwieniecki, M. A., Passioura, J. B., Randall, L. B., Holbrook, N. M., & ST. Clair, D. A. (2004). Water relations under root chilling in a sensitive and tolerant tomato species. *Plant, Cell & Environment, 27*(8), 971–979. doi:10.1111/j.1365-3040.2004.01200.x.

Bornman, J. F., Reuber, S., Cen, Y. P., & Weissenböck, G. (2010). Ultraviolet radiation as a stress factor and the role of protective pigments. *Plant Ecology, 128*(1–2), 50–61.

Brundrett, M. C., & Tedersoo, L. (2018). Evolutionary history of mycorrhizal symbioses and global host plant diversity. *New Phytologist, 220*(4), 1108–1115. doi:10.1111/nph.14976.

Buttery, B. R., Buzzell, R. I., & Findlay, W. I. (1977). Relationships among photosynthetic rate, bean yield and other characters in field-grown cultivars of soybean. *Canadian Journal of Plant Science, 57*(1), 111–121.

Buttery, B. R., Sharma, P., & Jha, A. (2024). Antioxidant defense mechanisms in plants under temperature stress. *Plant Stress, 12*, 100271.

Caradonia, F., Francia, E., Morcia, C., Ghizzoni, R., Moulin, L., Terzi, V., & Ronga, D. (2019). Arbuscular mycorrhizal fungi and plant growth-promoting rhizobacteria avoid processing tomato leaf damage during chilling stress. *Agronomy, 9*(6), 299. https://doi.org/10.3390/agronomy9060299.

Chae, S. Y., Kim, Y. J., & Kim, H. J. (2021). Low temperature stress increases proline accumulation and antioxidant responses in tomato plants. *Horticulturae, 7*(12), 560.

Chugunova, O. B., Koroleva, O. A., & Vakulenko, V. V. (1980). Effect of low temperature on photosynthesis in cucumber plants. *Soviet Plant Physiology, 27*, 631–636.

Chen, S., Jin, W., Liu, A., Zhang, S., Liu, D., Wang, F., Lin, X., & Chaoxing, H. (2013). Arbuscular mycorrhizal fungi (AMF) increase growth and secondary metabolism in cucumber subjected to low temperature stress. *Scientia Horticulturae, 160*, 222–229. doi:10.1016/j.scienta.2013.05.039.

Corradi, N., & Bonfante, P. (2012). The arbuscular mycorrhizal symbiosis: Origin and evolution of a beneficial plant infection. *PLoS Pathogens, 8*(4), e1002600. doi:10.1371/journal.ppat.1002600.

Das, S., & Sarkar, S. (2024). Arbuscular mycorrhizal fungal contribution towards plant resilience to drought conditions. *Frontiers in Fungal Biology, 5*, 1355999. doi:10.3389/ffunb.2024.1355999.

de Oliveira, J. G., da Costa Aguiar Alves, P. L., & Vitória, A. P. (2009). Alterations in chlorophyll *a* fluorescence, pigment concentrations and lipid peroxidation to chilling temperature in coffee seedlings. *Environmental and Experimental Botany, 67*(1), 71–76. https://doi.org/10.1016/j.envexpbot.2009.03.005.

de Vries, F. T., Griffiths, R. I., Bailey, M., Craig, H., Girlanda, M., Gweon, H. S., Hallin, S., Kaisermann, A., Keith, A. M., Kretzschmar, M., Lemanceau, P., Lumini, E., Mason, K. E., Oliver, A., Ostle, N., Prosser, J. I., Thion, C., Thomson, B., & Bardgett, R. D. (2020). Soil bacterial networks are less stable under drought than fungal networks. *Nature Communications, 9*, 3033.

El-Yazied, A., A. , A., F. , K., El-Sayed, S. A., & El-Wahed, M. S. A. (2016). Effect of temperature stress on osmotic adjustment and antioxidant system in plants. *Middle East Journal of Agriculture Research, 5*(3), 321–330.

Dhalaria, R., Kumar, D., Kumar, H., Nepovimova, E., Kuča, K., Islam, M. T., & Verma, R. (2020). Arbuscular mycorrhizal fungi as potential agents in ameliorating heavy metal stress in plants. *Agronomy, 10*(6), 815. doi:10.3390/agronomy10060815.

Ditta, A., Mehmood, S., Imtiaz, M., & Tu, M. S. (2024). *Bio-organic amendments for heavy metal remediation: Water, soil and plant approaches and technologies (plant biology, sustainability and climate change)* (1st ed.). Elsevier Inc.

Ditta, A. (2018). Physiological and phenological responses of crop plants under heat stress. In A. Ramakrishna, & S. S. Gill (Eds.), *Metabolic adaptations in plants during abiotic stress, Taylor and Francis* (pp. 55–63). CRC Press USA.

Ditta, A., & Ullah, N. (2023). Perspectives of using plant growth-promoting rhizobacteria under salinity stress for sustainable crop production. In *Plant stress mitigators: Types, techniques and functions* (pp. 231–247). Elsevier. https://www.sciencedirect.com/book/9780323898713. doi:10.1016/B978-0-323-89871-3.00024-0.

Dolatabadian, A. (2021). Plant–microbe interaction. *Biology, 10*(1), 15. doi:10.3390/biology10010015.

Drevon, J. J., Alkama, N., Bargaz, A., Paula Rodiño, A., Sungthongwises, K., & Zaman-Allah, M. (2015). The legume-rhizobia symbiosis. In *Grain legumes* (pp. 267–290). Springer. http://www.springer.com/in/book/9781493927968. doi:10.1007/978-1-4939-2797-5_9.

Frew, A., Powell, J. R., Allsopp, P. G., Sallam, N., & Johnson, S. N. (2022). Arbuscular mycorrhizal fungi promote plant growth and resilience under climate stress. *Current Opinion in Plant Biology, 68*, 102243.

Gavito, M. E., Olsson, P. A., Rouhier, H., Medina-Peñafiel, A., Jakobsen, I., Bago, A., & Azcón-Aguilar, C. (2005). Temperature constraints on the growth and functioning of root organ cultures with arbuscular mycorrhizal fungi. *New Phytologist, 168*(1), 179–188. doi:10.1111/j.1469-8137.2005.01481.x.

Guo, H., Nazim, H., Liang, Z., & Yang, D. (2019). Regulatory mechanisms of arbuscular mycorrhizal fungi on plant growth and stress resistance. *Frontiers in Plant Science, 10*, 1069. https://doi.org/10.3389/fpls.2019.01069.

Haagenson, D. M., Cunningham, S. M., Joern, B. C., Volenec, J. J., & Wiersma, D. W. (2003). Autumn defoliation effects on alfalfa winter survival, root physiology, and gene expression. *Crop Science, 43*(1), 134–150.

Harper, C. J., Walker, C., Schwendemann, A. B., Kerp, H., & Krings, M. (2020). *Archaeosporites rhyniensis* gen. et sp. nov. (Glomeromycota, Archaeosporaceae) from the Lower Devonian Rhynie chert: A fungal lineage morphologically unchanged for more than 400 million years. *Annals of Botany, 126*(5), 915–928.

Heflish, A. A., Abdel-Fattah, G. M., & Elhindi, K. M. (2021). Arbuscular mycorrhizal fungi mitigate drought stress in plants through improved nutrient and water uptake. *Saudi Journal of Biological Sciences, 28*(11), 6435–6445.

Hendrickson, L., Ball, M. C., Wood, J. T., Chow, W. S., & Furbank, R. T. (2010). Low temperature effects on photosynthesis and growth of grapevine. *Plant, Cell & Environment, 27*(7), 795–809.

Hong, E., Xia, X., Ji, W., Li, T., Xu, X., Chen, J., Chen, X., & Zhu, X. (2023). Effects of high temperature stress on the physiological and biochemical characteristics of *Paeonia ostii*. *International Journal of Molecular Sciences, 24*(13), 11180. https://doi.org/10.3390/ijms241311180.

Hossain, M. A., Bhattacharjee, S., Armin, S. M., Qian, P., Xin, W., Li, H. Y., Burritt, D. J., Fujita, M., & Tran, L. S. P. (2015). Hydrogen peroxide priming modulates abiotic oxidative stress tolerance: Insights from ROS detoxification and scavenging. *Frontiers in Plant Science,, 6*, 420.

Jha, S. S., & Songachan, L. S. (2022). Research on diversity and community composition of arbuscular mycorrhizal fungi species in India: A review. *Plant Archives, 20*(2), 4201–4226.

Jiang, Y., Yang, B., Deyholos, M. K., & Zhu, Y. (2019). Physiological and molecular responses of plants to high temperature stress. *Plant Physiology and Biochemistry, 141*, 1–9.

Jian, P., Zha, Q., Hui, X., Tong, C., & Zhang, D. (2024). Research progress of arbuscular mycorrhizal fungi improving plant resistance to temperature stress. *Horticulturae, 10*(8), 855. doi:10.3390/horticulturae10080855.

Joe, P., Han, E. J., Choi, Y. R., & Lee, Y. B. (1998). Air temperature and relative humidity affect the growth of chrysanthemum plantlets in a microponic system. *Horticulture, Environment, and Biotechnology, 39*, 625–628.

Kaur, H., & Suseela, V. (2020). AMF-induced reprogramming of plant metabolome and transcriptome under nutrient and biotic stress. *Plant and Soil, 456*, 115–131.

Khatib, K., Paulsen, G. M., & Patterson, T. G. (1999). High temperature effects on photosynthetic processes in wheat and maize. *Plant Physiology, 121*(3), 827–835.

Khaliq, A., Perveen, S., Alamer, K. H., Zia Ul Haq, M., Rafique, Z., Alsudays, I. M., Althobaiti, A. T., Saleh, M. A., Hussain, S., & Attia, H. (2022). Arbuscular mycorrhizal fungi symbiosis to enhance plant–soil interaction. *Sustainability, 14*(13), 7840. doi:10.3390/su14137840.

Khattak, A. M., & Pearson, S. (2006). Spectral filters and temperature effects on the growth and development of chrysanthemums under low light integral. *Plant Growth Regulation, 49*, 61–68.

Kumar, N., Krishnamoorthy, V., & Nalina, L. (2006). Low temperature induced chlorophyll degradation in arecanut seedlings. *Indian Journal of Plant Physiology, 11*(4), 412–416.

Lei, A. Q., Zhou, J. H., Rong, Z. Y., Alqahtani, M. D., Gao, X. B., & Wu, Q. S. (2024). Mycorrhiza-triggered changes in leaf food quality and secondary metabolite profile in tea at low temperatures. *Rhizosphere, 29*. https://www.sciencedirect.com/science/journal/24522198. doi:10.1016/j.rhisph.2023.100840.

Li, Y., Matsubara, Y., Miyawaki, C., & Liu, Y. (2008). Temperature stress tolerance and increase in antioxidative enzyme activities in mycorrhizal strawberry plants. *Acta Horticulturae, 774*, 391–396.

Li, X., Jiang, H., Liu, F., Cai, J., Dai, T., Cao, W., & Jiang, D. (2013). Induction of chilling tolerance in wheat during germination by pre-soaking seed with nitric oxide and gibberellin. *Plant Growth Regulation*, 71(1), 31-40.

Liu, A., Chen, S., Chang, R., Liu, D., Chen, H., Ahammed, G. J., Lin, X., & He, C. (2014). Arbuscular mycorrhizae improve low temperature tolerance in cucumber via alterations in H_2O_2 accumulation and ATPase activity. *Journal of Plant Research, 127*(6), 775–785. doi:10.1007/s10265-014-0657-8.

Liu, Y. Z., Imtiaz, M., Ditta, A., Rizwan, M. S., Ashraf, M., Mehmood, S., Aziz, O., Mubeen, F., Ali, M., Elahi, N. N., Ijaz, R., S, L., Shuang, C., & Tu, S. (2020). Response of growth, antioxidant enzymes, and root exudates production towards as stress in *Pteris vittata* and *Astragalus sinicus* colonized by arbuscular mycorrhizal fungi. *Environmental Science and Pollution Research, 27*, 2340–2352.

Lugo, M. A., Ontivero, R. E., Iriarte, H. J., Yelikbayev, B., & Pagano, M. C. (2023). The diversity of arbuscular mycorrhizal fungi and their associations in south America: A case study of Argentinean and Brazilian cattle raising productive ecosystems: A review. *Diversity, 15*(9). http://www.mdpi.com/journal/diversity/. doi:10.3390/d15091006.

Luo, N., Wei, S., Li, J., Gu, W., He, D., Qu, T., Qiao, T., & Yang, Z. (2014). Effects of low-temperature stress on root system characteristics and electric conductivity of maize seedlings. *Chinese Journal of Ecology, 33*, 2694–2699.

Ma, X., Zhu, Y., Li, C., & He, C. (2015). Arbuscular mycorrhizal fungi improve phosphorus uptake and gene expression in cucumber seedlings under low temperature stress. *Plant Physiology and Biochemistry, 87*, 38–45.

Ma, Y., Qin, H., & Li, Z. (2018). Arbuscular mycorrhizal fungi improve seedling quality and photosynthetic performance of cucumber under low temperature stress. *Scientia Horticulturae, 235*, 333–340.

Ma, Q., Luo, Y., Wen, Y., Hill, P. W., Chadwick, D. R., Wu, L., & Jones, D. L. (2020). Carbon and sulphur tracing from soil organic sulphur in plants and soil microorganisms. *Soil Biology and Biochemistry*, 150, 107971.

Mathur, S., Sharma, M. P., & Jajoo, A. (2018). Improved photosynthetic efficacy of maize plants with arbuscular mycorrhizal fungi under high temperature stress. *Journal of Photochemistry and Photobiology B: Biology, 180*, 149–154.

Mwampashi, L. L., Magubika, A. J., Ringo, J. F., Theonest, D. J., Tryphone, G. M., Chilagane, L. A., & Nassary, E. K. (2024). Exploring agro-ecological significance, knowledge gaps, and research priorities in arbuscular mycorrhizal fungi. *Frontiers in Microbiology, 15*. https://www.frontiersin.org/journals/microbiology#. doi:10.3389/fmicb.2024.1491861.

Naveed, M., Ditta, A., Ahmad, M., Mustafa, A., Ahmad, Z., Conde-Cid, M., Tahir, S., Shah, S. A. A., Abrar, M. M., & Fahad, S. (2021). Processed animal manure improves morpho-physiological and biochemical characteristics of *Brassica napus* L. under nickel and salinity stress. *Environmental Science and Pollution Research, 28*(33), 45629–45645. https://link.springer.com/journal/11356. doi:10.1007/s11356-021-14004-3.

Nazir, A., Sarfraz, W., Allah, D., Khalid, N., Farid, M., Shafiq, M., Firdaus-E-Bareen, Rizvi, Z. F., & Naeem, N. (2023). Synergistic impact of two autochthonous saprobic fungi (*A. niger* and *T. pseudokoningii*) on the growth, ionic contents, and metals uptake in *Brassica juncea L.* and *Vigna radiata* L. under tannery solid waste contaminated soil. *International Journal of Phytoremediation, 25*(11), 1488–1500. doi:10.1080/15226514.2023.2166457.

Paradis, R. Y., Dalpé, Y., & Charest, C. (1995). The combined effect of arbuscular mycorrhizas and short-term cold exposure on wheat. *New Phytologist, 129*(4), 637–642. https://doi.org/10.1111/j.1469-8137.1995.tb03033.x.

Pasbani, B., Salimi, A., Aliasgharzad, N., & Hajiboland, R. (2020). Colonization with arbuscular mycorrhizal fungi mitigates cold stress through improvement of antioxidant defense and accumulation of protecting molecules in eggplants. *Scientia Horticulturae, 272*, 109575. doi:10.1016/j.scienta.2020.109575.

Qu, L., Wang, M., & Biere, A. (2021). Interactive effects of mycorrhizae, soil phosphorus, and light on growth and induction and priming of defense in *Plantago lanceolata. Frontiers in Plant Science, 12*, 647372.

Raza, M. A. S., Ibrahim, M. A., Ditta, A., Iqbal, R., Aslam, M. U., Muhammad, F., Ali, S., Çiğ, F., Ali, B., Muhammad Ikram, R., Muzamil, M. N., Rahman, M. H., Alwahibi, M. S., & Elshikh, M. S. (2023). Exploring the recuperative potential of brassinosteroids and nano-biochar on growth, physiology, and yield of wheat under drought stress. *Scientific Reports, 13*(1). https://www.nature.com/srep/. doi:10.1038/s41598-023-42007-2.

Reddy, A. S. N., Ali, G. S., Celesnik, H., & Day, I. S. (2010). Coping with stresses: Roles of calcium- and calcium/calmodulin-regulated gene expression. *The Plant Cell, 23*(6), 2010–2032.

Ristic, Z., Bukovnik, U., & Prasad, P. V. V. (2007). Correlation between heat stability of thylakoid membranes and loss of chlorophyll in winter wheat under heat stress. *Crop Science, 47*(5), 2067–2073.

Rotoni, C., Leite, M. F. A., Pijl, A., Kowalchuk, G. A., & Kuramae, E. E. (2025). Synergy between AMF and accompanying microbiome enriched with PGPB enhances root development and microbiome dynamics. *npj Sustainable Agriculture, 3*(1). doi:10.1038/s44264-025-00081-1.

Sailaja, B., Voleti, S. R., Subrahmanyam, D., Sarla, N., & Vishnukiran, T. (2015). Prediction and expression analysis of heat shock transcription factors in rice under heat stress. *PLOS ONE, 10*(3), e0120717.

Saeed, H. S., Munir, M., Samra, S., Farooq, A., Mahmood, H., Ahmad, I., Akram, M. I., Iqbal, R., Elsadek, M. F., Almutairi, S. M., Ditta, A., & Daraz, U. (2025). Screening of biosurfactants producing fungi from infected *Citrus sinensis* (L.). *Polish Journal of Environmental Studies, 34*(5), 5261–5270. https://www.pjoes.com/pdf-190588-115894?filename=Screening%20of.pdf. doi:10.15244/pjoes/190588.

Saleem, A., Raza, M. A. S., Tahir, M. A., Iqbal, R., Aslam, M. U., Toleikiene, M., Khan, M. S., Alwahibi, M. S., Elshikh, M. S., & Ditta, A. (2025). Impact of biogas slurry on physiological and antioxidant mechanisms of wheat under drought stress. *Polish Journal of Environmental Studies, 34*(2), 1721–1731.

Saleem, A., Raza, M. A. S., Khan, I. H., Tahir, M. A., Iqbal, R., Aslam, M. U., Ejaz, Z., Ditta, A., Almunqedhi, B. M., & Al Farraj, D. A. (2024). Amelioration of drought stress in wheat by using plant growth-promoting rhizobacteria and biogas slurry. *Global Nest Journal, 26*(8). https://journal.gnest.org/system/files/2024-10/gnest_05890_final.pdf. doi:10.30955/gnj.005890.

Scharf, K. D., Berberich, T., Ebersberger, I., & Nover, L. (2012). The plant heat stress transcription factor (Hsf) family: Structure, function and evolution. *Biochimica et Biophysica Acta (BBA) - Gene Regulatory Mechanisms, 1819*(2), 104–119.

Singh, P., Singh, R. K., Li, H. B., Guo, D. J., Sharma, A., Lakshmanan, P., & Malviya, M. K. (2020). The role of beneficial microbes in sustainable agriculture and environmental management. *Climate Change and Environmental Sustainability,, 8*(2), 137–149.

Smith, S. E., Jakobsen, I., Grønlund, M., & Smith, F. A. (2011). Roles of arbuscular mycorrhizas in plant phosphorus nutrition: Interactions between pathways of phosphorus uptake in arbuscular mycorrhizal roots have important implications for understanding and manipulating plant phosphorus acquisition. *Plant Physiology, 156*(3), 1050–1057.

Smith, S. E., & Read, D. J. (2008). *Mycorrhizal symbiosis* (3rd ed.). Academic Press.

Strullu-Derrien, C., Selosse, M. A., Kenrick, P., & Martin, F. M. (2018a). The origin and evolution of mycorrhizal symbioses: From palaeomycology to phylogenomics. *New Phytologist, 220*(4), 1012–1030. https://onlinelibrary.wiley.com/journal/10.1111/(ISSN)1469-8137. doi:10.1111/nph.15076.

Strullu-Derrien, C., Spencer, A. R. T., Goral, T., Dee, J., Honegger, R., Kenrick, P., Longcore, J. E., & Berbee, M. L. (2018b). New insights into the evolutionary history of Fungi from a 407 Ma Blastocladiomycota fossil showing a complex hyphal thallus. *Philosophical Transactions of the Royal Society of London. Series B, Biological Sciences, 373*(1739), 20160502. doi:10.1098/rstb.2016.0502.

Taylor, T. N., Remy, W., Hass, H., & Kerp, H. (2005). Fossil arbuscular mycorrhizae from the Early Devonian. *Mycologia, 87*(4), 560–573.

Tang, H., Hassan, M. U., Feng, L., Nawaz, M., Shah, A. N., Qari, S. H., Liu, Y., & Miao, J. (2022). The critical role of arbuscular mycorrhizal fungi to improve drought tolerance and nitrogen use efficiency in crops. *Frontiers in Plant Science, 13*. https://www.frontiersin.org/journals/plant-science. doi:10.3389/fpls.2022.919166.

Tian F., Wang J.C., Bai X.X., Yang Y.B., Huang L., Liao X.F. (2023) Symbiotic seed germination and seedling growth of mycorrhizal fungi in *Paphiopedilum hirsutissimun* (Lindl. Ex Hook.) Stein from China. *Plant Signal Behav* 18(1):2293405.

Tharanath, A. C., Upendra, R. S., & Rajendra, K. (2024). Soil symphony: A comprehensive overview of plant–microbe interactions in agricultural systems. *Applied Microbiology, 4*(4), 1549–1567. doi:10.3390/applmicrobiol4040106.

Thind, S., Chaudhary, M. S., Ditta, A., Hussain, I., Parveen, A., Ullah, N., Mahmood, Q., Al-Ashkar, I., & El-Sabagh, A. (2022). Impact of mycorrhizal fungi from different rhizospheric soils on fungal colonization, growth, and chlorophyll contents of *Cenchrus ciliaris*. *Agronomy, 12*(11), 2644. doi:10.3390/agronomy12112644.

Tian, X. (2008). Effect of high temperature stress on some physiological characteristics of buckwheat. *Journal of Anhui Agricultural Sciences, 36*, 13519–13520.

Tobin, W. T., Bennett, A. E., & Bever, J. D. (2022). Resource limitation and mycorrhizal symbiosis under global environmental change. *Trends in Plant Science, 27*(10), 1012–1024.

van der Heijden, M. G. A., Martin, F. M., Selosse, M. A., & Sanders, I. R. (2015). Mycorrhizal ecology and evolution: The past, the present, and the future. *New Phytologist, 205*(4), 1406–1423. https://doi.org/10.1111/nph.13288.

Wang, H., Hao, Z., Zhang, X., Xie, W., & Chen, B. (2022). Arbuscular mycorrhizal fungi induced plant resistance against *Fusarium* wilt in jasmonate biosynthesis defective mutant and wild type of tomato. *Journal of Fungi, 8*(5), 422. doi:10.3390/jof8050422.

Wang, X., Feng, H., Wang, Y., Wang, M., Xie, X., Chang, H., & Wang, E. (2021). Mycorrhizal symbiosis modulates the rhizosphere microbiota to promote rhizobia–legume symbiosis. *Molecular Plant, 14*(3), 503–516.

Wasaya, A., Yaqoob, S., Ditta, A., Yasir, T. A., Sarwar, N., Javaid, M. M., Al-Ashkar, I., Sabagh, A. E. L., & Farooq, M. U. (2024). Exogenous application of β-aminobutyric acid improved water relations, membrane stability index, and achene yield in sunflower hybrids under terminal drought stress. *Polish Journal of Environmental Studies, 33*(5), 5367–5479. https://www.pjoes.com/pdf-177182-111658?filename=Exogenous%20Application%20of.pdf. doi:10.15244/pjoes/177182.

Wille, L., Messmer, M. M., Studer, B., & Hohmann, P. (2019). Insights to plant–microbe interactions provide opportunities to improve resistance breeding against root diseases in grain legumes. *Plant, Cell & Environment, 42*(1), 20–40. doi:10.1111/pce.13214.

Wu, Q. S., Xia, R. X., & Zou, Y. N. (2010). Improved soil structure and citrus growth after inoculation with arbuscular mycorrhizal fungi under low temperature stress. *Scientia Horticulturae, 124*(3), 289–293.

Xiang, Y., Zhu, X. C., Song, F. B., & Xu, H. W. (2010). Arbuscular mycorrhiza improves photosynthesis and water status in maize under low temperature stress. *Plant and Soil, 331*(1–2), 129–137.

Xiao, J., Li, H., Wang, R., & Zhang, F. (2023). Arbuscular mycorrhizal fungi regulate heat shock protein expression and aquaporin activity under temperature stress. *Plant Physiology and Biochemistry, 197*, 107655.

Xing, X., Zhang, J., Li, Y., & Wang, C. (2018). Arbuscular mycorrhizal fungi alleviate high temperature stress in lily plants by improving physiological characteristics. *Scientia Horticulturae, 234*, 84–91.

Xu, Q., Paulsen, A. Q., & Guikema, J. A. (1995). Functional and ultrastructural injury to photosynthesis in wheat by high temperature during maturation. *Environmental and Experimental Botany, 35*(1), 43–54.

Yano-Melo, A. M., Orivaldo, J. S., Jr., & Maia, L. C. (2003). Tolerance of mycorrhized banana (*Musa* sp. cv. Pacovan) plantlets to saline stress. *Agriculture, Ecosystems & Environment, 95*(1), 343–348. https://doi.org/10.1016/S0167-8809(02)00195-0.

Yan, Z., Ma, T., Guo, S., Liu, R., & Li, M. (2021). Leaf anatomy, photosynthesis, and chlorophyll fluorescence of lettuce as influenced by arbuscular mycorrhizal fungi under high temperature stress. *Scientia Horticulturae, 280*, 109933.

Zha, L., Liu, W., Yang, Q., & Zhang, Y. (2020). Regulation of cold stress response by CBF transcription factors in plants. *Plant Cell Reports, 39*(4), 489–503.

Zhang, L. H., Zhou, G., Sun, B. T., Li, X. H., Wang, S. S., & Shan, W. (2011). Physiological and biochemical effects of high temperature stress on the seedlings of two *Rhododendron* species of subgenus *Hymenanthes*. *Journal of Plant Sciences, 29*, 362–369.

Zhang, X., Wang, L., Meng, H., Wen, H., Fan, Y., & Zhao, J. (2020). Maize responses to low temperature stress: Physiological and biochemical changes. *Journal of Plant Growth Regulation, 39*(2), 639–650.

Zheng, C., Jiang, D., Liu, F., Dai, T., Jing, Q., & Cao, W. (2011). Effects of heat stress on photosynthesis, antioxidant enzyme activity and gene expression in wheat leaves. *Biologia Plantarum, 55*(3), 529–533.

Zhou, J., Wang, Y., & Li, X. (2012). Arbuscular mycorrhizal fungi improve cold resistance in teak seedlings by enhancing antioxidant enzyme activities. *African Journal of Biotechnology, 11*(63), 12750–12757.

Zou, M. Q., Yuan, L. Y., Zhu, S. D., Liu, S., Ge, J., & Wang, C. (2017). Effects of heat stress on photosynthetic characteristics and chloroplast ultrastructure of a heat-sensitive and heat-tolerant cultivar of wucai (*Brassica campestris* L.). *Acta Physiologiae Plantarum, 39*, 30. https://doi.org/10.1007/s11738-016-2333-7.

Zhu, X., Song, F., & Liu, F. (2017). Arbuscular mycorrhizal fungi and tolerance of temperature stress in plants. In Q. S. Wu (Ed.), *Arbuscular mycorrhizas and stress tolerance of plants* (pp. 163–193). Springer.

Zhu, X., Song, F., & Xu, H. (2010a). Influence of arbuscular mycorrhiza on lipid peroxidation and antioxidant enzyme activity of maize plants under temperature stress. *Mycorrhiza, 20*, 325–332.

Zhu, X. C., Song, F. B., & Xu, H. W. (2010b). Arbuscular mycorrhizae improve low temperature stress in maize via alterations in host water status and photosynthesis. *Plant and Soil, 331*, 129–137.

Chapter 15

Arbuscular mycorrhizal fungi and biotic stress tolerance in plants

Sohaib Mehmood[a], Muhammad Rashid[b], Hamida Akbar[a] and Natasha Amjad[b]
[a] *Institute of Botany, Bahauddin Zakariya University, Multan, Pakistan,* [b] *University of Management and Technology Lahore Sub campus Sialkot, Sialkot, Pakistan*

15.1 Introduction

Crop production has been in constant demand since the inception of agriculture, approximately 10,000 years ago, because of the large number of destructive organisms such as insect pests, nematodes, rodents, plant pathogens, viruses, bacteria, fungi, and competitive weeds. These biotic stressors, in combination with other abiotic factors that include drought, extreme temperatures, and nutrient restraints, can lead to significant agricultural productivity drops. Although physical, biological, and chemical methods of control have been developed, preharvest losses of crops to pests have been one of the significant constraints to food production globally. These organisms continue to present a huge pressure on the environment, underscoring the importance of effective and sustainable crop protection strategies (Oerke, 2006).

Although synthetic pesticides are still one of the main methods used to control crop pests and diseases, they have caused severe environmental and public health effects because of their massive and repetitive application. Pesticides can get into the human body by consumption, breathing, or skin contact and stay, accumulate, or disrupt vital body functions. Many classes of pesticides, such as organochlorines, organophosphates, carbamates, triazines, pyrethroids, and neonicotinoids, have been linked to endocrine disruption, reproductive toxicity, neurobehavioral impairments, genotoxic effects, and carcinogenicity. The food products, water sources, and even the breast milk of human beings have been found to contain these residues, raising concerns about chronic and synergistic exposure. In addition, agricultural systems that use agrochemicals have a negative effect on the environment and carry a high amount of chemicals to the ecosystem. These restrictions support the necessity to decrease the use of synthetic pesticides, as well as to think about safer and more sustainable methods of crop protection (Nicolopoulou-Stamati et al., 2016).

Because of the environmental and health constraints of the intensive use of pesticides, the use of biologically based approaches has become a more popular approach in sustainable agriculture. The members of the phylum Glomeromycota are the arbuscular mycorrhizal fungi (AMF), which establish symbiosis with the roots of the majority of the terrestrial plants and help to increase nutrient uptake, especially phosphorus uptake. In addition to their nutritional action, AMFs have been demonstrated to enhance plant sensitivity to various types of biotic stressors by altering plant immune responses, controlling rhizosphere microbial communities, and decreasing pathogen colonization. All these features make AMF a versatile product that can be used to enhance the well-being of plants without overrelying on synthetic agrochemicals to achieve such outcomes (Begum et al., 2019).

This chapter discusses the development of arbuscular mycorrhizal symbiosis and elucidates the molecular, physiological, and ecological processes whereby AMFs increase plant tolerance to biotic stress. It also explains the application of AMF in safeguarding plants against certain pathogens and pests and the possible implementation of AMF in sustainable agriculture.

15.1.1 Overview of arbuscular mycorrhizal fungi

The AMFs belong to the phylum Glomeromycota, which had been classified as part of the previous phylum, Zygomycota, because of their morphological traits. Further molecular phylogenetic studies, however, showed that Zygomycota was not

Advancing Sustainable Agriculture Through Arbuscular Mycorrhizal Fungi. DOI: https://doi.org/10.1016/B978-0-443-45130-0.00006-0

monophyletic, and it was abandoned and officially named Glomeromycota as a separate fungal lineage (Spatafora et al., 2016).

Mycorrhizas are symbiotic relationships between specialized root soil fungi and plant roots, where both parties can functionally benefit from an interface. In vesicular arbuscular mycorrhizae (VAM), fungi colonize the cortical tissues of plant roots and form specialized intracellular structures that facilitate bidirectional nutrient exchange between the plant and the fungus. This symbiotic association represents a continuum of plant–fungus interactions and reflects a highly conserved mutualistic relationship that has persisted throughout plant evolution. (Brundrett, 2002).

Arbuscular mycorrhizal symbiosis is a symbiotic biological relationship between plant roots and fungi of the phylum Glomeromycota in which both sides benefit. The fungus infects the root cortex and develops special internal tubes known as arbuscules, which act as a place of nutrient exchange. The fungus improves the absorption of mineral nutrients and water to the plant, and the water, in turn, through its wide hyphal network in the soil, through which the fungus absorbs the mineral nutrients to the plant, and the plant supplies the carbohydrates formed in the process of photosynthesis to the fungus. This mutual exchange of resources characterizes AM as a symbiotic and mutualistic cooperative type of relationship that is important in the nutrition of plants and in the functioning of the ecosystem (Parniske, 2008).

After symbiosis, AMFs acquire specific morphological forms that help in the process of nutrient exchange and survival in the host. The arbuscule, a highly branched, tree-like intracellular structure that forms within root cortical cells and is the main point of bidirectional nutrient exchange, is the most typical structure. The periarbuscular membrane allows the exchange of phosphate, nitrogen, and carbon compounds through the regulation of exchanges with the surrounding arbuscule, which is surrounded by a plant-derived periarbuscular membrane that ensures cellular integrity. Besides arbuscules, several AM fungi develop vesicles, storage organs of lipids, and other reserves between or inside cells of the root. The fungus outside the root forms an expansive extraradicular hyphal system, which hunts the soil around the plant and greatly expands the surface area of absorption among plants, greatly increasing the uptake of mineral nutrients and water. All these structures allow AM fungi to serve as a powerful connection between the soil resources and the plant metabolism (Parniske, 2008).

15.1.2 Biotic stress in plants: A major challenge

Plant diseases have been considered a significant challenge to global food security, with huge losses in yields reported in leading staple crops. The recent world evaluations show that average yield losses in wheat, rice, maize, potato, and soybean are 21.5, 30.3, 22.6, 17.2, and 21.4, respectively, in the presence of crop pests and pathogens. The plant diseases also lead to massive preharvest losses for smallholder farmers in most of the developing regions, further increasing food insecurity and economic instability. In addition to yield loss, plant illnesses impact species variety and cost of production because of management efforts and have downstream effects on human health and livelihood. The burden on agricultural systems through biotic stress is further escalated with the rising frequency and geographic distribution of emerging plant pathogens that are brought about by climate change and global trade networks (Ristaino et al., 2021).

Fungal pathogen biotic stress is a serious limitation to crop production, especially of economically significant cereals like wheat. *Zymoseptoria tritici* causes *Septoria tritici* blotch, which is one of the most devastating foliar infections affecting wheat in temperate parts of Europe. The pathogen is adapted to wet conditions and is now haunting large grain staple regions of production, with almost 70% of the fungicides used in the European Union annually. In susceptible cultivars the loss in yield has been reported to be as high as 50 detrimental in extreme cases of epidemics, whereas the average loss on untreated crops is approximately 20%. High reproductive ability, mixed sexual and asexual life cycle, high spore production (estimated up to 10^{10}–10^{11} spores per hectare per season), and genetic plasticity greatly contribute to the high susceptibility of rapid adaptation and development of fungicide resistance. These attributes outline how the consistency of biotic factors may jeopardize crop stability, monetary yields, and long-term food security (Fones & Gurr, 2015).

15.2 Establishment of arbuscular mycorrhizal fungi symbiosis: A foundation for defense

The arbuscular mycorrhizal symbioses have a strong effect on the interactions of plants with other organisms, and augmented resistance to soil-borne pathogens is commonly recorded in mycorrhizal plants. In the formation of mycorrhizas the plant does not merely switch off defense; rather, plant defense responses are modulated through crosstalk between the jasmonate-dependent and salicylic acid (SA)-dependent signaling pathways. This defense modulation can affect the response of the plant to the attackers and make it more efficient by making the tissues more responsive to the challenge with an efficient activation of defense mechanisms instead of keeping the constitutive defense activation expensive (Fig. 15.1). As such

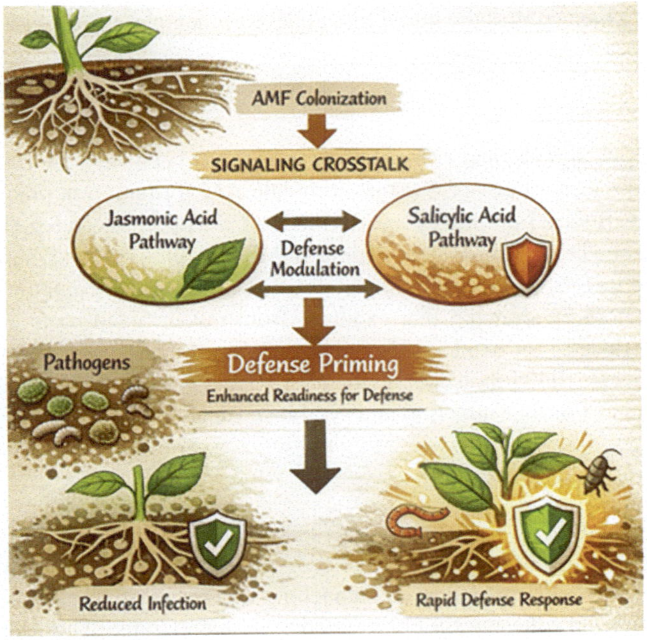

FIGURE 15.1 Plant AMF symbiosis and defense system via hormonal signaling. *AMF*, Arbuscular mycorrhizal fungi.

the formation of AMF symbiosis is a useful platform of defense because it correlates the development of symbiosis with system-wide processes of plant resistance and stress responses interaction (Pozo & Azcón-Aguilar, 2007).

15.2.1 Molecular dialogue and recognition

The process of aerial mycorrhizal development is triggered by mutual communication between plant roots and fungal symbionts via a complex of molecular dialogues. In the presymbiotic stage, plant roots secrete signaling molecules, including strigolactones, which induce fungal metabolism, hyphal growth, and branching, hence leading to the transition to active symbiotic interaction. AM fungi respond by the release of diffusible signaling molecules called "myc factors," which induce symbiosis-specific root responses and transcriptional activation of plant genes needed during AM development. This is a two-way chemical communication to guarantee the recognition of each other, the symbiotic genetic program, and the preparation of the two partners to be colonized and form arbuscules successfully in the cells of the roots (Parniske, 2008).

15.2.1.1 Presymbiotic signaling

The initial stage of arbuscular mycorrhizal development is the presymbiotic stage and is characterized by the early chemical interactions between plant roots and fungal partners, which precede physical contact. At this phase the plant roots secrete signaling molecules into the rhizosphere, which trigger fungal development and branching to stimulate the shift in growth development between exploration hyphae and active host-directed growth. In turn, AM fungi release diffusible signaling molecules, which correspond to the activation of symbiosis-related genes in the host root. This bi-directional communication between the two organisms gears the two toward compatibility so that the fungal penetration and the consequent colonization take place in a synchronized and regulated way.

15.2.1.1.1 Strigolactones

The AMFs have mutualistic relationships with the roots of over 80% of terrestrial plants, although they do not grow massively in the absence of a host. In the early stages of host recognition, AM fungal hyphae undergo extensive branching around the host roots before the creation of the appressorium, which is needed to penetrate the root. This is a dichotomous reaction that is elicited by host root signaling molecules. The spectroscopic analysis and chemical production of a branching factor of isolated strigolactone 5-deoxy-strigol of Lotus japonicus were determined. It was demonstrated that natural strigolactones, namely, 5-deoxy-strigol, sorgolactone, and strigol, as well as the synthetic analogue GR24, caused extensive hypertrophy of the germinating spores of *Gigaspora margarita* at extremely low concentrations. Such results indicate that strigolactones are important presymbiosis signals, which control the initial fungal developmental responses required to initiate AM symbiosis (Akiyama et al., 2005).

15.2.1.1.2 Myc factors

AMFs produce diffusible symbiotic signals, and *Glomus intraradices* has been shown to secrete a mixture of sulfated and nonsulfated simple lipochitooligosaccharides (LCOs) that stimulate AM formation in plant species from diverse families (Fabaceae, Asteraceae, and Umbelliferae). In the legume *Medicago truncatula*, these fungal signals stimulate root growth and branching through the symbiotic Does Not Make Infections (DMI) signaling pathway, supporting the model that rhizobia Nod factors and AM fungal Myc factors activate signaling pathways with common components. The study further supports the diffusible nature of Myc factors by showing that exudates from mycorrhized roots and germinating spores contain these LCOs, whereas nonmycorrhized root exudates do not, indicating fungal origin. Overall, identifying Myc-LCOs as genuine mycorrhizal signals provides direct molecular evidence for early fungal to plant communication that helps trigger the symbiotic program and facilitates further infection and symbiosis development (Maillet et al., 2011).

15.2.1.2 Root colonization process

15.2.1.2.1 Appressorium formation

The presymbiotic period leads to the physical contact of a fungal hypha with the root surface at the stage of root colonization. Upon further growth of the hyphae because of numerous branching as a response to plant cues, the hyphae tip touches the epidermis, where it forms a specialized adhesion structure called a "hyphopodium." This structure attaches to the epidermal cell wall, flattening against the root wall, which is the exact site of entry. The host cellular reorganization that is provoked by the formation of the hypha podium consists of the production of a prepenetration apparatus inside the cell that was contacted by the fungi and determines the intracellular path of fungal entry. This synchronized structuring and cellular reaction make penetration controlled as opposed to invasive growth (Bonfante & Genre, 2010).

15.2.1.2.2 Intracellular hyphal growth

The arbuscular mycorrhizal fungal hyphae do not grow much in the absence of host-derived signals and extend only a few centimeters in the form of asymbiotic mycelium. When the roots release exudates like strigolactones, the development of fungi changes to a presymbiotic stage dominated by more branching of the hyphae and an increase in metabolic rate. After touching the epidermis of the root, the fungus forms a hyphopodium, which sticks to the plant cell wall tightly and usually produces one penetrating hypha. Notably the host cell has firm control over intracellular growth. The invagination of the plant plasma membrane to take in the penetrating hypha keeps the fungal and plant cytoplasm apart. At the same time the host constructs an apparatus of prepenetration, whereby the cytoplasm and the nucleus rearrange at the contact point, and a column of cytoplasmic material predicts the intracellular route of the hypha. The prepenetration apparatus (PPA) foliates components of the secretory pathway, such as endoplasmic reticulum, Golgi bodies, and secretory vesicles, and severe exocytosis steadily accretes the perifungal membrane in front of the protruding haler tip. This synchronous action produces controlled intracellular accommodation and not invasion (Genre & Bonfante, 2016).

15.2.2 Physiological and architectural changes in host roots

When arbuscular mycorrhizal symbiosis is established, there are also coordinated physiological and structural changes in host roots in which fungi are accommodated. The AM colonization has effects on root architecture and changes the lateral formation of roots and the branching pattern, and the cortical cells rearrange their membrane system to build the periarbuscular membrane of the arbuscules developing around. The changes will be necessary to enable the creation of a highly specialized symbiotic interface that enables management of nutrient exchange without interfering with host cellular integrity. These forms of architectural and metabolic reprogramming are the structural basis of functional symbiosis (Parniske, 2008).

15.2.2.1 Changes in root morphology and exudation

The colonization of AMF has a notably important effect on the architecture of the root system, as it prompts lateral growth of roots and alters root branching phenotypes, which leads to exploration of the soil and nutrient uptake. Such morphological changes are strongly associated with hormonal signaling pathways, especially auxin and strigolactone, during phosphorus starvation. Moreover, AM symbiosis changes the root exudation patterns, such as the release of strigolactones and other signaling molecules into the rhizosphere, which control fungal growth and the establishment of symbiosis. These highly organized structural and biochemical changes enhance the communication of the root fungus and enhance the process of foraging for nutrients (Gutjahr & Paszkowski, 2013).

15.2.2.2 Establishment of arbuscular mycorrhizal symbiosis

Molecular communication between the roots of the plants and fungal hyphae in the rhizosphere triggers the formation of arbuscular mycorrhizal symbiosis, which is further followed by fungal settlement on the epidermis of the root and the creation of a specialized structure of penetration. When entering the fungus proliferates in the root cortex under rigid host control, and it differentiates into greatly branched arbuscules within the cells of the cortex. This is achieved by a well-coordinated programmed reorganization of the host cells to host the fungus and also to produce a functional symbiotic interface. The effective differentiation of arbuscules is the initiation of a stable endosymbiotic relationship that allows nutrient exchange and creates the structural basis of further physiological and defense-related reactions (Parniske, 2008).

15.2.2.3 Development of symbiotic interface

In the process of arbuscular mycorrhizal colonization, there is the formation of specialized symbiosis interfaces that allow interaction between the fungus and the host plant. A plant-derived membrane encloses intracellular fungal structures like arbuscules, forming a perifungal interface that provides physical isolation between fungal and plant cytoplasm. This interface is an apoptotic compartment that is controlled and allows the exchange of nutrients without damaging the host cells (Bonfante & Genre, 2010).

15.2.2.3.1 Periarbuscular membrane

In arbuscular mycorrhizal symbiosis, every growing arbuscule is encased by a plant-derived membrane called the periarbuscular membrane, creating a large plant-fungal interface that specializes in nutrient exchange. Phosphate that is released by the fungus into the periarbuscular space is taken into the cortical cell via certain plant transporters like MtPT4, which is localized in the periarbuscular membrane. MtPT4 depletion was shown to cause preterm arbuscular degeneration and symbiosis end, which is vital to the stability of the nutrient transfer and symbiosis, with functional analysis of the periarbuscular membrane in this process (Javot et al., 2007).

15.2.2.3.2 Periarbuscular space

Periarbuscular space is the apoplastic space that exists between the fungal plasma membrane and the plant-derived periarbuscular membrane. Phosphate is translocated during symbiosis by fungal hyphae as polyphosphate, and once hydrolyzed in the arbuscule, inorganic phosphate is released into the periarbuscular space. It is then through this compartment that phosphate is transported across the periarbuscular membrane to the cortical cell via MtPT4. The periarbuscular space is thus the immediate exchange zone that allows the transfer of nutrients to be controlled and yet separation between the fungal and plant cytoplasm to occur (Javot et al., 2007).

15.3 Mechanisms of arbuscular mycorrhizal fungi–mediated biotic stress tolerance

The amylose-mycorrhizal types of fungi increase plant resistance to biotic stresses in a multilayered pathway that can be generally referred to as mycorrhiza-induced resistance. Cameron et al. (2013) suggest that mycorrhiza-induced resistance (MIR) is a result of the mutual interaction between the direct plant–fungus immune signaling and indirect effects through the mycorrhizosphere microbiome. Preliminary colonization of roots by AMF initiates a transient activation of the inherent immunity of plants through fungal microbe associated molecular patterns (MAMPs) perception, resulting in systemic priming of defenses reliant on salicylic acid (SA) and consequently resulting in ultimate colonization by fungi. The AMF then inhibits local immune responses by acting through effector action and hormonal regulation and, at the same time, changes metabolic processes and exudation patterns of the roots. These alterations reorganize the mycorrhizosphere, which enhances the recruitment of induced systemic resistance (ISR)-stimulating rhizobacteria, which further primes jasmonic acid (JA) and ethylene (ET)-dependent pathways of defense. Therefore AMF-induced biotic stress resistance cannot be simply described as the positive effect of better nutrition but as a systemic and coordinated defense mechanism that involves immune priming, hormonal reprogramming, and resistance via the microbiome (Cameron et al., 2013).

15.3.1 Direct mechanisms

AMFs have been known to help in plant protection by direct ecological interaction at the rhizosphere, and competition among microbes is the determinant of pathogen success. Pathogens that are soil-borne must first grow in the rhizosphere before infecting tissues of the host, and their settlement is highly dependent on the microbial community. The nature of disease-suppressive soils shows that microbial consortia cause pathogens to be blocked by competition for nutrients and sites of infection, antibiotic compounds, secretion of lytic enzymes, and consumption of stimulatory pathogen molecules. Moreover, plants are active controllers of the rhizosphere microbiome by using root exudation, which specifically recruits beneficial microorganisms that confer superior antagonist function and the establishment of a protective microenvironment of the roots. These are direct microbiome-mediated processes that decrease pathogen pressure at the root–soil interface before activating the systemic plant defense responses and constitute an imperative initial guard line in the interaction of plants and microbes (Berendsen et al., 2012).

15.3.1.1 Physical barriers and competition

AMFs help in direct pathogen suppression by means of competition interaction and structural occupancy in the root–soil interface. The growth of extensive extraradical hyphal networks helps in increasing the surface coverage of the root surfaces and alters the rhizosphere microenvironment, thereby reducing the available sites of infection for soil-borne pathogens. The evidence reviewed shows that the presence of AMF colonization leads to a decrease in the growth of pathogens, as it competes with them in space and nutrients in the rhizosphere and can alter the dynamics of the microbial community in favor of antagonistic ones. Moreover, AMF-related alterations in soil microbial structure favor inhibitory environments that inhibit the growth of pathogens and the development of diseases. These are local direct ecological processes that act at the root interface and form a significant part of AMF-mediated disease control in sustainable agricultural systems (Umer et al., 2025).

15.3.1.1.1 The hyphal network as a barrier

AMFs assist in the suppression of pathogens by occupying the root tissues spatially and competing with the colonization sites and host-derived resources. There are several mechanisms happening in parallel in mycorrhiza-induced resistance, such as competition for colonization sites or photosynthates, root architecture changes, and changes in the rhizosphere microbial population. The support of good bioprotection usually involves an existing AM symbiosis, which means that the presence of fungal colonization will decrease the chances of pathogen entry and restrict its successful establishment. This is further supported by experimental evidence employing split root systems, which has shown that AMF can affect the success of pathogens in areas other than the actual point of infection, which supports both local and systemic effects. Hence, instead of a strict mechanical barrier, the AMF hyphal network is a spatial and ecological barrier anticipating infection niches and limiting the growth of pathogens at the initial stage of contact (Pozo & Azcón-Aguilar, 2007).

15.3.1.1.2 Competition for infection sites and nutrients

The AMFs also increase nutrient uptake of the plant, especially phosphorus and nitrogen, whereby the extended extraradical hyphae networks reach past the root depletion zone. This enhanced nutrient uptake changes nutrient distribution and availability in the rhizosphere and root tissues, which affects the interaction between soil microorganisms. AMFs indirectly influence the environment of resources that facilitate the proliferation of pathogens by improving the host nutritional condition and altering the carbon allocation patterns. As most of the soil-borne pathogens require localized nutrient availability and host-derived assimilates to be successful colonizers, AMF-mediated nutrient capture can be used to reduce pathogen growth. Nutrient competition in AMF-related systems is therefore achieved by altering the resource dynamics and does not take place exclusively by physical exclusion (Begum et al., 2019).

15.3.1.2 Mycorrhizosphere modifications

AMFs create a separate ecological space called the mycorrhizosphere in which root tissues and extraradical hyphae collaborate to determine the structure of the microbial community in soil. The ubiquitous AMF hyphal system alters carbon movement, nutrient gradients, and microsite structure within the surrounding soil and, in this manner, provides niches that are selective to bacteria and fungi. The attached hyphae and localized nutrient enrichment also favor populations of microbes that can be useful to increase nutrient mobilization and pathogen control. It has been shown experimentally that the AMF colonization is capable of modulating rhizosphere microbial communities and selectively recruiting beneficial microorganisms, such as antagonistic or growth-promoting bacteria. So AMFs have an essential role in the development of the microbiome as nutrient gainers but are also able to suppress ecologically by indirectly shaping the mycorrhizosphere ecology via ecological restructuring (Frey-Klett et al., 2007).

15.3.1.2.1 Changes in rhizosphere microbiome

AMFs alter the structure and composition of rhizosphere microbial communities through their ability to alter belowground biotic interactions and nutrient processes. The composition of microbial communities in soil has been demonstrated as having a direct control on the processes that govern the ecosystem, and that community structure change can have a major impact on multifunctionality in the ecosystem. As AMFs have much contact with root systems and soil biota, their presence leads to changes in microbial assemblage, which dictate nutrient cycling efficiency and biological stability. It has been shown that soil biodiversity and certain community structures are important determinants of ecosystem functioning, and thus, AMF-correlated alterations in the rhizosphere microbiome structure can alter both the functional capabilities of soils and affect the growth of plants (Bender et al., 2016).

15.3.1.2.2 Recruitment of mycorrhiza helper bacteria

The interaction of AMF with rhizosphere bacteria, collectively known as mycorrhiza helper bacteria, is responsible for the formation and operation of the mycorrhizal symbiosis. These bacteria can increase the efficiency of germination of fungal spores, highly developed hyphae, and colonization of the rhizosphere by producing signaling molecules, phytohormones, and enzymes, which change the rhizosphere environment. It can change the patterns of root exudation and alter fungal gene expression, thus enhancing successful symbiotic colonization. Notably, recruitment of mycorrhizal helper bacteria not only results in enhanced colonization of plants by mycorrhizae but also the promotion of plant health and suppression of plant disease in a synergistic manner between plant, fungus, and bacteria. This three-part interaction underscores the fact that selective recruitment of beneficial bacterial partners in the mycorrhizosphere is partially a driver of AMF-mediated stress tolerance (Frey-Klett et al., 2007).

15.3.2 Indirect (host-mediated) mechanisms

The indirect AMF-mediated biotic stress tolerance mechanisms are mediated by systemic host immunity reprogramming instead of pathogen inhibitory actions. In mycorrhizal plants the mechanism of resistance is largely linked to a process called defense priming, where the plant is in a sensitized physiological process that allows the plant to activate immune reactions much faster and more intensely than otherwise when the pathogen attacks (Dejana et al., 2022). Instead, it has been shown that AMF colonization does not lead to the constitutive overexpression of defense-related genes. Rather, mycorrhizal plants exhibit a stronger transcriptional response upon pathogen or stress challenge, indicating that priming reduces metabolic costs while maintaining defensive preparedness. Notably the research reveals that this immune reprogramming is highly dependent on the availability of phosphorus, which connects nutrient status to systemic defense regulation and brings about the incorporation of nutritional cues with immune pathways. The authors also describe the changes in the defense-associated transcriptional networks and hormonal signaling elements, which show that AMFs reorganize the plant defense system via orchestrated regulatory changes and not region-specific mechanical defense. Taken together, these results prove that AMF-mediated tolerance is host-mediated and context-sensitive modulation of immune signaling, growth-defense balance, and nutrient sensing (Dejana et al., 2022).

15.3.2.1 Priming of plant immune responses

Mycorrhiza-induced resistance: Mycorrhizal fungi induce a systemic host-mediated defense mechanism whereby AMFs prime plant immune responses without eliciting constitutive defense activation. The colonization of AMF instead of stimulating the basal defense gene expression leads to a state of physiological hypersensitivity that facilitates faster and enhanced transcriptional activation when pathogens are challenged. This chilled state is defined by augmented responsiveness of defense-associated genes, augmented amassing of antimicrobial metabolites, and fortified cell wall-associated defenses in the wake of infection. Notably the systemic root colonization through the AMF is capable of increasing the resistance in the distal tissues, such as leaves, suggesting long-distance signaling pathways in the host (Jung et al., 2012).

MIR is, at the molecular level, the modulation of certain major hormonal signaling networks, such as jasmonic acid and ET pathways, which are usually linked with resistance to necrotrophic pathogens and herbivorous insects. These signaling cascades may be fine-tuned by AMF colonization, which results in a more robust JA/ET-dependent defense response at attack. At the same time, there is interaction with SA-reliant pathways, which are indicative of an active hormonal cross-talk that dictates the specificity and strength of the immune response (Cameron et al., 2013). This regulatory integration will make mycorrhizal plants stay immune-ready but without wasteful allocation of resources to defense under nonstress conditions.

15.3.2.1.1 Relationship with systemic acquired resistance and induced systemic resistance pathways

MIR has similar functionality to ISR and systemic acquired resistance (SAR) but is mechanistically different. SAR is usually activated because of infection by pathogens, and its activation is largely conditioned by the presence of SA and pathogenesis-related (PR) genes. On the contrary, beneficial rhizobacteria induce ISR that is mostly controlled by JA and ET signaling, with negligible accumulation of SA. MIR is like ISR in that it depends on JA/ET signaling and priming-based defense activation, but it also has interactions with SA-dependent pathways depending on the pathogen and environmental conditions. The antagonism and balance between the SA and JA signaling pathways are one of the main regulatory nodes that determine the outcome of MIR (Pieterse et al., 2014).

15.3.2.2 Enhanced nutritional status

Amelioration of the nutritional status of plants is one of the most elementary host-mediated mechanisms underlying AMF-related biotic stress tolerance. AMFs are important to enhance the absorption of phosphorus (P), nitrogen (N), and other micronutrients by developing extensive extraradical hyaline networks, which extend outside the root depletion zone. These hyphae obtain immobile phosphate pools and organic sources of nitrogen that would not otherwise be utilized by the nonmycorrhizal plants, where they relocate nutrients to the cortical cells through special symbiotic interfaces (Smith & Smith, 2011). Besides phosphorus, AMFs also help in the absorption of nitrogen and aid in the procurement of micronutrients like zinc and copper and thus promote the overall metabolic ability and photosynthesis (Wang et al., 2010).

Better nutrient uptake does not act directly against pathogens, but as a result, it not only increases host vigor and tolerance to infection. Properly nourished plants can afford the resources to defend against compounds, synthesize defense compounds, and strengthen cell walls as well as repair damaged tissues. According to this notion of resource-based tolerance a higher level of nutritional conditions implies that plants can suffer attacks by pathogens with less penalty on yield, even in instances where infection has taken place. There is metaanalytical evidence to suggest that better phosphorus nutrition mediated by AMF can be associated with a lower level of disease severity, though the mechanism can be in terms of better host fitness but not necessarily by a direct effect on host immunity (Veresoglou & Rillig, 2012).

15.3.2.3 Modulation of defense pathways

It can dynamically reprogram interlinked immunoregulatory signaling pathways by modulating plant defense pathways. The core of this regulatory system consists of the crosstalk of SA, jasmonic acid (JA), and ET and organizes resistance against different categories of pathogens. The SA-dependent signaling is generally linked to the defense of biotrophic pathogens and SAR. JA and ET signaling are generally linked with the defense of necrotrophic pathogens and herbivorous insects, often in the context of ISR. It is not independent interactions but rather an elaborate regulatory web consisting of antagonistic and synergistic relationships defining the specific strength of defense (Pieterse et al., 2014).

The immune tolerance of plants during AMF colonization should not be lost, but the ability to respond to pathogens must be preserved. This is done by a fine-tuned modulation and nonconstitutive activation of the defense pathways. Mycorrhizal plants normally show a decreased expression of some SA-dependent responses at early colonization to enable the symbiotic population and systemic priming that increases JA/ET-dependent defense responsiveness when attacked by pathogens (Jung et al., 2012).

At the molecular scale, AMF-induced priming leads to an increase in the activation of defense-related genes after infection. These consist of PR proteins genes, phenylpropanoid pathway enzymes, lipoxygenases associated with the production of JA, and stress-responsive network transcription factors. Such genes exhibit enhanced transcriptional activation in response to pathogen attack in experimental conditions in mycorrhizal plants relative to nonmycorrhizal controls and thus confirm the remodeling of defense gene regulatory pathways by AMF (Cameron et al., 2013).

15.4 Arbuscular mycorrhizal fungi protection against specific biotic stressors

15.4.1 Soil-borne pathogens

AMF protection has maximum chances of success in *Fusarium* wilt pathosystems when research indicates lower wilt severity and a specific host signaling need (Fig. 15.2). An example is preinoculation of tomato with *Rhizophagus irregularis*, which improves resistance to *Fusarium oxysporum*, and the experiment specifically tested the involvement of jasmonates

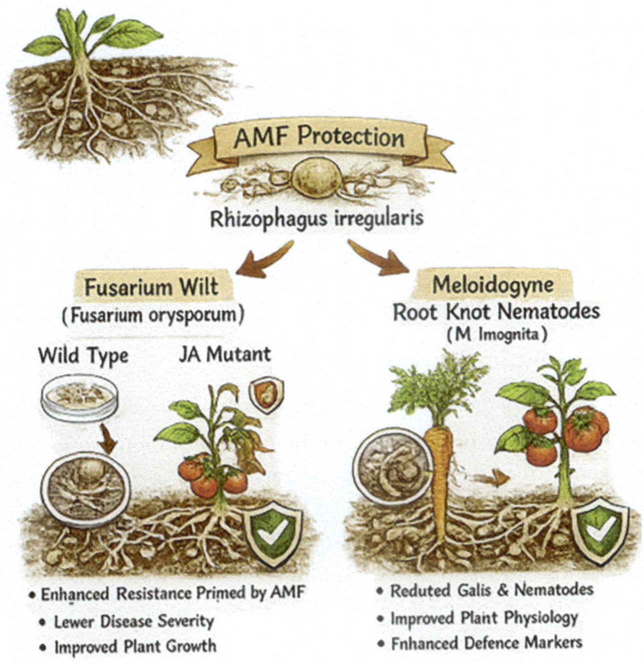

FIGURE 15.2 Plant protection through AMF against soil-borne pathogens. *AMF*, Arbuscular mycorrhizal fungi.

(JA) biosynthesis by comparing wild type to a JA-biosynthesis mutant, supporting the conclusion that the AMF-mediated protection can rely on host JA-related defense capacity. Recent greenhouse studies in tomato demonstrate AMF can minimize the effect of *Fusarium* wilt and also enhance the growth and health characteristics of plants, which supports the fact that AMF can be used as a disease management tool in controlled conditions of infection (Wang et al., 2022).

15.4.2 Meloidogyne (root-knot nematodes; e.g., *M. incognita*)

In the case of root-knot nematodes, AMF effects are commonly indicated in the form of lower infestation measures, galling, and reproduction, and in a high-quality study, both are quantified. AMF-based therapies alleviated the effects of *Meloidogyne* incognita in the presence of measuring plant growth and physiology, as well as defense-relevant biochemical markers in the presence of nematode pressure, which supports a composite model of resource-based tolerance and induced/primed defense responses. Species comparative studies in tomato show that the extent of benefit can be AMF species selective, nematode suppressive effects are not species-wide, and different AMF isolates and host interactions exist (Sulieman et al., 2025).

15.5 Factors influencing biocontrol efficacy

AMFs are not effective in inhibiting plant diseases, and the efficacy of this effect differs between experimental and field situations. The section will look at biological, environmental, and ecological variables, which dictate the permanence and intensity of AMF-mediated biocontrol.

15.5.1 Fungal identity and diversity

The biocontrol efficacy of AMF is strongly influenced by fungal identity, as different AM isolates vary significantly in their ability to suppress plant pathogens. A metaanalysis of 106 studies demonstrated that while pathogen identity had a limited influence on the magnitude of disease suppression, the identity of the AM fungal isolate had a dramatic effect on the level of protection conferred to host plants. In particular, changes in *Glomeraceae* isolate identity were sufficient to alter suppression of fungal pathogens, and differences in AM species richness also affected nematode control outcomes. These findings indicate that AM fungi are not functionally equivalent and that fungal identity and diversity are critical determinants of consistent and effective biocontrol performance (Veresoglou & Rillig, 2012).

15.5.2 Host plant genotype

The efficacy of AMF-mediated biocontrol is highly influenced by the genotype of host plants since the genotype dissimilarity in the essential defense signaling capacities is needed to mediate mycorrhiza-primed resistance. FN in tomato, *Funneliformis mosseae,* also minimized the severity of early blight and induced more PR genes (PR1-PR3) in response to *Alternaria solani* infection and enhanced a faster induction of CTE-type (PR1–PR3) and defense enzymes (PAL, LOX, chitinase, 0-1,3-glucanase). It was found that JA-enhanced plants (35S::PS) were more resistant to it and that the JA-biosynthesis mutant spr2 could not induce these defenses, and as such, AMF-primed systemic resistance requires functional JA signaling. This furnishes the exact mechanistic support that the inconsistent results of AMF biocontrol in cultivars and in the field can be attributed to host genetic variation in hormone pathways (Song et al., 2015).

15.5.3 Environmental and agronomic context

Environmental factors and agronomic practices have a potent influence on the biocontrol efficacy of AMF, as they affect the colonization of the fungus as well as its responsiveness to the host. Special attention should be paid to the availability of soil phosphorus because the presence of high levels of phosphorus inhibits the colonization of AMF and the activation of defense mechanisms based on symbiosis. Moreover, soil moisture, temperature, tillage level, fertilization plan, and the usage of pesticides may also modify AMF colonization and the rhizosphere microbial community organization, therefore, changing the effects of diseases. As a result, field variability tends to be the reason behind differences in AMF-mediated biocontrol found in controlled conditions not being consistently reproduced in agricultural systems (Smith & Smith, 2011).

15.5.4 Interaction with other soil microbes

The AMF interactions with other soil microorganisms are very strong, as the biological control of diseases in the rhizosphere is caused by a more complex network of microorganisms, not by AMF alone. The alterations in rhizosphere microbial community composition can be achieved by AMF shifts in root exudation and hyphae-mediate nutrient fluxes, which favor useful bacteria and repress pathogenic populations. Induced resistance and antagonism to pathogens can be promoted by synergistic interactions between rhizobacteria and mycorrhiza helper bacteria and plant growth, but competitive or antagonistic microbial interactions could decrease AMF activity. The local soil microbiome is therefore crucial toward conditioning the presence or absence of the AMF-mediated biocontrol in the field conditions (Berendsen et al., 2012).

15.6 Ecological and evolutionary perspectives

The ecology of arbuscular mycorrhizal-mediated biocontrol and the long-term evolutionary interactions among plants, fungi, pathogens, and soil microbial communities are decribed as follows.

15.6.1 Natural versus agroecosystems

AM fungal communities can influence ecosystem functioning at multiple hierarchical levels. At a direct level the extraradical mycelium interacts with the abiotic environment and contributes to key soil processes. For example, AM fungi play an important role in soil aggregation, thereby improving soil structure and stability. They also influence soil chemistry through nutrient uptake and redistribution, particularly phosphorus and other mineral elements.

Indirectly, AM fungi affect ecosystem functioning through their impacts on individual plants, which subsequently scale up to influence broader ecological processes. By altering plant growth, nutrient acquisition, and physiological performance, AM fungi can modify ecosystem productivity and nutrient cycling. Furthermore, AM fungi can influence plant community composition and microbial community structure. They interact with other soil microorganisms, including those involved in nitrogen cycling, and can shape plant, plant pathogen, and plant herbivore interactions. These multilayered interactions highlight the complex and far-reaching ecological roles of AM fungi in both natural and managed ecosystems (Powell & Rillig, 2018).

15.6.2 Tripartite evolutionary dynamics

Plant AMF pathogen interactions are a tripartite system of evolution where there is a mutualistic symbiosis and pathogenic infection, with each other under reciprocal selection pressure on each other. AMFs tune the signaling of host defense and resource allocation, and pathogens have developed mechanisms of surmounting or exploiting symbiosis-related plant physiology alterations. Coevolved compatibility between plant genotype, AMF identity, and pathogen lifestyle is thus a determinant of disease suppression, which has been the result of dynamic ecological and evolutionary trade-offs and not necessarily established protective benefits (Pozo & Azcón-Aguilar, 2007).

15.7 Agricultural applications

Agricultural AMF used as biocontrol agents rely on the availability of high-quality inoculum and on the ability to control quality. Large-scale production of obligate biotrophs requires host-dependent systems (substrate-based trap culture or in vitro root organ culture methods). Although trap cultures are very popular, they commonly produce heterogeneous inoculum with variable propagule density, but in vitro systems produce standardized and contaminant-free inoculum and are more reproducible. Nonetheless, commercial scalability, formulation stability, and storage viability are significant limitations affecting performance and consistency of fields (Berruti et al., 2016).

The AMF in agriculture can only be successfully implemented when incorporated into integrated pest management structures and not as an independent concept. AMFs act mostly as preventive agents by setting up early colonization of the root and priming of systemic defense before the pressure of pathogens increases. Agronomic practices, especially phosphorus fertilization, application of fungicides, and soil disturbance, can have a strong impact on their efficacy and can suppress colonization and symbiotic functionality. Combination with compatible biological agents such as growth-promoting rhizobacteria in plants and reduced input management systems is also a strategy that improves disease suppression and sustainability (Begum et al., 2019).

AMF has been proven by field validation research to alleviate the intensity of soil-borne pathogens such as *Fusarium, Phytophthora*, and root-knot nematodes; enhance nutrient uptake; and yield stability. Disease suppression levels, however, differ across environments because of the differences in AMF isolate identity, host genotype, soil microbial communities, and environmental stress conditions. Metaanalytical data have confirmed that AMF-mediated pathogen suppression is context-specific and is usually most effective in low-input or moderate stress systems, but stressed AMF-based biocontrol is not applicable universally as a substitute for chemical control (Gleason et al., 2012).

15.8 Challenges and future research

Although it is experimentally well known that AMFs have mediated biocontrol, there are a few scientific and practical issues that prevent regular application to the field. The biocontrol efficacy of a particular fungus differs with fungal isolate identity, host genotype, soil fertility, and environmental contexts, which is one of the key limitations. Specifically, phosphorus fertilizations and extensive use of agrochemicals may inhibit AMF colonization and symbiosis-dependent defense activation, which is why the effects of AMF remain inconsistent with field conditions (Smith & Smith, 2011).

15.8.1 Limitations

The existing constraints are the absence of a standardized inoculum quality assay, incomplete knowledge of tripartite signaling networks of plant–AMF pathogen interaction, and a lack of long-term field validation of the interaction. Numerous studies are performed in a controlled greenhouse environment, whereas in a real agricultural setting, the microbiomes in soil and heterogeneity in the environment can change the AMF activity. The specificity between AMF isolates and host genotypes is not fully comprehended by the molecular processes that regulate the specificity, which limits its use as a precision tool.

15.8.2 Advanced molecular tools

It is these gaps in knowledge that can be overcome by emerging molecular technologies. The systems level of interaction between plant AMF pathogens is beginning to be characterized in detail through high-throughput transcriptomics, metagenomics, and microbiome profiling. RNA-seq analyses have identified extensive transcriptional reprogramming of mycorrhiza-induced resistance, with defense priming and hormonal crosstalk being the key regulators (Jung et al., 2012).

15.8.3 Climate-resilient agriculture

In climatic change conditions, AMF can also be very vital in strengthening crop resistance to combined abiotic and biotic stresses. AMF enhances water utilization efficiency, nutrient uptake during times of drought, and stress endurance, whose effect may indirectly be resistance to pathogens in climate-stressed systems. Future studies are needed to help find climate-adapted strains of AMF and to determine the effects of increasing temperature, changing precipitation patterns, and soil erosion on symbiosis stability and biocontrol activity.

15.9 Conclusion

AMFs are multifunctional symbionts that enrich the nutrition of plants, immune signaling, and tolerance to diverse biotic stress factors. AMFs led to reduced disease severity in the soil-borne and foliar pathosystem through pathogen interference, mycorrhizosphere remodeling, and systemic defense priming mechanisms. Nevertheless, the extent of biocontrol effect is very site-specific, and it is influenced by fungal identity, host genotype, environmental factors, and interactions in complex microbial communities in soils.

One important thing about this chapter that can be learnt is that AMF-mediated disease suppression is not one direct antagonistic process but a concerted ecological activity, containing nutrient-mediated tolerance, hormonal crosstalk, and immune priming. The variability seen in the field conditions is, however, not to be seen as failure but as a sign of ecological sensitivity and complexity at the system level.

In the future, the further development of the molecular tool, microbiome engineering, and climate-resilient agriculture can provide positive prospects in optimizing the AMF application strategies. By better isolating selection, standard inoculum generation, and implementation into ecologically conscious management systems, AMFs have significant promise as biocontrol agents in the low-input and regenerative agricultural systems of the future.

References

Akiyama, K., Matsuzaki, K.-I., & Hayashi, H. (2005). Plant sesquiterpenes induce hyphal branching in arbuscular mycorrhizal fungi. *Nature, 435*, 824–827.

Begum, N., Qin, C., Ahanger, M. A., Raza, S., Khan, M. I., Ashraf, M., Ahmed, N., & Zhang, L. (2019). Role of arbuscular mycorrhizal fungi in plant growth regulation: Implications in abiotic stress tolerance. *Frontiers in Plant Science, 10*, 1068.

Bender, S. F., Wagg, C., & van der Heijden, M. G. (2016). An underground revolution: Biodiversity and soil ecological engineering for agricultural sustainability. *Trends in Ecology & Evolution, 31*, 440–452.

Berendsen, R. L., Pieterse, C. M., & Bakker, P. A. (2012). The rhizosphere microbiome and plant health. *Trends in Plant Science, 17*, 478–486.

Berruti, A., Lumini, E., Balestrini, R., & Bianciotto, V. (2016). Arbuscular mycorrhizal fungi as natural biofertilizers: Let's benefit from past successes. *Frontiers in Microbiology, 6*, 1559.

Bonfante, P., & Genre, A. (2010). Mechanisms underlying beneficial plant–fungus interactions in mycorrhizal symbiosis. *Nature Communications, 1*, 48.

Brundrett, M. C. (2002). Coevolution of roots and mycorrhizas of land plants. *New Phytologist, 154*, 275–304.

Cameron, D. D., Neal, A. L., van Wees, S. C., & Ton, J. (2013). Mycorrhiza-induced resistance: More than the sum of its parts? *Trends in Plant Science, 18*, 539–545.

Dejana, L., Ramírez-Serrano, B., Rivero, J., Gamir, J., Lopez-Raez, J. A., & Pozo, M. J. (2022). Phosphorus availability drives mycorrhiza-induced resistance in tomato. *Frontiers in Plant Science, 13*, 1060926.

Fones, H., & Gurr, S. (2015). The impact of *Septoria tritici* Blotch disease on wheat: An EU perspective. *Fungal Genetics and Biology, 79*, 3–7.

Frey-Klett, P., Garbaye, J., & Tarkka, M. (2007). The mycorrhiza helper bacteria revisited. *New Phytologist, 176*, 22–36.

Genre, A., & Bonfante, P. (2016). The structure of arbuscular mycorrhizas: A cell biologist's view. In *Molecular mycorrhizal symbiosis* (pp. 33–45). UNI-FIND.

Gleason, F. H., Crawford, J. W., Neuhauser, S., Henderson, L. E., & Lilje, O. (2012). Resource-seeking strategies of zoosporic true fungi in heterogeneous soil habitats at the microscale level. *Soil Biology and Biochemistry, 45*, 79–88.

Gutjahr, C., & Paszkowski, U. (2013). Multiple control levels of root system remodeling in arbuscular mycorrhizal symbiosis. *Frontiers in Plant Science, 4*, 204.

Javot, H., Penmetsa, R. V., Terzaghi, N., Cook, D. R., & Harrison, M. J. (2007). A *Medicago truncatula* phosphate transporter indispensable for the arbuscular mycorrhizal symbiosis. *Proceedings of the National Academy of Sciences, 104*, 1720–1725.

Jung, S. C., Martinez-Medina, A., Lopez-Raez, J. A., & Pozo, M. J. (2012). Mycorrhiza-induced resistance and priming of plant defenses. *Journal of Chemical Ecology, 38*, 651–664.

Maillet, F., Poinsot, V., André, O., Puech-Pagès, V., Haouy, A., Gueunier, M., Cromer, L., Giraudet, D., Formey, D., & Niebel, A. (2011). Fungal lipochitooligosaccharide symbiotic signals in arbuscular mycorrhiza. *Nature, 469*, 58–63.

Nicolopoulou-Stamati, P., Maipas, S., Kotampasi, C., Stamatis, P., & Hens, L. (2016). Chemical pesticides and human health: The urgent need for a new concept in agriculture. *Frontiers in Public Health, 4*, 148.

Oerke, E.-C. (2006). Crop losses to pests. *The Journal of Agricultural Science, 144*, 31–43.

Parniske, M. (2008). Arbuscular mycorrhiza: The mother of plant root endosymbioses. *Nature Reviews Microbiology, 6*, 763–775.

Pieterse, C. M., Zamioudis, C., Berendsen, R. L., Weller, D. M., Van Wees, S. C., & Bakker, P. A. (2014). Induced systemic resistance by beneficial microbes. *Annual Review of Phytopathology, 52*, 347–375.

Powell, J. R., & Rillig, M. C. (2018). Biodiversity of arbuscular mycorrhizal fungi and ecosystem function. *New Phytologist, 220*, 1059–1075.

Pozo, M. J., & Azcón-Aguilar, C. (2007). Unraveling mycorrhiza-induced resistance. *Current Opinion in Plant Biology, 10*, 393–398.

Ristaino, J. B., Anderson, P. K., Bebber, D. P., Brauman, K. A., Cunniffe, N. J., Fedoroff, N. V., Finegold, C., Garrett, K. A., Gilligan, C. A., & Jones, C. M. (2021). The persistent threat of emerging plant disease pandemics to global food security. *Proceedings of the National Academy of Sciences, 118*, E2022239118.

Smith, S. E., & Smith, F. A. (2011). Roles of arbuscular mycorrhizas in plant nutrition and growth: New paradigms from cellular to ecosystem scales. *Annual Review of Plant Biology, 62*, 227–250.

Song, Y., Chen, D., Lu, K., Sun, Z., & Zeng, R. (2015). Enhanced tomato disease resistance primed by arbuscular mycorrhizal fungus. *Frontiers in Plant Science, 6*, 163569.

Spatafora, J. W., Chang, Y., Benny, G. L., Lazarus, K., Smith, M. E., Berbee, M. L., Bonito, G., Corradi, N., Grigoriev, I., & Gryganskyi, A. (2016). A phylum-level phylogenetic classification of zygomycete fungi based on genome-scale data. *Mycologia, 108*, 1028–1046.

Sulieman, A. M. E., Elbadri, G. M., Alanazi, N. A., Alazmi, M., Alrashidi, A., & Alothman, N. S. (2025). Mitigation of root-knot nematode infestation in tomato plants by mycorrhizal fungi. *Frontiers in Sustainable Food Systems, 9*, 1609286.

Umer, M., Anwar, N., Mubeen, M., Li, Y., Ali, A., Alshaharni, M. O., & Liu, P. (2025). Roles of arbuscular mycorrhizal fungi in plant growth and disease management for sustainable agriculture. *Frontiers in Microbiology, 16*, 1616273.

Veresoglou, S. D., & Rillig, M. C. (2012). Suppression of fungal and nematode plant pathogens through arbuscular mycorrhizal fungi. *Biology Letters, 8*, 214–217.

Wang, B., Yeun, L. H., Xue, J. Y., Liu, Y., Ané, J. M., & Qiu, Y. L. (2010). The presence of three mycorrhizal genes in the common ancestor of land plants suggests a key role of mycorrhizas in the colonization of land by plants. *New Phytologist, 186*, 514–525.

Wang, H., Hao, Z., Zhang, X., Xie, W., & Chen, B. (2022). Arbuscular mycorrhizal fungi induced plant resistance against *Fusarium* wilt in jasmonate biosynthesis defective mutant and wild type of tomato. *Journal of Fungi, 8*, 422.

Chapter 16

The complex network of signals and responses during interactions between plants and arbuscular mycorrhizal fungi

Kamran Akbar[a], Tabassum Yaseen[b], Allah Ditta[c], Javed Iqbal[b], Banzeer Ahsan Abbasi[d], Zakir Ullah[e], Salma Noureen[a], Madiha Imtiaz[a], Zahid Ullah[a] and Shumaila Ijaz[f]

[a] *Department of Botany, Bacha Khan University Charsadda, Charsadda, Pakistan,* [b] *Department of Botany, Bacha Khan University, Charsada, Pakistan,* [c] *Department of Environmental Sciences, Shaheed Benazir Bhutto University Sheringal, Sheringal, Pakistan,* [d] *Department of Botany, Rawalpindi Women University, Rawalpindi, Pakistan,* [e] *Department of Plant Sciences, Faculty of Biological Sciences, Quaid-i-Azam University Islamabad, Islamabad, Pakistan,* [f] *Medical School, Shenzhen University, Shenzhen, China*

16.1 Introduction

Arbuscular mycorrhizae fungi (AMF) species like *Rhizophagus irregularis* possess advanced RNA interference (RNAi) systems, harboring numerous AGO and RNA-dependent RNA polymerase (RdRp) genes. Epigenetic modifications, such as 5-methylcytosine (5mC) and *N*6-methyladenine (6mA), also play a crucial role in fungal genome control and the success of symbiosis. The signaling from AMF can either activate or suppress plant immune responses, depending on the length and context of the CO chain. Recent research suggests that immune and symbiotic signaling pathways overlap, indicating a continuum of responses rather than a strict binary system. Central to this entire process is calcium signaling, which acts as an integrator of immune and symbiotic responses. In addition to enhancing plant growth, AMF contribute to improved soil structure, nutrient cycling, and climate regulation (Nazir et al., 2023). Their ancient symbiotic mechanisms share evolutionary roots with nitrogen-fixing rhizobia, underscoring their profound ecological and evolutionary importance (Saeed et al., 2025). Mycorrhizal symbiosis involves the occupation of plant roots by fungi, facilitating nutrient exchange. AMF penetrate root cells, forming structures called arbuscules that enhance nutrient absorption (Ahammed & Hajiboland, 2023), for carbon storage and climate regulation (Xiaoyu & Zhu, 2024). AMF represent the most widespread type, forming associations with over 90% of terrestrial plant species, such as angiosperms and gymnosperms (Lanfranco et al., 2016). In return for carbon-rich compounds from plants, AMF provide essential nutrients and improve water retention in soils, thus enhancing plant growth and survival (Ahammed & Hajiboland, 2023. AMF contribute to soil health by improving nutrient cycling, enhancing soil structure, and acting as a carbon sink, which is vital for carbon storage and climate regulation (Xiaoyu & Zhu, 2024). AMF form widespread symbiotic associations with plants, enhancing nutrient uptake, water retention, and stress tolerance. This mutualism significantly contributes to plant growth, ecological resilience, and sustainable agriculture, highlighting its importance in plant evolution and ecosystem dynamics (Nadeem et al., 2017; Thind et al., 2022).

AMF enhances phosphorus and nitrogen in plants, significantly influencing plant evolution and ecology. They improve soil health, promote carbon storage, and affect soil weathering, showcasing their vital role in ecosystems and improving yields and growth of various crops (Akbar et al., 2025; Noreen et al., 2023; Xiaoyu & Zhu, 2024; Yaseen et al., 2016a, 2016b). Root exudates play a crucial role in establishing symbiosis by releasing signals that activate fungal responses. Myc factors and fungal signaling molecules facilitate chemoperception, enabling fungi to recognize and respond to host plants, initiating the symbiotic relationship (Yizhu et al., 2020; Tahat & Sijam, 2012). Root exudates, particularly the active

Advancing Sustainable Agriculture Through Arbuscular Mycorrhizal Fungi. DOI: https://doi.org/10.1016/B978-0-443-45130-0.00003-5

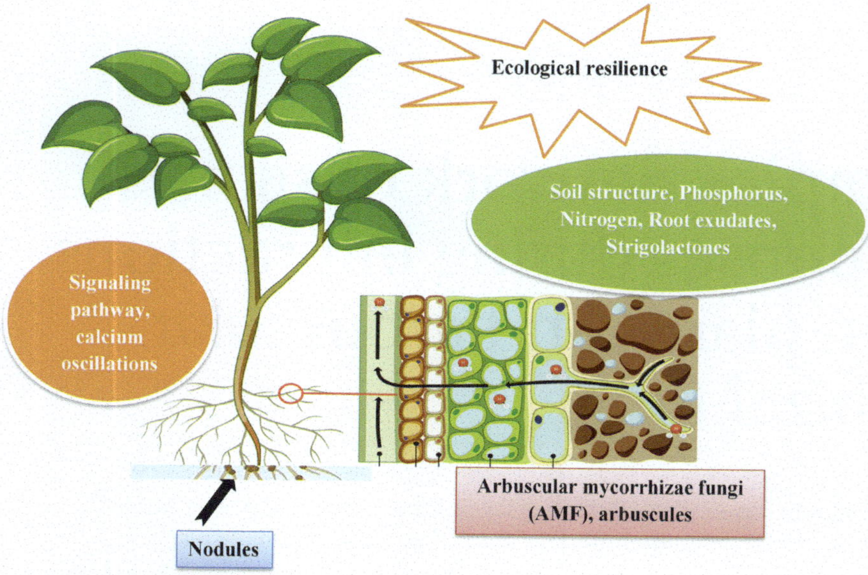

FIGURE 16.1 The role of arbuscular mycorrhizal fungi in plant symbiosis, soil health, and ecosystem functioning.

factor from wild-type tomato, influence a significant function in presymbiotic communication by stimulating fungal growth and branching, whereas Myc factors" from fungi trigger host responses, facilitating effective establishment of mycorrhizal symbiosis (López-Ráez & Pozo, 2013; Yaseen et al., 2011, 2013).

16.2 Overview of arbuscular mycorrhizal fungi symbiosis

Mycorrhizal fungi release signaling molecules like Myc factors and indole-3-acetic acid, which facilitate presymbiotic communication and host recognition. These molecules enhance fungal activation and improve the establishment efficiency of mycorrhizal symbiosis through molecular crosstalk with host plants (Wu et al., 2021). The AM fungi activate the common symbiosis (SYM) signaling pathway, involving calcium spiking regulated by LysM domain proteins. GRAS factors regulate transcriptional pathways essential for arbuscule growth, whereas short RNAs (sRNAs) modulate gene expression during symbiosis (Díaz et al., 2025). The dynamic relationship between plants and microorganisms in the rhizosphere plays a critical role in shaping plant physiology and improving agricultural productivity. Among these microbes, AMF are particularly important for their ability to enhance nutrient availability, bolster stress resilience, and stimulate plant growth, offering a pathway toward more sustainable farming practices (Lutz et al., 2023). AMF, classified within the Glomeromycota phylum, are obligate biotrophs that establish one of the most ancient and widespread forms of root symbiosis with terrestrial plants (Smith & Read, 2008). This mutualistic association involves intricate molecular signaling that leads to the development of specialized root structures called arbuscules, where essential nutrient exchange occurs (Parniske, 2008). AMF significantly improve the uptake of key nutrients, especially phosphorus and nitrogen, whereas plants, in turn, supply carbohydrates derived from photosynthesis to the fungi (Fig. 16.1).

16.3 Root exudates and presymbiotic communication

The formation of this symbiosis begins with a finely tuned molecular dialogue between the plant and the fungal associate. During the initial phase, plant roots secrete strigolactones (SLs) into the surrounding soil, which trigger spore germination and branching of fungal hyphae. In response, AM fungi emit signaling molecules such as lipochitooligosaccharides (LCOs) and chitooligosaccharides (COs) that are detected by specific receptors on the plant cell surface (Gough & Bécard, 2016). This acknowledgment triggers a cascade of intracellular signals that prepare the plant for colonization and activate widespread changes in gene expression. A key cascade included in coordinating this response is the common Sym (symbiosis signaling pathway), which plays a central function in both AMF symbiosis and root nodule formation in legumes (Debellé, 2020).

In legumes like peas, beans, and lucerne, nitrogen-fixing bacteria (like *Rhizobium* spp.) colonize root tissues, resulting in nodulation, a different type of beneficial association. The molecular underpinnings of this nitrogen-fixing symbiosis are investigated using the legume *Medicago truncatula* as a model organism (Oldroyd, 2013). The creation of calcium oscillations surrounding plant cell nuclei is a crucial step in the Sym pathway. A calcium/calmodulin-dependent protein kinase is activated by these calcium signals, and it is this kinase that mediates the alterations in gene expression required to create the arbuscular interface.

These arbuscules, known as the peri-arbuscular membrane, surround the symbiotic association and act as a specialized place for nutrient exchange and molecular communication. Constructing and maintaining this interface requires extensive reprogramming structure and function of the host cell, emphasizing the deep coordination between the genomes of the plant and the fungus. The PAM is biochemically different from the standard PM (plasma membrane), enriched with specific transport proteins and lipid components, such as phosphate and ammonium transporters, which ensure efficient and regulated nutrient exchange. This selective composition helps maintain a balanced symbiosis and prevents the fungal partner from exploiting the host plant (Wang et al., 2017).

AMF not only influence the immediate root environment but also trigger broader physiological changes throughout the plant. These systemic alterations affect plant nutrition, growth, and the ability to withstand both abiotic and biotic stresses, implying a complex communication pathway from roots to shoots. The presence of AMF is frequently associated with noticeable metabolic shifts in aerial parts of the plant, pointing to a highly coordinated, whole-plant response (Fiorilli et al., 2018). Moreover, this systemic impact enhances the plant's resilience to environmental challenges. For instance, AMF colonization improves withstanding loss of water by facilitating better water absorption and stimulating the synthesis of stress-responsive transcription and metabolites (Das & Sarkar, 2024). AMF also support the uptake of poorly mobile micronutrients, including copper and zinc, which are essential for different physiological functions (Founoune-Mboup et al., 2024).

Recent discoveries have also revealed the regulatory role of sRNAs in the AMF–plant relationship. These compounds modulate gene expression in both the fungal and plant partners and can travel between them, influencing the symbiosis on both systemic and local levels. This phenomenon of cross-kingdom RNAi opens new research avenues for enhancing AMF-related benefits in agriculture. A deeper understanding of the molecular dialogue between plants and AMF may help develop strategies to improve crop productivity and stress tolerance, especially under adverse conditions.

Chitin and its derived chitosan—key mechanism of fungal cell walls—play crucial roles in plant–fungus communication. In legume–rhizobium symbioses, plant-derived compounds such as flavonoids and betaines stimulate rhizobial nod gene expression, leading to the production of LCOs, which are central to establishing symbiosis (Gutjahr & Parniske, 2013). Similarly, plants recognize chitin-derived molecules known as chitooligosaccharides (COs) as microbial signatures that activate defense responses. These include changes in gene expression, mitogen-activated transcription kinase activation, ROS (reactive oxygen species) production, and calcium signaling (Cope et al., 2019). Eight-unit COs are particularly potent in eliciting these defenses and are detected by LysM domain-containing receptor-like kinases. In *M. truncatula*, for example, a receptor complex composed of MtLYK9, MtCERK1, and MtLYR4 binds to these COs to initiate immune signaling (Bozsoki et al., 2017). Similar mechanisms are observed in species such as rice and *Arabidopsis thaliana* (Desaki et al., 2019).

Interestingly, beneficial fungi, such as AMF, also produce COs, compelling the host plant to balance defense responses with the initiation of symbiosis (Desaki et al., 2019). Although it was once thought that longer CO chains triggered immune responses and shorter ones activated symbiosis, newer findings suggest that longer COs can also stimulate both symbiotic and immune signaling pathways. In *M. truncatula*, for instance, 8-U COs can promote Ca (calcium) oscillations and activate genes connected to symbiosis, even though they are perceived by receptors traditionally associated with immune responses (Feng et al., 2019).

AMF also release both COs (chitooligosaccharides) and LCOs that help establish symbiosis. Among these, sulfated LCOs are particularly effective at triggering calcium signaling and promoting fungal colonization. In *M. truncatula*, these LCOs induce Ca oscillations in root hair cells and atrichoblasts, prime sites for fungal entry (Genre et al., 2013). Even in nonlegume plants like French carrot and marigold, treatments that combine nonsulfated and sulfated LCOs improve root migration and calcium signaling, highlighting the universal role of LCOs in fostering AM symbiosis (Genre et al., 2013). Furthermore, LCOs can suppress plant defense responses triggered by COs, facilitating a more favorable environment for fungal colonization. Interestingly, LCOs produced by non-AM fungi can also activate untimely symbiotic signals, including root hair proliferation in *Vicia sativa* and *M. truncatula*, signifying that LCO production is a widespread trait among fungi (Genre et al., 2013). Structural studies of fungal secretions have shown that these LCOs closely resemble those produced by AMF, highlighting their conserved function in enabling plant–fungal communication. This dual role of LCOs and COs in orchestrating protection and partnership illustrates the complication of biochemical signaling in plant–microbe interactions.

A complex system distinguishes plants between harmful and mutualistic fungi, and their adjusted reactions to maximize defense and mutualistic effectiveness are demonstrated by the interplay between COs and LCOs. AMF can establish colonization and encourage nutrient exchange, for example, when LCOs are present and decrease immunological signaling that is triggered by eight-residue COs (Feng et al., 2019). Ca (calcium) signaling is essential in both second messenger pathways, and receptor assemblies that may integrate various chemical signals mediate this to devote stability between symbiosis and defense response. Chitin derivatives in plant–fungal interactions have evolutionary and functional significance that can be better understood by comprehending these processes (Desaki et al., 2019; Genre et al., 2013; Maillet et al., 2011). Their vital significance in fostering plant health and production is highlighted by AMF's capacity to modify plant responses through COs and LCOs. These chemicals facilitate fungal colonization by influencing root development, including lateral root formation, and improving nutrient uptake by promoting AM symbiosis (Khan et al., 2011).

Plants create important compounds known as SLs, whereas fungi produce LCOs and COs to form symbiotic relationships. Plant roots release SLs, a class of phytohormones produced from carotenoids. These molecules have two functions: They regulate plant development and operate as signaling molecules that facilitate interactions with other soil organisms. SLs were first identified as stimulants for germination of parasitic weeds like those in the genus *Striga* (Cook et al., 1966). Since then, they have been identified as significant regulators of plant architecture and promoters of the development of AMF (Ruyter-Spira et al., 2012). To create symbiotic interactions with host plants, AMF depend on SLs to stimulate branching and hyphal development. Members of the Glomeraceae and Gigasporaceae families are especially well-documented for this relationship, as SLs promote hyphal growth, mitochondrial size increase, and higher ATP levels (Akiyama et al., 2005; Besserer et al., 2006, 2008). It is interesting to note that hydroxy fatty acids and other substances released by roots can also cause hyphal extension, showing SLs are a part of a bigger signaling pathway (Nagahashi & Douds, 2011).

16.4 Molecular signaling pathways in arbuscular mycorrhizal fungi–plant interactions

A key factor in SLs' bioactivity is their structural makeup. These molecules are made up of an enol ether link between a tricyclic lactone and a methylbutenolide, both of which are necessary for their operation (Fig. 16.2). Their action is greatly impacted by changes to the A and B rings, like the addition of acetyl, keto, or hydroxyl groups (Akiyama et al., 2010). Important elements in SL production and secretion have been found through plant genetic research. For example, orobanchol, a well-known SL, is exuded by the G-subfamily ABC transporter PDR1 in *Petunia*. PDR1 mutations cause decreased SL release, which hinders AMF colonization by reducing intraradical fungal dissemination and hyphopodium initiation (Kretzschmar et al., 2012). The targeted nature of SL signaling is shown by the localized expression in hypodermal passage cells, which enables AMF to enter the root cortex (Sharda & Koide, 2008).

The production of SL is strictly controlled and sensitive to environmental factors, especially the availability of phosphate. Since the symbiosis facilitates phosphate acquisition, plants increase SL synthesis in response to phosphate deprivation to enhance AMF colonization (Yoneyama et al., 2012). GRAS-type transcription factors, particularly NSP1 and NSP2, coordinate this process by controlling the expression of D27 (DWARF27), which codes for a β-carotene isomerase essential for the first stages of the SL biosynthesis pathway (Alder et al., 2012; Liu et al., 2011). AMF colonization is hampered in mutants lacking NSP1 or NSP2 because of abnormal SL production and exudation. Interestingly, in *Medicago*, NSP2 is regulated by microRNA (miRNA) miR171h, which locally inhibits SL production to limit colonization to particular root zones. Colonization of normally uncolonized root elongation zones is made possible by overexpressing a miR171h-resistant form of NSP2 exposes a spatial mechanism of regulation in SL-mediated interactions (Lauressergues et al., 2012).

16.5 Strigolactones as key plant signals

The F-box protein D3/MAX2/RMS4 relates with the α/β-hydrolase receptor D14/DAD2/HTD2 to reveal SLs. When SLs bind to D14 a conformational shift occurs that makes it easier for it to engage with D3 and activate an E3 ligase. This enzyme influences particular polypeptides for destruction to facilitate downstream signaling (Hamiaux et al., 2012). Mutants of rice and pea that lack the F-box protein D3 demonstrate how important this protein is for AMF colonization. Despite exhibiting increased hyphal branching as a result of heightened SL exudation, these mutants are unable to maintain AMF accommodation and the expression of genes linked to symbiosis (Foo et al., 2012; Yoshida et al., 2012). Interestingly, rice D14 mutants show higher colonization levels, perhaps because of enhanced SL release, indicating that SL signaling is not necessarily necessary for AMF colonization. This implies that D3 might be involved in symbiotic-promoting pathways other than SL signaling (Mashiguchi et al., 2023).

Calcium spiking, which entails dynamic and recurring spikes in calcium levels inside the nucleus and perinuclear cytoplasm, is an essential part of plant cellular signaling. The endoplasmic reticulum is probably involved in these calcium oscillations, which come from calcium reserves close to the nuclear envelope (Luginbuehl & Oldroyd, 2016). Core molecular

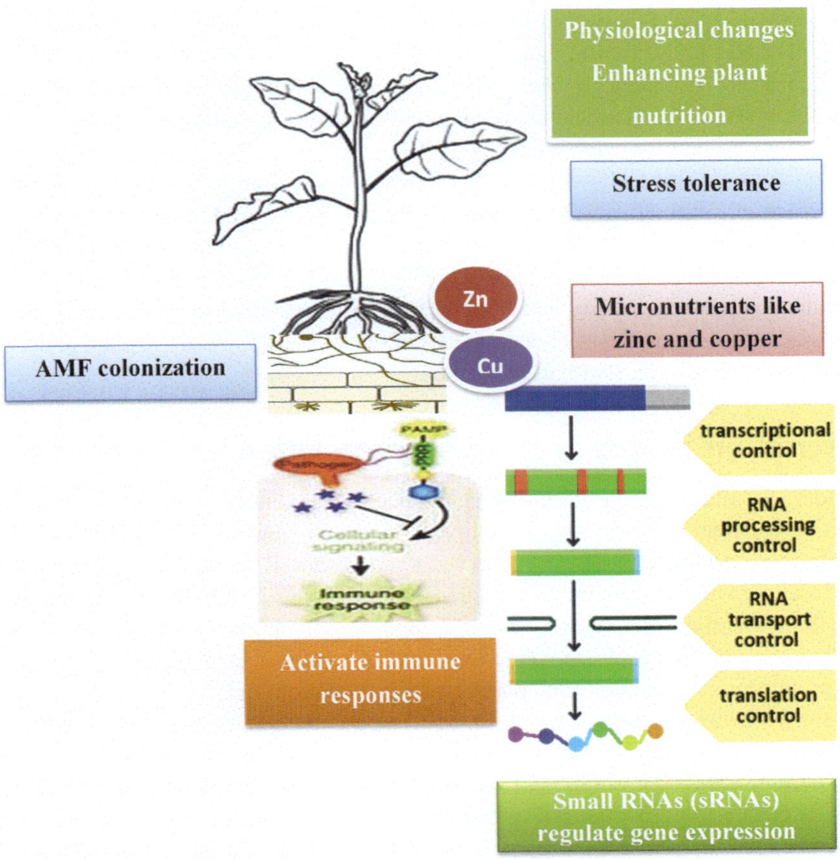

FIGURE 16.2 Systemic signaling and molecular crosstalk in arbuscular mycorrhizal fungi–plant symbiosis. *AMF*, Arbuscular mycorrhizal fungi.

machinery encompassing calcium channels mediating release, counter-ion channels like potassium or anion channels to balance charges, and calcium ATPases that pump calcium back into the store for replenishment is thought to be required to enable calcium spikes.

The CASTOR and POLLUX channels, also referred to as MtDMI1, are two of these essential elements that have been recognized as counter-ion channels. Similarly, it is suggested that the nuclear-contained calcium pump that restores calcium levels in these supplies is the SERCA-type ATPase MCA8 (Chabaud et al., 2011). Further information reveals that calcium spiking is disrupted by alterations in the nucleoporin genes NENA, NUP85, and NUP133. These genes are members of the NUP107 subcomplex, a collection of nuclear pore enzymes that may be involved in transporting vital splicing machinery elements, like pumps and channels, to their inner nuclear envelope functional areas (Luginbuehl & Oldroyd, 2016). The exact mechanisms controlling this transfer are still unknown, though.

Calcium signaling is a crucial component in the establishment of symbiotic relationships between microorganisms and plants like rhizobia and AMF. One of the earliest physiological responses observed in plants during such interactions is the generation of calcium oscillations within the nucleus (Luginbuehl & Oldroyd, 2016). In the case of rhizobia, this signaling cascade is initiated shortly after the detection of Nod factors, which leads to calcium spikes in root hair cells within minutes (Ehrhardt et al., 1996). This process involves a complex network of signaling elements, including receptors located on the PM and proteins associated with the nuclear envelope.

Similar calcium signaling activity is triggered by AMF in plant roots, consistent in the absence of direct hyphal contact, indicating that dispersible signals like Myc factors are capable of inducing this response (Chabaud et al., 2011). These Myc factors include molecules like chitooligosaccharides (COs) and LCOs, which have been revealed to provoke nuclear calcium oscillations in the root epidermis. Both sulfated and nonsulfated fungal-derived LCOs, as well as compact-chain COs, have demonstrated the ability to activate calcium spiking in root hairs.

Although the calcium signaling reactions initiated by rhizobia and AMF share many resemblances, there are subtle distinctions in the characteristics of these signals. For example, the calcium spikes elicited by fungal signals often last longer and are more variable than those triggered by Nod factors or LCOs (Sun et al., 2015). These observations suggest that the nature and concentration of the signaling molecules can influence the profile of the calcium response. Furthermore, the pattern of calcium oscillations changes throughout the symbiotic process. Early stages, such as the preinfection phase in cortical cells, are marked by low-frequency spikes, whereas increased spiking frequency is observed during symbiont entry into host cells (Yuan et al., 2022). This dynamic regulation of calcium signaling reflects its role in coordinating gene expression and cellular changes needed for symbiosis.

This calcium-mediated communication allows plants to recognize and differentiate between various microbial partners at the earliest stages of interaction. Although it was originally thought that AMF and rhizobia strength elicit distinct calcium-peaking patterns, research has shown that LCO-induced responses are largely similar across these organisms (Czaja et al., 2012). Nevertheless, fungal signals, especially those involving COs, tend to produce more prolonged and diverse calcium patterns, pointing to a certain degree of specificity. These signals also display cell-type specificity. For example, in rice roots, calcium signaling is only detected in trichoblasts when both LCOs and COs are applied together, underlining the importance of signal combination and concentration (Sieberer et al., 2012).

16.6 GRAS transcription factors in symbiotic regulation

GRAS transcription factors constitute a large group of plant-specific regulatory proteins involved in diverse biological functions, ranging from development and hormone signaling to environmental responses. Typically ranging from 360 to 850 amino acids in length, these proteins possess a conserved C-terminal GRAS domain characterized by motifs such as PFYRE, SAW, and VHIID, along with two leucine heptad repeats (Pysh et al., 1999). The N-terminal section, by contrast, is highly changeable and often essentially disordered, enabling flexible interactions with other proteins (Sun et al., 2011). A notable subgroup within the GRAS family includes the DELLA proteins, which carry a DELLA motif in their N-terminus that facilitates protein-protein interactions, particularly in gibberellin signaling pathways (la Rosa et al., 2014). The acronym GRAS originates from the names of early-discovered members: GA insensitive, RGA (Repressor of GA1), and SCL (SCARECROW-like) proteins (Pysh et al., 1999).

GRAS proteins are integral to various physiological and developmental processes, including shoot and root architecture, hormonal regulation, and responses to environmental stress. Their roles in symbiosis are also well documented, particularly in the context of gibberellin signaling, where DELLA proteins act as negative regulators by forming complexes with other transcription factors such as those in the indeterminate domain family (Yoshida et al., 2014).

Numerous species have 29 orthologous groups found through genome-wide investigations (Cenci et al., 2017). According to Xue et al. (2015), these groups have been divided into 17 different subfamilies, including 5 recently identified ones that are connected to mycorrhizal signaling, such as RAD1 and RAM1. With several GRAS subfamilies missing from the Brassicales, these subfamilies notably exhibit taxonomic uniqueness, underscoring the evolutionary diversity of GRAS proteins and their distinct functions in different plant ancestors. GRAS proteins play a variety of roles in complex regulatory networks that go beyond specific biological pathways. The relationship between the transcription factors SHORT-ROOT (SHR) and SCARECROW (SCR), which are well-known for their role in root radial organization, is a notable example (Cui et al., 2014). In addition, SCL3 bridges the GA/DELLA and SCR/SHR pathways to facilitate gibberellin-promoted root cell elongation (Heo et al., 2011; Ito & Fukazawa, 2021). These illustrations highlight how GRAS proteins can combine several hormonal and developmental cues to control the growth and differentiation of plants.

The role of GRAS proteins in symbiotic activities, particularly mycorrhizal connections, is a very fascinating feature of their activity. It has been demonstrated that GRAS proteins like RAD1 and RAM1 are key regulators of arbuscular mycorrhiza (AM) (Floss et al., 2013). These transcription factors help integrate fungal structures into host root cells and are crucial for the development and upkeep of AM symbiosis, like RAM1, which stimulates the production of membrane proteins necessary for arbuscule function and genes concerned in the metabolism of fats and carbohydrates (Pimprikar et al., 2016). Furthermore, RAM1 forms regulatory networks that govern arbuscule growth by interacting with other GRAS-domain proteins, such as RAD1 (Xu et al., 2018).

In addition, in AM symbiosis, DELLA proteins have two roles, aiding in arbuscule degeneration through certain regulatory complexes in addition to encouraging arbuscule development. For instance, DELLA activates RAM1 transcription by forming a composite with CYCLOPS and other proteins, which starts the creation of arbuscules (Floss et al., 2016). The dynamic and intricate roles of DELLA proteins across several symbiotic phases are highlighted by the fact that DELLA also relates with MYB1 and NSP1 to control arbuscule disintegration (Floss et al., 2017).

GRAS proteins, such as NSP1 and NSP2, are essential for mycorrhization and nodulation in legumes. These transcription factors work together to produce SL biosynthesis and early nodulation gene expression, both of which are essential for AM fungal colonization (Delaux et al., 2013). Mycorrhizal colonization is decreased in mutants lacking NSP1 or NSP2, highlighting the critical function of these proteins in symbiotic signaling (Gobbato et al., 2012). To control mycorrhizal gene expression, NSP2 also interacts with RAM1, underscoring the interdependent functions of GRAS proteins in symbiotic interaction (Gobbato et al., 2012).

The capacity of GRAS proteins to form heterodimers, which is frequently crucial for their action, is an example of their regulatory plasticity (Gobbato et al., 2012). Cooperative interactions between GRAS proteins interact and regulators allow them to take part in a variety of activities without directly binding DNA, according to analysis of the GRAS gene family at the genome level (Cenci et al., 2017). GRAS proteins can function as transcriptional controllers by building complexes that alter gene transcription in a context-dependent way thanks to their ability to interact with other proteins. GRAS proteins have been linked to light signaling and stress responses in addition to their functions in symbiosis and gibberellin signaling. They help plants adapt to environmental difficulties by controlling root architecture under abiotic stress (Cenci et al., 2017). Their ability to integrate environmental cues to optimize plant development and resource allocation is further demonstrated by their involvement in light signaling. Notwithstanding these developments, little is known about the molecular processes that underlie posttranscriptional and posttranslational modifications and their relations to control proteins. miRNAs like miR171h have been found in recent studies to target GRAS genes, including NSP2, during mycorrhization, which further complicates the regulation of these genes (Bajczyk et al., 2023).

16.7 Small RNAs and RNA interference in symbiosis

Plant growth and stress responses are significantly influenced by RNAi, a common control process seen in all eukaryotes (Castel & Martienssen, 2013). RNAi is mediated by sRNAs, which, depending on sequence complementarity, direct gene silencing either posttranscriptionally (PTGS) or transcriptionally (TGS). miRNAs and short interfering RNAs (siRNAs) are two types of plant sRNAs that are formed from distinct biogenesis pathways. They typically have a length of 20–28 nucleotides (Song et al., 2012). When single-stranded RNAs form constant hairpin structures the DICER-like enzyme DCL1 breaks them down to produce miRNAs. By cleaving target mRNAs or suppressing their translation, these miRNAs mainly aid in PTGS (Bajczyk et al., 2023). On the other hand, RdRps process dsRNAs (double-stranded RNAs), which can be exogenous or endogenous. siRNAs, including 24-nucleotide heterochromatic (hc-siRNAs), which direct RNA-directed DNA methylation, siRNAs aid in TGS and PTGS through 21–22 nucleotide siRNAs (Fei et al., 2013).

After being synthesized, sRNA duplexes enter RISC with AGO (ARGONAUTE) proteins executing RNAi stabilized by HEN1 (HUA ENHANCER1) methylation. According to sRNA length, 5′ nucleotide, and double-stranded structure, AGO proteins, which are conserved throughout eukaryotes, display functional specialization (Li et al., 2022). In addition to their functions inside cells, sRNAs and AGO proteins frequently have extracellular vesicles (EVs) or apoplastic gaps that allow them to travel outside of their original cells. Some plant species also have distinct RNAi components, such as *Chlamydomonas reinhardtii* AGO1's abnormally lengthy sRNAs, which show evolutionary adaptations, and the monocot-specific DCL5, which contributes to temperature-sensitive male sterility (Teng et al., 2020).

To control plant–biotic interactions, such as symbiosis and pathogen defense, plant sRNAs are essential. They control endogenous genes, guard against viral infections, and shield the genome against transposable elements (TEs). The host RNAi machinery converts viral dsRNAs into viral sRNAs in response to viral infections, either by targeting viral genomes to reduce infections or, in certain situations, by altering host transcripts to increase viral persistence (Smith et al., 2011). In addition, plant sRNAs can react to elicitors from bacteria and fungi, such as chitin and flagellin, to activate antimicrobial defenses (Lopez-Marquez et al., 2021). Moreover, sRNAs generated from plants can enter diseases and alter their pathogenicity. For example, cotton-derived miR159 and miR166 target genes that make *Verticillium dahliae* virulence genes (Zhang et al., 2016). This phenomenon, called trans-kingdom RNAi, has sparked research into plant-derived sRNAs for antimicrobial applications (Cai et al., 2018).

Across various species, AMF colonization causes a substantial reprogramming of sRNA expression in plants (Mendoza-Soto et al., 2022). According to Devers et al. (2011), miRNAs affected by AM fungi in *M. truncatula* target transcription factors implicated in immune suppression and AM formation. Likewise, in *Nicotiana attenuata*, AGO7 controls AM-induced sRNAs that affect colonization via food metabolism and phytohormone signaling (Pradhan et al., 2023). Similar to the enhanced nodulation observed in *M. truncatula* mutants (Hobecker et al., 2017), loss of AGO7 promotes mycorrhizal colonization, indicating shared regulatory mechanisms in symbiotic relationships. The purpose of miRNAs in AM symbiosis is still not fully understood, despite advances in knowledge. For example, miR399, a key competitor in phosphate starvation signaling, suppresses PHO2 to control AM colonization; yet, overexpression of this gene does not counteract the suppression of AM symbiosis brought on by elevated phosphate levels (Branscheid et al., 2010). Additional data suggest that AM colonization is orchestrated by a larger regulatory network that includes elements such as PHR (Das et al., 2022). Auxin signaling contributes positively to AM symbiosis, in contrast to miRNA regulation's repression of pathogen contacts. For instance, overexpression of miR396b decreases mycorrhizal colonization by targeting transcription factors in the bHLH and GRF families, but downregulation of miR393 increases auxin signaling, which is essential for arbuscule production (Bazin et al., 2013; Etemadi et al., 2014).

Fungal colonization is regulated by spatially constrained expression patterns of certain miRNAs. Fungal colonization in *M. truncatula* is restricted to particular root areas by miR171 isoforms, such as mtr-miR171h, which target NSP2, a GRAS transcription factor essential for AM gene expression (Hofferek et al., 2014). To ensure compatibility with AM fungus, miR171b shields important regulators such as LOM1 from silencing in AM-compatible plants (Couzigou et al., 2017). *Populus trichocarpa* also exhibits significant miRNA reprogramming following AM colonization, with maintained expression patterns across different mycorrhizal species. A priming effect against foliar pathogens may be shown by systemic responses like changes in miRNA in leaves (Couzigou et al., 2017). The intricate regulatory networks that underlie AM symbiosis may be clarified by future studies that include a variety of RNA types, including small and circular RNAs, as well as additional investigations on the DCL and AGO families (Pradhan & Requena, 2022).

DCL (dicer-like) proteins, AGO (Argonaute) proteins, and RdRps are essential elements of the RNAi machinery found in nearly all fungal lineages, like AM fungus (Lax et al., 2020). Their genomes frequently contain numerous genes that encode these components (Silvestri et al., 2019). In contrast to other fungi, AM fungi have distinct gene repertoires associated with RNAi (Lanfranco & Bonfante, 2023). The growth of AGO genes in *R. irregularis*, which has 26 full-length AGO genes and 14 incomplete AGO genes with the PIWI domain but no other domains, is a noteworthy example (Lee et al., 2018). In addition, 21 genes make up the RdRp family in *R. irregularis*, some of which encode peptides with atypically short and distinctive RdRp domains (Silvestri et al., 2020). One or two DCL genes and a prokaryotic-type class I ribonuclease III enzyme, which was probably obtained through horizontal gene transfer from cyanobacteria, coexist with these gene expansions and may aid in the construction of sRNAs (Dallaire et al., 2021). The sRNA methyltransferase HEN1, which stabilizes sRNAs, is also retained by AM fungi; this gene is lost in a number of other fungal taxa (Chaturvedi et al., 2021). RNAi-related genes are actively expressed and regulated throughout the AM fungal life cycle, according to transcriptomic and proteomic studies (Dallaire et al., 2021). The activity of a single DCL enzyme is compatible with the characteristic sRNA lengths of 24–26 nucleotides seen in the sequencing of sRNAs from AM fungi, such as *R. irregularis* and *Gigaspora margarita* (Silvestri et al., 2019). Some of these sRNAs are 2′-*O*-methylated at the 3′ end, a change made possible by HEN1 homologs, and they exhibit a 5′ uracil bias (Dallaire et al., 2021). Two main sRNA-generating loci have been found through genomic analyses: those tied to protein-coding genes, which create sRNAs of different lengths, and those associated with TEs, which create sRNAs that are 22–26 nt long (Silvestri et al., 2020). Furthermore, sRNA generation in AM fungi may possibly be aided by a noncanonical, DCL-independent pathway that was earlier noted in the nonmycorrhizal fungus *Mucor circinelloides*. The RNase III enzyme, which is derived from prokaryotes, is essential to this route (Perez-Arques et al., 2020).

As both basis and targets of RNAi-mediated stillness, TEs are crucial to the genome dynamics of AM fungi. Repetitive elements make up almost half of the genome in *R. irregularis*, whereas TEs account for 49% of the entire pool of sRNA. The remaining sRNAs come from protein-coding genes (10%) or unannotated genomic areas (41%), many of which are close to TEs, indicating potential control process connections (Dallaire et al., 2021). Given their entangled evolutionary processes, it is believed that the proliferation of AGO genes in *R. irregularis* coevolved with TEs. 5mC marks at CG sites are a crucial epigenetic characteristic that AM fungi use to control TEs through DNA methylation as well (Silvestri et al., 2020). Furthermore, AM fungi have a distinct epigenetic signature: 6mA at ApT motifs, which is mostly linked to symbiotic genes and enhanced transcription (Chaturvedi et al., 2021).

R. irregularis arranges its nuclei into separate euchromatic and heterochromatic compartments at the chromosomal level. Core activities are located in gene-rich euchromatic regions, but repetitive sequences and genes encoding concealed proteins that are strongly expressed during symbiosis are found in heterochromatic regions (Yildirir et al., 2022). The dynamic connections between host plants and fungal genomes are reflected in this chromatin organization, and there is evidence that host plants can alter DNA methylation and fungal chromatin. On the other hand the effector protein RiNEL1 from *R. irregularis*, which interacts with plant histone H2B to inhibit resistance gene expression and promote symbiosis, demonstrates how AM fungi can affect host epigenetics (Wang et al., 2021). The existence of sRNAs that map to viral genomes within AM fungi suggests that these fungi may also use RNAi as a defense mechanism against viruses (Silvestri et al., 2019). Nevertheless, little is known about the functions of mycoviruses in AM fungal biology (Silvestri et al., 2019). In addition to self-defense, AM fungi may participate in cross-kingdom RNAi, a process in which species exchange sRNAs to control gene expression. Fungal sRNAs produced from TEs can inhibit host defense genes in harmful interactions, hence promoting infection (Weiberg et al., 2013). Similarly, fungal sRNAs target plant genes to maintain symbiosis in mutualistic interactions like ectomycorrhizal symbiosis (Wong-Bajracharya et al., 2022). Although there is currently little concrete proof of cross-kingdom RNAi in AM symbiosis, related research raises the possibility. For instance, fungal gene expression in AM fungus has been manipulated by the use of host-induced gene silencing. When the fungus *R. irregularis* is silenced in *M. truncatula* roots, fungal colonization decreases. The monosaccharide transporter gene

MST2 (Helber et al., 2011) and the fungal genes implicated in phosphate transport, effectors, and MAPK cascades have been studied using comparable techniques (Tsuzuki et al., 2016). Fungal sRNAs that target host genes have also been found through computational and experimental research. For example, an sRNA from *R. irregularis* affects the transcription factors of *M. truncatula* WRKY to control colonization. In mutualistic AM relationships, fungal sRNAs may modify host defense responses and membrane remodeling to promote arbuscule formation, but they also frequently target host immunological genes such as WRKY transcription factors and MAPK cascade components. According to recent research the scope of interorganism communication goes beyond traditional cross-kingdom boundaries. These additional complexities include the mRNA transfer between organisms, the role of host AGO4 proteins in gene silencing through methylation, and sRNAs in EVs (Qiao et al., 2023).

Intracellular calcium ion concentration ([Ca^{2+}]) has increasingly been recognized as a fundamental and adaptable signaling mechanism across a wide range of organisms, from prokaryotes to eukaryotes (Luan & Wang, 2021; Plattner & Verkhratsky, 2015). In the plant kingdom, calcium signaling is integral to interpreting a variety of environmental cues, including both abiotic stressors and biotic interactions like herbivore attacks and the detection of microbe-associated molecular patterns (MAMPs; Edel et al., 2017). One reason this single ion might mediate such a diverse array of pathways is because of its chemical reactivity; Ca^{2+} can form precipitates with key anions like phosphate (PO_4^{3-}), potentially disrupting cellular metabolism if not tightly regulated (Edel et al., 2017). Therefore plant cells maintain low cytosolic and nuclear [Ca^{2+}]—typically around 100 nM, about 10,000 times lower than extracellular concentrations—through the active transport via ATP-dependent Ca^{2+} pumps and Ca^{2+}/H^+ exchangers (Luan & Wang, 2021). Temporary elevations in this concentration are sufficient to activate a range of structural and signaling proteins. Indeed, several Ca^{2+}-responsive sensor proteins have been identified, including calcium-dependent protein kinases, calmodulin, calmodulin-like proteins, and calcineurin B-like proteins (Ravi et al., 2023).

16.8 Epigenetic regulation and cross-kingdom RNA signaling

One of the key challenges in plant cell signaling is to interpret the information encoded in these calcium fluctuations. Evidence suggests that downstream responses are finely tuned based on the specific characteristics of the Ca^{2+} signal—commonly referred to as the "calcium signature." This includes the spatial and temporal dynamics of Ca^{2+} changes, such as their duration, intensity, gradient steepness, and oscillation frequency (Jiang & Ding, 2023).

Research into plant calcium signaling has evolved in tandem with advancements in molecular tools and imaging technologies. Early studies relied on the use of synthetic fluorescent calcium dyes like Calcium Green-1 and Fura-2, or on Ca^{2+}-sensitive bioluminescent proteins like aequorin. These were later supplemented by genetically encoded calcium indicators (GECIs), starting with recombinant aequorin and expanding to include fluorescent probes like Cameleon, GECO, and GCaMP proteins (Grenzi et al., 2021). Because of the progressive nature of these technological developments, different experimental approaches have been employed over time depending on the available tools. Initial work often involved direct microinjection of dyes like Calcium Green-1 into plant cells before stimulation with infectious (Xu & Heath, 1998) or symbiotic activators (Ehrhardt et al., 1996). Notably, Knight et al. (1991) were among the first to show that aequorin-expressing *Nicotiana plumbaginifolia* plants exhibited light emissions in response to mechanical stimuli, cold exposure, and pathogen-derived signals, thus confirming the occurrence of cytosolic Ca^{2+} elevations.

In the study of plant–microbe relations, early research highlighted the central role of calcium (Ca^{2+}) signaling in plant responses to pathogen-associated molecules. These investigations uncovered various Ca^{2+}-mediated responses, such as transient calcium increases in tobacco cell suspensions treated with oligogalacturonic acid (Chandra et al., 1997), sustained intracellular Ca^{2+} elevations triggered by β-glucan perception (Mithöfer et al., 1999), and biphasic calcium dynamics in response to *Phytophthora*-derived elicitors (Lecourieux et al., 2002). With time the advent of ratiometric genetically encoded calcium indicators (GECIs), particularly the Yellow Cameleon series developed by Miyawaki et al. (1999), enabled high-resolution Ca^{2+} imaging at the subcellular level, albeit at the cost of the quantitative precision previously offered by aequorin-based systems. Research using these probes has increasingly focused on nuclear calcium spiking events associated with beneficial microbial interactions (Barker et al., 2017). More recently, the development of intensiometric GECIs, like GCaMP and GECO, has further enhanced imaging capabilities by offering improved brightness, thereby facilitating detailed analysis of calcium oscillations during both defense and symbiosis signaling (Keinath et al., 2015; Kelner et al., 2018).

16.9 Integration of immune and symbiotic signaling

The use of diverse calcium indicators and experimental systems has led to a broad but fragmented body of data, making comparisons across studies difficult (see Fig. 16.1), though recent efforts are beginning to address this issue

(Binci et al., 2024). Among various MAMPs, chitin—a polymer of *N*-acetyl-glucosamine (GlcNAc) and a major component of fungal cell walls—has been extensively studied. Its fibrillar form has limited solubility and is generally thought to act at the immediate site of fungal-plant contact. In contrast, soluble chitin oligomers (COs) can diffuse in aqueous environments and trigger plant immune responses. Long-chain COs (with ≥8 GlcNAc units) rapidly elevate cytosolic Ca^{2+} levels, initiating defense pathways (Cao et al., 2014), whereas shorter COs (4–5 GlcNAc units), typically released by AMF, elicit nuclear Ca^{2+} spiking, a hallmark of symbiotic signaling (Barker et al., 2017). This functional dichotomy has led to the proposal that plants distinguish between short and long COs by assembling distinct receptor complexes (Feng et al., 2019; Zhang et al., 2021; see Fig. 16.2). However, because fungi—both symbiotic and pathogenic—can release COs of varying lengths, and given the technical difficulty of profiling the exact composition of fungal exudates, establishing a clear link between CO length and fungal identity remains challenging.

One potential solution lies in the identification of LCOs in AM fungal exudates. These modified, short-chain chitin derivatives, which structurally resemble Nod factors from rhizobia, also induce nuclear calcium spiking and are thought to enhance or even override the symbiotic signaling effects of short COs (Oldroyd, 2013). However, LCOs have also been detected in nonsymbiotic fungal exudates (Rush et al., 2020), adding complexity to their signaling role. Moreover, other signaling compounds unique to AM fungi or pathogens have long been suspected based on experimental observations (Bonfante & Requena, 2011). Notably, much of the existing research on fungal pathogenesis has focused on leaf tissues, while studies comparing root responses to both symbiotic and pathogenic fungi are scarce (Genre et al., 2009), complicating efforts to draw general conclusions because of anatomical and physiological differences between tissues (Tehrani & Mitra, 2023; Zhang & Kong, 2021).

Nevertheless, functional evidence supports the idea that plants consistently interpret CO signals. For example, Volpe et al. (2020) showed that short-chain COs alone can promote AM symbiosis, while exogenously applied long-chain COs suppress AM colonization (Zhang et al., 2021). This suggests that plants can selectively decode microbial MAMPs despite the complexity of their signaling environment. In addition, recent findings by Yu et al. (2023) have revealed a role for plant-secreted lysin motif (LysM) proteins in selectively binding long-chain COs during arbuscule formation, thereby attenuating defense responses. Whether similar mechanisms function earlier during AM colonization remains an open but intriguing question.

Two additional factors help clarify the significance of short-chain chitin oligomers (COs) as symbiotic signals. First, the diffusion properties of chitin-derived molecules in soil are crucial. Short-chain COs, like many similar oligosaccharides, are highly water-soluble, whereas this solubility decreases markedly once the chitin chain exceeds six *N*-acetylglucosamine (GlcNAc) units (Lodhi et al., 2014). This indicates that evolutionary pressure may favor short-chain COs as signaling molecules because their higher mobility allows them to reach plant roots more efficiently than their longer counterparts.

Secondly, fungal pathogens often adopt stealth tactics to avoid detection by plant immune systems. Well-documented mechanisms include the secretion of LysM domain proteins, which bind to long-chain COs released from fungal cell walls by plant chitinases. These proteins compete with plant receptors, effectively dampening defense responses (Volk et al., 2019). Symbiotic fungi, by contrast, benefit from revealing their presence early in the interaction, whereas pathogens are more successful by remaining undetected. Consistent with this, AMF possess multiple chitin synthases and chitinases (Sun et al., 2019; Tisserant et al., 2013), enabling them to produce short-chain COs independently of plant activity. These fungi release short-chain COs even in the absence of a plant host (Genre et al., 2013), suggesting an active role in early symbiotic signaling.

However, recent studies suggest that the distinction between immune and symbiotic signaling pathways in plants may not be as clear-cut as previously assumed. Both short- and long-chain COs can induce a temporary rise in cytosolic calcium levels ($[Ca^{2+}]$) (Binci et al., 2024), and nuclear Ca^{2+} spiking is triggered by both forms in species such as *M. truncatula*, rice, and barley (Feng et al., 2019; Li et al., 2022; Zhang et al., 2021). Despite this the required genetic components differ between these responses (Binci et al., 2024; Feng et al., 2019). Interestingly, when COs are applied to plants already colonized by AM fungi, short-chain COs tend to enhance symbiosis, whereas long-chain COs appear to suppress it (Volpe et al., 2020; Zhang et al., 2021).

These findings suggest a far more nuanced plant perception of fungal signals than previously thought. Factors include the concentration of COs, the composition and acetylation level of the molecules, the nutritional state of the plant, redundancy among receptor proteins, and concurrent activation of alternative signaling pathways, which may all influence the outcome (see Fig. 16.2). Notably, although many studies have involved exposing whole plant organs to aqueous CO solutions, in nature, signal exchange is likely to be highly localized. It occurs in proximity, if not direct contact, between fungal hyphae and plant cells. In such conditions, the plant likely integrates multiple inputs, including both soluble and surface-bound MAMPs. This complex signal integration likely shapes the Ca^{2+}-mediated responses and their downstream decoding.

Overall, the long-standing model separating fungal signals into either pathogenic or symbiotic categories warrants reconsideration. Rather than a strict dichotomy a more accurate representation may be a continuum of interactions or overlapping identities. Similarly, the idea of unique calcium signatures precisely encoding distinct responses is being challenged. The new perspective positions calcium as a central node in the plant's interpretation of fungal signals—both beneficial and harmful—acting as a shared language in the broader context of plant–fungus communication. We aim to delineate the intricate interplay between plant responses to symbiotic and pathogenic fungi, focusing on the central role of calcium (Ca^{2+}) signaling, starting from the well-established aspects.

16.10 Technological advances in calcium signaling research

Advances in calcium monitoring and imaging technologies have played a crucial role in our understanding of Ca^{2+} dynamics. Early discoveries of Ca^{2+} spiking were made in root hairs during the rhizobium–legume interaction (Ehrhardt et al., 1996; Miwa et al., 2006). The development of genetically encoded calcium indicators (GECIs) later enabled more precise observations, revealing synchronous nuclear and perinuclear Ca^{2+} oscillations (Kelner et al., 2018; Sieberer et al., 2009). These oscillations were also found in epidermal cells lacking root hairs (atrichoblasts) during early AM colonization (Chabaud et al., 2011), as well as in inner root tissues colonized by either rhizobia or AM fungi (Sieberer et al., 2012). Similar Ca^{2+} signaling patterns have been reported in actinorhizal associations (Chabaud et al., 2016) and even during interactions with the endophytic fungus *Fusarium* (Skiada et al., 2020).

In the two most well-characterized symbioses—rhizobia-induced nodulation and AM symbiosis—the eliciting molecules (Nod factors and Myc factors, respectively) have been shown to consistently induce Ca^{2+} spiking. This oscillatory Ca^{2+} signaling is considered a signature event within the so-called common symbiotic signaling pathway (Oldroyd, 2013), marking it as a critical early response to beneficial microbes.

Conversely, transient increases in cytosolic Ca^{2+} concentration ($[Ca^{2+}]$cyt) are also hallmark events during plant immune responses to pathogenic fungi (Zipfel & Oldroyd, 2017). Plant immunity is broadly divided into two branches: pattern-triggered immunity (PTI) and effector-triggered immunity (ETI; Yuan et al., 2021). PTI is initiated at the cell surface via pattern recognition receptors (PRRs) that detect conserved microbial features, whereas ETI is activated intracellularly through NLR proteins that sense pathogen effectors. This binary classification is now under reconsideration because of findings that show PTI and ETI are functionally intertwined and enhance each other in mounting effective immune responses (Ngou et al., 2022). Regardless of this reevaluation, both PTI and ETI rely on Ca^{2+} signals to propagate downstream responses, facilitated by a Ca^{2}-regulated phosphocode involving calcium-dependent protein kinases and potentially epigenetic modifications (Erickson et al., 2022; Hannan Parker et al., 2022).

PTI-associated Ca^{2+} signals are typically rapid and steep, forming part of a broader cascade, including ROS production, MAP kinase activation, and immune gene expression (Bjornson et al., 2021; Köster et al., 2022). Both microbes and purified triggers, including chitin-based compounds, can initiate such PTI responses (Cao et al., 2014; Keinath et al., 2015; Ranf et al., 2011). Historically, these responses were characterized using aequorin-based calcium reporters, which helped define the classic PTI Ca^{2+} profile. More recent studies using fluorescent GECIs have shown that these signals may also include oscillatory patterns at the level of individual cells (Keinath et al., 2015; Thor & Peiter, 2014), regulated by specific Ca^{2+} channels and pumps (Köster et al., 2022). In contrast, ETI-linked Ca^{2+} responses are delayed in onset and more prolonged. This slower cytosolic influx is dependent on specific PM Ca^{2+} channels (Kim et al., 2022) and the formation of resistosomes—NLR protein complexes essential for triggering hypersensitive cell death (Ngou et al., 2022).

Although symbiotic and immune signaling pathways are typically localized to different cellular compartments and exhibit distinct Ca^{2+} profiles, growing evidence supports substantial cross-talk between them. Initial clues came from microscopic studies during AM symbiosis (Gianinazzi-Pearson, 1996), followed by transcriptomic data revealing that AM fungal colonization activates numerous defense-related genes across plant species (Rey & Jacquet, 2018), with expression patterns varying by developmental stage (Giovannetti et al., 2015).

This coregulation may arise from the ability of AM-derived signals to trigger canonical defense responses such as ROS bursts and MAP kinase cascades (Bozsoki et al., 2017). Interestingly, molecules previously thought to be purely symbiotic, such as the tetrameric chitin oligomer CO_4—which induces nuclear Ca^{2+} spiking at nanomolar concentrations (Genre et al., 2013)—can also activate ROS and MAPK signaling via the same receptors involved in immune responses to octameric COs (CO_8) and chitin (Bozsoki et al., 2017). This indicates that the signaling responses to these molecules are not strictly exclusive to either symbiosis or immunity.

Both CO_4 and CO_8 have been shown to trigger a pronounced cytosolic Ca^{2+} influx, with the magnitude and nature of the response strongly influenced by the activator's concentration (Binci et al., 2024). AM fungal exudates can also elicit PTI-like Ca^{2+} responses; for example, germinating spore secretions from *G. margarita* stimulate rapid cytosolic Ca^{2+} elevations in aequorin-expressing soybean cell cultures (Navazio et al., 2007), resembling classical PTI signaling. It remains an open question whether and to what extent cellular mechanisms traditionally linked to plant immunity are also functionally involved in signaling during the perception of symbiotic fungal cues and, consequently, in the modulation of arbuscule formation and degradation. A notable example includes certain LysM-type receptors, like *M. truncatula* LYK9 (Bozsoki et al., 2017; Feng et al., 2019; Gibelin-Viala et al., 2019) and *Lotus japonicus* CERK6 (Binci et al., 2024; Bozsoki et al., 2017). These receptors, which display high affinity for chitin molecules, are implicated in triggering defense-associated responses, including ROS production, MAPK activation, and cytosolic Ca^{2+} influx. At the same time, they appear to contribute variably to the control of mycorrhizal colonization.

Proteins from the LYRIIIA family further highlight this dual functionality. They are involved in plant immune responses upon detection of LCOs, yet they also promote AM fungal colonization in species that form mycorrhizal associations. Intriguingly, their role in defense signaling seems to be conserved even in nonmycotrophic plants (Wang et al., 2023a). These findings suggest a model in which an early rise in cytosolic [Ca^{2+}] may act as a signaling intermediary, helping to either initiate or fine-tune the symbiotic interaction. This cytosolic response might be linked to downstream messengers such as mevalonate, which has been implicated in signaling pathways supporting symbiosis (Venkateshwaran et al., 2015).

The complexity of these interactions increases when considering that both CO_4 and Nod factors have demonstrated an ability to suppress immune responses in plants that do not naturally engage in mycorrhizal associations (Liang et al., 2013). This highlights their broader immunomodulatory potential beyond classical symbiotic contexts. Adding to this complexity is the emerging contrast between plant lineages—particularly Fabaceae and monocots—regarding nuclear Ca^{2+} spiking regulation. In barley, for instance, nutrient status, especially levels of phosphorus and nitrogen, plays a significant role in determining cellular responsiveness to short-chain COs. Under nutrient-deficient conditions a higher proportion of nuclei exhibit Ca^{2+} spiking in response to LCOs and, to a lesser extent, CO4. This enhanced responsiveness is associated with increased colonization by AM fungi and is mediated through transcription factors NSP1 and NSP2 (Li et al., 2022). It remains to be determined whether the plant's nutritional state also influences the interplay between immune and symbiotic signaling components, such as the competitive dynamics between OsCERK1 and its partners OsMYR1 and OsCeBiP, which differentially regulate defense and symbiosis in rice (Zhang et al., 2021). This represents a compelling direction for future research.

References

Ahammed, G. J., & Hajiboland, R. (2024). Introduction to arbuscular mycorrhizal fungi and higher plant symbiosis: Characteristic features, functions, and applications. In *Arbuscular mycorrhizal fungi and higher plants: Fundamentals and applications* (pp. 1–17). Springer Nature Singapore.

Akbar, K., Yaseen, T., Abbasi, B. A., Iqbal, J., Ullah, Z., Ijaz, S., Mustafa, A. E.-Z. M. A., Elshikh, M. S., Ercisli, S., Harsonowati, W., & Iqbal, R. (2025). Role and genetic diversity of arbuscular mycorrhizal fungi in improving productivity and enhancing nutrient absorption in chickpea *(Cicer arietinum* L.) genotypes. *Genetic Resources and Crop Evolution, 72*, 2459–2473. https://doi.org/10.1007/s10722-024-02132.

Akiyama, K., Matsuzaki, K., & Hayashi, H. (2005). Plant sesquiterpenes induce hyphal branching in arbuscular mycorrhizal fungi. *Nature, 435*, 824–827.

Akiyama, K., Ogasawara, S., Ito, S., & Hayashi, H. (2010). Structural requirements of strigolactones for hyphal branching in AM fungi. *Plant & Cell Physiology, 51*, 1104–1117.

Alder, A., Jamil, M., Marzorati, M., Bruno, M., Vermathen, M., Bigler, P., Ghisla, S., Bouwmeester, H., Beyer, P., & Al-Babili, S. (2012). The path from β-carotene to carlactone, a strigolactone–like plant hormone. *Science, 335*, 1348–1351.

Bajczyk, M., Jarmolowski, A., Jozwiak, M., Pacak, A., Pietrykowska, H., Sierocka, I., Swida-Barteczka, A., Szewc, L., & Szweykowska-Kulinska, Z. (2023). Recent insights into plant miRNA biogenesis: Multiple layers of miRNA level regulation. *Plants, 12*, 342.

Barker, D. G, Chabaud, M., Russo, G., & Genre, A. (2017). Nuclear Ca^{2+} signalling in arbuscular mycorrhizal and actinorhizal endosymbioses: On the trail of novel underground signals. *New Phytologist, 214*, 533–538.

Bazin, J., Khan, G. A., Combier, J. P., Bustos-Sanmamed, P., Debernardi, J. M., Rodriguez, R., Sorin, C., Palatnik, J., Hartmann, C., Crespi, M., & Lelandais-Brière, C. (2013). miR396 affects mycorrhization and root meristem activity in the legume *Medicago truncatula*. *Plant Journal, 74*, 920–934.

Besserer, A., Becard, G., Jauneau, A., Roux, C., & Sejalon-Delmas, N. (2008). GR24, a synthetic analog of strigolactones, stimulates the mitosis and growth of the arbuscular mycorrhizal fungus *Gigaspora rosea* by boosting its energy metabolism. *Plant Physiology, 148*, 402–413.

Besserer, A., Puech-Pagès, V., Kiefer, P., Gomez-Roldan, V., Jauneau, A., Roy, S., Portais, J.-C., Roux, C., Bécard, G., & Séjalon-Delmas, N. (2006). Strigolactones stimulate arbuscular mycorrhizal fungi by activating mitochondria. *PLoS Biology, 4*, 226.

Binci, F., Offer, E., Crosino, A., Sciascia, I., Kleine-Vehn, J., Genre, A., Giovannetti, M., & Navazio, L. (2024). Spatially and temporally distinct Ca^{2+} changes in *Lotus japonicus* roots orient fungal-triggered signalling pathways towards symbiosis or immunity. *Journal of Experimental Botany, 75*, 605–619.

Bjornson, M., Pimprikar, P., Nürnberger, T., & Zipfel, C. (2021). The transcriptional landscape of *Arabidopsis thaliana* pattern-triggered immunity. *Nature Plants, 7*, 579–586.

Bonfante, P., & Requena, N. (2011). Dating in the dark: How roots respond to fungal signals to establish arbuscular mycorrhizal symbiosis. *Current Opinion in Plant Biology, 14*, 1–7.

Bozsoki, Z., Cheng, J., Feng, F., Gysel, K., Vinther, M., Andersen, K. R, Oldroyd, G., Blaise, M., Radutoiu, S., & Stougaard, J. (2017). Receptor-mediated chitin perception in legume roots is functionally separable from Nod factor perception. *Proceedings of the National Academy of Sciences of the United States of America, 114*, E8118–E8127.

Branscheid, A., Sieh, D., Pant, B. D., May, P., Devers, E. A., Elkrog, A., & Krajinski, F. (2010). Expression pattern suggests a role of miR399 in the regulation of the cellular response to local Pi increase during arbuscular mycorrhizal symbiosis. *Molecular Plant-Microbe Interactions, 23*, 915–926.

Cai, Q., He, B., Kogel, K. H., & Jin, H. (2018). Cross-kingdom RNA trafficking and environmental RNAi—nature's blueprint for modern crop protection strategies. *Current Opinion in Microbiology, 46*, 58–64.

Cao, Y., Liang, Y., Tanaka, K., Nguyen, C. T, Jedrzejczak, R. P, Joachimiak, A., & Stacey, G. (2014). The kinase LYK5 is a major chitin receptor in *Arabidopsis* and forms a chitin-induced complex with related kinase CERK1. *Elife, 3*, e03766.

Castel, S., & Martienssen, R. (2013). RNA interference in the nucleus: Roles for small RNAs in transcription, epigenetics and beyond. *Nature Reviews Genetics, 14*, 100–112.

Cenci, A., & Rouard, M. (2017). Evolutionary analyses of GRAS transcription factors in angiosperms. *Frontiers in Plant Science, 8*, 273.

Chabaud, M., Genre, A., Sieberer, B. J, Faccio, A., Fournier, J., Novero, M., Barker, D. G, & Bonfante, P. (2011). Arbuscular mycorrhizal hyphopodia and germinated spore exudates trigger Ca^{2+} spiking in the legume and nonlegume root epidermis. *New Phytologist, 189*, 347–355.

Chabaud, M., Gherbi, H., Pirolles, E., Vaissayre, V., Fournier, J., Moukouanga, D., Franche, C., Bogusz, D., Tisa, L. S, Barker, D. G., & Svistoonoff, S. (2016). Chitinase-resistant hydrophilic symbiotic factors secreted by *Frankia* activate both Ca^{2+} spiking and NIN gene expression in the actinorhizal plant *Casuarina glauca*. *New Phytologist, 209*, 86–93.

Chandra, S., Stennis, M., & Low, P. S. (1997). Measurement of Ca^{2+} fluxes during elicitation of the oxidative burst in aequorin-transformed tobacco cells. *Journal of Biological Chemistry, 272*, 28274–28280.

Chaturvedi, A., Cruz Corella, J., Robbins, C., Loha, A., Menin, L., Gasilova, N., Masclaux, F. G., Lee, S. J., & Sanders, I. R. (2021). The methylome of the model arbuscular mycorrhizal fungus, *Rhizophagus irregularis*, shares characteristics with early diverging fungi and *Dikarya*. *Communications Biology, 4*, 901.

Cook, C. E., Whichard, L. P., Turner, B., Wall, M. E., & Egley, G. H. (1966). Germination of witchweed (*Striga lutea* Lour.): Isolation and properties of a potent stimulant. *Science, 154*, 1189–1190.

Cope, K. R., Bascaules, A., Irving, T. B., Venkateshwaran, M., Maeda, J., Garcia, K., Rush, T. A., Ma, C., Labbé, J., Jawdy, S., Steigerwald, E., Setzke, J., Fung, E., Schnell, K. G., Wang, Y., Schlief, N., Bücking, H., Strauss, S. H., Maillet, F., … Expand, A. (2019). The ectomycorrhizal fungus *Laccaria bicolor* produces lipochitooligosaccharides and uses the common symbiosis pathway to colonize *Populus* roots. *Plant Cell, 31*, 2386–2410.

Couzigou, J. M., Lauressergues, D., Andre, O., Gutjahr, C., Guillotin, B., Becard, G., & Combier, J. P. (2017). Positive gene regulation by a natural protective miRNA enables arbuscular mycorrhizal symbiosis. *Cell Reports, 20*, 1339–1350.

Cui, H., Kong, D., Liu, X., & Hao, Y. S. (2014). SCR-LIKE 23 and SHORT-ROOT control bundle sheath cell fate and function in *Arabidopsis thaliana*. *Plant Journal, 78*, 319–327.

Czaja, L. F., Hogekamp, C., Lamm, P., Maillet, F., Martinez, E. A., Samain, E., Dénarié, J., Küster, H., & Hohnjec, N. (2012). Transcriptional responses toward diffusible signals from symbiotic microbes reveal MtNFP- and MtDMI3-dependent reprogramming of host gene expression by arbuscular mycorrhizal fungal lipochitooligosaccharides. *Plant Physiology, 159*, 1671–1685.

Dallaire, A., Manley, B. F., Wilkens, M., Bista, I., Quan, C., Evangelisti, E., Bradshaw, C. R., Ramakrishna, N. B., Schornack, S., Butter, F., Paszkowski, U., & Miska, E. A. (2021). Transcriptional activity and epigenetic regulation of transposable elements in the symbiotic fungus *Rhizophagus irregularis*. *Genome Research, 31*, 2290–2302.

Das, D., Paries, M., Hobecker, K., Gigl, M., Dawid, C., Lam, H., Zhang, J., Chen, M., & Gutjahr, C. (2022). Phosphate starvation response transcription factors enable arbuscular mycorrhiza symbiosis. *Nature Communications, 13*, 477.

Das, S., & Sarkar, S. (2024). Arbuscular mycorrhizal fungal contribution towards plant resilience to drought conditions. *Frontiers in Fungal Biology, 5*, 1355999.

Debellé, F. (2020). The common symbiotic signaling pathway. In F. J. de Bruijn (Ed.), *The model legume* Medicago truncatula (pp. 523–528). John Wiley & Sons, Ltd.

Delaux, P. M., Bécard, G., & Combier, J. P. (2013). NSP1 is a component of the Myc signaling pathway. *New Phytologist, 199*, 59–65.

Desaki, Y., Kohari, M., Shibuya, N., & Kaku, H. (2019). MAMP-triggered plant immunity mediated by the LysM-receptor kinase CERK1. *Journal of General Plant Pathology, 85*, 1–11.

Devers, E. A., Branscheid, A., May, P., Krajinski, F., & Bäumlein, H. (2011). Stars and symbiosis: MicroRNA- and microRNA*-mediated transcript cleavage involved in arbuscular mycorrhizal symbiosis. *Plant Physiology, 156*, 1990–2010.

Díaz, V., Villalobos, M., Arriaza, K., Flores, K., Hernández-Saravia, L. P., & Velásquez, A. (2025). Decoding the dialog between plants and arbuscular mycorrhizal fungi: A molecular genetic perspective. *Genes, 16*(2), 143.

Edel, K. H, Marchadier, E., Brownlee, C., Kudla, J., & Hetherington, A. M. (2017). The evolution of calcium-based signalling in plants. *Current Biology, 27*, R667–R679.

Ehrhardt, D. W., Wais, R., & Long, S. R. (1996). Calcium spiking in plant root hairs responding to rhizobium nodulation signals. *Cell, 85*, 673–681.

Erickson, J., Weckwerth, P., Romeis, T., & Lee, J. (2022). What's new in protein kinase/phosphatase signalling in the control of plant immunity? *Essays in Biochemistry, 66*, 621–634.

Etemadi, M., Gutjahr, C., Couzigou, J. M., Zouine, M., Lauressergues, D., Timmers, A., Audran, C., Bouzayen, M., Becard, G., & Combier, J. P. (2014). Auxin perception is required for arbuscule development in arbuscular mycorrhizal symbiosis. *Plant Physiology, 166*, 281–292.

Fei, Q., Xia, R., & Meyers, B. C. (2013). Phased, secondary, small interfering RNAs in posttranscriptional regulatory networks. *Plant Cell, 25*, 2400–2415.

Feng, F., Sun, J., Radhakrishnan, G. V., Lee, T., Bozsóki, Z., Fort, S., Gavrin, A., Gysel, K., Thygesen, M. B., Andersen, K. R., Radutoiu, S., Stougaard, J., & Oldroyd, G. E. D. (2019). A combination of chitooligosaccharide and lipochitooligosaccharide recognition promotes arbuscular mycorrhizal associations in *Medicago truncatula*. *Nature Communications, 10*, 5047.

Fiorilli, V., Vannini, C., Ortolani, F., Garcia-Seco, D., Chiapello, M., Novero, M., Domingo, G., Terzi, V., Morcia, C., Bagnaresi, P., Moulin, L., Bracale, M., & Bonfante, P. (2018). Omics approaches revealed how arbuscular mycorrhizal symbiosis enhances yield and resistance to leaf pathogen in wheat. *Scientific Reports, 8*, 9625.

Floss, D. S., Gomez, S. K., Park, H.-J., MacLean, A. M., Müller, L. M., Bhattarai, K. K., Lévesque-Tremblay, V., Maldonado-Mendoza, I. E., & Harrison, M. J. (2017). A transcriptional program for arbuscule degeneration during AM symbiosis is regulated by MYB1. *Current Biology, 27*, 1206–1212.

Floss, D. S., Levy, J. G., Lévesque-Tremblay, V., Pumplin, N., & Harrison, M. J. (2013). DELLA proteins regulate arbuscule formation in arbuscular mycorrhizal symbiosis. *Proceedings of the National Academy of Sciences of the United States of America, 110*, E5025–E5034.

Foo, E., Yoneyama, K., Hugill, C., Quittenden, L., & Reid, J. (2012). Strigolactones and the regulation of pea symbioses in response to nitrate and phosphate deficiency. *Molecular Plant, 6*, 76–87.

Founoune-Mboup, H., Diallo, B., Adigoun, R. F. R., Kane, A., & Fall, A. F. (2024). Contribution of arbuscular mycorrhizal fungi to the bioavailability of micronutrients (iron and zinc) in millet accessions. *Frontiers in Plant Science, 15*, 1364469.

Genre, A., Chabaud, M., Balzergue, C., Puech-Pagès, V., Novero, M., Rey, T., Fournier, J., Rochange, S., Bécard, G., Bonfante, P., & Barker, D. G. (2013). Short-chain chitin oligomers from arbuscular mycorrhizal fungi trigger nuclear Ca^{2+} spiking in *Medicago truncatula* roots and their production is enhanced by strigolactone. *New Phytologist, 198*, 190–202.

Genre, A., Ortu, G., Bertoldo, C., Martino, E., & Bonfante, P. (2009). Biotic and abiotic stimulation of root epidermal cells reveals common and specific responses to arbuscular mycorrhizal fungi. *Plant Physiology, 149*, 1424–1434.

Gianinazzi-Pearson, V. (1996). Plant cell responses to arbuscular mycorrhizal fungi: Getting to the roots of the symbiosis. *Plant Cell, 8*, 1871–1883.

Gibelin-Viala, C., Amblard, E., Puech-Pages, V., Bonhomme, M., Garcia, M., Bascaules-Bedin, A., Fliegmann, J., Wen, J., Mysore, K. S., le Signor, C., Jacquet, C., & Gough, C. (2019). The *Medicago truncatula* LysM receptor-like kinase LYK9 plays a dual role in immunity and the arbuscular mycorrhizal symbiosis. *New Phytologist, 223*, 1516–1529.

Giovannetti, M., Mari, A., Novero, M., & Bonfante, P. (2015). Early *Lotus japonicus* root transcriptomic responses to symbiotic and pathogenic fungal exudates. *Frontiers in Plant Science, 6*, 480.

Gobbato, E., Marsh, J. F., Vernié, T., Wang, E., Maillet, F., Kim, J., Miller, J. B., Sun, J., Bano, S. A., Ratet, P., Mysore, K. S., Dénarié, J., Schultze, M., & Oldroyd, G. E. D. (2012). A GRAS-type transcription factor with a specific function in mycorrhizal signaling. *Current Biology, 22*, 2236–2241.

Gough, C., & Bécard, G. (2016). Strigolactones and lipo-chitooligosaccharides as molecular communication signals in the arbuscular mycorrhizal symbiosis. In F. Martin (Ed.), *Molecular mycorrhizal symbiosis* (pp. 107–124). John Wiley & Sons.

Grenzi, M., Resentini, F., Vanneste, S., Zottini, M., Bassi, A., & Costa, A. (2021). Illuminating the hidden world of calcium ions in plants with a universe of indicators. *Plant Physiology, 187*, 550–571.

Gutjahr, C., & Parniske, M. (2013). Cell and developmental biology of arbuscular mycorrhiza symbiosis. *Annual Review of Cell and Developmental Biology, 29*, 593–617.

Hamiaux, C., Drummond, R. S. M., Janssen, B. J., Ledger, S. E., Cooney, J. M., Newcomb, R. D., & Snowden, K. C. (2012). DAD2 is an α/β hydrolase likely to be involved in the perception of the plant branching hormone, strigolactone. *Current Biology, 22*, 2032–2036.

Hannan Parker, A., Wilkinson, S. W, & Ton, J. (2022). Epigenetics: A catalyst of plant immunity against pathogens. *New Phytologist, 233*, 66–83.

Helber, N., Wippel, K., Sauer, N., Schaarschmidt, S., Hause, B., & Requena, N. (2011). A versatile monosaccharide transporter that operates in the arbuscular mycorrhizal fungus *Glomus* sp. is crucial for the symbiotic relationship with plants. *Plant Cell, 23*, 3812–3823.

Heo, J.-O., Chang, K. S., A Kim, I., Lee, M.-H., Lee, S. A., Song, S.-K., Lee, M. M., & Lim, J. (2011). Funneling of gibberellin signaling by the GRAS transcription regulator scarecrow-like 3 in the *Arabidopsis* root. *Proceedings of the National Academy of Sciences of the United States of America, 108*, 2166–2171.

Hobecker, K. V., Reynoso, M. A., Bustos-Sanmamed, P., Wen, J., Mysore, K. S., Crespi, M., Blanco, F. A., & Zanetti, M. E. (2017). The microRNA390/TAS3 pathway mediates symbiotic nodulation and lateral root growth. *Plant Physiology, 174*, 2469–2486.

Hofferek, V., Mendrinna, A., Gaude, N., Krajinski, F., & Devers, E. A. (2014). MiR171h restricts root symbioses and shows like its target NSP2 a complex transcriptional regulation in *Medicago truncatula*. *BMC Plant Biology, 14*, 199.

Ito, T., & Fukazawa, J. (2021). SCARECROW-LIKE3 regulates the transcription of gibberellin-related genes by acting as a transcriptional co-repressor of GAI-ASSOCIATED FACTOR1. *Plant Molecular Biology, 105*, 463–482.

Jiang, Y., & Ding, P. (2023). Calcium signaling in plant immunity: A spatiotemporally controlled symphony. *Trends in Plant Science, 28*, 74–89.

Keinath, N. F, Waadt, R., Brugman, R., Schroeder, J. I, Grossmann, G., Schumacher, K., & Krebs, M. (2015). Live cell imaging with R-GECO1 sheds light on flg22- and chitin-induced transient $[Ca^{2+}]$cyt patterns in *Arabidopsis*. *Molecular Plant, 8*, 1188–1200.

Kelner, A., Leitão, N., Chabaud, M., Charpentier, M., & De Carvalho-Niebel, F. (2018). Dual color sensors for simultaneous analysis of calcium signal dynamics in the nuclear and cytoplasmic compartments of plant cells. *Frontiers in Plant Science, 9*, 245.

Khan, W., Costa, C., Souleimanov, A., Prithiviraj, B., & Smith, D. L. (2011). Response of *Arabidopsis thaliana* roots to lipo-chitooligosaccharide from *Bradyrhizobium japonicum* and other chitin-like compounds. *Plant Growth Regulation, 63*, 243–249.

Kim, N. H, Jacob, P., & Dangl, J. L. (2022). Con-Ca^{2+}-tenating plant immune responses via calcium-permeable cation channels. *New Phytologist, 234*, 813–818.

Knight, M. R, Campbell, A. K, Smith, S. M, & Trewavas, A. J. (1991). Transgenic plant aequorin reports the effects of touch and cold-shock and elicitors on cytoplasmic calcium. *Nature, 352*, 524–526.

Köster, P., DeFalco, T. A, & Zipfel, C. (2022). Ca^{2+} signals in plant immunity. *EMBO Journal, 41*, e11074.

Kretzschmar, T., Kohlen, W., Sasse, J., Borghi, L., Schlegel, M., Bachelier, J. B., Reinhardt, D., Bours, R., Bouwmeester, H. J., & Martinoia, E. (2012). A petunia ABC protein controls strigolactone-dependent symbiotic signalling and branching. *Nature, 483*, 341–344.

Lanfranco, L., & Bonfante, P. (2023). Lessons from arbuscular mycorrhizal fungal genomes. *Current Opinion in Microbiology, 75*, 102357.

Lanfranco, L., Bonfante, P., & Genre, A. (2016). The mutualistic interaction between plants and arbuscular mycorrhizal fungi. *Microbiology spectrum, 4*(6), 10–1128.

la Rosa, N. M.-D., Sotillo, B., Miskolczi, P., Gibbs, D. J., Vicente, J., Carbonero, P., Oñate-Sánchez, L., Holdsworth, M. J., Bhalerao, R., Alabadí, D., & Blázquez, M. A. (2014). Large-scale identification of gibberellin-related transcription factors defines group VII ETHYLENE RESPONSE FACTORS as functional DELLA partners. *Plant Physiology, 166*, 1022–1032.

Lauressergues, D., Delaux, P. M., Formey, D., Lelandais-Brière, C., Fort, S., Cottaz, S., Bécard, G., Niebel, A., Roux, C., & Combier, J. (2012). The microRNA miR171h modulates arbuscular mycorrhizal colonization of *Medicago truncatula* by targeting NSP2. *Plant Journal, 72*, 512–522.

Lax, C., Tahiri, G., Patiño-Medina, J. A., Canovas-Marquez, J. T., Perez-Ruiz, J. A., Osorio Concepcion, M., Navarro, E., & Calo, S. (2020). The evolutionary significance of RNAi in the fungal kingdom. *International Journal of Molecular Sciences, 21*, 9348.

Lecourieux, D., Mazars, C., Pauly, N., Ranjeva, R., & Pugin, A. (2002). Analysis and effects of cytosolic free calcium increases in response to elicitors in *Nicotiana plumbaginifolia* cells. *Plant Cell, 14*, 2627–2641.

Lee, S., Kong, M., Harrison, P., & Hijri, M. (2018). Conserved proteins of the RNA interference system in the arbuscular mycorrhizal fungus *Rhizoglomus irregulare* provide new insight into the evolutionary history of *Glomeromycota*. *Genome Biology and Evolution, 10*, 328–343.

Li, X.-R., Sun, J., Albinsky, D., Zarrabian, D., Hull, R., Lee, T., Jarratt-Barnham, E., Chiu, C. H., Jacobsen, A., Soumpourou, E., Albanese, A., Kohlen, W., Luginbuehl, L. H., Guillotin, B., Lawrensen, T., Lin, H., Murray, J., Wallington, E., Harwood, W., … Oldroyd, G. E. H. (2022). Nutrient regulation of lipochitooligosaccharide recognition in plants via NSP1 and NSP2. *Nature Communications, 13*, 6421.

Li, Z., Li, W., Guo, M., Liu, S., Liu, L., Yu, Y., Mo, B., Chen, X., & Gao, L. (2022). Origin, evolution and diversification of plant ARGONAUTE proteins. *Plant Journal, 109*, 1086–1097.

Liang, Y., Cao, Y., Tanaka, K., Thibivilliers, S., Wan, J., Choi, J., Kang, C., Qiu, J., & Stacey, G. (2013). Nonlegumes respond to rhizobial Nod factors by suppressing the innate immune response. *Science, 341*, 1384–1387.

Liu, W., Kohlen, W., Lillo, A., Op den Camp, R., Ivanov, S., Hartog, M., Limpens, E., Jamil, M., Smaczniak, C., Kaufmann, K., Yang, W.-C., Hooiveld, G. J. E. J., Charnikhova, T., Bouwmeester, H. J., Bisseling, T., & Geurts, R. (2011). Strigolactone biosynthesis in *Medicago truncatula* and rice requires the symbiotic GRAS-type transcription factors NSP1 and NSP2. *Plant Cell, 23*, 3853–3865.

Lodhi, G., Kim, Y.-S., Hwang, J.-W., Kim, S.-K., Jeon, Y.-J., Je, J.-Y., Ahn, C.-B., Moon, S.-H., Jeon, B.-T., & Park, P.-J. (2014). Chitooligosaccharide and its derivatives: Preparation and biological applications. *BioMed Research International, 2014*, 1–13.

Lopez-Marquez, D., Del-Espino, A., Lopez-Pagan, N., Rodríguez-Negrete, E. A., Rubio-Somoza, I., Ruiz-Albert, J., Bejarano, E. R., & Beuzon, C. R. (2021). miR825-5p targets the TIR-NBS-LRR gene MIST1 and down-regulates basal immunity against Pseudomonas syringae in *Arabidopsis*. *Journal of Experimental Botany, 72*, 7316–7334.

López-Ráez, J. A., & Pozo, M. J. (2013). *Chemical signalling in the arbuscular mycorrhizal symbiosis: Biotechnological applications* (pp. 215–232). Springer. https://doi.org/10.1007/978-3-642-39317-4_11.

Luan, S., & Wang, C. (2021). Calcium signaling mechanisms across kingdoms. *Annual Review of Cell and Developmental Biology, 37*, 311–340.

Luginbuehl, L., & Oldroyd, G. E. (2016). Calcium signaling and transcriptional regulation in arbuscular mycorrhizal symbiosis. *Molecular Mycorrhizal Symbiosis, 8*, 125–140.

Lutz, S., Bodenhausen, N., Hess, J., Valzano-Held, A., Waelchli, J., Deslandes-Hérold, G., Schlaeppi, K., & van der Heijden, M. G. A. (2023). Soil microbiome indicators can predict crop growth response to large-scale inoculation with arbuscular mycorrhizal fungi. *Nature Microbiology, 8*, 2277–2289.

Maillet, F., Poinsot, V., André, O., Puech-Pagès, V., Haouy, A., Gueunier, M., Cromer, L., Giraudet, D., Formey, D., Niebel, A., Martinez, E. A., Driguez, H., Bécard, G., & Dénarié, J. (2011). Fungal lipochitooligosaccharide symbiotic signals in arbuscular mycorrhiza. *Nature, 469*, 58–63.

Mashiguchi, K., Morita, R., Tanaka, K., Kodama, K., Kameoka, H., Kyozuka, J., Seto, Y., & Yamaguchi, S. (2023). Activation of strigolactone biosynthesis by the DWARF14-LIKE/KARRIKIN-INSENSITIVE2 pathway in mycorrhizal angiosperms, but not in *Arabidopsis*, a non-mycorrhizal plant. *Plant & Cell Physiology, 64*, 1066–1078.

Mendoza-Soto, A. B., Rodríguez-Corral, A. Z., Bojórquez-López, A., Cervantes-Rojo, M., Castro-Martínez, C., & Lopez-Meyer, M. (2022). Arbuscular mycorrhizal symbiosis leads to differential regulation of genes and miRNAs associated with the cell wall in tomato leaves. *Biology, 11*, 854.

Mithöfer, A., Ebel, J., Bhagwat, A. A, Boller, T., & Neuhaus-Url, G. (1999). Transgenic aequorin monitors cytosolic calcium transients in soybean cells challenged with β-glucan or chitin elicitors. *Planta, 207*, 566–574.

Miwa, H., Sun, J., Oldroyd, G. ED., & Downie, J. A. (2006). Analysis of Nod-factor-induced calcium signaling in root hairs of symbiotically defective mutants of *Lotus japonicus*. *Molecular Plant–Microbe Interactions, 19*, 914–923.

Miyawaki, A., Griesbeck, O., Heim, R., & Tsien, R. Y. (1999). Dynamic and quantitative Ca^{2+} measurements using improved cameleons. *Proceedings of the National Academy of Sciences of the United States of America, 96*, 2135–2140.

Nadeem, S. M., Khan, M. Y., Waqas, M. R., Binyamin, R., Akhtar, S., & Zahir, Z. A. (2017). Arbuscular mycorrhizas: An overview. *Arbuscular mycorrhizas and stress tolerance of plants*, 1–24.

Nagahashi, G., & Douds, D. D., Jr. (2011). The effects of hydroxy fatty acids on the hyphal branching of germinated spores of AM fungi. *Fungal Biology, 115*, 351–358.

Navazio, L., Moscatiello, R., Genre, A., Novero, M., Baldan, B., Bonfante, P., & Mariani, P. (2007). A diffusible signal from arbuscular mycorrhizal fungi elicits a transient cytosolic calcium elevation in host plant cells. *Plant Physiology, 144*, 673–681.

Nazir, A., Sarfraz, W., Allah, D., Khalid, N., Farid, M., Shafiq, M., Bareen, F.-e-, Rizvi, Z. F., & Naeem, N. (2023). Synergistic impact of two autochthonous saprobic fungi (*A. niger* and *T. pseudokoningii*) on the growth, ionic contents, and metals uptake in *Brassica juncea* L. and *Vigna radiata* L. under tannery solid waste contaminated soil. *International Journal of Phytoremediation, 25*(11), 1488–1500. 10.1080/15226514.2023.2166457.

Ngou, B. P. M., Ding, P., & Jones, J. D. G. (2022). Thirty years of resistance: Zig-zag through the plant immune system. *Plant Cell, 34*, 1447–1478.

Noreen, S., Yaseen, T., Iqbal, J., Abbasi, B. A., Ijaz, S., Ali, I., & Chalgham, W. (2023). Morphological and molecular characterization of arbuscular mycorrhizal fungi and its influence on soil physiochemical properties and plant nutrition. *ACS Omega, 8*, 32468–32482. http://pubs.acs.org/journal/acsodf.

Oldroyd, G. E. D. (2013). Speak, friend, and enter: Signalling systems that promote beneficial symbiotic associations in plants. *Nature Reviews Microbiology, 11*, 252–263.

Parniske, M. (2008). Arbuscular mycorrhiza: The mother of plant root endosymbioses. *Nature Reviews Microbiology, 6*, 763–775.

Perez-Arques, C., Navarro-Mendoza, M. I., Murcia, L., Navarro, E., Garre, V., & Nicolas, F. E. (2020). A non-canonical RNAi pathway controls virulence and genome stability in *Mucorales*. *PLoS Genetics, 16*, 1008611.

Pimprikar, P., Carbonnel, S., Paries, M., Katzer, K., Klingl, V., Bohmer, M. J., Karl, L., Floss, D. S., Harrison, M. J., Parniske, M., & Gutjahr, C. (2016). A CCaMK-CYCLOPS-DELLA complex activates transcription of RAM1 to regulate arbuscule branching. *Current Biology, 26*, 987–998.

Plattner, H., & Verkhratsky, A. (2015). The ancient roots of calcium signalling evolutionary tree. *Cell Calcium, 57*, 123–132.

Pradhan, M., Baldwin, I. T., & Pandey, S. P. (2023). Argonaute7 (AGO7) optimizes arbuscular mycorrhizal fungal associations and enhances competitive growth in *Nicotiana attenuata*. *New Phytologist, 40*, 382–398.

Pradhan, M., & Requena, N. (2022). Distinguishing friends from foes: Can smRNAs modulate plant interactions with beneficial and pathogenic organisms? *Current Opinion in Plant Biology, 69*, 102259.

Pysh, L. D., Wysocka-Diller, J. W., Camilleri, C., Bouchez, D., & Benfey, P. N. (1999). The GRAS gene family in *Arabidopsis*: Sequence characterization and basic expression analysis of the SCARECROW-LIKE genes. *Plant Journal, 18*, 111–119.

Qiao, A., Gao, Z., & Roth, R. (2023). A perspective on cross-kingdom RNA interference in mutualistic symbioses. *New Phytologist, 240*, 68–79.

Ranf, S., Eschen-Lippold, L., Pecher, P., Lee, J., & Scheel, D. (2011). Interplay between calcium signalling and early signalling elements during defence responses to microbe- or damage-associated molecular patterns. *The Plant Journal, 68*, 100–113.

Ravi, B., Sanyal, S. K, & Pandey, G. K. (2023). Calcium decoders and their targets: The holy alliance that regulate cellular responses in stress signaling. *Advances in Protein Chemistry and Structural Biology, 134*, 371–439.

Rey, T., & Jacquet, C. (2018). Symbiosis genes for immunity and vice versa. *Current Opinion in Plant Biology, 44*, 64–71.

Rush, T. A., Puech-Pagès, V., Bascaules, A., Jargeat, P., Maillet, F., Haouy, A., Maës, A. Q., Carriel, C. C., Khokhani, D., Keller-Pearson, M., Tannous, J., Cope, K. R., Garcia, K., Maeda, J., Johnson, C., Kleven, B., Choudhury, Q. J., Labbé, J., Swift, C., … Ané, J. M. (2020). Lipo-chitooligosaccharides as regulatory signals of fungal growth and development. *Nature Communications, 11*, 3897.

Ruyter-Spira, C., Al-Babili, S., van der Krol, S., & Bouwmeester, H. (2012). The biology of strigolactones. *Trends in Plant Science, 18*, 72–83.

Saeed, H. S., Munir, M. aì, Samra, S., Farooq, A., Mahmood, H., Ahmad, I., Irfan Akram, M., Iqbal, R., Elsadek, M. F., Almutairi, S. M., Ditta, A., & Daraz, U. (2025). Screening of biosurfactants producing fungi from infected *Citrus sinensis* (L.). *Polish Journal of Environmental Studies, 34*(5), 5261–5270. 10.15244/pjoes/190588.

Sharda, J. N., & Koide, R. T. (2008). Can hypodermal passage cell distribution limit root penetration by mycorrhizal fungi? *New Phytologist, 180*, 696–701.

Sieberer, B. J., Chabaud, M., Fournier, J., Timmers, A. C. J., & Barker, D. G. (2012). A switch in Ca^{2+} spiking signature is concomitant with endosymbiotic microbe entry into cortical root cells of *Medicago truncatula*. *Plant Journal, 69*, 822–830.

Sieberer, B. J, Chabaud, M., Timmers, A. C, Monin, A., Fournier, J., & Barker, D. G. (2009). A nuclear-targeted cameleon demonstrates intranuclear Ca^{2+} spiking in *Medicago truncatula* root hairs in response to rhizobial nodulation factors. *Plant Physiology, 151*, 1197–1206.

Silvestri, A., Fiorilli, V., Miozzi, L., Accotto, G. P., Turina, M., & Lanfranco, L. (2019). In silico analysis of fungal small RNA accumulation reveals putative plant mRNA targets in the symbiosis between an arbuscular mycorrhizal fungus and its host plant. *BMC Genomics, 20*, 169.

Silvestri, A., Turina, M., Fiorilli, V., Miozzi, L., Venice, F., Bonfante, P., & Lanfranco, L. (2020). Different genetic sources contribute to the small RNA population in the arbuscular mycorrhizal fungus *Gigaspora margarita*. *Frontiers in Microbiology, 11*, 395.

Skiada, V., Avramidou, M., Bonfante, P., Genre, A., & Papadopoulou, K. K. (2020). An endophytic *Fusarium*–legume association is partially dependent on the common symbiotic signalling pathway. *New Phytologist, 226*, 1429–1444.

Smith, N. A., Eamens, A. L., & Wang, M. B. (2011). Viral small interfering RNAs target host genes to mediate disease symptoms in plants. *PLoS Pathogens, 7*, 1002022.

Smith, S. E., & Read, D. J. (2008). *Mycorrhizal symbiosis* (4th ed.). London Academic Publishing.

Song, X., Li, P., Zhai, J., Zhou, M., Ma, L., Liu, B., Jeong, D. H., Nakano, M., Cao, S., Liu, C., Chu, C., Wang, X.-J., Green, P. J., Meyers, B. C., & Cao, X. (2012). Roles of DCL4 and DCL3b in rice phased small RNA biogenesis. *The Plant Journal, 69*, 462–474.

Sun, J., Miller, J. B., Granqvist, E., Wiley-Kalil, A., Gobbato, E., Maillet, F., Cottaz, S., Samain, E., Venkateshwaran, M., Fort, S., Morris, R. J., Ané, J.-M., & Dénarié, J. (2015). Activation of symbiosis signaling by arbuscular mycorrhizal fungi in legumes and rice. *Plant Cell, 27*, 823–838.

Sun, X., Chen, W., Ivanov, S., MacLean, A. M, Wight, H., Ramaraj, T., Mudge, J., Harrison, M. J, & Fei, Z. (2019). Genome and evolution of the arbuscular mycorrhizal fungus *Diversispora epigaea* (formerly *Glomus versiforme*) and its bacterial endosymbionts. *New Phytologist, 221*, 1556–1573.

Sun, X., Xue, B., Jones, W. T., Rikkerink, E., Dunker, A. K., & Uversky, V. N. (2011). A functionally required unfoldome from the plant kingdom: Intrinsically disordered N-terminal domains of GRAS proteins are involved in molecular recognition during plant development. *Plant Molecular Biology, 77*, 205–223.

Tahat, M. M., & Sijam, K. (2012). Arbuscular mycorrhizal fungi and plant root exudates bio-communications in the rhizosphere. *African Journal of Microbiology Research, 6*(46), 7295–7301. https://doi.org/10.5897/AJMR12.2250.

Tehrani, N., & Mitra, R. M. (2023). Plant pathogens and symbionts target the plant nucleus. *Current Opinion in Microbiology, 72*, 102284.

Teng, C., Zhang, H., Hammond, R., Huang, K., Meyers, B. C., & Walbot, V. (2020). Dicer like 5 deficiency confers temperature-sensitive male sterility in maize. *Nature Communications, 11*, 2912.

Thind, S., Chaudhary, M. S, Ditta, A., Hussain, I., Parveen, A., Ullah, N., Mahmood, Q., Al-ashkar, I., & El-Sabagh, A. (2022). Impact of mycorrhizal fungi from different rhizospheric soils on fungal colonization, growth, and chlorophyll contents of *Cenchrus ciliaris*. *Agronomy, 12*(11). 10.3390/agronomy12112644.

Thor, K., & Peiter, E. (2014). Cytosolic calcium signals elicited by the pathogen-associated molecular pattern flg22 in stomatal guard cells are of an oscillatory nature. *New Phytologist, 204*, 873–881.

Tisserant, E., Malbreil, M., Kuo, A., Kohler, A., Symeonidi, A., Balestrini, R., Charron, P., Duensing, N., Dit, F. N. F., Gianinazzi-Pearson, V., Gilbert, L. B., Handa, Y., Herr, J. R., Hijri, M., Koul, R., Kawaguchi, M., Krajinski, F., Lammers, P. J., Masclaux, F. G., … Martin, F. (2013). Genome of an arbuscular mycorrhizal fungus provides insight into the oldest plant symbiosis. *Proceedings of the National Academy of Sciences of the United States of America, 110*, 20117–20122.

Tsuzuki, S., Handa, Y., Takeda, N., & Kawaguchi, M. (2016). Strigolactone-induced putative secreted protein 1 is required for the establishment of symbiosis by the arbuscular mycorrhizal fungus *Rhizophagus irregularis*. *Molecular Plant-Microbe Interactions, 4*, 277–286.

Venkateshwaran, M., Jayaraman, D., Chabaud, M., Genre, A., Balloon, A. J., Maeda, J., Forshey, K., den Os, D., Kwiecien, N. W., Coon, J. J., Barker, D. G., & Ané, J.-M. (2015). A role for the mevalonate pathway in early plant symbiotic signaling. *Proceedings of the National Academy of Sciences of the United States of America, 112*, 9781–9786.

Volk, H., Marton, K., Flajšman, M., Radišek, S., Tian, H., Hein, I., Podlipnik, Č, Thomma, B., Košmelj, K., Javornik, B., & Berne, S. (2019). Chitin-binding protein of Verticillium nonalfalfae disguises fungus from plant chitinases and suppresses chitin-triggered host immunity. *Molecular Plant–Microbe Interactions, 32*, 1378–1390.

Volpe, V., Carotenuto, G., Berzero, C., Cagnina, L., Puech-Pagès, V., & Genre, A. (2020). Short chain chito-oligosaccharides promote arbuscular mycorrhizal colonization in *Medicago truncatula*. *Carbohydrate Polymers, 229*, 115505.

Wang, P., Jiang, H., Boeren, S., Dings, H., Kulikova, O., Bisseling, T., & Limpens, E. (2021). A nuclear-targeted effector of *Rhizophagus irregularis* interferes with histone 2B mono-ubiquitination to promote arbuscular mycorrhisation. *New Phytologist, 230*, 1142–1155.

Wang, S., Xie, X., Che, X., Lai, W., Ren, Y., Fan, X., Hu, W., Tang, M., & Chen, H. (2023a). Host and virus-induced gene silencing of HOG1-MAPK cascade genes in Rhizophagus irregularis inhibit arbuscule development and reduce resistance of plants to drought stress. *Plant Biotechnology Journal, 4*, 866–883.

Wang, T., Gasciolli, V., Gaston, M., Medioni, L., Cumener, M., Buendia, L., Yang, B., Bono, J. J, He, G., & Lefebvre, B. (2023b). LysM receptor-like kinases involved in immunity perceive lipo-chitooligosaccharides in mycotrophic plants. *Plant Physiology, 192*, 1435–1448.

Wang, W., Shi, J., Xie, Q., Jiang, Y., Yu, N., & Wang, E. (2017). Nutrient exchange and regulation in arbuscular mycorrhizal symbiosis. *Molecular Plant, 10*, 1147–1158.

Weiberg, A., Wang, M., Lin, F. M., Zhao, H., Zhang, Z., Kaloshian, I., Huang, H. D., & Jin, H. (2013). Fungal small RNAs suppress plant immunity by hijacking host RNA interference pathways. *Science, 342*, 118–123.

Wong-Bajracharya, J., Singan, V. R., Monti, R., Plett, K. L., Ng, V., Grigoriev, I. V., Martin, F. M., Anderson, I. C., & Plett, J. M. (2022). The ectomycorrhizal fungus Pisolithus microcarpus encodes a microRNA involved in cross-kingdom gene silencing during symbiosis. *Proceedings of the National Academy of Sciences of the United States of America, 119*, 2103527119.

Wu, C., Wu, C., Qu, J., Liu, L., Kang, H., Sun, H., Zhang, Y., Ghorbani, A., & Pehlivan, N. (2021). *Quo vadis*: Signaling molecules and small secreted proteins from mycorrhizal fungi at the early stage of mycorrhiza formation. *Symbiosis, 85*(2), 123–143. https://doi.org/10.1007/S13199-021-00793-1.

Xiaoyu, A., & Zhu, Y. (2024). The role of arbuscular mycorrhizal fungi for plant from the symbiotic perspective. *Medscien, 1*(10). https://doi.org/10.61173/gc7mmm83.

Xu, H., & Heath, M. C. (1998). Role of calcium in signal transduction during the hypersensitive response caused by basidiospore-derived infection of the cowpea rust fungus. *Plant Cell, 10*, 585–598.

Xu, Y., Zhu, S., Liu, F., Wang, W., Wang, X., Han, G., & Cheng, B. (2018). Identification of arbuscular mycorrhiza fungi responsive microRNAs and their regulatory network in maize. *International Journal of Molecular Sciences, 19*, 3201.

Xue, L., Cui, H., Buer, B., Vijayakumar, V., Delaux, P.-M., Junkermann, S., & Bucher, M. (2015). Network of GRAS transcription factors involved in the control of arbuscule development in *Lotus japonicus*. *Plant Physiology, 167*, 854–871.

Yaseen, T., Ali, K., Munsif, F., Rab, A., Ahmad, M., Israr, M., & Baraich, A. K. (2016a). Influence of arbuscular mycorrhizal fungi, *Rhizobium* inoculation, and rock phosphate on growth and quality of lentil. *Pakistan Journal of Botany, 48*(5), 2101–2107.

Yaseen, T., Burni, T., & Hussain, F. (2011). Effect of arbuscular mycorrhizal inoculation on nutrient uptake, growth and productivity of cowpea (*Vigna unguiculata*) varieties. *African Journal of Biotechnology, 10*(43), 8593–8598. https://doi.org/10.5897/ajb10.1494.

Yaseen, T., Hussain, F., & Noor, M. (2013). Change in growth and productivity of Burgundy due to rock phosphate, VAM, and *Rhizobium* inoculation. *Sarhad Journal of Agriculture, 29*(4), 537–542.

Yaseen, T., Hussain, F., & Shakeel, M. (2016b). Effect of *Rhizobium*, VAM, and rock phosphate inoculation on growth and productivity and isolation of VAMF spores of *Pisum sativum*. *Pure and Applied Biology, 5*(3), 564–572. https://dx.doi.org/10.19045/bspab.2016.50072.

Yildirir, G., Sperschneider, J., Malar, C. M., Chen, E. C. H., Iwasaki, W., Cornell, C., & Corradi, N. (2022). Long reads and Hi-C sequencing illuminate the two-compartment genome of the model arbuscular mycorrhizal symbiont *Rhizophagus irregularis*. *New Phytologist, 233*, 1097–1107.

Yizhu, L, Imtiaz, M., Ditta, A., Shahid Rizwan, M., Ashraf, M., Mehmood, S., … Tu, S. (2020). Response of growth, antioxidant enzymes and root exudates production towards As stress in *Pteris vittata* and in *Astragalus sinicus* colonized by arbuscular mycorrhizal fungi. *Environmental Science and Pollution Research, 27*(2), 2340–2352. 10.1007/s11356-019-06785-5.

Yoneyama, K., Xie, X., Kim, H. I., Kisugi, T., Nomura, T., Sekimoto, H., Yokota, T., & Yoneyama, K. (2012). How do nitrogen and phosphorus deficiencies affect strigolactone production and exudation? *Planta, 235*, 1197–1207.

Yoshida, H., Hirano, K., Sato, T., Mitsuda, N., Nomoto, M., Maeo, K., Koketsu, E., Mitani, R., Kawamura, M., Ishiguro, S., Tada, Y., Ohme-Takagi, M., Matsuoka, M., & Ueguchi-Tanaka, M. (2014). DELLA protein functions as a transcriptional activator through the DNA binding of the indeterminate domain family proteins. *Proceedings of the National Academy of Sciences of the United States of America, 111*, 7861–7866.

Yoshida, S., Kameoka, H., Tempo, M., Akiyama, K., Umehara, M., Yamaguchi, S., Hayashi, H., Kyozuka, J., & Shirasu, K. (2012). The D3 F-box protein is a key component in host strigolactone responses essential for arbuscular mycorrhizal symbiosis. *New Phytologist, 196*, 1208–1216.

Yu, H., Bai, F., Ji, C., Fan, Z., Luo, J., Ouyang, B., Deng, X., Xiao, S., Bisseling, T., Limpens, E., & Pan, Z. (2023). Plant lysin motif extracellular proteins are required for arbuscular mycorrhizal symbiosis. *Proceedings of the National Academy of Sciences of the United States of America, 120*, e2301884120.

Yuan, M., Ngou, B. PM., Ding, P., & Xin, X. F. (2021). PTI-ETI crosstalk: An integrative view of plant immunity. *Current Opinion in Plant Biology, 62*, 102030.

Yuan, P., Luo, F., Gleason, C., & Poovaiah, B. W. (2022). Calcium/calmodulin-mediated microbial symbiotic interactions in plants. *Frontiers in Plant Science, 13*, 984909.

Zhang, C., He, J., Dai, H., Wang, G., Zhang, X., Wang, C., Shi, J., Chen, X., Wang, D., & Wang, E. (2021). Discriminating symbiosis and immunity signals by receptor competition in rice. *Proceedings of the National Academy of Sciences of the United States of America, 118*, e2023738118.

Zhang, M., & Kong, X. (2021). How plants discern friends from foes. *Trends in Plant Science, 27*, 107–109.

Zhang, T., Zhao, Y. L., Zhao, J. H., Wang, S., Jin, Y., Chen, Z. Q., Fang, Y. Y., Hua, C. L., Ding, S. W., & Guo, H. S. (2016). Cotton plants export microRNAs to inhibit virulence gene expression in a fungal pathogen. *Nature Plants, 2*, 16153.

Zipfel, C., & Oldroyd, G. E. D. (2017). Plant signalling in symbiosis and immunity. *Nature, 543*, 328–336.

Chapter 17

Arbuscular mycorrhizal fungi and plant secondary metabolite production

Saliha Rukhsar[a], Ayesha Noor Awan[a], Samra saeed[a], Muhammad Usman Jamshaid[b] and Nousheen Yousaf[a]

[a] *Department of Botany, Government College University, Lahore, Pakistan,* [b] *School of Agriculture and Biology, Shanghai Jiao Tong University, Shanghai, China*

17.1 Introduction to arbuscular mycorrhizal fungi

The structures known as arbuscules, which are present in the cortical cells of many plant roots, are the source of the term "arbuscular." For arbuscular mycorrhizal fungi, arbuscules and storage vesicles are regarded as symbiosis markers (Schüßler et al., 2001). The term "vesicular-arbuscular mycorrhiza" was used for many decades but has been abandoned because it was found that only 80% of arbuscular mycorrhizal fungi form vesicles. However, the term "arbuscular" is still kept even though there is a greater appreciation for the variety of fungal structures (Smith & Read, 2008). Arbuscular mycorrhizal fungi exhibit a biotrophic life cycle, meaning that they require a living host plant for their reproductive processes. Unlike most fungi, AMFs lack a known sexual reproductive cycle. Their reproduction is primarily asexual through spores. Arbuscular mycorrhizal fungi is classified based on DNA sequence under the subphylum Glomeromycotina within the phylum Mucoromycota, which was previously classified under Glomeromycota (Spatafora et al., 2016).

They form obligately symbiotic associations with the roots of nearly 80% of terrestrial plants. As for the plants, they include the angiosperms and gymnosperms as well as the sporophytes of the pteridophytes, which all possess true roots, along with the gametophytes of some liverworts and pteridophytes, which do not (Read et al., 2000). Such relationships facilitate an exchange of nutrients between the fungus and the host plant (Smith & Read, 2008). The life cycle encompasses six essential stages, each characterized by distinct morphological and physiological attributes. This form of ancient symbiosis is over 450 million years old and exists in 80% of terrestrial plant species (Klironomos et al., 2000). In addition, it has been proposed that AM fungi increase plant diversity. Without AM fungi, many plants may find it difficult to coexist (Van der Heijden et al., 2008). According to a recent study, AM fungi have an impact on a plant community's stability over time (Lei et al., 2014). By influencing various plant species and reducing productivity variability over time, they achieve this.

17.1.1 Characteristics of arbuscular mycorrhizal fungi

The presence of branching structures called arbuscules, which form in the root cortical cells of host plants and facilitate symbiotic exchange, is a characteristic of arbuscular mycorrhizas. Carbon compounds can move from the plant to the fungal cell and nutrients from the fungal cell to the plant, thanks to this exchange system (Hock & Verma, 1995; Smith & Read, 1997). Types of soil's propagules that initiate new colonization include spores, infected root fragments, and mycelium. Formation of roots, which consist of some morphologically distinct types of arbuscular mycorrhiza, serves as an interface between the symbionts for the cross-talk of signal molecules and other nutrients activities of extraradical mycelia in the soil and their sporulation. The active processes of root development, coupled with fungal colonization, result in the formation of AM-root systems (Brundrett & Tedersoo, 2018).

Advancing Sustainable Agriculture Through Arbuscular Mycorrhizal Fungi. DOI: https://doi.org/10.1016/B978-0-443-45130-0.00004-7

17.1.2 Diversity of arbuscular mycorrhizal fungi

About 300,000 plant species and 200 plant families, comprising 1000 genera, have been found to exhibit the arbuscular mycorrhizal association (Bagyaraj, 1991). Mycorrhizal roots are as common in plants as photosynthetic leaves (Mosse, 1986). The AM fungi belong to the order Glomales and phylum Zygomycota (Redecker et al., 2000). However, they were recently assigned to the phylum "Glomeromycota" (Schussler et al., 2001). There are 150 species, 10 genera, 8 families, and 4 orders in the Glomeromycota. *Gigaspora, Glomus, Scutellospora*, and *Acaulospora* are the common genera (Schüßler, 2005). Genetic diversity: the high variability in ribosomal DNA (SSU, ITS, and LSU regions) aids in species identification (Öpik et al., 2013).

17.1.2.1 Evolution of arbuscular mycorrhizal fungi

The late 1800s saw the discovery and description of arbuscular mycorrhizas. Their widespread presence in many plant groups, especially in tropical regions, became apparent quickly (Gallaud, 1905). But until the middle of the 1950s, little was known about their purpose. Except for the work of Peyronel's, (1923) the majority of writings concerning the identity of the fungi before 1953 can be ignored. He demonstrated that the hyphae of the endophyte could be linked to the soil-dwelling sporocarps of fungi belonging to the Endogonaceae family. In a later important review, Butler (1939) concurred that the fungi referred to as *Rhizophagus* were probably imperfect members of the Endogonaceae, which at the time comprised the majority of fungi currently categorized under Glomeromycota. A species of *Endogone*, later identified as *Glomus*, colonized mycorrhizal strawberry plants, as Mosse's work (1953) convincingly showed. The modern era of AM research began with this. Mosse quickly built on these preliminary discoveries, as did Baylis, Germann, Nicolson, and Daft. Through inoculation, they demonstrated how fungi belonging to the Endogonaceae family developed symbiotic relationships with a wide variety of plant species (Smith & Read, 2008). The easily readable book "A History of Research on Arbuscular Mycorrhiza" (Koide & Mosse, 2004) provides additional details on the history of arbuscular mycorrhizal research. A major turning point was reached in 1974 when a number of important concepts were first presented at a successful symposium on endomycorrhizas.

AMFs are among the oldest known plant partners, with fossil evidence tracing their relationship with plants back to at least 460 million years ago, concurring with the time plants began to colonize land. Their long asexual evolutionary history has resulted in unique genetic traits, including high genetic diversity within individuals and a lack of host specificity. This is different from the host specialization found in other types of mycorrhizae (Smith & Read, 2008). Considering their wide evolutionary range and the finding of fossils of mycorrhizal fungal-like structures dating back 450 million years (Redecker et al., 2000), it is believed that the AM symbiosis is ancestral to land plants and probably helped them transition from aquatic to terrestrial environments (Selosse & Le Tacon, 1998). Whether arbuscular mycorrhizal fungi originated from biotrophic fungi that parasitized early land plants or from soil saprotrophs is still unknown. Symbiotic fungal associations in plant groups that are closest to land plants, such as Charo- and Coleochaetophyta, require further study (Rosling et al., 2024).

17.1.3 Morphological features of arbuscular mycorrhizal fungi

The majority of plant communities contain arbuscular mycorrhizas (AM) (Schwab et al., 1991) and are nature's most common symbiotic partnership. Two primary types have been identified by researchers, the *Arum* type and the *Paris* type, despite their widespread distribution both geographically and among plant groups. *Arum maculatum* and *Paris quadrifolia*, the plants from which they were initially described, are the names of these varieties (Gallaud, 1905). Fungal species of the *Arum* type produce arbuscules within cortical cells and intercellular hyphae between them. Arbusculate coils in the root cortex and extensive intracellular hyphal coils are characteristics of the *Paris* type. The *Paris* type almost completely lacks the intercellular phase of colonization. A comprehensive analysis of the literature (Smith & Smith, 1997) showed that the morphology of AM associations in angiosperms can be linked to plant groups. Smith and Smith (1997) discovered that 21 families had either an intermediate form or combined both the *Arum* and *Paris* types, whereas 30 families formed the *Arum* type and 41 families formed the *Paris* type. Generally speaking, only one type was produced by members of a single plant genus. Research on different plant–fungus combinations indicates that the plant species has a significant influence on the arbuscular mycorrhizal morphological type (Cavagnaro et al., 2001; Smith & Smith, 1997).

17.1.3.1 Hyphae

Many plant roots and certain mycorrhizal fungi have arbuscules inside their cortical cells. Nonetheless a wide variety of intraradical structures produced by arbuscular mycorrhizal fungi are identified (Dickson, 2004). Among these are fully

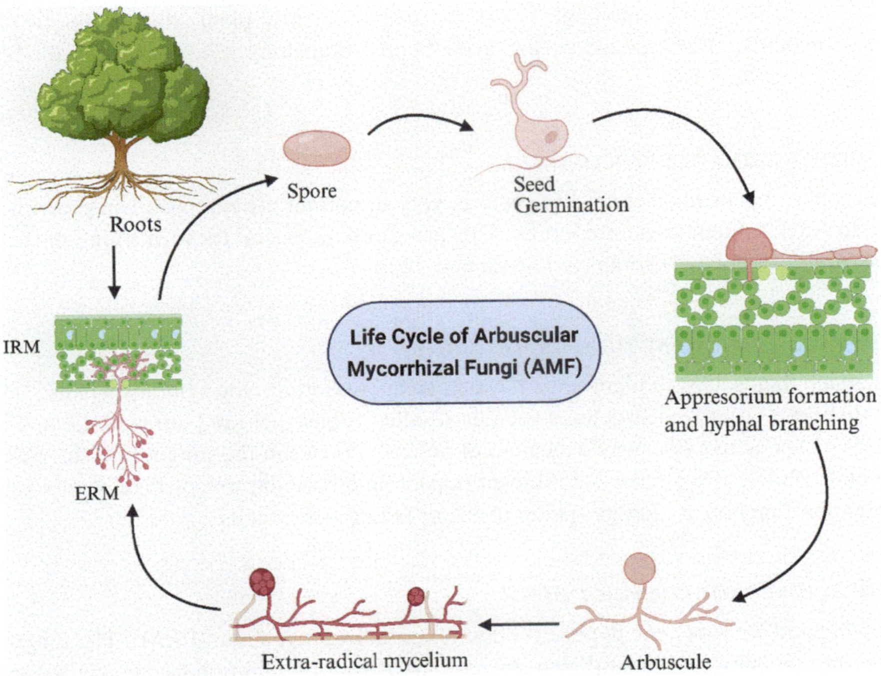

FIGURE 17.1 AMF's life cycle starts when fungal spores germinate within the soil, then hyphae can grow in the direction of the host plant by developing appressoria and penetrating the root cortex. After penetration, highly branched structures, arbuscules, are formed intensively inside the roots that help in nutrient exchange through symbiosis processes. Extraradical mycelium (ERM) is formed by fungus surrounding the colonized roots, producing new spores and completing the life cycle. *AMF*, Arbuscular mycorrhizal fungi.

formed intracellular hyphal coils, which occasionally show up without any arbuscules. Both plant and fungal partners influence the variations in development patterns. This makes it more difficult to identify a symbiosis as "AM" based solely on the intraradical fungal shape. The root, the fungus in the soil, and the vast network of fungi are the three main components of an arbuscular mycorrhiza. Yet, this network does not produce any vegetative, nonsymbiotic structures resembling ectomycorrhizas or intricate mycelial strands like rhizomorphs (Dickson et al., 2007; Smith & Read, 2008).

17.1.3.2 Spores

AM fungi can produce spores as large as 500 μm in diameter. They have thick, resilient walls, a lot of storage lipids, and some carbohydrates. Chitin and occasionally β (1–3) glucan are found in these walls (Gianinazzi-Pearson et al., 1994). Numerous nuclei are present in each spore; estimates vary by species, ranging from 800 to roughly 35,000 (Hosny et al., 1998). According to recent research, all of the nuclei in *Glomus intraradices, Glomus etunicatum*, and probably *Scutellospora castanea* are haploid, like other fungi, and show no indications of ploidy variation (Bianciotto et al., 1995). Although more details are required, it is assumed that this holds true for all AM fungi. The amounts of DNA per spore and estimates of the number of nuclei show that *Glomus versiforme* and *Scutellospora persica* have DNA contents of 1.7 and 3.4 pg, respectively (Viera & Glenn, 1990). On the other hand, fluorimetry of the stained nuclei reveals lower values for different fungi from four genera, ranging from 0.26 to 1.65 pg (Hosny et al., 1998). Although this is not true for all AM fungi, *G. intraradices's* genome size of roughly 16.54 Mb (0.017 pg) is small in comparison to other fungi.

17.1.4 Life cycle of arbuscular mycorrhizal fungi

The life cycle of AMFs includes different stages that occur at different times. This process is complex and revolves around their necessary partnership with plant roots (Fig. 17.1).

17.1.4.1 Spore germination (presymbiosis)

AMF's life cycle starts when fungal spores germinate in the soil under the right circumstances. Without the help of the host plant, this occurs naturally (Gianinazzi-Pearson, 1996; Requena et al., 2007). Fungal colonies develop into unique structures as they grow several centimeters (Giovannetti et al. 1994). Because of the host plant, this asymbiotic phase transitions into the presymbiotic phase, which is characterized by extensive hyphal branching (Giovannetti et al., 1993; Piliarová et al., 2019).

17.1.4.2 *Hyphal growth and host recognition*

This phase of the arbuscular mycorrhizal fungi life cycle is very important. The hyphae can grow in the direction of the host plant's roots due to AMF's chemotaxic properties. This procedure is crucial for identifying the host roots, developing appressoria, and establishing symbiosis (Sbrana & Giovannetti, 2005).

17.1.4.3 *Root colonization and symbiosis formation*

When fungal hyphae attach themselves to plant roots through an appressorium, the symbiosis phase starts. The fungi enter the cortex and create distinct, specialized structures such as vesicles, arbuscules, and intracellular and intercellular hyphae (Giovannetti et al., 1993). The active two-way exchange of nutrients between the fungus and the plant takes place in the arbuscules (Requena et al., 2007). They have a significant impact on arbuscular mycorrhizal symbiosis. In addition, some AMFs produce vesicles that function as storage spaces inside or between root cells.

17.1.4.4 *Extraradical mycelium development*

By penetrating the soil beyond the roots, the hyphae produce extraradical mycelium (ERM). The fungus spreads a network of extraradical hyphae into the surrounding soil from the colonized root. In addition to connecting several plants to form common mycorrhizal networks, this mycelium is in charge of nutrient search (Piliarová et al., 2019).

17.1.4.5 *Asexual spore formation*

The formation of new asexual spores in the ERM marks the end of the life cycle (Requena & Breuninger, 2004). These new spores in the soil are produced by the extraradical hyphae. The spores can endure for a long time in the soil because of their sturdy walls. They serve as the main means of AMF's survival and dissemination (Piliarová et al., 2019).

17.1.4.6 *Survival and propagation*

According to this connection the AM fungi increase plant growth and reproduction by facilitating better access to nutrients such as nitrogen and phosphorus. Through their mycelium, which is found both inside and outside of the roots, they also assist plants in absorbing more water from the soil (Bago et al., 2001). The plants give carbon in exchange for producing sugars through photosynthesis (Pfeffer et al., 1999). The fungal structures inside the roots can either actively or passively transfer these sugars to the fungi (Doidy et al., 2012). After colonizing plants, arbuscular mycorrhizal fungi remain attached to the root systems and can spread to other soil types (Mishra et al., 2018). In addition, AMF detoxifies plant environments with elevated heavy metal concentrations (Hildebrandt et al., 1999) and triggers the production of several phytohormones. Danneberg et al. (1993) observed that several compounds, including gibberellins, auxins, abscisic acid, and compounds resembling cytokinins, were present in higher concentrations in the roots and shoots of plants colonized by AMFs.

17.2 Plant secondary metabolites

17.2.1 Definition

Secondary metabolites (SMs) are multipurpose metabolites that typically contribute to environmental communication and plant defense (Hartmann, 2007). Research on plant SMs has grown over the last 50 years. A common characteristic of plants is their wide variety of SMs. Currently, over 100,000 substances have been identified. These comprise a variety of nitrogen-containing compounds, including cyanogenic glycosides, alkaloids, amines, nonprotein amino acids, alkamides, glucosinolates, and peptides, as well as numerous nitrogen-free compounds, including phenolics, polyketides, polyacetylenes, saponins, and terpenes (Wink, 2009). In plants, dozens of minor components are frequently accompanied by a few major SMs, which typically originate from various classes and biochemical pathways. Plant SMs have been used in traditional medicine for centuries because of their important biological activities. These days, they play a significant role in the production of valuable goods such as fine chemicals, pharmaceuticals, cosmetics, and, more recently, nutraceuticals (Wink, 2009; Kumar et al. 2021; Zhu et al. 2023).

17.2.2 Description

Over the past 50 years, there has been an increase in research on SMs found in plants. In addition to being a major source of active medications, these molecules are crucial for how plants adapt to their surroundings. To produce and study plant SMs, plant cell culture technologies emerged at the end of the 1960s. To increase the production of these plant compounds, researchers have investigated a number of tactics utilizing in vitro systems (Bourgaud et al., 2001).

17.2.2.1 Classification

The chemical structure of SMs, such as whether they contain a sugar or rings, can be used to classify them. Others are grouped according to their chemical makeup, such as whether or not they contain nitrogen. Classification is also influenced by their solubility in various solvents and the pathway by which they are produced, such as the phenylpropanoid pathway, which yields tannins. Typically, their biosynthetic pathways are used to classify them (Harborne et al., 1996). It is common to identify three major families of molecules: alkaloids and flavonoids, terpenes and steroids, and phenolics (Bourgaud et al., 2001; Tiwari & Rana, 2015). Some of these can lead to serious effects. Alkaloids are defined as pharmacologically active, nitrogen-containing compounds from plants. They can block ion channels, inhibit enzymes, or interfere with neurotransmission. This can lead to hallucinations, loss of coordination, convulsions, vomiting, and even death. Phenolics can disrupt digestion, slow growth, block enzyme activity and cell division, or simply taste bad (Tiwari & Rana, 2015). One of the most prevalent and chemically varied classes of natural products is terpenes. They belong to a unique class of hydrocarbon-based substances that are made from isoprene. The number of five-carbon units that terpenes contain determines their classification.

They play a physiological as well as an ecological role in plants. They can support insect pollinators, act as insecticides, play a role in allelopathy, or act as plant hormones such as gibberellin and abscisic acid (Kabera et al., 2014). Over 4500 distinct types of flavonoids have been identified to date, making them a large class of phenolic compounds. The majority of plant tissues contain them, and they are frequently found in vacuoles. They can be larger oligomers, dimers, or monomers. Plant metabolites that fall under the category of flavonoids include leucoanthocyanidins, catechins, anthocyanins, chalcones, aurones, flavanones, isoflavonoids, flavones, and flavonols (Tiwari & Rana, 2015). Benzopyranones are a broad family of plant metabolites that includes coumarins and stilbenes. More than 800 species contain more than 1500 different types of coumarins. These substances are present in the seed coats, fruits, flowers, roots, leaves, and stems of plants, with the fruits and flowers typically containing the highest concentrations (Kabera et al., 2014; Tiwari & Rana, 2015).

17.2.3 Mechanisms of their production

17.2.3.1 Biosynthesis

The pathways of biosynthesis produce both primary and SMs (Herbert, 1989; Kabera et al., 2014). Energy is needed for biosynthetic reactions and is obtained from the citric acid cycle and the energy released during the glycolysis of carbohydrates. ATP (adenosine triphosphate), a high-energy molecule produced by the breakdown of primary compounds, is formed by the oxidation of glucose, fatty acids, and amino acids. Fuel-building reactions involving intermediate molecules along the pathways recycle ATP. Biosynthesis, also known as anabolism, is a reduction reaction, whereas catabolism entails the oxidation of starting molecules. Because of this a reducing agent or hydrogen donor is required, and nicotinamide adenine dinucleotide phosphate is typically used. These catalysts are called coenzymes, and the most prevalent one is CoA (coenzyme A), which is composed of pantetheine phosphate and adenosine diphosphate (Goura et al., 2025; Michal & Schomburg, 2013).

Fig. 17.2 shows the biosynthesis of SMs. The primary pathways for biosynthesis are mevalonic acid for terpenes, steroids, and alkaloids; acetate-malonate for phenols and alkaloids; shikimic acid for phenols, tannins, and aromatic alkaloids; and the pentose pathway for glycosides and polysaccharides (Dewick, 2002). Biosynthetic intermediates are made from metabolites from the energy-generating process, such as photosynthesis, glycolysis, and the Krebs cycle. The primary building blocks for the biosynthesis of SMs are 1-deoxylulose 5-phosphate, shikimic acid, mevalonic acid, and acetyl-CoA (acetyl coenzyme A) (Giweli et al., 2013).

17.3 Ecological importance of arbuscular mycorrhizal fungi in the environment

17.3.1 Increase soil fertility

Numerous human activities have caused soil degradation and, in the case of agricultural soils, a gradual decline in soil fertility. The diversity and activity of the soil's biota are just as important to its quality as its physical or chemical

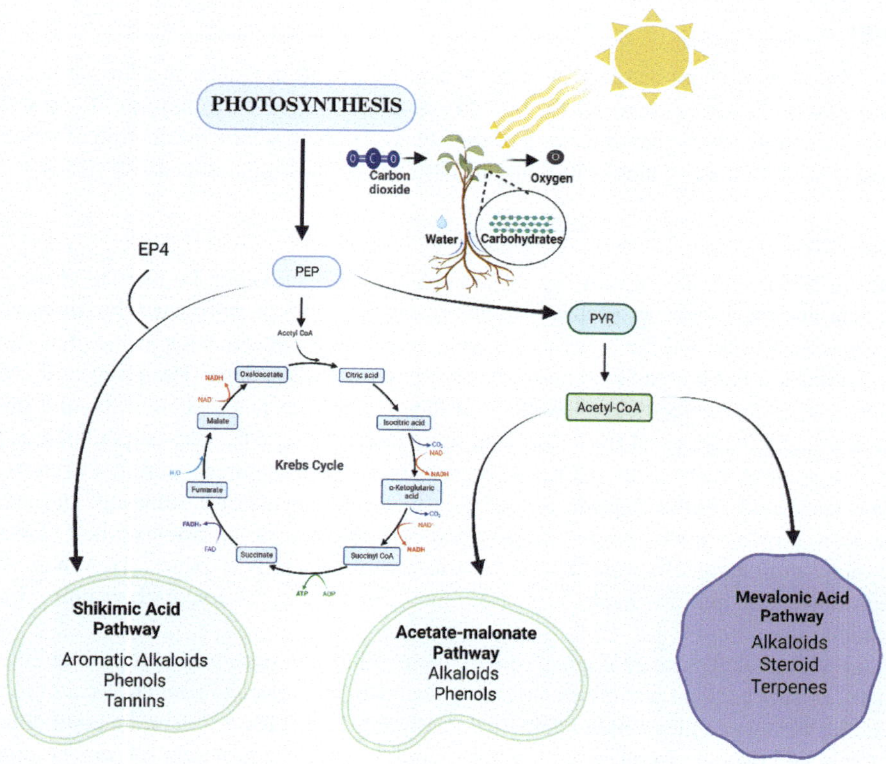

FIGURE 17.2 Biosynthesis of secondary metabolites.

characteristics (Doran & Linn 1994). AMFs are vital parts of soil biota and are present in almost every ecological setting, including natural ecosystems (especially those that support plant communities with a high species diversity) and conventional cropping systems (especially when administered using sustainable practices) (Gianinazzi & Schüepp, 1994). Research on soil fungal pathogens, such as *Aphanomyces, Fusarium, Phytophthora*, and *Verticillium* (Azcon-Aguilar & Barea, 1996), and nematodes that cause root rots, lesions, and galls, respectively, has primarily assessed the effect of AMF on the decrease of soil-borne diseases (Guillemin et al., 1993; Pinochet et al., 1996). According to certain research, mycorrhizal defense may also be effective against *Pseudomonas syringae* and *Erwinia carotovora* (Garcia-Garrido & Ocampo, 1989). As obligatory symbionts, AMFs rely on plant roots for their life cycle. In exchange, they lessen disease in the plants and lower the number of harmful microorganisms in the soil, particularly in areas with a limited supply of P (Linderman, 1994). This begs the question of how arbuscular mycorrhizal fungi can be controlled to maximize the advantages these microorganisms can offer for plant health through management strategies that increase their capacity to lower soil-borne illnesses. But doing so will largely depend on our comprehension of the functional processes that AMF perform and the creation of methods to increase their activity in soils (Jeffries et al., 2003). Arbuscular mycorrhizae fungi enhance the physical, chemical, and biological health of soil (Fall et al., 2022).

17.3.2 Nutrient recycling

A very common type of soil fungus, AMFs are becoming more and more recognized for their role in nutrient cycling processes (Hodge & Storer, 2014; Smith & Smith, 2011; Veresoglou et al., 2012). Most land plants develop symbiotic relationships with these fungi, and it has been shown that they can improve plant P nutrition and thus increase plant growth (Clark & Zeto, 2000). In addition, it has been demonstrated that arbuscular mycorrhizal fungi transfers nitrogen from soil to plants in certain cases (Atul-Nayyar et al., 2009; Bender et al., 2015; Cavagnaro et al., 2012; George et al., 1995).

17.3.3 Ecosystem maintenance

More than 90% of the higher plants surveyed in terrestrial ecosystems have symbiotic relationships with AMFs, which are known as arbuscular mycorrhizae (AM) (Smith & Read, 2008). Actually, arbuscular mycorrhizal fungi is distributed all over the world (Kivlin et al., 2011; Treseder & Cross, 2006). They were widespread in a variety of settings, including delicate and deteriorated ecosystems. It is well known that AMFs help host plants acquire mineral nutrients (specifically P), aid in soil aggregation processes, and preserve the stability, productivity, and ecosystem functioning of plant communities (Van der Heijden et al., 1998a). Numerous studies have shown that the AMFs are one of the primary mediators of ecosystem succession and disturbed communities; as such, they may have potential roles in ecosystem restoration at various scales (Wang, 2017; Van der Heijden et al., 1998b).

17.3.4 Benefits in phytoremediation

The process of bioremediation involves treating contaminated soil with microorganisms. AMFs and plants work together in a process called root-colonizer symbiosis, which is a component of the general phytoremediation cycle. By slowing down the rate of root-to-shoot translocation, AMFs have been shown to lessen metal toxicity to plants (Fan & Liu, 2011; Tomer et al., 2021). Among the most prevalent soil microorganisms, arbuscular mycorrhizal fungi play a crucial role in the soil–plant system and can be found in practically every type of habitat and climate. More precisely, it has been demonstrated that heavy metal toxicity can impact AMF; however, arbuscular mycorrhizal fungi frequently colonize mycotrophic plants that grow in heavy metal-contaminated soils (Leyval et al., 1997). Numerous factors, including fungal genotype; root length density; competition among AMF communities; plant uptake of metal via arbuscular mycorrhizal symbiosis; soil contamination conditions; the metal itself; concentrations of offered metals; the rhizosphere (pH, CEC, etc.); experimental conditions (plant growth stage, light intensity, available N and P); interactions between P and metals; plant species; plant size; and litter inputs, influence how much arbuscular mycorrhizal fungi affects metal plant uptake. Gadd (1993) asserts that free amino, hydroxyl, carboxyl, and other groups can aid in HM binding in both living and dead fungal cell wall components. AMF enters cortical root cells through the formation of intraradical hyphae and ERM that pierce intercellular gaps. Fungal vacuoles appear to play a significant role in retention, binding, and immobilization when HM uptake is reduced. They play a role in detoxifying potentially harmful metal ions and controlling the concentrations of metal ions in the cytosol. Because the fungal cell wall contains free amino, hydroxyl, carboxyl, and other functional groups, its chitin and glomalin are also significant (Christie et al. 2004; Shaker-Koohi, 2014).

17.3.5 Crop protection against disease

Plant pathogen damage is lessened by AM fungi and the interactions they have with plants (Harrier & Watson, 2004). Numerous plant species have been shown to interact in these ways, such as AM fungus sp. *G. intraradices* affects the pathogenic fungi *Fusarium oxysporum* f. sp. *radicis-lycopersici*, and then arbuscular mycorrhizal fungus significantly reduced *Fusarium* root rot on tomato (Caron et al., 1985). Another species, *Glomus mosseae,* affects the *Meloidogyne incognita* nematode development, which was reduced by mycorrhizal roots on tomato (Sikora, 1978). Given the rising expense of inorganic fertilizers and the risks to the environment and public health posed by pesticides and chemically resistant pathogens, AM fungi might offer a more appropriate and environmentally acceptable substitute for sustainable agriculture. Arbuscular mycorrhizal fungi play a significant role in the rhizosphere of plants and have the potential to influence the prevalence and severity of root diseases (Linderman, 1992). The main findings from the different observations are as follows: (1) AM associations lessen the harm that plant pathogens, particularly fungi and nematodes, cause; (2) arbuscular mycorrhizal symbiosis increases root resistance or tolerance, though this varies depending on the crop; (3) protection is not always effective against all pathogens; and (4) soil and other environmental factors influence disease protection. Therefore, depending on the host plant and the cultural system, different AM fungi and plant pathogens interact differently (Akhtar et al., 2008).

17.4 Collaboration between arbuscular mycorrhizal fungi and plant secondary metabolites

The plant–soil microbial relationship serves as an environment-friendly method for decreasing agricultural dependence on chemical fertilizers in recent times. AMFs establish beneficial symbiotic relationships between themselves and flowering plants, bryophytes, or ferns (Begum et al., 2019b). Selosse et al. (2015) identified AMF-host plant symbiosis as the most ancient plant-mutualistic relationship, which existed more than 400 million years ago. Multiple benefits arise from the association between AMF symbiosis and plant roots, but their effects differ in magnitude (Amani Machiani et al., 2022; Rahmat & Soheilikhah, 2024). The main advantage of arbuscular mycorrhizal fungi for plants involves enhancing nutrient absorption, especially phosphorus and iron and zinc, through their large hyphal structures (Amani Machiani et al., 2021;

Battini et al., 2017; Grümberg et al., 2014). Fall et al. (2022) reported that AMF obtains P from less bioavailable P minerals by boosting soil biological activity and enzyme production, as well as acidifying through H^+ release. AMF enhances water absorption by increasing root hydraulic conductivity through its extended mycelial surface area (Li et al., 2019; Ostadi et al., 2022). The mutualistic relationship between arbuscular mycorrhizal fungi and host plants delivers another advantage through enhanced photosynthesis by controlling chloroplast enzyme functions and decreasing chlorophyll breakdown while stimulating chlorophyll development (Begum et al., 2019b).

The mechanisms through which AMF enhance plant tolerance against environmental stressors lead to improved antioxidant defense responses (Al-Arjani et al., 2020). Amani Machiani et al. found that soybean plants inoculated with AMF indicated increased antioxidant enzyme activities, which included Superoxide Dismutase (SOD) along with Ascorbate Peroxidase (APX) and Glutathione Peroxidase (GPX), leading to better drought stress performance. The metabolic pathways of plants undergo reprogramming because of AMF, which produces alterations in their primary metabolites, such as sugars and organic acids, and SMs such as terpenoids, alkaloids, and flavonoids. The performance of plants under stress depends significantly on the availability and production of these compounds. The degree to which AMF modifies primary and SMs in plants varies according to plant species, AMF species, environmental factors, and their combined interactions (Amani Machiani et al., 2022; Wu et al., 2023).

17.5 Characteristic events occurring during the manufacturing of secondary metabolites and reinforcement of arbuscular mycorrhizal fungi

17.5.1 Symbiosis in plant roots (direct or indirect)

AMFs form a symbiotic relationship by colonizing plant roots, creating structures such as arbuscules, vesicles, and hyphae within root cortical cells. This direct colonization enhances nutrient uptake (especially phosphorus), improves plant growth, and can alter root architecture (Bi et al., 2019; Druille et al., 2013; Thind et al., 2022; Yizhu et al., 2020). AMF hyphae widen beyond the root zone, directly increasing the root system's ability to access water and nutrients from the soil. Direct competition with soil pathogens for root colonization sites is another mechanism, providing protection against root diseases (Weng et al., 2022). The role of arbuscular mycorrhizal fungi in plant–water relations is most commonly attributed to indirect mechanisms such as enhancement of plant nutrition and osmoregulation in the host plants. AMFs can improve plant–water relations via several indirect mechanisms (Augé et al., 2015; Mo et al., 2016). AMF can indirectly influence plant metabolism and defense by altering the plant's nutritional status, which in turn affects the allocation of resources to defensive compounds. They can modulate plant hormone levels (e.g., auxins, gibberellins, and cytokinins), thereby influencing plant growth and stress responses. Indirectly, arbuscular mycorrhizal fungi can contribute to soil structure stability by affecting the biochemical and morphological properties of host plants (Bi et al., 2019; Fontana et al., 2009; Wu et al., 2013).

17.5.2 Enhancing photosynthetic capacity

Apart from enhancing growth, plant nutrition, resistance against stresses, and enhancing biomass, AMF also enhance photosynthetic characters of a plant by improving light energy utilization and increasing pigments, gas exchange, stomatal conductance, nutrient and water uptake, and other factors involved in photosynthesis (Carranca et al., 2015; Peng et al., 2024; Tang et al., 2022). These studies have been observed in *Medicago sativa*, citrus, cotton seedlings, *Oryza sativa, Antirrhinum majus, Panicum virgatum, Triticum aestivum,* and *Camellia sinensis* cv. and *Vitis vinifera* L. cv. under both optimal and stress conditions (Basyal & Walker, 2023; Peng et al., 2024; Ye et al., 2022).

17.5.3 Phytochrome concentration

This ancient symbiosis, dating from 460 million years ago, is not directly related to the concentration of phytochromes in a plant but, however, affects plant morphological and gene expression responses to light and arbuscular mycorrhizal fungi colonization (Heckman et al., 2001; Saha et al., 2023). However, under a low R:FR ratio defense mechanism was suppressed by AMF, although it also counteracts plants' shade avoidance syndrome, which suppresses the defense mechanism of the plant as it prioritizes growth over defense, but AMF cannot initiate defense gene expression (Saha et al., 2023).

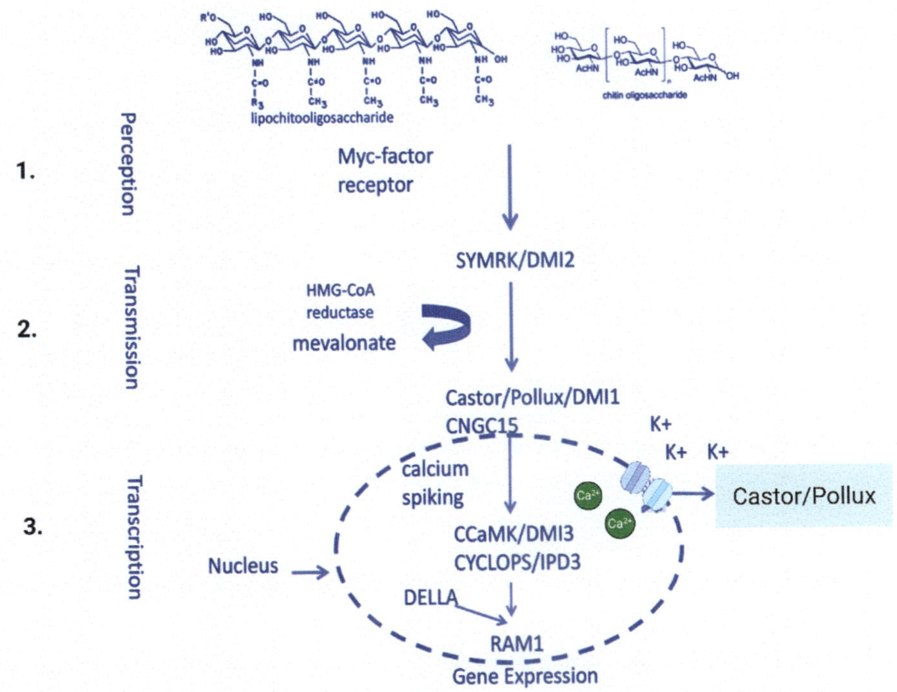

FIGURE 17.3 Myc factors (Myc-LCOs and Myc-COs) through LysM receptors of roots detected. HMG-CoA reductase interaction with SYMRK/DMI2 results in calcium oscillation and mevalonate generation. Afterward, there is a calcium spike with the help of DMI1; the calcium oscillations are decoded by a CCaMK/DMI3 and activate CYCLOPS/IPD3 by phosphorylation. Then, CCaMK–CYCLOPS complex interacts with DELLA proteins, and RAM1 encodes key transcriptional factors initiating the expression of genes.

17.5.4 Signaling pathway

AMF maintains a complex signaling pathway between plants and AMFs. Three main steps in signaling are perception and signal transmission, followed by transcriptional response.

The first step in the symbiotic signaling cascade is molecular crosstalk between strigolactones from the plant, which initiate hyphopodium formation and Myc factors of fungi (Fig. 17.3). AM fungul presence is initiated by the detection of diffusible signals such as lipochitooligosaccharide (Myc-LCOs) and short-chain chitin oligosaccharides (Myc-COs) and N-acetylglucosamine via plant roots through LysM receptor-like kinases, that is, OsCERK1, SlLYK10, LjNFR1, MtLYK3, and the receptor-like kinase SYMRK/DMI2 (Boyno et al., 2023; Ho-Plágaro & García-Garrido, 2022a; Sun et al., 2015). After Myc factors' recognition, calcium oscillation in the nucleus is spiking by HMG-CoA reductase interacts with SYMRK/DMI2 and generates mevalonate. A product of HMGR1 activity responsible for symbiotic gene expression and both nuclear and perinuclear calcium spiking with the help of DMI1, which transports signals to the nucleus via the plasma membrane (MacLean et al., 2022, 2017; Vankateshwaran et al., 2015).

The calcium oscillations are decoded by a CCaMK/DMI3, a calcium and calmodulin-dependent kinase, which acts as a master regulator and activates CYCLOPS/IPD3 by phosphorylation. The CCaMK–CYCLOPS complex then interacts with GRAS family target promoters such as DELLA proteins and RAM1, which encode key transcriptional factors initiating the expression of genes necessary for fungal settlement, arbuscules establishment, and key nutrient exchange. This transcriptional response enhances the symbiosis as it narrows a pathway to cellular reprogramming in the root cortex (Ho-Plágaro & García-Garrido, 2022a, 2022b; Pimprikar & Gutjahr, 2018). Postsymbiosis development sustenance is done with the alliance of ARK1 and ARK2 (Montero et al., 2021).

17.6 Effect of arbuscular mycorrhizal fungi in plant stress tolerance

Among land plants, arbuscular mycorrhizas (AMs) are the most common symbiosis. The related fungus lives inside the tissues of their host plant after colonizing the plant roots. This mutualistic relationship shields the plant from unfavorable soil conditions in addition to being crucial for improving plant growth by promoting the uptake of water and vital nutrients

(Hashem et al., 2018). In addition to all of these positive effects, AMF has a significant impact on plants' ability to withstand stress, which is covered below.

17.6.1 Tolerance to temperature stress

One of the most significant environmental variables influencing plant development and productivity worldwide is temperature. Stresses from low and high temperatures have an impact on a variety of physiological and biochemical systems and activities. It has been demonstrated that arbuscular mycorrhizal (AM) symbiosis increases plants' resistance to temperature stress (Anjum et al., 2011; Pant et al., 2021). Photosynthesis (photosynthetic rate, chlorophyll, and chlorophyll fluorescence), plasma membrane permeability (malondialdehyde and ATPase), reactive oxygen species and antioxidants, osmotic adjustment, carbohydrate metabolism, nutrient acquisition, and secondary metabolism under low or high temperature stress are all covered in this chapter, along with the impact of AM symbiosis on plant growth and biomass production (Anschutz et al., 2014; Auge et al., 2014; Pant et al., 2021). Through improved water and nutrient intake, increased photosynthetic capacity and efficiency, protection from oxidative damage, and an increase in osmolyte accumulation, the mechanisms of AM symbiosis also improve the host plants' resistance to temperature stress (Auge et al., 2014; Hashem et al., 2018; Pant et al., 2021; Wu & Zou, 2017).

17.6.2 Tolerance of salt stress

In many parts of the world, especially in arid and semiarid regions, soil salinity is continuously rising and has turned into a significant land degradation issue. A 30% loss of land within the next 25 years and a 50% loss by the middle of the 21st century are predicted as a result of increased salinization of arable land, which is predicted to have catastrophic worldwide implications. Saline-growing plants often withstand three different types of stress: oxidative, osmotic, and ionic (Bojorquez-Quintal et al., 2014; Bothe, 2012; Pant et al., 2021). Arbuscular mycorrhizal fungi are common soil microorganisms that form symbiotic relationships with most higher plants, create a physical connection between soil and plant roots, are essential to natural ecosystems, and are primarily found in salty environments. AMF's extraradical hyphae extend several meters beyond the depletion zone, boosting root surface area and assisting the plant in absorbing nutrients. In fact, AMF enhances the plant's overall metabolic activity and physiological processes while assisting in the reduction of physiological drought, which is frequently brought on by salty environments (Bojorquez-Quintal et al., 2014; Bothe, 2012; Pant et al., 2021). As a result, using arbuscular mycorrhizal fungi could provide a less expensive and practical way to address the salinity issue. Through enhanced growth and nutritional balance, water homeostasis, ion maintenance, homeostasis, and mitigation of the harmful effects of ion toxicity, regulation of the photosynthesis process, osmotic homeostasis, mitigation of oxidative damage/stress, control over ultrastructural changes, and molecular approaches for mitigating salinity stress, AMFs help to reduce salinity stress (Bothe, 2012; Campanelli et al., 2013; Pant et al., 2021; Wu & Zou, 2017).

17.6.3 Tolerance of waterlogging

One environmental element that has a detrimental impact on plant survival, growth, and development is waterlogging. The same thing happens when fungi, including AMFs, are present in aquatic environments. It has been found that AMF and a variety of aquatic and wetland plant species coexist in harmony. Numerous permanent and seasonal wetlands, such as lake and stream flooding, mangroves, salt marshes, rivers, riparian and floodplains, peat swamp forests, and other wetlands, have also been discovered to contain AMF (Dodd et al., 2012; Dvorak & Ross, 1986; Estrada et al., 2013; Gao et al., 2025; Pant et al., 2021; Tuheteru & Wu, 2017). The majority of the species in the Glomeraceae family, including *Rhizophagus fasciculatus, F. geosporus*, and *Funneliformis mosseae*, are found in all kinds of marshes. Numerous parameters, such as oxygen availability, seasonal variations, P availability, water depth, AMF type, and other vegetation kinds, affect colonization, spore density, species richness, and variety of AMF in a puddle environment. Through enhanced nutritional status and possible adaptation, the presence of AMF can increase plant growth and biomass. In certain aquatic habitat types, it may also hasten the succession of early pioneer vegetation (Cheng et al., 2021; Estrada et al., 2013; Liu et al., 2024; Tuheteru & Wu, 2017; Zhang et al., 2024). Crop resistance to puddle conditions can be enhanced by AMF. AMF improves the biochemistry and physiology, increases crop tolerance through nitrogen absorption, increases plant growth and biomass, and speeds up the area's aquatic vegetation succession (Dodd et al., 2012; Estrada et al., 2013; Pant et al., 2021; Tuheteru & Wu, 2017; Xiang et al., 2024; Wang et al., 2024; Wu & Zou, 2017).

17.6.4 Tolerance of drought stress

Plant development and crop productivity are significantly inhibited by drought stress. Overall, 80% of plants on land can have AMFs invade their roots and form an arbuscular mycorrhizal symbiosis. Although a long-term soil drought severity significantly reduced root colonization and hyphal development in the soil, a relatively short-term soil drought did not seem to deter root AMF colonization. Plant growth was improved, and plant survival under drought stress was raised by such a shift in mycorrhizal development (Al-Arjani et al., 2020; Bahadur et al., 2019). AMF has been found to improve a variety of plants' resistance to drought. First the morphology of mycorrhizal plants, particularly the root and leaf epicuticular wax, could adjust to drought stress (Begum et al., 2021). In addition, mycorrhizal plants had a direct water-uptake channel through extraradical hyphae. Furthermore, by means of physiological processes in nutrient uptake and biochemical processes involving hormones, osmotic adjustment, and antioxidant systems, AMF improved the host plant's resistance to drought. To improve soil structure and control plant–soil water interactions, AMF also released glomalin, a glomalin-related soil protein, into the soil (Al-Arjani et al., 2020; Begum et al., 2019a; Zhao et al., 2022). More information was provided regarding the molecular mechanisms underlying the expression of pertinent stress genes. Prospects for the future in this area are given (Bahadur et al., 2019; Chen et al., 2013; Pant et al., 2021; Wu & Zou, 2017; Zhu et al., 2024).

17.6.5 Tolerance of heavy metals

The majority of terrestrial plants form symbiotic relationships with soil fungi known as arbuscular mycorrhizal (AM) fungi. In this type of symbiosis the fungi exchange carbon for nutrients and water from the host plant. Research has shown that AM fungi have positive benefits on the host plant's growth when it is stressed by things such as drought, salinity, and heavy metals (Nazir et al., 2023; Saeed et al., 2025). The processes that could improve both the host plants and the fungal tolerance under stress are still unclear, though (Abdelmoneim et al., 2014; Bothe et al., 2009; Pant et al., 2021; Wu & Zou, 2017). The fungi's exceptional capacity to establish physiological and morphological mechanisms is primarily responsible for their mitigating effects on plant growth and the environment under stress. One of the most significant pressures that negatively impacts plant growth and the environment is the stress caused by heavy metals. One of the most efficient ways to lessen the negative impacts of stress on plant growth and the environment is to employ biological techniques, such as using soil microbes, such as AM fungi, plant growth-promoting rhizobacteria, and endophytic bacteria (Adeyemi et al., 2021; Pant et al., 2021; Wu & Zou, 2017).

17.7 Synthesis of secondary metabolite compounds from medicinal plants

Over 80% of individuals in poor nations now get their primary medical care from herbal remedies. Terpenoids, flavonoids, phenolics, alkaloids, glycosides, tannins, and other bioactive components of SMs are thought to be responsible for the therapeutic benefits of aromatic and medicinal plants (Ran et al., 2022). Terpenes, alkaloids, and flavonoids constitute the most abundant group of SMs among the substances named (Table 17.1). The AMF–plant symbiotic relationship has beneficial effects on the quantity and quality of medicinal and aromatic plants as well as on the uptake of nutrients and water, crop productivity and quality, soil fertility, and stress tolerance (Roaa, 2020).

17.7.1 Alkaloids

Alkaloids are compounds with one or more nitrogen atoms in the heterocyclic ring that resemble alkaloids (Zhao et al., 2022). These metabolites in plants come from terpenes, aromatic chemicals, and amino acids. They are divided into pyridine, quinolizidine, indole-quinoline, benzylisoquinoline, pyrrolizidine, pyrrolidine, and tropane groups according to their amino acid precursors (Dey et al., 2020, 2024; Pant et al., 2021; Roaa, 2020). Out of the 20,000 alkaloid compounds that have been found so far, 600 bioactive alkaloid compounds have been identified (mostly from plants). Alkaloids have both harmful and beneficial effects on bolstering the plant's defenses against biotic stressors (Borges et al., 2017; Roaa, 2020; Zhao et al., 2022). Certain alkaloids have therapeutic qualities and are used extensively in the pharmaceutical sector. We can list morphine, cocaine, nicotine, colchicine, quinine, strychnine, and caffeine in this instance. Numerous investigations documented increased levels of alkaloids following plant inoculation with AMF. For instance, *Gloriosa superba* L. plant colonization with various AMF species raised the concentration of the alkaloid colchicine in the plant's tuber, seed, and aerial organ (Borges et al., 2017; Pandey et al., 2019; Zhao et al., 2022). Furthermore, *Prosopis laevigata* inoculation with *Gigaspora rosea* was linked to an increase in the alkaloid trigonelline, which has medicinal qualities. Plants colonized with

TABLE 17.1 Comparison between AMF fungi along secondary metabolites and host's plants.

Sr. no.	Secondary metabolites	AMF species	Plant's name	References
1	**Alkaloids**	*Schizonepeta tenuifolia*	*Curcuma longa*	Andrade-Linares et al. (2016, 2013)
		Rhizophagus fasciculatus	*Arnica montana*	Jurkiewicz et al. (2010)
		Rhizophagus irregularis	*Satureja macrostema* (Benth.) Briq	Rydlová et al. (2016)
2	**Terpenoids**	*Rhizophagus intraradices, Funneliformis mosseae*	*Angelica archangelica*	Zitterl-Eglseer et al. (2015)
		Rhizophagus clarus	*Inula ensifolia*	Zubek et al. (2010)
		Pseudomonas fluorescens	*Catharanthus roseus*	Jaleel et al. (2007)
		Trichoderma harzianum, Bacillus subtilis	*Ocimum tenuiflorum* L.	Rydlová et al. (2016)
		F. mosseae	*Thymus vulgaris* L.	Golubkina et al. (2020)
		Glomus hoi	*Coriandrum sativum* L.	
3	**Phenolies**	*R. intraradices*	*Mentha spicata*	Bagheri et al. (2014)
		Glomus spp.	*Anadenanthera colubrina*	Pedone-Bonfim et al. (2018)
		Gigaspora albida	*Passiflora alata*	de Oliveira et al. (2015)
		Glomus, Gigaspora, Acaulospora sp.	*C. longa*	Dutta and Neog (2016)
		Funneliformis geosporum, R. intraradices	*A. montana*	Jurkiewicz et al. (2010)
		F. mosseae	*Echinacea purpurea* (L.)	Araim et al. (2009)
4	**Saponins**	*Claroideoglomus lamellosum*	*Origanum onites*	Karagiannidis et al. (2011)
		Bacillus megaterium, R. intraradices	*Bacopa monnieri*	Gupta et al. (2017)
		R. clarus, C. etunicatum, P. aduncu	*Piper aduncum* L.	de Oliveira et al. (2020)
		R. fasciculatus	*Mentha piperita*	Gupta et al. (2011)
		F. mosseae	*Salvia miltiorrhiza*	Meng et al. (2015)
5	**Other bioactive (oil)**	*R. clarus, C. etunicatum, P. aduncum*	*P. aduncum* L.	Ferreira et al. (2019)
		Glomus macrocarpum	*Anethum graveolens*	Neag et al. (2018)
		C. etunicatum	*Mentha crispa*	Urcoviche et al. (2015)
		R. clarus	*Cymbopogon citratus*	Lermen et al. (2015)
		R. intraradices, F. mosseae	*A. archangelica*	Zitterl-Eglseer et al. (2015)

AMF, arbuscular mycorrhizal fungi.

various AMF species also exhibited increased accumulation of the alkaloids trigonelline, castanospermine, and camptothecin (Amani Machiani et al., 2022; Pandey et al., 2018; Zhao et al., 2022).

17.7.2 Terpenoids

With about 27,000 molecules, terpenoids (also known as isoprenoids) are the most varied class of plant metabolites (Pandey et al., 2018; Sacchettini & Poulter, 1997). The fusing of isoprene units results in the ring or chain structure found in terpenoids. Terpenoids can be categorized as hemiterpenes (C5), monoterpenes (C10, geraniol, linalool, limonene, etc.), sesquiterpenes (C15, abscisic acid, humulene, germacrene, bisabolene, etc.), diterpenes (C20 and gibberellic acid), triterpenes (C30, ursolic acid, oleanolic acid, betulinic acid, etc.), tetraterpenes (C40, lupeol, diosgenin, stigmasterol, lanosterol, etc.), and polyterpenes ($C > 40$) based on the number of isoprene units used in their structure (Pandey et al., 2018;

Pusztahelyi et al., 2015). Plants produce and store terpenoids in their trichomes and epidermal structures (Covello et al., 2007; Pandey et al., 2018). Some of these chemicals (such as gibberellic acid and abscisic acid) control plant development and stress responses, and they are crucial for plants to adapt to harsh environmental conditions (Pusztahelyi et al., 2015). Certain monoterpenes, such as linalool, and sesquiterpenes, such as caryophyllene, are poisonous to diseases and herbivorous insects. They can also serve as signaling molecules to draw in insect natural enemies (Bakkali et al., 2008; Kappers et al., 2005; Pandey et al., 2018; Sharma et al., 2017).

Furthermore, when viruses or animals that consume fungi attack plants, AMFs cause the host plant to produce more terpenoids, such as catalpol. Transferred to the soil or exterior hyphae, catalpol serves as a signal or defensive agent to keep nearby plants or symbiotic fungi safe (Babikova et al., 2014; Duhamel et al., 2013; Pandey et al., 2020). Numerous studies have demonstrated that the symbiotic relationship between AMF and plants can stimulate the synthesis of terpenoids and affect the composition and content of terpenoids, despite some reports indicating neutral (Nell et al., 2010) or even negative effects (Karagiannidis et al., 2011; Khaosaad et al., 2006; Lermen et al., 2015; Sharma et al., 2017; Weisany et al., 2015). Furthermore, some terpenoids, such as the sesquiterpene artemisinin in Artemisia annua, are bioactive and the primary active ingredient in therapeutic herbs (Amani Machiani et al., 2022; Pandey et al., 2020; Zhao et al., 2022).

17.7.3 Phenolics

Plant tissues create a broad class of SMs called phenolics, also known as phenylpropanoids, to ward off diseases, parasites, predators, and Ultraviolet (UV) rays (Singla et al., 2024). They are produced via the phenylpropanoid pathway by the pentose phosphate and shikimate processes (Michalak, 2006; Zaynab et al., 2018). Phenolic acids, flavonoids, coumarins, stilbenes, and monolignols are the five primary categories of phenolics. In addition to their positive effects under stress the majority of the chemicals have antioxidant activity, which gives the plant important therapeutic qualities (Krishnaiah et al., 2011. They are also crucial signaling molecules in plant development and the interactions between microbes and plants. AMFs induce variations in the quantity and quality of distinct SMs by reprogramming the phenylpropanoid pathway to build a functioning symbiosis (Schweiger & Muller, 2015). Flavonoids and phenolic acids are two significant classes of chemicals in the phenylpropanoid pathway (Amani Machiani et al., 2022; Zhao et al., 2022).

17.7.4 Falovenoids

Flavonoids, curcuminoids, coumarins, tannins, stilbenes, lignans, phenolic acids, and quinones are among the many substances that are classified as phenolics because they all share one or more phenol groups (Cosme et al., 2020). It has been demonstrated that arbuscular mycorrhizal fungus raises the phenol content of therapeutic plants. For example, the synthesis of formononetin, a molecule that has antibacterial, antioxidant, antilipidemic, antidiabetic, anticancer, and neuroprotective properties , was elevated in *M. sativa* plants cultivated with *Rhizophagus intraradices* present. Under greenhouse conditions, *Curcuma longa* colonized by AMF species from the genera *Glomus/Rhizophagus, Gigaspora*, and *Acaulospora* sp. produced about 26% more curcumin, an antiinflammatory, antioxidant, anticancer, antiseptic, antiplasmodial, astringent, digestive, and diuretic compound (Pant et al., 2021; Zhao et al., 2022).

17.7.5 Saponins

Saponins are a significant family of SMs that are generated by plants. A carbohydrate component (an oligo or monosaccharide chain) joined to a steroid or triterpene is what defines their structure (Hussein & El-Anssary, 2019). The various biological functions of saponins explain their wide application in pharmaceutical, cosmetic, and food industries (Isah, 2019). An increase in saponin concentration has been reported in mycorrhizal plants. For example, inoculation of *Glycyrrhiza glabra* plants with *Claroideoglomus etunicatum* (Johny et al., 2021) and *F. mosseae* and/or *Diversispora epigaea* (Liu et al., 2009) was associated with the accumulation of triterpenoid saponin glycyrrhizic acid. Similarly an accumulation of saponins was observed in the aerial parts of *Anchusa officinalis* L. (Cartabia et al., 2021; Pant et al., 2021) and *Passiflora alata* (Muniz et al., 2021) plants inoculated with *Rhizophagus irregularis* and *Acaulospora longula*, respectively (Pant et al., 2021; Zhao et al., 2022).

17.7.6 Other bioactive compound production

The naphthodianthrones (anthraquinone derivatives) hypericin and pseudohypericin are mostly derived from *Hypericum* species. Among their various medicinal qualities are sedatives, antiseptics, and antispasmodics. *Hypericum perforatum* colonized by *R. intraradices* alone or by a combination of *F. constrictum, F. geosporum, F. mosseae*, and *R. intraradices* under greenhouse conditions had higher levels of hypericin and pseudohypericin, according to Zubek et al. (2012) and Pant et al. (2021). A steroidal lactone with a variety of pharmacological activities, such as cardioprotective, antiinflammatory, immunomodulatory, antiangiogenesis, antimetastasis, and anticarcinogenic qualities, withaferin-A has long been utilized in ayurvedic medicine (Mirjalili et al., 2009). According to Johny et al. (2021), in greenhouse conditions, the combination of the medicinal plant *Withania somnifera* and *R. irregularis* raised the concentration of withaferin-A in comparison to noninoculated plants (Pant et al., 2021; Zhao et al., 2022).

17.8 Participation of arbuscular mycorrhizal fungi for essential nutrient acquisition

More than 60% of terrestrial plants, including numerous significant crop species such as wheat, corn, and soybean, depend on the arbuscular mycorrhizal (AM) symbiosis for nutrient uptake. AM fungi are common and can make up as much as half of the soil's microbial biomass (Bhandari & Garg, 2017; Li et al., 2019; Neumann & George, 2010; Wang et al., 2006). As an extension of the root system, the fungus's ERM enhances the intake of trace elements, including copper and zinc as well as phosphate (P), nitrogen (N), sulfur, and magnesium. In addition, AM fungi offer their host plant nonnutritional advantages, such as increased plant resistance, survival, and fitness as ecosystem engineers, as well as advantages over disadvantages and mutualism (Bhandari & Garg, 2017; Bucking & Kafle, 2015; Li et al., 2019). In addition to diffusion-limited cations Zn, Cu, Fe, Mn, K, and NH_4^+, the positive effects are particularly noticeable with P and N. Although they have been documented, beneficial effects on the uptake of other plant nutrients, such as (NO_3^-), Ca, and Mg, are negligible. Munkvold et al. (2004) compared several strains of the same mycorrhizal fungus and observed a significant linear correlation between the hyphal length density and P acquisition. They also found a positive correlation between the extent of the extraradical network and the uptake of immobile nutrients (Kuyper et al., 2021; Li et al., 2019).

17.8.1 Phosphate and nitrogen fixation

A growing amount of evidence, however, indicates that AM fungi are capable of actively transferring nitrogen to their host, that the host plant's carbon supply promotes this transport, and that the host's periarbuscular membrane can help the mycorrhizal interface actively absorb nitrogen (Bucking & Arjun Kafle, 2015). The level to which plants prosper from the mycorrhizal relationship relies on the kind of fungus, features of the root system, environmental context, and maintenance. These elements do not operate independently, and greater focus should be placed on their interconnections, notably the most intricate one: fungus × plant × environment × management. The benefits that plants receive from the symbiosis are mostly determined by the properties of the root system, such as the diameter and density of the root hair (Bhandari & Garg, 2017; Kuyper et al., 2021; Li et al., 2019). Although ammonium (NH_4^+) predominates in many undisturbed or extremely acidic soils, nitrate (NO_3^-) is the predominant form of N that is accessible to plants and fungi in the majority of agricultural soils. Although NH_4^+ is preferred by fungi, Hawkins and George (2001) described that root and shoot biomass, hyphal length densities, and N transfer to the plant through the hyphae were all lower when NH_4^+ was the only source of nitrogen for mycorrhizal plants than they were following a NO_3^- supply (Bhandari & Garg, 2017; Bucking & Arjun Kafle, 2015). The obligate biotrophic life cycle of AM fungi has led to the general belief that AM fungi cannot use organic N sources, even though organic N can make up a sizable portion of total soil N in many soils. However, several investigations have shown that AM fungus hyphae develop into organic patches of various complexity and transfer ^{15}N to their host plant from these organic patches (Hodge & Fitte, 2010; Leigh et al., 2009), and that greater plant N concentrations may result from this (Bhandari & Garg, 2017; Bucking & Arjun Kafle, 2015; Li et al., 2019).

17.9 Application of arbuscular mycorrhizal fungi for sustainable agricultural systems

17.9.1 Effect of arbuscular mycorrhizal fungi on the plant soil system

AMF helps to improve the biological, chemical, and physical composition of soil. AMF, by forming a complex network of hyphae with the plant's rooting system, not only improves plant growth and root system but also protects the soil against aeolian and water erosion (Gutjahr et al., 2013). It seems to amplify global food production by making soil a storehouse of many essential nutrients by preserving them from leaching and by improving soil condition (Cavagnaro et al., 2015; Khaliq et al., 2022).

17.9.2 Arbuscular mycorrhizal fungi application techniques in agriculture

17.9.2.1 Seed coating with arbuscular mycorrhizal fungi

By coating or inoculating the seeds with a formulation containing AMF the seeds are ready for increased root establishment for mycorrhizal associations (Boyno et al., 2023). Research showing soybean plants that have been grown from a 1:4 ratio of previous soybean planting soil and AMF-coated seeds showed the highest number of roots with increased size. However, root elongation was most observed in the plants growing from seeds that had been coated with the formula containing lime with the highest AMF composition. The highest soybean yield was 2.046 t ha^{-1}, resulting from the seed coat treatment (2:4) without liming (Edi et al., 2023). AMF and nanoparticle seed coating of the maize plant improve its physiological performance under drought stress by enhancing photosynthate (Fatma et al., 2025).

17.9.2.2 Soil drenching

Soil drenching with AMF inoculants (introducing beneficial AMF directly into the soil) can promote plant growth, enhance nutrient uptake, and improve resistance to drought and soil-borne diseases. A series of experiments on drought tolerance and disease resistance of cowpea in which soil drenched with AMF before sowing grew improved cowpea plants then noninoculated soil before sowing. The improved growth is because of the readily available immobile phosphorus in soil. However, soil drenching with fungicides or pesticides has negative consequences on AMF by reducing root colonization, deteriorating hyphae and spores, or destroying the overall fungal webbing (Hage-Ahmed et al., 2019).

17.9.2.3 Production of biofertilizers

AMF could be used as biofertilizer, as it directly promotes plant growth by translocating water and nutrients, obtaining more nitrogen, phosphorus, and different less mobile nutrients; crop protection; provides resistance against root-related and soil-borne diseases; reduces soil erosion; increases soil productivity by improving soil texture; makes humus more porous; and facilitates root penetration deep in the soil (Fossalunga & Bonfante, 2023; Kuila & Ghosh, 2022). Because of increasing cost, environmental and health hazardous consequences, and its negative impact on climate, chemical fertilizers are possibly replaced by AMF, which is potentially recognized globally as a natural biofertilizer (Madawala, 2021). As AMF is a lead actor in mobilizing phosphorus, it can enhance P uptake 3.1–4.7 times higher than why AMF is the best biofertilizer to use over chemical fertilizers (Anand et al., 2022; Osorio et al., 2001). AMF fertilizer can be applied to seed, seedlings, roots, and soil by either a pot culture, substrate-based bioreactor system, mycorrhizal donor plant system, or in vitro propagation. AMF biofertilizer is a major need of agriculture for both the enhancement of crop productivity and sustainable alternatives to chemical fertilizers, as now sustainability is one of the main criteria of the scientists. Moreover, the effectual practical application of AMF biofertilizers is necessarily in harmony with the overall field management patterns (Fossalunga & Bonfante, 2023; Ghorui et al., 2025; Symanczik et al., 2018).

17.9.2.4 Implementation of the hydroponic system

The microbial revolution is a sustainable tool to revolutionize agriculture in this growing population. The use of AMF in a hydroponic system can improve crop yield, quality, antioxidant compounds, and ecofriendly as there is less dependence on chemical fertilizers. However, AMF has the potential to enhance growth and productivity in hydroponic culture, but its use is still controversial (Cela et al., 2022; Ijdo et al., 2011; Othman et al., 2024). Recent research on the growth of tomato plant seedlings showed that high doses of the AM fungus *R. irregularis* in hydroponic culture lead to a significant improvement in the quality and growth of tomato seedlings (Roussis et al., 2022). Another research study was done on the impact of selenium and AMF inoculation on strawberries grown in hydroponic culture. AMF cannot only enhance the nutritional value of the strawberry but can also nullify the stress induced by Se on the plants and may help in future Se biofortification in strawberry (Antoniou et al., 2021). Integrating arbuscular mycorrhiza into hydroponic systems is feasible and can substantially enhance plant growth, nutrient uptake, and stress tolerance, benefits comparable to those seen in soil-based cultivation. Success, however, depends on thoughtful system design, effective inoculation strategies, careful nutrient management, and a clear assessment of economic viability (Roussis et al., 2022).

17.10 Conclusion and future perspectives

Over the past 20 years, numerous studies on AMF have demonstrated their numerous advantages for crop productivity and soil health. Because mycorrhizal application can successfully lower the quantitative use of chemical fertilizer input, particularly of phosphorus, it is generally accepted that AMF could be considered a replacement for inorganic fertilizers in the near future (Ortas, 2012). Because of their detrimental effects on food product quality, soil health, and air and water systems, the ongoing use of inorganic fertilizers, herbicides, and fungicides has led to a number of issues for plants, soil, and human health.

Author's contribution

S.R., A.N.A., S.A.: writing, conceptualization, analysis, visualization; M.U., N.Y.: supervision, review, and project administration.

Declaration of competing interest

The authors declare no competing interests.

References

Abdelmoneim, T. S., Moussa, T. A. A., Almaghrabi, O. A., & Abdelbagi, I. (2014). Investigation the effect of arbuscular mycorrhizal fungi on the tolerance of maize plant to heavy metals stress. *Life Science Journal, 11*(4), 255–263. http://www.lifesciencesite.com/lsj/life1104/037_23346life110414_255_263.pdf.

Adeyemi, N. O., Atayese, M. O., Sakariyawo, O. S., Azeez, J. O., Abayomi Sobowale, S. P., Olubode, A., Mudathir, R., Adebayo, R., & Adeoye, S. (2021). Alleviation of heavy metal stress by arbuscular mycorrhizal symbiosis in *Glycine max* (L.) grown in copper, lead and zinc contaminated soils. *Rhizosphere, 18.* http://www.journals.elsevier.com/rhizosphere. doi:10.1016/j.rhisph.2021.100325.

Akhtar, M. S., & Siddiqui, Z. A. (2008). Arbuscular mycorrhizal fungi as potential bioprotectants against plant pathogens. In *Mycorrhizae: Sustainable agriculture and forestry* (pp. 61–97). Springer Netherlands.

Al-Arjani, A. B. F., Hashem, A., & AbdAllah, E. F. (2020). Arbuscular mycorrhizal fungi modulates dynamics tolerance expression to mitigate drought stress in *Ephedra foliata* Boiss. *Saudi Journal of Biological Sciences, 27*(1), 380–394.

Amani Machiani, M., Javanmard, A., Habibi Machiani, R., & Sadeghpour, A. (2022). Arbuscular mycorrhizal fungi and changes in primary and secondary metabolites. *Plants, 11*(17), 2183.

Amani Machiani, M., Javanmard, A., Morshedloo, M. R., Aghaee, A., & Maggi, F. (2021). *Funneliformis mosseae* inoculation under water deficit stress improves the yield and phytochemical characteristics of thyme in intercropping with soybean. *Scientific Reports, 11*(1). www.nature.com/srep/index.html. doi:10.1038/s41598-021-94681-9.

Anand, K., Pandey, G. K., Kaur, T., Pericak, O., Olson, C., Mohan, R., Akansha, K., Yadav, A., Devi, R., Kour, D., Rai, A. K., Kumar, M., & Yadav, A. N. (2022). Arbuscular mycorrhizal fungi as a potential biofertilizers for agricultural sustainability. *Journal of Applied Biology and Biotechnology, 10*, 90–107. jabonline.in/index.php. doi:10.7324/JABB.2022.10s111.

Andrade-Linares, D. R., & Franken, P. (2013). Fungal endophytes in plant roots: taxonomy, colonization patterns, and functions. In Symbiotic endophytes (pp. 311–334). Berlin, Heidelberg: Springer Berlin Heidelberg.

Andrade-Linares, D. R., Veresoglou, S. D., & Rillig, M. C. (2016). Temperature priming and memory in soil filamentous fungi. *Fungal Ecology, 21*, 10–15.

Anjum, S. A., Xie, X. y., Wang, L. C., Saleem, M. F., Man, C., & Lei, W. (2011). Morphological, physiological and biochemical responses of plants to drought stress. *African Journal of Agricultural Research, 6*(9), 2026–2032. http://www.academicjournals.org/AJAR/PDF/pdf2011/4%20May/Anjum%20et%20al.pdf.

Anschutz, U., Becker, D., & Shabala, S. (2014). Going beyond nutrition: Regulation of potassium homoeostasis as a common denominator of plant adaptive responses to environment. *Journal of Plant Physiology, 171*(9), 670–687.

Antoniou, O., Chrysargyris, A., Xylia, P., & Tzortzakis, N. (2021). Effects of selenium and/or arbuscular mycorrhizal fungal inoculation on strawberry grown in hydroponic trial. *Agronomy, 11*(4). https://www.mdpi.com/2073-4395/11/4/720/pdf. doi:10.3390/agronomy11040721.

Araim, G., Saleem, A., Arnason, J. T., & Charest, C. (2009). Root colonization by an arbuscular mycorrhizal (AM) fungus increases growth and secondary metabolism of purple coneflower, *Echinacea purpurea* (L.) Moench. *Journal of Agricultural and Food Chemistry, 57*(6), 2255–2258. http://pubs.acs.org/doi/pdfplus/10.1021/jf803173x. doi:10.1021/jf803173x.

Atul-Nayyar, A., Hamel, C., Hanson, K., & Germida, J. (2009). The arbuscular mycorrhizal symbiosis links N mineralization to plant demand. *Mycorrhiza, 19*(4), 239–246. doi:10.1007/s00572-008-0215-0.

Augé, R. M., Toler, H. D., & Saxton, A. M. (2014). Arbuscular mycorrhizal symbiosis and osmotic adjustment in response to NaCl stress: A meta-analysis. *Frontiers in Plant Science, 5*, 562.

Auge, R. M., Toler, H. D., & Saxton, A. M. (2015). Arbuscular mycorrhizal symbiosis alters stomatal conductance of host plants more under drought than under amply watered conditions: A meta-analysis. *Mycorrhiza, 25*, 13–24.

Azcón-Aguilar, C., & Barea, J. M. (1996). Arbuscular mycorrhizas and biological control of soil-borne plant pathogens: An overview of the mechanisms involved. *Mycorrhiza, 6*, 457–464.

Babikova, Z., Gilbert, L., Bruce, T., Dewhirst, S. Y., Pickett, J. A., & Johnson, D. (2014). Arbuscular mycorrhizal fungi and aphids interact by changing host plant quality and volatile emission. *Functional Ecology, 28*(2), 375–385. http://www.blackwellpublishing.com/journals/FEC. doi:10.1111/1365-2435.12181.

Bagheri, S., Ebrahimi, M. A., Davazdahemami, S., & Moghadam, J. M. (2014). Terpenoids and phenolic compounds production of mint genotypes in response to mycorrhizal bio-elicitors. *Technical Journal of Engineering and Applied Sciences, 4*, 339–348.

Bago, B., Pfeffer, P., & Shachar-Hill, Y. (2001). Could the urea cycle be translocating nitrogen in the arbuscular mycorrhizal symbiosis? *New Phytologist, 149*(1), 4–8. doi:10.1046/j.1469-8137.2001.00016.x.

Bagyaraj, D. J. (1991). Ecology of vesicular arbuscular mycorrhizae. In D. K. Arora, B. Rai, K. G. Mukerji, & G. R. Knudsen (Eds.). In *Handbook of applied mycology: i* (pp. 3–34). Marcel Dekker, Inc.

Bahadur, A., Batool, A., Nasir, F., Jiang, S., Mingsen, Q., Zhang, Q., Pan, J., Liu, Y., & Feng, H. (2019). Mechanistic insights into arbuscular mycorrhizal fungi-mediated drought stress tolerance in plants. *International Journal of Molecular Sciences, 20*(17), 4199. doi:10.3390/ijms20174199.

Bakkali, F., Averbeck, S., Averbeck, D., & Idaomar, M. (2008). Biological effects of essential oils—A review. *Food and Chemical Toxicology, 46*(2), 446–475. doi:10.1016/j.fct.2007.09.106.

Basyal, B., & Walker, B. J. (2023). Arbuscular mycorrhizal fungi enhance yield and photosynthesis of switchgrass (*Panicum virgatum* L.) under extreme drought and alters the biomass composition of the host plant. *Biomass and Bioenergy, 177*. http://www.journals.elsevier.com/biomass-and-bioenergy/. doi:10.1016/j.biombioe.2023.106936.

Battini, F., Grønlund, M., Agnolucci, M., Giovannetti, M., & Jakobsen, I. (2017). Facilitation of phosphorus uptake in maize plants by mycorrhizosphere bacteria article. *Scientific Reports, 7*(1). www.nature.com/srep/index.html. doi:10.1038/s41598-017-04959–0.

Begum, N., Ahanger, M. A., Su, Y., Lei, Y., Mustafa, N. S. A., Ahmad, P., & Zhang, L. (2019a). Improved drought tolerance by AMF inoculation in maize (*Zea mays*) involves physiological and biochemical implications. *Plants, 8*(12), 579.

Begum, N., Akhtar, K., Ahanger, M. A., Iqbal, M., Wang, P., Mustafa, N. S., & Zhang, L. (2021). Arbuscular mycorrhizal fungi improve growth, essential oil, secondary metabolism, and yield of tobacco (*Nicotiana tabacum* L.) under drought stress conditions. *Environmental Science and Pollution Research, 28*(33), 45276–45295.

Begum, N., Qin, C., Ahanger, M. A., Raza, S., Khan, M. I., Ashraf, M., Ahmed, N., & Zhang, L. (2019b). Role of Arbuscular mycorrhizal fungi in plant growth regulation: Implications in abiotic stress tolerance. *Frontiers in Plant Science, 10*, 1068.

Bender, S. F., Conen, F., & Heijden, M. G. (2015). Mycorrhizal effects on nutrient cycling, nutrient leaching and N_2O production in experimental grassland. *Soil Biology and Biochemistry, 80*, 283–292.

Bhandari, P., & Garg, N. (2017). Dynamics of arbuscular mycorrhizal symbiosis and its role in nutrient acquisition: An overview. In *Mycorrhiza—Nutrient uptake, biocontrol, ecorestoration* (pp. 21–43). Springer Science and Business Media LLC. doi:10.1007/978-3-319-68867-1_2.

Bi, Y., Xiao, L., & Sun, J. (2019). An arbuscular mycorrhizal fungus ameliorates plant growth and hormones after moderate root damage due to simulated coal mining subsidence: A microcosm study. *Environmental Science and Pollution Research, 26*(11), 11053–11061.

Bianciotto, V., Barbiero, G., & Bonfante, P. (1995). Analysis of the cell cycle in an arbuscular mycorrhizal fungus by flow cytometry and bromodeoxyuridine labelling. *Protoplasma, 188*(3), 161–169.

Bojórquez-Quintal, E., Velarde-Buendía, A., Ku-González, A., Carillo-Pech, M., Ortega-Camacho, D., Echevarría-Machado, I., & Martínez-Estévez, M. (2014). Mechanisms of salt tolerance in habanero pepper plants (*Capsicum chinense* Jacq.): Proline accumulation, ions dynamics and sodium root-shoot partition and compartmentation. *Frontiers in Plant Science, 5*, 605.

Borges, C. V., Minatel, I. O., Gomez-Gomez, H. A., & Lima, G. P. P. (2017). Medicinal plants: Influence of environmental factors on the content of secondary metabolites. In *Medicinal plants and environmental challenges* (pp. 259–277). Springer International Publishing. http://www.springer.com/in/book/9783319687162. doi:10.1007/978-3-319-68717-9_15.

Bothe, H., Regvar, M., & Turnau, K. (2009). Arbuscular mycorrhiza, heavy metal, and salt tolerance. In *Soil heavy metals* (pp. 87–111). *Springer Berlin Heidelberg*.

Bothe, H. (2012). Arbuscular mycorrhiza and salt tolerance of plants. *Symbiosis, 58*(1–3), 7–16. doi:10.1007/s13199-012-0196-9.

Bourgaud, F., Gravot, A., Milesi, S., & Gontier, E. (2001). Production of plant secondary metabolites: A historical perspective. *Plant Science, 161*(5), 839–851. doi:10.1016/s0168-9452(01)00490-3.

Boyno, G., Danesh, Y. R., Demir, S., Teniz, N., Mulet, J. M., & Porcel, R. (2023). The complex interplay between arbuscular mycorrhizal fungi and strigolactone: Mechanisms, sinergies, applications and future directions. *International Journal of Molecular Sciences, 24*(23), 16774.

Brundrett, M. C., & Tedersoo, L. (2018). Evolutionary history of mycorrhizal symbioses and global host plant diversity. *New Phytologist, 220*(4), 1108–1115. http://onlinelibrary.wiley.com/journal/10.1111/(ISSN)1469-8137. doi:10.1111/nph.14976.

Bücking, H., & Kafle, A. (2015). Role of arbuscular mycorrhizal fungi in the nitrogen uptake of plants: Current knowledge and research gaps. *Agronomy, 5*(4), 587–612.

Butler, P. M. (1939). Studies of the mammalian dentition—Differentiation of the post-canine dentition. *Proceedings of the Zoological Society of London, 109*(1), 1–36. doi:10.1111/j.1469-7998.1939.tb00021.x.

Campanelli, A., Ruta, C., Mastro, G. D., & Morone-Fortunato, I. (2013). The role of arbuscular mycorrhizal fungi in alleviating salt stress in *Medicago sativa* L. var. icon. *Symbiosis, 59*(2), 65–76. doi:10.1007/s13199-012-0191-1.

Caron, M., Fortin, J. A., & Richard, C. (1985). Influence of substrate on the interaction of *Glomus intraradices* and *Fusarium oxysporum* f. sp. *radicis-lycopersici* on tomatoes. *Plant Soil, 87*, 233–239.

Carranca, C., Torres, M. O., & Madeira, M. (2015). Underestimated role of legume roots for soil N fertility. *Agronomy for Sustainable Development, 35*, 1095–1102.

Cartabia, M., Girometta, C. E., Milanese, C., Baiguera, R. M., Buratti, S., Branciforti, D. S., Vadivel, D., Girella, A., Babbini, S., Savino, E., & Dondi, D. (2021). Collection and characterization of wood decay fungal strains for developing pure mycelium mats. *Journal of Fungi, 7*(12). https://www.mdpi.com/2309-608X/7/12/1008/pdf. doi:10.3390/jof7121008.

Cavagnaro, T. R., Barrios-Masias, F. H., & Jackson, L. E. (2012). Arbuscular mycorrhizas and their role in plant growth, nitrogen interception and soil gas efflux in an organic production system. *Plant and Soil, 353*, 181–194.

Cavagnaro, T. R., Bender, S. F., Asghari, H. R., & van der Heijden, M. G. A. (2015). The role of arbuscular mycorrhizas in reducing soil nutrient loss. *Trends in Plant Science, 20*(5), 283–290. www.elsevier.com/inca/publications/store/3/0/9/6/0/index.htt. doi:10.1016/j.tplants.2015.03.004.

Cavagnaro, T. R., Gao, L.-L., Smith, F. A., & Smith, S. E. (2001). Morphology of arbuscular mycorrhizas is influenced by fungal identity. *New Phytologist, 151*(2), 469–475. doi:10.1046/j.0028-646x.2001.00191.x.

Cela, F., Avio, L., Giordani, T., Vangelisti, A., Cavallini, A., Turrini, A., Sbrana, C., Pardossi, A., & Incrocci, L. (2022). Arbuscular mycorrhizal fungi increase nutritional quality of soilless grown lettuce while overcoming low phosphorus supply. *Foods, 11*(22). www.mdpi.com/journal/foods. doi:10.3390/foods11223612.

Chen, X. W., Wu, F. Y., Li, H., Chan, W. F., Wu, C., Wu, S. C., & Wong, M. H. (2013). Phosphate transporters expression in rice (*Oryza sativa* L.) associated with arbuscular mycorrhizal fungi (AMF) colonization under different levels of arsenate stress. *Environmental and Experimental Botany, 87*, 92–99. doi:10.1016/j.envexpbot.2012.08.002.

Cheng, X. F., Wu, H. H., Zou, Y. N., Wu, Q. S., & Kuča, K. (2021). Mycorrhizal response strategies of trifoliate orange under well-watered, salt stress, and waterlogging stress by regulating leaf aquaporin expression. *Plant Physiology and Biochemistry, 162*, 27–35. http://www.journals.elsevier.com/plant-physiology-and-biochemistry/. doi:10.1016/j.plaphy.2021.02.026.

Christie, P., Li, X., & Chen, B. (2004). Arbuscular mycorrhiza can depress translocation of zinc to shoots of host plants in soils moderately polluted with zinc. *Plant and Soil, 261*(1), 209–217.

Clark, R. B., & Zeto, S. K. (2000). Mineral acquisition by arbuscular mycorrhizal plants. *Journal of Plant Nutrition, 23*(7), 867–902. doi:10.1080/01904160009382068.

Cosme, P., Rodríguez, A. B., Espino, J., & Garrido, M. (2020). Plant phenolics: Bioavailability as a key determinant of their potential health-promoting applications. *Antioxidants, 9*(12), 1–20. https://www.mdpi.com/2076-3921/9/12/1263. doi:10.3390/antiox9121263.

Covello, P. S., Teoh, K. H., Polichuk, D. R., Reed, D. W., & Nowak, G. (2007). Functional genomics and the biosynthesis of artemisinin. *Phytochemistry, 68*(14), 1864–1871. doi:10.1016/j.phytochem.2007.02.016.

Danneberg, G., Latus, C., Zimmer, W., Hundeshagen, B., Schneider-Poetsch, H. j., & Bothe, H. (1993). Influence of vesicular-arbuscular mycorrhiza on phytohormone balances in maize (*Zea mays* L.). *Journal of Plant Physiology, 141*(1), 33–39. doi:10.1016/s0176-1617(11)80848-5.

de Oliveira, T. B., de Lucas, R. C., de Almeida Scarcella, A. S., Contato, A. G., Pasin, T. M., Martinez, C. A., & de Lourdes Teixeira de Moraes Polizeli, M. (2020). Fungal communities differentially respond to warming and drought in tropical grassland soil. *Molecular Ecology, 29*(8), 1550–1559. http://onlinelibrary.wiley.com/journal/10.1111/(ISSN)1365-294X. doi:10.1111/mec.15423.

de Oliveira, T. B., Gomes, E., & Rodrigues, A. (2015). Thermophilic fungi in the new age of fungal taxonomy. *Extremophiles, 19*(1), 31–37.

Dewick, P. M. (2002). *Medicinal Natural Products: A Biosynthetic Approach* (2nd ed.). Chichester, UK: John Wiley & Sons Ltd.

Dey, D., Habibovic, A., Löcken, A., Wintersberger, P., Pfleging, B., Riener, A., … Terken, J. (2020). Taming the eHMI jungle: A classification taxonomy to guide, compare, and assess the design principles of automated vehicles' external human-machine interfaces. *Transportation Research Interdisciplinary Perspectives, 7*, 100174.

Dey, S., Choudhury, D., Mukherjee, C., & Dutta, S. (2024). Applications of plant growth promoting bacteria (PGPB) and vesicular arbuscular mycorrhizal (VAM) fungi on *Aloe barbadensis* Mill., a medicinally important succulent herb. *Agricultural Science Digest, 44*(2), 201–211.

Dickson, D. W. (2004). Apoptotic mechanisms in Alzheimer neurofibrillary degeneration: Cause or effect? *Journal of Clinical Investigation, 114*(1), 23–27. http://www.jci.org. doi:10.1172/JCI22317.

Dickson, S., Smith, F. A., & Smith, S. E. (2007). Structural differences in arbuscular mycorrhizal symbioses: More than 100 years after Gallaud, where next? *Mycorrhiza, 17*(5), 375–393.

Dodd, I. C., & Perez-Alfocea, F. (2012). Microbial amelioration of crop salinity stress. *Journal of Experimental Botany, 63*(9), 3415–3428.

Doidy, J., Grace, E., Kühn, C., Simon-Plas, F., Casieri, L., & Wipf, D. (2012). Sugar transporters in plants and in their interactions with fungi. *Trends in Plant Science, 17*(7), 413–422. doi:10.1016/j.tplants.2012.03.009.

Doran, J. W., & Linn, D. M. (1994). Microbial ecology of conservation management systems. In J. L. Hatfield, & B. A. Stewart (Eds.), *Soil biology: Effects on soil quality. (Advances in soil science)* (pp. 1–27). Florida: Lewis Publishers.

Druille, M., Cabello, M. N., Omacini, M., & Golluscio, R. A. (2013). Glyphosate reduces spore viability and root colonization of arbuscular mycorrhizal fungi. *Applied Soil Ecology, 64*, 99–103.

Duhamel, M., & Vandenkoornhuyse, P. (2013). Sustainable agriculture: Possible trajectories from mutualistic symbiosis and plant neodomestication. *Trends in Plant Science, 18*(11), 597–600.

Dutta, S. C., & Neog, B. (2016). Accumulation of secondary metabolites in response to antioxidant activity of turmeric rhizomes co-inoculated with native arbuscular mycorrhizal fungi and plant growth promoting rhizobacteria. *Scientia Horticulturae, 204*, 179–184. www.elsevier.com/inca/publications/store/5/0/3/3/1/6. doi:10.1016/j.scienta.2016.03.028.

Dvorak, J., & Ross, K. (1986). Expression of tolerance of Na+, K+, Mg+, Cl− and SO4+ ions and sea water in the amphiploid of *Triticum aestivum×Elytrigia elongata*. *Crop Science, 26*, 658–660.

Edi, A., Jamilah, J., & Ernita, M. (2023). Arbuscular mycorrhizal fungi as seed coat of soybean grown on ultisol applied with various doses of lime. *Planta Tropika, 11*(1), 25–32.

Estrada, B., Aroca, R., Maathuis, F. J. M., Barea, J. M., & Ruiz-Lozano, J. M. (2013). Arbuscular mycorrhizal fungi native from a Mediterranean saline area enhance maize tolerance to salinity through improved ion homeostasis. *Plant, Cell and Environment, 36*(10), 1771–1782. doi:10.1111/pce.12082.

Fall, A. F., Nakabonge, G., Ssekandi, J., Founoune-Mboup, H., Apori, S. O., Ndiaye, A., Badji, A., & Ngom, K. (2022). Roles of arbuscular mycorrhizal fungi on soil fertility: Contribution in the improvement of physical, chemical, and biological properties of the soil. *Frontiers in Fungal Biology, 3*, 723892.

Fan, Q. J., & Liu, J. H. (2011). Colonization with arbuscular mycorrhizal fungus affects growth, drought tolerance and expression of stress-responsive genes in *Poncirus trifoliata*. *Acta Physiologiae Plantarum, 33*(4), 1533–1542. doi:10.1007/s11738-011-0789-6.

Fatma, W. S., Ilyas, S., Budi, S. W., & Widowati, L. R. (2025). Synergistic effects of iron oxide nanoparticles and arbuscular mycorrhizal fungi through seed coating on maize growth and yield under drought stress. *Journal of Ecological Engineering, 26*(9), 400–419. http://www.jeeng.net. doi:10.12911/22998993/205120.

Ferreira, A. I., da Costa Ferreira, P., Cooper, C. L., & Oliveira, D. (2019). How daily negative affect and emotional exhaustion correlates with work engagement and presenteeism-constrained productivity. *International Journal of Stress Management, 26*(3), 261–271. http://www.apa.org/pubs/journals/str/. doi:10.1037/str0000114.

Fontana, A., Reichelt, M., Hempel, S., Gershenzon, J., & Unsicker, S. B. (2009). The effects of arbuscular mycorrhizal fungi on direct and indirect defense metabolites of *Plantago lanceolata* L. *Journal of Chemical Ecology, 35*(7), 833–843. doi:10.1007/s10886-009-9654-0.

Fossalunga, A. S. D., & Bonfante, P. (2023). Arbuscular mycorrhizal fungi as biofertilisers. *Current Biology, 33*(11), 460–463.

Gadd, G. M. (1993). Interactions of fungi with toxic metals. *New Phytologist, 124*(1), 25–60. doi:10.1111/j.1469-8137.1993.tb03796.x.

Gallaud, I. (1905). Etudes sur les mycorrhizes endotrophes. *Revue générale de botanique, 17*, 479–500.

Gao, D., Fu, G., Wan, X., & Memon, F. A. (2025). Future water stress in China: Projections under socioeconomic and climate change scenarios. *Journal of Environmental Management, 391*. https://www.sciencedirect.com/science/journal/03014797. doi:10.1016/j.jenvman.2025.126486.

Garcia-Garrido, J. M., & Ocampo, J. A. (1989). Effect of VA mycorrhizal infection of tomato on damage caused by *Pseudomonas syringae*. *Soil Biology and Biochemistry, 21*(1), 165–167. doi:10.1016/0038-0717(89)90027-8.

George, E., Marschner, H., & Jakobsen, I. (1995). Role of arbuscular mycorrhizal fungi in uptake of phosphorus and nitrogen from soil. *Critical Reviews in Biotechnology, 15*(3–4), 257–270.

Ghorui, M., Chowdhury, S., & Burla, S. (2025). Recent advances in the commercial formulation of arbuscular mycorrhizal inoculants. *Frontier in Industrial Microbiology, 3*, 1553472.

Gianinazzi, S., & Schüepp, H. (1994). *Impact of arbuscular mycorrhizas on sustainable agriculture and natural ecosystems*. Birkhäuser Basel. doi:10.1007/978-3-0348-8504-1.

Gianinazzi-Pearson, V. (1996). Plant cell responses to arbuscular mycorrhizal fungi: Getting to the roots of the symbiosis. *The Plant Cell, 8*(10), 1871–1883. doi:10.1105/tpc.8.10.1871.

Gianinazzi-Pearson, V., Gollotte, A., Dumas-Gaudot, E., Franken, P., & Gianinazzi, S. (1994). *Gene expression and molecular modifications associated with plant responses to infection by arbuscular mycorrhizal fungi* (pp. 179–186). Springer Nature. doi:10.1007/978-94-011-0177-6_27.

Giovannetti, M., Sbrana, C., & Logi, C. (1994). Early processes involved in host recognition by arbuscular mycorrhizal fungi. *The New Phytologist, 127*(4), 703–709. doi:10.1111/j.1469-8137.1994.tb02973.x.

Giovannetti, M., Ayio, L., Sbrana, C., & Silvia Citernesi, A. (1993). Factors affecting appressorium development in the vesicular–arbuscular mycorrhizal fungus *Glomus mosseae* (Nicol. & Gerd.) Gerd. & Trappe. *New Phytologist, 123*(1), 115–122. doi:10.1111/j.1469-8137.1993.tb04537.x.

Giweli, A. A., Džamić, A. M., Soković, M., Ristić, M. S., Janaćković, P., & Marin, P. D. (2013). The chemical composition, antimicrobial and antioxidant activities of the essential oil of salvia fruticosa growing wild in Libya. *Archives of Biological Sciences, 65*(1), 321–329. http://archonline.bio.bg.ac.rs/VOL65/SVESKA1/37%20-%20Giweli.pdf. doi:10.2298/ABS1301321GSerbia.

Golubkina, N., Krivenkov, L., Sekara, A., Vasileva, V., Tallarita, A., & Caruso, G. (2020). Prospects of arbuscular mycorrhizal fungi utilization in production of *Allium* plants. *Plants, 9*(2). https://www.mdpi.com/2223-7747/9/2/279/pdf. doi:10.3390/plants9020279.

Goura, K., Legrifi, I., Kallali, N. S., Taoussi, M., Kenfaoui, J., Meddich, A., & Lahlali, R. (2025). Beyond survival: The role of secondary metabolites in plant defense mechanisms. *Journal of Crop Health, 77*(4), 121. https://doi.org/10.1007/s10343-025-01183-z.

Grümberg, B. C., Urcelay, C., Shroeder, M. A., Vargas-Gil, S., & Luna, C. M. (2014). The role of inoculum identity in drought stress mitigation by arbuscular mycorrhizal fungi in soybean. *Biology and Fertility of Soils, 51*(1), 1–10. link.springer.de/link/service/journals/00374/index.htm. doi:10.1007/s00374-014-0942-7.

Guillemin, J. P., Abdel-Fattah, G. M., Trouvelot, A., Gianinazzi, S., & Gianinazzi-Pearson, V. (1993). Interactions between soil-applied fungicides, endomycorrhiza fungal activity and plant growth. *Trends in Agriculture Science, 1*, 161–172.

Gupta, R., Mehta, G., Khasa, Y. P., & Kuhad, R. C. (2011). Fungal delignification of lignocellulosic biomass improves the saccharification of cellulosics. *Biodegradation, 22*(4), 797–804. doi:10.1007/s10532-010-9404-6.

Gupta, S., Wali, A., Gupta, M., & Annepu, S. K. (2017). Fungi: An effective tool for bioremediation. In *Plant-microbe interactions in agro-ecological perspectives: 2* (pp. 593–606). http://www.springer.com/in/book/9789811065927. doi:10.1007/978-981-10-6593-4_24.

Gutjahr, C., & Paszkowski, U. (2013). Multiple control levels of root system remodeling in arbuscular mycorrhizal symbiosis. *Frontier in Plant Sciences, 4*, 204.

Hage-Ahmed, K., Rosner, K., & Steinkellner, S. (2019). Arbuscular mycorrhizal fungi and their response to pesticides. *Pest Management Science, 75*(3), 583–590.

Harborne, J. B. (1996). Chemotaxonomy of anthocyanins and phytoalexins in the Compositae. In D. J. N. Hind, & H. J. Boentje (Eds.), *Compositae: Systematics* (pp. 207–218). *Royal Botanic Gardens, Kew*.

Harrier, L. A., & Watson, C. A. (2004). The potential role of arbuscular mycorrhizal (AM) fungi in the bioprotection of plants against soil-borne pathogens in organic and/or other sustainable farming systems. *Pest Management Science, 60*(2), 149–157. doi:10.1002/ps.820.

Hartmann, T. (2007). From waste products to ecochemicals: Fifty years research of plant secondary metabolism. *Phytochemistry, 68*(22–24), 2831–2846. doi:10.1016/j.phytochem.2007.09.017.

Hashem, A., Alqarawi, A. A., Radhakrishnan, R., Al-Arjani, A. B. F., Aldehaish, H. A., Egamberdieva, D., & Abd_Allah, E. F. (2018). Arbuscular mycorrhizal fungi regulate the oxidative system, hormones and ionic equilibrium to trigger salt stress tolerance in *Cucumis sativus* L. *Saudi Journal of Biological Sciences, 25*(6), 1102–1114.

Hawkins, H. J., & George, E. (2001). Reduced 15N-nitrogen transport through arbuscular mycorrhizal hyphae to *Triticum aestivum* L. supplied with ammonium vs. nitrate nutrition. *Annals of Botany, 87*(3), 303–311. http://aob.oxfordjournals.org/. doi:10.1006/anbo.2000.1305.

Heckman, D. S., Geiser, D. M., Eidell, B. R., Stauffer, R. L., Kardos, N. L., & Hedges, S. B. (2001). Molecular evidence for the early colonization of land by fungi and plants. *Science, 293*(5532), 1129–1133. doi:10.1126/science.1061457.

Herbert, R. B. (1989). *The biosynthesis of secondary metabolites* (2nd, p. 200). London: Chapman and Hall.

Hildebrandt, U., Kaldorf, M., & Bothe, H. (1999). The zinc violet and its colonization by arbuscular mycorrhizal fungi. *Journal of Plant Physiology, 154*(5–6), 709–717. doi:10.1016/S0176-1617(99)80249-1.

Hock, B., & Verma, A. (1995). *Mycorrhiza structure, function, molecular biology and biotechnology*. Berlin, Germany.: Springer-Verlag.

Hodge, A., & Fitter, A. H. (2010). Substantial nitrogen acquisition by arbuscular mycorrhizal fungi from organic material has implications for N cycling. *Proceedings of the National Academy of Sciences of the United States, 107*, 13754–13759.

Hodge, A., & Storer, K. (2014). Arbuscular mycorrhiza and nitrogen: Implications for individual plants through to ecosystems. *Plant and Soil, 386*(1–2). http://www.wkap.nl/journalhome.htm/0032-079X. doi:10.1007/s11104-014-2162-1.

Ho-Plágaro, T., & García-Garrido, J. M. (2022a). Molecular regulation of arbuscular mycorrhizal symbiosis. *International Journal of Molecular Sciences, 23*(11), 5960.

Ho-Plágaro, T., & García-Garrido, J. M. (2022b). Multifarious and interactive roles of GRAS transcription factors during arbuscular mycorrhiza development. *Frontier in Plant Science, 13*, 908.

Hosny, M., Gianinazzi-Pearson, V., & Dulieu, H. (1998). Nuclear DNA content of 11 fungal species in *Glomales*. *Genome, 41*(3), 422–428.

Hussein, R. A., & El-Anssary, A. A. (2019). Plants secondary metabolites: The key drivers of the pharmacological actions of medicinal plants. *Herbal Medicine, 1*(3), 11–30.

IJdo, M., Cranenbrouck, S., & Declerck, S. (2011). Methods for large-scale production of AM fungi: Past, present, and future. *Mycorrhiza, 21*, 1–16.

Isah, T. (2019). Stress and defense responses in plant secondary metabolites production. *Biological Research, 52*(1). doi:10.1186/s40659-019-0246-3.

Jaleel, C. A., Manivannan, P., Sankar, B., Kishorekumar, A., Sankari, S., & Panneerselvam, R. (2007). Paclobutrazol enhances photosynthesis and ajmalicine production in *Catharanthus roseus*. *Process Biochemistry, 42*(11), 1566–1570.

Jeffries, P., Gianinazzi, S., Perotto, S., Turnau, K., & Barea, J. M. (2003). The contribution of arbuscular mycorrhizal fungi in sustainable maintenance of plant health and soil fertility. *Biology and Fertility of Soils, 37*(1), 1–16. link.springer.de/link/service/journals/00374/index.htm. doi:10.1007/s00374-002-0546-5.

Johny, L., Cahill, D. M., & Adholeya, A. (2021). AMF enhance secondary metabolite production in ashwagandha, licorice, and marigold in a fungi-host specific manner. *Rhizosphere, 17*, 100314.

Jurkiewicz, A., Ryszka, P., Anielska, T., Waligórski, P., Białońska, D., Góralska, K., & Turnau, K. (2010). Optimization of culture conditions of *Arnica montana* L.: Effects of mycorrhizal fungi and competing plants. *Mycorrhiza, 20*(5), 293–306.

Kabera, J. N., Semana, E., Mussa, A. R., & He, X. (2014). Plant secondary metabolites: Biosynthesis, classification, function and pharmacological properties. *Journal of Pharmacy and Pharmacology, 2*(7), 377–392.

Kappers, I. F., Aharoni, A., Herpen, T. W. J. M., Luckerhoff, L. L. P., Dicke, M., & Bouwmeester, H. J. (2005). Genetic engineering of terpenoid metabolism attracts bodyguards to arabidopsis. *Science, 309*(5743), 2070–2072. doi:10.1126/science.1116232.

Karagiannidis, N., Thomidis, T., Lazari, D., Panou-Filotheou, E., & Karagiannidou, C. (2011). Effect of three Greek arbuscular mycorrhizal fungi in improving the growth, nutrient concentration, and production of essential oils of oregano and mint plants. *Scientia Horticulturae, 129*(2), 329–334. doi:10.1016/j.scienta.2011.03.043.

Khaliq, A., Perveen, S., Alamer, K. H., Zia Ul Haq, M., Rafique, Z., Alsudays, I. M., Althobaiti, A. T., Saleh, M. A., Hussain, S., & Attia, H. (2022). Arbuscular mycorrhizal fungi symbiosis to enhance plant–soil interaction. *Sustainability, 14*(13), 7840. doi:10.3390/su14137840.

Khaosaad, T., Vierheilig, H., Nell, M., Zitterl-Eglseer, K., & Novak, J. (2006). Arbuscular mycorrhiza alter the concentration of essential oils in oregano (*Origanum* sp., Lamiaceae). *Mycorrhiza, 16*(6), 443–446. doi:10.1007/s00572-006-0062-9.

Kivlin, S. N., Hawkes, C. V., & Treseder, K. K. (2011). Global diversity and distribution of arbuscular mycorrhizal fungi. *Soil Biology and Biochemistry, 43*(11), 2294–2303.

Klironomos, J. N. (2000). Host-specificity and functional diversity among arbuscular mycorrhizal fungi. In C. R. Bell, M. Brylinsky, & P. Johnson-Green (Eds.), Microbial biosystems: New f*rontiers*. Proceedings of the 8th International Symposium on Microbial ecology*: 1* (pp. 845–851). Halifax, Canada: Atlantic Canada Society for Microbial Ecology.

Koide, R. T., & Mosse, B. (2004). A history of research on arbuscular mycorrhiza. *Mycorrhiza, 14*(3), 145–163. link.springer.de/link/service/journals/00572/index.htm. doi:10.1007/s00572-004-0307-4.

Krishnaiah, D., Sarbatly, R., & Nithyanandam, R. (2011). A review of the antioxidant potential of medicinal plant species. *Food and Bioproducts Processing, 89*(3), 217–233.

Kuila, D., & Ghosh, S. (2022). Aspects, problems and utilization of arbuscular mycorrhizal (AM) application as bio-fertilizer in sustainable agriculture. *Current Research in Microbial Sciences, 3*, 100107. doi:10.1016/j.crmicr.2022.100107.

Kumar, S., Arora, N., & Upadhyay, H. (2021). Arbuscular mycorrhizal fungi: Source of secondary metabolite production in medicinal plants. In *New and future developments in microbial biotechnology and bioengineering* (pp. 155–164). Elsevier.

Kuyper, T. W., Wang, X., & Muchane, M. N. (2021). The interplay between roots and arbuscular mycorrhizal fungi influencing water and nutrient acquisition and use efficiency. In *The root systems in sustainable agricultural intensification* (pp. 193–220). Hoboken, NJ, USA: Wiley.

Lei, Y., Yang, K., Wang, B., Sheng, Y., Bird, B. W., Zhang, G., & Tian, L. (2014). Response of inland lake dynamics over the Tibetan Plateau to climate change. *Climatic Change, 125*(2), 281–290. www.wkap.nl/journalhome.htm/0165-0009. doi:10.1007/s10584-014-1175-3.

Leigh, J., Hodge, A., & Fitter, A. H. (2009). Arbuscular mycorrhizal fungi can transfer substantial amounts of nitrogen to their host plant from organic material. *New Phytologist, 181*, 199–207.

Lermen, C., Morelli, F., Gazim, Z. C., da Silva, A. P., Goncalves, J. E., Dragunski, D. C., & Alberton, O. (2015). Essential oil content and chemical composition of *Cymbopogon citratus* inoculated with arbuscular mycorrhizal fungi under different levels of lead. *Industrial Crops and Products, 76*, 734–738.

Leyval, C., Tumau, K., & Haselwandter, K. (1997). Effect of heavy metal pollution on mycorrhizal colonization and function: Physiological, ecological and applied aspects. Interactions between heavy metals and mycorrhizal fungi in polluted soils: Physiological, ecological and applied aspects. *Mycorrhiza, 7*, 139–153.

Li, J., Meng, B., Chai, H., Yang, X., Song, W., Li, S., Lu, A., Zhang, T., & Sun, W. (2019). Arbuscular mycorrhizal fungi alleviate drought stress in C3 (*Leymus chinensis*) and C4 (*Hemarthria altissima*) grasses via altering antioxidant enzyme activities and photosynthesis. *Frontiers in Plant Science, 10*. https://www.frontiersin.org/articles/10.3389/fpls.2019.00499/pdf. doi:10.3389/fpls.2019.00499.

Li, L., McCormack, M. L., Chen, F., Wang, H., Ma, Z., & Guo, D. (2019). Different responses of absorptive roots and arbuscular mycorrhizal fungi to fertilization provide diverse nutrient acquisition strategies in Chinese fir. *Forest Ecology and Management, 433*, 64–72.

Li, X., Sun, X., Meng, Y., Liang, J., Wu, F., & Li, J. (2019). Dice loss for data-imbalanced NLP tasks. *arXiv*. preprint arXiv:1911.02855 https://doi.org/10.48550/arXiv.1911.02855 .

Linderman, R. G. (1992). VA mycorrhizae and soil microbial interactions. In G. J. Bethelenfalvay, & R. G. Linderman (Eds.), *Mycorrhizae in sustainable agriculture, ASA Special Publication No. 54* (pp. 45–70). ASA.

Linderman, R. G. (1994). Role of VAM fungi in biocontrol. In FL Pfleger, & RG Linderman (Eds.), *Mycorrhizae and plant health* (pp. 1–26). APS Press.

Liu, K., Ding, X., Deng, B., & Chen, W. (2009). Isolation and characterization of endophytic taxol-producing fungi from Taxus chinensis. *Journal of Industrial Microbiology and Biotechnology, 36*(9), 1171–1177. doi:10.1007/s10295-009-0598-8.

Liu, M., Wang, H., Lin, Z., Ke, J., Zhang, P., Zhang, F., Ru, D., Zhang, L., Xiao, Y., & Liu, X. (2024). Arbuscular mycorrhizal fungi inhibit necrotrophic, but not biotrophic, aboveground plant pathogens: A meta-analysis and experimental study. *New Phytologist, 241*(3), 1308–1320. http://onlinelibrary.wiley.com/journal/10.1111/(ISSN)1469-8137. doi:10.1111/nph.19392.

MacLean, A. M., Bravo, A., & Harrison, M. J. (2017). Plant signaling and metabolic pathways enabling arbuscular mycorrhizal symbiosis. *The Plant Cell, 29*(10), 2319–2335.

MacLean, A. M., Bravo, A., & Harrison, M. J. (2022). Plant signaling and metabolic pathways enabling arbuscular mycorrhizal symbiosis. *The Plant Cell, 29*(10), 2319–2335.

Madawala, H. M. S. P. (2021). Arbuscular mycorrhizal fungi as biofertilizers: Current trends, challenges, and future prospects. In *Biofertilizers: Volume 1: Advances in bio-inoculants* (pp. 83–93). Elsevier BV. doi:10.1016/b978-0-12-821667-5.00029-4.

Meng, L., Zhang, A., Wang, F., Han, X., Wang, D., & Li, S. (2015). Arbuscular mycorrhizal fungi and rhizobium facilitate nitrogen uptake and transfer in soybean/maize intercropping system. *Frontiers in Plant Science, 6*, 1–10. http://journal.frontiersin.org/article/10.3389/fpls.2015.00339/pdf. doi:10.3389/fpls.2015.00339.

Michal, G., & Schomburg, D. (2013). *Biochemical pathways: An atlas of biochemistry and molecular biology: Second edition biochemical pathways: An atlas of biochemistry and molecular biology: Second edition*. John Wiley and Sons http://onlinelibrary.wiley.com/book/10.1002/9781118657072. doi:10.1002/9781118657072.

Michalak, A. (2006). Phenolic compounds and their antioxidant activity in plants growing under heavy metal stress. *Polish Journal of Environmental Studies, 15*(4), 523–530.

Mirjalili, M. H., Moyano, E., Bonfill, M., Cusido, R. M., & Palazón, J. (2009). Steroidal lactones from *Withania somnifera*, an ancient plant for novel medicine. *Molecules, 14*(7), 2373–2393. http://www.mdpi.com/1420-3049/14/7/2373/pdf. doi:10.3390/molecules14072373Iran.

Mishra, V., Ellouze, W., & Howard, R. J. (2018). Utility of arbuscular mycorrhizal fungi for improved production and disease mitigation in organic and hydroponic greenhouse crops. *Journal of Horticulture, 5*(03), 1–10. doi:10.4172/2376-0354.1000237.

Mo, Y., Wang, Y., Yang, R., Zheng, J., Liu, C., Li, H., Ma, J., Zhang, Y., Wei, C., & Zhang, X. (2016). Regulation of plant growth, photosynthesis, antioxidation and osmosis by an arbuscular mycorrhizal fungus in watermelon seedlings under well-watered and drought conditions. *Frontiers in Plant Science, 7*, 1–15.

Montero, H., Lee, T., Pucker, B., Ferreras-Garrucho, G., Oldroyd, G., Brockington, S. F., Miyao, A., & Paszkowski, U. (2021). A mycorrhiza-associated receptor-like kinase with an ancient origin in the green lineage. *Proceedings of the National Academy of Sciences, 118*(25). doi:10.1073/pnas.2105281118.

Mosse, B. (1953). Fructifications associated with mycorrhizal strawberry roots. *Nature, 171*, 974.

Mosse, B. (1986). Mycorrhiza in a sustainable agriculture. *Biological Agriculture & Horticulture, 3*(2-3), 191–209. doi:10.1080/01448765.1986.9754471.

Muniz, C. R., Leugi, G. B., de Marco Pereira, C., Przeybilovicz, E., & Alves, A. M. (2021). Uma analise sobre exclusao digital durante a pandemia de COVID-19 no Brasil: Quem tem direito as cidades inteligentes? *Revista de Direito da Cidade, 13*(2), 700–728.

Munkvold, L., Kjøller, R., Vestberg, M., Rosendahl, S., & Jakobsen, I. (2004). High functional diversity within species of arbuscular mycorrhizal fungi. *New Phytologist, 164*(2), 357–364. doi:10.1111/j.1469-8137.2004.01169.x.

Nazir, A., Sarfraz, W., Allah, D., Khalid, N., Farid, M., Shafiq, M., Firdaus-E-Bareen, Rizvi, Z. F, & Naeem, N. (2023). Synergistic impact of two autochthonous saprobic fungi (*A. niger* and *T. pseudokoningii*) on the growth, ionic contents, and metals uptake in *Brassica juncea* L. and *Vigna radiata* L. under tannery solid waste contaminated soil. *International Journal of Phytoremediation, 25*(11), 1488–1500. doi:10.1080/15226514.2023.2166457.

Neag, M. A., Mocan, A., Echeverría, J., Pop, R. M., Bocsan, C. I., Crisan, G., & Buzoianu, A. D. (2018). Berberine: Botanical occurrence, traditional uses, extraction methods, and relevance in cardiovascular, metabolic, hepatic, and renal disorders. *Frontiers in Pharmacology, 9*. https://www.frontiersin.org/articles/10.3389/fphar.2018.00557/full. doi:10.3389/fphar.2018.00557.

Nell, M., Wawrosch, C., Steinkellner, S., Vierheilig, H., Kopp, B., Lössl, A., Franz, C., Novak, J., & Zitterl-Eglseer, K. (2010). Root colonization by symbiotic arbuscular mycorrhizal fungi increases sesquiterpenic acid concentrations in *Valeriana officinalis* L. *Planta Medica, 76*(4), 393–398. doi:10.1055/s-0029-1186180.

Neumann, E., & George, E. (2010). Nutrient uptake: The arbuscular mycorrhiza fungal symbiosis as a plant nutrient acquisition strategy. In *Arbuscular mycorrhizas: Physiology and function* (pp. 137–167). Springer Netherlands. http://www.springerlink.com/openurl.asp?genre=book&isbn=978-90-481-9488-9. doi:10.1007/978-90-481-9489-6_7.

Öpik, M., Zobel, M., Cantero, J. J., Davison, J., Facelli, J. M., Hiiesalu, I., Jairus, T., Kalwij, J. M., Koorem, K., Leal, M. E., Liira, J., Metsis, M., Neshataeva, V., Paal, J., Phosri, C., Põlme, S., Reier, U., Saks, U., Schimann, H., … Moora, M. (2013). Global sampling of plant roots expands the described molecular diversity of arbuscular mycorrhizal fungi. *Mycorrhiza, 23*(5), 411–430. doi:10.1007/s00572-013-0482-2.

Ortas, I. (2012). The effect of mycorrhizal fungal inoculation on plant yield, nutrient uptake and inoculation effectiveness under long-term field conditions. *Field crops research, 125*, 35–48.

Osorio, N., & Habte, M. (2001). Synergistic influence of an arbuscular mycorrhizal fungus and a P solubilizing fungus on growth and P uptake of *Leucaena leucocephala* in an Oxisol. *Arid Land Research and Management, 15*(3), 263–274.

Ostadi, A., Javanmard, A., Machiani, M. A, Sadeghpour, A., Maggi, F., Nouraein, M., Morshedloo, M. R, Hano, C., & Lorenzo, J. M. (2022). Co-application of TiO_2 nanoparticles and arbuscular mycorrhizal fungi improves essential oil quantity and quality of sage (*Salvia officinalis* L.) in Drought Stress conditions. *Plants, 11*(13), 1659. doi:10.3390/plants11131659.

Othman, Y. A., Alananbeh, K. M., & Tahat, M. M. (2024). Can arbuscular mycorrhizal fungi enhance crop productivity and quality in hydroponics? A meta-analysis. *Sustainability, 16*(9), 3662.

Pandey, D. K., Kaur, P., & Dey, A. (2018). Arbuscular mycorrhizal fungi: Effects on secondary metabolite production in medicinal plants. In *Fungi and their role in sustainable development: Current perspectives* (pp. 507–538). *Springer Singapore*.

Pandey, D., Kaur Kehri, H., Zoomi, I., Akhtar, O., & Singh, A. K. (2019). *Mycorrhizal fungi: Biodiversity, ecological significance, and industrial applications* (pp. 181–199). Springer Science and Business Media LLC. doi:10.1007/978-3-030-10480-1_5.

Pandey, R. R., Loushambam, S., & Srivastava, A. K. (2020). Arbuscular mycorrhizal and dark septate endophyte fungal associations in two dominant ginger species of Northeast India. *Proceedings of the National Academy of Sciences, India Section B: Biological Sciences, 90*(4), 885–894.

Pant, P., Pandey, S., & Dall'Acqua, S. (2021). The influence of environmental conditions on secondary metabolites in medicinal plants: A literature review. *Chemistry & Biodiversity, 18*(11), e2100345.

Pedone-Bonfim, M. V. L., da Silva, D. K. A., Maia, L. C., & Yano-Melo, A. M. (2018). Mycorrhizal benefits on native plants of the Caatinga, a Brazilian dry tropical forest. *Symbiosis, 74*(2), 79–88. http://www.springer.com/life+sci/microbiology/journal/13199. doi:10.1007/s13199-017-0510-7.

Peng, Z., Zulfiqar, T., Yang, H., Wang, M., & Zhang, F. (2024). Effect of arbuscular mycorrhizal fungi (AMF) on photosynthetic characteristics of cotton seedlings under saline-alkali stress. *Scientific Reports, 14*(1). https://www.nature.com/srep/. doi:10.1038/s41598-024-58979-8.

Peyronel's, B. (1923). *Preliminary researches on the endotrophic mycorrhiza and on the normal fungous flora of the roots of phanerogams. Rivista di Biologia, 5*, 463–485.

Pfeffer, P. E., Douds, D. D., Bécard, G., & Shachar-Hill, Y. (1999). Carbon uptake and the metabolism and transport of lipids in an arbuscular mycorrhiza. *Plant Physiology, 120*(2), 587–598. http://www.plantphysiol.org/. doi:10.1104/pp.120.2.587.

Piliarová, M., Ondreičková, K., Hudcovicová, M., Mihálik, D., & Kraic, J. (2019). Arbuscular mycorrhizal fungi—Their life and function in ecosystem. *Agriculture (Pol'nohospodarstvo), 65*(1), 3–15. https://content.sciendo.com/view/journals/agri/agri-overview.xml. doi:10.2478/agri-2019-0001.

Pimprikar, P., & Gutjahr, C. (2018). Transcriptional regulation of arbuscular mycorrhiza development. *Plant and Cell Physiology, 59*(4), 673–690. http://pcp.oxfordjournals.org/. doi:10.1093/pcp/pcy024.

Pinochet, J., Calvet, C., Camprubí, A., & Fernández, C. (1996). Interactions between migratory endoparasitic nematodes and arbuscular mycorrhizal fungi in perennial crops: A review. *Plant and Soil, 185*(2), 183–190. doi:10.1007/bf02257523.

Pusztahelyi, T., Holb, I. J., & Pocsi, I. (2015). Secondary metabolites in fungus-plant interactions. *Frontiers in Plant Science, 6*, 573.

Rahmat, S., & Soheilikhah, Z. (2024). Arbuscular mycorrhizal fungi and plant secondary metabolism. In *Arbuscular mycorrhizal fungi and higher plants: Fundamentals and applications* (pp. 99–121). *Springer Nature Singapore*.

Ran, Z., Ding, W., Cao, S., Fang, L., Zhou, J., & Zhang, Y. (2022). Arbuscular mycorrhizal fungi: Effects on secondary metabolite accumulation of traditional Chinese medicines. *Plant Biology, 24*(6), 932–938.

Read, R. C., Camp, N. J., Di Giovine, F. S., Borrow, R., Kaczmarski, E. B., Chaudhary, A. G. A., Fox, A. J., & Duff, G. W. (2000). An interleukin-1 genotype is associated with fatal outcome of meningococcal disease. *Journal of Infectious Diseases, 182*(5), 1557–1560. doi:10.1086/315889.

Redecker, D. (2000). Specific PCR primers to identify arbuscular mycorrhizal fungi within colonized roots. *Mycorrhiza, 10*(2), 73–80.

Redecker, D., Morton, J. B., & Bruns, T. D. (2000). Ancestral lineages of arbuscular mycorrhizal fungi (Glomales). *Molecular Phylogenetics and Evolution, 14*, 276–284.

Requena, N., Serrano, E., Ocón, A., & Breuninger, M. (2007). Plant signals and fungal perception during arbuscular mycorrhiza establishment. *Phytochemistry, 68*(1), 33–40. doi:10.1016/j.phytochem.2006.09.036.

Requena, N., & Breuninger, M. (2004). *The old arbuscular mycorrhizal symbiosis in the light of the molecular era* (pp. 323–356). Springer Science and Business Media LLC. doi:10.1007/978-3-642-18819-0_13.

Roaa, M. H. (2020). A review article: The importance of the major groups of plants secondary metabolism phenols, alkaloids, and terpenes. *International Journal for Research in Applied Sciences and Biotechnology (IJRASB), 7*(5), 354–358.

Rosling, A., Sahraei, S. E., Kalsoom Khan, F., Desirò, A., Bryson, A. E., Mondo, S. J., Grigoriev, I. V., Bonito, G., & Sánchez-García, M. (2024). Evolutionary history of arbuscular mycorrhizal fungi and genomic signatures of obligate symbiosis. *BMC Genomics, 25*(1). https://bmcgenomics.biomedcentral.com/. doi:10.1186/s12864-024-10391-2.

Roussis, I., Beslemes, D., Kosma, C., Triantafyllidis, V., Zotos, A., Tigka, E., Mavroeidis, A., Karydogianni, S., Kouneli, V., Travlos, I., & Kakabouki, I. (2022). The influence of arbuscular mycorrhizal fungus *Rhizophagus irregularis* on the growth and quality of processing tomato (*Lycopersicon esculentum* Mill.) Seedlings. *Sustainability, 14*(15), 9001. doi:10.3390/su14159001.

Rydlová, J., Jelínková, M., Dušek, K., Dušková, E., Vosátka, M., & Püschel, D. (2016). Arbuscular mycorrhiza differentially affects synthesis of essential oils in coriander and dill. *Mycorrhiza, 26*(2), 123–131. link.springer.de/link/service/journals/00572/index.htm. doi:10.1007/s00572-015-0652-5.

Sacchettini, J. C., & Poulter, C. D. (1997). Creating isoprenoid diversity. *Science, 277*(5333), 1788–1789. doi:10.1126/science.277.5333.1788.

Saeed, H. S., Munir, M., Samra, S., Farooq, A., Mahmood, H., Ahmad, I., Akram, M. I., Iqbal, R., Elsadek, M. F., Almutairi, S. M., Ditta, A., & Daraz, U. (2025). Screening of biosurfactants producing fungi from infected *Citrus sinensis* (L.). *Polish Journal of Environmental Studies, 34*(5), 5261–5270. https://www.pjoes.com/pdf-190588-115894?filename=Screening%20of.pdf. doi:10.15244/pjoes/190588.

Saha, H., Kaloterakis, N., Harvey, J. A., Putten, W. V. D., & Biere, A. (2023). Effects of light quality on colonization of tomato roots by arbuscular mycorrhizal fungus (AMF) and implications for growth and defense. *Plants, 11*(7), 861.

Sbrana, C., & Giovannetti, M. (2005). Chemotropism in the arbuscular mycorrhizal fungus *Glomus mosseae*. *Mycorrhiza, 15*(7), 539–545. doi:10.1007/s00572-005-0362-5.

Schussler, A., Schwarzott, D., & Walker, C. (2001). A new fungal phylum, the Glomeromycota, phylogeny and evolution. *Mycological Research, 105*, 1413–1421.

Schwab, M. E., & Schnell, L. (1991). Channeling of developing rat corticospinal tract axons by myelin-associated neurite growth inhibitors. *Journal of Neuroscience, 11*(3), 709–721.

Schweiger, R., & Muller, C. (2015). Leaf metabolome in arbuscular mycorrhizal symbiosis. *Current Opinion in Plant Biology, 26*, 120–126.

Selosse, M. A., Strullu-Derrien, C., Martin, F. M., Kamoun, S., & Kenrick, P. (2015). Plants, fungi and oomycetes: A 400-million year affair that shapes the biosphere. *New Phytologist, 206*(2), 501–506.

Selosse, M.-A., & Le Tacon, F. (1998). The land flora: A phototroph-fungus partnership? *Trends in Ecology & Evolution, 13*(1), 15–20. doi:10.1016/s0169-5347(97)01230-5.

Shaker-Koohi, S. (2014). Role of arbuscular mycorrhizal (AM) fungi in phytoremediation of soils contaminated: A review. *International journal of Advanced Biological and Biomedical Research, 2*(5), 1854–1864.

Sharma, R., Magotra, A., Manhas, R. S., & Chaubey, A. (2017). Antagonistic potential of a psychrotrophic fungus: Trichoderma velutinum ACR-P1. *Biological Control, 115*, 12–17. http://www.elsevier.com/inca/publications/store/6/2/2/7/9/1/index.htt. doi:10.1016/j.biocontrol.2017.08.024.

Sikora, R. A. (1978). Einfluß der endotrophen Mykorrhiza (*Glomus mosseae*) auf das Wirt-Parasit-Verhältnis von *Meloidogyne incognita* in Tomaten/Effect of the endotrophic mycorrhizal fungus, *Glomus mosseae*, on the host-parasite relationship of *Meloidogyne incognita* in tomato. *Zeitschrift für Pflanzenkrankheiten und Pflanzenschutz/Journal of Plant Diseases and Protection, 85*(3/4), 197–202.

Singla, P., Bhardwaj, R. D., Sharma, S., & Sunidhi (2024). Plant–fungus interaction: A stimulus–response theory. *Journal of Plant Growth Regulation, 43*(2), 369–381. https://www.springer.com/journal/344. doi:10.1007/s00344-023-11100-1.

Smith, F. A., & Smith, S. E. (1997). Tansley review no. 96 structural diversity in (vesicular)-arbuscular mycorrhizal symbioses. *The New Phytologist, 137*(3), 373–388.

Schüßler, A. (2005). Glomeromycota phylogeny and taxonomy. Department of Biology, Technical University of Darmstadt. Retrieved from http://www.tu-darmstadt.de/fb/bio/bot/schuessler/amphylo/amphylogeny.html

Smith, S. E., & Read, D. J. (1997). *Mycorrhizal Symbiosis* (2nd ed.). San Diego, California: Academic Press.

Smith, S. E., & Read, D. J. (2008). *Mycorrhizal symbiosis* (3rd edn.). Amsterdam: Academic Press, Elsevier.

Smith, S. E., & Smith, F. A. (2011). Roles of arbuscular mycorrhizas in plant nutrition and growth: New paradigms from cellular to ecosystem scales. *Annual Review of Plant Biology, 62*, 227–250. doi:10.1146/annurev-arplant-042110-103846.

Spatafora, J. W., Chang, Y., Benny, G. L., Lazarus, K., Smith, M. E., Berbee, M. L., Bonito, G., Corradi, N., Grigoriev, I., Gryganskyi, A., James, T. Y., O'Donnell, K., Roberson, R. W., Taylor, T. N., Uehling, J., Vilgalys, R., White, M. M., & Stajich, J. E. (2016). A phylum-level phylogenetic classification of zygomycete fungi based on genome-scale data. *Mycologia, 108*(5), 1028–1046. http://www.mycologia.org/content/108/5/1028.full.pdf. doi:10.3852/16-042.

Sun, M., Gadad, S. S., Kim, D. S., & Kraus, W. L. (2015). Discovery, annotation, and functional analysis of long noncoding RNAs controlling cell-cycle gene expression and proliferation in breast cancer cells. *Molecular cell, 59*(4), 698–711.

Symanczik, S., Bouamri, R., Rachidi, F., & Mäder, P. (2018). Mycorrhizal fungi as natural bio-fertilizers: How to produce and use (Technical Handbook). *Research Institute of Organic Agriculture (FiBL) & Ecole Nationale d'Agriculture (ENA).*

Tang, H., Hassan, M. U., Feng, L., Nawaz, M., Shah, A. N., Qari, S. H., Liu, Y., & Miao, J. (2022). The critical role of arbuscular mycorrhizal fungi to improve drought tolerance and nitrogen use efficiency in crops. *Frontiers in Plant Science, 13*. https://www.frontiersin.org/journals/plant-science. doi:10.3389/fpls.2022.919166.

Thind, S., Chaudhary, M. S, Ditta, A., Hussain, I., Parveen, A., Ullah, N., Mahmood, Q., Al-Ashkar, I., & El-Sabagh, A. (2022). Impact of mycorrhizal fungi from different rhizospheric soils on fungal colonization, growth, and chlorophyll contents of *Cenchrus ciliaris*. *Agronomy, 12*(11), 2644. doi:10.3390/agronomy12112644.

Tiwari, R., & Rana, C. S. (2015). Plant secondary metabolites: A review. *International Journal of Engineering Research and General Science, 3*(5), 661–670.

Tomer, A., Singh, R., Singh, S. K., Dwivedi, S. A., Reddy, C. U, Amanthra Keloth, M. R., & Rachel, R. (2021). *Role of fungi in bioremediation and environmental sustainability* (pp. 187–200). Springer Science and Business Media LLC. doi:10.1007/978-3-030-54422-5_8.

Treseder, K. K., & Cross, A. (2006). Global distributions of arbuscular mycorrhizal fungi. *Ecosystems, 9*(2), 305–316. doi:10.1007/s10021-005-0110-x.

Tuheteru, F. D., & Wu, Q. S. (2017). Arbuscular mycorrhizal fungi and tolerance of waterlogging stress in plants. In *Arbuscular mycorrhizas and stress tolerance of plants* (pp. 43–66). Springer Singapore.

Urcoviche, R. C., Gazim, Z. C., Dragunski, D. C., Barcellos, F. G., & Alberton, O. (2015). Plant growth and essential oil content of *Mentha crispa* inoculated with arbuscular mycorrhizal fungi under different levels of phosphorus. *Industrial Crops and Products, 67*, 103–107. www.elsevier.com/inca/publications/store/5/2/2/8/2/5. doi:10.1016/j.indcrop.2015.01.016.

Van der Heijden, M. G., Bardgett, R. D., & Van Straalen, N. M. (2008). The unseen majority: Soil microbes as drivers of plant diversity and productivity in terrestrial ecosystems. *Ecology Letters, 11*(3), 296–310.

Van der Heijden, M. G., Klironomos, J. K., Ursic, M., Moutoglis, P., Streitwolf-Engel, R., Boller, T., Wiemken, A., & Sanders, I. R. (1998a). Mycorrhizal fungal diversity determines plant biodiversity, ecosystem variability and productivity. *Nature, 396*(6706), 69–72.

Van der Heijden, M. G. A., Boller, T., Wiemken, A., & Sanders, I. R. (1998b). Different arbuscular mycorrhizal fungal species are potential determinants of plant community structure. *Ecology, 79*(6), 2082–2091. http://esajournals.onlinelibrary.wiley.com/hub/journal/10.1002/%28ISSN%291939-9170/issues/. doi:10.1890/0012-9658(1998)079[2082:DAMFSA]2.0.CO;2.

Venkateshwaran, M., Jayaraman, D., Chabaud, M., Genre, A., Balloon, A. J., Maeda, J., Forshey, K., Os, D., Kwiecien, N. W., Coon, J. J., Barker, D. G., & Ané JM, J. M. (2015). A role for the mevalonate pathway in early plant symbiotic signaling. *Proceedings of the National Academy of Sciences, 112*(31), 9781–9786.

Veresoglou, S. D., Chen, B. D., & Rillig, M. C. (2012). Arbuscular mycorrhiza and soil nitrogen cycling. *Soil Biology & Biochemistry, 46*, 53–62.

Viera, A., & Glenn, M. G. (1990). DNA content of Vesicular-arbuscular mycorrhizal fungal spores. *Mycologia, 82*(2), 263. doi:10.2307/3759857.

Wang, B. Q. I. U., & Qiu, Y. L. (2006). Phylogenetic distribution and evolution of mycorrhizas in land plants. *Mycorrhiza, 16*(5), 299–363.

Wang, D., Wu, W. J., Tian, X., Xiang, N., Hashem, A., Abd_Allah, E. F., Wu, Q. S., & Zou, Y. N. (2024). AMF improves response to waterlogging stress in cucumber. *Rhizosphere, 30*. https://www.sciencedirect.com/science/journal/24522198. doi:10.1016/j.rhisph.2024.100891.

Wang, F. (2017). Arbuscular mycorrhizas and ecosystem restoration. In *Arbuscular mycorrhizas and stress tolerance of plants* (pp. 245–292). Springer Singapore. http://www.springer.com/us/book/9789811041143. doi:10.1007/978-981-10-4115-0_11.

Weisany, W., Raei, Y., & Pertot, I. (2015). Changes in the essential oil yield and composition of dill (*Anethum graveolens* L.) as response to arbuscular mycorrhiza colonization and cropping system. *Industrial Crops and Products, 77*, 295–306.

Weng, W., Yan, J., Zhou, M., Yao, X., Gao, A., Ma, C., Cheng, J., & Ruan, J. (2022). Roles of arbuscular mycorrhizal fungi as a biocontrol agent in the control of plant diseases. *Microorganisms, 10*(7). https://www.mdpi.com/2076-2607/10/7/1266/pdf?version=1655879237. doi:10.3390/microorganisms10071266.

Wink, M. (2009). Mode of action and toxicology of plant toxins and poisonous plants. *Mitteilungen aus dem Julius Kühn-Institut, 421*, 93–112.

Wu, Q. S., Silva, F. S. B., Hijri, M., & Kapoor, R. (2023). Editorial: Arbuscular mycorrhiza-mediated augmentation of plant secondary metabolite production. *Frontiers in Plant Science, 14*, 1150900.

Wu, Q. S., Srivastava, A. K., & Zou, Y. N. (2013). AMF-induced tolerance to drought stress in citrus: A review. *Scientia Horticulturae, 164*, 77–87.

Wu, Q. S., & Zou, Y. N. (2017). Arbuscular mycorrhizal fungi and tolerance of drought stress in plants. In *Arbuscular mycorrhizas and stress tolerance of plants* (pp. 25–41). Springer Singapore. http://www.springer.com/us/book/9789811041143. doi:10.1007/978-981-10-4115-0_2.

Xiang, N., Liu, Z., Tian, X., Wang, D., Hashem, A., Abd-Allah, E. F., & Zou, Y. N. (2024). Improved waterlogging tolerance in roots of cucumber plants after inoculation with arbuscular mycorrhizal fungi. *Horticulturae, 10*(5), 478.

Ye, Q., Wang, H., & Li, H. (2022). Arbuscular mycorrhizal fungi improve growth, photosynthetic activity, and chlorophyll fluorescence of *Vitis vinifera* L. cv. ecolly under drought stress. *Agronomy, 12*(7), 1563.

Yizhu, L., Imtiaz, M., Ditta, A., Rizwan, M. S., Ashraf, M., Mehmood, S., Aziz, O., Mubeen, F., Ali, M., Elahi, N. N., Ijaz, R., Lele, S., Shuang, C., & Tu, S. (2020). Response of growth, antioxidant enzymes and root exudates production towards as stress in *Pteris vittata* and in *Astragalus sinicus* colonized by arbuscular mycorrhizal fungi. *Environmental Science and Pollution Research, 27*(2), 2340–2352. https://link.springer.com/journal/11356. doi:10.1007/s11356-019-06785-5.

Zaynab, M., Fatima, M., Abbas, S., Sharif, Y., Umair, M., Zafar, M. H., & Bahadar, K. (2018). Role of secondary metabolites in plant defense against pathogens. *Microbial Pathogenesis, 124*, 198–202. http://www.elsevier.com/inca/publications/store/6/2/2/9/1/5/index.htt. doi:10.1016/j.micpath.2018.08.034.

Zhang, C., Xiang, X., Yang, T., Liu, X., Ma, Y., Zhang, K., Liu, X., & Chu, H. (2024). Nitrogen fertilization reduces plant diversity by changing the diversity and stability of arbuscular mycorrhizal fungal community in a temperate steppe. *Science of the Total Environment, 918*. https://www.sciencedirect.com/science/journal/00489697. doi:10.1016/j.scitotenv.2024.170775.

Zhao, Y. Y., Cartabia, A., Lalaymia, I., & Declerck, S. (2022). Arbuscular mycorrhizal fungi and production of secondary metabolites in medicinal plants. *Mycorrhiza, 32*(3–4), 221–256. link.springer.de/link/service/journals/00572/index.htm. doi:10.1007/s00572-022-01079-0.

Zhu, F., Wen, W., Cheng, Y., Alseekh, S., & Fernie, A. R. (2023). Integrating multiomics data accelerates elucidation of plant primary and secondary metabolic pathways. *Biotech, 4*(1), 47–56.

Zhu, G., Qiao, Y., Miao, C., Miao, L., Qin, S., & Zhang, Y. (2024). Arbuscular mycorrhizal fungi promote soil ammonia but not nitrate assimilation in the Mu Us Desert. *Applied Soil Ecology, 204*, 105726. doi:10.1016/j.apsoil.2024.105726.

Zitterl-Eglseer, K., Nell, M., Lamien-Meda, A., Steinkellner, S., Wawrosch, C., Kopp, B., & Novak, J. (2015). Effects of root colonization by symbiotic arbuscular mycorrhizal fungi on the yield of pharmacologically active compounds in *Angelica archangelica* L. *Acta Physiologiae Plantarum, 37*(2), 21.

Zubek, S., Mielcarek, S., & Turnau, K. (2012). Hypericin and pseudohypericin concentrations of a valuable medicinal plant *Hypericum perforatum* L. are enhanced by arbuscular mycorrhizal fungi. *Mycorrhiza, 22*(2), 149–156.

Zubek, S., Stojakowska, A., Anielska, T., & Turnau, K. (2010). Arbuscular mycorrhizal fungi alter thymol derivative contents of *Inula ensifolia* L. *Mycorrhiza, 20*(7), 497–504. doi:10.1007/s00572-010-0306-6.

Chapter 18

Inoculum production, formulation, and quality control of arbuscular mycorrhizal fungi

Shakeel Ahmad[a], Ayesha Ali[a], Waseem Hassan[a], Hasan Riaz[b], Muhammad Usman Jamshaid[a], Wazir Ahmed[a] and Muhammad Imran[a]

[a] *Department of Soil and Environmental Sciences, Muhammad Nawaz Shareef University of Agriculture, Multan, Pakistan,*

[b] *Department of Plant Pathology, Muhammad Nawaz Shareef University of Agriculture, Multan, Pakistan*

18.1 Introduction

18.1.1 Overview of arbuscular mycorrhizal fungi

Arbuscular mycorrhizal fungi (AMFs) are a vital group of soil microorganisms that form symbiotic associations with the roots of approximately 80% of terrestrial plant species (Smith & Read, 2008). These fungi belong to the phylum Glomeromycota and are characterized by their ability to penetrate the cortical cells of plant roots, forming specialized structures such as arbuscules and vesicles, which facilitate the exchange of nutrients (Berruti et al., 2016). The mutualistic relationship enhances the uptake of essential nutrients, particularly phosphorus (P), and improves plant tolerance to biotic and abiotic stresses.

18.1.2 Importance of arbuscular mycorrhizal fungi in sustainable agriculture

With increasing global concerns over soil degradation, chemical fertilizer dependency, and declining crop productivity, the role of AMF in sustainable agriculture has gained renewed attention. AMF contribute significantly to nutrient cycling, soil structure stabilization, and plant health, making them key components in low-input and organic farming systems (Gianinazzi et al., 2010). Their contribution to plant growth and soil health can reduce the need for synthetic fertilizers and pesticides, thereby promoting environmentally sound agricultural practices (Rillig et al., 2019). The benefits of AMF extend beyond plant nutrition; they also enhance soil aggregation through the production of glomalin—a glycoprotein that binds soil particles—and influence microbial community dynamics in the rhizosphere (Rillig, 2004). Moreover, their symbiotic efficiency makes them suitable candidates for integrated nutrient management strategies in degraded or nutrient-poor soils.

18.1.3 Need for inoculum production and quality control

The increasing global emphasis on sustainable and ecofriendly agriculture has led to a growing interest in biological alternatives to chemical fertilizers, particularly biofertilizers such as AMF. However, for AMF to be effectively applied at the field level, reliable and scalable inoculum production systems and rigorous quality control (QC) measures are essential.

18.1.3.1 Challenges in natural arbuscular mycorrhizal fungi populations

Under natural field conditions, the population density and diversity of AMF are often insufficient to support optimal plant growth, especially in degraded, intensively farmed, or disturbed soils (Brundrett & Abbott, 2002). Practices such as excessive tillage, monocropping, and overuse of synthetic inputs reduce native AMF propagule density and infectivity, leading to the need for external inoculation (Verbruggen & Kiers, 2010).

Advancing Sustainable Agriculture Through Arbuscular Mycorrhizal Fungi. DOI: https://doi.org/10.1016/B978-0-443-45130-0.00002-3

18.1.3.2 Importance of inoculum production

Mass production of AMF inoculum allows for the reintroduction or enhancement of symbiotic fungi in soils where their natural presence is limited or functionally compromised. Effective inoculum must contain a sufficient number of viable propagules (spores, hyphal fragments, and colonized root pieces) capable of colonizing plant roots under field conditions (Ijdo et al., 2011). Furthermore, it must be adaptable to diverse soil and climatic conditions and compatible with common cropping systems.

Three main methods are used for AMF inoculum production: (1) soil-based pot cultures using host plants, (2) substrate-based systems using sterilized carrier materials such as vermiculite, and (3) in vitro root-organ cultures (ROCs) using transformed root systems (Fortin et al., 2002). Each method has specific advantages and limitations related to scalability, contamination risk, and cost-effectiveness. Developing standardized protocols for AMF multiplication remains a critical research area.

18.1.3.3 Need for formulation and storage protocols

AMF are obligate biotrophs, which makes their formulation and storage more complex compared to free-living microorganisms. An ideal formulation should maintain the viability and infectivity of AMF propagules over time, be easy to apply, cost-effective, and resistant to environmental stress during transportation and storage (Chandrasekaran et al., 2014). Poor formulation can lead to the death of spores or reduced colonization efficiency, rendering the inoculum ineffective in the field.

18.1.3.4 Critical role of quality control

The absence of standardized QC frameworks in many developing countries has resulted in the commercialization of AMF inocula with uncertain or substandard efficacy. This not only diminishes farmer confidence but can also cause inconsistencies in research outcomes and practical field applications (Barea et al., 2005). Reliable QC should include

- Spore count and viability tests (e.g., tetrazolium staining and germination assays),
- Colonization potential using trap plants and root staining,
- Microbial purity assessments to ensure the absence of pathogens or unwanted organisms,
- Shelf life studies under different environmental conditions.

Countries, such as India and Brazil, have initiated efforts to regulate biofertilizer quality through government guidelines (INSA, 2009; MAPA, 2011), but global harmonization is still lacking. Development of international standards and accreditation systems could enhance confidence in AMF technologies and stimulate their broader adoption.

18.1.3.5 Implications for sustainable agriculture

High-quality AMF inoculum can play a transformative role in sustainable agriculture by improving nutrient use efficiency, reducing dependence on synthetic inputs, enhancing drought resistance, and supporting carbon sequestration (Lehmann et al., 2014). Ensuring its consistent quality and efficacy through proper production and control measures is crucial for realizing these benefits at scale.

18.2 Biology and ecology of arbuscular mycorrhizal fungi relevant to inoculum production

Understanding the biology and ecology of AMF is fundamental for optimizing their mass multiplication, formulating effective inocula, and ensuring their field performance. The obligate symbiotic lifestyle, slow growth rate, and specific environmental requirements of AMF necessitate tailored approaches for successful inoculum production.

18.2.1 Taxonomy and morphology

AMF are classified under the phylum Glomeromycota, with prominent genera, including Glomus, Acaulospora, Gigaspora, Scutellospora, and Rhizophagus (Schüßler et al., 2001). Morphologically, AMF produce large multinucleate spores, branched hyphae, vesicles (storage structures), and arbuscules (site of nutrient exchange within roots). These structures are crucial for both plant colonization and inoculum viability. Spores are the primary units used in inoculum production. Their size, wall structure, and viability influence their shelf life and infectivity. For instance, *Glomus intraradices*

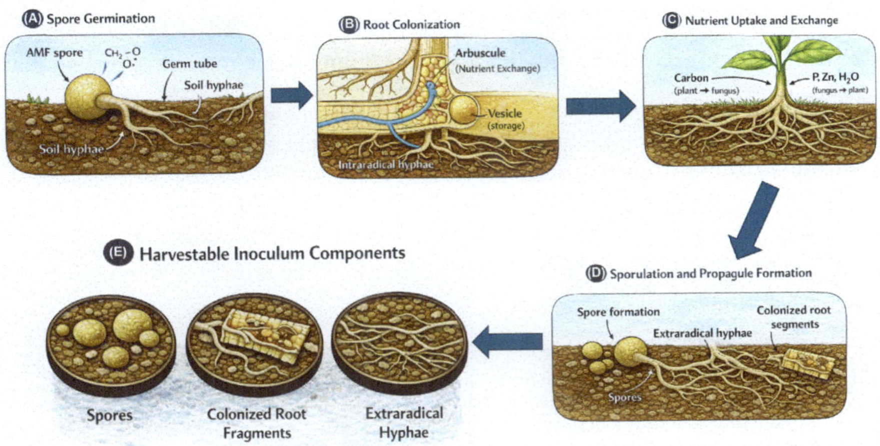

FIGURE 18.1 Generalized life cycle of arbuscular mycorrhizal fungi (AMF) illustrating sequential developmental stages: (A) spore germination stimulated by host root exudates; (B) root penetration and cortical colonization with formation of arbuscules and vesicles; (C) bidirectional nutrient exchange between host plant and fungus; (D) sporulation and propagule development in soil and roots; and (E) biologically active harvestable inoculum components, including spores, colonized root fragments, and extraradical hyphae used for AMF inoculum production.

(now *Rhizophagus irregularis*) produces abundant and relatively fast-germinating spores, making it ideal for commercial production (Ijdo et al., 2011).

18.2.2 Life cycle and reproduction

AMF exhibit a life cycle that is tightly linked to their host plants (Fig. 18.1). Colonization begins with spore germination and hyphal growth toward plant roots in response to root exudates (strigolactones) (Akiyama et al., 2005). Once in contact with the root surface, AMF penetrate the cortex and form arbuscules, where bidirectional nutrient exchange occurs—plants provide carbohydrates, whereas fungi supply phosphorus, zinc, and other immobile nutrients (Smith & Smith, 2011).

The obligate biotrophic nature of AMF (they cannot complete their life cycle without a host plant) presents a major challenge for in vitro or axenic culture methods. This biological constraint mandates the use of host plants, such as maize or sorghum, or transformed ROCs (e.g., carrot roots) for propagation.

18.2.3 Ecological adaptation and environmental influences

AMF are widely distributed across terrestrial ecosystems and exhibit significant ecological plasticity. However, their distribution, community composition, and symbiotic performance are influenced by soil type, pH, temperature, moisture, cropping system, and anthropogenic disturbances (Öpik et al., 2006; van der Heijden et al., 2008). Ecologically, AMF contribute to nutrient cycling, soil aggregation, and drought resilience, making them essential in both natural ecosystems and agricultural systems. The diversity and abundance of AMF in soil influence the effectiveness of inocula, as native communities can outcompete introduced strains or synergize with them (Rodriguez & Sanders, 2015). Understanding these ecological interactions helps in selecting appropriate AMF strains for inoculum production based on target soil and climatic conditions. For instance, strains adapted to saline or acidic soils may perform better in such environments (Gai et al., 2009).

18.2.4 Host specificity and symbiotic efficiency

Although AMFs form associations with a broad range of host plants, the degree of symbiotic efficiency can vary based on host-fungus compatibility (Helgason et al., 2002). Some AMF species show a preference toward certain plant taxa, affecting nutrient transfer efficiency and colonization rates. For effective inoculum production, it is critical to match AMF strains with compatible and responsive host plants that support rapid and abundant fungal propagation. Host plants, such as *Sorghum bicolor, Zea mays*, and *Allium cepa*, are commonly used in trap cultures because of their high root biomass and susceptibility to AMF colonization (Gerdemann & Nicolson, 1963).

18.2.5 Functional traits relevant to inoculum design

Key functional traits of AMF that influence their suitability for inoculum production and application include

- Sporulation rate: Species like *R. irregularis* sporulate prolifically and are amenable to bulk production.
- Root colonization ability: Some species rapidly colonize roots and form extensive hyphal networks, which is important for early plant establishment.
- Environmental tolerance: Heat, drought, or salinity tolerance determines the survival and performance of inocula in stress-prone environments.
- Nutrient transport efficiency: Fungal strains vary in their capacity to enhance P or N uptake in plants, which impacts field-level benefits (Smith & Smith, 2011).

Selecting AMF species with desired functional traits ensures better compatibility with the target crop and environment, increasing inoculum efficacy and consistency.

18.3 Inoculum production techniques

The successful application of AMF in sustainable agriculture hinges on the availability of effective and standardized inoculum. Inoculum production involves the propagation of viable fungal propagules—spores, hyphae, and colonized root fragments—that can be used to establish symbiosis with host plants. Several techniques, including soil-based, substrate-based, and in vitro ROCs, are employed depending on the target crop, application method, and desired scale of production (Declerck et al., 2005; Ijdo et al., 2011).

18.3.1 Overview of inoculum types

AMF inoculum can be broadly categorized based on the production method into three types:

18.3.1.1 Soil-based inoculum

Soil-based inoculum is one of the oldest and most straightforward methods, involving the use of live host plants grown in sterilized or semisterile soil inoculated with AMF spores. After a growth period of 8–12 weeks, the soil containing spores, hyphal fragments, and colonized root pieces is harvested and used as inoculum (Gerdemann & Nicolson, 1963).

This method is inexpensive and does not require specialized equipment, making it suitable for on-farm production. However, it is labor-intensive, susceptible to contamination, and provides inconsistent results because of environmental variability (Salomon, Watthana, Vosátka, & Albrechtová, 2022).

18.3.1.2 Substrate-based inoculum

Substrate-based inoculum involves growing AMF with host plants in inert substrates such as perlite, vermiculite, sand, or a mixture thereof. The use of inert substrates improves control over microbial contamination and facilitates easier harvesting of inoculum components (Crush, 1974). This method is widely adopted for commercial AMF production and allows better standardization than soil-based methods. The final product contains spores, colonized root segments, and extraradical mycelium embedded in the substrate, which can be applied directly to seeds or roots.

18.3.1.3 Root-organ culture (in vitro) inoculum

ROC systems, also known as in vitro or axenic cultures, use transformed roots (usually *Agrobacterium rhizogenes*–induced carrot roots) grown on a nutrient medium under sterile conditions. This method enables the growth of AMF without whole plants and facilitates the production of clean, high-quality inoculum (Bécard & Fortin, 1988). ROC systems are ideal for research and high-value inoculum production because they offer high levels of purity, scalability, and reproducibility. However, they are limited by high cost, technical complexity, and the fact that not all AMF species can be easily cultured in vitro (Declerck et al., 2005).

18.3.2 Host plant selection and growth conditions

The choice of host plant greatly influences AMF propagation efficiency. Ideal host plants produce abundant, fine roots that are highly susceptible to AMF colonization. Commonly used hosts include *S. bicolor, Z. mays, A. cepa*, and *Stylosanthes* spp. because of their rapid root growth and high colonization rates (Ijdo et al., 2011). Environmental conditions such as temperature (20°C–28°C), light (16-hour photoperiod), moisture, and nutrient availability (particularly low phosphorus levels) must be optimized to favor mycorrhizal colonization and sporulation (Smith & Smith, 2011). Excessive phosphorus suppresses colonization, so careful nutrient management is essential (Nouri et al., 2014).

18.3.3 Cultivation in controlled versus open systems

Inoculum production can be carried out in either open (field-based or greenhouse) or controlled (growth chamber or lab) systems:

- Open systems, including greenhouse or field beds, are cost-effective and suitable for large-scale production. However, they are vulnerable to contamination by pathogens and nontarget organisms, and environmental variables can affect quality (Crush, 1974).
- Controlled systems, such as ROC or sterile substrate cultures, provide uniform conditions and allow for the production of contamination-free inocula. These systems are ideal for producing research-grade or commercial bioinoculants, though they involve higher capital and operational costs (Declerck et al., 2005).

18.3.4 Scale-up considerations for mass production

Scaling up AMF inoculum production from lab to commercial scale requires careful planning to maintain consistency, viability, and cost-effectiveness (Table 18.1).

Key considerations include

- Selection of high-performing AMF strains with proven efficiency across diverse environments (Rodriguez & Sanders, 2015).
- Optimization of growth systems that balance cost and contamination control—for example, using modular substrate-based systems with automated irrigation and environmental control.
- Harvesting and processing techniques such as air-drying, sieving, and QC measures (e.g., spore viability and colonization potential tests).
- Storage and packaging that preserve spore viability and ensure ease of application, such as encapsulation in beads or granules for seed coating (Salomon et al., 2012).
- Developing partnerships with agricultural extension systems, regulatory bodies, and private sector stakeholders is essential for creating sustainable inoculum production chains.

18.4 Formulation and storage of arbuscular mycorrhizal fungi inoculum

The successful use of AMFs in agriculture depends not only on efficient production but also on proper formulation, packaging, and storage of the inoculum. Formulation refers to the process of mixing the propagules (spores, colonized root fragments, and hyphae) with carriers or additives that improve their delivery, shelf life, and ease of application. Effective formulation ensures that the inoculum remains viable and infective until it reaches the plant rhizosphere (Bashan et al., 2014).

18.4.1 Objectives and challenges in formulation

The primary objectives of AMF inoculum formulation are

- To ensure uniform distribution of propagules during application.
- To protect the propagules from environmental stressors (e.g., desiccation, UV radiation, temperature extremes).
- To extend shelf life and retain propagule viability during storage and transport.
- To enhance compatibility with farming practices, including seed coating, drip irrigation, and transplant dipping.

However, formulating AMF inoculum presents unique challenges. AMF are obligate biotrophs that cannot be cultured without host plants, and their spores are sensitive to desiccation, temperature fluctuations, and microbial competition. Moreover, maintaining the viability of multiple propagule types (spores, hyphae, and root fragments) within a single formulation adds complexity (Cranenbrouck et al., 2005; Ijdo et al., 2011).

TABLE 18.1 Major arbuscular mycorrhizal fungi (AMF) inoculum production techniques.

Inoculum production technique	Host requirement	Major propagule types	Key advantages	Major limitations	Typical application	Reference(s)
Soil-based pot culture	Whole host plant (e.g., maize, sorghum)	Spores, hyphae, and colonized root fragment	Simple, low cost, suitable for on-farm production; mimics natural conditions	High contamination risk; low standardization; bulky product	Small-scale, on-farm or experimental use	Gerdemann and Nicolson (1963), Salomon, Watthana, Vosátka, & Albrechtová, 2022
Substrate-based culture (sand/ vermiculite/ perlite)	Whole host plant	Spores, hyphae, colonized roots	Better control of contaminants; easier harvesting; moderate scalability	Requires sterilized substrates; still host-dependent	Commercial inoculum production	Crush (1974), Ijdo et al. (2011)
ROC/in vitro culture	Transformed roots (e.g., carrot hairy roots)	Spores, extraradical hyphae	High purity; reproducible; pathogen-free inoculum	High cost; technical expertise required; limited AMF species	Research and high-quality commercial products	Bécard and Fortin (1988), Declerck et al. (2005)
Aeroponic culture systems	Whole host plant	Spores, hyphae	Enhanced sporulation; easy root access; reduced soil contamination	High setup cost; limited commercial adoption	Advanced research and pilot-scale production	Hung and Sylvia (1988), Dalpé and Monreal (2004)
Hydroponic/ nutrient film systems	Whole host plant	Hyphae, spores	Controlled nutrient supply; clean inoculum	Suppressed sporulation in some AMF; system complexity	Experimental studies	Sylvia and Hubbel (1986)
Field-based multiplication plots	A whole host plant in the field soil	Spores, hyphae, roots	Very low cost; suitable for local adaptation	Extremely variable quality; pathogen risk	Large-area soil inoculation	Brundrett et al. (1996), Schwartz et al. (2006)
Bioreactor-assisted ROC	Transformed roots	Spores, hyphae	Scalable, automated, high propagule density	Capital-intensive; still under development	Industrial-scale inoculum production	, Ijdo et al. (2011)
Trap culture technique	Whole host plant (AMF-responsive species)	Spores, colonized roots	Effective for isolating native AMF strains	Time-consuming; not fully standardized	Strain isolation and premultiplication	Brundrett, Bougher, Dell, Grove, & Malajczuk, 1996, INVAMINVAM (n.d.)

ROC, root-organ culture.

18.4.2 Types of formulations

Different types of AMF formulations have been developed to cater to various application modes, scalability, and environmental conditions.

18.4.2.1 Solid carriers (e.g., vermiculite, peat, soil mixes)

Solid carriers are the most common formulation type and include materials such as peat, vermiculite, perlite, zeolite, talc, or sterilized soil. These carriers support the survival of AMF propagules by maintaining moisture and shielding them from environmental stresses (Rillig & Mummey, 2006). Peat-based formulations are widely used because of their high organic content and moisture-holding capacity, which favors spore longevity. Vermiculite and perlite provide aeration and physical support and are often used in substrate-based inocula. Carrier materials must be nontoxic, inert, and free of antagonistic microorganisms (Xavier & Germida, 2003).

18.4.2.2 Liquid formulations

Liquid formulations involve suspending AMF spores or fragmented colonized roots in sterile water, saline, or nutrient buffers. These formulations are convenient for application through fertigation, seed coating, or dipping of seedlings. Liquid forms may include additives such as humic acids, surfactants, or protective agents to enhance stability (Adholeya et al., 2005). Although easy to handle, liquid formulations generally have shorter shelf lives because of the risk of microbial contamination and limited water activity, which can compromise spore viability if not properly managed (Gianinazzi et al., 2010).

18.4.2.3 Encapsulated formulations (e.g., beads, capsules)

Encapsulation involves embedding AMF propagules in biodegradable polymeric matrices, such as alginate, to form beads or capsules. This technique provides physical protection, moisture retention, and controlled release of inoculum upon application (Bashan et al., 2002). Encapsulated formulations are highly desirable for precision agriculture, seed coatings, or nursery transplants because of their targeted delivery and improved shelf life. However, this technology can be costly and technically complex to scale up for commercial use (Cai et al., 2017).

18.4.3 Packaging and storage conditions

Proper packaging is essential to maintain the biological integrity of AMF inoculum. Key considerations include

- Moisture-proof containers (e.g., laminated pouches, vacuum-sealed bags) to prevent desiccation.
- Opaque packaging to protect against UV-induced spore degradation.
- Sterile or aseptic packaging to minimize microbial contamination in liquid or encapsulated formulations.

Storage should ideally be in cool, dark, and dry environments. Temperatures between 4°C and 15°C are generally suitable for most formulations, especially solid carriers and encapsulated forms (Gianinazzi et al., 2010). Avoiding freeze–thaw cycles is critical to preserving spore viability.

18.4.4 Shelf life and viability considerations

Shelf life refers to the period during which the inoculum retains acceptable levels of viability and colonization potential. Several factors influence this:

- Formulation type: Solid carriers and encapsulated forms typically have longer shelf lives (up to 6–12 months) than liquid formulations (2–6 months).
- Spore maturity and concentration: More mature spores are often more resilient and retain infectivity longer.
- Storage conditions: Consistent temperature and humidity levels improve shelf life and reduce propagule degradation.

Periodic testing of viability—such as spore germination rates, root colonization assays, or quantitative Polymerase Chain Reaction (qPCR)-based detection—is crucial for QC and assurance before field application (Cranenbrouck et al., 2005; Ijdo et al., 2011).

18.5 Quality control of arbuscular mycorrhizal fungi inoculum

The consistent performance and reliability of AMF inoculants in agricultural applications depend heavily on rigorous QC measures. Given the biological complexity of AMFs and their dependence on living hosts, maintaining standardized quality across batches is both a scientific and commercial challenge. QC is essential for ensuring that AMF products meet the expectations of farmers, researchers, and regulatory authorities in terms of viability, purity, and effectiveness (Cranenbrouck et al., 2005; Ijdo et al., 2011).

18.5.1 Importance of quality control in arbuscular mycorrhizal fungi products

The lack of consistent regulatory frameworks and QC procedures has resulted in a wide variability in the effectiveness of commercial AMF products. Some products on the market may contain inadequate or nonviable propagules, contaminating microbes, or insufficient infectivity levels (Schwartz et al., 2006; Salomon, Watthana, Vosátka, & Albrechtová, 2022). QC ensures the reproducibility of AMF effects across different environments, the credibility of inoculant producers, compliance with national and international standards, and protection of ecosystems from the introduction of pathogens or invasive organisms.

18.5.2 Parameters for quality assessment

Several key parameters are routinely assessed to evaluate the quality of AMF inoculum:

18.5.2.1 Spore count and viability

Spore density is a fundamental measure of inoculum quality, usually expressed as the number of spores per gram or milliliter of product (Table 18.2). Spore viability is tested using staining methods (e.g., tetrazolium salts such as MTT (3-(4,5-Dimethylthiazol-2-yl)-2,5-iphenyltetrazolium bromide) or trypan blue) to distinguish live from dead spores (Phillips & Hayman, 1970). However, viable spores may not always be infective, so viability assessments must be complemented with infectivity tests (Schenck & Perez, 1990).

18.5.2.2 Infectivity and colonization potential

Infectivity is the ability of the inoculum to colonize host roots under controlled conditions. Standard bioassays using trap plants such as *Z. mays* or *A. cepa* in sterilized substrates are used to assess root colonization rates, typically measured after 4–8 weeks. Root samples are stained and examined microscopically to determine the percentage of root length colonized (Trouvelot et al., 1986). This is considered one of the most important indicators of product efficacy.

18.5.2.3 Microbial purity and contamination check

AMF inoculants may inadvertently harbor other microorganisms, including pathogens or competing fungi. Microbial purity is assessed through serial dilution plating on selective media and by using molecular tools (e.g., qPCR, Denaturing Gradient Gel Electrophoresis (DGGE)) to detect nontarget DNA (Xavier & Germida, 2003). Ensuring purity is essential to prevent disease introduction or microbial competition that can impair AMF functionality.

18.5.3 Standard protocols and guidelines

Several research groups and international organizations have proposed standard operating procedures for AMF inoculum quality testing. These include

- ISO/TR 24276:2006—Guidance on the detection of genetically modified organisms, which can be adapted for molecular screening.
- Mycorrhizae Quality Standards (proposed by the International Mycorrhiza Society and the European Mycorrhizal Products Quality Assurance project).
- Guidelines by International Culture Collection of (Vesicular) Arbuscular Mycorrhizal Fungi INVAM (n.d.)—including protocols for spore isolation, quantification, and trap culturing.

Although not universally adopted, these protocols provide a baseline for harmonizing AMF QC across different regions and industries (Ijdo et al., 2011).

18.5.4 Accreditation and certification systems

To ensure quality assurance at the industry level, some countries and organizations have initiated **certification systems** for AMF products:

- In Europe, some countries have adopted or are developing AMF certification schemes based on minimum spore count, colonization ability, and purity.
- In India the Department of Fertilizers and the Indian Council of Agricultural Research (ICAR) have guidelines for biofertilizer standards, including AMFs (Government of India, 2021).
- Third-party testing laboratories and academic institutions are sometimes engaged for neutral verification of product quality.
- A recognized certification system ensures farmer trust and facilitates product acceptance in export markets, thereby promoting the sustainable use of AMF technology in agriculture (Adholeya et al., 2005).

TABLE 18.2 Quality control parameters for AMF inoculum.

QC parameter (what to verify)	Recommended method(s)	How to report (units/outputs)	Practical interpretation (what "good" means)	Reference(s)
Product identity (AMF present and correctly labeled)	Morphological Identification (ID) of spores + (where feasible) DNA-based verification (species/consortium confirmation)	Species/consortium name; method used; voucher images or sequence IDs (if available)	Label should match actual AMF taxa; this improves reproducibility and trust	Basiru et al. (2021), Salomon et al. (2022)
Propagule density (dose strength)	Wet sieving/decanting for spores and microscopic counting; for root-based inoculum, quantify colonized root fragments	Spores g^{-1} (or spores mL^{-1}); % colonized root pieces; propagules per dose	Higher and consistently reported propagule density supports consistent field performance	Phillips and Hayman (1970), Basiru et al. (2021)
Spore viability (living vs. dead propagules)	Vital stains (e.g., MTT) and/or germination tests under controlled conditions	% viable spores; germination %; time-to-germination	Viability should be reported at **manufacture** and ideally after **storage** to support shelf life claims	An and Hendrix (1988), Koziol et al. (2024)
Infectivity/inoculum potential (ability to colonize roots)	Standardized greenhouse bioassay using a responsive host to quantify colonization after a set time	% root colonization; frequency/intensity indices; inoculum potential metric	Product must *actually* colonize roots under controlled conditions before field claims	Giovannetti and Mosse (1980), Liu and Luo (1994), Salomon et al. (2022)
Colonization quantification method standardization (comparability)	Clearing + staining of roots, gridline intersect, or standardized scoring system	Method named, number of root intersects; % colonization	Using standardized quantification prevents over/underreporting of colonization	Phillips and Hayman (1970), Giovannetti and Mosse (1980)
Contaminant/pathogen screening (biosafety)	Culture-based screening (where appropriate) + targeted pathogen checks: assess the presence of known harmful contaminants	Presence/ absence of pathogens; contaminant taxa detected; sterility statement (for in vitro products)	Absence of plant pathogens is critical; contamination is a common cause of inconsistent outcomes	Koziol et al. (2024), Salomon et al. (2022)
Formulation/carrier quality (delivery and survival)	Moisture content; carrier compatibility tests; particle size suitability (if granular/powder)	Moisture %; carrier type; storage stability notes	The carrier must support propagule survival and easy application without damaging propagules	Malusá et al. (2012), Basiru et al. (2021)
Shelf life/stability (performance after storage)	Repeat viability + infectivity assays at planned shelf-life points	Viability % and infectivity after X months; storage temperature	Claims should be supported by data at the *end of shelf life* (not only at production)	Salomon et al. (2022), Koziol et al. (2024)
Minimum functional dose guidance (label accuracy)	Dose–response bioassays (controlled) to support the recommended field dose	Propagules per plant/per kg seed/per ha; application method	Label dose should be justified by infectivity/effectiveness evidence	Salomon et al. (2022), Basiru et al. (2021)
Inoculum composition (propagule types included)	Microscopy-based confirmation of spores, hyphae, and colonized root fragments; (optional) quantify each component	Presence/ absence + relative abundance of propagule types	Mixed propagule composition can improve infectivity (especially with root fragments/vesicles)	Biermann and Linderman (1983), Ijdo et al. (2011)

AMF, arbuscular mycorrhizal fungi; *QC*, quality control.

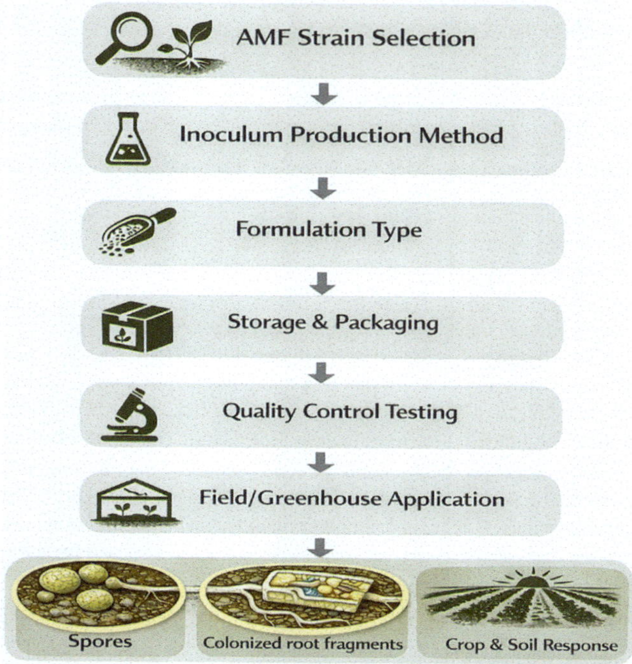

FIGURE 18.2 Flowchart of AMF inoculum development from production to field application. *AMF*, Arbuscular mycorrhizal fungi.

18.6 Applications and performance evaluation

The ultimate goal of AMF inoculum production is its application in agricultural and ecological systems to enhance plant productivity, nutrient use efficiency, and soil health (Fig. 18.2). Evaluating the performance of AMF under field and controlled conditions is essential for validating the efficacy of formulations and determining their integration with other agronomic practices.

18.6.1 Field trials and greenhouse testing

Field trials and greenhouse experiments serve as critical platforms for testing AMF inoculum efficacy across different crop species, soils, and environmental conditions. In greenhouse settings, variables, such as soil sterilization, nutrient regimes, and host specificity, can be tightly controlled to isolate the effect of AMF colonization on plant growth (Smith & Read, 2008). Greenhouse tests are typically the first step, involving trap cultures to verify infectivity, biomass measurement, nutrient uptake, root colonization assessment (Plenchette et al., 1983), and stress response evaluation, especially under drought or salinity.

Field trials extend these observations under realistic agronomic conditions. Multilocation trials are preferred to assess performance across agroecological zones. Performance metrics include

- Plant height, biomass, and yield
- Root colonization rates
- Phosphorus uptake and soil nutrient dynamics (Mäder et al., 2000)

Challenges in field evaluation include interactions with native AMF populations, variability in soil fertility, and environmental stressors, which can obscure treatment effects (Ryan & Graham, 2002).

18.6.2 Crop responses and soil health indicators

AMF applications have been shown to significantly influence crop productivity and key soil health indicators. These benefits, however, are context-dependent. Crop responses include

- Enhanced phosphorus, nitrogen, and micronutrient uptake (Smith & Smith, 2011)

- Improved water use efficiency
- Better resistance to abiotic and biotic stresses
- Soil health indicators impacted by AMF include
  - Increased microbial biomass and activity
  - Enhanced soil aggregation via glomalin production (Rillig, 2004)
  - Improvement in nutrient cycling and organic matter stabilization

AMF also fosters plant biodiversity in mixed cropping systems and supports sustainable agriculture by reducing chemical fertilizer dependence (Van der Heijden et al., 2008).

18.6.3 Integration with other biofertilizers

For holistic soil fertility management, AMF inoculants are often combined with other beneficial microorganisms such as Rhizobia for nitrogen fixation in legumes, phosphate-solubilizing bacteria, *Trichoderma*, and *Bacillus* spp. for pathogen control and growth promotion. Synergistic effects have been reported when AMFs are coinoculated with other biofertilizers, leading to enhanced plant growth and yield (Azcón-Aguilar & Barea, 1997; Berruti et al., 2016). However, compatibility among microbial inoculants must be assessed in advance, as antagonistic interactions may suppress colonization or activity (Koide & Mosse, 2004). Integrated inoculant strategies also require appropriate carrier formulations and application timing to ensure survival and performance.

18.7 Challenges and future perspectives

Despite the ecological and agronomic benefits of AMF the widespread adoption and commercialization of AMF-based products face several technical, economic, and regulatory hurdles. Addressing these challenges requires integration of scientific advances, supportive policies, and market development strategies.

18.7.1 Technical and economic challenges

One of the major technical limitations in AMF inoculum production is their obligate symbiotic nature, which necessitates a host plant for growth, making pure culture production more complex and costly compared to free-living microorganisms (Ijdo et al., 2011). Moreover, inconsistencies in inoculum quality, colonization efficiency, and field performance across diverse soils and crops reduce farmer confidence and market reliability (Schwartz et al., 2006). Economic challenges include

- High production and labor costs because of slow growth rates and the need for controlled conditions.
- Limited shelf life and viability of inocula under ambient storage conditions.
- Lack of economies of scale in many regions because of nascent commercial infrastructure.
- In addition, there is often limited awareness among farmers and extension workers regarding the benefits of AMF, resulting in a low demand for such biofertilizers (Rai & Varma, 2005).

18.7.2 Advances in biotechnology and genomic tools

Recent breakthroughs in molecular biology and biotechnology offer promising avenues to enhance AMF research and application. Metagenomics and amplicon sequencing now enable the identification and monitoring of AMF communities in soil and plant roots without traditional microscopy (Öpik et al., 2010). Transcriptomics and proteomics are being applied to understand gene expression and metabolic pathways involved in AMF–host symbiosis (Tisserant et al., 2012). Genomic editing tools like CRISPR/Cas9 may potentially be used in host plants to enhance their symbiotic compatibility and colonization efficiency, although ethical and regulatory concerns remain. Bioreactor-based ROCs, synthetic microbial communities, and nanotechnology-assisted formulations are additional areas gaining attention for scalable and effective AMF inoculum development (Gianinazzi et al., 2010).

18.7.3 Policy and regulatory issues

Currently the regulation of AMF-based biofertilizers varies greatly across countries and lacks harmonization. There is often no standard protocol for inoculum quality assessment, formulation registration, or efficacy verification (Malusá et al., 2012).

As a result, market flooding with substandard products is a recurring issue that undermines trust and investment. Key policy gaps include

- Absence of certification systems or third-party verification.
- Limited inclusion of AMF in national soil fertility or organic farming schemes.
- Insufficient government funding for AMF-related R&D and awareness programs.
- Policy reforms should promote
  - Incentives for AMF use in sustainable farming.
  - Capacity building in AMF production and application.
  - Collaborative networks among research institutions, industries, and regulatory bodies.

18.7.4 Potential for commercialization and global markets

The global biofertilizer market is expanding, and AMF inoculants represent a growing segment because of the demand for sustainable agricultural inputs. The commercialization potential of AMF hinges on

- Developing cost-effective mass production methods.
- Standardizing formulations with long shelf lives.
- Educating end users and integrating AMF use with conventional farming systems.

Emerging markets in Asia, Africa, and Latin America—characterized by nutrient-poor soils and smallholder farming—can particularly benefit from AMF-based products (Singh et al., 2019). Public–private partnerships and international collaborations are essential to scale up production and application, supported by robust supply chains and policy support. With proper regulation, innovation, and outreach, AMF inoculants have the potential to contribute significantly to climate-resilient agriculture, organic farming, and soil restoration globally.

18.8 Conclusion and recommendations

18.8.1 Conclusion

AMF are indispensable for sustainable agriculture because of their role in enhancing nutrient uptake, improving soil structure, and increasing plant resilience to stress. Over the past few decades, substantial progress has been made in understanding AMF biology, ecology, and their application in crop production systems. However, challenges remain in the large-scale, cost-effective, and standardized production of high-quality AMF inoculum. This chapter has provided a comprehensive overview of the key elements involved in AMF inoculum development:

- **Biological and ecological insights** into AMF that guide inoculum production strategies.
- **Techniques and systems** used for AMF cultivation, including soil-based, substrate-based, and in vitro methods.
- **Formulation types**, storage requirements, and viability constraints.
- **QC protocols** to ensure spore viability, colonization potential, and microbial purity.
- **Applications and performance evaluation**, highlighting the beneficial outcomes in crop productivity and soil health.
- **Current limitations**, including technical, regulatory, and commercialization challenges.

Overall, AMF inoculants hold great potential for integration into ecofriendly agricultural practices and can play a pivotal role in reducing chemical fertilizer dependency.

18.8.2 Recommendations

Based on the current knowledge and practical experiences discussed throughout the chapter, the following recommendations are proposed for researchers, industry stakeholders, policymakers, and practitioners:

- **Enhance research on native AMF strains:** Focus on the identification, isolation, and mass multiplication of indigenous AMF strains adapted to local agroecological conditions to improve inoculum efficacy and field performance.
- **Develop cost-effective production protocols:** Optimize low-input cultivation systems (e.g., trap culture using regionally available substrates) to reduce production costs without compromising quality.
- **Promote quality assurance standards:** National and international regulatory bodies should formulate and enforce **standardized guidelines** for AMF inoculum quality assessment, labeling, and certification.

- **Support infrastructure and training:** Establish decentralized AMF production units, particularly in rural and periurban areas, and train technicians and farmers in inoculum application and handling.
- **Foster public–private partnerships:** Encourage collaboration between research institutions, private biotech firms, and extension services to scale up inoculum production and market development.
- **Integrate AMF with other sustainable inputs:** Develop synergistic formulations combining AMF with nitrogen-fixing bacteria, phosphate-solubilizing microbes, or organic amendments to enhance overall biofertilizer performance.
- **Policy inclusion and financial incentives:** Governments should include AMF-based biofertilizers in soil health and organic agriculture schemes and provide subsidies or financial incentives for their adoption.
- **Advance biotechnological tools:** Support interdisciplinary research using genomics, metabolomics, and precision agriculture tools to improve AMF functional traits, monitor field dynamics, and tailor site-specific recommendations.

The successful incorporation of AMF into mainstream agriculture depends on bridging the gap between scientific knowledge and practical implementation. With appropriate investment, policy support, and awareness-building, AMF inoculants can become a cornerstone of resilient and regenerative farming systems worldwide.

References

Adholeya, A., Goswami, S., & Singh, R. (2005). Practical manual on mycorrhizal technology. *Centre for mycorrhizal research. TERI Press.*

Akiyama, K., Matsuzaki, K., & Hayashi, H. (2005). Plant sesquiterpenes induce hyphal branching in arbuscular mycorrhizal fungi. *Nature, 435*(7043), 824–827. https://doi.org/10.1038/nature03608.

An, Z.-Q., & Hendrix, J. W. (1988). Determining the viability of endogonaceous spores with a vital stain. *Mycologia, 80*(2), 259–261. https://doi.org/10.1080/00275514.1988.12025532.

Azcón-Aguilar, C., & Barea, J. M. (1997). Arbuscular mycorrhizas and biological control of soil-borne plant pathogens—An overview of the mechanisms involved. *Mycorrhiza, 6*(6), 457–464. https://doi.org/10.1007/s005720050147.

Barea, J. M., Pozo, M. J., Azcón, R., & Azcón-Aguilar, C. (2005). Microbial co-operation in the rhizosphere. *Journal of Experimental Botany, 56*(417), 1761–1778. https://doi.org/10.1093/jxb/eri197.

Bashan, Y., de-Bashan, L. E., Prabhu, S. R., & Hernandez, J. P. (2014). Advances in plant growth-promoting bacterial inoculant technology: Formulations and practical perspectives (1998–2013). *Plant and Soil, 378*(1–2), 1–33. https://doi.org/10.1007/s11104-013-1956-x.

Bashan, Y., Hernandez, J. P., Leyva, L. A., & Bacilio, M. (2002). Alginate microbeads as inoculant carriers for plant growth-promoting bacteria. *Biology and Fertility of Soils, 35*(5), 359–368. https://doi.org/10.1007/s00374-002-0481-5.

Basiru, S., Mwanza, H. P., & Hijri, M. (2021). Analysis of arbuscular mycorrhizal fungal inoculant benchmarks. *Microorganisms, 9*(1), 81. https://doi.org/10.3390/microorganisms9010081.

Bécard, G., & Fortin, J. A. (1988). Early events of vesicular–arbuscular mycorrhiza formation on Ri T-DNA transformed roots. *New Phytologist, 108*(2), 211–218. https://doi.org/10.1111/j.1469-8137.1988.tb03698.x.

Berruti, A., Lumini, E., Balestrini, R., & Bianciotto, V. (2016). Arbuscular mycorrhizal fungi as natural biofertilizers: Let's benefit from past successes. *Frontiers in Microbiology, 6*, 1559. https://doi.org/10.3389/fmicb.2015.01559.

Biermann, B., & Linderman, R. G. (1983). Use of vesicular–arbuscular mycorrhizal roots, intraradical vesicles, and extraradical vesicles as inoculum. *New Phytologist, 95*, 97–105. https://doi.org/10.1111/j.1469-8137.1983.tb03472.x.

Brundrett, M., & Abbott, L. K. (2002). Arbuscular mycorrhizas in plant communities. In K. Sivasithamparam, K. W. Dixon, & R. L. Barrett (Eds.), *Microorganisms in Plant Conservation and Diversity* (pp. 151–194). Kluwer Academic Publishers.

Brundrett, M., Bougher, N., Dell, B., Grove, T., & Malajczuk, N. (1996). Working with Mycorrhizas in Forestry and Agriculture. *ACIAR Monograph No. 32*. Canberra, Australia: Australian Centre for International Agricultural Research.

Brundrett, M., Melville, L., & Peterson, L. (1996). *Practical methods in mycorrhiza research*. Mycologue Publications.

Cai, F., Yu, G., Wang, P., Wei, Z., Fu, L., Shen, Q., & Huang, Q. (2017). Alginate bead encapsulation improves survival of *Bacillus amyloliquefaciens* and its biocontrol efficacy against tomato bacterial wilt. *Biological Control, 110*, 98–105. https://doi.org/10.1016/j.biocontrol.2017.04.005.

Chandrasekaran, M., Krishnamoorthy, R., Paulraj, M. G., & Ignacimuthu, S. (2014). Development and formulation of arbuscular mycorrhizal fungi for sustainable agriculture. *Journal of Biofertilizers & Biopesticides, 5*(2), 1–7. https://doi.org/10.4172/2155-6202.1000154.

Cranenbrouck, S., Voets, L., Bivort, C., Renard, L., Strullu, D. G., & Declerck, S. (2005). Methodologies for in vitro cultivation of arbuscular mycorrhizal fungi with root organs. *In Vitro Cellular & Developmental Biology—Plant, 41*(3), 259–267. https://doi.org/10.1079/IVP2005648.

Crush, J. R. (1974). Plant growth responses to vesicular-arbuscular mycorrhiza IV. The relative efficiency of inocula. *New Phytologist, 73*(3), 743–749. https://doi.org/10.1111/j.1469-8137.1974.tb02153.x.

Declerck, S., Dufey, M., & Strullu, D. G. (2005). In vitro culture of arbuscular mycorrhizal fungi: An overview. *Mycorrhiza, 15*(3), 163–175. https://doi.org/10.1007/s00572-004-0324-7.

Dalpé, Y., & Monreal, M. (2004). Arbuscular Mycorrhiza Inoculum to Support Sustainable Cropping Systems. *Crop Management, 3*, 1–11. https://doi.org/10.1094/CM-2004-0301-09-RV.

Fortin, J. A., Bécard, G., Declerck, S., Dalpé, Y., St-Arnaud, M., Coughlan, A. P., & Piché, Y. (2002). Arbuscular mycorrhiza on root-organ cultures. *Canadian Journal of Botany, 80*(1), 1–20. https://doi.org/10.1139/b01-138.

Gai, J., Christie, P., Feng, G., Li, X., & Zhang, J. (2009). A preliminary survey of arbuscular mycorrhizal fungal spore density and diversity in saline soils of the Yellow River Delta. *Mycorrhiza, 19*(8), 613–622. https://doi.org/10.1007/s00572-009-0259-y.

Gerdemann, J. W., & Nicolson, T. H. (1963). Spores of mycorrhizal *Endogone* species were extracted from soil by wet sieving and decanting. *Transactions of the British Mycological Society, 46*(2), 235–244. https://doi.org/10.1016/S0007-1536(63)80079-0.

Gianinazzi, S., Gollotte, A., Binet, M. N., van Tuinen, D., Redecker, D., & Wipf, D. (2010). Agroecology: The key role of arbuscular mycorrhizas in ecosystem services. *Mycorrhiza, 20*, 519–530. https://doi.org/10.1007/s00572-010-0333-3.

Giovannetti, M., & Mosse, B. (1980). An evaluation of techniques for measuring vesicular–arbuscular mycorrhizal infection in roots. *New Phytologist, 84*, 489–500. https://doi.org/10.1111/j.1469-8137.1980.tb04556.x.

Government of India. (2021). Fertiliser (Inorganic, Organic or Mixed) (Control) Fifth Amendment Order, 2021 (S.O. 2671(E), July 1, 2021). Ministry of Agriculture and Farmers Welfare, Department of Agriculture, Cooperation and Farmers Welfare.

Helgason, T., Daniell, T. J., Husband, R., Fitter, A. H., & Young, J. P. W. (2002). Diversity and specificity of arbuscular mycorrhizal fungi associated with two species of *Plantago*. *New Phytologist, 154*(3), 741–749. https://doi.org/10.1046/j.1469-8137.2002.00497.x.

Hung, L. L., & Sylvia, D. M. (1988). Production of vesicular–arbuscular mycorrhizal fungus inoculum in aeroponic culture. *Applied and Environmental Microbiology, 54*, 353–357. https://doi.org/10.1128/AEM.54.2.353-357.1988.

International Culture Collection of (Vesicular) Arbuscular Mycorrhizal Fungi (INVAM). (n.d.). Protocols for arbuscular mycorrhizal fungi (AMF) isolation and trap culture. West Virginia University. Retrieved August 4, 2026, from https://invam.wvu.edu

Ijdo, M., Cranenbrouck, S., & Declerck, S. (2011). Methods for large-scale production of AM fungi: Past, present, and future. *Mycorrhiza, 21*, 1–16. https://doi.org/10.1007/s00572-010-0337-x.

INSA. (2009). *Biofertilizer quality control order, 2009*. Indian National Science Academy.

Koide, R. T., & Mosse, B. (2004). A history of research on arbuscular mycorrhiza. *Mycorrhiza, 14*(3), 145–163. https://doi.org/10.1007/s00572-004-0307-4.

Koziol, L., Lubin, T., & Bever, J. D. (2024). An assessment of twenty-three mycorrhizal inoculants reveals limited viability of AM fungi, pathogen contamination, and negative microbial effects on crop growth for commercial products. *Applied Soil Ecology, 202*, 105559. https://doi.org/10.1016/j.apsoil.2024.105559.

Lehmann, A., Veresoglou, S. D., Leifheit, E. F., & Rillig, M. C. (2014). Arbuscular mycorrhizal influence on zinc nutrition in crop plants—A meta-analysis. *Soil Biology and Biochemistry, 69*, 123–131. https://doi.org/10.1016/j.soilbio.2013.11.001.

Liu, R.-J., & Luo, X.-S. (1994). A new method to quantify the inoculum potential of arbuscular mycorrhizal fungi. *New Phytologist, 128*(1), 89–92. https://doi.org/10.1111/j.1469-8137.1994.tb03990.x.

Mäder, P., Edenhofer, S., Boller, T., Wiemken, A., & Niggli, U. (2000). Arbuscular mycorrhizae in a long-term field trial comparing low-input (organic, biological) and high-input (conventional) farming systems in a crop rotation. *Biology and Fertility of Soils, 31*(2), 150–156. https://doi.org/10.1007/s003740050638.

Malusá, E., Sas-Paszt, L., & Ciesielska, J. (2012). Technologies for beneficial microorganisms inocula used as biofertilizers. *The Scientific World Journal, 2012*, 491206. https://doi.org/10.1100/2012/491206.

MAPA. (2011). Pecuária e Abastecimento—Instrução Normativa No. *30. Ministério da Agricultura*.

Nouri, E., Breuillin-Sessoms, F., Feller, U., & Reinhardt, D. (2014). Phosphorus and nitrogen regulate arbuscular mycorrhizal symbiosis in *Petunia hybrida*. *PLoS ONE, 9*(3), e90841. https://doi.org/10.1371/journal.pone.0090841.

Öpik, M., Moora, M., Liira, J., & Zobel, M. (2006). Composition of root-colonizing arbuscular mycorrhizal fungal communities in different ecosystems around the globe. *Journal of Ecology, 94*(4), 778–790. https://doi.org/10.1111/j.1365-2745.2006.01136.x.

Öpik, M., Vanatoa, A., Vanatoa, E., Moora, M., Davison, J., Kalwij, J. M., & Zobel, M. (2010). The online database MaarjAM reveals global and ecosystemic distribution patterns in arbuscular mycorrhizal fungi (*Glomeromycota*). *New Phytologist, 188*(1), 223–241. https://doi.org/10.1111/j.1469-8137.2010.03334.x.

Phillips, J. M., & Hayman, D. S. (1970). Improved procedures for clearing roots and staining parasitic and vesicular-arbuscular mycorrhizal fungi for rapid assessment of infection. *Transactions of the British Mycological Society, 55*(1), 158–161. https://doi.org/10.1016/S0007-1536(70)80110-3. IN16–IN18.

Plenchette, C., Fortin, J. A., & Furlan, V. (1983). Growth responses of several plant species to mycorrhizae in a soil of moderate P-fertility. *Mycologia, 75*(5), 879–885. https://doi.org/10.1080/00275514.1983.12023868.

Rai, M. K., & Varma, A. (Eds.). (2005). *Mycorrhization in plants: Molecular biology, ecological studies and practical applications*. Springer.

Rillig, M. C. (2004). Arbuscular mycorrhizae, glomalin, and soil aggregation. *Canadian Journal of Soil Science, 84*(4), 355–363. https://doi.org/10.4141/S04-003.

Rillig, M. C., & Mummey, D. L. (2006). Mycorrhizas and soil structure. *New Phytologist, 171*(1), 41–53. https://doi.org/10.1111/j.1469-8137.2006.01750.x.

Rillig, M. C., Aguilar-Trigueros, C. A., Lehmann, A., Kohler, J., & Wulf, A. (2019). Mycorrhizal fungal diversity regulates soil carbon cycling. *The ISME Journal, 13*, 1188–1199. https://doi.org/10.1038/s41396-018-0340-4.

Rodriguez, A., & Sanders, I. R. (2015). The role of community and population ecology in applying mycorrhizal fungi for improved food security. *ISME Journal, 9*, 1053–1061. https://doi.org/10.1038/ismej.2014.207.

Ryan, M. H., & Graham, J. H. (2002). Is there a role for arbuscular mycorrhizal fungi in production agriculture? *Plant and Soil, 244*(1–2), 263–271. https://doi.org/10.1023/A:1020207631893.

Salomon, M. J., Watthana, S., Vosátka, M., Albrechtová, J., et al. (2022). Establishing a quality management framework for commercial inoculants containing arbuscular mycorrhizal fungi. *iScience, 25*(7), 104636. doi:10.1016/j.isci.2022.104636.

Salomon, M. J., Watts-Williams, S. J., McLaughlin, M. J., Bücking, H., Singh, B. K., Hutter, I., Schneider, C., Martin, M. F., Vosátka, M., Guo, L., Ezawa, T., Saito, M., Declerck, S., Zhu, Y.-G., Bowles, T., Abbott, L. K., Smith, F. A., Cavagnaro, T. R., & van der Heijden, M. G. A. (2022). Establishing a quality management framework for commercial inoculants containing arbuscular mycorrhizal fungi. *iScience, 25*(7), 104636. https://doi.org/10.1016/j.isci.2022.104636.

Schenck, N. C., & Perez, Y. (1990). *Manual for the identification of VA mycorrhizal fungi* (2nd ed.). Synergistic Publications.

Schüßler, A., Schwarzott, D., & Walker, C. (2001). A new fungal phylum, the Glomeromycota: Phylogeny and evolution. *Mycological Research, 105*(12), 1413–1421. https://doi.org/10.1017/S0953756201005196.

Schwartz, M. W., Hoeksema, J. D., Gehring, C. A., Johnson, N. C., Klironomos, J. N., Abbott, L. K., & Pringle, A. (2006). The promise and the potential consequences of the global transport of mycorrhizal fungal inoculum. *Ecology Letters, 9*(5), 501–515. https://doi.org/10.1111/j.1461-0248.2006.00910.x.

Singh, R., Sharma, R., & Manhas, R. K. (2019). Biofertilizers: Current status and future perspectives. In A. Rakshit, H. B. Singh, & A. Sen (Eds.), *Biofertilizers: Technology, marketing and usage* (pp. 1–28). Springer.

Smith, F. A., & Smith, S. E. (2011). Roles of arbuscular mycorrhizas in plant nutrition and growth: New paradigms from cellular to ecosystem scales. *Annual Review of Plant Biology, 62*, 227–250. https://doi.org/10.1146/annurev-arplant-042110-103846.

Smith, S. E., & Read, D. J. (2008). *Mycorrhizal symbiosis* (3rd ed.). Academic Press.

Sylvia, D. M., & Hubbell, D. H. (1986). Growth and sporulation of vesicular–arbuscular mycorrhizal fungi in hydroponic culture. *Mycologia, 78*, 429–435. https://doi.org/10.1080/00275514.1986.12025364.

Tisserant, E., Kohler, A., Dozolme-Seddas, P., Balestrini, R., Benabdellah, K., Colard, A., & Martin, F. (2012). The transcriptome of the arbuscular mycorrhizal fungus *Glomus intraradices* (DAOM 197198) reveals functional tradeoffs in an obligate symbiont. *New Phytologist, 193*(3), 755–769. https://doi.org/10.1111/j.1469-8137.2011.03948.x.

Trouvelot, A., Kough, J. L., & Gianinazzi-Pearson, V. (1986). Mesure du taux de mycorhization VA d'un système radiculaire. Recherche de méthodes d'estimation ayant une signification fonctionnelle. *Physiological and genetical aspects of mycorrhizae* (pp. 217–221). INRA Press.

van der Heijden, M. G. A., Bardgett, R. D., & van Straalen, N. M. (2008). The unseen majority: Soil microbes as drivers of plant diversity and productivity in terrestrial ecosystems. *Ecology Letters, 11*(3), 296–310. https://doi.org/10.1111/j.1461-0248.2007.01139.x.

Verbruggen, E., & Kiers, E. T. (2010). Evolutionary ecology of mycorrhizal functional diversity in agricultural systems. *Evolutionary Applications, 3*(5–6), 547–560. https://doi.org/10.1111/j.1752-4571.2010.00145.x.

Vosátka, M., Albrechtová, J., Patten, C. L., & Glick, B. R. (2012). Manipulating the rhizosphere microbiome to enhance the growth and health of crops. In M. K. Rai (Ed.), *Microbial bioprospecting for sustainable development* (pp. 215–234). Springer.

Xavier, L. J. C., & Germida, J. J. (2003). Selective interactions between arbuscular mycorrhizal fungi and rhizosphere bacteria and their potential application for improving plant growth. *Canadian Journal of Microbiology, 49*(11), 827–837. https://doi.org/10.1139/w03-108.

Chapter 19

Arbuscular mycorrhizal fungi as biofertilizers for sustainable agriculture

Anoo Gunasekera [a], **Dinoo Gunasekera** [b], **W.J.A.P.M. Jayasooriya** [c] **and Disna Ratnasekera** [d]

[a] *Department of Biomedical Sciences, Faculty of Health Sciences, CINEC Campus, Malabe, Sri Lanka,* [b] *Department of Information and Communication Technology, Faculty of Technology, University of Ruhuna, Matara, Sri Lanka,* [c] *Faculty of Graduate Studies, University of Ruhuna, Matara, Sri Lanka,* [d] *Department of Agricultural Biology, Faculty of Agriculture, University of Ruhuna, Matara, Sri Lanka*

19.1 Introduction

Modern agriculture faces unprecedented challenges in the face of global food demands amidst economic instability, climate change, and ecosystem degradation (Food and Agriculture Organization of the United Nations (FAO), 2020). Although chemical fertilizers temporarily address soil nutrient deficiencies, their application requires precise calibration to both soil deficits and crop-specific needs (Li et al., 2020). The present agricultural scenario presents some serious contradictions: (1) price escalation of global fertilizers with increasing usage of agrochemicals; (2) increasing cost of production and energy; and (3) collateral damage to soil health, water regimes, and human well-being, to name a few, by chemical-rich agro-practices (Ahmad et al., 2023; Ditta et al., 2022; Ullah et al., 2020). Exorbitant use of fertilizers, beyond the crop requirements in many cases, is a direct consequence of the high nutrient demand of the modern cultivars grown (Singh & Singh, 2017). Low nutrient-use efficiency is the immediate consequence of enormous fertilization, much of which ends up in the degradation of soil, imbalance in microbial populations, and in the contamination of water bodies via leaching/runoff, jeopardizing both ecosystem and human health (Meena et al., 2020). Such an unsustainable development trajectory warrants urgent consideration of alternative fertilization measures balancing production and environmental needs.

Biofertilizers, which are complexes of beneficial microorganisms for increasing nutrient availability to plants, are an ecofriendly alternative to chemical fertilizers in sustainable agriculture. Research studies evidence the potential for increasing harvest fertilization by 25%–30% when combined with chemical fertilizers (Pal et al., 2015). Arbuscular mycorrhizal fungi (AMFs) appear to be a promising bioinoculant for sustainable agriculture in the context of long-term soil fertility, plant nutrition, and pathogen protection (Thirkell et al., 2017). Symbiotic AMFs exist with almost 90% of the land plants (Davison et al., 2015), forming special arbuscules for nutrient exchange and vesicles for storage within root cortical cells (Pepe et al., 2016). AMFs act as a biocontrol agent against soil-borne pathogens and provide an amelioration to stress and toxicity because of heavy metals while ensuring the sustainability of soil health.

Dominating agricultural soils, AMFs provide essential ecosystem services in organic farming. Their symbioses offer several agronomical benefits, namely, (1) enhanced nutrient acquisition by extraradical hyphae extending past root depletion zones to absorb immobile nutrients (especially P and N) via specific transporters and transport these as polyphosphate granules (Sato et al., 2019); (2) improved water relations as the very extensive mycelial network can retain moisture and access soil micropores unavailable to roots (Augé et al., 2015); and (3) the ability of the hyphal surface area to provide an absorptive capacity 100 times more than that of root systems so that the nutrients can be translocated at a very rapid rate (Püschel et al., 2020). These mechanisms justify AMF's role in sustainable agriculture when hyphae penetrate soil crevices to confront nutrient pools while forming a mat of moisture retention. AMF symbiosis compensates for root zones of depletion that enhance better soil moisture retention, ultimately enhancing plant growth, hence reducing fertilizer dependence as one of the bases of ecological intensification.

Advancing Sustainable Agriculture Through Arbuscular Mycorrhizal Fungi. DOI: https://doi.org/10.1016/B978-0-443-45130-0.00008-4

19.2 Arbuscular mycorrhizal fungi-mediated nutrient absorption

AMFs greatly help plants in acquiring nutrients, primarily phosphorus (P) and nitrogen (N) (Fu et al., 2023). Their extraradical hyphae, being one to two times smaller than roots in diameter, can penetrate deeper layers of soil, hence furthering the absorption zones for nutrients as well as enhancing root growth and nutrient uptake. Through their extensive hyphal network, AMFs create an enormous root surface area for the absorption of nutrients, whereby AMF assist in the mineral nutrition of plants, especially in nutrient-challenged environments.

Mineral nutrient transporters (MNTs) control plant–microbe interactions by regulating nutrient exchange and the dynamics of microbial colonization. These transporters transfer essential nutrients in both directions between hosts and symbionts, yet they are still selective about which microbes are allowed to enter the plant tissues (Sun et al., 2021). By preventing overcolonization, MNTs maintain symbiotic balance, promoting beneficial associations while inhibiting pathogenic invasion. Plants also use MNTs for pathogen defense in two ways: depriving pathogens of nutrients and triggering immune responses in the plant (Banasiak et al., 2021). Hence their implication in nutrient partitioning and microbial selectivity places MNTs at the forefront of symbiotic optimization. MNTs affect plant–microbe interactions directly and indirectly. Direct regulation happens through managing nutrient exchanges and controlling microbial populations in the rhizosphere. Indirect effects emerge from pathways involving pathogen competitiveness and activation of host defense. Said in another way, in multiplicities, MNTs define the critical nexus for productively harnessing symbiotic plant–microbe interactions while also affording resistance to pathogens.

19.2.1 Mechanisms of nutrient absorption: Phosphorus (P) and nitrogen (N)

Mainly, AMFs increase P and N uptake by plants through the luxuriant hyphal network that facilitates absorption and transport of these important nutrients. AMFs secrete phosphatases that mineralize organic phosphorus into plant-available forms (Jindo et al., 2023), overcoming soil phosphorus limitations and promoting plant growth (Cheng et al., 2023).

Nutrient uptake in AMF symbiosis occurs through specialized membrane transporters that enhance phosphorus and ammonium solubility. AMFs overcome soil phosphate limitations by absorbing and transferring it to host plants via mycorrhiza-induced phosphate transporters, which show upregulated expression in both fungal hyphae and roots during symbiosis (Menge, 2023). Arbuscular mycorrhizal (AM) symbiosis activates specific plant Pi transporters, including MtPT4 and LjPT4, which function in noncolonized root tips as potential Pi sensors (Volpe et al., 2015). Tomato demonstrates this through upregulation of LePT3-5 during colonization, with LePT4 playing a crucial symbiotic role (Xu et al., 2007). The P transporter, *RiPT7* in *Rhizophagus irregularis*, mediates phosphate (Pi) transport across membranes depending on Pi gradients (Xie et al., 2022). In addition, PHR transcription factors, particularly *PHR2*, regulate mycorrhizal symbiosis genes via P1BS motifs, further optimizing P uptake (Shi et al., 2021). These findings reveal complex Pi transport regulation, where AM-induced transporters operate beyond strict symbiosis contexts while maintaining essential fungal-mediated nutrient transfer functions. Increase phosphorus acquisition by plants via secretion of acid phosphatase to mineralize organic phosphates and include proton extrusion and organic acid exudation to solubilize soil minerals (Francis et al., 2023). Further symbiotic association also induces anthocyanin accumulation as a stress–responsive pigment and reactive oxygen species (ROS) production, indicating that defense pathways are activated against biotic/abiotic stressors (Courty et al., 2015).

AMF similarly facilitate nitrogen acquisition by promoting ammonium and nitrate uptake through extraradical hyphae (Hornstein and Sederoff, 2024) while enhancing nitrogen assimilation enzymes in host plants. AMF ingresses into carbon partitioning: this symbiosis enhances photosynthesis because of nutrient status enhancement (Bell et al., 2021), which comes with abundant carbohydrates being translocated to roots or fungal structures like arbuscules. The tripartite nutrient exchange (P, N, and C) highlights the role of AMF in improving plant resource use efficiency under nutrient constraints. A better understanding of such mechanisms will enable us to exploit AMF symbiosis for sustainable agriculture.

AMF regulate ammonium transporters (AMTs) and thus increase nitrogen uptake by their hosts. In so doing, they optimize nutrient exchange to support plant growth under nutrient stress. The mycorrhiza-inducible AMTs localize to the peri-arbuscular membrane, which is the critical interface for symbiotic nutrient transfer in the arbuscular mycorrhizae. AMTs, akin to the phosphate transporter MtPT4 in *Medicago truncatula*, take up ammonium into cortical cells in soybean and *Medicago* (Breuillin-Sessoms et al., 2015). Studies of *Medicago* mutants reveal that the transporters of Pi (PT4) and ammonium (AMT2;3) control the lifespan of the arbuscule (Breuillin-Sessoms et al., 2015; Javot et al., 2007). Hence these transporters might also be signaling pathways that stimulate maintenance of arbuscules, which suggests a dual function for these transporters in nutrient delivery and symbiotic maintenance.

19.2.2 Mechanisms of nutrient absorption: Other macro/micro nutrients

AMFs transfer sulfur (S) to the host plants and enhance S nutrition and regulation of sulfate transporter expression (Giovannetti et al., 2013). One specific arbuscular sulfate transporter has been identified in this symbiotic pathway (Giovannetti et al., 2013). Potassium (K^+) accumulates in fungal structures (spores, hyphae, and vesicles) (Olsson et al., 2008), whereas AM-colonized *Lotus japonicus* roots upregulate K^+ transporters (Guether et al., 2009). This AM-mediated K^+ uptake correlates with increased plant tolerance to abiotic stress factors, such as salinity and drought, thereby showing another facet of AM involvement in nutrient acquisition and stress alleviation.

Moreover, AMFs offer a sustainable biofortification strategy to enhance crop micronutrient content, complementing genetic and agronomic approaches (Pellegrino & Bedini, 2014). Beyond yield improvement, AM symbiosis elevates macro- and micronutrient concentrations in edible tissues, significantly impacting food quality. For zinc (Zn), AMF consistently increases plant tissue concentrations across diverse environments, though soil properties (texture and pH) and nutrient availability (particularly Zn/P deficiency) modulate this effect (Lehmann et al., 2014). Similar benefits extend to copper (Cu) and iron (Fe), with Fe enrichment most pronounced in intermediate-duration experiments, whereas manganese (Mn) improvements appear herb-specific. Field studies demonstrate AMF's practical value in chickpea inoculation, boosting both yield and grain nutritional quality, with elevated protein, Fe, and Zn levels (Pellegrino & Bedini, 2014). A zinc transporter (GintZnT1) was first characterized in AMF (*Glomus intraradices*). Subsequently, genomic analysis of *R. irregularis* (DAOM-197198) identified additional candidate transporters for copper (Cu), iron (Fe), and zinc (Zn) (Tamayo et al., 2014). The findings position AMF as a multifunctional tool that simultaneously addresses agricultural productivity and nutritional security through enhanced micronutrient acquisition (Zn, Cu, and Fe), improved protein content, and reduced dependence on synthetic inputs, particularly in marginal soils where conventional biofortification proves challenging.

These coordinated responses demonstrate plants' sophisticated adaptation strategies through microbial symbioses, optimizing nutrient uptake while reinforcing stress resilience. The observed physiological changes collectively improve mineral bioavailability and stress tolerance, underscoring the evolutionary advantage of plant–microbe interactions in nutrient-limited environments. Understanding the mechanisms of MNTs is of much potential for agriculture, mainly for microbial-mediated crop production systems. Their capacity to accelerate nutrient use efficiency vis-à-vis the control of microbial populations provides newer possibilities for sustainable cultivation conditions. In addition to their pronounced role in mycorrhizal associations for enhancing uptake of immobile nutrients (P, Zn, and Cu) through a wide array of hyphal networks, they also operate in water and nutrient uptake under water-limited conditions (Wang et al., 2024).

Plants actively control AM colonization with nutrient-dependent feedback mechanisms, where fungal associations are adjusted according to their nutritional requirements. Phosphorus is a strong inhibitor of symbiosis; in line with several studies, elevated Pi levels inhibit AM colonization (Balzergue et al., 2013; Bonneau et al., 2013). Systematic nutrient screening in *Petunia hybrida* and *R. irregularis* AMF systems has shown that, other than Pi, nitrate also inhibits AM development, whereas sulfate, Mg^{2+}, Ca^{2+}, and Fe^{3+} do not show a significant effect (Nouri et al., 2014). Interestingly, starvation, especially for nitrate, can override Pi-mediated suppression, thus signaling that starvation may override Pi inhibition to conserve the relationship under nutrient-limited conditions.

Beyond merely nutrient responses, AM regulation has many levels of intricacy. AM symbiosis depends ultimately on partner identity, environmental context, resource access, and functional diversity within the interacting organisms (Walder & van der Heijden, 2015). This outlines why nutrient effects on AM symbiosis appear to differ for various host-fungus combinations and environmental scenarios. Collectively the findings suggest that plants maintain a sophisticated, multilayered regulatory network whereby plants balance nutritional status with the costs/benefits of symbiosis: (1) direct nutrient changes to colonization, (2) simultaneous stimulatory and inhibitory signals, and (3) interactions with partners modulated by context. Such a mechanistic viewpoint intersects with managing AM associations in agricultural systems, particularly in maintaining beneficial symbioses and maximizing fertilizer use based on crop nutritional demands.

19.3 Inoculum production and application methods

19.3.1 Inoculation methods

In large-scale commercial inoculum production, trap-plant cultivation dominates AMF propagation methods, whereas alternative approaches remain underutilized. Soilless culture systems, particularly aeroponics and hydroponics, offer distinct advantages by producing contaminant-free spores while optimizing host plant growth conditions (IJdo et al., 2011). These methods show significant promise for large-scale commercial production. Moreover, root-organ monoxenic culture enables efficient mass propagation of select AMF species through transformed root cultures.

The monoxenic method involves culturing excised roots (hairy-root culture) transformed by *Agrobacterium rhizogenes*, which contains the Ri plasmid to induce prolific root growth without shoot development. A single Petri dish can generate substantial quantities of spores, hyphae, and colonized root material within months. However, this protocol has only been successfully adapted for limited AMF species.

Inoculum effectiveness varies significantly depending on propagule type, as different AMF structures exhibit varying colonization efficiencies (Hart & Klironomos, 2002). Key propagules include extraradical hyphae, intraradical vesicles (particularly important for fungal regrowth in certain species), and root fragments. The selection of appropriate inoculum sources represents a critical factor determining successful root colonization and subsequent symbiotic establishment. Different AM fungal families exhibit distinct propagation preferences. For example, species from the Glomeraceae family propagate mainly through hyphal fragments, whereas Gigasporaceae, Acaulosporaceae, and Scutellosporaceae families rely more on spore germination (Berruti et al., 2016). For multispecies inoculum production using the common trap-plant method, processing should include substrate sieving and root fragmentation to recover all propagule types (spores, hyphae, and colonized roots). Although some applications use purified spores or specific propagule combinations, the crude inoculum approach ensures comprehensive fungal diversity.

19.3.2 Selection and composition of arbuscular mycorrhizal fungal inoculants

The success of inoculation depends on selecting appropriate AM fungal strains. Inoculants are typically sourced from culture collections or isolated from soils similar to those in the study, though some experiments use commercial AMF products. Certain AMF species exhibit greater stress tolerance and thrive in polluted or harsh environments (Hildebrandt et al., 2007). Native AMF from osmotically stressed regions often handle salt stress more effectively than nonnative strains, inferring stress resistance of AMF native to stressed environments. Thus selecting stress-adapted AMF improves revegetation success in degraded soils. Indigenous AMF outperform commercial inoculants in enhancing plant resistance to root-knot nematodes, promoting growth in Mn-contaminated soil, and increasing shoot biomass in calcareous soils, demonstrating their superior efficacy (Briccoli Bati et al., 2015; Labidi et al., 2015). Under saline conditions, native AMF inoculation enhanced plant performance more effectively than cultured AMF, increasing shoot biomass, photosystem II efficiency, stomatal conductance, and glutathione levels. Native AMF consistently outperforms commercial strains, despite manufacturers' claims of broad adaptability. Commercial inoculants often fail to deliver promised results under field conditions.

Moreover, current practices favor single-species AMF inoculations, which typically enhance shoot biomass more effectively than multispecies applications. Application of functionally diverse AMF communities showed no growth benefits in host plants (Gosling et al. 2015). Under controlled single-stress conditions (greenhouse trials), host plants benefit most from specific stress-adapted fungi, whereas field conditions with multiple stressors may require more diverse AMF communities. Research indicates that AMF partnership composition, not just diversity, drives functional outcomes (Wagg et al., 2015). Most studies focus on three generalist species (*Rhizophagus intraradices, Funneliformis mosseae*, and *R. irregularis*) because of their broad host compatibility, global distribution, and cultivation ease (Öpik et al., 2010). Notably, intraspecific variation often affects plant response more than interspecific differences (Angelard et al., 2010), making strain selection crucial for optimal inoculation.

Common use of single AMF species (*R. intraradices, R. irregularis*, and *F. mosseae*) remains valid because of their intrinsic functional diversity. With *R. irregularis'* genome sequenced, comparing isolates could reveal the genetic bases for AMF effectiveness, enabling targeted strain development (Lin et al., 2014; Tisserant et al., 2013). However, plants show highly variable AMF responses, suggesting crop breeding programs should incorporate AMF compatibility as a key selection criterion to optimize symbiotic benefits.

The demonstrated superiority of native AMF in ecological restoration highlights the importance of inoculant composition, where locally adapted strains consistently outperform commercial alternatives in stress resilience. For agricultural systems, integrating genomic insights to design tailored AMF inoculants while breeding crops for enhanced symbiosis will maximize the benefits of these critical plant-fungal partnerships across both natural and managed ecosystems.

19.3.3 Arbuscular mycorrhizal fungi as a low-cost biofertilizer

Biofertilizers enhance soil fertility through natural processes, improving both soil health and plant growth. AMFs significantly boost soil quality and crop yields as a sustainable alternative to synthetic fertilizers. AMF can reduce phosphorus requirements by up to 50%, depending on plant species and environmental conditions (Ortas, 2012). AMF inoculation is a nature-based approach that aligns with global goals for sustainable intensification of agriculture while maintaining productivity. Research confirms that locally sourced AMF strains often outperform commercial inoculants. Encouraging

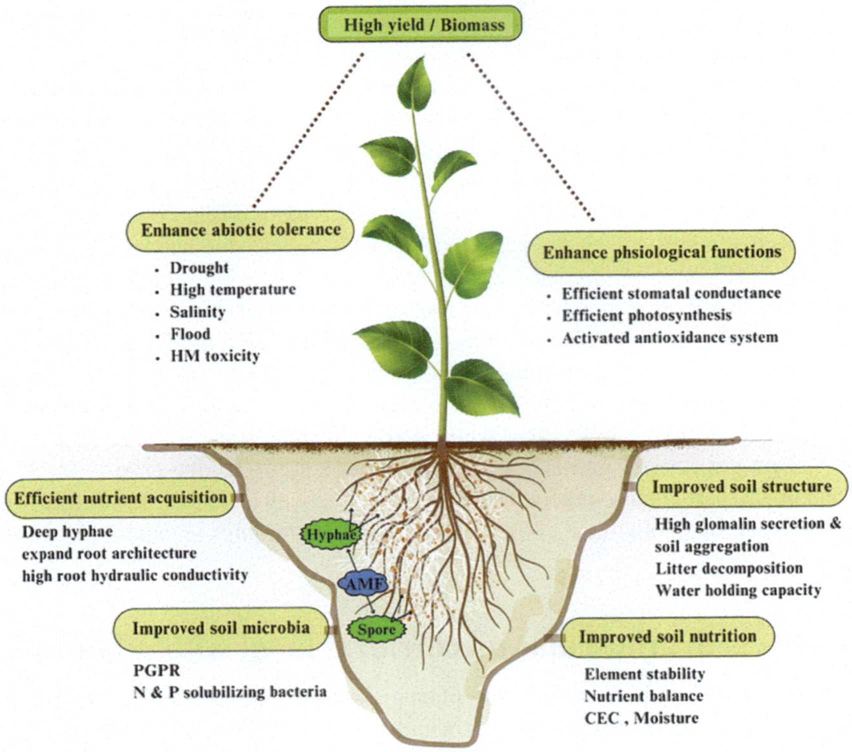

FIGURE 19.1 Mechanisms of AMF in improving soil health, nutrient uptake, and crop productivity. *AMF*, Arbuscular mycorrhizal fungi; *PGPR*, plant growth-promoting rhizobacteria.

farmers to cultivate native AMF provides a sustainable, cost-effective biofertilizer solution, particularly valuable for developing regions where chemical fertilizers strain budgets and ecosystems. Unlike synthetic inputs, AMFs provide a reduction in dependence on nonrenewable phosphorus, enhance soil structure and water retention, and lower production costs in the long term.

Chemical fertilizers impose serious threats to ecosystems and human health through soil degradation, water pollution, and food contamination (Yang et al., 2004). In contrast, AMF applications offer an ecofriendly solution through maintaining productivity while cutting fertilizer use, improving soil structure and nutrient cycling, and reducing agriculture's environmental footprint. With proper species-specific adaptation, AMF biofertilizers could substantially replace synthetic inputs, particularly in stress-prone environments. Their widespread adoption represents a critical step toward sustainable agriculture.

AMF inoculation enhances plant growth in both controlled and field environments through improved nutrient acquisition by host plants. Field studies demonstrate comparable efficacy to greenhouse trials, even where native AMF populations exist (Ceballos et al., 2013). To promote widespread AMF adoption, large-scale field validations and cost-benefit analyses are essential for demonstrating their agricultural advantages. Research confirms that locally sourced AMF strains often outperform commercial inoculants. On-farm inoculant production from locally sourced AMF strains offers a sustainable, cost-effective biofertilization strategy. This approach leverages adapted native AMF strains, reduces dependence on synthetic inputs, and lowers production costs for smallholder farmers (Jha & Songachan, 2023).

19.4 Arbuscular mycorrhizal fungi and soil health

AMFs serve as vital soil ecosystem engineers, enhancing soil health through multiple interconnected mechanisms (Fig. 19.1). AMFs significantly improve soil aggregation through multiple mechanisms, enhancing key ecosystem services, including carbon storage, water regulation, and erosion control (Leifheit et al., 2015). Their hyphal networks physically bind soil particles, whereas glomalin, a glycoprotein secreted by AMFs, acts as a biological glue, stabilizing both macroaggregates (>250 μm) and microaggregates (<250 μm) (Gou et al., 2023). These aggregates create optimal soil conditions by maintaining aerobic environments that prevent nutrient loss, providing microhabitats for beneficial microbial communities, and reducing decomposition rates of organic matter (Merino-Martín et al., 2021).

The combined effects of hyphal enmeshment and glomalin deposition not only improve soil structure but also enhance nutrient cycling efficiency. This dual function makes AMFs crucial for sustainable soil management, particularly in systems vulnerable to degradation. Their ability to simultaneously modify soil physical properties and biological activity underscores their unique role as ecosystem engineers in terrestrial habitats (Fig. 19.1). AMFs enhance soil stability by improving aggregate formation, thereby reducing erosion and increasing nutrient retention (Guo et al., 2023). This mechanism may significantly mitigate erosion-driven nutrient losses, a relationship yet to be tested in natural agricultural systems. Although existing studies demonstrate AMF's ability to reduce nutrient leaching, most evidence comes from sterilized soil systems where native soil structure and microbiome interactions are artificially disrupted (Guo et al., 2023).

Recently reported that AMF symbiosis enhanced carbon allocation to maize roots and soil, particularly in low-SOC (soil organic carbon) conditions, by increasing macroaggregate formation (+28%) and microbial biomass carbon (+19%), improving SOC chemical composition, and elevating glomalin-related proteins in high-SOC soil (Siyao et al., 2024). AMF can decompose litter (Fig. 19.1). Reports showed that while AMFs enhance soil carbon storage, they may concurrently stimulate organic matter decomposition despite lacking saprotrophic ability (Hodge et al., 2001). Studies demonstrate AMF accelerate litter breakdown, particularly under elevated CO_2/N, likely by activating hyphosphere bacteria that liberate nitrogen while reducing patch carbon (Koller et al., 2013). Validating these findings in intact field soils remains an essential next step for practical applications.

19.5 Synergistic effects of arbuscular mycorrhizal fungi with other soil microbes

Among the synergistic effects of AMF with other microbes, plant growth-promoting rhizobacteria (PGPR) have particularly wider usage. Recent studies demonstrated that combined inoculation with AMF and PGPR enhances plant growth and productivity more effectively than single applications. This synergistic effect has been documented in diverse crops, including tomato (*Solanum lycopersicum*), maize (*Zea mays*), cereal crops, pigeon pea (*Cajanus cajan*), rice (*Oryza sativa*), and wheat (*Triticum aestivum*) (Ji et al., 2021; Mathimaran et al., 2020; Sagar et al., 2021).

The chemical changes in the rhizosphere of mycorrhizal plants influence bacterial community composition, leading to the mycorrhizosphere effect. PGPR contribute to AM symbiosis and enhance host plant growth. Bacteria can attach to AM hyphae or reside within AM spore walls. Hyphae-associated bacteria utilize fungal exudates as nutrients and may rely on the mycelium for rhizosphere colonization. Mycorrhizal plants exhibit increased populations of beneficial bacteria, including nitrogen fixers and Gram-negative strains. PGPR stimulate AM fungal development by enhancing spore germination and hyphal elongation. PGPR and AMFs engage in nutrient exchange via direct contact, classifying these bacteria as mycorrhiza helper bacteria, boosting AM propagule germination, hyphal proliferation, and root colonization (Ramasamy et al., 2011). Coinoculation of tobacco seedlings with the AMF *F. mosseae* and two PGPR strains (*Bacillus* sp. and *Pseudomonas* sp.) significantly enhanced multiple growth parameters: aboveground and root dry biomass, nitrogen/phosphorus/potassium uptake, plant height, stem diameter, maximum leaf area, chlorophyll content, and root morphology (Zeng et al., 2024). This coinoculation further showed improved rhizosphere microbial community structure and function, demonstrating that AMF-PGPR synergism offers an effective, sustainable approach for optimizing tobacco seedling production through microbial cocultivation (Zeng et al., 2024).

Although synergistic interactions between AMF and PGPR are well-documented, antagonistic or neutral effects have also been observed (Diagne et al., 2020; Savastano & Bais, 2024). These variable outcomes depend on differences in inoculum strains, host plant species, environmental factors, and native rhizosphere microbiomes (Cai et al., 2021; Savastano & Bais, 2024), underscoring the complexity of these microbial interactions. Notably, research on AMF–PGPR in tobacco (*Nicotiana tabacum*) shows the seedling stage is the ideal inoculation stage, as early root contact maximizes symbiotic establishment (Angúlo-Castro et al., 2021). The rhizosphere microbiome, serving as a critical plant–soil interface, dynamically responds to microbial inoculants. However, most studies have overlooked how these microbial community interactions may mediate AMF–PGPR effects on plant growth (Savastano & Bais, 2024).

19.6 Efficacy of arbuscular mycorrhizal fungi inoculation in agricultural systems

19.6.1 Soil nutrient status

AMFs perform well in low-nutrient environments compared to well-fertilized or highly fertilized soils. Slow-release fertilizers like rock phosphate enhance AMF colonization and activity more effectively than conventional P fertilizers, with

AMF-colonized plants outperforming nonmycorrhizal plants receiving double the superphosphate dose (Thirkell et al., 2017). Optimal nitrogen levels (low-medium) further boost AMF establishment, sporulation, and plant growth, particularly when supplied early (Getman-Pickering et al., 2021). Arbuscular mycorrhizae enhance plant growth under reduced fertilizer inputs (Jarosz et al., 2021). Organic fertilizers, compared to inorganic ones, further improve AM-mediated growth and biocontrol. Elevated phosphorus (P) suppresses AM colonization, whereas low P stimulates root branching for enhanced uptake (García-Caparrós et al., 2021). AM-driven biomass gains peak under moderate fertilization but decline at high doses. Herbivore resistance is also strengthened by AM at intermediate fertilization levels, with no effects under low or excessive inputs (Kuila & Ghosh, 2022). Notably, phosphorus-rich organic fertilizers maximize AM-induced resistance.

Beyond expanding nutrient absorption through their extensive hyphal networks, mycorrhizae enhance phosphorus availability through biochemical processes. They solubilize insoluble phosphates and primary minerals via organic acid secretion and directly mineralize organic phosphorus through acid phosphatase activity (Li et al., 2020; Sato et al., 2015). Certain species additionally hydrolyze organic phosphorus compounds (Li et al., 2020). Mycorrhizae foster symbiotic relationships with rhizosphere bacteria by providing plant-derived photosynthates (Azcón-Aguilar & Barea, 2015), forming microbial consortia that collectively improve phosphate mineralization (Battini et al., 2017; Schneider et al., 2019). These interactions facilitate the mobilization of both inorganic and organic phosphorus into the soil solution.

Adopting integrated soil management strategies, such as optimized fertilization, inoculation with phosphate-solubilizing microorganisms, and mycorrhizae-assisted bacteria, can significantly improve phosphorus use efficiency. Such approaches not only reduce phosphorus toxicity risks but also minimize leaching losses (Battini et al., 2017), supporting sustainable agricultural systems.

19.6.2 Arbuscular mycorrhizal fungi–host–soil affinity

Host-fungal compatibility significantly influences AM efficiency, with intraspecific strain diversity and host cultivar-specific responses playing key roles (Table 19.1). Crop diversity shapes AMF community composition, bolstering ecological resilience and symbiotic efficiency. Polyculture systems, such as intercropping and rotational schemes, foster diverse fungal networks that optimize nutrient transfer and soil ecosystem services through enhanced microbial coexistence (Chaudhary et al., 2025). Such biodiversity maintains agroecosystem stability amid environmental fluctuations (Lugo et al., 2023). Cultivar genotypes exhibit substantial variation in their interactions with environmental conditions (Chu et al., 2013). The molecular basis of plant-mediated fungal performance remains poorly understood but may involve carbohydrate and lipid allocation. For instance, host-dependent expression of monosaccharide transporter genes (*RiMSTs*) in *R. irregularis* differs between intraradical and extraradical hyphae, alongside variations in phosphate uptake (Ait Lahmidi et al., 2016). In addition, biomass accumulation correlates with extraradical mycelial growth, underscoring genotype-environment complexity (Sawers et al., 2017). Moreover, upland ecosystems harbor native AMF that demonstrate high efficiency in symbiotic associations with upland rice. Crop rotation with rice enhances soil phosphorus levels and native AMF populations (Maiti et al., 2012). Although most AMF species are unsuitable for wetland conditions, certain native strains, including *Glomus etunicatum, Glomus mosseae*, and *G. intraradices,* exhibit remarkable adaptability, improving phosphorus uptake, root colonization, and yield in flooded rice systems across varying soil fertility levels (Watanarojanaporn et al., 2013). Notably, *G. intraradices* enhances rice growth, photosynthesis, and antioxidant activity under drought stress (Ruíz-Sánchez et al., 2010).

In addition, AMF performance depends on environmental conditions and farming practices. Soil moisture, temperature, pH, and nutrient levels critically regulate fungal activity and root colonization. Tillage damages hyphal networks, impairing colonization efficiency, whereas agrochemical overuse disrupts AMF populations and microbial interactions (Wahab et al., 2023). AMF further engage in cross-kingdom partnerships with soil microbiota, forming dynamic consortia that drive biogeochemical cycles. These mutualisms critically sustain pedospheric health, carbon storage, and plant–microbe synergies (Frąc et al., 2018). AMFs demonstrate remarkable ecological plasticity, forming adaptable communities essential for agricultural sustainability (Table 19.2). Studies reveal successional progression in AMF assemblages, with temporal specialization and modified host interactions (Bennett et al., 2013). Environmental fluctuations, seasonal dynamics, and anthropogenic activities collectively shape AMF diversity patterns. Intensive monocropping systems particularly diminish fungal richness, underscoring the necessity for conservation-oriented practices that maintain functional AMF diversity (Higo et al., 2019).

19.6.3 Native arbuscular mycorrhizal fungi overintroduced arbuscular mycorrhizal fungi

Native AMFs demonstrate exceptional potential as biofertilizers, with studies documenting significant yield improvements in diverse cropping systems, including rice paddies (Kalamulla et al., 2021). The ecological advantage of indigenous AMF

TABLE 19.1 Host root-AMF compatibility: Root exudates released by diverse plant species recruit arbuscular mycorrhizal fungi to establish symbiotic associations.

Crop species	AMF species	Chemical compound	References
Rice (*Oryza sativa*)	*Gigaspora rosea*	Orobanchol	Cardoso et al. (2014)
Wheat (*Triticum aestivum* L.)	*Glomus mosseae, Glomus intraradices, Glomus etunicatum*	Strigolactone (GR24)	Moosavi et al. (2024)
Maize (*Zea mays* L.)	Glomeraceae, Gigasporaceae	5-Dexoystrigol, sorgomol	Yoneyama et al. (2015)
Sorghum bicolor	*G. rosea*	Strigolactones	Besserer et al. (2006)
Carrot (*Daucus carota*)	*Gigaspora gigantea*	1-Hydroxy fatty acid, 2-hydroxytetradecanoic acid	Nagahashi et al. (2010)
Lotus japonicus	*Gigaspora margarita*	Strigolactones	Akiyama et al. (2005)
White clover (*Trifolium repens*)	*Glomus intraradix*	Flavonoids	Siqueira et al. (1991)
Alfalfa (*Medicago sativa* L.)	*G. etunicatum, Glomus macrocarpum*	Flavonoids	Tsai et al. (1991)
Tomato (*Solanum lycopersicum*)	*Gigaspora* spp., *Glomus* spp.	Strigolactones, flavones, flavonols	Akiyama et al. (2005), Steinkellner et al. (2007)
Soybean (*Glycine max*)	*Rhizophagus intraradices*	Flavonoids, isoflavones	Antunes et al. (2006)
Potato (*Solanum tuberosum*)	*G. intraradices, G. mosseae*	Strigolactones, glycoalkaloids	Lingua et al. (2013)
Sunflower (*Helianthus annuus*)	*G. intraradices, G. mosseae*	Strigolactones, phenolic compounds	Sarwar et al. (2020)
Bean (*Phaseolus vulgaris*)	*G. intraradices, Gigaspora* spp.	Flavonoids, isoflavones	Aisyah (2015)
Onion (*Allium cepa*)	*G. mosseae, G. intraradices*	Flavonoids, organosulfur compounds	Golubkina et al. (2020)
Lettuce (*Lactuca sativa*)	*G. intraradices, G. mosseae*	Phenolic compounds, sesquiterpene lactones	Stojanović et al. (2023)
Cucumber (*Cucumis sativus*)	*G. intraradices, Gigaspora* spp.	Strigolactones, cucurbitacins	Etri (2022)
Pepper (*Capsicum annuum*)	*G. intraradices, G. mosseae*	Capsaicinoids, flavonoids	Kebu et al. (2024)
Apple (*Malus domestica*)	*G. intraradices, G. mosseae*	Phenolic compounds, flavonoids	Arnold et al. (2024)

Chaudhary et al. (2025).
AMF, arbuscular mycorrhizal fungi.

strains lies in their evolutionary adaptation to local pedoclimatic conditions, resulting in superior colonization efficiency and stress resilience compared to introduced species. This local adaptation enhances nutrient cycling efficiency and promotes more stable plant–microbe interactions, leading to improved ecosystem functioning across various environments from forest ecosystems to semiarid agroecosystems (Djotan et al., 2024; Vieira et al., 2020).

The ecological plasticity of AMF communities positions them as critical components in developing climate-smart agricultural systems. Recent research emphasizes their multifunctional roles in: (1) improving soil carbon storage through glomalin production, (2) increasing plant water relations during drought stress, and (3) assessing complex belowground interactions that support biodiversity (Enebe & Erasmus, 2023). Field applications in agroforestry systems demonstrate how native AMF inoculants can simultaneously address food production challenges while delivering ecosystem services, increasing crop yields by 15%–40% while improving soil organic matter content (Hu et al., 2024).

Further, studies reveal novel mechanisms underlying AMF–plant–environment interactions, particularly their role in mitigating abiotic stresses through enhanced antioxidant production, modulating rhizosphere microbiome composition, and

TABLE 19.2 AMF-mediated enhanced productivity in some crop species, soil health, and carbon stabilization.

AMF species	Crop species	Mechanism	References
Rhizophagus intraradices	Wheat (*Triticum aestivum*	Increased grain production	Sharma et al. (2011)
Claroideoglomus etunicatum	Soybean (*Glycine max*)	Increased root biomass	Vieira et al. (2020)
Gigaspora margarita	Maize (*Zea mays*)	Increased plant height	Malembaka et al. (2021)
Rhizophagus clarus	Tomato (*Solanum lycopersicum*	Improve phosphorus absorption	Higo et al. (2019)
Funneliformis mosseae	Olive (*Olea europaea*)	Enhanced glomalin secretion and soil aggregation	Montes-Borrego et al. (2014)
Glomus mosseae	Rice (*Oryza sativa*)	Increased carbon sequestration	Suriyagoda (2022)
Glomus etunicatum, G. mosseae, Glomus intraradices	Rice (*O. sativa*)	Improved P uptake, photosynthesis, and yield, enhanced stress resistance	Watanarojanaporn et al. (2013), Ruíz-Sánchez et al. (2010)
Funneliformis caledonium	Barley (*Hordeum vulgare*)	Enhanced soil microbial biomass	Mwafulirwa et al. (2021)
Rhizophagus irregularis	Potato (*Solanum tuberosum*)	Improved nutrient uptake and tuber yield	Deja-Sikora et al. (2023)
Gigaspora gigantea	Sunflower (*Helianthus annuus*)	Enhanced phosphate solubilization	Sarah et al. (2016)
Claroideoglomus claroideum	Bean (*Phaseolus vulgaris*)	Increased nitrogen fixation efficiency	Khoso et al. (2025)
Rhizophagus fasciculatus	Pepper (*Capsicum annuum*)	Improved fruit quality and antioxidant content	Padmavathi et al. (2015)
Scutellospora heterogama	Lettuce (*Lactuca sativa*)	Enhanced mineral nutrition and growth	Baslam et al. (2013)
Septoglomus constrictum	Strawberry (*Fragaria × ananassa*)	Enhanced fruit production and sugar content	Todeschini et al. (2018), Lingua et al. (2013)
Claroideoglomus lamellosum	Apple (*Malus domestica*)	Enhanced root system development	Cook et al. (2024)
Funneliformis coronatus	Citrus (*Citrus sinensis*)	Improved phosphorus and zinc uptake	Liu et al. (2024)
Gigaspora albida	Coffee (*Coffea arabica*)	Enhanced plant growth and bean quality	Urgiles-Gómez et al. (2021)
Acaulospora mellea	Tea (*Camellia sinensis*)	Improved leaf quality and antioxidant levels	Liu et al. (2021)
Rhizophagus custos	Cassava (*Manihot esculenta*)	Enhanced starch production and root yield	da Silva et al. (2021), Ceballos et al. (2013)
Scutellospora pellucida	Banana (*Musa acuminata*)	Improved disease resistance and fruit quality	Jefwa et al. (2012)
Entrophospora infrequens	Sweet potato (*Ipomoea batatas*)	Improved tuber formation and beta-carotene content	Meenakshisundaram (2011)
Rhizophagus diaphanus	Sugarcane (*Saccharum officinarum*)	Improved sugar yield and stress tolerance	Prasad (2024)
Pacispora scintillans	Yam (*Dioscorea alata*)	Enhanced tuber development	Tchabi et al. (2010)
Ambispora leptoticha	Pineapple (*Ananas comosus*)	Enhanced fruit development and sugar content	Guillemin et al. (1994)

Mwampashi et al. (2024).
AMF, arbuscular mycorrhizal fungi.

improving nutrient use efficiency in low-input systems. The integration of indigenous AMF into cropping systems represents a viable pathway toward developing more resilient and productive agroecosystems in the face of climate change.

19.7 Molecular mechanisms of arbuscular mycorrhizal fungi–plant interactions

The establishment of AM symbiosis requires coordinated molecular recognition between fungi and host plants, mediated by reciprocal genetic regulation (Birhanu & Negussie, 2024). This interaction progresses through stage-specific signaling events, beginning with plant-derived biochemical cues. Genetic modification of AMF strains offers potential for augmenting stress resilience and nutrient acquisition capabilities, enabling the development of tailored inoculants for precision agriculture (Reinhart et al., 2024). Modern molecular tools have transformed AMF research, with next-generation sequencing enabling precise identification of cryptic and low-abundance species (Bainard et al., 2017; Liu et al., 2014). Metagenomic analyses further elucidate functional gene networks governing nutrient cycling and stress adaptation in AMF communities (Wahab et al., 2024; Hyde et al., 2024). Pioneering studies employing high-throughput sequencing have revealed novel AMF-bacterial synergisms that enhance plant nutrition and phytoprotection (Young & Gillung, 2020), demonstrating the power of these approaches to uncover complex soil microbiome interactions.

Under environmental stress, root exudates, including sugars, amino acids, and secondary metabolites, modulate rhizosphere microbial dynamics. Notably, strigolactones, while serving as germination signals for parasitic plants, also critically promote AM fungal development (Abdelhalim et al., 2019; Nazir et al., 2023; Saeed et al., 2025; Thind et al., 2022; Yizhu et al., 2020). Stress environments further trigger plant secretion of hyphal branching factors, which enhance AMF growth and morphological differentiation during early symbiosis (Yoneyama et al., 2007). In response, AMFs release Myc factors that induce plant symbiotic pathways (Navazio et al., 2007). Successful colonization ends in arbuscule formation, specialized structures facilitating bidirectional nutrient exchange (particularly phosphate and nitrogen) between symbionts. Further, advanced molecular tools, including high-throughput sequencing, have enabled the tracking of drought-tolerant AMF strains that enhance soil aggregation (Lahlali et al., 2021). Targeted inoculation with these strains fosters crop resilience against abiotic stresses, particularly water deficit and soil deterioration.

Understanding the metabolic pathways and gene functions of AMF can facilitate the development of novel biofertilizers and biopesticides. Genetic modification of AMF to synthesize plant-beneficial compounds or pathogen-resistant metabolites presents a sustainable alternative to conventional agrochemicals. Cutting-edge molecular and imaging tools, such as metagenomics and stable isotope probing, enable accurate assessment of soil health and AMF interactions by elucidating microbial community structure and functional dynamics. These analytical approaches support evidence-based decision-making for optimizing AMF inoculant efficacy and monitoring long-term soil quality improvements.

19.7.1 Limitations to scalable arbuscular mycorrhizal fungi inoculant production and effective field utilization

AMFs are obligate biotrophs, requiring host plants compulsorily for growth, which complicates large-scale inoculum production. Commonly, three primary inoculum types exist: **soil-based** (containing spores, hyphae, and root fragments but potentially contaminated with weeds/pathogens), **crude inoculum** (produced via trap plants in sterile media, offering higher propagule purity), and **root fragments** (isolated from colonized hosts). However, crude inoculum is most widely used because of its standardized propagules. However, mass production remains challenging despite advances in cultivation and seed-coating techniques (van der Heijden et al., 2015). The key constraints in large-scale inoculum production are as follows: (1) **Host dependence**: AMF require time- and space-intensive cocultivation with plants; (2) **Quality control**: No rapid methods exist to verify colonization rates preapplication; and (3) **Logistical demands**: Large-scale inoculum handling is laborious, though simplified for transplanted crops (lower quantities needed). These limitations necessitate specialized protocols for AMF culture collections, differing markedly from those for free-living microbes. The lack of axenic cultivation methods further restricts industrial scalability, underscoring the need for innovation in propagation and application technologies to harness AMF's agricultural potential.

Large-scale AMF inoculation may appear economically and logistically challenging, but restored AMF communities persist under conservation practices such as cover cropping and reduced tillage (Säle et al., 2015). A cost-effective alternative is targeted "fertility islands," where localized AMF inoculation aids gradual mycelial network establishment. This approach is particularly effective for revegetating degraded soils, as inoculated patches enhance native plant recovery in nutrient-depleted areas. Once established, AMF networks remain intact if no disruptive practices are employed, offering long-term benefits without repeated applications (Berruti et al., 2016).

Thus AMF restoration incurs only an initial cost, which can be amortized over time if fungal persistence is maintained. Studies show that AMF inoculation can be more cost-effective than conventional fertilization, offering significant savings for growers and land restoration. Commercial interest in AMF has led to the production of inoculants, typically containing a limited number of generalist fungal species. These species are selected for ease of propagation and broad host compatibility across diverse environments. Although some manufacturers offer single-formula products, others provide specialized blends for different crops and conditions. However, despite claims of universal suitability, the actual benefits of commercial inoculants vary and are not always guaranteed (Faye et al., 2013).

19.8 Research gaps in arbuscular mycorrhizal fungi application

Despite advances in modern techniques, critical knowledge gaps remain in AMF taxonomy and biogeography. A significant proportion of AMF species are still unclassified, and their global distribution patterns remain poorly documented, limiting insights into their ecological functions and agricultural potential (Yurkov et al., 2023). High-throughput sequencing and phylogenomic approaches could resolve these uncertainties by elucidating species diversity, evolutionary linkages, and habitat specialization.

Functional diversity in AMF needs deeper investigation to explore the interspecific variation that exists in nutrient mobilization, stress adaptation, and host compatibility; however, the broader ecological consequences of these traits remain unclear (Bainard et al., 2017). Systematic studies are needed to evaluate how AMF taxa differentially influence crop productivity, soil nutrient cycling, and ecosystem stability as key considerations for sustainable land management (Martin & van der Heijden, 2024).

The mechanistic understanding of AMF–plant symbiosis, such as phosphorus exchange and strigolactone signaling, is established; the underlying genetic and metabolic networks demand further characterization (Qi et al., 2022; Shen & Duan, 2024). Deciphering these molecular mechanisms could unlock novel strategies for improving crop resilience and yield (Boyno et al., 2023).

The interplay between AMF and abiotic factors, including soil properties (pH, nutrient gradients, and hydrology), climate change, and pollutants, remains insufficiently characterized. These variables critically regulate AMF colonization efficiency and symbiotic performance, with anthropogenic disturbances potentially destabilizing AMF–plant networks (Boorboori & Zhang, 2022; Rillig et al., 2024). Elucidating these relationships is vital for developing AMF-based climate resilience strategies.

Although the agricultural benefits of AMF are established, the differential effects of farming practices on AMF communities require deeper investigation. Organic farming systems, characterized by minimal synthetic inputs and elevated organic matter, consistently demonstrate higher AMF colonization rates and diversity than conventional systems (Wang et al., 2023). Conversely, intensive tillage practices degrade AMF populations through physical disruption of hyphal networks and reduced spore viability (Kuila & Ghosh, 2022). Emerging evidence suggests that conservation practices like no-till systems with cover crops can significantly improve AMF abundance and soil functionality through sustained organic inputs (Tian et al., 2024).

The limitations of potential applications of AMF in sustainable agriculture include insufficient data on species-specific efficacy, optimal inoculation protocols, and long-term effects on soil ecosystems (Wahab et al., 2023). Furthermore, AMF function within complex soil microbiomes, where interactions with bacteria, fungi, and soil fauna collectively influence plant health and ecosystem processes (Fall et al., 2022; Hnini et al., 2024). A system-level understanding of these multitrophic relationships is essential for developing effective microbial management strategies. Addressing these research priorities will significantly advance both fundamental knowledge of AMF ecology and their practical implementation in environmental stewardship.

19.9 Conclusion and recommendations

AMFs play a diversified role in sustainable agriculture, performing as biofertilizers that enhance nutrient uptake (particularly nitrogen, phosphorus, and micronutrients) and also act as bioprotectants, improving plant resilience to biotic and abiotic stresses while maintaining soil health and ecosystem sustainability. The effectiveness of AMF biofertilizers depends on fungal and crop compatibility, soil conditions, and inoculation techniques. However, critical knowledge gaps remain in AMF taxonomy, symbiotic mechanisms, and environmental interactions, with traditional identification methods proving limited and molecular pathways in plant-fungal symbiosis still incompletely understood. To advance AMF research and applications, it is recommended to adopt advanced molecular techniques such as high-throughput sequencing and metagenomics for improved species identification and functional characterization, whereas mechanistic studies should focus on deciphering symbiotic signaling across diverse plant-fungal combinations. In addition, integrated lab and field studies are needed to assess AMF responses to environmental factors, and holistic approaches must be employed to understand AMF interactions within the broader soil microbiome using multiomics tools and ecological modeling. Long-term field trials should validate AMF efficacy in real-world agricultural and restoration scenarios, with site-specific inoculation protocols developed to optimize performance across different agroecological conditions. Overcoming these challenges will position AMF biofertilizers as a cornerstone of sustainable crop production, ensuring food security in the face of climate change and environmental degradation. This chapter underscores the multifaceted benefits of AMF in agroecosystems, emphasizing their role in advancing ecoconscious agriculture for a resilient future.

References

Abdelhalim, T., Jannoura, R., & Joergensen, R. G. (2019). Mycorrhiza response and phosphorus acquisition efficiency of sorghum cultivars differing in strigolactone composition. *Plant and Soil, 437*(1–2), 55–63. http://www.wkap.nl/journalhome.htm/0032-079X. doi:10.1007/s11104-019-03960-y.

Ahmad, M., Hussain, A., Dar, A., Luqman, M., Ditta, A., Iqbal, Z., Ahmad, H. T., Nazli, F., Soufan, W., Almutairi, K., & Sabagh, A. E. (2023). Combating iron and zinc malnutrition through mineral biofortification in maize through plant growth promoting *Bacillus* and *Paenibacillus* species. *Frontiers in Plant Science, 13*. https://www.frontiersin.org/journals/plant-science. doi:10.3389/fpls.2022.1094551.

Aisyah, S. (2015). *Induction of prenylated isoflavonoids and stilbenoids in legumes (Doctoral dissertation)*, Wageningen University.

Ait Lahmidi, N., Courty, P. E., Brulé, D., Chatagnier, O., Arnould, C., Doidy, J., Berta, G., Lingua, G., Wipf, D., & Bonneau, L. (2016). Sugar exchanges in arbuscular mycorrhiza: RiMST5 and RiMST6, two novel *Rhizophagus irregularis* monosaccharide transporters, are involved in both sugar uptake from the soil and from the plant partner. *Plant Physiology and Biochemistry, 107*, 354–363. http://www.journals.elsevier.com/plant-physiology-and-biochemistry/. doi:10.1016/j.plaphy.2016.06.023.

Akiyama, K., Matsuzaki, K. I., & Hayashi, H. (2005). Plant sesquiterpenes induce hyphal branching in arbuscular mycorrhizal fungi. *Nature, 435*(7043), 824–827. doi:10.1038/nature03608.

Angelard, C., Colard, A., Niculita-Hirzel, H., Croll, D., & Sanders, I. R. (2010). Segregation in a mycorrhizal fungus alters rice growth and symbiosis-specific gene transcription. *Current Biology, 20*(13), 1216–1221. doi:10.1016/j.cub.2010.05.031.

Angulo-Castro, A., Ferrera-Cerrato, R., Alarcón, A., Almaraz-Suárez, J. J., Delgadillo-Martínez, J., Jiménez-Fernández, M., & García-Barradas, O. (2021). Improved growth of bell pepper (*Capsicum annuum*) plants by inoculating arbuscular mycorrhizal fungi and beneficial rhizobacteria. *Scientia Fungorum*, 51. doi:10.33885/sf.2021.51.1299.

Antunes, P. M., Rajcan, I., & Goss, M. J. (2006). Specific flavonoids as interconnecting signals in the tripartite symbiosis formed by arbuscular mycorrhizal fungi, *Bradyrhizobium japonicum* (Kirchner) Jordan and soybean (*Glycine max* (L.) Merr.). *Soil Biology and Biochemistry, 38*(3), 533–543. doi:10.1016/j.soilbio.2005.06.008.

Arnold, M., & Gramza-Michalowska, A. (2024). Recent development on the chemical composition and phenolic extraction methods of apple (*Malus domestica*)—A review. *Food and Bioprocess Technology, 17*(9), 2519–2560.

Augé, R. M., Toler, H. D., & Saxton, A. M. (2015). Arbuscular mycorrhizal symbiosis alters stomatal conductance of host plants more under drought than under amply watered conditions: A meta-analysis. *Mycorrhiza, 25*(1), 13–24. link.springer.de/link/service/journals/00572/index.htm. doi:10.1007/s00572-014-0585-4.

Azcón-Aguilar, C., & Barea, J. M. (2015). Nutrient cycling in the mycorrhizosphere. *Journal of Soil Science and Plant Nutrition, 15*(ahead). doi:10.4067/s0718-95162015005000035.

Balzergue, C., Chabaud, M., Barker, D. G., Bécard, G., & Rochange, S. F. (2013). High phosphate reduces host ability to develop arbuscular mycorrhizal symbiosis without affecting root calcium spiking responses to the fungus. *Frontiers in Plant Science, 4*. http://journal.frontiersin.org/Journal/10.3389/fpls.2013.00426/pdf. doi:10.3389/fpls.2013.00426.

Banasiak, J., Jamruszka, T., Murray, J. D, & Jasiński, M. (2021). A roadmap of plant membrane transporters in arbuscular mycorrhizal and legume–rhizobium symbioses. *Plant Physiology, 187*(4), 2071–2091. doi:10.1093/plphys/kiab280.

Bainard, L. D., Chagnon, P. L., Cade-Menun, B. J., LambEG, LaForge, K., Schellenberg, M., & Hamel, C. (2017). Plant communities and soil properties mediate agricultural land use impacts on arbuscular my-corrhizal fungi in the Mixed Prairie ecoregion ofthe North American Great Plains. *Agriculture, Ecosystems and Environment, 249*, 187–195. doi:10.1016/j.agee.2017.08.010.

Baslam, M., Garmendia, I., & Goicoechea, N. (2013). Enhanced accumulation of vitamins, nutraceuticals and minerals in lettuces associated with arbuscular mycorrhizal fungi (AMF): A question of interest for both vegetables and humans. *Agriculture, 3*(1), 188–209. doi:10.3390/agriculture3010188.

Battini, F., Grønlund, M., Agnolucci, M., Giovannetti, M., & Jakobsen, I. (2017). Facilitation of phosphorus uptake in maize plants by mycorrhizosphere bacteria. *Scientific Reports, 7*(1). doi:10.1038/s41598-017-04959-0.

Bell, C. A., Magkourilou, E., Urwin, P. E., & Field, K. J. (2021). The influence of competing root symbionts on below-ground plant resource allocation. *Ecology and Evolution, 11*(7), 2997–3003. http://onlinelibrary.wiley.com/journal/10.1002/(ISSN)2045-7758. doi:10.1002/ece3.7292.

Bennett, A. E., Daniell, T. J., Öpik, M., Davison, J., Moora, M., Zobel, M., Selosse, M. A., & Evans, D. (2013). Arbuscular mycorrhizal fungal networks vary throughout the growing season and between successional stages. *PLoS One, 8*(12). http://www.plosone.org/article/fetchObject.action?uri=info%3Adoi%2F10.1371%2Fjournal.pone.0083241&representation=PDF. doi:10.1371/journal.pone.0083241. United Kingdom.

Berruti, A., Lumini, E., Balestrini, R., & Bianciotto, V. (2016). Arbuscular mycorrhizal fungi as natural biofertilizers: Let's benefit from past successes. *Frontiers in Microbiology, 6*. doi:10.3389/fmicb.2015.01559.

Besserer, A., Puech-Pagès, V., Kiefer, P., Gomez-Roldan, V., Jauneau, A., Roy, S., Portais, J.-C., Roux, C., Bécard, G., Séjalon-Delmas, N., & Chory, J. (2006). Strigolactones stimulate arbuscular mycorrhizal fungi by activating mitochondria. *PLoS Biology, 4*(7), e226. doi:10.1371/journal.pbio.0040226.

Birhanu, M. W., & Negussie, Z. T. (2024). Arbuscular mycorrhizal (AM) fungi symbiosis in sustainable production of sorghum (*Sorghum bicolor* L. Moench) under drought stress: An emerging biofertilizer in dryland areas. *Advances in Agriculture, 2024*(1). https://onlinelibrary.wiley.com/journal/9403. doi:10.1155/2024/8814543.

Bonneau, L., Huguet, S., Wipf, D., Pauly, N., & Truong, H. N. (2013). Combined phosphate and nitrogen limitation generates a nutrient stress transcriptome favorable for arbuscular mycorrhizal symbiosis in *Medicago truncatula*. *New Phytologist, 199*(1), 188–202. doi:10.1111/nph.12234.

Boorboori, M. R., & Zhang, H. Y. (2022). Arbuscular mycorrhizal fungi are an influential factor in improving the phytoremediation of arsenic, cadmium, lead, and chromium. *Journal of Fungi, 8*(2). https://www.mdpi.com/2309-608X/8/2/176/pdf. doi:10.3390/jof8020176.

Boyno, G., Danesh, Y. R., Demir, S., Teniz, N., Mulet, J. M., & Porcel, R. (2023). The complex interplay between arbuscular mycorrhizal fungi and strigolactone: Mechanisms, sinergies, applications and future directions. *International Journal of Molecular Sciences, 24*(23). http://www.mdpi.com/journal/ijms. doi:10.3390/ijms242316774.

Breuillin-Sessoms, F., Floss, D. S., Gomez, S. K., Pumplin, N., Ding, Y., Levesque-Tremblay, V., Noar, R. D., Daniels, D. A., Bravo, A., Eaglesham, J. B., Benedito, V. A., Udvardi, M. K., & Harrison, M. J. (2015). Suppression of arbuscule degeneration in *Medicago truncatula* phosphate transporter4 mutants is dependent on the ammonium transporter 2 family protein AMT2;3. *The Plant Cell, 27*(4), 1352–1366. doi:10.1105/tpc.114.131144.

Briccoli Bati, C., Santilli, E., & Lombardo, L. (2015). Effect of arbuscular mycorrhizal fungi on growth and on micronutrient and macronutrient uptake and allocation in olive plantlets growing under high total Mn levels. *Mycorrhiza, 25*(2), 97–108. doi:10.1007/s00572-014-0589-0.

Cai, X., Zhao, H., Liang, C., Li, M., & Liu, R. (2021). Effects and mechanisms of symbiotic microbial combination agents to control tomato *Fusarium* crown and root rot disease. *Frontiers in Microbiology, 12*. doi:10.3389/fmicb.2021.629793.

Cardoso, C., Charnikhova, T., Jamil, M., Delaux, P.-M., Verstappen, F., Amini, M., Lauressergues, D., Ruyter-Spira, C., Bouwmeester, H., & Öpik, M. (2014). Differential activity of *Striga hermonthica* seed germination stimulants and *Gigaspora rosea* Hyphal branching factors in rice and their contribution to underground communication. *PLoS One, 9*(8), e104201. doi:10.1371/journal.pone.0104201.

Ceballos, I., Ruiz, M., Fernández, C., Peña, R., Rodríguez, A., Sanders, I. R., & Rillig, M. (2013). The in vitro mass-produced model mycorrhizal fungus, *Rhizophagus irregularis*, significantly increases yields of the globally important food security crop cassava. *PLoS One, 8*(8), e70633. doi:10.1371/journal.pone.0070633.

Chaudhary, A., Poudyal, S., & Kaundal, A. (2025). Role of arbuscular mycorrhizal fungi in maintaining sustainable agroecosystems. *Applied Microbiology, 5*(1), 6. doi:10.3390/applmicrobiol5010006.

Cheng, Z., Wu, S., Du, J., Liu, Y., Sui, X., & Yang, L. (2023). Reduced arbuscular mycorrhizal fungi (AMF) diversity in light and moderate fire sites in Taiga Forests, Northeast China. *Microorganisms, 11*(7), 1836. doi:10.3390/microorganisms11071836.

Chu, Q., Wang, X., Yang, Y., Chen, F., Zhang, F., & Feng, G. (2013). Mycorrhizal responsiveness of maize (*Zea mays* L.) genotypes as related to releasing date and available P content in soil. *Mycorrhiza, 23*(6), 497–505. doi:10.1007/s00572-013-0492-0.

Cook, C., Huskey, D., Mazzola, M., & Somera, T. (2024). Effect of rootstock genotype and arbuscular mycorrhizal fungal (AMF) species on early colonization of apple. *Plants, 13*(10), 1388. doi:10.3390/plants13101388.

Courty, P. E., Smith, P., Koegel, S., Redecker, D., & Wipf, D. (2015). Inorganic nitrogen uptake and transport in beneficial plant root-microbe interactions. *Critical Reviews in Plant Sciences, 34*, 4–16. www.tandf.co.uk/journals/titles/07352689.asp. doi:10.1080/07352689.2014.897897.

da Silva, W. F., da Silva Lima, A. M, & Lopes, E. A. P (2021). Levantamento de fungos micorrízicos arbusculares em associação a espécie vegetal *Manihot esculenta* Crantz. *Diversitas Journal, 6*(4), 3779–3795. doi:10.48017/dj.v6i4.1813.

Davison, J., Moora, M., Öpik, M., Adholeya, A., Ainsaar, L., Bâ, A., Burla, S., Diedhiou, A. G., Hiiesalu, I., Jairus, T., Johnson, N. C., Kane, A., Koorem, K., Kochar, M., Ndiaye, C., Pärtel, M., Reier, Ü., Saks, Ü., Singh, R., … Zobel, M. (2015). Global assessment of arbuscular mycorrhizal fungus diversity reveals very low endemism. *Science, 349*(6251), 970–973. doi:10.1126/science.aab1161.

Deja-Sikora, E., Werner, K., & Hrynkiewicz, K. (2023). AMF species do matter: *Rhizophagus irregularis* and *Funneliformis mosseae* affect healthy and PVY-infected *Solanum tuberosum* L. in a different way. *Frontiers in Microbiology, 14*. doi:10.3389/fmicb.2023.1127278.

Diagne, N., Ndour, M., Djighaly, P. I., Ngom, D., Ngom, M. C. N., Ndong, G., Svistoonoff, S., & Cherif-Silini, H. (2020). Effect of plant growth promoting rhizobacteria (PGPR) and arbuscular mycorrhizal fungi (AMF) on salt stress tolerance of *Casuarina obesa* (Miq.). *Frontiers in Sustainable Food Systems, 4*. www.frontiersin.org/journals/sustainable-food-systems#. doi:10.3389/fsufs.2020.601004.

Ditta, A., Ullah, N., Imtiaz, M., Li, X., Jan, A. U., Mehmood, S., Rizwan, M. S., & Rizwan, M. (2022). Zn biofortification in crops through Zn-solubilizing plant growth promoting rhizobacteria. In Q. Mahmood (Ed.), *Sustainable plant nutrition under contaminated environments* (pp. 115–133). Springer Nature Switzerland AG.

Djotan, A. K. G., Matsushita, N., & Fukuda, K. (2024). Year-round dynamics of arbuscular mycorrhizal fungi communities in the roots and surrounding soils of *Cryptomeria japonica*. *Mycorrhiza, 34*(1–2), 119–130. https://www.springer.com/journal/572. doi:10.1007/s00572-024-01143-x.

Enebe, M. C., & Erasmus, M. (2023). Symbiosis—A perspective on the effects of host traits and environmental parameters in arbuscular mycorrhizal fungal richness, colonization and ecological functions. *Agriculture, 13*(10), 1899. doi:10.3390/agriculture13101899.

Etri, K. (2022). Influence of arbuscular mycorrhizal inoculation on growth and stress resistance of maize and cucumber *[Master's thesis]*. doi:10.13140/RG.2.2.20808.53767.

Fall, A. F., Nakabonge, G., Ssekandi, J., Founoune-Mboup, H., Apori, S. O., Ndiaye, A., Badji, A., & Ngom, K. (2022). Roles of arbuscular mycorrhizal fungi on soil fertility: Contribution in the improvement of physical, chemical, and biological properties of the soil. *Frontiers in Fungal Biology, 3*. www.frontiersin.org/journals/fungal-biology. doi:10.3389/ffunb.2022.723892.

Faye, A., Dalpé, Y., Ndung'u-Magiroi, K., Jefwa, J., Ndoye, I., Diouf, M., & Lesueur, D. (2013). Evaluation of commercial arbuscular mycorrhizal inoculants. *Canadian Journal of Plant Science, 93*(6), 1201–1208. doi:10.4141/cjps2013-326.

Food and Agriculture Organization of the United Nations (FAO). (2020). Crop prospects and food situation—quarterly global report No. *1*. FAO. doi:10.4060/ca8032en.

Frąc, M., Hannula, S. E., Bełka, M., & Jędryczka, M. (2018). Fungal biodiversity and their role in soil health. *Frontiers in Microbiology, 9*. doi:10.3389/fmicb.2018.00707.

Francis, B., Aravindakumar, C. T., Brewer, P. B., & Simon, S. (2023). Plant nutrient stress adaptation: A prospect for fertilizer limited agriculture. *Environmental and Experimental Botany, 213*, 105431. doi:10.1016/j.envexpbot.2023.105431.

Fu, W., Yan, M., Zhao, L., Zeng, X., Cai, B., Qu, S., & Wang, S. (2023). Inoculation with arbuscular mycorrhizal fungi increase calcium uptake in *Malus robusta*. *Scientia Horticulturae, 321*, 112295. doi:10.1016/j.scienta.2023.112295.

García-Caparrós, P., Lao, M. T., Preciado-Rangel, P., & Sanchez, E. (2021). Phosphorus and carbohydrate metabolism in green bean plants subjected to increasing phosphorus concentration in the nutrient solution. *Agronomy, 11*(2), 245. doi:10.3390/agronomy11020245.

Getman-Pickering, Z. L., Stack, G. M., & Thaler, J. S. (2021). Fertilizer quantity and type alter mycorrhizae-conferred growth and resistance to herbivores. *Journal of Applied Ecology, 58*(5), 931–940. http://onlinelibrary.wiley.com/journal/10.1111/(ISSN)1365-2664. doi:10.1111/1365-2664.13833.

Giovannetti, M., Avio, L., & Sbrana, C. (2013). Improvement of nutraceutical value of food by plant symbionts. In *Natural products* (pp. 2641–2662). Springer.

Golubkina, N., & Caruso, G. (2020). Nutritional and antioxiant properties of *Allium cepa* L. In A. M. Jaiswal (Ed.), *Nutritional composition and antioxidant properties of fruits and vegetables* (pp. 73–89). Academic Press.

Gosling, P., Jones, J., & Bending, G. D. (2015). Evidence for functional redundancy in arbuscular mycorrhizal fungi and implications for agroecosystem management. *Mycorrhiza, 25*(6), 529–537. doi:10.1007/s00572-015-0651-6.

Gou, X., Ni, H., Sadowsky, M. J., Chang, X., Liu, W., & Wei, X. (2023). Arbuscular mycorrhizal fungi alleviate erosion-induced soil nutrient losses in experimental agro-ecosystems. *Catena, 220*, 106687. doi:10.1016/j.catena.2022.106687.

Guether, M., Balestrini, R., Hannah, M., He, J., Udvardi, M. K., & Bonfante, P. (2009). Genome-wide reprogramming of regulatory networks, transport, cell wall and membrane biogenesis during arbuscular mycorrhizal symbiosis in *Lotus japonicus*. *New Phytologist, 182*(1), 200–212. doi:10.1111/j.1469-8137.2008.02725.x.

Guillemin, J.-P., Gianinazzi, S., Gianinazzi-Pearson, V., & Marchal, J. (1994). Contribution of arbuscular mycorrhizas to biological protection of micropropagated pineapple (*Ananas comosus* (L.) Merr) against *Phytophthora cinnamomi* Rands. *Agricultural and Food Science, 3*(3), 241–251. doi:10.23986/afsci.72702.

Hart, M., & Klironomos, J. (2002). Colonization of roots by arbuscular mycorrhizal fungi using different sources of inoculum. *Mycorrhiza, 12*(4), 181–184. doi:10.1007/s00572-002-0169-6.

Higo, M., Kang, D.-J., & Isobe, K. (2019). First report of community dynamics of arbuscular mycorrhizal fungi in radiocesium degradation lands after the Fukushima-Daiichi Nuclear disaster in Japan. *Scientific Reports, 9*(1). doi:10.1038/s41598-019-44665-7.

Hildebrandt, U., Regvar, M., & Bothe, H. (2007). Arbuscular mycorrhiza and heavy metal tolerance. *Phytochemistry, 68*(1), 139–146. doi:10.1016/j.phytochem.2006.09.023.

Hnini, M., Rabeh, K., & Oubohssaine, M. (2024). Interactions between beneficial soil microorganisms (PGPR and AMF) and host plants for environmental restoration: A systematic review. *Plant Stress, 11*, 100391. doi:10.1016/j.stress.2024.100391.

Hodge, A., Campbell, C. D., & Fitter, A. H. (2001). An arbuscular mycorrhizal fungus accelerates decomposition and acquires nitrogen directly from organic material. *Nature, 413*(6853), 297–299. doi:10.1038/35095041.

Hornstein, E. D., & Sederoff, H. (2024). Back to the future: Re-engineering the evolutionarily lost arbuscular mycorrhiza host trait to improve climate resilience for agriculture. *Critical Reviews in Plant Sciences, 43*(1), 1–33.

Hu, X., Chen, D., Yan, F., Zheng, X., Fang, X., Bai, Y., Zhao, J., Ma, X., Ma, C., Cai, X., Deng, D., Sun, G., Sun, F., Zhou, J., & Liu, L. (2024). Global research trends on the effects of arbuscular mycorrhizal fungi on the soil carbon cycle: A bibliometric analysis. *Ecological Indicators, 158*, 111543. doi:10.1016/j.ecolind.2023.111543.

Hyde, K. D., Baldrian, P., Chen, Y., Thilini Chethana, K. W., De Hoog, S., Doilom, M., de Farias, A. R. G., Gonçalves, M. F. M., Gonkhom, D., Gui, H., Hilário, S., Hu, Y., Jayawardena, R. S., Khyaju, S., Kirk, P. M., Kohout, P., Luangharn, T., Maharachchikumbura, S. S. N., Manawasinghe, I. S., … Walker, A. (2024). Current trends, limitations and future research in the fungi? *Fungal Diversity, 125*(1), 1–71. https://www.springer.com/journal/13225. doi:10.1007/s13225-023-00532-5.

IJdo, M., Cranenbrouck, S., & Declerck, S. (2011). Methods for large-scale production of AM fungi: Past, present, and future. *Mycorrhiza, 21*(1), 1–16. doi:10.1007/s00572-010-0337-z.

Jarosz, Z., Michałojć, Z., Pitura, K., Dzida, K., & Koter, M. (2021). Influence of fertilization and mycorrhizae on the nutritional status of rhododendron (*Rhododendron hybridum*) in a nursery. *Agriculture, 11*(6), 538. doi:10.3390/agriculture11060538.

Javot, H., Penmetsa, R. V., Terzaghi, N., Cook, D. R., & Harrison, M. J. (2007). A *Medicago truncatula* phosphate transporter indispensable for the arbuscular mycorrhizal symbiosis. *Proceedings of the National Academy of Sciences of the United States of America, 104*(5), 1720–1725. doi:10.1073/pnas.0608136104.

Jefwa, J. M., Kahangi, E., Losenge, T., Mung'atu, J., Ngului, W., Ichami, S. M., Sanginga, N., & Vanluawe, B. (2012). Arbuscular mycorrhizal fungi in the rhizosphere of banana and plantain and the growth of tissue culture cultivars. *Agriculture, Ecosystems and Environment, 157*, 24–31. doi:10.1016/j.agee.2012.03.014.

Jha, S. S., & Songachan, L. S. (2023). The usage of arbuscular mycorrhizal fungi (AMF) as a biofertilizer. *Research Square*. https://www.researchsquare.com/browse. doi:10.21203/rs.3.rs-2559546/v1.

Ji, L., Yang, Y., Yang, N., Khan, A., & Yang, L. (2021). Seasonal variation of diversity and co-occurrence patterns of arbuscular mycorrhizal fungal communities in mixed broadleaf-conifer forests. *Applied Soil Ecology, 158*, 103782. doi:10.1016/j.apsoil.2020.103782.

Jindo, K., Audette, Y., Olivares, F. L., Canellas, L. P., Smith, D. S., & Voroney, R. P. (2023). Biotic and abiotic effects of soil organic matter on the phytoavailable phosphorus in soils: A review. *Chemical and Biological Technologies in Agriculture, 10*(1). https://chembioagro.springeropen.com. doi:10.1186/s40538-023-00401-y.

Kalamulla, R., Karunarathna, S. C., Tibpromma, S., Galappaththi, M. C. A., Suwannarach, N., Stephenson, S. L., Asad, S., Salem, Z. S., & Yapa, N. (2021). Arbuscular mycorrhizal fungi in sustainable agriculture. *Sustainability, 14*, 12250. doi:10.3390/su141912250.

Kebu, Z., Gure, A., & Jilo Molole, G. (2024). Total phenolic and flavonoid content, lipophilic components, and antioxidant activities of *Capsicum annuum* varieties grown in Omo Nada, Jimma, Ethiopia. *Natural Product Communications, 19*(12). doi:10.1177/1934578x241306244.

Khoso, A. G., Li, Y., & Liu, D. (2025). The arbuscular mycorrhizal fungus *Claroideoglomus etunicatum* (Glomerales: Claroideoglomeraceae) inoculated wheat plants mediated responses of *Sitobion avenae* (Fabricius), (Hemiptera: Aphididae) to water deficit. *Journal of Economic Entomology, 118*(1), 203–218. https://academic.oup.com/jee/issue. doi:10.1093/jee/toae293.

Koller, R., Rodriguez, A., Robin, C., Scheu, S., & Bonkowski, M. (2013). Protozoa enhance foraging efficiency of arbuscular mycorrhizal fungi for mineral nitrogen from organic matter in soil to the benefit of host plants. *New Phytologist, 199*(1), 203–211. doi:10.1111/nph.12249.

Kuila, D., & Ghosh, S. (2022). Aspects, problems and utilization of arbuscular mycorrhizal (AM) application as bio-fertilizer in sustainable agriculture. *Current Research in Microbial Sciences, 3*, 100107. doi:10.1016/j.crmicr.2022.100107.

Labidi, S., Jeddi, F. B., Tisserant, B., Yousfi, M., Sanaa, M., Dalpé, Y., & Sahraoui, A. L. H. (2015). Field application of mycorrhizal bio-inoculants affects the mineral uptake of a forage legume (*Hedysarum coronarium* L.) on a highly calcareous soil. *Mycorrhiza, 25*(4), 297–309. link.springer.de/link/service/journals/00572/index.htm. doi:10.1007/s00572-014-0609-0.

Lahlali, R., Ibrahim, D. S. S., Belabess, Z., Roni, M. Z. K., Radouane, N., Vicente, C. S. L., Menéndez, E., Mokrini, F., Barka, E. A, de Melo e Mota, M. G., & Peng, G. (2021). High-throughput molecular technologies for unraveling the mystery of soil microbial community: Challenges and future prospects. *Heliyon, 7*(10), e08142. doi:10.1016/j.heliyon.2021.e08142.

Lehmann, A., Veresoglou, S. D., Leifheit, E. F., & Rillig, M. C. (2014). Arbuscular mycorrhizal influence on zinc nutrition in crop plants—A meta-analysis. *Soil Biology and Biochemistry, 69*, 123–131. www.elsevier.com/inca/publications/store/3/3/2. doi:10.1016/j.soilbio.2013.11.001.

Leifheit, E. F., Verbruggen, E., & Rillig, M. C. (2015). Arbuscular mycorrhizal fungi reduce decomposition of woody plant litter while increasing soil aggregation. *Soil Biology and Biochemistry, 81*, 323–328. doi:10.1016/j.soilbio.2014.12.003.

Li, Z. F., Lü, P. P., Wang, Y. L., Yao, H., Maitra, P., Sun, X., Zheng, Y., & Guo, L. D. (2020). Response of arbuscular mycorrhizal fungal community in soil and roots to grazing differs in a wetland on the Qinghai-Tibet plateau. *PeerJ, 2020*(6). https://peerj.com/archives/?journal=peerj. doi:10.7717/peerj.9375.

Lin, K., Limpens, E., Zhang, Z., Ivanov, S., Saunders, D. G. O., Mu, D., Pang, E., Cao, H., Cha, H., Lin, T., Zhou, Q., Shang, Y., Li, Y., Sharma, T., van Velzen, R., de Ruijter, N., Aanen, D. K., Win, J., Kamoun, S., … Butler, G. (2014). Single nucleus genome sequencing reveals high similarity among nuclei of an endomycorrhizal fungus. *PLoS Genetics, 10*(1), e1004078. doi:10.1371/journal.pgen.1004078.

Lingua, G., Bona, E., Manassero, P., Marsano, F., Todeschini, V., Cantamessa, S., Copetta, A., D'Agostino, G., Gamalero, E., & Berta, G. (2013). Arbuscular mycorrhizal fungi and plant growth-promoting pseudomonads increases anthocyanin concentration in strawberry fruits (*Fragaria x ananassa* var. Selva) in conditions of reduced fertilization. *International Journal of Molecular Sciences, 14*(8), 16207–16225. doi:10.3390/ijms140816207.

Liu, C. Y., Guo, X. N., Dai, F. J., & Wu, Q. S. (2024). Mycorrhizal symbiosis enhances P uptake and indole-3-acetic acid accumulation to improve root morphology in different citrus genotypes. *Horticulturae, 10*(4). www.mdpi.com/journal/horticulturae. doi:10.3390/horticulturae10040339.

Liu, R. C., Xiao, Z. Y., Hashem, A., Abd_Allah, E. F., Xu, Y. J., & Wu, Q. S. (2021). Unraveling the interaction between arbuscular mycorrhizal fungi and *Camellia* plants. *Horticulturae, 7*(9), 322.

Liu, W., Jiang, S., Zhang, Y., Yue, S., Christie, P., Murray, P. J., Li, X., & Zhang, J. (2014). Spatiotemporal changes in arbuscular mycorrhizal fungal communities under different nitrogen inputs over a 5-year period in intensive agricultural ecosystems on the North China Plain. *FEMS Microbiology Ecology, 90*(2), 436–453. http://femsec.oxfordjournals.org/. doi:10.1111/1574-6941.12405.

Lugo, M. A., Ontivero, R. E., Iriarte, H. J., Yelikbayev, B., & Pagano, M. C. (2023). The diversity of arbuscular mycorrhizal fungi and their associations in South America: A case study of Argentinean and Brazilian cattle raising productive ecosystems: A review. *Diversity, 15*(9). http://www.mdpi.com/journal/diversity/. doi:10.3390/d15091006.

Maiti, D., Singh, R. K., & Variar, M. (2012). Rice-based crop rotation for enhancing native arbuscular mycorrhizal (AM) activity to improve phosphorus nutrition of upland rice (*Oryza sativa* L.). *Biology and Fertility of Soils, 48*(1), 67–73. doi:10.1007/s00374-011-0609-6.

Malembaka, R., Onwonga, R., Jefwa, J., Ayuke, F., & Nabahungu, L. (2021). Role of native arbuscular mycorrhizal fungi on maize (*Zea mays*) growth and nutrient uptake in acidic soils under controlled conditions. *International Journal of Agronomy and Agricultural Research, 18*, 20–32.

Martin, F. M., & van der Heijden, M. G. A. (2024). The mycorrhizal symbiosis: Research frontiers in genomics, ecology, and agricultural application. *New Phytologist, 242*(4), 1486–1506. http://onlinelibrary.wiley.com/journal/10.1111/(ISSN)1469-8137. doi:10.1111/nph.19541.

Mathimaran, N., Jegan, S., Thimmegowda, M. N., Prabavathy, V. R., Yuvaraj, P., Kathiravan, R., Sivakumar, M. N., Manjunatha, B. N., Bhavitha, N. C., Sathish, A., Shashidhar, G. C., Bagyaraj, D. J., Ashok, E. G., Singh, D., Kahmen, A., Boller, T., & Mäder, P. (2020). Intercropping transplanted pigeon pea with finger millet: Arbuscular mycorrhizal fungi and plant growth promoting rhizobacteria boost yield while reducing fertilizer input. *Frontiers in Sustainable Food Systems, 4*. www.frontiersin.org/journals/sustainable-food-systems#. doi:10.3389/fsufs.2020.00088.

Meena, R., Kumar, S., Datta, R., Lal, R., Vijayakumar, V., Brtnicky, M., Sharma, M., Yadav, G., Jhariya, M., Jangir, C., Pathan, S., Dokulilova, T., Pecina, V., & Marfo, T. (2020). Impact of agrochemicals on soil microbiota and management: A review. *Land, 9*(2), 34. doi:10.3390/land9020034.

Meenakshisundaram, M. (2011). Studies on association of arbuscular mycorrhizal fungi with *Gluconacetobacter diazotrophicus* and its effect on improvement of *Sorghum bicolor* (L.). *International Journal of Current Scientific Research (IJCSR), 1*(2), 23–30.

Menge, E. M. (2023). Understanding the mechanisms of nutrient transfer between arbuscular mycorrhizal fungi (AMF) and host plants. *International Journal of Science and Research Archive, 10*(2), 557–567. doi:10.30574/ijsra.2023.10.2.1013.

Merino-Martín, L., Stokes, A., Gweon, H. S., Moragues-Saitua, L., Staunton, S., Plassard, C., Oliver, A., Bissonnais, Y. L., & Griffiths, R. I. (2021). Interacting effects of land use type, microbes and plant traits on soil aggregate stability. *Soil Biology and Biochemistry, 154*. www.elsevier.com/inca/publications/store/3/3/2. doi:10.1016/j.soilbio.2020.108072.

Montes-Borrego, M., Metsis, M., & Landa, B. B. (2014). Arbuscular mycorhizal fungi associated with the olive crop across the Andalusian landscape: Factors driving community differentiation. *PLoS One, 9*(5). http://www.plosone.org/article/fetchObject.action?uri=info%3Adoi%2F10.1371%2Fjournal.pone.0096397&representation=PDF. doi:10.1371/journal.pone.0096397.

Moosavi, M., Khorassani, R., & Tavakkol Afshari, R. (2024). Strigolactone (GR24) and abscisic acid induced drought tolerance in wheat by ameliorating nutrient uptake. *Journal of Plant Growth Regulation*, 1–16. doi:10.1007/s00344-024-11223-w.

Mwafulirwa, L., Baggs, E. M., Russell, J., Hackett, C. A., Morley, N., de la Fuente Cantó, C., & Paterson, E. (2021). Identification of barley genetic regions influencing plant–microbe interactions and carbon cycling in soil. *Plant and Soil, 468*(1–2), 165–182. http://www.wkap.nl/journalhome.htm/0032-079X. doi:10.1007/s11104-021-05113-6.

Mwampashi, L. L., Magubika, A. J., Ringo, J. F., Theonest, D. J., Tryphone, G. M., Chilagane, L. A., & Nassary, E. K. (2024). Exploring agro-ecological significance, knowledge gaps, and research priorities in arbuscular mycorrhizal fungi. *Front. Microbiol. 15*, 1491861. doi:10.3389/fmicb.2024.1491861.

Nagahashi, G., Douds, D. D., & Ferhatoglu, Y. (2010). Functional categories of root exudate compounds and their relevance to AM fungal growth. In H. Koltai, & Y. Kapulnik (Eds.), *Arbuscular mycorrhizas: Physiology and function* (pp. 33–56). Springer.

Navazio, L., Moscatiello, R., Genre, A., Novero, M., Baldan, B., Bonfante, P., & Mariani, P. (2007). A diffusible signal from arbuscular mycorrhizal fungi elicits a transient cytosolic calcium elevation in host plant cells. *Plant Physiology, 144*(2), 673–681. doi:10.1104/pp.106.086959.

Nazir, A., Sarfraz, W., Allah, D., Khalid, N., Farid, M., Shafiq, M., Firdaus-E-Bareen, Rizvi, Z. F, & Naeem, N. (2023). Synergistic impact of two autochthonous saprobic fungi (*A. niger* and *T. pseudokoningii*) on the growth, ionic contents, and metals uptake in *Brassica juncea* L. and *Vigna radiata* L. under tannery solid waste contaminated soil. *International Journal of Phytoremediation, 25*(11), 1488–1500. doi:10.1080/15226514.2023.2166457.

Nouri, E., Breuillin-Sessoms, F., Feller, U., Reinhardt, D., & Dutilh, B. E. (2014). Phosphorus and nitrogen regulate arbuscular mycorrhizal symbiosis in *Petunia hybrida. PLoS One, 9*(3), e90841. doi:10.1371/journal.pone.0090841.

Olsson, P. A., Hammer, E. C., Wallander, H., & Pallon, J. (2008). Phosphorus availability influences elemental uptake in the mycorrhizal fungus *Glomus intraradices*, as revealed by particle-induced X-ray emission analysis. *Applied and Environmental Microbiology, 74*(13), 4144–4148. doi:10.1128/AEM.00376-08.

Öpik, M., Vanatoa, A., Vanatoa, E., Moora, M., Davison, J., Kalwij, J. M., Reier, Ü., & Zobel, M. (2010). The online database Maarj AM reveals global and ecosystemic distribution patterns in arbuscular mycorrhizal fungi (*Glomeromycota*). *New Phytologist, 188*(1), 223–241. doi:10.1111/j.1469-8137.2010.03334.x.

Ortas, I. (2012). The effect of mycorrhizal fungal inoculation on plant yield, nutrient uptake and inoculation effectiveness under long-term field conditions. *Field Crops Research, 125*, 35–48. doi:10.1016/j.fcr.2011.08.005.

Padmavathi, T., Dikshit, R., & Seshagiri, S. (2015). Effect of *Rhizophagus* spp. and plant growth-promoting *Acinetobacter junii* on *Solanum lycopersicum* and *Capsicum annuum. Brazilian Journal of Botany, 38*(2), 273–280. doi:10.1007/s40415-015-0144-z.

Pal, S., Singh, H., Farooqui, A., & Rakshit, A. (2015). Fungal biofertilizers in Indian agriculture: Perception, demand and promotion. *Journal of Eco-Friendly Agriculture, 10*(2), 101–113.

Pellegrino, E., & Bedini, S. (2014). Enhancing ecosystem services in sustainable agriculture: Biofertilization and biofortification of chickpea (*Cicer arietinum* L.) by arbuscular mycorrhizal fungi. *Soil Biology and Biochemistry, 68*, 429–439. doi:10.1016/j.soilbio.2013.09.030.

Pepe, A., Giovannetti, M., & Sbrana, C. (2016). Different levels of hyphal self-incompatibility modulate interconnectedness of mycorrhizal networks in three arbuscular mycorrhizal fungi within the Glomeraceae. *Mycorrhiza, 26*(4), 325–332. doi:10.1007/s00572-015-0671-2.

Prasad, K. (2024). Occurrence and interactions of arbuscular mycorrhizal fungi (*Rhizophagus fasciculatus*) and rhizospheric fungi in *Saccharum officinarum* L. *Tropical Plants, 3*(1). doi:10.48130/tp-0024-0035.

Püschel, D., Bitterlich, M., Rydlová, J., & Jansa, J. (2020). Facilitation of plant water uptake by an arbuscular mycorrhizal fungus: A Gordian knot of roots and hyphae. *Mycorrhiza, 30*(2–3), 299–313. doi:10.1007/s00572-020-00949-9.

Qi, S., Wang, J., Wan, L., Dai, Z., da Silva Matos, D. M., Du, D., Egan, S., Bonser, S. P., Thomas, T., & Moles, A. T. (2022). Arbuscular mycorrhizal fungi contribute to phosphorous uptake and allocation strategies of *Solidago canadensis* in a phosphorous-deficient environment. *Frontiers in Plant Science, 13*. https://www.frontiersin.org/journals/plant-science. doi:10.3389/fpls.2022.831654.

Ramasamy, K., Joe, M. M, Kim, K.-Y., Lee, S.-M., Shagol, C., Rangasamy, A., Chung, J.-B., Islam, M. R., & Sa, T.-M. (2011). Synergistic effects of arbuscular mycorrhizal fungi and plant growth promoting rhizobacteria for sustainable agricultural production. *Korean Journal of Soil Science and Fertilizer, 44*(4), 637–649. doi:10.7745/kjssf.2011.44.4.637.

Reinhart, K. O., Vermeire, L. T., Penn, C. J., & Lekberg, Y. (2024). Experimental evidence that poor soil phosphorus (P) solubility typical of drylands due to calcium co-precipitation favors autonomous plant P acquisition over collaboration with mycorrhizal fungi. *Soil Biology and Biochemistry, 199*. https://www.sciencedirect.com/science/journal/00380717. doi:10.1016/j.soilbio.2024.109605.

Rillig, M. C., Lehmann, A., Orr, J. A., & Rongstock, R. (2024). Factors of global change affecting plants act at different levels of the ecological hierarchy. *Plant Journal, 117*(6), 1781–1785. http://onlinelibrary.wiley.com/journal/10.1111/(ISSN)1365-313X. doi:10.1111/tpj.16509.

Ruiz-Sánchez, M., Aroca, R., Muñoz, Y., Polón, R., & Ruiz-Lozano, J. M. (2010). The arbuscular mycorrhizal symbiosis enhances the photosynthetic efficiency and the antioxidative response of rice plants subjected to drought stress. *Journal of Plant Physiology, 167*(11), 862–869. doi:10.1016/j.jplph.2010.01.018.

Saeed, H. S., Munir, M., Samra, S., Farooq, A., Mahmood, H., Ahmad, I., Akram, M. I., Iqbal, R., Elsadek, M. F., Almutairi, S. M., Ditta, A., & Daraz, U. (2025). Screening of biosurfactants producing fungi from infected *Citrus sinensis* (L.). *Polish Journal of Environmental Studies, 34*(5), 5261–5270. https://www.pjoes.com/pdf-190588-115894?filename=Screening%20of.pdf. doi:10.15244/pjoes/190588.

Sagar, A., Rathore, P., Ramteke, P. W., Ramakrishna, W., Reddy, M. S., & Pecoraro, L. (2021). Plant growth promoting rhizobacteria, arbuscular mycorrhizal fungi and their synergistic interactions to counteract the negative effects of saline soil on agriculture: Key macromolecules and mechanisms. *Microorganisms, 9*(7). https://www.mdpi.com/2076-2607/9/7/1491/pdf. doi:10.3390/microorganisms9071491.

Säle, V., Aguilera, P., Laczko, E., Mäder, P., Berner, A., Zihlmann, U., van der Heijden, M. G. A., & Oehl, F. (2015). Impact of conservation tillage and organic farming on the diversity of arbuscular mycorrhizal fungi. *Soil Biology and Biochemistry, 84*, 38–52. www.elsevier.com/inca/publications/store/3/3/2. doi:10.1016/j.soilbio.2015.02.005.

Sarah, S., & Ibrar, M. (2016). Effects of arbuscular mycorrhizal fungi on spores density and root colonization of four hybrids of sunflower (*Helianthus annuus* L.) at different rock phosphate levels. *Sarhad Journal of Agriculture, 32*(4), 258–266.

Sarwar, Y., & Shahbaz, M. (2020). Modulation in growth, photosynthetic pigments, gas exchange attributes and inorganic ions in sunflower (*Helianthus annuus* L.) by strigolactones (GR24) achene priming under saline conditions. *Pakistan Journal of Botany, 52*(1), 23–31.

Sato, T., Ezawa, T., Cheng, W., & Tawaraya, K. (2015). Release of acid phosphatase from extraradical hyphae of arbuscular mycorrhizal fungus *Rhizophagus clarus*. *Soil Science and Plant Nutrition, 61*(2), 269–274. http://www.tandfonline.com/toc/tssp20/current. doi:10.1080/00380768.2014.993298.

Sato, T., Hachiya, S., Inamura, N., Ezawa, T., Cheng, W., & Tawaraya, K. (2019). Secretion of acid phosphatase from extraradical hyphae of the arbuscular mycorrhizal fungus *Rhizophagus clarus* is regulated in response to phosphate availability. *Mycorrhiza, 29*(6), 599–605. doi:10.1007/s00572-019-00923-0.

Savastano, N., & Bais, H. (2024). Synergism or antagonism: Do arbuscular mycorrhizal fungi and plant growth-promoting rhizobacteria work together to benefit plants? *International Journal of Plant Biology, 15*(4), 944–958. doi:10.3390/ijpb15040067.

Sawers, R. J. H., Svane, S. F., Quan, C., Grønlund, M., Wozniak, B., Gebreselassie, M. N., González-Muñoz, E., Chávez Montes, R. A., Baxter, I., Goudet, J., Jakobsen, I., & Paszkowski, U. (2017). Phosphorus acquisition efficiency in arbuscular mycorrhizal maize is correlated with the abundance of root-external hyphae and the accumulation of transcripts encoding PHT1 phosphate transporters. *New Phytologist, 214*(2), 632–643. http://www.wiley.com/bw/editors.asp?ref=0028-646X&site=1. doi:10.1111/nph.14403.

Schneider, K. D., Thiessen Martens, J. R., Zvomuya, F., Keith Reid, D., Fraser, T. D., Lynch, D. H., O'Halloran, I. P., & Wilson, H. F. (2019). Options for improved phosphorus cycling and use in agriculture at the field and regional scales. *Journal of Environmental Quality, 48*, 1247–1264. https://www.agronomy.org/publications/jeq48. doi:10.2134/jeq2019.02.0070.

Sharma, M. P., Reddy, U. G., & Adholeya, A. (2011). Response of arbuscular mycorrhizal fungi on Wheat (*Triticum aestivum* L.) grown conventionally and on beds in a sandy loam soil. *Indian Journal of Microbiology, 51*(3), 384–389. doi:10.1007/s12088-011-0134-1.

Shen, Y., & Duan, T. (2024). The interaction between arbuscular mycorrhizal fungi (AMF) and grass endophyte (Epichloë) on host plants: A review. *Journal of Fungi, 10*(3), 174. doi:10.3390/jof10030174.

Shi, J., Zhao, B., Zheng, S., Zhang, X., Wang, X., Dong, W., Xie, Q., Wang, G., Xiao, Y., Chen, F., Yu, N., & Wang, E. (2021). A phosphate starvation response-centered network regulates mycorrhizal symbiosis. *Cell, 184*(22), 5527. doi:10.1016/j.cell.2021.09.030.

Singh, B., & Singh, V. K. (2017). Fertilizer management in rice. In *Rice production worldwide* (pp. 217–253). Springer International Publishing. http://www.springer.com/in/book/9783319475141. doi:10.1007/978-3-319-47516-5_10.

Siqueira, J. O., Safir, G. R., & Nair, M. G. (1991). Stimulation of vesicular-arbuscular mycorrhiza formation and growth of white clover by flavonoid compounds. *New Phytologist, 118*(1), 87–93. doi:10.1111/j.1469-8137.1991.tb00568.x.

Siyao, C., Jing, K. T., Yangle, S., Yee, H. C., Zihao, D., Hao, L., Hui, S., & Yao, L. (2024). The integration of Internet of Things (IoT) technology into fire evacuation plan in high-rise buildings. *Journal of Advanced Research in Applied Sciences and Engineering Technology, 42*(1), 239–253. https://semarakilmu.com.my/journals/index.php/applied_sciences_eng_tech/article/download/4289/4321. doi:10.37934/araset.42.1.239253.

Steinkellner, S., Lendzemo, V., Langer, I., Schweiger, P., Khaosaad, T., Toussaint, J. P., & Vierheilig, H. (2007). Flavonoids and strigolactones in root exudates as signals in symbiotic and pathogenic plant-fungus interactions. *Molecules, 12*(7), 1290–1306. http://www.mdpi.org/molecules/papers/12071290.pdf. doi:10.3390/12071290Austria.

Stojanović, M., Savić, S., Delcourt, A., Hilbert, J.-L., Hance, P., Maksimović, J. D., & Maksimović, V. (2023). Phenolics and sesquiterpene lactones profile of red and green lettuce: Combined effect of cultivar, microbiological fertiliser, and season. *Plants, 12*(14), 2616. doi:10.3390/plants12142616.

Sun, Y., Wang, M., Mur, L. A. J., Shen, Q., & Guo, S. (2021). The cross-kingdom roles of mineral nutrient transporters in plant-microbe relations. *Physiologia Plantarum, 171*(4), 771–784. http://onlinelibrary.wiley.com/journal/10.1111/(ISSN)1399-3054. doi:10.1111/ppl.13318.

Suriyagoda, L. B. D. (2022). Rice production in nutrient-limited soils: Strategies for improving crop productivity and land sustainability. *Journal of the National Science Foundation of Sri Lanka, 50*(3), 521. doi:10.4038/jnsfsr.v50i3.10601.

Tamayo, E., Gómez-Gallego, T., Azcón-Aguilar, C., & Ferrol, N. (2014). Genome-wide analysis of copper, iron and zinc transporters in the arbuscular mycorrhizal fungus *Rhizophagus irregularis*. *Frontiers in Plant Science, 5*. http://journal.frontiersin.org/Journal/10.3389/fpls.2014.00547/pdf. doi:10.3389/fpls.2014.00547.

Tchabi, A., Coyne, D., Hountondji, F., Lawouin, L., Wiemken, A., & Oehl, F. (2010). Efficacy of indigenous arbuscular mycorrhizal fungi for promoting white yam (*Dioscorea rotundata*) growth in West Africa. *Applied Soil Ecology, 45*(2), 92–100. doi:10.1016/j.apsoil.2010.03.001.

Thind, S., Chaudhary, M. S., Ditta, A., Hussain, I., Parveen, A., Ullah, N., Mahmood, Q., Al-Ashkar, I., & El-Sabagh, A. (2022). Impact of mycorrhizal fungi from different rhizospheric soils on fungal colonization, growth, and chlorophyll contents of *Cenchrus ciliaris*. *Agronomy, 12*(11), 2644. doi:10.3390/agronomy12112644.

Thirkell, T. J., Charters, M. D., Elliott, A. J., Sait, S. M., & Field, K. J. (2017). Are mycorrhizal fungi our sustainable saviours? Considerations for achieving food security. *Journal of Ecology, 105*(4), 921–929. http://www.blackwellpublishing.com/journals/JEC. doi:10.1111/1365-2745.12788.

Tian, L., Wang, T., Cui, S., Li, Y., Gui, W., Yang, F., Chen, J., Dong, R., Gu, X., Zhao, X., Zhang, M., Chen, C., & Li, Z. (2024). Diversified cover crops and no-till enhanced soil total nitrogen and arbuscular mycorrhizal fungi diversity: A case study from the Karst Area of Southwest China. *Agriculture, 14*(7), 1103. doi:10.3390/agriculture14071103.

Tisserant, E., Malbreil, M., Kuo, A., Kohler, A., Symeonidi, A., Balestrini, R., Charron, P., Duensing, N., Frei Dit Frey, N., Gianinazzi-Pearson, V., Gilbert, L. B., Handa, Y., Herr, J. R., Hijri, M., Koul, R., Kawaguchi, M., Krajinski, F., Lammers, P. J., Masclaux, F. G., … Martin, F. (2013). Genome of an arbuscular mycorrhizal fungus provides insight into the oldest plant symbiosis. *Proceedings of the National Academy of Sciences of the United States of America, 110*(50), 20117–20122. http://www.pnas.org/content/110/50/20117.full.pdf+html. doi:10.1073/pnas.1313452110.

Todeschini, V., AitLahmidi, N., Mazzucco, E., Marsano, F., Gosetti, F., Robotti, E., Bona, E., Massa, N., Bonneau, L., Marengo, E., Wipf, D., Berta, G., & Lingua, G. (2018). Impact of beneficial microorganisms on strawberry growth, fruit production, nutritional quality, and volatilome. *Frontiers in Plant Science, 9*. doi:10.3389/fpls.2018.01611.

Tsai, S. M., & Phillips, D. A. (1991). Flavonoids released naturally from alfalfa promote development of symbiotic *Glomus* spores *in vitro*. *Applied and Environmental Microbiology, 57*, 1485–1488.

Ullah, N., Ditta, A., Khalid, A., Mehmood, S., Rizwan, M. S., Ashraf, M., Mubeen, F., Imtiaz, M., & Iqbal, M. M. (2020). Integrated effect of algal biochar and plant growth promoting rhizobacteria on physiology and growth of maize under deficit irrigations. *Journal of Soil Science and Plant Nutrition, 20*(2), 346–356. https://www.springer.com/life+sciences/plant+sciences/journal/42729. doi:10.1007/s42729-019-00112-0.

Urgiles-Gómez, N., Avila-Salem, M. E., Loján, P., Encalada, M., Hurtado, L., Araujo, S., Collahuazo, Y., Guachanamá, J., Poma, N., Granda, K., Robles, A., Senés, C., & Cornejo, P. (2021). Plant growth-promoting microorganisms in coffee production: From isolation to field application. *Agronomy, 11*(8), 1531. doi:10.3390/agronomy11081531.

van der Heijden, M. G. A., Martin, F. M., Selosse, M. A., & Sanders, I. R. (2015). Mycorrhizal ecology and evolution: The past, the present, and the future. *New Phytologist, 205*(4), 1406–1423. http://onlinelibrary.wiley.com/journal/10.1111/(ISSN)1469-8137. doi:10.1111/nph.13288.

Vieira, L. C., Silva, D. K. A., Escobar, I. E. C., Silva, J. M., Moura, I. A., Oehl, F., & da Silva, G. A. (2020). Changes in an arbuscular mycorrhizal fungi community along an environmental gradient. *Plants, 9*(1), 52. doi:10.3390/plants9010052.

Volpe, V., Giovannetti, M., Sun, X. G., Fiorilli, V., & Bonfante, P. (2015). The phosphate transporters *LjPT4* and *MtPT4* mediate early root responses to phosphate status in non-mycorrhizal roots. *Plant, Cell & Environment*. doi:10.1111/pce.12659.

Wagg, C., Barendregt, C., Jansa, J., & van der Heijden, M. G. A. (2015). Complementarity in both plant and mycorrhizal fungal communities are not necessarily increased by diversity in the other. *Journal of Ecology, 103*(5), 1233–1244. http://onlinelibrary.wiley.com/journal/10.1111/(ISSN)1365-2745. doi:10.1111/1365-2745.12452. Switzerland.

Wahab, A., Batool, F., Muhammad, M., Zaman, W., Mikhlef, R. M., Qaddoori, S. M., Ullah, S., Abdi, G., & Saqib, S. (2024). Unveiling the complex molecular dynamics of arbuscular mycorrhizae: A comprehensive exploration and future perspectives in harnessing phosphate-solubilizing microorganisms for sustainable progress. *Environmental and Experimental Botany, 219*. https://www.sciencedirect.com/science/journal/00988472. doi:10.1016/j.envexpbot.2023.105633.

Wahab, A., Muhammad, M., Munir, A., Abdi, G., Zaman, W., Ayaz, A., Khizar, C., & Reddy, S. P. P. (2023). Role of arbuscular mycorrhizal fungi in regulating growth, enhancing productivity, and potentially influencing ecosystems under abiotic and biotic stresses. *Plants, 12*, 3102. doi:10.3390/plants12173102.

Walder, F., & van der Heijden, M. G. A. (2015). Regulation of resource exchange in the arbuscular mycorrhizal symbiosis. *Nature Plants, 1*, 15159. doi:10.1038/nplants.2015.159.

Wang, Q., Liu, M., Wang, Z., Li, J., Liu, K., & Huang, D. (2024). The role of arbuscular mycorrhizal symbiosis in plant abiotic stress. *Frontiers in Microbiology, 14*, 1323881. doi:10.3389/fmicb.2023.1323881.

Wang, Y., Zou, Y., Shu, B., & Wu, Q. (2023). Deciphering molecular mechanisms regarding enhanced drought tolerance in plants by arbuscular mycorrhizal fungi. *Scientia Horticulturae, 308*, 111591. doi:10.1016/j.scienta.2022.111591.

Watanarojanaporn, N., Boonkerd, N., Tittabutr, P., Longtonglang, A., Young, J. P. W., & Teaumroong, N. (2013). Effect of rice cultivation systems on indigenous arbuscular mycorrhizal fungal community structure. *Microbes and Environments, 28*(3), 316–324. https://www.jstage.jst.go.jp/article/jsme2/28/3/28_ME13011/_pdf. doi:10.1264/jsme2.ME13011Thailand.

Xie, X., Lai, W., Che, X., Wang, S., Ren, Y., Hu, W., Chen, H., & Tang, M. (2022). A SPX domain-containing phosphate transporter from *Rhizophagus irregularis* handles phosphate homeostasis at symbiotic interface of arbuscular mycorrhizas. *New Phytologist, 234*(2), 650–671. doi:10.1111/nph.17973.

Xu, G.-h., Chague, V., Melamed-Bessudo, C., Kapulnik, Y., Jain, A., Raghothama, K. G., Levy, A. A., & Silber, A. (2007). Functional characterization of LePT4: A phosphate transporter in tomato with mycorrhiza-enhanced expression. *Journal of Experimental Botany, 58*(10), 2491–2501. doi:10.1093/jxb/erm096.

Yang, S., Li, F., Malhi, S. S., Wang, P., Suo, D., & Wang, J. (2004). Long-term fertilization effects on crop yield and nitrate nitrogen accumulation in soil in northwestern China. *Agronomy Journal, 96*(4), 1039–1049. https://acsess.onlinelibrary.wiley.com/journal/14350645. doi:10.2134/agronj2004.1039.

Yizhu, L., Imtiaz, M., Ditta, A., Rizwan, M. S., Ashraf, M., Mehmood, S., Aziz, O., Mubeen, F., Ali, M., Elahi, N. N., Ijaz, R., Lele, S., Shuang, C., & Tu, S. (2020). Response of growth, antioxidant enzymes and root exudates production towards as stress in *Pteris vittata* and in *Astragalus sinicus* colonized by arbuscular mycorrhizal fungi. *Environmental Science and Pollution Research, 27*(2), 2340–2352. https://link.springer.com/journal/11356. doi:10.1007/s11356-019-06785-5.

Yoneyama, K., Arakawa, R., Ishimoto, K., Kim, H. I., Kisugi, T., Xie, X., Nomura, T., Kanampiu, F., Yokota, T., Ezawa, T., & Yoneyama, K. (2015). Difference in striga-susceptibility is reflected in strigolactone secretion profile, but not in compatibility and host preference in arbuscular mycorrhizal symbiosis in two maize cultivars. *New Phytologist, 206*(3), 983–989. http://onlinelibrary.wiley.com/journal/10.1111/(ISSN)1469-8137. doi:10.1111/nph.13375Japan.

Yoneyama, K., Xie, X., Kusumoto, D., Sekimoto, H., Sugimoto, Y., Takeuchi, Y., & Yoneyama, K. (2007). Nitrogen deficiency as well as phosphorus deficiency in sorghum promotes the production and exudation of 5-deoxystrigol, the host recognition signal for arbuscular mycorrhizal fungi and root parasites. *Planta, 227*(1), 125–132. doi:10.1007/s00425-007-0600-5.

Young, A. D., & Gillung, J. P. (2020). Phylogenomics—Principles, opportunities and pitfalls of big-data phylogenetics. *Systematic Entomology, 45*(2), 225–247. http://onlinelibrary.wiley.com/journal/10.1111/(ISSN)1365-3113. doi:10.1111/syen.12406.

Yurkov, A. P., Kryukov, A. A., Gorbunova, A. O., Kudriashova, T. R., Kovalchuk, A. I., Gorenkova, A. I., Bogdanova, E. M., Laktionov, Y. V., Zhurbenko, P. M., Mikhaylova, Y. V., Puzanskiy, R. K., Bagrova, T. N., Yakhin, O. I., Rodionov, A. V., & Shishova, M. F. (2023). Diversity of arbuscular mycorrhizal fungi in distinct ecosystems of the North Caucasus, a temperate biodiversity hotspot. *Journal of Fungi, 10*, 11. doi:10.3390/jof10010011.

Zeng, W., Xiang, D., Li, X., Gao, Q., Chen, Y., Wang, K., Qian, Y., Wang, L., Li, J., Mi, Q., Huang, H., Xu, L., Zhao, M., Zhang, Y., & Xiang, H. (2024). Effects of combined inoculation of arbuscular mycorrhizal fungi and plant growth-promoting rhizosphere bacteria on seedling growth and rhizosphere microecology. *Frontiers in Microbiology, 15*. https://www.frontiersin.org/journals/microbiology#. doi:10.3389/fmicb.2024.1475485.

Chapter 20

Enhancing crop production through arbuscular mycorrhizal fungi

Muhammad Shehzad[a], Muhammad Mazhar Iqbal[b], Aqeel Ahmad[a], Khalid Mehmood[a], Nosheen Zahid[c], Mehdi Maqbool[c], Hafiz Muhammad Rashad Javeed[d], Khuram Mubeen[e], Muhammad Abdul Qayyum[f] and Tayyaba Naz[g]

[a]*Department of Agronomy, Faculty of Agriculture, University of Poonch, Rawalakot, Pakistan,* [b]*Department of Soil and Environmental Sciences, College of Agriculture, University of Sargodha, Sargodha, Pakistan,* [c]*Department of Horticulture, Faculty of Agriculture, University of Poonch, Rawalakot, Pakistan,* [d]*Department of Environmental Sciences, COMSATS University Islamabad, Vehari Campus, Vehari, Pakistan,* [e]*Department of Agronomy, Muhammad Nawaz Shareef University of Agriculture, Multan, Pakistan,* [f]*Department of Soil and Environmental Sciences, Ghazi University, Dera Ghazi Khan, Pakistan,* [g]*Institute of Soil and Environmental Sciences, University of Agriculture Faisalabad, Faisalabad, Pakistan*

20.1 Introduction

Agricultural production increased in the 20th century primarily because of advances in plant breeding and genetics, the development and application of pesticides, and the use of phosphorus- and nitrogen-based fertilizers. The rhizosphere is home to countless organisms that support plant growth, nutrient uptake, stress tolerance, disease prevention, carbon sequestration, and other aspects of the ecosystem. These microorganisms, which assist plants in absorbing nutrients through their roots, include actinomycetes, bacteria, and mycorrhizal fungi. Arbuscular mycorrhizal fungi (AMFs) are vital to natural ecosystems, but their role is declining because of the widespread use of chemical fertilizers in agriculture (Nazir et al., 2023; Saeed et al., 2025; Thind et al., 2022; Yizhu et al., 2020). Most terrestrial plants live in symbiosis with either AMFs or vesicular AMFs. This symbiotic relationship promotes the growth and development of the host plant because AMFs can absorb phosphorus (P) and other ecosystem functions mineral nutrients from the soil. These also contribute to increased plant diversity and enhanced disease resistance. This process is accomplished by AMF mycelia, which can grow both inside and outside the host root system. Targeted management of AMFs can improve the sustainability and profitability of agricultural systems. AMFs are particularly effective when nutrient availability is low, and nutrients are bound to organic matter and soil particles. Therefore these play an important role in sustainable agricultural systems (Devi et al., 2021). Agricultural productivity has steadily increased in temperate regions since the mid-20th century and, in recent years, has also steadily increased in tropical regions, largely because of improvements in varieties, equipment, fertilizers, and pesticides (Craswell & Karjalainen, 1990). Nevertheless, agricultural yields have stagnated recently, primarily because of soil degradation, which is sometimes allied with unmanageable farming practices (Bindraban et al., 2012; Gomiero, 2016; Grassini et al., 2013). Food production must increase dramatically this century, on ever-decreasing arable land, to sustainably meet future demand (Godfray et al., 2010). Numerous factors, such as global population growth, potential competition between food and nonfood production, the market value of agricultural goods in several countries, competition between agriculture and urbanization for land, and food shortages in developing countries, necessitate this increase in crop yields (Bruinsma, 2003). However, the methods producers can use to raise yields often lead to higher input charges, more fieldwork, and a greater risk of financial failure. At the same time, environmental contamination from the use of synthetic substances (such as pesticides, phosphates, and nitrates in waterways) is increasing and is sometimes irreversible. Furthermore, cultivation systems have become increasingly dependent on chemicals because of side effects on biogenesis and soil depletion (Clermont-Dauphin et al., 1997). Consequently the demand for high-quality food and clean agriculture is rising, and growing knowledge about

Advancing Sustainable Agriculture Through Arbuscular Mycorrhizal Fungi. DOI: https://doi.org/10.1016/B978-0-443-45130-0.00015-1

food production is ultimately contributing to a reduction in chemical use in industrialized countries (Boiffin et al., 2001). However, the high demand for food in developing countries indicates a trend toward intensification, primarily through increased fertilizer use (Norse, 2003), as plant nutrition has been crucial for the sharp rise in both food supply and demand. The use of phosphate (P) fertilizer has more than quadrupled, and that of nitrogen (N) fertilizer has almost tripled. These advances were made possible by the massive expansion of nitrogen and phosphorus fertilizer use and the introduction of highly productive agricultural systems at comparatively low costs (Schultz et al., 1995). However, the amplified use of manures and extremely productive systems has also led to ecological glitches such as air pollution, declining biodiversity, and impaired ecosystem functions, as well as a deterioration of soil, surface, and groundwater quality (Socolow, 1999). Increased nutrient availability can lead to direct or indirect pollution. Leaching, volatilization, acidification, and denitrification can be caused directly by improper application, excessive fertilizer use, or inadequate fertilizer management. Air pollution with carbon dioxide and nitrogen arises indirectly from production (using fossil fuels in the Haber–Bosch process).

20.2 Attributes of arbuscular mycorrhizal fungus

AMFs are specialized microorganisms characterized by considerable morphological and physiological diversity and live in the plant rhizosphere (Smith & Read, 2008b). Symbiosis through AMFs is one of the primordial and most extensive forms of mycorrhizal association (Finlay, 2008). It belongs to the phylum Glomeromycota, specifically to *Glomus* spp., which reproduce asexually and form symbiotic relationships with over 80% of terrestrial plant species (Bonfante & Gene, 2010), together with forest trees, wild grasses, and various crop plants (Harrison, 2005).

AMFs belong to the phylum Glomeromycota, specifically to *Glomus* spp. AMF refers to the endosymbiotic relationship among plant roots and fungi of the order Glomeromycota. These are characterized by two specific structures: arbuscules, complex branched structures in the root cells that are essential for symbiotic nutrient exchange, and vesicles, balloon-like structures that store nutrients in the cortical cells of the host root (Pepe et al., 2016). As obligatory biotrophic endophytes (Harrison, 2005), to obtain carbon and energy, these totally rely on a host plant (Finlay, 2008), as they cannot assimilate carbohydrates from other sources. Because of AMF's beneficial impacts on plant development and soil health, the carbon loss is balanced out (Hamel & Strullu, 2006). In AMF symbiosis, two types of root colonization exist (Hodge et al., 2010): (1) The arum type is categorized by the intercellular spread of the symbiont between the root's cortical cells, leading to the formation of mortal arbuscules at intracellular hyphal branches; and (2) in the Paris type the fungus reproduces directly from cell to cell in the cortex, creating intracellular hyphal spirals and intercalary arbuscules (Bücking et al., 2012; Giovannini et al., 2020).

AMFs create a significant number of spores that demonstrate rapid growth under a variety of stress circumstances (Abdelhameed & Metwally, 2019). In addition, these develop resistant hyphae that can penetrate the root of the host and extend into the soil, where they interact with the particles of the soil (Javaid, 2010). After that, these establish a network of extraradical mycelial structures, which increases the surface area of the fungus that can absorb nutrients and makes it easier for the fungus to transport mineral nutrients from the soil to the plants that are feeding on it (Javaid, 2010; Smith & Read, 2008b).

20.3 Importance of arbuscular mycorrhizal fungi in agriculture

Biofertilizers are compounds containing microorganisms that, when supplied to the soil, improve its fertility and promote plant growth. These are known as biofertilizers because they deliver the host plant with essential nutrients, water, and protection from harmful pathogens, receiving photosynthetic products in return (Smith & Read, 2008b). Therefore it is expected that inorganic fertilizers could be replaced by AMF in the coming years, offering benefits for plants and the environment. The widespread use of chemical fertilizers, herbicides, and insecticides leads to significant problems for health, soil, and plants. These degrade food quality and disrupt soil and water systems. The efficacy of AMF as a biofertilizer has been investigated by Lekberg and Koide (2005), McGonigle (1988), and Berruti et al. (2016). Researchers advocate the use of AMF as a biofertilizer to increase sustainable crop yields. The use of AMF is assessed to reduce the use of biochemical fertilizers by approximately 50%. However, this estimate varies depending on plant species and the type of abiotic stress present.

AMFs help to uptake the soil nutrients, particularly N and P, thereby promoting the development of host plants (Smith et al., 2011). Numerous studies have demonstrated that AMF can assimilate nitrogen and transfer it to neighboring plants (Battini et al., 2017; Hodge & Storer, 2015; Turrini et al., 2018). The AMF symbiosis is crucial for regulating nitrogen (N) and phosphorus (P) concentrations. According to Wang et al. (2018) the connection between AMF and salt stress has a substantial impact on the amount of phosphorus (P) and nitrogen (N) that is present in the aboveground sections of plants, as well as the ratio of nitrogen to phosphorus.

Currently, increased crop yields in many regions worldwide are attributed to the excessive use of chemical fertilizers (Ju et al., 2009). Soil organisms carry out the majority of nutrient transformations in the soil. Their activities significantly influence the nutrient cycle. According to Robertson and Groffman (2007) and Van der Heijden et al. (2008), this can be used to determine the availability of nutrients to plants, the storage of those nutrients in the soil, and the possible loss of those nutrients from the plant–soil system. In addition to improving the nutrient status of plants, beneficial bacteria in the rhizosphere also enhance crop quality. Baslam et al. (2011) showed that mycorrhizal compounds increase the enrichment of various minerals, chlorophyll, carotenoids, anthocyanins, and tocopherols. AMFs are already being used in the large-scale cultivation of potatoes (Hijri, 2016) and maize (Sabia et al., 2015).

20.4 Scope and objectives

20.4.1 Scope

This chapter examines the role of AMFs in increasing crop production in modern and sustainable agricultural systems. It covers the biological and functional aspects of AMFs, their interactions with crops and the soil, and the mechanisms by which these enhance nutrient uptake, plant growth, and stress tolerance. This chapter explores both ecological and agronomic aspects, linking fundamental mycorrhizal processes with practical field applications. The focus is on AMF's contribution to soil health, the reduction of chemical fertilizer use, and the improvement of crop productivity under varying soil and climatic conditions. Recent advances in molecular biology and biotechnology related to AMF diversity, function, and inoculum development are also addressed. The discussion extends to the challenges associated with large-scale applications and the variability of AMF performance across different agroecosystems.

20.4.2 Objectives

The main objectives of this chapter are to:

1. Explain the biological basis and mechanisms of AMF–plant symbiosis in relation to nutrient acquisition and crop growth.
2. Highlight the role of AMF in improving soil structure, microbial activity, and overall soil fertility.
3. Examine the contribution of AMF in enhancing crop tolerance to abiotic and biotic stresses.
4. Evaluate the potential of AMF as biofertilizers for reducing dependency on chemical inputs.
5. Discuss recent biotechnological and molecular advancements in AMF research.
6. Provide practical recommendations for the effective integration of AMF into sustainable crop production systems.

20.4.3 Arbuscular mycorrhizal fungi: Concept and biology

According to Smith and Read (2008b), AMF is a member of the subphylum Glomeromycotina and forms the most common and basic symbiotic interactions with the roots of roughly 72% of all terrestrial plant species. These associations include a wide variety of crops, trees, and herbaceous plants within the category. This mutualistic symbiosis, known as arbuscular mycorrhiza, is characterized by the fungus forming distinctive, tree-like erections called arbuscules inside the root cortex. These arbuscules are the primary locations of nutrient shift between the host plant and the fungus. The relationship is based primarily on bidirectional nutrient transfer: The fungus provides the plant with minerals (especially N and P) that it takes up from the soil, whereas the plant supplies the obligatory biotrophic fungus with fixed carbon in the form of sugars and fats; the plant is the source of nitrogen (Luginbuehl & Oldroyd, 2017; Parniske, 2008).

20.4.4 Taxonomy and life cycle

AMF belongs to the monophyletic phylum Mucoromycota and subphylum Glomeromycotina. These are obligate biotrophs, meaning they are unable to complete their life cycle without a living host plant. Reproduction occurs primarily asexually through the formation of large, multinucleated spores in the soil. These spores can germinate spontaneously; however, hyphal growth and proliferation are significantly promoted by chemical signals from the roots, particularly strigolactones (Akiyama et al., 2005).

20.4.5 Colonization process

The formation of symbiosis follows a carefully regulated dialogue.

Presymbiotic signaling: The root of the plant produces strigolactones, which are substances that promote the germination of spores and the branching of hyphae. Mycorrhizal factors, which are a mixture of chito-oligosaccharides and lipo-chitooligosaccharides, are released by the fungus, which in turn activates a symbiotic signaling pathway that the plant also shares with rhizobia bacteria. Mycorrhizal factors are meant to be released by the plant (Maillet et al., 2011).
Hyphal penetration: An appressorium is formed on the surface of the root by the fungal hypha, which then enters the root and develops either intercellularly or intracellularly after passing through the cortex.
Arbuscule formation: Within inner cortical cells the fungus initiates the development of the **arbuscule**. The plant cell membrane invaginates and proliferates to envelop the finely branched arbuscule, creating an extensive interfacial matrix for exchange. Arbuscules have a short lifespan (a few days), after which these degenerate (Smith & Smith, 2011).

20.4.6 Structures and functional integration

Intraradical structures: Inside the root the fungus forms **hyphae, arbuscules**, and sometimes **vesicles** (lipid-storage organs).

Extraradical mycelium: From the root a huge network of hyphae penetrates the earth below. As a result, this serves as an essential extension of the root system, which investigates a higher volume of soil. This mycelium is responsible for nutrient uptake and can also interconnect different plants, forming a common mycorrhizal network.

Nutrient exchange: The extraradical hyphae absorb inorganic phosphate (Pi) and ammonium, converting them into polyphosphates and arginine, respectively, for transport to the arbuscule. Here, these are released to the plant. In return the plant transports sugars (hexoses) to the arbusculated cell. Recent research highlights that lipids are also a major and essential carbon source delivered from the plant to AMF (Keymer et al., 2017).

20.4.7 Functional benefits

Enhanced plant nutrition: The most documented benefit is the **improved uptake of immobile soil nutrients**, particularly phosphate (P) and zinc (Zn). The fungal hyphae can access pores too small for roots and exploit P beyond the plant's depletion zone.

Water relations: The extraradical mycelium can enhance plant water uptake and drought tolerance, though the mechanism is often indirect via improved plant phosphorus status.

Soil structure: The extensive hyphal network and the production of a sticky glycoprotein, **glomalin**, help to bind soil particles, improving soil aggregation and stability (Rillig, 2004).

Biotic stress resistance: AMF colonization can bring systemic resistance in plants, providing enhanced protection against certain root pathogens and nematodes.

20.4.8 Arbuscular mycorrhizae symbiosis

To characterize the various types of collaboration between soil-dwelling fungi and the roots or rhizomes of higher plants, Frank coined the term "mycorrhiza" in 1985. Allen (1991), who took a more objective or microbiologically oriented view of the relationship between plants and fungi, defined mycorrhiza as "a mutualistic symbiosis between plant and fungus localized in a root or root-like structure, with energy flowing primarily from the plant to the fungus and inorganic resources flowing from the fungus to the plant." So, mycorrhiza is a way of saying that the two species are in a mutually beneficial association. According to Molina et al. (1992) the type of mycorrhiza that results is entirely determined by the specific fungi and plants involved in the relationship. Kohout et al. (2014)and Saito (2000) both note that there has been significant progress in understanding how these fungi function and in classifying them. As these are regarded as obligate symbionts, they are classified as belonging to the phylum Glomeromycota . Abbott and Gazey (1994) as well as Pellegrino and Bedini (2014) provide extensive evidence for their effectiveness in agriculture. Ectomycorrhizal (ECM) fungi, also called ECM fungi, occur in many different habitats; however, these only form symbioses with 3% of terrestrial plant families (Smith & Read, 2008). Both the phyla Ascomycota and Basidiomycota are responsible for the classification of these extracellular matrix fungi, also known as ECM fungi (Hibbett et al., 2000; Siddiqui et al., 2008). AMFs, on the other hand, are completely dependent on plants for their carbon supply. Within the framework of symbiosis, both ECM and arbuscular mycorrhizal (AM) fungi can consume 20%–40% of the carbon fixed in the plant through photosynthesis

(McNear, 2013). The microorganisms that are present in the soil have a substantial impact on the physical health of plants as well as the fertility of the soil. AMFs, which are often commonly referred to as AMFs, are an important part of the community of microorganisms that live in the soil. Plant reproduction and development are both significantly impacted by their presence. According to Brundrett (2002), over 80% of land plants exhibit a symbiotic association with AMF. Arbuscular mycorrhizae are a synergetic association between certain fungi in the soil and the roots of higher plants. According to Redecker (2002), these associations have existed for 460 million years. These are widespread and ancient. Through the formation of aspersoria, fungal hyphae in ambrosia (AM) symbioses can penetrate the root cortex cell walls. This leads to the formation of arbuscules, or clusters, which interact with the host's cytoplasm and establish intraradical hyphal colonization (Smith, 1997). The plant and fungus can exchange nutrients via these highly branched arbuscules. AMFs are also capable of producing vesicles, which serve as storage organs (Smith, 1997). According to Johnson et al. (2002), researchers assume that plants in natural environments with AMFs donate 10%–20% of the carbon fixed by photosynthesis to their fungal partners instead of retaining it.

According to Van der Heijden et al. (2008), AMFs can improve soil quality and suppress organisms that reduce agricultural yield. AMFs intermingle directly with the soil by forming extraradical hyphae that can penetrate several centimeters deep. According to this facilitates nutrient uptake by host plants, particularly phosphorus. Hodge et al. (2001) suggest that the extraradical mycelium of AMFs supports the transport of nitrogen compounds (N) bound to organic matter in plant litter. Tisdall and Oades (1979) state that AM fungal hyphae are responsible for soil aggregate establishment by the production of glomalin, which significantly contributes to soil compaction. A glycoprotein called "globulin" is formed in the soil by AMF, specifically on their hyphae. Glomalin was discovered by Wright and Upadhyaya (1996) and named after the organism that produces it, which belongs to the phylum Glomeromycota. Glomalin is so insoluble in water that it cannot be dissolved in water. The extremely persistent glycoprotein-like compound (Wright & Upadhyaya, 1998), thus, remains in the soil even after the hyphae die (Driver et al., 2005). It improves the physical properties of the soil, increases carbon sequestration, enriches the mineral content, promotes microbial activity, soothes pollutants, and finally contributes to the restoration of the soil ecosystem. Glomalin can positively influence the soil by acting as a substrate for microbial communities, as a binding agent for aggregate formation, as a chelating agent for heavy metals and pollutants, and as a means of long-term carbon storage (Singh et al., 2020). The indirect or "secondary" effects of glomalin on the development and stabilization of soil aggregates improve symbiotic relationships and growth conditions (Andrade et al., 1998; Rillig & Mummey, 2006).

Extraradical hyphae of AMF increased the surface area of the root systems of the host plant, which enhanced the potential for absorption, facilitated interaction with other microorganisms, and served as a key transport mechanism for energy-rich plant assimilates into the soil. This was accomplished by increasing the surface area of the root systems. The primary characteristics of the symbiosis are the uptake of phosphorus (P) by the host plants and the development of extraradical fungal hyphae networks, which increase the amount of water that the plants take in. This symbiosis can trigger physiological and chemical signals at the cellular level, modify the structure of plant communities, and increase plant resistance to various biotic and abiotic stressors. The underground networks of mycorrhizal hyphae connect plants of the same species as well as plants of other kinds. As well as secreting defensive proteins, lipo-chitooligosaccharides, and strigolactones, these are responsible for facilitating the exchange of resources. The fungi can discover their host plants with the assistance of strigolactones and other chemicals that are secreted by the roots. When AMFs are exposed to strigolactones, these experiences increase growth and branching. Several chemical substances that are collectively referred to as mycorrhizal factors are released by the fungi in response to this signal. These factors significantly influence the interaction between AMFs and nitrogen-fixing bacteria. The AM compounds are further enhanced by the activation of seven genes identified as SYM genes.

20.5 The importance of arbuscular mycorrhizal fungi in improving abiotic stress tolerance

20.5.1 Drought stress mitigation

Drought stress is among the most damaging abiotic growth inhibitors for plants, as it reduces water availability for cell growth, impairs nutrient uptake, inhibits photosynthesis, and increases oxidative stress with the accumulation of reactive oxygen species (ROS). In most cropping systems, drought also leads to lower fertilizer efficiency and destabilizes yields. One potential solution to drought stress, which has proven to be a promising application of biotechnological research, is the use of AMF—soil fungi that form a symbiosis with plant roots, thereby improving their survival in water-scarce environments. The authors reiterate that AMF symbiosis is widespread and of great agricultural importance, as it colonizes a significant proportion of soil plants and has a substantial influence on plant water balance and stress physiology (Das & Sarkar, 2024; Tang et al., 2022).

One of the most significant contributions of AMF to drought tolerance is the improvement of the plant's water balance. AMFs form extraradical hyphae that extend the root depletion zone, thereby increasing the depth penetrating the soil. This also allows plants access to water trapped in drying soil microhabitats. Further research shows that AMF can improve the physical properties of the hydraulic pathway between soil and plant by increasing water transport to the roots and the effective absorption length of the root system (Abdalla et al., 2023). Modeling and experiments suggest that AMF can delay the development of a water crisis—that is, plants can transpire for longer periods and tolerate the loss of water potential for a longer time despite progressive soil desiccation (Abdalla et al., 2023). AMFs also influence root structure (e.g., increased branching and altered root hair growth) and can improve soil structure through hyphal networks and glomalin-mediated support of aggregation, thereby indirectly promoting water retention and infiltration. These physical soil improvements are particularly important in sandy or degraded soils, where water runs off easily, and drought progresses rapidly (Madouh & Quoreshi, 2023).

Drought typically restricts the diffusion and mass transport of nutrients, mainly phosphorus (P) and nitrogen (N), which in turn exacerbates stress through reduced photosynthetic performance and growth inhibition. AMFs boost nutrient uptake (especially P) through hyphal uptake and increased root surface area. This can improve water use efficiency and stabilize growth even in dry soils. A crop-specific review details how AMFs maintain membrane integrity, improve plant water content and nutrient uptake, and protect the photosynthetic apparatus from oxidative damage caused by drought—all contributing to better performance under drought stress (Tang et al., 2022). Drought tolerance depends not only on the amount of water present in the soil but also on the efficiency of plants in transpiring water into the soil via roots and tissues. The expression and activity of aquaporins (proteins and water channels) can be regulated by AMFs, influencing root hydraulic conductivity and leaf gas exchange. Current literature describes aquaporin regulation as one of the mechanisms that explains why mycorrhizal plants, compared to nonmycorrhizal plants, can maintain higher stomatal conductivity and better water balance under drought stress conditions (Batool et al., 2026; Das & Sarkar, 2024). Drought stress often leads to an oxidative upheaval that damages lipids, proteins, and nucleic acids. Antioxidant systems (enzymatic and nonenzymatic) that strengthen AMF-colonized plants and suppress the formation of ROS and cell damage are frequently observed. Furthermore, AMFs can support osmotic adaptation by modulating compatible solutes (e.g., proline and soluble sugars), enabling cells to maintain their turgor and enzyme activity during dehydration. Recent summaries of their effects, such as antioxidant protection, osmotic stabilization, and enhanced physiological stability, describe their biochemical benefits as consistent, drought-mitigating effects of AMFs in numerous plant species (Batool et al., 2026; Das & Sarkar, 2024). Abscisic acid (ABA), jasmonates, auxins, cytokinins, gibberellins, and strigolactones are also potent hormones that regulate plant behavior under drought stress. AMFs interact with this signaling network to influence the mechanism by which plants regulate the balance between growth and stress survival. According to a recent literature review by Plant and Soil, AMF-mediated drought resistance is related to hormonal interactions, such as the regulation of ABA-related responses (the key to stomata control and drought signaling) and the interplay between these and other hormones that regulate root development, mycorrhization, and drought acclimatization (Chandrasekaran et al., 2025).

20.5.2 Salinity stress alleviation

Saline soils inhibit plant growth by osmotic stress (impaired water uptake) and ion toxicity, predominantly because of the storage of large amounts of sodium and chloride ions in plant tissue. AMF can support plants in surviving saline environments through a combination of correlated physiological and biochemical processes. These processes mitigate salt damage and improve overall plant performance. Recent studies show that AMF-colonized plants tend to be healthier in saline soils compared to nonmycorrhizal plants in terms of growth, photosynthesis, and nutrient uptake. Improved ion homeostasis is one of the main mechanisms AMF uses to cope with salt damage. AMF can help plants maintain a better potassium–sodium balance by either restricting the uptake and transport of sodium ions to the aerial parts of the plant or by enhancing the selective uptake and utilization of potassium ions, which are necessary for enzyme function and stomatal function. This improves ion homeostasis, minimizing toxicity and promoting the metabolic stability of salt-stressed plants. In addition, AMF makes osmotic adaptation easier to achieve. To survive in salty environments with a low soil solution water potential, plants need to keep their cell pressure (turgor) stable. It has been found that AMFs are related to greater concentrations of suitable solutes, such as proline, glycine betaine, and soluble carbohydrates. These solutes assist plants in retaining water and stabilizing proteins and membranes during the effects of stress. Furthermore, AMFs enhance the plant's antioxidant defense system and help eliminate ROS generated during salt stress. This reduces lipid peroxidation and membrane damage and can protect the photosynthetic apparatus. The result is improved chlorophyll retention and carbon assimilation. Studies also show that AMFs can improve soil structure and nutrient uptake (especially of phosphorus), indirectly increasing plant vigor in saline soils.

20.5.3 Heavy metal toxicity and soil contamination

Soils contaminated with heavy metals, such as cadmium (Cd), lead (Pb), arsenic (As), zinc (Zn), mercury (Hg), and chromium (Cr), as well as other pollutants, hinder plant growth, impair nutrient uptake, damage cell membranes, and cause oxidative stress. AMFs help reduce these effects by forming symbiotic relationships with roots and restructuring the flow and effects of pollutants within the soil–plant system. Extensive reviews and metaanalyses demonstrate that AMFs represent valuable phytonutrients in phytoremediation, as these increase plant biomass and stress tolerance and influence the behavior of metals in the rhizosphere. The most important process is the immobilization and sequestration of metals. AMF hyphae can bind metal ions to the cell wall and other structures. AMF-associated compounds (especially glomalin-related soil proteins) also promote soil aggregation and can facilitate the immobilization of metals, thereby reducing their bioavailability. AMF can mitigate the toxicity of metals in the roots by restricting their transport to the aboveground parts of the plant and inducing compartmentalization within the root tissue. The latter is typical of AMF-colonized plants and contributes to the mitigation of cadmium toxicity.

Furthermore, AMFs indirectly strengthen plant defenses: These improve the uptake of phosphorus and micronutrients, increase the activity of antioxidant enzymes, and reduce oxidative damage from metals. These physiological benefits help plants maintain photosynthesis and growth even in contaminated soils, which would otherwise severely impair their performance. In addition to heavy metals, AMF can also contribute to improving plant resistance in soils contaminated with organic pollutants (e.g., polycyclic aromatic hydrocarbons (PAHs), petroleum derivatives, and pesticides). Inoculation with AMFs can support the phytoremediation of organic and inorganic pollutants. Studies show that it promotes plant growth and modifies the rhizosphere environment in a way that favors pollutant conversion and microbial activity. However, the results vary depending on the type of pollutant and its impact on AMF colonization. In general, AMF should be considered biological buffers that reduce pollutant stress and promote plant growth, especially in remediation programs based on integrated approaches with tolerant plant species and soil management strategies.

20.6 Role of arbuscular mycorrhizal fungi in managing biotic stresses

20.6.1 Suppression of soilborne pathogens

AMF can reduce the damage caused by most soilborne pathogens and pests, both through direct action of the fungi in the rhizosphere and indirectly through the plants themselves. Recent studies describe AMFs as natural partners in biological pest control, protecting plants from fungi, bacteria, viruses, and nematodes and supporting the soil microbiota, which inhibits pathogen colonization (Delaeter et al., 2024; Farhaoui et al., 2025). Immune priming/induced systemic resistance (ISR) is one of the most important mechanisms. Root colonization by AMF can activate defense mechanisms, enabling plants to respond more strongly and rapidly to pathogen attack by activating defense genes and metabolites (Farhaoui et al., 2025; Weng et al., 2022). This activation is often attributed to hormonal signaling pathways (typically jasmonate/ethylene-mediated, ISR) and can counteract most pathological soil infections such as *Fusarium, Rhizoctonia*, and other root rots (Delaeter et al., 2024; Weng et al., 2022). AMFs also utilize competition and rhizosphere modification—the occupation of root infection sites, competition for available resources, the alteration of root exudates, and the restructuring of the root/soil microbiome toward disease-suppressing communities (Farhaoui et al., 2025). There is also evidence that shared mycorrhizal networks have the potential to transmit defense signals between plants. Recent studies report on the transfer of jasmonic acid (JA) through shared mycorrhizal networks, which promotes the recruitment of disease-suppressing bacteria and the protection of neighboring plants (Zhang et al., 2025). Regarding soil pests the role of AMF in controlling plant-parasitic nematodes is being intensively discussed. Both direct (space/nutrient competition) and indirect mechanisms (ISR, tolerance, and altered interactions in the rhizosphere) are involved (Schouteden et al., 2015). AMF can also influence root-feeding insects/larvae and other herbivores by altering plant nutrition and defense compounds. The intensity and direction of this protection depend on AMF diversity and field conditions (Frew et al., 2022).

20.6.2 Induced systemic resistance in crops

AMFs not only improve plant nutrient uptake but also induce systemic resistance—a plant-wide defense mechanism in which tissues distant from the colonized roots become more susceptible to attacks from pathogens and pests. This so-called mycorrhizal-induced resistance (MIR) is considered as effective as other systemic defense mechanisms, such as ISR and (in certain cases) systemically acquired resistance, as it is effective against a broad spectrum of attackers (Pozo & Azcon-Aguilar, 2013). Activation of defense mechanisms is a characteristic feature of AMF-stimulated systemic resistance. Instead of maintaining defenses at a permanently elevated level of activity (which would be a costly cultivation approach), AMF-colonized plants can be kept in a dormant state and respond more quickly and strongly to infections. It has been found that pretreatment with AMF tends to increase the speed and intensity of defense responses (enzymes; antimicrobial metabolites; and cell wall reinforcement, e.g., lignification and callose deposition) following pathogen infection (Farhaoui et al., 2025; Weng et al., 2022).

Mechanistically, there is a strong relationship between MIR and hormonal signaling. Numerous reports and studies highlight the importance of JA and ethylene (ET) signaling pathways (which are typical of ISR defense mechanisms), with salicylic acid (SA) also potentially playing a role, depending on the host plant, AMF species, and pathogen type (biotrophic or necrotrophic). To illustrate, it has been shown that preinoculation with AMF can increase the expression of defense genes and activate genes of the JA and SA signaling pathways after pathogen infestation, which is consistent with systemic priming.

20.7 Biotechnological and molecular advances in arbuscular mycorrhizal fungi research

20.7.1 Molecular diversity and functional genomics of arbuscular mycorrhizal fungi

AMFs are difficult to cultivate and identify based solely on morphological characteristics. Therefore molecular markers form the basis for describing AMF diversity in roots and soils. Ribosomal DNA regions, particularly the small subunit of ribosomal (SSU) (18S) rRNA (with or without ITS and large subunit of ribosomal (LSU)), amplified with AMF-specific primers and sequenced using amplicons, serve as the target in most collaborative studies. However, AMF rDNA can exhibit high intraindividual variability, and reference databases are incomplete for many taxa. This limits species-level identification and renders exact approaches unreliable without the use of curated frameworks. To standardize molecular diversity, researchers frequently use the MaarjAM database to systematize collected global AMF-SSU rDNA sequences into virtual taxa (VT)—a practical operational taxonomy for comparing species richness and biogeography in studies and ecosystems. MaarjAM enables navigation and searching by VT, species name, geography, biome, and host metadata and is an important resource for molecular ecology and mapping AMF biodiversity. Combined with barcoding strategies, AMF genomics has advanced rapidly with long sequencing, Hi-C scaffolding, and single-nucleus techniques. A persistent challenge is that AMFs are often multinucleated and can exhibit considerable genetic diversity between nuclei, complicating genome assembly, haplotype separation, and the interpretation of individual genotypes. Recent assemblies using long sequencing and Hi-C from *Rhizophagus irregularis* (one of the AMF model organisms) provide evidence of increased contiguity and can reveal genome organization, including repetitive and transposable elements of enriched regions, both of which are thought to play a role in genome evolution and the emergence of functional innovations. AMF genomes are considered a source for a better understanding of symbiotic genes, genome dynamics, and the adaptation of AMF to different environments, linking patterns of molecular diversity to the processes of evolution.

20.7.2 Development of arbuscular mycorrhizal fungi-based inoculants

AMF-based inoculation refers to the production and application of viable propagules (spores, colonized root fragments, and/or hyphal pieces) to achieve early colonization of target plants and associated benefits such as improved nutrient uptake and stress tolerance. As AMFs are obligate biotrophs (these require a living host root for their life cycle), inoculum production typically involves (1) isolation/selection, (2) mass propagation, (3) formulation, (4) quality control (QC), and (5) application.

Production (mass propagation) is usually carried out in substrate cultures (host plants are grown in soil/sand/substrate and inoculated with AMF). This method is relatively simple and widely used for large-scale propagation. Modern production methods also include direct-to-field cultivation systems, hydroponics/aeroponics, and in vitro root organ culture. In vitro systems can produce cleaner inoculum (low contamination) but require specialized facilities and careful strain maintenance.

After propagation the inoculum is processed into user-friendly products (powders, granules, seed coatings, or liquid suspensions). Current formulation research focuses on improving shelf life and stability through appropriate carriers/fillers and storage methods while maintaining the viability and performance of the propagation units under field conditions. QC is essential, as market studies show that some commercial products contain few or no viable propagation units and may not promote root colonization. QC typically measures spore density, the number of infectious propagation units (e.g., MPN), and colonization/infection potential, while also checking for contamination and genetic/functional consistency.

References

Abbott, L. K., & Gazey, C. (1994). An ecological view of the formation of VA mycorrhizas. *Plant and Soil, 159*, 69–78.

Abdalla, M., Bitterlich, M., Jansa, J., Püschel, D., & Ahmed, M. A. (2023). The role of arbuscular mycorrhizal symbiosis in improving plant water status under drought. *Journal of Experimental Botany, 74*(16), 4808–4824. https://doi.org/10.1093/jxb/erad249.

Abdelhameed, R. E., & Metwally, R. A. (2019). Alleviation of cadmium stress by arbuscular mycorrhizal symbiosis. *International Journal of Phytoremediation, 21*(7), 663–671.

Akiyama, K., Matsuoka, H., & Hayashi, H. (2005). Plant sesquiterpenes induce hyphal branching in arbuscular mycorrhizal fungi. *Nature, 435*(7043), 824–827.

Allen, M. F. (1991). *The ecology of mycorrhiza* p. 184. Cambridge University Press.

Andrade, G., Mihara, K. L., Linderman, R. G., & Bethlenfalvay, G. J. (1998). Soil aggregation status and rhizobacteria in the mycorrhizosphere. *Plant and Soil, 202*, 89–96.

Baslam, M., Garmendia, I., & Goicoechea, N. (2011). Arbuscular mycorrhizal fungi (AMF) improved the growth and nutritional quality of greenhouse-grown lettuce. *Journal of Agricultural and Food Chemistry, 59*, C5504–C5515. https://doi.org/10.1021/jf200501c.

Batool, A., Li, S.-S., Dong, H.-J., Bahadur, A., Tu, W., Zhang, Y., Xiao, Y., Feng, S.-Y., Wang, M., Zhang, J., Sheng, H.-B., He, S., Li, Z.-Y., Kang, H.-R., Lan, D.-Y., He, X.-Y., & Xiao, Y.-L. (2026). Battle of arbuscular mycorrhizal fungi against drought stress: A gateway to sustainable agriculture. *Journal of Fungi, 12*(1), 20. https://doi.org/10.3390/jof12010020.

Battini, F., Grønlund, M., Agnolucci, M., Giovannetti, M., & Jakobsen, I. (2017). Facilitation of phosphorus uptake in maize plants by mycorrhizosphere bacteria. *Scientific Reports, 7*, 4686.

Berruti, A., Lumini, E., Balestrini, R., & Bianciotto, V. (2016). Arbuscular mycorrhizal fungi as natural biofertilizers: Let's benefit from past successes. *Frontiers in Microbiology, 6*, 1–13.

Bindraban, P. S., van der Velde, M., Ye, L., van den Berg, M., Materechera, S., Kiba, D. I., Tamene, L., Ragnarsdottir, K. V., Jongschaap, R. E. E., Hoogmoed, M., Hoogmoed, W. B., Beek, C. L., & van Lynden, G. W. J. (2012). Assessing the impact of soil degradation on food production. *Current Opinion in Environmental Sustainability, 4*, 478–488.

Boiffin, J., Malézieux, E., & Picard, D. (2001). Cropping systems for the future. In J. Nosberger, H. H. Geiger, & P. C. Struik (Eds.), *Crop science: Progress and prospects* (pp. 261–279). CAB International.

Bonfante, P., & Genre, A. (2010). Mechanisms underlying beneficial plant—Fungus interactions in mycorrhizal symbiosis. *Nature Communications, 1*, 48.

Bruinsma, J. (2003). *World agriculture towards 2015/2030: An FAO perspective* p. 432. Earthscan Publications Ltd.

Brundrett, M. C. (2002). Coevolution of roots and mycorrhizas of land plants. *New Phytologist, 154*, 275–304.

Bücking, H., Liepold, E., & Ambilwade, P. (2012). The role of the mycorrhizal symbiosis in nutrient uptake of plants and the regulatory mechanisms underlying these transport processes. In N. K. Dhal, & S. C. Sahu (Eds.), *Plant science* (pp. 107–138). Intechopen.

Chandrasekaran, M., Ponmani, T., & Ahmad, S. (2025). Arbuscular mycorrhizal fungi mediated hormonal interplay under drought stress conditions: Current insights and future directions. *Plant and Soil, 517*, 2041–2058. https://doi.org/10.1007/s11104-025-07972-9.

Clermont-Dauphin, C., & Meynard, J. M. (1997). L'emploi des pesticides et des engrais en agriculture. In P. Stengel, & S. Gelin (Eds.), *Sols: Interfaces fragiles* (pp. 175–186). INRA.

Craswell, E. T., & Karjalainen, U. (1990). Recent research on fertilizer problems in Asian agriculture. *Fertilizer Research, 26*, 243–248.

Das, S., & Sarkar, S. (2024). Arbuscular mycorrhizal fungal contribution towards plant resilience to drought conditions. *Frontiers in Fungal Biology, 5*. https://doi.org/10.3389/ffunb.2024.1355999.

Delaeter, M., Magnin-Robert, M., Randoux, B., & Lounès-Hadj Sahraoui, A. (2024). Arbuscular mycorrhizal fungi as biostimulant and biocontrol agents: A review. *Microorganisms, 12*(7), 1281.

Devi, S. H., Bhupenchandra, I., Sinyorita, S., Chongtham, S. K., & Lamalakshmi Devi, E. (2021). *Mycorrhizal fungi and sustainable agriculture*. IntechOpen 10.5772/intechopen.99262.

Driver, J. D., Holben, W. E., & Rillig, M. C. (2005). Characterization of glomalin as a hyphal wall component of arbuscular mycorrhizal fungi. *Soil Biology and Biochemistry, 37*(1), 101–106. https://doi.org/10.1016/j.soilbio.2004.06.011.

Farhaoui, A., Taoussi, M., Laasli, S.-E., Legrifi, I., El Mazouni, N., Meddich, A., Hijri, M., & Lahlali, R. (2025). Arbuscular mycorrhizal fungi and their role in plant disease control: A state-of-the-art. *The Microbe, 8*, 100438. https://doi.org/10.1016/j.microb.2025.100438.

Finlay, R. D. (2008). Ecological aspects of mycorrhizal symbiosis: With special emphasis on the functional diversity of interactions involving the extraradical mycelium. *Journal of Experimental Botany, 59*(5), 1115–1126.

Frew, A., Antunes, P. M., Cameron, D. D., Hartley, S. E., Johnson, S. N., Rillig, M. C., & Bennett, A. E. (2022). Plant herbivore protection by arbuscular mycorrhizas: A role for fungal diversity? *New Phytologist, 233*(3), 1022–1031. https://doi.org/10.1111/nph.17781.

Giovannini, L., Palla, M., Agnolucci, M., Avio, L., Sbrana, C., Turrini, A., & Giovannetti, M. (2020). Arbuscular mycorrhizal fungi and associated microbiota as plant biostimulants: Research strategies for the selection of the best performing Inocula. *Agronomy, 10*, 106.

Godfray, H. C. J., Beddington, J. R., Crute, I. R., Haddad, L., Lawrence, D., Muir, J. F., Pretty, J., Robinson, S., Thomas, S. M., & Toulmin, C. (2010). Food security: The challenge of feeding 9 billion people. *Science, 327*, 812–818.

Gomiero, T. (2016). Soil degradation, land scarcity and food security: Reviewing a complex challenge. *Sustainability, 8*, 281.

Grassini, P., Eskridge, K. M., & Cassman, K. G. (2013). Distinguishing between yield advances and yield plateaus in historical crop production trends. *Nature Communications, 4*, 2918.

Hamel, C., & Strullu, D. (2006). Arbuscular mycorrhizal fungi in field crop production: Potential and new direction. *Canadian Journal of Plant Science, 86*, 941–950.

Harrison, M. J. (2005). Signaling in the arbuscular mycorrhizal symbiosis. *Annual Review of Microbiology, 59*, 19–42.

Hibbett, D. S., Gilbert, L. B., & Donoghue, M. J. (2000). Evolutionary instability of ectomycorrhizal symbioses in basidiomycetes. *Nature, 407*, 506–508.

Hijri, M. (2016). Analysis of a large dataset from field mycorrhizal inoculation trials on potato showed a highly significant increase in yield. *Mycorrhiza, 2*, 209–214.

Hodge, A., Campbell, C. D., & Fitter, A. H. (2001). An arbuscular mycorrhizal fungus accelerates decomposition and acquires nitrogen directly from organic matter. *Nature, 413*, 297–299.

Hodge, A., Helgason, T., & Fitter, A. H. (2010). Nutritional ecology of arbuscular mycorrhizal fungi. *Fungal Ecology, 3*(4), 267–273.

Hodge, A., & Storer, K. (2015). Arbuscular mycorrhiza and nitrogen: Implications for individual plants through to ecosystems. *Plant and Soil, 386*, 1–19.

Javaid, A. (2010). Role of arbuscular mycorrhizal fungi in nitrogen fixation in legumes. In M. S. Khan, J. Musarrat, & A. Zaidi (Eds.), *Microbes for legume improvement* (pp. 409–426). Springer Wien New York.

Johnson, D., Leake, J. R., Ostle, N., Ineson, P., & Read, D. J. (2002). *In situ* 13CO2 pulse- labelling of upland grasslands demonstrates a rapid pathway of carbon flux from arbuscular mycorrhizal mycelia to the soil. *New Phytologist, 153*, 327–334.

Ju, X.-T., Xing, G.-X., Chen, X.-P., Zhang, S.-L., Zhang, L.-J., Liu, X.-J., Cui, Z.-L., Yin, B., Christie, P., Zhu, Z. L., & Zhang, F.-S. (2009). Reducing environmental risk by improving N management in intensive Chinese agricultural systems. *Proceedings of the National Academy of Sciences of the United States of America, 106*, 3041–3046.

Keymer, A., Pimprikar, P., Wewer, V., Huber, C., Brands, M., Bucerius, S. L., & Gutjahr, C. (2017). Lipid transfer from plants to arbuscular mycorrhiza fungi. *Elife, 6*, e29107.

Kohout, P., Sudova´, R., Janouskova´, M., Ctvrtlı'kova´, M., Hejda, M., Pa'nková´, H., Slavı'kova´, R., Stajerova´, K., Vosa'tka, M., & Sy'korova´, Z. (2014). Comparison of commonly used primer sets for evaluating arbuscular mycorrhizal fungal communities: Is there a universal solution? *Soil Biology and Biochemistry, 68*, 482–493.

Lekberg, Y., & Koide, R. T. (2005). Is plant performance limited by the abundance of arbuscular mycorrhizal fungi? A meta-analysis of studies published between 1988 and 2003. *New Phytologist, 168*, 189–204.

Luginbuehl, L. H., & Oldroyd, G. E. (2017). Understanding the arbuscule at the heart of endomycorrhizal symbioses in plants. *Current Biology, 27*(17), R952–R963.

Madouh, T. A., & Quoreshi, A. M. (2023). The function of arbuscular mycorrhizal fungi associated with drought stress resistance in native plants of arid desert ecosystems: A review. *Diversity, 15*(3), 391. https://doi.org/10.3390/d15030391.

Maillet, F., Poinsot, V., André, O., Puech-Pagès, V., Haouy, A., Gueunier, M., & Dénarié, J. (2011). Fungal lipochitooligosaccharide symbiotic signals in arbuscular mycorrhiza. *Nature, 469*(7328), 58–63.

McGonigle, T. P. (1988). A numerical analysis of published field trials with vesicular-arbuscular mycorrhizal fungi. *Functional Ecology, 2*, 473–478.

McNear, D. H., Jr (2013). The rhizosphere—Roots, soil and everything in between. *Nature Education Knowledge, 4*(3), 1.

Molina, R., Massicotte, H., & Trappe, J. M. (1992). Specificity phenomena in mycorrhizal symbiosis: Community-ecological consequences and practical implications. In M. F. Allen (Ed.), *Mycorrhizal functioning: An integrative plant-fungal process* (pp. 357–423). Chapman and Hall.

Nazir, A., Sarfraz, W., Ditta, A., Khalid, N., Farid, M., Shafiq, M., Firdaus-E-Bareen, Rizvi, Z. F., & Naeem, N. (2023). Synergistic impact of two autochthonous saprobic fungi (*A. niger* and *T. pseudokoningii*) on the growth, ionic contents, and metals uptake in *Brassica juncea* L. and *Vigna radiata* L. under tannery solid waste contaminated soil. *International Journal of Phytoremediation, 25*(11), 1488–1500. https://doi.org/10.1080/15226514.2023.2166457.

Norse, D. (2003) Fertilizer and World Food Demand-Implication for Environmental Stresses. *IFA-FAO Agriculture Conference "Global Food Security and the Role of Sustainable Fertilization", International Fertilizer Industry Association (IFA), Rome*.

Parniske, M. (2008). Arbuscular mycorrhiza: The mother of plant root endosymbioses. *Nature Reviews Microbiology, 6*(10), 763–775.

Pellegrino, E., & Bedini, S. (2014). Enhancing ecosystem services in sustainable agriculture: Biofertilization and biofortification of chickpea (*Cicer arietinum* L.) by arbuscular mycorrhizal fungi. *Soil Biology and Biochemistry, 68*, 429–439.

Pepe, A., Giovannetti, M., & Sbrana, C. (2016). Different levels of hyphal self-incompatibility modulate interconnectedness of mycorrhizal networks in three arbuscular mycorrhizal fungi within the Glomeraceae. *Mycorrhiza, 26*, 325–332.

Pozo, M. J., & Azcón-Aguilar, C. (2013). Mycorrhiza-induced resistance: More than the sum of its parts? *Trends in Plant Science, 18*(10), 539–545.

Rillig, M. C. (2004). Arbuscular mycorrhizae, glomalin, and soil aggregation. *Canadian Journal of Soil Science, 84*(4), 355–363.

Rillig, M. C., & Mummey, D. L. (2006). Mycorrhizas and soil structure. *New Phytologist, 171*, 41–53.

Robertson, G. P., & Groffman, P. M. (2007). Nitrogen transformations. In E. A. Paul (Ed.), *Soil microbiology, biochemistry, and ecology* (pp. 341–364). Springer.

Sabia, E., Claps, S., Morone, G., Bruno, A., Sepe, L., & Aleandri, R. (2015). Field inoculation of arbuscular mycorrhiza on maize (*Zea mays* L.) under low inputs: Preliminary study on quantitative and qualitative aspects. *Italian Journal of Agronomy, 10*, 30–33.

Saeed, H. S., Munir, M., Samra, S., Farooq, A., Mahmood, H., Ahmad, I., Akram, M. I., Iqbal, R., Elsadek, M. F., Almutairi, S. M., Ditta, A., & Daraz, U. (2025). Screening of biosurfactants producing fungi from infected *Citrus sinensis* (L.). *Polish Journal of Environmental Studies, 34*(5), 5261–5270. https://doi.org/10.15244/pjoes/190588.

Saito, M. (2000). Symbiotic exchange of nutrients in arbuscular mycorrhizas: Transport and transfer of phosphorus. In D. D. Douds, & Y. Kapunik (Eds.), *Arbuscular mycorrhizas: Physiology and function* (pp. 85–106). Kluwer.

Schouteden, N., De Waele, D., Panis, B., & Vos, C. M. (2015). Arbuscular mycorrhizal fungi for the biocontrol of plant-parasitic nematodes: A review of the mechanisms involved. *Frontiers in Microbiology, 6*, 1280. https://doi.org/10.3389/fmicb.2015.01280.

Schultz, R. C., Colletti, J. P., Isenhart, T. M., Simkins, W. W., Mize, C. W., & Thompson, M. L. (1995). Design and placement of a multi-species riparian buffer strip system. *Agroforestry System, 29*, 1–16.

Siddiqui, Z. A., & Pichtel, J. (2008). Mycorrhizae: An overview. In Z. A. Siddiqui, M. S. Akhter, & K. Futai (Eds.), *Mycorrhizae: Sustainable agriculture and forestry* (pp. 1–35). Springer.

Singh, A. K., Zhu, X., Chen, C., Wu, J., Yang, B., Zakari, S., & Liu, W. (2020). The role of glomalin in mitigation of multiple soil degradation problems. *Critical Reviews in Environmental Science and Technology, 52*(9), 1604–1638.

Smith, F. A., & Smith, S. E. (2011). Roles of arbuscular mycorrhizas in plant nutrition and growth: New paradigms from cellular to ecosystem scales. *Annual Review of Plant Biology., 62*, 227–250.

Smith, S. E., Jakobsen, I., Grnlund, M., & Smith, F. A. (2011). Roles of arbuscular mycorrhizas in plant phosphorus nutrition: Interactions between pathways of phosphorus uptake in arbuscular mycorrhizal roots have important implications for understanding and manipulating plant phosphorus acquisition. *Plant Physiology, 156*, 1050–1057.

Smith, S. E., & Read, D. J. (2008a). *Mycorrhizal symbiosis* (3rd ed.). Academic Press.

Smith, S. E., & Read, D. J. (1997). *Mycorrhizal symbiosis* p. 605. Academic Press.

Smith, S. E., & Read, D. J. (2008b). *Mycorrhizal symbiosis* p. 800. Academic Press.

Socolow, R. H. (1999). Nitrogen management and the future of food: Lessons from the management of energy and carbon. *Proceedings of the National Academy of Sciences USA, 96*, 6001–6008.

Tang, H., Hassan, M. U., Feng, L., Nawaz, M., Shah, A. N., Qari, S. H., Liu, Y., & Miao, J. (2022). The critical role of arbuscular mycorrhizal fungi to improve drought tolerance and nitrogen use efficiency in crops. *Frontiers in Plant Science, 13*. https://doi.org/10.3389/fpls.2022.919166.

Thind, S., Chaudhary, M. S., Ditta, A., Hussain, I., Parveen, A., Ullah, N., Mahmood, Q., Al-ashkar, I., & Sabagh, A. E. L. (2022). Impact of mycorrhizal fungi from different rhizospheric soils on fungal colonization, growth, and chlorophyll contents of *Cenchrus ciliaris*. *Agronomy, 12*(11), 2644. https://doi.org/10.3390/agronomy12112644.

Tisdall, J. M., & Oades, J. M. (1979). Stabilization of soil aggregates by the root systems of rye grass. *Australian Journal of Soil Research, 17*, 429–441.

Turrini, A., Bedini, A., Loor, M. B., Santini, G., Sbrana, C., Giovannetti, M., & Avio, L. (2018). Local diversity of native arbuscular mycorrhizal symbionts differentially affects the growth and nutrition of three crop plant species. *Biology and Fertility of Soils, 54*, 203–217.

Van der Heijden, M. G. A., Bardgett, R. D., & van Straalen, N. M. (2008). The unseen majority: Soil microbes as drivers of plant diversity and productivity in terrestrial ecosystems. *Ecology Letters, 11*, 296–310.

Wang, Y., Wang, M., Li, Y., Wu, A., & Huang, J. (2018). Effects of arbuscular mycorrhizal fungi on growth and nitrogen uptake of *Chrysanthemum morifolium* under salt stress. *PLoS ONE, 13*(4), e0196408.

Weng, W., Yan, J., Zhou, M., Yao, X., Gao, A., Ma, C., Cheng, J., & Ruan, J. (2022). Roles of arbuscular mycorrhizal fungi as a biocontrol agent in the control of plant diseases. *Microorganisms, 10*(7), 1266. https://doi.org/10.3390/microorganisms10071266.

Wright, S. F., & Upadhyaya, A. (1996). Extraction of an abundant and unusual protein from soil and comparison with hyphal protein from arbuscular mycorrhizal fungi. *Soil Science, 161*, 575–586.

Wright, S. F., & Upadhyaya, A. (1998). A survey of soils for aggregate stability and glomalin, a glycoprotein produced by hyphae of arbuscular mycorrhizal fungi. *Plant and Soil, 198*, 97–107.

Yizhu, L., Imtiaz, M., Ditta, A., Rizwan, M. S., Ashraf, M., Mehmood, S., Aziz, O., Mubeen, F., Ali, M., Elahi, N. N., Ijaz, R., S, L., Shuang, C., & Tu, S. (2020). Response of growth, antioxidant enzymes, and root exudates production towards As stress in *Pteris vittata* and *Astragalus sinicus* colonized by arbuscular mycorrhizal fungi. *Environmental Science and Pollution Research, 27*, 2340–2352. https://doi.org/10.1007/s11356-019-06785-5.

Zhang, et al. (2025). Common mycorrhizal networks facilitate plant disease resistance by chemical communication and recruitment of disease-suppressive bacteria (abstract page). *Cell Host & Microbe*.

Chapter 21

Arbuscular mycorrhizal fungi and organic farming

Khalid Mehmood[a], Muhammad Shehzad[a], Muhammad Mazhar Iqbal[b], Majid Mahmood Tahir[c], Hafiz Muhammad Rashad Javeed[d], Ansa Tamkeen[a], Omer Farooq[e] and Tayyaba Naz[f]

[a]*Department of Agronomy, Faculty of Agriculture, University of Poonch, Rawalakot, Pakistan,* [b]*Department of Soil and Environmental Sciences, College of Agriculture, University of Sargodha, Sargodha, Pakistan,* [c]*Department of Soil and Environmental Sciences, Faculty of Agriculture, University of Poonch, Rawalakot, Pakistan,* [d]*Department of Environmental Sciences, COMSATS University Islamabad, Vehari Campus, Vehari, Pakistan,* [e]*Department of Agronomy, Institute of Agronomy, Bahauddin Zakariya University, Multan, Pakistan,* [f]*Department of Soil and Environmental Sciences, Institute of Soil and Environmental Sciences, University of Agriculture Faisalabad, Faisalabad, Pakistan*

21.1 Introduction

Arbuscular mycorrhizal fungi (AMFs) are the fungal species found in the rhizosphere. These fungi make a symbiotic association with 70%–90%**terrestrial** plant roots and enhance water and nutrient uptake through plant carbon exchange. Arbuscular mycorrhizae are tree-like structures produced inside the root cells, connected through fungal hyphae extending in the soil volume, and are responsible for nutrient exchange. In a fungal soil plant association, plants provide sugars to fungi, and in return, fungi regulate water and nutrients like nitrogen, phosphorus, and also provide defense against plant stresses. AMFs also act as natural biofertilizers and improve the soil characteristics by promoting soil aggregation, defining the structure of the soil. These factors facilitate carbon sequestration and minimize the dependence on chemical fertilizers, favoring the growth and yield of crops (Wahdan et al., 2024).

21.1.1 Organic farming

Organic farming is the practice of producing crops without synthetic chemicals, that is, pesticides, herbicides, fertilizers, or genetically modified organisms. It is a sustainable agricultural practice that totally relies on natural processes to produce nutritious food and build soil health sustainably. Crop farming involves practices such as crop rotations, composting, and maintaining stress through biological pest protection measures to enhance optimum crop productivity. Organic agriculture involves practices such as crop rotation, composting, and the use of organic manures, which benefit the crop through nutrient regulation. Agro-ecosystems managed by soil microorganisms could perform several functions. The specific species of bacteria, fungi, and archaea can improve soil structure and promote crop growth by enhancing nutrient cycling.

The microorganisms, such as bacteria, fungi, and algae, play a crucial role in the organic farming system by minimizing the use of soluble biomolecules and biocides. Many crops are associated with plants and play a substantial role in plant nutrient regulation and host plant resistance. The role of fungi like mycorrhizae as natural fertilizer for plants improved the plant nutrient uptake and increased the crop yield by improving the diversity by improving soil structure, enhancing water absorption, and controlling disease attack.

21.1.2 Role of microorganisms in organic farming

In agricultural systems, microorganisms act as a living engine to regulate plant nutrition and protect the plants from pests and diseases. The microorganisms contribute significantly to improving crop productivity and soil health. There is a need to fully

Advancing Sustainable Agriculture Through Arbuscular Mycorrhizal Fungi. DOI: https://doi.org/10.1016/B978-0-443-45130-0.00022-9

harness the potential role of microorganisms by advancing microbial genomics using biotechnological applications. The role of microorganisms in maintaining soil health through water and nutrient regulation needs detailed investigation to understand the degradation of organic matter in soil, which ultimately affects nitrogen fixation and the solubilization of phosphorus. Some microbes, such as mycorrhizal fungi and plant growth–promoting rhizobacteria (PGPR), provide resistance to plants against pathogens for achieving resilient agricultural systems (Das & Sengupta, 2024). There is an evident role of microbes as microbial inoculants and biofertilizers could reduce reliance on synthetic inputs like chemical fertilizers and pesticides to promote environmentally sustainable agriculture. Employing microbial diversity in organic farming could perform a wide range of functions for sustainable agriculture. Among the wide range of microbial species that can play a role in organic farming, some important ones are discussed below.

21.1.3 Actinomycetes

Actinomycetes are essential filamentous soil bacteria and play a vital role in breaking down complex organic matter to substances that can boost soil fertility. The substances also act as nontoxic biocontrol agents to control pathogens. The key components, such as compost and humus, are the substrates for plant growth–promoting substances through nutrient cycling. The actinomycetes can produce natural antibiotics against pathogens and improve the structure of the soil in favor of plant growth.

21.1.4 Bacteria

The role of bacteria is essential in crop production, as they enhance soil fertility, protect the crops from pests, and boost growth and yield. Specific strains of bacteria are involved in nitrogen fixation to convert atmospheric nitrogen into usable nitrogen, such as *Rhizobium* and *Azotobacter*. Some bacterial species, such as *Bacillus* and *Pseudomonas*, can release nutrients like phosphate-solubilizing bacteria (PSB) that make phosphorus available to plants. In organic farming, these bacteria are the primary agents involved in the cycling of nutrients to fix nitrogen fixation through decomposition and suppress the pathogen. In organic systems, crop productivity can be increased through practices like crop rotation, the use of compost to boost microbial diversity and abundance, and the reduction of chemical use compared to conventional methods. Recently, commercial PGPR that produce hormones, such as indole-3-acetic acid and gibberellins, have been used for plant growth enhancement.

21.1.5 Fungi

The fungi have widespread influence on abiotic and biotic microflora in various agro-ecosystems of organic farming. The agro-microflora regulates the cycling of organic and inorganic inputs in agricultural systems, minimizing the need for agrochemicals for food production. The microorganism that acts as biopesticides can naturally suppress pests and diseases. The species of *Bacillus thuringiensis* are used for controlling insects, *Beauveria bassiana* for fungal insect pathogens, and *Trichoderma* for fungal disease control. Similarly, certain AMF can facilitate water and nutrient (phosphorus and zinc) uptake through extending the root systems. Similarly, *Trichoderma* species suppress soil-borne diseases.

21.2 Perspectives of arbuscular mycorrhizal fungi in organic farming

21.2.1 Improving soil organic matter

The application of organic residues, such as farmyard manure, compost, and green manures, improves the soil organic matter status that favors the diversity of microbial populations through enhanced activity. The increase in organic matter through carbon sequestration to produce glomalin-related soil protein could improve the soil structure by promoting aggregation, which stabilizes soil particles, ensuring a more stable soil structure for reducing erosion. AMFs play a major role in contributing to the soil carbon pool, and their role ranges between 25% and 85% of the total organic soil pool (Fall et al., 2022). Thus increasing the diversity of AMF is expected to increase the nutrient cycling in soil, which will promote an increase in organic matter content.

21.2.2 Nutrient cycling and acquisition

AMF significantly increases the effective root surface area, enhancing the decomposition of organic matter through mineralization to sparingly assess available phosphorus forms that have restricted mobility in soil (Nazir et al., 2023; Saeed et al., 2025; Thind et al., 2022; Yizhu et al., 2020). The decomposition and mineralization processes in organic farming also facilitate nitrogen uptake by interacting with nitrogen-fixing bacteria. In nutrient cycling the nutrients move in the plant body through the exchange of water and nutrients among biotic and abiotic inputs. Hence it can become an indication of nutrients absorbed by plants and consumed by soil use and thereby enhance the facility. Thus an effective system that focuses on absorption of crops, transferred through livestock or residues, and returned to the soil for reuse, maintaining fertility over time.

21.2.3 Crop diversification through rotations

Crop diversification involves alternating different plant species by carrying out practices that minimize the dependence on chemical fertilizers and attaining higher yields through better crop management. AMFs need specific hosts for long-term survivals, such as intercropping and crop rotations, which enhance the diversity and stability of crops by lowering the environmental impacts and promoting sustainable farming. These rotations and diversification could increase yield up to 25% by improving water use and enhancing the soil health through soil fertility improvement and pest management measures.

21.2.4 Tillage practices and soil structure

Tillage enhances AMF populations, and disruption of fungal networks in intensive practices (conventional tillage) reduces biomass and stability. Conservation or reduced tillage promotes the growth of AMF that improves water infiltration and increases carbon sequestration, leading to better soil structure. Intensive tillage practices reduce AMF colonization and growth of fungal mycelium. Reduced or conservation tillage practices should be adopted for the proliferation of hyphal networks. Improved AMF hyphae growth produces glycoprotein named glomalin that improves soil structure by creating better aggregation. It will favor stability of microbial habitat through enhanced infiltration and reduce losses through erosion control in an organic farming system. Thus better colonization improves the growth of plants through better biomass accumulation under nutrient-limited soil environments (Wang et al., 2025).

21.2.5 Applications of arbuscular mycorrhizal fungi to organic farming

The beneficial microorganisms, such as AMF, are beneficial soil microbes that are recognized as important symbionts to colonize with more than 80% agronomic crops and terrestrial plant species (Posta & Hong Duc, 2019). AMFs have attained popularity in sustainable agriculture because of their capability to increase the host plant's access to water and nutrients and induce tolerance to stresses such as drought, salinity, and pathogens to improve soil fertility (Begum et al., 2019). AMF have attained interest among farmers for their commercial use as microbial inoculants with a mixture of naturally occurring beneficial soil microorganisms. In conventional farming, intensification of crops and deep plowing under mono-cropping systems make high use of chemical inputs and minimize the diversity of AMF (Kazeeroni & Al-Sadi, 2016). AMFs play a vital role in organic farming by developing symbiotic relationships with the plant roots. The AMF market has grown significantly as the fastest growing biofertilizer business (Soti et al., 2023). In organic farming the diversity of AMF species creates richness in agro-ecosystems. Therefore the use of organic manures, composts, and slow-release fertilizers encourages species diversity to promote soil health on a sustainable basis (Mullath et al., 2019). The cycling of biosynthetic inputs like biofertilizers (fixing nitrogen and solubilizing phosphorus) is greatly influenced by AMF. The interaction of AMF with vascular plant species can minimize the dependence on synthetic inputs through nutrient acquisition, thereby improving soil health, crop productivity, and ecosystem sustainability in organic agricultural systems. AMFs also act as biocontrol agents to suppress the pests and diseases. They minimize the dependency on external inputs in agricultural ecosystems by maintaining ecological balance in low-input farming systems (Al-Hinai et al., 2025). These organisms form a complex web to improve soil health in conventional farming systems. In organic farming the use of cover cropping and composting under reduced tillage boosts the crop resilience to improve soil structure and support a sustainable food system, unlike conventional farming, which often harms these beneficial microbes (Heuck et al., 2025).

21.3 Microbial inputs in organic farming

21.3.1 Biofertilizers

Biofertilizers are vital microbial inoculants like bacteria and fungi that can convert raw organic nutrients into plant-available forms, improving the structure of soil and increasing the yields by 5%–7% by minimizing the need for synthetic chemicals.

These living microbial formulations of *Rhizobium, Azotobacter, Mycorrhiza*, and PSB are applied to seeds or soil to improve the fertility of the soil. AMF inoculants are often used when the native population is low or in those soils that have previously been fumigated or disturbed through tillage, etc. In organic farming, these inoculations are applied to degraded soils of nurseries, especially in horticultural crops. Thus the use of these biofertilizers is highly effective to maximize crop production organically on low-input agricultural systems.

21.3.2 Biopesticides

Biopesticides are natural, microbial-based pesticides such as *B. thuringiensis* and *B. bassiana* that are effective in controlling insects. They could be used against fungal insect pathogens, as *Trichoderma* species are capable of controlling fungal disease infestations. These are natural materials of plant and animal origin that are environment-friendly safe alternatives. The compounds are extracts of different plant species made from microorganisms and have toxic effects on plant pathogens. The *B. thuringiensis* is predominantly used for the control of harmful insects, whereas the bacterium *Bacillus subtilis* is a bacterial species used in biological protection.

21.3.3 Compost and vermicompost

Compost is made of nonsynthetic substances, including approved synthetics used as plant or soil amendments. The feedstock is piled or mixed to an exposure at a minimum heat of 55°C and stored at a 2:1 or 3:1 brown-to-green ratio. The pile is kept moist (like a damp sponge) and turned weekly for aeration, which speeds up decomposition. Vermicompost is made from allowed feedstock materials, either nonsynthetic substances or synthetics approved for use as plant or soil amendments. Aerobic conditions are maintained at 70°C–90°C for organic matter through forced-air pipes, and turning is done till production of the finished product. For organic use, contamination of crops, soil, water, pathogenic organisms, and heavy metals is prohibited. When applied to soil, amendments act as beneficial microbial species that act as slow-release fertilizers to improve organic carbon for improving physical and chemical properties of soil in favor of plant growth.

21.3.4 Liquid microbial cultures

Liquid microbial cultures contain living microorganisms in the form of concentrated suspensions of beneficial bacteria, fungi, or actinomycetes such as *B. subtilis* or nutrient-fixing bacteria. They can be applied as foliar spray, as seed treatment, or applied to soil in drenching to enhance soil fertility, manage pests, or promote plant growth. They include effective microorganisms and Jadam microbial solution and improve nutrient availability by enhancing uptake of nutrients such as N, P, and K. Liquid formulations are easier to apply and have higher cell concentrations and prolonged shelf life.

21.4 Application of arbuscular mycorrhizal fungi in organic crops

21.4.1 Resistance to abiotic stress

Abiotic stresses negatively affect the growth and yield of crops because of nonliving factors in extreme conditions, such as heat and cold stress, salinity, drought, nutrient deficiency, and heavy metal toxicity. These stress factors cause disturbance in ion balance and immobilize toxic elements in the soil. AMF extends fungal hyphae to develop an effective root system to extract water from deeper soil layers to improve crop water use efficiency. This helps to maintain leaf turgor, protecting the crops from wilting and sustaining growth under water scarcity periods.

AMF can regulate the Na^+ and K^+ ion balance through osmotic adjustments, enhancing the chlorophyll content and improving crop growth and yield under saline conditions that are induced by salinity stress.

In nutrient-deficient organic soils, AMF can cope with nutrient deficiency by improving nutrient use efficiency by enhancing the uptake of nitrogen, phosphorus, copper, and zinc, thus reducing dependence on external sources.

AMFs, in association with plant cells, provide cell membrane stability by regulating the metabolic processes and improving antioxidant enzymes such as superoxide dismutase (SOD), catalase (CAT), and peroxidase (POD), and protect cells from damage in extreme heat or cold conditions.

AMFs minimize the translocation of heavy metals to plant shoots by immobilizing them in the soil.

21.4.2 Arbuscular mycorrhizal fungi and biotic stresses in organic crops

Biotic stress is induced through living factors that affect the growth and yield of crops. These living organisms could be the insect pests, weeds, pathogens (bacterial, fungal, and viral), and nematodes (Wahab, Muhammad et al., 2023). AMF, in an organic farming system, could develop a mutualistic association, develop a biological shield against pathogens, and minimize the damage caused by living pests. AMF acts as a biofertilizer and biopesticides, thereby strengthening the defense system and reducing the use of synthetic inputs. These fungi could physically block the invaders and compete for resources such as nutrients and space. AMFs produce protective compounds by altering root exudates through hormonal signaling. The symbiotic relationship would regulate nutrient/water uptake and improve the overall health of plants by triggering defense systems as a natural tool for sustainable agriculture.

21.5 Role of arbuscular mycorrhizal fungi in managing biotic stresses

21.5.1 Soil-borne disease suppression

Soil-borne disease can be managed by using microorganisms. Some techniques, such as soil solarization, are effective for disinfesting soil or growing media; similarly, the use of heat, soil fumigants, fungicides, and seed treatment can suppress soil-borne disease. The use of AMFs that compete for nutrients (phosphorus) and space by developing root health and thickness and alter unfavorable root exudates against the diseases caused by *Rhizoctonia, Fusarium,* and *Phytophthora* species.

21.5.2 Induced systemic resistance against pests and pathogens

Induced systemic resistance is beneficial microbes (e.g., *Pseudomonas* and *Bacillus*) that prime a plant's immune system, applied before stress to withstand pests and pathogens. Acid/ethylene pathways could boost resistance across the plant without direct infection. It is a strategy to control insects/pathogens through gene expression or improved enzyme activity. As AMFs produce phenolics, pathogenesis-related proteins, and phytoalexins that activate plant defense pathways (jasmonic acid and salicylic acid) and create resistance against pathogens and pests.

21.5.3 Nematode damage control

In organic farming, integrated nematode damage techniques like crop rotation with resistant varieties, adding organic matter, and soil solarization could be used. However, for severe infestations, the use of nematicides could be effective. But it is preferable to use nematode-resistant seeds or rootstocks and soil amendments containing biocontrol agents, such as the use of beneficial fungi such as *Paecilomyces lilacinus* to combat nematodes. The application of AMF increases the lignification of the root cell wall and improves the tolerance against nematodes by reducing the penetration of nematodes into the plant body.

21.5.4 Reducing insect pest severity

AMFs are sustainable biocontrol agents that reduce insect pest severity and plant pathogen infestation by enhancing host plant defense mechanisms. AMF for over 80% of terrestrial plants bolster plant growth and nutrient uptake and induce systemic resistance to biotic stresses. They also increase the release of compounds, reducing feeding and reproduction to improve plant vigor and tolerance against pests. AMF could employ barriers like mycorrhiza-induced resistance, enhancing defense-related enzyme activities, such as polyphenol oxidase, POD, and SOD. Similarly, they create physical barriers in the roots and alter root exudates to reduce profiles, reducing infestation of soil-borne pathogens and root-knot nematodes. Moreover, changing the composition of volatile compounds released by plants also repels insects and pests.

21.5.5 Suppression of parasitic weed infestations

AMFs are recognized as an ecofriendly biological control effective for suppressing root parasitic weeds such as *Striga* (witchweed) and *Orobanche* (broomrape). AMF enhances competitiveness to host plants through reducing germination signals for parasitic weeds such as *Striga* and *Orobanche* and suppresses the infestation of parasitic weeds in organic farming systems by promoting host growth and nutrient uptake. The mechanisms for suppression involve the production of strigolactones that cause a reduction of germination stimulants in host roots. Similarly, improved plant nutrition reduces susceptibility to parasitic attack and induces systematic resistance through rhizosphere modifications that are less suited to parasitic development.

21.5.6 Arbuscular mycorrhizal fungi biotic stress tolerance mechanisms

AMFs boost plant resilience to biotic stress (like pests/diseases) by building osmoprotectants (proline and polyamines), cross-tolerance response, and maintaining redox homeostasis through hormonal shifts (abscisic acid (ABA), salicylic acid (SA), jasmonic acid (JA)) and antioxidant defenses [reactive oxygen species (ROS) scavenging] systems.

AMF involves the following physiological and biochemical processes:

AMF boosts enzymes and molecules to neutralize ROS caused by stress, preventing oxidative damage by regulating the antioxidant system by initiating a cascade of molecular events.

AMFs maintain the cell turgor and scavenge ROS, especially under water stress by increasing proline and polyamines (spermidine and spermine) through osmolyte accumulation. They enhance the plant resistance against fungi, viruses, and herbivores by triggering systemic defenses.

AMFs effectively compete with pathogens by limiting the space in the root cortex, thereby starving invading organisms; they also regulate water balance through aquaporins that prevent dehydration.

AMFs prevent dehydration through efficient water regulation and modulate stress signals against drought, pathogens, and insects by regulating plant hormones such as ABA, SA, and JA, respectively, to optimize growth through hormonal cross-talks.

It triggers plant genes and mitogen-activated protein kinase (MAPK) cascades, priming the plant's immune system and strengthening general stress tolerance through gene expression and signaling. The activation of defense systems (hormones, antioxidants, and osmolytes) has priming effect on the defense system to withstand water and improve nutrient stress.

AMFs extend through its hyphae, extending them to extract water and minerals, significantly (P, N, etc.) uptake through carbon exchange for managing stress and maintaining growth through water and nutrient acquisition.

AMF stabilizes the heavy metals by reducing their toxicity by reshaping the rhizosphere with beneficial microbes to enhance plant health by creating stress resistance.

Fungal hyphae form a dense network around roots, creating a physical barrier that hinders pathogen entry and colonization. AMF alter root exudates and favor beneficial microbes (like rhizobia and *Pseudomonas*) that suppress pathogens, improving soil microbial community structure through rhizosphere modifications (Israel et al., 2022).

21.5.7 Arbuscular mycorrhizal fungi biotic stress management in organic farming

The cereal crops such as maize and wheat; pluses such as chickpea and pigeon pea; horticultural fruits such as banana and citrus; vegetable such as brinjal, chili, and tomato; and other plants such as tea and coffee are treated either as seed treatment or soil inoculum by incorporating AMF. Application of farmyard manure (FYM) or compost to soil or root dipping before transplanting in AMF suspension is beneficial to these crops against biotic stresses. These microbes could be applied as biofertilizer to seeds before sowing or mixed with soil during land preparation to enhance the germination process. Liquid microbial solutions are also available that can be sprayed on foliage for disease resistance and improved nutrient absorption.

References

Al-Hinai, A., Janke, R., Sieverding, E., Farooq, M., & Menezes-Blackburn, D. (2025). Identification and characterization of native arbuscular mycorrhizal fungi in plants growing under organic and conventional farming conditions in Oman. *Soil & Environmental Health, 3*(2), 100140.

Begum, N., Qin, C., Ahanger, M. A., Raza, S., Khan, M. I., Ashraf, M., Ahmed, N., & Zhang, L. (2019). Role of arbuscular mycorrhizal fungi in plant growth regulation: Implications in abiotic stress tolerance. *Frontiers in Plant Science, 10*, 1068.

Das, D., & Sengupta, S. (2024). The role of microorganisms in agriculture: Enhancing soil health, crop productivity, and sustainable farming practices. *International Journal of Agriculture and Food Science, 6*(2), 163–165.

Fall, A. F., Nakabonge, G., Ssekandi, J., Founoune-Mboup, H., Apori, S. O., Ndiaye, A., Badji, A., & Ngom, K. (2022). Roles of arbuscular mycorrhizal fungi on soil fertility: Contribution to the improvement of physical, chemical, and biological properties of the soil. *Frontiers in Fungal Biology, 3*, 723892.

Heuck, M. K., Powell, J. R., Kath, J., Birnbaum, C., & Frew, A. (2025). Organic management shapes AM fungal community structure and function, partially mitigating the negative effects of conventional agriculture. *Functional Ecology, 39*, 1328–1342.

Israel, A., Langrand, J., Fontaine, J., & Sahraoui, A. L.-H. (2022). Significance of arbuscular mycorrhizal fungi in mitigating abiotic environmental stress in medicinal and aromatic plants: A review. *Foods, 11*, 2591.

Kazeeroni, E. A., & Al-Sadi, A. M. (2016). 454-pyrosequencing reveals variable fungal diversity across farming systems. *Frontiers in Plant Science, 7*, 314.

Mullath, S. K., Błaszkowski, J., Govindan, B. N., Dhaheri, L. A., Symanczik, S., & Al-Yahya'ei, M. N. (2019). Organic farming practices in a desert habitat increased the abundance, richness, and diversity of arbuscular mycorrhizal fungi. *Emirates Journal of Food and Agriculture, 31*(12), 969–979.

Nazir, A., Sarfraz, W., Ditta, A., Khalid, N., Farid, M., Shafiq, M., Firdaus-e-Bareen, Z. F. R., & Naeem, N. (2023). Synergistic impact of two autochthonous saprobic fungi (*A. niger* and *T. pseudokoningii*) on the growth, ionic contents, and metals uptake in *Brassica juncea* L. and *Vigna radiata* L. under tannery solid waste contaminated soil. *International Journal of Phytoremediation, 25*(11), 1488–1500.

Posta, K., & Hong Duc, N. (2019). Benefits of arbuscular mycorrhizal fungi application to crop production under water scarcity. *Drought-detection and solutions.*

Saeed, H. S., Munir, M., Samra, S., Farooq, A., Mahmood, H., Ahmad, I., Akram, M. I., Iqbal, R., Elsadek, M. F., Almutairi, S. M., Ditta, A., & Daraz, U. (2025). Screening of biosurfactants producing fungi from infected *Citrus sinensis* (L.). *Polish Journal of Environmental Studies, 34*(5), 5261–5270.

Soti, P., Kariyat, R., & Racelis, A. (2023). Effective farm management promotes native AMF and benefit organic farming systems. *Agriculture, Ecosystems & Environment, 342*, 108240.

Thind, S., Chaudhary, M. S., Ditta, A., Hussain, I., Parveen, A., Ullah, N., Mahmood, Q., Al-ashkar, I., & Sabagh, A. E. L. (2022). Impact of mycorrhizal fungi from different rhizospheric soils on fungal colonization, growth, and chlorophyll contents of *Cenchrus ciliaris*. *Agronomy, 12*(11), 2644.

Wahab, A., Muhammad, M., Munir, A., Abdi, G., Zaman, W., Ayaz, A., Khizar, C., & Reddy, S. P. P. (2023). Role of arbuscular mycorrhizal fungi in regulating growth, enhancing productivity, and potentially influencing ecosystems under abiotic and biotic stresses. *Plants, 12*, 3102. https://doi.org/10.3390/plants12173102.

Wahdan, S. F. M., Asran, A. G. A., Abdellatef, M., Atia, M. A. M., & Ji, L. (2024). Arbuscular mycorrhizal fungi in organic versus conventional farming. *Arbuscular* mycorrhizal fungi and higher plants*: Fundamentals and* applications, 259–286.

Wang, H., Guo, R., Xu, J., Yang, T., Wang, M., Zhang, L., Jiang, L., Gao, Y., Guo, J., Shi, L., Gao, Y., Sun, M., & Zhang, T. (2025). Organic fertilizer enhances crop yield and soil health by increasing arbuscular mycorrhizal fungi species richness. *Functional Ecology, 39*(11), 3210–3224.

Yizhu, L., Imtiaz, M., Ditta, A., Rizwan, M. S., Ashraf, M., Mehmood, S., Aziz, O., Mubeen, F., Ali, M., Elahi, N. N., Ijaz, R., Lelel, S., Shuang, C., & Tu, S. (2020). Response of growth, antioxidant enzymes, and root exudates production towards As stress in *Pteris vittata* and *Astragalus sinicus* colonized by arbuscular mycorrhizal fungi. *Environmental Science and Pollution Research, 27*, 2340–2352.

Chapter 22

Arbuscular mycorrhizal fungi and urban agriculture

Sana Ashraf[a], Meshal Saeed[a], Muhammad Kashif Irshad[b,c], Sajid Rashid Ahmad[a], Sobia Ashraf[b], Umer Farooq[d], Maimoona Shahbaz[a], Noor Fatima[a] and Humra Imran[a]

[a] *College of Earth and Environmental Sciences, University of the Punjab, Quaid-e-Azam Campus, Lahore, Pakistan,* [b] *Department of Environmental Sciences, Government College University, Faisalabad, Pakistan,* [c] *Department of Environmental and Energy Engineering, Yonsei University, Wonju, South Korea,* [d] *Department of Food Engineering, University of Agriculture Faisalabad, Faisalabad, Pakistan*

22.1 Introduction

Urban agriculture (UA) is a permanent and dynamic factor of the urban socioeconomic and ecological system (Dona et al., 2021). Yuan et al. (2022) stated UA as "an industry that produces, processes, and markets food and fuel, primarily in response to the daily demand of consumers within a town, city, or metropolis, on land and water dispersed throughout the urban and peri-urban area, applying intensive production methods, using and reusing natural resources and urban wastes to yield a diversity of crops and livestock." UA promotes urban social and economic development by providing an alternative to available urban resources, as the world relies on common urban resources, including land and water, with other urban functions resulting in their depletion.

Arbuscular mycorrhizal fungi (AMF) are considered essential components of the soil microbiome as they play a major role in the cycling of nutrients and help plants adapt to different environmental conditions (Liu et al., 2020; Nazir et al., 2023; Thind et al., 2022). They are essential for reducing soil erosion, enhancing phytoremediation, eliminating harmful microorganisms, and protecting agroecosystems (Saeed et al., 2025). Achieving future food security in more sustainable agricultural systems depends on how AMFs affect nutrient bioavailability and ecosystem services. It is crucial to understand the connection between AMF and soil–plant nutrient dynamics for enhancing crop production in a sustainable way.

Depending on social circumstances, economic conditions, and spatial availability, UA has several different forms, such as rooftops, vertical farms, community gardens, and urban horticulture (Fig. 22.1). Rooftop greenhouses and outdoor gardens and farms are examples of rooftop agriculture, an urban agricultural practice carried out on buildings that includes both protected and unprotected techniques. It accounts for one-fourth of all the urban surfaces (Appolloni et al., 2021) and is considered a feasible futuristic choice (Drottberger et al., 2023). In addition to enhanced spatial resolution, using rooftops has many benefits for cities and their residents, such as reducing urban heat islands, conserving energy, and managing water resources (Chen et al., 2025).

Rooftop UA is a unique approach for refugees who have been displaced for a long time to take charge of their lives and improve their health and well-being, whereas they live in the same place for an indefinite amount of time, which for Palestinians can last a lifetime (Tarchahani et al., 2023). Vertical gardens are gardens that use different types of plants and specific methods to decorate the walls of buildings (Ekren, 2017). Vertical greening uses plants to cover buildings, which benefits the environment, supports the economy, and positively influences human health (Wang et al., 2022). Another type of UA includes vertical greening systems, which are widely used in wastewater mitigation, as they are suitable for the environment and very good at getting rid of pollutants such as nitrogen, phosphorus, and organic matter (Wang et al., 2024).

Community gardens are specific areas in cities where people work together for crop production. It provides several advantages by reducing food insecurity issues conserving biodiversity in agriculture, and lessening pollution and transportation emissions (Raneng et al., 2023). The outcomes included the acquisition of horticultural knowledge, a connection to nature,

Advancing Sustainable Agriculture Through Arbuscular Mycorrhizal Fungi. DOI: https://doi.org/10.1016/B978-0-443-45130-0.00016-3

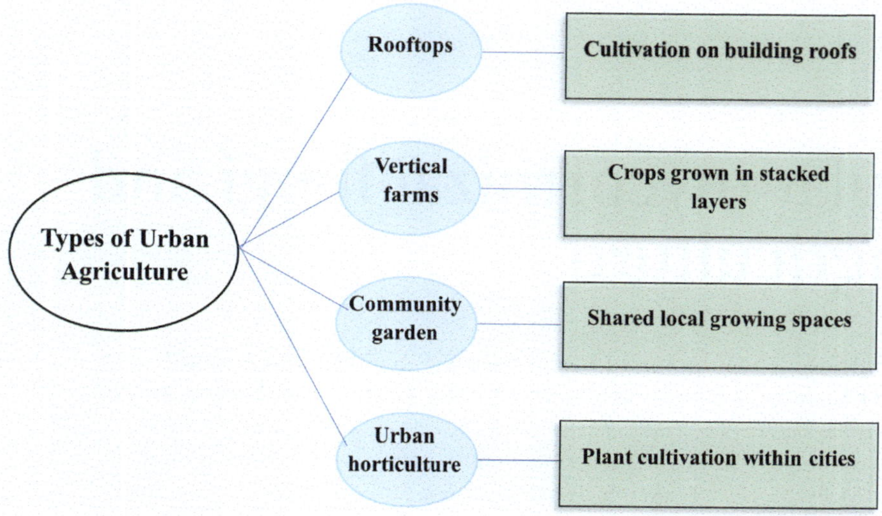

FIGURE 22.1 Forms of urban agriculture.

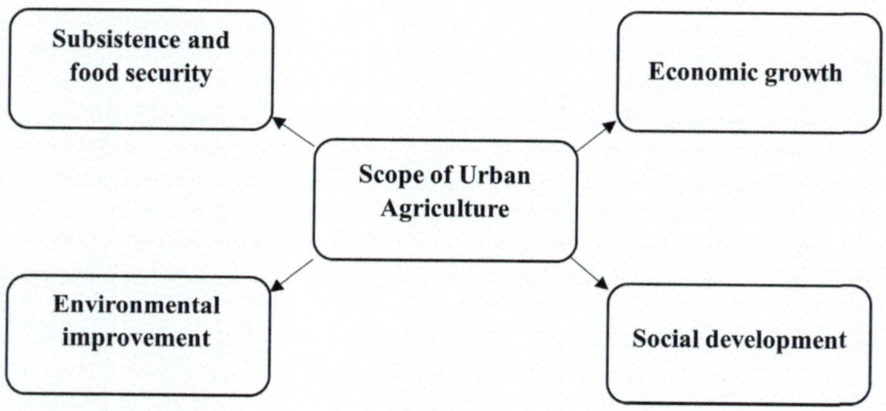

FIGURE 22.2 Scope of urban agriculture.

social relationships, community involvement through cooperative endeavors, and a sense of collective pride (Truong et al., 2022).

Urban horticulture involves cultivating a variety of crops and plants in densely populated areas, such as cities and towns, thereby reevaluating the relationship between urban environments and food production (Sashika et al., 2024). Singapore has been known as a Garden City for many years. The idea is now that it will be a "city in nature," where urban gardening will help restore nature (Sia et al., 2023). UA is a broader field that includes food security, environmental improvement, social progress, and economic growth. In the last ten years, more and more research has been done on UA and peri-UA and the need to come up with new ways to ensure sufficient food supply for urban populations (Lucertini et al., 2021). As a result, vertical farming is quickly becoming popular around the world because it can provide reliable and long-lasting food sources (Oh et al., 2023). Urbanization, population shifts, and climate change are the factors that are responsible for hunger, food insecurity, malnutrition, and the degradation of the environment. To deal with these problems, we need a plan that focuses on food systems to ensure everyone has enough food and nutrition, even for future generations. The United Nations has also suggested urban farming as one way to make the food system more sustainable. There has been a lot of growth in commercial UA (Alves et al., 2022). UA can help lessen the environmental damage caused by traditional farming methods and global food supply chains. This is because it can effectively close off nutrient and water cycles in cities and eradicates farming on less fertile soils (Langemeyer et al., 2021).

Better food production and environmental conditions in a region ultimately lead to economic growth. UA improves both national and local communities by creating job opportunities, making money, saving money on food transportation, encouraging entrepreneurship, helping small businesses grow, and adding to the urban economy (Fig. 22.2).

TABLE 22.1 Factors governing the agriculture sector in developed and developing nations.

Parameter	Developed countries	Developing countries
Land availability	Limited and expensive land restricts large-scale farming	Often cultivated on public or open vacant land without proper ownership rights
Governance	Strict laws and regulations for zoning and health reduce food production	Weak and absent policy frameworks
Infrastructure	Advanced, costly infrastructure dependent on technology	Poor infrastructure with limited access to water, storage
Cost	High input and labor costs	Low-cost but less efficient methods
Environmental conditions	Urban pollution, limited sunlight, dense cities	Contaminated soil, polluted water, heavy metal pollution

UA helps reduce obesity and health disparities related to diet by increasing the availability of fresh fruits and vegetables, which, in turn, reduces food poverty and improves nutritional health. Urban farms and gardens give people and nonprofits the power to grow their own food, which has led researchers to look into food security as a result. Urban farms and gardens are proposed as effective settings and activities for science, technology, engineering, and mathematics education, alongside the intergenerational transmission of skills and ecological knowledge (Ilieva et al., 2022). Globally the challenges faced by the agriculture sector include land availability, poor economic conditions of farmers, laws and regulations, and water availability.

Research shows that urban farming is harder in developing countries than in developed ones (Dona et al., 2021). Some authors reported that UA applies different strategies in the Global North and the Global South. This classification of the world order divides the world into two groups. The first group includes countries from North America, Europe, some Asian countries, and some Southern Hemisphere countries, such as Australia and New Zealand. The second group includes Latin American countries, the African continent, the Middle East, and parts of Asia. The separation between developed and developing nations is shown by the way these borders are drawn, which is based on economic, social, and political effects (Bennedetti et al., 2023).

In industrialized countries, UA is driven by the need for self-sufficiency and to build community ties that are shaped by a variety of social, economic, environmental, and technological factors. Developed countries focus on UA as a method of urban planning to enhance the social and health well-being of residents. The current analysis indicates that industrialized nations substantially recognize the potential of UA as a preventive health care strategy in urban environments. One of the main purposes of UA in industrialized countries is to encourage social interactions and fun activities. It is widely known that spending time in nature regularly is good for one's physical and mental health. In the Global North, these production areas often have a short-term view because they are threatened by more profitable land uses and mostly by market-driven efforts.

The main goal of UA in developing countries is to create an extra source of income and make sure that people have enough food. But they still face issues regarding technology utilization, getting support from policies, and dealing with social and cultural issues. We can learn a lot from how urban farming is done in developing countries. Developing countries often put the economic benefits of UA first. A study conducted by Bennidetti et al. (2023) in Quito, Ecuador, showed that having a garden at home increased the value of the property. In the Global South, production usually takes place in places that are dirty and dangerous to health because of poor management and pollution. In many cases, UA is not legally recognized, and activities like leisure or recreation are rarely reported. Some factors that are responsible for constraining agricultural growth in developing and developed countries are demonstrated in Table 22.1.

22.2 Mechanisms of arbuscular mycorrhizal fungi that enhance urban agriculture

Most land plants can form symbiotic relationships with AMF. This type of symbiosis influences plant growth and yield, their nutrient absorption capability (especially phosphorus and other nutrients that do not move), the ability to resist biotic and abiotic stressors (like salt and drought), soil aggregation, and carbon sequestration. AMF can change how the microbial community is organized and make it easier for other microorganisms to grow in the rhizosphere. This can affect a number of biological processes, such as biological nitrogen fixation (Singh et al., 2024).

22.2.1 Soil nutrient acquisition

AMF are a nearly universal group of soil microorganisms that interact directly with the roots and root-like structures of most plants in almost all major terrestrial biomes, including cities. When extraradical fungal hyphae colonize the roots of a suitable host plant, they extend into the soil and extract and assimilate nutrients from places that plant roots cannot reach or from sources such as tiny soil pores or soil minerals. Mycorrhizal fungi transport nutrients from the soil to the host plant via intracellular structures in the plant root cortex, receiving in return a portion of carbon derived from the plant in the form of carbohydrates and lipids. AMF plays a major role in providing plants with phosphorus (P) and helps break down other important nutrients such as nitrogen (N), water, and other micronutrients (Bate-Weldon et al., 2024). A metaanalysis showed that the average nutrient absorption of plants that were inoculated with AMF increased by 49%. The AMF has increased the uptake of macronutrients like P, N, and potassium (K) by 86%, 35%, and 26%, respectively, indicating the efficacy of AMF inoculation in nutrient absorption (Chandrasekaran, 2022).

In karst shrubland, AMF and roots are essential for plant nutrient uptake, especially in soils that lack nutrients with diverse plant varieties. However, it is still unclear whether an accumulation strategy of AMF and roots exists in the soils of karst shrubland habitats. Root traits that are associated with nutrient absorption, such as root biomass, arbuscular mycorrhizal colonization, root acid phosphatase activity, and nitrogen fixation, in two species that fix nitrogen: *Albizia odoratissima* (Linn. f.) Benth. and *Cajanus cajan* (Linn.) Millsp., grown in ammonium-rich soil that was either mixed or not mixed, with and without AMF inoculation. Both plants showed higher levels of AMF colonization, root biomass, and relative growth rate. However, they also showed lower levels of N_2 fixation and root acid phosphatase activity in the rhizosphere when grown in heterogeneous soil compared to homogeneous soil. Plants grown in heterogeneous soil that had AM fungi added to it had more root biomass and more active root acid phosphatase than plants grown in soil that did not have AMF added to it. Arbuscular mycorrhizal symbiosis resulted in increased root biomass that influences nutrient uptake in plants. This shows how arbuscular mycorrhizae and root characteristics help plants to get nutrients in the nutrient-poor and varied soils of the karst shrubland ecosystem (Liang et al., 2022).

There are many different types of urban green spaces, each with its own set of environmental conditions and management strategies. Park green areas usually have more diverse plants and better soil nutrients because they are well-managed, with regular trimming, watering, fertilizing, and soil amendments. On the other hand, green spaces along roadsides are more affected by human activities, such as pollution from traffic, soil compaction, and water stress. These factors often cause soil conditions to get worse, which means that management strategies need to be more comprehensive and less focused (Guo et al., 2025).

22.2.2 Pollution mitigation

Pollution caused by anthropogenic sources is becoming a worldwide concern. The physicochemical remediation strategies are, nevertheless, expensive and environmentally damaging. An improved method of cleaning up contaminated regions is phytoremediation, which utilizes microbes and plants. Better outcomes are achieved by using AMF global organisms that require a host to survive since they benefit both the fungi and the plants. Numerous studies have shown that AMF increases plant resistance to heavy metal and persistent organic pollutant pollution by affecting the uptake and dispersion of heavy metals inside host plants. They either neutralize the pollutants in roots or transfer them to the aerial parts of plants, leading to phytostabilization and phytoextraction. AMF as a phytoremediation agent enhances the concentration of chlorophyll pigments, amino acids, carbohydrates, total sugars, and macronutrient content such as P and N. It results in decreased oxidative stress and increases soil stability, making plant bodies more resilient to soil stress. The synthesis of glomalin shields hyphae and AMF from pollution-induced stress. Contaminants accumulate in AMF structures (spores, extraradical and intraradical hyphae, vesicles, and arbuscules) or within plant structures (root cells, shoots, leaves, or stems) as a result of mycorrhizal interactions, where metabolic activity slows down and the plant suffers less damage.

22.2.3 Role of arbuscular mycorrhizal fungi in enhancing stress tolerance

AMF is a beneficial symbiont that makes host plants immune to high levels of salt stress through mycorrhizal symbiosis. Recent studies have shown that AMF can improve plant growth and nutrient uptake, reduce loss of tomato production in saline conditions, and increase the salt tolerance in different plant species. The beneficial role of AMF symbiosis in reducing the accumulation of reactive oxygen species caused by simultaneous drought and heat, as well as by combined drought and heat shock stress. Fungi can survive efficiently even in polluted and harsh environments. AMF colonizes plant roots and

TABLE 22.2 Role of AMF as biostimulant and biofertilizer in urban agriculture.

Feature	AMF as biostimulant	AMF as biofertilizer
Primary role	Nutrient acquisition (P, N, micronutrients)	Stress tolerance (drought, salinity, disease)
Mechanism	Solubilizes nutrients, extends root systems	Triggers plant defense/hormonal pathways
Main benefit	Improved plant nutrition and growth	Enhanced plant vigor and resilience
Target function	Nutrient availability/uptake	Stress response/plant fitness
Impact on yield	Directly increases because of better feeding	Indirectly improves by reducing losses

AMF, arbuscular mycorrhizal fungi.

alters root structure and the nutritional makeup of the soil. This improves mycelial growth and heavy metal absorption, and it also lowers heavy metal concentration by producing secondary metabolites (Chandrasekaran et al., 2021).

22.3 Role of arbuscular mycorrhizal fungi in sustainable agriculture

22.3.1 Arbuscular mycorrhizal fungi as biofertilizers and biostimulants

Biostimulants enhance plant metabolism, leading to increased crop yield and improved quality; they also protect plants from environmental stresses such as water scarcity, soil salinization, and suboptimal growth temperatures; and they promote plant growth by facilitating greater nutrient absorption (Sun et al., 2023). Biostimulants can be either microbial or nonmicrobial. Microbial biostimulants include AMF, plant growth–promoting rhizobacteria (PGPR), nonpathogenic fungi, protozoa, and nematodes. AMF are among the most prominent microbial biostimulants and have an important role in cultivating better, healthier, and more functional foods in sustainable agriculture. AMF helps plants get nutrients and water, makes them more resistant to salinity, drought, and heavy metals, and slows down soil erosion. It is also a proven source of crop supplements that are beneficial for the environment (Sun et al., 2023).

AMF also improves soil health for better plant growth by changing its physicochemical properties. AMF makes the structure of the soil better by releasing glomalin, a glycoprotein that acts like a biological glue to hold soil particles together and make larger soil aggregates. This makes the soil more resistant to water and wind erosion. AMF also improves soil water retention properties and prevents nutrients from leaching out, especially nitrogen (N) and phosphate (P), which are responsible for promoting plant growth (Delaeter et al., 2024).

AMF inocula are now being used more and more as biostimulants to help plants take up minerals. But their potential as a viable alternative to chemical phytosanitary products in the biocontrol market is not well understood or talked about (Delaeter et al., 2024). Choosing the right parts for a microbial biostimulant product is very important for how well it works. The inherent properties of the chosen inoculants, their relationships with host plants, and the interactions among PGPM strains and organic matrices can significantly influence the biostimulant activity of each formulation. Biostimulants can meet a wider range of needs, such as improving soil fertility, restoring degraded soil, and maintaining yield, because their effects and effectiveness vary so much (Castiglione et al., 2021).

AMF acts as a biofertilizer as it enhances the nutrient absorption ability of the host plant, playing a key role in sustainable agriculture. AMF makes a large network of hyphae (extraradical mycelium) that goes far beyond the roots of the plant. This greatly increases the surface area for nutrient absorption, which lets the plant get nutrients from a larger area of soil. The role of AMF in UA as a biostimulant and biofertilizer is depicted in Table 22.2.

22.3.2 Role in organic and low-input farming systems

AMF receives between 5% and 20% of the total carbon intake by plants, which increases the accumulation of soil organic carbon (SOC) below ground. Because biotic interactions are more important in organic farming techniques than chemical inputs, the functional roles of AMF in soil health highlight their potential to mitigate the effects of climate change and improve agricultural sustainability (Park et al., 2024).

AMF's contribution to improving crop productivity and maintaining soil fertility is one of agronomy's most studied and well-known topics. According to Wahdan et al. (2024) the AMF is a specific type of organism that helps plants quench their desire for water and minerals while also favorably influencing other soil bacteria. Through microbial-assisted molecular pathway-triggered immunity, they help plants acquire resistance to diseases, improve heavy metal tolerance in plants, and

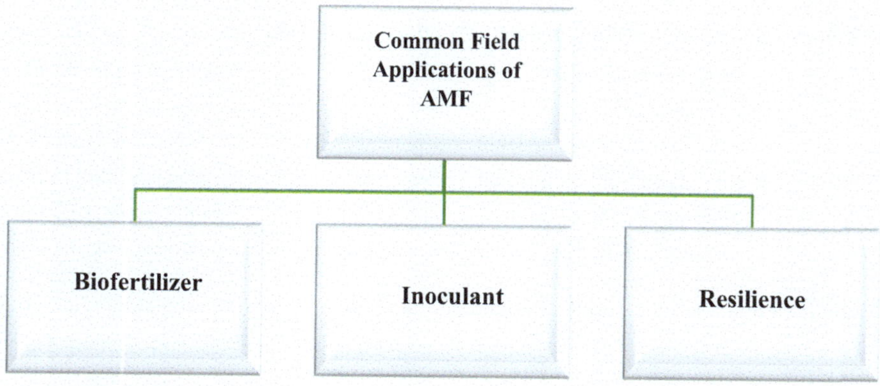

FIGURE 22.3 Common field applications of AMF. *AMF*, Arbuscular mycorrhizal fungi.

improve soil structure. AMF is becoming inevitable for the maintenance of agricultural fields in a sustainable agroecosystem (Chen et al., 2022).

22.3.3 Field applications, inoculum production, and formulation challenges

Fig. 22.3 indicates the application of AMF in the field. Inoculants of AMF are long-lasting biological materials that can support plants in many ways, especially in disturbed agroecosystems and when phytomanagement interventions are used. Most studies show that introduced AMF can live in the target field for a few months to several years, but their population will go down (60%), or they will completely vanish (30%). Further analysis reveals that AMF inoculation has both beneficial and detrimental effects on indigenous AMF species, manifesting as suppression (33%), stimulation (38%), exclusion (19%), and neutral effects (10% of cases examined) (Basiru et al., 2020). Because AMF are obligatory symbionts, their commercial or on-farm production must use a living host plant. The following are the methods applied for AMF inoculation into different plant species.

In vivo methods (pot/trap culture and on-farm production) include the most common approach. AMF spores or colonized roots are introduced to a host plant (like corn, millet, or grasses) grown in a low-nutrient, sterile, or fumigated substrate (often sand, vermiculite, or a soil/compost mix). The plant roots serve as the "factory" where the AMF multiplies, and the resulting mixture of spores, hyphae, and colonized roots is harvested as the inoculum (Dobo & Popovic, 2022). Another technique is in vitro/root organ culture (ROC). This technique is used for cleaner and high-density inoculum, which relies on genetically modified (Ri T-DNA-transformed) plant roots grown in a nutrient medium without the rest of the plant, producing contamination-free AMF. By being the core component of these production processes, AMF provides a viable, cost-effective, and environment-friendly alternative to traditional chemical fertilization, directly supporting the transition to more sustainable agricultural practices (Wahab et al., 2023).

22.3.3.1 *Formulation challenges*

The bioformulations based on mycorrhizae are good for the environment, with extended shelf life. The commercialization of AMF inoculants has made a lot of progress, such as better production methods, strain selection, and bioformulation strategies that make AMF-based products more stable, effective, and scalable. The factors that are responsible for reducing AMF growth are listed below.

- High production costs make it hard for many people to use, so costs need to be reduced so that everyone can use it.
- Even though they take a lot of time and money, thorough studies should be part of the strain acquisition process to help with registration.
- To make sure that microbial consortia work, it is important to optimize and control production and do field validation. Regular feedback from agronomic technicians can help find things that are going well and things that need upgradation (Boyno et al., 2024).
- Fake and low-quality goods influence consumer trust and slow market growth, especially in areas with weak rules.
- Performance inconsistencies across different environments, limited knowledge of optimal application methods, and competition with native soil microbes make it even harder to be sure that AMF inoculation will work.

- Another big problem is that farmers do not know much about biofertilizers because they cannot get the information, resources, or money they need, which makes it hard for them to use them widely. Maintaining genetic stability, quality control, and regulatory consistency are still a major environmental challenge (Ghorui et al., 2025).

22.3.4 Inoculation with other microbial inoculants (consortia approach)

The efficacy of the AMF consortia approach stems from the principles of ecological redundancy and functional complementarity. A diverse microbial consortium is generally more adaptable to varying field conditions (soil type, pH, and moisture) and can deliver a wider range of services, effectively buffering the plant against environmental uncertainties. For example, although AMF excel at phosphorus mobilization, a coinoculated PGPR might specialize in producing phytohormones to boost root development or solubilizing other trace minerals.

Research indicates that the combination can lead to a "multiplex" biofertilizer effect, which is critical for restoring microbial diversity in degraded agricultural soils and establishing a sustainable, low-input farming system. However, the successful development of consortia inoculants requires careful screening and formulation to ensure that the chosen strains are mutually compatible, competitive against native soil microbes, and stable within the commercial carrier product to maintain viability and functional synergy until field application (Perez-Moncada et al., 2024).

22.4 Arbuscular mycorrhizal fungi in urban agriculture: Difficulties and limitations in urban environments

22.4.1 Soil substrate limitations

The physical and chemical properties of the soil substrate make it very hard for AMF to grow and work well in cities. UA often relies on disturbed, compacted, or artificial soils and container systems, which present considerable abiotic challenges to AMF symbiosis. In urban agricultural systems, substrates often undergo compaction because of spatial limitations, containerized cultivation, and consistent irrigation cycles. This compaction significantly modifies the pore structure of the growth medium, a crucial factor influencing AMF activity and the overall dynamics of soil–water interactions. Püschel et al. (2024) conducted research that shows how AMF affects water retention depends on the physical state of the substrate. In highly compacted substrates the presence of AMF may not enhance and could potentially reduce water-retention capacity. On the other hand, AMF hyphae can make capillary water transport better on less compacted surfaces by making the pores more connected. The results show that the structure of the substrate has a big effect on how well AMF works in cities. This is because compaction reduces pore space and aeration, which limits hyphal growth and makes it harder for the fungi to transport water and nutrients.

Along with compaction the widespread loss of SOC is a major abiotic problem in urban container systems. Urbanization consistently leads to reduced SOC stocks, with highly urbanized areas displaying markedly lower levels in comparison to less disturbed soils (Zhang et al., 2024). SOC is essential for maintaining a functional soil microbiome, as it feeds many microbial communities, such as AMF, and helps important enzymatic reactions that are important for nutrient cycling and soil health. Zhang et al. (2024) determined that the structure of the microbial community accounts for approximately 25% of the variation in aggregate-related SOC, exerting a more substantial influence than soil enzymes. So, less organic carbon in urban soils weakens the biological foundation that is necessary for efficient AMF–plant symbiosis, which makes it harder for nutrients to move between plants and fungi and makes container-based production systems less resilient overall.

High levels of pollution, in addition to low carbon content, also make AMF less effective. The high levels of pollution in urban container substrates, especially heavy metals, are another issue regarding the effectiveness of AMF. This makes the soil microbial population less functional and less diverse. Higher amounts of elements like selenium (Se) put a lot of stress on the body, which makes microbial diversity and ecosystem stability go down a lot (Zhang et al., 2024).

In polluted soils, all major diversity indicators show big drops, which means that microbes are less able to survive. The loss of diversity leads to an ecological network that is broken and unstable, with less complexity and more modularity—signs of disrupted microbial connectivity and interaction (Zhang et al., 2024). The abundance of heavy metals imposes a deterministic selection mechanism that constrains the community's adaptive capacity and obstructs the ecological complexity essential for critical soil processes, including the successful colonization and effectiveness of AMF.

In the end, urbanization hurts the functioning microbial community, which hurts soil health and makes it harder for AMF to work together. There is a big drop in both microbial biomass and enzymatic activity along the urbanization gradient. The biomass of critical microbial groups, such as gram-positive bacteria and Actinomycota, is significantly reduced in highly urbanized soils, indicating a considerable decrease in active microbial populations (Zhang et al., 2024). This reduction corresponds with substantial decreases in enzyme activity, specifically *N*-acetylglucosamine hydrolase and acid phosphatase, which show declines of about 21% and 33%, respectively, signifying impaired nutrient cycling capacity (Zhang et al., 2024). Structural equation modeling demonstrated that urbanization indirectly reduces SOC by altering soil physicochemical properties and diminishing microbial biomass and enzyme activity, which subsequently lowers aggregate-related carbon content (Zhang et al., 2024). The decrease in microbial functioning disrupts plant–microbe interactions essential for nutrient uptake, posing a considerable challenge to sustaining productive and biologically active soils in urban agricultural systems.

22.4.2 Problems with inoculants

There are a lot of problems with inoculants in urban container-substrate systems. They often face a constrained supply of suitable AMF inoculants, with production concentrated in a few countries and companies (Basiru et al, 2020). Liquid formulations, featuring a single strain (*Rhizophagus irregularis*, DAOM 197198), facilitate fertigation but demonstrate restricted functional diversity (Basiru et al., 2020). Powdered (>60%) and granular (~25%) products include a wider range of species, but they can be contaminated, mixed unevenly, and require more work, which is a problem in crowded urban areas (Basiru et al., 2020). There are not many AMF species available; most of them are *R. irregularis* and *Funneliformis mosseae*, which limits strain–host compatibility in callous soils (Basiru et al., 2020). Propagule counts are high, but they do not always match up with dosage recommendations (Basiru et al., 2020). Seed-coating methods (~26%) improve efficiency, but they do not work as well because the coating is not uniform and does not last long (Basiru et al., 2020). The performance of AMF is significantly influenced by substrate chemistry, phosphorus concentration, and stress conditions. This makes single-strain solutions easy to use but limited in terms of biology, whereas multistrain consortia are more complicated but provide more biological diversity (Basiru et al., 2020).

However, AMF inoculants do not always fit well with practical container systems. Some possible reasons are that strains grown in very controlled in vitro conditions may not work well in more complicated field or container substrates (Rosa et al., 2020). Persistence presents a risk of the inoculant spreading and changing local microbial communities, which could have an impact on system performance over time (Rosa et al., 2020). Suitability depends on how well the strain, host, and substrate fit together. Using multiisolate or local strains may be helpful, but each case needs to be looked at separately (Rosa et al., 2020).

22.4.3 Physical limitations

Urban container systems make it even harder to set up AMF and make plants work well. The limited root volume at establishment is a real problem for container-grown perennials. It reduces early vegetative vigor and may even slow down structural canopy growth (Yuri et al., 2024). According to Yuri et al. (2024) the initial growth-related issues are caused by the physical ability of the roots to grow, not by a lack of macronutrients. The initial yields were similar across treatments, but in the second year, the quality of the fruit varied by nursery type. This suggests that the quality of the fruit affected sink development and resource allocation (Yuri et al., 2024). The authors contend that nursery type and planting conditions can either obscure or amplify the effects of root volume; however, extensive root systems are not universally requisite for short-term survival or yield, indicating that water availability, site microclimate, and training system function as colimiting factors (Yuri et al., 2024).

The size of the container is another important physical factor that affects plant growth in urban and containerized systems. Seedlings grown in larger containers had better survival rates after 2 years and grew faster in the beginning than seedlings grown in smaller containers (Siqueira et al., 2025). The quality of sites impacted outcomes; high-quality sites improved survival rates for all species, whereas low-quality sites reduced them overall, although some species still required larger containers to endure stress (Siqueira et al., 2025). Larger volumes mechanically reduce root deformation and increase the uptake of water and nutrients, thereby improving stress resistance. Under ideal circumstances or prolonged nursery cultivation, inferior grades may attain acceptable performance; however, they carry an increased likelihood of failure (Siqueira et al., 2025). Consequently, for urban container systems, medium to large containers (~150–310 cm^3) offer a cost-effective solution, whereas smaller containers (~90–125 cm^3) are suitable exclusively for low-stress environments (Siqueira et al., 2025).

Püschel et al. (2024) discovered that the impact of AMF on substrate water-holding capacity (WHC) is largely dependent on substrate structure and compaction. In their investigation, AMF colonization was significantly established in both studies; however, WHC responses exhibited variability between the settings. A study conducted by Püschel et al. (2024) concluded that pore configuration and bulk density affect the role of AMF hyphae in either hindering or promoting water retention. The author also proposed that in compacted substrates, fungal structures may occupy limited pore space and hinder aeration;

conversely, in looser substrates, AMF hyphae interlace larger pores and facilitate capillary water transport. So the structure and level of compaction of the substrate are important physical factors that affect the role of AMF in how water moves in container systems (Püschel et al., 2024).

22.4.4 Interaction with pollutants in cities

In cities, AMF are exposed to several pollutants that make their symbiotic function less effective. Heavy metals are quite hazardous to the environment. Urban horticulture soils often obtain zinc (Zn) from fertilizers, manures, and industrial sources, which may interfere with the symbiotic relationship between plants and AMF. Bate-Weldon et al. (2024) found that pea plants grown in soils with high Zn levels (250–400 mg kg^{-1}) had less AMF root colonization and a much lower transfer of phosphorus from the fungus to the plant roots. Even though fungal hyphae stayed the same, the process of exchanging nutrients got worse. Higher levels of Zn made it much harder for plants to transfer carbon to fungi. This suggests that the stress caused their mutually beneficial relationship to break down (Bate-Weldon et al., 2024).

The research suggested that AMF might persist in sequestering Zn within their hyphae, indicating a potential buffering role; however, the overall nutrient and carbon flow was obstructed. Mechanistically, elevated Zn likely interferes with fungal P-transport proteins and undermines root–fungal interfaces, reducing symbiotic benefits. Moderate Zn contamination common in urban soils may impede AMF functionality and reduce nutrient uptake in crops. Urban soil management must regulate zinc inputs and encourage the use of tolerant AMF strains to preserve soil functionality and food production (Bate-Weldon et al., 2024).

Salinity, in addition to heavy metals, represents a considerable abiotic stress affecting plant growth and productivity, particularly in urban and container-based agricultural systems where salt accumulation commonly arises from limited leaching and frequent irrigation. Kakabouki et al. (2023) examined the role of AMF in mitigating the adverse effects of salt stress on *Linum usitatissimum* (flax). Under saline conditions, AMF-inoculated plants exhibited significantly enhanced height, root biomass, and chlorophyll content compared to noninoculated controls (Kakabouki et al., 2023). The researchers observed enhanced physiological responses, including elevated relative water content and diminished electrolyte leakage, indicating that AMF symbiosis contributes to the maintenance of membrane stability and water balance under salt stress (Kakabouki et al., 2023). These benefits were attributed to the fungi's ability to improve nutrient uptake, particularly phosphorus and potassium, and to sustain ion homeostasis by limiting sodium accumulation in plant tissues. Kakabouki et al. (2023) determined that AMF substantially improve plant tolerance to salt by enhancing nutrient efficiency, osmotic regulation, and overall physiological function. The data suggest that AMF inoculation could be an effective and long-lasting way to improve plant growth and productivity in urban container systems that are affected by salt.

Lammel et al. (2025) examined the influence of various microplastic (MP) types and shapes on the community structure of AMF across three distinct soil types. The overall diversity of AMF remained relatively stable; however, significant changes occurred at the genus level. Fiber-shaped MPs caused a decrease in *Diversispora*, whereas fragment-shaped MPs led to an increase in *Claroideoglomus* and *Rhizophagus*, genera that are more resilient to disturbance (Lammel et al., 2025). The changes were greatly affected by the type of soil and the properties of the mycorrhizae, which shows that particle pollution can change the composition of AMF apart from growth retardation. The researchers concluded that MPs act as both physical and chemical stressors, modifying soil pore structure and nutrient dynamics, which indirectly influences AMF colonization and symbiotic effectiveness (Lammel et al., 2025). This shows that particle integration in urban soils can change AMF communities, which can change their role in nutrient cycling and soil stability.

22.4.5 Problems with maintenance: Saving media, contamination, shading, and watering

Problems with maintenance, such as media contamination, substrate degradation, and deficient light and water resources, influence how reliable and effective AMF are in urban and contained systems. Salomon et al. (2022) noted that many commercial AMF inoculants failed to facilitate adequate root colonization, mainly because of contaminated carrier media, diminished spore viability, and inferior substrate quality. The study showed that in vivo AMF products often bring in unwanted bacteria or weed seeds when sterilization and quality control are not done. The authors advocated for standardized testing, sterile carriers, and clear product labeling as essential measures to maintain inoculum quality and ensure consistent AMF efficacy in controlled environments.

Environmental factors make it even harder to keep AMF in these systems. Zhang et al. (2022) indicate that shade and excessive irrigation (flooding) markedly diminished AMF colonization and phosphate transfer efficiency in rice plants. The research revealed that low light and waterlogging diminished extraradical hyphal density and downregulated P-transport genes (GintPT and OsPT11), signifying decreased carbon allocation from host plants to fungus. The findings indicate that inconsistent watering and inadequate light exposure, prevalent in UA, can reduce the advantages of AMF and perhaps negate their symbiotic effects.

Maintaining pristine media, regulated watering, and adequate light exposure is crucial for preserving AMF functionality in container and urban systems. Inadequate management procedures diminish inoculant efficacy and compromise the ecological and nutritional advantages conferred by AMF symbiosis in these limited growing conditions.

22.4.6 Conclusion

To sum up, AMFs are an essential and sustainable approach to improving UA. Their function as microbial biofertilizers and biostimulants is essential for boosting crop quality and yield, particularly in urban settings where frequent stressors are present. By expanding the effective absorption region beyond the plant roots, AMF networks of hyphae greatly enhance nutrient intake, especially for nitrogen and phosphate. In addition, AMF improves the structure and WHC of the soil by producing glomalin, which is essential for resistance to nutrient runoff and erosion. AMF bioinoculants are a proven source of crop supplements that provide a biologically sound and ecologically sound substitute for conventional agrochemicals, thereby bolstering the shift to more resilient and sustainable food systems in urban environments.

List of abbreviations

AMF	Arbuscular mycorrhizal fungi
AMFs	Arbuscular mycorrhizal fungi (plural use)
AMF PGPR	Arbuscular mycorrhizal fungi—plant growth–promoting rhizobacteria
BR	Root ball volume class (BR25, BR50, BR100) (urban horticulture rootstock study)
DAOM 197198	Designation of a commercial AMF strain (*Rhizophagus irregularis* culture code)
DNA	Deoxyribonucleic acid
F. mosseae	*Funneliformis mosseae* (AMF species)
GintPT	Glomus intraradices phosphate transporter
HM	Heavy metals
H_2O_2	Hydrogen peroxide
MDA	Malondialdehyde
M	Molar concentration
NPK	Nitrogen, phosphorus, potassium (foliar nutrient concentrations)
OsPT11	Oryza sativa phosphate transporter 11
OM	Organic matter
PGPR	Plant growth–promoting rhizobacteria
PGPM	Plant growth–promoting microorganisms
POP	Persistent organic pollutants
R. irregularis	*Rhizophagus irregularis* (AMF species)
RA	Rooftop agriculture
RGR	Relative growth rate
ROC	Root organ culture
ROS	Reactive oxygen species
SOC	Soil organic carbon
TCSA	Trunk cross-sectional area
T-DNA	Transfer DNA
UA	Urban agriculture
VGS/VGSs	Vertical greening system(s)
WHC	Water holding capacity

References

Appolloni, E., Orsini, F., Specht, K., Thomaier, S., Sanyé-Mengual, E., Pennisi, G., & Gianquinto, G. (2021). The global rise of urban rooftop agriculture: A review of worldwide cases. *Journal of Cleaner Production, 296*, 126556. https://doi.org/10.1016/j.jclepro.2021.126556.

Basiru, S., Mwanza, H. P., & Hijri, M. (2020). Analysis of arbuscular mycorrhizal fungal inoculant benchmarks. *Microorganisms, 9*(1), 81. https://doi.org/10.3390/microorganisms9010081.

Bate-Weldon, M. P. A., Edmondson, J. L., & Field, K. J. (2024). Impact of zinc on arbuscular mycorrhizal-mediated nutrient exchange in urban horticulture. *iScience, 27*(8), 110580. https://doi.org/10.1016/j.isci.2024.110580.

Bennedetti, L. V., de Almeida Sinisgalli, P. A., Ferreira, M. L., & Lemes de Oliveira, F. (2023). Challenges to promote sustainability in urban agriculture models: A review. *International Journal of Environmental Research and Public Health, 20*(3). https://doi.org/10.3390/ijerph20032110.

Boyno, G., Ansari, R. A., Durak, E. D., Güneş, H., Çevik, R., & Demir, S. (2024). Arbuscular mycorrhizal technology in sustainable agriculture: Current knowledge and challenges in agroforestry. In *Mycorrhizal symbiosis and agroecosystem restoration* (pp. 173–195). Springer Nature. https://doi.org/10.1007/978-981-99-5030-0_8.

Castiglione, A. M., Mannino, G., Contartese, V., Bertea, C. M., & Ertani, A. (2021). Microbial biostimulants as response to modern agriculture needs: Composition, role and application of these innovative products. *Plants, 10*(8). https://doi.org/10.3390/plants10081533.

Chandrasekaran, M. (2022). Arbuscular mycorrhizal fungi mediated enhanced biomass, root morphological traits and nutrient uptake under drought stress: A meta-analysis. *Journal of Fungi, 8*(7). https://doi.org/10.3390/jof8070660.

Chandrasekaran, M., Boopathi, T., & Manivannan, P. (2021). Comprehensive assessment of ameliorative effects of AMF in alleviating abiotic stress in tomato plants. *Journal of Fungi, 7*(4), 303.

Chen, J., Li, J., Yang, Y., Wang, Y., Zhang, Y., & Wang, P. (2022). Effects of conventional and organic agriculture on soil arbuscular mycorrhizal fungal community in low-quality farmland. *Frontiers in Microbiology, 13*. https://doi.org/10.3389/fmicb.2022.914627.

Chen, Y., Liu, Y., Slootweg, M., Hu, M., Tukker, A., & Chen, W. Q. (2025). Unlocking rooftop potential for sustainable cities: A systematic review. *Frontiers of Engineering Management, 12*(2), 223–237. https://doi.org/10.1007/s42524-024-4053-3.

Püschel, D., Jana, R., Radka, S., Jan, J., & Michael, B. (2024). Soil compaction reversed the effect of arbuscular mycorrhizal fungi on soil hydraulic properties. *Mycorrhiza, 34*(4), 361–368. https://doi.org/10.1007/s00572-024-01153-9.

Delaeter, M., Magnin-Robert, M., Randoux, B. é, & Sahraoui, A. Lès-H. (2024). Arbuscular mycorrhizal fungi as biostimulant and biocontrol agents: A review. *Microorganisms, 12*(7), 1281. https://doi.org/10.3390/microorganisms12071281.

de Oliveira Alves, D., & de Oliveira, L. (2022). Commercial urban agriculture: A review for sustainable development. *Sustainable Cities and Society, 87*, 104185.

Dobo, B., & Popovic, V. (2022). Effect of arbuscular mycorrhizal fungi (AMF) and rhizobium inoculation on growth and yield of *Glycine max* L. varieties. *International Journal of Agronomy, 2022*, 1–10. https://doi.org/10.1155/2022/9520091.

Drottberger, A., Zhang, Y., Yong, J. W. H., & Dubois, M. C. (2023). Urban farming with rooftop greenhouses: A systematic literature review. *Renewable and Sustainable Energy Reviews, 188*. https://doi.org/10.1016/j.rser.2023.113884.

Ghorui, M., Chowdhury, S., & Burla, S. (2025). 2 Recent advances in the commercial formulation of arbuscular mycorrhizal inoculants. *Frontiers in Industrial Microbiology, 3*, 1553472.

Guo, J., Xin, Y., Li, X., Sun, Y., Hu, Y., & Wang, J. (2025). The assembly mechanisms of arbuscular mycorrhizal fungi in urban green spaces and their response to environmental factors. *Diversity, 17*(6), 425. https://doi.org/10.3390/d17060425.

Ilieva, R. T., Cohen, N., Israel, M., Specht, K., Fox-Kämper, R., Fargue-Lelièvre, A. ès, Poniży, L., Schoen, V., Caputo, S., Kirby, C. K., Goldstein, B., Newell, J. P., & Blythe, C. (2022). The socio-cultural benefits of urban agriculture: A review of the literature. *Land, 11*(5), 622. https://doi.org/10.3390/land11050622.

Kakabouki, I., Stavropoulos, P., Roussis, I., Mavroeidis, A., & Bilalis, D. (2023). Contribution of arbuscular mycorrhizal fungi (AMF) in improving the growth and yield performances of flax (*Linum usitatissimum* L.) to salinity stress. *Agronomy, 13*(9), 2416. https://doi.org/10.3390/agronomy13092416.

Lammel, D. R., Kim, S. W., Rong, L., Chen, H., Ingraffia, R., & Rillig, M. C. (2025). Effects of microplastic types and shapes on the community structure of arbuscular mycorrhizal fungi in different soil types. *Environmental Science and Pollution Research, 32*(20), 12504–12512. https://doi.org/10.1007/s11356-025-36408-1.

Langemeyer, J., Madrid-Lopez, C., Beltran, A. M., & Mendez, G. V. (2021). Urban agriculture—A necessary pathway towards urban resilience and global sustainability? *Landscape and Urban Planning, 210*, 104055.

Liang, Y., Pan, F., Jiang, Z., Li, Q., Pu, J., & Liu, K. (2022). Accumulation in nutrient acquisition strategies of arbuscular mycorrhizal fungi and plant roots in poor and heterogeneous soils of karst shrub ecosystems. *BMC Plant Biology, 22*(1). https://doi.org/10.1186/s12870-022-03514-y.

Liu, Y. Z., Imtiaz, M., Ditta, A., Rizwan, M. S., Ashraf, M., Mehmood, S., Aziz, O., Mubeen, F., Ali, M., Elahi, N. N., Ijaz, R., S, L., Shuang, C., & Tu, S. (2020). Response of growth, antioxidant enzymes, and root exudates production towards As stress in *Pteris vittata* and *Astragalus sinicus* colonized by arbuscular mycorrhizal fungi. *Environmental Science and Pollution Research, 27*, 2340–2352.

Lucertini, G., & Di Giustino, G. (2021). Urban and peri-urban agriculture as a tool for food security and climate change mitigation, and adaptation: The case of Mestre. *Sustainability, 13*(11), 5999.

Nazir, A., Sarfraz, W., Allah, D., Khalid, N., Farid, M., Shafiq, M., Firdaus-E-Bareen, Rizvi, Z. F., & Naeem, N. (2023). Synergistic impact of two autochthonous saprobic fungi (*A. niger* and *T. pseudokoningii*) on the growth, ionic contents, and metals uptake in *Brassica juncea* L. and *Vigna radiata* L. under tannery solid waste contaminated soil. *International Journal of Phytoremediation, 25*(11), 1488–1500. https://doi.org/10.1080/15226514.2023.2166457.

Oh, S., & Lu, C. (2023). Vertical farming-smart urban agriculture for enhancing resilience and sustainability in food security. *The Journal of Horticultural Science and Biotechnology, 98*(2), 133–140.

Park, S. H., Kang, B. R., Kim, J., Lee, Y., Nam, H. S., & Lee, T. K. (2024). Enhanced soil fertility and carbon dynamics in organic farming systems: The role of arbuscular mycorrhizal fungal abundance. *Journal of Fungi, 10*(9). https://doi.org/10.3390/jof10090598.

Pérez-Moncada, U. A., Santander, C., Ruiz, A., Vidal, C., Santos, C., & Cornejo, P. (2024). Design of microbial consortia based on arbuscular mycorrhizal fungi, yeasts, and bacteria to improve the biochemical, nutritional, and physiological status of strawberry plants growing under water deficits.3. *Plants, 13*(11), 1556.

Raneng, J., Howes, M., & Pickering, C. M. (2023). *Current and future directions in research on community gardens*: 79. Urban Forestry & Urban Greening.

Rosa, D., Pogiatzis, A., Bowen, P., Kokkoris, V., Richards, A., Holland, T., & Hart, M. (2020). Performance and establishment of a commercial mycorrhizal inoculant in viticulture. *Agriculture, 10*(11), 539. https://doi.org/10.3390/agriculture10110539.

Saeed, H. S., Munir, M., Samra, S., Farooq, A., Mahmood, H., Ahmad, I., … Daraz, U. (2025). Screening of biosurfactants producing fungi from infected *Citrus sinensis* (L.). *Polish Journal of Environmental Studies, 34*(5), 5261–5270. https://doi.org/10.15244/pjoes/190588.

Salomon, M. J., Watts-Williams, S. J., McLaughlin, M. J., Bücking, H., Singh, B. K., Hutter, I., Schneider, C., Martin, F. M., Vosatka, M., Guo, L., Ezawa, T., Saito, M., Declerck, S., Zhu, Y. G., Bowles, T., Abbott, L. K., Smith, F. A., Cavagnaro, T. R., & van der Heijden, M. G. A. (2022). Establishing a quality management framework for commercial inoculants containing arbuscular mycorrhizal fungi. *iScience, 25*(7). https://doi.org/10.1016/j.isci.2022.104636.

Sashika, M. N., Gammanpila, H. W., & Priyadarshani, S. V. G. N. (2024). Exploring the evolving landscape: Urban horticulture cropping systems–trends and challenges. *Scientia Horticulturae, 327*, 112870.

Sia, A., Tan, P. Y., & Er, K. B. (2023). The contributions of urban horticulture to cities' liveability and resilience: Insights from Singapore. *Plants, People, Planet, 5*(6), 828–841.

Singh, A., Pandey, A. K., Dodmani, B. A., Swati, R. J., Wongamthing, R., Mishra, S., & Karanwal, R. (2024). Arbuscular mycorrhizal fungi efficiency on plant growth and nutrient acquisition: A comprehensive review. *Microbiology Research Journal International, 34*(8), 13–22. https://doi.org/10.9734/mrji/2024/v34i81466.

Siqueira, D. P., Ford, C., Lloyd, A., White, D., Salvatierra, G., & Dungey, H. (2025). Container size and site quality affect survival and early growth performance of New Zealand native tree species. *Journal of Forestry Research, 36*(1). https://doi.org/10.1007/s11676-025-01851-w.

Sun, W., & Shahrajabian, M. H. (2023). The application of arbuscular mycorrhizal fungi as microbial biostimulant, sustainable approaches in modern agriculture. *Plants, 12*(17), 3101.

Tarchahani, A., & Loos, J. (2023). Flowers in the dark: The contribution of rooftop urban agriculture to human well-being in the Ein El-Hilweh Refugee Camp. *Lebanon. World Development Sustainability, 2*, 100057.

Thind, S., Chaudhary, M. S., Ditta, A., Hussain, I., Parveen, A., Ullah, N., Mahmood, Q., Al-Ashkar, I., & El-Sabagh, A. (2022). Impact of mycorrhizal fungi from different rhizospheric soils on fungal colonization, growth, and chlorophyll contents of *Cenchrus ciliaris*. *Agronomy, 12*(11), 2644. https://doi.org/10.3390/agronomy12112644.

Truong, S., Gray, T., & Ward, K. (2022). *Enhancing urban nature and place-making in social housing through community gardening*: 72. Urban Forestry & Urban Greening.

Wadumestrige Dona, C. G., Mohan, G., & Fukushi, K. (2021). Promoting urban agriculture and its opportunities and challenges—A global review. *Sustainability, 13*(17), 9609. https://doi.org/10.3390/su13179609.

Wahab, A., Muhammad, M., Munir, A., Abdi, G., Zaman, W., Ayaz, A., Khizar, C., & Reddy, S. P. P. (2023). Role of arbuscular mycorrhizal fungi in regulating growth, enhancing productivity, and potentially influencing ecosystems under abiotic and biotic stresses. *Plants, 12*(17). https://doi.org/10.3390/plants12173102.

Wahdan, S. F. M., Asran, A. G. A., Abdellatef, M., Atia, M. A. M., & Ji, L. (2024). Arbuscular mycorrhizal fungi in organic versus conventional farming. In *Arbuscular mycorrhizal fungi and higher plants: Fundamentals and applications* (pp. 259–286). Springer Nature. https://doi.org/10.1007/978-981-99-8220-2_12.

Wang, P., Yew, H., Wong, C., Yong Tan, S., Li, W., & Tong, C. (2022). Vertical greening systems: Technological benefits, progresses and prospects. *Sustainability, 14*(20), 12997. https://doi.org/10.3390/su142012997.

Wang, W., Zhou, X., Wu, S., Zhao, M., Jin, Z., Bei, K. e, Zheng, X., & Fan, C. (2024). Vertical green wall systems for rainwater and sewage treatment. *Sustainability, 16*(17), 7593. https://doi.org/10.3390/su16177593.

Ekren, E. (2017). Advantages and Risks of Vertical Gardens. Journal of Bartin Faculty of Forestry, 19 (1): 51-57.

Yuan, G. N., Marquez, G. P. B., Deng, H., Iu, A., Fabella, M., Salonga, R. B., Ashardiono, F., & Cartagena, J. A. (2022). A review on urban agriculture: Technology, socio-economy, and policy. *Heliyon, 8*(11). https://doi.org/10.1016/j.heliyon.2022.e11583.

Yuri, J. éA., Simeone, D., Fuentes, M., Sepúlveda, Á, Palma, M., Moya, M., & Sánchez-Contreras, J. (2024). Reduced root volume at establishment, canopy growth and fruit production in 'Lapins'/'Colt' and 'Regina'/'Gisela 12' sweet cherry trees. *Horticulturae, 10*(6), 579. https://doi.org/10.3390/horticulturae10060579.

Zhang, S., Yun, W., Xia, Y., Wu, S., You, Z., & Rillig, M.C. (2022). Arbuscular mycorrhiza reduced nitrogen loss via runoff, leaching, and emission of N_2O and NH_3 from microcosms of paddy fields. Water, Air, & Soil Pollution, 233(1), 1–14. https://doi.org/10.1007/s11270-021-05429-0.

Zhang, F., Zhong, J., Zhao, Y., Cai, C., Liu, W., Wang, Q., Wang, W., Wang, H., Jiang, X., & Yuan, R. (2024). Urbanization-induced soil organic carbon loss and microbial-enzymatic drivers: Insights from aggregate size classes in Nanchang city, China. *Frontiers in Microbiology, 15*. https://doi.org/10.3389/fmicb.2024.1367725.

Chapter 23

Arbuscular mycorrhizal fungi and rehabilitation in degraded lands: Supporting plant growth potential in a degraded environment

Muhammad Usama[a], Muniba Farhad[b], Muhammad Iqbal[a], Shahbaz Ali Khan[c], Hina Rizvi[a], Hafiz Syed Tanzeem-ul-Haq[a], Hadia Ahsan[a], Hafiz Muhammad Tauqeer Khan[d], Alisha Amdad[a] and Zartasha[a]

[a] *Department of Environmental Sciences, Government College University Faisalabad, Faisalabad, Pakistan,* [b] *Department of Chemistry, Government College University Faisalabad, Faisalabad, Pakistan,* [c] *National Compliance Centre, Ministry of Commerce, Islamabad, Pakistan,* [d] *Department of Environmental Sciences, University of Gujrat, Gujrat, Pakistan*

23.1 Introduction

Globally, soil degradation, referred to as less productive/fertile soil, is a leading environmental issue affecting food security and overall functionality of ecosystems and sustainability. Primary reasons for degraded soils, such as anthropogenic factors like pollution, various agricultural activities, and deforestation, diminish soil productivity and ecological functionality (Solomon et al., 2024). Unfortunately, degraded soils represent an adverse situation in terms of low microbial diversity, nutrient depletion, compaction, erosion, and accumulation of toxic pollutants (Singh et al., 2022). According to a global estimate, approximately 50% soils are moderately degraded, and 25% soils are severely affected (Srivastava et al., 2019).

Arbuscular mycorrhizal fungi (AMF), a symbiotic fungi belong to the phylum Glomeromycota and have the potential to develop a beneficial relationship with almost 80% of terrestrial plants (Kashyap et al., 2024). The AMF develop extensive extraradical hyphae and colonize with plant roots, which stabilize soil aggregates and improve root architecture for better uptake of water and nutrients (Wang et al., 2022). Moreover, the presence of AMF generates glomalin that binds soil particles, elevates aggregate stability, and participates in soil carbon (C) sequestration (Gerami et al., 2021; Kashyap et al., 2024). The AMF perform a pivotal role in restoring degraded soils through boosting soil fertility, microbial diversity, and nutrient cycling. For instance, AMF inoculation in soil mobilizes essential nutrients, that is, nitrogen (N), phosphorus (P), potassium (K), and zinc (Zn), and provide them to the plants while diminishing the requirement of synthetic fertilizers (Ahmed et al., 2025). The symbiotic features of AMF in soil boost the accumulation of organic matter and microbial biomass, which help to improve soil productivity and structure (Zhang et al., 2024; Zhou et al., 2024). In addition, the interactions of AMF with advantageous rhizobacteria such as *Bacillus* and *Pseudomonas* upgraded microbial colonization along with enzymatic activities, which further accelerate soil restoration (Herawati et al., 2024; Sagar et al., 2021).

Besides fertility rehabilitation, the AMF help to develop resistance in plants against different stress conditions such as salinity, compaction, drought, and contamination (Ditta & Ullah, 2023; Saleem et al., 2024, 2025). Such stresses regulate ion homeostasis, osmotic balance, and antioxidant enzymatic activity, assisting plants to flourish in severe environments (Navarro et al., 2024; Xu et al., 2024). Moreover, AMF participate in bioremediation through immobilizing heavy metals (HMs) via the secretion of extracellular polymeric substances (EPS) and organic acids, diminishing their bioavailability and phytotoxic impacts (Patel et al., 2022; Riaz et al., 2021). The AMF hyphal networks, along with the production of glomalin, further support the stabilization of pollutants and soil decontamination (Chen et al., 2019).

Advancing Sustainable Agriculture Through Arbuscular Mycorrhizal Fungi. DOI: https://doi.org/10.1016/B978-0-443-45130-0.00039-4

Among environmentally friendly and sustainable biological solutions for the rehabilitation of soils, the usage of AMF as a biological tool has become famous in scientific research (Matos et al., 2021). The AMF exhibit a natural and sustainable biological system that can improve soil structure and quality as well as plant resistance against degraded conditions. The dynamic role of AMF for soil stabilization, nutrient mobilization, and bioremediation highlights its ecological and agricultural importance. Being a novel bioecological strategy, AMF-based technologies grasp vast potential for the future management and rehabilitation of degraded soils.

23.2 Soil degradation

Land degradation is a process that declines land's ability to perform essential functions and deliver ecosystem services (Solomon et al., 2024). Moreover the degraded soils often harbor deleterious environmental conditions, such as the absence of topsoil and/or low fertility because of wind- and water-driven soil erosion, elevated pH, soil compaction, limited soil microbial status, and a high proportion of hazardous substances like HMs (Nazir et al., 2023; Saeed et al., 2025; Singh et al., 2022). Globally, 25% of the soil is intensely degraded, whereas 50% of the soil is partially degraded because of natural and man-made activities during recent decades (Srivastava et al., 2019). In the 21st century, the issue of soil degradation and its negative influences on the environment has been a leading challenge worldwide. Fortunately, AMF inoculum can play an effective role in restoring degraded soils (Matos et al., 2021).

23.3 Arbuscular mycorrhizal fungi and the mycorrhizosphere ecology

The rhizosphere is the notable compartment of soil (thin layer) enclosing the living roots, which is a pivotal and vigorous section because of variation in physical, chemical, and biological characteristics through roots. Interestingly the plant roots are not the only reason for this active compartment of the soil (Wang et al., 2022). AMF are fundamental soil microorganisms that develop mutualistic symbiotic linkages with plant roots (Mulyadi & Jiang, 2023), since plants initially colonized land around 450 million years ago (Delaux & Schornack, 2021). Likewise, the mycorrhizosphere is known as a portion of soil enclosing the living roots colonized through mycorrhizal fungi. It is constituted of the roots and hyphal surface, as well as the soil connecting the roots and hyphae, which is directly customized via mycorrhizal fungi interfacing hyphal and root exudates (Wang et al., 2022).

23.3.1 Classification and importance

The AMF have the ability to develop symbiotic relations with the majority of terrestrial plant species (almost 80%). Such characteristics of AMF species exhibit their ecological significance within the soil environments (Kashyap et al., 2024). A total of 340 AMF species have already been discovered worldwide, and new AMF species are still being explored (Sedaghati et al., 2021). The AMF may be categorized depending on spore morphology, spore wall structure, and formation of spores (Boyno et al., 2023). They are categorized as *Ascomycota, Basidiomycota, Glomeromycota*, and *Mucoromycota*. AMF may also be classified into different groups (such as arbutoid, ectomycorrhiza, ectendomycorrhiza, endomycorrhiza, ericoid, monotropoid, and orchid) based on various aspects such as fungus type, lineages with host plant, and structure between fungus and plant (Sedaghati et al., 2021).

The AMF have the ability to grow extensive mycorrhizal extraradical hyphal structures that proliferate within the peripheral soil matrix, exceeding the root boundary, therefore expanding the surface area of the root and its functional interface with the rhizosphere/soil (Wang et al., 2022). The spores generated by AMF species are separate or in clusters or glomerocarps (sporocarps) (Błaszkowski et al., 2021). Several other features of AMF are as follows: (1) The hyphal length density in soil (greater than 108 m/m^3) is significantly greater than roots (c.104 m/m^3) and (2) the diameter of AMF hyphae is 2–10 μm, which is two orders of magnitude thinner than fine roots (about 0.2 mm or less) (Wang et al., 2022).

23.4 Roles of arbuscular mycorrhizal fungi in the restoration of degraded soils

23.4.1 Restoration of soil degraded by erosion

Soil erosion is a leading environmental concern worldwide. It is a phenomenon in which topsoil is detached or excavated by natural forces like water flow and wind (Sharma, 2024). In addition, this phenomenon is held responsible for land degradation and desertification, which, as a result, displaces the most fertile layer of soil containing a large amount of nutrients, leading to deterioration of soil productivity (Abdel Rahman, 2023). According to Lal (2020) the losses of soil organic carbon (SOC) worldwide are estimated as 1.0 and 1.3 billion t/year by wind erosion and water flow, respectively. Hence, the AMF provide an exponential eco-friendly biological medium for stabilizing soil structure, enhancing nutrient cycling, and raising soil water-holding capacity and productivity while shrinking soil erosion (Ansari et al., 2024).

Interestingly, AMF inoculation substantially enhanced the content of polysaccharide in both root and hyphae zones as well as raised total organic carbon (OC) in the compacted soil where varieties of grass, that is, Bermuda (*Cynodon dactylon*) and Vetiver (*Chrysopogon zizanioides*), were grown, thus lowering the risk of soil erosion. Furthermore, a high content of glomalin in AMF-inoculated treatments compared to controls was reported (Li et al., 2025). Likewise, Chen et al. (2021) found that the AMF inoculation increased the formation of water-stable macroaggregates (2–5 mm) in highly compacted inorganic silt (up to 95% of maximum dry density).

Applying AMF can improve soil structure and reduce soil erosion. According to Muthukumar and Karthikeyan (2024) the extraradical mycelial network of AMF develops a three-dimensional matrix enmeshing the soil particles. A high secretion of glomalin in the soil treated with AMF inoculation results in soil binding and therefore facilitates controlling soil erosion (Kashyap et al., 2024). Besides, the hyphal network developed by AMF considerably strengthened root development and plant growth, which prevents plants from wind and soil erosion (Bello & Fabiyi, 2024).

23.4.2 Rehabilitation of soil degraded by water

23.4.2.1 Drought

Droughts are the leading root cause of land degradation, which may cause negative impacts on ecosystems (Ahmad et al., 2022). Globally, the land (around 1.9 billion) has been degraded because of drought and desertification (da Silva et al., 2022). Fall et al. (2022) have reported that the hyphal networks of AMF raise the water-holding capacity of soil and support plants to regulate hydration and flourish even under drought.

Xu et al. (2024) reported that AMF inoculation supported root development, increased gas exchange attributes, and enhanced the biomass of kenaf and the content of abscisic acid (ABA) and indole-3-acetic acid (IAA) in the roots by 6.69% and 26.45%, respectively, under drought stress. Moreover, AMF regulate the expression of aquaporin-related genes (such as HcPIP1;2 and HcPIP2;7) and raise the tolerance of kenaf against drought. AMF during drought stress, induced tolerance in kenaf and regulated the expression of aquaporin-related genes (such as HcPIP1;2 and HcPIP2;7). Similarly, the AMF inoculation in drought-affected soil enhanced 28.5% of grain yield in wheat, exhibiting the ability of AMF to facilitate water-use efficiency and uptake of nutrients (Duan et al., 2024).

The association of AMF with plant roots leads to altered root architecture via developing finer and wider root systems, which supports better water uptake from the soil (Ahmed et al., 2025; Thind et al., 2022). Improved root absorption was achieved because of a wider root system, promoted by AMF hyphae, which can expand the area outside the root zone, eventually increasing the available volume of soil solution (Chandrasekaran, 2022; Zhang et al., 2018). Colonization of AMF also stimulates the plant hormone, that is, IAA, which plays its role in the functioning of stomata and enhances stomatal density, which improves the potential of plants to regulate water loss during transpiration (Chareesri et al., 2020). Under water-stressed conditions, the AMF treatment enhances the capacity of plants to gather osmolytes such as melatonin, trehalose, and proline, which facilitate sustaining cellular turgor and stabilizing proteins (Ahmad et al., 2024; Ghani et al., 2022).

23.4.2.2 Waterlogged

Waterlogging in soil is predominant in various arable regions, which obviously deteriorates the quality and productivity of crops (Yang et al., 2024). All pores in waterlogged soil become filled, which affects soil air volume ratios and ideal solid pore material, thereby affecting the physiological efficacy of plants (Pais et al., 2022). Furthermore, the surplus water in soils may result in variation of soil properties like diminished soil oxygen availability, disturbing its diffusion to plant tissues and causing hypoxic/anoxic conditions (Pais et al., 2022; Sasidharan et al., 2017).

AMF markedly enhanced the adaptability of crops under waterlogged stress via producing high biomass and increasing nutrient absorption (Nie et al., 2024). AMF inoculation in waterlogged soil significantly promoted the colonization of the roots of rice. In addition, the AMF species, that is, *Funneliformis mosseae* and *Rhizophagus irregularis*, have performed a supportive role for improving nutrients such as calcium (Ca), P, Zn, N, and magnesium (Mg). Compared to the control treatment, both AMF species performed excellently to enhance stomatal conductance and transpiration rate in rice plants (Iqbal et al., 2021).

During an intermittent flooding situation, the occurrence of AMF uplifted the expression of defense marker genes like OsPBZ1 and OsPR1 in rice plants (Wang et al., 2021). Nevertheless, AMF induces the synthesis of proline and glycine betaine during waterlogged stress, which improves leaf osmotic potential for regulating cell turgidity through osmoregulation. Interestingly the functionality of the cell is not interrupted by these compounds (Das et al., 2022).

23.4.3 Restoration of soil fertility

23.4.3.1 Nutrient mobilization

Globally, nutrient depletion is a leading issue that deteriorates the overall quality of soil because of aggressive traditional cropping practices to meet the demand for food/agriculture products (Bhattacharyya et al., 2023). The efficacy of AMF to ameliorate the soil deterioration is evident from the information provided by various studies (Singh et al., 2022). Likewise, the interactive role of AMF species in soil with diverse microorganisms participates in improving soil fertility. Being a microbial agent in soil, the scientists have confirmed that AMF enhance the microecological conditions of rhizosphere soil, which causes enrichment of soil nutrients (Xing et al., 2025). As confirmed by Dedysh et al. (2022), AMF improved the abundance of *Actinobacteria* and *Acidobacteria,* thus supporting soil structure and nutrient cycling.

Furthermore, the presence of AMF in nutrient-deficient soil expands its hyphal network, which enhances the surface area of roots for absorption/uptake of water and nutrients such as P, K, N, and Zn, thus promoting plant growth while lowering the requirement of chemical fertilizers (Ahmed et al., 2025). The incorporation of AMF (i.e., *R. irregularis* CD1) markedly enhanced the yield of seed of cotton (28.54%) and P acquisition (13.65%–43.27%) as well as improved photosynthesis and plant health, compared to the control (Marwanto et al., 2024). Mycorrhizal inoculation (consortia) effectively upgraded grain yield (21.2%), levels of OC, nutrients [P, K, iron (Fe), Zn], and protein in wheat plants (Akbar et al., 2023).

The AMF fungal hypha produces glomalin, which is a nondegradable and stable glycoproteinaceous substance that can precipitate high quantities of C and N within soil. Consequently, the existence of AMF-secreted glomalin in soil plays its role in storing and taking up C and N as well as the formation of stable complexes, leading to soil fertility (Germai et al., 2021). Moreover, AMF support carbon sequestration by organic acids and glomalin production to shelter fungal hyphae, responsible for nutrient mobility between soil, AMF, and plants (Koshila Ravi et al., 2019; Yang et al., 2022).

Besides this microbial connectivity, the AMF species in nutrient-deficient soil enhance nutrient availability through forming extensive hyphal networks that promote the surface area for nutrient absorption, predominantly for N and P (Ahmed et al., 2025). Plants and their associated mycorrhizae have evolved specialized mechanisms to chelate insoluble Fe via the secretion of siderophores. These siderophores facilitate plant growth through enhancing Fe availability in rhizosphere soil (Upadhayay et al., 2019).

23.4.3.2 Soil organic matter addition

According to Ahammed and Hajiboland (2024), soil organic matter (SOM) characterizes an extensive C reservoir that is directly affected by several factors, such as plant growth, fundamental development, and photosynthetic C transmission into the biomass of plants. However, the degradation of SOM adversely influences soil health because it may diminish the filtering of HMs (Ferreira et al., 2022). The SOM content in soil directly promotes soil aggregate stability, which performs a vital role in impeding erosion and stabilizing soil structure, improves water retention, and enhances nutrient availability critical for crop productivity (Omer et al., 2024). Hawkins et al. (2023) reported that AMF characterize a universally significant pathway for the flow of C from plants to soil, which is converted into SOM via many kinds of chemical and microbial processes.

Under AMF symbiosis, findings of Zhou et al. (2024) revealed that the SOC content was enhanced by 4.5% and microbial necromass carbon (MNC) by 47.5% in soil, compared to the control. Furthermore the inoculation of AMF in soil obviously promoted the content of SOM over the control treatment (Mulyadi & Jiang, 2023). Likewise, Galazka et al. (2020) reported that the occurrence of AMF in soil produces a significant amount of glomalin, which is directly linked with higher content of OC and stabilized soil structure.

In soil, the AMF contribute substantially to SOM by creating demand for sequestering plant-derived C and transferring it to subsurface hyphal networks (Javeed et al., 2023). The AMF-induced glomalin represents a possible soil ecosystem function by facilitating storage of OC and improving soil aggregate structure (Ahammed et al., 2023; Riaz et al., 2021). In soil, the AMF inoculation influenced the soil C because it enhances the decomposition rate of SOM through interaction with other biota (Bi et al., 2020). The accumulation of AMF-induced glomalin and microbial necromass in soil was linked with the abundance of AMF and mineral-associated organic C. The elevated soil microbial biomass observed in AMF-symbiotic soils originated predominantly from AMF biomass, with fungal necromass carbon significantly leading to SOC accumulation (Zhou et al., 2024).

23.4.3.3 *Microbial abundance*

Soil microorganisms perform a vital role in sustaining the structure and functionality of soil ecosystems. Moreover, soil biodiversity can develop complex linkages with plants, exhibiting interrelationships of plant species with soil (Pedrinho et al., 2024). Zhang et al. (2024) confirmed that the AMF in soil interact with other soil microorganisms like rhizosphere fungi (i.e., *Trichoderma* and *Aspergillus*) and bacteria (i.e., *Bacillus* and *Pseudomonas*), which significantly alter the diversity, composition, and functionality of the microbial community.

According to Sagar et al. (2021) and Yu et al. (2022), the participation of AMF in the enrichment of microbial communities within the rhizosphere is through bridging symbiotic relationships with plant growth–promoting rhizobacteria (PGPR) such as *Bacillus* and *Pseudomonas* species, which promote nutrient uptake, leading to better root health. Furthermore a positive correlation of AMF-induced glomalin with microbial respiration and biomass C in various soils has been reported (Balik et al., 2022).

Nevertheless, the AMF enhance the microbial diversity (fungi and bacteria) by providing glomalin (containing nutrients and C) (Farhad et al., 2024). Being a substrate of microbial decomposition, AMF-induced glomalin improves microbial catabolic functions by releasing extracellular enzymes (Singh et al., 2022). Hyphae of AMF play their role as a transport agent or provide passage for the bacterial community (Ujvári et al., 2021). Furthermore the release of exudates from AMF can attract bacteria and promote their growth (Jiang et al., 2021). AMF hyphae supported the mobility of bacteria between two interdependent plants through the same mycorrhizal network (de Novais et al., 2020). The AMF hyphae could participate as a pathway for phosphate-solubilizing bacteria that promotes the heterogeneous distribution of bacteria (Wang et al., 2023). According to Faghihinia et al. (2023), the AMF hyphosphere furnishes an energy-rich compartment for microbes, helps the distribution of microbes in soil matrix through water film-coated hyphae, supports interaction between predators and prey, and selectively recruits advantageous bacteria and archaea.

Plants assign 4%–20% of C to AMF in the rhizosphere, which stimulates enzymatic activity in the soil for the microbial community (Gujre et al., 2021). Findings of Carteron et al. (2021) show that the AMF enhance the biomass of saprotrophic fungi through the dissolution of organic matter into mineral matter in the soil. A symbiotic relationship between *Artemisia annua* roots and AMF regulates the utilization of several C resources through rhizospheric microbes, influencing the functionality of microbial community within the rhizosphere, thereby enhancing uptake of nutrients by plants (Ma et al., 2021). Moreover, the AMF efficiently provide a suitable and adaptable environment for microbes (i.e., rhizobacteria) via the provision of nutrients in the rhizosphere, which, as a result, upgrades overall soil fertility (Herawati et al., 2024).

23.4.4 Reclamation of soil degraded with salinity

23.4.4.1 *Saline soil*

Soil salinization is one of the leading factors of soil degradation, in which a high level of water-soluble salts is accumulated in the soil, causing adverse influences on crop productivity and ecological health (Ahmed et al., 2024; Ali et al., 2024; Negacz et al., 2022). According to Ferreira et al. (2022), the soil is categorized as saline if the electrical conductivity (EC) of the saturation extract is greater than 4 dS/m. Globally, Mandal et al. (2025) reported that 833 million hectares of land were considered 62% saline soil (at 30–100 cm depth) and 424 million hectares of land were 85% saline soil (at 0–30 cm depth). AMF can support plant growth and resilience in salt-affected soil (Pooja et al., 2025).

Under saline conditions, various AMF species (i.e., *Glomus deserticola, Pasmodium brasilianum, F. mosseae, Glomus etunicatum, Glomus aggregatum, Gigaspora margarita, Glomus monosporum, Glomus clarum, and R. irregularis*) improved plant productivity and antioxidant enzymatic activity (Yilmaz et al., 2023; Yizhu et al., 2020). Findings of Liu et al. (2023) revealed that AMF colonization in salt-affected soil upregulates the expression of salt sensitive genes (i.e., S1SOS1 and S1SOS2), promoting the potential of tomatoes to resist saline stress. Similarly the AMF promote the scavenging of reactive oxygen species (ROS) in rice because of enhanced expression of salt-resistant genes (i.e., OsPRX, OsHBP1b, Os10g, and OsNCX) in inoculated plants over non-inoculated ones (Zhang et al., 2023). Likewise, Navarro et al. (2024) demonstrated that AMF species, that is, *R. irregularis*, can precisely take up elements (Ca^{2+}, Mg^{2+}, and K^+) to sustain internal K^+:Na^+ and Ca^{2+}:Na^+ ratios within limits compared to noninoculated plants. Consequently the AMF in salt-affected soils can promote the development of native flora and support the restoration and stabilization of soil and ecological recovery (Pooja et al., 2025).

AMF also increase uptake of K^+ from saline or acidic soils through enhancing K^+ availability via interaction of AMF with K-solubilizing microorganisms, which help to release K^+ from mineral forms (Olaniyan et al., 2022; Singh et al., 2022). Under stress conditions, the AMF sustain osmoregulation by improving nutrient translocation, allowing plants to boost turgor pressure and avoid cellular dehydration, which is responsible for plant decay (Evelin et al., 2019;

Shi et al., 2023). Moreover, under saline-stressed plants, the AMF-regulated expression of aquaporin genes prevailing in plant parts (leaves and roots), which may improve water and nutritional status in AMF-inoculated plants because aquaporins perform their roles as important pathways to transport water and nutrients (Cheng et al., 2021). Moreover, the root morphology in AMF-inoculated plants is altered, which supports the AMF to discover the soil efficiently, causing improvement of water uptake in plants withstanding saline stress (Navarro et al., 2024).

23.4.4.2 Sodic soil

According to Sree et al. (2025), sodium (Na) saturated soils exhibit a high accumulation of Na in cation exchange complexes, causing reduced permeability and structural degradation. Based on terrain attributes, the soils are categorized as Na-saturated as exchangeable Na is 5%–40% (Ferreira et al., 2022). A high concentration of Na^+ disrupts soil aggregate stability by replacing divalent cations such as Mg^{2+} and Ca^{2+}, causing soil dispersion, diminished pore connectivity, and raised bulk density. Such soil situations exhibit low water infiltration and high risk of soil erosion and surface runoff (Choudhary & Kharche, 2018). Globally a total of approximately 434 million hectares, or 52.2%, in 100 countries is considered sodic (Sree et al., 2025).

AMF inoculation improved stress-enduring capability by uplifting Na^+ concentrations in shoots (13.46%) and roots (47.6%), respectively, in contrast to noninoculated ones (Peng et al., 2024). Under salt stress, the mycorrhizal colonization of wheat plants supports the acquisition of nutrient balance, which regulates the uptake of Na^+ and chloride (Cl^-), lessens membrane permeability along with proline, and boosts photosynthetic pigments, resulting in better growth and productivity of plants (Elgharably et al., 2021).

Under AMF treatment, the plants have the ability to hold Na^+ within intraradical fungal hyphae and root cell vacuoles, restricting mobility of toxic ions from roots to shoots (Parvin et al., 2020). Furthermore, in alkaline soil, by increasing hairy root and root development, plants developed mycorrhiza for K^+ mobile building, which, in turn, helped mycorrhizal fungal chains to move H^+ ions into the soil, leading to adjusted soil pH (Peng et al., 2024).

23.4.4.3 Saline-sodic soil

Saline-sodic soils represent a high concentration of soluble salt along with exchangeable Na^+, which deteriorates their productivity. Several basic characteristics of saline-sodic soil are as follows: Na adsorption ratio (SAR) > 13, EC > 4 dS/m, exchangeable Na percentage (ESP) > 15 (Sree et al., 2025). Globally a total area (3793,123 hectares) is covered by sodic (85.6%) and saline-sodic (13.3%) soils (Mandal et al., 2025).

Nawaz et al. (2023) confirmed that the inoculation of AMF in saline–alkaline soil develops resilience by upgrading the K^+/Na^+ ratio, which supports the development of tolerance against salt stress. Similarly the presence of AMF in saline soil helps to manage Cl^- homeostasis through increasing its sequestration in vacuoles, diminishing the toxicity of Cl^-, and facilitating plant growth. For instance, under saline–alkali stress, the inoculation of *F. mosseae* in *Populus simonii* × *Pinus nigra* upgraded the distribution of Cl^- and plant biomass, as well as regulated Cl^- and Na^+ homeostasis. Moreover, the AMF inoculation supports better uptake of water and nutrients, which further develops resistance against salt stress in plants (Dong et al., 2023; Evelin et al., 2019). Fan et al. (2024) also reported that the inoculation of AMF in soil under saline–alkali conditions improves soil quality by enrichment of nutrients (N, P, K), which, in turn, improves the yield of maize by 18.63%. Scientists have reported low aggregate stability in sodic-saline soils, but the occurrence of AMF in such soils produces glomalin, which helps to develope stable aggregates. Despite reduced glomalin secretion in soil facing salinity, the positive variation in its chemical nature and response behavior could ruminate on its effects on soil aggregation (Aliasgharzad & Malekzadeh, 2024).

23.4.5 Remediation of polluted soil

23.4.5.1 Organic pollutants

Organic pollutants [such as pharmaceuticals, pesticides, and petroleum-based hydrocarbons, i.e., polycyclic aromatic hydrocarbons (PAHs)] are composed of C-containing substances that negatively impact the environment (Iyiola et al., 2024; Zeb et al., 2022). Such toxicants are discharged into the surroundings because of anthropogenic activities (such as agriculture and mining), which causes soil degradation and eventually leads to pollutant deposition in agricultural commodities (Bakhshayeshan-Agdam et al., 2024). Furthermore, according to Fecih & Baoune (2019), the utilization of AMF in bioremediation is a leading biological source that could diminish persistent organic pollutants (POPs), which causes the elimination of the pollutants from soils and promotes the quality and productivity of plants under contaminated soils.

Findings of research by Li et al. (2023a) revealed that the inoculation of AMF (*Rhizophagus intraradices, F. mosseae*, and *Laroideoglomus etunicatum*) significantly improved organic acids (α-linolenic acid, octadecanedioic acid, arachidonic acid, 5-methoxysalicylic acid, and 10,12,14-octadecarachidonic acid). Moreover, AMF species elevated the levels of dehydrogenase, laccase, and polyphenol oxidase, which performed their vital roles in the biodegradation of PAHs. The AMF inoculation obviously diminished the residual concentration of PAHs in soil by 48.5%. Most obvious findings in terms of plant biomass, shoot PAHs concentration, root PAHs concentration, soil bacterial biomass (51.7%, 53.1%, 565%, and 141%, respectively), as well as enzymatic activities (polyphenol oxidase and soil catalase activity by 51.9% and 100%), were influenced by AMF inoculation (Shi et al., 2022).

Glomalin promoted phenanthrene sorption by Fe-montmorillonites via elevating their hydrophobic potential and surface binding (Chen et al., 2019). Likewise the presence of glomalin boosts the deposition of PAHs in roots of ryegrass (*Lolium multiflorum* Lam.) and therefore diminishes translocation of PAHs into aboveground parts of the plant (Chen et al., 2018). Several mechanisms of AMF are held responsible for the development of resistance in plants against POPs as follows: (1) Plant biomass improves because of better water availability and nutrient provision; (2) stimulated microbial biomass and promoted activities of pollutant-degrading enzymes; (3) lowered toxic influences of pollutants in terms of oxidative stress; (4) reduced usage of herbicides and pesticides; and (5) accumulated pollutants into AMF structures (Bakhshayeshan-Agdam et al., 2024).

23.4.5.2 *Inorganic pollutants*

23.4.5.2.1 Heavy metals/Metalloids

HMs, that is, nickel (Ni), lead (Pb), cadmium (Cd), chromium (Cr), copper (Cu), Zn, mercury (Hg), and arsenic (As) pollution cause a negative influence on the soil because of their low degradability (Angon et al., 2024). HM release for a prolonged period may alter the overall functionality of agricultural soil, deteriorate the quality of crops, and ultimately cause health issues in humans via the food chain (Yan & Yang, 2025). The AMF have exceptional potential to decontaminate such HM-polluted soils and improve soil fertility and develop resistance of plants against HMs stress (Silva-Castro et al., 2022).

Furthermore, Khan et al. (2020) revealed that the inoculation of AMF in HM-polluted soil enhanced plant productivity. Moreover, the AMF develop stress tolerance in plants, demonstrating an obvious role in ecological restoration. Similarly, AMF also form a symbiotic relationship with *Medicago truncatula* that further supports reducing the concentration of Cd in plant parts (Li et al., 2023b). In particular, Cr stress tolerance in plants is enhanced directly by AMF through stabilization and indirectly through the symbiotic relationship between host plants and AMF, which, in turn, improves physiological processes and nutrient acquisition (Ahammed et al., 2023; Riaz et al., 2021).

AMF have the ability to form complexes with HMs, thus supporting the lowering of HMs toxicity to the plants while immobilizing HMs' within the soil (Fabiyi et al., 2024). Interestingly the AMF also participated in reducing bioavailability of HMs in soil and their translocation in the parts of pea plant through the following mechanism: (1) keeping HM ions in hyphae, spores, arbuscules, mycelia, and vesicles; (2) forming a defensive barrier around roots of plant to restrict uptake of HMs; and (3) releasing of EPS from cell wall containing amine, imidazole, carboxyl, phosphoric, and free hydroxyl functional groups, which may adsorb HMs (Farhad et al., 2024; Patel et al., 2022; Riaz et al., 2021). In addition, the production of organic acids (such as citric, malic, and oxalic acids) as well as other compounds (such as metallothioneins, phytochelatins, and glomalin) by AMF effectively bind HMs ions (Farhad et al., 2024).

23.4.5.2.2 Combating soil compaction

Soil compaction is a phenomenon in which distortion and densification of soil occur under different stresses above its threshold, exerted by livestock and machinery (i.e., tracks, wheels, and rollers) (Alaoui & Diserens, 2018). It alters soil structure, together with aggregates and their stability, lowering porosity (including microporosity), permeability, and water infiltration potential (Ferreira et al., 2022). According to Yin et al. (2020), such weak soil structure develops resistance to root penetration and disturbs water fluxes and nutrient and gaseous exchanges and harms soil biodiversity.

The inoculation of AMF, in compacted soils, effectively promoted root morphology, building a structural base for root-driven pore formation (Du et al., 2024). Applying AMF species (*G. etunicatum, Glomus claroideum, F. mosseae*, and *Glomus fasciculatum*) at various levels (15, 30, and 45 g/polybag) exhibited obvious improvement in soil porosity and reduction in soil bulk density, compared to the control (Mulyadi & Jiang, 2023).

Furthermore, AMF result in the betterment of soil structure by producing glomalin. It is a glycoprotein that may facilitate soil aggregate stability, upgrading aeration and water retention, thereby contributing to minimizing soil erosion (Ji et al., 2019; Singh et al., 2022). Although during soil aggregate development, glomalin is released from the spore wall (in and around soil particles), AMF external hyphae, which, in turn, support soil particles to link with partially decomposed particulate organic matter, diminish bulk density, and increase porosity, available water content, and OC (Singh et al., 2022). Soil physical structure is indirectly altered by the interaction of AMF with plant roots (Aminzadeh et al., 2025). Furthermore, according to Bi et al. (2025), the AMF secrete extracellular polysaccharides that perform their role for soil aggregation, strengthen structural integrity, and develop resistance to erosion, consequently diminishing the risk of soil degradation.

23.5 Conclusion

Soil degradation has emerged as a major environmental issue with the ongoing anthropogenic activities. Globally a considerable portion of soil is experiencing land degradation, resulting in a severe decline in the productivity and quality of crops. Underdeveloped countries facing food security issues are on the verge of malnutrition and economic crises because of reduced crop production from degraded lands. Various physical, chemical, and biological approaches have been proposed over the period to address the degradation issue of soil. AMF presence in soil exhibited extraordinary impacts on the restoration of these degraded soils because of several beneficial functions. The extraordinary potential of AMF for soil improvement, mobilizing nutrients, and bioremediating such affected soils by numerous mechanisms confirms their ecological importance. AMF induce tolerance in plants growing on affected soils by stimulating microbial communities and soil enzymes. The production of glomalin offers carbon sequestration, soil aggregation, and a nutrient reservoir to affected soils. In addition, the scalability and effectiveness of soil remediation by AMF is improving day by day. As research and innovation continue to evolve, the integration of AMF-based approaches into broader soil remediation strategies offers hope for sustainable soil rehabilitation and improved crop production.

Abbreviations

ABA	abscisic acid
AMF	arbuscular mycorrhizal fungi
As	arsenic
Cd	cadmium
Ca	calcium
C	carbon
Cl	chloride
Cr	chromium
Cu	copper
EC	electrical conductivity
EPS	extracellular polymeric substances
IAA	indole-3-acetic acid
Fe	iron
Pb	lead
Mg	magnesium
Hg	mercury
MNC	microbial necromass carbon
Ni	nickel
N	nitrogen
OC	organic carbon
POPs	persistent organic pollutants
P	phosphorus
PGPR	plant growth-promoting rhizobacteria
PAHs	polycyclic aromatic hydrocarbons
K	potassium
ROS	reactive oxygen species
Na	sodium
SOC	soil organic carbon
SOM	soil organic matter
Zn	zinc

References

AbdelRahman, M. A. (2023). An overview of land degradation, desertification, and sustainable land management using GIS and remote sensing applications. *Rendiconti Lincei. Scienze Fisiche e Naturali, 34*(3), 767–808.

Ahammed, G. J., & Hajiboland, R. (2024). Introduction to arbuscular mycorrhizal fungi and higher plant symbiosis: Characteristic features, functions, and applications. *Arbuscular mycorrhizal fungi and higher plants: Fundamentals and applications* (pp. 1–17). Springer Nature Singapore.

Ahammed, G. J., Shamsy, R., Liu, A., & Chen, S. (2023). Arbuscular mycorrhizal fungi-induced tolerance to chromium stress in plants. *Environmental Pollution, 327*, 121597.

Ahmad, H. T., Hussain, A., Aimen, A., Jamshaid, M. U., Ditta, A., Asghar, H. N., & Zahir, Z. A. (2022). Improving resilience against drought stress among crop plants through inoculation of plant growth-promoting rhizobacteria. In Azamal Husen, & Mohammad Jawaid (Eds.), *Harsh environment and plant resilience: Molecular and functional aspects* (pp. 387–408). Springer Cham.

Ahmad, J, Hayat, F, Khan, U, Ahmed, N, Li, J, Ercisli, S, Iqbal, S, Javed, HU, Alyas, T, Tu, P, & Chen, J (2024). Melatonin: A promising approach to enhance abiotic stress tolerance in horticultural plants. *South African Journal of Botany, 164*, 66–76.

Ahmed, AM, Wais, AH, Ditta, A, Islam, MR, Chowdhury, MK, Hosen, MM, Islam, HN, Soufan, W, Sabagh, AE, & Islam, MS (2024). Seed germination and early seedling growth of sorghum (*Sorghum bicolor* L. Moench) genotypes under salinity stress. *Polish Journal of Environmental Studies, 33*(3), 3019–3032.

Ahmed, N., Li, J., Li, Y., Deng, L., Deng, L., Chachar, M., & Tu, P. (2025). Symbiotic synergy: How arbuscular mycorrhizal fungi enhance nutrient uptake, stress tolerance, and soil health through molecular mechanisms and hormonal regulation. *IMA Fungus, 16*, e144989.

Akbar, M, Chohan, SA, Yasin, NA, Ahmad, A, Akram, W, & Nazir, A (2023). Mycorrhizal inoculation enhanced tillering in field grown wheat, nutritional enrichment and soil properties. *PeerJ, 11*, e15686.

Alaoui, A., & Diserens, E. (2018). Mapping soil compaction – A review. *Current Opinion in Environmental Science & Health, 5*, 60–66.

Ali, MY, Manj, IA, Ahmed, I, Iqbal, R, Ditta, A, Kumarasamy, V, Eldin, SM, Ali, I, Alwahibi, MS, & Elshikh, MS (2024). Characterization of salt-tolerant cultivars of date palm based on morphological and biochemical responses under salinity stress. *Polish Journal of Environmental Studies, 33*(4), 4019–4029.

Aliasgharzad, N., & Malekzadeh, E. (2024). Glomalin and carbon sequestration in terrestrial ecosystems. *Arbuscular mycorrhizal fungi and higher plants: Fundamentals and applications* (pp. 239–258). Springer Nature Singapore.

Aminzadeh, A., Dorostkar, V., & Asghari, H. R. (2025). Soil structural stability improvement using arbuscular mycorrhizal fungi and biochar in water repellent and non-water repellent soil. *Soil Use and Management, 41*(1), e70024.

Angon, P. B., Islam, M. S., Das, A., Anjum, N., Poudel, A., & Suchi, S. A. (2024). RETRACTED: Sources, effects and present perspectives of heavy metals contamination: Soil, plants and human food chain. *Heliyon, 10*(7), e28357.

Ansari, R. A., Sumbul, A., Rizvi, R., & Mahmood, I. (2024). Impact of environmental gases on mycorrhizal symbiosis and its influence on ecosystem functioning under the current climate change scenario. *Mycorrhizal symbiosis and agroecosystem restoration* (pp. 51–76). Springer Nature Singapore.

Bakhshayeshan-Agdam, H., Houshani, M., & Salehi-Lisar, S. Y. (2024). Arbuscular mycorrhizal fungi in plant tolerance to organic pollutants and associated food safety. *Arbuscular mycorrhizal fungi and higher plants: Fundamentals and applications* (pp. 177–191). Springer Nature Singapore.

Balík, J., Kulhánek, M., Černý, J., Sedlář, O., Suran, P., & Asrade, D. A. (2022). The influence of organic and mineral fertilizers on the quality of soil organic matter and glomalin content. *Agronomy, 12*(6), 1375.

Bello, T. T., & Fabiyi, O. A. (2024). Advancement in mycorrhizal fungi-based sustainable plant disease management. *Mycorrhizal symbiosis and agroecosystem restoration* (pp. 125–145). Springer Nature Singapore.

Bhattacharyya, R., Bhatia, A., Ghosh, B. N., Santra, P., Mandal, D., Kumar, G., & Chaudhari, S. K. (2023). Soil degradation and mitigation in agricultural lands in the Indian Anthropocene. *European Journal of Soil Science, 74*(4), e13388.

Bi, Y., Tian, L., Du, X., & Yin, K. (2025). Optimizing soil pore structure in mined land: Integrating arbuscular mycorrhizal fungi and mixed planting for ecological restoration. *Plant and Soil, 515*, 1405–1422.

Bi, Y., Xiao, L., Guo, C., & Christie, P. (2020). Revegetation type drives rhizosphere arbuscular mycorrhizal fungi and soil organic carbon fractions in the mining subsidence area of northwest China. *Catena, 195*, 104791.

Błaszkowski, J., Niezgoda, P., Zubek, S., Meller, E., Milczarski, P., Malicka, M., & Magurno, F. (2021). *Dominikia bonfanteae* and *Glomus atlanticum*, two new species in the *Glomeraceae* (phylum Glomeromycota) with molecular phylogenies reconstructed from two unlinked loci. *Mycological Progress, 20*, 131–148.

Boyno, G., Demir, S., Rezaee Danesh, Y., Durak, E. D., Çevik, R., Farda, B., & Pellegrini, M. (2023). A new technique for the extraction of arbuscular mycorrhizae fungal spores from rhizosphere. *Journal of Fungi, 9*(8), 845.

Carteron, A., Beigas, M., Joly, S., Turner, B. L., & Laliberté, E. (2021). Temperate forests dominated by arbuscular or ectomycorrhizal fungi are characterized by strong shifts from saprotrophic to mycorrhizal fungi with increasing soil depth. *Microbial Ecology, 82*, 377–390. doi:10.1007/s00248-020-01540-7.

Chandrasekaran, M. (2022). Arbuscular mycorrhizal fungi mediated enhanced biomass, root morphological traits and nutrient uptake under drought stress: A meta-analysis. *Journal of Fungi, 8*(7), 660.

Chareesri, A., De Deyn, G. B., Sergeeva, L., Polthanee, A., & Kuyper, T. W. (2020). Increased arbuscular mycorrhizal fungal colonization reduces yield loss of rice (*Oryza sativa* L.) under drought. *Mycorrhiza, 30*(2), 315–328.

Chen, XW, Wong, JTF, Wang, JJ, et al. (2021). Effects of mycor rhizal Bermuda grass on low-range soil matric suction. *Journal of Soils and Sediments, 21*, 990–1000.

Chen, S., Sheng, X., Qin, C., Waigi, M. G., & Gao, Y. (2019). Glomalin-related soil protein enhances the sorption of polycyclic aromatic hydrocarbons on cation-modified mont morillonite. *Environment International, 132*, 105093.

Chen, S., Wang, J., Waigi, M. G., & Gao, Y. (2018). Glomalin-related soil protein influences the accumulation of polycyclic aromatic hydrocarbons by plant roots. *Science of the Total Environment, 644*, 465–473.

Cheng, X. F., Wu, H. H., Zou, Y. N., Wu, Q. S., & Kuča, K. (2021). Mycorrhizal response strategies of trifoliate orange under well-watered, salt stress, and waterlogging stress by regulating leaf aquaporin expression. *Plant Physiology and Biochemistry, 162*, 27–35.

Choudhary, O. P., & Kharche, V. K. (2018). Soil salinity and sodicity, *Soil science: An introduction* (12, pp. 353–384). New Delhi: Indian Society of Soil Science.

da Silva, D. F., Moreira, J. V., de Sousa, L. I. S., Santana, M. C., Mota, J. C. A., dos Santos Queiroz, A., & de Araujo Pereira, A. P (2022). Arbuscular mycorrhizal fungi community in soils under desertification and restoration in the Brazilian semiarid. *Microbiological Research, 264*, 127161.

Das, D., Ullah, H., Himanshu, S. K., Tisarum, R., Cha-Um, S., & Datta, A. (2022). Arbuscular mycorrhizal fungi inoculation and phosphorus application improve growth, physiological traits, and grain yield of rice under alternate wetting and drying irrigation. *Journal of Plant Physiology, 278*, 153829.

de Novais, C. B., Sbrana, C., da Conceição Jesus, E., Rouws, L. F. M., Giovannetti, M., Avio, L., & de Faria, S. M. (2020). Mycorrhizal networks facilitate the colonization of legume roots by a symbiotic nitrogen-fixing bacterium. *Mycorrhiza, 30*(2), 389–396.

Dedysh, S. N., Ivanova, A. A., Begmatov, S. A., Beletsky, A. V., Rakitin, A. L., Mardanov, A. V., & Ravin, N. V. (2022). Acidobacteria in fens: Phylogenetic diversity and genome analysis of the key representatives. *Microbiology, 91*(6), 662–670.

Delaux, PM, & Schornack, S (2021). Plant evolution driven by interactions with symbiotic and pathogenic microbes. *Science, 371*, eaba6605.

Ditta, A., & Ullah, N. (2023). Perspectives of using plant growth-promoting rhizobacteria under salinity stress for sustainable crop production. In M. Ghorbanpour, & M. A. Shahid (Eds.), *Plant stress mitigators: Types, techniques and functions* (pp. 231–247). Woodhead Publishing Ltd., Elsevier UK ISBN 978-0-323-89871-3.

Dong, F., Wang, Y., Tao, J., Xu, T., & Tang, M. (2023). Arbuscular mycorrhizal fungi affect the expression of PxNHX gene family, improve photosynthesis and promote *Populus simonii×P. nigra* growth under saline-alkali stress. *Frontiers in Plant Science, 14*, 1104095.

Du, X., Bi, Y., Tian, L., Li, M., & Yin, K. (2024). Enhancing infiltration characteristics of compact soil in open-pit dumps through arbuscular mycorrhizal fungi inoculation in *Amorpha fruticosa*: Mechanisms and effects. *Catena, 247*, 108515.

Duan, H. X., Luo, C. L., Zhou, R., Zhao, L., Zhu, S. G., Chen, Y., & Xiong, Y. C. (2024). AM fungus promotes wheat grain filling via improving rhizospheric water & nutrient availability under drought and low density. *Applied Soil Ecology, 193*, 105159.

Elgharably, A., & Nafady, N. A. (2021). Inoculation with *Arbuscular mycorrhizae, Penicillium funiculosum* and *Fusarium oxysporum* enhanced wheat growth and nutrient uptake in the saline soil. *Rhizosphere, 18*, 100345.

Evelin, H., Devi, T. S., Gupta, S., & Kapoor, R. (2019). Mitigation of salinity stress in plants by arbuscular mycorrhizal symbiosis: Current understanding and new challenges. *Frontiers in Plant Science, 10*, 470.

Fabiyi, O. A., Ansari, R. A., Rizvi, R., Sumbul, A., & Mahmood, I. (2024). Influence of arbuscular mycorrhizal fungi on soil health amelioration and plant fitness under hostile environment. *Mycorrhizal symbiosis and agroecosystem restoration* (pp. 227–248). Springer Nature Singapore.

Faghihinia, M., Jansa, J., Halverson, L. J., & Staddon, P. L. (2023). Hyphosphere microbiome of arbuscular mycorrhizal fungi: A realm of unknowns. *Biology and Fertility of Soils, 59*(1), 17–34.

Fall, A. F., Nakabonge, G., Ssekandi, J., Founoune-Mboup, H., Apori, S. O., Ndiaye, A., & Ngom, K. (2022). Roles of arbuscular mycorrhizal fungi on soil fertility: contribution in the improvement of physical, chemical, and biological properties of the soil. *Frontiers in Fungal Biology, 3*, 723892.

Fan, L., Zhang, P., Cao, F., Liu, X., Ji, M., & Xie, M. (2024). Effects of AMF on maize yield and soil microbial community in sandy and saline soils. *Plants, 13*(15), 2056.

Farhad, M., Noor, M., Yasin, M. Z., Nizamani, M. H., Turan, V., & Iqbal, M. (2024). Interactive suitability of rice stubble biochar and arbuscular mycorrhizal fungi for improving wastewater-polluted soil health and reducing heavy metals in peas. *Sustainability, 16*(2), 634.

Fecih, T., & Baoune, H. (2019). Arbuscular mycorrhizal fungi remediation potential of organic and inorganic compounds. *Microbial technology for the welfare of society* (pp. 247–257). Springer Singapore.

Ferreira, C. S., Seifollahi-Aghmiuni, S., Destouni, G., Ghajarnia, N., & Kalantari, Z. (2022). Soil degradation in the European Mediterranean region: Processes, status and consequences. *Science of the Total Environment, 805*, 150106.

Gałązka, A., Grządziel, J., Gałązka, R., Gawryjołek, K., Ukalska-Jaruga, A., & Smreczak, B. (2020). Fungal community, metabolic diversity, and glomalin-related soil proteins (GRSP) content in soil contaminated with crude oil after long-term natural bioremediation. *Frontiers in Microbiology, 11*, 572314.

Gerami, Z., Lakzian, A., Hemati, A., Amirifar, A., Asgari Lajayer, B., & Van Hullebusch, E. D. (2021). Effect of cadmium on sorghum root colonization by glomeral fungi and its impact on total and easily extractable glomalin production. *Environmental Science and Pollution Research, 28*, 34570–34583.

Ghani, M. I., Ali, A., Atif, M. J., Ali, M., Amin, B., & Cheng, Z. (2022). Arbuscular mycorrhizal fungi and dry raw garlic stalk amendment alleviate continuous monocropping growth and photosynthetic declines in eggplant by bolstering its antioxidant system and accumulation of osmolytes and secondary metabolites. *Frontiers in Plant Science, 13*, 849521.

Gujre, N., Agnihotri, R., Rangan, L., Sharma, M. P., & Mitra, S. (2021). Deciphering the dynamics of glomalin and heavy metals in soils contaminated with hazardous municipal solid wastes. *Journal of Hazardous Materials, 416*, 125869.

Hawkins, H. J., Cargill, R. I., Van Nuland, M. E., Hagen, S. C., Field, K. J., Sheldrake, M., & Kiers, E. T. (2023). Mycorrhizal mycelium as a global carbon pool. *Current Biology, 33*(11), R560–R573.

Herawati, A., Mujiyo, M., Dewi, W. S., Syamsiyah, J., & Romadhon, M. R. (2024). Improving microbial properties in Psamments with mycorrhizal fungi, amendments, and fertilizer. *Eurasian Journal of Soil Science, 13*(1), 59–69.

Iqbal, M. T., Ahmed, I. A., Isik, M., Sultana, F., & Ortaş, I. (2021). Role of mycorrhizae inoculations on nutrient uptake in rice grown under aerobic and anaerobic water management. *Journal of Plant Nutrition, 44*(4), 550–568.

Iyiola, A. O., Ipinmoroti, M. O., Akingba, O. O., Ewutanure, J. S., Setufe, S. B., Bilikoni, J., & Wangboje, O. M. (2024). Organic chemical pollutants within water systems and sustainable management strategies. *Water crises and sustainable management in the global south* (pp. 211–251). Springer Nature Singapore.

Javeed, H. M. R., Ali, M., Zamir, M. S. I., Qamar, R., Kanwal, S., Andleeb, H., & Masood, N. (2023). Biochar and arbuscular mycorrhizae fungi to improve soil organic matter and fertility. *Sustainable agriculture reviews 61: Biochar to improve crop production and decrease plant stress under a changing climate* (pp. 331–354). Springer International Publishing.

Ji, L., Tan, W., & Chen, X. (2019). Arbuscular mycorrhizal mycelial networks and glomalin-related soil protein increase soil aggregation in Calcaric Regosol under well-watered and drought stress conditions. *Soil and Tillage Research, 185*, 1–8.

Jiang, F., Zhang, L., Zhou, J., George, T. S., & Feng, G. (2021). Arbuscular mycorrhizal fungi enhance mineralisation of organic phosphorus by carrying bacteria along their extraradical hyphae. *New Phytologist, 230*(1), 304–315.

Kashyap, P., Sharma, I., Kashyap, S., & Agarwala, N. (2024). Arbuscular mycorrhizal fungi (AMF)-mediated control of foliar fungal diseases. *Arbuscular mycorrhizal fungi and higher plants: Fundamentals and applications* (pp. 193–223). Springer Nature Singapore.

Khan, M. A., Ramzani, P. M. A., Zubair, M., Rasool, B., Khan, M. K., Ahmed, A., & Iqbal, M. (2020). Associative effects of lignin-derived biochar and arbuscular mycorrhizal fungi applied to soil polluted from Pb-acid batteries effluents on barley grain safety. *Science of the Total Environment, 710*, 136294.

Koshila Ravi, R., Anusuya, S., Balachandar, M., & Muthukumar, T (2019). Microbial interactions in soil formation and nutrient cycling. *Mycorrhizosphere and pedogenesis* (pp. 363–382). Springer Singapore.

Lal, R. (2020). Soil erosion and gaseous emissions. *Applied Sciences, 10*(8), 2784.

Li, M., Chen, X. W., & Leung, A. K. (2025). Grass species and mycorrhizal fungi improved aggregate stability of compacted and vegetated soils. *Plant and Soil, 511*(1), 1081–1099.

Li, W., Chen, K., Li, Q., Tang, Y., Jiang, Y., & Su, Y. (2023b). Effects of arbuscular mycorrhizal fungi on alleviating cadmium stress in *Medicago truncatula* Gaertn. *Plants, 12*(3), 547.

Li, X., Kang, X., Zou, J., Yin, J., Wang, Y., Li, A., & Ma, X. (2023a). Allochthonous arbuscular mycorrhizal fungi promote *Salix viminalis* L.–mediated phytoremediation of polycyclic aromatic hydrocarbons characterized by increasing the release of organic acids and enzymes in soils. *Ecotoxicology and Environmental Safety, 249*, 114461.

Liu, Y., Lu, J., Cui, L., Tang, Z., Ci, D., Zou, X., & Si, T. (2023). The multifaceted roles of arbuscular mycorrhizal fungi in peanut responses to salt, drought, and cold stress. *BMC Plant Biology, 23*(1), 36.

Ma, J., Ma, Y., Wei, Z., Wu, J., Sun, C., Yang, J., & Huang, J. (2021). Effects of arbuscular mycorrhizal fungi symbiosis on microbial diversity and enzyme activities in the rhizosphere soil of *Artemisia annua*. *Soil Science Society of America Journal, 85*(3), 703–716.

Mandal, A. K., Arora, S., Sharma, P. C., & Yadav, R. K. (2025). Spatial assessment, mapping, and characterization of salt-affected soils in Uttar Pradesh state of the Gangetic plain (IGP), India, for planning reclamation and management. *Environmental Monitoring and Assessment, 197*(7), 739.

Marwanto, M, Bustamam, H, Handajaningsih, M, et al. (2024). Co-application of arbuscular mycorrhizae via seed coating and phosphorus fertilizer for enhancing growth, yield, and nutrient uptake in ultisols for maize. *TERRA: Journal of Land Restoration, 7*, 8–13.

Matos, P. S., Silva, C. F. D., Damian, J. M., Cerri, C. E. P., Pereira, M. G., & Zonta, E. (2021). Beneficial services of glomalin and arbuscular mycorrhizal fungi in degraded soils in Brazil. *Scientia Agricola, 79*(5), e20210064.

Mulyadi, & Jiang, L (2023). Combined application of arbuscular mycorrhizal fungi (AMF) and nitrogen fertilizer alters the physicochemical soil properties, nitrogen uptake, and rice yield in a polybag experiment. *Agriculture, 13*(7), 1364.

Muthukumar, T., & Karthikeyan, A. (2024). Diversity of arbuscular mycorrhizal fungi in mined land: Distribution and function in reclamation of mined land ecosystems. *Mycorrhizal symbiosis and agroecosystem restoration* (pp. 3–30). Springer Nature Singapore.

Navarro, J. M., Pérez-Tornero, O., & Morte, A. (2024). Effects of salinity on 'Fino 95' lemon trees inoculated with arbuscular mycorrhizal fungi. *Horticulturae, 10*(6), 612.

Nawaz, A., Qamar, Z. U., Marghoob, M. U., Imtiaz, M., Imran, A., & Mubeen, F. (2023). Contribution of potassium solubilizing bacteria in improved potassium assimilation and cytosolic K+/Na+ ratio in rice (*Oryza sativa* L.) under saline-sodic conditions. *Frontiers in Microbiology, 14*, 1196024.

Nazir, A, Sarfraz, W, Ditta, A, Khalid, N, Farid, M, Shafiq, M, Firdaus-e-Bareen, ZF Rizvi, & Naeem, N (2023). Synergistic impact of two autochthonous saprobic fungi (*A. niger* and *T. pseudokoningii*) on the growth, ionic contents, and metals uptake in *Brassica juncea* L. and *Vigna radiata* L. under tannery solid waste contaminated soil. *International Journal of Phytoremediation, 25*(11), 1488–1500.

Negacz, K., Malek, Ž., de Vos, A., & Vellinga, P. (2022). Saline soils worldwide: Identifying the most promising areas for saline agriculture. *Journal of Arid Environments, 203*, 104775.

Nie, W., He, Q., Guo, H., Zhang, W., Ma, L., Li, J., & Wen, D. (2024). Arbuscular mycorrhizal fungi: Boosting crop resilience to environmental stresses. *Microorganisms,, 12*(12), 2448.

Olaniyan, F. T., Alori, E. T., Adekiya, A. O., Ayorinde, B. B., Daramola, F. Y., Osemwegie, O. O., & Babalola, O. O. (2022). The use of soil microbial potassium solubilizers in potassium nutrient availability in soil and its dynamics. *Annals of Microbiology, 72*(1), 45.

Omer, E., Szlatenyi, D., Csenki, S., Alrwashdeh, J., Czako, I., & Láng, V. (2024). Farming practice variability and its implications for soil health in agriculture: A review. *Agriculture, 14*(12), 2114.

Pais, I. P., Moreira, R., Semedo, J. N., Ramalho, J. C., Lidon, F. C., Coutinho, J., & Scotti-Campos, P. (2022). Wheat crop under waterlogging: Potential soil and plant effects. *Plants, 12*(1), 149.

Parvin, S., Van Geel, M., Yeasmin, T., Verbruggen, E., & Honnay, O. (2020). Effects of single and multiple species inocula of arbuscular mycorrhizal fungi on the salinity tolerance of a Bangladeshi rice (*Oryza sativa* L.) cultivar. *Mycorrhiza, 30*(4), 431–444.

Patel, S., Sharma, A., & Batra, N. G. (2022). Arbuscular Mycorrhizal fungi-Assisted Bioremediation of heavy metals: A revaluation. *Innovations in environmental biotechnology* (pp. 785–804). Springer Nature Singapore.

Pedrinho, A., Mendes, L. W., de Araujo Pereira, A. P., Araujo, A. S. F., Vaishnav, A., Karpouzas, D. G., & Singh, B. K. (2024). Soil microbial diversity plays an important role in resisting and restoring degraded ecosystems. *Plant and Soil, 500*(1), 325–349.

Peng, Z., Zulfiqar, T., Yang, H., Wang, M., & Zhang, F. (2024). Effect of arbuscular mycorrhizal fungi (AMF) on photosynthetic characteristics of cotton seedlings under saline-alkali stress. *Scientific Reports, 14*(1), 8633.

Pooja, P., Tallapragada, S., Lamba, A., & Punia, S. (2025). Role played by arbuscular mycorrhizal fungi in amelioration of salinity stress: A review. *Plant and Soil, 511*(1), 27–42.

Riaz, M, Kamran, M, Fang, Y, Wang, Q, Cao, H, Yang, G, Deng, L, Wang, Y, Zhou, Y, Anastopoulos, I, & Wang, X (2021). Arbuscular mycorrhizal fungi-induced mitigation of heavy metal phytotoxicity in metal contaminated soils: a critical review. *Journal of Hazardous Materials, 402*, 123919.

Saeed, HS, Munir, M, Samra, S, Farooq, A, Mahmood, H, Ahmad, I, Akram, MI, Iqbal, R, Elsadek, MF, Almutairi, SM, Ditta, A, & Daraz, U (2025). Screening of biosurfactants producing fungi from infected *Citrus sinensis* (L.). *Polish Journal of Environmental Studies, 34*(5), 5261–5270.

Sagar, A., Rathore, P., Ramteke, P. W., Ramakrishna, W., Reddy, M. S., & Pecoraro, L. (2021). Plant growth promoting rhizobacteria, arbuscular mycorrhizal fungi and their synergistic interactions to counteract the negative effects of saline soil on agriculture: Key macromolecules and mechanisms. *Microorganisms, 9*(7), 1491.

Saleem, A, Raza, MAS, Tahir, MA, Iqbal, R, Aslam, MU, Toleikiene, M, Khan, MS, Alwahibi, MS, Elshikh, MS, & Ditta, A (2025). Impact of biogas slurry on physiological and antioxidant mechanisms of wheat under drought stress. *Polish Journal of Environmental Studies, 34*(2), 1721–1731.

Saleem, A., Raza, M. A. S., Khan, I. H., Tahir, M. A., Iqbal, R., Aslam, M. U., Ejaz, Z., Ditta, A., AlMunqedhi, B. M., & Al Farraj, D. A. (2024). Amelioration of drought stress in wheat by using plant growth-promoting rhizobacteria and biogas slurry. *Global NEST Journal, 26*(8), 05890.

Sasidharan, R., Bailey-Serres, J., Ashikari, M., Atwell, B. J., Colmer, T. D., Fagerstedt, K., & Voesenek, L. A. (2017). Community recommendations on terminology and procedures used in flooding and low oxygen stress research. *New Phytologist, 214*(4), 1403–1407.

Sedaghati, E., Yazdanpanah, M., & Nadi, M. (2021). A review of taxonomic studies of arbuscular mycorrhizal fungi in Iran. *Mycologia Iranica, 8*(2), 15–30.

Sharma, P. (2024). Biochar application for sustainable soil erosion control: a review of current research and future perspectives. *Frontiers in Environmental Science, 12*, 1373287.

Shi, J., Wang, X., & Wang, E. (2023). Mycorrhizal symbiosis in plant growth and stress adaptation: From genes to ecosystems. *Annual Review of Plant Biology, 74*(1), 569–607.

Shi, Y., Wang, S., Guo, J., Xu, Z., Wang, S., & Sang, Y. (2022). Effects of arbuscular mycorrhizal inoculation on the phytoremediation of PAH-contaminated soil: A meta-analysis. *Chemosphere, 307*, 136033.

Silva-Castro, G. A., Cano, C., Moreno-Morillas, S., Bago, A., & García-Romera, I. (2022). Inoculation of indigenous arbuscular mycorrhizal fungi as a strategy for the recovery of long-term heavy metal-contaminated soils in a mine-spill area. *Journal of Fungi, 9*(1), 56.

Singh, A. K., Zhu, X., Chen, C., Wu, J., Yang, B., Zakari, S., & Liu, W. (2022). The role of glomalin in mitigation of multiple soil degradation problems. *Critical Reviews in Environmental Science and Technology, 52*(9), 1604–1638.

Solomon, N., Birhane, E., Tilahun, M., Schauer, M., Gebremedhin, M. A., Gebremariam, F. T., & Newete, S. W. (2024). Revitalizing Ethiopia's highland soil degradation: A comprehensive review on land degradation and effective management interventions. *Discover Sustainability, 5*(1), 106.

Sree, S. S., Al-zharani, M., Nasr, F. A., Alneghery, L. M., Kumar, T. T., Sureshkumar, B. T., & Dineshkumar, R. (2025). Innovative bio-amelioration strategies for sustainable reclamation of salt-affected agroecosystems: A review. *Plant and Soil, 516*, 67–114.

Srivastava, P., Giri, N., & Mandal, D. (2019). 137 Cs technology for soil erosion and soil carbon redistribution. *Current Science, 116*(6), 888–889.

Thind, S, Chaudhary, MS, Ditta, A, Hussain, I, Parveen, A, Ullah, N, Mahmood, Q, Al-ashkar, I, & Sabagh, AEL (2022). Impact of mycorrhizal fungi from different rhizospheric soils on fungal colonization, growth, and chlorophyll contents of *Cenchrus ciliaris. Agronomy, 12*(11), 2644.

Ujvári, G., Turrini, A., Avio, L., & Agnolucci, M. (2021). Possible role of arbuscular mycorrhizal fungi and associated bacteria in the recruitment of endophytic bacterial communities by plant roots. *Mycorrhiza, 31*(5), 527–544.

Upadhayay, V. K., Singh, J., Khan, A., Lohani, S., & Singh, A. V. (2019). Mycorrhizal mediated micronutrients transportation in food based plants: A biofortification strategy. *Mycorrhizosphere and pedogenesis* (pp. 1–24). Springer Singapore.

Wang, F., Zhang, L., Zhou, J., Rengel, Z., George, T. S., & Feng, G. (2022). Exploring the secrets of hyphosphere of arbuscular mycorrhizal fungi: Processes and ecological functions. *Plant and Soil, 481*(1), 1–22.

Wang, G., Jin, Z., George, T. S., Feng, G., & Zhang, L. (2023). Arbuscular mycorrhizal fungi enhance plant phosphorus uptake through stimulating hyphosphere soil microbiome functional profiles for phosphorus turnover. *New Phytologist, 238*(6), 2578–2593.

Wang, Y., Bao, X., & Li, S. (2021). Effects of arbuscular mycorrhizal fungi on rice growth under different flooding and shading regimes. *Frontiers in Microbiology, 12*, 756752.

Xing, D., Zhang, H., Cheng, H., Zhang, F., Shi, Y., Peng, S., & Twagirayezu, G. (2025). Arbuscular mycorrhizal fungi decrease soil nitrate by accelerating mulberry uptake and dissimilatory nitrate reduction to ammonium. *Environmental Technology & Innovation, 38*, 104139.

Xu, G., Pan, J., Rehman, M., Li, X., Cao, S., Wang, C., & Chen, P. (2024). Arbuscular mycorrhizal fungi-mediated drought stress tolerance in kenaf (*Hibiscus cannabinus* L.): A mechanistic approach. *Plant Growth Regulation, 103*(3), 803–824.

Yan, Y., & Yang, Y. (2025). Revealing the synergistic spatial effects in soil heavy metal pollution with explainable machine learning models. *Journal of Hazardous Materials, 482*, 136578.

Yang, Y, Luo, W, Xu, J, et al. (2022). Fallow land enhances carbon sequestration in glomalin and soil aggregates through regulating diversity and network complexity of Arbuscu lar Mycorrhizal Fungi under climate change in relatively high-latitude regions. *Frontiers in Microbiology, 13*, 930622.

Yang, R., Wang, C., Yang, Y., Harrison, M. T., Zhou, M., & Liu, K. (2024). Implications of soil waterlogging for crop quality: A meta-analysis. *European Journal of Agronomy, 161*, 127395.

Yilmaz, A., Yildirim, E., Yilmaz, H., Soydemir, H. E., Güler, E., Ciftci, V., & Yaman, M. (2023). Use of arbuscular mycorrhizal fungi for boosting antioxidant enzyme metabolism and mitigating saline stress in Sweet Basil (*Ocimum basilicum* L.). *Sustainability, 15*(7), 5982.

Yin, R., Kardol, P., Thakur, M. P., Gruss, I., Wu, G. L., Eisenhauer, N., & Schädler, M. (2020). Soil functional biodiversity and biological quality under threat: Intensive land use outweighs climate change. *Soil Biology and Biochemistry, 147*, 107847.

Yizhu, L., Imtiaz, M., Ditta, A., Rizwan, M. S., Ashraf, M., Mehmood, S., Aziz, O., Mubeen, F., Ali, M., Elahi, N. N., Ijaz, R., S, Lelel, Shuang, C., & Tu, S. (2020). Response of growth, antioxidant enzymes, and root exudates production towards As stress in *Pteris vittata* and *Astragalus sinicus* colonized by arbuscular mycorrhizal fungi. *Environmental Science and Pollution Research, 27*, 2340–2352.

Yu, L., Zhang, H., Zhang, W., Liu, K., Liu, M., & Shao, X. (2022). Cooperation between arbuscular mycorrhizal fungi and plant growth-promoting bacteria and their effects on plant growth and soil quality. *PeerJ, 10*, e13080.

Zeb, B. S., Hayat, M. T., Zeb, T., Khan, F. Y., Abbasi, H. Z., Nawaz, I., & Ebadi, A. (2022). Uptake of organic pollutants and the effects on plants. *Sustainable plant nutrition under contaminated environments* (pp. 209–234). Springer International Publishing.

Zhang, F., Zou, Y. N., & Wu, Q. S. (2018). Quantitative estimation of water uptake by mycorrhizal extraradical hyphae in citrus under drought stress. *Scientia Horticulturae, 229*, 132–136.

Zhang, M., Che, R., Cheng, Z., Zhao, H., Wu, C., Hu, J., & Wu, Y. (2023). Decades of reforestation significantly change microbial necromass, glomalin, and their contributions to soil organic carbon. *Agriculture, Ecosystems & Environment, 346*, 108362.

Zhang, Q., Jin, C. H. E. N., Yingyue, L. I., Huang, T., Kailing, X. I. E., Jing, Z. H. O. U., & Xiaoyu, L. I. (2024). Effects of interactions between arbuscular mycorrhizal fungi and bacteria on the growth of *Lotus corniculatus* L.: From the perspective of regulating rhizosphere fungal community. *Pedosphere, 34*(2), 411–423.

Zhou, J., Bilyera, N., Guillaume, T., Yang, H., Li, F. M., & Shi, L. (2024). Microbial necromass and glycoproteins for determining soil carbon formation under arbuscular mycorrhiza symbiosis. *Science of The Total Environment, 955*, 176732.

Chapter 24

Arbuscular mycorrhizal fungi and climate-smart agriculture

Muhammad Talha Aslam[a], Saira Ghafoor[b], Imran Khan[a], Muhammad Umer Chattha[a], Rizwan Maqbool[a], Muhammad Zia Ul Haq[a], Muneeba Saleem[c] and Hossam S. El-Beltagi[d]

[a] *Department of Agronomy, University of Agriculture Faisalabad, Faisalabad, Pakistan,* [b] *Key Laboratory of Crop Physiology, Ecology, and Genetic Breeding, Ministry of Education/College of Agronomy, Jiangxi Agricultural University, Nanchang, China,* [c] *Research Center for Plant Functional Genes and Tissue Culture Technology, Jiangxi Agricultural University, Nanchang, China,* [d] *Agricultural Biotechnology Department, College of Agriculture and Food Sciences, King Faisal University, Al Hofuf, Saudi Arabia*

24.1 Introduction

Food security is a global challenge, and crop vulnerability intensifies it as observed in developing countries (Noya et al., 2018). This insecurity is compounded by three major concerns: anthropogenic climate change, a steady growth of the global population, and the accidental penalties of prevailing high-input agricultural systems. The recent intensive practices that include genetically improved crops, high reliance on chemical fertilizers, and unsustainable groundwater extraction have boosted crop production (Abdul Aziz et al., 2022; Shah & Wu, 2019). Nonetheless, such optimum gains are at a major environmental cost, including degrading the agricultural land, soil erosion, biodiversity loss, and the depletion of freshwater resources (Raza et al., 2019). Altogether, these undermine the environmental foundations pivotal for long-term sustainability in the food production system.

The global population is expected to reach 9.7 billion by 2050 (United Nations, 2015), and therefore, food production must increase substantially from present levels. Simultaneously, constant urbanization and increasing average incomes are shifting consumption patterns, which catalyze the profound dietary transition (He, 2024). This "nutrition transition" not only worsens the agricultural carbon footprint but also exacerbates the inequities in food access. Likewise, Deepika et al. (2023) reported 12% of the global population in a state of hunger. Hence achieving food security necessitates addressing production along with dietary sustainability and food distribution patterns.

The Green Revolution safeguarded the large-scale famines by a historic increase in cereal yields because of high-yielding cultivars and synthetic agrochemical applications. This significantly augmented (tripled per capita) the cereal production without a major expansion of arable land (Alam & Roy, 2024). However, its legacy is dual-edged. The environmental costs, that is, degradation of soil, water pollution, and loss of agrobiodiversity, are now apparent (Prasad, 2016). In addition, the yield of staple cereals increased in different areas, whereas the yield stagnation is linked to poor soil fertility, less genetic potential of monoculture systems, and the ecological fallout of input-intensive practices (John & Babu, 2021). This dependency on the genetically uniform crop also amplifies the systemic vulnerability to tragic pest or disease epidemics.

Critically, agriculture exists in a feedback loop with climate change, and the agricultural sector majorly contributes (24%–35%) to greenhouse gas (GHG) emissions in the world (Lynch et al., 2021). Previous studies also mentioned that agriculture is a major anthropogenic source of methane (CH_4) and nitrous oxide (N_2O) (Ahmed et al., 2025), which also affects agricultural productivity (Ali et al., 2023; Ditta et al., 2024; Rahman & Ditta, 2025; Rahman et al., 2025). Therefore it is imperative to develop an integrated, systems-based approach to safeguard future food security. For this the implementation of climate-smart agricultural practices is imperative to diminish the agricultural GHG emissions, improve the resilience of food systems to climate shocks, and sustainably promote crop productivity, thereby breaking the cycle between agricultural intensification and environmental deterioration.

Advancing Sustainable Agriculture Through Arbuscular Mycorrhizal Fungi. DOI: https://doi.org/10.1016/B978-0-443-45130-0.00031-X

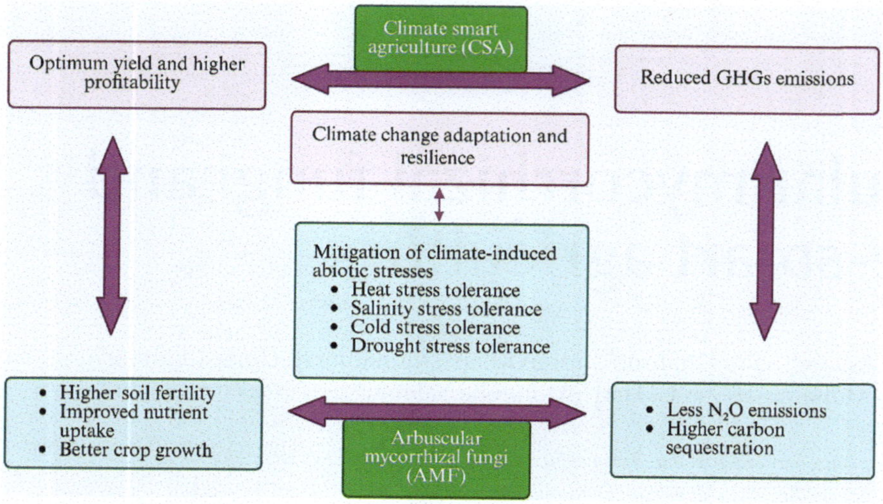

FIGURE 24.1 Arbuscular mycorrhizal fungi as a key constituent of climate-smart agriculture. *GHGs*, Greenhouse gases.

24.2 Climate-smart agriculture

Climate-smart agriculture (CSA) is the integrated framework designed to cope with the interlinked priorities of food security and climate change (Fig. 24.1). It also follows the key principles of sustainable agriculture (balancing the long-term ecological stewardship with near-term economic viability). CSA incorporates climate change adaptation and mitigation into the planning and development of agricultural systems (Majhi et al., 2023). Its objectives are threefold: to sustainably promote the agricultural productivity and farmer profits; to improve resilience and adaptive capacity to climate change; and to decrease and/or eliminate GHG emissions (Kabato et al., 2025; Mesfin & Markos, 2019).

24.3 Arbuscular mycorrhizal fungi as a tool to practice climate-smart agriculture

Arbuscular mycorrhizal fungi (AMF) are obligate symbionts within the phylum Glomeromycota and produce a mutual relationship with ~80% of terrestrial plant species, including main agricultural field crops (Deepika et al., 2023). Moreover, AMF mutualism with the members of the Brassicaceae and Chenopodiaceae families develops specialized structures (arbuscules and vesicles) responsible for symbiosis within the root cortex, and they serve as a primary interface for resource exchange (Liu & Chen, 2024). For example, fungi deliver the host plant with immobile soil nutrients (P, Zn, etc.) while receiving ~20% of the plant's ATP (adenosine triphosphate, the primary energy currency of the cell) (Deepika et al., 2023). This exchange augments the nutrient acquisition and tolerance to different biotic and abiotic stresses (Nazir et al., 2023; Saeed et al., 2025; Wahab et al., 2023).

AMF also affect the ecosystem diversity and productivity by their critical function between plant roots and the soil matrix (Lenoir et al., 2016). For example, AMF plays a dynamic role in nutrient cycling that reduces leaching losses and improves soil structure with an agroecosystem (Cavagnaro et al., 2015; Liu et al., 2020; Thind et al., 2022). In addition, AMF hyphae help in soil carbon sequestration by synthesizing the glomalin-related soil proteins (GRSPs), directly linking their activity to climate change mitigation (Martínez-Garcia et al., 2017). The low host specificity of AMFs provides an added advantage for their integration into diverse and sustainable agricultural systems across varied environmental conditions. Furthermore, given the abovementioned multifunctional benefits, AMFs represent a supreme biological input for CSA.

24.4 Role of arbuscular mycorrhizal fungi in the productivity of agricultural systems

AMF improves soil health and plant productivity. Their multifunctional role makes them prime candidates for use as bioinoculants within a climate-smart agricultural system. Specifically, biofertilization with AMF substantially decreased the 50% reliance on the inorganic fertilizers (Begum et al., 2019b); however, crop species and prevailing environmental conditions played a crucial role in decreasing the inorganic fertilizer use (Begum et al., 2019b). This decrease in input use directly aligns with the CSA goals of sustainable intensification and mitigation.

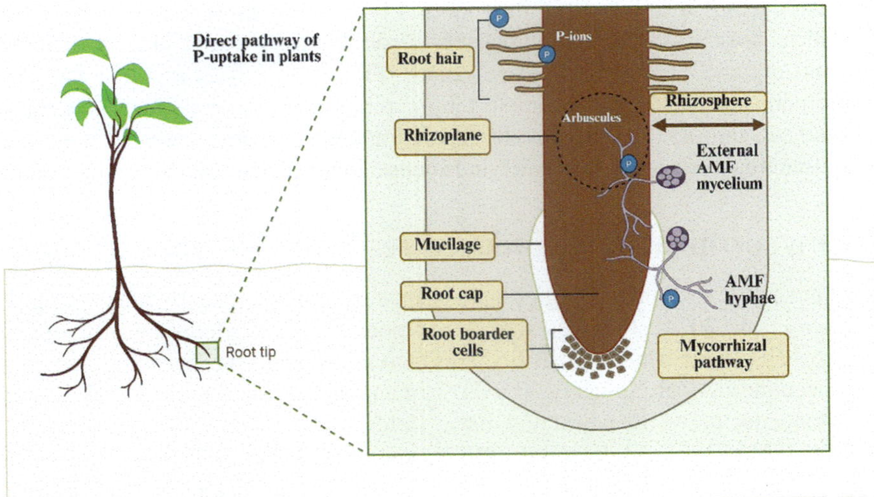

FIGURE 24.2 Phosphorus uptake pathway through arbuscular mycorrhizal fungi (AMF) association.

24.5 Plant nutrition and growth

The growth and productivity of crops are fundamentally constrained by nutrient phytoavailability within the rhizosphere, which is governed by complex soil physicochemical properties and environmental conditions. AMFs form mutualistic symbioses with most major crops, directly enhancing nutrient acquisition. The extraradical hyphae of AMF, constituting a common mycorrhizal network, possess a far greater surface-area-to-volume ratio than root hairs, enabling exploration of a larger soil volume and improved uptake of water and immobile nutrients (Deepika et al., 2023). In exchange for plant-derived photoassimilates, this network facilitates the absorption and translocation of essential macro- and micronutrients, promoting plant growth even under abiotic stress.

The increased acquisition of phosphorus (P) is a paramount promise of the AMF symbiotic relationship. Soil P is predominantly immobile and insoluble, which limits the root uptake efficiency (Zhu et al., 2018). Moreover, plants use two main pathways for P acquisition: the root epidermis and hairs (direct pathway) and the AMF-mediated indirect pathway (Smith et al., 2003). Here, fungal hyphae absorb inorganic phosphate (Pi) by specialized symporters and solubilize and transport the fixed soil P by exudates to the plant root interface (Fig. 24.2). Thus it bypasses the diffusion limitations inherent to the direct pathway (Chiu & Paszkowski, 2019).

In addition, AMFs considerably improve the plant nitrogen (N) nutrition. Fungi acquire N from both inorganic and organic sources, subsequently stimulating the biological N_2 fixation in legumes through increasing nodulation and nitrate reductase activities (Devi et al., 2021). The fungi mainly absorb ammonium (NH_4^+) ion, thereby decreasing the nitrate (NO_3^-) ion's vulnerability to leaching losses, which is a major economic and environmental cost in agriculture (Jansa et al., 2019). Many researchers documented that AMF associations decreased the N and P leaching by approximately 40% and 50%, respectively (Martínez-Garcia et al., 2017). Furthermore, AMFs also contribute 50% of a plant's N requirement (Miransari, 2011). Thus AMF biofertilization signifies an essential strategy for sustainable nutrient cycling within CSA. The AMF inoculation presents a viable option to promote agricultural sustainability and system resilience through increasing the fertilizer use efficiency and decreasing the reliance on synthetic inputs (Chaudhary et al., 2025).

24.6 Improved soil structure and fertility status

Climate-induced threats to agricultural sustainability, that is, soil erosion, fertility loss, organic carbon transformation, soil compaction, and nutrient leaching, significantly undermine food security (Hamidov et al., 2018). Among these, soil structure is a fundamental factor of soil fertility status. AMFs are imperative architects of soil structure, affecting the physical, chemical, and biological properties of soils. Extraradical hyphae secrete the GRSPs that act as binding agents, stabilizing micro- into macroaggregates (Deepika et al., 2023). This augments the soil water retention, decreases soil erosion and surface runoff, and reduces the leaching losses, which subsequently improve soil quality and fertility status. Moreover, studies highlighted that the stabilized root zone improves the plant resistance to erosive forces (Gutjahr & Paszkowski, 2013). Furthermore, GRSPs considerably contribute to soil organic carbon (SOC), acting as a durable carbon sink. AMFs also improve the biogeochemical cycling in soil by alleviating the soil P limitation through extending hyphal networks beyond root depletion zones. Besides, AMF also interact with phosphate-solubilizing bacteria and their own exudates to solubilize the insoluble inorganic and organic P contents (Kalayu, 2019). Fall et al. (2022) reported that AMF mycelium itself constitutes a considerable pool of SOC. AMFs promote the microbial interactions, which increase nutrient acquisition, stress tolerance, and overall soil fertility. It is possible by regulating the rhizosphere microbiome via exudates and altered root deposits. These multifunctional benefits make AMF indispensable for climate-resilient soil management.

24.7 Arbuscular mycorrhizal fungi–induced resilience in crops to climate change

Human-based anthropogenic activity is intensifying climate change, which significantly alters the global temperature and precipitation regimes (Savo et al., 2024). Previous studies mentioned that the global mean temperature is projected to increase up to 2.4°C–4°C by 2100 (Gonzalez et al., 2010). This increase worsens the key abiotic stresses (drought, heat, and salinity) that collectively account for approximately 50%–80% of annual yield loss in the world (Oshunsanya et al., 2019). Notably, drought stress, alone, decreased the wheat and maize yields up to 21% and 40%, respectively, between 1980 and 2015 (Daryanto et al., 2016). Moreover, Pandey et al. (2017) reported that the probability of compound stress (drought and heat or salinity and heat) events is increasing and presents a more complex challenge than individual stressors. In this context, AMF inoculation arises as a strategic biological tool to increase crop resilience. AMFs alleviate the harmful effects of multiple abiotic stresses via increasing the plant water relations, nutrient acquisition, and osmotic adjustment. AMF shows symbiosis to sustain crop productivity under challenging conditions (Table 24.1). AMFs perform as an effective bioprotectant through bolstering crop vigor and stress tolerance simultaneously, making AMF an integral part of a climate-smart agricultural system to adapt to a changing climate.

24.8 Arbuscular mycorrhizal fungi and salinity stress

Climate change–induced changes in soil's physical, chemical, and biological properties augment the salinization, which is a critical hydrological threat to global agriculture (Ahmed et al., 2025). According to an estimate, 20% of irrigated land (~120 countries) is already affected in the world (Pareek et al., 2020). The arid, semiarid, and coastal regions are facing increased salinization because of the altered precipitation patterns, high evapotranspiration, and increased sea level that hinder the hydrological balances and intensify the soluble salt accumulation (Hassani et al., 2021). Moreover, intensive irrigation practices in the region of global warming majorly contribute to secondary salinization (Okur & Örçen, 2020).

The colonization of AMF promotes salinity tolerance through integrated plant physiological mechanisms. It includes the restoration of ionic homeostasis, where AMFs alleviate the harmful Na^+/K^+ and Na^+/Ca^{2+} ratios caused by salinity stress, thereby protecting cellular function (Evelin et al., 2019). Moreover, AMFs stimulate the compatible osmolyte (trehalose, glycine betaine, and proline) accumulation that helps in osmotic adjustment and cellular protection (Evelin et al., 2019). In oxidative stress, AMF promote the host's antioxidant defense system by increasing the superoxide dismutase and catalase activity (Wang et al., 2021). Enhanced root exploration and regulation of aquaporins promise improved water uptake, lessening the physiological drought imposed by salt ions (Evelin et al., 2019). Together, these adaptations promise better photosynthetic efficiency and overall plant growth under salinity stress (Table 24.1).

24.9 Arbuscular mycorrhizal fungi and drought stress

Drought represents a major abiotic stress limiting global crop productivity. Global warming and erratic rainfall patterns are increasing the frequency and severity of drought events in arid and semiarid regions (Ahmed et al., 2019). This worsens the plant water deficit by promoting soil evaporation and plant transpiration, potentially adverse to any theoretical benefits from atmospheric CO_2 fertilization because of drought-induced stomatal closure (Bhattacharya, 2021). The deficit water condition disrupts the basic physiological processes, that is, photosynthesis, enzyme functioning, and nutrient transport, which leads to hormonal and nutritional imbalances (Bhattacharya, 2021; Siddique et al., 2023). Osmotic stress decreases turgor pressure, hindering plant growth, whereas the overproduction of reactive oxygen species (ROS) causes oxidative damage to cellular components (Khan et al., 2024). However, AMF inoculation mitigates drought stress by an integrated mechanism. Physically the extra radial hyphal network improves water uptake from a larger soil volume and regulates host aquaporins (Zhang et al., 2018). Physiologically, AMF increase the plant water status, osmotic adjustment, and stomatal regulation (Table 24.1), although these modulate stress-related phytohormones (Bahadur et al., 2019). Thus AMFs boost the plant's antioxidant defense system, decreasing oxidative injury because of ROS activity in drought stress conditions (El-Beltagi et al., 2025; Zou et al., 2021).

TABLE 24.1 Effects of arbuscular mycorrhizal fungi association with different crops under salt and drought stress.

Host plant species	AMF species	Effects	References
Salt stress			
Tree cotton	*Funneliformis mosseae*	Improved phosphorus contents and biomass production	Tian et al. (2004)
Wheat	Inoculum of the following AMF species: *Entrophospora etunicata, F. mosseae, Rhizophagus intraradices*	Decreased salt ions, that is, Na^+ and Cl^-, increased nutrient uptake, plant growth, and grain yield	Daei et al. (2009)
Tomato	Inoculum of the following AMF species: *R. intraradices, F. mosseae, Rhizophagus aggregatus, Rhizophagus clarus, Funneliformis monosporus, Saxinis deserticola, Plasmodium brasilianum, E. etunicata, Gigaspora margarita*	Increased chlorophyll contents, fresh stem weight, plant growth	Basak et al. (2011)
Eggplant	*S. deserticola*	Improved growth and colonization of roots	Mohammad and Mittra (2013)
Rice	*Claroideoglomus etunicatum*	Increased plant biomass, stomatal conductance, soluble sugar contents, and free α-amino acid activity	Porcel et al. (2015)
Cucumber	Inoculum of the following AMF species: *E. etunicata, R. intraradices, F. mosseae*	Improved chlorophyll synthesis; uptake of mineral ions, that is, K, Ca, Zn, Fe, Mg, Mn, and Cu; and antioxidant activities. Reduced lipid peroxidation, electrolyte leakage, and Na^+ accumulation	Hashem et al. (2018)
Soybean	Inoculum of the following AMF species: *F. mosseae, R. intraradices, C. etunicatum*	Reduced H_2O_2 and lipid peroxidation. Improved auxin production, membrane protection, root functions, nutrient acquisition, and plant growth	Hashem et al. (2019a)
Tomato	*Rhizophagus fasciculatus*	Improved phosphorus (P) uptake and activity of the phosphatase enzyme. Reduced pH	Sivakumar et al. (2019)
Lettuce	*Claroideoglomus claroideum*	Decreased oxidative damage and improved plant growth and development and crop yield	Santander et al. (2020)
Pea	Single species (*R. intraradices*), formulated species (*F. mosseae* and *R. intraradices*), and multispecies (*R. fasciculatum* and Gi. sp.)	Improved nutrient uptake, chlorophyll synthesis, osmolyte accumulation, crop growth, biomass, and yield	Parihar et al. (2020)
Drought stress			
Maize	*Rhizophagus irregularis*	Improved apoplastic water flow	Bárzana et al. (2014)
Tomato			
Barley	*R. intraradices*	Improved root volume, phosphorous concentration, and phosphatase activity	Bayani et al. (2015)
Common beans	Inoculum of the following AMF species: *E. etunicata, R. intraradices, F. monosporus*	Improved vegetative growth, crop yield, total protein, total sugar, and essential nutrients (N, P, K, Mg, Ca, Fe, Zn, Mn, and Cu) uptake. Reduced proline content	Salim and Abou El-Yazied (2015)
Tomato	Inoculum of the following AMF species: *F. mosseae, E. etunicata, E. claroidea, Rhizoglomus microaggregatum, Funneliformis geosporus, R. irregularis*	Increased fruit yield, WUE, crop biomass, and leaf water potential	Bakr et al. (2017)

(continued on next page)

TABLE 24.1 Effects of arbuscular mycorrhizal fungi association with different crops under salt and drought stress—cont'd

Host plant species	AMF species	Effects	References
Tomato	Inoculum of the following AMF species: *F. mosseae* and *R. irregularis*	Enhanced plant height, crop biomass, WUE (intrinsic), stomatal density, CO_2 absorption, and proline contents. Reduced H_2O_2 and hormonal (ABA) synthesis in plant parts (leaf and root)	Chitarra et al. (2016)
Tomato	*R. irregularis*	Improved growth rate, photosystem II efficiency, and strigolactone production	Ruiz-Lozano et al. (2016)
Melon	Inoculum of the following AMF species: *R. intraradices* and *Glomus* spp.	Improved photosynthetic efficiency, physiological attributes, and chlorophyll production	Cakmakci et al. (2017)
Chickpea	Inoculum of the following AMF species: *C. etunicatum, R. irregularis,* and *F. mosseae*	Increased leghemoglobin concentration, nitrate reductase activity, and N fixation	Hashem et al. (2019b)
Wheat	Inoculum of the following AMF species: *R. intraradices, F. mosseae,* and *Funneliformis geosporum*	Enhanced PS-I and PS-II activities by reducing photosystem damage	Mathur et al. (2019)
Maize	*Glomus versiforme*	Reduced oxidative damage and enhanced plant growth, photosynthetic rate, chlorophyll concentration, nutrient uptake, and mineral assimilation	Begum et al. (2019a)
Soybean	*R. clarus*	Enhanced water status, chlorophyll pigments, photosynthesis, and crop yield	Oliveira et al. (2022)

AMF, arbuscular mycorrhizal fungi; *WUE*, water use efficiency; *ABA*, abscisic acid.

24.10 Arbuscular mycorrhizal fungi and heat stress

The increase in average temperature poses a major threat to global agricultural productivity. The report mentions that a 1.5°C rise is expected between 2030 and 2052 under current emissions trajectories (Deepika et al., 2023). Likewise, a 3°C–4°C increase impairs plant growth and development and, subsequently, 15%–35% yield loss across Africa and Asia (Hassan et al., 2021). Heat stress significantly damages the reproductive processes and seed setting, causing catastrophic production losses; each 1°C rise in temperature can decrease the wheat yield up to 4%–6% in the world (Asseng et al., 2015).

In heat stress the physiological injuries range from visible foliar scorching and growth inhibition to cellular and metabolic dysfunctions. Critically, high temperatures disrupt photosynthesis through damaging chloroplast structures and hindering Rubisco activity (Bhattacharya, 2022). Subsequently an overproduction of ROS triggers complex molecular adaptations, that is, osmolyte accumulation and stress-related phytohormones (Rasheed et al., 2022). Inoculation with AMF improves the plant's thermo-tolerance by various interconnected mechanisms (Table 24.2). For example, AMFs promote the root water uptake and hydraulic conductivity via regulation of aquaporin genes, thus mitigating the heat-induced water deficiency (Oyebamiji et al., 2024). AMFs also promise the improved photosynthetic efficiency and enhanced antioxidant synthesis to lower the oxidative damage (Begum et al., 2022). Moreover, AMFs increase the osmolyte and cryoprotectant accumulation, contributing to improved cellular homeostasis under high-temperature stress (Jian et al., 2024).

24.11 Arbuscular mycorrhizal fungi and cold stress

Though anthropogenic activity-based climate change is linked to increasing global temperature, it also increases the frequency and persistence of extreme cold events over complex atmospheric teleconnections, that is, disruptions to the stratospheric polar vortex (Cohen et al., 2021). Such cold extremes comprise chilling (0°C–15°C) and freezing (<0°C) stress, which imposes a major threat to global agriculture by lowering plant growth and development (Jha et al., 2017). Studies highlight that staple cereals are more vulnerable; for example, a substantial loss in rice yield occurs below 15°C–20°C, and spring frosts annually influence ~85% of the global wheat production at the booting stage (Yue et al., 2016).

TABLE 24.2 Effects of arbuscular mycorrhizal fungi association with different crops in temperature stress.

Host plant	AMF species	Effects	References
High temperature stress			
Maize	*Entrophospora etunicata*	Increased net photosynthetic rate, stomatal conductance, transpiration rate, fluorescence, quantum efficiency of PSII photochemistry, photochemical efficiency, chlorophyll content, water use efficiency, water-holding capacity, and relative water content	Zhu et al. (2011)
Cucumber	*Funneliformis mosseae*	Enhanced fresh (root + shoot) biomass, antioxidant activities, and phenolic concentration	Haghighi et al. (2015)
Wheat	Inoculum of the following AMF species: *Rhizophagus irregularis, F. mosseae, Funneliformis geosporum,* and *Claroideoglomus claroideum*	Improved nutrient uptake and yield	Cabral et al. (2016)
Cucumber	*R. irregularis*	Increased photosynthesis by protecting the photosynthetic apparatus	Ma et al. (2019)
Maize	*Funneliformis* sp.	Regulate photosystem (PS-II) heterogeneity	Mathur and Jajoo (2020)
Tomato Chili pepper Cucumber	*R. irregularis*	Increased seed vigor, productivity, and fruit quality	Reva et al. (2021)
Low temperature stress			
Cucumber	*F. mosseae*	Improved biomass (fresh and dry), secondary metabolites, and enzymatic activity. Reduced oxidative species	Chen et al. (2013)
Tomato	*F. mosseae*	Reduced cell membrane injury and enhanced photosystem (PS-II) efficiency and seedling growth	Caradonia et al. (2019)
Barley	Inoculum of the following AMF species: *Glomus versiforme* and *R. irregularis*	Improved survival rate and cold stress resilience	Hajiboland et al. (2019)
Cucumber	*R. irregularis*	Enhanced photosynthetic efficiency, carbon sink, and sugar content in leaves	Ma et al. (2019)
Eggplant	Inoculum of the following AMF species: *F. mosseae, Claroideoglomus etunicatum, R. irregularis,* and *Diversispora versiformis*	Improved photochemical reactions and antioxidant production. Reduced membrane injury	Pasbani et al. (2020)
Water melon	*Rhizophagus intraradices*	Reduced H_2O_2 activity, MDA level, and electrolyte leakage. Enhanced photosynthesis	Shirani Bidabadi and Mehralian (2020)
Barley	Inoculum of the following AMF species: *D. versiformis* and *R. irregularis*	Enhanced antioxidant defense system and phenolic contents. Reduced H_2O_2 and line accumulation	Joudmand et al. (2021)

AMF, arbuscular mycorrhizal fungi; *MDA*, malondialdehyde.

The damage induced by cold stress includes chlorosis, necrosis, and wilting and induces cellular dysfunction by membrane damage, electrolyte leakage, and the accumulation of ROS (Manasa et al., 2022). Cold injury disrupts the reproductive processes that lead to pollen sterility and ovule abortion, which drastically lessen the crop yield (Jahed et al., 2023). However, AMFs promote plant resilience to low-temperature injury by integrating physio-biochemical and molecular adaptations (Table 24.2). For example, AMF colonization enhances the root hydraulic conductivity that promotes the water uptake under chilling conditions, thereby mitigating the cold-induced water deficiency (Liu et al., 2023). AMF also enhances nutrient acquisition and photosynthesis by triggering specific gene transporters and the antioxidant defense system for

ROS scavenging (Ma et al., 2018). Moreover, AMF promote the accumulation of cryoprotective osmolytes (proline and soluble sugars) and modulate the gene expressions involved in metabolic and hormonal functioning for systemic protection (Devi et al., 2019).

24.12 Arbuscular mycorrhizal fungi–mediated mitigation of climate change

The global food production system is a significant source of anthropogenic GHG emissions (Lynch et al., 2021). It contributes ~21%–37% of the annual total, with direct agricultural activities accounting for ~12% (Lynch et al., 2021). Among GHGs, CO_2 and N_2O are major pollutants, whereas N_2O is a potent GHG with a global warming potential 298 times that of CO_2 (Deepika et al., 2023). Moreover, high use of N-based fertilizers significantly increases the N losses via microbial nitrification and denitrification (Aslam et al., 2026). The use of AMF reduces N_2O emissions by increasing the plant N uptake, thereby lowering the soil nitrate availability for denitrifying microbes. Martínez-Garcia et al. (2017) mentioned that AMF hyphae compete with ammonia-oxidizing bacteria for ammonium, which suppresses the nitrification-derived N_2O production (Storer et al., 2018). CSA bestows a robust framework to address this dual challenge, and the inoculation of crops using AMF offers a promising biological strategy to improve resilience alongside emission mitigation. However, direct agricultural CO_2 emissions are comparatively reduced; agricultural systems influence the carbon cycle by land-use change and soil management. For this, enhanced SOC sequestration is a key mitigation strategy where AMFs significantly hinder the negative emissions by transferring the plant-derived carbon into the soil through the hyphal network. This network further promotes the formation of stable soil aggregates by GRPs (Wang et al., 2016). This process builds a long-term carbon sink, directly aiding climate change mitigation.

24.13 Agricultural practices and arbuscular mycorrhizal fungi symbiosis–crop rotations, tillage, and agrochemicals

AMF are the key to sustainable agroecosystems; however, conventional agricultural practices compromise the abundance and diversity of AMF in soil (Liu et al., 2024). Likewise, intensive tillage, monocropping, and high use of agrochemicals boost the short-term productivity but negatively impact soil health and decrease the mutualistic benefits of AMF symbiosis (Fallah et al., 2023). Research showed that intensive tillage application disturbs the soil structure (Aslam et al., 2025) and breaks the hyphal networks of AMF (Thomopoulos et al., 2025). Likewise, monocropping reduces AMF diversity by a limited number of host plants, while crop rotations along with polycultures strengthen the AMF communities, leading to better nutrient cycling (Guzman et al., 2021). In addition, ample fertilization (N and P) changes the cost–benefit balance of the symbiosis. Because plants become less dependent on AMF in high nutrient availability, fewer mutualistic fungal strains provide fewer benefits (Thirkell et al., 2017). Besides, these fertilizers and the fungicide application acidify the soil, which limits the AMF propagule production and colonization (Hage-Ahmed et al., 2019). Collectively, such practices limit the ecosystem services, emphasizing the dire need to revisit the management system to conserve native AMF communities for harnessing their complete potential.

24.14 Arbuscular mycorrhizal fungi symbiosis and climate change

The multifunctional benefits of AMF enable them as a key bioinoculant for CSA. Although the efficiency of AMF is influenced by various climatic factors, as plants can modify their root exudation and belowground carbon allocation patterns under climate change. This further affects the composition and function of associated AMF communities (Ahmed et al., 2025). Cotton et al. (2018) described the impact of elevated atmospheric CO_2 on AMF communities. They mentioned that the overall richness (AMF colonization) often remains unchanged; however, their compositional shifts were obvious, favoring taxa within the *Glomeraceae* over the *Gigasporaceae* (Deepika et al., 2023). In contrast, N deposition consistently reduced the AMF richness and diversity; however, nitrophilous species like *Rhizophagus irregularis* may proliferate (Williams et al., 2017). Moreover, no clear pattern was noticed under the effect of high temperature and change in precipitation regimes (Shi et al., 2017).

This variation highlights the emerging state of knowledge about the impact of climate change on AMF symbiosis. The obligate symbiotic nature of AMF poses substantial methodological challenges, making it difficult to isolate fungal responses from those of the host plant (Liu & Chen, 2024). Thus current literature emphasizes the vulnerability of AMF communities to global change but highlights the need for further research to predict outcomes reliably.

24.15 Conclusion and future perspectives

Climate change results in a critical impairment to global food security via destabilizing agricultural productivity. Efforts to increase crop production are heavily reliant on the natural resource-intensive practice, which intensifies environmental degradation. However, CSA promises the transformation of food systems with better productivity, sustainability, and decreased carbon footprint. For this, AMFs are a pivotal biological tool. The dual role of AMF as biofertilizers and bioprotectants promises increased nutrient acquisition, better soil health, and crop resilience to different abiotic stresses, which perfectly align with the objectives of CSA.

Conversely the effective integration of AMF into extensive agricultural practice faces substantial challenges, including the context-dependent nature of the symbiosis, plant genotype- and fungal-dependent benefits, soil type, and management practices. There is a significant research gap to cover these challenges. This knowledge gap makes it difficult to predict whether AMF symbioses will consistently function as a climate mitigation tool under future conditions. So, future research must prioritize addressing the above-given challenges, enabling the development of robust, sustainable, and climate-resilient agricultural systems for future food security.

Acknowledgment

We thank the Deanship of Scientific Research, Vice Presidency for Graduate Studies and Scientific Research, King Faisal University, Saudi Arabia (KFU260124), for supporting this research work.

Funding

Supported by the Deanship of Scientific Research, Vice Presidency for Graduate Studies and Scientific Research, King Faisal University, Saudi Arabia (KFU260124), for supporting this research.

References

Abdul Aziz, M., Brini, F., Rouached, H., & Masmoudi, K (2022). Genetically engineered crops for sustainably enhanced food production systems. *Frontiers in Plant Science, 13*, 1027828.

Ahmed, K., Shahid, S., Chung, E. S., Wang, X. J., & Harun, S. B. (2019). Climate change uncertainties in seasonal drought severity-area-frequency curves: Case of the arid region of Pakistan. *Journal of Hydrology, 570*, 473–485.

Ahmed, Z., Gui, D., Yunfei, L., Murtaza, G., Aslam, T., Ali, S., & Raza, M. Y. (2025). Climate smart agriculture. In *Climate resilient and sustainable agriculture: Volume 1: Adaptation and mitigation strategies* (pp. 407–440). Springer Nature Switzerland.

Alam, M. Z., & Roy, M. D. (2024). The reduction of abiotic stress in food crops through climate-smart mycorrhiza-enriched biofertilizer. *AIMS Microbiology, 10*(3), 674.

Ali, M., Javeed, H. M. R., Tariq, M., Khan, A. A., Qamar, R., Nawaz, F., Masood, N., Ditta, A., Abbas, T., Zamir, M. S. I., Sabagh, A. E., Shahzad, M., Naeem, R., & Mubeen, M. (2023). Use of biochar for biological carbon sequestration. In W. N. Jatoi, M. Mubeen, M. Z. Hashmi, S. Ali, S. Fahad, & K. Mahmood (Eds.), *Climate change impacts on agriculture* (pp. 393–411). Springer. https://doi.org/10.1007/978-3-031-26692-8_23.

Aslam, M. T., Khan, I., Chattha, M. U., & Murtaza, G. (2026). Coated urea improves productivity and nitrogen use efficiency in wheat. *Journal of Soil Science and Plant Nutrition, 26*, 865–887.

Aslam, M. T., Khan, I., Chattha, M. U., Raza, A., Chattha, M. B., Ali, M., & Fatima, I. (2025). Perspective chapter: Redesigning agroecological practices for enhanced resource use efficiency in agroecosystems. *Sustainable agroecosystems-principles and practices*. IntechOpen.

Asseng, S., Ewert, F., Martre, P., Rötter, R. P., Lobell, D. B., Cammarano, D., & Zhu, Y. (2015). Rising temperatures reduce global wheat production. *Nature Climate Change, 5*(2), 143–147.

Bahadur, A., Batool, A., Nasir, F., Jiang, S., Mingsen, Q., Zhang, Q., & Feng, H. (2019). Mechanistic insights into arbuscular mycorrhizal fungi-mediated drought stress tolerance in plants. *International Journal of Molecular Sciences, 20*(17), 4199.

Bakr, J., Daood, H. G., Pék, Z., Helyes, L., & Posta, K. (2017). Yield and quality of mycorrhized processing tomato under water scarcity. *Applied Ecology and Environmental Research, 15*(1), 401–413.

Bárzana, G., Aroca, R., Bienert, G. P., Chaumont, F., & Ruiz-Lozano, J. M. (2014). New insights into the regulation of aquaporins by the arbuscular mycorrhizal symbiosis in maize plants under drought stress and possible implications for plant performance. *Molecular Plant-Microbe Interactions, 27*(4), 349–363.

Basak, H., Demir, K., Kasim, R., & Okay, F. Y. (2011). The effect of endo-mycorrhiza (VAM) treatment on the growth of tomato seedlings grown under saline conditions. *African Journal of Agricultural Research, 6*(11), 2532–2538.

Bayani, R., Saateyi, A., & Faghani, E. (2015). Influence of arbuscular mycorrhiza in phosphorus acquisition efficiency and drought-tolerance mechanisms in barley (*Hordeum vulgare* L.). *International Journal of Biosciences, 7*, 86–94.

Begum, N., Ahanger, M. A., Su, Y., Lei, Y., Mustafa, N. S. A., Ahmad, P., & Zhang, L. (2019a). Improved drought tolerance by AMF inoculation in maize (*Zea mays*) involves physiological and biochemical implications. *Plants, 8*(12), 579.

Begum, N., Qin, C., Ahanger, M. A., Raza, S., Khan, M. I., Ashraf, M., & Zhang, L. (2019b). Role of arbuscular mycorrhizal fungi in plant growth regulation: implications in abiotic stress tolerance. *Frontiers in Plant Science, 10*, 1068.

Begum, N., Wang, L., Ahmad, H., Akhtar, K., Roy, R., Khan, M. I., & Zhao, T. (2022). Co-inoculation of arbuscular mycorrhizal fungi and the plant growth-promoting rhizobacteria improves growth and photosynthesis in tobacco under drought stress by up-regulating antioxidant and mineral nutrition metabolism. *Microbial Ecology, 83*(4), 971–988.

Bhattacharya, A. (2021). Effect of soil water deficit on growth and development of plants: A review. In *Soil water deficit and physiological issues in plants* (pp. 393–488). Springer.

Bhattacharya, A. (2022). Effect of low temperature stress on photosynthesis and allied traits: A review. *Physiological processes in plants under low temperature stress*, 199–297.

Cabral, C., Ravnskov, S., Tringovska, I., & Wollenweber, B. (2016). Arbuscular mycorrhizal fungi modify nutrient allocation and composition in wheat (*Triticum aestivum* L.) subjected to heat stress. *Plant and Soil, 408*(1), 385–399.

Cakmakci, O., Cakmakci, T., Durak, E. D., Demir, S., & Sensoy, S. (2017). Effects of arbuscular mycorrhizal fungi in melon (*Cucumis melo* L.) seedling under deficit irrigation. *Fresenius Environmental Bulletin, 26*(12), 7513–7520.

Caradonia, F., Francia, E., Morcia, C., Ghizzoni, R., Moulin, L., Terzi, V., & Ronga, D. (2019). Arbuscular mycorrhizal fungi and plant growth-promoting rhizobacteria avoid processing tomato leaf damage during chilling stress. *Agronomy, 9*(6), 299.

Cavagnaro, T. R., Bender, S. F., Asghari, H. R., & van der Heijden, M. G. (2015). The role of arbuscular mycorrhizas in reducing soil nutrient loss. *Trends in Plant Science, 20*(5), 283–290.

Chaudhary, A., Poudyal, S., & Kaundal, A. (2025). Role of arbuscular mycorrhizal fungi in maintaining sustainable agroecosystems. *Applied Microbiology, 5*(1), 6.

Chen, S., Jin, W., Liu, A., Zhang, S., Liu, D., Wang, F., & He, C. (2013). Arbuscular mycorrhizal fungi (AMF) increase growth and secondary metabolism in cucumber subjected to low temperature stress. *Scientia Horticulturae, 160*, 222–229.

Chitarra, W., Pagliarani, C., Maserti, B., Lumini, E., Siciliano, I., Cascone, P., & Guerrieri, E. (2016). Insights on the impact of arbuscular mycorrhizal symbiosis on tomato tolerance to water stress. *Plant Physiology, 171*(2), 1009–1023.

Chiu, C. H., & Paszkowski, U. (2019). Mechanisms and impact of symbiotic phosphate acquisition. *Cold Spring Harbor Perspectives in Biology, 11*(6), a034603.

Cohen, J., Agel, L., Barlow, M., Garfinkel, C. I., & White, I. (2021). Linking Arctic variability and change with extreme winter weather in the United States. *Science, 373*(6559), 1116–1121.

Cotton, T. A. (2018). Arbuscular mycorrhizal fungal communities and global change: An uncertain future. *FEMS Microbiology Ecology, 94*(11), fiy179.

Daei, G., Ardekani, M. R., Rejali, F., Teimuri, S., & Miransari, M. (2009). Alleviation of salinity stress on wheat yield, yield components, and nutrient uptake using arbuscular mycorrhizal fungi under field conditions. *Journal of Plant Physiology, 166*(6), 617–625.

Daryanto, S., Wang, L., & Jacinthe, P. A. (2016). Global synthesis of drought effects on maize and wheat production. *PLoS One, 11*(5), e0156362.

Deepika, S., Goswami, V., & Kothamasi, D. (2023). Arbuscular mycorrhizal fungi (AMF) and climate-smart agriculture: Prospects and challenges. In M. W. Ansari, A. K. Singh, & N. Tuteja (Eds.), *Global* climate change and plant stress management (pp. 175–200). Wiley.

Devi, S. H., Bhupenchandra, I., Sinyorita, S., Chongtham, S. K., & Devi, E. L. (2021). Mycorrhizal fungi and sustainable agriculture. *Nitrogen in agriculture – Physiological, agricultural and ecological aspects*, 1–19.

Devi, T. S., Gupta, S., & Kapoor, R. (2019). Arbuscular mycorrhizal fungi in alleviation of cold stress in plants. In *Advancing frontiers in mycology & mycotechnology: Basic and applied aspects of fungi* (pp. 435–455). Springer Singapore.

Ditta, A, Mehmood, S, Imtiaz, M, & Tu, MS (2024). *Bio-organic Amendments for heavy metal remediation: Water, soil and plant approaches and technologies (plant biology, sustainability and climate change* (1st edition). Elsevier Inc. https://doi.org/10.1016/C2022-0-03004-X.

El-Beltagi, H. S., El-Waraky, E. A., El-Mogy, M. M., El-Nady, M. F., Ismail, A. M., Belal, E. B., & Metwaly, M. M. S. (2025). Microbial-priming of seeds with plant growth-promoting rhizobacteria and arbuscular mycorrhizal fungi for improving cotton (*Gossypium barbadense* L.) growth, yield, and water productivity under drought stress. *Biologia, 80*(4), 779–803.

Evelin, H., Devi, T. S., Gupta, S., & Kapoor, R. (2019). Mitigation of salinity stress in plants by arbuscular mycorrhizal symbiosis: Current understanding and new challenges. *Frontiers in Plant Science, 10*, 470.

Fall, A. F., Nakabonge, G., Ssekandi, J., Founoune-Mboup, H., Apori, S. O., Ndiaye, A., & Ngom, K. (2022). Roles of arbuscular mycorrhizal fungi on soil fertility: Contribution to the improvement of physical, chemical, and biological properties of the soil. *Frontiers in Fungal Biology, 3*, 723892.

Fallah, N., Pang, Z., Zhou, Y., Yuan, Z., & Lin, W. (2023). Sugarcane-peanut intercropping stimulatory effects on arbuscular mycorrhizal and entomopathogenic fungi promote soil health and crop productivity. *Field Crops Research, 303*, 109110.

Gonzalez, P., Neilson, R. P., Lenihan, J. M., & Drapek, R. J. (2010). Global patterns in the vulnerability of ecosystems to vegetation shifts due to climate change. *Global Ecology and Biogeography, 19*(6), 755–768.

Gutjahr, C., & Paszkowski, U. (2013). Multiple control levels of root system remodeling in arbuscular mycorrhizal symbiosis. *Frontiers in Plant Science, 4*, 204.

Guzman, A., Montes, M., Hutchins, L., DeLaCerda, G., Yang, P., Kakouridis, A., & Kremen, C. (2021). Crop diversity enriches arbuscular mycorrhizal fungal communities in an intensive agricultural landscape. *New Phytologist, 231*(1), 447–459.

Hage-Ahmed, K., Rosner, K., & Steinkellner, S. (2019). Arbuscular mycorrhizal fungi and their response to pesticides. *Pest Management Science, 75*(3), 583–590.

Haghighi, M., Mozafariyan, M., & Abdolahipour, B. (2015). Effect of cucumber mycorrhiza inoculation under low and high root temperature grown on hydroponic conditions. *Journal of Crop Science and Biotechnology, 18*(2), 89–96.

Hajiboland, R., Joudmand, A., Aliasgharzad, N., Tolrá, R., & Poschenrieder, C. (2019). Arbuscular mycorrhizal fungi alleviate low-temperature stress and increase freezing resistance as a substitute for acclimation treatment in barley. *Crop and Pasture Science, 70*(3), 218–233.

Hamidov, A., Helming, K., Bellocchi, G., Bojar, W., Dalgaard, T., Ghaley, B. B., & Schönhart, M. (2018). Impacts of climate change adaptation options on soil functions: A review of European case-studies. *Land Degradation & Development, 29*(8), 2378–2389.

Hashem, A., Abd_Allah, E. F., Alqarawi, A. A., Wirth, S., & Egamberdieva, D. (2019a). Comparing symbiotic performance and physiological responses of two soybean cultivars to arbuscular mycorrhizal fungi under salt stress. *Saudi Journal of Biological Sciences, 26*(1), 38–48.

Hashem, A., Alqarawi, A. A., Radhakrishnan, R., Al-Arjani, A. B. F., Aldehaish, H. A., Egamberdieva, D., & Abd_Allah, E. F. (2018). Arbuscular mycorrhizal fungi regulate the oxidative system, hormones, and ionic equilibrium to trigger salt stress tolerance in *Cucumis sativus* L. *Saudi Journal of Biological Sciences, 25*(6), 1102–1114.

Hashem, A., Kumar, A., Al-Dbass, A. M., Alqarawi, A. A., Al-Arjani, A. B. F., Singh, G., & Abd_Allah, E. F. (2019b). Arbuscular mycorrhizal fungi and biochar improve drought tolerance in chickpea. *Saudi Journal of Biological Sciences, 26*(3), 614–624.

Hassan, M. U., Chattha, M. U., Khan, I., Chattha, M. B., Barbanti, L., Aamer, M., & Aslam, M. T. (2021). Heat stress in cultivated plants: Nature, impact, mechanisms, and mitigation strategies—A review. *Plant Biosystems—An International Journal Dealing With All Aspects of Plant Biology, 155*(2), 211–234.

Hassani, A., Azapagic, A., & Shokri, N. (2021). Global predictions of primary soil salinization under a changing climate in the 21st century. *Nature Communications, 12*(1), 6663.

He, P. (2024). Navigating the global dietary transition and its impact on planetary and human health: The role of middle-income countries. *Food Science and Nutrition Cases*, fsncases20240015.

Jahed, K. R., Saini, A. K., & Sherif, S. M. (2023). Coping with the cold: Unveiling cryoprotectants, molecular signaling pathways, and strategies for cold stress resilience. *Frontiers in Plant Science, 14*, 1246093.

Jansa, J., Forczek, S. T., Rozmoš, M., Püschel, D., Bukovská, P., & Hršelová, H. (2019). Arbuscular mycorrhiza and soil organic nitrogen: Network of players and interactions. *Chemical and Biological Technologies in Agriculture, 6*(1), 1–10.

Jha, U. C., Bohra, A., & Jha, R. (2017). Breeding approaches and genomics technologies to increase crop yield under low-temperature stress. *Plant Cell Reports, 36*(1), 1–35.

Jian, P., Zha, Q., Hui, X., Tong, C., & Zhang, D. (2024). Research progress of arbuscular mycorrhizal fungi improving plant resistance to temperature stress. *Horticulturae, 10*(8), 855.

John, D. A., & Babu, G. R. (2021). Lessons from the aftermath of the green revolution on the food system and health. *Frontiers in Sustainable Food Systems, 5*, 644559.

Joudmand, A., Hajiboland, R., & Aliasgharzad, N. (2021). Mitigation of chilling and freezing stresses through colonization with arbuscular mycorrhizal fungi in spring barley. *Journal of Plant Process Function, 9*(40), 13–24.

Kabato, W., Getnet, G. T., Sinore, T., Nemeth, A., & Molnár, Z. (2025). Towards climate-smart agriculture: Strategies for sustainable agricultural production, food security, and greenhouse gas reduction. *Agronomy, 15*(3), 565.

Kalayu, G. (2019). Phosphate-solubilizing microorganisms: promising approach as biofertilizers. *International Journal of Agronomy, 2019*(1), 4917256.

Khan, I., Chattha, M. U., Maqbool, R., Ali, M., Asif, M., Hassan, M. U., & Aslam, M. T. (2024). Perspectives on phytohormone application to enhance salinity tolerance in plants. *New insights into phytohormones*. IntechOpen.

Lenoir, I., Fontaine, J., & Sahraoui, A. L. H. (2016). Arbuscular mycorrhizal fungal responses to abiotic stresses: A review. *Phytochemistry, 123*, 4–15.

Liu, Y. Z., Imtiaz, M., Ditta, A., Rizwan, M. S., Ashraf, M., Mehmood, S., Aziz, O., Mubeen, F., Ali, M., Elahi, N. N., Ijaz, R., S, Lelel, Shuang, C., & Tu, S. (2020). Response of growth, antioxidant enzymes, and root exudates production towards as stress in *Pteris vittata* and *Astragalus sinicus* colonized by arbuscular mycorrhizal fungi. *Environmental Science and Pollution Research, 27*, 2340–2352. https://doi.org/10.1007/s11356-019-06785-5.

Liu, R., & Chen, Y. (2024). Mycorrhizal symbiosis: Evolution, opportunities, challenges, and prospects. *Arbuscular mycorrhizal fungi in sustainable agriculture: Inoculum production and application*, 1–35.

Liu, R., Li, M., Guo, S., & Chen, Y. (2024). The role of AMF community composition, diversity, and distribution in sustainable agroecosystems. In *Arbuscular mycorrhizal fungi in sustainable agriculture: Inoculum production and application* (pp. 281–317). Springer Nature Singapore.

Liu, Y., Lu, J., Cui, L., Tang, Z., Ci, D., Zou, X., & Si, T. (2023). The multifaceted roles of arbuscular mycorrhizal fungi in peanut responses to salt, drought, and cold stress. *BMC Plant Biology, 23*(1), 36.

Lynch, J., Cain, M., Frame, D., & Pierrehumbert, R. (2021). Agriculture's contribution to climate change and role in mitigation are distinct from predominantly fossil CO_2-emitting sectors. *Frontiers in Sustainable Food Systems, 4*, 518039.

Ma, J., Janoušková, M., Ye, L., Bai, L. Q., Dong, R. R., Yan, Y., & He, C. X. (2019). Role of arbuscular mycorrhiza in alleviating the effect of cold on the photosynthesis of cucumber seedlings. *Photosynthetica, 57*(1), 86–95.

Ma, J., Sun, C., Bai, L., Dong, R., Yan, Y., Yu, X., & Li, Y. (2018). Transcriptome analysis of cucumber roots reveals key cold-resistance genes induced by AM fungi. *Plant Molecular Biology Reporter, 36*(1), 135–148.

Majhi, P. K., Samal, I., Bhoi, T. K., Pattnaik, P., Pradhan, C., Gupta, A., & Senapati, S. K. (2023). Climate-smart agriculture: An integrated approach to address climate change and food security. In *Climate change and insect biodiversity* (pp. 221–245). CRC Press.

Manasa, S. L., Panigrahy, M., Panigrahi, K. C., & Rout, G. R. (2022). Overview of cold stress regulation in plants. *The Botanical Review, 88*(3), 359–387.

Martínez-García, L. B., De Deyn, G. B., Pugnaire, F. I., Kothamasi, D., & Van der Heijden, M. G. (2017). Symbiotic soil fungi enhance ecosystem resilience to climate change. *Global Change Biology, 23*(12), 5228–5236.

Mathur, S., & Jajoo, A. (2020). Arbuscular mycorrhizal fungi protect maize plants from high temperature stress by regulating photosystem II heterogeneity. *Industrial Crops and Products, 143*, 111934.

Mathur, S., Tomar, R. S., & Jajoo, A. (2019). Arbuscular mycorrhizal fungi (AMF) protect the photosynthetic apparatus of wheat under drought stress. *Photosynthesis Research, 139*(1), 227–238.

Mesfin, S. G., & Markos, E. (2019). A review of the role of climate-smart agriculture for sustainable crop production. *Journal of Biology, Agriculture and Healthcare, 9*, 21–29.

Miransari, M. (2011). Arbuscular mycorrhizal fungi and nitrogen uptake. *Archives of Microbiology, 193*(2), 77–81.

Mohammad, A., & Mittra, B. (2013). Effects of inoculation with stress-adapted arbuscular mycorrhizal fungus *Glomus deserticola* on growth of *Solanum melogena* L. and *Sorghum sudanese* Staph. Seedlings under salinity and heavy metal stress conditions. *Archives of Agronomy and Soil Science, 59*(2), 173–183.

Nazir, A, Sarfraz, W, Ditta, A, Khalid, N, Farid, M, Shafiq, M, Firdaus-e-Bareen, ZF Rizvi, & Naeem, N (2023). Synergistic impact of two autochthonous saprobic fungi (*A. niger* and *T. pseudokoningii*) on the growth, ionic contents, and metals uptake in *Brassica juncea* L. and *Vigna radiata* L. under tannery solid waste contaminated soil. *International Journal of Phytoremediation, 25*(11), 1488–1500. https://doi.org/10.1080/15226514.2023.2166457.

Noya, I., Gonzalez-Garcia, S., Bacenetti, J., Fiala, M., & Moreira, M. T. (2018). Environmental impacts of the cultivation phase associated with agricultural crops for feed production. *Journal of Cleaner Production, 172*, 3721–3733.

Okur, B., & Örçen, N. (2020). Soil salinization and climate change. In *Climate change and soil interactions* (pp. 331–350). Elsevier.

Oliveira, T. C., Cabral, J. S. R., Santana, L. R., Tavares, G. G., Santos, L. D. S., Paim, T. P., & Mendes, G. C. (2022). The arbuscular mycorrhizal fungus *Rhizophagus clarus* improves physiological tolerance to drought stress in soybean plants. *Scientific Reports, 12*(1), 9044.

Oshunsanya, S. O., Nwosu, N. J., & Li, Y. (2019). Abiotic stress in crops under climatic conditions. In *Sustainable agriculture, forest and environmental management* (pp. 71–100). Springer Singapore.

Oyebamiji, Y. O., Nahar, S., Shamsudin, N. A., Salisu, M. A., Akinola, S. A., Adebayo, I. A., & Arsad, H. (2024). The role of arbuscular mycorrhizal fungi in alleviating heat and drought stress in vegetables. *Egyptian Journal of Botany, 64*(2), 497–505.

Pandey, G., Yadav, C. B., Sahu, P. P., Muthamilarasan, M., & Prasad, M. (2017). Salinity induced differential methylation patterns in contrasting cultivars of foxtail millet (*Setaria italica* L.). *Plant Cell Reports, 36*(5), 759–772.

Pareek, A., Dhankher, O. P., & Foyer, C. H. (2020). Mitigating the impact of climate change on plant productivity and ecosystem sustainability. *Journal of Experimental Botany, 71*(2), 451–456.

Parihar, M., Rakshit, A., Rana, K., Tiwari, G., & Jatav, S. S. (2020). The effect of arbuscular mycorrhizal fungi inoculation in mitigating salt stress of pea (*Pisum Sativum* L.). *Communications in Soil Science and Plant Analysis, 51*(11), 1545–1559.

Pasbani, B., Salimi, A., Aliasgharzad, N., & Hajiboland, R. (2020). Colonization with arbuscular mycorrhizal fungi mitigates cold stress through improvement of antioxidant defense and accumulation of protective molecules in eggplants. *Scientia Horticulturae, 272*, 109575.

Porcel, R., Redondo-Gómez, S., Mateos-Naranjo, E., Aroca, R., Garcia, R., & Ruiz-Lozano, J. M. (2015). Arbuscular mycorrhizal symbiosis ameliorates the optimum quantum yield of photosystem II and reduces non-photochemical quenching in rice plants subjected to salt stress. *Journal of Plant Physiology, 185*, 75–83.

Prasad, S. C. (2016). Innovating at the margins: The system of rice intensification in India and transformative social innovation. *Ecology and Society, 21*(4), 9.

Rahman, H. U., & Ditta, A. (2025). Mine remediation and climate changes. In A. A. H. A. Latef, E. M. Zayed, & A. A. Omar (Eds.), *Remediation of pollution and climatic changes to avoid environmental risks*. Springer Singapore https://doi.org/10.1007/978-981-96-5674-5_5.

Rahman, H. U., Khan, M., & Ditta, A. (2025). Recent advances in sustainable waste management practices. In A. A. H. Abdel Latef, E. M. Zayed, & A. A. Omar (Eds.), *Bioremediation and nanotechnology for climate change mitigation* (pp. 103–136). Springer. https://doi.org/10.1007/978-981-96-3069-1_5.

Rasheed, A., Jie, Y., Nawaz, M., Jie, H., Ma, Y., Shah, A. N., & Qari, S. H. (2022). Improving drought stress tolerance in ramie (*Boehmeria nivea* L.) using molecular techniques. *Frontiers in Plant Science, 13*, 911610.

Raza, A., Razzaq, A., Mehmood, S. S., Zou, X., Zhang, X., Lv, Y., & Xu, J. (2019). Impact of climate change on crop adaptation and strategies to tackle its outcome: A review. *Plants, 8*(2), 34.

Reva, M., Cano, C., Herrera, M. A., & Bago, A. (2021). Arbuscular mycorrhizal inoculation enhances endurance to severe heat stress in three horticultural crops. *HortScience, 56*(4), 396–406.

Ruiz-Lozano, J. M., Aroca, R., Zamarreño, Á. M., Molina, S., Andreo-Jiménez, B., Porcel, R., & López-Ráez, J. A. (2016). Arbuscular mycorrhizal symbiosis induces strigolactone biosynthesis under drought and improves drought tolerance in lettuce and tomato. *Plant, Cell & Environment, 39*(2), 441–452.

Saeed, H. S., Munir, M., Samra, S., Farooq, A., Mahmood, H., Ahmad, I., Akram, M. I., Iqbal, R., Elsadek, M. F., Almutairi, S. M., Ditta, A., & Daraz, U. (2025). Screening of biosurfactants producing fungi from infected *Citrus sinensis* (L.). *Polish Journal of Environmental Studies, 34*(5), 5261–5270. https://doi.org/10.15244/pjoes/190588.

Salim, B. B. M., & Abou El-Yazied, A. (2015). Effect of mycorrhiza on growth, biochemical constituents, and yield of snap bean plants under water deficit conditions. *Journal of Horticultural Science & Ornamental Plants (JHSOP), 7*(3), 131–140.

Santander, C., Ruiz, A., García, S., Aroca, R., Cumming, J., & Cornejo, P. (2020). Efficiency of two arbuscular mycorrhizal fungal inocula to improve saline stress tolerance in lettuce plants by changes in antioxidant defense mechanisms. *Journal of the Science of Food and Agriculture, 100*(4), 1577–1587.

Savo, V., Kohfeld, K. E., Sillmann, J., Morton, C., Bailey, J., Haslerud, A. S., & Lepofsky, D. (2024). Using human observations with instrument-based metrics to understand changing rainfall patterns. *Nature Communications, 15*(1), 9563.

Shah, F., & Wu, W. (2019). Soil and crop management strategies to ensure higher crop productivity within sustainable environments. *Sustainability, 11*(5), 1485.

Shi, G., Yao, B., Liu, Y., Jiang, S., Wang, W., Pan, J., & Zhou, H. (2017). The phylogenetic structure of AMF communities shifts in response to gradient warming with and without winter grazing on the Qinghai–Tibet Plateau. *Applied Soil Ecology, 121*, 31–40.

Shirani Bidabadi, S., & Mehralian, M. (2020). Arbuscular mycorrhizal fungi inoculation to enhance chilling stress tolerance of watermelon. *Gesunde Pflanzen, 72*(2), 171–179.

Siddique, MBA, Khalid, A, Ditta, A, Mahmood, S, Alataway, A, Dewidar, AZ, & Mattar, MA (2023). Climate change variables modify microbial community structure and soil enzymes involved in nitrogen and phosphorus metabolism. *Rhizosphere, 28*, 100793. https://doi.org/10.1016/j.rhisph.2023.100793.

Sivakumar, K., Kumaresan, G., & Sugapriya, N. (2019). Arbuscular mycorrhizal fungi (AM fungi) and phosphate-solubilizing bacteria (PSB) on tolerance of tomato under salt stress. *Journal of Pharmacognosy and Phytochemistry, 8*, 4717–4721.

Smith, S. E., Smith, F. A., & Jakobsen, I. (2003). Mycorrhizal fungi can dominate phosphate supply to plants irrespective of growth responses. *Plant Physiology, 133*(1), 16–20.

Storer, K., Coggan, A., Ineson, P., & Hodge, A. (2018). Arbuscular mycorrhizal fungi reduce nitrous oxide emissions from N_2O hotspots. *New Phytologist, 220*(4), 1285–1295.

Thind, S., Chaudhary, M. S., Ditta, A., Hussain, I., Parveen, A., Ullah, N., Mahmood, Q., Al-Ashkar, I., & Sabagh, A. E. L. (2022). Impact of mycorrhizal fungi from different rhizospheric soils on fungal colonization, growth, and chlorophyll contents of *Cenchrus ciliaris*. *Agronomy, 12*(11), 2644. https://doi.org/10.3390/agronomy12112644.

Thirkell, T. J., Charters, M. D., Elliott, A. J., Sait, S. M., & Field, K. J. (2017). Are mycorrhizal fungi our sustainable saviours? Considerations for achieving food security. *Journal of Ecology, 105*(4), 921–929.

Thomopoulos, S., Munkholm, L. J., Elsgaard, L., & Ravnskov, S. (2025). Tillage and residue management modulate the links between soil physical signatures and arbuscular mycorrhizal fungal biomarkers. *Geoderma, 454*, 117204.

Tian, C. Y., Feng, G., Li, X. L., & Zhang, F. S. (2004). Different effects of arbuscular mycorrhizal fungal isolates from saline or non-saline soil on salinity tolerance of plants. *Applied Soil Ecology, 26*(2), 143–148.

United Nations Framework Convention on Climate Change. (2015). Adoption of the Paris Agreement: Draft decision -/CP.21: Proposal /by the President. *Conference of the Parties to the United Nations Framework Convention on Climate Change. President*. Geneva: UN 12 Dec 2015.

Wahab, A., Muhammad, M., Munir, A., Abdi, G., Zaman, W., Ayaz, A., & Reddy, S. P. P. (2023). Role of arbuscular mycorrhizal fungi in regulating growth, enhancing productivity, and potentially influencing ecosystems under abiotic and biotic stresses. *Plants, 12*(17), 3102.

Wang, G., Wang, L., Ma, F., Yang, D., & You, Y. (2021). Earthworm and arbuscular mycorrhiza interactions: Strategies to motivate antioxidant responses and improve soil functionality. *Environmental Pollution, 272*, 115980.

Wang, Z. G., Bi, Y. L., Jiang, B., Zhakypbek, Y., Peng, S. P., Liu, W. W., & Liu, H. (2016). Arbuscular mycorrhizal fungi enhance soil carbon sequestration in the coalfields, northwest China. *Scientific Reports, 6*(1), 34336.

Williams, A., Manoharan, L., Rosenstock, N. P., Olsson, P. A., & Hedlund, K. (2017). Long-term agricultural fertilization alters arbuscular mycorrhizal fungal community composition and barley (*Hordeum vulgare*) mycorrhizal carbon and phosphorus exchange. *New Phytologist, 213*(2), 874–885.

Yue, Y., Zhou, Y., Wang, J. A., & Ye, X. (2016). Assessing wheat frost risk with the support of GIS: An approach coupling a growing season meteorological index and a hybrid fuzzy neural network model. *Sustainability, 8*(12), 1308.

Zhang, F., Zou, Y. N., & Wu, Q. S. (2018). Quantitative estimation of water uptake by mycorrhizal extraradical hyphae in citrus under drought stress. *Scientia Horticulturae, 229*, 132–136.

Zhu, J., Li, M., & Whelan, M. (2018). Phosphorus activators contribute to legacy phosphorus availability in agricultural soils: A review. *Science of the Total Environment, 612*, 522–537.

Zhu, X. C., Song, F. B., Liu, S. Q., & Liu, T. D. (2011). Effects of arbuscular mycorrhizal fungus on photosynthesis and water status of maize under high temperature stress. *Plant and Soil, 346*(1), 189–199.

Zou, Y. N., Wu, Q. S., & Kuča, K. (2021). Unravelling the role of arbuscular mycorrhizal fungi in mitigating the oxidative burst of plants under drought stress. *Plant Biology, 23*, 50–57.

Chapter 25

New frontiers for sustainable agriculture through nanotechnology and arbuscular mycorrhizal fungi

Nasir Assad[a], Muhammad Ali[a], Rehan Ullah[a], Muskan Zahra[a], Parmal Arshad[a], Minahil Shehzad[a], Memoona Mukhtar[a], Muhammad Nauman Khan[b,c] and Barkat Ullah[b]

[a] *Institute of Chemistry, University of Sargodha, Sargodha, Pakistan,* [b] *Biology Laboratory, University Public School, University of Peshawar, Peshawar, Pakistan,* [c] *Department of Botany, University of Chakwal, Punjab, Pakistan*

25.1 Introduction

Agriculture is one of the pillars on which humans have flourished and existed at the lowest level of civilization. The experience of farming and taming animals has made a great contribution to the success and development of society since the dawn of human civilization (Bellwood, 2004). Nevertheless, contemporary agricultural systems have to cope with new challenges like never before. The world population is moving toward the 10-billion mark in 2050, and hence, the pressure to produce food is expected to rise steeply. But agricultural land is not growing; actually, in most parts, it is shrinking because of the urban sprawl and landscape destruction through climatic changes (Habib et al., 2025; Hussain et al., 2024; Kaplan et al., 2024). This has necessitated the immediate need to grow more food using fewer resources and also with less effect on the environment. Being able to attain sustainable agriculture in this kind of environment is an intimidating yet fundamental undertaking.

Soil degradation is one of the most severe problems in agriculture nowadays. Constant monoculture replacement, heavy use of agrochemical fertilizers and pesticides, and heavy mechanization of the topsoil have resulted in erosion, compacting, salinization, and loss of organic matter. Such modifications drastically affect the capacities of the soil to promote the growth of healthy plants (Telo da Gama, 2023). Due to almost one-third of the soils are either moderately to strongly degraded in the whole world, the agricultural productivity is directly endangered. Healthy soil, which is the basis of agriculture, is becoming an endangered resource in the quest to attain agricultural resources. The problem is worsened by a lack of water. Farming consumes almost 70% of all the available fresh water in the world, most of which is squandered in the process of irrigation and bad management. Water shortages are increasingly occurring in the areas that have already been hit by environmental changes (Mishra et al., 2021). Unfit rainfalls and droughts contribute to the farm pressure, as not all farmers can always irrigate their crops, which can make them lose their production and stay behind financially. At the same time, excessive use and misuse of water cause environmental deterioration and water contamination.

Climate change further compounds these problems. The rise in global temperatures shortens crop cycles, reduces productivity, and increases the vulnerability of plants and animals to diseases and environmental stress. Floods, storms, and other extreme weather patterns like long droughts are increasing and destroying the crops and infrastructure (Jha & Dev, 2024; Kaplan et al., 2025). In addition, agriculture is not only a victim of climate change but also a significant contributor to it. The sector emits major greenhouse gases: methane from livestock, carbon dioxide from deforestation and land-use changes, and nitrous oxide from fertilizer application. Therefore both climate change mitigation and agricultural adaptation must be integral components of any sustainability strategy.

Advancing Sustainable Agriculture Through Arbuscular Mycorrhizal Fungi. DOI: https://doi.org/10.1016/B978-0-443-45130-0.00024-2

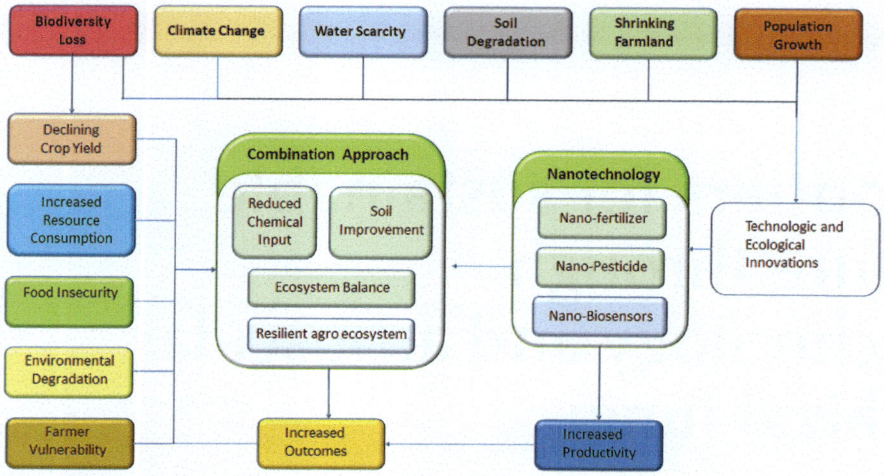

FIGURE 25.1 Conceptual flowchart: path to sustainable agriculture through challenges and technological solutions. This flowchart presents an overview of the main challenges facing sustainable agriculture, such as resource scarcity, climate change, and soil degradation, and maps them to emerging technological solutions, including nanotechnology, precision farming, and biological inputs like arbuscular mycorrhizal fungi (AMF). The flow demonstrates how these innovations converge to build a more resilient and ecofriendly agricultural system.

The other key issue is the extinction of biodiversity, which is also commonly known as a silent crisis. The destruction of their habitats, as well as the diffusion of pesticides and disease, is causing declines in the population of crucial pollinators like bees, which are crucial to the reproduction of many crops. There are also the soil microbes, such as good bacteria and fungi, which cycle many of the soil nutrients and which promote plant health as well. Such biodiversity loss weakens the resilience of agroecosystems, making them more dependent on chemical inputs and susceptible to pests, diseases, and other environmental perturbations.

Use of synthetic manure and pesticides in the long term has reduced crop production, yet in the short term, this has increased yields, which has led to a variety of environmental and health-related issues. Such nutrient runoff into the water bodies results in eutrophication, dead zones, and loss of aquatic ecosystems. Moreover, in case these chemicals are left in the food and water supply, they can cause severe health hazards. Pests and weeds have become resistant over the years, and it forced the use of increased doses or the invention of new chemicals in a continuous, costly, and unsustainable cycle (Khan et al., 2023). Moreover, such inputs are usually based on fossil fuels, which, under further exploitation, pose a risk to environmental security and energy security.

Socioeconomic dynamics are also deciding the sustainability terrain. Numerous small-holding farmers in the developing countries, a group that constitutes the majority of food producers in the world, lack access to modern technology, extension, credit, and market. These limits make them especially vulnerable to environmental and monetary shocks. Such obstacles as unstable land tenure systems, sex discrimination, and poor access also make the potential of their sustainable practices even more difficult (Balas et al., 2022). At the same time, food security is increased by inefficiencies and losses throughout the global food supply chain, which runs between harvest and consumption.

As these challenges are so large and interconnected, it is not tenable to make minor adjustments in the current agricultural systems. It requires a paradigm change, meaning a rise in productivity and a strengthening of resistance and less damage to the environment. The concept of sustainable agriculture is intended to support the demands of the current generation without jeopardizing the potential of having the need to be met in future generations. It needs resource-saving farming practices that are more productive in terms of biodiversity and help people in rural families sustain their socioeconomic livelihood (Fig. 25.1).

The technology innovation is critical to this change. Two of the most likely solutions include nanotechnology and the incorporation of biological agents, including the arbuscular mycorrhizal fungi (AMF). Such developments present opportunities to address the shortcomings of traditional agriculture, such as nutrient delivery, soil health, enhanced stress absorption, and decreased dependence on man-made inputs (Faheem et al., 2024). Nanotechnology works at the atomic and molecular level, where it is possible to come up with nanoparticles (NPs) that have unique properties that can be used in agriculture (Ditta, 2012; Ditta & Arshad, 2016; Ditta et al., 2015, 2020). As an example, nanofertilizers may slowly liberate nutrients in plant roots where they are absorbed more effectively, and less is lost by leaching or run-off (Aslam et al., 2025; Ditta & Ullah, 2022; Gul et al., 2024; Khan et al., 2024; Naz et al., 2025). Pesticides in nanoform can be used to attack particular pests using a smaller amount, thereby containing environmental pollution together with health hazards. Moreover, nanotechnology enables precision agriculture by use of nanosensors, which provide real-time monitoring of soil properties, nutrient status, and plant health (Parameswari et al., 2024). These information-based insights support smarter decisions regarding irrigation, fertilization, and pest control, helping maximize resource efficiency and crop yields. Nanomaterials can also contribute to water purification and soil remediation, addressing some of the most critical environmental challenges in agriculture (Batool et al., 2025; Rahman et al., 2025; Usman et al., 2025).

Conversely, AMF have symbiotic interactions with the roots of plants, and they stretch into the ground and increase the intake of water and other essential plant nutrients, especially phosphorus, which is usually deficient. AMF also enhance soil structure by producing glomalin, a protein that binds soil aggregates, and this was one factor that helped in promoting plant growth as well as resisting stress (such as drought, salinity, and pathogens). This results in better aeration, retention of water, and carbon sequestration (Wahab et al., 2023). Being one of the biological techniques, AMF present an ecofriendly way of enhancing crop yields and reviving the health of the soil.

A combination of nanotechnology and AMF can become a synergistic approach (Sayed et al., 2024). Although NPs can increase the speed of delivery and effectiveness of the agrochemicals and nutrients in the soil, AMF can give an ecological balance and can alleviate the possible toxicity of the NPs to give a stable and healthy soil ecosystem. This combination can lower the amount of fertilizer or pesticides, as well as increase crop output and decrease environmental stresses. In addition, the innovations are aligned with the world's agendas, such as climate-smart agriculture and United Nations Sustainable Development Goals (Ma & Rahut, 2024). They can be integrated into various farming systems and offer benefits to both large-scale commercial agriculture and smallholder farms. They offer an avenue where efficiency is boosted, greenhouse gas emissions are decreased, and climatic variations are able to cope with. This indicates a conceptual framework of the challenges and proposed technological solutions in sustainable agriculture.

In light of these challenges and opportunities, adopting sustainable agricultural practices is no longer a choice but a necessity—especially through the use of advanced technologies such as nanotechnology and AMF. Their combined application has the potential to revolutionize food production, making it more efficient, resilient, and ecologically sound. Policymakers, scientists, and agricultural stakeholders are increasingly recognizing this potential and are investing in research, development, and regulatory frameworks to ensure safe and effective implementation (Lescrauwaet et al., 2022).

Overall the current global agricultural crisis demands a comprehensive and forward-thinking response. The growing pressures on food systems underscore the urgent need for innovative approaches. Nanotechnology and AMF, used individually or in combination, offer promising avenues for achieving sustainable agriculture. They address the critical challenge of feeding a growing population while preserving the planet's natural resources, making them essential tools for the future of farming.

25.2 Nanotechnology in modern agriculture

Nanotechnology is a rapidly developing discipline in agriculture, promising novel and sustainable solutions to numerous problems faced by conventional agricultural systems (Assad et al., 2025e). Primarily, nanotechnology involves the design, manipulation, and use of materials with nanoscale dimensions, typically ranging from 1 to 100 nm (Hashmi et al., 2024; Jabbar et al., 2023). At this scale, materials would have enhanced physical, chemical, and biological properties than their bulk counterparts (Assad et al., 2025d; Ghaffar et al., 2023). Such unique properties are an increase in reactivity, surface area, and solubility and can be customized to perform particular functions. This quality of nanomaterials is very appealing as a way to improve the productivity of agriculture, enhance sustainability, and safeguard the environment (Assad et al., 2025a).

Nanomaterials that are used in agriculture vary in their purposes and functions and also vary concerning types and composition. Nanomaterials can be synthesized from a variety of metals, including silver (Khan et al., 2023), copper (Thiruvengadam et al., 2019), zinc (Assad et al., 2024), and gold (Assad et al., 2025b), as well as metal oxides such as titanium dioxide (Shi et al., 2013) and fullerenes and carbon nanotubes as carbon-based nanomaterials and polymer-based nanocarriers (Gangwar et al., 2023). Nanomaterials possess unique properties that allow for diverse applications in agriculture. As an example, metal oxide NPs have antimicrobial properties (Gul et al., 2025; Kanwal et al., 2025) and can be useful when controlling pests and diseases in the agricultural field. Carbon-based nanomaterials, with their amazing mechanical strength and ability to conduct electricity, can be used as a delivery medium and in sensor technologies. Targeted nutrient delivery is among the most promising applications of nanotechnology in agriculture (Mubeen et al., 2023; Munir et al., 2024; Yin et al., 2021). Even though traditional fertilizers are essential in crop growth, they are unproductive because large amounts of nutrients are washed, volatilized, or run off before being taken up by crops. This poor efficiency not only causes waste but also causes environmental problems such as greenhouse emissions and water pollution. This dilemma is solved by nanotechnology because it has developed nanofertilizers and nanocarriers with nutrients that release nutrients in specific sites in a slow manner and in controlled ways. Such intelligent delivery systems could be designed to react to environmental factors or throughout needs; this makes their nutrient-use efficiency significantly higher and losses very minimal (Sargazi et al., 2021). The technology of nano also improves the performance of the agrochemicals that include pesticides and herbicides. Traditional formulations tend to be used at high application rates and are therefore prone to pollute the environment and contribute to the growth of resistance by pests and weeds. Nanoformulations, conversely, facilitate pinpoint delivery of active elements toward concentrating destinations as lesser doses and thus decrease off-target and ecological effects (Xin et al., 2020). NPs have been shown to increase the solubility, stability, and bioavailability of these chemicals so they can be more effective at lower concentrations, as well as have antimicrobial potential (Noor et al., 2025). This accuracy lowers the environmental impact of crop protection with minimal loss in efficacy.

Other than enhancing the delivery of nutrients and chemicals, nanotechnology has more to offer in terms of the desired impacts on the sustainability of the environment in line with sustainable agriculture (An et al., 2022). The decreased runoff and leaching perform the roles of conserving the water quality and preventing the destruction of aquatic life. Reduced input usages lessen the amount of energy used in producing and shipping fertilizers and pesticides and the overall carbon footprint. Moreover, some types of nanomaterials show adsorption or degradation properties of environmental stressors, and this has raised hopes for different kinds of nanomaterials in the remediation and rejuvenation of soils (Dhanapal et al., 2024; Naz et al., 2022).

Recent empirical studies further illustrate the practical benefits of NP-based applications in crop systems. For instance, TiO_2-NPs applied to spinach (*Spinacia oleracea*) enhanced enzymatic activities and nitrate absorption, leading to improved biomass (Bijali & Acharya, 2020; Vega-Vásquez et al., 2020). In cadmium (Cd)-stressed wheat, foliar application of TiO_2-NPs improved plant height, photosynthesis, spike length, and grain yield (Irshad et al., 2021). Likewise, hydroponic application of TiO_2-NPs in rice (*Oryza sativa*) lowered arsenic accumulation, showing remediation potential (Wu et al., 2021).

ZnO-NPs, on the other hand, alleviate zinc deficiency and enhance dehydrogenase enzyme activity critical for metabolism (Paramo et al., 2020). Their antimicrobial activity spans various pathogens such as *Pseudomonas aeruginosa, Bacillus subtilis*, and *Aspergillus niger* (Keerthana et al., 2021; Silva et al., 2022; Tripathi et al., 2023). ZnO-NPs have also promoted yield and root growth in ladyfinger (okra) and wheat when synthesized using *Aloe vera*, especially under Cd and drought stress (Keerthana et al., 2021; Singh et al., 2019).

In addition to enhancing nutrient delivery, NPs also serve as potent antimicrobial agents, offering a sustainable alternative to traditional crop protection methods. Metal oxide NPs, especially ZnO, Ag, and CuO, show broad-spectrum inhibition of plant pathogens via mechanisms such as membrane disruption, ROS generation, and DNA interference. ZnO-NPs have demonstrated efficacy against *Salmonella enterica, Candida albicans, A. niger*, and others. Their use in foliar sprays enhances resistance in crops such as wheat, rice, and okra, reducing both disease incidence and agrochemical load. Compared to synthetic pesticides, NP-based antimicrobials operate at lower dosages, limit ecotoxicity, and show reduced resistance development, making them ideal for integrated pest management (IPM) strategies.

Nanotechnology makes this possible by providing nanosensors that can keep a record of the real-time soil and crop status. These sensors exhibit high sensitivity for monitoring nutrient deficiencies, soil moisture status, pest infestations, and disease outbreaks, thereby providing real-time information that enables farmers to optimize input application and improve resource-use efficiency. This is what precision agriculture is all about: high productivity but low input waste and pollution. The use of nanotechnology allows improving the efficiency and automation of farm management to an even greater extent by integrating it with digital agriculture tools and the Internet of Things (Dhanapal et al., 2024). Besides all the potential, nanotechnology in agriculture is also facing some risks and challenges. The fate, toxicity, and behavior of nanomaterials in the soil–plant–environment continuum have yet to be comprehended. Inappropriate management or overconcentration of NPs could be detrimental to the positive soil microbiome, nontarget organisms, or even human health (Mubeen et al., 2023). To ensure responsible implementation, further research is needed to understand NPs' interactions with biological systems, develop safe-by-design nanomaterials, and establish appropriate regulatory frameworks.

Conclusively, nanotechnology has immense potential in the modernization of agriculture that includes an upgrade in the use of nutrients, better delivery of agrochemicals, promotion of less pollution to the environment, and facilitation of precision farming. With various types of materials and flexible functionalities, it can be used to tackle numerous agricultural problems. Responsible use of nanotechnology can greatly help in sustainable agriculture in meeting the expanding food demand without degrading the environment.

25.3 Role of arbuscular mycorrhizal fungi

AMF are the symbiotic, ancient, and widespread microorganisms that have been coevolving with plants in the terrestrial ecosystems for more than 400 million years (Saeed et al., 2025; Schüßler & Walker, 2011). More than 80% of vascular plant species form mutualistic associations with roots of this kind of fungi, which are key components of natural and agricultural ecosystems (Brundrett, 2004). Unlike other synthetic technologies of this kind, AMF are living companions of plants, which exist in an intricate and mutually dependent interaction with their hosts, with the exchange of nutrients facilitated and signaling molecules between the two partners (Liu & Chen, 2024). Association starts with the release of signaling compounds by plant roots into the rhizosphere, and this triggers the germination of AMF spores (Ellouze et al., 2012). The fungal hyphae, in their turn, grow toward the root and enter its cortical cells, causing the development of special fungal structures, the arbuscules, which, because of their shape, with a highly branched, tree-like structure, become the main location of exchange of nutrients (Garg & Chandel, 2011). The plant provides the fungus with photosynthetically generated carbohydrates through these arbuscules, and in turn, the fungus results in acquiring soil nutrients that are required but may be limited (Khan et al., 2022).

The influence of AMF is one of the most important agricultural values involving the increased uptake of insoluble mineral nutrients, chiefly phosphorus, zinc, and copper (Bhantana et al., 2021; Liu et al., 2020). The fungal extraradical mycelium, a wide network of hyphae that wend far beyond the range of the root system, serves as an expansion of the root system of the plant and represents a massive expansion in root soil volume that currently searches for nutrients (Martin et al., 2000). The result of this increased availability is an increase in nutrient status and an increase in biomass produced, with a corresponding increase in crop yields, especially in nutrient-deficient or degraded soils. AMF also play significant roles in the resilience of the plants against abiotic stress (Begum et al., 2019; Nazir et al., 2023; Thind et al., 2022). Their hyphal net functions to increase water uptake, hence making plants more tolerant to drought (Bahadur et al., 2019). They are also protective in salinity and heavy metal-contaminated soils, as they regulate the intake of ions and sequester the toxic metal within fungal cells, and thus hinder the translocation of harmful compounds into the plant organs (Dhalaria et al., 2020). All these stress-relieving activities occur by way of stimulating plant defensive responses, such as the modulation of phytohormones and the boosting of the antioxidant apparatus, both enzymatic and nonenzymatic in character, that counteract oxidative stress (Assad et al., 2025c).

In addition to plant and level advantages, AMF are crucial soil health and ecosystem structural designers. They secrete glomalin, the sole glycoprotein that is powerful in the binding of soil. Glomalin is involved in the formation of durable soil aggregates, which makes soil have a good structure, porosity, and water retention, as well as less erosion. These shifts not only create a friendlier root environment but are also in favor of microbial biodiversity and soil fertility in general. Somehow, AMF are the essential elements of sustainable agriculture, providing a superior and ecologically safe alternative to chemical fertilizers, increasing the long-term efficiency and endurance of the agroecosystems.

25.4 Synergistic potential of nanotechnology and arbuscular mycorrhizal fungi

Nanotechnology combined with the use of AMF in the agricultural sector is a revolutionary solution that enhances the yield and stress tolerance of plants and minimizes reliance on chemical fertilizing (Joel et al., 2023). Although each technology individually enhances crop productivity and soil health, their integration produces synergistic effects that far exceed their isolated applications. This synergism is attributed to the complementary nature of nanomaterials and the AMF in enhancing the uptake of nutrients to achieve ecological stability and maximize the efficiency of agricultural inputs. NPs can be used to deliver nutrients, pesticides, and even growth regulators in a highly targeted fashion, which reduces waste because of volatilization/leaching, whereas AMF develop extensive hyphal networks, which enhance water and nutrient intake by plant-fungal association.

The most important benefit of such a combination is the improved nutrient distribution system. Slow-release NPs have been designed to slow down the rate of nutrient release to meet the requirements of the plant and prevent wastage and increase the uptake rate (Wahab et al., 2023). Simultaneously the AMF hyphae access the deeper micropores of the soil to deliver such nanonutrients to the roots of the plant and enhance growth and yield. Further, AMF work as innate sitters, curbing the toxicity of NPs. They convert and trap NPs in the soil matrix; this minimizes the bioavailability of the NPs and safeguards the useful organisms in the soil (Adeleke et al., 2024). Such biological buffering sustains the biodiversity of soil microbes and favors the safe use of nanotechnology in agricultural practices. The synergy also improves plant tolerance to biotic and abiotic stresses; AMF have a stimulating role in defense via systemic resistance possibilities and enhanced physiological conditions because of nanomaterials' defense-related agents (Faryal et al., 2022; Shah et al., 2021).

The other significant advantage of this integrated method is the minimized use of synthetic agrochemicals. The efficiency of nutrient use and resistance to pests that is provided by the dual application reduces the requirement for fertilizer and pesticides. It leads to affordable agriculture and less environmental pollution. The feasibility of such synergy is proven in the real world, with the use of nanofertilizers and AMF inoculants resulting in the increased phosphorus uptake in maize, wheat, and legumes. Likewise the pesticides nanoencapsulated during the AMF colonization process produce better

TABLE 25.1 Synergistic roles of nanotechnology and AMF in sustainable agriculture.

Aspect	Nanotechnology	AMF	Synergistic effect
Nutrient uptake	Delivers nutrients in nanoform for controlled release	Enhances nutrient absorption through hyphal extension	Maximized nutrient-use efficiency and reduced leaching
Soil health	May pose toxicity risks at high concentrations	Improves soil structure and microbial diversity	AMF buffer NPs' toxicity and maintain microbial balance
Stress tolerance	Delivers stress-mitigating agents (e.g., antioxidants)	Activates plant stress pathways and enhances water uptake	Dual protection from abiotic stresses such as drought and salinity
Plant growth promotion	Enhances germination and growth via nanocarriers	Produces phytohormones and expands nutrient access	Increased biomass and crop yield
Pest and disease resistance	Nanoscale pesticides target pathogens effectively	Boosts plant immune responses through symbiosis	Stronger plant defenses and reduced disease load
Environmental impact	Reduces chemical input via precision application	Natural, biodegradable symbiotic fungi	Environment-friendly farming with lower runoff and pollution

AMF, arbuscular mycorrhizal fungi; *NPs*, nanoparticles.

control of pests with little or no effect on the environment. The integrated system favors the sustenance of the intensified activities on diverse crops and soils, which translates to flexibility across diverse agro-provisioning conditions. Ultimately the complementary activity of nanotechnology and AMF provides a universal gateway to sustainable farming, as it promises productivity, conservation of resources, and environmental integrity (Adeleke et al., 2024; Joel et al., 2023; Wahab et al., 2023) (Table 25.1).

25.5 Advancements in application and integration

The introduction of nanotechnology coupled with AMF in agriculture can be termed a landmark toward sustainable agriculture with a new level of realization on the importance of the plant–fungi relationship and evolution in nanoscience. Recent developments have highlighted synergistic opportunities of these technologies in terms of enhanced crop productivity, reduced resource use during output, and reduction of environmental impacts. Fungi-based biofilms are based on nanomaterial synthesis, fungal biology, and agronomy, multiple lines of multidisciplinary investigation that have resulted in the development of proficient, realistic, ecofriendlier formulations (Ullah et al., 2024).

Innovations have also embraced the application of biodegradable nanoformulations, like those made of chitosan, silica, and particles made of lipid, in the manufacture of fertilizers, pesticides, and stimulants. The nanocarriers are more stable; release and deliver to a specific area; and respond to soil moisture, pH, or root exudates, reducing nutrient loss and maximizing uptake. At the same time, methods of AMF formulation, such as spore encapsulation and loading in nanocarriers, have increased fungal survival and coordinated administration with nanoagrochemicals, increasing synergistic effects in the rhizosphere (Ghorui et al., 2025). Field and greenhouse tests demonstrate the good perspectives. As an example, maize and wheat were fed on nanophosphorus and AMF to show even triple improvement in yields in comparison to controls, and similar effects were recorded in legumes and other vegetables, such as tomatoes and strawberries; pest resistance and nutrient uptakes were particularly bolstered (Channab et al., 2024).

These technologies prove especially beneficial in degraded, saline, or drought-prone soils, where AMF-induced glomalin production and nanobased delivery of osmoprotectants enhance fertility and resilience (Almotairy, 2024). Despite these advancements, further research is needed to optimize dosages, compatibility, and long-term ecological effects. Efforts are also underway to develop affordable, farmer-friendly delivery systems for broader adoption. Ultimately the combination of nanotechnology and AMF holds transformative potential for global agriculture, offering sustainable solutions across diverse environments. Continued interdisciplinary research, stakeholder engagement, and regulatory support are essential to realizing its full promise.

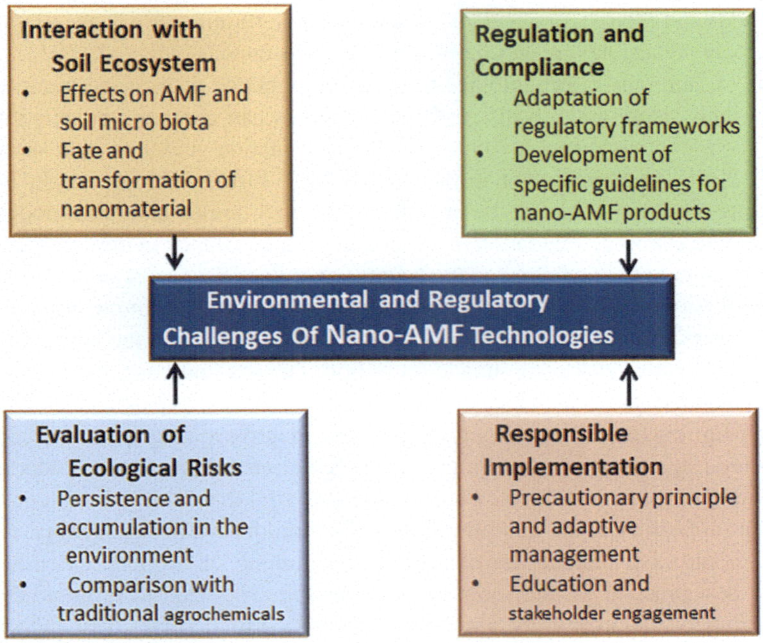

FIGURE 25.2 Environmental and regulatory considerations in the integration of nanotechnology and arbuscular mycorrhizal fungi (AMF) in agriculture. This figure illustrates the key environmental and regulatory challenges involved in the agricultural application of nanotechnology in conjunction with AMF. It highlights areas such as nanoparticle toxicity, regulatory frameworks, risk assessment procedures, biosafety protocols, and the role of AMF in mitigating environmental stress and buffering nanoparticle impacts.

25.6 Environmental and regulatory considerations

The inculcation of nanotechnology and AMF in rearing agriculture offers solutions to enhance crop productivity and sustainability in the agricultural systems, but it also exhibits significant environmental and regulatory factors that should be critically taken into consideration to ensure safe and responsible exploitation (Fig. 25.2). Identifying the interaction dynamics of these technologies with microbiota of the soil and overall ecosystem stability, determining ecological risks and issues of toxicity, and creating suitable policy and compliance systems are important steps toward enabling adoption of these technologies at the same time as protecting the health of the environment and people.

Nanotechnology deals with the application of man-made nanomaterials (ENMs), which display exotic physicochemical properties because of their microscopic size and large surface area (Barhoum et al., 2022; Hussain et al., 2025). Although these properties give such agricultural input (nanofertilizers and nanopesticides) an added efficacy, they also cause certain concerns regarding unwanted effects on soil life forms and other ecological processes in the ecosystem. The NPs can also react with useful soil microorganisms, such as bacteria, fungi, and archaea, which take crucial roles in the cycling of nutrients, breakdown of organic materials, and fitness of plants (Dinesh et al., 2012). NPs' exposure to microbial communities may have variable consequences based on the type, concentration, size, and surface chemistry of the nanomaterials, soil properties, and other factors.

Under circumstances involving the presence of AMF the most important symbiotic fungi on which growth depends in terms of nutrient acquisition and soil support, the situation becomes yet more complicated. AMF not only provide soil aggregation and glomalin biomass, but they also affect the structure and activity of other soil microbes (Li et al., 2022). The effect of nanomaterials in the soil could influence the rate of AMF colonization, spore germination, and hyphal growth. It has also been found in some studies that some NPs are harmful to AMF in high doses and even kill the AMFs, hindering the symbiotic associations, causing the loss of positive impacts associated with the AMFs (Ebbisa, 2022). Other studies, however, indicate that AMF may counter the toxicity effect of NPs by adsorbing and sequestering NPs, making them less bioavailable to soil organisms.

Considering this complex interaction, careful evaluation of the environmental fate, transport, and transformation of nanomaterials in the soil environment is urgent. This covers their persistence, their accumulation, their bioaccumulation potential in food chains, and their effects on the enzymatic activity of the soil and soil microbial diversity. Application of sophisticated analysis methods and ecotoxicological tests is useful in determining safe concentration levels and in informing the formulation of nanoagrochemicals that are less harmful to the environment (Ali et al., 2023).

The ecological hazards of nano-AMF technologies have to be compared against the traditional agricultural inputs to comprehend the degree of their safety and advantages. Nanomaterials can contain innovative threats unlike the harp of agrochemicals because of their length and receptivity to novel risk evaluation guidelines that should prescribe nanospecific descriptors (Roshan et al., 2024). Further the combination with AMF further adds variables, including fungal-mediated transformations and buffering effects, that have to be included in the environmental impact models.

In terms of regulation, nanotechnology regulation in agriculture is in its infancy in most countries around the world, with some greater regulations and guidelines depending on the country or region (Kumari et al., 2023). Regulatory authorities are also struggling to set up regulatory measures such as standardized test protocols, defining criteria of classification, labeling of such products, and even the postmarket surveillance of nanoenabled products in the agricultural sector. Innovation and access to markets, as well as effective risk communication to both farmers and consumers, can be hampered because of uncoordinated international regulations.

Regulatory strategies are turning to precautionary principles and adaptive management procedures as ways of managing these challenges. These include stepwise product testing, environmental surveillance, stakeholder involvement, and constant revision of guidelines in view of new scientific knowledge. It is also vital that AMF-related considerations be established in these frameworks so that biofertilizers and combined nano-AMF products can pass quality, efficacy, and safety levels. Responsible use policies should also include the education and training of farmers, extension workers, and industry stakeholders to spread the best practice of informed decision-making. The need to promote data and researchers to be transparent with other researchers, manufacturers, regulators, and users also increases trust and technology transfer.

Finally, although the combination of nanotechnology and AMF has prospects both significant and direct to sustainable agriculture, emphasis is more on environmental and regulatory factors that should be taken into account to prevent such innovations from further destroying the soil ecosystems and human health without the intent of doing so. Strict scientific assessment of ecological effects, development of specific risk tools, development of the overall regulatory system, and active stakeholder involvement will be the key to ensuring that the benefits of these technologies are seen in a responsible and sustainable way. With these considerations, nano-AMF solutions should effectively be incorporated into the agricultural system, and it should lead to food security and environmental conservation, maintain a balance, and be safe (Joel et al., 2023).

This flowchart illustrates the key ecological interactions, potential risks, and regulatory needs associated with nano-AMF technologies. It emphasizes soil microbiome impact, NPs' behavior, AMF compatibility, and environmental fate and outlines regulatory and stakeholder strategies for safe and sustainable deployment.

25.7 Future directions and research needs

As the world agriculture sector is working to support the increased food requirements of a rising population without compromising on environmental sustainability, interdisciplinary approaches are becoming more critical to sustainable farming. The aspects of nanotechnology and the use of AMF, such as future, may indeed be the dawn of a new era of greater productivity and resource use and the ability to resist biotic and abiotic stresses (Wahab et al., 2023). But to be able to bring out the full potential of this, we still need to do numerous research and development to mitigate the current shortcomings and increase practical applications.

One of the most urgent vectors of further development is the enhancement of interdisciplinary cooperation between material science, microbiology, plant physiology, soil science, and agronomy. The line of relationships between nanomaterials, AMF, plants, and soil ecosystems is too complicated and requires interdisciplinary research that will overcome the boundaries of these areas (Ahmad & Ahmad, 2024). This type of synergy is crucial to develop nanomaterials that are not only functional but also biocompatible and environmentally salubrious and, at the same time, clarifies how the AMF affect the functionality of the NPs and, in consequence, the well-being of the plants under different environmental conditions. With the help of established bioanalytical technologies at the molecular and nanoscale, including genomics, transcriptomics, and metabolomics, these connections will be revealed at the level of their complex cellular and biochemical pathways (Wang et al., 2025).

Inclusion of nano-AMF technologies into precision agriculture is another area of research that is essential. The most advanced advancement in data-driven agriculture (remote sensing, geographic information systems, and machine learning) allows optimizing crop and resource management at the field level (Bhat & Huang, 2021).

The real-time version of nanosensors can assist in real-time surveying of the soil nutrient status, soil moisture levels, plant stress indicators, and AMF colonization rates, which would allow selective and appropriate application of nanofertilizers and bioinoculant applications. This will be more efficient in terms of input as well as mitigating environmental effects because it does not result in overapplication. The future advancements of this vision will be building smart delivery systems that react to the environmental stimuli and the signals of the plants.

The implementation of the nano-AMF technologies on a large scale is one of the biggest bottlenecks; it is hard to roll out technologies under laboratory conditions to real and ever-changing agricultural operations. Despite this encouraging outcome in small-scale experimentation, there is a lot of need to replicate these gains in the real world of commerce, with the smallholder farmers in the developing world being at one end of the spectrum and the large-scale, industrial complex being on the other end of the spectrum. This necessitates serious field tests in varying climatic, soil, and management settings. These involve optimizing the stability of formulation, dose, application window, and delivery mode to provide cost-effective results, repeated effects, and acceptability by the users. The forms have to be shelf-stable, simple to apply without specific tools, and fit into IPM and organic farming systems (Stejskal et al., 2021). Regulatory and safety aspects continue to be paramount to the nano-AMF research in the future. To evaluate the fate, persistence, and ecological effects of nanomaterials, especially the interaction of nanomaterials with AMF and other soil organisms, long-term monitoring and field trials on the environment are required (Rasheela et al., 2024).

25.8 Conclusion

The fusion of nanotechnology and AMF represents a transformative stride toward sustainable, high-efficiency agriculture. By uniting the precision and innovation of nanomaterials with the evolutionary resilience of AMF, this integrated approach offers a pathway to address the most pressing agricultural challenges of the 21st century. Nanotechnology enables the targeted and controlled delivery of agrochemicals, thereby reducing input losses, environmental contamination, and resource dependency. In parallel, AMF operate as living soil engineers, extending root systems, enhancing nutrient and water absorption, and improving plant tolerance to drought, salinity, and toxic elements. This synergy is not merely complementary but profoundly synergistic. When applied together, nanomaterials and AMF can amplify each other's benefits, leading to improved plant growth, enhanced soil health, and reduced reliance on synthetic fertilizers and pesticides. The result is a model of agriculture that is both productive and ecologically regenerative, restoring the vitality of soils, promoting biodiversity, and minimizing the ecological footprint of food production. Yet, this promise must be approached with caution and scientific responsibility. The environmental fate, bioavailability, and long-term impacts of NPs in complex soil ecosystems remain underexplored, particularly regarding their interactions with beneficial microbes like AMF. It is essential to adopt a "safe-by-design" framework grounded in rigorous research, risk assessment, and transparent regulation to ensure that these technologies contribute positively without unintended ecological consequences. Looking ahead the convergence of nanotechnology and AMF offers more than a technological fix; it embodies a paradigm shift toward systems-level thinking in agriculture. It encourages innovation that respects ecological boundaries and leverages biological partnerships. With continued interdisciplinary collaboration, policy support, and investment in sustainable innovation, this integrated approach can play a critical role in achieving global food security while safeguarding the planet for future generations.

References

Adeleke, B. S., Olowe, O. M., Ayilara, M. S., Fasusi, O. A., Omotayo, O. P., Fadiji, A. E., Onwudiwe, D. C., & Babalola, O. O. (2024). Biosynthesis of nanoparticles using microorganisms: A focus on endophytic fungi. *Heliyon, 10*(21), e39636.

Ahmad, F., & Ahmad, S. (2024). NPs and soil microorganisms interactions in crop management-current status and future prospects. In: S. Jain, A. Gupta, & N. Verma (Eds.), *Industrial applications of soil microbes* (Vol. 4, pp. 361–403). Bentham Science Publishers. https://doi.org/10.2174/9789815124996124040021.

Ali, S., Ahmad, N., Dar, M. A., Manan, S., Rani, A., Alghanem, S. M. S., & Mostafa, Y. S. (2023). Nano-agrochemicals as substitutes for pesticides: Prospects and risks. *Plants, 13*(1), 109.

Almotairy, H. (2024). Mitigating metal/metalloid stress in crops: Strategies for sustainable agricultural resilience. *Abiotic stress in crop plants—Ecophysiological responses and molecular approaches*. IntechOpen.

An, C., Sun, C., Li, N., Huang, B., Jiang, J., Shen, Y., & Wang, C. (2022). Nanomaterials and nanotechnology for the delivery of agrochemicals: Strategies towards sustainable agriculture. *Journal of Nanobiotechnology, 20*(1), 11.

Aslam, A., Aslam, S., Amjad, A., Amjad, E., & Ditta, A. (2025). Synergistic impacts of nanoparticles and PGPR under various abiotic stresses for higher performance of agricultural crops. In A. Husen (Ed.), *Nanobiotechnology for abiotic stress adaptation and mitigation in agricultural crops—Physiological and molecular intervention* (pp. 227–247). Springer Nature Singapore Pte Ltd.

Assad, N., Abbas, A., ur Rehman, M. F., & Naeem-ul-Hassan, M. (2024). Photo-catalytic and biological applications of phyto-functionalized zinc oxide nanoparticles synthesized using a polar extract of *Equisetum diffusum* D. *RSC Advances, 14*(31), 22344–22358.

Assad, N., Ahmad, H., Abbas, A., Fayyaz Ur Rehman, M., Naeem-ul-Hassan, M., Sher, M., & Assad, Y. (2025a). Green synthesis of selenium nanoparticles using *Equisetum diffusum*: Characterization, antibacterial potential, effects on plant growth. *IEEE Transactions on NanoBioscience*, 1. https://doi.org/10.1109/TNB.2025.3567327.

Assad, N., Ansa, M. B. L., Sher Ullah, M. A., Wahab, S., Kaplan, A., Khan, M. N., & Ullah, B. (2025e). Nanotechnology in oilseed crops: A New Frontier for abiotic stress adaptation. In *Oilseed crops under abiotic stress* (p. 507). https://doi.org/10.1007/978-981-96-8346-8_16.

Assad, N., Laila, M. B., Naeem Ul Hassan, M., Fayyaz ur Rehman, M., Ali, L., Mustaqeem, M., Ullah, B., Khan, M. N., Iqbal, M., Ercişli, S., Alarfaj, A. A., Ansari, M. J., & Malik, T. (2025b). Eco-friendly synthesis of gold nanoparticles using *Equisetum diffusum* D. Don. with broad-spectrum antibacterial, anticancer, antidiabetic, and antioxidant potentials. *Scientific Reports, 15*(1), 19246.

Assad, N., Laila, M. B., Naeem-Ul-Hassan, M., Ali, B., Kaplan, A., Khan, M. N., & Ozdemir, F. A. (2025c). Phytohormones as key players: Boosting oilseed crop tolerance to nutrient deficiency stress. *Oilseed crops under abiotic stress: Mitigation strategies and future perspectives*, 415. https://doi.org/10.1007/978-981-96-8346-8_13.

Assad, N., Naeem-Ul-Hassan, M., Ajaz Hussain, M., Abbas, A., Sher, M., Muhammad, G., & Farid-ul-Haq, M. (2025d). Diffused sunlight assisted green synthesis of silver nanoparticles using *Cotoneaster nummularia* polar extract for antimicrobial and wound healing applications. *Natural Product Research, 39*(8), 2203–2217.

Bahadur, A., Batool, A., Nasir, F., Jiang, S., Mingsen, Q., Zhang, Q., & Feng, H. (2019). Mechanistic insights into arbuscular mycorrhizal fungi-mediated drought stress tolerance in plants. *International Journal of Molecular Sciences, 20*(17), 4199.

Balas, M., Lemmen, C., & Casimiro, I. M. (2022). Factors contributing to gender inequality in land access and land tenure security in Sub-Saharan Africa: A literature review. In *Paper presented at the 27th FIG Congress 2022: Volunteering for the future-Geospatial excellence for a better living* (pp. 11–15). Poland: Warsaw.

Barhoum, A., García-Betancourt, M. L., Jeevanandam, J., Hussien, E. A., Mekkawy, S. A., Mostafa, M., & Bechelany, M. (2022). Review on natural, incidental, bioinspired, and engineered nanomaterials: History, definitions, classifications, synthesis, properties, market, toxicities, risks, and regulations. *Nanomaterials, 12*(2), 177.

Batool, F., Tahira, M., Gul, M., Qadir, R., Akhtar, T., Ghumman, S. A., Amin, M., Sajjad, N., Rehman, M. A., & Ditta, A. (2025). Bioremediation of Nigrosine anionic dye from wastewater employing copper nanocomposite-modified *Cucumis melo* seeds powder: A comparison of performance evaluation between defatted and surface modified adsorbent. *Biomass Conversion and Biorefinery, 15*, 23523–23542.

Begum, N., Qin, C., Ahanger, M. A., Raza, S., Khan, M. I., Ashraf, M., & Zhang, L. (2019). Role of arbuscular mycorrhizal fungi in plant growth regulation: Implications in abiotic stress tolerance. *Frontiers in Plant Science, 10*, 1068.

Bellwood, P. (2004). *First farmers: The origins of agricultural societies*. John Wiley & Sons.

Bhantana, P., Rana, M. S., Sun, X.-c., Moussa, M. G., Saleem, M. H., Syaifudin, M., & Bhat, M. A. (2021). Arbuscular mycorrhizal fungi and its major role in plant growth, zinc nutrition, phosphorous regulation and phytoremediation. *Symbiosis, 84*(1), 19–37.

Bhat, S. A., & Huang, N.-F. (2021). Big data and AI revolution in precision agriculture: Survey and challenges. In *IEEE Access: 9* (pp. 110209–110222).

Bijali, J., & Acharya, K. (2020). Current trends in nano-technological interventions on plant growth and development: A review. *IET Nanobiotechnology, 14*(2), 113–119.

Brundrett, M. (2004). Diversity and classification of mycorrhizal associations. *Biological Reviews, 79*(3), 473–495.

Channab, B.-E., Idrissi, A. E., Ammar, A., Dardari, O., Marrane, S. E., El Gharrak, A., & Zahouily, M. (2024). Recent advances in nano-fertilizers: Synthesis, crop yield impact, and economic analysis. *Nanoscale, 16*(9), 4484–4513.

Dhalaria, R., Kumar, D., Kumar, H., Nepovimova, E., Kuča, K., Torequl Islam, M., & Verma, R. (2020). Arbuscular mycorrhizal fungi as potential agents in ameliorating heavy metal stress in plants. *Agronomy, 10*(6), 815.

Dhanapal, A. R., Thiruvengadam, M., Vairavanathan, J., Venkidasamy, B., Easwaran, M., & Ghorbanpour, M. (2024). Nanotechnology approaches for the remediation of agricultural polluted soils. *ACS Omega, 9*(12), 13522–13533.

Dinesh, R., Anandaraj, M., Srinivasan, V., & Hamza, S. (2012). Engineered nanoparticles in the soil and their potential implications to microbial activity. *Geoderma, 173*, 19–27.

Ditta, A., & Ullah, N. (2022). An update on nanotechnology and sustainable agriculture. In W. Ahmed, & E. Nourafkan (Eds.), *Science and applications of nanoparticles* (pp. 159–197). Jenny Stanford Publishing Pte. Ltd.

Ditta, A. (2012). How helpful is nanotechnology in agriculture? *Advances in Natural Science: Nanoscience and Nanotechnology, 3*, 033002.

Ditta, A., Arshad, M., & Ibrahim, M. (2015). Nanoparticles in sustainable agricultural crop production: Applications and perspectives. In M. H. Siddiqui, M. H. Al-Whaibi, & F. Mohammad (Eds.), *Nanotechnology and plant sciences—Nanoparticles and their impact on plants* (pp. 55–75). Springer.

Ditta, A., & Arshad, M. (2016). Applications and perspectives of using nanomaterials for sustainable plant nutrition. *Nanotechnology Reviews, 2*(5), 209–229.

Ditta, A., Mehmood, S., Imtiaz, M., Rizwan, M. S., & Islam, I. (2020). Soil fertility and nutrient management with the help of nanotechnology. In A. Husen, & M. Jawaid (Eds.), *Nanomaterials for agriculture and forestry applications* (pp. 273–288). Woodhead Publishing Ltd., Elsevier UK.

Ebbisa, A. (2022). Arbuscular mycorrhizal fungi (AMF) in optimizing nutrient bioavailability and reducing agrochemicals for maintaining sustainable agroecosystems. *Arbuscular mycorrhizal fungi in agriculture-new insights*. IntechOpen.

Ellouze, W., Hamel, C., Cruz, A. F., Ishii, T., Gan, Y., Bouzid, S., & St-Arnaud, M. (2012). Phytochemicals and spore germination: At the root of AMF host preference? *Applied Soil Ecology, 60*, 98–104.

Faheem, M., Ditta, A., & Du, J. (2024). *Nanomaterials in industrial chemistry* (1st ed.). CRC Press, Taylor & Francis Group.

Faryal, S., Ullah, R., Khan, M. N., Ali, B., Hafeez, A., Jaremko, M., & Qureshi, K. A. (2022). Thiourea-capped nanoapatites amplify osmotic stress tolerance in *Zea mays* L. by conserving photosynthetic pigments, osmolytes biosynthesis and antioxidant biosystems. *Molecules, 27*(18), 5744.

Gangwar, J., Kadanthottu Sebastian, J., Puthukulangara Jaison, J., & Kurian, J. T. (2023). Nano-technological interventions in crop production—A review. *Physiology and Molecular Biology of Plants, 29*(1), 93–107.

Garg, N., & Chandel, S. (2011). Arbuscular mycorrhizal networks: Process and functions. In Sustainable agriculture *volume 2* (pp. 907–930). Springer.

Ghaffar, S., Abbas, A., Naeem-ul-Hassan, M., Assad, N., Sher, M., Ullah, S., & Al Bratty, M. (2023). Improved photocatalytic and antioxidant activity of olive fruit extract-mediated ZnO nanoparticles. *Antioxidants, 12*(6), 1201.

Ghorui, M., Chowdhury, S., & Burla, S. (2025). Recent advances in the commercial formulation of arbuscular mycorrhizal inoculants. *Frontiers in Industrial Microbiology, 3*, 1553472.

Gul, M., Khan, R. S., Ul Islam, Z., Khan, S., Shumaila, A., Umar, S., Khan, S., Brekhna, Zahoor, M., & Ditta, A. (2024). Nanoparticles in plant resistance against bacterial pathogens: Current status and future prospects. *Molecular Biology Reports, 51*, 92.

Gul, H., Khan, H. A., Nasreen, Z., Assad, N., Turab, S. A., & Hanif, M. (2025). Exploring the antibacterial, anticoagulant, and hemolytic potential of green-synthesized Fe_2O_3 nanoparticles by *Cucurbita pepo* pulp. *IEEE Transactions on NanoBioscience*, 1. https://doi.org/10.1109/TNB.2025.3563307.

Habib, M., Singh, S., Jan, S., Jan, K., & Bashir, K. (2025). The future of the future foods: Understandings from the past towards SDG-2. *npj Science of Food, 9*(1), 138.

Hashmi, S., Ibrahim, M., Adnan, M., Ullah, A., Khan, M., Kamal, A., & Zaman, W. (2024). Green synthesis of silver nanoparticles from *Olea europaea* L. extracted polysaccharides, characterization, and its assessment as an antimicrobial agent against multiple pathogenic microbes. *Open Chemistry*, 22. https://doi.org/10.1515/chem-2024-0016.

Hussain, A., Elkarmout, A. F., Mansour, E. Z., Awais, M., Usman, M., Ahmad, H., & Ahmad, T. (2024). An environment friendly practice, the climate smart agriculture crop production and soil management systems: A review. *Journal of Sustainable Agricultural and Environmental Sciences, 3*(3), 101–124.

Hussain, S. B., Zahra, S., Hussain, S., Mahmood, H., Haq, A. u., Shafique, M., & Khan, M. N. (2025). Characterization and comparative antibacterial activities of gold nanoparticles synthesized by *Tagetes patula* L. (Asteraceae) flower extract. *Microscopy Research and Technique, 88*(6), 1869–1880.

Irshad, M. A., Nawaz, R., Ur Rehman, M. Z., Adrees, M., Rizwan, M., Ali, S., & Tasleem, S. (2021). Synthesis, characterization and advanced sustainable applications of titanium dioxide nanoparticles: A review. *Ecotoxicology and Environmental Safety, 212*, 111978.

Jabbar, A., Abbas, A., Assad, N., Naeem-ul-Hassan, M., Alhazmi, H. A., Najmi, A., & Amin, H. M. (2023). A highly selective Hg^{2+} colorimetric sensor and antimicrobial agent based on green synthesized silver nanoparticles using *Equisetum diffusum* extract. *RSC Advances, 13*(41), 28666–28675.

Jha, M. K., & Dev, M. (2024). Impacts of climate change. In *Smart internet of things for environment and healthcare* (pp. 139–159). Springer.

Joel, J. M., Johnson, R., & Puthur, J. T. (2023). Co-application of arbuscular mycorrhizal fungi and engineered nanomaterials: A promising strategy for crop resilience against abiotic stresses. *South African Journal of Botany, 162*, 314–323.

Kanwal, M., Sher, M., Abbas, A., Akhtar, S., Siddique, A. B., ul Hasan, M. N., Assad, N., Alhazmi, H. A., & Amin, H. M. A. (2025). Dual colorimetric sensing of Hg(II) and Fe(III) using sulfanilamide-stabilized silver nanoparticles and evaluating their photodegradation and antibacterial properties. *Journal of Water Process Engineering, 75*, 107981. https://doi.org/10.1016/j.jwpe.2025.107981.

Kaplan, A., Khan, M. N., Hayat, K., Iqbal, M., Ali, B., Wahab, S., Wahid, N., & Kanwal (2024). Coupling environmental factors and climate change: Impacts on plants and vegetation growth patterns in ecologically sensitive regions. In *Environment, climate, plant and vegetation growth* (pp. 307–358). Springer.

Kaplan, A., Khan, M. N., Wahab, S., Assad, N., Adnan, M., & Iqbal, M. (2025). Thallium in soil environments and its biological availability. In *Beneficial elements for remediation of heavy metals in polluted soil* (pp. 373–398). Elsevier.

Keerthana, P., Vijayakumar, S., Vidhya, E., Punitha, V., Nilavukkarasi, M., & Praseetha, P. (2021). Biogenesis of ZnO nanoparticles for revolutionizing agriculture: A step towards anti-infection and growth promotion in plants. *Industrial Crops and Products, 170*, 113762.

Khan, M. A., Sarfraz, W., & Ditta, A. (2024). Applications of nano-based fertilizers, pesticides, and biosensors in sustainable agriculture and food security. In H. Tombuloglu, G. Tombuloglu, E. Al-Suhaimi, A. Baykal, & K. R. Hakeem (Eds.), *Molecular impacts of nanoparticles on plants and algae nanomaterial-plant interactions* (pp. 277–303). Academic Press, Elsevier.

Khan, B. A., Nadeem, M. A., Nawaz, H., Amin, M. M., Abbasi, G. H., Nadeem, M., & Maqbool, R. (2023). Pesticides: Impacts on agriculture productivity, environment, and management strategies. In *Emerging contaminants and plants: Interactions, adaptations and remediation technologies* (pp. 109–134). Springer.

Khan, Y., Shah, S., & Tian, H. (2022). The roles of arbuscular mycorrhizal fungi in influencing plant nutrients, photosynthesis, and metabolites of cereal crops—A review. *Agronomy, 12*(9), 2191.

Khan, Z. U. R., Assad, N., Naeem-Ul-Hassan, M., Sher, M., Alatawi, F. S., Alatawi, M. S., Omran, A. M. E., Jame, R. M. A., Adnan, M., Khan, M. N., Ali, B., Wahab, S., Razak, S. A., Javed, M. A., Kaplan, A., & Rahimi, M. (2023). *Aconitum lycoctonum* L. (Ranunculaceae) mediated biogenic synthesis of silver nanoparticles as potential antioxidant, anti-inflammatory, antimicrobial and antidiabetic agents. *BMC Chemistry, 17*(1), 128.

Kumari, R., Suman, K., Karmakar, S., Mishra, V., Lakra, S. G., Saurav, G. K., & Mahto, B. K. (2023). Regulation and safety measures for nanotechnology-based agri-products. *Frontiers in Genome Editing, 5*, 1200987.

Lescrauwaet, L., Wagner, H., Yoon, C., & Shukla, S. (2022). Adaptive legal frameworks and economic dynamics in emerging tech-nologies: Navigating the intersection for responsible innovation. *Law and Economics, 16*(3), 202–220.

Li, Y., Xu, J., Hu, J., Zhang, T., Wu, X., & Yang, Y. (2022). Arbuscular mycorrhizal fungi and glomalin play a crucial role in soil aggregate stability in Pb-contaminated soil. *International Journal of Environmental Research and Public Health, 19*(9), 5029.

Liu, Y. Z., Imtiaz, M., Ditta, A., Rizwan, M. S., Ashraf, M., Mehmood, S., Aziz, O., Mubeen, F., Ali, M., Elahi, N. N., Ijaz, R., S, Lelel, Shuang, C., & Tu, S. (2020). Response of growth, antioxidant enzymes, and root exudates production towards as stress in *Pteris vittata* and *Astragalus sinicus* colonized by arbuscular mycorrhizal fungi. *Environmental Science and Pollution Research, 27*, 2340–2352.

Liu, R., & Chen, Y. (2024). Mycorrhizal symbiosis: Evolution, opportunities, challenges, and prospects. In *Arbuscular mycorrhizal fungi in sustainable agriculture: Inoculum production and application* (pp. 1–35). https://doi.org/10.1007/978-981-97-0296-1_1.

Ma, W., & Rahut, D. B. (2024). Climate-smart agriculture: Adoption, impacts, and implications for sustainable development. *Mitigation and Adaptation Strategies for Global Change, 29*(5), 44.

Martin, F. M., Perotto, S., & Bonfante, P. (2000). Mycorrhizal fungi: A fungal community at the interface between soil and roots. In *The rhizosphere* (pp. 279–312). CRC Press.

Mishra, B. K., Kumar, P., Saraswat, C., Chakraborty, S., & Gautam, A. (2021). Water security in a changing environment: Concept, challenges and solutions. *Water, 13*(4), 490.

Mubeen, B., Hasnain, A., Wang, J., Zheng, H., Naqvi, S. A. H., Prasad, R., & Farhan, M. (2023). Current progress and open challenges for combined toxic effects of manufactured nano-sized objects (MNO's) on soil biota and microbial community. *Coatings, 13*(1), 212.

Munir, A., Salah, M. A., Ali, M., Ali, B., Saleem, M. H., Samarasinghe, K., & Anas, M. (2024). Advancing agriculture: Harnessing smart nanoparticles for precision fertilization. *BioNanoScience, 14*(4), 3846–3863.

Naz, T., Iqbal, M. M., Ullah, I., Mubeen, M. A., & Ditta, A. (2025). Nano-silicon application enhances salt stress tolerance in *Lens culinaris* L.: A comparative study with bulk silicon. *Journal of Plant Nutrition, 48*(19), 3377–3393.

Naz, R., Khan, M., Hafeez, A., Fazil, M., Khan, M., Ali, B., & Alfaifi, M. (2022). Assessment of phytoremediation potential of native plant species naturally growing in a heavy metal-polluted industrial soils. *Brazilian Journal of Biology, 84*, e264473.

Nazir, A., Sarfraz, W., Allah, D., Khalid, N., Farid, M., Shafiq, M., Firdaus-E-Bareen, Rizvi, Z. F., & Naeem, N. (2023). Synergistic impact of two autochthonous saprobic fungi (*A. niger* and *T. pseudokoningii*) on the growth, ionic contents, and metals uptake in *Brassica juncea* L. and *Vigna radiata* L. under tannery solid waste contaminated soil. *International Journal of Phytoremediation, 25*(11), 1488–1500.

Noor, A., Ahmad, N., Ali, A., Ali, M., Iqbal, M., Khan, M. N., & Ercişli, S. (2025). Green synthesis of silver-doped ZnO nanoparticles from *Adiantum venustum* D. Don (Pteridaceae): Antimicrobial and antioxidant evaluation. *Journal of Basic Microbiology, 65*(4), e2400543.

Parameswari, P., Belagalla, N., Singh, B. V., Abhishek, G., Rajesh, G., Katiyar, D., & Paul, S. (2024). Nanotechnology-based sensors for real-time monitoring and assessment of soil health and quality: A review. *Asian Journal of Soil Science and Plant Nutrition, 10*(2), 157–173.

Paramo, L. A., Feregrino-Pérez, A. A., Guevara, R., Mendoza, S., & Esquivel, K. (2020). Nanoparticles in agroindustry: Applications, toxicity, challenges, and trends. *Nanomaterials, 10*(9), 1654.

Rahman, H. U., Khan, M., & Ditta, A. (2025). Recent advances in sustainable waste management practices. In A. A. H. Abdel Latef, E. M. Zayed, & A. A. Omar (Eds.), *Bioremediation and nanotechnology for climate change mitigation* (pp. 103–136). Springer.

Rasheela, A. R. P., Khalid, M. F., Abumaali, D. A., Alatalo, J. M., & Ahmed, T. (2024). Impact of abiotic stressors on soil microbial communities: A focus on antibiotics and their interactions with emerging pollutants. *Soil Systems, 9*(1), 2.

Roshan, M., Singh, I., Vats, A., Behera, M., Singh, D. P., Gautam, D., & De, S. J. I. M. (2024). Antimicrobial and antibiofilm effect of cannabinoids from Cannabis sativa against methicillin-resistant Staphylococcus aureus (MRSA) causing bovine mastitis. *Journal of Immunology and Microbiology*, 1–14.

Saeed, H. S., Munir, M., Samra, S., Farooq, A., Mahmood, H., Ahmad, I., Akram, M. I., Iqbal, R., Elsadek, M. F., Almutairi, S. M., Ditta, A., & Daraz, U. (2025). Screening of biosurfactants producing fungi from infected *Citrus sinensis* (L.). *Polish Journal of Environmental Studies, 34*(5), 5261–5270.

Sargazi, S., Mukhtar, M., Rahdar, A., Barani, M., Pandey, S., & Díez-Pascual, A. M. (2021). Active targeted nanoparticles for delivery of poly (ADP-ribose) polymerase (PARP) inhibitors: A preliminary review. *International Journal of Molecular Sciences, 22*(19), 10319.

Sayed, E. G., Desoukey, S., Desouky, A. F., Farag, M. F., El-Kholy, R. I., & Azoz, S. N. (2024). Synergistic influence of arbuscular mycorrhizal fungi inoculation with nanoparticle foliar application enhances chili (*Capsicum annuum* L.) antioxidant enzymes, anatomical characteristics, and productivity under cold-stress conditions. *Plants, 13*(4), 517.

Schüßler, A., & Walker, C. (2011). The Fungal Kingdom: Evolution, Morphology, and Ecology. In D. T. Wicklow, & G. M. Carroll (Eds.), *Evolution of the 'plant-symbiotic' fungal phylum, Glomeromycota* (pp. 163–185). Springer.

Shah, S. N., Ahmad, M., Zafar, M., Hadi, F., Khan, M. N., Noor, A., & Iqbal, M. (2021). Application of spore morphology to solve identification problems in certain species of family Dryopteridaceae from Malakand Division, Pakistan. *Microscopy Research and Technique, 84*(8), 1897–1910.

Shi, H., Magaye, R., Castranova, V., & Zhao, J. (2013). Titanium dioxide nanoparticles: A review of current toxicological data. *Particle and Fibre Toxicology, 10*(1), 15.

Silva, S., Dias, M. C., & Silva, A. M. (2022). Titanium and zinc based nanomaterials in agriculture: A promising approach to deal with (A) biotic stresses? *Toxics, 10*(4), 172.

Singh, J., Kumar, S., Alok, A., Upadhyay, S. K., Rawat, M., Tsang, D. C., & Kim, K.-H. (2019). The potential of green synthesized zinc oxide nanoparticles as nutrient source for plant growth. *Journal of Cleaner Production, 214*, 1061–1070.

Stejskal, V., Vendl, T., Aulicky, R., & Athanassiou, C. (2021). Synthetic and natural insecticides: Gas, liquid, gel and solid formulations for stored-product and food-industry pest control. *Insects, 12*(7), 590.

Telo da Gama, J. (2023). The role of soils in sustainability, climate change, and ecosystem services: Challenges and opportunities. *Ecologies, 4*(3), 552–567.

Thind, S., Chaudhary, M. S., Ditta, A., Hussain, I., Parveen, A., Ullah, N., Mahmood, Q., Al-ashkar, I., & Sabagh, A. E. L. (2022). Impact of mycorrhizal fungi from different rhizospheric soils on fungal colonization, growth, and chlorophyll contents of *Cenchrus ciliaris*. *Agronomy, 12*(11), 2644.

Thiruvengadam, M., Chung, I.-M., Gomathi, T., Ansari, M. A., Gopiesh Khanna, V., Babu, V., & Rajakumar, G. (2019). Synthesis, characterization and pharmacological potential of green synthesized copper nanoparticles. *Bioprocess and Biosystems Engineering, 42*(11), 1769–1777.

Tripathi, S., Mahra, S., Tiwari, K., Rana, S., Tripathi, D. K., Sharma, S., & Sahi, S. (2023). Recent advances and perspectives of nanomaterials in agricultural management and associated environmental risk: A review. *Nanomaterials, 13*(10), 1604.

Ullah, Z., Iqbal, J., Abbasi, B. A., Ijaz, S., Ahmad, S., Khan, S., & Mehmood, Y. (2024). Nano-formulation for agriculture applicability. In *Revolutionizing agriculture: A comprehensive exploration of agri-nanotechnology* (pp. 325–367). Springer.

Usman, M., Batool, F., Iqbal, T., Noreen, S., Gondal, H. Y., Roheen, T., Qadir, R., Amin, M., Sajid, S., & Ditta, A. (2025). Harnessing de-oiled *Glycine max* seeds-anchored-CuO nanoparticles for adsorptive removal of crystal violet dye with comprehensive mechanistic insights. *RSC Advances, 15*, 24406–24423.

Vega-Vásquez, P., Mosier, N. S., & Irudayaraj, J. (2020). Nanoscale drug delivery systems: From medicine to agriculture. *Frontiers in Bioengineering and Biotechnology, 8*, 79.

Wahab, A., Muhammad, M., Munir, A., Abdi, G., Zaman, W., Ayaz, A., & Reddy, S. P. P. (2023). Role of arbuscular mycorrhizal fungi in regulating growth, enhancing productivity, and potentially influencing ecosystems under abiotic and biotic stresses. *Plants, 12*(17), 3102.

Wang, Y., Xiao, Z., Wang, Z., Lee, D., Ma, Y., Wilhelm, S., & Jiang, W. (2025). Multi-omics approaches to decipher the interactions of nanoparticles and biological systems. *Nature Reviews Bioengineering, 3*(4), 333–348.

Wu, X., Hu, J., Wu, F., Zhang, X., Wang, B., Yang, Y., & Wang, X. (2021). Application of TiO_2 nanoparticles to reduce bioaccumulation of arsenic in rice seedlings (*Oryza sativa* L.): A mechanistic study. *Journal of Hazardous Materials, 405*, 124047.

Xin, X., Judy, J. D., Sumerlin, B. B., & He, Z. (2020). Nano-enabled agriculture: From nanoparticles to smart nanodelivery systems. *Environmental Chemistry, 17*(6), 413–425.

Yin, H., Cao, Y., Marelli, B., Zeng, X., Mason, A. J., & Cao, C. (2021). Soil sensors and plant wearables for smart and precision agriculture. *Advanced Materials, 33*(20), 2007764.

Chapter 26

Perceptions, knowledge, and practices among farmers regarding social impacts of mycorrhizal technology adoption

Samah Ait Benichou[a], Mounia Cherki[a] and Karima El Bouqdaoui[a,b]

[a] *Laboratory of Health, Environment and Biotechnologies, Ain Chock's Sciences Faculty, Hassan II University of Casablanca, Casablanca, Morocco,* [b] *Laboratory of Care, Health and Sustainable Development, Higher Institute of Nursing Professions and Health Techniques, Casablanca, Morocco*

26.1 Introduction

Currently, many agricultural soils, especially in arid and semiarid environments, are fragile and subject to profound biological and chemical disturbances, leading to lower crop yields and, as a result, numerous difficulties in meeting the food needs of local populations in vulnerable regions (FAO, 2020a; Shokri et al., 2025). Therefore, to meet this challenge and increase agricultural productivity and sustainability without any risk to human health, the search for new alternative green strategies is becoming a global priority (FAO, 2020b). Agroecology plays a central role in restoring degraded ecosystems. It proposes environmentally friendly practices, such as the use of plant cover, water management, and the restoration of soil fertility through phytoremediation and bioremediation (Naeem et al., 2024; Rahman & Ditta, 2025; Rahman et al., 2025; Wezel & Rath, 2002). The latter is an increasingly popular approach that relies on microorganisms (bacteria and soil fungi) to break down or transform pollutants, including hydrocarbons, pesticides, and other organic contaminants (FAO, 2020b).

Microorganisms, particularly mycorrhizal fungi, play a crucial role in restoring soil structure. Without the use of aggressive chemical treatments, symbiotic microorganisms improve soil health and establish mutually beneficial interactions with plants, contributing to their nutrition. In this context, arbuscular mycorrhizal fungi (AMF) in the soil are of particular interest because of their ability to establish a natural and beneficial symbiotic association with plant roots (Cao et al., 2025; Wahab et al., 2023). These fungi form a network of filaments, called hyphae, which extend into the soil and come into contact with plant roots. This fungal relationship promotes plant growth and health and allows for better absorption of nutrients, particularly nitrogen, potassium, phosphorus, and even heavy metals. In addition, AMF facilitate access to water by increasing the soil's water retention capacity and reducing the effects of water stress in plants. They help to strengthen their resistance to abiotic (erosion and salinity) and biotic (pathogens) stresses (Kalamulla et al., 2022). In return, plants provide these fungi with sugars and other organic carbon compounds produced by photosynthesis (Salmeron-Santiago et al., 2022; Smith & Read, 2008). This relationship is therefore mutually beneficial and constitutes a key ecological mechanism for optimizing the use of soil resources, particularly in nutrient-poor environments. This symbiosis not only improves plant growth but also soil structure by promoting the formation of stable aggregates (Smith & Read, 2008).

Thus the use of AMF represents a promising innovation in agroecology. It has the potential to reduce the need for chemical fertilizers and pesticides (which seriously and permanently affect soil biofunctioning and pose risks to human health); improve root growth, soil health, fertility, and biodiversity of soils; and increase agricultural yields and farmers' incomes in a sustainable manner (Dalgaard et al., 2003). However, farmers' adoption of this biotechnology depends on several factors, including their level of knowledge, practical experience, perception of benefits, and constraints specific to their production system, which is known as the knowledge, application, and attitude method. In this context, participatory

Advancing Sustainable Agriculture Through Arbuscular Mycorrhizal Fungi. DOI: https://doi.org/10.1016/B978-0-443-45130-0.00030-8

TABLE 26.1 Classification of methods for evaluating sustainable agricultural techniques.

Classification criteria	Methods	Objective	Agricultural technology	Data collected
Quantitative approaches	Standardized surveys, semistructured interviews, participatory groups, and direct observation	Understanding farmers' perceptions, expectations, and motivations	All techniques (AMF, PGPR, *Trichoderma*, compost, mixed inoculum, etc.)	Opinion and representation, local knowledge, constraints of use, and social acceptability
Comparative tests	Experimental plots, controlled trials, and agronomic protocols	Assessing the agronomic, ecological, and economic effectiveness of techniques	All techniques (AMF, PGPR, *Trichoderma*, compost, mixed inoculum, etc.)	Yield, root colonization, nutrient content, biomass, and pH

AMF, arbuscular mycorrhizal fungi; *GRSP*, glomalin-related soil protein.

approaches play an essential role in assessing the acceptability and conditions for integrating these fungi into local agricultural practices.

This chapter provides an in-depth analysis of the adoption of AMF as a sustainable technology. It is based on a review of international experiences, supplemented by surveys and field trials conducted with farmers to assess their perceptions, needs, and skills. This study aims to identify the technical, economic, and social benefits of using AMF techniques, while highlighting the key factors that facilitate their adoption in different agroecological contexts.

26.1.1 Methodology for assessing farmers' perceptions and practices

Several case studies in multiple countries provide a range of innovative, organic, and sustainable agricultural techniques with specific results in terms of adaptive management, soil nutrition and health assessment, and strengthening soil biodiversity and its ecological functions (Davis & Nkonya, 2012; Njaramanana et al., 2022). In addition to arbuscular mycorrhizal fungi (AMF) technology, there are several agricultural techniques that use plant or algae extracts, bacteria, or other fungi, which also seek to promote crop productivity while respecting the environment and health. Among these techniques the most used are microbial biofertilization plant growth promoting rhizobacteria (PGPR), *Trichoderma* (beneficial antagonistic fungi), mixed inoculum (AMF + PGPR + *Trichoderma*), enriched microbiological compost, fungal or microbial endophytes, and natural biostimulants (algae, plant extracts, etc.). These techniques are similar or complementary, providing services related to crops and their productivity, as well as other services related to the ecosystem. However, their effectiveness depends on soil type, farming practices (irrigation, plowing, crop rotation, etc.), climatic conditions, and economic circumstances (Gianinazzi et al., 2010; Ndiko et al., 2024).

The evaluation of these different agricultural and sustainable techniques can be classified according to two criteria: qualitative approaches (standardized surveys and questionnaires, interviews, participatory workshops, and direct observation) and comparative trials under real conditions (demonstration plots) (Table 26.1).

16.1.2 Qualitative approaches: Participatory vision

Qualitative approaches focus on farmers' perceptions and practices (attitudes, knowledge, and skills) and are based on surveys, questionnaires, interviews, and observations. Several types of surveys are standardized and used in agricultural research. They involve targeted sampling represented by farmers selected (size, location, and crops) by "snowball" or proportional sampling, who respond to a questionnaire structured with closed and open questions measuring their knowledge, attitudes, experiences, and crop needs according to proven protocols (Jha et al., 2020). Semistructured interviews may also be conducted to explore responses and local perceptions in greater depth, achieve data saturation, and identify recurring themes. The analysis of surveys and interviews is based on the use of qualitative methods inspired by Minayo (2012) and Yin (2018): thematic coding, categorization, and cross-interpretation with the theoretical framework. Multiple case studies are used to explore the diversity of production contexts and agroecological systems. This qualitative approach can be collective, bringing together several farmers (codesign and collective analysis) in workshops to share trial results, benefits, constraints, and conditions for adoption.

26.1.3 Comparative trials: Experimental plots

Comparative trials are experimental evaluation methods that aim to test the agronomic effectiveness of the technique used in real or controlled conditions (plots and greenhouses). It is based on measurable data such as yield and colonization. For example, the comparison of biologically treated plots (inoculation with indigenous or commercial AMF) with control plots (noninoculated) in date palm experiments in Morocco (Gagou et al., 2025). These tests are conducted with enough repetitions (≥4), randomly distributed to limit environmental bias. Data collection is carried out by measuring yield (weight and number of fruits), quality (Brix and antioxidant content), root colonization and spore density, and soil characteristics (pH, nutrients, and carbon content) (Liu et al., 2016) and by standard diagnostic methods (root staining and spore analysis by sieving and centrifugation) (Boyno et al., 2023).

26.2 Application of the knowledge, attitudes, and practices method to participatory assessment

AMF adoption remains marginal among farmers worldwide. This situation is explained by a lack of knowledge, cautious perceptions, and farming practices that are sometimes incompatible with their development. International KAP (knowledge, attitudes, and practices) studies show that most farmers do not understand the biological mechanisms of mycorrhizal symbiosis and frequently consider AMF as simple organic inputs, which limits their voluntary use (Ebekine-Onah & Dio, 2025). Many farmers, as well as some technical advisors, still consider AMF as organic amendments, fast-acting biofertilizers, or direct substitutes for mineral fertilizers. This confusion stems primarily from a lack of understanding of their living, symbiotic nature. Unlike composts, manures, or biochars, which act independently of the plant by directly improving nutrient availability or soil structure, AMF only function in the presence of a compatible host plant. Their effectiveness is therefore conditional, gradual, and rarely immediate. Root colonization takes time, and the observed benefits often relate more to improved nutrient efficiency and crop resilience than to an instant yield increase. This lack of a rapid response contributes to an underestimation of their agronomic value (Ryan & Graham, 2018).

KAP experiments on the use of agronomic diseases in agriculture have been conducted worldwide. These studies highlight significant variations depending on climatic, genetic, and socio-agronomic contexts. We will present a synthesis of some studies and farmer surveys carried out in Europe, Asia, Africa, and the Americas.

26.2.1 Case study 1: America–French Guiana: Perceptions, practices, and adoption of mycorrhizal symbioses

French Guiana is located in northeastern South America, with a humid equatorial climate. The **soils are predominantly ferralitic, acidic, and nutrient-poor**, which severely constrains local agriculture, dominated by small farms and subsistence crops. In this context, **arbuscular mycorrhizal symbioses** constitute a key agroecological lever for improving phosphate nutrition and crop resilience. However, their adoption remains limited by **incomplete knowledge, cautious attitudes toward risk, and disruptive agricultural practices, which** justify the use of the **KAP** to support the agroecological transition (Dessart et al., 2019).

In this context, participatory approaches involving farmers, aimed at strengthening their knowledge of controlled materials framework, appear as essential levers for promoting the adoption of agroecological innovations. The **MYMYX (Mimic program mycorrhizal networks)** is an *interactive serious game. The game* developed by INRAE (2020a, 2020b) aims to support the agroecological transition by focusing on the role of soil mycorrhizal networks. It is an educational and participatory tool that is part of a coconstruction approach, enabling the analysis of the impact of agricultural practices on soil mycorrhizal networks and the trade-offs between yield, sustainability, and crops, as well as agroecological strategies adapted to local contexts (Corbeels et al., 2019; INRAE, 2020a, 2020b). The approach aims to support the agroecological transition through collective learning and the design of sustainable systems, using the KAP method as an analytical and pedagogical framework to structure farmers' thinking, objectify their perceptions, and support the evolution of their practices (Chave et al., 2018).

26.2.1.1 Knowledge of Guyanese farmers about arbuscular mycorrhizal fungi

In French Guiana, farmers' knowledge of AMF is primarily based on a hybrid of empirical knowledge and scientific input. The work of Chave and Angeon (2014) shows that farmers spontaneously associate mycorrhizae with soil fertility, crop vigor, and the reduction of soil-borne diseases, without, however, fully understanding the biological mechanisms, such as

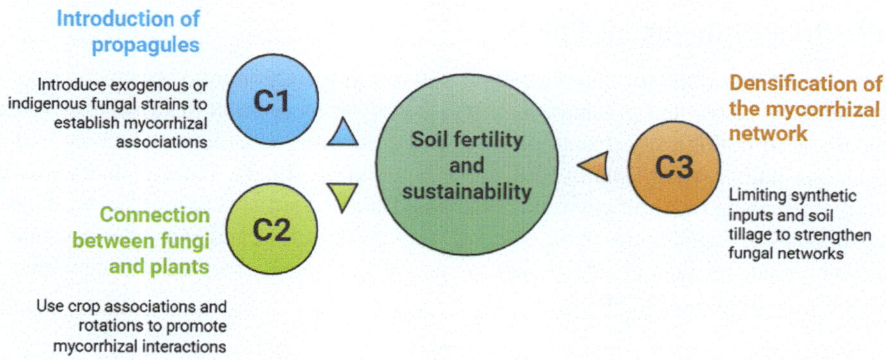

FIGURE 26.1 Distribution of management concepts favorable to mycorrhizal symbiosis (AMF) in French Guiana. *AMF*, Arbuscular mycorrhizal fungi.

phosphorus uptake and the functioning of common mycorrhizal networks (Chave et al., 2018). These perceptions are largely constructed through field observation in challenging pedoclimatic contexts characterized by poor soils.

Studies conducted in French Guiana also highlight that access to scientific knowledge on agroclimatic management remains limited outside of training and action research programs. The lack of targeted outreach and the absence of technical guidelines specifically adapted to agroclimatic conditions constitute major obstacles to the individual appropriation of this knowledge (Chave & Angeon, 2014; Chave et al., 2018).

The MYMYX system, involving 14 farmers (average age 46 years, areas 1–2 ha), consolidated the K space (knowledge) through training and surveys, promoting the appropriation of three concepts: C1 (introduction of propagules), C2 (plant–AMF connection), and C3 (network densification) (Fig. 26.1) (Chave et al., 2018).

The first phase relies on individual surveys conducted directly on farms, simultaneously aiming to assess the knowledge acquired by farmers following the first workshop (MYMYX); to identify and prioritize internal (perceptions, technical constraints, etc.) and external (economic, institutional, and land-related constraints) barriers to the adoption of mycorrhizotrophic practices; and to observe the practices actually implemented in the field. This phase is fully aligned with the theoretical frameworks of learning that consider knowledge as the basis for action, taking into account the socio-technical context of the farm (Chave et al., 2018). The reflective exchange undertaken with each farmer makes it possible to link formal knowledge and experiential knowledge, promoting a progressive awareness of the levers and constraints specific to each situation proposed and thus supporting the adjustment of agroecological agricultural practices (Chave et al., 2018; Ingram, 2018; Klerkx et al., 2012).

26.2.1.2 Practices carried out by Guyanese farmers

Current agroecological approaches increasingly emphasize indigenous mycorrhizal fungi (MF) already present in agricultural soils, rather than the systematic use of external inoculants. These indigenous MF are generally well-established and functional, but they are rarely used intentionally. Farmers increasingly favor intercropping systems capable of maintaining and strengthening existing mycorrhizal networks. This choice is based on a precautionary approach, given the environmental risks associated with introducing exogenous inoculants, which potentially threaten the microbiological balance of soils (Ijdo et al., 2011; Verbruggen et al., 2012). From an agronomic perspective, intercropping systems, such as cereal–legume combinations or the use of mycotrophic relay plants, ensure the continuity of shared mycorrhizal networks. They thus contribute to improved mineral nutrition and increased crop protection, while preserving the soil's biological balance (Smith & Read, 2008). These systems promote more efficient mobilization of phosphorus and nitrogen and enhance crop tolerance to environmental stresses, exceeding those achieved through artificial inoculation (Pellegrino et al., 2011).

In French Guiana, agricultural practices favorable to mycorrhizal fungi are mostly part of agroecological systems, often developed to address local constraints such as high pest pressure or limited access to inputs. Without explicitly targeting mycorrhization, farmers adopt practices such as reduced tillage, the application of compost and manure, intercropping, and agroforestry (Chave et al., 2018).

The study by Chave et al. (2018) includes multisite experiments conducted between April and August 2016 on contrasting soils (pH 6–8). Agricultural practices promoting mycorrhizal symbiosis were grouped into three main categories, corresponding to increasing levels of intervention on soil fungal communities. Approximately 50% of the strategies fall under C3-type practices, characterized by indirect stimulation of mycorrhizal fungi through organic matter management and reduced soil disturbance (Verbruggen et al., 2013). Nearly 42% of the strategies correspond to category C2, which aims for

TABLE 26.2 Agricultural strategies promoting mycorrhizal symbiosis in French Guiana.

Strategy category	Share of practices (%)	Some practices used	State of the mycorrhizal networks	Ecological mechanisms
C3	≈50	Organic mulch, fragmented ramial wood (BRF), and limitations of phosphate fertilizers and pesticides	Significant densification of fungal and mycorrhizal networks; increased microbial biomass; and improved soil structural stability	Supply of organic carbon, maintenance of moisture, stimulation of extra root hyphae, and glomalin; improvement of the soil's biological habitat
C2	≈42	Intercropping (e.g., sorghum–cucumber), agroforestry, introduction of perennial or woody species (e.g., *Inga* spp.)	Maintenance and continuity of common mycorrhizal networks; improvement of nutrient sharing (phosphorus and nitrate)	Connectivity of hyphal networks, root complementarity, and mycotrophic relay plants ensure the persistence of AMF
C1	≈8	Spot application of commercial AMF inocula	Variable and often limited effects in open fields; effectiveness is dependent on the soil, climate, and biological context	Possible competition with native AMFs; dependent on host–fungus compatibility and soil conditions

AMF, arbuscular mycorrhizal fungi.

plant diversification to ensure the continuity of host plants and the connectivity of mycorrhizal networks (Cardinale et al., 2012; Van Der Heijden et al., 2015). Only 8% are strategies in category C1, corresponding to the direct application of exogenous mycorrhizal inocula (Hart et al., 2018; Ryan & Graham, 2018).

This distribution (Table 26.2) shows a clear orientation toward C2 and C3 strategies, based on comprehensive agroecological management of the soil–plant system, ensured by agroforestry and intercropping, which play a crucial role in the sustainable maintenance of functional mycorrhizal networks (Van Der Heijden et al., 2015). The frequent use of mulch and chipped branch wood contributes to the intensification of soil biological activity, particularly fungal activity, through enrichment with organic carbon and stimulation of existing hyphal networks (Lehmann & Rillig, 2015). Conversely the limited use of exogenous inoculants (C1) reveals that, in soils with biological activity, the mobilization of indigenous mycorrhizal fungi proves more reliable than direct inoculation (Hart et al., 2018).

C2-type strategies show significantly higher frequencies and intensities of mycorrhization than those observed with exogenous inocula (C1) or controls, as early as 3 weeks after cucumber planting. Root colonization of sorghum reaches nearly 100% after 8 weeks, highlighting the effectiveness of native mycorrhizal fungi in the absence of external propagule inputs (Chave, 2015; Harter, 2016). Experimental results obtained in French Guiana also show that utilizing local mycorrhizal fungi, particularly through the association of donor mycorrhizal plants, leads to faster and more intense mycorrhization than that achieved with exogenous commercial products (Chave et al., 2018; Pellegrino et al., 2011). This approach is particularly well received by farmers, as it aligns with a logic of self-sufficiency, cost control, and the development of local biological resources. This confirms that the functionality of mycorrhizal fungi depends primarily on farming practices and plant diversity, far more than on the application of external inoculants (Chave et al., 2018).

26.2.1.3 *Attitudes and perceptions of Guyanese farmers*

Farmers in French Guiana generally demonstrate a favorable attitude toward agroecologically modified organisms (AMF), particularly those committed to reducing chemical inputs. Mycorrhizae are largely perceived as a relevant lever for agroecological transition, capable of strengthening the resilience of cropping systems to soil degradation and water stress (Chave & Angeon, 2018).

This positive perception, however, remains contingent on a strong expectation of concrete results in the short term, particularly in terms of yield. Many farmers express reservations, believing that the agronomic effects of agroforestry systems (AFSs) are difficult to observe directly in the field. This caution aligns with analyses from the social sciences, which emphasize that uncertainty is the major obstacle to the adoption of agroecological innovations (Angeon & Chave, 2014; Duru et al., 2014).

Despite everything the attitudes observed among Guyanese farmers reflect a conditional openness among the nine prospective farmers interviewed. Five of them stated they were willing to use hypothetical local inoculants called "GuyaMyc," provided they were economically accessible and effective on degraded soils. The other four favored the self-multiplication of indigenous AMF (assistance fibrosa microorganisms), perceived as better suited to local conditions and less ecologically risky. This opposition illustrates the contrast between a logic of substituting commercial biological inputs and a logic of innovation, particularly in the sorghum–cucumber intercropping system (Chave & Angeon, 2018). Farmers' attitudes are also part of a broader context of concerns related to reducing inputs and adapting to climate change.

In conclusion the main levers identified to encourage the adoption of mycorrhizal bioinputs (MBIs) rely on collective training, participatory mechanisms, and field trials directly involving farmers. The MYMYX project, in particular, demonstrated that bringing scientific and empirical knowledge into dialogue improves the understanding of mycorrhizal processes and strengthens farmers' decision-making autonomy (Chave et al., 2018). Conversely, several obstacles remain, including a lack of local technical expertise, the absence of a clearly defined institutional framework for MBIs, and the perceived risk associated with the introduction of nonnative strains that could compete with indigenous communities (Verbruggen et al., 2012). These factors explain the strong preference of Guyanese farmers for strategies based on the valorization and ecological management of local mycorrhizae.

26.2.2 Case study 2: Europe–southwest France: Perceptions, practices, and adoption of mycorrhizal symbioses

In Europe, farmers use AMF to compensate for low phosphorus availability in soils, maintain productivity in intensive and regulated agricultural systems, increase yields, and stabilize crop growth in the face of seasonal climate variations and the continent's specific conditions (Panagos et al., 2022). In contrast, France, with its calcareous or silty soils and temperate climate, presents an ideal model for introducing these fungi into its agriculture.

The Mycoagra project makes a significant contribution to understanding the role of arbuscular mycorrhizal symbioses in the sustainability of agricultural systems, in-line with the Sustainable Development Goals (SDGs) defined by the United Nations in the 2030 Agenda (ONU, 2015). In particular, these results directly align with SDG 2, which aims to promote sustainable agriculture capable of ensuring food security while preserving natural resources. This functional biodiversity is a key factor in the resilience of agroecosystems to climate stresses and in ensuring the sustainability of food production (FAO, 2018).

The **Mycoagra project** adopts an approach that is not limited to experimental laboratory trials but is carried out **directly on agricultural plots cultivated by farmers** (RAIN, 2022). This project aims to produce and develop **agroecological indicators** with **practical recommendations** easily usable by farmers who integrate mycorrhizal symbioses into their crops, while contributing to the **agroecological transition** (INRAE, 2022; RAIN, 2022). This project was carried out in southwestern France (French departments: Dordogne, Lot, and Corrèze). It aims to study the natural mycorrhization of intercropped plants (particularly maize and walnut) and to identify agricultural practices that are favorable or unfavorable to these symbioses in soil conservation contexts (Hirissou et al., 2022).

This project was conducted in partnership between the Dordogne Chamber of Agriculture, research institutes (INRAE and University of Burgundy), agricultural high schools, the CTIFL (Interprofessional Technical Centre for Fruits and Vegetables), and other local stakeholders. The project adopts an action research approach involving farmers, researchers, and institutions to evaluate the effects of mycorrhization using conventional methods: inputs/weed control (M1), organic farming (M2), interrow cover crops (M3), walnut–maize agroforestry (M4), and untreated/fungicide-treated maize seeds (M5/M6) (Hirissou et al., 2022). Before implementing experimental treatments, **farmers' KAP must be systematically assessed** (Hirissou et al., 2022; Schneider et al., 2017). Thus the KAP approach ensures that experimentation is not limited to soil biodynamics but integrates the socio-technical dimension, enabling the development of tailored local recommendations for farmers to enhance mycorrhizal processes.

26.2.2.1 Knowledge of farmers

Farmers interviewed in Dordogne, Lot, and Corrèze exhibited highly variable levels of knowledge regarding AMF. A significant proportion of them still equate AMF with conventional agricultural inputs, such as organic amendments or biofertilizers, leading them to expect rapid and direct effects on yields. However, this perception does not reflect the actual functioning of mycorrhizal symbiosis, whose benefits are more readily apparent in the medium and long term (Schneider et al., 2017; Smith & Read, 2008). However, farmers who participated in the Mycoagra workshops or field demonstrations showed a significantly improved understanding of the role of mycorrhizal fungi (MFGs). They more clearly

TABLE 26.3 Mycoagra project—AMF in France: results of the six AMF modalities.

Treatment	Covered/Culture	Arbuscular intensity (AI) (%)	AMF diversity
M1	Wild grasses	15	Witness
M2	Grasses and wild legumes	20	Slightly higher than M1
M3	Broad bean planted Oct.–Apr.	30–60	Twice compared to M1 and M2
M4	Agroforestry: Fernor/Fernette + corn	50–93	93 UTO (Territorial Observation Unit) common glomerulus
M5	Untreated corn	60	Depends on the site
M6	Thiram-treated corn*	15	Depends on the site

AMF, arbuscular mycorrhizal fungi.
b*(UTO: Territorial Observation Unit)

identified their contribution to phosphorus and water uptake, as well as their role in improving crop tolerance to water stress, which is well-documented in the scientific literature (Wang & Qiu, 2006). Participants also recognized the negative impact of fungicides and excessive phosphate fertilization on the functioning of mycorrhizal symbiosis (Brito et al., 2012; Jin et al., 2013).

Furthermore, farmers' knowledge of nutrient-rich media (MRAM) remains generally limited, as these are often perceived as conventional agricultural inputs, with the expectation of rapid and visible effects on yields (Hirissou et al., 2022). This narrow view is partly explained by a lack of information and concrete examples. Nevertheless, awareness-raising activities, such as participatory workshops and demonstrations in real-world conditions, have contributed to improving understanding of the fundamental role of MRAM in phosphorus and water absorption, as well as in improving soil quality (Schneider et al., 2017). In the Mycoagra project the knowledge acquired clearly shows that AMFs are very sensitive to intensive agricultural practices, particularly the use of fungicides, intensive tillage, and overfertilization with phosphorus and nitrogen (Brito et al., 2012; Druille et al., 2013; Hirissou et al., 2022; Kjøller & Rosendahl, 2000).

26.2.2.2 Practices of farmers

In Southwest France, for maize and walnut crops, six treatments (Table 26.3) were chosen. The selected plots were distributed over an area of approximately 500 km^2, located in the three departments: Dordogne (two plots), Lot (nine plots), and Corrèze (one plot). The results of the Mycoagra project show that cultivation practices directly and significantly influence the mycorrhization of crops. Walnut orchards managed using conventional farming methods without cover crops (M1) exhibit lower levels of mycorrhization than those observed in organic farming (M2), highlighting the negative effect of synthetic inputs on mycorrhizal fungi (Gianinazzi et al., 2010; Hirissou et al., 2022). However, the most pronounced effect is associated with the introduction of legume-based cover crops (M3), particularly fava beans, which act as a biological relay by maintaining mycorrhizal activity during the winter and promoting rapid root colonization in the spring (Hirissou et al., 2022; Van Der Heijden et al., 2015).

In maize cultivation, fungicide seed coating (M6) significantly reduces the frequency and intensity of mycorrhization compared to untreated seeds (M5), confirming the inhibitory effect of fungicides on the establishment of mycorrhizal fungi (Jin et al., 2013). Finally the walnut–maize AFS (M4) highlights the structuring role of perennial crops in maintaining soil mycorrhizal networks and suggests the existence of functional interactions between annual and perennial crops via a shared mycorrhizal network (Van Tuinen et al., 2020). These results confirm the value of agroecological practices for preserving mycorrhizal symbioses and strengthening the sustainability of agricultural systems (Hirissou et al., 2022).

Experimental setups show that a field bean cover crop (M3), sown from October to April, doubles the intensity of mycorrhization and increases arbuscular arbuscule abundance two- to four folds compared to spontaneous grass-dominated cover crops (M1 and M2). The number of MFG species observed on walnut and field bean roots is also doubled, with field bean acting as a highly mycorrhizal relay plant during the winter. In maize the use of untreated seeds (M5) results in a significantly higher frequency and intensity of mycorrhization than that observed with thiram-treated seeds (M6), confirming the negative effect of fungicides despite maize's strong dependence on MGFs (Hirissou et al., 2022). Finally, in agroforestry (M4), the presence of a common mycorrhizal network between walnut trees and maize highlights a functional mutualism between perennial and annual plants, with a sharing of resources within the system (Van Tuinen et al., 2020).

Overall, favorable practices (M3/M4/M5) are more readily adopted by farmers who have received training, whereas conventional practices (M1/M2/M6) persist in unsupported plots. Observations were made on faba beans in the M3 trial, identified as a highly mycorrhizal plant, as well as on maize grown from untreated seed (M5) (Hirissou et al., 2022). These awareness-raising activities also highlight, in a concrete way, the benefits of relay crops and winter cover crops, particularly legumes such as broad beans or alfalfa, for maintaining vascular tissue during periods of low activity in main crops. This is consistent with the agroecological principles promoted by the FAO and with the sustainable soil management objectives set out in the 2030 Agenda (FAO, 2019; ONU, 2015).

26.2.2.3 Attitudes and perceptions of farmers

Farmers' attitudes toward integrated pest management reflect a gradual evolution in their perception of the benefits and risks associated with this symbiosis. Initial skepticism is frequently observed because of the lack of immediate economic returns (Gianinazzi et al., 2010). However, field demonstrations and participatory workshops conducted within the Mycoagra project have significantly altered these perceptions. Farmers' who have observed visible effects on relay crops, particularly highly mycorrhizal faba beans, or on maize grown from untreated seed, are showing increased interest in AFSs and developing a more favorable attitude. This shift is reflected in greater acceptance of fungicide reduction, the planting of relay crops, and the adoption of agroecological principles (Hirissou et al., 2022; Jin et al., 2013).

In AFSs combining walnut and maize the observed attitudes are more proactive. Farmers positively perceive the existence of a shared mycorrhizal network between the trees and annual crops, recognized as a factor in long-term soil fertility. This perception is consistent with experimental results on the walnut–maize association studied in Mycoagra, which, through carbon isotope analysis, highlighted the probable existence of a common mycorrhizal network ensuring carbon exchange between a perennial and an annual crop (Van Tuinen et al., 2020), and with the conceptual frameworks of sustainable agroforestry (FAO, 2019). These approaches resonate with walnut growers, for whom leveraging mycorrhizal symbioses is a relevant strategy for improving the establishment and profitability of young orchards, which are characterized by still limited root systems (Ryan & Graham, 2018). In France, approximately 30% of walnut farms were managed organically in 2015, reflecting a trend toward the adoption of agroecological practices based on reduced inputs and soil activity (Agence française pour le développement et la promotion de l'agriculture biologique, 2016).

26.2.3 Case study 3: Asia—Perceptions, practices, and adoption of mycorrhizal symbiosis

Cereals, particularly rice, are of paramount importance for human nutrition worldwide because of their high starch, lipid, and protein content (Sarwar et al., 2013). Asian culinary culture is heavily dependent on rice consumption as the main source of the diet. Indeed, Asia is known for rice cultivation and is considered the world's leading rice producer. Farmers use a wide variety of techniques such as flooded rice fields, terraces, dry farming, and irrigated or rain-fed systems. However, rice cultivation is often subject to several constraints, mainly related to climatic conditions, the presence of pathogens (fungi and bacteria) or pests, and the degradation of soil structure and nutrient content, particularly phosphorus (P), iron (Fe), zinc (Zn), and nitrogen (N) (Wilson & Talbot, 2009). Thus, to reduce the effects of these constraints and avoid crop losses to achieve high yields that meet the needs of populations, farmers are increasingly interested in intercropping or combining rice with cereals or legumes such as corn, fava beans, soybeans, or beans. This intercropping can increase AMF colonization and provide benefits to rice cultivation based on its specific nutrient requirements (Ma et al., 2025). Indeed, AMF colonization of the host plant's roots improves fertility, protects against infections and pests, combats water stress, and promotes phosphorus absorption (Liu et al., 2016; Pozo de la Hoz et al., 2021).

China, the world's leading rice producer, is a country known for its tropical, fragile, and intensively cultivated soils, which have led to their nutrient depletion and degradation. Therefore, to support sustainable agriculture with high yields, Chinese farmers are increasingly interested in alternating rice cultivation with cereals or legumes and combining it with AMF (Ma et al., 2025). Recently, several studies have highlighted the central role of AMF, especially in optimizing phosphate nutrition in rice (Mitra et al., 2023; Wang et al., 2021). However, despite these significant scientific advances, the effective transfer of knowledge about the use of these mycorrhizal fungi to all farmers remains uneven; hence the need to make considerable efforts to improve farmers' technical knowledge, strengthen their training resources, and transform scientific research results into widespread agricultural practices (Sun et al., 2022).

26.2.3.1 Knowledge of Chinese farmers about arbuscular mycorrhizal fungi

Throughout the world, farmers' level of technical knowledge is a key factor in the adoption of new, more efficient agricultural practices that could increase production. In China, farmers' knowledge of AMF fungi appears to be generally heterogeneous

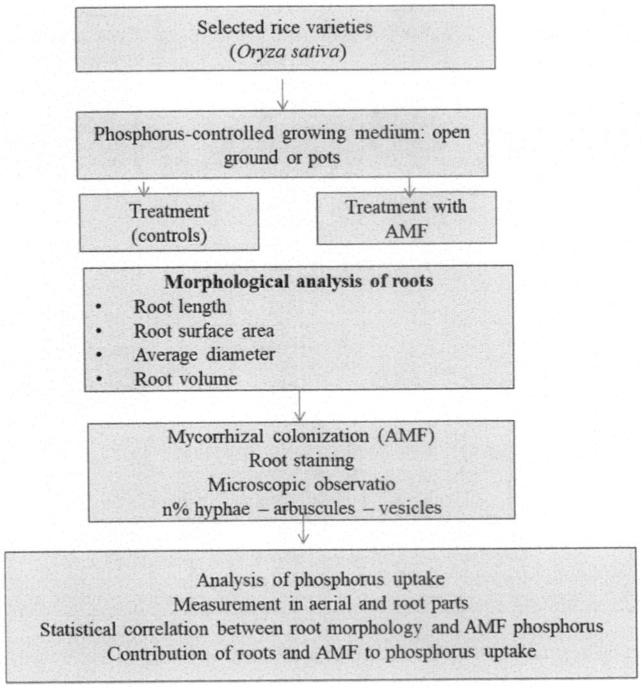

FIGURE 26.2 Strategy cultivation treatment procedures using AMF (Changchun, China) and analyses performed. *AMF*, Arbuscular mycorrhizal fungi.

and highly dependent on the socio-agronomic context, level of academic education, and degree of integration into national and international agricultural extension programs. In many intensive rice-growing regions, farmers traditionally have empirical knowledge of the beneficial effects of soil biological fertility, without identifying the role of AMF as key elements in P uptake (Mitra et al., 2023; Wang et al., 2021).

This knowledge often remains indirect, associated with practices such as organic matter management, crop rotation of rice with other cereals or legumes, and reduced tillage, which promote mycorrhizal symbiosis. However, using farmers' educational attainment to assess their knowledge of agricultural production has two limitations. First, educational attainment and knowledge of agricultural production are not necessarily correlated. Second, even if it is difficult to improve farmers' educational attainment, it is possible to improve their knowledge through theoretical and practical training (workshops, interactive games, simulations, etc.). Thus, in areas benefiting from applied research and technological development programs, particularly those involved in university or government pilot projects, Chinese farmers are developing a more formalized understanding of mycorrhizae, which are perceived as biological material that can improve phosphorus use efficiency and reduce dependence on mineral and chemical fertilizers. This evolution in knowledge is closely linked to the efforts of research institutes and agricultural extension services, which play a central role in disseminating scientific concepts to the field. However, in China, farmers' knowledge of AMF lies at the interface between traditional knowledge, scientific innovations, and agricultural policies geared toward sustainable intensification, especially for rice farming systems (Sun et al., 2022).

26.2.3.2 Practices carried out by Chinese farmers

Several Chinese experimental studies, conducted under controlled conditions or in experimental rice fields, describe the practical methods of using AMF in rice cultivation systems, their interactions with the root characteristics of rice, and their effectiveness in absorbing nutrients and increasing production. These studies vary in their objectives, ranging from understanding the role of AMF in phosphorus absorption to evaluating agronomic techniques applied directly by farmers with a view to optimizing yields and the sustainability of rice systems.

In Changchun, in northeastern China, Ma et al. (2025) conducted field and pot experiments on intercropping rice (*Oryza sativa* L.) and soybeans (*Glycine max* L. Merr.) in dry cultivation with AMF association (Fig. 26.2). They sought to deepen knowledge of the biological mechanisms of phosphorus uptake by rice roots and to assess the relative contribution of root morphological characteristics and AMF colonization to phosphorus uptake. They demonstrate that AMFs do not replace the plant's intrinsic strategies but complement them. Indeed the results showed that intercropping rice and soybeans led to a

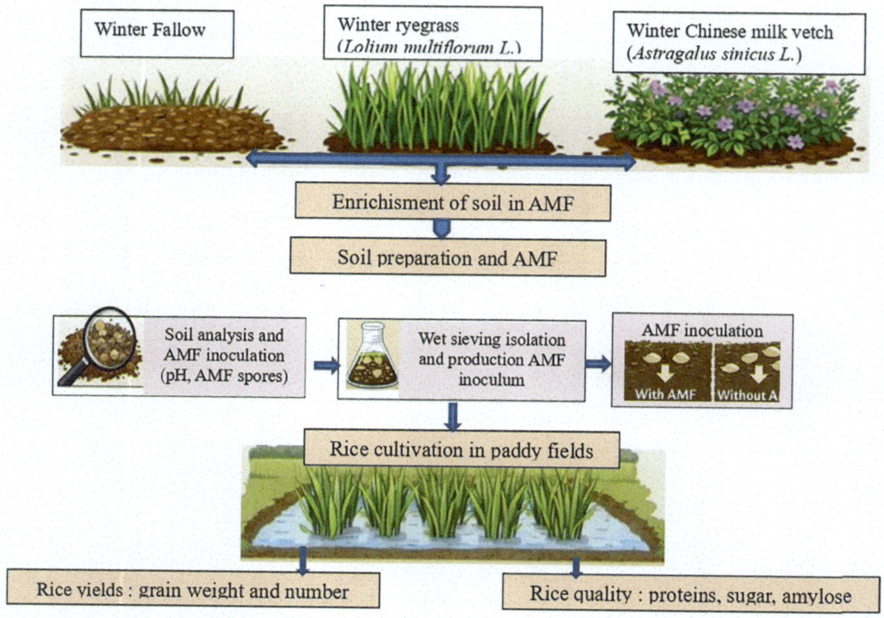

FIGURE 26.3 Strategy for rice cultivation treatments using AMF (Changzhou, China) and analyses performed. *AMF*, Arbuscular mycorrhizal fungi.

significant increase in total biomass and yield compared to monoculture in open fields and that FMAs complement the roots to improve phosphorus uptake. These practices may directly influence Chinese farmers' attitudes toward the use of FMAs in rice cultivation.

However, it has long been believed that rice (*O. sativa* L.) grown in paddy fields was colonized by AMF to a limited extent, if at all, under the anaerobic and flooded conditions of the rhizosphere (Vallino et al., 2009, 2013). Recently, however, several studies have shown that symbiosis with AMF also occurs in flooded rice under certain cultivation conditions (Cao et al., 2025; Wang et al., 2021).

In Guangdong, southern China, Wang et al. (2021) tested the functionality of inoculated AMF on the growth of rice grown in soil samples taken from local rice fields. The study was conducted under different flooding conditions (wet soil, soil with alternate irrigation, and submerged soil) and light conditions (full light or shade). The results indicate that in wet soils, AMF inoculation can be beneficial for rice growth and yield, whereas under flooded and reduced light conditions, these effects become negative, probably because of lower colonization and reduced carbon–phosphorus exchange in the symbiosis between the plant and the fungus. Thus prolonged irrigation or flooding conditions constitute an unfavorable environment for these fungi. For this reason, it is necessary to maintain mycorrhizal colonization through early inoculation of plants, which allows symbiosis to be established first in an aerial environment and then maintained after transplantation into flooded fields, thanks to partial oxygenation of the rhizosphere induced by the aeriferous tissues of rice.

In this context, in Guangzhou in southern China, studies by Cao et al. (2025) were conducted in pots under flooded rice field conditions. They adopted three treatments representing typical winter soil management situations in rice fields: soil left bare without cultivation (control), soil with winter ryegrass (*Lolium multiflorum* L.) cultivation, and soil with Chinese astragalus cultivation (Fig. 26.3). After preparing and enriching the soil with AMF spores from the winter treatments, they grew rice (*O. sativa*) under paddy conditions, keeping certain cultivation parameters constant (rice type, water, and standard fertilization) to ensure that the differences observed were mainly because of the mycorrhizal community. Comparisons of rice grown in different winter cropping systems allow the specific effect of mycorrhizal communities on rice growth and production to be determined. This experiment can be adopted in the field under conventional conditions by Chinese farmers.

26.2.3.3 *Attitudes and perceptions of Chinese farmers*

Even though AMFs are scientifically recognized for their beneficial effects, particularly on improving the mineral nutrition of crops, few farmers, including those in China, are familiar with, accept, or apply this technique in their crops (Berruti et al., 2016). The transition from scientific research on the use of AMF to its actual application in rice cultivation by farmers can yield favorable results in terms of rice quantity and quality. Thus the use of AMFs is part of a systemic dynamic, where scientific research, farmers' perceptions, and field practices evolve in an interdependent manner. This changes their attitude toward this agroecological innovation, which works in rice farming systems by reinforcing its credibility and promoting a positive attitude toward its integration into agricultural practices.

Despite the lack of sociological analysis in the literature on the adoption of AMF by Chinese farmers, Chinese scientific research on these fungi is very rich in terms of mechanisms and beneficial effects on plants, including rice, such as the study by Cao et al. (2025), which shows that AMF developed via winter green manure improves rice yield and quality. This result may strongly encourage Chinese farmers seeking to increase their rice yields to adopt green manure and AMF inoculum in their sustainable rice farming practices.

26.2.4 Case study 4: Africa–Nigeria: Perceptions, practices, and adoption of mycorrhizal symbioses

Nigeria, located in West Africa under a contrasting tropical climate, has predominantly rainfed agricultural systems based on food crops (maize, cassava, and legumes) planted on ferruginous and ferralitic soils, generally poor in available phosphorus, acidic to neutral, and subject to strong degradation. In this context, arbuscular mycorrhizal symbioses (AMF) constitute an important agronomic lever, but their adoption remains limited by insufficient knowledge, disruptive agricultural practices, and short-term-oriented attitudes, highlighting the need for a KAP approach to support their sustainable integration (Dessart et al., 2019; Smith & Read, 2008).

The analyzed study is part of an integrated approach aimed at evaluating the role of integrated farming practices (IFPs) not only on crop agronomic performance but also on farmers' knowledge, practices, and attitudes in different agroecological contexts in Nigeria. The methodology is based on a cross-analysis of several case studies conducted between 2018 and 2023, combining field trials, observations, and feedback from farmers. Four contexts were studied: maize and wheat cultivation in areas with poor soils and prone to drought, cassava production in degraded areas of southern Nigeria, rice cultivation in hydromorphic environments, and organic farming and soil rehabilitation systems. This approach allows us to understand IFPs as a technical, socio-economic, and environmental lever, linked to the principles of sustainable agriculture and soil conservation.

The integration of AFSs into the studied agricultural systems is fully aligned with the principles of agroforestry, contributing to soil conservation, improved soil structure, and reduced erosion. By promoting soil particle aggregation, water retention, and microbial activity, AFS enhances the sustainability of production systems while maintaining, or even increasing, agricultural yields (Sarwade et al., 2024; Zhang et al., 2024).

26.2.4.1 Farmers' knowledge about arbuscular mycorrhizal fungi

The results show that farmers' initial knowledge of mycorrhizal fungi remains generally limited and empirical. In most of the areas studied, producers lacked a scientific understanding of the mechanisms of mycorrhizal symbiosis but intuitively associated the presence of MF with improved soil fertility and better crop vigor. This perception is consistent with studies showing that farmers more readily recognize the visible effects of soil microorganisms than their specific biological functions (Antoszewski et al., 2022; Ortiz & Sansinenea, 2022). However, in areas receiving technical support (organic farming and soil restoration projects), farmers have gradually acquired a functional knowledge of AMFs, including their role in phosphorus absorption, improving soil structure, and crop tolerance to water stress (Basiru et al., 2023; Fall et al., 2022).

26.2.4.2 Farmers' practices on the three crops studied

In the case of maize, grown mainly in north-central Nigeria, farmers have adopted the inoculation of AFSs at planting time, often in combination with a reduction in mineral fertilizers. This practice has led to a significant increase in yields, estimated at between 20% and 30%, particularly under water-deficit conditions, thanks to improved soil exploration and phosphorus and water uptake (Adebowale et al., 2023; Olorunfemi et al., 2020). These results confirm studies highlighting the role of AFS in increasing nutrient use efficiency and the resilience of cereal crops to drought (Kakouridis et al., 2022; Das & Sarkar, 2024).

For cassava, a major food crop in southern Nigeria, AMFs were applied during the planting of cuttings. Farmers have progressively integrated this practice into integrated soil fertility management, in conjunction with organic amendments. The results show an increase of approximately 25% in root yield, improved drought tolerance, and a reduction in the incidence of soil-borne diseases, reflecting an overall improvement in soil quality (Adebowale et al., 2023; Ibrahim et al., 2019). In Anambra State, mycorrhizal inoculation has been carried out in nurseries or at transplanting. This practice has promoted better root system structure and more efficient nitrogen and phosphorus uptake, resulting in an average yield increase of 18% and a reduced reliance on chemical fertilizers (Opara-Nadi et al., 2020). These results align with the findings of several studies indicating that mycorrhizal fungi significantly improve crop performance in hydromorphic environments with limited fertility (Bhantana et al., 2021; Liu et al., 2024).

26.2.4.3 *Farmers' attitudes after applications of arbuscular mycorrhizal fungi*

One of the major contributions of the study lies in the analysis of how farmers' attitudes changed after experimentation with AFSs. Across all the case studies the observation of positive agronomic effects led to increased producer confidence in these biological solutions. This confidence translated into a voluntary reduction in chemical inputs, motivated both by lower production costs and the long-term preservation of soil fertility. Furthermore, farmers who achieved good results played a key role in the horizontal dissemination of practices within rural communities, thus promoting the collective adoption of AFS (Bello et al., 2024; Ogunyemi et al., 2021).

With a positive attitude, those who are environmentally conscious view agroecologically modified (AMF) as profitable (reducing chemical fertilizers and increasing yields) and resilient to water stress (Thind et al., 2022). Despite this the intensive use of synthetic fertilizers generates skepticism among other farmers who increasingly recognize their role in soil microbial biodiversity, thus encouraging a transition to agroecological practices. Nigerian farmers perceive AFSs as a natural solution to combat soil degradation and drought, with an implicit understanding gained from trials showing yield increases (maize: 20%–30%, cassava: 25%, and rice: 18%). However, gaps remain in AFS taxonomy, strain diversity, and molecular mechanisms, limiting informed adoption. Recommendations emphasize training and demonstrations to educate on these benefits, aligned with SDGs 2 and 13 (FAO, 2020).

26.3 Socioeconomic and environmental impacts of adopting arbuscular mycorrhizal fungi in agriculture

It should be noted that mycorrhizae form a mutual symbiotic association between plant roots and mycorrhizal fungi, in which each partner derives a direct benefit. The plant provides the fungus with carbon compounds produced by photosynthesis, whereas the fungus improves the absorption of water and minerals, particularly phosphorus, nitrogen, and certain trace elements. This association plays a key role in the ecosystem services provided by nature to human society. It contributes to carbon sequestration and climate change mitigation and plays a part in the biogeochemical cycle of nutrients, plant biodiversity, and increased productivity of natural and agricultural ecosystems.

In the context of sustainable and agroecological agriculture the use of mycorrhizae, and particularly AMF, significantly reduces the use of chemical inputs (mineral fertilizers and chemical pesticides). This approach promotes qualitative and quantitative optimization of agricultural yields and good management and remediation of degraded or nutrient-depleted soils (Sarfraz et al., 2024a, 2024b, 2024c). It leads to better growth and greater resistance of plants to biotic stresses (soil pathogens and parasites) and abiotic stresses (drought, salinity, nutritional deficiencies, and heavy metals). These benefits will have lasting positive environmental impacts on ecosystems, particularly those with polluted or degraded soils. They also contribute to a net improvement in the socioeconomic conditions of farmers, particularly those with small agricultural fields.

26.3.1 Environmental impacts

The use of AMFs provides numerous ecosystem benefits (for plants and the environment). Mycorrhizal activity helps to stabilize degraded soils, improve the effectiveness of phytoremediation, biofertilization, and bioprotection, and increase agricultural yields (Aslam et al., 2024; Ditta et al., 2024). These benefits are illustrated in Fig. 26.4 by Son et al. (2024) in their review of the main roles of glomalin-related soil protein (GRSP) secreted by AMF hyphae.

26.3.1.1 *Soil stabilization*

In this section, it is important to recall certain properties of soil to understand the role of AMFs in its stabilization. Soil is an essential component of the terrestrial ecosystem. It is the result of a complex combination of geological, biological, and chemical processes. It is a matrix in which various components interact: a mineral phase consisting mainly of fine particles resulting from the weathering of rocks (the mineral phase), an organic phase composed of decomposing organic matter, and living organisms, which, through their biological activities, enrich the environment (Brady & Weil, 2008). Soil, therefore, plays a crucial role in plant growth as a physical support but also as a reservoir of water and nutrients necessary for plant development (Giller et al., 2011). It is considered a key element in the integrated management of agricultural ecosystems. Indeed, soil fertility and its ability to support plant development depend on two fundamental characteristics: its texture and

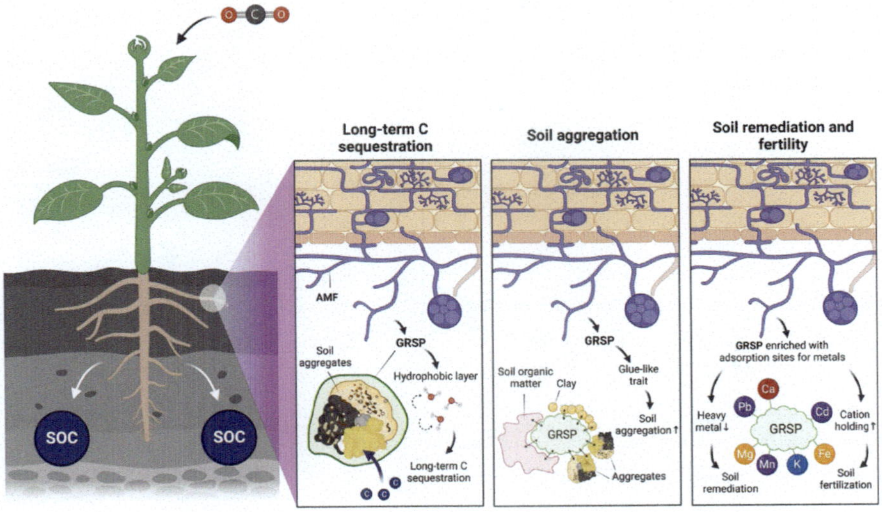

FIGURE 26.4 The main benefits of glomalin-related soil protein (GRSP)—long-term carbon (C) sequestration, soil aggregation, and soil remediation and fertility. Atmospheric CO_2 is fixed by plants via photosynthesis, and then the plant C is incorporated into the soil as soil organic carbon (SOC). Arbuscular mycorrhizal fungi (AMF) inoculate the roots of the plant and grow their extraradical hyphal network into the soil, releasing GRSP to the soil. GRSP can promote long-term C sequestration by forming a hydrophobic layer around aggregates. GRSP can increase soil aggregation by binding soil organic matter and clays, thus forming microaggregates and macroaggregates using its glue-like trait. GRSP can bind and sequester metals, thereby decreasing their bioavailability and toxicity in the soil. (*From Son et al. (2024).*)

its structure. Soil texture refers to the relative proportion of three types of mineral particles: sand, silt, and clay. It plays an important role in soil properties: water retention, permeability, and nutrient storage capacity. When the soil has a large particle size (sandy soils), water retention capacity is low, and nutrients are quickly drained away, which can lead to nutrient deficiencies. Conversely, soil with small particles (clay) has a high water retention capacity, but it can become compacted, making it difficult for roots to oxygenate and for plants to metabolize (Hillel, 2004).

Soil structure is the arrangement of particles into aggregates that vary in size and shape, directly affecting soil porosity and aeration (Brady & Weil, 2008). Good structure, with sufficient pore spaces, promotes air and water circulation, which are essential for root growth and nutrient availability. AMFs play an important role in regulating soil structure and stabilization. Through the hyphae present in large quantities in the soil, they act on the aggregation of soil glomerules, promoting its stability (Liu et al., 2020b; Tisdall, 1991). The hyphae of AMF secrete glomalin, which is a hydrophobic and thermotolerant extracellular glycoprotein (Fig. 26.5). Glomalin acts as a glue that adheres to small particles (<250 μm) to form stable macroaggregates comprising clay, silt, and fine sand (>250 μm) (Wright & Upadhyaya, 1998). This clumping ensures soil stability and helps combat erosion and the leaching of nutrients and organic matter, thereby increasing soil porosity and plant productivity. This aggregation also helps preserve soil organic carbon (Liu et al., 2020b).

26.3.1.2 *Improving phytoremediation*

Phytoremediation is an innovative method of soil remediation and decontamination. It is based on the use of hyperaccumulator plants to absorb and concentrate heavy metals such as lead, cadmium, and mercury and toxic chemicals (Baker et al., 2020; Saeed et al., 2025). This method is consistent with the SDGs. It is environmentally sustainable because it does not require the use of chemicals and allows for gradual and less invasive soil rehabilitation. It is economical (less expensive) and can be used over large areas, making it particularly suitable for heavily polluted land.

However, the use of phytoremediation remains limited and depends primarily on the bioavailability of pollutants. One way to improve the performance of this method is to inoculate soils with AMFs. The GSRPs produced by hyphae contribute to the immobilization of trace metals (Vodnik et al., 2007), polycyclic aromatic hydrocarbons (Joner et al., 2001), and polychlorinated biphenyls (Teng et al., 2010). These proteins increase bioprotection and stimulate plants' defenses against abiotic stress (Liu et al., 2020a; Nazir et al., 2023). They increase soil fertility and therefore plants' access to nutrients in the rhizosphere. This results in a clear improvement in the growth of mycorrhizal plants.

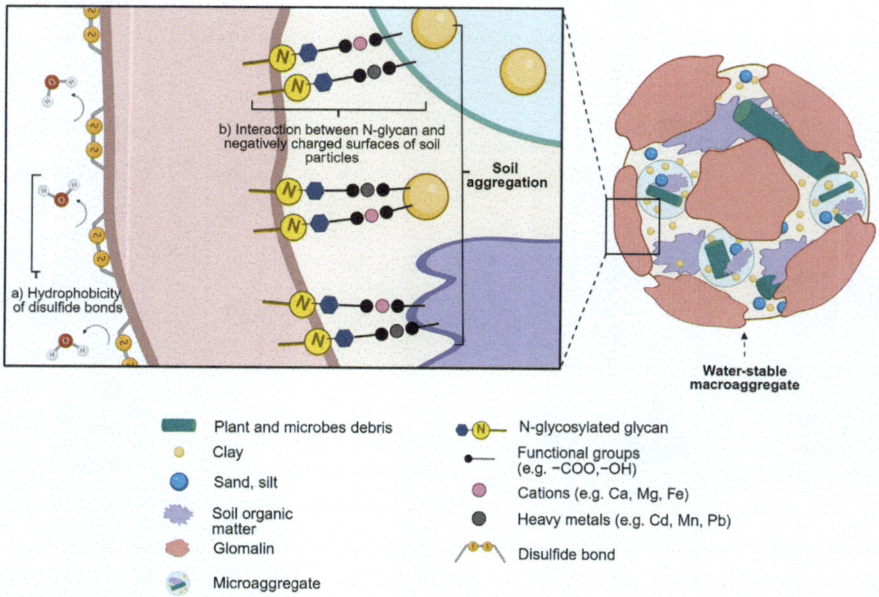

FIGURE 26.5 The outline of the formation of a water-stable macroaggregate facilitated by GRSP: (A) The disulfide bonds within glomalin can help form a sticky, insoluble, hydrophobic layer around aggregates; (b) the *N*-glycosylation within glomalin not only expands the reaction sites for cation and heavy metal adsorption but also helps bind soil organic matter, microaggregates, and soil particles, producing a water-stable aggregate. *GRSP*, Glomalin-related soil protein. (*From Son et al. (2024).*)

26.3.1.3. Biofertilization

Microorganisms actively participate in biogeochemical cycles and the degradation of organic matter, thereby releasing essential nutrients necessary for plant growth (such as nitrogen, phosphorus, and certain trace elements). Agricultural practices involving mycorrhizal soils significantly improve plant growth. Specifically for AMF, the elongation of extraroot hyphae increases the contact surface between soil minerals and plant roots to extract water and nutrients from more distant areas and transfer them to the host plant (Khalvati et al., 2008). AMFs enable the mobilization of organic phosphorus and hydrolyze it into inorganic phosphorus that can be assimilated by the plant. In exchange, carbohydrates are transferred from the host plant to the AMF (Bucking & Shachar-Hill, 2005; Smith et al., 2011).

As with phosphorus, AMF mycelium is also capable of extracting nitrogen in the form of ammonium ions, nitrates, and amino acids (Hawkins et al., 2000). It can also accelerate the breakdown of organic matter to increase its bioavailability to plants.

The potential applications of AMFs in agriculture and their role in increasing crop yields and reducing inputs such as phosphate and nitrogen fertilizers were discussed in the previous case studies and are yielding positive results for farmers.

26.3.1.4 Bioprotection

Plants face challenges from biotic stresses (pathogens causing root and leaf diseases in plants) and abiotic stresses (drought, salinity, pollution, radioactivity, etc.) that can affect their growth and fruit development.

The tolerance of mycorrhizal plants to biotic and abiotic stress is primarily attributed to improved mineral and water nutrition, leading to better plant development (Tisdal et al., 1991). The mycelial network improves water absorption, reduces osmotic stress, promotes better stomatal conductance, and increases photosynthesis. Investigations by Zhang et al. (2021) into the negative impacts of soil cracks on maize growth showed that these effects were mitigated by AMF inoculation and led to an increase in aboveground and root biomass. The results suggest that the beneficial effects of AMF on root stress may be related to the regulation of hormone levels and an increase in the percentage of rootlets. This promotes root and leaf growth and increases the plant's resistance to environmental stress.

AMFs appear to reduce the incidence and/or severity of harmful effects caused by certain root phytopathogenic fungi, such as *Rhizoctonia, Fusarium, Verticillium, Phytophthora, Pythium*, and *Aphanomyces*; certain bacteria such as *Xanthomonas campestris*; and certain nematodes such as *Meloidogyne incognita* and *Pratylenchus penetrans* (Lounes-Hadj Sahraoui, 2013).

TABLE 26.4 Effects of combining AMFs on the yield of different crops.

Culture	AMF species	Yield	References
Cotton (*Gossypium hirsutum* L.)	*Rhizophagus irregularis*	28.54% increase of yield	Gao et al. (2020)
Onion (*Allium cepa* L.)	*Glomus iranicum*	5% increase in yield when using only AMF (92% increase in yield when using AMF + compost)	Bouamir et al. (2025)
Maïs (*Zea mays* L.)	*Funneliformis mosseae*	18.63%–25.77% for yield; 3.72%–7.56% for effective ear number; 10.56%–11.22% for grain number per ear	Fan et al. (2024)
Agrumes *Citrus* trees	–	17.4% increase of yield	Yao, Lin, Chen, Lei, & Zhu, 2006
Leguminous herb (*Stylosanthes gracilis*)	–	71.0% increase of yield	Yao, Lin, Chen, Lei, & Zhu, 2006
Sweet potato varieties	A mixture (*R. irregularis, F. mosseae, Claroideoglomus claroideum*, and *Claroideoglomus etunicatum*)	28.68% increase in shoot biomass in week 20	Sakha et al. (2019)

AMF, arbuscular mycorrhizal fungi.

26.3.2 Benefits for farmers

Very few studies have been conducted on the economic benefits of using AMFs. This is largely because of the difficulty of setting up and monitoring experimental trials. Nevertheless, we can mention two benefits: increased agricultural yields and reduced use of chemical inputs.

26.3.2.1 Increased agricultural yield

The potential applications of AMFs, which stem from the environmental benefits mentioned before, offer several advantages. They increase agricultural yields, which directly translates into higher income when crops are sold. Some of the benefits include (1) facilitating plant hydration in dry soil conditions and preventing losses because of drought and unfavorable soil conditions, as is often the case for rice crops in paddy fields or temporarily dry soils; (2) particle aggregation by GSRPs improves soil structure, increases carbon retention, and protects it from microbial degradation. It promotes biofertilization and therefore ensures better nutrient supply and sustained long-term productivity; and (3) increases plant bioprotection, creating a soil and rhizosphere environment that is more resilient to the development of pathogens and root diseases.

Table 26.4 provides a nonexhaustive summary of bibliographic data on the effect of AMFs on agricultural yields for different crops. For example, a study conducted by Gao et al. (2020) showed a beneficial effect of AMF use on the quantity and quality of cotton fibers, with a 28.54% increase in production. The effect of fungi can be boosted by the use of natural organic amendments such as compost. Tests carried out by Bouamir et al. (2025) showed that inoculating onion roots with 3 kg/ha of commercial AMF granules (containing 120 propagules/mL of *Glomus iranicum* spores) with 40 T/ha of compost resulted in a 67% increase in plant height, whereas yield increased by 92%. However, inoculation with AMF alone (3 kg/ha) increased plant height and onion yield by 29% and 5%, respectively.

26.3.2.2 Reduction in the use of chemical inputs

Soil inoculation with AMF significantly increases the bioavailability of nutrients. The advantage provided by mycorrhizal soils in terms of nitrogen and phosphorus availability has been discussed previously. This significantly reduces the need for chemical fertilizers, particularly phosphorus, while maintaining or increasing agricultural yields. A study by Qian et al. (2024), conducted in maize fields using conventional farming methods in China, highlighted that AMF inoculation, closely linked to fertilization regimes, can effectively reduce fertilizer use while maintaining or even improving agricultural yields. That said, using less fertilizer means lower costs for farmers, with yields maintained or increased, thus having a positive economic impact.

26.3.3 Social effects

According to our research, no studies on the social nature of AMFs have been published. It follows from what has been discussed in this chapter that the use of AMF in sustainable agriculture is an important lever for farmers' financial and social autonomy. Indeed the benefits of the symbiotic association of AMF with plants are numerous: fertilizing and stabilizing the soil and increasing the bioprotection of plants. This results reduced the costs of chemical inputs (fertilizers and pesticides) and soil treatment costs and increased agricultural yields and, consequently, improved income from their production. This practice therefore strengthens farmers' social and economic autonomy by providing them with stable incomes, reducing their vulnerability, and boosting their self-confidence. The promotion of AMF use in agricultural practices was initiated by scientific research conducted in partnership with individuals or NGOs (associations or cooperatives). This practice is often linked to pilot projects, institutional extension, and rural development programs. This collaboration is a positive social lever that strengthens the exchange of knowledge (trials, results, and practices) between partners and leads to collective learning. It highlights the farming profession and helps to build trust and mutual support. It could lead to exchanges between local leaders in the field, who will ensure technology transfer. Bringing farmers closer to institutional actors promotes their inclusion in collective projects and strengthens cooperation between farmers. In the long term the effects will be social resilience in rural communities, collective adoption of sustainable practices, and strengthening of the link between farmers and institutions.

26.4 Conclusion

AMFs are microorganisms that reside in the rhizosphere of various cultivated plants. They can effectively increase the nutritional components of crops by improving soil properties and secreting enzymes. They also constitute one of the effective and environmentally friendly derivatives for inhibiting or eradicating plant diseases by increasing bioprotection and resistance to biotic and abiotic stresses. To survive, AMFs depend on soluble carbohydrates and lipids provided by their hosts. In turn, they also act as a bridge between the soil and plants, promoting the survival of host plants by improving their nutrient uptake capacity.

Comparing several case studies conducted across four continents reveals a common trajectory: the gradual shift from empirical knowledge to genuine technical mastery of AMF, supported by concrete agronomic results. These symbioses are emerging as a cross-cutting lever for the agroecological transition, simultaneously promoting improved yields, soil preservation, and a reduction in chemical inputs, regardless of soil and climate conditions or cropping systems.

Analyses conducted in several regions, particularly in southwestern France and French Guiana, reveal widespread obstacles to the development of mycorrhizal technologies: a strong reliance on synthetic inputs and a desire for simplified farming practices. However, these constraints tend to lessen when farmers engage in collaborative action research initiatives, such as the Mimic Mycorrhizal Networks (MYMYX) and Mycoagra programs. These approaches foster the emergence of common strategies based on the establishment of mycorrhizal cover crops, the absence of seed treatments, and the integration of agroforestry. Observations in Asia and Africa corroborate these findings, highlighting similar obstacles to the adoption of mycorrhizal microorganisms (MMM) but also a comparable potential for utilizing natural mycorrhizae within multifunctional agricultural systems that reconcile productivity, biodiversity, and ecosystem services.

Awareness campaigns play a crucial role in the adoption of farming practices that support mycorrhizal fungi, particularly the use of legume-rich cover crops. These cover crops ensure the biological continuity of mycorrhizal networks and a regular supply of organic carbon, essential conditions for maintaining fungal activity. Furthermore the development of AFSs appears to be a major lever for strengthening plant diversity and the resilience of agroecosystems by sustainably stabilizing mycorrhizal networks.

Despite these advances, several factors still hinder the wider adoption of these practices: Insufficient specific training, agronomic and economic benefits that are sometimes barely perceptible in the short term, and uncertainties related to changes in farming practices. These observations underscore the importance of strengthened technical and economic support, as well as long-term demonstration programs, to ensure a smooth transition to agricultural systems that fully integrate mycorrhizal symbioses as a key driver of sustainability.

References

Adebowale, A. A., Olorunfemi, I. E., & Ibrahim, M. (2023). Arbuscular mycorrhizal fungi as a tool for improving crop productivity under drought stress. *Journal of Sustainable Agriculture, 15*(3), 215–228.

Agence française pour le développement et la promotion de l'agriculture biologique. (2016). *Les chiffres clés de l'agriculture biologique en France*. Agence française pour le développement et la promotion de l'agriculture biologique https://www.agencebio.org.

Angeon, V., & Chave, M. (2014). Farmers' perceptions and uncertainty in agroecological innovation processes. *Agronomy for Sustainable Development, 34*(4), 787–797. https://doi.org/10.1007/s13593-014-0217-1.

Antoszewski, M., Kiers, E. T., & van der Heijden, M. G. A. (2022). Farmers' knowledge of soil microbial diversity and ecosystem services. *Agriculture and Human Values, 39*(2), 545–559. https://doi.org/10.1007/s10460-021-10253-7.

Aslam, Naveed, M., Aslam, S., Amjad, E., Amjad, A., & Ditta, A. (2024). Remediation approaches for heavy metals contaminated soils. In A. Ditta, S. Mehmood, M. Imtiaz, & M. S. Tu (Eds.), *Bio-organic amendments for heavy metal remediation: Water, soil and plant approaches and technologies (plant biology, sustainability and climate change)* (pp. 459–477). Elsevier. https://doi.org/10.1016/B978-0-443-21610-7.00029-X.

Baker, A. J. M., Brooks, R. R., Reeves, R. D., & Jaffré, T. (2020). *Phytoremediation: Transformation of contaminants in soil and water*. Springer https://doi.org/10.1007/978-3-030-00000-0.

Basiru, A., Fall, D., & Ndiaye, M. (2023). Functional knowledge of arbuscular mycorrhizal fungi in tropical farming systems. *Soil Systems, 7*(1), 12. https://doi.org/10.3390/soilsystems7010012.

Bello, T. O., Ogunyemi, O., & Adekunle, V. (2024). Adoption dynamics of biofertilizers among smallholder farmers in West Africa. *Outlook on Agriculture, 53*(1), 45–56. https://doi.org/10.1177/00307270231234567.

Berruti, A., Lumini, E., Balestrini, R., & Bianciotto, V. (2016). Arbuscular mycorrhizal fungi as natural biofertilizers: Let's benefit from past successes. *Frontiers in Microbiology, 6*, 1559. https://doi.org/10.3389/fmicb.2015.01559.

Bhantana, P., Lazarovits, G., & Paulitz, T. (2021). Arbuscular mycorrhizal fungi in flooded rice systems. *Applied Soil Ecology, 168*, 104124. https://doi.org/10.1016/j.apsoil.2021.104124.

Bouamri, R., Ait Jillali, S., Ou-zine, M., Midhat, L., Meziani, R., & El Kinany, S. (2025). Effect of mycorrhizal fungi and compost as agroecological practices on onion (*Allium cepa* L.) growth and yield under field conditions. *African and Mediterranean Agricultural Journal – Al Awamia, 146*, 91–100. https://doi.org/10.34874/IMIST.PRSM/afrimed-i146.52772.

Boyno, G., Demir, S., Rezaee Danesh, Y., Durak, E. D., Çevik, R., Farda, B., & Pellegrini, M. (2023). Une nouvelle technique d'extraction des spores de champignons mycorhiziens arbusculaires de la rhizosphère. *Journal of Fungi, 9*(8), 845. https://doi.org/10.3390/jof9080845.

Brady, N. C., & Weil, R. R. (2008). *The nature and properties of soils* (14th ed.). Pearson Education.

Brito, I., Goss, M. J., & de Carvalho, M. (2012). Impact of tillage and fertilizer management on arbuscular mycorrhiza. *Applied Soil Ecology, 61*, 44–51. https://doi.org/10.1016/j.apsoil.2012.04.010.

Bucking, H., & Shachar-Hill, Y. (2005). Phosphate uptake, transport, and transfer by the arbuscular mycorrhizal fungus *Glomus intraradices* is stimulated by increased carbohydrate availability. *New Phytologist, 165*, 899–912. https://doi.org/10.1111/j.1469-8137.2004.01274.x.

Cao, M., Ye, S., Jin, C., Cheng, J., Xiang, Y., Song, Y., Xin, G., & He, C. (2025). The communities of arbuscular mycorrhizal fungi established by different winter green manures in paddy fields promote post-cropping rice production. *Journal of Integrative Agriculture, 24*(4), 1588–1605.

Cardinale, B. J., Duffy, J. E., Gonzalez, A., Hooper, D. U., Perrings, C., Venail, P., Narwani, A., Mace, G. M., Tilman, D., Wardle, D. A., Kinzig, A. P., Daily, G. C., Loreau, M., Grace, J. B., Larigauderie, A., Srivastava, D. S., & Naeem, S. (2012). Biodiversity loss and its impact on humanity. *Nature, 486*(7401), 59–67. https://doi.org/10.1038/nature11148.

Chave, M. (2015). *Dynamiques des réseaux mycorhiziens en systèmes agroécologiques tropicaux. [Thèse de doctorat*. Université de Guyane.

Chave, M., & Angeon, V. (2014). Agroecological knowledge and farmers' practices in French Guiana. *Ecology and Society, 19*(3), 12. https://doi.org/10.5751/ES-06586-190312.

Chave, M., & Angeon, V. (2018). Participatory approaches and farmers' learning processes in agroecology. *Sustainability, 10*(9), 3204. https://doi.org/10.3390/su10093204.

Chave, M., Angeon, V., & Corbeels, M. (2018). Co-designing agroecological systems using participatory tools: The MYMYX serious game. *Agricultural Systems, 165*, 188–199. https://doi.org/10.1016/j.agsy.2018.06.012.

Corbeels, M., Cardinael, R., & Chave, M. (2019). Agroecological transitions and soil biodiversity management. *Agronomy for Sustainable Development, 39*, 1–14. https://doi.org/10.1007/s13593-019-0590-3.

Dalgaard, T., Hutchings, N. J., & Porter, J. R. (2003). Agroecology, scaling, and interdisciplinarity. *Agriculture, Ecosystems & Environment, 100*(1), 39–51. https://doi.org/10.1016/S0167-8809(03)00152-X.

Das, S., & Sarkar, S. (2024). Mycorrhizal-assisted nutrient uptake under climate stress. *Journal of Plant Nutrition and Soil Science, 187*(1), 65–78. https://doi.org/10.1002/jpln.202300321.

Davis, K., Nkonya, E., Kato, E., Mekonnen, D. A., Odendo, M., Miiro, R., & Nkuba, J. (2012). Impact of farmer field schools on agricultural productivity and poverty in East Africa. *World Development, 40*(2), 402–413. https://doi.org/10.1016/j.worlddev.2011.05.019.

Dessart, F. J., Barreiro-Hurlé, J., & van Bavel, R. (2019). Behavioural factors affecting the adoption of sustainable farming practices. *Journal of Environmental Management, 234*, 141–148. https://doi.org/10.1016/j.jenvman.2018.12.101.

Ditta, A., Mehmood, S., Imtiaz, M., & Tu, M. S. (2024). *Bio-organic amendments for heavy metal remediation: Water, soil and plant approaches and technologies (plant biology, sustainability and climate change)* (1st ed.). Elsevier Inc. https://doi.org/10.1016/C2022-0-03004-X.

Druille, M., Cabello, M., & Omacini, M. (2013). Arbuscular mycorrhizal fungi diversity under agricultural intensification. *Agriculture, Ecosystems & Environment, 164*, 36–44. https://doi.org/10.1016/j.agee.2012.09.017.

Duru, M., Therond, O., & Fares, M. (2014). Designing agroecological transitions. *Agronomy for Sustainable Development, 35*, 1237–1257. https://doi.org/10.1007/s13593-014-0258-5.

Ebekine-Onah, S., & Dio, M. (2025). Knowledge, attitudes, and practices of farmers toward mycorrhizal biofertilizers. *Sustainability, 17*(2), 455. https://doi.org/10.3390/su17020455.

Fall, D., Ndiaye, M., & Basiru, A. (2022). Impact of arbuscular mycorrhizal fungi on crop resilience under water stress conditions. *Soil Systems, 6*(3), 45. https://doi.org/10.3390/soilsystems6030045.

Fan, L., Zhang, P., Cao, F., Liu, X., Ji, M., & Xie, M. (2024). Effects of AMF on maize yield and soil microbial community in sandy and saline soils. *Plants, 13*, 2056. https://doi.org/10.3390/plants13152056.

FAO. (2018). *Scaling up agroecology to achieve the sustainable development goals*. FAO.

FAO. (2019). *Agroforestry for sustainable agriculture*. FAO.

FAO. (2020a). *Agroecology to reverse soil degradation and sustain food production*. FAO https://openknowledge.fao.org/server/api/core/bitstreams/bb2a86db-7f53-4e70-91ca-35ceb9d777db/content.

FAO. (2020b). *Soil biodiversity: State of knowledge*. FAO https://openknowledge.fao.org/handle/20.500.14283/cb1928en.

Gagou, E., Guérin, C., Chakroune, K., Abbas, M., Lamkami, T., El Jaziri, M., & Hakkou, A. (2025). Morphological and molecular characterization of arbuscular mycorrhizal fungi from the rhizosphere of date palm (*Phoenix dactylifera* L.) in the oasis of Figuig, Morocco. *Diversity, 17*(10), 710. https://doi.org/10.3390/d17100710.

Gao, X., Guo, H., Zhang, Q., Zhang, L., Gou, Z., Liu, Y., Wei, J., Chen, A., Chu, Z., & Zeng, F. (2020). Arbuscular mycorrhizal fungi enhanced the growth, yield, fiber quality, and phosphorus regulation in upland cotton (*Gossypium hirsutum* L.). *Scientific Reports, 10*, 2084. https://doi.org/10.1038/s41598-020-59180-3.

Gianinazzi, S., Gollotte, A., Binet, M. N., van Tuinen, D., Redecker, D., & Wipf, D. (2010). Agroecology: The key role of arbuscular mycorrhizas. *Mycorrhiza, 20*(8), 519–530. https://doi.org/10.1007/s00572-010-0333-3.

Giller, K. E., Witter, E., Corbeels, M., & Tittonell, P. (2011). The role of soil organic matter in maintaining sustainable agriculture. *Soil Use and Management, 27*(2), 215–220. https://doi.org/10.1111/j.1475-2743.2011.00343.x.

Hart, M. M., Antunes, P. M., & Abbott, L. K. (2018). Challenges in the commercialization of arbuscular mycorrhizal fungi. *Trends in Plant Science, 23*(9), 830–834. https://doi.org/10.1016/j.tplants.2018.06.004.

Harter, R. D. (2016). *Soil chemical processes* (2nd ed.). Wiley.

Hawkins, H.-J., Johansen, A., & George, E. (2000). Uptake and transport of organic and inorganic nitrogen by arbuscular mycorrhizal fungi. *Plant and Soil, 226*, 275–285. https://doi.org/10.1023/A:1026500810385.

Hillel, D. (2004). *Introduction to environmental soil physics*. Academic Press.

Hirissou, F., Corbeels, M., & Gianinazzi, S. (2022). Mycorrhizal functioning in agroforestry systems: Results from the Mycoagra project. *Innovations Agronomiques, 85*, 383–391.

Ibrahim, M., Adebowale, A. A., & Olorunfemi, I. E. (2019). The role of arbuscular mycorrhizal fungi in improving soil fertility and crop productivity: A case study with cassava. *Soil Biology & Biochemistry, 134*, 45–55.

Ijdo, M., Cranenbrouck, S., & Declerck, S. (2011). Methods for large-scale production of arbuscular mycorrhizal fungi inoculum. *Mycorrhiza, 21*(1), 1–16. https://doi.org/10.1007/s00572-010-0337-z.

Ingram, J. (2018). Agricultural knowledge systems and innovation. *NJAS—Wageningen Journal of Life Sciences, 84*, 1–5. https://doi.org/10.1016/j.njas.2017.12.001.

INRAE. (2022). *MycoAgra: Fonctionnement des mycorhizes et services écosystémiques en agroforesterie—Rapport final de projet*. INRAE.

INRAE. (2020a). *MYMYX: Mimic mycorrhizal networks*. INRAE.

INRAE. (2020b). *MYMYX serious game: Educational guide*. INRAE.

Jha, S., Kaechele, H., Lana, M., Amjath-Babu, T. S., & Sieber, S. (2020). Exploring farmers' perceptions of agricultural technologies: A case study from Tanzania. *Sustainability, 12*(3), 998. https://doi.org/10.3390/su12030998.

Jin, H., Germida, J. J., & Walley, F. L. (2013). Effects of fungicides on arbuscular mycorrhizal fungi. *Biology and Fertility of Soils, 49*(3), 329–337. https://doi.org/10.1007/s00374-012-0731-1.

Joner, E. J., Johansen, A., Loibner, A. P., De La Cruz, M. A., Szolar, O. H. J., Portal, J.-M., & Leyval, C. (2001). Rhizosphere effects on microbial community structure, dissipation, and toxicity of polycyclic aromatic hydrocarbons (PAHs) in spiked soil. *Environmental Science & Technology, 35*, 2773–2777. https://doi.org/10.1021/es001843h.

Kakouridis, K., Giannopoulos, G., & Papadopoulos, A. (2022). Enhancing cereal crop resilience to drought through agroforestry systems. *Agricultural Water Management, 261*, 107402. https://doi.org/10.1016/j.agwat.2022.107402.

Kalamulla, R., Karunarathna, S. C., Tibpromma, S., Galappaththi, M. C. A., Suwannarach, N., Stephenson, S. L., Asad, S., Salem, Z. S., & Yapa, N. (2022). Arbuscular mycorrhizal fungi in sustainable agriculture. *Sustainability, 14*, 12250. https://doi.org/10.3390/su141912250.

Khalvati, M. A., Hu, Y., Mozafar, A., & Schmidthalter, U. (2008). Quantification of water uptake by arbuscular mycorrhizal hyphae and its significance for leaf growth, water relations, and gas exchange of barley subjected to drought stress. *Plant Biology, 7*, 706–712. https://doi.org/10.1055/s-2005-873043.

Kjøller, R., & Rosendahl, S. (2000). Effects of fungicides on arbuscular mycorrhizal fungi. *Mycorrhiza, 10*(3), 127–134. https://doi.org/10.1007/s005720000070.

Klerkx, L., van Mierlo, B., & Leeuwis, C. (2012). Evolution of agricultural innovation systems. *Agricultural Systems, 108*, 1–10. https://doi.org/10.1016/j.agsy.2012.03.006.

Lehmann, A., & Rillig, M. C. (2015). Arbuscular mycorrhizal contribution to soil aggregation. *Soil Biology and Biochemistry, 89*, 1–12. https://doi.org/10.1016/j.soilbio.2015.06.017.

Liu, Y. Z., Imtiaz, M., Ditta, A., Rizwan, M. S., Ashraf, M., Mehmood, S., Aziz, O., Mubeen, F., Ali, M., Elahi, N. N., Ijaz, R., Lele, S., Shuang, C., & Tu, S. (2020a). Response of growth, antioxidant enzymes, and root exudates production towards As stress in *Pteris vittata* and *Astragalus sinicus* colonized by arbuscular mycorrhizal fungi. *Environmental Science and Pollution Research, 27*, 2340–2352. https://doi.org/10.1007/s11356-019-06785-5.

Liu, H., Wang, X., Liang, C., Ai, Z., Wu, Y., Xu, H., Xue, S., & Liu, G. (2020b). Glomalin-related soil protein affects soil aggregation and recovery of soil nutrients following natural revegetation on the Loess Plateau. *Geoderma, 357*, 113921. https://doi.org/10.1016/j.geoderma.2019.113921.

Liu, W., Zhang, Y., Jiang, S., Deng, Y., Christie, P., Murray, P. J., & Zhang, J. (2016). Arbuscular mycorrhizal fungi in soil and roots respond differently to phosphorus inputs. *Scientific Reports, 6*, 24902. https://doi.org/10.1038/srep24902.

Liu, W., Zhang, Y., Jiang, S., Deng, Y., Christie, P., Murray, P. J., & Zhang, J. (2024). Mycorrhizal contribution to nutrient efficiency in low-fertility soils. *Scientific Reports, 14*, 2156.

Lounes-Hadj Sahraoui, A. (2013). La mycorhize à arbuscules: Quels bénéfices pour l'homme et son environnement dans un contexte de développement durable ? *Revue Sciences et Technologies—Synthèse, 26*, 6–19.

Ma, H., Zhang, H., Gao, Q., Li, S., Yu, Y., Ma, J., & Zhang, H. (2025). The relative contribution of root morphology and arbuscular mycorrhizal fungal colonization on phosphorus uptake in rice/soybean intercropping under dry cultivation. *Plants, 14*(1), 106.

Minayo, M. C. D. S. (2012). Qualitative analysis: Theory, steps, and reliability. *Ciência & Saúde Coletiva, 17*, 621–626. https://doi.org/10.1590/S1413-81232012000300007.

Mitra, D., Panneerselvam, P., Senapati, A., Chidambaranathan, P., Nayak, A. K., & Mohapatra, P. K. D. (2023). Arbuscular mycorrhizal fungi response to soil phosphorus utilization and enzyme activities in aerobic rice under phosphorus-deficient conditions. *Life, 13*(5), 1118.

Naeem, N., Sarfraz, W., Khalid, N., Rizvi, Z. F, Nazir, A., Ejaz, U., Amjad, N., Safdar, L., Amanat, R., Akhtar, S., & Ditta, A. (2024). Factors affecting the remediation of wastewater worldwide through eco-technologies. In A. Ditta, S. Mehmood, M. Imtiaz, & M. S. Tu (Eds.), *Bio-organic amendments for heavy metal remediation: Water, soil and plant approaches and technologies (plant biology, sustainability and climate change)* (pp. 21–31). Elsevier. https://doi.org/10.1016/B978-0-443-21610-7.00022-7.

Nazir, A., Sarfraz, W., Ditta, A., Khalid, N., Farid, M., Shafiq, M., Firdaus-e-Bareen, Z. F. R., & Naeem, N. (2023). Synergistic impact of two autochthonous saprobic fungi (*A. niger* and *T. pseudokoningii*) on the growth, ionic contents, and metals uptake in *Brassica juncea* L. and *Vigna radiata* L. under tannery solid waste contaminated soil. *International Journal of Phytoremediation, 25*(11), 1488–1500. https://doi.org/10.1080/15226514.2023.2166457.

Ndeko, A. B., Diedhiou, A. G., Founoune-Mboup, H., Chuma, G. B., Mugumaarhahama, Y., Diouf, D., Fails, S., Mushagalusan, N., & Kane, A. (2024). Site climate, more than soil properties and topography shape the natural arbuscular mycorrhizal symbiosis in maize and spore density within rainfed maize (*Zea mays* L.) cropland in the eastern DR Congo. *PLoS One, 19*(12), e0312581. https://doi.org/10.1371/journal.pone.0312581.

Njaramanana, N. M. R., Rahetlah, V. B., Trap, J., & Autfray, P. (2022). Field arbuscular mycorrhizal inoculation increased plant performance without phosphorus fertilizer supply in four promoted upland rice varieties in Madagascar. *Experimental Agriculture, 58*, e57. https://doi.org/10.1017/S0014479722000419.

Ogunyemi, O., Adekunle, V., & Bello, T. O. (2021). Promoting collective adoption of agroforestry practices in rural communities. *Agriculture and Human Values, 38*(5), 1307–1320. https://doi.org/10.1007/s10460-021-10247-1.

Olorunfemi, I. E., Adebowale, A. A., & Ibrahim, M. (2020). Mycorrhizal inoculation in agroforestry systems under water-deficit conditions: Effects on nutrient uptake and soil exploration. *Journal of Sustainable Agriculture, 14*(2), 120–134.

ONU. (2015). *Transformer notre monde: Le Programme de développement durable à l'horizon 2030*. Nations Unies.

Opara-Nadi, E., Bello, T. O., & Ogunyemi, O. (2020). Cassava yield enhancement and fertilizer reduction through mycorrhizal inoculation in Anambra State, Nigeria. *Outlook on Agriculture, 49*(2), 123–133.

Ortiz, D., & Sansinenea, E. (2022). Microbial diversity in agroecosystems: Beyond specific biological functions. *Agronomy, 12*(4), 897. https://doi.org/10.3390/agronomy12040897.

Panagos, P., Ballabio, C., Meusburger, K., Spinoni, J., Alewell, C., & Borrelli, P. (2022). Soil erosion in Europe: Climate change, land use and policy perspectives. *Science of the Total Environment, 810*, 152258. https://doi.org/10.1016/j.scitotenv.2021.152258.

Pellegrino, E., Bedini, S., Avio, L., Bonari, E., & Giovannetti, M. (2011). Field inoculation effectiveness of native arbuscular mycorrhizal fungi in a Mediterranean agroecosystem. *Soil Biology and Biochemistry, 43*(2), 367–376. https://doi.org/10.1016/j.soilbio.2010.11.002.

Pozo de la Hoz, J., Rivero, J., Azcón-Aguilar, C., Urrestarazu, M., & Pozo, M. J. (2021). Mycorrhiza-induced resistance against foliar pathogens is uncoupled from nutritional effects under different light intensities. *Journal of Fungi, 7*(6), 402.

Qian, S., Xu, Y., Zhang, Y., Wang, X., Niu, X., & Wang, P. (2024). Effect of AMF inoculation on reducing excessive fertilizer use. *Microorganisms, 12*(8), 1550. https://doi.org/10.3390/microorganisms12081550.

Rahman, H. U., & Ditta, A. (2025). Mine remediation and climate changes. In A. A. H. A. Latef, E. M. Zayed, & A. A. Omar (Eds.), *Remediation of pollution and climatic changes to avoid environmental risks*. Springer https://doi.org/10.1007/978-981-96-5674-5_5.

Rahman, H. U., Khan, M., & Ditta, A. (2025). Recent advances in sustainable waste management practices. In A. A. H. AbdelLatef, E. M. Zayed, & A. A. Omar (Eds.), *Bioremediation and nanotechnology for climate change mitigation* (pp. 103–136). Springer. https://doi.org/10.1007/978-981-96-3069-1_5.

RAIN. (2022). *RAIN: Réseau pour l'agroécologie et l'innovation naturelle—Rapport de projet*. RAIN.

Ryan, M. H., & Graham, J. H. (2018). Little evidence that arbuscular mycorrhizal fungi increase crop yields. *New Phytologist, 219*(1), 4–8. https://doi.org/10.1111/nph.15080.

Saeed, H. S, Munir, M., Samra, S., Farooq, A., Mahmood, H., Ahmad, I., Akram, M. I., Iqbal, R., Elsadek, M. F., Almutairi, S. M., Ditta, A., & Daraz, U. (2025). Screening of biosurfactants producing fungi from infected *Citrus sinensis* (L.). *Polish Journal of Environmental Studies, 34*(5), 5261–5270. https://doi.org/10.15244/pjoes/190588.

Sakhaet, M. A., Jefwa, J., & Gweyi-Onyango, J. P. (2019). Effects of arbuscular mycorrhizal fungal inoculation on growth and yield of two sweet potato varieties. *Journal of Agriculture and Ecology Research International, 18*(3), 1–8. https://doi.org/10.9734/JAERI/2019/v18i330063.

Salmerón-Santiago, I. A., Martínez-Trujillo, M., Valdez-Alarcón, J. J., Pedraza-Santos, M. E., Santoyo, G., Pozo, M. J., & Chávez-Bárcenas, A. T. (2022). An updated review on the modulation of carbon partitioning and allocation in arbuscular mycorrhizal plants. *Microorganisms, 10*(1), 75. https://doi.org/10.3390/microorganisms10010075.

Sarfraz, W., Farid, M., Khalid, N., Sheeren, H., Zarnab, R., Nazir, A., Naeem, N., Jabeen, F., Goraya, A., Naseer, A., & Ditta, A. (2024a). Bioremediation of metal(loid) contaminated soils using organic amendments. In A. Ditta, S. Mehmood, M. Imtiaz, & M. S. Tu (Eds.), *Bio-organic amendments for heavy metal remediation: Water, soil and plant approaches and technologies (plant biology, sustainability and climate change)* (pp. 641–658). Elsevier. https://doi.org/10.1016/B978-0-443-21610-7.00023-9.

Sarfraz, W., Khalid, N., Naeem, N., Latif, A., Rizvi, Z. F, Jahangir, A., Shehbaz, A., Amanat, M., Shabbir, F., Ejaz, U., & Ditta, A. (2024b). Toxicity and bioremediation of heavy metals contaminated tannery wastewater. In A. Ditta, S. Mehmood, M. Imtiaz, & M. S. Tu (Eds.), *Bio-organic amendments for heavy metal remediation: Water, soil and plant approaches and technologies (plant biology, sustainability and climate change)* (pp. 257–269). Elsevier. https://doi.org/10.1016/B978-0-443-21610-7.00013-6.

Sarfraz, W., Naeem, N., Farid, M., Khalid, N., Gillani, S. A, Zainab, S. Z, Basharat, M., Safa, S., Ejaz, U., & Ditta, A. (2024c). Factors affecting the remediation of soil through ecotechnologies. In A. Ditta, S. Mehmood, M. Imtiaz, & M. S. Tu (Eds.), *Bio-organic amendments for heavy metal remediation: Water, soil and plant approaches and technologies (plant biology, sustainability and climate change)* (pp. 629–639). Elsevier. https://doi.org/10.1016/B978-0-443-21610-7.00013-6.

Sarwade, P., Singh, R., & Kumar, A. (2024). Enhancing crop productivity through integrated mycorrhizal management under changing climate scenarios. *Agricultural Systems, 223*, 104863. https://doi.org/10.1016/j.agsy.2024.104863.

Sarwar, M. H., Sarwar, M. F., Sarwar, M., Qadri, N. A., & Moghal, S. (2013). The importance of cereals (Poaceae: Gramineae) nutrition in human health: A review. *Journal of Cereals and Oilseeds, 4*(3), 32–35.

Schneider, K. D., Lynch, J. P., & Voroney, R. P. (2017). Arbuscular mycorrhizal fungi and nutrient acquisition strategies in agroecosystems. *Agronomy for Sustainable Development, 37*, 44. https://doi.org/10.1007/s13593-017-0454-2.

Shokri, N., Robinson, D. A., Afshar, M., Alewell, C., Aminzadeh, M., Arthur, E., Broothaerts, N., Campbell, G. A., Eklund, L., Gupta, S., Harper, R., Hassani, A., Hohenegger, C., Keller, T., Kiener, M., Lebron, I., Madani, K., Marwala, T., Matthews, F., … Or, D. (2025). Rethinking global soil degradation: Drivers, impacts, and solutions. *Reviews of Geophysics, 63*(4), e2025RG000883. https://doi.org/10.1029/2025RG000883.

Smith, S. E., & Read, D. J. (2008). *Mycorrhizal symbiosis* (3rd ed.). Academic Press.

Smith, S. E., Jakobsen, I., Grønlund, M., & Smith, F. A. (2011). Roles of arbuscular mycorrhizas in plant phosphorus nutrition: Interactions between pathways of phosphorus uptake in arbuscular mycorrhizal roots. *Plant Physiology, 156*, 1050–1057. https://doi.org/10.1104/pp.111.174581.

Son, Y., Martinez, C. E., & Kao-Kniffin, J. (2024). Three important roles and chemical properties of glomalin-related soil protein. *Frontiers in Soil Science, 4*, 1418072. https://doi.org/10.3389/fsoil.2024.1418072.

Sun, X., Lyu, J., & Ge, C. (2022). Knowledge and farmers' adoption of green production technologies: An empirical study on IPM adoption intention in major indica-rice-producing areas in the Anhui province of China. *International Journal of Environmental Research and Public Health, 19*(21), 14292.

Teng, Y., Luo, Y., Sun, X., Tu, C., Xu, L., Liu, W., Li, Z., & Christie, P. (2010). Influence of arbuscular mycorrhiza and *Rhizobium* on phytoremediation by alfalfa of an agricultural soil contaminated with weathered PCBs: A field study. *International Journal of Phytoremediation, 12*(5), 516–533. https://doi.org/10.1080/15226510903390475.

Thind, S., Chaudhary, M. S, Ditta, A., Hussain, I., Parveen, A., Ullah, N., Mahmood, Q., Al-ashkar, I., & Sabagh, A. E. L. (2022). Impact of mycorrhizal fungi from different rhizospheric soils on fungal colonization, growth, and chlorophyll contents of *Cenchrus ciliaris*. *Agronomy, 12*(11), 2644. https://doi.org/10.3390/agronomy12112644.

Tisdall, J. M. (1991). Fungal hyphae and structural stability of soil. *Australian Journal of Soil Research, 29*, 729–743. https://doi.org/10.1071/SR9910729.

Vallino, M., Fiorilli, V., & Bonfante, P. (2014). Rice flooding negatively impacts root branching and arbuscular mycorrhizal colonization, but not fungal viability. *Plant, Cell & Environment, 37*(3), 557–572.

Vallino, M., Greppi, D., Novero, M., Bonfante, P., & Lupotto, E. (2009). Rice root colonisation by mycorrhizal and endophytic fungi in aerobic soil. *Annals of Applied Biology, 154*(2), 195–204.

Van der Heijden, M. G. A., Martin, F. M., Selosse, M.-A., & Sanders, I. R. (2015). Mycorrhizal ecology and evolution. *New Phytologist, 205*(4), 1406–1423. https://doi.org/10.1111/nph.13288.

Van Tuinen, D., Jacquot, E., & Gianinazzi, S. (2020). Functional mutualism between perennial and annual plants mediated by arbuscular mycorrhizal fungi: Resource sharing in agroecosystems. *Mycorrhiza, 30*(2), 123–136. https://doi.org/10.1007/s00572-019-00919-2.

Verbruggen, E., Röling, W. F. M., Gamper, H. A., Kowalchuk, G. A., & van der Heijden, M. G. A. (2012). Arbuscular mycorrhizal fungi and soil sustainability. *Ecology Letters, 15*(5), 505–514. https://doi.org/10.1111/j.1461-0248.2012.01770.x.

Vodnik, D., Grčman, H., Maček, I., van Elteren, J. T., & Kovačevič, M. (2007). The contribution of glomalin-related soil protein to Pb and Zn sequestration in polluted soil. *Science of the Total Environment, 392*, 130–136. https://doi.org/10.1016/j.scitotenv.2007.11.016.

Wahab, A., Muhammad, M., Munir, A., Abdi, G., Zaman, W., Ayaz, A., Khizar, C., & Reddy, S. P. P. (2023). Role of arbuscular mycorrhizal fungi in regulating growth, enhancing productivity, and potentially influencing ecosystems under abiotic and biotic stresses. *Plants, 12*(17), 3102. https://doi.org/10.3390/plants12173102.

Wang, B., & Qiu, Y.-L. (2006). Phylogenetic distribution of mycorrhizas. *Mycorrhiza, 16*(5), 299–363. https://doi.org/10.1007/s00572-005-0030-x.

Wang, Y., Bao, X., & Li, S. (2021). Effects of arbuscular mycorrhizal fungi on rice growth under different flooding and shading regimes. *Frontiers in Microbiology, 12*, 756752.

Wezel, A., & Rath, T. (2002). Resource conservation strategies in agro-ecosystems of semi-arid West Africa. *Journal of Arid Environments, 51*(3), 383–400. https://doi.org/10.1006/jare.2001.0953.

Wilson, R. A., & Talbot, N. J. (2009). Under pressure: Investigating the biology of plant infection by *Magnaporthe oryzae*. *Nature Reviews Microbiology, 7*, 185–195. https://doi.org/10.1038/nrmicro2032.

Wright, S. F., & Upadhyaya, A. (1998). A survey of soils for aggregate stability and glomalin, a glycoprotein produced by hyphae of arbuscular mycorrhizal fungi. *Plant and Soil, 198*, 97–107. https://doi.org/10.1023/A:1004347701584.

Yao, Q., Lin, F. X., Chen, J. Z., Lei, X. T., & Zhu, H. H. (2006). *Réponses des semis d'agrumes et d'une herbe légumineuse,* Stylosanthes gracilis, *à l'inoculation fongique mycorhizienne arbusculaire* (pp. 63–67). International Society for Horticultural Science.Seoul, Korea.

Yin, R. K. (2018). *Case study research and applications: Design and methods* (6th ed.). Sage Publications.

Zhang, H., Liu, X., & Chen, J. (2024). AMF-mediated drought tolerance in crops. *Agricultural Water Management, 286*, 108388. https://doi.org/10.1016/j.agwat.2023.108388.

Zhang, J., Bi, Y., Song, Z., Xiao, L., & Christie, P. (2021). Arbuscular mycorrhizal fungi alter root and foliar responses to fissure-induced root damage stress. *Ecological Indicators, 127*, 107800. https://doi.org/10.1016/j.ecolind.2021.107800.

Chapter 27

Economic analysis of arbuscular mycorrhizal fungi in crop production

Muhammad Rashid and Natasha Amjad
University of Management and Technology Lahore Sub Campus Sialkot, Sialkot, Pakistan

27.1 Introduction

27.1.1 Overview of arbuscular mycorrhizal fungi

Mycorrhizal symbiosis is a unique association between the plant root and specific taxa of fungi. This is a mutualistic interaction in which both partners play a role and get benefits. The plant root gets protection, food, water, and nutrition through the extensive network of fungal hyphae, whereas the fungi get the root exudates for their nutrition (sugars, amino acids, proteins, and peptides). There are six types of mycorrhizal associations. Among these types the arbuscular mycorrhizal fungi (AMFs) are most important, established by the phylum Glomeromycota. It is estimated that about 80% aquatic and terrestrial plant species are AMF-associated. Examples of these fungi are *Funneliformis mosseae, Rhizoglomus intraradices, Rhizophagus fasciculatus, Acaulospora laevis, Acaulospora scrobiculata, Gigaspora margarita, Gigaspora rosea, Scutellaspora calospora*, and *Dentiscutata heterogama* (Bennett & Groten, 2022; Tedersoo et al., 2020).

AMFs are commonly found in agricultural fields, grassland or savannas, and tropical and subtropical forests. The fungal arbuscules are present within the root cortical cells that facilitate the resource exchange between them. AMFs produce spores in soil that germinate in soil in the presence of moisture and infect the host plant root system. These produce the arbuscules in the root and become a source of nutrition exchanges between the plant and fungi, that is, carbon, nitrogen, phosphate, potassium, and sulfate. Furthermore, they formulated the mycorrhizal networks that spread from root to surrounding soil (Khoza et al., 2025; Zeleke et al., 2024).

AMFs are obligate biotrophs that get essential carbon compounds, primarily carbohydrates and sugars, from their host plants' roots. In turn the plant root gets protection from soil nematodes and nutrition from an extended network of hyphae. Phosphate uptake is a major issue and limiting factor for plant growth and is enhanced through the AMF extended network of hyphae. In turn the plant increases up to 30% photosynthetically fixed carbon to its AMF partners. The current research exhibited the importance of lipids as a significant source that increases the AMF colonization and growth. AMF also increases the resistance and tolerance against abiotic and biotic stress, such as drought, salinity, metal toxicity, intraspecific competition, and soil-borne diseases. The stress tolerance through AMF is gained by improving the water relations, osmotic adjustment, and antioxidant enzyme activity (Liu & Chen et al., 2024; Parihar et al., 2020; Popoola et al., 2025).

27.1.2 Role of arbuscular mycorrhizal fungi in modern agriculture

AMF has emerged as an important component of sustainable agricultural practices because of having various plant health benefits, including nutrient uptake, stress tolerance, and plant growth. It has a significant effect on plant physiology and plant ecology (Wahab et al., 2023). One of the most important roles is the phosphorus acquisition from the root zone because phosphorus is an immobile nutrient present in soil and unavailable to plants, and plants suffer from phosphorus deficiency. The AMFs extend their network through the hyphae and make the availability of phosphorus to the root in phosphorus-depleted zones of the root. This increased level of phosphorus reduces the need for synthetic phosphate fertilizer, as a result, in environmental savings and to negative impact of synthetic fertilizer (Fall et al., 2022).

Advancing Sustainable Agriculture Through Arbuscular Mycorrhizal Fungi. DOI: https://doi.org/10.1016/B978-0-443-45130-0.00023-0

AMF not only increases the phosphorus, but it also enhances the uptake of other essential nutrients such as nitrogen (N), copper (Cu), iron (Fe), and zinc (Zn) that increase the plant growth and productivity. The AMF also increases the soil aggregation property by producing glomalin proteins (a glycoprotein that binds soil particles into aggregates) and improves the soil aeration, water retention, and erosion resistance (Sun & Hesam Shahrajabian, 2023).

The biotic and abiotic stresses in modern agriculture are mitigated by using the AMF. The AMF colonization improves the plant tolerance against salinity, drought, heavy metal toxicity, and extreme temperature. It improves the tolerance by modulation of plant physiological responses by increasing antioxidant enzymes, osmolytes, and stress-related hormones. It protects the plant against the plant pathogen either by physically obstructing pathogen entry or by inducing systemic resistance in host plants (Chauhan et al., 2022).

AMFs contribute to agroecosystem biodiversity and stability by increasing microbial activity and facilitating plant–plant interactions. Their presence can improve crop quality, such as enhancing essential oil content in medicinal plants or increasing micronutrient concentrations in edible parts (Katarzyna et al., 2021). With reference to climate-smart agriculture, AMFs are increasingly viewed as essential in decreasing greenhouse gas emissions and enhancing carbon sequestration. Their ability to increase root biomass and soil aggregation leads to higher organic carbon input into soils, playing a role in climate change mitigation (Sangwan et al., 2023).

27.2 Mechanisms of arbuscular mycorrhizal fungi in enhancing crop productivity

27.2.1 Nutrient uptake and use efficiency

The AMF significantly enhanced the nutrition uptake, particularly phosphorus, nitrogen, zinc, copper, and other micronutrients, which led to improved plant growth, development, and productivity, as well as nutrient use efficiency (Thind et al., 2022). Phosphorus is an essential macronutrient for the growth of plants, but because of its low mobility in soil, it hinders its availability for the plant root. AMF improves the phosphorus by hyphal extension, phosphatase enzyme, and P transporters. AMF hyphae spread their network long distance from the root surface area and access the phosphorus that the plant root cannot obtain directly. It is also investigated that AMFs secrete phosphatase enzyme that directly hydrolyzes the organic phosphorus into plant-available inorganic phosphorus (Begum et al., 2019; Li et al., 2021). AMF also regulated the P transporter in *Medicago truncatula* (*MtPT4*) that directly facilitates the uptake of phosphorus from fungal hyphae. Various studies showed that AMF increases the uptake of phosphorus by 30%–50%, which reduces the need for phosphorus fertilizer (Wang et al., 2024).

The nitrogen uptake in plants is also improved by AMF through the fungal hyphal absorption, organic decomposition of nitrogen, and nitrogen-fixing bacteria (Saleem et al., 2024; Sarfraz et al., 2019; Ullah et al., 2020). With the help of AMF, plants absorb ammonium (NH_4^+) and nitrate (NO_3^-) from the soil. In the role of decomposer, fungal hyphae break down the organic nitrogen and make it available for the plants. It has been observed that some AMFs help the bacteria to promote nodulation in legumes with the interaction of rhizobia (Gorgia & Tsikou, 2026; Xiao et al., 2025).

Micronutrients, such as Zn, Cu, and Fe, are also improved by AMF. This improvement is carried out by producing siderophores and organic acids, which both solubilize the Zn, Cu, and Fe. The efficiency of uptake is improved by the hyphal transporter. AMF also improves nutrient use efficiency by reducing fertilizer dependency and preventing nutrient loss. It also modifies the root morphology by increasing the absorptive surface area (Bhantana et al., 2021).

27.2.2 Improvement in plant stress tolerance

The AMFs significantly improve the plant tolerance and resistance against the different abiotic and biotic stress, including the water deficit, flooding, heavy metal toxicity, salinity, extreme temperature, and pathogen attacks (Fig. 27.1). Protection is provided by changing plant morphology, physiology, biochemistry, and molecular mechanisms. Drought stress is mitigated by enhancing the water uptake, osmotic adjustment, and stomatal regulation. The fungal hyphae extend their network to the deep soil layer to access the water. Under drought stress the AMF-inoculated plant showed better accumulation of osmolytes (glycine betaine and proline) for the cell turgor pressure. AMF also improves stomatal conductance and reduces water loss. An AMF-inoculated wheat plant showed better enzyme activity (superoxide dismutase (SOD) and catalase (CAT)) that reduced the oxidative damage under drought stress (Begum et al., 2019; Hashem et al., 2018).

AMF reduces the Na^+ uptake and increases the K^+/Na^+ ratio under salinity stress. It also increases the jasmonic acid (JA) and Abscisic acid (ABA) to increase the salinity tolerance. It enhanced the antioxidant enzymes (CAT, POD, and SOD) to mitigate salt stress. In the tomato plant, AMF (*F. mosseae*) decreases the Na^+ accumulation by 40% under salt stress. The AMF-inoculated rice plant showed an increased level of proline under 100 mM NaCl. AMF detoxified the metal by

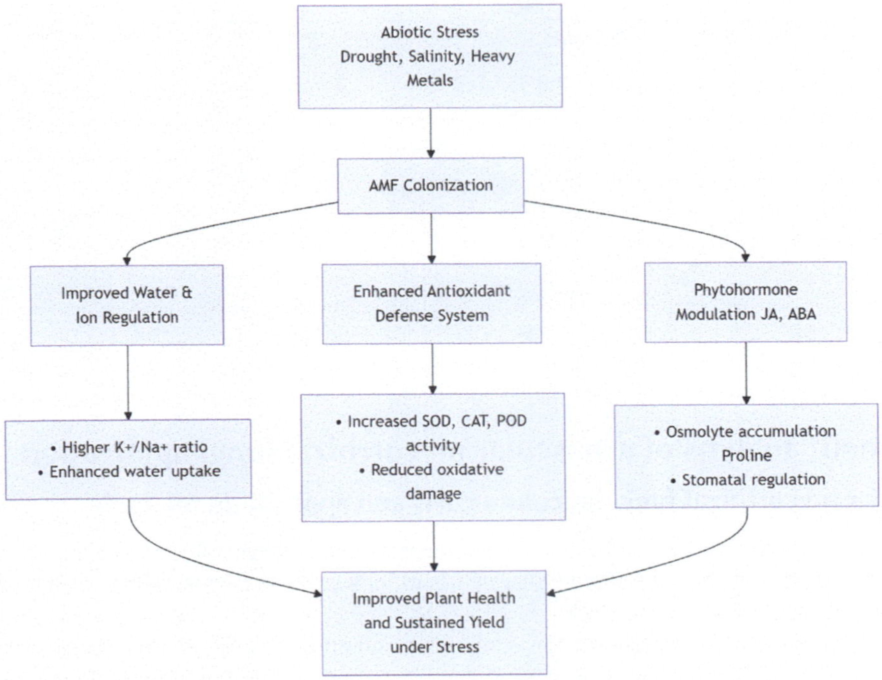

FIGURE 27.1 Schematic illustration of the role of arbuscular mycorrhizal fungi (AMF) in enhancing plant tolerance to abiotic stresses (drought, salinity, and heavy metals) through improved water and ion regulation, strengthened antioxidant defense, and phytohormone modulation. *JA*, Jasmonic acid.

immobilization, phytochelatin synthesis, and rhizosphere modification. The fungal cell wall (glomalin) holds the Cd, Pb, and As for the prevention of damage. AMFs increase metallothionein production for HM sequestration. It also changes the pH and secretes organic acids to reduce HM bioavailability. Cd uptake is reduced by 50% in soybean by the inoculation of AMF. In sunflowers, Pb tolerance is increased by the synthesis of glutathione by the application of AMF (Chandrasekaran et al., 2021; Zhang et al., 2005; Hashem et al., 2018; Pan et al., 2020).

Extreme temperature is mitigated by the production of heat shock proteins in AMF-inoculated plants. AMF upregulated HSP70 and HSP90 under thermal stress (Chandrasekaran et al., 2021). AMF also increases unsaturated fatty acids to maintain membrane fluidity. The cold stress tolerance in AMF-inoculated wheat plants is achieved by producing soluble sugar (Chauhan et al., 2022). The inoculated AMF grapevine mitigated heat stress (40°C) by increasing the antioxidant capacity (He et al., 2021). The AMF protects against pathogens through JA and salicylic acid pathways. It also outcompetes pathogens such as *Fusarium* and *Phytophthora* for space and nutrients. It produces chitinases and β-1,3-glucanases to inhibit pathogens (Salvioli & Bonfante, 2019).

27.2.3 Enhancement of soil health and structure

AMFs play an important role in soil structure, fertility, and health by improving the stability of aggregates, decomposition of organic matter, and microbial activity (Saeed et al., 2025). The fungal hyphae bind the soil particles and form the macroaggregates. In addition, AMFs produce and secrete a glomalin protein that holds the soil particle, it also called biological glue and is water stable. AMF also triggers the plant root to produce polysaccharides that help with soil aggregation. It also prevents soil erosion. A field study on AMF-inoculated soybean showed a 40% reduction in soil erosion because of soil-binding properties of AMF (Bedini et al., 2020; Zhang et al., 2023).

Soil organic matter and carbon sequestration are improved by AMF inoculation. Long-term carbon sequestration is achieved by hyphal biomass and increasing the GRPS accumulation that resists decomposition. AMFs stimulate saprophytic microbes for litter decomposition. Leaf decomposition is increased by 30% in agroforestry by the application of AMF. AMF enhanced the rhizosphere microbiome diversity, fostering symbiotic bacteria Plant Growth-Promoting Rhizobacteria (PGPR) and fungi. The AMF improved the soil porosity and water infiltration in eroded and compacted land. It reduced the bulk density in heavy clay soil (Hannula et al., 2021; Kohl et al., 2022; Han et al., 2025).

TABLE 27.1 Economic and environmental benefits.

Parameter	Impact of AMF	References
Phosphorus fertilizer savings	20%–50% reduction	Pellegrino et al. (2015)
N fertilizer savings	10%–30% reduction	Bender et al. (2015)
Micronutrient savings	30%–40% reduction	Lehmann et al. (2014)
Reduced green house gas (GHG) emissions	Lower N_2O from urea	Bender et al. (2014)
Soil health improvement	Less acidification	Guo et al. (2010)

AMF, arbuscular mycorrhizal fungi.

27.3 Cost–benefit analysis of arbuscular mycorrhizal fungi application

27.3.1 Arbuscular mycorrhizal fungi inoculum costs and application methods

The inoculum cost, application methods, and crop improvement by AMF determined the economic feasibility of AMF. The cost of inoculum is dependent on formulation, that is, liquid, powered, and granulation. It also depends on the strain specificity and production scale (Basu et al., 2023).

The different application methods influence labor, equipment, and efficiency. The AMF can be used with seed coating, soil drenching/spraying, nursery inoculation, and in-furrow application. Seed coating of maize increases the yield by 15%–20%. Using the drip irrigation with AMF reduced the phosphorus fertilizers by 30% in a tomato farm. Similarly the soybean crop exhibited a 12% increase in yield by the application of AMF in the furrow. The pepper seedling inoculated with AMF showed better drought resistance (Begum et al., 2019; Biel et al., 2021; Pellegrino et al., 2015).

27.3.2 Reduction in chemical fertilizer use

AMF reduces the uptake of phosphorus, nitrogen, and other macronutrients, thus reducing the synthetic fertilizers (Table 27.1). It has been investigated that the application of *Rhizophagus irregularis* (granular AMF inoculant) in maize plants reduces the 30% phosphorus fertilizer (60–42 kg/ha) without affecting yield. The application of *Claroideoglomus etunicatum* (AMF) in soybean increased nitrogen fixation, reducing urea use by 35% and boosting yield by 15% (Bender et al., 2014; Soretire et al., 2020).

27.3.3 Yield and quality improvements

AMF increased water absorption, nutrient uptake, and stress resistance. These attributes increase the production and yield of plants. It increases the nitrogen, potassium, phosphorus, zinc, and copper through the extended network of hyphae in the root zone. The maize inoculated with AMF increases phosphorus by 30%–50%, and this yield increases by 15%–20%. Plants inoculated with AMF have improved the efficiency for drought resistance, thus improving the yield under drought stress. For example, the wheat inoculated with *R. irregularis* exhibited 20% higher yield under water deficit conditions (Begum et al., 2019; Zhang et al., 2018, 2019).

The AMF provides relief against oxidative stress. The tomato plant with AMF had 40% less damage because of salt stress and an increased level of sugar content in fruit under salinity stress. AMF increases the secondary metabolites' production. It increases the flavonoids, antioxidants, phenolics, and essential oils in crops. The AMF-inoculated basil exhibited 25% higher essential oil contents. The maize yield (18%) and the protein content (12%) were increased with AMF inoculation. AMF reduced the arsenic toxicity and increased the grain weight by 10% with the inoculation of AMF (Hart et al., 2015; Rivero et al., 2015).

Nutritional quality and postharvest are also increased in fruit, vegetables, and legumes by the application of AMF. In tomato the lycopene (20%) contents increase by the inoculation of AMF. The AMF increases the sugar content (15%) and shelf life of strawberries. In legumes the nodulation and nitrogen fixation are increased (30%), which raises the seed protein content. AMF increase the iron and zinc bioavailability in chickpeas (Chen et al., 2019; Lehmann et al., 2021; Zou et al., 2021).

27.4 Arbuscular mycorrhizal fungi integration with other biofertilizers

27.4.1 Coapplication with phosphate-solubilizing bacteria

AMF and phosphate-solubilizing bacteria (PSB) are two key biofertilizers that enhance phosphorus (P) availability in soil (Ahmad et al., 2022; Ditta et al., 2018; Ditta & Ullah, 2023; Ullah et al., 2021). However, AMFs extend their hyphal networks to access and transport immobilized P, and PSB secrete organic acids and enzymes to solubilize inorganic phosphates. Their combined application can lead to synergistic effects, improving plant P uptake, growth, and yield more effectively than their individual use (Taktek et al., 2015; Zhang et al., 2019).

PSB solubilizes the insoluble phosphates and converts them into plant-available forms ($H_2PO_4^-$ and HPO_4^{2-}) (Hussain et al., 2020; Khan et al., 2022; Naveed et al., 2020). AMFs absorb and translocate these solubilized P forms through their extensive hyphae, increasing P for the plants (Liu et al., 2020; Nazir et al., 2023). Some PSBs produce the iodole acetic acid (phytohormone) that boosts the root growth and increases the space for AMF colonization (Perveen et al., 2023). AMF improves the soil structure and makes it a favorable environment for the activity of PSB. AMF spore germination and hyphal growth are also supported by the PSB. AMF also releases the exudates that help the PSB proliferation. It has been investigated that coinoculation of *R. irregularis* (AMF) and *Bacillus megaterium* (PSB) increased the uptake of P by 45% in maize growth in P-deficient soil. The other study on soybean indicated that the combined application of *F. mosseae* (AMF) and *Pseudomonas fluorescens* (PSB) increased 50% grain yield and reduced the P (25%) (Jiang et al., 2017; Nacoon et al., 2020).

27.4.2 Synergies with organic and mineral fertilizers

The use of AMF with organic or mineral fertilizers not only increases nutrient use efficiency but also increases soil health, plant growth, development, and yield. It also enhanced the resistance against abiotic and biotic stress. The study indicated that AMFs interact with fertilizer by increasing the soil root interface area, secreting various enzymes (phosphatases), and modulating root structure and function by producing different phytohormones (auxin). The extent is influenced by the fertilizer (mineral vs. organic), plant species, and soil condition (pH, texture, and organic matter) (Thirkell et al., 2020a; Zhang et al., 2019).

It has been reported that wheat grown with biochar and AMF exhibited 20% higher nitrogen retention. Similarly, soybean intercropped with clover (a green manure) and inoculated with AMF showed a 15% increase in biological nitrogen fixation, highlighting the role of leguminous cover crops in promoting rhizosphere synergy. It has been investigated that rice grown with AMF and only 50% of the suggested nitrogen dose maintained yields equivalent to full-dose treatments while reducing nitrogen leaching. This was attributed to enhanced nitrogen-use efficiency driven by AMF-induced upregulation of glutamine synthase in roots. It is also observed that tomato plants fertilized with rock phosphate and AMF exhibited a 40% increase in phosphorus uptake compared to rock phosphate alone, especially in alkaline soils where AMF helped solubilize fixed phosphorus (Chun et al., 2021; Hammer et al., 2016; Pellegrino et al., 2021; Xie et al., 2022).

27.5 Environmental and sustainability impacts

27.5.1 Reduction in fertilizer runoff and pollution

Nitrogen and phosphorus have positive points in the improvement of plant growth, whereas these pose negative and serious threats to environmental sustainability, contributing to problems, for example, groundwater contamination, eutrophication, and soil degradation. The unnecessary use of synthetic fertilizers leads to the leaching of nitrates into water bodies, promoting algal blooms and creating hypoxic "dead zones," whereas eutrophication in fresh water is caused because of phosphorus runoff. AMF is an ecological solution to these problems to mitigate the impact on environment. By increasing the nutrient uptake efficiency and promoting soil health, AMFs decrease the dependence on chemical fertilizers, thereby lowering the risk of nutrient runoff and pollution (Cavagnaro, 2022; Rillig et al., 2019).

AMFs increase nutrient use efficiency by extending their hyphae network 10–100 times outside the root zone and increase the soil volume explored for nutrients. The extended network increases the acquisition of more nutrients, such as NH_4^+, NO_3^-, and $H_2PO_4^-$, before they are lost to leaching. The AMF releases phosphatases and organic acids that act as biological phosphorus solubilizers. It has been observed that with AMF application in soybean fields, runoff of phosphorus is reduced, which leads to an increase in internal cycling of soil P and reduces the need for excess fertilizers. It has been investigated that AMF reduced the nitrate by 50%. AMF increases the accumulation of nitrogen into plant tissue, thus reducing the nitrate in the soil. It has been studied that the combination of AMF and cover crops reduced nitrate leaching by 60% in agroecosystems, emphasizing the value of integrated soil management strategies. For phosphorus, AMFs convert insoluble calcium- or iron-bound phosphate into plant-available orthophosphate ($H_2PO_4^-$), thereby minimizing the accumulation of unused P in soil, which is prone to runoff. Recent studies reinforce the environmental benefits of AMF integration. In maize, AMF with a 50% reduced nitrogen application led to a 25% decrease in nitrogen leaching. In wheat, AMF combined with no-till farming practices achieved a 50% reduction in nitrate leaching. In soybean and tomato systems, AMF coapplied with rock phosphate and compost, respectively, reduced phosphorus runoff by 30%–35% (Chen et al., 2023; Liu et al., 2022; Xie et al., 2022).

27.5.2 Long-term soil fertility and carbon sequestration

AMF increases the nutrient uptake through their extended hyphae network in the plant root zone. They increase the uptake of phosphorus, nitrogen, copper, zinc, and iron. These nutrients increase the plant growth, development, and yield. The AMF remains in the soil for a long time if the soil is not disturbed by extensive plowing. AMF protect the soil surface by producing glomalin-related proteins. This protein binds the soil particles in the form of aggregates. The soil aggregates prevent the leaching of nutrients from deep layer of soil. It reduces soil erosion and increases the water retention capacity. AMFs play an important role in stabilizing organic carbon in soils through various pathways. Glomalin-related soil proteins contribute 5%–15% of the total organic carbon in soil. These proteins are highly resistant to microbial decomposition and provide long-term carbon storage. AMF hyphae play a significant role in microbial biomass carbon production. Their exudates facilitate the formation of stable soil organic matter. It colonized plants that allocate a higher proportion of photosynthates to roots, supporting belowground carbon inputs. It has been studied that soils associated with AMF often contain 10%–30% higher soil organic carbon than nonmycorrhizal systems. Agricultural practices such as reduced tillage and organic amendments further promote AMF activity and carbon sequestration. It reduced the carbon loss through oxidation. The application of compost and animal manure promotes the AMF diversity and colonization, along with the improvement of nutrients (Lehmann et al., 2020; Wilson et al., 2020).

27.5.3 Alignment with Sustainable Development Goals

The AMF plays an important role in promoting agroecological sustainability, with direct implications for achieving several United Nations Sustainable Development Goals. It also plays an important role in zero hunger. Zero hunger is also achieved by increasing nutrition uptake and resistance to stress. It is also achieved through reducing the cost by decreasing the dependency on chemical fertilizers. Maize field treated with AMF showed 40% greater increase in drought tolerance. AMFs protect the environment by soil carbon sequestration through stabilization of organic matter and improve nitrogen use efficiency, resulting in a decrease in nitrous oxide emissions from agricultural fields.

Life on land sustains because of AMF. AMFs improve the soil structure and water retention and facilitate the tree seedling establishment. The seedling establishment helped the reforestation. AMFs help in water cleaning and sanitation; for example, the AMF increases nutrient retention, reducing the fertilizer runoff into water bodies. It prevents soil erosion and improves the watershed's health. The natural nutrients improve the plant health against the environmental diseases (Cavagnaro, 2023; Rillig et al., 2023; Yang, 2023).

27.6 Challenges and limitations in arbuscular mycorrhizal fungi adoption

27.6.1 Variability in arbuscular mycorrhizal fungi effectiveness

The AMF effectiveness for the improvement of plant growth depends on various factors, such as environmental, biological, and management factors. Its effectiveness also depends upon the plant variety, genotypes, species, and cultivars. Sawers et al. (2017) observed significant genotypic variation in maize responsiveness to AMF, suggesting that breeding programs have inadvertently selected for lower mycorrhizal dependency. It also depends on the AMF species and its functional diversity. There is functional diversity that depends on the AMF species and strains. For example, for the use of phosphorus uptake, *R. irregularis* is used, whereas *C. etunicatum* is used for drought tolerance. It is investigated that single inocula generally perform better than the combination because of complementary functional traits (Pellegrino et al., 2021; Thirkell et al., 2020b).

Soil conditions also influence the growth and positive effects, such as pH, soil texture, salinity, and phosphorus content. It is also demonstrated that a high level of soil phosphorus inhibits the AMF formation because of a decrease in plant carbon allocation. In saline soils, some AMF species increase salt tolerance, but colonization may be halted at high salt levels.

Temperature and water availability directly influence the AMF activity and growth. Waterlogging and extreme drought cause a negative impact on hyphae growth and spore germination. However, in moderate drought, AMF can increase plant drought tolerance through improved root water uptake and osmotic adjustment (Evelin et al., 2019; He et al., 2020; Zhao et al., 2021).

Agricultural management practices such as crop rotation, tillage, and the use of chemicals also impact the AMF. Tillage disrupts mycorrhizal networks, whereas long-term monoculture can reduce AMF diversity (Verzeaux et al., 2017). Conversely, conservation agriculture and reduced input systems promote AMF proliferation and activity (Thirkell et al., 2020a, 2020b).

27.6.2 Inoculum availability and standardization

The availability and standardization of AMF inoculants are essential for their successful application in agriculture, forestry, and the restoration of land. However, different challenges exist, including inconsistent production methods, variability in inoculum quality, and a lack of global regulatory frameworks. AMF inoculum is produced by various production methods, that is, in vitro system, pot culture, and large-scale production. Pot culture is the most popular method, but it is labor-intensive and very slow. Although the in vitro system (root organ culture) is high purity and sterile production, it is costly and limited to specific species. Liquid fermentation is commercial scale using bioreactors, but it is still experimental (Cruz-Paredes et al., 2021; Dejaifve et al., 2023).

Different challenges are faced during the inoculation; for example, the nontarget microbes (pathogenic fungi) are produced, which decreases the efficiency of AMF. Dry or high temperatures reduce the spore survival during storage. To standardize, efforts are made to improve reproducibility; researchers and industry have proposed different methods, for example, the molecular quantification to conform and verify the inoculum composition and ISO/TS 20196:2022 guidelines for AMF inoculum testing (International Organization for Standardization (ISO), 2022). Bioassay protocols to assess colonization potential before field application (Pellegrino & Bedini, 2022).

27.6.3 Farmer awareness and knowledge gaps

The farmers are unaware of the AMF's function for growth improvement and production. Surveys across low- and middle-income countries consistently show that many farmers are unaware of the existence or function of AMF. For instance a study in sub-Saharan Africa reported that fewer than 20% of surveyed farmers had any knowledge of soil microbial inoculants, including AMF (Vanek et al., 2016). Similarly, Muriuki et al. (2020) found that although farmers recognized improved plant health when using compost or manure (often containing native AMF), they lacked understanding of the underlying microbial mechanisms.

The integration of AMF into farm practices requires technical understanding—especially regarding inoculum application, compatibility with crop types, and expected benefits under various soil conditions. However, extension programs often focus on fertilizers, pesticides, and improved seed varieties, with minimal emphasis on microbial biofertilizers (Pretty et al., 2018).

Farmers often perceive AMF technology as complex, especially when benefits are not immediate or visually observable. Unlike fertilizers that provide rapid greening or growth, the effects of AMF are gradual and context-specific (Schwartz et al., 2021). Furthermore, variability in field performance because of environmental or soil factors adds to the skepticism among farmers regarding the reliability of AMF inoculants.

Even when farmers are informed, the lack of access to reliable AMF inoculants remains a barrier. Most commercial inoculants are expensive, imported, or poorly adapted to local soil conditions. A study in India (Singh et al., 2023) showed that locally produced AMF inocula were more affordable and effective, but their distribution was limited to a few pilot regions because of weak supply chains and regulatory hurdles.

27.7 Conclusion and future outlook

27.7.1 Summary of economic impacts

AMFs is a low-cost application having high potential to decrease the chemical fertilizer dependence, improve crop yields, and increase resource use efficiency—particularly in smallholder and resource-limited farming systems (Fig. 27.2). Economic evaluations across diverse agroecological zones specify significant cost savings and yield improvements. The AMF application reduces the phosphorus fertilizer to 50% without changing the yield of crops, which reduces the financial

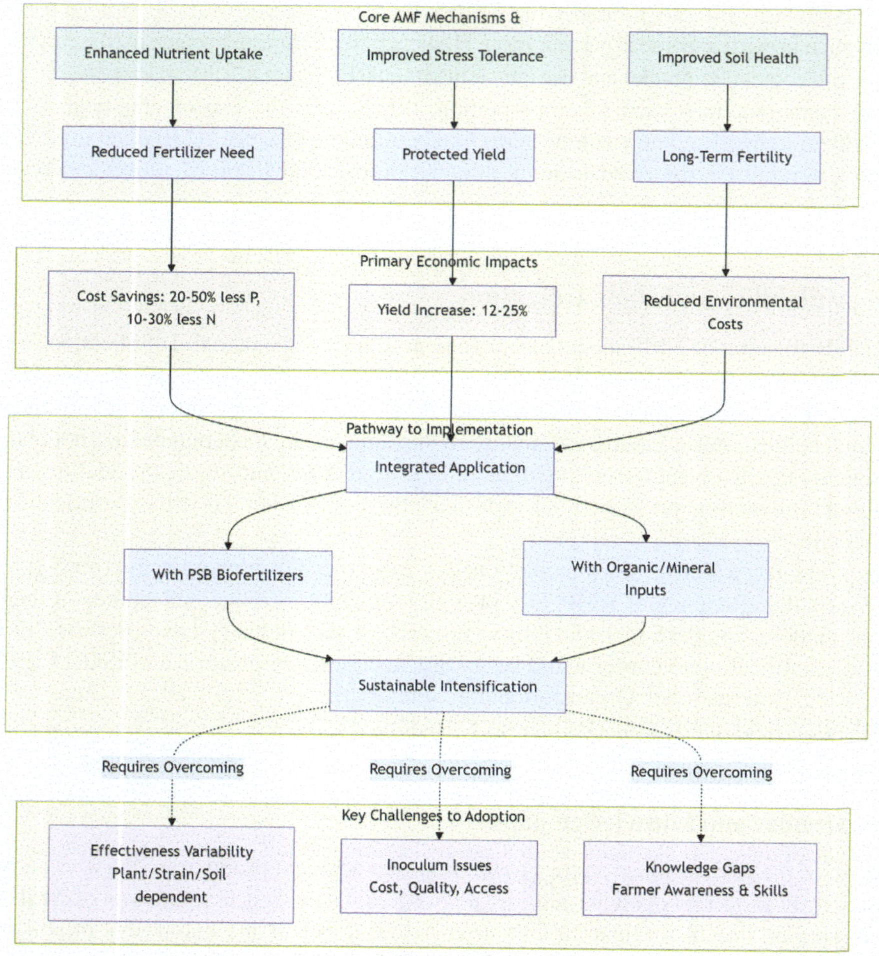

FIGURE 27.2 Mechanisms, impacts, and implementation pathway for arbuscular mycorrhizal fungi (AMF), including key benefits, economic outcomes, and the primary challenges to widespread adoption. *PSB*, Phosphate-solubilizing bacteria.

burden on farmers. An inoculated plant increases its yield. It reduced the risk of drought and nutrient stress that led to an increase in yield and economic benefit (El Malahi et al., 2025).

27.7.2 Research and innovation needs

AMF remains disturbed in agricultural land. To deal with this disturbance, research and innovation are needed to minimize the disturbance and inoculation problems. Strain specificity can be improved by the continuous exploration of local AMF in native soil. Climates also play a significant role in increasing symbiotic efficiency. Plant breeding should be adopted to develop varieties that have high potential for symbiosis with AMF under low input costs. Application methods should be improved, such as seed coating, dipping, and in-furrow inoculation. AMF effectiveness is often influenced by interactions with other soil organisms (e.g., rhizobacteria and pathogens). Integrated soil microbiome research can lead to synergistic bioformulations. Metagenomics, isotopic tracing, and real-time colonization monitoring are helping advance in understanding of AMF dynamics in complex soil ecosystems.

27.7.3 Roadmap for scaling arbuscular mycorrhizal fungi in crop production

To mainstream AMF into sustainable agriculture a multidimensional roadmap is required—linking science, policy, farmer education, and market mechanisms (Fig. 27.3). The government should provide support for the AMF uses as a biofertilizer and for environmental suitability. The farmer should be given subsidies to use AMF-based fertilizers. Farmers and agronomists should be trained for the well-skilled use of AMF in agricultural fields.

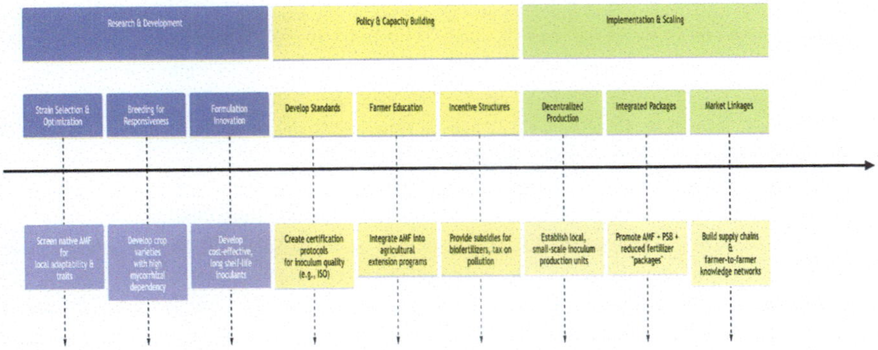

FIGURE 27.3 Roadmap for scaling the adoption of arbuscular mycorrhizal fungi (AMF) in agriculture, integrating research and development, policy and capacity building, and implementation and scaling strategies.

For the local production of AMF, decentralized production units should be used for local AMF strain production for cost-effectiveness. There should be a regular mechanism for certification and approval for AMF inoculants. Codevelop AMF technologies with farmers to ensure context-relevant practices. Develop metrics to track AMF performance in real-world settings, including yield gains, input savings, and environmental benefits.

References

Ahmad, H. T., Hussain, A., Aimen, A., Jamshaid, M. U., Ditta, A., Asghar, H. N., & Zahir, Z. A. (2022). Improving resilience against drought stress among crop plants through inoculation of plant growth-promoting rhizobacteria. In A Husen, & M Jawaid (Eds.), *Harsh environment and plant resilience: Molecular and functional aspects* (pp. 387–408). Springer. https://doi.org/10.1007/978-3-030-65912-7_16.

Basu, S., Rabara, R. C., & Negi, S. (2023). Towards a sustainable future: Elucidating the economic viability and scalability of arbuscular mycorrhizal fungi (AMF) application in modern agriculture. *Applied Soil Ecology, 192*, 105074. https://doi.org/10.1016/j.apsoil.2023.105074.

Bedini, S., Pellegrino, E., Avio, L., Pellegrini, S., Bazzoffi, P., Argese, E., & Giovannetti, M. (2020). Changes in soil aggregation and glomalin-related soil protein content as affected by the arbuscular mycorrhizal fungal species *Glomus mosseae* and *Glomus intraradices*. *Soil Biology and Biochemistry, 142*, 107707. https://doi.org/10.1016/j.soilbio.2020.107707.

Begum, N., Qin, C., Ahanger, M. A., Raza, S., Khan, M. I., Ashraf, M., Ahmed, N., & Zhang, L. (2019). Role of arbuscular mycorrhizal fungi in plant growth regulation: Implications in abiotic stress tolerance. *Frontiers in Plant Science, 10*. https://doi.org/10.3389/fpls.2019.01068.

Bender, S. F., Plantenga, F., Neftel, A., Jocher, M., Oberholzer, H. R., Köhl, L., Giles, M., Daniell, T. J., & Van Der Heijden, M. G. A. (2014). Symbiotic relationships between soil fungi and plants reduce N_2O emissions from soil. *ISME Journal, 8*(6), 1336–1345. 10.1038/ismej.2013.224.

Bender, S. F., & Van der Heijden, M. G. A. (2015). Soil biota enhances agricultural sustainability by improving crop yield, nutrient uptake, and reducing nitrogen leaching losses. *Journal of Applied Ecology, 52*(1), 228–239. https://doi.org/10.1111/1365-2664.12351.

Bennett, A. E., & Groten, K. (2022). The costs and benefits of plant-arbuscular mycorrhizal fungal interactions. *Annual Review of Plant Biology, 73*, 649–672. https://doi.org/10.1146/annurev-arplant-102820-124504.

Bhantana, P., Rana, M. S., Sun, X. C., Moussa, M. G., Saleem, M. H., Syaifudin, M., Shah, A., Poudel, A., Pun, A. B., Bhat, M. A., Mandal, D. L., Shah, S., Zhihao, D., Tan, Q., & Hu, C. X. (2021). Arbuscular mycorrhizal fungi and its major role in plant growth, zinc nutrition, phosphorous regulation and phytoremediation. *Symbiosis, 84*(1), 19–37. https://doi.org/10.1007/s13199-021-00756-6.

Biel, C., Camprubí, A., Lovato, P. E., & Calvet, C. (2021). On-farm reduced irrigation and fertilizer doses, and arbuscular mycorrhizal fungal inoculation improve water productivity in tomato production. *Scientia Horticulturae, 288*. https://doi.org/10.1016/j.scienta.2021.110337.

Cavagnaro, T. R. (2022). The role of arbuscular mycorrhizas in reducing soil nutrient loss. *Trends in Plant Science*. https://doi.org/10.1016/j.tplants.2021.11.002.

Cavagnaro, T. R., et al. (2023). Mycorrhizas in climate-smart agriculture: Carbon sequestration and ecosystem resilience. *Global Change Biology, 29*(2). https://doi.org/10.1111/gcb.16491.

Chandrasekaran, M., Boopathi, T., & Manivannan, P. (2021). Comprehensive assessment of ameliorative effects of AMF in alleviating abiotic stress in tomato plants. *Journal of Fungi, 7*(4), 303.

Chandrasekaran, M., Boopathi, T., & Manivannan, P. (2021). Comprehensive assessment of ameliorative effects of AMF in alleviating abiotic stress in tomato plants. *Journal of Fungi, 7*(4), 303.

Chauhan, S., Mahawar, S., Jain, D., Udpadhay, S. K., Mohanty, S. R, Singh, A., Maharjan, E., & Yadav, A. N. (2022). Boosting sustainable agriculture by arbuscular mycorrhiza under stress condition: Mechanism and future prospective. *BioMed Research International, 2022*(1). https://doi.org/10.1155/2022/5275449.

Chen, S., Zhao, H., Zou, C., Li, Y., Chen, Y., Wang, Z., & Zhao, P. (2019). Arbuscular mycorrhizal fungi improve the quality and shelf life of strawberry fruits. *Scientia Horticulturae, 256*. https://doi.org/10.1016/j.scienta.2019.108589.

Chen, S., Zhao, P., Li, J., & Zhu, Y. (2023). AMF and organic amendments reduce phosphorus leaching in vegetable systems under field conditions. *Science of the Total Environment, 856*, 159122. https://doi.org/10.1016/j.scitotenv.2022.159122.

Chun, S. C., Paramasivan, M., & Chandrasekaran, M. (2021). Arbuscular mycorrhizal fungi and reduced nitrogen doses enhance nitrogen use efficiency in rice. *Journal of Plant Nutrition, 44*(7), 1013–1026. https://doi.org/10.1080/01904167.2020.1860213.

Cruz-Paredes, C., Davey, M., Jakobsen, I., & Nybroe, O. (2021). Challenges in commercializing arbuscular mycorrhizal inoculants: Quality control and consistency. *Frontiers in Microbiology, 12*. https://doi.org/10.3389/fmicb.2021.723382.

Dejaifve, P. A., Lecomte, J., & Declerck, S. (2023). Scaling up arbuscular mycorrhizal fungal inoculant production via liquid fermentation: Challenges and opportunities. *Biotechnology Advances, 62*, 108075. https://doi.org/10.1016/j.biotechadv.2022.108075.

Ditta, A., Imtiaz, M., Mehmood, S., Rizwan, M. S., Mubeen, F., Aziz, O., Qian, Z., Ijaz, R., & Tu, S. (2018). Rock phosphate-enriched organic fertilizer with phosphate-solubilizing microorganisms improves nodulation, growth, and yield of legumes. *Communications in Soil Science and Plant Analysis, 49*(21), 2715–2725. https://doi.org/10.1080/00103624.2018.1538374.

Ditta, A., & Ullah, N. (2023). *Perspectives of using plant growth-promoting rhizobacteria under salinity stress for sustainable crop production plant stress mitigators: Types, techniques and functions* (pp. 231–247). Elsevier. https://doi.org/10.1016/B978-0-323-89871-3.00024-0.

Evelin, H., Devi, T. S., Gupta, S., & Kapoor, R. (2019). Mitigation of salinity stress in plants by arbuscular mycorrhizal symbiosis: Current understanding and new challenges. *Frontiers in Plant Science, 10*. https://doi.org/10.3389/fpls.2019.00470.

Fall, A. F., Nakabonge, G., Ssekandi, J., Founoune-Mboup, H., Apori, S. O., Ndiaye, A., Badji, A., & Ngom, K. (2022). Roles of arbuscular mycorrhizal fungi on soil fertility: Contribution in the improvement of physical, chemical, and biological properties of the soil. *Frontiers in Fungal Biology, 3*. https://doi.org/10.3389/ffunb.2022.723892.

Gorgia, P., & Tsikou, D. (2026). Tripartite Symbiosis Between Legumes, Arbuscular Mycorrhizal Fungi and Nitrogen Fixing Rhizobia: Interactions and Regulation. *Plant, Cell & Environment*. https://doi.org/10.1111/pce.15341.

Guo, J. H., Liu, X. J., Zhang, Y., Shen, J. L., Han, W. X., Zhang, W. F., Christie, P., Goulding, K. W. T., Vitousek, P. M., & Zhang, F. S. (2010). Significant acidification in major chinese croplands. *Science, 327*(5968), 1008–1010. https://doi.org/10.1126/science.1182570.

Hammer, E. C., Balogh-Brunstad, Z., Jakobsen, I., Olsson, P. A., Stipp, S. L. S., & Rillig, M. C. (2016). Biochar increases arbuscular mycorrhizal plant growth enhancement and ameliorates salinity stress. *Applied Soil Ecology, 96*, 114–121. https://doi.org/10.1016/j.apsoil.2015.07.014.

Han, Y., Yuan, G., Yang, X., Fang, L., Liang, Y., Zhou, B., & Wei, Z. (2025). Arbuscular mycorrhizal fungi enhance soil nutrient cycling by regulating soil bacterial community structures in mango orchards with different soil fertility rates. *Frontiers in Microbiology, 16*, 1615694. https://doi.org/10.3389/fmicb.2025.1615694.

Hannula, S. E., Heinen, R., Huberty, M., Steinauer, K., De Long, … Bezemer, T. M. (2021). Persistence of plant-mediated microbial soil legacy effects in soil and inside roots. *Nature communications, 12*(1), 5686.

Hart, M., Ehret, D. L., Krumbein, A., Leung, C., Murch, S., Turi, C., & Franken, P. (2015). Inoculation with arbuscular mycorrhizal fungi improves the nutritional value of tomatoes. *Mycorrhiza, 25*(5), 359–376. https://doi.org/10.1007/s00572-014-0617-0.

Hashem, A, Alqarawi, A. A., Radhakrishnan, R., Al-Arjani, A. B. F., Aldehaish, H. A., & Egamberdieva, D. (2018). Arbuscular mycorrhizal fungi regulate the oxidative system, hormones and ionic equilibrium to trigger salt stress tolerance in *Cucumis sativus* L. *Saudi Journal of Biological Sciences, 25*, 1102–1114.

Hashem, A., Alqarawi, A. A., Radhakrishnan, R., Al-Arjani, A. B. F., Aldehaish, H. A., Egamberdieva, D., & Abd Allah, E. F. (2018). Arbuscular mycorrhizal fungi regulate the oxidative system, hormones and ionic equilibrium to trigger salt stress tolerance in *Cucumis sativus* L. *Saudi Journal of Biological Sciences, 25*(6), 1102–1114. https://doi.org/10.1016/j.sjbs.2018.03.009.

He, F., Zhang, H., & Tang, M. (2021). Aquaporin-mediated improvement in root hydraulic conductance is involved in arbuscular mycorrhizal fungi-induced heat tolerance in grapevine. *Journal of Plant Physiology, 264*, 153477.

He, J. D., Dong, T., Wu, H. H., Zou, Y. N., Wu, Q. S., & Kuča, K. (2020). Mycorrhizas enhance drought tolerance by regulating aquaporins and antioxidant enzymes in trifoliate orange. *Environmental and Experimental Botany, 171*, 103925. https://doi.org/10.1016/j.envexpbot.2019.103925.

Hussain, A., Zahir, Z. A., Ditta, A., Tahir, M. U., Ahmad, M., Mumtaz, M. Z., Hayat, K., & Hussain, S. (2020). Production and implication of bio-activated organic fertilizer enriched with zinc-solubilizing bacteria to boost up maize (*Zea mays* L.) production and biofortification under two cropping seasons. *Agronomy, 10*(1). https://doi.org/10.3390/agronomy10010039.

International Organization for Standardization (ISO). (2022). *ISO/TS 20196:2022: In vitro techniques for arbuscular mycorrhizal (AM) fungal inoculants—Requirements for production and quality control* https://www.iso.org/standard/79767.html.

Jiang, Y., Wang, W., Xie, Q., Liu, N. a, Liu, L., Wang, D., Zhang, X., Yang, C., Chen, X., Tang, D., & Wang, E. (2017). Plants transfer lipids to sustain colonization by mutualistic mycorrhizal and parasitic fungi. *Science, 356*(6343), 1172–1175. https://doi.org/10.1126/science.aam9970.

Katarzyna, T., Guido, L., & Silvio, G. (2021). Arbuscular mycorrhizal fungi as plant biostimulants. In *Microbiome stimulants for crops* (pp. 333–348). Woodhead Publishing.

Khan, K. S., Qadir, M. F., Ahmad, A., Naveed, M., Raza, T., & Ditta, A. (2022). Efficacy of different endophytic bacterial strains in enhancing growth, yield, and physiological and biochemical attributes of *Linum usitatissimum* L. *Journal of Soil Science and Plant Nutrition, 22*(4), 4365–4376. https://doi.org/10.1007/s42729-022-01035-z.

Khoza, T., Masenya, A., Khanyile, N., & Thosago, S. (2025). Alleviating plant density and salinity stress in *Moringa oleifera* using arbuscular mycorrhizal fungi: A review. *Journal of Fungi, 11*(4), 328. https://doi.org/10.3390/jof11040328.

Kohl, L., Lukasiewicz, C. E., & van der Heijden, M. G. A. (2022). Arbuscular mycorrhizal fungi reduce decomposition of woody plant litter while increasing soil aggregation. *Soil Biology and Biochemistry, 169*, 108689. https://doi.org/10.1016/j.soilbio.2022.108689.

Lehmann, A., Veresoglou, S. D., Leifheit, E. F., & Rillig, M. C. (2014). Arbuscular mycorrhizal influence on zinc nutrition in crop plants—A meta-analysis. *Soil Biology and Biochemistry, 69*, 123–131. https://doi.org/10.1016/j.soilbio.2013.11.001.

Lehmann, A., Veresoglou, S. D., & Rillig, M. C. (2021). Mycorrhizas and soil aggregation increase zinc and iron bioavailability in chickpea. *Plant and Soil, 460*(1), 281–294. https://doi.org/10.1007/s11104-020-04796-7.

Lehmann, A., Zheng, W., Ryo, M., Soutschek, K., Roy, J., & Rillig, M. C. (2020). Farmyard manure improves phosphorus uptake by wheat from phosphorus-fixing soils through microbial interventions. *Soil Biology and Biochemistry, 142*, 107722. https://doi.org/10.1016/j.soilbio.2020.107722.

Li, G. H., Cheng, Q., Li, L., Lu, D. l., & Lu, W. p. (2021). N, P and K use efficiency and maize yield responses to fertilization modes and densities. *Journal of Integrative Agriculture, 20*(1), 78–86. https://doi.org/10.1016/S2095-3119(20)63214-2.

Liu, R., & Chen, Y. (2024). Mycorrhizal symbiosis: Evolution, opportunities, challenges, and prospects. In *Arbuscular mycorrhizal fungi in sustainable agriculture: Inoculum production and application* (pp. 1–35). https://doi.org/10.1007/978-981-97-0296-1_1.

Liu, Y., Li, X., Cai, K., Luo, X., & Zhang, Y. (2022). Arbuscular mycorrhizal fungi with reduced chemical fertilizers improve phosphorus retention in soil and mitigate runoff. *Agriculture, Ecosystems & Environment, 324*. https://doi.org/10.1016/j.agee.2021.107718.

Liu, Y. Z., Imtiaz, M., Ditta, A., Rizwan, M. S., Ashraf, M., Mehmood, S., Aziz, O., Mubeen, F., Ali, M., Elahi, N. N., Ijaz, R., Lelel, S., Shuang, C., & Tu, S. (2020). Response of growth, antioxidant enzymes, and root exudates production towards As stress in *Pteris vittata* and *Astragalus sinicus* colonized by arbuscular mycorrhizal fungi. *Environmental Science and Pollution Research, 27*, 2340–2352.

El Malahi, S. E., Touhami, D., Ganoudi, M., Ennami, M., Maach, M., Hassani, L. M. I., & Jauregui, I. (2025). Harnessing the power of microbial allies: AMF and PGPR as biostimulants for Sustainable Bioeconomy Development in the Global South. In *Sustainable Bioeconomy Development in the Global South: Volume I Status and perspectives: 1* (pp. 261–291). https://doi.org/10.1007/978-981-96-0640-5_10.

Muriuki, A. W., Okalebo, J. R., Othieno, C. O., Njoroge, R., & Odeny, D. A. (2020). Smallholder farmers' perceptions and knowledge on soil fertility management practices in Central Kenya. *Journal of Agricultural Extension and Rural Development, 12*(4). https://doi.org/10.5897/JAERD2020.1166.

Nacoon, S., Jogloy, S., Riddech, N., Mongkolthanaruk, W., Kuyper, T. W., & Boonlue, S. (2020). Interaction between phosphate-solubilizing bacteria and arbuscular mycorrhizal fungi on growth promotion and tuber inulin content of *Helianthus tuberosus* L. *Scientific Reports, 10*(1), 4916.

Naveed, M., Bukhari, S. S., Mustafa, A., Ditta, A., Alamri, S., El-Esawi, M. A., Rafique, M., Ashraf, S., & Siddiqui, M. H. (2020). Mitigation of nickel toxicity and growth promotion in sesame through the application of a bacterial endophyte and zeolite in nickel contaminated soil. *International Journal of Environmental Research and Public Health, 17*(23), 1–27. https://doi.org/10.3390/ijerph17238859.

Nazir, A., Sarfraz, W., Allah, D., Khalid, N., Farid, M., Shafiq, M., Firdaus-E-Bareen, Rizvi, Z. F., & Naeem, N. (2023). Synergistic impact of two autochthonous saprobic fungi (*A. niger* and *T. pseudokoningii*) on the growth, ionic contents, and metals uptake in *Brassica juncea* L. and *Vigna radiata* L. under tannery solid waste contaminated soil. *International Journal of Phytoremediation, 25*(11), 1488–1500. https://doi.org/10.1080/15226514.2023.2166457.

Pan, J., Huang, C., Peng, F., Zhang, W., Luo, J., Ma, S., & Xue, X. (2020). Effect of arbuscular mycorrhizal fungi (AMF) and plant growth-promoting bacteria (PGPR) inoculations on *Elaeagnus angustifolia* L. in saline soil. *Applied Sciences, 10*(3), 945. https://doi.org/10.3390/app10030945.

Parihar, M., Rakshit, A., Meena, V. S., Gupta, V. K., Rana, K., Choudhary, M., Tiwari, G., Mishra, P. K., Pattanayak, A., Bisht, J. K., Jatav, S. S., Khati, P., & Jatav, H. S. (2020). The potential of arbuscular mycorrhizal fungi in C cycling: A review. *Archives of Microbiology, 202*(7), 1581–1596. https://doi.org/10.1007/s00203-020-01915-x.

Pellegrino, E., & Bedini, S. (2022). AMF and rock phosphate improve phosphorus nutrition in tomato under low-P stress. *Frontiers in Plant Science, 12*, 725541. https://doi.org/10.3389/fpls.2021.725541.

Pellegrino, E., Öpik, M., & Bonari, E. (2021). Trait-based assembly of arbuscular mycorrhizal fungal communities determines soil carbon formation and plant nutrition. *New Phytologist, 230*(1). https://doi.org/10.1111/nph.17082.

Pellegrino, E., Öpik, M., Bonari, E., & Ercoli, L. (2015). Responses of wheat to arbuscular mycorrhizal fungi: A meta-analysis of field studies from 1975 to 2013. *Soil Biology and Biochemistry, 84*, 210–217. https://doi.org/10.1016/j.soilbio.2015.02.020.

Perveen, R., Hussain, A., Ditta, A., Dar, A., Aimen, A., Ahmad, M., Alataway, A., Dewidar, A. Z., & Mattar, M. A. (2023). Growth and yield of okra exposed to a consortium of rhizobacteria with different organic carriers under controlled and natural field conditions. *Horticulturae, 9*(1), 8. https://doi.org/10.3390/horticulturae9010008.

Popoola, B. M., Olanbiwoninu, A. A., Adeyemi, O. A., & Adebayo, E. A. (2025). Physiological mechanisms behind the plants and microbes induced soil remediation. In *Soil bioremediation: 256* (pp. 243–271). Springer Science and Business Media LLC. https://doi.org/10.1007/978-3-031-76886-6_10.

Pretty, J., Benton, T. G., Bharucha, Z. P., Dicks, L. V., Flora, C. B., Godfray, H. C. J., Goulson, D., Hartley, S., Lampkin, N., Morris, C., Pierzynski, G., Prasad, P. V. V., Reganold, J., Rockström, J., Smith, P., Thorne, P., & Wratten, S. (2018). Global assessment of agricultural system redesign for sustainable intensification. *Nature Sustainability, 1*(8), 441–446. https://doi.org/10.1038/s41893-018-0114-0.

Rillig, M. C., Aguilar-Trigueros, C. A., Camenzind, T., Cavagnaro, T. R., Degrune, F., Hohmann, P., Lammel, D. R., Mansour, I., Roy, J., Van der Heijden, M. G. A., & Yang, G. (2019). Why farmers should manage the arbuscular mycorrhizal symbiosis. *New Phytologist, 222*(3), 1171–1175. https://doi.org/10.1111/nph.15602.

Rivero, J., Gamir, J., Aroca, R., Pozo, M. J., & Flors, V. (2015). Metabolic transition in mycorrhizal tomato roots under drought and salinity: Increased antioxidant activity and changes in sugar and polyamine metabolism. *Plant Journal, 107*(3), 729–748. https://doi.org/10.1111/tpj.15311.

Saeed, H. S., Munir, M., Samra, S., Farooq, A., Mahmood, H., Ahmad, I., Akram, M. I., Iqbal, R., Elsadek, M. F., Almutairi, S. M., Ditta, A., & Daraz, U. (2025). Screening of biosurfactants producing fungi from infected *Citrus sinensis* (L.). *Polish Journal of Environmental Studies, 34*(5), 5261–5270. https://doi.org/10.15244/pjoes/190588.

Saleem, A., Raza, M. A. S., Khan, I. H., Tahir, M. A., Iqbal, R., Aslam, M. U., Ejaz, Z., Ditta, A., Almunqedhi, B. M., & Al Farraj, D. A. (2024). Amelioration of drought stress in wheat by using plant growth-promoting rhizobacteria and biogas slurry. *Global Nest Journal, 26*(8). https://doi.org/10.30955/gnj.005890.

Salvioli, A., & Bonfante, P. (2019). The impact of the mycorrhizosphere on plant nutrition and growth: Insights from multi-omics approaches. *Current Opinion in Biotechnology, 62*, 77–85.

Sangwan, S., Saxena, G., Barik, P., & Bana, R. S. (2023). *Bioinoculants: Biological option for mitigating global climate change arbuscular mycorrhizal fungi: A keystone to climate-smart agriculture*. Springer Nature.

Sarfraz, R., Hussain, A., Sabir, A., Ben Fekih, I., Ditta, A., & Xing, S. (2019). Role of biochar and plant growth promoting rhizobacteria to enhance soil carbon sequestration—A review. *Environmental Monitoring and Assessment, 191*(4). https://doi.org/10.1007/s10661-019-7400-9.

Sawers, R. J. H., Svane, S. F., Quan, C., Grønlund, M., Wozniak, B., Gebreselassie, M.-N., & Varoquaux, N. (2017). Variation in arbuscular mycorrhizal fungal colonization and phosphorus acquisition among maize cultivars: A field study in Denmark and the USA. *Journal of Plant Nutrition and Soil Science, 180*(3). https://doi.org/10.1002/jpln.201600621.

Schwartz, M. W., Hoeksema, J. D., Gehring, C. A., Johnson, N. C., Klironomos, J. N., & Koide, R. T. (2021). The promise and the potential consequences of the global transport of mycorrhizal inoculum. *Ecology Letters, 24*(1). https://doi.org/10.1111/ele.13613.

Singh, A. K., Rai, A., & Singh, N. (2023). Scaling constraints of indigenous arbuscular mycorrhizal fungal inoculants in Indian agriculture: Supply chain and policy gaps. *Frontiers in Sustainable Food Systems, 7*. https://doi.org/10.3389/fsufs.2023.1125027.

Soretire, A. A, Adeyemi, N. O, Atayese, M. O, Sakariyawo, O. S, & Adewunmi, A. (2020). Nodulation and biological nitrogen fixation in soybean (*Glycine max* L.) as influenced by phosphorus fertilization and arbuscular mycorrhizal inoculation. *Acta Universitatis Sapientiae, Agriculture and Environment, 12*(1), 22–44. https://doi.org/10.2478/ausae-2020-0003.

Sun, W., & Hesam Shahrajabian, M. (2023). The application of arbuscular mycorrhizal fungi as microbial biostimulant, sustainable approaches in modern agriculture. *Plants, 12*(17), 3101. https://doi.org/10.3390/plants12173101.

Taktek, S., Trépanier, M., Servin, P. M., St-Arnaud, M., Piché, Y., Fortin, J. A., & Antoun, H. (2015). Trapping of phosphate solubilizing bacteria on hyphae of the arbuscular mycorrhizal fungus *Rhizophagus irregularis* DAOM 197198. *Soil Biology and Biochemistry, 90*, 1–9. https://doi.org/10.1016/j.soilbio.2015.07.016.

Tedersoo, L., Bahram, M., & Zobel, M. (2020). How mycorrhizal associations drive plant population and community biology. *Science, 367*(6480), eaba1223.

Thind, S., Chaudhary, M. S, Ditta, A., Hussain, I., Parveen, A., Ullah, N., Mahmood, Q., Al-Ashkar, I., & El-Sabagh, A. (2022). Impact of mycorrhizal fungi from different rhizospheric soils on fungal colonization, growth, and chlorophyll contents of *Cenchrus ciliaris*. *Agronomy, 12*(11), 2644. https://doi.org/10.3390/agronomy12112644.

Thirkell, T. J., Cameron, D. D., & Hodge, A. (2020a). Contrasting nitrogen fertilisation treatments alter AMF colonisation and root exudate composition in wheat. *Plant and Soil, 451*(1–2), 259–272. https://doi.org/10.1007/s11104-020-04518-z.

Thirkell, T. J., Pastok, D., & Field, K. J. (2020b). Functional diversity in arbuscular mycorrhiza: A trait-based framework to understand mycorrhizal symbiosis. *Plant and Soil, 452*(1–2), 1–16. https://doi.org/10.1007/s11104-020-04529-w.

Ullah, N., Ditta, A., Imtiaz, M., Li, X., Jan, A. U., Mehmood, S., Rizwan, M. S., & Rizwan, M. (2021). Appraisal for organic amendments and plant growth-promoting rhizobacteria to enhance crop productivity under drought stress: A review. *Journal of Agronomy and Crop Science, 207*(5), 783–802. https://doi.org/10.1111/jac.12502.

Ullah, N., Ditta, A., Khalid, A., Mehmood, S., Rizwan, M. S., Ashraf, M., Mubeen, F., Imtiaz, M., & Iqbal, M. M. (2020). Integrated effect of algal biochar and plant growth promoting rhizobacteria on physiology and growth of maize under deficit irrigations. *Journal of Soil Science and Plant Nutrition, 20*(2), 346–356. https://doi.org/10.1007/s42729-019-00112-0.

Vanek, S. J., Thies, J., & Drinkwater, L. E. (2016). Lack of knowledge, not resources, limits farmer adoption of microbial inoculants in sub-Saharan Africa. *Agronomy for Sustainable Development, 36*(3), 44. https://doi.org/10.1007/s13593-016-0383-9.

Verzeaux, J., Hirel, B., Dubois, F., Lea, P. J., & Tétu, T. (2017). Agricultural practices to improve nitrogen use efficiency through the use of arbuscular mycorrhizae: Basic and agronomic aspects. *Plant Science, 264*, 48–56. https://doi.org/10.1016/j.plantsci.2017.08.004.

Wahab, A., Muhammad, M., Munir, A., Abdi, G., Zaman, W., Ayaz, A., Khizar, C., & Reddy, S. P. P. (2023). Role of arbuscular mycorrhizal fungi in regulating growth, enhancing productivity, and potentially influencing ecosystems under abiotic and biotic stresses. *Plants, 12*(17). https://doi.org/10.3390/plants12173102.

Wang, Q., Liu, M., Wang, Z., Li, J., Liu, K. E., & Huang, D. (2024). The role of arbuscular mycorrhizal symbiosis in plant abiotic stress. *Frontiers in Microbiology, 14*. https://doi.org/10.3389/fmicb.2023.1323881.

Wilson, G. W. T., Rice, C. W., Rillig, M. C., Springer, A., & Hartnett, D. C. (2020). Soil aggregation and carbon sequestration are tightly correlated with the abundance of arbuscular mycorrhizal fungi: Results from long-term field experiments. *New Phytologist, 225*(3). https://doi.org/10.1111/nph.16279.

Xiao, J., Kao-Kniffin, J., & Zhu, B. (2025). Arbuscular mycorrhizal fungi enhance nitrogen acquisition from, but not carbon loss of, organic matter in soil. *New Phytologist, 247*(3), 1415–1425.

Xie, K., Deng, Y., Zhang, W., Chen, N., & Zhang, X. (2022). Intercropping with clover enhances soybean nitrogen fixation through mycorrhiza-mediated rhizosphere acidification. *Agriculture, Ecosystems & Environment, 324*, 107713. https://doi.org/10.1016/j.agee.2021.107713.

Yang, G., et al. (2023). AMF enhances agroecological sustainability through soil structure and biodiversity. *Agriculture, Ecosystems & Environment, 345*, 108321. https://doi.org/10.1016/j.agee.2023.108321.

Zeleke, G., Dobo, B., & Asefa, F. (2024). Diversity of arbuscular mycorrhizal fungi under the rhizosphere soil of different cropping systems. *Journal of BioScience and Biotechnology, 13*(1), 87–101.

Zhang, F., Wang, P., Zou, Y. N., Wu, Q. S., & Kuca, K. (2018). Effects of mycorrhizal fungi on growth and drought tolerance of wheat seedlings under drought stress. *Plant and Soil, 426*. https://doi.org/10.1007/s11104-018-3618-5.

Zhang, S., Lehmann, A., Zheng, W., You, Z., & Rillig, M. C. (2019). Arbuscular mycorrhizal fungi increase grain yields: A meta-analysis. *New Phytologist, 222*(1), 543–555. https://doi.org/10.1111/nph.15570.

Zhang, S., Lehmann, A., Zheng, W., You, Z., & Rillig, M. C. (2023). Arbuscular mycorrhizal fungi increase soil aggregation and stabilisation of organic carbon: A meta-analysis. *Soil Biology and Biochemistry, 186*. https://doi.org/10.1016/j.soilbio.2023.109182.

Zhang, X. H., Zhu, Y. G., Chen, B. D., Lin, A. J., Smith, S. E., & Smith, F. A. (2005). Arbuscular Mycorrhizal Fungi Contribute to Resistance of Upland Rice to Combined Metal Contamination of Soil. *Journal of Plant Nutrition, 28*(12), 2065–2077. https://doi.org/10.1080/01904160500320871.

Zhao, R., Guo, W., Bi, Q., Guo, J., Zhang, L., & Christie, P. (2021). Temperature-driven shifts in soil microbial communities reduce AMF colonization and nutrient uptake. *Soil Biology and Biochemistry, 160*, 108350. https://doi.org/10.1016/j.soilbio.2021.108350.

Zou, Y. N., Wu, Q. S., & Kuča, K. (2021). Unravelling the role of arbuscular mycorrhizal fungi in mitigating plant oxidative stress. *Environmental Science and Pollution Research, 28*(24), 30460–30470. https://doi.org/10.1007/s11356-021-13992-6.

Chapter 28

Arbuscular mycorrhizal fungi in agriculture: Policy and regulatory frameworks

Sana Yaqoob[a], Saboor Muarij Bunny[b], Alyan Ashraf[b] and Usman Khalid Chaudhry[b]
[a] *Institute of Soil & Environmental Sciences, University of Agriculture, Faisalabad, Pakistan,* [b] *Pakistan Environmental Protection Agency, Ministry of Climate Change & Environmental Coordination, Islamabad, Pakistan*

28.1 Role of arbuscular mycorrhizal fungi in agriculture

Symbiotic soil microbes, such as AMF, form associations with the roots of terrestrial plants and play a vital role in mitigating various plant stress conditions, including drought, salinity, and soil degradation. This microbial partnership is vital for promoting sustainable agricultural practices under stress, as it enhances nutrient cycling, alleviates abiotic stresses, and improves soil structure. As a key component of integrated agroecosystem management, AMF contributes to more resilient and productive farming systems. The association of AMF with plant roots enhances nutrient uptake efficiency, particularly for nitrogen (N), phosphorus (P), iron (Fe), and zinc (Zn). By extending their extraradical hyphae into the soil, AMFs effectively increase the absorptive surface area for nutrient acquisition (Fig. 28.1). This symbiotic relationship reduces reliance on chemical fertilizers while enhancing plant resilience to abiotic stresses through improved antioxidant defense mechanisms, osmotic regulation, and soil health. In addition, AMFs contribute to better soil aeration and water retention by forming extensive hyphal networks that interact with plant root systems (Begum et al., 2019).

Arbuscular mycorrhizal fungi (AMF) effectively mitigate abiotic stress conditions and enhance crop yield. Research demonstrates that the application of *Acaulospora longula* in melon hybrids from arid regions significantly improves biomass production and nutritional content (Banerjee et al., 2023). The mechanisms by which AMF counteract salinity stress in plants are multifaceted, including enhanced photosynthetic rates, improved antioxidant functioning, and increased ion homeostasis (Evelin et al., 2019). These adaptive strategies critically support crop productivity, particularly in regions characterized by poor soil health conditions. By boosting plant resilience through complex physiological processes, AMFs play a vital role in agricultural sustainability, offering a promising approach to mitigating environmental challenges in crop production.

The application of AMF in crop plants not only mitigates abiotic stress conditions but also significantly enhances the nutritional and quality parameters of vegetables, grains, and fruits. AMF inoculation improves the uptake efficiency of essential minerals such as zinc and iron, as well as critical phytochemicals, such as phenols and carotenoids, thereby substantially benefiting the nutritional profile of food crops (Rouphael et al., 2015). These biochemical improvements directly contribute to superior food quality characteristics, including enhanced taste, better texture, appealing color, and extended shelf life. The impact of AMF is particularly noteworthy, as these improvements meet both domestic consumer preferences and stringent international food quality standards (Pascual et al., 2018).

Beyond mitigating abiotic stress and contributing to crop yield, AMFs play a crucial role in soil health improvement by supporting nitrogen-fixing bacteria in plant root nodules and collaborating with potassium-solubilizing bacteria (Akyol et al., 2019). The AMF significantly contributes to sustainable agroecosystems through enhanced soil aggregation, which improves soil fertility and reduces soil erosion. However, the effectiveness of AMF can be limited by practices, such as heavy tillage and monoculture. To optimize AMF network efficiency in plant roots, agricultural experts recommend conservation practices, including organic farming, crop rotation, and reduced tillage, which support the intricate symbiotic relationships between fungi and plant roots while promoting overall soil health and agricultural sustainability (Lehmann et al., 2017).

Advancing Sustainable Agriculture Through Arbuscular Mycorrhizal Fungi. DOI: https://doi.org/10.1016/B978-0-443-45130-0.00034-5

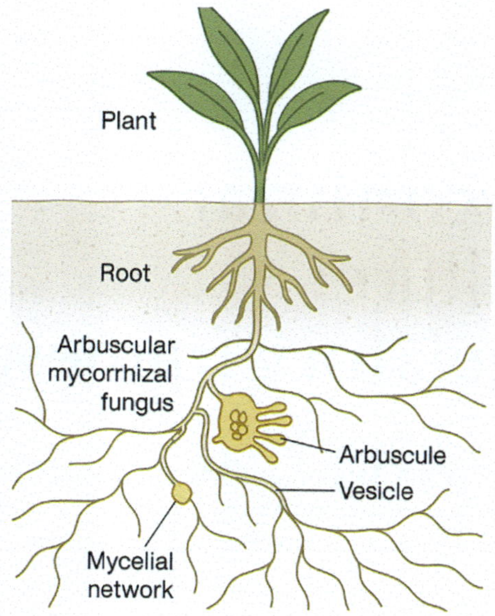

FIGURE 28.1 AMF–plant symbiotic network. *AMF*, Arbuscular mycorrhizal fungi.

TABLE 28.1 Key benefits of AMF in agriculture.

Sr. no.	Benefit	Mechanism	Example crop
1.	Improved nutrient uptake	Hyphal extension and solubilization of P, Zn	Wheat, maize
2.	Drought tolerance	Osmotic regulation and water-use efficiency	Melon, tomato
3.	Salinity stress mitigation	Ion homeostasis and antioxidant production	Rice, barley
4.	Enriched food quality	Increased mineral and phytochemical content	Tomato, spinach
5.	Soil health improvement	Soil aggregation and microbial synergy	Multiple crops

AMF, arbuscular mycorrhizal fungi.

AMF emerged as a unique microbial solution for sustainable agriculture, offering many benefits as discussed in Table 28.1.

AMFs exhibit optimal performance only within specific ranges of soil types, crop species, and environmental conditions, thereby inherently limiting their potential benefits (Kokkoris et al., 2019). Consequently, there is a critical need to investigate the precise compatibility requirements for AMF to function efficiently, including a comprehensive understanding of host–plant interactions and local soil microbiome dynamics. The efficiency of AMF varies significantly across different ecological conditions, with its performance being particularly sensitive to agricultural practices. Moreover, subsequent tillage operations and the chemical application of fertilizers, pesticides, and weedicides can substantially disrupt the delicate fungal networks within host plant roots, further compromising the effectiveness of AMF symbiosis (Ryan & Graham, 2018). Beyond these specific challenges, producing, storing, colonizing, and applying AMF inoculants presents a complex process that spans from laboratory to field conditions. The large-scale adoption of AMF applications remains challenging because of the lack of standardized protocols and technical complexity. To maximize the potential benefits of AMF in agricultural settings a comprehensive approach is crucial, encompassing targeted research, farmer training programs, and practical field applications. These strategic interventions are essential for bridging the gap between scientific understanding and practical implementation, ultimately facilitating the effective integration of AMF technologies in modern agricultural practices (Díaz et al., 2025).

The applications of AMF for sustainable and beneficial crop production are expected to expand globally, driven by sudden climatic changes and the urgent need to address food security challenges. Innovative improvements are critical for advancing AMF production, understanding its complex symbiotic relationships, and establishing direct connections with sustainable agricultural practices. These developments promise significant benefits for crop production and environmental preservation, positioning AMF as a key strategy in mitigating the impacts of environmental uncertainty and supporting global

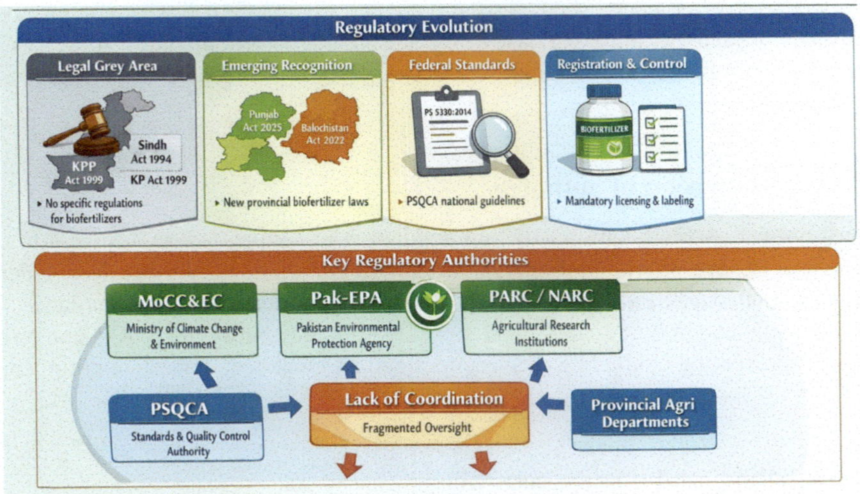

FIGURE 28.2 Overview of the current regulatory framework for biofertilizers in Pakistan, illustrating the evolution from a legal grey area to emerging provincial recognition and federal standardization, alongside the roles of key federal and provincial authorities. The figure highlights existing gaps in coordination, enforcement, and the absence of AMF-specific quality and biosafety criteria. *AMF*, Arbuscular mycorrhizal fungi; *MoCC&EC*, Ministry of Climate Change and Environmental Coordination; *NARC*, National Agricultural Research Centre; *Pak-EPA*, Pakistan Environmental Protection Agency; *PARC*, Pakistan Agricultural Research Council; *PSQCA*, Pakistan Standards and Quality Control Authority.

food systems. The integration of AMF is crucial for improved crop production, as it enhances nutrient uptake efficiency, improves food quality, and significantly reduces reliance on chemical fertilizers. By providing a sustainable agricultural solution, AMF offers a promising approach to addressing global food security challenges under increasingly unpredictable climatic conditions. This innovative biological approach not only optimizes crop performance but also contributes to more environmentally resilient and sustainable agricultural practices.

28.2 Current status of biofertilizer regulation in Pakistan

The regulatory landscape for biofertilizers in Pakistan has evolved from a prolonged legal vacuum to a fragmented but gradually formalizing framework. Historically the absence of a clear legal definition and dedicated standards allowed unregulated commercialization of microbial products, including AMF. In recent years, provincial legislative amendments and the introduction of national quality standards have signaled growing recognition of biofertilizers as critical inputs for sustainable agriculture. However, regulatory authority remains divided between federal and provincial institutions, resulting in uneven implementation, limited AMF-specific quality criteria, and weak coordination among oversight bodies. Fig. 28.2 summarizes the evolution of biofertilizer regulation in Pakistan and highlights the roles of key regulatory authorities and existing coordination gaps.

28.2.1 Legal recognition and early frameworks

For many years, Pakistan did not have any official legal definition or national regulation for biofertilizers, resulting in the commercialization of unstandardized microbial inoculants (Saha et al., 2023). Provincial legislation, such as Sindh's Fertilizer (Control) Act of 1994 and KP's Fertilizer Act of 1999, addressed solely chemical fertilizers and did not include microbial or AMF-based products, thereby leaving them in a legal grey area. Consequently, AMF inoculants were disseminated without the requirement of specific licensing or quality assessment, provided they adhered to general fertilizer regulations (Itamah et al., 2025).

28.2.2 Emerging recognition of biofertilizers

Recently, provinces have initiated amendments to fertilizer regulations to acknowledge microbial products. Punjab's Fertilizer Control Act 2025 defines "biofertilizer" as any carrier-based product that contains nutrient-mobilizing microorganisms, implicitly encompassing AMF (Malusá & Vassilev, 2014). Similarly the 2022 Act of Balochistan emphasizes the enhancement of soil quality through the regulation of fertilizer application. Federally the Pakistan Standards and

Quality Control Authority (PSQCA) has established Pakistan Standard PS 5330:2014, which delineates quality and testing specifications for biofertilizers (Santos et al., 2024). However, the standard predominantly addresses bacterial inoculants (e.g., nitrogen-fixers and phosphorus-solubilizers) and excludes AMF-specific criteria, such as propagule density or root colonization efficacy (Mishra & Barolia, 2020).

28.2.3 Federal–provincial division of authority

Following the 18th Constitutional Amendment of 2010 the regulation of fertilizers was devolved to the provincial level, leading to the development of diverse provincial frameworks (Sultan et al., 2025). Nevertheless, federal authorities, such as PSQCA, establish national standards and certification protocols for "biofertilizer" in accordance with PS 5330-2014, thereby impacting interprovincial and import quality regulation. Coordination between PSQCA and provincial departments continues to be essential for preventing conflicting dual regimes and guaranteeing consistent enforcement (Shahzad et al., 2025).

28.2.4 Registration, labeling, and quality control

Historically, provincial registration systems were established primarily for chemical fertilizers rather than microbial formulations, resulting in inadequate labeling and insufficient strain information (Saha et al., 2023). The PSQCA standard now requires the inclusion of information, such as microbial count ($\leq 10^7$ CFU/g), carrier sterilization, and microorganism type on product labels, although adherence to these requirements remains variable (Mishra & Barolia, 2020). Licensing is now mandatory under Punjab's 2025 legislation, encompassing all biofertilizer manufacturers. However, the absence of AMF-specific efficacy testing protocols highlights the necessity for enhanced laboratory and surveillance capabilities (Malusá & Vassilev, 2014).

28.3 Regulatory authorities involved

Pakistan's biofertilizer sector, including AMF, is governed by numerous federal and provincial agencies that frequently operate independently without an integrated regulatory framework (Naveed et al., 2015).

28.3.1 Ministry of Climate Change and Environmental Coordination

The Ministry of Climate Change and Environment Control, via the Pakistan Environmental Protection Agency (Pak-EPA), supervises environmental legislation and biosafety measures. It functions as the national focal point for the Cartagena Protocol on Biosafety; however, its mandate has predominantly centered on genetically modified organism (GMO) rather than natural microbial inoculants, such as AMF (Casey, 2022). Although Ministry of Climate Change and Environmental Coordination (MoCC&EC)'s climate policies promote sustainable agriculture and the utilization of biofertilizers, there is no established biosafety review mechanism for AMF strains (Chaudhry, 2017).

28.3.2 Pakistan Environmental Protection Agency

Pak-EPA administers the Pakistan Environmental Protection Act (1997) and the National Biosafety Rules (2005). Although authorized to regulate ecological hazards, it has not conducted reviews or monitoring of AMF-based biofertilizers, as they involve non-GMO organisms (Casey, 2022). This results in a void where the potential environmental impacts of AMF releases are not assessed within any biosafety framework.

28.3.3 Pakistan Agricultural Research Council and National Agricultural Research Centre

Pakistan Agricultural Research Council (PARC), operating under the Ministry of National Food Security & Research, oversees research and technological advancement in microbial inoculants, such as Biozote. It has performed AMF field trials and provides technical expertise in the development of standards (Mallmann et al., 2026. The National Agricultural Research Centre (NARC), as PARC's premier center, facilitates testing and provides training for farmers. Nevertheless, their contribution to regulatory decision-making remains informal, resulting in a sluggish translation of research findings into enforceable standards (Naveed et al., 2015).

28.3.4 Pakistan Standards and Quality Control Authority

Under the Ministry of Science and Technology, PSQCA establishes national standards and enforces the Pakistan Standard PS 5330:2014 for biofertilizers (Ali et al., 2015). The standard acknowledges microbial inoculants but omits specific benchmarks for AMF, such as spore count or colonization efficacy. Enforcement is constrained by limited testing capacity for live microbial products, resulting in only partial compliance (Rana et al., 2022).

28.3.5 Provincial Department of Agriculture

Following the 18th Amendment, provincial departments are responsible for enforcing fertilizer regulations and conducting product testing. Punjab's 2025 Act now authorizes biofertilizer producers; however, provinces still lack AMF-specific testing protocols and adequately trained inspectors (Ali et al., 2017). Coordination difficulties among provinces may lead to inconsistent implementation.

28.3.6 Coordination deficiencies

There is no centralized coordinating authority connecting MoCC&EC, PSQCA, PARC, and provincial regulators, leading to regulatory overlap and fragmented enforcement (Casey, 2022). Effective AMF governance necessitates an interministerial framework that consolidates biosafety, standardization, and agricultural input regulation.

28.4 Challenges in implementation and enforcement

Pakistan's fertilizer regulations have developed unevenly, leading to regulatory fragmentation: Sindh and Khyber Pakhtunkhwa continue to operate under outmoded frameworks, whereas Punjab and Balochistan have implemented more recent laws that encompass biofertilizers (Itamah et al., 2025). This absence of standardization leads to ambiguity regarding definitions, testing procedures, and registration processes, thereby allowing noncompliant entities to take advantage of legal loopholes (Yadav & Yadav, 2024). Stakeholders caution that inconsistent provincial testing protocols undermine confidence in AMF-based products (Saha et al., 2023). Extensive adulteration and counterfeit fertilizers reveal systemic deficiencies in monitoring and enforcement (Ashraf et al., 2020). Biofertilizers, especially AMF inoculants, frequently exhibit low viable counts and contamination; however, such deficiencies often remain undetected because of the lack of established testing protocols (Mishra & Barolia, 2020). Enforcement continues to be predominantly reactive; while periodic raids in Punjab have enhanced supervision, they have not resolved the more extensive issue of the absence of systematic surveillance across the nation (Narayana et al., 2025). No Pakistani regulation specifies quality standards for AMF, such as propagule density, spore viability, or root colonization efficacy. Existing standards depend on bacterial CFU counts, which are unsuitable for fungi (Mishra & Barolia, 2020). This regulatory gap hinders the validation of product claims, enabling substandard AMF inoculants to proliferate without oversight (Songachan et al., 2023). Most provincial laboratories do not possess microbiological facilities or adequately trained personnel for biofertilizer analysis. Testing infrastructure primarily addresses chemical fertilizers, rather than living microbial products that necessitate cold-chain management and viability evaluation (Itamah et al., 2025). Absent formal collaboration with PARC and university laboratories, the capacity for regulatory testing remains limited. Despite the implementation of new laws, procedural obstacles, such as prolonged testing processes, a limited number of inspectors, and the presence of informal biofertilizer producers, hamper consistent enforcement. The sector's informality results in numerous producers operating without licenses, whereas penalties and inspections occur intermittently, thereby undermining compliance incentives (Ntsomboh-Ntsefong et al., 2025). Limited awareness and previous encounters with ineffective biofertilizers have undermined farmers' confidence. Insufficient training on the correct application of AMF, including timing, compatibility with fungicides, and storage, results in improper use and suboptimal outcomes. Extension initiatives continue to be constrained, thereby perpetuating the cycle of minimal adoption and widespread mistrust (Hussain et al., 2025).

28.5 Recommendations for improvement

Pakistan should develop a cohesive national biofertilizer framework to synchronize provincial legislation and guarantee uniform quality standards. A national biofertilizer coordination committee, established under Ministry of National Food Security & Research (MNFSR) or the Council of Common Interests, could standardize registration, licensing, and testing procedures across provinces (Das et al., 2025). Drawing from the 2023 biopesticide harmonization initiative, Pakistan should establish comprehensive national guidelines for biofertilizers, encompassing efficacy trial data, strain verification procedures, and environmental safety standards (Haider et al., 2025). Legal updates should modernize outdated statutes (e.g., Sindh 1994) while maintaining federal consistency through PSQCA standards for interprovincial commerce, thereby minimizing regulatory loopholes and enhancing compliance (Channa et al., 2025). Formulating national standards specific to AMF is of paramount importance. PSQCA, PARC, and academic institutions should work together to amend PS 5330:2014 by incorporating metrics, such as minimum viable propagule quantities and limits on spore purity and contamination. International assays, such as most probable number or root staining, can serve as benchmarks for local methodologies (Aziz et al., 2025). Provincial authorities are required to broaden inspection efforts to encompass biofertilizer manufacturers and suppliers. A Biofertilizer Regulatory Inter-Agency Council, consisting of PSQCA, provincial agriculture departments, PARC/NARC, and Pak-EPA, should collaborate on data-sharing, inspection procedures, and revisions to standards.

A comprehensive framework should explicitly specify biofertilizers and AMF inoculants within legal statutes, eliminating ambiguity and ensuring uniform regulation across all provinces (Sultan et al., 2025). Punjab's 2025 Fertilizer Act provides a comprehensive framework by incorporating both "biofertilizers" and "biostimulants" within its legal scope. AMF, as an indigenous and low-risk entity, should be subject to regulation emphasizing efficacy rather than hazard assessment, whereas GM bioinputs should be governed by more stringent biosafety regulations (Rachna et al., 2024). An expert technical advisory panel could conduct an annual review of emerging research and consequently revise the list of permitted species and formulations. A national reference standard established by PSQCA should serve as the foundation for provincial enforcement. Federal agencies concentrate on standardization and certification, whereas provincial authorities are responsible for implementation and inspection, ensuring subsidiarity while maintaining consistency (Channa et al., 2025).

28.6 Collaborative framework for arbuscular mycorrhizal fungi biofertilizer regulation in Pakistan

Establishing a comprehensive regulatory framework for AMF and biofertilizers in Pakistan necessitates the integration of quality assurance, biosafety protocols, and mechanisms for collaborative innovation. A comprehensive Pakistan biofertilizer standard established by PSQCA should specify microbial counts, purity levels, shelf life, and particular AMF parameters (e.g., viable propagule density, species verification through DNA barcoding) to ensure consistent quality control across the country (Mishra & Barolia, 2020). Product registration should be centralized via a Biofertilizer Registration Committee comprising PARC, PSQCA, and provincial departments, ensuring that only products with validated local efficacy trials are authorized (Santos et al., 2024). Mandatory licensing of producers and dealers based on technical capacity and adherence to good manufacturing practices will foster uniform standards and reduce the risk of adulteration. Randomized quality inspections and digital traceability through QR codes or barcodes can establish accountability and discourage the sale of counterfeit products (Mishra & Barolia, 2020). A nationally approved list of strains should direct the safe application of microbes, emphasizing indigenous AMF species, such as *Rhizophagus intraradices* and *Funneliformis mosseae* (Sudha Samuel & Veeramani, 2021). Import controls and quarantine inspections for exotic strains should be coordinated under the supervision of Pak-EPA to reduce ecological risks (Arjjumend & Koutouki, 2021). Safety protocols must guarantee the absence of human and animal pathogens and require labeling to inform users of necessary precautions. A unified approval process can integrate PSQCA certification with provincial registration to enhance compliance efficiency. Renewal cycles, occurring every 3–5 years, should be contingent upon performance evaluations and feedback from farmers, thereby facilitating the removal of underperforming products (Atieno et al., 2020). Products qualifying for government subsidies or public procurement are required to possess a valid certification, thereby connecting regulatory compliance with incentive mechanisms.

Collaborative networks among PARC, NARC, universities, and private enterprises are vital for advancing and scaling innovation. Public laboratories should spearhead AMF strain screening, culture preservation, and training initiatives for provincial inspectors, whereas private enterprises contribute capital investment and extensive distribution networks (Ali, 2025; Santos et al., 2024). Extension services should implement demonstration sites and farmer field schools within each agro-ecological zone to foster trust and enhance awareness (Hussain et al., 2025). An Inter-Agency Biofertilizer Council comprising PSQCA, MoCC&EC, PARC, and provincial regulators should oversee the coordination of standards updates, biosafety evaluations, and data exchange. Evidence from local trials should serve as a direct basis for amendments to standards and subsidy policies. This dynamic, "science-led" feedback model reflects best practices from international biofertilizer governance (Shahzad et al., 2025).

References

Akyol, T. Y., Niwa, R., Hirakawa, H., Maruyama, H., Sato, T., Suzuki, T., & Sato, S. (2019). Impact of the introduction of *Arbuscular mycorrhizal* fungi on the root microbial community in agricultural fields. *Microbes and Environments, 34*(1), 23–32.

Ali, M. (2025). Strategies for transforming agro-commodity into agro-industrial clusters in Pakistan. *Pakistan Development Review, 64*, 53–72.

edited by Ali, M., Ahmed, F., Channa, H., & Davies, S. (2017). Chapter 6. Pakistan's Fertilizer Sector: Structure, policies, performance, and impacts. In D. J. Spielman, J. Sohail, P. D. Malik, & N. Ahmad (Eds.), *Agriculture and the rural economy in Pakistan* University of Pennsylvania Press. edited by.

Ali, M., Ahmed, F., Channa, H., Davies, S., Ali, M., Ahmed, F., Channa, H., & Davies, S. (2015). The role of regulations in the Fertilizer Sector of Pakistan. *2015 Conference August 9-14, 2015* https://doi.org/10.22004/AG.ECON.211559 .

Arjjumend, H., & Koutouki, K. (2021). Analysis of Indian and Canadian laws on biofertilizers. *Agrár- És Környezetjog, 16*(30), 7–23.

Ashraf, M. T, Mufti, S., Jan, U., Anayat, R., Nisar, F., & Ahmad, N. (2020). Role of biofertilizers in vegetable crop production: A review. *International Journal of Chemical Studies, 8*(6), 2810–2814.

Atieno, M., Herrmann, L., Nguyen, H. T., Thi Phan, H., Khoi Nguyen, N., Srean, P., Maw Than, M., Zhiyong, R., Tittabutr, P., Shutsrirung, A., Bräu, L., & Lesueuret, D. (2020). Assessment of biofertilizer use for sustainable agriculture in the great mekong region. *Journal of Environmental Management, 275*(111300), 111300.

Aziz, Z. N., Hussein, A. G., & Kadhim, Z. J. (2025). Effect of biofertilizers (mycorrhizal fungi, *Trichoderma, Bacillus subtilis* bacteria) and seaweed extracts on the growth and yield of mint (*Mentha spicata*): An integrative study of vegetative growth and secondary compounds. *International Journal of Advanced Biochemistry Research, 9*(5), 100–106.

Banerjee, A., Chakraborty, N., Acharya, K., & Chakraborty, N. (2023). Role of endophytes, plant growth promoting rhizobacteria, and *arbuscular mycorrhizal* fungi in stress tolerance. In *Biology and biotechnology of environmental stress tolerance in plants* (pp. 57–96). Apple Academic Press.

Begum, N., Qin, C., Ahanger, M. A., Raza, S., Khan, M. I., Ashraf, M., & Zhang, L. (2019). Role of *arbuscular mycorrhizal* fungi in plant growth regulation: Implications in abiotic stress tolerance. *Frontiers in Plant Science, 10*, 1068.

Casey, J. P. (2022). Policy coherence for national climate change adaptation and invasive species management in four countries. *CABI Agriculture and Bioscience, 3*(1). https://doi.org/10.1186/s43170-022-00077-8.

Channa, K. A, Khoso, D. H, & Malik, A. H (2025). Anti-money laundering regulations in Pakistan: A comparative analysis with global best practices. *Journal of Money Laundering Control, 28*(3), 518–532.

Chaudhry, Q. U. Z. (2017). *Climate change profile of Pakistan*. Asian Development Bank.

Das, A. K, Lee, D. a-S., Woo, Y.-J. i, Sultana, S., Mahmud, A., & Yun, B.-W. (2025). The impact of flooding on soil microbial communities and their functions: A review. *Stresses, 5*(2), 30.

Díaz, V., Villalobos, M., Arriaza, K., Flores, K., Hernández-Saravia, L. P., & Velásquez, A. (2025). Decoding the dialog between plants and *arbuscular mycorrhizal* fungi: A molecular genetic perspective. *Genes, 16*(2), 143.

Evelin, H., Devi, T. S., Gupta, S., & Kapoor, R. (2019). Mitigation of salinity stress in plants by *arbuscular mycorrhizal* symbiosis: Current understanding and new challenges. *Frontiers in Plant Science, 10*, 470.

Haider, A., Tansif, U. R., Dr., & Anees, A. (2025). Legal challenges in regulating sports governance and athletes' rights in Pakistan. *Al-Aasar, 2*(3), 48–58.

Hussain, K., Talha Azeem, M., Raees, S., & Iqbal, H. (2025). From waste to wealth: Assessing awareness and adoption of organic compost/biofertilizer in the context of urban horticulture and food safety in Karachi City. *The Critical Review of Social Sciences Studies, 3*(3), 2001–2020.

Itamah, E., Bello, T. K., Muhammed Waziri, S., & Ugwueke, S. (2025). Crimsonpublishers. *Progress in Petrochemical Science, 7*(1), 768–784.

Kokkoris, V., Hamel, C., & Hart, M. M. (2019). Mycorrhizal response in crop versus wild plants. *PLoS One, 14*(8), e0221037.

Lehmann, A., Zheng, W., & Rillig, M. C. (2017). Soil biota contributions to soil aggregation. *Nature Ecology & Evolution, 1*(12), 1828–1835.

Malusá, E., & Vassilev, N. (2014). A contribution to set a legal framework for biofertilisers. *Applied Microbiology and Biotechnology, 98*(15), 6599–6607.

Mishra, B. K., & Barolia, S. K. (2020). Quality assessment of microbial inoculants as biofertilizer. *International Journal of Current Microbiology and Applied Sciences, 9*(10), 3715–3729.

Narayana, S., Jagetiya, P., Mathur, R., Mishra, R., & Chauhan, S. (2025). Biofertilizer commercialization: Future prospects and challenges. *Agricultural Research Journal, 62*(1), 1–15.

Naveed, M., Mehboob, I., Shaker, M. A., Hussain, M. B., & Farooq, M. (2015). Biofertilizers in Pakistan: Initiatives and limitations. *International Journal of Agriculture and Biology, 17*(3), 411–420.

Ntsomboh-Ntsefong, G., Gabriel, M. S. T., Namuene, K. S., Mélanie, D. T. C., & Kingsley, T. M. (2025). Biofertilizers: An integrated approach to improving soil fertility, plant nutrition, forest and environmental sustainability. *OALib, 12*(08), 1–26.

Pascual, J. A., Ceglie, F., Tuzel, Y., Koller, M., Koren, A., Hitchings, R., & Tittarelli, F. (2018). Organic substrate for transplant production in organic nurseries. A review. *Agronomy for Sustainable Development, 38*, 1–23.

Rachna, Thakur, V., Sharma, P., & Nelson, R. (2024). Agri-financing and rural development. In *Modern horizons in agriculture: 1* (pp. 250–269). Stella International Publication. Retrieved January 7, 2026, from. https://www.researchgate.net/publication/382563395_Agri-financing_and_Rural_Development .

Rana, A. W., Gill, S., & Akram, I. (2022). *Seed system of Pakistan: Policy challenges and prospects*. International Food Policy Research Institute https://doi.org/10.2499/p15738coll2.136381.

Rouphael, Y., Franken, P., Schneider, C., Schwarz, D., Giovannetti, M., Agnolucci, M., De Pascale, S., Bonini, P., & Colla, G. (2015). *Arbuscular mycorrhizal* fungi act as biostimulants in horticultural crops. *Scientia Horticulturae, 196*, 91–108.

Ryan, M. H., & Graham, J. H. (2018). Little evidence that farmers should consider the abundance or diversity of *arbuscular mycorrhizal* fungi when managing crops. *New Phytologist, 220*(4), 1092–1107.

Saha, S., Paul, D., Poudel, T. R., Basunia, N. M., Hasan, T., Hasan, M., Li, B., Reza, R., Haque, A. R., Hanif, M. A., Sarker, M., Roberts, N. J., Khoso, M. A., Wu, H., & Shen, H.-l. (2023). Biofertilizer science and practice for agriculture and forestry: A review. *Journal of Applied Biology & Biotechnology*. https://doi.org/10.7324/jabb.2023.148741.

Santos, F., Melkani, S., Oliveira-Paiva, C., Bini, D., Pavuluri, K., Gatiboni, L., Mahmud, A., Torres, M., McLamore, E., & Bhadha, J. H. (2024). Biofertilizer use in the United States: Definition, regulation, and prospects. *Applied Microbiology and Biotechnology, 108*(1), 511.

Shahzad, M., Hayat, R., Mujtaba, G., Waseem, U. rR., & Muhammad, N. (2025). Biofertilizers in sustainable agriculture: Mechanisms, applications, and future prospects. *Discover Agriculture, 3*(1). https://doi.org/10.1007/s44279-025-00318-0.

Songachan, L. S., Deka, A., & Jha, S. S. (2023). Diversity and distribution of arbuscular mycorrhizal fungi in selected medicinal plants. *Agricultural Science Digest—A Research Journal*. https://doi.org/10.18805/ag.d-5834.

Sudha Samuel, S., & Veeramani, A. (2021). Advantages of arbuscular mycorrhizal fungi (AMF) production for the profitability of agriculture and biofertilizer industry. *Mycorrhizal fungi—Utilization in agriculture and industry*. IntechOpen.

Sultan, M. T., Noman, A. M., & Raza, H. (2025). Food laws of Pakistan: Current flaws and strategic future reforms. *Legal Research & Analysis, 3*(1), 80–88.

Yadav, A., & Yadav, K. (2024). Challenges and opportunities in biofertilizer commercialization. *SVOA Microbiology, 5*(1), 1–14.

Chapter 29

Case studies on agricultural applications of arbuscular mycorrhizal fungi in the field: Pros and cons

Kashif Tanwir[a], Muhammad Zulqurnain Haider[a], Muhammad Shahid[b] and Muhammad Tariq Javed[a]

[a] *Department of Botany, Faculty of Life Sciences, Government College University, Faisalabad, Pakistan,* [b] *Department of Bioinformatics and Biotechnology, Government College University Faisalabad, Faisalabad, Pakistan*

29.1 Introduction

The term "mycorrhiza" was derived from the Greek word "mushrooms" for the first time by Frank (1885), which factually means "fungus roots". Mycorrhizal fungi are capable of forming symbiotic associations with the roots of plants in a similar pattern to the mutualistic relationship that exists between the roots of legume plants and nitrogen-fixing bacteria. Seven different types of mycorrhizal association (arbuscular, ectendo, ericoid, ecto, monotropoid, orchidaceous, and arbutoid mycorrhizae) exist in nature; however, ectomycorrhizae and arbuscular mycorrhizae are the two most abundant and prevalent types (Allen et al., 2003; Martin et al., 2017). An arbuscular mycorrhizal fungus (AMF) is the most abundant association and capable of forming root and hyphal associations with over 90% of vascular plants that exist naturally (Begum et al., 2019; Chen et al., 2017; Teste et al., 2020). AMFs exist globally as a coerce mutualist organism that belongs to the Glomeromycota phylum (Montesinos-Navarro et al., 2019; Redecker et al., 2000). Plants growing in different kinds of stresses also get nonnutritional benefits from AMF associations, alongside the nutritional benefits (Püschel et al., 2021). In some plants, it is also observed that the antioxidant enzyme activity was also enhanced by some AMF species applications (Malicka et al., 2021) and are also involved in the maintenance of ionic homeostasis under salinity stress (Evelin et al., 2019).

The current chapter compiles various field-based case studies to assess the practical pros and cons associated with AMF applications in diverse agroecosystems. Eminent pros of AMF field applications include improved nutrients and water use efficiency (WUE), improved stress tolerance, and decreased need for synthetic fertilizers (Nazir et al., 2023; Saeed et al., 2025; Thind et al., 2022; Yizhu et al., 2020). In addition, AMF promotes long-term soil health through carbon sequestration, pathogen resistance, and improved microbial diversity. But, along with all these pros, the AMF application also faces some cons, including inconsistent efficacy in terms of soil type, fungal and plant species, high costs of inoculants, and incompatibility with intensive tillage or chemical inputs, restricting broader adoption. Accessibility for farmers to quality inoculants and gaps in knowledge regarding AMF–crop interactions further impede implementation (Fig. 29.1). Although AMF shows significant potential for regions facing resource limitations and environmental stress, its success relies on customized strategies that consider local soil conditions, crop rotations, and socio-economic factors. Further, this chapter also highlights the importance of AMF in sustainable agriculture while stressing the necessity for practical, context-specific strategies to optimize its advantages.

29.2 Arbuscular mycorrhizal fungi

AMFs are commonly present and first evolved among all the other types of mycorrhizas (Ahmad et al., 2020). In the plant rhizosphere, soil and plant roots interrelate with several beneficial microorganisms, including AMF, which is highly effective

Advancing Sustainable Agriculture Through Arbuscular Mycorrhizal Fungi. DOI: https://doi.org/10.1016/B978-0-443-45130-0.00021-7

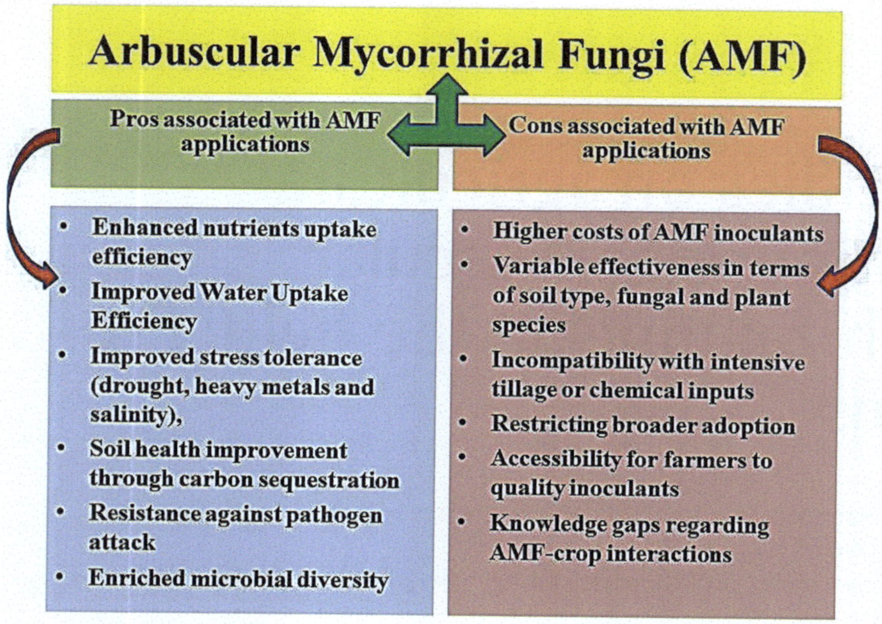

FIGURE 29.1 Pros and cons associated with the symbiotic associations between plants and field applications of AMF. *AMF*, Arbuscular mycorrhizal fungi.

for ecosystem sustainability (Jeffries & Barea, 2012; Sosa-Hernández et al., 2019). This interaction is most primeval and evolved about 460 million years ago (Smith & Smith, 2011). The term "ecosystem engineers" is also used for AMF because they enrich the accretion of vital minerals from soil (Bücking & Kafle, 2015). They control the accumulation of vital nutrients in plants by forming a vital network of extraradical fungal mycelium (ERM) (Diagne et al., 2020). This transport network is working through the reciprocal transport of phosphate and ammonium ions from fungal hyphae to plant tissue in response to carbon (C) uptake from plant roots (Begum et al., 2019). These fungal hyphae form specialized structures termed "arbuscules," which are subcellular, root-curricular structures and derived from the Latin word "arbusculum," which means "bush" or "tree" (Gutjahr & Parniske, 2013; Montesinos-Navarro et al., 2019). These AMF have diverse forms and types of origin of numerous fungal clades and plant types (Van der Heijden et al., 2015). The application of AMF significantly enhanced the accumulation and utilization of soil nutrients by *Artemisia annua* L. and augmented its relative artemisinin content (Ma et al., 2021).

29.3 Pros of arbuscular mycorrhizal fungi field applications

AMF field applications have various pros, and the most common are described in detail below.

1. Enhanced nutrient uptake efficiency
2. Improved water uptake efficiency
3. Improved stress tolerance (drought, heavy metals, and salinity)
4. Soil health improvement through carbon sequestration
5. Resistance against pathogen attack
6. Enriched microbial diversity

29.3.1 Enhanced nutrient uptake efficiency

The well-documented benefit of AMF colonization in field settings is its profound impact on nutrient acquisition, particularly for immobile nutrients like phosphorus (P). Soil P often exists in forms unavailable to plants or is fixed rapidly (AMFs overcome this limitation through several mechanisms). The extensive network of AMF extraradical hyphae, significantly finer and more widespread than plant root hairs, can explore a larger volume of soil. This allows access to nutrient pools beyond the rhizosphere depletion zone, which are otherwise inaccessible to the host root system (Smith & Read, 2008). This extended reach is crucial for scavenging P from distant soil pores. AMF can release organic acids and enzymes (e.g., phosphatases) into the soil, which can solubilize recalcitrant forms of P, making them available for uptake. This direct enzymatic activity contributes to the mobilization of otherwise unavailable P compounds (Ortas et al., 2021).

The fungal hyphae possess highly efficient P transporters, which actively absorb inorganic phosphate (Pi) from the soil solution and translocate it to the plant roots. Within the plant roots, specialized symbiotic P transporters facilitate the uptake of Pi from the arbuscules into the host cells (Yang et al., 2023). This "mycorrhizal pathway" significantly supplements the plant's own direct uptake pathway. Recent field trials have consistently shown increased grain yield and enhanced P content in crops inoculated with AMF, especially under reduced fertilization regimes, demonstrating improved phosphorus fertilizer use efficiency (Hu et al., 2024).

Although phosphorus uptake is the most prominent, AMFs also play a role in nitrogen cycling and uptake. They can facilitate the uptake of various N forms, including ammonium and nitrate, and contribute to the mineralization of organic N in the soil (Hodge & Fitter, 2010). Studies have indicated that AMF can enhance nitrogen content in crop tissues and improve overall nitrogen use efficiency in field conditions (Mohseni et al., 2024).

The mycelia of AMF form a complex network surrounding the plant roots and can not only facilitate the uptake of minerals bound with the litter of the soil but also enhance the area of absorption by roots. Further, these mycelia transfer these absorbed minerals to their host plants. This phenomenon of mineral absorption by mycelia is termed "direct mineral cycling." This phenomenon is composed of two major components: one is fungi, which act as primary decomposers in forests and form mycorrhizal associations with plant roots that enable the accumulation of mineral nutrients to their plant partner (Shah et al., 2025).

29.3.2 Improved water uptake efficiency

Beyond nutrient benefits, AMF symbiosis significantly improves plant water relations, particularly under drought stress conditions, a common challenge in many agricultural regions. Similar to nutrient uptake the widespread extraradical hyphae of AMF extend far beyond the root depletion zone, enabling the plant to access water from a larger soil volume, including micropores that are too small for root penetration. This effectively expands the plant's access to soil moisture (Auge, 2004; Birhane et al., 2023). The fungal hyphae can act as hydraulic conduits, directly transporting water from the soil to the plant roots. This direct transfer pathway, coupled with improvements in root hydraulic conductivity in mycorrhizal plants, contributes to better water acquisition (Ruiz-Lozano, 2003). AMF colonization induces physiological changes in the host plant that enhance WUE. This includes better stomatal regulation, increased photosynthetic rates, and accumulation of compatible osmolytes (e.g., proline and sugars) that help maintain cell turgor and protect cellular structures under water deficit (Al-Arjani et al., 2020; Seutra et al., 2021). These adaptations allow mycorrhizal plants to maintain growth and productivity even under reduced water availability.

29.3.3 Improved stress tolerance (drought, salinity, and heavy metals)

29.3.3.1 Drought and water stress

AMFs significantly improve plant water relations. Beyond enhanced water uptake efficiency because of the expansive hyphal network (as discussed previously), AMF-colonized plants exhibit better stomatal regulation, higher hydraulic conductivity, and accumulate more compatible osmolytes (e.g., proline and soluble sugars) to maintain cell turgor and protect cellular structures under water deficit (Fig. 29.2). This leads to improved photosynthetic efficiency and sustained growth under drought conditions, as evidenced in numerous field studies with various crops (Al-Arjani et al., 2020; Seutra et al., 2021).

29.3.3.2 Salinity stress

High soil salinity induces both osmotic and ionic stress in plants. AMFs mitigate these effects by enhancing nutrient uptake (diluting the concentration of toxic ions), regulating ion transport (e.g., reducing Na^+ uptake and increasing K^+/Na^+ ratio), and improving antioxidant defenses to combat salt-induced oxidative stress (Evelin et al., 2019; Roy et al., 2021). Mycorrhizal plants often maintain better water status and photosynthetic rates in saline environments (Fig. 29.2).

29.3.3.3 Heavy metal toxicity

AMFs enhance plant tolerance to heavy metals by reducing their uptake, facilitating their chelation and compartmentalization within fungal or plant tissues, and strengthening the plant's antioxidant machinery (Goh et al., 2021). Conversely, AMF can also alleviate nutrient deficiency stresses by improving the acquisition of essential, often immobile, nutrients such as phosphorus, zinc, and copper, which are critical for various metabolic and defense processes (Ortas et al., 2021).

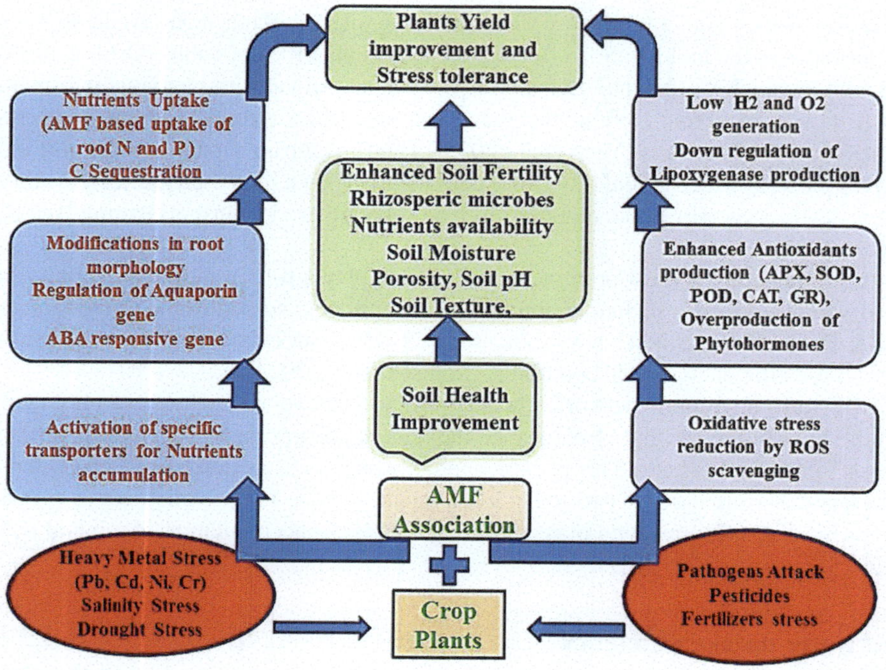

FIGURE 29.2 Showing the different mechanisms of stress tolerance associated with the symbiotic associations between plants and field applications of AMF. *AMF*, Arbuscular mycorrhizal fungi; *ROS*, reactive oxygen species.

29.3.4 Mechanisms of arbuscular mycorrhizal fungi–mediated carbon sequestration

AMF contribute to soil carbon sequestration through several direct and indirect mechanisms that enhance the quantity and stability of soil organic matter.

29.3.4.1 Direct carbon input through fungal biomass

The extraradical mycelial network of AMF represents a significant input of carbon into the soil. As hyphae grow, senesce, and turnover, they contribute directly to the particulate organic matter fraction of SOC (Chen et al., 2017). The fungal biomass itself, rich in carbon, serves as a substrate for soil microbial communities, contributing to the overall carbon pool. Field studies have shown that AMF hyphal biomass can constitute a substantial portion of the microbial biomass carbon in agricultural soils, especially under conditions favoring mycorrhizal development (Chen et al., 2017).

29.3.4.2 Production of glomalin-related soil protein

Perhaps the most significant direct contribution of AMF to stable SOC is the production of glomalin-related soil protein (GRSP). Glomalin, a glycoprotein produced by AMF hyphae, is highly recalcitrant to decomposition, with a slow turnover rate, making it a stable form of carbon in the soil (Rillig & Mummey, 2006). GRSP acts as "super glue," binding soil particles together into stable macroaggregates. These aggregates protect occluded organic matter from microbial decomposition by physically isolating it, thereby enhancing the long-term sequestration of carbon within the soil structure (Singh et al., 2024). Field applications of AMF have consistently demonstrated an increase in GRSP content, leading to enhanced soil aggregation and, consequently, greater SOC stability (Singh et al., 2024).

29.3.5 Mechanisms of arbuscular mycorrhizal fungi–mediated resistance against pathogen attack

The enhanced disease resistance observed in AMF-colonized plants in field settings is multifaceted, encompassing direct and indirect mechanisms that collectively bolster the plant's defense capabilities (Table 29.1).

29.3.5.1 Mycorrhiza-induced resistance and priming

The most prominent mechanism is mycorrhiza-induced resistance (MIR), a form of systemic resistance where AMF colonization primes the plant for a faster and stronger defense response upon subsequent pathogen challenge

TABLE 29.1 Examples of AMF field applications on plants facing pathogen attacks and their modes of action.

AMF species	Host plant	Disease/ Responsible pathogen	Effect	Mode of action	References
Funneliformis mosseae	*Lycopersicum esculentum*	Root rot/*Phytophthora nicotianae* var. parasitica 201	Disease symptoms were decreased	Enhanced shoot phosphorus (P) uptake	Trotta et al. (1996)
Rhizophagus intraradices	*Linum usitatissimum*	*Oidium lini*	Disease severity levels were increased	Enriched shoot growth	Dugassa et al. (1996)
R. intraradices	*Pisum sativum* cv. Solara	Root rot/*Aphanomyces euteiches*	Disease symptoms and severity were decreased	Shoot P accumulation level increased	Bødker et al. (1998)
F. mosseae	Tomato and eggplant	Verticillium wilt/*Verticillium dahliae* Kleb	Abridged effect of pathogen	Enhanced shoot phosphorus (P) and N levels and growth	Karagiannidis et al. (2002)
F. mosseae	*Zea mays*	*Rhizoctonia solani* AG1-IA	Condensed intensities of disease severity and symptoms	Enhanced production of root phytoalexin 2,4-dihydroxy-7-methoxy-2H-1,4-benzoxazin-3(4*H*) and systemic resistance	Song et al. (2011)
R. intraradices	*Glycine max* cv. NEAU-D50	Root rot/*P. sojae*	Disease symptoms were decreased	Amended jasmonic acid, systemic resistance in shoot and root N accumulation	Li et al. (2013)
R. irregularis	*Vitis* spp.	*Plasmopara viticola*	Not determined	Increase of shoot stilbenoids	Bruisson et al. (2016)
R. irregularis	*Solanum lycopersicum* cv. Better Boy	*Botrytis cinerea* CECT2100	Disease symptoms were decreased	Increase in lignan production	Sanmartín et al. (2020)
F. mosseae	*G. max* HN48	Root rot/*Fusarium oxysporum*	Abridged effect of pathogen	Downregulated genes involved in protein production responsible for components of fungal cell wall	Zhang et al. (2020)
R. intraradices and *F. mosseae*	*S. lycopersicum* cv. Red Rock	Fusarium wilt/*F. oxysporum* f. sp. lycopersici	Abridged effect of pathogen	Increased plant growth, activities of glutathione reductase, and dehydroascorbate reductase	Hashem et al. (2021)
Glomus versiforme	*Salvia miltiorrhiza*	Fusarium wilt/*F. oxysporum*	Disease incidence levels were decreased	Upregulated the plant growth, defense genes, and β-1,3-glucanase, chitinase, and PAL activities in root	Pu et al. (2022)
Mycorrhizal consortium (five species)	*Coffea arabica*	Coffee leaf rust/*Hemileia vastatrix*	Abridged levels of disease severity and frequency	Not determined	Vallejos-Torres et al. (2023)
An unidentified native AM fungal species	*Cajanus cajan* cv. GT101	Fusarium wilt/*F. udum*	Disease symptoms were decreased	Upregulated the plant growth, defense genes, and β-1,3-glucanase, chitinase, and PAL activities in root	Jha and Mohamed (2024)

AMF, arbuscular mycorrhizal fungi; *PAL*, phenylalanine ammonia lyase.

(Bhupenchandra et al., 2024). This "priming" means that defense-related genes are not constitutively upregulated to high, energy costly levels but rather are poised for rapid and robust activation only when a pathogen attack occurs. This leads to accelerated and amplified defense responses: Mycorrhizal plants typically exhibit quicker and more intense bursts of reactive oxygen species, faster lignification, and earlier accumulation of pathogenesis-related (PR) proteins upon pathogen infection compared to nonmycorrhizal plants (Hao et al., 2023).

29.3.5.2 Direct interference with pathogen growth

AMF can directly inhibit the growth or proliferation of certain pathogens through various means: competition for resources: AMFs compete with root pathogens for colonization sites and essential nutrients, thereby reducing pathogen establishment and spread (Mishra et al., 2024). Physical barrier: The dense network of AMF hyphae around and within the root cortex can act as a physical barrier, making it more difficult for pathogens to penetrate and colonize root tissues. Antimicrobial metabolites: Although less commonly cited than MIR, some AMF strains may produce compounds with mild antimicrobial properties that could directly impact pathogen viability (Chaudhary et al., 2025).

29.3.6 Enrichment of microbial diversity

AMF field application contributes to an enriched soil microbial diversity through several interconnected mechanisms, primarily centered on the rhizosphere, the narrow zone of soil influenced by plant roots and their exudates.

29.3.6.1 Altered root exudation and rhizosphere priming

AMF colonization significantly modifies the quantity and quality of root exudates (e.g., sugars, organic acids, and amino acids). Although AMF themselves receive carbon from the plant, their presence can trigger shifts in the plant's metabolic pathways, leading to the release of novel or altered compounds into the rhizosphere (Fan et al., 2022). These exudates act as carbon and energy sources, selectively attracting and stimulating the growth of specific microbial taxa that can utilize these compounds. This "rhizosphere priming" effect by AMF creates a more favorable and diverse nutritional environment for a wider range of beneficial soil microorganisms, including plant growth–promoting rhizobacteria and other beneficial fungi.

29.4 Cons associated with arbuscular mycorrhizal fungi applications

Cons are the different aspects that can alter the activity of AMF application in plants and are also involved in the decrease of its influence on plants. Important ones are described here:

1. Higher costs of AMF inoculants
2. Variable effectiveness in terms of soil type, fungal, and plant species
3. Incompatibility with intensive tillage or chemical inputs
4. Restricting broader adoption
5. Accessibility for farmers to quality inoculants
6. Knowledge gaps regarding AMF–crop interactions

29.4.1 The high cost of arbuscular mycorrhizal fungi inoculants: A barrier to widespread adoption in sustainable agriculture

AMFs are crucial components of healthy soil ecosystems, forming symbiotic relationships with the roots of over 80% of terrestrial plants, including many important crops (Jeffries & Barea, 2012). This symbiosis provides numerous benefits to plants, such as enhanced nutrient and water uptake, increased tolerance to various abiotic and biotic stresses, and improved soil structure (Fall et al., 2022). Consequently, AMF inoculants are increasingly recognized as a vital tool for sustainable agriculture, offering an ecofriendly alternative to conventional chemical fertilizers and pesticides. However, despite their demonstrable advantages, the widespread adoption of AMF inoculants is significantly hampered by their relatively high production costs, posing a major challenge for both manufacturers and end-users.

Furthermore the quality control and standardization of AMF inoculants contribute to their overall expense. Ensuring a consistent concentration of viable propagules (spores, colonized root fragments, or extraradical mycelium) and maintaining the genetic purity and efficacy of specific AMF strains are critical for product performance (Chaudhary et al., 2025). This often involves rigorous testing, which adds to the manufacturing overhead. The diversity of AMF species and their varying effectiveness across different soil types and plant species also present a challenge, requiring producers to offer a range of products or to invest in identifying broad-spectrum strains, further impacting research and development costs (Wahab et al., 2023). The challenges associated with storage and transportation also play a role in the final price of AMF inoculants. Maintaining the viability of live fungal propagules during storage and transit requires specific conditions, such as controlled temperature and humidity, which can be expensive to maintain, particularly for sensitive formulations. Although advancements in carrier substrates and formulation techniques are being made to enhance shelf life and stability, these innovations often come with their own associated development and material costs (Chaudhary et al., 2025).

From an economic perspective, the high initial investment for AMF inoculants can be a deterrent for farmers, especially in regions with limited financial resources. Although studies demonstrate significant benefits such as increased crop yield, reduced reliance on chemical fertilizers, and improved plant resilience (Delavaux et al., 2025), quantifying these long-term economic returns can be challenging and may not always be immediately apparent in short-term seasonal analyses. The lack of widespread awareness among farmers about the comprehensive benefits and cost-effectiveness of AMF in the long run also hinders market penetration.

29.4.2 Variable effectiveness of arbuscular mycorrhizal fungi: A multifaceted

29.4.2.1 Interaction of soil, fungal, and plant species

AMFs are ubiquitous soil microorganisms forming mutualistic symbioses with the roots of most terrestrial plants, playing a vital role in nutrient cycling, plant health, and ecosystem stability (Aggarwal et al., 2011). Although their general benefits, such as enhanced nutrient uptake (particularly phosphorus), improved water relations, and increased stress tolerance, are well-established, the practical application of AMF inoculants in agriculture often encounters significant variability in their effectiveness. This variability is largely attributed to complex interactions among the specific soil characteristics, the particular AMF species or strains, and the host–plant species, necessitating a nuanced understanding for successful implementation (Khaliq et al., 2022).

29.4.2.2 Influence of soil type on arbuscular mycorrhizal fungi effectiveness

Soil characteristics exert a profound influence on AMF community composition, colonization rates, and ultimately, their symbiotic efficacy. Key soil factors include the following sections.

29.4.2.2.1 Nutrient availability

High levels of readily available nutrients, especially phosphorus (P), can significantly reduce AMF colonization and the plant's reliance on the fungal symbiosis. When P is abundant, plants may prioritize direct nutrient uptake, diminishing the need for the fungal intermediary. Conversely, in nutrient-poor soils, AMF demonstrate their greatest benefits, acting as critical conduits for nutrient acquisition (Bhantana et al., 2021).

Here we are trying to clarify the various challenges that are responsible for obstructing the large-scale acceptance of AMF inoculants and trying to suggest different practices to endorse their application for sustainable agriculture. Higher application of phosphorus fertilizers in crops has negatively affected the mycorrhizal symbiosis by taking up higher concentrations of P without providing carbohydrates to AMF (William et al., 2013). Higher concentration of soil P in plants reduced the formation and exudation of strigolactones, which ultimately alter the formation of AMF symbiosis in the roots of plants. Higher accumulation of root P fertilization significantly reduced the supply of root soluble carbohydrates, which act as AMF food and, in response, limited the formation of appressoria and fresh infection (García-Caparros et al., 2021; López-Ráez et al., 2017). The increase in phosphorus accumulation also retards the arbuscule formation in AMF colonies (Ghorui et al., 2024; Kobae et al., 2016). Furthermore the existence of functional plant and fungal Pi: H+ symporter (PT) genes is accountable for P accumulation in extraradical mycelia, which is delayed in higher P concentrations in soils, leading to a shift toward AMF's parasitic nature (Getman-Pickering et al., 2021). Higher potassium (K) accumulation in plant roots also decreases root exudation, which in response also limits the carbohydrate supply for AMF colonization (Getman-Pickering et al., 2021).

29.4.2.2.2 Soil pH

Soil pH is a major determinant shaping AMF species composition. Different AMF species exhibit distinct pH optima, and extreme pH values can negatively impact fungal growth, spore germination, and the establishment of symbiosis (Jamiołkowska et al., 2018).

29.4.2.2.3 Soil texture and structure

Soil texture influences pore space, aeration, and water movement, which, in turn, affect fungal hyphal growth and spread. Well-structured soils with good aeration generally support healthier AMF populations. Practices like intensive tillage can disrupt soil structure and negatively impact AMF diversity and abundance.

29.4.2.2.4 Fungal species/strain specificity

Although AMFs are considered generalist symbionts, associating with a wide range of plant species, significant functional diversity exists among different AMF species and even within isolates of the same species. This functional variation can manifest in the following:

Colonization rates and patterns: Different AMF species may vary in their ability to colonize host roots, their root colonization intensity, and the development of extraradical hyphae.

Nutrient acquisition efficiency: The efficiency with which different AMF species acquire and transfer specific nutrients (e.g., phosphorus, nitrogen, water) to the host plant can vary substantially (Tanwir et al., 2023). For example, some strains might be more effective at mobilizing tightly bound soil phosphorus, whereas others might excel in water uptake under drought conditions.

Stress amelioration: The capacity of AMF to confer tolerance to abiotic stresses (e.g., drought, salinity, and heavy metals) or biotic stresses (e.g., pathogens) is often strain-specific.

29.4.2.2.5 Host specificity (functional and phylogenetic)

Although not strictly host-specific, AMF can exhibit preferences or varying degrees of effectiveness with different plant phylogenetic groups or even individual genotypes within a species. This "functional specificity" means that a particular AMF strain might perform optimally with one crop species but show limited benefits with another. Studies have shown that AMF communities beneath different host plants can significantly differ.

29.4.2.2.6 Plant species/genotype responsiveness

The plant partner also plays a crucial role in determining the outcome of the mycorrhizal symbiosis. Factors related to the host plant include the following:

Mycorrhizal responsiveness: Plants vary in their inherent mycorrhizal responsiveness, which is the degree to which they benefit from AMF colonization. Some plant species are highly dependent on AMF for nutrient uptake, especially in low-fertility soils, whereas others may be less responsive or even exhibit a parasitic relationship under certain conditions (Cardoso et al., 2017).

Root architecture and physiology: The root architecture, exudation patterns, and inherent nutrient acquisition strategies of different plant species can influence AMF colonization and effectiveness. For instance, plant phosphorus acquisition strategies can influence AMF diversity and community composition in roots (Chaudhary et al., 2025).

Phenology and life cycle: The timing of AMF colonization and the plant's growth stage can influence the benefits derived. Annual crops may exhibit different mycorrhizal associations compared to perennials.

Genotypic variation: Even within a single plant species, different genotypes can exhibit varying degrees of AMF colonization and responsiveness (Berger, 2021). This highlights the potential for breeding programs to select for plants that are more compatible with beneficial AMF.

29.4.3 Incompatibility with intensive tillage or chemical inputs

The utilization of agriculture-based chemicals, containing both systematic and nonsystematic fungicides, has variable effects on AMF. Mostly spray of fungicides adversely influences spore formation and their germination, hyphal growth, colonization, hyphal phosphatase activity, and sporulation (Channabasava et al., 2015). Fungicides' effects on AMF are always dose-dependent and also retard the AMF properties even in lesser quantities (Rodriguez-Morelos et al., 2021). Pencycuron, flutolanil, azoxystrobin, and fenpropimorph are different types of chemical fungicides that can affect AMF contrarily depending on their quantities (Ghorui et al., 2024).

Another factor that physically affects the root colonization of AMF through limiting hyphal networks and spore formation is tillage practices (Bito et al., 2012). Soil instabilities through tillage not only hamper AM-mediated water and nutrient uptake but also disorder extraradical myceliums and glomalin-based formation of soil aggregates (Sale et al., 2015; Zinati et al., 2025). On the other hand a decline in soil tillage leads to an upsurge in soil organic carbon accumulation together with an increase in soil aggregation, reducing erosion and moisture retention, which eventually encourages AMF colonization (Higo et al., 2020; Wang et al., 2018). Various other agricultural practices, such as crop rotation with non-AMF crops and long fallow periods, can significantly increase AMF propagule density, communities, and activities (Higo et al., 2020). Waterlogging situations in paddy rice cultivation can also hamper AMF growth and activities. Generally, these agrarian practices also play an active role in determining the connections between AMF and plants, influencing soil health and nutrient accumulation (Ghorui et al., 2024). All these tillage aspects are responsible for hampering the implementation of AMF inoculants for agricultural sustainability.

29.4.4 Restricting broader adoption

Despite the well-documented ecological and agricultural benefits of AMF, including enhanced nutrient uptake, improved plant resilience to stress, and reduced reliance on chemical inputs, their widespread adoption in mainstream agriculture remains limited (Ghorui et al., 2025). Numerous interconnected factors restrict the broader adsorption and integration of AMF inoculants into conventional farming practices, posing significant challenges for both producers and end-users. Addressing these barriers is crucial for unlocking the full potential of AMF as a cornerstone of sustainable agriculture (Marro et al., 2022).

One of the most significant limiting factors is the inconsistent and unpredictable field performance of AMF inoculants (Basiru & Hijri, 2022). Although greenhouse and controlled laboratory experiments consistently demonstrate positive effects, these often fail to translate consistently to diverse field conditions.

29.4.5 Accessibility of farmers to quality arbuscular mycorrhizal fungi inoculants

The increasing recognition of AMF as vital bioenhancers for sustainable agriculture underscores the importance of their widespread adoption (Conti et al., 2025). However, a significant bottleneck limiting the integration of AMF into conventional farming practices, particularly for smallholder and resource-limited farmers, is the inadequate accessibility to high-quality inoculants. This multifaceted challenge encompasses issues of availability, affordability, awareness, and reliability, collectively hindering the broader realization of AMF's ecological and economic benefits (Liu & Chen, 2024).

One of the primary barriers to accessibility is the limited availability of high-quality AMF inoculants in many regions, particularly in developing countries (Ghorui et al., 2024). Unlike synthetic fertilizers or pesticides that are widely distributed through established supply chains, the market for AMF inoculants is relatively nascent and unevenly developed.

29.4.6 Knowledge gaps regarding arbuscular mycorrhizal fungi–crop interactions

AMFs are acknowledged as critical components of sustainable agricultural systems, fostering enhanced nutrient acquisition, improved stress tolerance, and overall plant health in a wide array of crops (Wahab et al., 2023). Despite extensive research demonstrating these benefits, a comprehensive understanding of the intricate interactions between specific AMF species/strains, diverse crop genotypes, and varying environmental conditions remains incomplete (Wille, 2020). This knowledge gap is a significant impediment to the consistent and predictable application of AMF inoculants in agricultural settings, thereby restricting their broader adoption (Kadowaki et al., 2018).

One of the most pressing knowledge gaps lies in elucidating the precise mechanisms of functional specificity between AMF and crop genotypes. Although AMF are broadly considered generalist symbionts, studies increasingly reveal that the benefits conferred are not uniform across all plant–fungal pairings (Claassens, 2024). This "functional specificity" means that a particular AMF strain might significantly enhance the growth of one crop variety, while offering minimal or no benefit, or even exhibiting a parasitic tendency.

Furthermore, there is a substantial knowledge gap regarding the complex interactions within the soil microbiome and how AMF mediate these interactions to benefit crops. Agricultural soils harbor diverse microbial communities, and AMFs do not operate in isolation.

29.5 Conclusion

Despite different cons the pros of AMF inoculation in crop plants were more important. We accomplish that AMF, along with other soil microbiota, can meaningfully increase the soil fertility. AMF application also copes with different types of plant stresses through enhancing nutrient uptake, carbon sequestration, and other metabolic changes in plants and the rhizosphere. Although AMF shows significant potential for regions facing resource limitations and environmental stress, its success relies on customized strategies that consider local soil conditions, crop rotations, and socio-economic factors.

Future adoption necessitates the development of cost-effective inoculant production, farmer education, and policies that support agro-ecological practices. Further, this chapter also highlights the importance of AMF in sustainable agriculture while stressing the necessity for practical, context-specific strategies to optimize its advantages.

References

Aggarwal, A., Kadian, N., Tanwar, A., Yadav, A., & Gupta, K. K. (2011). Role of arbuscular mycorrhizal fungi (AMF) in global sustainable development. *Journal of Applied and Natural Science, 3*(2), 340.

Ahmad, M., Summuna, B., Gupta, S., & Sheikh, P. A. (2020). Mycorrhizal association: An important tool for the management of root diseases. *International Journal of Current Microbiology and Applied Sciences, 9*(9), 1958–1966.

Al-Arjani, A. B., Hashem, A., Abd Allah, E. F., & Ahmad, P. (2020). Arbuscular mycorrhizal fungi: A critical role in enhancing plant drought tolerance. *Journal of Plant Interactions, 15*(1), 127–138.

Allen, M. F., Swenson, W., Querejeta, J. I., Egerton-Warburton, L. M., & Treseder, K. K. (2003). Ecology of mycorrhizae: A conceptual framework for complex interactions among plants and fungi. *Annual Review of Phytopathology, 41*, 271–303.

Auge, R. M. (2004). Arbuscular mycorrhizae and soil water in the context of plant hydraulic and nonhydraulic signaling. *Plant and Soil, 264*(1–2), 1–11.

Basiru, S., & Hijri, M. (2022). The potential applications of commercial arbuscular mycorrhizal fungal inoculants and their ecological consequences. *Microorganisms, 10*(10), 1897.

Begum, N., Ahanger, M. A., Raza, S., Nazir, A., & Ashraf, M. (2019). Arbuscular mycorrhizal fungi and plant interactions under abiotic stress: A review. *Plant Physiology and Biochemistry, 142*, 190–202.

Begum, N., Qin, C., Ahanger, M. A., Raza, S., Khan, M. I., Ashraf, M., & Zhang, L. (2019). Role of arbuscular mycorrhizal fungi in plant growth regulation: Implications in abiotic stress tolerance. *Frontiers in Plant Science, 10*, 1068.

Berger, F., & Gutjahr, C. (2021). Factors affecting plant responsiveness to arbuscular mycorrhiza. *Current Opinion in Plant Biology, 59*, 101994.

Bhantana, P., Rana, M. S., Sun, X. C., Moussa, M. G., Saleem, M. H., Syaifudin, M., & Hu, C. X. (2021). Arbuscular mycorrhizal fungi and its major role in plant growth, zinc nutrition, phosphorous regulation and phytoremediation. *Symbiosis, 84*(1), 19–37.

Bhupenchandra, I., Chongtham, S. K., Devi, A. G., Dutta, P., Sahoo, M. R., Mohanty, S., & Swapnil, P. (2024). Unlocking the potential of arbuscular mycorrhizal fungi: Exploring role in plant growth promotion, nutrient uptake mechanisms, biotic stress alleviation, and sustaining agricultural production systems. *Journal of Plant Growth Regulation, 44*(12), 6802–6840.

Birhane, E., Assegie, A., & Yitaferu, B. (2023). Water stress changes on AMF colonization, stomatal conductance and photosynthesis of *Dalbergia sissoo* seedlings grown in entisol soil under nursery condition. *Journal of Agricultural and Environmental Sciences, 8*(1), 47–58.

Bødker, L., Kjøller, R., & Rosendahl, S. (1998). Effect of phosphate and the arbuscular mycorrhizal fungus *Glomus intraradices* on disease severity of root rot of peas (*Pisum sativum*) caused by *Aphanomyces euteiches*. *Mycorrhiza, 8*(3), 169–174.

Brito, I., Goss, M. J., & De Carvalho, M. (2012). Effect of tillage and crop on arbuscular mycorrhiza colonization of winter wheat and triticale under Mediterranean conditions. *Soil Use and Management, 28*(2), 202–208.

Bruisson, S., Maillot, P., Schellenbaum, P., Walter, B., Gindro, K., & Deglène-Benbrahim, L. (2016). Arbuscular mycorrhizal symbiosis stimulates key genes of the phenylpropanoid biosynthesis and stilbenoid production in grapevine leaves in response to downy mildew and grey mould infection. *Phytochemistry, 131*, 92–99.

Bücking, H., & Kafle, A. (2015). Role of arbuscular mycorrhizal fungi in the nitrogen uptake of plants: Current knowledge and research gaps. *Agronomy, 5*(4), 587–612.

Cardoso, E. J., Nogueira, M. A., & Zangaro, W. (2017). Importance of mycorrhizae in tropical soils. In *Diversity and benefits of microorganisms from the tropics* (pp. 245–267). Springer International Publishing.

Channabasava, A., Lakshman, H. C., & Jorquera, M. A. (2015). Effect of fungicides on association of arbuscular mycorrhiza fungus *Rhizophagus fasciculatus* and growth of Proso millet (*Panicum miliaceum* L.). *Journal of Soil Science and Plant Nutrition, 15*(1), 35–45.

Chaudhary, A., Poudyal, S., & Kaundal, A. (2025). Role of arbuscular mycorrhizal fungi in maintaining sustainable agroecosystems. *Applied Microbiology, 5*(1), 6.

Chen, S., Zhao, H., Zou, C., Li, Y., Chen, Y., Wang, Z., Jiang, Y., Liu, A., Zhao, P., Wang, M., & Ahammed, G. J. (2017). Combined inoculation with multiple arbuscular mycorrhizal fungi improves growth, nutrient uptake and photosynthesis in cucumber seedlings. *Frontiers in Microbiology, 8*, 2516.

Claassens, A. (2024). *Mycorrhizal gateways: Advancing interpretations of fungal symbiosis in sugarcane agriculture*. Southern Cross University.

Conti, G., Urcelay, C., Gundel, P. E., & Piñeiro, G. (2025). The potential of arbuscular mycorrhizal fungi to improve soil organic carbon in agricultural ecosystems: A meta-analytical approach. *Functional Ecology, 39*(4), 1016–1030.

Delavaux, C. S., Burrill, H., Menning, R., Duell, E. B., Bryant, R. L., Lubin, T., & Bever, J. D. (2025). Origin matters: Mycorrhizal growth response and induced resistance to pathogens depend on mycorrhizal and pathogen source. *New Phytologist, 248*(3), 1516–1526.

Diagne, N., Ngom, M., Djighaly, P. I., Fall, D., Hocher, V., & Svistoonoff, S. (2020). Roles of arbuscular mycorrhizal fungi on plant growth and performance: Importance in biotic and abiotic stressed regulation. *Diversity, 12*(10), 370.

Dugassa, G. D., Von Alten, H., & Schönbeck, F. (1996). Effects of arbuscular mycorrhiza (AM) on health of *Linum usitatissimum* L. infected by fungal pathogens. *Plant and Soil, 185*(2), 173–182.

Evelin, H., Devi, T. S., Gupta, S., & Kapoor, R. (2019). Mitigation of salinity stress in plants by arbuscular mycorrhizal symbiosis: Current understanding and new challenges. *Frontiers in Plant Science, 10*, 470.

Evelin, H., Kapoor, R., & Giri, B. (2019). Mycorrhizal symbiosis alleviates salt stress: Recent advances and challenges. *Journal of Plant Growth Regulation, 38*(4), 1157–1167.

Fall, A. F., Nakabonge, G., Ssekandi, J., Founoune-Mboup, H., Apori, S. O., Ndiaye, A., & Ngom, K. (2022). Roles of arbuscular mycorrhizal fungi on soil fertility: Contribution in the improvement of physical, chemical, and biological properties of the soil. *Frontiers in Fungal Biology, 3*, 723892.

Fan, Y., Zhu, Y., Li, G., Li, Z., & Li, R. (2022). Arbuscular mycorrhizal fungi alter root exudation to shape rhizosphere bacterial communities: A review. *Frontiers in Microbiology, 13*, 965432.

Frank, A. B. (1885). Uber di auf werzelsymbiose beruhende Ernahrung gewisser Baume durch unterirdischeplize. *Berichte der Deutschen Botanischen Gesellschaft, 3*, 128–145.

Garcia-Caparros, P., Lao, M. T., Preciado-Rangel, P., & Sanchez, E. (2021). Phosphorus and carbohydrate metabolism in green bean plants subjected to increasing phosphorus concentration in the nutrient solution. *Agronomy, 11*(2), 245.

Getman-Pickering, Z. L., Stack, G. M., & Thaler, J. S. (2021). Fertilizer quantity and type alter mycorrhizae-conferred growth and resistance to herbivores. *Journal of Applied Ecology, 58*(5), 931–940.

Ghorui, M., Chowdhury, S., Balu, P., & Burla, S. (2024). Arbuscular mycorrhizal inoculants and its regulatory landscape. *Heliyon, 10*(9), e30359.

Ghorui, M., Chowdhury, S., & Burla, S. (2025). Recent advances in the commercial formulation of arbuscular mycorrhizal inoculants. *Frontiers in Industrial Microbiology, 3*, 1553472.

Goh, T. B., Thami, K., Ariffin, A., & Hamid, K. S. (2021). Role of arbuscular mycorrhizal fungi in heavy metal tolerance and phytoremediation of contaminated soils. *Frontiers in Plant Science, 12*, 689045.

Gutjahr, C., & Parniske, M. (2013). Cell and developmental biology of arbuscular mycorrhiza symbiosis. *Annual Review of Cell and Developmental Biology, 29*(1), 593–617.

Hao, Z., Yang, W., & Wei, M. (2023). Arbuscular mycorrhizal fungi promote crop disease resistance: Mechanisms and applications. *Journal of Agricultural Science, 15*(2), 1–15.

Hashem, A., Akhter, A., Alqarawi, A. A., Singh, G., Almutairi, K. F., & Abd_Allah, E. F. (2021). Mycorrhizal fungi induced activation of tomato defense system mitigates *Fusarium* wilt stress. *Saudi Journal of Biological Sciences, 28*(10), 5442–5450.

Higo, M., Tatewaki, Y., Iida, K., Yokota, K., & Isobe, K. (2020). Amplicon sequencing analysis of arbuscular mycorrhizal fungal communities colonizing maize roots in different cover cropping and tillage systems. *Scientific Reports, 10*(1), 6039.

Hodge, A., & Fitter, A. H. (2010). Mycorrhizal networks. *Annual Review of Plant Biology, 61*, 305–333.

Hu, M., Hu, H., Li, S., Wang, K., & Qin, S. (2024). Effect of AMF inoculation on reducing excessive fertilizer use. *Agriculture, 14*(3), 543.

Jamiołkowska, A., Księżniak, A., Gałązka, A., Hetman, B., Kopacki, M., & Skwaryło-Bednarz, B. (2018). Impact of abiotic factors on development of the community of arbuscular mycorrhizal fungi in the soil: A review. *International Agrophysics, 32*(1), 133–140.

Jeffries, P., & Barea, J. M. (2012). 4 Arbuscular Mycorrhiza: A key component of sustainable plant–soil ecosystems. In B. Hock (Ed.), *Fungal associations* (pp. 51–75). Springer.

Jha, Y., & Mohamed, H. I. (2024). Endophytic *Pseudomonas pseudoalcaligenes* and arbuscular mycorrhizal fungi mediated anti-autophagy and induction of catalase and antioxidant enzymes in pigeon pea against fungal pathogen. *Journal of Plant Pathology, 106*(1), 225–240.

Kadowaki, K., Yamamoto, S., Sato, H., Tanabe, A. S., Hidaka, A., & Toju, H. (2018). Mycorrhizal fungi mediate the direction and strength of plant–soil feedbacks differently between arbuscular mycorrhizal and ectomycorrhizal communities. *Communications Biology, 1*(1), 196.

Karagiannidis, N., Bletsos, F., & Stavropoulos, N. (2002). Effect of Verticillium wilt (*Verticillium dahliae* Kleb.) and mycorrhiza (*Glomus mosseae*) on root colonization, growth and nutrient uptake in tomato and eggplant seedlings. *Scientia Horticulturae, 94*(1–2), 145–156.

Khaliq, A., Perveen, S., Alamer, K. H., Ul Haq, M. Z., Rafique, Z., Alsudays, I. M., Althobaiti, A. T., Saleh, M. A., Hussain, S., & Attia, H. (2022). Arbuscular mycorrhizal fungi symbiosis to enhance plant–soil interaction. *Sustainability, 14*(13), 7840.

Kobae, Y., Ohmori, Y., Saito, C., Yano, K., Ohtomo, R., & Fujiwara, T. (2016). Phosphate treatment strongly inhibits new arbuscule development but not the maintenance of arbuscule in mycorrhizal rice roots. *Plant Physiology, 171*(1), 566–579.

Li, T., Lin, G., Zhang, X., Chen, Y., Zhang, S., & Chen, B. (2014). Relative importance of an arbuscular mycorrhizal fungus (*Rhizophagus intraradices*) and root hairs in plant drought tolerance. *Mycorrhiza, 24*(8), 595–602.

Liu, R., & Chen, Y. (2024). Mycorrhizal symbiosis: Evolution, opportunities, challenges, and prospects. In M. Parihar, A. Rakshit, A. Adholeya, & Y. Chen (Eds.), *Arbuscular mycorrhizal fungi in sustainable agriculture: Inoculum production and application* (pp. 1–35). Springer.

López-Ráez, J. A., Shirasu, K., & Foo, E. (2017). Strigolactones in plant interactions with beneficial and detrimental organisms: The Yin and Yang. *Trends in plant Science, 22*(6), 527–537.

Ma, J., Ma, Y., Wei, Z., Wu, J., Sun, C., Yang, J., & Huang, J. (2021). Effects of arbuscular mycorrhizal fungi symbiosis on microbial diversity and enzyme activities in the rhizosphere soil of *Artemisia annua*. *Soil Science Society of America Journal, 85*(3), 703–716.

Malicka, M., Magurno, F., Posta, K., Chmura, D., & Piotrowska-Seget, Z. (2021). Differences in the effects of single and mixed species of AMF on the growth and oxidative stress defense in *Lolium perenne* exposed to hydrocarbons. *Ecotoxicology and Environmental Safety, 217*, 112252.

Marro, N., Grilli, G., Soteras, F., Caccia, M., Longo, S., Cofré, N., & Urcelay, C. (2022). The effects of arbuscular mycorrhizal fungal species and taxonomic groups on stressed and unstressed plants: A global meta-analysis. *New Phytologist, 235*(1), 320–332.

Martin, F. M., Uroz, S., & Barker, D. G. (2017). Ancestral alliances: Plant mutualistic symbioses with fungi and bacteria. *Science, 356*(6340), Eaad4501.

Mishra, A. K., Bhowmick, M. K., Das, R., Pidikiti, P., Maurya, P. K., Sharma, S., & Singh, S. (2024). Potentials and prospects of AMF for soil carbon sequestration and nutrient cycling in rice-based cropping system. In *Arbuscular mycorrhizal fungi in sustainable agriculture: Nutrient and crop management* (pp. 113–129). Springer Nature Singapore.

Mohseni, F., Rouhi, H., & Moradi, S. (2024). The critical role of arbuscular mycorrhizal fungi to improve drought tolerance and nitrogen use efficiency in crops. *Frontiers in Plant Science, 15*, 919166.

Montesinos-Navarro, A., Valiente-Banuet, A., & Verdú, M. (2019). Mycorrhizal symbiosis increases the benefits of plant facilitative interactions. *Ecography, 42*(3), 447–455.

Nazir, A., Sarfraz, W., Allah, D., Khalid, N., Farid, M., Shafiq, M., Firdaus-E-Bareen, Rizvi, Z. F., & Naeem, N. (2023). Synergistic impact of two autochthonous saprobic fungi (*A. niger* and *T. pseudokoningii*) on the growth, ionic contents, and metals uptake in *Brassica juncea* L. and *Vigna radiata* L. under tannery solid waste contaminated soil. *International Journal of Phytoremediation, 25*(11), 1488–1500. doi:10.1080/15226514.2023.2166457.

Ortas, I., Al-Arjani, A. B., & Abd Allah, E. F. (2021). The role of arbuscular mycorrhizal fungi in improving nutrient uptake and plant growth: A review. *Journal of King Saud University—Science, 33*(4), 101569.

Pu, C., Ge, Y., Yang, G., Zheng, H., Guan, W., Chao, Z., & Huang, L. (2022). Arbuscular mycorrhizal fungi enhance disease resistance of *Salvia miltiorrhiza* to *Fusarium* wilt. *Frontiers in Plant Science, 13*, 975558.

Püschel, D., Bitterlich, M., Rydlová, J., & Jansa, J. (2021). Drought accentuates the role of mycorrhiza in phosphorus uptake. *Soil Biology and Biochemistry, 157*, 108243.

Redecker, D., Morton, J. B., & Bruns, T. D. (2000). Ancestral lineages of arbuscular mycorrhizal fungi (Glomales). *Molecular Phylogenetics and Evolution, 14*, 276–284.

Rillig, M. C., & Mummey, D. L. (2006). Mycorrhizae and soil structure. *New Phytologist, 171*(1), 41–53.

Rodriguez-Morelos, V. H., Calonne-Salmon, M., Bremhorst, V., Garcés-Ruiz, M., & Declerck, S. (2021). Fungicides with contrasting mode of action differentially affect hyphal healing mechanism in *Gigaspora* sp. and *Rhizophagus irregularis*. *Frontiers in Plant Science, 12*, 642094.

Roy, J., Saha, B., & Roy, S. (2021). Arbuscular mycorrhizal fungi: Potential bio-inoculants for enhancing crop tolerance against salinity stress. *Environmental Technology & Innovation, 24*, 102045.

Ruiz-Lozano, J. M. (2003). Arbuscular mycorrhizal symbiosis and drought tolerance: An update. *Mycorrhiza, 13*(6), 329–335.

Saeed, H. S., Munir, M., Samra, S., Farooq, A., Mahmood, H., Ahmad, I., Akram, M. I., Iqbal, R., Elsadek, M. F., Almutairi, S. M., Ditta, A., & Daraz, U. (2025). Screening of biosurfactants producing fungi from infected *Citrus sinensis* (L.). *Polish Journal of Environmental Studies, 34*(5), 5261–5270. https://www.pjoes.com/pdf-190588-115894?filename=Screening%20of.pdf. doi:10.15244/pjoes/190588.

Säle, V., Aguilera, P., Laczko, E., Mäder, P., Berner, A., Zihlmann, U., & Oehl, F. (2015). Impact of conservation tillage and organic farming on the diversity of arbuscular mycorrhizal fungi. *Soil Biology and Biochemistry, 84*, 38–52.

Sanmartín, N., Pastor, V., Pastor-Fernández, J., Flors, V., Pozo, M. J., & Sánchez-Bel, P. (2020). Role and mechanisms of callose priming in mycorrhiza-induced resistance. *Journal of Experimental Botany, 71*(9), 2769–2781.

Seutra, M. P., Zaid, A., & Alqarawi, A. A. (2021). Arbuscular mycorrhizal fungi: Enhancing plant stress tolerance and production under global climate change. *Applied Soil Ecology, 166*, 104033.

Shah, S., Nawaz, T., & Fahad, S. (2025). Arbuscular mycorrhizal fungi role in plant beneficial elements uptake and nutrient acquisition. *Rhizosphere, 35*, 101141.

Singh, A., Pandey, A. K., Dodmani, B. A., Joshi, R., Wongamthing, R., Mishra, S., & Karanwal, R. (2024). Arbuscular mycorrhizal fungi efficiency on plant growth and nutrient acquisition: A comprehensive review. *Microbiology Research Journal International, 34*(8), 13–22.

Smith, S. E., & Read, D. J. (2008). *Mycorrhizal symbiosis* (3rd ed.). Academic Press.

Smith, S. E., & Smith, F. A. (2011). Roles of arbuscular mycorrhizas in plant nutrition and growth: New paradigms from cellular to ecosystem scales. *Annual Review of Plant Biology, 62*(1), 227–250.

Song, Y. Y., Cao, M., Xie, L. J., Liang, X. T., Zeng, R. S., Su, Y. J., & Luo, S. M. (2011). Induction of DIMBOA accumulation and systemic defense responses as a mechanism of enhanced resistance of mycorrhizal corn (*Zea mays* L.) to sheath blight. *Mycorrhiza, 21*(8), 721–731.

Sosa-Hernández, M. A., Leifheit, E. F., Ingraffia, R., & Rillig, M. C. (2019). Subsoil arbuscular mycorrhizal fungi for sustainability and climate-smart agriculture: A solution right under our feet? *Frontiers in Microbiology, 10*, 744.

Tanwir, K., Abbas, S., Shahid, M., Chaudhary, H. J., & Javed, M. T. (2023). Mycorrhizal symbiosis and nutrients uptake in plants. In *Plant ionomics: Sensing, signaling, and regulation* (pp. 73–95). John Wiley & Sons.

Teste, F. P., Jones, M. D., & Dickie, I. A. (2020). Dual-mycorrhizal plants: Their ecology and relevance. *New Phytologist, 225*, 1835–1851.

Thind, S., Chaudhary, M. S., Ditta, A., Hussain, I., Parveen, A., Ullah, N., Mahmood, Q., Al-Ashkar, I., & El-Sabagh, A. (2022). Impact of mycorrhizal fungi from different rhizospheric soils on fungal colonization, growth, and chlorophyll contents of *Cenchrus ciliaris*. *Agronomy, 12*(11), 2644. doi:10.3390/agronomy12112644.

Trotta, A., Varese, G. C., Gnavi, E., Fusconi, A., Sampo, S., & Berta, G. (1996). Interactions between the soilborne root pathogen *Phytophthora nicotianae* var. parasitica and the arbuscular mycorrhizal fungus *Glomus mosseae* in tomato plants. *Plant and Soil, 185*(2), 199–209.

Vallejos-Torres, G., Gaona-Jimenez, N., Ordoñez-Sánchez, L., Lozano, C., Arevalo, A. A., Lozano, A., & Marín, C. (2023). Effects of arbuscular mycorrhizal fungi, poultry manure compost, and cadmium on plant growth and nutrient absorption of *Oryza sativa*. *Chilean Journal of Agricultural Research, 83*(6), 656–665.

van Der Heijden, M. G., Martin, F. M., Selosse, M. A., & Sanders, I. R. (2015). Mycorrhizal ecology and evolution: The past, the present, and the future. *New Phytologist, 205*(4), 1406–1423.

Wang, Y., Wang, M., Li, Y., Wu, A., & Huang, J. (2018). Effects of arbuscular mycorrhizal fungi on growth and nitrogen uptake of *Chrysanthemum morifolium* under salt stress. *PLoS One, 13*(4), e0196408.

Wille, L. (2020). *The root rot complex of pea-Screening for resistance and quantification of microbial key players in the rhizosphere*. ETH Zurich.

Willmann, M., Gerlach, N., Buer, B., Polatajko, A., Nagy, R., Koebke, E., & Bucher, M. (2013). Mycorrhizal phosphate uptake pathway in maize: Vital for growth and cob development on nutrient poor agricultural and greenhouse soils. *Frontiers in Plant Science, 4*, 533.

Yang, G. B., Ding, P., & Yang, B. (2023). Understanding the mechanisms of nutrient transfer between arbuscular mycorrhizal fungi (AMF) and host plants. *International Journal of Scientific Research in Agricultural Sciences, 7*(1), 1–10.

Yizhu, L., Imtiaz, M., Ditta, A., Rizwan, M. S., Ashraf, M., Mehmood, S., Aziz, O., Mubeen, F., Ali, M., Elahi, N. N., Ijaz, R., Lele, S., Shuang, C., & Tu, S. (2020). Response of growth, antioxidant enzymes and root exudates production towards as stress in *Pteris vittata* and in *Astragalus sinicus* colonized by arbuscular mycorrhizal fungi. *Environmental Science and Pollution Research, 27*(2), 2340–2352. https://link.springer.com/journal/11356. doi:10.1007/s11356-019-06785-5.

Zhang, R., Mu, Y., Li, X., Li, S., Sang, P., Wang, X., & Xu, N. (2020). Response of the arbuscular mycorrhizal fungi diversity and community in maize and soybean rhizosphere soil and roots to intercropping systems with different nitrogen application rates. *Science of the Total Environment, 740*, 139810.

Zinati, G., Carrara, J. E., Das, S., Caetani, R., Kalra, A., Carr, E. A., & Heller, W. P. (2025). Impact of tillage practices and arbuscular mycorrhizal fungi inoculation on organic sweet corn yield and nutritional quality. *Soil and Tillage Research, 251*, 106545.

Chapter 30

Enhancing metal tolerance in plants through arbuscular mycorrhizal fungi

Sana Akram[a], **Muhammad Sohail Akram**[a], **Muhammad Tariq Javed**[a] and **Hassan Javed Chaudhary**[b]

[a] *Department of Botany, Faculty of Life Sciences, Government College University Faisalabad, Faisalabad, Pakistan,*

[b] *Department of Plant Sciences, Faculty of Biological Sciences, Quaid-i-Azam University, Islamabad, Pakistan*

30.1 Introduction

Heavy metals (HMs) effluence is a global environmental problem. The HMs have metallic properties and are hazardous, even at low concentrations (Rehman & Singh, 2020). Some HMs are nonessential and toxic for plant growth, such as mercury (Hg), arsenic (As), cadmium (Cd), and lead (Pb), which can easily move from soil to plants because of their mobile properties (Gautam et al., 2025). However, other HMs (such as Cu, Zn, Se, Mo, and Ni) are less toxic and cause health risk and environmental contamination (Rehman & Singh, 2020). Usually the concentration of HMs is evaluated by the presence of HMs in bedrock and their testimony from the environment (Zhang et al., 2022). The slight content of HMs naturally exists in the earth's crust and thus in soils through processes such as weathering of parent rock, volcanic activity, erosion, and atmospheric deposition. However, the current environmental concern largely stems from anthropogenic activities that dramatically elevate these HM levels in the environment. Industrialization, mining, smelting, improper waste disposal, agricultural practices (e.g., use of contaminated fertilizers and pesticides), and urban runoff are major contributors to HM pollution in the environment. Therefore levels of HMs exceed the natural concentrations, which caused detrimental effects on soil fertility, plants, and human health, and it has been recognized as one of the most critical threats to the agro-ecosystems' sustainability (Gautam et al., 2025; Kumar et al., 2022; Padhiary & Kumar, 2024). HMs can accumulate in grains and integrate into the food chain, potentially causing various health issues in humans once they enter the diet (Angon et al., 2024; Chang et al., 2018; Vanisree et al., 2022). After entering HMs into the environment, they remain there for several years (Verma et al., 2021). The HMs negatively influence plant growth and productivity through several mechanisms, which include (1) interruption with the activity of essential nutrients, (2) changes in plant morphophysiological attributes, (3) harmful effects on soil microbial growth, particularly symbiotic microbes like rhizobia, and (4) alterations of soil structures (Goyal et al., 2020). Though most HMs are naturally found on the soil surface, their harmful effects are intensified when these metals are introduced into the environment in excessive amounts (Sharafi & Salehi, 2025). Augmented meditations of HMs in plant tissues result in toxicity, disrupt the integrity of cell membranes, hinder photosynthesis and mineral uptake, and negatively influence various morphophysiological and biological processes, ultimately impeding growth rates. Besides, HMs promote the production of reactive oxygen species (ROS) that harm cellular organelles and DNA and hinder ATP production, leading to lipid peroxidation (Madkour, 2019). The degree of damage imposed by HMs is affected by the prevailing stress level, exposure duration, HM concentration, and their bioavailability within plant tissues (Kosakivska et al., 2021). Plants have natural defense mechanisms, including antioxidant enzymes that assist in alleviating HM stress; however, at higher concentrations of HMs, these enzymes lose their effectiveness (Chandrakar et al., 2020; Rasheed et al., 2021).

Various physical and chemical practices have been exploited to lower HMs' levels in soil, including soil washing, acidification, excavation, landfilling, and vitrification. These methods are exclusive and may compromise soil quality, making it unsuitable for rhizospheric microbial communities. Contrariwise, biological strategies that include fungi and microbes have emerged as viable and economical alternatives, as they are simpler to implement and contribute to the

Advancing Sustainable Agriculture Through Arbuscular Mycorrhizal Fungi. DOI: https://doi.org/10.1016/B978-0-443-45130-0.00013-8

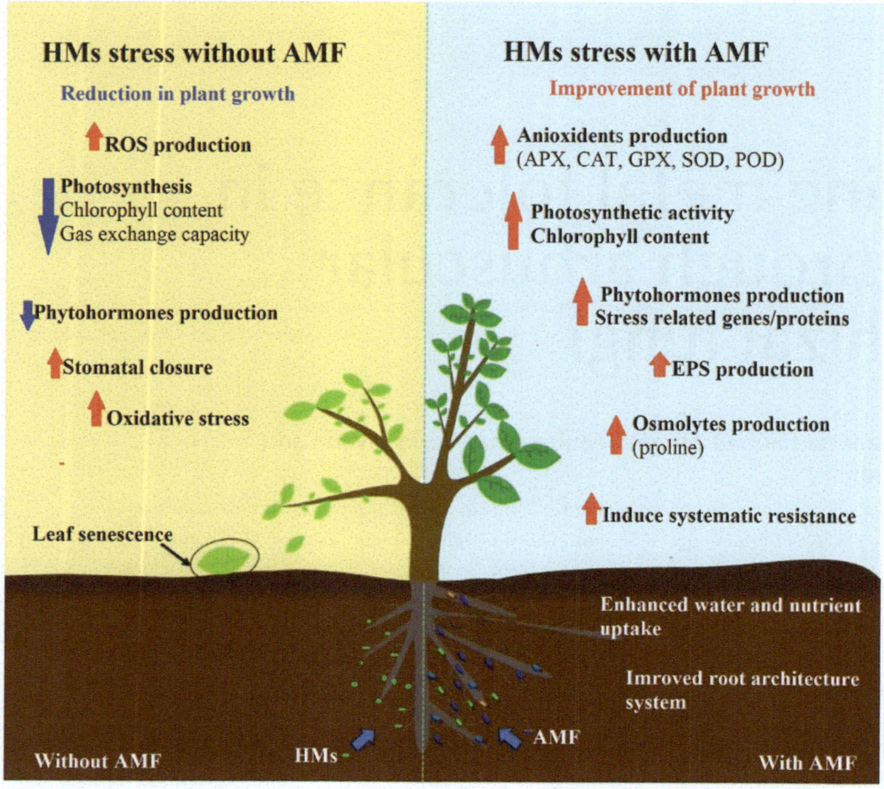

FIGURE 30.1 Role of arbuscular mycorrhizal fungi in plants under heavy metal stress. Graphical representation to highlight the role of arbuscular mycorrhizal fungi in plants under heavy metal stress. *AMF*, Arbuscular mycorrhizal fungi; *CAT*, catalase; *EPS*, extracellular polymeric substances; *HMs*, heavy metals; *POD*, peroxidase; *ROS*, reactive oxygen species; *SOD*, superoxide dismutase.

preservation of the natural ecosystem (Medfu Tarekegn et al., 2020; Sommer et al., 2025). The arbuscular mycorrhizal fungi (AMFs) are symbiotic entities that form beneficial associations with plants' roots, providing a natural and sustainable method to improve growth and yield of plants when exposed to abiotic stresses (Budak et al., 2025; Sommer et al., 2025).

AMFs boost plant growth by establishing a symbiotic association with plant roots, which significantly enhance the effective root surface area through their extensive hyphal networks (Budak et al., 2025; Sommer et al., 2025), thereby improving the uptake of essential nutrients such as phosphorus (P), nitrogen (N), and micronutrients as well as water (Ahmed et al., 2025). This acquisition of nutrients and water improves plant physiological functions, including biomass accumulation and photosynthesis, particularly under abiotic stresses such as drought, HMs, and salinity (Wahab et al., 2023). The AMF symbiotic association induces transcriptional reprogramming in plants by triggering genes related to nutrient uptake, hormonal signaling, and stress management (Zhao et al., 2024). Besides, AMFs affect plant hormonal regulation and activate antioxidant defense mechanisms, leading to improved stress tolerance and metabolic efficiency. These combined effects intensify plant development, elevate productivity, and promote sustainable growth in nutrient-poor or challenging soil environments, making AMF an important component for ecofriendly agricultural and ecosystem management (Gao et al., 2023; Gong et al., 2023). Therefore the main aim of this chapter is to study the function of AMF in sustaining an ecofriendly environment within ecosystems that are contaminated with HMs, while also highlighting the various mechanisms through which AMFs mitigate the harmful impacts of HMs on plants (Fig. 30.1).

30.2 Indirect impact of arbuscular mycorrhizal fungi on metal tolerance

The association among host plants and AMF is enhanced by rapidly growing hyphae that are capable of thriving in challenging conditions characterized by metal toxicity and environmental stress (Riaz et al., 2021; Wahab et al., 2023). Throughout the symbiotic process, AMF can trigger various adaptations in the host plants to better cope with these toxic environments. The mycorrhizal fungus promotes tolerance in the plants indirectly by enhancing the uptake of water and

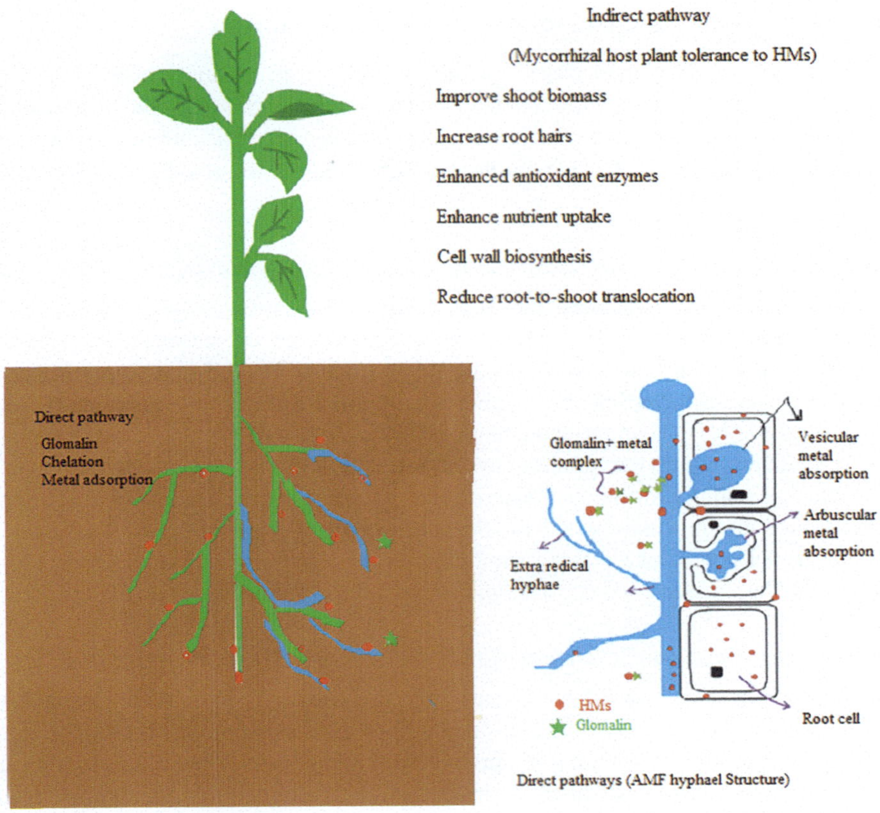

FIGURE 30.2 Arbuscular mycorrhizal fungi–mediated direct and indirect phytotoxicity of heavy metals. *AMF*, Arbuscular mycorrhizal fungi; *HMs*, heavy metals.

vital mineral nutrients, boosting shoot biomass, altering root architecture, and reducing oxidative stress caused by HMs (Fig. 30.2).

30.2.1 Acquisition of nutrients mediated by arbuscular mycorrhizal fungi

AMFs, classified in the phylum Glomeromycota, establish an extensive and ancient mutualistic symbiosis with almost 80% of terrestrial plants (Sportes et al., 2021). In this symbiotic association the fungi colonize plant roots intracellularly, creating distinctive structures such as arbuscules (for nutrient exchange) and vesicles (for storage) while spreading a widespread extraradical mycelial network into the surrounding soil (Wang et al., 2022). This symbiotic association is well-known for improving host plant nutrition, predominantly phosphorus (P) uptake, but its role in extenuating several abiotic stresses, including HM toxicity gained significant attention (Nazir et al., 2023; Singhal et al., 2023; Thind et al., 2022; Yizhu et al., 2020).

The most fundamental mechanism conferred by AMF is the expansion of an extensive extraradical hyphal network that can travel far greater soil volume than plant roots (Ahammed et al., 2023). These fine hyphae can penetrate soil pores distant from roots and efficiently scavenge for immobile nutrients such as phosphorus (P), zinc (Zn), and copper (Cu), which often become less accessible under HMs stress because of precipitation or competitive inhibition (Bhupenchandra et al., 2024).

A phosphate transporter exists in mycorrhizal fungi, facilitating inorganic phosphate movement from the soil to plants (Johri et al., 2015). In the extraradial mycelium of *Glomus mosseae, G. versiforme*, and *G. intraradices*, multiple genes that code for transporter proteins, including GiPT, GmosPT, and GvPT, have been exposed, which indicate a role in phosphate uptake (Bhantana et al., 2021; Dhalaria et al., 2020). The colonization by mycorrhizae significantly improved the phosphorus (P) nutrition of plants, limited As uptake, and retained a greater amount of As in the roots by upregulating the expression of the AM-induced P transporter gene *MsPT4*. Elevated levels of As in the soil led to a downregulation of MsPT4 expression (Li et al., 2018). Recently, Chang et al. (2018) investigated the potential role of the AMF against metal stress of cadmium and lanthanum on maize in a pot experiment. The study proved that the fungus treatment eliminates the hazardous effects of metals by improving the nutrient uptake and biomass accumulation.

30.2.2 Arbuscular mycorrhizal fungi–mediated defense mechanisms

Under HMs' toxicity, plants produce ROS in excessive quantities; higher concentrations of ROS in cells result in oxidative stress results, which can damage cellular organelles and may lead to cell death (Sharma et al., 2019). Plants counteract these increased ROS levels through a variety of antioxidant mechanisms (both enzymatic and nonenzymatic), involving a range of enzymes that function synergistically to manage the increased ROS levels triggered by HMs in plants (Mansoor et al., 2023).

30.2.2.1 Activation of enzymatic antioxidants by arbuscular mycorrhizal fungi

The research exposed that these AMFs enhanced the function of antioxidative enzymes such as superoxide dismutase (SOD), peroxidase (POD), and catalase (CAT), raising the concentrations of antioxidative substances (with proline, carotenoids, and α-tocopherol) in plants exposed to arsenic stress when compared to their nonmycorrhizal counterparts. Moreover, the mycorrhizal plants displayed increased ratios of glutathione:glutathione disulfide (GSH:GSSG) and ascorbate:dehydroascorbate (AsA:DHA), which contributed to the neutralization of H_2O_2 and the reduction of lipid peroxidation (Cheema & Garg, 2024). In addition, previous studies have shown that AMF can reduce ROS levels in mycorrhizal plants (*Cajanus cajan*) under Cd and salinity stress by bolstering the antioxidative defense system (Garg & Chandel, 2015). Recently a study demonstrated that inoculation of AM *Rhizophagus intraradices* in Cd-stressed *Glycine max* diminished oxidative stress and reduced the uptake of Cd. Like another study demonstrated that inoculation of AMF stimulates the enzymatic defense system of Cu-stressed *Dysosma versipellis* and reduces oxidative stress (Adeyemi et al., 2021).

30.2.2.2 Activation of nonenzymatic antioxidants by arbuscular mycorrhizal fungi

Nonenzymatic antioxidants comprise compounds such as ascorbic acid (AsA), GSH, carotenoids, and phenolic compounds. GSH reductase plays a vital role in scavenging ROS and safeguarding against membrane damage triggered by free radicals (Rao et al., 2025). GSH is an essential intracellular nonprotein thiol that preserves the redox equilibrium within cells and scavenges excess ROS (Georgiou-Siafis et al., 2023). Jiang et al. (2016) reported that *Funneliformis mosseae* application in Cd-stressed *Solanum nigrum* enhanced the activity of GR in leaves of the host plant, which reduces the oxidative stress induced by Cd. As a consequence the antioxidative system activated by mycorrhizae can assist in preserving reactive ROS homeostasis in plants under HMs, thereby promoting better HM accretion in mycorrhizal plants.

30.2.3 Influence on root morphology

Mycorrhizal fungi facilitate morphological changes in their host plants, permitting them to improve endurance in harsh environments, such as those caused by metal toxicity (Branco et al., 2022). Fungi that form associations with host plants enhance the development of strong root systems and promote root hair growth and surface area, which in turn increases the ability to absorb nutrients under abiotic stress (Wahab et al., 2023). Previous studies indicated that changes in root morphological structure aid in tolerance to HMs. Mycorrhizal plants usually display lower concentrations of metals in the rhizosphere compared to those lacking mycorrhizal associations. The reduction in HMs' concentrations leads to a decline in their bioavailability within host plants. This decline in metal levels in the mycorrhizosphere is linked to the improvement of the rhizosphere through high soil pH (Janeeshma & Puthur, 2020). As a consequence, AMFs play a critical role in altering the soil environment and microbial communities by releasing specific substances (Jamiołkowska et al., 2018).

A current investigation directed by Singh et al. (2019) showed *Zea mays* inoculation with AMF frequently enhanced soil enzyme activity, including β-glucosidase, dehydrogenase, and both acid and alkaline phosphatase, when compared to noninoculated actions. Fungi that form close associations with the roots of plants establish a physical barrier that limits the uptake of HMs and proliferation of pathogens and enhances rhizosphere microbial activity. Besides, AMF reduces the population of pathogenic fungi by competing for both space and nutrients.

30.2.4 Heavy metal sequestration by arbuscular mycorrhizal fungi in vacuoles

The sequestration/compartmentalization of toxic substances in cellular vacuoles is an effective strategy used by AMF to contribute to the host plant in mitigating the detrimental effects of HMs (Nedjimi, 2021). Numerous studies indicated that AMF can actively participate in this sequestration, either directly within their hyphal vacuoles or indirectly by affecting the host plant's vacuolar transport systems. This procedure often includes binding of HMs to chelating agents such as phytochelatins (PCs) and metallothioneins (MTs), which are then transported into the vacuole, thereby reducing their bioavailability and toxic effects on both the fungal and plant cells (Dhalaria et al., 2020). This compartmentalization reduces the risk of phytotoxicity in the aboveground parts of the plant and enables the host plant to survive in a metal-polluted soil. Overall the AMF-assisted sequestration enhances the phytoremediation capacity of host plants.

30.2.5 Arbuscular mycorrhizal fungi mediates in binding of metal ions to their hyphal cell wall

The fungal cell wall, composed of biopolymers like chitin, acts as a physical barrier and a binding site for metal ions, effectively sequestering them before they can enter the plant root. This mycelial network acts as a physical barrier that can sequester metal ions, making them less bioavailable to plant roots (Chakraborty et al., 2020). The AMF executes a vigorous role in mitigating the phytotoxicity of HMs by sequestering these contaminants within their hyphal structures. The fungal cell wall, composed of functional groups, such as free hydroxyl groups, imidazole carboxyl groups, and amino groups that present numerous binding sites for HMs, effectively immobilizes HMs and hinders their mobility to the host plants (Kumar & Dwivedi, 2021). These interactions involve mechanisms such as biosorption, complexation, and precipitation. Moreover, AMFs are recognized for sequestering metals within their spores (Cornejo et al., 2013). Multiple studies have demonstrated that enduring metal concentrations prompt fungal cell walls to act as a primary defense mechanism by obstructing the ingress of HMs (Emamverdian et al., 2015). AMF also improved cell wall biosynthesis and reduced the metal uptake by enhancing the surface area and thickness of host plant roots. However, at elevated concentrations of metal ions, they infiltrate the cell wall, accumulate in the fungal cell membrane, or be translocated into the fungal cytoplasm, consequently reducing the concentration of metal ions within the plants (Zhang et al., 2019). This mediation by AMF signifies a vigorous natural attenuation process, contributing significantly to phytostabilization efforts in metal-contaminated environments.

30.2.6 Arbuscular mycorrhizal fungi–mediated heavy metal chelation

An important phenomenon used by AMF to remove HMs from the polluted environment is the chelation of metal ions by a ligand that shows strong affinity against metal ions within the cytosol. Chelating compounds are formed when the concentration of HMs exceeds the threshold level within the cell, which reduces the solubility and reactivity of metal ions by making them less available in the cytoplasm (Mishra et al., 2024; Nazir et al., 2023; Saeed et al., 2025; Thind et al., 2022; Yizhu et al., 2020). A highly significant molecular mechanism by which AMFs enhance HM tolerance is through the stimulation of PC and MT synthesis in both the host plant and the mycorrhizal structures (Dhalaria et al., 2020). PCs and MTs are small, cysteine-rich peptides and proteins, respectively, that play a central role in metal detoxification by binding to HMs, forming stable complexes that reduce their toxicity. Mostly the HMs enter the cytoplasm via Zrt–Irt protein transporters, stress tolerance mechanism linked with thiol peptide (phytochelation) production, which forms the complex with metal ions, whereas AMFs boost the production of phytochelation (Begum et al., 2019). Like another study demonstrated that AMFs improve the biosynthesis of PCs, subsequently, a greater number of PC-metal complexes are produced and transported to the vacuole, consequently reducing their detrimental impact of Cd stress (Jiang et al., 2016). In mycorrhizal plants and AMFs, MTs activate in response to HMs. Several studies showed that in response to HMs, numerous MT genes are activated that diminish hazardous effects in the host plant (Liu et al., 2009).

30.2.7 Arbuscular mycorrhizal fungi mediate the protection of the PSII reaction center and also improve gas exchange capacity

A substantial concentration of metals in the soil is transferred to various parts of the plant, inducing the enzyme activity associated with the biosynthesis of chlorophyll, which subsequently impacts the photosynthetic development, predominantly the PS II reaction center. The AMF application improves the photosynthetic content, with a distinguished upsurge in chlorophyll *a*, which plays a critical role in host photosynthesis. Several studies indicated plants inoculated with AMF exhibited high chlorophyll synthesis, leading to improved photosynthesis (Hashem et al., 2016). Similarly, earlier research has also demonstrated that AMF facilitated biosynthesis of chlorophyll by promoting the de novo protein synthesis, which directly influences the absorption of vital elements such as magnesium (Mg) and iron (Fe), both of which are critical components of the chlorophyll molecule (Ef et al., 2015). The synergetic relationship of HMs and AMF enhances the photosynthetic capacity and productivity, as well as gas exchange aptitudes and improved transcriptional fluxes.

30.3 Arbuscular mycorrhizal fungi directly mediated the remediation of heavy metals

Fungal hyphae enhance the absorption of HMs via various physiological mechanisms, which include (1) synthesis of glomalin protein, (2) AMF can actively sequester metals within their own fungal structures, (3) chelation of HMs within the cellular membrane, and (4) activation of specific genes that initiate the phytostabilization process.

30.3.1 Arbuscular mycorrhizal fungi secrete various extracellular biomolecules (glomalin) into the rhizosphere

A glycoprotein produced by AMF hyphae and spores, called glomalin, after association with plants that reduces the bioavailability of HMs (Son et al., 2024). Glomalin, produced by AMF via their spore walls and mycelium, is a glycoprotein that shows resistance to both water and temperature (Agnihotri et al., 2022) as well, and it plays an essential role in environmental conservation. The GRSP has recalcitrance and long-term stability in soil, making a significant contribution to the storage of carbon and nitrogen over prolonged periods (Wang et al., 2018). Moreover, various searches demonstrated that glomalin has substantial potential for chelating/speciation of HMs and diminishing their availability in the rhizosphere of the host plant (Wang et al., 2017).

Another study indicates 1 g of glomalin from specific AMF strains can bind Pb (1.12 mg), Cd (0.08 mg), and Cu (4.3 mg) from contaminated soil, which underscores the abilities of glomalin to interact with various metals and their concentrations (Gonzalez-Chavez et al., 2004). However, the AMF also enhances glomalin levels as a defense mechanism in Cd-polluted soil, preventing its movement to sorghum roots, which results in less Cd accumulation in plants (Babadi et al., 2019). Furthermore, it has been recognized that MF inoculation, particularly *Glomus aggregatum*, reduces HMs (including Cd, Pb, Zn, and Cu) in plant shoots. This might be attributed to glomalin secretion by the fungus that chelates the harmful metals (Nafady & Elgharably, 2018).

Lately, Hu et al. (2019) proposed that *R. intraradices, G. versiforme*, and *F. mosseae* formed a symbiotic relationship that boosted GRSP in the soil and reduced Pb and Cd content in maize plants, which was attributed to a rise in soil pH. The GRSP's adhesive nature acts like glue; it links small soil particles into aggregates of varying size, thus providing stability by reforming strategy (Son et al., 2024). However, prolonged exposure to HMs considerably impacted soil AMF characteristics, potentially hindering the GRSP development. Nevertheless, there is a lack of comprehensive information on how AMF-produced GRSP impacts soil quality, particle size, and aggregate structure in areas polluted with HMs. Therefore a deeper knowledge of GRSP's link to soil traits and HM pollution is vital (Zhang et al., 2024). This understanding will help in using plants to fix polluted sites. It also aids in rebuilding natural systems. Overall, AMF may act as an effective approach to facilitate the recovery of soils polluted by metals.

30.3.2 Accretion of heavy metals by fungal structures

30.3.2.1 *Accretion of heavy metals on the surface of hyphae*

The AMF sequester significant amounts of HMs through their fungal structures, which in turn diminish the absorption of HMs by host plants, consequences in reduced noxiousness and improving host plant tolerance. The HMs accumulation process by AMF occurs because of hyphal extension both intracellularly and extracellularly (Wu et al., 2019). The accumulation of HMs onto fungal structures takes place through adsorption, subsequently leading to the transformation of HMs from one form to another (Dwivedi, 2023). Consequently, AMFs diminish the translocation of HMs into the cytoplasm of host plants by accumulating HMs on their hyphal surfaces (Wu et al., 2019). These fungi are recognized as potent biological agents that remediate HMs within ecosystems.

30.3.2.2 *Arbuscular mycorrhizal fungi–mediated production of extracellular polymeric substances*

The AMFs produce EPS (extracellular polymeric substances), a complex mixture of high molecular weight polymers that appear as small particles on their surfaces. The EPS encompasses functional groups such as carboxyl, phosphoric, hydroxyl and amine groups, which reveal a robust tendency for metal adsorption through mechanisms such as surface settling, chelation, and swapping ions (Wang et al., 2022). The EPS production by AMF is often upregulated under environmental stresses such as drought, HM contamination, and nutrient deficit. Previous studies revealed that AMFs improve the EPS secretion in response to Cr(VI) stress, helping the reduction and immobilization of toxic metals through the formation of EPS-metal complexes on the fungal surface. This defensive mechanism is further supported by the occurrence of functional groups such as phosphate and carboxyl within the EPS, which bind and sequester metal ions (Bhatia et al., 2024).

Moreover, the chemical composition of AMF contains EPS (low-molecular-weight compounds) such as carbohydrates like glucose and oligosaccharides, as well as amorphous polymers, which establish valuable relations with soil bacteria and support the development of the microbial community. The AMF mycelial exudates (rich in EPS) promote the growth and activity of useful soil microbes by altering the surrounding microbial community. These relations lead to enhanced nutrient cycling, improved plant nutrient acquisition (particularly phosphorus), and boosted plant growth (Wang et al., 2022).

30.3.2.3 Compartmentalization of metals through their intraradical structures

The compartmentalization of metals within the intraradical structures of AMF implies vigorous physiological mechanisms for metal detoxification and homeostasis in plants under HMs stress. AMF symbiotically colonize plant roots, establishing specific intraradical structures such as vesicles, arbuscules, and hyphae, which assist not only in the exchange of nutrients but also in immobilizing and sequestering potentially noxious metal ions (Chen et al., 2018). Most of HMs, including Cd, Pb, and Cr, are often compartmentalized within fungal cell walls, vacuoles, or bound to EPS in the fungal structures, thereby reducing their bioavailability and translocation to aerial plant tissues. This intracellular metal sequestration is facilitated by metal-binding ligands (such as MT, PC, and GSH), which chelate metal ions within the AMF cells (Jaroszuk-Ściseł et al., 2025). The strategic sequestering and localization within the intraradical compartments efficiently alleviate metal-induced oxidative stress and cytotoxicity, enhancing plant tolerance to contaminated soils. As well the dynamic modulation of metal compartmentalization within these fungal structures underscores the role of AMF as biological buffers in metal-polluted environments, indicating their potential efficacy in phytoremediation and sustainable agriculture under HMs (Mishra et al., 2024; Zhao et al., 2024).

30.3.2.4 Activation of specific genes that initiate the phytostabilization process

Beyond specific metal transporters and chelators, AMFs also influence the broader regulation of stress–responsive genes in plants under HMs stress. This involves a complex interplay of gene expression that enhances the plant's ability to cope with the adverse conditions. AMF inoculation can lead to the augmented expression of genes linked to cell wall modification and strengthening. Such as some studies indicated that AMF inoculation in Pb-contaminated media augmented gene expression such as *MtPrx05* and *MtPrx10*, which are intricate in cell wall polysaccharide cross-linking, enhancing Pb fixation in the root cell wall of host plants, making them more resistant to Pb toxicity by binding the metal. Besides, various transcriptional genes within cells of fungus-activated genes *GmarMT* "cDNA-encoding MT-like functional polypeptide" in *Gigaspora margarita* spores were reported which exhibited tolerance against HMs (Dhalaria et al., 2022). Numerous genes contribute to the maintenance of cellular homeostasis under HMs stress, including *GmarMT*1, *GintABC*1, *GrosMT*1, and *RintZnT*1 (Azcon et al., 2013). The gene *GmarMT1* encodes MTs in *G. margarita* (BEG34), which help in maintaining the redox potential of the fungus and reducing the oxidative stress, whereas *GintABC1* enables the detoxification of HMs (González-Guerrero et al., 2007).

For instance, genes encoding natural resistance-associated macrophage proteins (NRAMPs) play a role in Cd uptake, and certain ABC-type xenobiotic transporters (e.g., PtABCC1) have been implicated in Hg sequestration (Tao & Lu, 2022). Besides, TFs are master regulators that bind to specific DNA sequences in gene promoters, thereby controlling gene expression. In response to HM stress, various TF families, including WRKY, NAC, MYB, and AP2/ERF, are activated to orchestrate the transcriptional reprogramming necessary for phytostabilization (Sun et al., 2025). These TFs can upregulate genes involved in metal transport, chelation, and stress tolerance. Overall the activation of specific genes is the cornerstone of the phytostabilization process, enabling plants to adapt to and mitigate the detrimental effects of environmental contaminants. A comprehensive understanding of these genetic and molecular underpinnings is crucial for developing sustainable and effective phytoremediation strategies.

30.3.2.5 Sequestration of metals by arbuscular mycorrhizal fungi's own fungal structure

AMF can actively sequester metals within their fungal structures, including deposition of HMs in the vacuoles of AMF and specialized fungal structures such as vesicles and hyphae, which help detoxify potentially harmful metal ions by regulating their concentrations in the cytosol (Mishra et al., 2024). Zhang et al. (2021) reported that AMF (*Rhizophagus irregularis*) has been shown to upsurge Pb uptake in both colonized and noncolonized roots, and to transfer and sequester Pb within the colonized root segments. This indicated the low mobility of the sequestered metal. By sequestering metals, AMF prevents them from reaching sensitive plant tissues, particularly the shoots.

30.4 Summary

AMFs significantly enhance metal tolerance in plants through a multifaceted symbiotic relationship that improves plant resilience in HM-contaminated soils. AMFs colonize plant roots, forming extensive hyphal networks that extend soil exploration and increase nutrient and water uptake, which is critical under metal stress conditions. Likewise the fungal hyphae secrete glomalin (a glycoprotein that binds HMs), limiting their movement toward plant tissues. In addition, AMFs regulate metal uptake by plants, often confining translocation to aerial parts, which aids plants in avoiding metal toxicity. At the molecular level, AMFs induce metal tolerance by modulating gene expression linked with metal transport, stress response, and antioxidative defense. For instance, AMF also enhances the plant antioxidant system, increasing activities of enzymes such as SOD, CAT, and APX, which alleviates oxidative stress caused by HMs, improves cellular redox balance, and stabilizes plant metabolism under metal-polluted soil. In addition, AMFs influence plant root morphology and physiology, promoting greater root surface area and altered root exudation, which improves nutrient acquisition and metal immobilization and reduces metal toxicity. Overall, AMFs boost metal tolerance in plants by immobilizing metals, modulating metal transport and detoxifying genes, and improving root function and growth. This symbiotic association enables plants to thrive in HM-polluted soils and can be used as a biological tool for the management of metal-polluted environments and phytoremediation applications.

References

Adeyemi, N. O., Atayese, M. O., Sakariyawo, O. S., Azeez, J. O., Olubode, A., Ridwan, M., & Adeoye, S. (2021). A Growth and phosphorus uptake of soybean (*Glycine max* L.) in response to arbuscular mycorrhizal fungus *Rhizophagus intraradices* inoculation in heavy metal-contaminated soils. *Soil and Sediment Contamination: An International Journal, 30*(6), 698–713.

Agnihotri, R., Sharma, M. P., Prakash, A., Ramesh, A., Bhattacharjya, S., Patra, A. K., & Kuzyakov, Y. (2022). Glycoproteins of arbuscular mycorrhiza for soil carbon sequestration: Review of mechanisms and controls. *Science of the Total Environment, 806*, 150571.

Ahammed, G. J., Shamsy, R., Liu, A., & Chen, S. (2023). Arbuscular mycorrhizal fungi-induced tolerance to chromium stress in plants. *Environmental Pollution, 327*, 121597.

Ahmed, N., Li, J., Li, Y., Deng, L., Deng, L., Chachar, M., & Tu, P. (2025). Symbiotic synergy: How arbuscular mycorrhizal fungi enhance nutrient uptake, stress tolerance, and soil health through molecular mechanisms and hormonal regulation. *IMA Fungus, 16*, e144989.

Angon, P. B., Islam, M. S., Das, A., Anjum, N., Poudel, A., & Suchi, S. A. (2024). Sources, effects and present perspectives of heavy metals contamination: Soil, plants and human food chain. *Heliyon, 10*(7), e28357.

Azcon, R., Medina, A., Aroca, R., & Ruiz-Lozano, J. M. (2013). Abiotic stress remediation by the arbuscular mycorrhizal symbiosis and rhizosphere bacteria/yeast interactions. *Molecular Microbial Ecology of the Rhizosphere, 1*, 991–1002.

Babadi, M., Zalaghi, R., & Taghavi, M. (2019). A non-toxic polymer enhances sorghum-mycorrhiza symbiosis for bioremediation of Cd. *Mycorrhiza, 29*, 375–387.

Begum, N., Qin, C., Ahanger, M. A., Raza, S., Khan, M. I., Ashraf, M., & Zhang, L. (2019). Role of arbuscular mycorrhizal fungi in plant growth regulation: Implications in abiotic stress tolerance. *Frontiers in Plant Science, 10*, 1068.

Bhantana, P., Rana, M. S., Sun, X. C., Moussa, M. G., Saleem, M. H., Syaifudin, M., & Hu, C. X. (2021). Arbuscular mycorrhizal fungi and its major role in plant growth, zinc nutrition, phosphorous regulation and phytoremediation. *Symbiosis, 84*, 19–37.

Bhatia, S. K., Panesar, P. S., & Mehariya, S. (Eds.). (2024). *Microbial exopolysaccharides: Production and applications*. CRC Press.

Bhupenchandra, I., Chongtham, S. K., Devi, A. G., Dutta, P., Sahoo, M. R., Mohanty, S., & Swapnil, P. (2024). Unlocking the potential of arbuscular mycorrhizal fungi: Exploring role in plant growth promotion, nutrient uptake mechanisms, biotic stress alleviation, and sustaining agricultural production systems. *Journal of Plant Growth Regulation, 44*(12), 6802–6840.

Branco, S., Schauster, A., Liao, H. L., & Ruytinx, J. (2022). Mechanisms of stress tolerance and their effects on the ecology and evolution of mycorrhizal fungi. *New Phytologist, 235*(6), 2158–2175.

Budak, B., Onay, T. T., Goren, A. Y., & Khalvati, A. (2025). Experimental study for recovery of heavy metals from contaminated soil using arbuscular mycorrhizal fungi. *International Journal of Global Warming, 35*(1), 1–15.

Chakraborty, M., Hasanuzzaman, M., Rahman, M., Khan, M. A. R., Bhowmik, P., Mahmud, N. U., & Islam, T. (2020). Mechanism of plant growth promotion and disease suppression by chitosan biopolymer. *Agriculture, 10*(12), 624.

Chandrakar, V., Yadu, B., Xalxo, R., Kumar, M., & Keshavkant, S. (2020). Mechanisms of plant adaptation and tolerance to metal/metalloid toxicity. In *In Hasanuzzaman, M., (Ed.), Plant* ecophysiology and adaptation under climate change: Mechanisms and perspectives II: Mechanisms of adaptation and stress amelioration (pp. 107–135). Springer Nature.

Chang, Q., Diao, F. W., Wang, Q. F., Pan, L., Dang, Z. H., & Guo, W. (2018). Effects of arbuscular mycorrhizal symbiosis on growth, nutrient and metal uptake by maize seedlings (*Zea mays* L.) grown in soils spiked with lanthanum and cadmium. *Environmental Pollution, 241*, 607–615.

Cheema, A., & Garg, N. (2024). Arbuscular mycorrhizae reduced arsenic induced oxidative stress by coordinating nutrient uptake and proline-glutathione levels in *Cicer arietinum* L. (chickpea). *Ecotoxicology, 33*(2), 205–225.

Cornejo, P., Pérez-Tienda, J., Meier, S., Valderas, A., Borie, F., Azcón-Aguilar, C., & Ferrol, N. (2013). Copper compartmentalization in spores as a survival strategy of arbuscular mycorrhizal fungi in cu-polluted environments. *Soil Biology and Biochemistry, 57*, 925–928.

Dhalaria, R., Kumar, D., Kumar, H., Nepovimova, E., Kuča, K., Torequl Islam, M., & Verma, R. (2020). Arbuscular mycorrhizal fungi as potential agents in ameliorating heavy metal stress in plants. *Agronomy, 10*(6), 815.

Dwivedi, S. K. (2023). Fungi mediated detoxification of heavy metals: Insights on mechanisms, influencing factors and recent developments. *Journal of Water Process Engineering, 53*, 103800.

Ef, A. A., Abeer, H., Alqarawi, A. A., & Hend, A. A. (2015). Alleviation of adverse impact of cadmium stress in sunflower (*Helianthus annuus* L.) by arbuscular mycorrhizal fungi. *Pakistan Journal of Botany, 47*(2), 785–795.

Emamverdian, A., Ding, Y., Mokhberdoran, F., & Xie, Y. (2015). Heavy metal stress and some mechanisms of plant defense response. *The Scientific World Journal, 2015*(1), 756120.

Gao, Y., Huang, S., Wang, Y., Lin, H., Pan, Z., Zhang, S., & Chen, Y. (2023). Analysis of the molecular and biochemical mechanisms involved in the symbiotic relationship between arbuscular mycorrhiza fungi and *Manihot esculenta* Crantz. *Frontiers in Plant Science, 14*, 1130924.

Garg, N., & Chandel, S. (2015). Role of arbuscular mycorrhiza in arresting reactive oxygen species (ROS) and strengthening antioxidant defense in *Cajanus cajan* (L.) Millsp. nodules under salinity (NaCl) and cadmium (Cd) stress. *Plant Growth Regulation, 75*, 521–534.

Gautam, A., Banerjee, B., Singh, A., & Bithel, N. (2025). Heavy metals (HMs) uptake by plants through bioremediation to restore damaged ecosystem and environment: A sustainable approach. In *In Kumar, P. , Singh, A., Rajput, V. D., Minkina, T. & Singh A. K., (Eds.), Smart technologies in sustainable agriculture* (pp. 123–145). Apple Academic Press.

Georgiou-Siafis, S. K., & Tsiftsoglou, A. S. (2023). The key role of GSH in keeping the redox balance in mammalian cells: Mechanisms and significance of GSH in detoxification via formation of conjugates. *Antioxidants, 12*(11), 1953.

Gong, M., Bai, N., Wang, P., Su, J., Chang, Q., & Zhang, Q. (2023). Co-inoculation with arbuscular mycorrhizal fungi and dark septate endophytes under drought stress: Synergistic or competitive effects on maize growth, photosynthesis, root hydraulic properties and aquaporins? *Plants, 12*, 2596. doi:10.3390/plants12142596.

Gonzalez-Chavez, M. C., Carrillo-Gonzalez, R., Wright, S. F., & Nichols, K. A. (2004). The role of glomalin, a protein produced by arbuscular mycorrhizal fungi, in sequestering potentially toxic elements. *Environmental Pollution, 130*(3), 317–323.

González-Guerrero, M., Cano, C., Azcón-Aguilar, C., & Ferrol, N. (2007). GintMT1 encodes a functional metallothionein in *Glomus intraradices* that responds to oxidative stress. *Mycorrhiza, 17*, 327–335.

Goyal, D., Yadav, A., Prasad, M., Singh, T. B., Shrivastav, P., Ali, A., & Mishra, S. (2020). Effect of heavy metals on plant growth: An overview. In *In Naeem, M., Ansari, A.A., Gill, S.S., (Eds.), Contaminants in agriculture: Sources, impacts and management* (pp. 79–101). Springer Nature.

Hashem, A., Abd_Allah, E. F., Alqarawi, A. A., Al Huqail, A. A., Egamberdieva, D., & Wirth, S. (2016). Alleviation of cadmium stress in *Solanum lycopersicum* L. by arbuscular mycorrhizal fungi via induction of acquired systemic tolerance. *Saudi Journal of Biological Sciences, 23*(2), 272–281.

Hu, Z. H., Zhuo, F., Jing, S. H., Li, X., Yan, T. X., Lei, L. L., & Jing, Y. X. (2019). Combined application of arbuscular mycorrhizal fungi and steel slag improves plant growth and reduces Cd, Pb accumulation in *Zea mays*. *International Journal of Phytoremediation, 21*(9), 857–865.

Jamiołkowska, A., Księżniak, A., Gałązka, A., Hetman, B., Kopacki, M., & Skwaryło-Bednarz, B. (2018). Impact of abiotic factors on development of the community of arbuscular mycorrhizal fungi in the soil: A review. *International Agrophysics, 32*(1), 133–140.

Janeeshma, E., & Puthur, J. T. (2020). Direct and indirect influence of arbuscular mycorrhizae on enhancing metal tolerance of plants. *Archives of Microbiology, 202*(1), 1–16.

Jaroszuk-Ściseł, J., Nowak, A., Pac-Sosińska, M., Kołodyńska, D., & Komaniecka, I. (2025). Heavy metal biosorption ability of EPS obtained from cultures of *Fusarium culmorum* strains with different effects on cereals. *Sustainability, 17*(8), 3744.

Jiang, Q. Y., Tan, S. Y., Zhuo, F., Yang, D. J., Ye, Z. H., & Jing, Y. X. (2016). Effect of *Funneliformis mosseae* on the growth, cadmium accumulation and antioxidant activities of *Solanum nigrum*. *Applied Soil Ecology, 98*, 112–120.

Johri, A. K., Oelmüller, R., Dua, M., Yadav, V., Kumar, M., Tuteja, N., & Stroud, R. M. (2015). Fungal association and utilization of phosphate by plants: Success, limitations, and future prospects. *Frontiers in Microbiology, 6*, 984.

Kosakivska, I. V., Babenko, L. M., Romanenko, K. O., Korotka, I. Y., & Potters, G. (2021). Molecular mechanisms of plant adaptive responses to heavy metals stress. *Cell Biology International, 45*(2), 258–272.

Kumar, P., Kumar, S., & Singh, R. P. (2022). Severe contamination of carcinogenic heavy metals and metalloid in agroecosystems and their associated health risk assessment. *Environmental Pollution, 301*, 118953.

Kumar, V., & Dwivedi, S. K. (2021). Mycoremediation of heavy metals: Processes, mechanisms, and affecting factors. *Environmental Science and Pollution Research, 28*(9), 10375–10412.

Li, J., Sun, Y., Jiang, X., Chen, B., & Zhang, X. (2018). Arbuscular mycorrhizal fungi alleviate arsenic toxicity to *Medicago sativa* by influencing arsenic speciation and partitioning. *Ecotoxicology and Environmental Safety, 157*, 235–243.

Liu, T., Liu, S., Guan, H., Ma, L., Chen, Z., Gu, H., & Qu, L. J. (2009). Transcriptional profiling of arabidopsis seedlings in response to heavy metal lead (Pb). *Environmental and Experimental Botany, 67*(2), 377–386.

Madkour, L. H. (2019). Function of reactive oxygen species (ROS) inside the living organisms and sources of oxidants. *Pharmaceutical Sciences & Analytical Research Journal, 2*, 180023.

Mansoor, S., Ali, A., Kour, N., Bornhorst, J., AlHarbi, K., Rinklebe, J., & Chung, Y. S. (2023). Heavy metal induced oxidative stress mitigation and ROS scavenging in plants. *Plants, 12*(16), 3003.

Medfu Tarekegn, M., Zewdu Salilih, F., & Ishetu, A. I. (2020). Microbes used as a tool for bioremediation of heavy metal from the environment. *Cogent Food & Agriculture, 6*(1), 1783174.

Mishra, P., Tiwari, A., Dubey, V. K., Kushwaha, A., Singh, S. K., Singh, V. K., & Singh, N. K. (2024). Arbuscular mycorrhizal fungi (AMF): A natural tool for phytoremediation of heavy metals (HMs). In *In Madhav, S., Gupta, G.P., Yadav, R.K., Mishra, R. & Hullebusch, E.V., (Eds), Phytoremediation: Biological treatment of environmental pollution* (pp. 225–243). Springer Nature Switzerland.

Nafady, N. A., & Elgharably, A. (2018). Mycorrhizal symbiosis and phosphorus fertilization effects on *Zea mays* growth and heavy metals uptake. *International Journal of Phytoremediation, 20*(9), 869–875.

Nazir, A., Sarfraz, W., Allah, D., Khalid, N., Farid, M., Shafiq, M., Firdaus-E-Bareen, Rizvi, Z. F., & Naeem, N. (2023). Synergistic impact of two autochthonous saprobic fungi (*A. niger* and *T. pseudokoningii*) on the growth, ionic contents, and metals uptake in *Brassica juncea* L. and *Vigna radiata* L. under tannery solid waste contaminated soil. *International Journal of Phytoremediation, 25*(11), 1488–1500. doi:10.1080/15226514.2023.2166457.

Nedjimi, B. (2021). Phytoremediation: A sustainable environmental technology for heavy metals decontamination. *SN Applied Sciences, 3*(3), 286.

Padhiary, M., & Kumar, R. (2024). Assessing the environmental impacts of agriculture, industrial operations, and mining on agro-ecosystems. In *In Azrour, M., Mabrouki, J., Alabdulatif, A., Guezzaz, A., Amounas, F., (Eds), Smart internet of things for environment and healthcare* (pp. 107–126). Springer Nature Switzerland.

Rahman, Z., & Singh, V. P. (2020). Bioremediation of toxic heavy metals (THMs) contaminated sites: Concepts, applications and challenges. *Environmental Science and Pollution Research, 27*(22), 27563–27581.

Rao, M. J., Duan, M., Zhou, C., Jiao, J., Cheng, P., Yang, L., & Zheng, B. (2025). Antioxidant defense system in plants: Reactive oxygen species production, signaling, and scavenging during abiotic stress-induced oxidative damage. *Horticulturae, 11*(5), 477.

Rasheed, A., Hassan, M. U., Fahad, S., Aamer, M., Batool, M., Ilyas, M., & Li, H. (2021). Heavy metals stress and plants defense responses. In *In Fahad, S., Sonmez, O., Saud, S., Wang, D., Wu, C., Adnan, M., Turan, V., (Eds.), Sustainable soil and land management and climate change* (pp. 57–82). CRC Press.

Riaz, M., Kamran, M., Fang, Y., Wang, Q., Cao, H., Yang, G., Deng, L., Wang, Y., Zhou, Y., Anastopoulos, I., & Wang, X. (2021). Arbuscular mycorrhizal fungi-induced mitigation of heavy metal phytotoxicity in metal contaminated soils: A critical review. *Journal of Hazardous Materials, 402*, 123919. doi:10.1016/j.jhazmat.2020.123919.

Saeed, H. S., Munir, M., Samra, S., Farooq, A., Mahmood, H., Ahmad, I., Akram, M. I., Iqbal, R., Elsadek, M. F., Almutairi, S. M., Ditta, A., & Daraz, U. (2025). Screening of biosurfactants producing fungi from infected *Citrus sinensis* (L.). *Polish Journal of Environmental Studies, 34*(5), 5261–5270. https://www.pjoes.com/pdf-190588-115894?filename=Screening%20of.pdf. doi:10.15244/pjoes/190588.

Sharafi, S., & Salehi, F. (2025). Comprehensive assessment of heavy metal (HMs) contamination and associated health risks in agricultural soils and groundwater proximal to industrial sites. *Scientific Reports, 15*(1), 7518. doi:10.1038/s41598-025-91453-7.

Sharma, P., Jha, A. B., & Dubey, R. S. (2019). Oxidative stress and antioxidative defense system in plants growing under abiotic stresses. In *In Pessarakli, M., (Ed.), Handbook of plant and crop stress* (pp. 93–136). CRC press.

Singh, G., Pankaj, U., Chand, S., & Verma, R. K. (2019). Arbuscular mycorrhizal fungi-assisted phytoextraction of toxic metals by *Zea mays* L. from tannery sludge. *Soil and Sediment Contamination, 28*(8), 729–746. http://www.tandfonline.com/toc/bssc20/current. doi:10.1080/15320383.2019.1657381.

Singhal, R. K., Fahad, S., Kumar, P., Choyal, P., Javed, T., Jinger, D., Singh, P., Saha, D., Md, P., Bose, B., Akash, H., Gupta, N. K., Sodani, R., Dev, D., Suthar, D. L., Liu, K., Harrison, M. T., Saud, S., Shah, A. N., & Nawaz, T. (2023). Beneficial elements: New players in improving nutrient use efficiency and abiotic stress tolerance. *Plant Growth Regulation, 100*(2), 237–265. https://www.springer.com/journal/10725. doi:10.1007/s10725-022-00843-8.

Sommer, N., Guo, Y., Rasche, F., Hagemann, M. H., & Zörb, C. (2025). The potential of four legume trees for mercury phytoremediation and the role of arbuscular mycorrhizal fungi. *Research Square*. https://www.researchsquare.com/browse. doi:10.21203/rs.3.rs-5336359/v1.

Son, Y., Martínez, C. E., & Kao-Kniffin, J. (2024). Three important roles and chemical properties of glomalin-related soil protein. *Frontiers in Soil Science, 4*, 1418072.

Sportes, A., Hériché, M., Boussageon, R., Noceto, P.-A., van Tuinen, D., & Wipf, D. (2021). Pierre emmanuel courty, a historical perspective on mycorrhizal mutualism emphasizing arbuscular mycorrhizas and their emerging challenges. *Mycorrhiza, 31*(6), 637–653. doi:10.1007/s00572-021-01053-2.

Sun, T., Delaplace, P., Li, G., James, A., Pan, J., & Zhang, J. (2025). Novel insights into the effect of arbuscular mycorrhizal fungi inoculation in soils under long-term biosolids application: Emphasis on antibiotic and metal resistance genes, and mobile genetic elements. *Environmental Pollution, 369*, 125846. doi:10.1016/j.envpol.2025.125846.

Tao, J., & Lu, L. (2022). Advances in genes-encoding transporters for cadmium uptake, translocation, and accumulation in plants. *Toxics, 10*(8), 411. doi:10.3390/toxics10080411.

Thind, S., Chaudhary, M. S., Ditta, A., Hussain, I., Parveen, A., Ullah, N., Mahmood, Q., Al-Ashkar, I., & El-Sabagh, A. (2022). Impact of mycorrhizal fungi from different rhizospheric soils on fungal colonization, growth, and chlorophyll contents of *Cenchrus ciliaris*. *Agronomy, 12*(11), 2644. doi:10.3390/agronomy12112644.

Vanisree, C. R., Sankhla, M. S., Singh, P., Jadhav, E. B., Verma, R. K., Awasthi, K. K., & Nagar, V. (2022). Heavy metal contamination of food crops: Transportation via food chain, human consumption, toxicity and management strategies. In *In Saleh, H. M. & Hassan, A. I., (Eds.), Heavy metals-new insights* (pp. 1–23). IntechOpen.

Verma, S., Bhatt, P., Verma, A., Mudila, H., Prasher, P., & Rene, E. R. (2021). Microbial technologies for heavy metal remediation: Effect of process conditions and current practices. *Clean Technologies and Environmental Policy, 25*(38), 1–23.

Wahab, A., Muhammad, M., Munir, A., Abdi, G., Zaman, W., Ayaz, A., Khizar, C., & Reddy, S. P. P. (2023). Role of arbuscular mycorrhizal fungi in regulating growth, enhancing productivity, and potentially influencing ecosystems under abiotic and biotic stresses. *Plants, 12*(17). www.mdpi.com/journal/plants. doi:10.3390/plants12173102.

Wang, F., Zhang, L., Zhou, J., Rengel, Z., George, T. S., & Feng, G. (2022). Exploring the secrets of hyphosphere of arbuscular mycorrhizal fungi: Processes and ecological functions. *Plant and Soil, 481*(1), 1–22.

Wang, Q., Lu, H., Chen, J., Hong, H., Liu, J., Li, J., & Yan, C. (2018). Spatial distribution of glomalin-related soil protein and its relationship with sediment carbon sequestration across a mangrove forest. *Science of the Total Environment, 613–614*, 548–556. doi:10.1016/j.scitotenv.2017.09.140.

Wang, W., Shi, J., Xie, Q., Jiang, Y., Yu, N., & Wang, E. (2017). Nutrient exchange and regulation in arbuscular mycorrhizal symbiosis. *Molecular Plant, 10*(9), 1147–1158. doi:10.1016/j.molp.2017.07.012.

Wu, S., Zhang, X., Huang, L., & Chen, B. (2019). Arbuscular mycorrhiza and plant chromium tolerance. *Soil Ecology Letters, 1*(3–4), 94–104. doi:10.1007/s42832-019-0015-9.

Yizhu, L., Imtiaz, M., Ditta, A., Rizwan, M. S., Ashraf, M., Mehmood, S., Aziz, O., Mubeen, F., Ali, M., Elahi, N. N., Ijaz, R., Lele, S., Shuang, C., & Tu, S. (2020). Response of growth, antioxidant enzymes and root exudates production towards as stress in *Pteris vittata* and in *Astragalus sinicus* colonized by arbuscular mycorrhizal fungi. *Environmental Science and Pollution Research, 27*(2), 2340–2352. https://link.springer.com/journal/11356. doi:10.1007/s11356-019-06785-5.

Zhang, H., Ren, W., Zheng, Y., Li, Y., Zhu, M., & Tang, M. (2021). Arbuscular mycorrhizal fungi increase Pb uptake of colonized and non-colonized *Medicago truncatula* root and deliver extra Pb to colonized root segment. *Microorganisms, 9*(6), 1203. doi:10.3390/microorganisms9061203.

Zhang, Q., Liu, H., Liu, F., Ju, X., Dinis, F., Yu, E., & Yu, Z. (2022). Source identification and superposition effect of heavy metals (HMs) in agricultural soils at a high geological background area of karst: A case study in a typical watershed. *International Journal of Environmental Research and Public Health, 19*(18), 11374. doi:10.3390/ijerph191811374.

Zhang, X., Wang, Z., Lu, Y., Wei, J., Qi, S., Wu, B., & Cheng, S. (2024). Sustainable remediation of soil and water utilizing arbuscular mycorrhizal fungi: A review. *Microorganisms, 12*(7), 1255. doi:10.3390/microorganisms12071255.

Zhang, X. F., Hu, Z. H., Yan, T. X., Lu, R. R., Peng, C. L., Li, S. S., & Jing, Y. X. (2019). Arbuscular mycorrhizal fungi alleviate Cd phytotoxicity by altering Cd subcellular distribution and chemical forms in *Zea mays*. *Ecotoxicology and Environmental Safety, 171*, 352–360. http://www.elsevier.com/inca/publications/store/6/2/2/8/1/9/index.htt. doi:10.1016/j.ecoenv.2018.12.097.

Zhao, S., Yan, L., Kamran, M., Liu, S., & Riaz, M. (2024). Arbuscular mycorrhizal fungi-assisted phytoremediation: A promising strategy for cadmium-contaminated soils. *Plants, 13*(23), 3289.

Chapter 31

Endophytic fungi and sustainable agricultural production

Muhammad Arshad[a], Muhammad Irfan Ullah[a], Muhammad Mustaqeem[b] and Allah Ditta[c,d]

[a] *Department of Entomology, University of Sargodha, Sargodha, Pakistan,* [b] *Institute of Chemistry, University of Sargodha, Sargodha, Pakistan,* [c] *Department of Environmental Sciences, Shaheed Benazir Bhutto University Sheringal, Sheringal, Pakistan,* [d] *School of Biological Sciences, The University of Western Australia, Perth, WA, Australia*

31.1 Introduction

31.1.1 Overview of microbial symbioses in sustainable agriculture

The interaction of microbes is a crucial element in sustainable agriculture, as it enhances soil and plant conditions. Symbiotic interactions between microorganisms and plants, as well as between plants and the environment, can be beneficial to crop production and sustainability (Perveen et al., 2023; Saeed et al., 2025; Saleem et al., 2024). Microbial symbiosis has drawn recent interest in the area of sustainable agriculture. The use of microbial biopesticides could replace the widespread use of conventional pesticides and fertilizers in the production of organic crops, thereby alleviating nontarget effects on the environment. Negative effects' microorganisms have beneficial effects on the health of plants in the form of endophytic entomopathogenic fungi, rhizobacteria, and growth regulators (Ahmad et al., 2021; Ditta & Ullah, 2023; Ditta et al., 2022; Hasnain et al., 2025; Sarfraz et al., 2019; Shah et al., 2022; Ullah et al., 2021). Such symbiotic relationships can increase crop production as they act as plant growth regulators and biofertilizers, and, at the same time, lower the population of pests (Khan et al., 2022; Muhammad et al., 2024; Naveed et al., 2020). Endophytic fungi are the only microbes with the highest level of application to defend their hosts. There is also the use of endophytic fungi as a biocontroller application to produce bioactive compounds and enhance the plant's immune system (Yan et al., 2019). Endophytic fungi are members of the phylum Ascomycota, and most commonly, they live and reproduce within the inner tissues of plants. Endophytic fungi have been shown to have mutualistic, commensal, or latent interactions with their host species of use.

Endophytic fungi infect plant tissues without causing disease and are considered a key component of sustainable agriculture because of their strong abilities to improve plant development, increase stress tolerance, and provide antipathogen defense (Nazir et al., 2023; Thind et al., 2022; Yizhu et al., 2020; Zhan et al., 2024). These endoplastic fungi live in different plant tissues such as leaves, stems, and roots and interact with host plants. Several taxa of endophytic fungi have the capacity to colonize plant tissues and therefore enhance the nutrition of plants. Endophytic fungi are categorized into various types, depending on the type of relationship that they have with the host plant (Rodriguez et al., 2019). The greatest group is that of mutualistic endophytes. In this type of relationship, fungi get carbohydrates and nutrients from plants, and endophytic fungi protect plants from herbivores and pathogens by producing bioactive compounds. In this type of relationship, fungi also assist plants in getting water and nutrients (phosphorus). The mutualistic relationship improves plant growth, develops resistance to diseases, and improves the overall plant health (Zhan et al., 2024). Another relationship is with commensal endophytic fungi that reside in plant tissues without harming plant enemies but do not give space to pathogens within plant tissues (Arnold et al., 2003). A type of endophytic fungus is reported as latent endophytic fungi. It is present inside the plant tissues and is activated when triggered by some abiotic stressors. When the latent endophytic fungi are triggered by nutrient deficiency or any other abiotic factor, they become pathogenic for plants.

Arbuscular mycorrhizal fungi (AMFs) create a symbiotic relationship with plants that improves plant growth and health. The endophytic fungi penetrate through plant tissues and can be found in different parts of the plant, and on the other

Advancing Sustainable Agriculture Through Arbuscular Mycorrhizal Fungi. DOI: https://doi.org/10.1016/B978-0-443-45130-0.00040-0

hand, the AMFs enter and grow within the root systems. The main role of the same fungi is to enhance the health of the plant, mobilizing soil nutrients and increasing the network of hyphae within the soil matrix. They also provide pest resistance, making plants resistant to a variety of herbivores; some AMFs have even been reported to have a use in weed suppression (Delavaux et al., 2017). In experiments with the invasive weed *Impatiens glandulifera*, AMF and endophytes, like *Colletotrichum acutatum* and *Cladosporium oxysporum*, had a suppressive effect on the growth of the weed and on its aphid predator. With this, endophytic fungi and AMF can be used together to combat the weeds and insect pests in sustainable farming (Razak & Gange, 2023). The AMF and endophytic fungi are, therefore, necessary in plant growth and productivity, although through different mechanisms. AMF are mainly associated with nutrient uptake and alleviation of soil, although the endophytic fungi provide an expanded range of ecosystem advantages, such as accelerated plant expansion, disease resistance, and stress tolerance. Together, these fungal taxa have the potential to balance each other, thus supporting sustainable agricultural systems and productivity.

In modern agricultural systems, both endophytic fungi and AMF could be helpful by improving crop production, enhancing the nutritional food quality, and reducing the hazardous environmental impacts. The hazardous chemicals and pesticides with negative impacts on nontarget organisms could be replaced with these fungi. These fungi also improve the plants by managing the challenges posed by climate change, including high temperatures, drought conditions, and increasing salinity (Zhang et al., 2024). As a result, microbial symbioses offer a sustainable, ecofriendly solution for increasing agricultural productivity while protecting natural resources and the environment (Sharma & Sharma, 2021). Endophytic fungi reside inside the plant tissues asymptomatically and belong to Ascomycota. There is a diverse number of an estimated 300,000 plant species present in temperate and tropical rainforest regions. Each plant species had two or more species of endophytic fungi.

31.1.2 Diversity and ecology of endophytic fungi

Plantae and endophytes evolved over millions of years. Endophytic fungi show significant diversification in their taxonomic composition, which can vary among host plant species, geographically separate individuals of the same species, and different tissues or organs of the host plant (Fang et al., 2019). Endophytic fungi show notable diversity and can colonize almost every known plant species in the aquatic and terrestrial habitats. It is anticipated that each plant species possesses at least one fungal endophyte; however, environmental and genetic factors drive the colonization patterns. It is estimated that there are approximately 1.2 million fungal endophytes globally, but about 84% of these are still unidentified. This gap highlights a vast potential for research and discovery. The medicinal plants disclose 2–58 fungal endophyte species per plant, which emphasize further about the biodiversity within this group (Prajapati et al., 2025). Studies on plants inhabiting extreme environments exposed 174 diverse fungal isolates from 9 wild plant species, which represent 13 different genera. Variation in species richness and evenness reveals that some host plants serve as reservoirs for varied fungal communities. High colonization frequencies within leaf tissues also indicate that endophytes play a major role in plant health (Sharma et al., 2025). Endophytic fungi consist of various groups in terms of their ecological nature, mode of relationship, host preference, and colonization. The major taxonomic groups of endophytic fungi include Zygomycota, Ascomycota, and Basidiomycota, entomopathogenic fungi.

Ascomycetes include genera, such as *Epichloë, Neotyphodium*, and *Claviceps*, which are mostly associated with grasses and other herbaceous plants. Ascomycetes are characterized by the production of ascospores and are known to form a wide range of mutualistic, commensal, and pathogenic relationships with plants. *Epichloë* species are considered beneficial endophytes for grasses, showing resistance to herbivory, drought, and diseases. Basidiomycetes have the *Corticium* and *Ganoderma* genera. These fungi are found in woody plants and trees, where they contribute to wood decay, lignin breakdown, and nutrient cycling. Although they are less common as endophytes than ascomycetes, some species can establish beneficial relationships with plants, improving plant health and resistance to environmental stresses (Arnold et al., 2003). The genus *Mucor* and *Rhizopus* belong to the Zygomycota, and some species of endophytic fungi belong to this order. Zygomycetes are mainly involved in saprotrophic and opportunistic relationships with plants, breaking down organic material in the soil and contributing to nutrient cycling. Some zygomycetes are known to form symbiotic relationships with plants, providing benefits such as nitrogen fixation.

Some entomopathogenic fungi can live inside plant tissues as endophytes, such as *Beauveria* and *Metarhizium*. These fungi are known for their ability to parasitize insects but can also enhance plant resistance to herbivores and pests by producing toxic compounds. Their role as plant endophytes is emerging as an important part of the plant's natural defense system (Vidal & Jaber, 2015). According to Hawksworth (1991), there are 2.3 million fungal species reported globally, including 1,000,000–1,250,000 species of endophytic fungi. Fungi are classified into 5 subkingdoms on the basis of molecular identification, whereas these suborders are further divided into 16 divisions. Most endophytic fungi belong to

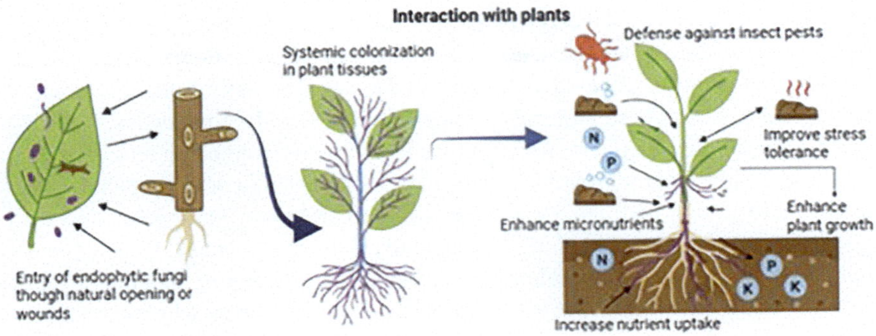

FIGURE 31.1 Interaction of endophytic entomopathogenic fungi with plants.

Ascomycota, whereas others belong to Basidiomycota, Mucoromycota, and so on. Bhunjun et al. (2023) estimated that there are more than 3 million species of endophytic fungi that have been reported. It is confusing to identify all the fungal species on the basis of morphological characteristics because of the limited characteristics, so it is highly recommended to identify fungal strains via molecular analysis (Göker et al., 2009).

Endophytic fungi showed a close relationship with their host plants. Based on the endophytic fungi's relationship with their host, they are divided into two major groups. The first group has clustered (Class 1) endophytic fungi that belong to the family Clavicipitaceae. The host plants in Group 1 belong to herbaceous species such as *Lolium* spp. There are limited host plants in Group 1 classification. The second group of endophytic fungi belongs to the Clavicipitaceae family and is further divided into three clusters (Classes 2–4) (Rodriguez et al., 2009). Endophytic fungi show a wide host range. Although most of the fungal species are host-specific, others can colonize several plant species. For example, the genus *Neotyphodium* typically associates with grasses, whereas *Penicillium* and *Aspergillus* species can inhabit a range of plants, including crops, trees, and herbaceous plants (Arnold et al., 2003). The ability of endophytic fungi to colonize multiple plant species is an adaptive trait that enables them to flourish in diverse habitat conditions and ecological niches. Endophytic fungi benefit plants in several ways. It helps the plants in nutrient uptake, helps from pathogens, and resists several abiotic factors. Endophytic fungi play a crucial role in controlling insect pests on various host plants. Endophytic fungi naturally occur in many forests, grasslands, and other plant species. Endophytic fungi can be found with aquatic plants, algae, and duckweed in the aquatic environment and help the plants in nutrient absorption, which eventually promotes plant growth (Makino et al., 2022). Endophytic fungi may also be found in the environment of plants and form symbiotic relationships in severe environmental conditions such as arid, saline, and desert areas, where the availability of water is restricted, and temperature variation is high (Georgousaki et al., 2022).

31.1.3 Interaction of endophytic fungi with plants

Endophytic fungi colonize plant tissues through inter- or intracellular ways by developing mutualistic or commensal interactions with their host plants (Barrow, 2003). After entering through specific pathways, it starts to colonize into plant tissues and then spreads throughout the plant. The whole process by which endophytic fungi enter the plant is strictly dependent on plant–fungus interaction and the morphology of the plant tissues (Fig. 31.1). The growth of fungus within roots could be wider than above-ground plant parts (Schulz & Boyle, 2005). Variation in colonization of fungi may depend on the group and class to which the endophyte belongs (Rodriguez et al., 2009). The clavicipitaceous endophytes, commonly known as C-endophytes or Class 1, with a limited host range, broadly colonize the plant shoots and rhizomes. The endophytes other than clavicipitaceous are commonly known as NC-endophytes, reported to have a diverse host range. These NC-endophytes can be classified further into functional classes depending on patterns of host colonization and the mechanism of transmission among host plants (Rodriguez et al., 2009).

Endophytic fungi primarily enter plants through wounds, stomata, lenticels, or natural openings. These opening points are easy-access entry points for fungi, which colonize in plant tissues more readily. After the fungal entry into plants, it starts to colonize and establish systemically in plant parts, including the leaves, stems, and roots. Endophytic fungi disperse in all plant parts through vascular tissues and establish in plant tissues. The endophytic fungi produce hyphae and connect one plant tissue to another, and this hyphal network helps in nutrient exchange and transports metabolites in all plant tissues. Systemic colonization confirms the plant and fungus relation, wherein the fungus can influence plant health and metabolism across the entire plant (Frey-Klett et al., 2007).

Endophytes use molecular signals released by both plants and endophytes, and these chemical signals are used to identify the host plants. The signaling molecules, phenolic acids, flavonoids, and strigolacetones are produced by plants and attract the suitable endophytic fungi. In addition, fungi also produce molecules that control plant processes to benefit the fungus and secrete them into the plant tissues. These molecules influence plant signaling pathways, primarily improving nutrient uptake and stimulating the plant's defense system, which protects the fungus from being recognized as a pathogen. In some cases, these molecules can also induce plant growth by increasing root and shoot development (Frey-Klett et al., 2007).

Plants usually balance their growth and defense system against herbivores, pathogens, and environmental stressors. When plants form symbiotic relationships with fungi, particularly endophytic fungi, they experience a growth-defense trade-off (Pangesti et al., 2013). The endophytic microorganisms produce auxins or alkaloids for plant growth, and these microorganisms improve the trade-off by inducing hormonal responses. The enhanced plant resistance to plant herbivores and pathogenic microorganisms is because of beneficial symbionts (Pieterse et al., 2009). The best example for a plant relation with symbionts describing both enhanced growth and increased resistance is the grass and *Epichloë* symbiosis relationship. Fungal endophytes of the genus *Epichloë* are beneficial symbionts for grasses because they produce bioactive alkaloids that show protection against plant herbivores (Schardl et al., 2013). Moreover, *Epichloë* endophytes can also positively adjust the hormonal pathways related to plant growth (Schmid et al., 2017). A few endophytic fungi reduce the plant's requirement for defense compound production for the plant's protection from pests and pathogenic microbes. In return, the plant distributes more resources to increase growth than defense, which ultimately helps to improve the overall productivity. This balance between growth and defense is context-dependent, with environmental conditions, plant genotype, and fungal species all influencing the outcome (Wang et al., 2017).

Endophytic association with plants can be affected by environmental factors. The significant differences were observed in the structures of endophytes among different plants because of varied environmental factors (Hardoim et al., 2015). Environmental factors, such as humidity, temperature, soil texture, and light availability, play a crucial role in determining endophytic structures. Biogeography adds another layer of variation, with tropical ecosystems generally supporting higher endophyte diversity than temperate regions because of their favorable climate and rich plant diversity. Plant stage and genotype are also important factors in evaluating the endophytic community, as older plants are considered to have more diverse fungal communities because of repeated colonization and structural variation in their tissues over time. Plant genotypes and species showed variation in the degrees of susceptibility to fungal colonization. Based on genotypes, some plants have advanced mechanisms to attract or repel specific endophytic fungal species, which leads to differences in colonization patterns. In addition, plant genotype can influence the types of interactions that occur, such as mutualism versus pathogenicity (Arnold et al., 2003). Some endophytes are adapted to extreme climates and can help plants survive such stresses by improving water retention and nutrient uptake (Aroca, 2013). The microorganisms, that is, mycorrhizal fungi, bacteria, viruses, and other microorganisms, also had a great impact on endophytic fungi. Soil properties such as soil structure, texture, pH level, nutrient availability, and moisture content play a significant role in endophyte colonization in plants. Endophytic fungi are more likely to flourish in soils with higher organic matter, as they depend on plant roots for nutrient exchange. Soils that are nutrient-poor or have low pH may limit the diversity of endophytic fungi, whereas well-aerated, fertile soils tend to support a broader range of fungal species (Stone et al., 2000). Endophytic fungi interact with other microorganisms and mycorrhizal fungi, which may have positive or negative interactions for plant health. These interactions can also affect the establishment and function of endophytes within the plant tissues. Sometimes, beneficial microorganisms may increase the endophytic colonization by improving nutrient availability or providing chemical signals that attract the fungi to plant roots (Frey-Klett et al., 2007).

31.2 Functional roles of endophytic fungi in sustainable agriculture

31.2.1 Plant and soil health improvement

Endophytic fungi significantly contribute to sustainable agriculture by improving nutrient acquisition for plants. The nutrient availability is naturally limited in soils, whereas plants rely on various mechanisms to obtain essential elements such as nitrogen, potassium, phosphorus, and micronutrients. Endophytic fungi, through their symbiotic relationships with plants, enhance nutrient uptake by acting as a facilitator between the soil and the plant roots (Verma et al., 2021).

Endophytic fungi capable of colonizing plant roots can alter the concentration of soil nutrients, thereby making them more accessible for absorption and use by plants. Endophytic fungi also help to dissolve phosphorus, making it available to the hosting plants. Some species of fungi, including *Trichoderma, Penicillium*, and *Aspergillus*, produce enzymes that break the complex organic compounds, releasing nutrients that the plant uses in plant growth (Tang et al., 2023). Plant growth could be affected by *Beauveria bassiana* seed treatment, and it depends on the nutrient availability of the soil, as the plant growth is directly proportional to soil fertility (Tall & Meyling, 2018). Inoculation of endophytic fungi, *Purpureocillium lavendulum, Purpureocillium lilacinum,* and *Metarhizium marquandii*, to maize, beans, and soybeans, produces indoleacetic acid (IAA) and solubilizes P that helps in growth promotion (Baron et al., 2020). A mutualistic association of mycorrhizal fungi forms with the plant roots, which increases nutrient uptake in plants, especially phosphorus, in exchange for carbohydrates from the plant (Shi et al., 2023). Entomopathogenic fungus, *Metarhizium robertsii*, can colonize endophytically in bean and switchgrass plants, and after killing *Galleria mellonella* larvae, it makes a source of nitrogen to the host plants (Behie et al., 2012). Various species of *Metarhizium* and *B. bassiana* are used to kill the insect pests and carry out the nitrogen from insects to host plants (Behie and & Bidochka, 2014). Several metabolic pathways and molecules are involved in the transfer of nutrients within the plant because of endophytic fungi. The production of spike and root length in wheat increased after inoculation of *B. bassiana* through seed inoculation and soil treatment.

Previously, various studies reported on the production of plant hormones by endophytic fungi. The most common phytohormones include auxins, gibberellins (GAs), and cytokinins, whereas these hormones act as chemical signals and messengers for plant growth. The major phytohormone is auxin, which is considered main regulator of plant development. Auxins are responsible for shoot and root development in plants. GAs are responsible for seed germination, flourishing, stem elongation, sexual expression, and fruit formation (Bömke & Tudzynski, 2009). GAs by endophytic fungi use the mevalonic acid pathway, and the final bioactive compounds are GA1 and GA3. Gibberellin production by *Aspergillus fumigatus* and *Cladosporium* sp. increased the growth of soybean and cucumber plants, respectively (Humayun et al., 2010). Endophytic fungi target plant development by direct regulation of endogenous hormonal pathways. This phenomenon increased the endogenous concentrations of the phytohormones auxins, GAs, and salicylic acid, which were accompanied by a decrease in the levels of abscisic acid and jasmonic acid.

31.2.2 Abiotic stress tolerance

During the life cycle of the crops, there are many abiotic stresses, which are drought, salinity, extreme temperatures, toxicity to heavy metals, and nutritional deficiency. Those stresses have significant adverse effects on plant growth and crop productivity, such as leading to ion toxicity, photosynthesis inhibition, impaired membrane integrity, and increased reactive oxygen species production. Endophytic fungi compensate for the negative impact of these environmental factors by either causing stress-resistant responses of the plant or by directly producing antistress biochemicals (Lata et al., 2018). Therefore the endophytic fungi may induce or trigger the plant synthesis of antioxidant enzymes, including catalase, superoxide dismutase, and lipoxygenase; antioxidant secondary metabolites, including phenols and flavonoids; organic solutes, including proline and aspartic acid; and phytohormones, including abscisic acid and GAs (Tyagi et al., 2022).

Endophytic fungi can deal with all abiotic stressors, including temperature fluctuation, drought, toxic heavy metals, and salinity. Endophytic fungi, *B. bassiana*, and *Isaria fumosorosea* Wize. (Ascomycota; Hypocreales), and *Metarhizium brunneum* help to promote cabbage plant growth under water stress, whereas *B. bassiana* improves the growth of strawberry plants. The saline stress in soybean crops was reduced by an endophytic fungus, *Metarhizium anisopliae* (Khan et al., 2012). Drought stress in plants is responsible for lower water availability, leading to a lower photosynthesis rate, and ultimately results in oxidative damage to cells. Salinity stress can lead to ion toxicity and osmotic stress, which decreases water uptake and causes cellular damage and nutrient imbalance. *Paecilomyces formosus* helps to resolve salinity problems in cucumber plants. Cucumber plants with inoculated endophytic fungi had higher growth as compared to noninoculated plants (Khan et al., 2012). The high-temperature stress in soybean and sunflower plants was managed by the inoculation of the endophytic fungi *Aspergillus niger* (Ismail et al., 2020).

31.2.3 Biocontrol of pathogens and pests

Over the past several decades, biological control methods using useful organisms have gained growing appreciation as a way of reducing dependency on synthetic pesticides (Fig. 31.2). The endophytic fungi, in this case, have potential because of their ability to form mutualistic associations with the host plants, regulate the plant metabolism, improve stress resilience, and defend against insect herbivory. Annual losses to crops run between 20% and 40% because of biotic stresses related to pathogens, herbivores, and weeds; about 10% and 15% of these losses are related to diseases (Mohammad-Razdari et al., 2022). There are several mechanisms of action of endophytic fungi in reducing pathogen-associated yield losses in crops. Endophytic fungi can elicit local and systemic plant defenses against crop pathogens indirectly (Poveda et al., 2020). Directly, endophytic fungi produce volatile and nonvolatile antimicrobial compounds (antibiosis), occupy rhizospheric, phyllospheric, and internal plant space and nutrients, and have direct antimicrobial interactions like mycoparasitism and nematoparasitism (Poveda & Baptista , 2021).

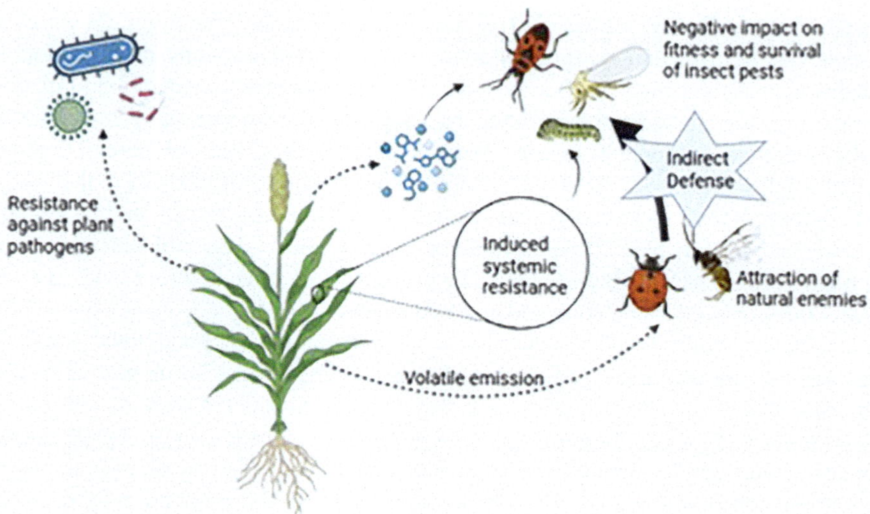

FIGURE 31.2 Role of endophytes as biocontrol agents in agricultural crops.

The endophytic fungi could have an influence on insect pests in a number of ways. They can directly infect insect tissue or produce insecticidal compounds, but indirectly, they can boost host plant defense responses and/or trigger the biosynthesis of plant-produced compounds with insecticidal/antifeedant activities (Poveda, 2021). Plants with endophytic fungi are also prolific producers of secondary metabolites, including alkaloids, steroids, terpenoids, peptides, polyketides, flavonoids, quinones, phenolics, chlorinated derivatives, and volatile organic compounds (VOCs) (Kaddes et al., 2019). These metabolites exhibit activity against a wide range of pathogens and herbivores, and some are reported to have antiviral, antibacterial, and antifungal properties (Latz et al., 2018). In addition, they offer defense against biotic stress, such as diseases and pests. Endophytic fungi use different mechanisms to establish colonization of host plants and cause damage caused by insect herbivores, which require crop species, the targeted fungal endophytes, the method of application, and target insect taxa. The initial step of endophytic colonization by endophytic fungi in insect pests encompasses colonization of the surface of the plant, cuticle penetration, and later growth in the body cavity, which forms the major entry route to most endophytic fungi (Vega et al., 2008).

In insect invasion, after efficient adhesion, conidia germinate to give rise to hyphae, and the conidial surface proteins, including the *Metarhizium* adhesin genes MAD1 and MAD2 of *M. anisopliae*, the hydrophobin proteins hyd1 and hyd2 of *B. bassiana*, support binding to plant and insect surfaces and to recognition of insect-specific compounds that in turn lead to cuticle degradation. It was reported that in *M. robertsii*, adhesive MAD1 is necessary in cuticle penetration, but that adhesion to the conidial spores of plants is mediated by MAD2 (Wang &d St Leger, 2007). The following insect cuticle and plant cell wall degradation allow the entry of the entomopathogenic fungus (EPF) into plant and insect cells, achieved by an enzymatic activity of various proteases and mechanical penetration of specific hyphal structures, including aspersoria (Barelli et al., 2016). To cause an insect's death, the EPF must evade the insect's immune system to successfully infect it as a parasite. After entering the cuticle, the EPF spreads throughout the hemolymph, where the spores further develop into blastospores (yeast-like asexual spores), feed on the hemocoel, and produce insecticidal metabolites, eventually causing death (Fan et al., 2017). These activities through which EPFs reduce herbivore damage in host plants include antibiosis or pest avoidance/feeding deterrent (Vega et al., 2008), insect fitness reduction, enhanced predation and parasitism risk (Shymanovich & Faeth, 2018), insect mortality (Rasool et al., 2021), slowed insect growth, and altered plant defense gene expression (Ahmad et al., 2020). One possible mechanism is direct defense or indirect plant defense by amplifying plant volatiles that attract more olfactory-foraging natural enemies (Fuchs & Krauss, 2019).

Fungal endophytes can also be utilized as biocontrol agents since they induce innate defense responses in host plants and also protect themselves (Huang et al., 2020). These microbes produce various bioactive metabolites with antibacterial and antifungal effects. Some of the compounds that are formed are alkaloids, terpenoids, flavonoids, chlorinated derivatives, quinols, peptides, steroids, polyketides, flavonoids, phenols, and a range of VOCs. Therefore it has been observed that crop plants inoculated with endophytic fungi tend to face the attacks of various phytopathogens. An example is the treatment of cucumber with 30 species of endophytes, especially *Penicillium* sp. and *Hypocrea* sp., which had a significant effect in reducing the growth of the fungi *Fusarium oxysporum* f. sp. *cucumerinum* by inhibiting the development of their mycelia and the exacerbation of the wilt disease (Abro et al., 2019). Likewise, endophytic *Trichoderma* isolates have been demonstrated to reduce the prevalence of red root rot of *Acacia mangium* seedlings caused by *Ganoderma philippii*, highlighting the potential of these biocontrol agents (Gafur, 2023). The mechanism through which endophytes suppress pathogens is not well comprehended, though it is likely to be very diverse. It is possible to directly kill pathogen cells with fungi, hinder their growth, invisibly introduce oneself into host tissues, paralyze target organisms, and interfere with resource uptake by the pathogen, and often a complex of these strategies is used. As an illustration, *Pochonia chlamydosporia* and *P. lilacinum* can adhere to, infect, colonize, and consequently degrade egg content in root-knot nematodes (*Meloidogyne* spp.), thereby exemplifying a versatile antagonistic behavior (Swarnakumari, 2021). The possible mechanisms of plant disease suppression vary depending on the species of EPFs. There are several mechanisms by which the association of these plant-EPFs benefits the host to resist pathogens. For example, direct inhibition of pathogens, including resource competition (space and nutrients), mycoparasitism, and antibiosis, and indirect inhibition by induced systemic resistance and the stimulation of secondary metabolites (Urooj et al., 2021). For instance, *B. bassiana* in an endophytic way can limit the growth and development of many soilborne plant diseases, such as *Pythium, Rhizoctonia*, and *Fusarium*. The *Beauveria* spp. are well-known to produce a range of bioactive metabolites that induce systemic resistance and stop the growth of fungal pathogens.

31.2.4 Entomopathogenic fungi as natural insecticides

Entomopathogenic fungi serve as biological control agents, used to manage various agricultural insect pests through direct infection or endophytically (Mubeen et al., 2022). To overcome excessive insecticide use, biocontrol agents are the best alternative for managing insect pests. Entomopathogenic fungi adhere to the insect body and penetrate through the cuticle. It breaks the insect's cuticle, whereas other microorganisms, bacteria, and viruses need to be ingested by the host. When fungi attach to the insect cuticle, different enzymes, including lipases, proteases, and chitinases, are produced to break down the cuticle components (protein, chitin, and lipids) (Akello et al., 2008).

Endophytic fungi, particularly entomopathogenic species, like *B. bassiana* and *M. anisopliae*, are emerging as natural insecticides through plant inoculation rather than direct spraying, providing systemic pest protection with reduced environmental impact, though their global commercial adoption remains limited compared to traditional microbial biopesticides. EPFs are able to colonize in different plant parts as well as reduce the pest populations in several economic crops like maize, sorghum, soybean, cotton, and tomato. *B. bassiana, I. fumosorosea*, and *Lecanicillium lecanii* were applied to *Phaseolus vulgaris* L. (Fabaceae) plants and tested against *Tetranychus urticae* Koch (Acari: Tetranychidae). Successfully colonized EPFs in plants increased the plant growth and had a negative impact on *T. urtica* reproduction (Dash et al., 2018). Major crops, wheat, tomato (*Solanum lycopersicon* L.), soybean (*Glycine max* L.), green bean, and switchgrass (*Panicum virgatum* L.), growth was increased as a result of the symbiotic relationship of EPF (Behie & Bidochka, 2014). In citrus, EPF colonization through the trunk resulted in a suppressed population of *Diaphorina citri* and improved the tree strength; EPFs may be integrated into integrated pest management (IPM) programs targeting vector-borne diseases such as Huanglongbing (Lezama-Gutiérrez et al., 2012).

Use of entomopathogenic fungi (EPFs) has cost-effectiveness, a pesticide residue-free environment, and enhanced biodiversity. However, several genera, including *Lecanicillium, Metarhizium, Hirsutella, Isaria*, and *Beauveria*, have been commercialized as commercially developed to manage insect pests in the field. Over 170 EPFs-based mycopesticides exist globally, primarily for direct contact but adaptable for endophytism in IPM, with regional availability in Latin America, India, and Europe. Challenges like variable colonization (20%–100%) limit dedicated endophytic branding, but ongoing trials promote seed coatings for enhanced persistence. Some of the commercially available entomopathogenic fungi are listed in Table 31.1.

Microbial insecticides play an important role in the integrated pest management programs without harming the environment and nontarget organisms. EPF *B. bassiana* has been used in China for the management of pine moths *Dendrolimus* spp. and is considered one of the major biocontrol agents over one million hectares. In Europe and the United States, commercially EPFs have been used to manage economic pests like European corn borer, Asiatic corn borer, whitefly, thrips, aphids, and mealybugs. An EPF strain, ATCC 74040, has been commercially registered and used to control many soft-bodied insects from the orders Homoptera, Heteroptera, and Coleoptera (Maina et al., 2018).

31.2.5 Challenges and limitations

Though microbial insecticides are being considered the best alternative for the management of insect pests, there are still some limitations for using these insecticides in the field. As microbes have a short life cycle, these insecticides require a high cost to mass culture and produce commercially. Further specific environment is required to germinate and cause

TABLE 31.1 List of microbial insecticides commercially available.

Sr. no.	Commercial name	Fungus	Company	References
1.	Naturalis	*Beauveria bassiana*	Biogard, CBC (Europe) S.r.L., Grassobbio, Italy	Scovero et al. (2025)
2.	Mycotal	*Akanthomyces muscarius*	Koppert Biological System	Scovero et al. (2025)
3.	BotaniGard	*B. bassiana*	Laverlam, International/Bioworks, United States	Stansly and Natwick (2010)
4.	PFR-97	*Isaria fumorosea*	Certis, United States	Stansly and Natwick (2010)
5.	Metatech WP	*Metarhizium anisopliae*	Dudutech Division, Flamingo Horticulture (K) Ltd., Naivasha, Kenya	Ndereyimana et al. (2019)
6.	Beauvitech WP	*B. bassiana*	Dudutech Division, Flamingo Horticulture (K) Ltd., Naivasha, Kenya	Ndereyimana et al. (2019)

infections (Maina et al., 2018). EPFs species can be isolated from soil and can be easily affected through soil factors such as PH, humidity, soil temperature, and other soil microbes. EPFs germination can be negatively affected by high soil temperature. Entomopathogenic endophytic fungi face noteworthy hurdles in practical application as biocontrol agents, including application method, fungal strain, inconsistent endophytic colonization rates (often 20%–80%) influenced by plant species, and inoculation technique, which restrict the reliable systemic protection against insect pests. Environmental factors can strictly impact the efficacy: High temperatures reduce viability, ultraviolet radiation degrades the spores, soil microbes compete or antagonize fungi, and humidity fluctuations deter persistence, resulting in poor field performance as compared to controlled conditions. There are also challenges in mass production that involve propagule stability and formulation for shelf life, whereas isolation of novel strains is very labor-intensive because of irregular soil distribution and geographical variability. Despite effective and promising laboratory or greenhouse results, the performance in the field often falls short because of these inconsistencies, leading to poor overall acceptance in IPM programs. Recent advances in biotechnology, such as genetic engineering, optimized delivery systems, and UV protectants, offer mitigation strategies, but extensive research is needed to reduce variability, enhance scalability, and integrate entomopathogenic endophytic fungi as sustainable alternatives to synthetic insecticides.

31.2.6 Future perspectives and research directions

Future perspectives for endophytic fungi highlight their integration into sustainable agriculture and biotechnology, exploiting advances in genomics, metagenomics, and high-throughput omics technologies to explain plant-endophyte interactions, identify biosynthetic gene clusters, and optimize metabolite production for applications in biocontrol, biofertilizers, and stress tolerance. Genetic engineering and improvement in strain strategies, including coculturing and optimized fermentation media, are helpful to overcome limitations in secondary metabolite yield during repeated subculturing, enabling scalable production of ecofriendly biopesticides that mimic host plant metabolism.

In agriculture, endophytic fungi hold a significant position to enhance crop resilience to abiotic and biotic stresses through induced systemic resistance, with field trials targeting seed coatings and inoculants for major crops such as cotton, maize, and rice to boost yield under climate change scenarios. Bioprospecting from diverse ecosystems, including medicinal plants, will expand the collection of bioactive compounds for nutraceuticals, bioremediation, and biobased manufacturing, supported by AI-driven predictive modeling for strain selection and ecological roles.

Commercialization also faces some regulatory and scalability hurdles, but trends point toward IPM formulations combining endophytes with microbial consortia, reducing synthetic pesticide reliance amid global biopesticide market growth to USD 36.86 billion by 2033, with emphasis on standardized protocols for consistent colonization and nontarget safety. Long-term research priorities include elucidating molecular colonization mechanisms, multitrophic interactions, and ecological sustainability to position endophytic fungi as key players in circular bioeconomies.

References

Abro, M. A., Sun, X., Li, X., Jatoi, G. H., & Guo, L. D. (2019). Biocontrol potential of fungal endophytes against *Fusarium oxysporum* f. sp. *cucumerinum* causing wilt in cucumber. *Plant Pathology Journal, 35*(6), 598–608. http://www.ppjonline.org/journal/download_pdf.php?doi=10.5423/PPJ.OA.05.2019.0129. doi:10.5423/PPJ.OA.05.2019.0129.

Ahmad, H. T., Hussain, A., Aimen, A., Jamshaid, M. U., Ditta, A., Asghar, H. N., & Zahir, Z. A. (2021). Improving resilience against drought stress among crop plants through inoculation of plant growth-promoting rhizobacteria. In *Harsh environment and plant resilience: Molecular and functional aspects* (pp. 387–408). Springer International Publishing. https://link.springer.com/book/10.1007/978-3-030-65912-7. doi:10.1007/978-3-030-65912-7_16.

Ahmad, I., Del Mar Jiménez-Gasco, M., Luthe, D. S., Shakeel, S. N., & Barbercheck, M. E. (2020). Endophytic *Metarhizium robertsii* promotes maize growth, suppresses insect growth, and alters plant defense gene expression. *Biological Control, 144*. http://www.elsevier.com/inca/publications/store/6/2/2/7/9/1/index.htt. doi:10.1016/j.biocontrol.2019.104167.

Akello, J., Dubois, T., Coyne, D., & Kyamanywa, S. (2008). Endophytic *Beauveria bassiana* in banana (*Musa* spp.) reduces banana weevil (*Cosmopolites sordidus*) fitness and damage. *Crop Protection, 27*(11), 1437–1441. doi:10.1016/j.cropro.2008.07.003.

Arnold, A. E., Mejía, L. C., Kyllo, D., Rojas, E. I., Maynard, Z., Robbins, N., & Herre, E. A. (2003). Fungal endophytes limit pathogen damage in a tropical tree. *Proceedings of the National Academy of Sciences of the United States of America, 100*(26), 15649–15654. doi:10.1073/pnas.2533483100.

Aroca, R. (2013). Symbiotic endophytes. *Soil Biology, 37*. doi:10.1007/978-3-642-39317-4.

Baron, N. C., de Souza Pollo, A., & Rigobelo, E. C. (2020). *Purpureocillium lilacinum* and *Metarhizium marquandii* as plant growth-promoting fungi. *PeerJ, 2020*(5). https://peerj.com/archives/?journal=peerj. doi:10.7717/peerj.9005.

Barrow, J. R. (2003). Atypical morphology of dark septate fungal root endophytes of *Bouteloua* in arid southwestern USA rangelands. *Mycorrhiza, 13*(5), 239–247. doi:10.1007/s00572-003-0222-0.

Barelli, L., Moonjely, S., Behie, S. W., & Bidochka, M. J. (2016). Fungi with multifunctional lifestyles: Endophytic insect pathogenic fungi. *Plant Molecular Biology, 90*, 657–664. doi:10.1007/s11103-015-0413-z.

Behie, S. W., & Bidochka, M. J. (2014). Ubiquity of insect-derived nitrogen transfer to plants by endophytic insect-pathogenic fungi: An additional branch of the soil nitrogen cycle. *Applied and Environmental Microbiology, 80*(5), 1553–1560. http://aem.asm.org/content/80/5/1553.full.pdf. doi:10.1128/AEM.03338-13Canada.

Behie, S. W., Zelisko, P. M., & Bidochka, M. J. (2012). Endophytic insect-parasitic fungi translocate nitrogen directly from insects to plants. *Science, 336*(6088), 1576–1577. http://www.sciencemag.org/content/336/6088/1576.full.pdf. doi:10.1126/science.1222289.

Bhunjun, C. S., Phukhamsakda, C., Hyde, K. D., McKenzie, E. H., Saxena, R. K., & Li, Q. (2023). Do all fungi have ancestors with endophytic lifestyles? *Fungal Divers, 16*, 1–26.

Bömke, C, & Tudzynski, B (2009). Diversity, regulation, and evolution of the gibberellin biosynthetic pathway in fungi compared to plants and bacteria. *Phytochemistry, 70*, 1876–1893. doi:10.1016/j.phytochem.2009.05.020.

Dash, C. K., Bamisile, B. S., Keppanan, R., Qasim, M., Lin, Y., Islam, S. U., Hussain, M., & Wang, L. (2018). Endophytic entomopathogenic fungi enhance the growth of *Phaseolus vulgaris* L. (Fabaceae) and negatively affect the development and reproduction of *Tetranychus urticae* Koch (Acari: Tetranychidae). *Microbial Pathogenesis, 125*, 385–392. http://www.elsevier.com/inca/publications/store/6/2/2/9/1/5/index.htt. doi:10.1016/j.micpath.2018.09.044.

Delavaux, C. S., Smith-Ramesh, L. M., & Kuebbing, S. E. (2017). Beyond nutrients: A meta-analysis of the diverse effects of arbuscular mycorrhizal fungi on plants and soils. *Ecology, 98*(8), 2111–2119. http://esajournals.onlinelibrary.wiley.com/hub/journal/10.1002/%28ISSN%291939-9170/issues/. doi:10.1002/ecy.1892.

Ditta, A., & Ullah, N. (2023). Perspectives of using plant growth-promoting rhizobacteria under salinity stress for sustainable crop production. In *Plant stress mitigators: Types, techniques and functions* (pp. 231–247). Elsevier. https://www.sciencedirect.com/book/9780323898713. doi:10.1016/B978-0-323-89871-3.00024-0.

Ditta, A., Ullah, N., Imtiaz, M., Li, X., Jan, A. U., Mehmood, S., Rizwan, M. S., & Rizwan, M. (2022). Zn biofortification in crops through Zn-solubilizing plant growth promoting rhizobacteria. In Q. Mahmood (Ed.), *Sustainable plant nutrition under contaminated environments*. Springer Nature Switzerland AG.

Fan, Y., Liu, X., Keyhani, N. O., Tang, G., Pei, Y., Zhang, W., & Tong, S. (2017). Regulatory cascade and biological activity of *Beauveria bassiana* oosporein that limits bacterial growth after host death. *Proceedings of the National Academy of Sciences of the United States of America, 114*(9), E1578. http://www.pnas.org/content/114/9/E1578.full.pdf. doi:10.1073/pnas.1616543114.

Fang, K., Miao, Y. F., Chen, L., Zhou, J., Yang, Z. P., Dong, X. F., & Zhang, H. B. (2019). Tissue-specific and geographical variation in endophytic fungi of *Ageratina adenophora* and fungal associations with the environment. *Frontiers in Microbiology, 10*. https://www.frontiersin.org/journals/microbiology#. doi:10.3389/fmicb.2019.02919.

Frey-Klett, P., Garbaye, J., & Tarkka, M. (2007). The mycorrhiza helper bacteria revisited. *New phytologist, 176*(1), 22–36.

Fuchs, B., & Krauss, J. (2019). Can *Epichloë* endophytes enhance direct and indirect plant defence? *Fungal Ecology, 38*, 98–103. doi:10.1016/j.funeco.2018.07.002.

Gafur, A (2023). Red root rot disease of tropical estate forests: pathogen identification, dispersal and management. In Detection, diagnosis and management of soil-borne phytopathogens (pp. 159–178). Singapore: Springer Nature. https://doi.org/10.1007/978-981-19-8307-8_7.

Georgousaki, K., González-Menéndez, V., Tormo, J. R., Tsafantakis, N., Mackenzie, T. A., Martín, J., Gumeni, S., Trougakos, I. P., Reyes, F., Fokialakis, N., & Genilloud, O. (2022). Comoclathrin, a novel potent skin-whitening agent produced by endophytic *Comoclathris strains* associated with Andalusia desert plants. *Scientific Reports, 12*(1). Www.nature.com/srep/index.html. doi:10.1038/s41598-022-05448-9.

Göker, M., García-Blázquez, G., Voglmayr, H., Tellería, M. T., Martín, M., & Litvintseva, A. P. (2009). Molecular taxonomy of phytopathogenic fungi: A case study in *Peronospora*. *PLoS One, 4*(7), e6319. doi:10.1371/journal.pone.0006319.

Hardoim, P. R., Van Overbeek, L. S., Berg, G., Pirttilä, A. M., Compant, S., Campisano, A., Döring, M., & Sessitsch, A. (2015). The hidden world within plants: Ecological and evolutionary considerations for defining functioning of microbial endophytes. *Microbiology and Molecular Biology Reviews, 79*(3), 293–320. http://mmbr.asm.org/content/79/3/293.full.pdf+html. doi:10.1128/MMBR.00050-14.

Hasnain, M., Nazir, Q., Dar, A., Hussain, A., Mahmood, S., Saleem, I., Ahmad, H. T., Shah, S. S. H., Ali, M. A., Naveed, M., Iqbal, R., Mustafa, A. E. Z. M. A., Elshikh, M. S., & Ditta, A. (2025). Improving maize physiological attributes by regulating urease activity and zinc availability in rhizosphere through bioactivated zinc-coated urea (Engro Zabardast Urea). *Polish Journal of Environmental Studies, 34*(3), 2687–2695. https://www.pjoes.com/pdf-188186-120597?filename=Improving%20Maize.pdf. doi:10.15244/pjoes/188186.

Hamayun, M., Khan, S. A., Khan, A. L., Rehman, G., Kim, Y. H., Iqbal, I., Hussain, J., Sohn, E.-Y., & Lee, I. J. (2010). Gibberellin production and plant growth promotion from pure cultures of *Cladosporium* sp. MH-6 isolated from cucumber (*Cucumis sativus* L.). *Mycologia, 102*(5), 989–995.

Hawksworth, D. L. (1991). The fungal dimension of biodiversity: Magnitude, significance, and conservation. *Mycological Research, 95*(6), 641–655.

Huang, L. Q., Niu, Y. C., Su, L., Deng, H., & Lyu, H. (2020). The potential of endophytic fungi isolated from cucurbit plants for biocontrol of soilborne fungal diseases of cucumber. *Microbiological Research, 231*. Www.urbanfischer.de/journals/microbiolres/microbio.htm. doi:10.1016/j.micres.2019.126369.

Ismail, M. H., Hussain, A., Iqbal, A., Khan, S. A., & Lee, I. J. (2020). *Aspergillus niger* boosted heat stress tolerance in sunflower and soybean via regulating their metabolic and antioxidant system. *Journal of Plant Interactions, 15*(1), 223–232. http://www.tandf.co.uk/journals/titles/17429145.asp. doi:10.1080/17429145.2020.1771444.

Kaddes, A., Fauconnier, M. L., Sassi, K., Nasraoui, B., & Jijakli, M. H. (2019). Endophytic fungal volatile compounds as solution for sustainable agriculture. *Molecules, 24*(6). https://www.mdpi.com/1420-3049/24/6/1065/pdf. doi:10.3390/molecules24061065.

Khan, A. L., Hamayun, M., Khan, S. A., Kang, S. M., Shinwari, Z. K., Kamran, M., ur Rehman, S., Kim, J. G., & Lee, I. J. (2012). Pure culture of *Metarhizium anisopliae* LHL07 reprograms soybean to higher growth and mitigates salt stress. *World Journal of Microbiology and Biotechnology, 28*(4), 1483–1494. doi:10.1007/s11274-011-0950-9.

Khan, K. S., Qadir, M. F., Ahmad, A., Naveed, M., Raza, T., & Ditta, A. (2022). Efficacy of different endophytic bacterial strains in enhancing growth, yield, and physiological and biochemical attributes of *Linum usitatissimum* L. *Journal of Soil Science and Plant Nutrition, 22*(4), 4365–4376. https://www.springer.com/journal/42729. doi:10.1007/s42729-022-01035-z.

Lata, R., Chowdhury, S., Gond, S. K., & White, J. F. (2018). Induction of abiotic stress tolerance in plants by endophytic microbes. *Letters in Applied Microbiology, 66*(4), 268–276. doi:10.1111/lam.12855.

Latz, M. A. C., Jensen, B., Collinge, D. B., & Jørgensen, H. J. L. (2018). Endophytic fungi as biocontrol agents: Elucidating mechanisms in disease suppression. *Plant Ecology and Diversity, 11*(5–6), 555–567. http://www.tandf.co.uk/journals/titles/17550874.asp. doi:10.1080/17550874.2018.1534146.

Lezama-Gutiérrez, R., Molina-Ochoa, J., Chávez-Flores, O., Ángel-Sahagún, C. A., Skoda, S. R., Reyes-Martínez, G., Barba-Reynoso, M., Rebolledo-Domínguez, O., Ruíz-Aguilar, G. M. L., & Foster, J. E. (2012). Use of the entomopathogenic fungi *Metarhizium anisopliae, Cordyceps bassiana* and *Isaria fumosorosea* to control *Diaphorina citri* (Hemiptera: Psyllidae) in Persian lime under field conditions. *International Journal of Tropical Insect Science, 32*(1), 39–44. doi:10.1017/S1742758412000069.

Maina, U. M., Galadima, I. B., Gambo, F. M., & Zakaria, D. (2018). A review on the use of entomopathogenic fungi in the management of insect pests of field crops. *Journal of Entomology and Zoology Studies, 6*(1), 27–32.

Makino, A., Nakai, R., Yoneda, Y., Toyama, T., Tanaka, Y., Meng, X.-Y., Mori, K., Ike, M., Morikawa, M., Kamagata, Y., & Tamaki, H. (2022). Isolation of aquatic plant growth-promoting bacteria for the floating plant duckweed (*Lemna minor*). *Microorganisms, 10*(8), 1564. doi:10.3390/microorganisms10081564.

Mohammad-Razdari, A., Rousseau, D., Bakhshipour, A., Taylor, S., Poveda, J., & Kiani, H. (2022). Recent advances in E-monitoring of plant diseases. *Biosensors and Bioelectronics, 201*, 113953. doi:10.1016/j.bios.2021.113953.

Mubeen, N., Khalid, A., Ullah, M. I., Altaf, N., Arshad, M., Amin, L., Talat, Q., Sadaf, A., & Farwa (2022). Effect of *Metarhizium anisopliae* on the nutritional physiology of the fall armyworm, *Spodoptera frugiperda* (J.E. Smith) (Lepidoptera: Noctuidae). *Egyptian Journal of Biological Pest Control, 32*(1). https://link.springer.com/journal/41938. doi:10.1186/s41938-022-00573-z.

Muhammad, M., Begum, S., Ditta, A., Liu, Y. H., Li, L., & Li, W. J. (2024). Progress in soil microbiome research. In *Progress in soil science. Metabolites treasure from endophytes: Advances, applications, and technological challenges* (pp. 171–206). Springer.

Naveed, M., Bukhari, S. S., Mustafa, A., Ditta, A., Alamri, S., El-Esawi, M. A., Rafique, M., Ashraf, S., & Siddiqui, M. H. (2020). Mitigation of nickel toxicity and growth promotion in sesame through the application of a bacterial endophyte and zeolite in nickel contaminated soil. *International Journal of Environmental Research and Public Health, 17*(23), 1–27. https://www.mdpi.com/1660-4601/17/23/8859/pdf. doi:10.3390/ijerph17238859.

Nazir, A., Sarfraz, W., Allah, D., Khalid, N., Farid, M., Shafiq, M., Firdaus-E-Bareen, Rizvi, Z. F., & Naeem, N. (2023). Synergistic impact of two autochthonous saprobic fungi (*A. niger* and *T. pseudokoningii*) on the growth, ionic contents, and metals uptake in *Brassica juncea* L. and *Vigna radiata* L. under tannery solid waste contaminated soil. *International Journal of Phytoremediation, 25*(11), 1488–1500. doi:10.1080/15226514.2023.2166457.

Ndereyimana, A., Nyalala, S., Murerwa, P., & Gaidashova, S. (2019). Pathogenicity of some commercial formulations of entomopathogenic fungi on the tomato leaf miner, *Tuta absoluta* (Meyrick) (Lepidoptera: Gelechiidae). *Egyptian Journal of Biological Pest Control, 29*(1). doi:10.1186/s41938-019-0184-y.

Pangesti, N., Pineda, A., Pieterse, C. M. J., Dicke, M., & Loon, J. J. A. (2013). Two-way plant mediated interactions between root-associated microbes and insects: From ecology to mechanisms. *Frontiers in Plant Science, 4*. doi:10.3389/fpls.2013.00414.

Pieterse, C. M. J., Leon-Reyes, A., Van der Ent, S., & Van Wees, S. C. M. (2009). Networking by small-molecule hormones in plant immunity. *Nature Chemical Biology, 5*, 308–316.

Perveen, R., Hussain, A., Ditta, A., Dar, A., Aimen, A., Ahmad, M., Alataway, A., Dewidar, A. Z., & Mattar, M. A. (2023). Growth and yield of okra exposed to a consortium of rhizobacteria with different organic carriers under controlled and natural field conditions. *Horticulturae, 9*(1). doi:10.3390/horticulturae9010008.

Poveda, J., & Baptista, P. (2021). Filamentous fungi as biocontrol agents in olive (*Olea europaea* L.) diseases: Mycorrhizal and endophytic fungi. *Crop Protection, 146*, 105672. doi:10.1016/j.cropro.2021.105672.

Poveda, J. (2021). *Trichoderma* as biocontrol agent against pests: New uses for a mycoparasite. *Biological Control, 159*, 104634. doi:10.1016/j.biocontrol.2021.104634.

Poveda, J., Abril-Urias, P., & Escobar, C. (2020). Biological control of plant-parasitic nematodes by filamentous fungi inducers of resistance: *Trichoderma*, mycorrhizal and endophytic fungi. *Frontiers in Microbiology, 11*. doi:10.3389/fmicb.2020.00992.

Prajapati, C., Rai, S. N., Singh, A. K., Chopade, B. A., Singh, Y., Singh, S. K., Haque, S., Prieto, M. A., & Ashraf, G. M. (2025). An update of fungal endophyte diversity and strategies for augmenting therapeutic potential of their potent metabolites: Recent advancement. *Applied Biochemistry and Biotechnology, 197*(5), 2799–2866. https://www.springer.com/journal/12010. doi:10.1007/s12010-024-05098-9.

Razak, N. A., & Gange, A. C. (2023). Multitrophic interactions between arbuscular mycorrhizal fungi, foliar endophytic fungi and aphids. *Microbial Ecology, 85*(1), 146–156.

Rasool, S., Vidkjær, N. H., Hooshmand, K., Jensen, B., Fomsgaard, I. S., & Meyling, N. V. (2021). Seed inoculations with entomopathogenic fungi affect aphid populations coinciding with modulation of plant secondary metabolite profiles across plant families. *New Phytologist, 229*(3), 1715–1727. http://onlinelibrary.wiley.com/journal/10.1111/. doi:10.1111/nph.16979. (ISSN)1469-8137.

Rodriguez, P. A., Rothballer, M., Chowdhury, S. P., Nussbaumer, T., Gutjahr, C., & Falter-Braun, P. (2019). Systems biology of plant-microbiome interactions. *Molecular Plant, 12*(6), 804–821. doi:10.1016/j.molp.2019.05.006.

Rodriguez, R. J., White Jr, J. F., Arnold, A. E., & Redman, R. S. (2009). Fungal endophytes: Diversity and functional roles. *New Phytologist, 182*(2), 314–330. doi:10.1111/j.1469-8137.2009.02773.x.

Saeed, H. S., Munir, M., Samra, S., Farooq, A., Mahmood, H., Ahmad, I., Akram, M. I., Iqbal, R., Elsadek, M. F., Almutairi, S. M., Ditta, A., & Daraz, U. (2025). Screening of biosurfactants producing fungi from infected *Citrus sinensis* (L.). *Polish Journal of Environmental Studies, 34*(5), 5261–5270. https://www.pjoes.com/pdf-190588-115894?filename=Screening%20of.pdf. doi:10.15244/pjoes/190588.

Saleem, A., Raza, M. A. S., Khan, I. H., Tahir, M. A., Iqbal, R., Aslam, M. U., Ejaz, Z., Ditta, A., Almunqedhi, B. M., & Al Farraj, D. A. (2024). Amelioration of drought stress in wheat by using plant growth-promoting rhizobacteria and biogas slurry. *Global Nest Journal, 26*(8). https://journal.gnest.org/system/files/2024-10/gnest_05890_final.pdf. doi:10.30955/gnj.005890.

Sarfraz, R., Hussain, A., Sabir, A., Ben Fekih, I., Ditta, A., & Xing, S. (2019). Role of biochar and plant growth promoting rhizobacteria to enhance soil carbon sequestration—A review. *Environmental Monitoring and Assessment, 191*(4). https://link.springer.com/journal/10661. doi:10.1007/s10661-019-7400-9.

Schardl, C. L., Florea, S., Pan, J., Nagabhyru, P., Bec, S., & Calie, P. J. (2013). The epichloae: Alkaloid diversity and roles in symbiosis with grasses. *Current Opinion in Plant Biology, 16*(4), 480–488. doi:10.1016/j.pbi.2013.06.012.

Schmid, J., Day, R., Zhang, N., Dupont, P. Y., Cox, M. P., Schardl, C. L., Minards, N., Truglio, M., Moore, N., Harris, D. R., & Zhou, Y. (2017). Host tissue environment directs activities of an *Epichloë* endophyte, while it induces systemic hormone and defense responses in its native perennial ryegrass host. *Molecular Plant-Microbe Interactions, 30*(2), 138–149. http://apsjournals.apsnet.org/doi/pdf/10.1094/MPMI-10-16-0215-R. doi:10.1094/MPMI-10-16-0215-R.

Schulz, B., & Boyle, C. (2005). The endophytic continuum. *Mycological Research, 109*(6), 661–686. doi:10.1017/s095375620500273x.

Scovero, S., Moraglio, S. T., & Tavella, L. (2025). Effects of tomato treatments with commercial formulations of entomopathogenic fungi on the pest *Tuta absoluta* and the predator *Macrolophus pygmaeus*. *Biological Control, 207*, 105832. doi:10.1016/j.biocontrol.2025.105832.

Shah, A. M. U.H., Ditta, A., Parveen, A., Thind, S., & Ebadi, A. G. (2022). Sustainable plant nutrition under contaminated environments. Transcription factors that scavenge reactive oxygen species in rhizobacteria. In: Q. Mahmood Springer Nature Switzerland AG.

Sharma, N., Murtaza, M., Ahmed, W., Goutam, U., & Jaglan, S. (2025). Isolation, characterization, and diversity of fungal endophytes from *Albizia lebbeck* in Jammu & Kashmir, India. *Journal of Applied Biology and Biotechnology, 13*(4), 89–99. https://www.jabonline.in/admin/php/uploads/1343_pdf.pdf. doi:10.7324/JABB.2025.232699.

Sharma, R., & Sharma, P. (2021). Fungal entomopathogens: A systematic review. *Egyptian Journal of Biological Pest Control, 31*(1). doi:10.1186/s41938-021-00404-7.

Shi, J., Wang, X., & Wang, E. (2023). Mycorrhizal symbiosis in plant growth and stress adaptation: From genes to ecosystems. *Annual Review of Plant Biology, 74*, 569–607. http://www.annualreviews.org/journal/arplant. doi:10.1146/annurev-arplant-061722-090342.

Shymanovich, T., & Faeth, S. H. (2018). Anti-insect defenses of *Achnatherum robustum* (sleepygrass) provided by two *Epichloë* endophyte species. *Entomologia Experimentalis et Applicata, 166*(6), 474–482. http://onlinelibrary.wiley.com/journal/10.1111/(ISSN)1570-7458. doi:10.1111/eea.12692.

Stone, J. K., Bacon, C. W., & White, J. F., Jr. (2000). An overview of endophytic microbes: Endophytism defined. In *Microbial endophytes* (pp. 17–44). CRC Press.

Stansly, P. A., & Natwick, E. T. (2010). Integrated systems for managing *Bemisia tabaci* in protected and open field agriculture. In *Bemisia: Bionomics and management of a global pest* (pp. 467–497). Springer Netherlands. http://www.springerlink.com/openurl.asp?genre=book&isbn=978-90-481-2459-6. doi:10.1007/978-90-481-2460-2_17.

Swarnakumari, N (2021). Effect of *Pochonia chlamydosporia* formulations on root-knot nematode, *Meloidogyne incognita* infestation in medicinal coleus. *Coleus Forskohlii PMHE, 27*, 270–273.

Tall, S, & Meyling, N. V. (2018). Probiotics for plants? Growth promotion by the entomopathogenic fungus *Beauveria bassiana* depends on nutrient availability. *Microbial Ecology, 76*, 1002–1008.

Tang, W., Gong, W., Xiao, R., Mao, W., Zhao, L., Song, J., Awais, M., Ji, X., & Li, H. (2023). Endophytic fungal community of *Stellera chamaejasme* L. and its possible role in improving host plants' ecological flexibility in degraded grasslands. *Journal of Fungi, 9*(4), 465. doi:10.3390/jof9040465.

Thind, S., Chaudhary, M. S., Ditta, A., Hussain, I., Parveen, A., Ullah, N., Mahmood, Q., Al-Ashkar, I., & El-Sabagh, A. (2022). Impact of mycorrhizal fungi from different rhizospheric soils on fungal colonization, growth, and chlorophyll contents of *Cenchrus ciliaris*. *Agronomy, 12*(11), 2644. doi:10.3390/agronomy12112644.

Tyagi, J., Chaudhary, P., Mishra, A., Khatwani, M., Dey, S., & Varma, A. (2022). Role of endophytes in abiotic stress tolerance: With special emphasis on *Serendipita indica*. *International Journal of Environmental Research, 16*(4). doi:10.1007/s41742-022-00439-0.

Ullah, N., Ditta, A., Imtiaz, M., Li, X., Jan, A. U., Mehmood, S., Rizwan, M. S., & Rizwan, M. (2021). Appraisal for organic amendments and plant growth-promoting rhizobacteria to enhance crop productivity under drought stress: A review. *Journal of Agronomy and Crop Science, 207*(5), 783–802. http://onlinelibrary.wiley.com/journal/10.1111/. doi:10.1111/jac.12502. (ISSN)1439-037X.

Urooj, F., Farhat, H., Tariq, A., Moin, S., Sohail, N., Sultana, V., Hameedi, S. F., Shams, Z. I., & Ehteshamul-Haque, S. (2021). Role of endophytic *Penicillium* species and *Pseudomonas monteilii* in inducing the systemic resistance in okra against root rotting fungi and their effect on some physiochemical properties of okra fruit. *Journal of Applied Microbiology, 130*(2), 604–616. doi:10.1111/jam.14894.

Vega, F. E., Posada, F., Catherine Aime, M., Pava-Ripoll, M., Infante, F., & Rehner, S. A. (2008). Entomopathogenic fungal endophytes. *Biological Control, 46*(1), 72–82. doi:10.1016/j.biocontrol.2008.01.008.

Verma, S. K., Sahu, P. K., Kumar, K., Pal, G., Gond, S. K., Kharwar, R. N., & White, J. F. (2021). Endophyte roles in nutrient acquisition, root system architecture development and oxidative stress tolerance. *Journal of Applied Microbiology, 131*(5), 2161–2177. doi:10.1111/jam.15111.

Vidal, S., & Jaber, L. R. (2015). Entomopathogenic fungi as endophytes: Plant-endophyte-herbivore interactions and prospects for use in biological control. *Current Science, 109*(1), 46–54. http://www.currentscience.ac.in/Volumes/109/01/0046.pdf.

Wang, C., & St Leger, R. J. (2007). A scorpion neurotoxin increases the potency of a fungal insecticide. *Nature Biotechnology, 25*(12), 1455–1456. doi:10.1038/nbt1357.

Wang, X., Li, Y., Song, X., Meng, Q., Zhu, J., Zhao, Y., & Yu, W. (2017). Influence of host tree species on isolation and communities of mycorrhizal and endophytic fungi from roots of a tropical epiphytic orchid, *Dendrobium sinense* (Orchidaceae). *Mycorrhiza, 27*(7), 709–718. doi:10.1007/s00572-017-0787-7.

Yan, L., Zhu, J., Zhao, X., Shi, J., Jiang, C., & Shao, D. (2019). Beneficial effects of endophytic fungi colonization on plants. *Applied Microbiology and Biotechnology, 103*(8), 3327–3340. Link.springer.de/link/service/journals/00253/index.htm. doi:10.1007/s00253-019-09713-2.

Yizhu, L., Imtiaz, M., Ditta, A., Rizwan, M. S., Ashraf, M., Mehmood, S., Aziz, O., Mubeen, F., Ali, M., Elahi, N. N., Ijaz, R., Lele, S., Shuang, C., & Tu, S. (2020). Response of growth, antioxidant enzymes and root exudates production towards as stress in *Pteris vittata* and in *Astragalus sinicus* colonized by arbuscular mycorrhizal fungi. *Environmental Science and Pollution Research, 27*(2), 2340–2352. https://link.springer.com/journal/11356. doi:10.1007/s11356-019-06785-5.

Zhang, W., Xia, K., Feng, Z., Qin, Y., Zhou, Y., Feng, G., Zhu, H., & Yao, Q. (2024). Tomato plant growth promotion and drought tolerance conferred by three arbuscular mycorrhizal fungi is mediated by lipid metabolism. *Plant Physiology and Biochemistry, 208*, 108478. https://doi.org/10.1016/j.plaphy.2024.108478.

Zhan, X., Wang, R., Zhang, M., Li, Y., Sun, T., Chen, J., Li, J., & Liu, T. (2024). *Trichoderma*-derived emodin competes with ExpR and ExpI of *Pectobacterium carotovorum* subsp. Carotovorum to biocontrol bacterial soft rot. *Pest Management Science, 80*(3), 1039–1052. https://doi.org/10.1002/ps.7835.

Chapter 32

Arbuscular mycorrhizal fungi as key players in soil health: Eco-physiological approaches toward sustainable agriculture

Elmostapha Outamamat, Anas Raklami and Loubna El Fels
Laboratory of Water Sciences, Microbial Biotechnologies, and Natural Resources Sustainability (AQUABIOTECH), Unit of Microbial Biotechnologies, Agrosciences, and Environment (BIOMAGE)-CNRST Labeled Research Unit N°4, Faculty of Sciences Semlalia, University Cadi Ayyad, Marrakech, Morocco

32.1 Introduction

In the context of climate change, ecosystems in arid and semiarid areas, in particular, are subject to the effects of specific climatic conditions characterized by infrequent or irregular rainfall and long, dry summer periods (Singh & Chudasama, 2021). These climatic constraints, combined with anthropogenic pressure, especially from input-intensive agricultural practices involving the use of synthetic inorganic fertilizers and agrochemicals, intensify the production processes needed to meet global food demand, without taking into account the impact of agrochemicals on the functioning of natural ecosystems (Liu et al., 2016; Singh & Chudasama, 2021).

This generally leads to the disruption of vegetation cover coupled with rapid soil erosion. As a result, gradual desertification sets in, resulting in soil depletion of nutrients and organic matter and the loss or reduction of soil microbial density and activity (Kumar et al., 2023). The degradation of soil quality and the disruption of natural ecosystem functioning are evident and therefore threaten environmental sustainability. Soil degradation is now one of the most pressing environmental and socioeconomic challenges worldwide. It could also serve as a warning for future food security (Timmis & Ramos, 2021). In recent years, agroecologists have come to realize that any program aimed at restoring fragile and degraded ecosystems must be thoroughly examined and implemented in a coherent approach (Wang et al., 2019). This is why every restoration initiative must incorporate components from different levels of the ecosystem, particularly soil microorganisms that are closely linked to plants (Nazir et al., 2023; Saeed et al., 2025; Singh Rawat et al., 2023; Thind et al., 2022).

Arbuscular mycorrhizal fungi (AMFs) are well known to be a key component of natural ecosystems around the world. They form mutualistic associations with the roots of more than 80% of terrestrial plants, thereby improving plant growth and survival in harsh environments, such as soils deficient in water and nutrients (Smith & Read, 2010). AMFs are known to improve plant growth and nutrient assimilation (Smith & Read, 2010) and increase their tolerance to biotic and abiotic stresses (Outamamat et al., 2021; Yizhu et al., 2020). At the same time, AMFs contribute to soil stability by promoting soil aggregation and water retention through the mycelial network and the production of glomalin (Singh et al., 2022). The ecological impact of AMF is particularly relevant in the extreme conditions of desert ecosystems (Xie et al., 2024). Today, they are considered an important component of the restoration and functioning of degraded ecosystems (Shen & Wang, 2011). AMF may be best suited to the soil and climatic conditions of the ecosystem concerned (Wang, 2017a, 2017b). In this context, this chapter aims to present the mechanisms by which AMFs influence soil health through their physical, chemical, and biological effects before exploring their role in resilience to abiotic stresses and their applications in agroecology and ecological restoration. Finally, we will discuss the current limitations to their use and the prospects for research and innovation in this field.

Advancing Sustainable Agriculture Through Arbuscular Mycorrhizal Fungi. DOI: https://doi.org/10.1016/B978-0-443-45130-0.00032-1

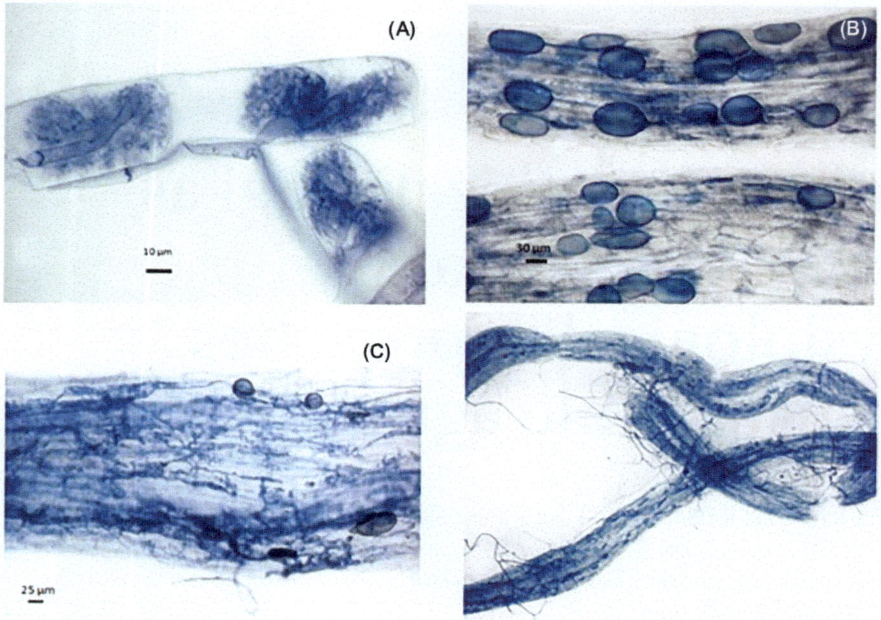

FIGURE 32.1 Characteristic structures of arbuscular mycorrhizal fungi: (A) intercellular arbuscules, (B) intraradicular vesicles, (C) intraradicular hyphae, and (D) extraradicular hyphae.

32.2 Biology and ecology of arbuscular mycorrhizal fungi

AMFs, also known as endomycorrhizae, are microorganisms found in the soil that establish a symbiotic relationship with 80%–90% of vascular plant species, including 90% of agricultural plants, particularly most crops such as cereals, vegetables, and horticultural plants (Fig. 32.1). These fungi are capable of forming mutualistic associations with the fine roots of nearly 80% of all terrestrial plants, whether woody, herbaceous, mosses, ferns, gymnosperms, angiosperms, several conifers, or the majority of flowering plants, both monocotyledons and dicotyledons (Smith & Read, 2010). Their distribution is ubiquitous in global ecosystems, mainly defined by the global distribution of known host plants (Stürmer et al., 2018).

AMFs, classified in the phylum Mucoromycota and the subphylum Glomeromycotina (Glomeromycetes), comprise three classes (Glomeromycetes, Archaeosporomycetes, and Paraglomeromycetes). They belong to 11 families and 25 genera and comprise nearly 250 species (Stürmer, 2012). Glomeromycota are obligate biotrophic symbionts that depend on carbon substrates provided by their host plants, accounting for up to 20% of the carbon fixed by the latter for their survival (Horsch, 2022). In this symbiotic interaction, fungi improve the supply of water and nutrients, such as phosphate and nitrogen, to the host plant (Bagyaraj et al., 2015; Püschel et al., 2020). This improvement occurs through extraradical and intraradical hyphae, arbuscules, and the root apoplast interface. Arbuscules, resulting from the intense branching of intracellular hyphae within the cells of the root cortex, produce characteristic branches. The extensive network of hyphae extends into the soil, forming a symbiotic interface for nutrient exchange (Cargill et al., 2025; Magallon-Servín et al., 2020). In addition, these fungi induce resistance to pathogens such as fungi, bacteria, viruses, and nematodes, while enhancing various biological processes governing ecosystem sustainability and functioning (Poveda et al., 2020; Vishwakarma et al., 2022; Zhang et al., 2020).

The taxonomy of endomycorrhizal fungi is diverse, encompassing different genera and species that establish specific associations with host plants. A thorough understanding of this taxonomy is crucial to understanding the diversity and functioning of these symbiotic partnerships (Brundrett & Tedersoo, 2018).

32.3 Structure of arbuscular mycorrhizal fungi

32.3.1 Spore

The spore serves as a storage and propagation organ for AMF. It germinates and gives rise to mycelial filaments. When the hyphae come into contact with a young root, they form an appressorium, enter, and spread rapidly, differentiating within the roots into arbuscules and, in some cases, vesicles (Fig. 32.1).

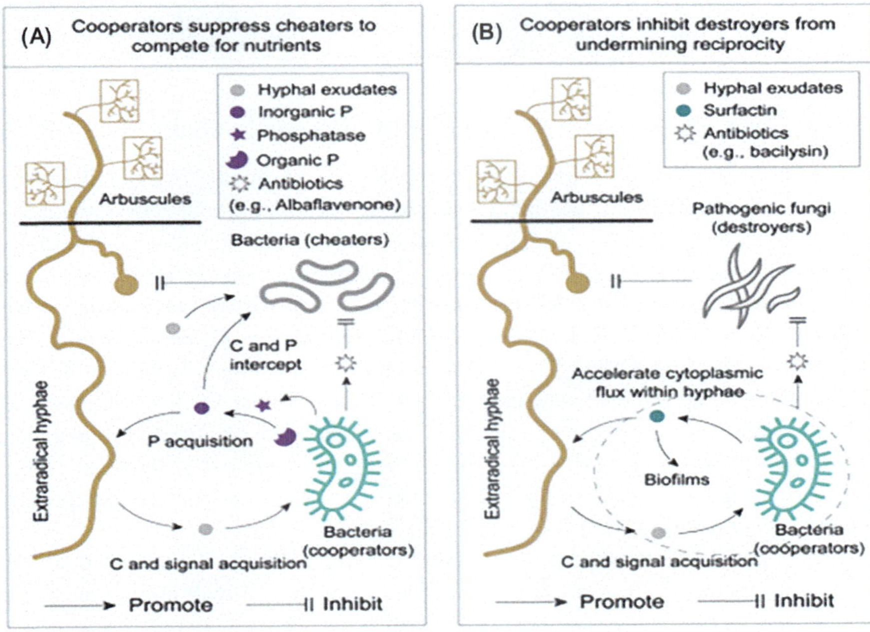

FIGURE 32.2 Interaction between AMF and solubilising bacteria in the mobilisation of organic and inorganic phosphorus. It shows AMF hyphae and PSBs working together to make P available to the plant (Duan et al., 2025). (A) Carbon and P exchange; (B) Carbon and growth-promoting factors feedback. *AMF*, Arbuscular mycorrhizal fungi; *PSBs*, phosphate-solubilizing bacteria.

32.3.2 Arbuscule

The arbuscule is the unit at which exchanges between the host and the fungus occur. It is a lateral branching of fungal hyphae in the root cortex cells, where the fungus penetrates and grows inside. The host cell membrane invaginates and envelops the fungus, and this new compartment provides direct contact between the fungus and the plant (Fig. 32.1).

32.3.3 Vesicle

The vesicle is a thin-walled storage structure with lipid content and generally appears in intercellular spaces (Fig. 32.1).

32.3.4 Extraradicular hypha

The extraradicular hypha produced by the arbuscular mycorrhizal fungus is one of the organs of propagation and can colonize a plant other than the plant from which it originated (Fig. 32.1).

32.4 Role of arbuscular mycorrhizal fungi in improving soil chemical properties

32.4.1 Improving phosphorus availability

Phosphorus is a vital element for plants, involved in energy molecules adenosine triphosphate (ATP), nucleic acids, and cell membranes (Malhotra et al., 2018). Although soils contain large amounts of organic and inorganic P, most of it is immobilized in an insoluble form, bound to iron, calcium, or aluminum oxides, which reduces its availability to less than 1%. Much of this P comes from past fertilizer inputs, forming what is known as legacy P (Jindo et al., 2023; Menezes-Blackburn et al., 2018).

AMFs play a key role in mobilizing these unavailable forms through several complementary mechanisms (Wahid et al., 2019). On the one hand, they associate with phosphate-solubilizing bacteria, which are capable of producing phosphatase enzymes and organic acids that release P from its mineral complexes (Duan et al., 2025; Xiaoxia et al., 2021) (Fig. 32.2). On the other hand, they induce local acidification, chelation, and ion exchange phenomena that solubilize insoluble mineral phosphates, including those derived from natural phosphates (Duan et al., 2025; Xiaoxia et al., 2021). The extramatrix hyphae then absorb the P thus released and transfer it directly to the plant, avoiding losses through slow diffusion. This dual biological and chemical action makes AMF major phosphorus activators, optimizing the use of soil resources and reducing dependence on phosphate fertilizers (Duan et al., 2025; Fall et al., 2022) (Table 32.1).

32.4.2 Contribution of arbuscular mycorrhizal fungi to nitrogen availability in the soil

Nitrogen is an essential component of amino acids, coenzymes, and phospholipids. In the soil, it is found in organic and mineral forms (nitrates and ammonium), but its availability is often limited by mineralization processes and losses through leaching (Shafreen et al., 2021). AMFs significantly improve plant access to nitrogen through direct absorption via specific transporters located in their hyphae (Rui et al., 2022). These are capable of taking up nitrogen in the form of nitrates (NO_3^-), ammonium (NH_4^+), and amino acids, then transferring it to arbuscules, intracellular structures specialized in nutrient exchange. In vitro studies have shown that 20%–50% of root nitrogen can come from the mycorrhizal pathway, with an overall contribution ranging from 21% to 75% depending on the experimental context (Govindarajulu et al., 2005; Jin et al., 2012). Ammonium (NH_4^+) appears to be the most energy-efficient source for AMF (Hawkins et al., 2000; Jin et al., 2012). Several mycorrhizal ammonium transporters have been identified, such as GintAMTs in *Rhizophagus irregularis*, LjAMT2;2 in *Lotus japonicus*, and MtAMT2;3/MtAMT2;4 in *Medicago truncatula* (Guether et al., 2009; Wang et al., 2017). More recently, it has been shown that mycorrhizal symbiosis not only activates ammonium transport but also stimulates its assimilation in plant tissues, as observed in rice (Wang et al., 2017, 2020).

In addition to this direct uptake, AMFs stimulate microbial decomposition activity, which accelerates the release of mineral nitrogen from organic matter (Wei et al., 2019). Although unable to break down complex organic compounds themselves (as they lack the necessary genes), AMFs cooperate with other soil microorganisms to mobilize organic nitrogen (Jansa et al., 2019) (Table 32.2). In this context, they can acquire approximately 31% of their total nitrogen from organic pockets, representing 3%–20% of the host plant's nitrogen (Hodge & Storer, 2015; Jansa et al., 2019).

Thus the mycelial network of AMF acts as a functional extension of the roots, capable of exploring distant areas that are poor in nutrients, capturing nitrogen there, and transporting it to the plant (Simard & Durall, 2004). Although the exact quantitative contribution of AMF to the overall nitrogen cycle remains to be determined, their role in optimizing and securing it is undeniable, particularly in poor, degraded soils or those subject to high losses through leaching (Ebbisa, 2022).

32.4.3 Contribution of arbuscular mycorrhizal fungi to the carbon cycle and sequestration

AMFs play a central role in the global carbon (C) cycle, acting as key players in its transfer, transformation, and stabilization in soils (Shukla et al., 2025). It is estimated that the annual amount of carbon allocated by plants to the underground mycelial networks of AMF is equivalent to nearly 36% of current carbon dioxide (CO_2) emissions from fossil fuels (Hawkins et al., 2023; Zhang et al., 2024). This massive contribution stems from the fact that host plants transfer between 4% and 20% of their photosynthetic carbon to their fungal partners, which is then redistributed in the soil (Luo et al., 2023; Parihar et al., 2020).

The carbon provided by AMF comes from several sources: root and fungal exudates, living hyphae, necromass from mycelial renewal, and hyphae-bound proteins such as glomalin. All of these flows contribute to the formation of mineral-associated organic matter, a particularly stable fraction of soil carbon (Kuyper & Jansa, 2023; Parihar et al., 2020). The hyphae of AMFs, whose density can reach more than 30 m g^{-1} of soil, are 10 times greater than root density (Košnar et al., 2025). Thus they represent a huge surface area for interaction with the mineral and microbial matrix of the soil. The carbon stored in these networks is estimated at between 54 and 900 kg ha^{-1}, constituting a carbon sink that is difficult to access by root systems alone (Zhu & Miller, 2003).

Fungal exudates, sugars, carboxylates, and amino acids readily link with soil minerals and contribute to the formation of stable aggregates (Zhou et al., 2024). In addition, glomalin and related proteins, produced at rates of more than 4 Mg ha^{-1} per year, accumulate in soils over long periods of up to 10 years (Steinberg & Rillig, 2003). These compounds enhance soil structural stability, protect organic matter from microbial degradation, and promote long-term carbon storage (Li et al., 2025). Furthermore the presence of AMF influences the decomposition dynamics of organic residues. In most cases, they accelerate the decomposition of organic inputs and stimulate nitrogen release (Wei et al., 2019). However, some studies show that AMF can slow down the decomposition of lignified substrates, such as wood, thereby contributing to carbon stabilization during decomposition (Leifheit et al., 2015; Verbruggen et al., 2016).

Finally, experimental observations confirm that CO_2 fluxes in hyphal compartments are higher in the presence of AMF, indicating increased respiration and a strong involvement in regulating soil C–N–P exchanges (Andrino de la Fuente, 2019; Parihar et al., 2020). A recent conceptual framework distinguishes four pathways of carbon stabilization mediated by AMF: regeneration, reprocessing, reorganization, and stabilization of organic matter (Wu et al., 2023) (Table 32.2). Nevertheless,

TABLE 32.1 Experimental data on AMF–bacteria interactions in the nitrogen and phosphorus cycles.

Nutrient	Nutrient source(s)	AMF involved	Associated bacteria	Plant(s) studied	Essential results	References
P	Phytin	*Rhizophagus irregularis*	*Pseudomonas alcaligenes* M20, *Bacillus megaterium* C4, *Rahnella aquatilis* HX2	Maize	Enhanced organic P mineralization, increased microbial biomass P	Wang et al. (2016)
P	Phytate	*R. irregularis*	*R. aquatilis* HX2	Carrot root organ cultures	Enhanced organic P mineralization	Zhang et al. (2018)
P	Phytin, KH_2PO_4	*R. irregularis*	Bacterial filtrate	*Medicago truncatula*	Enhanced P turnover	Wang et al. (2023)
N	Maize leaves, agricultural soil, NH_4NO_3	*R. irregularis*	–	Maize	Reduced net N_2O emissions	Storer et al. (2018)
N	Faba bean residues	*Funneliformis mosseae*	*Pseudomonas fluorescens* JL1	Maize	Upregulated *nosZ* expression, reduced N_2O emissions	Li et al. (2023)
N	Ryegrass leaves	*R. irregularis*	Live soil microbial inocula	*Brachypodium distachyon*	Increased N acquisition	Hestrin et al. (2019)
N	^{15}N-labeled chitin	*R. irregularis*	Various soil bacteria + protist (*Polysphondylium pallidum*)	–	Increased organic N utilization	Rozmoš et al. (2022)
N	–	*R. irregularis*	Soil microbial filtrate	*Plantago lanceolata*	Altered community structure of ammonia oxidizers	Veresoglou et al. (2019)

AMF, arbuscular mycorrhizal fungi.

TABLE 32.2 Role of AMF and associated bacteria in the carbon cycle and sequestration.

Carbon process	Carbon source(s)	AMF species	Associated bacteria	Plant(s)	Essential results	References
Soil carbon sequestration via glomalin production	Plant-derived C allocated to roots	*Rhizophagus irregularis*	–	Maize, wheat	Increased glomalin-related soil protein, enhanced soil C storage	Sellappan et al. (2022)
Stabilization of soil organic matter	Root exudates, litter inputs	AMF consortia	Soil microbial community	Grassland species	AMF hyphae contribute to soil aggregate formation and C stabilization	Wilson et al. (2009)
Priming of soil organic matter decomposition	Root-derived C	*Glomus hoi*	Diverse bacterial inocula	*Plantago lanceolata*	AMF hyphae and bacteria synergistically enhance litter decomposition and nutrient release	Nuccio et al. (2013)
Carbon fluxes within common mycorrhizal networks	Plant photosynthates	*Funneliformis mosseae*	Soil microbiome	Multiple plant species	AMFs mediate C allocation between plants and shape microbial communities	Bücking and Kafle (2015)
Enhanced C use efficiency	Plant-derived C	*Rhizoglomus intraradices*	Phosphate-solubilizing bacteria	Maize	Increased microbial biomass C and higher C use efficiency in soil	Zhang et al. (2018)

AMF, arbuscular mycorrhizal fungi.

accurately quantifying the proportion of carbon fixed and stabilized by AMF remains a challenge because of the temporal and spatial variability of plant–AMF–soil exchanges (Parihar et al., 2020). These results highlight the importance of integrating AMF into sustainable management strategies for agricultural and natural soils. Promoting their activity could not only improve soil fertility and structure but also enhance their role in carbon sequestration, contributing to climate change mitigation.

32.4.4 Improved absorption of trace elements

In addition to their role in acquiring macronutrients, AMFs significantly improve the absorption of essential trace elements such as potassium (K), calcium (Ca), magnesium (Mg), copper (Cu), zinc (Zn), iron (Fe), manganese (Mn), cobalt (Co), and sulfur (S). These elements, which are often immobile in the soil, become more accessible thanks to the vast extramatricial network of hyphae that explores a much larger volume of soil than that reached by roots alone (Lehmann & Rillig, 2015; Wahab et al., 2023). The hyphae also secrete chelating compounds, such as organic acids and siderophores, which increase the solubility and availability of these trace elements to the plant (Bhantana et al., 2021; Lehmann & Rillig, 2015). This strategy compensates for the low mobility of these nutrients in the soil and ensures a better nutritional balance. Furthermore, in polluted or heavy metal-rich environments, AMF can perform a protective function. They immobilize some of the excess elements in their fungal structures, thereby reducing their toxicity and protecting the host plant (Kumar & Saxena, 2019; Raklami et al., 2019). This buffering role, combined with their ability to mobilize nutrients in the event of deficiency, gives plants greater tolerance to nutritional and environmental stresses.

32.4.5 Role of arbuscular mycorrhizal fungi in soil structure and stability

AMFs play a key role in the formation, stability, and functioning of soil aggregates, which are essential for maintaining the fertility and sustainability of ecosystems. Soil aggregates provide fundamental ecosystem services: maintaining porosity, regulating gas exchange, water infiltration and storage, erosion resistance, and organic carbon sequestration (Chaudhary et al., 2025; Fall et al., 2022).

32.4.6 Direct mechanisms: hyphae and glomalin secretion

The extraradicular hyphae of AMF act as a biological net that entangles and connects primary soil particles, organic matter, and microaggregates to form stable macroaggregates (Morris et al., 2019). At the same time, these hyphae secrete exudates (sugars, organic acids, and proteins) and, above all, glomalin, a glycoprotein specific to AMF. Glomalin acts as a biological cement by binding mineral and organic particles, thereby increasing the cohesion and mechanical strength of aggregates (Prasad et al., 2018). These processes slow down the disintegration of macroaggregates and stimulate the renewal dynamics of microaggregates, thus contributing to sustainable soil structure.

32.4.7 Indirect mechanisms: Interaction with plants and microflora

Indirectly, AMFs influence soil structure through their effects on host plants: increased root biomass, altered fine root morphology, and stimulation of rhizodeposition (Jones et al., 2004). Mycorrhizal roots exude more organic compounds that serve as substrates for soil microorganisms. These microbial communities associated with AMF actively participate in biochemical processes (decomposition and production of extracellular polymers) that reinforce the structural stability of soils. The microsites within the aggregates thus constitute privileged niches for biophysical and biochemical interactions mediated by AMF (Agnolucci et al., 2015; Scheublin et al., 2010; Trabelsi & Mhamdi, 2013).

32.4.8 Effects on soil physical properties

By promoting aggregation, AMFs improve soil porosity and water retention capacity, resulting in better infiltration and increased resistance to compaction (Pauwels et al., 2023). This structuring also optimizes air and nutrient circulation, thereby reducing the risk of root asphyxia. In addition, the stabilization of aggregates limits soil disintegration under the effect of rain or wind, greatly reducing water and wind erosion (Fall et al., 2022; Zheng et al., 2023).

32.4.9 Influence of soil management practices

The benefits of AMF on soil structure and stability depend heavily on farming practices. Excessive mineral fertilization, particularly with phosphorus, reduces plants' dependence on AMF, leading to a decline in root colonization and fungal diversity (Jansa et al., 2006; Manoharan et al., 2017). Similarly, intensive use of pesticides or herbicides (e.g., glyphosate and azoxystrobin) can inhibit spore germination and limit mycelial expansion (Pagano et al., 2023). Conversely, moderate organic fertilizations or no-till/soil conservation practices promote the abundance and diversity of AMF, enhancing their positive effects on soil structure and aggregate stability (Fall et al., 2022; Gu et al., 2020).

32.5 Agroecological and restorative applications

32.5.1 Use of arbuscular mycorrhizal fungi in sustainable agriculture

AMFs are an essential component of beneficial plant microbiomes and represent a promising solution for improving soil health, ensuring sustainable agriculture, and enhancing food security (Nadeem et al., 2014; Suman et al., 2022). Their integration as biofertilizers reduces the use of chemical fertilizers and pesticides, while enhancing crop resilience to abiotic stresses (Berruti et al., 2016; Wahab et al., 2023). They also play a key role in agroforestry systems, intercropping, and organic farming, where they promote nutrient exchange and soil structuring (Fall et al., 2022). The following AMF-friendly management strategies are now being developed to maximize their benefits (Rillig et al., 2016; Volpe et al., 2023b):

1. Reduced or no tillage to preserve mycelial networks.
2. Diversification and crop rotation to increase AMF diversity and population.
3. Reduction in synthetic inputs (fertilizers and pesticides) to limit negative effects on colonization and mycorrhizal functions.
4. Introduction of cover crops to increase carbon flow to the soil and stimulate AMF activity.
5. Selection and genetic improvement of cultivars capable of exploiting indigenous AMF more efficiently.
6. Inoculation with effective AMF strains to enhance their functions.
7. Addition of biostimulants or mycorrhizal-auxiliary bacteria that stimulate hyphal growth and efficiency.

Furthermore, recent advances in molecular biology have highlighted the importance of chemical signals in plant–AMF dialogue. Strigolactones exuded by roots stimulate hyphal branching and the activation of symbiotic responses, whereas certain fungal molecules, such as chitooligosaccharides and lipo-chitooligosaccharides, trigger recognition reactions in the plant (Volpe et al., 2023a). Thus combining these sustainable agricultural practices with a better understanding of molecular signals opens up new perspectives for maximizing the effectiveness of AMF in agroecosystems.

32.5.2 Restoration of degraded soils

AMFs also play an essential role in the rehabilitation of poor or desertified soils, particularly in arid and semiarid areas (Elmostapha, Hanane, Rachid, & Lahcen, 2021; Wang, 2017a, 2017b). Their colonization of pioneer plants improves rooting, drought tolerance, and nutrient availability, creating a virtuous cycle of ecological restoration. They are also involved in assisted phytoremediation, where their ability to immobilize or redistribute certain pollutants (heavy metals and excess nutrients) helps to stabilize contaminated soils and limit the transfer of contaminants into food chains. Several case studies in arid areas have shown that mycorrhizal inoculation, combined with adapted plants, promotes the return of vegetation cover and improves the ecological resilience of soils (Raklami et al., 2019; Yan et al., 2020).

32.5.3 Limitations and prospects

Although AMFs have major potential, several obstacles hinder their widespread application. On the one hand the production of inoculants remains technically complex, costly, and dependent on the presence of host plants (Madawala, 2021). Their effectiveness varies according to soil and climate conditions, local diversity, and the plant species cultivated (Alguacil et al., 2014). Furthermore the variability of responses observed across crops and habitats is a practical limitation: some species benefit greatly from symbiosis, whereas others respond poorly, complicating the standardization of AMF-based biofertilizers (Berruti et al., 2016; Ghorui et al., 2024). Several approaches are emerging to overcome these limitations:

- The use of AMF-friendly management strategies (reduced tillage, crop diversification, reduced chemical inputs, cover crops, etc.) to preserve and strengthen native mycorrhizal networks (Rillig et al., 2016).

- Developing genomics, bioengineering, and understanding of molecular signals to improve symbiotic efficiency (Volpe et al., 2023a).
- The exploration of complex microbial partnerships, combining AMF, phosphorus-solubilizing bacteria, and other beneficial microorganisms, to create more effective microbial consortia that are adapted to the real conditions of agroecosystems (Wahid et al., 2016).

In short, future success will depend on an integrated approach combining fundamental research, technological innovation, and sustainable agricultural practices to make AMF an essential lever in agroecological transition and soil restoration.

32.6 Conclusions

AMFs play a key role in soil functioning and agricultural sustainability. Their biology and ecology give them a unique ability to interact with most terrestrial plants and directly influence soil fertility. Through various mechanisms, they improve phosphorus availability, promote nitrogen mobilization and transfer, participate in carbon cycling and sequestration, and optimize trace element uptake. At the same time, their role in aggregate stabilization and glomalin production contributes to soil structure and resilience. These ecosystem services have practical applications in agroecology, whether through the use of AMF as natural biofertilizers, their integration into sustainable agricultural systems, or their mobilization for the restoration of degraded soils. However, limitations remain because of variability in their effectiveness depending on species, soil conditions, and farming practices. Prospects lie in the selection of suitable strains, the development of more effective bioinoculants, and their integration into comprehensive sustainable soil management strategies. In short, AMFs appear to be strategic allies in reconciling **fertility, productivity, and sustainability**, paving the way for agriculture that is more resilient in the face of global change.

References

Agnolucci, M., Battini, F., Cristani, C., & Giovannetti, M. (2015). Diverse bacterial communities are recruited on spores of different arbuscular mycorrhizal fungal isolates. *Biology and Fertility of Soils, 51*(3), 379–389. doi:10.1007/s00374-014-0989-5.

Alguacil, M. M., Torrecillas, E., García-Orenes, F., & Roldán, A. (2014). Changes in the composition and diversity of AMF communities mediated by management practices in a Mediterranean soil are related with increases in soil biological activity. *Soil Biology and Biochemistry, 76*, 34–44. doi:10.1016/j.soilbio.2014.05.002.

Andrino de la Fuente, A. (2019). *Carbon and phosphorus trading in the arbuscular mycorrhizal symbiosis*. Institutionelles Repositorium der Leibniz Universität Hannover.

Bagyaraj, D. J., Sharma, M. P., & Maiti, D. (2015). Phosphorus nutrition of crops through arbuscular mycorrhizal fungi. *Current Science, 108*, 1288–1293.

Berruti, A., Lumini, E., Balestrini, R., & Bianciotto, V. (2016). Arbuscular mycorrhizal fungi as natural biofertilizers: Let's benefit from past successes. *Frontiers in Microbiology, 6*. doi:10.3389/fmicb.2015.01559.

Bhantana, P., Rana, M. S., Sun, X. c., Moussa, M. G., Saleem, M. H., Syaifudin, M., Shah, A., Poudel, A., Pun, A. B., Bhat, M. A., Mandal, D. L., Shah, S., Zhihao, D., Tan, Q., & Hu, C. X. (2021). Arbuscular mycorrhizal fungi and its major role in plant growth, zinc nutrition, phosphorous regulation and phytoremediation. *Symbiosis, 84*(1), 19–37. https://rd.springer.com/journal/13199. doi:10.1007/s13199-021-00756-6.

Brundrett, M. C., & Tedersoo, L. (2018). Evolutionary history of mycorrhizal symbioses and global host plant diversity. *New Phytologist, 220*(4), 1108–1115. http://onlinelibrary.wiley.com/journal/10.1111/(ISSN)1469-8137. doi:10.1111/nph.14976.

Bücking, H., & Kafle, A. (2015). Role of arbuscular mycorrhizal fungi in the nitrogen uptake of plants: Current knowledge and research gaps. *Agronomy, 5*(4), 587–612. doi:10.3390/agronomy5040587.

Cargill, R. I. M., Shimizu, T. S., Kiers, E. T., & Kokkoris, V. (2025). Cellular anatomy of arbuscular mycorrhizal fungi. *Current Biology, 35*(11), R545. https://www.sciencedirect.com/science/journal/09609822. doi:10.1016/j.cub.2025.03.053.

Chaudhary, A., Poudyal, S., & Kaundal, A. (2025). Role of arbuscular mycorrhizal fungi in maintaining sustainable agroecosystems. *Applied Microbiology, 5*, 6.

Duan, S., Jin, Z., Zhang, L., & Declerck, S. (2025). Mechanisms of cooperation in the plants-arbuscular mycorrhizal fungi-bacteria continuum. *The ISME Journal, 19*(1). doi:10.1093/ismejo/wraf023.

Ebbisa, A. (2022). Arbuscular mycorrhizal fungi in agriculture-new insights. In A. Ebbisa (Ed.), *Arbuscular mycorrhizal fungi (AMF) in optimizing nutrient bioavailability and reducing agrochemicals for maintaining sustainable agroecosystems*. IntechOpen.

Fall, A. F., Nakabonge, G., Ssekandi, J., Founoune-Mboup, H., Apori, S. O., Ndiaye, A., Badji, A., & Ngom, K. (2022). Roles of arbuscular mycorrhizal fungi on soil fertility: Contribution in the improvement of physical, chemical, and biological properties of the soil. *Frontiers in Fungal Biology, 3*. Www.frontiersin.org/journals/fungal-biology. doi:10.3389/ffunb.2022.723892.

Ghorui, M., Chowdhury, S., Balu, P., & Burla, S. (2024). Arbuscular mycorrhizal inoculants and its regulatory landscape. *Heliyon, 10*(9), e30359. doi:10.1016/j.heliyon.2024.e30359.

Govindarajulu, M., Pfeffer, P. E., Jin, H., Abubaker, J., Douds, D. D., Allen, J. W., Bücking, H., Lammers, P. J., & Shachar-Hill, Y. (2005). Nitrogen transfer in the arbuscular mycorrhizal symbiosis. *Nature, 435*(7043), 819–823. doi:10.1038/nature03610.

Gu, S., Wu, S., Guan, Y., Zhai, C., Zhang, Z., Bello, A., Guo, X., & Yang, W. (2020). Arbuscular mycorrhizal fungal community was affected by tillage practices rather than residue management in black soil of northeast China. *Soil and Tillage Research, 198*, 104552. doi:10.1016/j.still.2019.104552.

Guether, M., Neuhäuser, B., Balestrini, R., Dynowski, M., Ludewig, U., & Bonfante, P. (2009). A mycorrhizal-specific ammonium transporter from *Lotus japonicus* acquires nitrogen released by arbuscular mycorrhizal fungi. *Plant Physiology, 150*(1), 73–83. http://www.plantphysiol.org/cgi/reprint/150/1/73. doi:10.1104/pp.109.136390.

Hawkins, H. J., Cargill, R. I. M., Van Nuland, M. E., Hagen, S. C., Field, K. J., Sheldrake, M., Soudzilovskaia, N. A., & Kiers, E. T. (2023). Mycorrhizal mycelium as a global carbon pool. *Current Biology, 33*(11), R560. http://www.elsevier.com/journals/current-biology/0960-9822. doi:10.1016/j.cub.2023.02.027.

Hawkins, H.-J., Johansen, A., & George, E. (2000). Uptake and transport of organic and inorganic nitrogen by arbuscular mycorrhizal fungi. *Plant Soil, 226*, 275–285.

Hestrin, R., Hammer, E. C., Mueller, C. W., & Lehmann, J. (2019). Synergies between mycorrhizal fungi and soil microbial communities increase plant nitrogen acquisition. *Communications Biology, 2*(1). Nature.com/commsbio/. doi:10.1038/s42003-019-0481-8.

Hodge, A., & Storer, K. (2015). Arbuscular mycorrhiza and nitrogen: Implications for individual plants through to ecosystems. *Plant and Soil, 386*(1–2), 1–19. doi:10.1007/s11104-014-2162-1.

Horsch, C. C. A. (2022). *Assessing the contributions of arbuscular mycorrhizal community life-history traits and trait diversity to soil carbon cycling and plant nutrition*. McGill University.

Jansa, J., Forczek, S. T., Rozmoš, M., Püschel, D., Bukovská, P., & Hršelová, H. (2019). Arbuscular mycorrhiza and soil organic nitrogen: Network of players and interactions. *Chemical and Biological Technologies in Agriculture, 6*(1). Chembioagro.springeropen.com/. doi:10.1186/s40538-019-0147-2.

Jansa, J., Wiemken, A., & Frossard, E. (2006). The effects of agricultural practices on arbuscular mycorrhizal fungi. *Geological Society, London, Special Publications, 266*, 89–115.

Jin, H., Liu, J., Liu, J., & Huang, X. W. (2012). Forms of nitrogen uptake, translocation, and transfer via arbuscular mycorrhizal fungi: A review. *Science China Life Sciences, 55*(6), 474–482. doi:10.1007/s11427-012-4330-y.

Jindo, K., Audette, Y., Olivares, F. L., Canellas, L. P., Smith, D. S., & Voroney, R. P. (2023). Biotic and abiotic effects of soil organic matter on the phytoavailable phosphorus in soils: A review. *Chemical and Biological Technologies in Agriculture, 10*(1). https://chembioagro.springeropen.com. doi:10.1186/s40538-023-00401-y.

Jones, D. L., Hodge, A., & Kuzyakov, Y. (2004). Plant and mycorrhizal regulation of rhizodeposition. *New Phytologist, 163*, 459–480.

Košnar, J., Šmilauer, P., & Šmilauerová, M. (2025). Arbuscular and fine root-endophytic mycorrhizal fungi forage differently for nutrients in a seminatural temperate grassland. *Biology and Fertility of Soils, 61*, 773–790.

Kumar, S., David Raj, A., Kalambukattu, J. G., & Chatterjee, U. (2023). Climate change impact on land degradation and soil erosion in hilly and mountainous landscape: Sustainability issues and adaptation strategies. *Ecological footprints of climate change: Adaptive approaches and sustainability*. Springer.

Kumar, S., & Saxena, S. (2019). Arbuscular mycorrhizal fungi (AMF) from heavy metal-contaminated soils: Molecular approach and application in phytoremediation. *Biofertilizers for sustainable agriculture and environment*. Springer.

Kuyper, T. W., & Jansa, J. (2023). Arbuscular mycorrhiza: Advances and retreats in our understanding of the ecological functioning of the mother of all root symbioses. *Plant and Soil, 489*(1–2), 41–88. https://www.springer.com/journal/11104. doi:10.1007/s11104-023-06045-z.

Lehmann, A., & Rillig, M. C. (2015). Arbuscular mycorrhizal contribution to copper, manganese and iron nutrient concentrations in crops—A meta-analysis. *Soil Biology and Biochemistry, 81*, 147–158. Www.elsevier.com/inca/publications/store/3/3/2. doi:10.1016/j.soilbio.2014.11.013.

Leifheit, E. F., Verbruggen, E., & Rillig, M. C. (2015). Arbuscular mycorrhizal fungi reduce decomposition of woody plant litter while increasing soil aggregation. *Soil Biology and Biochemistry, 81*, 323–328.

Li, M. Y., Wang, W., Yin, H. H., Chen, Y., Ashraf, M., Tao, H. Y., Li, S. S., Wang, W. Y., Yang, C. L., Xiao, Y. L., Zhu, L., & Xiong, Y. C. (2025). The functional role of arbuscular mycorrhizal fungi in enhancing soil organic carbon stocks and stability in dryland. *Soil and Tillage Research, 248*. https://www.sciencedirect.com/science/journal/01671987. doi:10.1016/j.still.2024.106443.

Li, X., Zhao, R., Li, D., Wang, G., Bei, S., Ju, X., An, R., Li, L., Kuyper, T. W., Christie, P., Bender, F. S., Veen, C., van der Heijden, M. G. A., van der Putten, W. H., Zhang, F., Butterbach-Bahl, K., & Zhang, J. (2023). Mycorrhiza-mediated recruitment of complete denitrifying *Pseudomonas* reduces N_2O emissions from soil. *Microbiome, 11*(1). https://microbiomejournal.biomedcentral.com/. doi:10.1186/s40168-023-01466-5.

Liu, H. L., Willems, P., Bao, A. M., Wang, L., & Chen, X. (2016). Effect of climate change on the vulnerability of a socio-ecological system in an arid area. *Global and Planetary Change, 137*, 1–9. http://www.sciencedirect.com/science/journal/09218181. doi:10.1016/j.gloplacha.2015.12.014.

Luo, X., Liu, Y., Li, S., & He, X. (2023). Interplant carbon and nitrogen transfers mediated by common arbuscular mycorrhizal networks: Beneficial pathways for system functionality. *Frontiers in Plant Science, 14*. doi:10.3389/fpls.2023.1169310.

Madawala, H. M. S. P. (2021). Arbuscular mycorrhizal fungi as biofertilizers: Current trends, challenges, and future prospects. In *Biofertilizers* (pp. 83–93). Elsevier BV. doi:10.1016/b978-0-12-821667-5.00029-4.

Magallon-Servín, P., Antoun, H., Taktek, S., Bashan, Y., & de-Bashan, L. (2020). The maize mycorrhizosphere as a source for isolation of arbuscular mycorrhizae-compatible phosphate rock-solubilizing bacteria. *Plant and Soil, 451*(1–2), 169–186. doi:10.1007/s11104-019-04226-3.

Malhotra, H., Vandana, S. S., & Pandey, R. (2018). Phosphorus nutrition: Plant growth in response to deficiency and excess. In *Plant nutrients and abiotic stress tolerance* (pp. 171–190). Springer Singapor. http://www.springer.com/in/book/9789811090431. doi:10.1007/978-981-10-9044-8_7.

Manoharan, L., Rosenstock, N. P., Williams, A., & Hedlund, K. (2017). Agricultural management practices influence AMF diversity and community composition with cascading effects on plant productivity. *Applied Soil Ecology, 115*, 53–59. Www.elsevier.com/inca/publications/store/5/2/4/5/1/8/index.htt. doi:10.1016/j.apsoil.2017.03.012.

Menezes-Blackburn, D., Giles, C., Darch, T., George, T. S., Blackwell, M., Stutter, M., Shand, C., Lumsdon, D., Cooper, P., Wendler, R., Brown, L., Almeida, D. S., Wearing, C., Zhang, H., & Haygarth, P. M. (2018). Opportunities for mobilizing recalcitrant phosphorus from agricultural soils: A review. *Plant and Soil, 427*(1–2), 5–16. http://www.wkap.nl/journalhome.htm/0032-079X. doi:10.1007/s11104-017-3362-2.

Morris, E. K., Morris, D. J. P., Vogt, S., Gleber, S. C., Bigalke, M., Wilcke, W., & Rillig, M. C. (2019). Visualizing the dynamics of soil aggregation as affected by arbuscular mycorrhizal fungi. *ISME Journal, 13*(7), 1639–1646. http://www.nature.com/ismej/marketing/aims_scope.html. doi:10.1038/s41396-019-0369-0.

Nadeem, S. M., Ahmad, M., Zahir, Z. A., Javaid, A., & Ashraf, M. (2014). The role of mycorrhizae and plant growth promoting rhizobacteria (PGPR) in improving crop productivity under stressful environments. *Biotechnology Advances, 32*(2), 429–448. Www.elsevier.com/inca/publications/store/5/2/5/4/5/5/index.htt. doi:10.1016/j.biotechadv.2013.12.005.

Nazir, A., Sarfraz, W., Allah, D., Khalid, N., Farid, M., Shafiq, M., Firdaus-E-Bareen, Rizvi, Z. F., & Naeem, N. (2023). Synergistic impact of two autochthonous saprobic fungi (*A. niger* and *T. pseudokoningii*) on the growth, ionic contents, and metals uptake in *Brassica juncea* L. and *Vigna radiata* L. under tannery solid waste contaminated soil. *International Journal of Phytoremediation, 25*(11), 1488–1500. doi:10.1080/15226514.2023.2166457.

Nuccio, E. E., Hodge, A., Pett-Ridge, J., Herman, D. J., Weber, P. K., & Firestone, M. K. (2013). An arbuscular mycorrhizal fungus significantly modifies the soil bacterial community and nitrogen cycling during litter decomposition. *Environmental Microbiology, 15*(6), 1870–1881. doi:10.1111/1462-2920.12081.

Outamamat, E., Bourhia, M., Dounas, H., Salamatullah, A. M., Alzahrani, A., Alyahya, H. K., Albadr, N. A., Al Feddy, M. N., Mnasri, B., & Ouahmane, L. (2021). Application of native or exotic arbuscular mycorrhizal fungi complexes and monospecific isolates from saline semi-arid Mediterranean ecosystems improved phoenix dactylifera's growth and mitigated salt stress negative effects. *Plants, 10*(11). https://www.mdpi.com/2223-7747/10/11/2501/pdf. doi:10.3390/plants10112501.

Pagano, M. C., Kyriakides, M., & Kuyper, T. W. (2023). Effects of pesticides on the arbuscular mycorrhizal symbiosis. *Agrochemicals, 2*, 337–354.

Parihar, M., Rakshit, A., Meena, V. S., Gupta, V. K., Rana, K., Choudhary, M., Tiwari, G., Mishra, P. K., Pattanayak, A., Bisht, J. K., Jatav, S. S., Khati, P., & Jatav, H. S. (2020). The potential of arbuscular mycorrhizal fungi in C cycling: A review. *Archives of Microbiology, 202*(7), 1581–1596. Link.springer.de/link/service/journals/00203/index.htm. doi:10.1007/s00203-020-01915-x.

Pauwels, R., Graefe, J., & Bitterlich, M. (2023). An arbuscular mycorrhizal fungus alters soil water retention and hydraulic conductivity in a soil texture specific way. *Mycorrhiza, 33*, 165–179.

Poveda, J., Abril-Urias, P., & Escobar, C. (2020). Biological control of plant-parasitic nematodes by filamentous fungi inducers of resistance: *Trichoderma*, mycorrhizal and endophytic fungi. *Frontiers in Microbiology, 11*, 992.

Prasad, M., Chaudhary, M., Ramakrishnan, S., & Mahawer, S.K. (2018). Glomalin: a miracle protein for soil sustainability. *Indian Farmer*, 2018, 5, (9), 1092-100.

Püschel, D., Bitterlich, M., Rydlová, J., & Jansa, J. (2020). Facilitation of plant water uptake by an arbuscular mycorrhizal fungus: A Gordian knot of roots and hyphae. *Mycorrhiza, 30*(2–3), 299–313. doi:10.1007/s00572-020-00949-9.

Raklami, A., Bechtaoui, N., Tahiri, A.-, Anli, M., Meddich, A., & Oufdou, K. (2019). Use of rhizobacteria and mycorrhizae consortium in the open field as a strategy for improving crop nutrition, productivity and soil fertility. *Frontiers in Microbiology, 10*. doi:10.3389/fmicb.2019.01106.

Rillig, M. C., Sosa-Hernández, M. A., Roy, J., Aguilar-Trigueros, C. A., Vályi, K., & Lehmann, A. (2016). Towards an integrated mycorrhizal technology: Harnessing mycorrhiza for sustainable intensification in agriculture. *Frontiers in Plant Science, 7*(2016). http://journal.frontiersin.org/article/10.3389/fpls.2016.01625/full. doi:10.3389/fpls.2016.01625.

Rozmoš, M., Bukovská, P., Hršelová, H., Kotianová, M., Dudáš, M., Gančarčíková, K., & Jansa, J. (2022). Organic nitrogen utilisation by an arbuscular mycorrhizal fungus is mediated by specific soil bacteria and a protist. *The ISME Journal, 16*(3), 676–685. doi:10.1038/s41396-021-01112-8.

Rui, W., Mao, Z., & Li, Z. (2022). The roles of phosphorus and nitrogen nutrient transporters in the arbuscular mycorrhizal symbiosis. *International Journal of Molecular Sciences, 23*, 11027.

Saeed, H. S., Munir, M., Samra, S., Farooq, A., Mahmood, H., Ahmad, I., Akram, M. I., Iqbal, R., Elsadek, M. F., Almutairi, S. M., Ditta, A., & Daraz, U. (2025). Screening of biosurfactants producing fungi from infected *Citrus sinensis* (L.). *Polish Journal of Environmental Studies, 34*(5), 5261–5270. https://www.pjoes.com/pdf-190588-115894?filename=Screening%20of.pdf. doi:10.15244/pjoes/190588.

Scheublin, T. R., Sanders, I. R., Keel, C., & Van Der Meer, J. R. (2010). Characterisation of microbial communities colonising the hyphal surfaces of arbuscular mycorrhizal fungi. *ISME Journal, 4*(6), 752–763. doi:10.1038/ismej.2010.5.

Sellappan, R., Krishnan, A., & Thangavel, K. (2022). Role of plant microbiome in carbon sequestration for sustainable agriculture. In *Core microbiome: Improving crop quality and productivity* (pp. 190–205). Wiley.

Shafreen, M., Vishwakarma, K., Shrivastava, N., & Kumar, N. (2021). *Physiology and distribution of nitrogen in soils* (pp. 3–31). Springer Science and Business Media LLC. doi:10.1007/978-3-030-71206-8_1.

Shen, S. K., & Wang, Y. H. (2011). Arbuscular mycorrhizal (AM) status and seedling growth response to indigenous AM colonisation of Euryodendron excelsum in China: implications for restoring an endemic and critically endangered tree. *Australian Journal of Botany, 59*(5), 460–467.

Shukla, S., Didwania, N., & Choudhary, R. (2025). Arbuscular mycorrhizal fungi (AMF): A pathway to sustainable soil health, carbon sequestration, and greenhouse gas mitigation. *Journal of the Saudi Society of Agricultural Sciences, 24*, 22.

Simard, S. W., & Durall, D. M. (2004). Mycorrhizal networks: A review of their extent, function, and importance. *Canadian Journal of Botany, 82*(8), 1140–1165. http://pubs.nrc-cnrc.gc.ca. doi:10.1139/B04-116.

Elmostapha, O., Hanane, D., Rachid, B., & Lahcen, O. (2021). Application of arbuscular mycorrhizal fungi isolates from semi-arid Mediterranean ecosystems as biofertilizers in Argan tree development. Journal of Soil Science and Plant Nutrition, 22, 944–955 (2022)

Singh, A. K., Zhu, X., Chen, C., Wu, J., Yang, B., Zakari, S., Jiang, X. J., Singh, N., & Liu, W. (2022). The role of glomalin in mitigation of multiple soil degradation problems. *Critical Reviews in Environmental Science and Technology, 52*(9), 1604–1638. Www.tandf.co.uk/journals/titles/10643389.asp. doi:10.1080/10643389.2020.1862561.

Singh, P. K., & Chudasama, H. (2021). Pathways for climate change adaptations in arid and semi-arid regions. *Journal of Cleaner Production, 284*, 124744. doi:10.1016/j.jclepro.2020.124744.

Singh Rawat, V., Kaur, J., Bhagwat, S., Pandit, M. A., & Rawat, C. D. (2023). Deploying microbes as drivers and indicators in ecological restoration. *Restoration Ecology, 31*(1). doi:10.1111/rec.13688.

Smith, S. E., & Read, D. J. (2010). Mycorrhizal symbiosis. Academic press.

Steinberg, P. D., & Rillig, M. C. (2003). Differential decomposition of arbuscular mycorrhizal fungal hyphae and glomalin. *Soil Biology and Biochemistry, 35*(1), 191–194. doi:10.1016/S0038-0717(02)00249-3.

Storer, K., Coggan, A., Ineson, P., & Hodge, A. (2018). Arbuscular mycorrhizal fungi reduce nitrous oxide emissions from N2O hotspots. *New Phytologist, 220*(4), 1285–1295.

Stürmer, S. L. (2012). A history of the taxonomy and systematics of arbuscular mycorrhizal fungi belonging to the phylum Glomeromycota. *Mycorrhiza, 22*(4), 247–258. doi:10.1007/s00572-012-0432-4.

Stürmer, S. L., Bever, J. D., & Morton, J. B. (2018). Biogeography of arbuscular mycorrhizal fungi (Glomeromycota): A phylogenetic perspective on species distribution patterns. *Mycorrhiza, 28*, 587–603.

Suman, J., Rakshit, A., Ogireddy, S. D., Singh, S., Gupta, C., & Chandrakala, J. (2022). Microbiome as a key player in sustainable agriculture and human health. *Frontiers in Soil Science, 2*. Www.frontiersin.org/journals/soil-science. doi:10.3389/fsoil.2022.821589.

Thind, S., Chaudhary, M. S., Ditta, A., Hussain, I., Parveen, A., Ullah, N., Mahmood, Q., Al-Ashkar, I., & El-Sabagh, A. (2022). Impact of mycorrhizal fungi from different rhizospheric soils on fungal colonization, growth, and chlorophyll contents of *Cenchrus ciliaris*. *Agronomy, 12*(11), 2644. doi:10.3390/agronomy12112644.

Timmis, K., & Ramos, J. L. (2021). The soil crisis: The need to treat as a global health problem and the pivotal role of microbes in prophylaxis and therapy. *Microbial Biotechnology, 14*(3), 769–797. http://onlinelibrary.wiley.com/journal/10.1111/(ISSN)1751-7915. doi:10.1111/1751-7915.13771.

Trabelsi, D., & Mhamdi, R. (2013). Microbial inoculants and their impact on soil microbial communities: A review. *BioMed Research International, 2013*, 1–11. doi:10.1155/2013/863240.

Verbruggen, E., Jansa, J., Hammer, E. C., & Rillig, M. C. (2016). Do arbuscular mycorrhizal fungi stabilize litter-derived carbon in soil? *Journal of Ecology, 104*(1), 261–269. http://www.blackwellpublishing.com/journals/JEC. doi:10.1111/1365-2745.12496.

Veresoglou, S. D., Verbruggen, E., Makarova, O., Mansour, I., Sen, R., & Rillig, M. C. (2019). Arbuscular mycorrhizal fungi alter the community structure of ammonia oxidizers at high fertility via competition for soil NH4+. *Microbial Ecology, 78*(1), 147–158. Link.springer.de/link/service/journals/00248/index.htm. doi:10.1007/s00248-018-1281-2.

Vishwakarma, S. K., Ilyas, T., Malviya, D., Shafi, Z., Shahid, M., Yadav, B., Singh, U. B., Rai, J. P., Singh, H. B., & Singh, H. V. (2022). Arbuscular mycorrhizal fungi (AMF) as potential biocontrol agents. *Microorganisms for Sustainability, 40*, 197–222. https://www.springer.com/series/14379. doi:10.1007/978-981-19-5872-4_10.

Volpe, V., Chialva, M., Mazzarella, T., Crosino, A., Capitanio, S., Costamagna, L., Kohlen, W., & Genre, A. (2023a). Long-lasting impact of chitooligosaccharide application on strigolactone biosynthesis and fungal accommodation promotes arbuscular mycorrhiza in *Medicago truncatula*. *New Phytologist, 237*, 2316–2331.

Volpe, V., Magurno, F., Bonfante, P., Ghignone, S., & Lumini, E. (2023b). Diversity of arbuscular mycorrhizal fungi associated with six rice cultivars in Italian agricultural ecosystem managed with alternate wetting and drying. *Rice Science, 30*, 348–358.

Wahab, A., Muhammad, M., Munir, A., Abdi, G., Zaman, W., Ayaz, A., Khizar, C., & Reddy, S. P. P. (2023). Role of arbuscular mycorrhizal fungi in regulating growth, enhancing productivity, and potentially influencing ecosystems under abiotic and biotic stresses. *Plants, 12*(17). Www.mdpi.com/journal/plants. doi:10.3390/plants12173102.

Wahid, F., Sharif, M., Fahad, S., Adnan, M., Khan, I. A., Aksoy, E., Ali, A., Sultan, T., Alam, M., Saeed, M., Ullah, H., Basir, A., Noor, M., & Khan, N. A. (2019). Arbuscular mycorrhizal fungi improve the growth and phosphorus uptake of mung bean plants fertilized with composted rock phosphate fed dung in alkaline soil environment. *Journal of Plant Nutrition, 42*(15), 1760–1769. Www.tandf.co.uk/journals/titles/01904167.asp. doi:10.1080/01904167.2019.1643371.

Wahid, F., Sharif, M., Steinkellner, S., Khan, M. A., Marwat, K. B., & Khan, S. A. (2016). Inoculation of arbuscular mycorrhizal fungi and phosphate solubilizing bacteria in the presence of rock phosphate improves phosphorus uptake and growth of maize. *Pakistan Journal of Botany, 48*(2), 739–747. http://www.pakbs.org/pjbot/PDFs/48(2)/43.pdf.

Wang, F. (2017a). Arbuscular mycorrhizas and ecosystem restoration. In *Arbuscular mycorrhizas and stress tolerance of plants* (pp. 245–292). Springer Singapore. http://www.springer.com/us/book/9789811041143. doi:10.1007/978-981-10-4115-0_11.

Wang, F. (2017b). Occurrence of arbuscular mycorrhizal fungi in mining-impacted sites and their contribution to ecological restoration: Mechanisms and applications. *Critical Reviews in Environmental Science and Technology, 47*, 1901–1957.

Wang, F., Shi, N., Jiang, R., Zhang, F., & Feng, G. (2016). In situ stable isotope probing of phosphate-solubilizing bacteria in the hyphosphere. *Journal of Experimental Botany, 67*(6), 1689–1701. doi:10.1093/jxb/erv561.

Wang, G., Jin, Z., George, T. S., Feng, G., & Zhang, L. (2023). Arbuscular mycorrhizal fungi enhance plant phosphorus uptake through stimulating hyphosphere soil microbiome functional profiles for phosphorus turnover. *New Phytologist, 238*(6), 2578–2593. http://onlinelibrary.wiley.com/journal/10.1111/(ISSN)1469-8137. doi:10.1111/nph.18772.

Wang, S., Chen, A., Xie, K., Yang, X., Luo, Z., Chen, J., Zeng, D., Ren, Y., Yang, C., Wang, L., Feng, H., López-Arredondo, D. L., Herrera-Estrella, L. R., & Xu, G. (2020). Functional analysis of the OsNPF4.5 nitrate transporter reveals a conserved mycorrhizal pathway of nitrogen acquisition in plants. *Proceedings of the National Academy of Sciences, 117*(28), 16649–16659. doi:10.1073/pnas.2000926117.

Wang, S. L., Caldwell, C. D., Kilyanek, S. L., & Smukler, S. M. (2019). Using agroecology to stimulate the greening of agriculture in China: A reflection on 15 years of teaching and curriculum development. *International Journal of Agricultural Sustainability, 17*(4), 298–311. http://www.tandfonline.com/toc/tags20/current. doi:10.1080/14735903.2019.1633901.

Wang, W., Shi, J., Xie, Q., Jiang, Y., Yu, N., & Wang, E. (2017). Nutrient exchange and regulation in arbuscular mycorrhizal symbiosis. *Molecular Plant, 10*(9), 1147–1158. doi:10.1016/j.molp.2017.07.012.

Wei, L., Vosátka, M., Cai, B., Ding, J., Lu, C., Xu, J., Yan, W., Li, Y., & Liu, C. (2019). The role of arbuscular mycorrhiza fungi in the decomposition of fresh residue and soil organic carbon: A mini-review. *Soil Science Society of America Journal, 83*(3), 511–517. doi:10.2136/sssaj2018.05.0205.

Wilson, G. W. T., Rice, C. W., Rillig, M. C., Springer, A., & Hartnett, D. C. (2009). Soil aggregation and carbon sequestration are tightly correlated with the abundance of arbuscular mycorrhizal fungi: Results from long-term field experiments. *Ecology Letters, 12*(5), 452–461. doi:10.1111/j.1461-0248.2009.01303.x.

Wu, W. J., Zou, Y. N., Hashem, A., Avila-Quezada, G. D., Abd_Allah, E. F., & Wu, Q. S. (2023). *Rhizoglomus intraradices* is more prominent in improving soil aggregate distribution and stability than in improving plant physiological activities. *Agronomy, 13*(5). Www.mdpi.com/journal/agronomy/. doi:10.3390/agronomy13051427.

Xiaoxia, A., Liu, J., Liu, X., Ma, C., & Zhang, Q. (2021). Optimizing phosphorus application rate and the mixed inoculation of arbuscular mycorrhizal fungi and phosphate-solubilizing bacteria can improve the phosphatase activity and organic acid content in alfalfa soil. *Research Square*. https://www.researchsquare.com/browse. doi:10.21203/rs.3.rs-892549/v1.

Xie, T., Lin, Y., & Li, X. (2024). Responses of the arbuscular mycorrhizal fungi community to warming coupled with increased drought in an arid desert region. *Geoderma, 441*, 116744.

Yan, A., Wang, Y., Tan, S. N., Mohd Yusof, M. L., Ghosh, S., & Chen, Z. (2020). Phytoremediation: A promising approach for revegetation of heavy metal-polluted land. *Frontiers in Plant Science, 11*, 359.

Yizhu, L., Imtiaz, M., Ditta, A., Rizwan, M. S., Ashraf, M., Mehmood, S., Aziz, O., Mubeen, F., Ali, M., Elahi, N. N., Ijaz, R., Lele, S., Shuang, C., & Tu, S. (2020). Response of growth, antioxidant enzymes and root exudates production towards as stress in *Pteris vittata* and in *Astragalus* sinicus colonized by arbuscular mycorrhizal fungi. *Environmental Science and Pollution Research, 27*(2), 2340–2352. https://link.springer.com/journal/11356. doi:10.1007/s11356-019-06785-5.

Zhang, J., Zhao, R., Li, X., & Zhang, J. (2024). Potential of arbuscular mycorrhizal fungi for soil health: A review. *Pedosphere, 34*(2), 279–288. doi:10.1016/j.pedsph.2024.02.002.

Zhang, L., Feng, G., & Declerck, S. (2018). Signal beyond nutrient, fructose, exuded by an arbuscular mycorrhizal fungus triggers phytate mineralization by a phosphate solubilizing bacterium. *ISME Journal, 12*, 2339–2351.

Zhang, Y., Li, S., Li, H., Wang, R., Zhang, K. e-Q., & Xu, J. (2020). Fungi–nematode interactions: Diversity, ecology, and biocontrol prospects in agriculture. *Journal of Fungi, 6*(4), 206. doi:10.3390/jof6040206.

Zheng, Y., Chen, N., Yu, K., & Zhao, C. (2023). The effects of fine roots and arbuscular mycorrhizal fungi on soil macropores. *Soil and Tillage Research, 225*, 105528. doi:10.1016/j.still.2022.105528.

Zhou, J., Bilyera, N., Guillaume, T., Yang, H., Li, F.-M., & Shi, L. (2024). Microbial necromass and glycoproteins for determining soil carbon formation under arbuscular mycorrhiza symbiosis. *Science of the Total Environment, 955*, 176732. doi:10.1016/j.scitotenv.2024.176732.

Zhu, Y. G., & Miller, R. M. (2003). Carbon cycling by arbuscular mycorrhizal fungi in soil-plant systems. *Trends in Plant Science, 8*(9), 407–409. Www.elsevier.com/inca/publications/store/3/0/9/6/0/index.htt. doi:10.1016/S1360-1385(03)00184-5.

Chapter 33

Arbuscular mycorrhizal fungi (AMF) as potential mitigators of microplastic toxicity in soil ecosystems

Naima Huma Naveed[a], Aamir Ali[a], Aima Iram Batool[b], Muhammad Fayyaz ur Rehman[c], Muhammad Mustaqeem[c] and Allah Ditta[d,e]

[a] *Department of Botany, University of Sargodha, Sargodha, Pakistan,* [b] *Department of Zoology, University of Sargodha, Sargodha, Pakistan,* [c] *Institute of Chemistry, University of Sargodha, Sargodha, Pakistan,* [d] *Department of Environmental Sciences, Shaheed Benazir Bhutto University Sheringal, Sheringal, Pakistan,* [e] *School of Biological Sciences, The University of Western Australia, Perth, WA, Australia*

33.1 Introduction

Plastic is considered a ubiquitous and essential material because of its feasibility, versatility, and ability to change into various shapes (Ali et al., 2025). It is a synthetic material manufactured from polymers derived from petroleum. There is an enormous increase in the production of plastic with respect to its demand, which has led to the persistence of plastic products (Abedsoltan, 2024). Nearly 80% of plastics that are manufactured in industries are ultimately released into the natural environment (Megha et al., 2025). These plastics break down through various biological and physical processes in which they first degrade into macroplastics, mesoplastics, and ultimately into microplastics (MPs). MPs are the tiny plastic fragments that have a diameter of less than 5 mm in size (Turner et al., 2025). Mechanical degradation is carried out with tire abrasion, weathering, and UV degradation, whereas biological degradation occurs through organisms that can consume and degrade plastics such as bacteria, fungi, and algae (Gilani et al., 2023). This degradation phenomenon can actually worsen the situation by reducing the particle size of plastics and increasing their potential for being oxidized (Kasmuri et al., 2022).

Plastic pollution is recognized as a serious ecological threat of global concern. Since the 1950s, plastics have been overutilized; their production has been enhanced exponentially, reaching tons in years, because their stability, resilience, and affordable cost have stimulated international demand. Disintegration of large plastic fragments because of photooxidation, hydrolysis, UV irradiation, and biodegradation tends to generate small plastic elements having a size $\leq$5 mm, which are recognized as MPs. MPs predominantly harm aquatic environments, which include freshwater and marine ecosystems. These particles end up contaminating drinking water and food supplies. They not only build up in sediments but also get swallowed by aquatic life and become an indispensable part of the food chain (Roy, 2024). That is why it is pertinent to control MP pollution.

MPs act as transporters of many hydrophobic organic contaminants and heavy metals (Ditta et al., 2024; Hossain et al., 2025; Sarfraz et al., 2024a, 2024b, 2024c). Absorption of MPs toward organic contaminants is because of high hydrophobicity, small size, and large surface area. The aging process of MPs increases their adsorption mechanism toward heavy metals. Bioaccumulation of heavy metals such as Pb, Ar, Ni, and Cr increased when they adsorbed on the surface of MPs. As compared to other environmental materials, MPs can accumulate huge amounts of heavy metals. It also offers a pathway for transmission of heavy metals from the environment to the body of an organism (Adeleye et al., 2024). Therefore it is necessary to focus on these emerging contaminants because of the toxic effects on aquatic and land habitats. In water, MPs having different configurations are recognized, including polystyrene (PS), polyethylene (PE), and polypropylene

Advancing Sustainable Agriculture Through Arbuscular Mycorrhizal Fungi. DOI: https://doi.org/10.1016/B978-0-443-45130-0.00005-9

(PP). Microbeads, agricultural plastics, and fibers derived from sewage-sludge transport in water via farmland runoff and agricultural drainage. Entry of MPs in the aquatic ecosystem causes an increase in the density of water because of photooxidation and biofouling (Xu et al., 2024). Although many studies focus on the occurrence of MPs in rivers, their abundance has also been identified in lakes and ponds. These MPs can move over long distances and have a wide range of lasting effects on the ecological environment (Li et al., 2023a).

The occurrence of MPs in terrestrial ecosystems is 4–23 times with respect than aquatic environment, making it a significant reservoir for MPs (Mutshekwa et al., 2025). The main source of MPs in soil involves irrigated water, plastic mulching, composting, and fragmentation. It is very difficult to separate and discover new MPs because of their difficult sorting from the organic matter present in soil (Li et al., 2024). However, researchers have recognized high concentrations of MP contamination in particular soils. MPs stay in soil, and their accumulation damages the soil ecosystem, biodiversity, and functionality. For instance an increase in the concentration of MPs affects soil physicochemical properties, microbial diversity, and enzyme activity (Li et al., 2022). In addition, MPMPs can change the bulk density of soil by improving the stability of soil particles. Such alterations in soil fertility produced by MP contamination have unintentional detrimental impacts on plant growth (Zhou et al., 2023).

The absorption of MPs in plants is carried out in the form of aggregates. A significant factor involved in the uptake and transfer of MPs in plants is transpiration pull (Wang et al., 2022a). MPs that are absorbed from soil into epidermal tissues of plants via the pericycle and eventually transported in the axial part of plants (Ghosh & Kumar, 2025).

MPs have the potential to severely affect a wide range of physiological processes in plants once they are absorbed, such as causing oxidative damage, causing genotoxicity, reducing biomass, interfering with photosynthesis, delaying seed germination, and inhibiting plant growth (Li et al., 2023c). MPs can not only affect the plants but also fish, algae, and other trophic levels within the aquatic ecosystem. Fish are considered an important source of food for organisms of high trophic levels and humans. Ingestion of MPs has been reported in various species of fish through trophic transfer that ultimately hinders their growth, disrupts physiological functions, and diminishes the yield of the aquatic system (Ma & You, 2025). Accumulation of PS beads in tissues of fish induces oxidative stress and inflammation mainly in the epithelium of the stomach (Chen et al., 2022). Herbivorous fish accumulate high amounts of MPs as compared to carnivores and omnivores because of their long intestines; herbivorous fish allow more retention time (Schutten et al., 2024). MPs, along with other additives such as bisphenol and phthalates, disrupt the endocrine system of aquatic organisms (Ullah et al., 2023).

Owing to the widespread presence of MPs in the food chain, they have greater potential to enter our food, water supply, and even our lungs. Different studies reported their presence in blood, feces, kidney, liver, placenta, and breast milk (Hasan & Tarannum, 2025). It is found that the annual consumption of MPs in people is above 50,000 particles, and their concentration is higher if inhalation is considered. This shows that MPs are abundant in target organs, enter the body through various routes, and travel throughout the body via the circulatory system (Chaudhary et al., 2025). MPs' uptake in the human body depends upon the size, functional groups, solubility, and chemistry of MPs (Menichetti et al., 2024). Negatively charged MPs can easily pass through the gastrointestinal tract, forming direct contact with epithelial cells. In addition, MPs follow endocytic and persorption pathways for their translocation and absorption (Ahmadi et al., 2025). The interaction of MPs with biological phenomena reduces the efficiency of mucus in the colon, which eventually enhances the chances of colorectal cancer (Jaafarzadeh Haghighi Fard et al., 2025). So a complete understanding of the distribution, absorption, metabolism, and excretion of MPs is necessary to identify in order to determine their toxic effects.

Agricultural soils are increasingly contaminated with MPs because of the widespread use of plastic mulching films, sewage sludge application, irrigation with wastewater, atmospheric deposition, and litter fragmentation. Once introduced into soils, MPs persist for long periods because of their resistance to biodegradation, posing chronic risks to soil health and ecosystem sustainability. Among the most dynamic and intricate ecosystems are the soil ecosystems. These are characterized by interlinked physical structures, chemical processes, and biological interactions. Different types of MP are added from a variety of sources, thus showcasing the multiple routes through which these pollutants can infiltrate the soil. The moisture retention ability of the soil is negatively affected by the increasing levels of MP, which also leads to degradation of soil structure and permanently results in diminished irrigation water and rainwater that can be absorbed by the soil (Lawanga et al., 2022).

Adding more insult to injury, MPs may also function as carriers for several detrimental hazardous additives, heavy metals, and numerous organic pollutants. This intensifies the hazardous ecological consequences of MPs and is also an indication of negative effects on soil microbiota, enzymatic activities, invertebrates, and plant life. As a result of this, soil fertility and agricultural productivity are jeopardized in a long-lasting manner. Hypoxic conditions in the soil–plant system can be induced by this. It has also been noted that leftover layers of PE can compromise the integrity of soil aggregates, leading to reduced aeration and water permeability (Jin et al., 2022). The density of soil, water-holding capacity, and structural integrity are severely affected by environmental MP pollution (Hanif et al., 2024). Physicochemical characteristics of soil

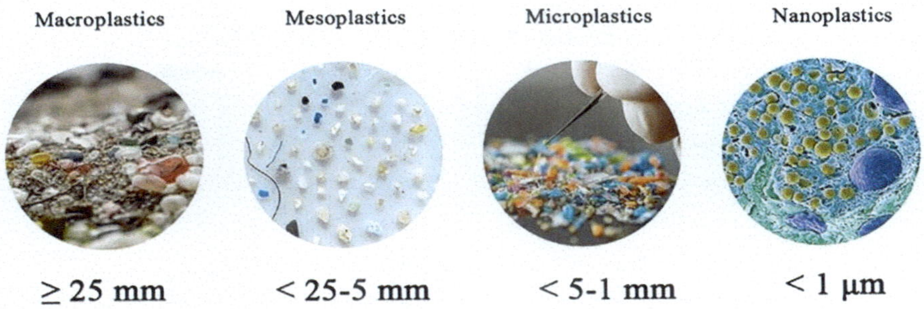

FIGURE 33.1 Types of microplastics.

are negatively impacted by the accumulation of MPs, leading to unsustainable farming practices and broader environmental decline (Lakhiar et al., 2024). Residues of insecticides that adhere to or accumulate within plastic particles can disrupt soil ecosystems, impair root growth, and reduce plant productivity in MP-contaminated soils (Fang et al., 2025).

Tolerance of soil against the stress induced by MPs can be enhanced by biological buffering systems, which is also a central focus point for researchers. The intermediate soil layer, which supports plant life, is also essential for microbial communities (Wu et al., 2024). Among these microorganisms, arbuscular mycorrhizal fungi (AMF), members of the phylum Glomeromycota, form symbiotic associations with approximately 80%–90% of terrestrial plants (Parniske, 2008; Schüßler et al., 2001). Through their extensive hyphal networks, AMF assist plants in absorbing key nutrients such as phosphorus, nitrogen, and various micronutrients, in exchange for carbon supplied by the host plants (Samanta et al., 2025). These fungi are integral components of rhizospheric microbial communities (Jajoo & Mathur, 2021), known for promoting plant growth even under adverse environmental conditions and forming associations with more than 90% of angiosperms (Lanfranco et al., 2016).

The sustainable nutrient exchange process between plant and AMF is based on their mutualism (Roth & Paszkowski, 2017). Plants provide AMF with energy in the form of carbohydrates and lipids; the fungi use their intricate mycelial networks to draw water and mineral nutrients from the soil, channeling them to plant roots through specialized structures (Liu et al., 2020). This partnership is recognized as one of the most efficient natural mechanisms for enhancing plant tolerance and reducing both biotic and abiotic stresses, including those resulting from soil contamination (Adeyemi et al., 2021; Ahmad et al., 2022a, 2022b, Ahmed et al., 2024; Ali et al., 2024a, 2024b; Diagne et al., 2020; Ditta & Ullah, 2023; Jan et al., 2022; Saleem et al., 2024, 2025; Ullah et al., 2021).

AMFs play an important role in soil aggregation, enhance water dynamics, regulate plant hormonal activity, improve tolerance to environmental stressors such as drought, salinity, heavy metals, and organic pollutants, and contribute to nutrient acquisition (Liu et al., 2020; Nazir et al., 2023; Thind et al., 2022). Because of their broad ecological functions, AMFs are increasingly being studied for their potential to counteract the harmful effects of emerging soil contaminants like MPs. Evidence suggests that AMF can influence the environmental fate and ecological behavior of MPs in soil–plant systems by affecting their accumulation, distribution, and migration patterns (Li et al., 2024; Zhang et al., 2024). This chapter explores the capacity of AMF to reduce MP toxicity within soil ecosystems. By integrating insights into soil science, plant physiology, microbiology, and environmental toxicology, it seeks to develop a comprehensive understanding of AMF-mediated interactions with MPs and their implications for sustainable soil management.

33.2 Microplastics in soil ecosystems: Sources, characteristics, and fate

33.2.1 Definition and classification of microplastics

MPs are commonly defined as plastic particles smaller than 5 mm in diameter, though some studies further subdivide them into mesoplastics (1–5 mm), MPs (1 μm to 1 mm), and nanoplastics (<1 μm) (Fig. 33.1) (Table 33.1). They vary widely in shape (fibers, fragments, films, pellets, and foams); polymer type [PE, PP, PS, polyvinyl chloride (PVC), PE terephthalate (PET), among others]; density; and surface chemistry. These characteristics strongly influence their behavior, transport, and biological interactions in soil environments.

33.2.2 Classification of MPs

According to Russo and Di Ciaccio (2024), MPs are categorized into two major types based on their emergence (Fig. 33.2).

TABLE 33.1 Categories of plastics based on particle size.

Category	Particle size
Megaplastics	Greater than 1 m
Large- or macroplastics	Smaller than 1 m
Mesoplastics	Smaller than 2.5 cm
Microplastics	Smaller than 5 mm
Nanoplastics	Smaller than 1000 nm

From Cui et al. (2025).

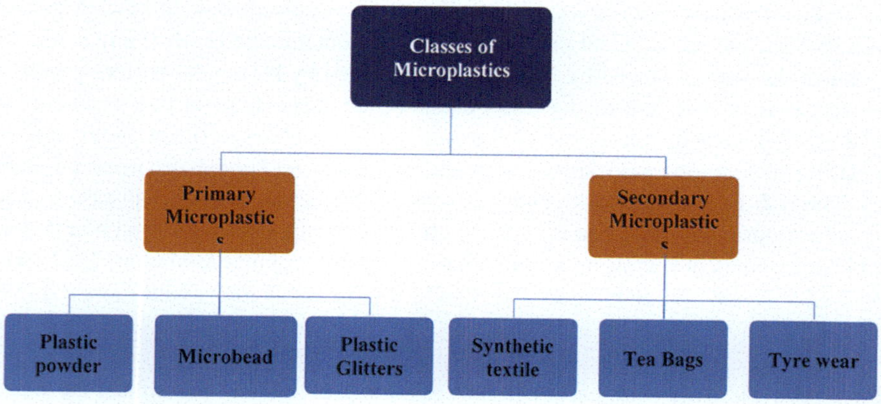

FIGURE 33.2 Classification of microplastics.

Primary and secondary MPs represent the two principal classes employed to differentiate between the sources of MPs (Song et al., 2024).

33.2.2.1 *Primary microplastics*

Primary MPs are made especially for applications in different technical and industrial contexts, as well as pharmaceutical vectors and abrasives in cosmetics (Patil et al., 2024). Primary MPs are specifically produced in microsize and utilized in numerous products of daily use, such as detergents, cleansers, facewash, moisturizers, fragrances, sunscreen, coloring agents, nail polishes, skin creams, and eyelashes (Alex et al., 2024; Chaudhari et al., 2024). Microbeads and resin pellets are two examples of primary MPs. Personal care and cosmetics may contain microbeads up to 5% with an average size of 250 μm to boost scrubbing and cleaning (Emenike et al., 2023; Gamage & Mahagamage, 2024). Polythene kinds of MPs are often beneficial for professional personal care applications. According to studies, hundreds of tons of microbeads are discharged every year in the United States by only utilizing detergents (Luzi et al., 2025). In facewash and toothpastes, microbeads are used as an abrasive to eliminate marks, and about 4000 microbeads can be emitted in a single use (Emenike et al., 2023). Therefore it becomes very difficult to remove these MPs with the sewage disposal methods, and they gradually accumulate in the ecosystems.

33.2.2.2 *Secondary microplastics*

In the environment, larger polymers slowly decompose into fragments as a result of a variety of complex environmental conditions, such as UV radiation, sunlight, higher temperature, wind, erosion, and waves (Jansen et al., 2024; Yousafzai et al., 2025). When plastic particles attain a size in microns, then these are known to be secondary MPs. These MPs also cannot be eliminated in sewage treatment. After a thorough treatment procedure, the elimination rate of MPs in wastewater can reach up to 88% (Bodzek & Pohl, 2022).

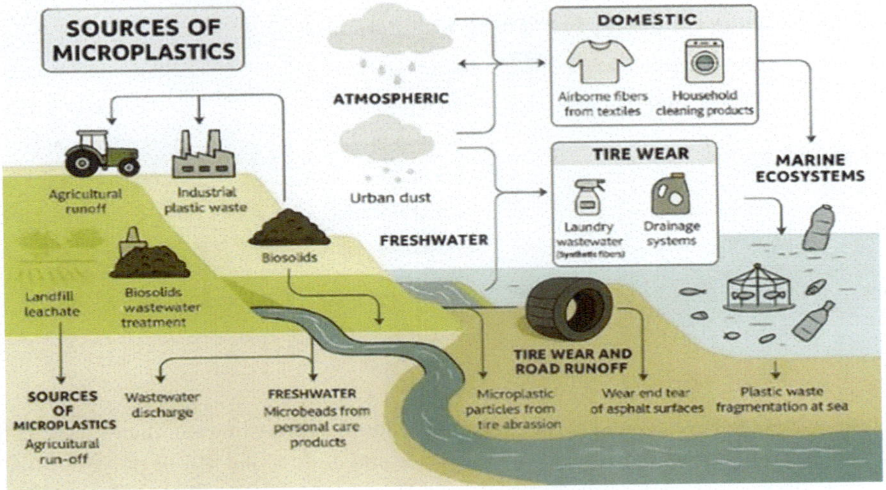

FIGURE 33.3 Sources of microplastics.

33.2.3 Sources of microplastics in soils

Soils receive MPs from multiple pathways, many of which are closely linked to human activities (Fig. 33.3). Agricultural practices represent a major source, particularly the use of plastic mulch films to improve crop yields and water use efficiency. Over time, these films fragment into smaller particles that accumulate in the soil. Sewage sludge and compost derived from municipal waste, often applied as fertilizers, contain substantial quantities of MPs originating from household products, textiles, and industrial effluents. Irrigation with treated or untreated wastewater further contributes to soil MP loads. Atmospheric deposition is another significant but often overlooked pathway. MP fibers and fragments can be transported over long distances by wind and deposited onto soil surfaces, even in remote regions. In addition, littering, tire wear particles, and degradation of construction materials contribute to localized soil contamination.

33.2.3.1 Microplastics on land

Uncontrolled management of garbage, direct littering, and industrial wastewater are the causes of MP emission into the environment. Many factories produced and disposed of these MPs on land (Amesho et al., 2023). Every year, Europe's agricultural soils receive approximately 120–800 t of MPs per million through sewage sludge management, contributing to the annual accumulation of 473,000–910,100 MT of plastic debris, including MPs, in its freshwater and terrestrial ecosystem (Milojevic & Cydzik-Kwiatkowska, 2021). Many facilities are utilized for the treatment of wastewater that eliminates MP fragments up to 99%, and some of them shift into sludge (Amesho et al., 2023).

33.2.3.2 Microplastics in freshwater

Larger plastic trash, fibers, microbeads, as well as MPs and macroplastics, can readily find their way into freshwater environments because of inadequate garbage disposal, agricultural drainage, urban runoff, untreated storms, and sewage debris (Gkanasos et al., 2021; Schell et al., 2022; Wang et al., 2022). The MPs follow a chain process for their dispersal from the terrestrial to the aquatic environment. As a result, consideration must be given to inputs that are the main source for the production of combined sewer overflow (CSOs), microplastics (MPs), and effluent outfalls (Gkanasos et al., 2021).

33.2.3.3 Microplastics in the atmosphere

MPs in the atmosphere are generated from the degradation of clothing materials because of washing and drying, erosion of synthetic rubber tires, breakdown of household furniture, pollutants from the synthetic textile sector, airborne pollutants from PVC production, and contamination from urban dust (Priyadarshini et al., 2024). In the air, high quantities of MPs exist in the air that travel as urban dust in the atmosphere because they possess low molecular weight. Debris present on roads, fragments of tires, synthetic fibers, and manufacturing paints are the well-known sources of these dust particles. Because of heavy rainfall, these airborne particles come to the ground and pollute the surface of land by various methods (Vlasov et al., 2022). At the end, these small particles fly and land on the water and sediment's surface (O'Brien et al., 2023).

TABLE 33.2 Types of microplastics commonly found in agricultural soils.

Microplastic type	Common polymer	Major sources	Key soil impacts
Fibers	Polyester, nylon	Textiles, sludge	Reduced aggregation, microbial shifts
Fragments	PE, PP	Plastic films, litter	Nutrient sorption, root interference
Films	LDPE	Mulching films	Reduced water infiltration
Pellets	PS	Industrial waste	Additive leaching
Nanoplastics	Various	Degradation products	Cellular uptake, oxidative stress

PE, polyethylene; *PP*, polypropylene; *PS*, polystyrene; *LDPE*, low density polyethylene.

33.2.3.4 *Microplastics from domestic discharge*

Natural fibers like cotton, mixes of polyester and rayon, decompose more quickly in water than synthetic fibers like polyester. Around 75% of cotton, 62% of rayon, 40% of blends of polyester and cotton, and 4% of polyester fibers had biodegraded (Nofal, 2022). As natural fibers possess a high degradation rate compared to synthetic fibers, they release more toxic chemicals and several dyes in the environment (Aldalbahi et al., 2021). Synthetic textile fibers from laundry have been recognized as the major source from International Union for Conservation of Nature (IUCN), resulting in 35% of MP contamination in the ocean worldwide (Hale et al., 2020). A 6-kg load of acrylic fabric could release almost 700,000 fibers during a house wash (Raguso, 2025). High concentrations of fibrous MPs enter the freshwater through washing of garments commonly manufactured from synthetic material (Ageel et al., 2022).

33.2.3.5 *Tire wear particles*

The rubber in tire treads is the primary source of MP particles in road dust. The frictional interaction between the rolling tire and the road surface creates MPs that cause tire wear. This contact results in tire wear, releasing particles of rubber and heat friction inside the tire, which could cause volatile tire components to evaporate (Wang et al., 2024). Tire wear particles are produced on the route and proliferate in the environment through surface runoff, wastewater effluents, and precipitation (Baensch-Baltruschat et al., 2021; Lindfors et al., 2025). Small particles are often released into the atmosphere and spread out, whereas large particles land on the surface of the road (Ahmad et al., 2025). Their concentration is reduced with an increase in the distance from the roads. This shows that the accumulation of suspended particles is a significant route to transport into the aquatic environment (Lindfors et al., 2025).

33.2.3.6 *Microplastics in the marine environment*

MPs have been steadily accumulating in the marine environment because of their high resistance and large-scale production. The primary sources of MPs entering the sea are atmospheric transportation, offshore operations, coastal land, and river intake (Bao et al., 2023). MPs can travel in many stages by vertical settlement or lateral movement after penetrating the ocean surface and reaching the deep ocean. The key element in regulating particle settlement is the density difference between the fluid and MP particles. The movement of MPs depends on their physical properties such as size, shape, and density (Shamskhany et al., 2021). PVC, polyamide, and PET MPs are heavier than water and mainly sink into the water column, whereas PP and PE float because of their lower densities (Yan et al., 2021).

33.2.4 Distribution and accumulation in soil profiles

It is essential to evaluate MPs' impact on the soil ecosystem to understand their toxicological properties (Ullah et al., 2025). Researchers considered soil as polluted soil that possesses 0.1% of MPs by soil weight (Table 33.2). MPs present in the soil damage the structure of the soil, reduce its infiltration capacity for precipitation and irrigation, and negatively affect its water retention ability (Liu et al., 2022). If these remains are present in huge numbers, they will obstruct the soil pores, disrupt the soil nitrogen cycle, and change microbial composition, ultimately impacting crop development (Yang et al., 2021). Because of their potentially harmful effects, together with their durability and persistence, the existence of MPs poses a danger to soil ecology and ecosystems (Paudel et al., 2024).

Once introduced into soil, MPs can be redistributed vertically and horizontally through tillage, bioturbation by soil fauna, water infiltration, and root growth. Furthermore, it has been reported that remaining PE mulch film weakens the structure of soil aggregates and reduces soil aeration and water permeability, which ultimately inhibits root growth and overall plant production (Gu et al., 2024). MPs can cause the top layer of soil to dry up and crack because they can interact with soil particles to produce chunks or erode the structural strength of soil. The overall density of the soil is a significant indicator of the amount of carbon stored in it, but the presence of MPs can result in an inaccurate measurement of this carbon storage (Huang et al., 2025). Studies have shown that MPs are not confined to surface layers but can penetrate deeper soil horizons, potentially reaching groundwater systems. The persistence of MPs in soil is influenced by polymer type, environmental conditions, and biological activity, but complete degradation is generally slow, leading to long-term accumulation.

33.2.5 Interaction of microplastics with soil properties

MPs interact with soil physical properties by modifying bulk density, porosity, and aggregation. Some MPs, like fibrous MPs, may increase soil porosity and water infiltration, while film-like particles can impede water movement and gas exchange. The bioavailability and absorption of nutrients, pesticides, and heavy metals may also become variable because of the presence of MPs in soil. Biologically, they provide novel surfaces for microbial colonization, forming so-called plastispheres that differ in composition from native soil microbial communities.

33.2.5.1 Physicochemical properties

MP accumulation negatively impacts the physicochemical properties of the soil, ultimately leading to unsustainable agricultural practices and environmental damage (Elbasiouny et al., 2023). As MPs are hydrophobic and have a large surface area, many dangerous substances may adsorb onto them and alter the reactions with other soil contaminants. The bulk densities of the soil containing fragments of High-density polyethylene (HDPE), PP, and PS decreased. The pH of alkaline soil increased after being contaminated with PP, whereas the pH of acidic soil declined (Fan et al., 2022).

33.2.5.2 Soil microorganisms

MPs can either directly or indirectly harm soil microorganisms by altering their microhabitats and nutrition availability by producing biotoxicity (Zhang et al., 2021). This biotoxicity in soil microorganisms can result from reactive oxygen species (ROS), facilitated oxidative stress, and harmful effects of leached MP additives, including heavy metals, plasticizing agents, antioxidants, flame retardants, and photostabilizers. MPs damage keystone bacterial species that are vital for providing important soil services, including nitrifiers, arbuscular mycorrhizas, and increase the number of disease-causing organisms that harm plants and animals (Csorba et al., 2024).

33.3 Toxic effects of microplastics in soil ecosystems

33.3.1 Effects on soil physical and chemical processes

MPs can disrupt soil structure by interfering with aggregate formation and stability. This has implications for erosion susceptibility, water retention, and root penetration. The presence of MPs may also influence soil pH and redox conditions, indirectly affecting nutrient cycling processes such as nitrification and phosphorus availability. Multiple studies have demonstrated that the carbon and nitrogen contents of microbial biomass in soil drastically decrease as the amount of MP film residues rises. Extrinsic factors like microbial growth and human activity have the ability to immediately integrate MPs into the soil matrix. This assimilation may cause modifications to the soil's structure and texture. Ordinary soil particles are often less compact than industrial polymers, which contributes to the visible changes in the structure and content of highly contaminated soils (Luo et al., 2024; Ma et al., 2024). By introducing polyester fibers the water-holding capacity, water table clumps, and bulk weight of soil vary according to concentration (Mangla et al., 2025).

The bulk density of the soil is reported to be decreased by contamination with fragments of high-density PE, PS, and PP. In soil contaminated with polyether sulfone and PS the generation of water-stable aggregates is significantly decreased (Qin et al., 2025). It has been reported that MPs alter the sizes and characteristics of soil's water-stable aggregates. As the concentration of MPs in the soil increased, more evaporation took place, reducing the soil's water content (Fang et al., 2024). PP MP exposure increased the pH of alkaline soil while lowering the pH of acidic soil, indicating that the effect of MPs on soil pH may vary depending on the type of soil (Li et al., 2023b). Variations in soil pH have been found to alter nutrient availability, which, in turn, affects plant development and agricultural output. The biogeochemical carbon cycle and the microbial accessibility to carbon in soils may be greatly impacted by the typically inert carbon present in MPs (Ma et al., 2024).

33.3.2 Phytotoxic effects of microplastics

Plants exposed to MP-contaminated soils often exhibit reduced germination, impaired root growth, altered nutrient uptake, and increased oxidative stress. The absorption of MPs by plants is an increasing matter of interest in ecological science and agriculture, as MPs are widespread contaminants in terrestrial and aquatic habitats. Although most studies have focused on their influence on aquatic ecosystems, there is a significant amount of data demonstrating that MPs can also get into terrestrial food chains through plants (Saeedi, 2023).

Vascular plants perform crucial functions through carbon and oxygen cycles while serving as essential foundations of ecological equilibrium. Vascular plants provide a variety of ecological services, which include preserving soil and water, regulating the climate, preventing wind and sand erosion, and creating biological habitats (Chowaniec et al., 2024). The relationship between MPs and vascular plants has garnered a lot of academic interest. Because of their size and rough surfaces, underwater vascular plants may potentially act as a trap for MPs. MPs carried by water currents can be captured by larger vascular plants and adsorbed on their large and rough body surfaces (Gallitelli et al., 2023). A study reported that even after thorough cleaning in the lab, there is still a significant attachment between the root surfaces and negatively charged MPs (He et al., 2023).

It is generally believed that the harmful effect of MPs in plants is associated with their size and dosage. The smaller sizes exhibit a more pronounced impact, and particles accumulate in the soil pore, enabling easy absorption by plants while also physically obstructing the pores in seeds and roots (Shi et al., 2024). The physical obstruction present in seeds and roots restricts the uptake of water and nutrients, leading to a decrease in relative seed germination as well as root and shoot growth. The detrimental impact on seed germination rises with the increasing size of plastic particles (De Silva et al., 2021). The MPMPs' polymer composition is a significant component in addition to their morphological features. Plant development may be impacted differently by plastics with varying polymer compositions. For instance, PVC MPs considerably reduce *Lepidium sativum* development, germination, and plant height as compared to PE and PP (Pignattelli et al., 2020). The impact of MPs on plant development, especially on leaves, is also reflected in photosynthesis. Typically the contents of photosynthetic pigments (such as chlorophyll *a*, carotenoids, and chlorophyll *b*) are detected to assess how MPs affect photosynthesis in plants. With rising PE MP concentrations, lettuce had a trend of lowering chlorophyll (Li et al., 2020).

MPs smaller in size have the highest inhibitory effects on chlorophyll, significantly reducing plant photosynthesis (Wang et al., 2021b). Moreover, smaller MPs can carry different pollutants across the plant root gap and accelerate the absorption of hazardous pollutants into plants. Research has shown that small-diameter MPs may be absorbed by plant roots and translocated to aerial tissues. According to the physical and chemical characteristics of the corresponding polymer and its additives, tiny lipophilic plastic particles may interact with organelles as well as membranes after being absorbed by an organism (Liese et al., 2024).

Larger MPs seem to attach to plant roots, inhibiting absorption of nutrients by plants (Li et al., 2024). MPs particles with a larger size cannot be absorbed by plant roots. They may influence the growth of susceptible plants by either indirect or direct mechanisms, such as harming root tissue, obstructing cell wall pores, or altering nutrient absorption. The magnitude and particulars of such impacts differ according to the properties of MPs (such as kind, size of particles, and shape) and the species of plants affected (Xu et al., 2024). Data shows that MPs adversely affect crop development and quality through decreasing plant biomass by 13% and levels of chlorophyll by 28% (Shi et al., 2024).

Oxidative stress can be caused by PE, PP, and PVC, which may have metabolic downstream consequences. *Brassica rapa* ssp. chinensis (Brassicaceae) showed a reduction in overall photosynthetic rate when exposed to PS beads (Liese et al., 2024). It was found that PE changed the biochemical makeup of many crop seedlings. A study revealed that glucosinolate concentrations in broccoli and radish sprouts (Brassicaceae) decreased when grown in a hydroponic solution with MPs (Bouaicha et al., 2022a, 2022b). Glucosinolates and the products of their hydrolysis protect against different pathogens and herbivores to plant species of the Brassicaceae family (Liese et al., 2024).

33.3.3 Impacts on soil microbial communities

Soil microorganisms are central to ecosystem functioning, driving organic matter decomposition and nutrient cycling. MPs can alter microbial biomass, diversity, and community composition, with effects varying depending on MP type, concentration, and exposure duration. Apart from the ability of MPs to act as vectors for chemicals, MPs can act as platforms for microbes (Zhao et al., 2024). The production of biofilms can impact relationships of MPs and biota on various scales. The colonization of MPs with parasites and hazardous algae/bacteria could induce infections as well as chemical toxicity. In addition, biofouling has been demonstrated to impact the destiny of MPs by modifying the particle features (e.g., density). When biofilms develop, the density of buoyant or floating MPs rises, causing these low-density particles to

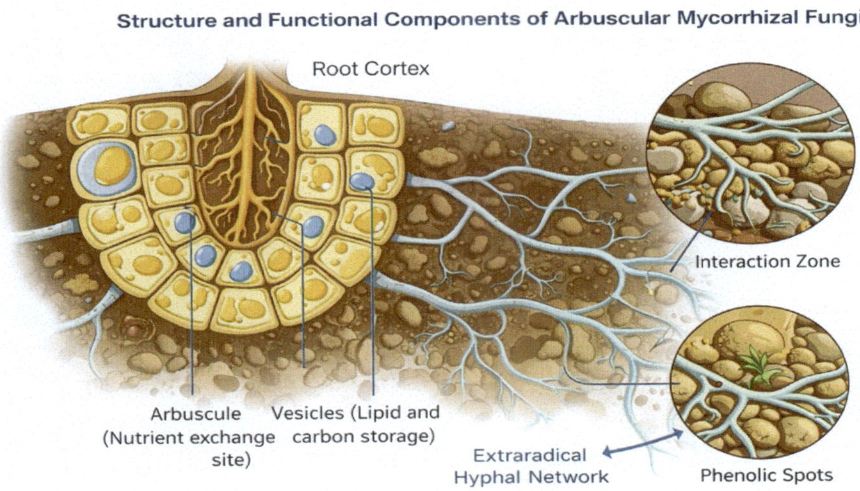

FIGURE 33.4 Structure and function of AMF. *AMF*, Arbuscular mycorrhizal fungi

settle. Moreover, MPs are probably integrated into heteroaggregates in the environment. Particulate matter (MPs and other suspended particles) and microorganisms (algae, protozoa, etc.) make up these aggregates, whereas biopolymers serve as binders (Chettri et al., 2023; Uthra et al., 2023; Yarahmadi et al., 2024).

Moreover, particulate organic matter and primary producers such as unicellular bacteria or algae supply nutrients to a variety of benthic as well as pelagic organisms (Sommer, 2024). Smaller MPs fall within the same size range as these natural foods. It was found that the zooplankton's nutritional intake and reproduction were significantly impacted by the PS particles. According to the research, zooplankton may survive on up to 40% less food than their actual food intake. MP beads not only have less nutritional value but also make it more difficult for zooplankton and other species to adjust to food shortages. The organisms' dietary consumption of MP particles is the cause of their incapacity to naturally lower their metabolic rate in response to periods of food scarcity (Gong et al., 2024; Meng et al., 2024; Saeedi, 2023). Some studies report reduced enzyme activities and microbial respiration, whereas others suggest shifts toward microbial taxa capable of colonizing plastic surfaces. These changes may have cascading effects on soil health and plant productivity.

33.4 Arbuscular mycorrhizal fungi: Biology and ecological functions

33.4.1 Taxonomy and life cycle

AMFs belong to the phylum Glomeromycota and are obligate symbionts, relying on host plants for carbon. This symbiotic association of AMF and plants is considered to be one of the most widespread and the most ancient types of plant–fungi associations, dating back to 460 million years, as per recent fossil evidence (Prasad et al., 2017). In the plant–AMF symbiotic relationship, an extensive and vast network of mycelia is formed by AMF that extends beneath the roots, creating the so-called common mycorrhizal network (CMN). CMN is found commonly in plants that are associated with AMF. There are basically two groups for the mycelial network formed, which are as below.

The name of the first one is **extraradical mycelia**. It makes the connection between the colonized root and the surrounding soil matrix. It also performs the primary function of uptake of water and nutrients like phosphorus from the soil and transporting them back to the plant (Fig. 33.4).

The second one is called the **intraradical mycelia** network. It forms a deep penetration into the root cortex. Their interaction results in the formation of structures called arbuscules (Fig. 33.4) and vesicles that promote nutrient exchange and enhance storage, respectively (Wang et al., 2024).

The **CMN** plays a significant role in terrestrial ecosystems. This influences plant communities as well as invasive species and facilitates nutrient transport, particularly phosphorus and nitrogen from fungi to plants (Pringle et al., 2009). Many vital processes in plants, such as the water potential of the leaf, stomatal conductance of the plant, relative water content (RWC), and assimilation of carbon dioxide, are also affected by the inoculation of AMF, as they successfully invade the host plant's root epidermal cells, and a symbiotic relationship is established. According to Hu et al. (2017) the mycelial network synthesizes glycoproteins that help the host plant absorb and transport minerals and water, thereby forming a favorable rhizosphere environment. This significantly improves the ability of plants to absorb micro- and macronutrients

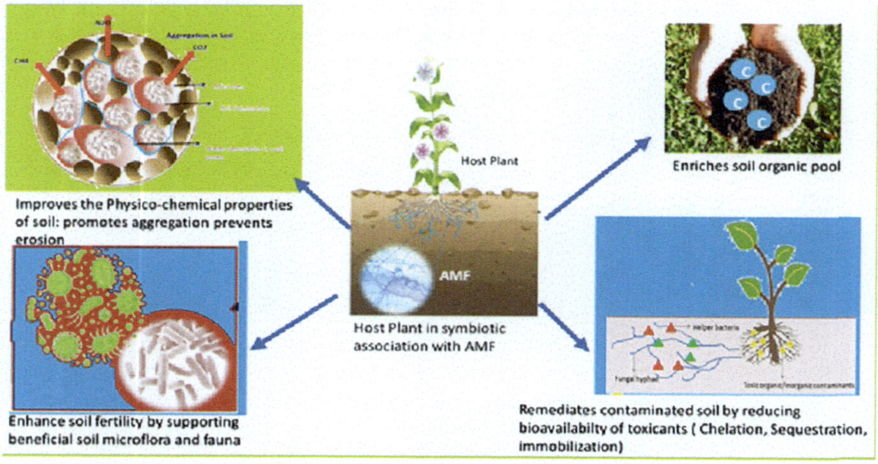

FIGURE 33.5 Role of AMF in soil health and maintenance. *AMF*, Arbuscular mycorrhizal fungi

and thus enhances their development. AMF can be used as biofertilizers and bioinoculants to increase crop production as a sustainable agricultural practice (Syamsiyah et al., 2018). They are known to improve soil structure, water-holding capacity (through substances like glomalin), and nutrient availability (Chandrasekaran et al., 2019; He et al., 2017). Their life cycle involves spore germination, hyphal growth, root colonization, and the formation of specialized structures such as arbuscules (sites of nutrient exchange) and vesicles (storage organs). AMF hyphae extend far beyond the root depletion zone, effectively increasing the absorptive surface area of plant roots.

33.4.2 Role of arbuscular mycorrhizal fungi in nutrient cycling

AMFs promote plant growth by boosting nutrient uptake, as their extramatrical hyphae penetrate deeper realms of soil nutrient reservoirs and acquire nutrients that are otherwise not accessible to the plant (Berruti et al., 2016; Pellegrino & Bedini, 2014; Smith & Smith, 2012) (Fig. 33.5). Hence these wonder fungi are increasingly gaining attention as biofertilizers. Characterization of high-affinity transporter proteins in AMF (Harrison & van Buuren, 1995) has led to extensive studies on nutritional aspects of acrylamide (AM) symbiosis from physiological as well as molecular perspectives (Bucher, 2007; Nagy et al., 2005; Paszkowski et al., 2002; Smith & Smith, 2012). One of the most well-established functions of AMF is the enhancement of plant phosphorus uptake, particularly in low-P soils. AMFs also contribute to nitrogen acquisition, either directly or indirectly through interactions with other soil microorganisms. In addition, AMF facilitate the uptake of micronutrients such as zinc, copper, and iron, which are often poorly mobile in soils. Several studies have reported the favorable effects of AMF on the uptake of nutrients such as phosphorus, nitrogen, iron, zinc, calcium, potassium, sulfur, and magnesium (Evelin et al., 2012; Lehmann & Rillig, 2015).

33.4.3 Arbuscular mycorrhizal fungi and soil structure

AMFs play a crucial role in soil aggregation through the physical entanglement of soil particles by hyphae and the production of glomalin-related soil proteins (GRSPs). The ability of AMF to regulate soil quality is ascribed to their beneficial effects on the maintenance of soil structure and stability and soil ecological interactions. The improved structure and stability of soil are correlated with glomalin production, which is a persistent glycoprotein released by AMF hyphae and spores. Besides Glomeromycota, no other fungal groups synthesize glomalin in copious amounts. The structural and functional properties of glomalin are akin to fungal hydrophobins, which are small, aggregating, and insoluble (in water) proteins released by various types of fungi (Nichols & Wright, 2004). Glomalin, being stable, hydrophobic, and impervious to heat damage, qualifies as an excellent soil-binding and stabilizing agent, as it forms stable soil aggregates and promotes organo-mineral complexes' development. This glycoprotein glues to the soil particles, forming homogenous aggregates comprising humus and minerals. In addition to conferring thermal stability to the soil, glomalin has been found to potentially sequester heavy metals, such as lead, zinc, iron, manganese, copper, and cadmium (Aslam et al., 2024; Naeem et al., 2024). This might reflect the capacity of AMF to sequester and immobilize heavy metals in the soil to preclude their entry into the plant system. Glomalin is purported to impart rigidity to the hypha and prevent its wear and tear as it grows between the soil particles. The protective

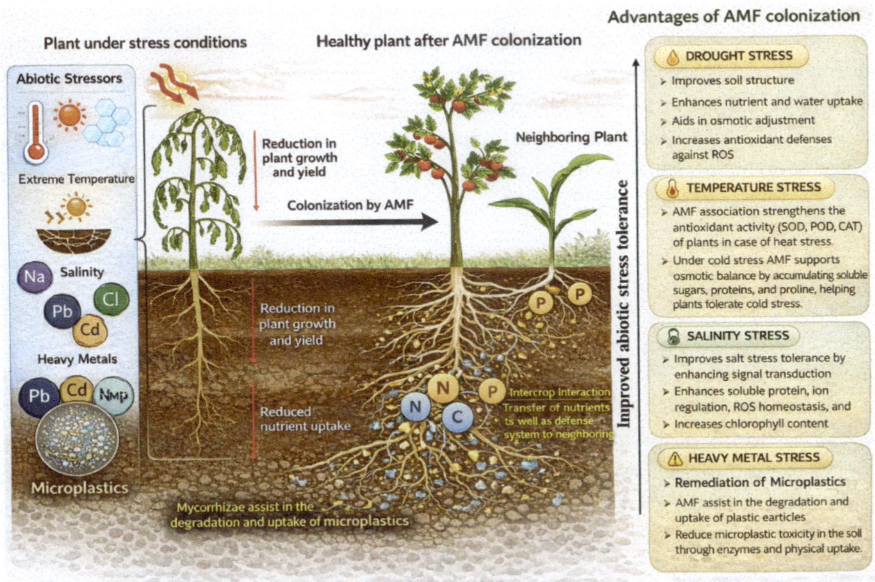

FIGURE 33.6 Overall attributes of AMF in protecting plants against stress conditions. *AMF*, Arbuscular mycorrhizal fungi; *CAT*, catalase; *POD*, peroxidases; *ROS*, reactive oxygen species; *SOD*, superoxide dismutase.

glomalin sheath of senescing hyphae is sloughed off into the soil, where it is associated with organic matter and mineral particles, assembling into microaggregates. Glomalin is also believed to improve the tilth and texture of soil to make it stable enough to withstand erosion (by wind and water) but permeable enough to allow movements of air and water and penetration of roots growing through it. These properties of glomalin make it an "essential guardian" of soil and the hyphae dwelling therein (Wright & Upadhyaya, 1999). These processes enhance aggregate stability, improve water infiltration, and reduce erosion, thereby contributing to overall soil resilience.

33.5 Role of arbuscular mycorrhizal fungi in imparting abiotic stress tolerance to plants

Abiotic stress refers to the negative impact of nonliving factors on plants; these factors include water shortage or drought stress, extremely low and high-temperature stress, and salinity stress (Ahmad et al., 2023; Elahi et al., 2023; Naveed et al., 2021; Raza et al., 2023; Shahzad et al., 2019; Wasaya et al., 2024). An overall reduction in agricultural productivity and impaired plant growth is the ultimate result of these abiotic stresses. Several research studies have been performed to date to develop strategies to combat stress conditions, which include physiological, biochemical, and molecular advancements (Parniske, 2008; Sharma et al., 2021; Smith & Smith, 2012). Although these biotechnological developments offer accuracy and the potential to introduce new features, the AMF association offers a more organic, sustainable, and ecologically friendly method of enhancing resistance to abiotic stress. Abiotic stresses—including drought, salinity, temperature extremes, and heavy metal toxicity—pose significant challenges to plant development and global agricultural sustainability. These stressors interfere with plant water relations, nutrient absorption, redox equilibrium, and overall metabolic stability, often resulting in reduced biomass and crop yield. Unlike chemical fertilizers or transgenic interventions that rely heavily on external inputs, AMF offer a natural, energy-efficient, and environmentally sound approach to strengthening plant tolerance against adverse conditions (Chakraborty et al. 2019; Samanta et al., 2025). AMFs form mutualistic associations with the roots of most terrestrial plants, developing intricate intra- and extraradical hyphal networks that considerably expand the plant's functional root zone. This extended root system enhances the uptake of essential but poorly mobile nutrients such as phosphorus, along with nitrogen and trace elements, particularly under nutrient-limited or stressful conditions. The resulting improvement in mineral nutrition sustains photosynthetic activity and supports plant growth when exposed to abiotic stresses (Fig. 33.6).

33.5.1 Drought stress tolerance

Under limited water availability, AMF-colonized plants consistently exhibit superior water-use efficiency, higher RWC, and enhanced physiological stability compared with nonmycorrhizal counterparts. These benefits are attributed to the capacity of

AMF hyphae to explore a larger soil volume, improving water extraction and root hydraulic conductivity while modulating the expression of aquaporin genes in root tissues (Augé et al., 2001, 2007). AMF also facilitates osmotic regulation through the buildup of compatible solutes such as proline, soluble sugars, and amino acids, which help preserve cell structure during dehydration (Porcel & Ruiz-Lozano, 2004). Moreover, AMFs contribute to hormonal balance—particularly by regulating abscisic acid (ABA)—to optimize stomatal behavior and minimize water loss. Enhanced antioxidant enzyme activity further protects mycorrhizal plants from oxidative damage and ensures efficient photosynthetic function during drought (Huang et al., 2017; Samanta et al., 2025).

33.5.2 Salinity stress tolerance

Excess soil salinity generates osmotic and ionic imbalances that restrict plant growth and disrupt metabolic processes. AMF symbiosis alleviates these adverse effects by improving phosphorus uptake, which stabilizes energy metabolism and cellular membranes in saline conditions (Evelin et al., 2013; Giri et al., 2007). Mycorrhizal plants maintain healthier ionic homeostasis by selectively absorbing potassium over sodium and compartmentalizing toxic ions in root vacuoles, reducing their accumulation in shoots (Wu et al., 2010). In addition, AMF enhance chlorophyll content, sustain photosynthetic efficiency, and promote stomatal conductance under salt stress, leading to greater biomass production and improved salinity tolerance (Chakraborty et al., 2019; Sheng et al., 2008).

33.5.3 Temperature stress (heat and cold)

Extreme temperatures destabilize cellular membranes, disrupt enzymatic functions, and intensify oxidative stress. Plants in AMF symbiosis exhibit fortified antioxidant defense systems because of increased activity of enzymes such as catalase (CAT), superoxide dismutase (SOD), and peroxidases (PODs), thereby reducing ROS accumulation during both heat and cold exposure (Samanta et al., 2025). Under cold conditions, AMF-colonized plants accumulate higher levels of soluble sugars and protective proteins, which help maintain cell integrity and membrane fluidity. Conversely, in hot environments, AMFs enhance nutrient uptake, maintain membrane structure, and protect photosynthetic systems from heat-induced injury (Teskey et al., 2015).

33.5.4 Heavy metal stress tolerance

AMFs significantly contribute to plant tolerance in heavy metal-contaminated soils. Their extraradical hyphae can immobilize metals such as cadmium, lead, mercury, and arsenic through mechanisms such as adsorption, chelation, and sequestration within fungal tissues, thereby minimizing their translocation to aerial parts of the plant (Gadd, 1993; Hildebrandt et al., 2007). The glycoprotein glomalin, secreted by AMF, plays an essential role in binding and stabilizing metals within soil aggregates, reducing metal solubility and bioavailability (Sharma et al., 2021; Wright & Upadhyaya, 1998). Mycorrhizal symbiosis also enhances the antioxidant potential of host plants, mitigating oxidative stress and promoting a biomass dilution effect that lessens overall metal toxicity (Dhalaria et al., 2020).

33.5.5 Microplastics stress and arbuscular mycorrhizal fungi–mediated mitigation

MPs have emerged as a new category of persistent abiotic pollutants that threaten terrestrial ecosystems. Their presence in agricultural soils alters physical structure, impairs nutrient cycles, disturbs microbial balance, and triggers oxidative stress in plants. In addition, MPs can interact directly with roots and act as carriers for toxic additives or heavy metals, intensifying plant stress responses. AMF mitigates these effects through multifaceted mechanisms. The extensive hyphal network acts as a biological filtration system, reducing root contact with MPs and restricting their mobility in soil. AMF-induced glomalin production enhances soil aggregation, trapping MPs within stable soil clusters and lowering their bioavailability (Leifheit et al., 2015; Rillig, 2004; Sharma et al., 2021). At the rhizosphere level, AMF modify microbial populations and enzyme activity, facilitating the transformation or immobilization of MP particles. Improved nutrient uptake and elevated antioxidant defenses in mycorrhizal plants further counteract oxidative stress and nutrient imbalance caused by MPs. Enhanced growth and biomass production under AMF colonization also dilute the physiological burden imposed by chronic MP exposure. Collectively, these processes position AMF as a nature-based solution for mitigating the harmful impacts of MPs and restoring soil ecological balance.

TABLE 33.3 Documented effects of microplastics on soil and plants.

Soil/Plant component	Observed effect	Ecological consequence
Soil aggregation	Decreased stability	Increased erosion
Enzyme activity	Reduced dehydrogenase, urease	Impaired nutrient cycling
Root growth	Shortened roots	Lower nutrient uptake
Oxidative stress	Increased ROS	Cellular damage
Yield	Reduced biomass	Food security risks

ROS, reactive oxygen species.

33.5.6 Integrative stress mitigation mechanisms

AMF-mediated stress tolerance involves a tightly coordinated network of physiological, biochemical, and molecular adjustments. These include activation of stress-responsive genes, regulation of polyamine metabolism, modulation of phytohormones such as indole-3-acetic acid and gibberellins, and enhancement of soil structure through increased aggregation and carbon stabilization. Through these interconnected pathways, AMFs strengthen plant adaptive capacity under multiple stress conditions, promoting sustainable growth and productivity in changing environments.

33.6 Interactions between arbuscular mycorrhizal fungi and microplastics in soil

33.6.1 Microplastics in the mycorrhizosphere

The mycorrhizosphere is defined as the zone of soil influenced by both plant roots and associated mycorrhizal fungi (Table 33.3). This is a highly dynamic interface where chemical gradients, microbial activity, and nutrient fluxes are intensified. MPs entering this zone encounter a biologically active environment characterized by high enzymatic activity, organic exudates, and dense microbial populations. These conditions can significantly alter the behavior, bioavailability, and ecological impacts of MPs. In addition to MPs, potentially toxic elements (PTEs) are also widely present in soil ecosystems (Khan et al., 2021). Because of their large specific surface area and inherent negative charge, MPs tend to adsorb positively charged toxic metals, thereby increasing the toxicity of PTEs in soil (Chen et al., 2024; Huang et al., 2021; Liu et al., 2024; Song et al., 2022). When MP invades the mycorrhizosphere, they encounter specific challenges and opportunities. Here, MP encounters a biologically active environment. This region is characterized by features like high enzymatic activity, a diverse array of organic exudates, and dense microbial populations. These conditions can significantly alter the behavior, bioavailability, and ecological impacts of MPs, potentially enhancing their persistence in the environment while also affecting soil health and plant growth. The presence of MP in the mycorrhizosphere can easily disturb the delicate balance between the microbial communities and the nutrient cycles. As so many previous researchers have reported the presence of MPs can inhibit microbial activity and alter the composition of microbial communities, leading to reduced nutrient cycling efficiency (Chen et al., 2021). This disruption can hinder plant growth and resilience, particularly in contaminated soils, emphasizing the need for further investigation into the interactions between MPs and mycorrhizal fungi.

33.6.2 Physical interactions between arbuscular mycorrhizal fungi hyphae and microplastics

As the AMF have the filamentous nature of hyphae, which enables them to physically interact with soil constituents and foreign particles, including MPs, by the penetration of their hyphae into the soil interface (Fig. 33.7). As earlier described in this chapter the MPs are of different sizes and shapes, so the fibrous and film-like MPs are particularly susceptible to entanglement within hyphal networks. This results in a significant reduction in their movement through soil layers, and this entrapment limits both vertical transport and the likelihood of root uptake (Moreno-Jiménez et al., 2022).

Furthermore, sensitive biological interfaces can be isolated from the MP by AMF-mediated soil aggregation that can encapsulate MPs within stable aggregates, isolating them. GRSPs are hydrophobic glycoproteins secreted by AMF. These proteins play a pivotal role in binding soil particles and forming stable aggregates. Owing to the hydrophobic nature of the MPs and these GRSPs, the conversion of MPs into soil aggregates is highly facilitated (Rillig, 2012a, 2012b). The same process of immobilization is mirrored in the sequestration mechanisms seen for heavy metals and organic pollutants in mycorrhizal soils. This highlights the potential of AMF as agents of soil remediation.

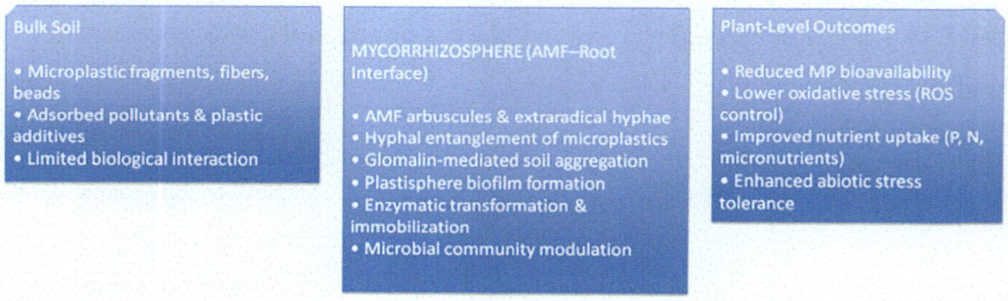

FIGURE 33.7 The mycorrhizosphere dynamics and microplastic interactions. *AMF*, Arbuscular mycorrhizal fungi; *ROS*, reactive oxygen species.

33.6.3 Chemical interactions and surface modification

MPs, particularly those in smaller fractions such as nanoplastics, possess large surface area-to-volume ratios, enabling them to adsorb organic matter, nutrients, and contaminants to form ecocoronas that influence their reactivity and interactions (Horton et al., 2017). In the mycorrhizosphere, AMF can alter soil chemical parameters, including pH, redox potential, and concentrations of organic acid, thereby influencing the surface properties of MPs. Root and fungal exudates that are rich in organic acids, polysaccharides, and phenolic compounds can form biogenic coatings on MP surfaces. Thereby reducing their chemical reactivity and potential toxicity. This surface modification can also decrease the adsorption of toxic additives and cocontaminants, minimizing ecological risks. In addition, the microbial consortia that are found associated with AMF may form biofilms on MPs, leading to further stabilization or even partial transformation of the plastic material.

33.6.4 Mechanisms of arbuscular mycorrhizal fungi–mediated mitigation of microplastic toxicity

33.6.4.1 Enhancement of plant nutritional status

Plant nutrient availability and uptake are hampered by the disruption in soil structure and nutrient dynamics because of the MP contamination (Zhou et al., 2021). AMFs counteract these effects by improving nutrient acquisition, particularly of phosphorus (PO_4^{3-}), nitrogen (NO_3), and micronutrients (particularly Fe and Zn), thereby maintaining essential metabolic processes under stress conditions. When there is adequate supply and availability of nutrients, support is provided in the synthesis of protective molecules such as osmolytes and antioxidants. For instance, adequate absorption of phosphorus enhances Adenosine triphosphate (ATP) synthesis and stabilizes cellular membranes. All these factors contribute toward stress tolerance and energy proficiency in stressful environments (Khan et al., 2023).

33.6.4.2 Modulation of plant stress physiology

MPs can either directly or indirectly harm soil microorganisms by altering their microhabitats and nutrition availability by producing biotoxicity (Zhang et al., 2021). This biotoxicity in soil microorganisms can result from ROS, facilitated oxidative stress, and harmful effects of leached MP additives, including heavy metals, plasticizing agents, antioxidants, flame retardants, and photostabilizers. Oxidative damage is also caused by exposure to MP stress in plants. This is marked by excessive production of ROS such as hydrogen peroxide and superoxide radicals (Wang et al., 2021a). The ROS are notorious for causing damage to proteins, lipids, and DNA, thus overall compromising the plant's health.

One of the ways pollutants affect plants is by directly or indirectly producing an excess of ROS (Mukherjee et al., 2019). A high level of ROS can impact the redox state of plant cells and cause oxidative damage (Wu et al., 2023). Plants develop both enzymatic and nonenzymatic antioxidant defense mechanisms to eliminate excess ROS and ensure their survival (Ahmad et al., 2010). When excess pollutants are transported to the photosynthetic organs of mycorrhizal plants, AMF promote the activities of antioxidant enzymes such as SOD, POD, CAT, ascorbate peroxidase (APX), glutathione POD, and glutathione *S*-transferase, which protect plants from the damage caused by these pollutants (Hu et al., 2020: Lenoir et al., 2017; Lu et al., 2020). On the other hand the regulation of the ascorbate–glutathione cycle in the cytoplasm and chloroplasts by AMF enhances plant tolerance to heavy metals (HM) stress, which resists oxidative stress and scavenges reactive oxygen radicals to maintain normal photosynthesis (Tan et al., 2022) (Fig. 33.8).

AMF symbiosis enhances antioxidant enzyme activities, including CAT, POD, SOD, and APX, thus helping plants neutralize ROS and maintain redox balance (Khan et al., 2025). In addition, AMF regulate hormonal pathways involving ABA, cytokinins, and auxins, which play key roles in stress signaling, stomatal regulation, and root development. This hormonal modulation allows plants to adjust morphologically and physiologically to MP-induced stress.

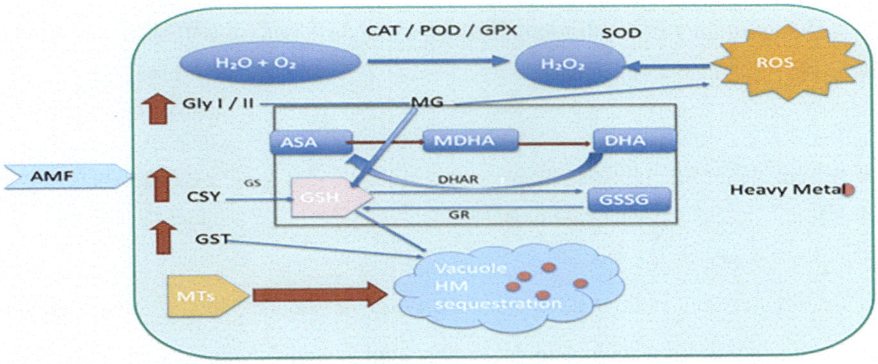

FIGURE 33.8 AMF-mediated regulation of antioxidant defenses and heavy-metal detoxification pathways in mycorrhizal plant cells. *AMF*, Arbuscular mycorrhizal fungi; *CAT*, catalase; *GST*, glutathione *S*-transferase; *POD*, peroxidases.

33.6.4.3 *Regulation of microplastic uptake and translocation*

The principal method of MP absorption by plants occurs via their roots. MPs present in the soil environment might cling to the root surface or be absorbed through the root structure owing to their diminutive size and electrostatic characteristics (Lin et al., 2024; Xu et al., 2022). MPs can enter plant cells by endocytosis through vesicle formation and are distributed within the cell (Liu et al., 2021; Rose et al., 2023). Knowing the mechanisms underlying this uptake is essential. Plants can absorb MPs through several mechanisms. As MPs exist in the soil, plant roots directly interact with them. Mycorrhizal fungi can attach to the root surface, and delicate root hairs can assist in their acquisition. Root cells lack selectivity, resulting in the unintended absorption of MPs alongside vital nutrients and water (Farow, 2023; Ramachandran et al., 2024). Smaller particles can infiltrate root cells with greater ease, but larger particles may persist on the root surface. A reduced molecular weight increases the likelihood of translocation into the plant (Oliveira & Ristroph, 2024; Roy et al., 2022). Positively charged microparticles may be repelled by negatively charged root surfaces, whereas negatively charged microparticles may be attracted to the roots (Wu et al., 2024). Once absorbed, these particles can move to aerial parts, potentially entering the food chain. AMF colonization can modify root anatomy—such as thickening the cortical layer, enhancing suberization, and strengthening apoplastic barriers—thereby reducing MP entry (Riaz et al., 2021). Moreover, AMF may influence the expression of membrane transporters and aquaporins, indirectly regulating particle movement within root tissues.

33.6.4.4 *Interaction with cocontaminants*

MPs can accumulate harmful compounds in soil systems locally by sorbing them because of their more substantial surface areas and hydrophobic characteristics. MPs can bind to both neutral and charged sites and adsorb onto hydrous oxides or participate with metals via direct physical adsorption (Narwal et al., 2024). In some habitats, MPs can function like a vector for pollutants that are absorbed by them as well as are subsequently transmitted through trophic stages, making the compounds available in soil (Shaji et al., 2025). Hydrophobic organic pollutants (such as polyaromatic hydrocarbons (PAH) and polychlorinated biphenyl), hydrophilic organic pollutants (like perfluoroalkyl acids), pharmaceuticals, and metals are among the environmental contaminants that MPs can absorb or adsorb (Gonçalves et al., 2023). Polyamide exhibited a higher affinity to adsorb antibiotics compared to PE, PVC, and PP because of its porous structure and interaction through hydrogen bonding between its amide groups and carbonyl groups of antibiotics.

Adsorption of pollutants by MPs may be influenced by the alterations in their diameter and polarities at the molecular level, when they start to degrade after exposure to UV light and microbes in an oxidative condition. MPs' aging can also pose a serious threat to ecosystems because of their increasing capacity to absorb various contaminants, including heavy metals and metalloids (Luo et al., 2020). In soils containing both MPs and other contaminants, AMF may alleviate compound stress by stabilizing soil chemistry, reducing contaminant uptake, and enhancing detoxification pathways. This multifunctional stress-buffering capacity underscores their role as biological shields in contaminated soils.

33.6.5 Arbuscular mycorrhizal fungi and mitigation strategies

33.6.5.1 *Arbuscular mycorrhizal fungi and reduction of microplastic uptake in crops*

Experimental evidence suggests that AMF can limit the translocation of MPs into above-ground plant tissues. For instance, crops colonized by *Rhizophagus irregularis* displayed reduced movement of polymethyl methacrylate particles from roots

TABLE 33.4 Summary of experimental evidence on AMF–microplastic interactions.

Study type	key findings	Implication
Laboratory	AMF reduces ROS in plants	Physiological protection
Greenhouse	Improved biomass under MP stress	Growth recovery
Field (limited)	Higher resilience in AMF-rich soils	Ecosystem buffering
Molecular	Altered gene expression	Stress adaptation

AMF, arbuscular mycorrhizal fungi; *MP*, microplastic; *ROS*, reactive oxygen species.

to shoots, largely because of sequestration within fungal hyphae and vesicles. Although complete prevention of MP entry was not achieved, AMF improved phosphorus uptake and modified plastic surface chemistry, enhancing plant resilience under contamination. Similarly, *Lactuca sativa* L. (lettuce) inoculated with AMF exhibited reduced uptake of PET by roughly 20% and an 11% increase in biomass relative to nonmycorrhizal plants. This improvement was linked to regulated hormone biosynthesis and nucleotide metabolism (Ali et al., 2022). However, the degree of mitigation varied with the type of polymer—PP and PS elicited different physiological responses—indicating that AMF effects are species- and plastic-specific.

33.6.5.2 *Effects of microplastics on arbuscular mycorrhizal fungi physiology and colonization*

Despite their benefits, high concentrations of MPs can impair AMF growth and colonization. Studies on maize rhizosphere fungi found that PE MPs significantly reduced AMF spore formation, root colonization, and metabolic activity, with colonization decreasing by up to 78% at elevated contamination levels (Gamage et al., 2025). Other research observed that polyester microfibers altered AMF community composition, influencing the abundance of genera such as *Diversispora, Claroideoglomus*, and *Rhizophagus* depending on soil type and MP form (Yin, Wang, & Zhou, 2021). These findings highlight the complex, species-dependent responses of AMF communities to different types of plastic contaminants.

33.6.5.3 *Microplastics, arbuscular mycorrhizal fungi, and soil carbon dynamics*

MPs can reshape microbial community structure and influence soil carbon cycling processes. Recent pot experiments demonstrated that both biodegradable and persistent MPs altered soil organic carbon (SOC) levels by affecting microbial necromass and GRSP content (Wang et al., 2021a). AMF genera, such as *Glomus* and *Paraglomus*, were identified as central nodes within soil microbial networks under MP influence, contributing to SOC stabilization through glomalin secretion. These findings suggest that while MPs disrupt microbial composition, AMF activity can partially restore carbon stability and maintain soil fertility. The interaction between MPs, AMF, and carbon dynamics may also have broader implications for climate regulation and carbon sequestration modeling.

33.6.5.4 *Broader impacts of microplastics on soil microbiomes*

MPs not only affect AMF but also disrupt the wider soil microbiome, altering community diversity, enzymatic functions, and trophic relationships. These changes impact key ecological processes such as nutrient cycling, organic matter turnover, and soil structural integrity. Plastic contamination shifts the abundance of dominant microbial taxa and modifies the activity of functional groups associated with carbon and nitrogen metabolism. As AMF responses occur within these complex networks, an integrated approach encompassing bacteria, fungi, plants, and soil chemistry is essential for understanding ecosystem-level impacts. Future research should focus on unraveling these interconnected dynamics to support sustainable soil management under growing pollution pressures (Table 33.4).

33.7 Knowledge gaps and future research priorities

Although work has been made in explaining the contribution of AMF to plant tolerance against conventional abiotic stresses, their functional role under emerging contaminants, particularly MPs, is still poorly articulated and supported by discontinuous empirical evidence (Table 33.5). Addressing these unresolved knowledge gaps is essential for positioning AMF-mediated mechanisms as credible, nature-based interventions for managing contamination in agricultural ecosystems.

First, empirical evidence linking AMF colonization to altered MP fate, transport, and sequestration in soils is still limited. Although AMF-mediated soil aggregation and hyphal networks are hypothesized to restrict MP mobility, direct quantitative

TABLE 33.5 Knowledge gaps and future research priorities.

Area	Knowledge gap	Research need
Detection	Nanoplastics in soil	Standardized methods
Biology	AMF species specificity	Functional screening
Time scale	Long-term impacts	Multiyear trials
Food safety	MP transfer to grains	Crop-specific studies
Management	AMF inoculation	Field validation

AMF, arbuscular mycorrhizal fungi; *MP*, microplastic.

measurements of MP retention within mycorrhizosphere soil fractions are scarce. Future work should focus on tracing labeled or fluorescent MPs across bulk soil, rhizosphere, and mycorrhizosphere compartments to resolve spatial partitioning driven by AMF activity.

Second the molecular and biochemical responses of AMF themselves to MPs remain largely unexplored. Most current studies evaluate plant responses in the presence of MPs, often overlooking whether MPs affect AMF physiology, hyphal integrity, or symbiotic efficiency. Elucidating AMF transcriptional, proteomic, and metabolic shifts under chronic MPs exposure will be essential to determine whether AMF act merely as passive barriers or as actively responsive biological agents.

Third, there is insufficient understanding of how MPs reshape the mycorrhizosphere microbiome and whether AMF can buffer or redirect these shifts. MPs are known to host distinct plastisphere communities and modify soil enzyme activities, yet their interaction with AMF-structured microbial networks is poorly resolved. Integrated multiomics approaches combining metagenomics, metatranscriptomics, and soil enzymology could clarify whether AMF stabilize microbial functions under MP stress or facilitate adaptive community reassembly.

Fourth, most available studies rely on short-term pot experiments and single polymer types, limiting ecological relevance. Agricultural soils are exposed to heterogeneous mixtures of MPs differing in size, shape, polymer chemistry, and aging status. Long-term, field-based studies across contrasting soil types and cropping systems are critically needed to assess the durability of AMF-mediated mitigation under realistic environmental conditions.

Fifth, interactions between MPs and cooccurring stressors—such as drought, salinity, heavy metals, and agrochemicals—remain largely unexplored in AMF systems. Given that MPs can act as carriers for metals and organic pollutants, future research should prioritize factorial experiments that examine whether AMF simultaneously mitigate MP stress and associated contaminant toxicity or whether trade-offs emerge under combined stress scenarios.

Finally the translation of mechanistic insights into applied frameworks is still nascent. Identifying AMF taxa or consortia with superior capacity to promote soil aggregation, microbial stability, or contaminant immobilization under MP exposure could enable targeted inoculation strategies. Coupling such biological interventions with sustainable soil management practices may accelerate the development of AMF-based remediation approaches tailored for MP-impacted agroecosystems.

Addressing these gaps will require interdisciplinary collaboration spanning soil physics, microbial ecology, plant physiology, and environmental chemistry. Progress in this direction will not only clarify the functional role of AMF in MP-contaminated soils but also position mycorrhizal symbiosis as a cornerstone of resilient and pollution-aware agricultural systems.

References

Abedsoltan, H. (2024). Applications of plastics in the automotive industry: Current trends and future perspectives. *Polymer Engineering & Science, 64*(3), 929–950.

Adeleye, A. T., Bahar, M. M., Megharaj, M., Fang, C., & Rahman, M. M. (2024). The unseen threat of the synergistic effects of microplastics and heavy metals in aquatic environments: A critical review. *Current Pollution Reports, 10*(3), 478–497.

Adeyemi, N. O., Atayese, M. O., Sakariyawo, O. S., Azeez, J. O., Olubode, A., Ridwan, M., Adebayo, R., & Adeoye, S. (2021). A growth and phosphorus uptake of soybean (*Glycine max* L.) in response to arbuscular mycorrhizal fungus *Rhizophagus* intraradices inoculation in heavy metal-contaminated soils. *Soil and Sediment Contamination: An International Journal, 30*, 1–16.

Ageel, H. K., Harrad, S., & Abdallah, M. A.-E. (2022). Occurrence, human exposure, and risk of microplastics in the indoor environment. *Environmental Science: Processes & Impacts, 24*(1), 17–31.

Ahmad, H. T., Hussain, A., Aimen, A., Jamshaid, M. U., Ditta, A., Asghar, H. N., & Zahir, Z. A. (2022a). Improving resilience against drought stress among crop plants through inoculation of plant growth-promoting rhizobacteria. In A. Husen, & M. Jawaid (Eds.), *Harsh environment and plant resilience: Molecular and functional aspects* (pp. 387–408). Springer.

Ahmad, M., Waraich, E. A., Shahid, H., Ahmad, Z., Zulfiqar, U., Mahmood, N., Al-Ashkar, I., Ditta, A., & Sabagh, A. E. (2023). Exogenously applied potassium enhanced morpho-physiological growth and drought tolerance of wheat by alleviating osmotic imbalance and oxidative damage. *Polish Journal of Environmental Studies, 32*(5), 4447–4459.

Ahmad, R., Hadi, F., Jan, A. U., & Ditta, A. (2022b). Straw incorporation in contaminated soil enhances drought tolerance but simultaneously increases the accumulation of heavy metals in rice. *Sustainability, 14*(17), 10578.

Ahmad, P., Jaleel, C. A., Salem, M. A., Nabi, G., & Sharma, S. (2010). Roles of enzymatic and nonenzymatic antioxidants in plants during abiotic stress. *Critical Reviews in Biotechnology, 30*, 161–175.

Ahmad, T., Gul, S., Peng, L., Mehmood, T., Huang, Q., Ahmad, A., Ali, H., Ali, W., Souissi, S., & Zinck, P. (2025). Microplastic mitigation in urban stormwater using green infrastructure: A review. *Environmental Chemistry Letters, 23*, 999–1024.

Ahmadi, P., Doyle, D., Mojarad, N., Taherkhani, S., Janzadeh, A., Honardoost, M., & Gholami, M. (2025). Effects of micro- and nanoplastic exposure on macrophages: A review of molecular and cellular mechanisms. *Toxicology Mechanisms and Methods, 35*, 823–846.

Ahmed, A. M., Wais, A. H., Ditta, A., Islam, M. R., Chowdhury, M. K., Pramanik, M. H., … Islam, M. S. (2024). Seed Germination and Early Seedling Growth of Sorghum (*Sorghum bicolor* L. Moench) Genotypes Under Salinity Stress. *Polish Journal of Environmental Studies, 33*(3), 3019–3032. https://doi.org/10.15244/pjoes/177180.

Aldalbahi, A., El-Naggar, M. E., El-Newehy, M. H., Rahaman, M., Hatshan, M. R., & Khattab, T. A. (2021). Effects of technical textiles and synthetic nanofibers on environmental pollution. *Polymers, 13*(1), 155.

Alex, R. K., Maes, T., & Devipriya, S. P. (2024). Clean, but not green: Emission assessment, forecast modelling and policy solutions for plastic microbeads from personal care products in India. *Emerging Contaminants, 10*(3), 100326.

Ali, M. Y., Manj, I. A., Ahmed, I., Iqbal, R., Ditta, A., Kumarasamy, V., Eldin, S. M., Ali, I., Alwahibi, M. S., & Elshikh, M. S. (2024a). Characterization of salt-tolerant cultivars of date palm based on morphological and biochemical responses under salinity stress. *Polish Journal of Environmental Studies, 33*(4), 4019–4029.

Ali, H., Khan, E., Saifullah, Farooq, M. A., & Ahmad, S. (2022). Impact of arbuscular mycorrhizal fungi on microplastic uptake in lettuce. Journal of Soil Science and Plant Nutrition, 22(3), 3145–3156

Ali, A. A. M., Khalid, A. A., Abd Razak, N. I., Maulana, N. S. M., Roslan, N. S., Razmi, R. S. B., Ruseli, W. M. A. W., Ibrahim, Y. S., Jaafar, M., & Shahrudin, R. (2024b). A review on the presence of microplastics in environmental matrices within Southeast Asia: Elucidating risk information through an analysis of microplastic characteristics such as size, shape, and type. *Water Emerging Contaminants & Nanoplastics, 3*(2), 12.

Ali, S., Khan, M. H., Zuhra, Z., & Wang, J. (2025). Innovative materials that behave like robots to combat plastic pollution. *Materials Horizons, 12*, 4042–4064.

Amesho, K. T., Chinglenthoiba, C., Samsudin, M. S., Lani, M. N., Pandey, A., Desa, M. N. M., & Suresh, V. (2023). Microplastics in the environment: An urgent need for coordinated waste management policies and strategies. *Journal of Environmental Management, 344*, 118713.

Aslam, M. N. aveed, Aslam, S., Amjad, E., Amjad, A., & Ditta, A. (2024). Remediation approaches for heavy metals contaminated soils. In A. Ditta, S. Mehmood, M. Imtiaz, & M. S. Tu (Eds.), *Bio-organic amendments for heavy metal remediation: Water, soil and plant approaches and technologies (plant biology, sustainability and climate change)* (pp. 459–477). Elsevier. https://doi.org/10.1016/B978-0-443-21610-7.00029-X.

Augé, R. M. (2001). Water relations, drought, and vesicular–arbuscular mycorrhizal symbiosis. *Mycorrhiza, 11*, 3–42.

Augé, R. M., Toler, H. D., Moore, J. L., Cho, K., & Saxton, A. M. (2007). Comparing contributions of soil versus root colonization to variations in stomatal behavior and soil drying in mycorrhizal *Sorghum bicolor* and *Cucurbita pepo*. *Journal of Plant Physiology, 164*, 1289–1299.

Baensch-Baltruschat, B., Kocher, B., Kochleus, C., Stock, F., & Reifferscheid, G. (2021). Tyre and road wear particles calculation of generation, transport, and release to water and soil with special regard to German roads. *Science of the Total Environment, 752*, 141939.

Bao, M., Xiang, X., Huang, J., Kong, L., Wu, J., & Cheng, S. (2023). Microplastics in the atmosphere and water bodies of coastal agglomerations: A mini-review. *International Journal of Environmental Research and Public Health, 20*(3), 2466.

Berruti, A., Lumini, E., Balestrini, R., & Bianciotto, V. (2016). Arbuscular mycorrhizal fungi as natural biofertilizers: Let's benefit from past successes. *Frontiers in Microbiology, 6*, 1559.

Bodzek, M., & Pohl, A. (2022). Removal of microplastics in unit processes used in water and wastewater treatment: A review. *Archives of Environmental Protection, 48*(4), 102–128.

Bouaicha, O., et al. (2022a). Microplastics make their way into the soil and rhizosphere. *Soil Security*.

Bouaicha, O., Tiziani, R., Maver, M., Lucini, L., Miras-Moreno, B., Zhang, L., Trevisan, M., Cesco, S., Borruso, L., & Mimmo, T. (2022b). Plant species-specific impact of polyethylene microspheres on seedling growth and the metabolome. *Science of the Total Environment, 840*, 156678.

Bucher, M. (2007). Functional biology of plant phosphate uptake at root and mycorrhiza interfaces. *New Phytologist, 173*, 11–26.

Chakraborty, K., Talapatra, K., Roy Das, A., Saha, A. K., & Das, P. (2019). Mycorrhizal colonization in *Drosera burmannii* Vahl, a carnivorous plant: Observation using light and fluorescence microscopy. *Ambient Science, 6*(1), 16–19.

Chandrasekaran, M., Chanratana, M., Kim, K., Seshadri, S., & Sa, T. (2019). Impact of arbuscular mycorrhizal fungi on photosynthesis, water status, and gas exchange of plants under salt stress—A meta-analysis. *Frontiers in Plant Science, 10*, 457. https://doi.org/10.3389/fpls.2019.00457.

Chaudhari, S., Parmar, H., & Samnani, P. (2024). Microplastics in personal care products and cosmetics. In *Microplastics pollution and its remediation* (pp. 171–188). Springer.

Chaudhary, H., Dwivedi, S., Sharma, D., & Dwivedi, N. (2025). A state-of-the-art review on toxicity and effect of micro/nano plastic on water life: Current statistics and bioremediation strategy. In M. P. Shah, & G. Yıldız Töre (Eds.), *Microplastics (MPs) in wastewater: Determination-treatment methods and effects on climate change* (pp. 273–303). Springer.

Chen, H., Liu, Y., & Wang, L. (2024). Combined toxicity of microplastics and co-existing inorganic pollutants in soil environments: A critical review of carrier effects and risks. *Science of The Total Environment, 906*, 167431.

Chen, M., Yue, Y., Bao, X., Feng, X., Ou, Z., Qiu, Y., … Yu, H. (2022). Effects of polystyrene nanoplastics on oxidative stress, histopathology, and intestinal microbiota in largemouth bass (Micropterus salmoides). *Aquaculture Reports, 27*, 101423.

Chettri, D., Pati, T., & Verma, A. K. (2023). Microbe-mediated biodegradation of microplastics from wastes. *Water and Environment Journal, 37*(4), 671–685.

Chen, Y., Leng, Y., Liu, X., & Wang, J. (2021). Microplastics acts as a vector for antibiotic resistance genes in landfill leachate. *Environmental Pollution, 284*, 117174.

Chowaniec, K., Zubek, S., Zalewska-Gałosz, J., Stanek, M., & Skubała, K. (2024). *A mosaic of biological soil crusts and vascular plants contributes to the spatial heterogeneity of key soil properties at different successional stages of restored inland sand dunes Mosaic of biological soil crusts and vascular plants contributes. Plant and Soil, 510*(1), 603–626. https://doi.org/10.1007/s11104-024-06941-y.

Csorba, C., Hackl, E., Reichenauer, T., van der Putten, W., & Sessitsch, A. (2024). *Sustainable use and conservation of soil microorganisms and invertebrates contributing to bioremediation and nutrient cycling*. Food and Agriculture Organization.

Cui, Y., Wu, Y., Shi, P., Ni, Y., Zeng, H., Zhang, Z., Zhao, C., Sun, W., & Yi, Q. 2025. Mitigating microplastic-induced organ damage: Mechanistic insights from the microplastic-macrophage axesRedox Biology84103688

De Silva, Y., Rajagopalan, U., & Kadono, H. (2021). Microplastics on the growth of plants and seed germination in aquatic and terrestrial ecosystems. *Global Journal of Environmental Science & Management (GJESM), 7*(3), 347–368.

Dhalaria, R., Kumar, D., Kumar, H., Nepovimova, E., Kuča, K., Islam, M. T., & Verma, R. (2020). Arbuscular mycorrhizal fungi as potential agents in ameliorating heavy metal stress in plants. *Agronomy, 10*(6), 815.

Diagne, N., Ngom, M., Djighaly, P. I., Fall, D., Hocher, V., & Svistoonoff, S. (2020). Roles of arbuscular mycorrhizal fungi on plant growth and performance: Importance in biotic and abiotic stressed regulation. *Diversity, 12*, 370.

Ditta, A., Mehmood, S., Imtiaz, M., & Tu, M. S. (2024). *Bio-organic amendments for heavy metal remediation: Water, soil and plant approaches and technologies (plant biology, sustainability and climate change* (1st ed.). Elsevier Inc.

Ditta, A., & Ullah, N. (2023). Perspectives of using plant growth-promoting rhizobacteria under salinity stress for sustainable crop production. In M. Ghorbanpour, & M. A. Shahid (Eds.), *Plant stress mitigators: Types, techniques and functions* (pp. 231–247). Woodhead Publishing Ltd.Elsevier.

Elahi, N. N., Raza, S., Rizwan, M. S., Albalawi, B. F. A., Ishaq, M. Z., Ahmed, H. M., Mehmood, S., Imtiaz, M., Farooq, U., Rashid, M., & Ditta, A. (2023). Foliar application of gibberellin alleviates the adverse impacts of drought stress and improves the growth, physiological, and biochemical attributes of canola (*Brassica napus* L.). *Sustainability, 15*(1), 78.

Elbasiouny, H., Mostafa, A. A., Zedan, A., Elbltagy, H. M., Dawoud, S. F., Elbanna, B. A., El-Shazly, S. A., El-Sadawy, A. A., Sharaf-Eldin, A. M., & Darweesh, M. (2023). Potential effect of biochar on soil properties, microbial activity, and *Vicia faba* properties affected by microplastics contamination. *Agronomy, 13*(1), 149.

Emenike, E. C., Okorie, C. J., Ojeyemi, T., Egbemhenghe, A., Iwuozor, K. O., Saliu, O. D., Okoro, H. K., & Adeniyi, A. G. (2023). From oceans to dinner plates: The impact of microplastics on human health. *Heliyon, 9*(10), e20440.

Evelin, H., Giri, B., & Kapoor, R. (2013). Ultrastructural evidence for AMF-mediated salt stress mitigation in *Trigonella foenum-graecum. Mycorrhiza, 23*(1), 71–86.

Evelin, H., Giri, B., & Kapoor, R. (2012). Contribution of *Glomus intraradices* inoculation to nutrient acquisition and mitigation of ionic imbalance in NaCl-stressed *Trigonella foenum-graecum. Mycorrhiza, 22*, 203–217.

Fan, P., Tan, W., & Yu, H. (2022). Effects of different concentrations and types of microplastics on bacteria and fungi in alkaline soil. *Ecotoxicology and Environmental Safety, 229*, 113045.

Fang, S., Hua, C., Yang, J., Liu, F., Wang, L., Wu, D., & Ren, L. (2025). Combined pollution of soil by heavy metals, microplastics, and pesticides: Mechanisms and anthropogenic drivers. *Journal of Hazardous Materials, 485*, 136812.

Fang, Z., Sallach, J., & Hodson, M. E. (2024). Size-and concentration-dependent effects of microplastics on soil aggregate formation and properties. *Journal of Hazardous Materials, 465*, 133395.

Farow, A. M. (2023). Unselective root absorption of emerging polymer contaminants during nutrient and water influx. *Terrestrial Environmental Toxicology, 14*(2), 88–97.

Gadd, G. M. (1993). Interactions of fungi with toxic metals. *New Phytologist, 124*(1), 25–60.

Gallitelli, L., Di Lollo, G., Adduce, C., Maggi, M., Trombetta, B., & Scalici, M. (2023). Aquatic plants entrap different sizes of plastics in indoor flume experiments. *Science of The Total Environment, 863*, 161051.

Gamage, G. G. H. M., Madanayake, N. H., Premarathna, M., & Madawala, H. MS. P. (2025). Effect of microplastics on rhizosphere and arbuscular mycorrhizal fungi of Zea mays. *Ceylon Journal of Science, 54*(3), 855–864. https://doi.org/10.4038/cjs.v54i3.8944856.

Gamage, S., & Mahagamage, Y. (2024). Microplastics in personal care products and cosmetics in Sri Lanka. *Heliyon, 10*(8), e29393.

Ghosh, S., & Kumar, Y. (2025). Effect of microplastics on seed storage proteins of plants. In *Microplastics* (pp. 403–428). Elsevier.

Giri, B., Kapoor, R., & Mukerji, K. G (2007). Improved tolerance of Acacia nilotica to salt stress by arbuscular mycorrhiza, *Glomus fasciculatum* may be partly related to elevated K/Na ratios in root and shoot tissues. *Microb Ecol, 54*, 753–760. doi:10.1007/s00248-007-9239-9.

Gilani, I. E., Sayadi, S., Zouari, N., & Al-Ghouti, M. A. (2023). Plastic waste impact and biotechnology: Exploring polymer degradation, microbial role, and sustainable development implications. *Bioresource Technology Reports, 24*, 101606.

Gkanasos, A., Tsiaras, K., Triantaphyllidis, G., Panagopoulos, A., Pantazakos, G., Owens, T., … Katsafados, N. (2021). Stopping macroplastic and microplastic pollution at source by installing novel technologies in river estuaries and wastewater treatment plants: The CLAIM project. *Frontiers in Marine Science, 8*, 738876.

Gonçalves, C. O., Santos, M. S., Silva, A. M., & Quinta-Ferreira, R. M. (2023). Interactions between microplastics and organic/inorganic pollutants in aquatic environments: Mechanisms, affecting factors, and ecotoxicological risks. *Environmental Research, 233*, 116413.

Gong, N., Wang, Z., Wang, X., & Shao, K. (2024). Uptake, removal, and trophic transfer of fluorescent polyethylene microplastics by freshwater model organisms: The impact of particle size and food availability. *Aquatic Toxicology, 277*, 107165.

Gu, X., Yin, R., Cai, W., Chen, P., Cui, K., Du, Y., Li, Y., & Cai, H. (2024). Residual plastic film decreases crop yield and water use efficiency through direct negative effects on soil physicochemical properties and root growth. *Science of the Total Environment, 946*, 174204.

Hale, R. C., Seeley, M. E., La Guardia, M. J., Mai, L., & Zeng, E. Y. (2020). A global perspective on microplastics. *Journal of Geophysical Research: Oceans, 125*(1), e2018JC014719.

Hanif, M., Aijaz, N., Azam, K., Akhtar, M., Laftah, W., Babur, M., … Benitez, I. (2024). Impact of microplastics on soil (physical and chemical) properties, soil biological properties/soil biota, and response of plants to it: A review. *International Journal of Environmental Science and Technology, 21*(16), 10277–10318.

Harrison, M. J., & Van Buuren, M. L. (1995). A phosphate transporter from the mycorrhizal fungus *Glomus versiforme*. *Nature, 378*, 626–629.

Hasan, M. M., & Tarannum, M. N. (2025). Adverse impacts of microplastics on soil physicochemical properties and crop health in agricultural systems. *Journal of Hazardous Materials Advances, 17*, 100528.

He, F., Sheng, M., & Tang, M. (2017). Effects of *Rhizophagus irregularis* on photosynthesis and antioxidative enzymatic system in *Robinia pseudoacacia* L. under drought stress. *Frontiers in Plant Science, 8*, 183. https://doi.org/10.3389/fpls.2017.00183.

He, D., Guo, T., Li, J., & Wang, F. (2023). Optimize lettuce washing methods to reduce the risk of microplastics ingestion: The evidence from microplastics residues on the surface of lettuce leaves and in the lettuce washing wastewater. *Science of the Total Environment, 868*, 161726.

Hildebrandt, U., Regvar, M., & Bothe, H. (2007). Arbuscular mycorrhiza and heavy metal tolerance. *Phytochemistry, 68*(1), 139–146.

Horton, A. A., Walton, A., Spurgeon, D. J., Lahive, E., & Svendsen, C. (2017). Microplastics in freshwater and terrestrial environments: Evaluating the current understanding to identify the knowledge gaps and future research priorities. *Science of the Total Environment, 586*, 127–141.

Hossain, A., Mondal, C. K., Hasan, M., Masud, M. A. A., Senapathi, V., Pal, S. C., Idris, A. M., & Towfiqul Islam, A. R. M. (2025). Navigating the hidden crisis: Microplastics and heavy metals in soil-transport, impact, and remediation. *Soil and Sediment Contamination: An International Journal, 34*, 2491–2532.

Hu, W., Zhang, H., Chen, H., & Tang, M. (2017). Arbuscular mycorrhizas influence *Lycium barbarum* tolerance of water stress in a hot environment. *Mycorrhiza, 27*(5), 451–463. https://doi.org/10.1007/s00572-017-0765-0.

Hu, S., Hu, B., Chen, Z., Vosátka, M., & Vymazal, J. (2020). Antioxidant response in arbuscular mycorrhizal fungi inoculated wetland plant under Cr stress. *Environmental Research, 191*, 110203.

Huang, D., Wang, Q., Jing, G., Ma, M., Li, C., Ma, F., & Luo, Z. (2021). Overexpression of MdIAA24 improves apple drought resistance by positively regulating strigolactone biosynthesis and mycorrhization. *Tree Physiology, 41*(1), 134–146. https://doi.org/10.1093/treephys/tpaa109.

Huang, Y. M., Zou, Y. N., & Wu, Q. S. (2017). Alleviation of drought stress by mycorrhizas is related to increased root H_2O_2 efflux in trifoliate orange. *Scientific Reports, 7*, 423–435.

Huang, Y., Yu, S., Zheng, Z., Xiao, X., Zhu, Z., Deng, L., Wei, H., Liang, J., Chen, S., & Holmer, M. (2025). Dual mobilization of buried microplastics and organic carbon driven by seagrass degradation: A case study from Swan Lake, China. *Frontiers in Marine Science, 12*, 1593776.

Jaafarzadeh Haghighi Fard, N., Ahmadi, B., Jahedi, F., Abolfazli, P., & Maniati, M. (2025). Microplastics as vectors for antibiotic resistance genes and their implications for gut health. *Discover Medicine, 2*(1), 6.

Jajoo, A., & Mathur, S. (2021). Role of arbuscular mycorrhizal fungi as an underground savior for protecting plants from abiotic stresses. *Physiology and Molecular Biology of Plants, 27*, 2589–2603.

Jan, A. U., Hadi, F., Ditta, A., Suleman, M., & Ullah, M. (2022). Zinc-induced anti-oxidative defense and osmotic adjustments to enhance drought stress tolerance in sunflower (*Helianthus annuus* L.). *Environmental and Experimental Botany, 193*, 104682.

Jansen, M., Andrady, A., Bornman, J., Aucamp, P., Bais, A., Banaszak, A., Barnes, P., Bernhard, G., Bruckman, L., Busquets, R., Häder, D.-P. ., Hanson, M. L., Heikkilä, A. M., Hylander, S., Lucas, R. M., Mackenzie, R., Madronich, S., Neale, P. J., Neale, R. E., … Zhu, L. (2024). Plastics in the environment in the context of UV radiation, climate change and the Montreal Protocol: UNEP Environmental Effects Assessment Panel, Update 2023. *Photochemical & Photobiological Sciences, 23*, 629–650.

Jin, T., Tang, J., Lyu, H., Wang, L., Gillmore, A. B., & Schaeffer, S. M. (2022). Activities of microplastics (MPs) in agricultural soil: A review of MPs pollution from the perspective of agricultural ecosystems. *Journal of Agricultural and Food Chemistry, 70*(14), 4182–4201.

Kasmuri, N., Tarmizi, N. A. A., & Mojiri, A. (2022). Occurrence, impact, toxicity, and degradation methods of microplastics in the environment—A review. *Environmental Science and Pollution Research, 29*(21), 30820–30836.

Khan, S., Shamshad, I., Khan, M. A., & Ahmad, L. (2021). Distribution and ecological risk assessment of potentially toxic elements in soil ecosystems: Anthropogenic inputs vs. geogenic origins. *Environmental Monitoring and Assessment, 193*(7), 415 ScienceDirect /Springer Link.

Khan, F., Siddique, A. B., Shabala, S., Zhou, M., & Zhao, C. (2023). Phosphorus plays key roles in regulating plants' physiological responses to abiotic stresses. *Plants, 12*(15), 2861. doi:10.3390/plants12152861.

Khan, Y., Shah, S., Li, D., & Yang, F. (2025). Arbuscular mycorrhizal fungi-mediated abiotic stress tolerance: Emerging roles in nutrient exchange, antioxidant defence, and hormonal crosstalk. *Plant Stress, 18*, 101068. doi:10.1016/j.stress.2025.101068.

Lakhiar, I. A., Yan, H., Zhang, J., Wang, G., Deng, S., Bao, R., Zhang, C., Syed, T. N., Wang, B., & Zhou, R. (2024). Plastic pollution in agriculture as a threat to food security, the ecosystem, and the environment: An overview. *Agronomy, 14*(3), 548.

Lanfranco, L., Bonfante, P., & Genre, A. (2016). The mutualistic interaction between plants and arbuscular mycorrhizal fungi. *Microbiology Spectrum, 4*, 727–747.

Lehmann, A., & Rillig, M. C. (2015). Arbuscular mycorrhizal contribution to copper, manganese, and iron nutrient concentrations in crops – A meta-analysis. *Soil Biology and Biochemistry, 81*, 147–158.

Leifheit, E. F., Verbruggen, E., & Rillig, M. C. (2015). Arbuscular mycorrhizal fungi reduce decomposition of woody plant litter while increasing soil aggregation. *Soil Biology and Biochemistry, 81*, 323–328.

Lenoir, I., Fontaine, J., Tisserant, B., Laruelle, F., & Sahraoui, A. L.-H. (2017). Beneficial contribution of the arbuscular mycorrhizal fungus, *Rhizophagus irregularis*, in the protection of *Medicago truncatula* roots against benzo[*a*]pyrene toxicity. *Mycorrhiza, 27*, 465–476.

Li, J., Shan, E., Zhao, J., Teng, J., & Wang, Q. (2023a). The factors influencing the vertical transport of microplastics in the marine environment: A review. *Science of the Total Environment, 870*, 161893.

Li, R., Tu, C., Li, L., Wang, X., Yang, J., Feng, Y., Zhu, X., Fan, Q., & Luo, Y. (2023b). Visual tracking of label-free microplastics in wheat seedlings and their effects on crop growth and physiology. *Journal of Hazardous Materials, 456*, 131675.

Li, T., Tao, S., Ma, M., Liu, S., Shen, M., & Zhang, H. (2024). Is the application of organic fertilizers becoming an undeniable source of microplastics and resistance genes in agricultural systems? *Science of the Total Environment, 912*, 169571.

Li, W., Wang, Z., Li, W., & Li, Z. (2022). Impacts of microplastics addition on sediment environmental properties, enzymatic activities, and bacterial diversity. *Chemosphere, 307*, 135836.

Li, X., Shi, F., Zhou, M., Wu, F., Su, H., Liu, X., Wei, Y., & Wang, F. (2024). Migration and accumulation of microplastics in soil-plant systems mediated by symbiotic microorganisms and their ecological effects. *Environment International, 191*, 108965.

Li, X., Wang, R., Dai, W., Luan, Y., & Li, J. (2023c). Impacts of micro (nano) plastics on terrestrial plants: Germination, growth, and litter. *Plants, 12*(20), 3554.

Li, Z., Li, Q., Li, R., Zhao, Y., Geng, J., & Wang, G. (2020). Physiological responses of lettuce (*Lactuca sativa* L.) to microplastic pollution. *Environmental Science and Pollution Research, 27*(24), 30306–30314.

Liese, B., Stock, N. L., Düwel, J., Pilger, C., Huser, T., & Müller, C. (2024). Uptake of microplastics and impacts on plant traits of savoy cabbage. *Ecotoxicology and Environmental Safety, 272*, 116086.

Lin, W., Jiang, R., Wu, J., Xiong, S., & Ouyang, G. (2024). Influencing mechanisms of microplastics existence on soil heavy metals accumulated by plants. *Journal of Hazardous Materials, 471*, 134293.

Lindfors, S., Österlund, H., Lorenz, C., Vianello, A., Nordqvist, K., Gopinath, K., Lykkemark, J., Lundy, L., Vollertsen, J., & Viklander, M. (2025). Microplastics and tyre wear particles in urban runoff from different urban surfaces. *Science of the Total Environment, 980*, 179527.

Liu, X., An, C., & Zhang, K. (2024). Interactive effects of microplastics and potentially toxic elements in soil-plant systems: Adsorption properties, bioavailability, and eco-toxicological outcomes. *Soil Biology and Biochemistry, 189*, 109270.

Liu, Y., Guo, R., Zhang, S., Yuan, X., & Wang, F. (2021). Endocytosis-mediated cellular uptake and distribution of nanoplastics in plant root cells. *Environmental Science & Technology, 55*(21), 14711–14721.

Liu, Y. Z., Imtiaz, M., Ditta, A., Rizwan, M. S., Ashraf, M., Mehmood, S., Aziz, O., Mubeen, F., Ali, M., Elahi, N. N., Ijaz, R., S, Lelel, Shuang, C., & Tu, S. (2020). Response of growth, antioxidant enzymes, and root exudates production towards As stress in *Pteris vittata* and *Astragalus sinicus* colonized by arbuscular mycorrhizal fungi. *Environmental Science and Pollution Research, 27*, 2340–2352.

Liu, J., Chen, J., Xie, K., Tian, Y., Yan, A., Liu, J., Huang, Y., Wang, S., Zhu, Y., Chen, A., & Xu, G. (2020). A mycorrhiza-specific H+-ATPase is essential for arbuscule development and symbiotic phosphate and nitrogen uptake. *Plant, Cell & Environment, 43*, 1069–1083.

Liu, Z., Cai, L., Dong, Q., Zhao, X., & Han, J. (2022). Effects of microplastics on water infiltration in agricultural soil on the Loess Plateau, China. *Agricultural Water Management, 271*, 107818.

Lu, F., Sheng, J., Hu, X., & Wang, Y. (2020). Effects of arbuscular mycorrhizal fungi Glomus mosseae on the growth and medicinal components of Dysosma versipellis under copper stress. *Ecotoxicology and Environmental Safety, 190*, 110115.

Luo, H., Chang, L., Ju, T., & Li, Y. (2024). Factors influencing the vertical migration of microplastics up and down the soil profile. *ACS Omega, 9*, 50064–50077.

Luo, J., Li, X., Jin, Y., Traore, I., Dong, L., Yang, G., & Wang, Y. (2020). Effects of arbuscular mycorrhizal fungi *Glomus mosseae* on the growth and medicinal components of *Dysosma versipellis* under copper stress. *Bulletin of Environmental Contamination and Toxicology, 107*, 924–930.

Luzi, B., Miino, M. C., Rada, E. C., Zullo, R., Baltrocchi, A. P. D., Torretta, V., & Galafassi, S. (2025). Critical review of microfiber release from textiles: Results, comparative challenges, mitigation strategies, and legislative perspectives. *Chemosphere, 378*, 144394.

Lwanga, E. H., Beriot, N., Corradini, F., Silva, V., Yang, X., Baartman, J., Rezaei, M., van Schaik, L., Riksen, M., & Geissen, V. (2022). Review of microplastic sources, transport pathways and correlations with other soil stressors: A journey from agricultural sites into the environment. *Chemical and Biological Technologies in Agriculture, 9*(1), 20.

Ma, Y., Yang, K., Yu, H., Tan, W., Gao, Y., & Lv, B. (2024). Effects and mechanism of microplastics on organic carbon and nitrogen cycling in agricultural soil: A review. *Soil Use and Management, 40*(1), e12971.

Ma, Y.-f., & You, X.-y (2025). Microplastics in freshwater ecosystems: A significant force of disrupting health and altering trophic transfer patterns by reduced assimilation efficiency of aquatic organisms. *Aquaculture, 594*, 741463.

Mangla, S., Pathak, G., & Bhan, V. (2025). Microplastics contamination in soil, water, air, and environmental health risk assessment. *Microplastics*, 3–30.

Megha, K., Anvitha, D., Parvathi, S., Neeraj, A., Sonia, J., & Mohanan, P. (2025). Environmental impact of microplastics and potential health hazards. *Critical Reviews in Biotechnology, 45*(1), 97–127.

Meng, L., Zheng, S., Zhao, Y., Liu, T., Liang, J., Zhu, M., & Sun, X. (2024). Seasonal microplastic ingestion by carnivorous chaetognaths in Jiaozhou Bay, China: Field evidence revealing microplastic trophic transfer. *Journal of Hazardous Materials, 478*, 135532.

Menichetti, A., Mordini, D., & Montalti, M. (2024). Penetration of microplastics and nanoparticles through skin: Effects of size, shape, and surface chemistry. *Journal of Xenobiotics, 15*(1), 6.

Milojevic, N., & Cydzik-Kwiatkowska, A. (2021). Agricultural use of sewage sludge as a threat to microplastic (MP) spread in the environment and the role of governance. *Energies, 14*(19), 6293.

Moreno-Jiménez, E., Leifheit, E. F., Plaza, C., Feng, L., Bergmann, J., Wulf, A., … Rillig, M. C. (2022). Effects of microplastics on crop nutrition in fertile soils and interaction with arbuscular mycorrhizal fungi. *Journal of Sustainable Agriculture and Environment, 1*(1), 66–72. https://doi.org/10.1002/sae2.12006.

Mukherjee, A., Agrawal, S. B., & Agrawal, M. (2019). Responses of tropical tree species to urban air pollutants: ROS/RNS formation and scavenging. *Science of the Total Environment, 710*, 136363.

Mutshekwa, T., Motitsoe, S. N., Maiyana, V. P., Mofu, L., Ajay, K., & Phophi, L. (2025). Microplastic occurrence and densities in small subtropical reservoirs of Southern Africa. *Discover Sustainability, 6*(1), 132.

Naeem, N., Sarfraz, W., Khalid, N., Rizvi, Z. F., Nazir, A., Ejaz, U., Amjad, N., Safdar, L., Amanat, R., Akhtar, S., & Ditta, A. (2024). Factors affecting the remediation of wastewater worldwide through eco-technologies. In A. Ditta, S. Mehmood, M. Imtiaz, & M. S. . Tu (Eds.), *Bio-organic amendments for heavy metal remediation: Water, soil and plant approaches and technologies (plant biology, sustainability and climate change)* (pp. 21–31). Elsevier. https://doi.org/10.1016/B978-0-443-21610-7.00022-7.

Nagy, R., Karandashov, V., Chague, V., Kalinkevich, K., Tamasloukht, M. B., Xu, G., & Bucher, M. (2005). The characterization of novel mycorrhiza-specific phosphate transporters from *Lycopersicon esculentum* and *Solanum tuberosum* uncovers functional redundancy in symbiotic phosphate transport in solanaceous species. *The Plant Journal, 42*, 236–250.

Narwal, N., Kakakhel, M. A., Katyal, D., Yadav, S., Rose, P. K., Rene, E. R., Rakib, M. R. J., Khoo, K. S., & Kataria, N. (2024). Interactions between microplastic and heavy metals in the aquatic environment: Implications for toxicity and mitigation strategies. *Water, Air, & Soil Pollution, 235*(9), 567. https://doi.org/10.1007/s11270-024-07343-7.

Naveed, M., Ditta, A., Ahmad, M., Mustafa, A., Ahmad, Z., Conde-Cid, M., Tahir, S., Shah, S. A. A., Abrar, M. M., & Fahad, S. (2021). Processed animal manure improves morpho-physiological and biochemical characteristics of *Brassica napus* L. under nickel and salinity stress. *Environmental Science and Pollution Research, 28*, 45629–45645.

Nazir, A., Sarfraz, W., Ditta, A., Khalid, N., Farid, M., Shafiq, M., Firdaus-e-Bareen, Rizvi, Z. F., & Naeem, N. (2023). Synergistic impact of two autochthonous saprobic fungi (*A. niger* and *T. pseudokoningii*) on the growth, ionic contents, and metals uptake in *Brassica juncea* L. and *Vigna radiata* L. under tannery solid waste contaminated soil. *International Journal of Phytoremediation, 25*(11), 1488–1500.

Nichols, K. A., & Wright, S. F. (2004). Contributions of fungi to soil organic matter in agro-ecosystems. In F. Magdoff, & R. R. Weil (Eds.), *Soil organic matter in sustainable agriculture* (pp. 179–198). CRC Press.

Nofal, R. M. (2022). Biodegradable textiles, recycling, and sustainability achievement. In *Handbook of biodegradable materials* (pp. 1–37). Springer Nature.

Oliveira, M. B., & Ristroph, K. D. (2024). Biophysical barriers in plant vascular systems: How surface characteristics and particle weight dictate microplastic movement. *Environmental Research, 242*, 117650 [1].

O'Brien, S., Rauert, C., Ribeiro, F., Okoffo, E. D., Burrows, S. D., O'Brien, J. W., Wang, X., Wright, S. L., & Thomas, K. V. (2023). There's something in the air: A review of sources, prevalence and behaviour of microplastics in the atmosphere. *Science of the Total Environment, 874*, 162193.

Parniske, M. (2008). Arbuscular mycorrhiza: The mother of plant root endosymbioses. *Nature Reviews Microbiology, 6*, 763–775.

Paszkowski, U., Kroken, S., Roux, C., & Briggs, S. P. (2002). Rice phosphate transporters include an evolutionarily divergent gene specifically activated in arbuscular mycorrhizal symbiosis. *Proceedings of the National Academy of Sciences, 99*, 13324–13329.

Patil, P. M., Mahamuni-Badiger, P., Ingavale, R. R., Patel, P. R., & Dhanavade, M. J. (2024). Usage of microplastic beads in pharmaceuticals and cosmetics industry: A review. In M. Shahnawaz, C. O. Adetunji, M. A. Dar, & D. Zhu (Eds.), *Microplastic pollution* (pp. 51–72). Springer.

Paudel, P., Kumar, R., Pandey, M. K., Paudel, P., & Subedi, M. (2024). Exploring the impact of microplastics on soil health and ecosystem dynamics: A comprehensive review. *Journal of Experimental Biology and Agricultural Sciences, 12*(2), 163–174.

Pellegrino, E., & Bedini, S. (2014). Enhancing ecosystem services in sustainable agriculture: Biofertilization and biofortification of chickpea (*Cicer arietinum* L.) by arbuscular mycorrhizal fungi. *Soil Biology and Biochemistry, 68*, 429–439.

Pignattelli, S., Broccoli, A., & Renzi, M. (2020). Physiological responses of garden cress (*L. sativum*) to different types of microplastics. *Science of The Total Environment, 727*, 138609.

Porcel, R., & Ruiz-Lozano, J. M. (2004). Arbuscular mycorrhizal influence on leaf water potential, solute accumulation, and oxidative stress in soybean plants subjected to drought stress. *J Exp Bot, 55*, 1743–1750. doi:10.1093/jxb/erh188.

Prasad, R., Bhola, D., Akdi, K., Cruz, C., Kvss, S., & Tuteja, N. (2017). Introduction to mycorrhiza: Historical development. In *Mycorrhiza-function, diversity, state of the art* (pp. 1–7). https://doi.org/10.1007/978-3-319-53064-2_1.

Pringle, A., Bever, J. D., Gardes, M., Parrent, J. L., Rillig, M. C., & Klironomos, J. N. (2009). Mycorrhizal symbioses and plant invasions. *Annual Review of Ecology, Evolution, and Systematics, 40*(1), 699–715. https://doi.org/10.1146/annurev.ecolsys.39.110707.173454.

Priyadarshini, S., Jagatee, S., & Das, A. P. (2024). Synthetic fabrics and microfiber pollution – An assessment of their global impact. In *Renewable energy generation and value addition from environmental microfiber pollution through advanced greener solution* (pp. 137–157). Springer.

Qin, Y., Tian, X., Zhang, J., Tu, Y., & Chen, C. (2025). Combined toxicity and ecological impacts of soil microplastics and heavy metals in rhizosphere microenvironments: A comprehensive review. *Journal of Soils and Sediments, 25*, 2551–2570.

Ramachandran, S., Kumar, A., & Singh, B. (2024). Root hair-assisted acquisition of microplastics and mycorrhizal attachment dynamics in contaminated soils. *Plant and Soil, 495*(1), 115–132.

Raguso, C. B. (2025). Microplastics and their associated contaminants in the marine environment: From detection to the mechanisms of interaction with marine biota (Doctoral dissertation, Università degli Studi di Milano-Bicocca). *Bicocca Open Archive*.

Raza, M. A. S., Ibrahim, M. A., Ditta, A., Iqbal, R., Aslam, M. U., Muhammad, F., Ali, S., Çiğ, F., Ali, B., Ikram, R. M., Muzamil, M. N., Rahman, M. H., Alwahibi, M. S., & Elshikh, M. S. (2023). Exploring the recuperative potential of brassinosteroids and nano-biochar on growth, physiology, and yield of wheat under drought stress. *Scientific Reports, 13*, 15015.

Riaz, M., Kamran, M., Fang, Y., Wang, Q., Cao, H., Yang, G., … Wang, X. (2021). Arbuscular mycorrhizal fungi-induced mitigation of heavy metal phytotoxicity in metal contaminated soils: A critical review. *Journal of Hazardous Materials, 402*, 123919. https://doi.org/10.1016/j.jhazmat.2020.123919.

Rillig, M. C. (2004). Arbuscular mycorrhizae, glomalin, and soil aggregation. *Canadian Journal of Soil Science, 84*, 355–363.

Rillig, M. C. (2012a). Microplastic in terrestrial ecosystems and the soil? *Environmental Science & Technology, 46*, 6453–6454.

Rillig, M. C. (2012b). Microplastics in terrestrial ecosystems and the soil environment: A review. *Science of the Total Environment*.

Rose, J., Kumar, M., & Thakur, S. (2023). Ecotoxicological impact of microplastics: Bioaccumulation patterns and trophic transfer through food chains. *Environmental Pollution, 320*, 121084.

Roy, S., Das, P., & Sengupta, S. (2022). Molecular weight and particle size as determining factors for the systemic translocation of microplastics in higher plants. *Chemosphere, 291*, 132945.

Roth, R., & Paszkowski, U. (2017). Plant carbon nourishment of arbuscular mycorrhizal fungi. *Current Opinion in Plant Biology, 39*, 50–56.

Roy, T. (2024). Aquatic ecosystem toxicity and food chain. In *Toxicity of aquatic system and remediation* (pp. 25–36). CRC Press.

Russo, P., & Di Ciaccio, F. (2024). Deep classification of microplastics through image fusion techniques. *IEEE Access, 12*, 134852–134861.

Saeedi, M. (2023). How microplastics interact with food chain: A short overview of fate and impacts. *Journal of Food Science and Technology, 61*, 1–11.

Saleem, A., Raza, M. A. S., Tahir, M. A., Iqbal, R., Aslam, M. U., Toleikiene, M., Khan, M. S., Alwahibi, M. S., Elshikh, M. S., & Ditta, A. (2025). Impact of biogas slurry on physiological and antioxidant mechanisms of wheat under drought stress. *Polish Journal of Environmental Studies, 34*(2), 1721–1731.

Saleem, A., Raza, M. A. S., Khan, I. H., Tahir, M. A., Iqbal, R., Aslam, M. U., Ejaz, Z., Ditta, A., AlMunqedhi, B. M., & Al Farraj, D. A. (2024). Amelioration of drought stress in wheat by using plant growth-promoting rhizobacteria and biogas slurry. *Global NEST Journal, 26*(8), 05890.

Samanta, I., Ghosh, K., Saikia, R., Savita, Maity, P. , J., & Chowdhary, G. (2025). Arbuscular mycorrhizal fungi—A natural tool to impart abiotic stress tolerance in plants. *Plant Signaling & Behavior, 20*(1), 2525843. https://doi.org/10.1080/15592324.2025.2525843.

Sarfraz, W., Farid, M., Khalid, N., Sheeren, H., Zarnab, R., Nazir, A., Naeem, N., Jabeen, F., Goraya, A., Naseer, A., & Ditta, A. (2024a). Bioremediation of metal(loid) contaminated soils using organic amendments. In A. Ditta, S. Mehmood, M. Imtiaz, & M. S. Tu (Eds.), *Bio-organic amendments for heavy metal remediation: Water, soil and plant approaches and technologies (plant biology, sustainability and climate change)* (pp. 641–658). Elsevier. https://doi.org/10.1016/B978-0-443-21610-7.00023-9.

Sarfraz, W., Khalid, N., Naeem, N., Latif, A., Rizvi, Z. F., Jahangir, A., Shehbaz, A., Amanat, M., Shabbir, F., Ejaz, U., & Ditta, A. (2024b). Toxicity and bioremediation of heavy metals contaminated tannery wastewater. In A. Ditta, S. Mehmood, M. Imtiaz, & M. S. Tu (Eds.), *Bio-organic amendments for heavy metal remediation: Water, soil and plant approaches and technologies (plant biology, sustainability and climate change)* (pp. 257–269). Elsevier. https://doi.org/10.1016/B978-0-443-21610-7.00013-6.

Sarfraz, W., Naeem, N., Farid, M., Khalid, N., Gillani, S. A., Zainab, S. Z., Basharat, M., Safa, S., Ejaz, U., & Ditta, A. (2024c). Factors affecting the remediation of soil through ecotechnologies. In A. Ditta, S. Mehmood, M. Imtiaz, & M. S. Tu (Eds.), *Bio-organic amendments for heavy metal remediation: Water, soil and plant approaches and technologies (plant biology, sustainability and climate change)* (pp. 629–639). Elsevier. https://doi.org/10.1016/B978-0-443-21610-7.00013-6.

Schell, T., Hurley, R., Buenaventura, N. T., Mauri, P. V., Nizzetto, L., Rico, A., & Vighi, M. (2022). Fate of microplastics in agricultural soils amended with sewage sludge: Is surface water runoff a relevant environmental pathway? *Environmental Pollution, 293*, 118520.

Schutten, K., Chandrashekar, A., Dougherty, L., Stevens, B., Parmley, E. J., Pearl, D., Provencher, J. F., & Jardine, C. M. (2024). How do life history and behaviour influence plastic ingestion risk in Canadian freshwater and terrestrial birds? *Environmental pollution, 347*, 123777.

Schüβler, A., Schwarzott, D., & Walker, C. (2001). A new fungal phylum, the Glomeromycota: Phylogeny and evolution. *Mycological Research, 105*, 1413–1421.

Shaji, S., Chellam, P. V., & Sundaram, B. (2025). Interactions of microplastics with co-occurring pollutants in soil environment. *Water, Air, & Soil Pollution, 236*, 212. https://doi.org/10.1007/s11270-025-07855-w.

Shahzad, H., Ullah, S., Iqbal, M., Bilal, H. M., Shah, G. M., Ahmad, S., Zakir, A., Ditta, A., Farooqi, M. A., & Ahmad, I. (2019). Salinity types and level-based effects on the growth, physiology, and nutrient contents of maize (*Zea mays*). *Italian Journal of Agronomy, 14*, 199–207.

Shamskhany, A., Li, Z., Patel, P., & Karimpour, S. (2021). Evidence of microplastic size impact on mobility and transport in the marine environment: A review and synthesis of recent research. *Frontiers in Marine Science, 8*, 760649.

Sharma, K., Gupta, S., Thokchom, S. D., Jangir, P., & Kapoor, R. (2021). Arbuscular mycorrhiza-mediated regulation of polyamines and aquaporins during abiotic stress: Deep insights into the recondite players. *Frontiers in Plant Science, 12*, 642101. https://doi.org/10.3389/fpls.2021.642101.

Sheng, M., Tang, M., Chan, H., Yang, B., Zhang, F., & Huang, Y. (2008). Influence of arbuscular mycorrhizae on photosynthesis and water status of maize plants under salt stress. *Mycorrhiza, 18*, 287–298.

Shi, X., Shi, R., Fu, X., Zhao, Y., Ge, Y., Liu, J., Chen, C., & Liu, W. (2024). Impact of microplastics on plant physiology: A meta-analysis of dose, particle size, and crop type interactions in agricultural ecosystems. *Science of The Total Environment, 955*, 177245.

Smith, S. E., & Smith, F. A. (2012). Fresh perspectives on the roles of arbuscular mycorrhizal fungi in plant nutrition and growth. *Mycologia, 104*, 1–13.

Sommer, U. (2024). Life forms of aquatic organisms. In *Freshwater and marine ecology* (pp. 53–113). Springer.

Song, Y., Wang, J., & Zhang, X. (2022). Adsorption behaviors and mechanisms of positively charged toxic metal ions on microplastics with large specific surface areas and negative inherent charges. *Journal of Hazardous Materials, 424*, 127392.

Song, J., Wang, C., & Li, G. (2024). Defining primary and secondary microplastics: A connotation analysis. *ACS ES&T Water, 4*(6), 2330–2332.

Syamsiyah, J., Herawati, A., & Mujiyo, M. (2018). The potential of arbuscular mycorrhizal fungi application on aggregate stability in Alfisol soil. *IOP Conference Series: Earth and Environmental Science, 142*, 012045. https://doi.org/10.1088/1755-1315/142/1/012045.

Tan, Z., Xuan, Z., Wu, C., Cheng, Y., Xu, C., Ma, X., & Wang, D. (2022). Effects of Selenium on the AsA-GSH system and photosynthesis of Pakchoi (*Brassica chinensis* L.) under lead stress. *Journal of Soil Science and Plant Nutrition, 22*, 5111–5122.

Teskey, R., Wertin, T., Bauweraerts, I., Ameye, M., McGuire, M. A., & Steppe, K. (2015). Responses of tree species to heat waves and extreme heat events. *Plant Cell Environ, 38*, 1699–1712.

Thind, S., Chaudhary, M. S., Ditta, A., Hussain, I., Parveen, A., Ullah, N., Mahmood, Q., Al-ashkar, I., & Sabagh, A. E. L. (2022). Impact of mycorrhizal fungi from different rhizospheric soils on fungal colonization, growth, and chlorophyll contents of *Cenchrus ciliaris*. *Agronomy, 12*(11), 2644.

Turner, A., Fenn, O. J., & Walklett, E. J. (2025). Relationships between sediment size distribution and microplastic abundance and characteristics along the strandline of a sandy embayment (Whitsand, Southwest England). *Marine Pollution Bulletin, 213*, 117686.

Ullah, N., Ditta, A., Imtiaz, M., Li, X., Jan, A. U., Mehmood, S., Rizwan, M. S., & Rizwan, M. (2021). Appraisal for organic amendments and plant growth-promoting rhizobacteria to enhance crop productivity under drought stress: A review. *Journal of Agronomy and Crop Science, 207*(5), 783–802.

Ullah, S., Ahmad, S., Guo, X., Ullah, S., Ullah, S., Nabi, G., & Wanghe, K. (2023). A review of the endocrine-disrupting effects of micro- and nano-plastics and their associated chemicals in mammals. *Frontiers in Endocrinology, 13*, 1084236.

Ullah, S., Janušas, G., & Hasan, Z. U. (2025). Optimization of hybrid polymer matrices for enhanced mechanical and impact properties. *AUTEX 2025 World Conference*. Dresden: Technische Universit¨at Dresden 11–13 June 2025, Dresden, Germany: Book of abstracts.

Uthra, K. T., Chitra, V., Damodharan, N., Devadoss, A., Kuehnel, M., Exposito, A. J., Nagarajan, S., Pitchaimuthu, S., & Pazhani, G. P. (2023). Microplastic emerging pollutants–impact on microbiological diversity, diarrhea, antibiotic resistance, and bioremediation. *Environmental Science: Advances, 2*(11), 1469–1487.

Vlasov, D., Ramírez, O., & Luhar, A. (2022). In *Road dust in urban and industrial environments: Sources, pollutants, impacts, and management: 13* (p. 607). MDPI.

Wang, Q., Liu, M., Wang, Z., Li, J., Liu, K., & Huang, D. (2024). The role of arbuscular mycorrhizal symbiosis in plant abiotic stress. *Frontiers in Microbiology, 14*, 1323881. https://doi.org/10.3389/fmicb.2023.1323881.

Wang, W., Yuan, W., Xu, E. G., Li, L., Zhang, H., & Yang, Y. (2022a). Uptake, translocation, and biological impacts of micro (nano) plastics in terrestrial plants: Progress and prospects. *Environmental Research, 203*, 111867.

Wang, F., Adams, C. A., Zhou, M., Sun, Y., Zhang, S., Wang, P., … Shcherbakov, A. (2021a). Microplastics in agricultural soils: An emerging threat to soil health, microbial ecology, crop productivity, and food safety Microplastics in Agricultural Soils. *An Emerging Threat .... Soil Biology and Biochemistry,* vol. 162 p. 108422.

Wang, Y., Zhou, B., Chen, H., Yuan, R., & Wang, F. (2022). Distribution, biological effects, and biofilms of microplastics in freshwater systems: A review. *Chemosphere, 299*, 134370.

Wasaya, A., Yaqoob, S., Ditta, A., Yasir, T. A., Sarwar, N., Javaid, M. M., Al-Ashkar, I., & Sabagh, A. E. (2024). Exogenous application of β-aminobutyric acid improved water relations, membrane stability index, and achene yield in sunflower hybrids under terminal drought stress. *Polish Journal of Environmental Studies, 33*(5), 5367–5379.

Wright, S. F., & Upadhyaya, A. (1998). A survey of soils for aggregate stability and glomalin, a glycoprotein produced by hyphae of arbuscular mycorrhizal fungi. *Plant and Soil, 198*, 97–107.

Wright, S. F., & Upadhyaya, A. (1999). Quantification of arbuscular mycorrhizal fungi activity by the glomalin concentration on hyphal traps. *Mycorrhiza, 8*, 283–285.

Wu, B., Qi, F., & Liang, Y. (2023). Fuels for ROS signaling in plant immunity. *Trends in Plant Science, 28*, 1124–1131.

Wu, Q.-S., Zou, Y.-N., & He, X.-H. (2010). Contributions of arbuscular mycorrhizal fungi to growth, photosynthesis, root morphology and ionic balance of citrus seedlings under salt stress. *Acta Physiologiae Plantarum, 32*, 297–304. https://doi.org/10.1007/s11738-009-0407-z.

Wu, H., Cui, H., Fu, C., Li, R., Qi, F., Liu, Z., Yang, G., Xiao, K., & Qiao, M. (2024). Unveiling the crucial role of soil microorganisms in carbon cycling: A review. *Science of The Total Environment, 909*, 168627.

Xu, G., Liu, Y., & Wang, Q. (2022). Microplastics in soil-plant systems: A review of source, phenotypic toxicity, and uptake mechanisms. *Science of The Total Environment, 843*, 157024.

Xu, G., Liu, Y., Song, X., & Wang, J. (2022). Effects of microplastics on root exudates and heavy metal accumulation by plants in co-contaminated soils. *Environmental Pollution, 306*, 119422.

Xu, H., Chen, C., Pang, Z., Zhang, G., Zhang, W., & Kan, H. (2024). Effects of microplastics concentration on plant root traits and biomass: Experiment and meta-analysis. *Ecotoxicology and Environmental Safety, 285*, 117038.

Xu, Y., Ou, Q., van der Hoek, J. P., Liu, G., & Lompe, K. M. (2024). Photo-oxidation of micro-and nanoplastics: Physical, chemical, and biological effects in environments. *Environmental Science & Technology, 58*(2), 991–1009.

Yan, M., Wang, L., Dai, Y., Sun, H., & Liu, C. (2021). Behavior of microplastics in inland waters: Aggregation, settlement, and transport. *Bulletin of Environmental Contamination and Toxicology, 107*, 700–709.

Yang, L., Zhang, Y., Kang, S., Wang, Z., & Wu, C. (2021). Microplastics in soil: A review on methods, occurrence, sources, and potential risk. *Science of the Total Environment, 780*, 146546.

Yarahmadi, A., Heidari, S., Sepahvand, P., Afkhami, H., & Kheradjoo, H. (2024). Microplastics and environmental effects: Investigating the effects of microplastics on aquatic habitats and their impact on human health. *Frontiers in Public Health, 12*, 1411389.

Yousafzai, S., Farid, M., Zubair, M., Naeem, N., Zafar, W., Asam, Z. U. Z., Farid, S., & Ali, S. (2025). Detection and degradation of microplastics in the environment: A review. *Environmental Science Advances, 4*, 1142–1165.

Yin, L., Wang, M., & Zhou, Q. (2021). Effects of microplastics on plant growth and arbuscular mycorrhizal fungal communities in a soil spiked with ZnO nanoparticles. *Soil Biology and Biochemistry, 155*, 108179.

Zhang, X., Li, Y., Ouyang, D., Lei, J., Tan, Q., Xie, L., … Farooq, T. H. (2021). Systematical review of interactions between microplastics and microorganisms in the soil environment. *Journal of Hazardous Materials, 418*, 126288.

Zhang, X., Wang, Z., Lu, Y., Wei, J., Qi, S., Wu, B., & Cheng, S. (2024). Sustainable remediation of soil and water utilizing arbuscular mycorrhizal fungi: A review. *Microorganisms, 12*(7), 1255.

Zhao, H., Hong, X., Chai, J., Wan, B., Zhao, K., Han, C., Zhang, W., & Huan, H. (2024). Interaction between microplastics and pathogens in subsurface systems: What we know so far. *Water, 16*(3), 499.

Zhou, J., Wen, Y., Marshall, M. R., Zhao, J., Gui, H., Yang, Y., Zeng, Z., Jones, D. L., & Zang, H. (2021). Microplastics as an emerging threat to plant and soil health in agroecosystems. *Science of the Total Environment, 787*, 147444. https://doi.org/10.1016/j.scitotenv.2021.147444.

Zhou, J., Jia, R., Brown, R. W., Yang, Y., Zeng, Z., Jones, D. L., & Zang, H. (2023). The long-term uncertainty of biodegradable mulch film residues and associated microplastics pollution on plant-soil health. *Journal of Hazardous Materials, 442*, 130055.

Chapter 34

Ectomycorrhizal fungi and sustainable agriculture

Munaza Batool[a], **Saqib Bashir**[a], **Sanober Gul**[b] **and Muhammad Amjad Bashir**[c]
[a] *Department of Soil and Environmental Sciences, Ghazi University, Dera Ghazi Khan, Pakistan,* [b] *Department of Plant Breeding and Genetics, Ghazi University, Dera Ghazi Khan, Pakistan,* [c] *Department of Plant Protection, Ghazi University, Dera Ghazi Khan, Pakistan*

34.1 Introduction

34.1.1 Importance of sustainable agriculture

Agriculture remains the backbone of human civilization, underpinning food security, rural livelihoods, and economic development worldwide (Aslam et al., 2025; Khan et al., 2024). However, the contemporary agricultural system is under intense pressure because of a convergence of global challenges, including rapid population growth, climate change, land degradation, water scarcity, and declining soil fertility (Table 34.1). The global population is projected to reach approximately 9.7 billion by 2050, necessitating a 60%–70% increase in food production compared to current levels (FAO, 2021a). Achieving this increase using conventional agricultural practices is increasingly untenable, as these methods have already pushed natural resources to critical limits.

Over the past century, agricultural intensification has relied heavily on synthetic fertilizers, chemical pesticides, monocropping, and intensive tillage to boost yields (Ali et al., 2023; Ditta & Ullah, 2023; Sarfraz et al., 2023). Although these practices initially resulted in remarkable productivity gains, their long-term environmental consequences have been severe. Excessive use of nitrogen and phosphorus fertilizers has disrupted global nutrient cycles, leading to soil acidification, nutrient imbalances, eutrophication of water bodies, and contamination of groundwater with nitrates (Ahmed et al., 2021). According to the Food and Agriculture Organization, approximately 33% of the world's soils are moderately to highly degraded, and if current trends continue, up to 90% of soils could be degraded by 2050 (FAO, 2015).

Soil degradation manifests through multiple processes, including erosion, salinization, compaction, loss of organic matter, and decline in biological activity (Ditta et al., 2022a, 2022b). These processes not only reduce crop productivity but also impair the soil's capacity to provide essential ecosystem services such as carbon sequestration, water filtration, and nutrient cycling. In South Asia, including Pakistan, soil salinity and sodicity have emerged as major constraints to agricultural sustainability, particularly in irrigated agroecosystems. Climate change further exacerbates these issues by increasing the frequency of droughts, floods, and temperature extremes, thereby intensifying stress on crops and soils alike (Ahmad et al., 2020). In response to these challenges, sustainable agriculture has emerged as a holistic paradigm that balances productivity with environmental stewardship and socioeconomic viability. Sustainable agriculture emphasizes resource-use efficiency, ecological integrity, resilience to climate variability, and intergenerational equity. Rather than relying solely on external inputs, it seeks to harness natural processes such as nutrient cycling, biological pest control, and symbiotic plant–microbe interactions to maintain soil fertility and crop productivity over the long term (Saeed et al., 2025).

A key pillar of sustainable agriculture is soil health, which encompasses the physical, chemical, and biological properties of soil. Among these, the biological component of soil health, particularly soil microbial diversity and activity, has gained increasing attention. Soil microorganisms play critical roles in organic matter decomposition, nutrient mobilization, aggregate formation, and suppression of soil-borne pathogens (Ditta & Khalid, 2016). Within this microbial community, mycorrhizal fungi have emerged as one of the most influential and ecologically significant groups.

Advancing Sustainable Agriculture Through Arbuscular Mycorrhizal Fungi. DOI: https://doi.org/10.1016/B978-0-443-45130-0.00017-5

TABLE 34.1 Major challenges in conventional agriculture and sustainable alternatives.

Conventional agriculture challenges	Environmental impacts	Sustainable agriculture solutions
Excessive chemical fertilizer use	Soil acidification, water pollution	Biological nutrient cycling, mycorrhizae
Intensive tillage	Soil erosion, carbon loss	Conservation tillage
Monocropping	Biodiversity loss, pest outbreaks	Crop diversification, agroforestry
Overirrigation	Salinity, waterlogging	Efficient irrigation, soil biology
Pesticide dependence	Nontarget toxicity	Biological control, microbial inoculants

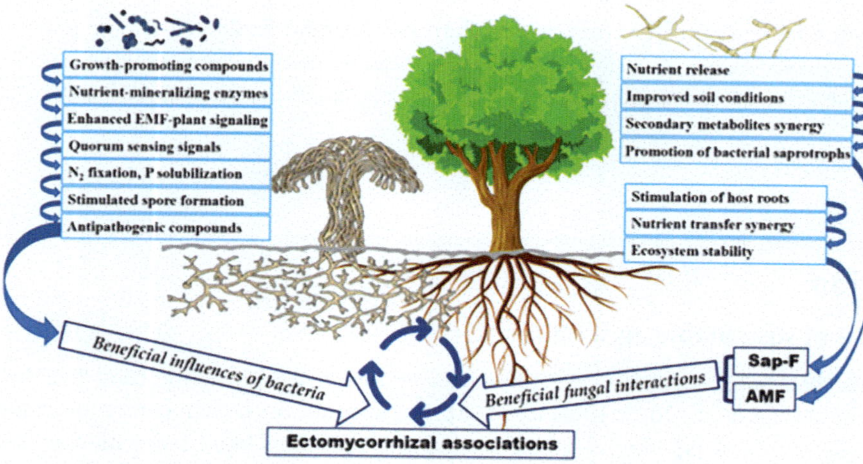

FIGURE 34.1 Illustration of the beneficial effects of bacteria and fungi on EM formation. *AMF*, Arbuscular mycorrhizal fungi; *EMF*, ectomycorrhizal fungi; *Sap-F*, saprotrophic fungi. (*From El Amrani (2024).*)

Recent advances in soil ecology and molecular biology have highlighted the central role of mycorrhizal symbioses in regulating nutrient dynamics, plant stress tolerance, and ecosystem stability. Mycorrhizal associations occur in approximately 90% of terrestrial plant species, underscoring their evolutionary and ecological importance. Among the different types of mycorrhizal fungi, ectomycorrhizal (ECM) fungi (EMF) are particularly notable for their functional versatility, enzymatic capacity, and strong influence on carbon and nutrient cycling.

Within the broader framework of climate-smart and regenerative agriculture, biological solutions based on beneficial soil microbes are increasingly viewed as viable alternatives to input-intensive farming systems. EMF, with their ability to enhance nutrient acquisition, improve drought tolerance, and support long-term carbon sequestration, represent a promising yet underutilized resource for sustainable land management. Their integration into agroforestry systems, perennial cropping systems, and rehabilitated degraded lands offers significant potential for restoring soil health while maintaining productivity.

34.1.2 Role of ectomycorrhizal fungi

EMF represent a specialized and evolutionarily advanced group of symbiotic fungi belonging primarily to the *Basidiomycota* and *Ascomycota*. These fungi form mutualistic associations with approximately 2%–3% of vascular plant species, predominantly woody perennials such as forest trees and shrubs (Fig. 34.1). Despite their limited host range relative to arbuscular mycorrhizal fungi, EMF dominate temperate and boreal forest ecosystems, where they play a pivotal role in ecosystem functioning and biogeochemical cycling. Economically and ecologically important plant families that commonly form ECM associations include Pinaceae (pines, spruces), Fagaceae (oaks, beeches), Betulaceae (birches), Dipterocarpaceae, and Myrtaceae. Many of these species are foundational components of forest ecosystems, timber industries, and agroforestry systems. The success and resilience of these plant species are closely linked to their association with EMF.

The defining anatomical features of ECM symbiosis include the formation of a hyphal mantle around fine root tips and the development of the Hartig net, a complex network of fungal hyphae that penetrates between epidermal and cortical root cells. Unlike arbuscular mycorrhizal fungi, EMF do not penetrate plant cell walls, thereby maintaining cellular integrity while establishing an extensive interface for nutrient exchange. One of the most remarkable attributes of EMF is their

TABLE 34.2 Comparison of ectomycorrhizal and arbuscular mycorrhizal fungi.

Characteristic	Ectomycorrhizal	Arbuscular mycorrhizal
Fungal phyla	*Basidiomycota, Ascomycota*	*Glomeromycota*
Host range	~2%–3% (mostly woody plants)	~80%–90% of plants
Root penetration	Intercellular (Hartig net)	Intracellular (arbuscules)
Enzymatic capacity	High (organic N & P)	Limited
Ecosystem dominance	Forests, agroforestry	Croplands

enzymatic versatility. These fungi produce a wide array of extracellular enzymes, including laccases, peroxidases, proteases, chitinases, and phosphatases, enabling them to access nutrients locked within complex organic matter. This ability allows EMF to mobilize organic nitrogen and phosphorus pools, which are often inaccessible to plant roots alone. Consequently, EMF are particularly important in nutrient-poor, acidic, or disturbed soils, where inorganic nutrient availability is limited.

In contrast to arbuscular mycorrhizal fungi, which primarily rely on inorganic nutrient uptake, EMF exhibit a saprotrophic-like capacity, blurring the boundary between symbiosis and decomposition (Table 34.2). This functional flexibility confers a competitive advantage in forest soils rich in organic matter and contributes to the accumulation and stabilization of soil carbon. Beyond nutrient acquisition, EMF enhance plant tolerance to abiotic stresses, including drought, salinity, heavy metal toxicity, and temperature extremes. By improving water uptake through extensive hyphal networks and regulating plant stress-response genes, EMF contribute significantly to plant survival under adverse environmental conditions. In addition, these play a role in pathogen suppression, either through competitive exclusion, production of antimicrobial compounds, or induction of systemic resistance in host plants. Given their multifunctional roles, EMF are increasingly recognized as key biological agents in ecosystem restoration, afforestation, reforestation, and sustainable land management. Their potential application extends beyond natural forests to include agroforestry systems, tree-based farming systems, and reclamation of degraded and saline soils.

34.2 Ectomycorrhizal fungi and plant symbiosis

34.2.1 Symbiotic relationship

The ECM symbiosis represents one of the most complex and evolutionarily sophisticated mutualistic relationships in terrestrial ecosystems. Fossil evidence and molecular clock analyses suggest that ECM symbioses evolved independently multiple times over the past 180–200 million years, coinciding with the diversification of woody plant lineages and the expansion of forest ecosystems (Brundrett, 2009). The establishment of ECM symbiosis begins with a finely regulated molecular dialogue between plant roots and fungal propagules. Plant roots release a diverse array of signaling molecules, including strigolactones, flavonoids, and phenolic compounds, which stimulate fungal spore germination and directional hyphal growth. In response, EMF secrete mycorrhizal (Myc) factors, such as lipochitooligosaccharides, which activate conserved symbiotic signaling pathways in plant roots.

Following recognition the fungus colonizes the root surface and initiates the formation of the mantle, a multilayered sheath of tightly packed hyphae. This mantle serves multiple functions, including protection of root tips from desiccation and pathogens, regulation of gas exchange, and mediation of nutrient transfer. In some associations the mantle can account for up to 40% of the total root biomass, highlighting its structural and metabolic significance.

Subsequently the fungus penetrates intercellular spaces of the root cortex to form the Hartig net, the primary site of nutrient exchange. Advanced imaging techniques, such as confocal microscopy and transmission electron microscopy, have revealed that the Hartig net dramatically increases the surface area of contact between fungal and plant tissues. This interface is enriched with specialized transport proteins, membrane-bound enzymes, and proton pumps that facilitate bidirectional nutrient flow. ECM associations exhibit varying degrees of host specificity. Generalist fungi, such as *Pisolithus tinctorius* and *Laccaria bicolor*, can associate with a wide range of host species, making them particularly valuable for inoculation in nurseries and restoration projects. In contrast, specialist fungi, such as *Tuber melanosporum* (black truffle), exhibit narrow host ranges and strict ecological requirements.

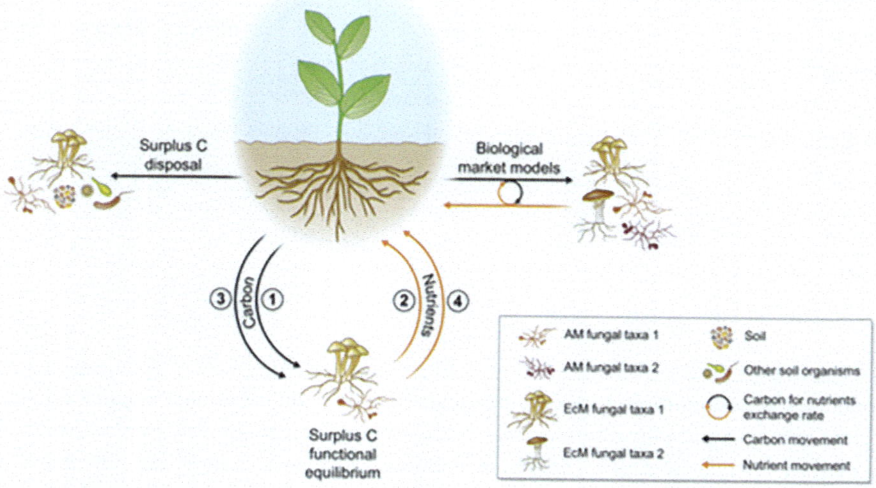

FIGURE 34.2 Nutrient and carbon exchange in ectomycorrhizal symbiosis. (*From Bunn et al. (2024).*)

34.2.2 Nutrient exchange

Nutrient exchange in ECM symbioses functions as a highly regulated biological marketplace, in which carbon, nitrogen, phosphorus, and micronutrients are traded based on supply and demand (Fig. 34.2). The extensive extraradical mycelium of EMF can explore soil volumes several orders of magnitude greater than plant roots alone, effectively extending the root system into nutrient-rich microsites. EMF deploy an impressive suite of extracellular enzymes to mobilize nutrients from complex organic substrates. Acid phosphatases liberate phosphate from organic compounds, whereas proteases and peptidases release amino acids from soil organic matter. In addition, the secretion of organic acids such as oxalate facilitates mineral weathering, releasing calcium, magnesium, and micronutrients.

Isotopic tracer studies using ^{15}N and ^{32}P have demonstrated that EMF can supply up to 80%–90% of host plant nitrogen and phosphorus requirements in nutrient-poor ecosystems. Nutrients absorbed by fungal hyphae are transported internally through cytoplasmic streaming and stored in vacuoles as polyphosphate granules, which act as dynamic nutrient reserves. In exchange the host plant allocates a substantial portion of its photosynthetically fixed carbon—typically 10%–20% of total photosynthate—to the fungal partner. Carbon is primarily transferred as sucrose, which is hydrolyzed by fungal invertases into glucose and fructose for metabolism. This carbon investment supports fungal growth, enzyme production, and soil exploration (Liu et al., 2020; Thind et al., 2022).

Recent advances in genomics and transcriptomics have revealed that nutrient exchange is tightly regulated at the molecular level. Genes encoding ammonium (AMT) and phosphate (PHT) transporters in EMF exhibit synchronized expression patterns with plant photosynthesis, indicating real-time coordination between partners. This dynamic regulation ensures mutual benefit and prevents exploitation by either symbiont. Importantly, ECM networks also function as major carbon sinks, contributing significantly to soil organic carbon pools. It is estimated that 5%–20% of global terrestrial carbon flux passes through mycorrhizal networks annually, underscoring their role in climate change mitigation.

34.3 Benefits of ectomycorrhizal fungi in sustainable agriculture

Sustainable agriculture seeks to balance high productivity with environmental stewardship, long-term soil health, and resilience to climatic variability. One of the most promising biological tools contributing to this goal is the symbiotic association between plants and EMF. EMF form mutualistic relationships with the roots of numerous woody and perennial plant species, particularly forest trees and agroforestry crops, where fungal hyphae envelop the root tips, forming a mantle, and penetrate the root cortex to create a Hartig net. Through this association, EMF facilitate nutrient acquisition, enhance soil structure, improve water relations, and increase plant tolerance to biotic and abiotic stresses. Unlike conventional agricultural practices that rely heavily on synthetic fertilizers and irrigation, ECM-based systems operate through biological efficiency. The fungal partner acquires nutrients and water from soil microsites inaccessible to roots, whereas the plant supplies carbohydrates derived from photosynthesis. This reciprocal exchange not only enhances plant growth but also improves ecosystem functioning at the soil–plant–microbe interface. As global agriculture faces challenges such as soil degradation,

TABLE 34.3 Effects of ectomycorrhizal fungi on soil fertility parameters.

Parameter	Nonmycorrhizal Soil	ECM-colonized soil	Improvement (%)
Aggregate stability	Low	High	40–60
Phosphatase activity	Baseline	2–3× higher	100–200
Nitrogen leaching	High	Reduced	40–70
Organic matter retention	Moderate	High	30–50

ECM, ectomycorrhizal.

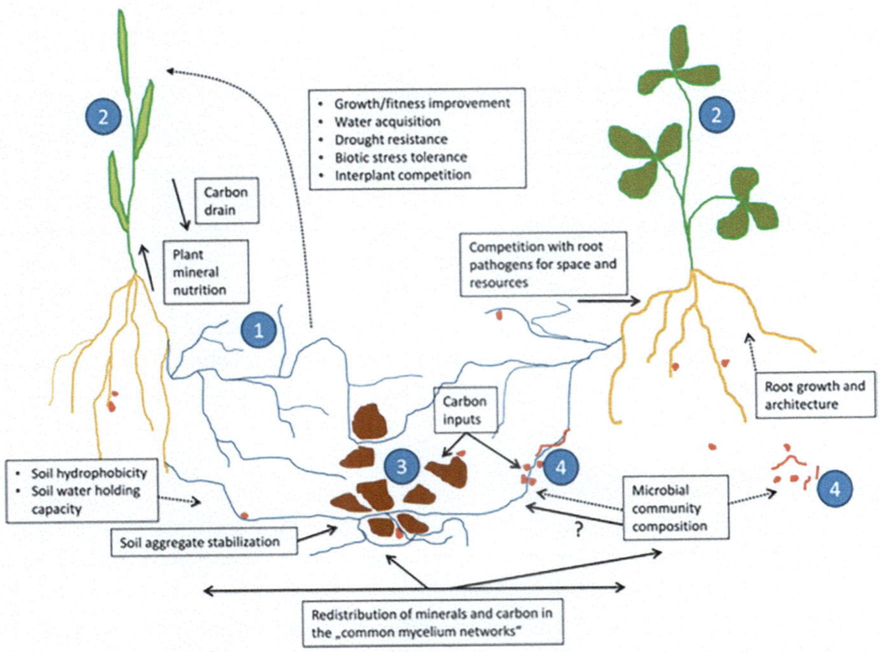

FIGURE 34.3 ECM hyphal networks binding soil particles into stable aggregates. *ECM*, Ectomycorrhizal. (*From Jansa et al. (2013).*)

nutrient depletion, climate-induced droughts, and increasing production demands, EMF are increasingly recognized as a cornerstone of sustainable and regenerative farming systems. The following sections examine in detail the major benefits of EMF in sustainable agriculture, with a particular focus on improved soil fertility, increased plant productivity, and enhanced drought tolerance.

34.3.1 Improved soil fertility

Soil fertility is a multidimensional concept encompassing physical structure, chemical nutrient availability, and biological activity. EMF significantly influence all three dimensions, making them powerful agents of soil restoration and long-term fertility enhancement. Their contribution extends beyond simple nutrient uptake to complex modifications of the rhizosphere that improve soil function and resilience (Table 34.3).

34.3.1.1 Soil structure and aggregation

One of the most critical contributions of EMF to soil fertility is their role in improving soil structure through aggregation (Fig. 34.3). ECM hyphae secrete glomalin-related soil proteins, sticky glycoproteins that bind soil particles into stable aggregates. These aggregates protect organic matter from rapid decomposition and enhance soil porosity, aeration, and water-holding capacity. Research indicates that soils colonized by mycorrhizal fungi exhibit up to 60% higher aggregate stability compared to nonmycorrhizal soils (Wilson et al., 2009). Improved aggregation reduces soil erosion by wind and

TABLE 34.4 Impact of ECM fungi on plant productivity.

Parameter	Non-ECM plants	ECM plants
Phosphorus uptake	Low–moderate	High
Root branching	Limited	3–5× increase
Biomass accumulation	Baseline	+30%–40%
Disease incidence	High	Reduced by 70%–90%

ECM, ectomycorrhizal.

water, particularly in degraded or sandy soils common in arid and semiarid regions. Enhanced porosity also facilitates deeper root penetration and greater microbial habitat diversity, reinforcing positive feedback loops within the soil ecosystem.

34.3.1.2 Enhancement of nutrient cycling and availability

EMF play a central role in nutrient cycling by mobilizing nutrients locked in organic and mineral forms. Through the secretion of organic acids such as oxalates, citrates, and malates, EMF solubilize phosphorus, calcium, magnesium, and micronutrients from otherwise unavailable mineral fractions. These acids chelate metal ions and lower pH in microsites, increasing nutrient bioavailability. Moreover, EMF actively stimulate the microbial loop, a process whereby fungal exudates promote bacterial growth, which, in turn, accelerates the mineralization of organic matter. This results in enhanced release of nitrogen, phosphorus, and sulfur in plant-available forms. Kaiser et al. (2015) demonstrated that ECM-colonized soils exhibit significantly higher microbial respiration and enzymatic activity compared to nonmycorrhizal soils. Metagenomic analyses further reveal that ECM-associated soils show two to three times higher activity of enzymes such as phosphatases, chitinases, and proteases, which are essential for the decomposition of organic residues and nutrient release (Courty et al., 2010).

34.3.1.3 Nitrogen acquisition and retention

Nitrogen is often the most limiting nutrient in agricultural systems. EMF exhibit a unique nitrogen acquisition strategy that enhances nitrogen use efficiency while minimizing losses. Unlike arbuscular mycorrhizal fungi, EMF can directly access organic nitrogen sources such as amino acids, peptides, and chitin through enzymatic degradation. EMF preferentially absorb ammonium (NH_4^+) rather than nitrate (NO_3^-), a strategy that significantly reduces nitrogen leaching and groundwater contamination. Studies show that ECM-dominated systems can reduce nitrogen losses by **40%–70%** compared to conventional fertilized systems (Talbot et al., 2008). In addition, fungal hyphae act as temporary nitrogen sinks, storing nitrogen within their biomass and releasing it gradually to the host plant. This buffering capacity ensures a steady nutrient supply even under fluctuating soil conditions.

34.3.1.4 Restoration of degraded soils

The cumulative effects of aggregation, nutrient mobilization, and microbial stimulation enable EMF to rehabilitate degraded soils. In tropical and subtropical agroforestry systems, ECM inoculation has been shown to restore fertility in highly weathered soils, leading to crop yield increases ranging from 25% to 300% (Ahmed et al., 2021). These benefits are particularly relevant in sustainable land management, reforestation programs, and low-input farming systems where chemical fertilizers are either unavailable or environmentally undesirable.

34.3.2 Increased plant productivity

Enhanced plant productivity is one of the most visible and economically significant benefits of ECM symbiosis. EMF improve plant growth through improved nutrition, altered root morphology, enhanced photosynthetic efficiency, and increased resistance to pests and diseases (Table 34.4).

34.3.2.1 Improved nutrient uptake efficiency

Stable isotope studies reveal that EMF supply up to 80% of host phosphorus and 60% of nitrogen requirements in mature plant systems (Hobbie & Högberg, 2012). The extensive hyphal network functions as an extension of the root system, exploring soil volumes far beyond the rhizosphere depletion zone. Phosphorus uptake is particularly enhanced, as ECM

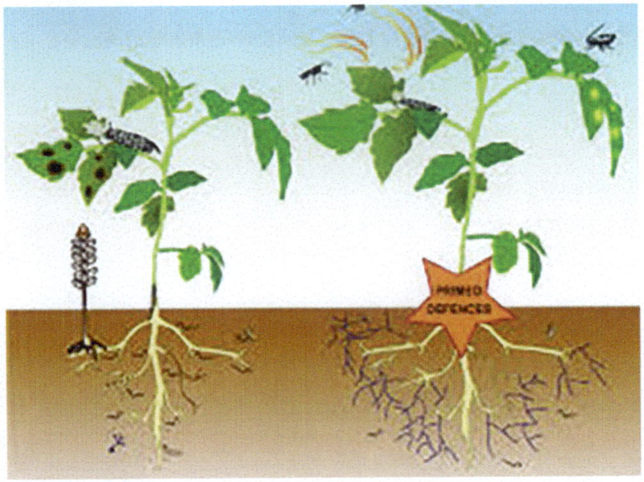

FIGURE 34.4 Comparison of root architecture in mycorrhizal versus nonmycorrhizal plants. *(From Jung & McCouch (2013).)*

hyphae access phosphorus bound to iron and aluminum oxides or present in organic forms inaccessible to plant roots. This significantly reduces dependence on phosphate fertilizers, a nonrenewable resource.

34.3.2.2 Root system architecture and hormonal regulation

EMF influence plant root development by modulating phytohormone signaling pathways (Fig. 34.4). EMF produce or stimulate the production of auxins e.g., Indole-3-acetic acid (IAA), cytokinins, and gibberellins, resulting in increased lateral root formation and fine root proliferation. Studies report a three- to fivefold increase in root branching density in ECM-colonized plants compared to controls (Felten et al., 2009). A more extensive root system enhances nutrient and water uptake while increasing root–soil contact.

34.3.2.3 Photosynthesis and carbon allocation

Transcriptomic and proteomic analyses demonstrate that ECM colonization induces the upregulation of genes associated with photosynthesis, carbon metabolism, and stress tolerance. Enzymes such as Rubisco activase and sucrose synthases show increased expression, leading to higher photosynthetic rates and improved carbohydrate transport (Plett & Martin, 2018). The increased photosynthate production compensates for carbon allocated to the fungal partner, resulting in a net gain in biomass. Field studies in ECM-inoculated *Eucalyptus* plantations report 35% higher biomass accumulation over 24 months (Bois et al., 2005).

34.3.2.4 Bioprotection and disease suppression

EMF provide a physical and biochemical barrier against root pathogens. The fungal mantle occupies up to 90% of root cortical space, preventing pathogen entry through competitive exclusion. In addition, EMF induce systemic resistance in host plants by activating jasmonic acid and salicylic acid signaling pathways. In vineyard systems, ECM inoculation reduced *Phytophthora* root infections by 70%–90% while simultaneously increasing grape yields by 22% (Schreiner, 2007). These protective effects reduce reliance on chemical pesticides, aligning with sustainable agriculture goals.

34.3.3 Drought tolerance

Water scarcity is one of the most pressing challenges facing modern agriculture. EMF significantly enhance plant drought tolerance through morphological, physiological, and molecular mechanisms, as clear from Table 34.5 (Shah et al., 2022).

34.3.3.1 Enhanced water uptake and hydraulic redistribution

ECM hyphae increase the effective absorptive surface area of roots by 10–100 times, enabling access to water held in soil micropores beyond the reach of plant roots (Auge et al., 2015). These micropores retain water at high tension levels, allowing plants to survive extended dry periods. ECM networks also facilitate hydraulic redistribution, moving water from wetter soil

TABLE 34.5 Drought tolerance traits in ECM-colonized plants.

Trait	Nonmycorrhizal	ECM-colonized
Root water uptake	Limited	High
Osmolyte accumulation	Low	2–4× higher
Survival under drought	Moderate	+40%
Stomatal regulation	Poor	Efficient

ECM, ectomycorrhizal.

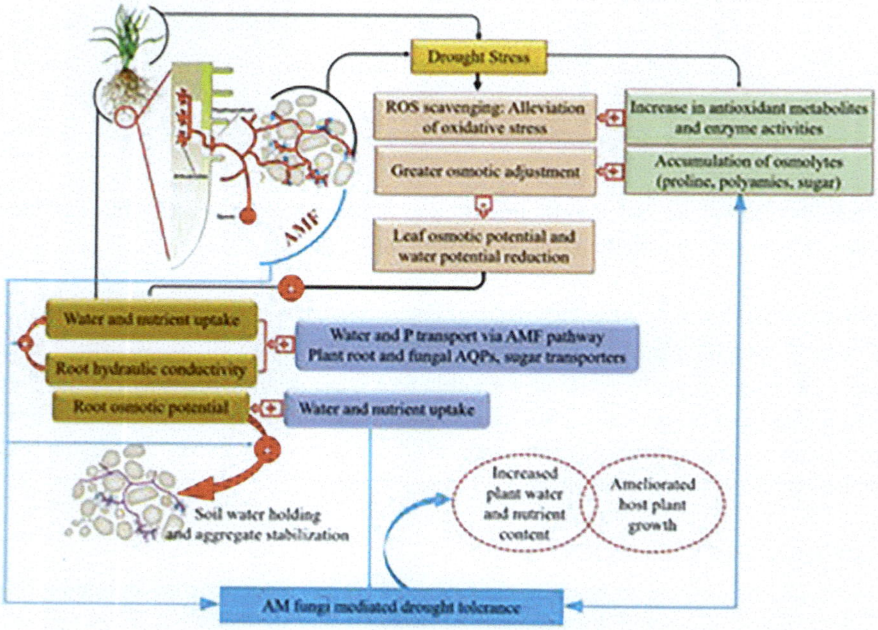

FIGURE 34.5 Mechanisms of drought tolerance mediated by ectomycorrhizal fungi. (*From Bahadur et al. (2019).*)

layers to drier zones at rates of 0.1–0.3 mm/day (Querejeta et al., 2007). This process not only benefits the host plant but also improves soil moisture availability for neighboring plants.

34.3.3.2 Cellular and molecular adaptations

At the cellular level, EMF stimulate the expression of aquaporins (PIP2 proteins), increasing membrane water permeability by three to five times (Barzana et al., 2012). This enhances root hydraulic conductivity under drought conditions. Metabolomic studies show that ECM-colonized plants accumulate significantly higher concentrations of osmoprotectants such as proline and glycine betaine. These compounds stabilize cellular structures and maintain turgor during water stress (Alvarez et al., 2009).

34.3.3.3 Field evidence and climate change relevance

Field trials with ECM-inoculated *Pinus taeda* demonstrate 40% higher survival rates under drought conditions compared to nonmycorrhizal plants (Warren et al., 2011). ECM plants maintained stomatal conductance at lower soil water potentials, indicating superior drought resilience. Given projections that drought-affected croplands may expand by 30% by 2100 (Dai, 2013), EMF represent a critical biological adaptation strategy for climate-resilient agriculture. Different mechanisms of drought tolerance mediated by EMF are illustrated in Fig. 34.5.

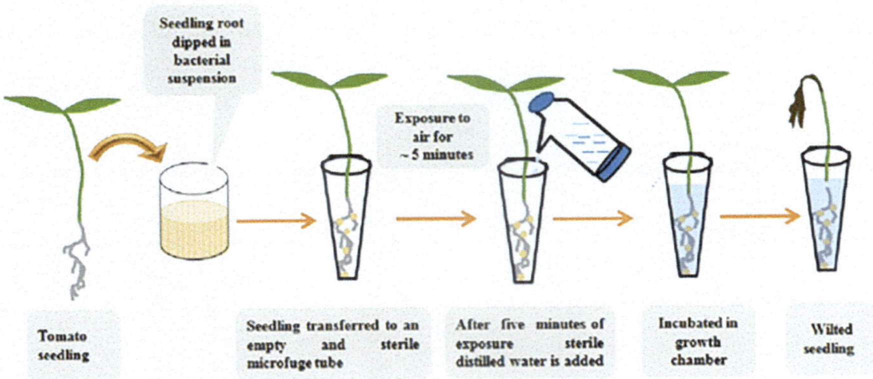

FIGURE 34.6 Schematic representation of ECM inoculation techniques: (A) seed coating, (B) root dipping, and (C) soil injection. *ECM*, Ectomycorrhizal. (*From Kumar et al. (2018).*)

34.4 Applications of ectomycorrhizal fungi in sustainable agriculture

34.4.1 Inoculation of crops

The application of EMF as bio-inoculants has emerged as a scientifically robust and economically viable strategy for advancing sustainable agriculture and forestry systems. EMF form mutualistic associations primarily with woody and perennial plant species, enhancing nutrient uptake, water acquisition, and resistance to biotic and abiotic stresses. Modern ECM inoculation protocols represent a significant evolution from early field trials, incorporating advances in fungal culturing, formulation chemistry, delivery technologies, and precision agriculture. These developments aim to maximize colonization efficiency, persistence in the field, and measurable agronomic benefits.

Historically, ECM inoculation was limited by difficulties in culturing obligate symbionts and by inconsistent field performance (Fig. 34.6). However, contemporary approaches utilize selected high-performance strains, optimized growth substrates, and protective carrier materials to ensure spore viability and compatibility with target hosts. Commercially successful ECM inoculants such as *P. tinctorius* (PT-270 strain) are now produced using submerged fermentation techniques, yielding inoculum densities exceeding 10^8 propagules per gram when formulated in lignocellulosic or peat-based carriers (Marx et al., 2002). These formulations provide physical protection, gradual release of propagules, and enhanced shelf life under storage and transport conditions. Precision application methods are central to the success of ECM inoculation programs. Three major delivery pathways—seed coating, root dipping, and soil injection—have been refined to suit different cropping systems and stages of plant development.

Seed coating: Technologies represent a cost-effective and scalable approach, particularly for nursery-raised seedlings and plantation forestry. Chitosan-encapsulated ECM spores, applied at rates of approximately 200–500 spores per seed, have demonstrated significant improvements in early seedling vigor. Chitosan not only protects spores from desiccation and ultraviolet radiation but also exhibits antimicrobial properties that suppress competing pathogens. Studies report germination rate increases of up to 25% and enhanced early root development compared to noninoculated controls (Taylor & Harrier, 2003).

Root dipping is widely employed in nursery and transplant production systems, especially for tree crops and agroforestry species. In this method, seedling roots are immersed in alginate-based hydrogel suspensions containing approximately 10^6 spores per milliliter. Alginate matrices provide a hydrated microenvironment that facilitates rapid fungal attachment and hyphal penetration. Colonization rates of 80%–90% have been consistently achieved under nursery conditions, translating into improved transplant survival and early field establishment (Kluber et al., 2012).

Soil injection techniques are increasingly used in field-scale applications where direct root contact at planting is not feasible. Subsurface injection of spore suspensions at depths of approximately 20 cm places propagules within the active rhizosphere zone while protecting them from surface-level temperature extremes and desiccation. Field trials indicate that application rates of 10^5 spores per milliliter maintain high viability and lead to stable colonization under diverse soil conditions (Teste et al., 2020).

Large-scale implementation of ECM inoculation has demonstrated substantial agronomic and economic benefits. In Brazil, ECM inoculation of *Eucalyptus* plantations across more than 600,000 ha has resulted in an average increase in mean annual increment of 3.7 m^3/ha while simultaneously reducing synthetic fertilizer inputs by approximately 40%

TABLE 34.6 Summary of the common ECM inoculation methods and their reported outcomes.

Application method	Inoculum type	Typical dose	Colonization efficiency (%)	Key benefits
Seed coating	Chitosan-encapsulated spores	200–500 spores/seed	60–75	Improved germination, low cost
Root dipping	Alginate hydrogel suspension	10^6 spores/mL	80–90	High survival, rapid establishment
Soil injection	Liquid spore suspension	10^5 spores/mL	50–70	Field-scale applicability

ECM, ectomycorrhizal.

TABLE 34.7 Key soil amendments and their effects on ECM performance.

Soil management practice	Application rate	Reported ECM benefit	Additional outcomes
Biochar	20 t/ha	+45% colonization	Improved water retention
Vermicompost tea	1:10 dilution	+30% hyphal growth	Enhanced microbial activity
No-till systems	Continuous	85% network preservation	Reduced erosion

ECM, ectomycorrhizal.

(Gomes et al., 2018). Similarly, in Europe, the deliberate inoculation of *Quercus ilex* seedlings with *T. melanosporum* has transformed marginal lands into high-value truffle orchards, generating revenues of up to €15,000/ha/year (Reyna & Garcia-Barreda, 2014). These examples illustrate the dual ecological and economic potential of ECM technologies. A summary of the common ECM inoculation methods and their reported outcomes is presented in Table 34.6.

34.4.2 Soil amendments and integrated soil management

The effectiveness of EMF in agricultural systems is strongly influenced by soil physical, chemical, and biological properties. Consequently, integrated soil management strategies that enhance soil structure, organic matter content, and microbial diversity can significantly amplify the benefits of ECM symbioses. Among these strategies, biochar amendments, compost teas, and conservation tillage practices have received considerable attention (Table 34.7).

Biochar amendments play a multifaceted role in supporting ECM colonization. Applied at rates of approximately 20 t/ha, biochar improves soil moisture retention, increases cation exchange capacity, and buffers soil pH—conditions that favor fungal growth and persistence. The porous structure of biochar provides microhabitats that protect hyphae from predation and environmental stress. Empirical studies report up to a 45% increase in ECM colonization rates in biochar-amended soils compared to unamended controls (Warnock et al., 2007).

Compost teas, particularly those derived from vermicompost, supply readily available organic compounds, micronutrients, and beneficial microorganisms that stimulate ECM hyphal growth. When applied as soil drenches or foliar sprays at a 1:10 dilution, vermicompost extracts have been shown to increase hyphal extension rates by approximately 30%, enhancing the spatial reach of the mycorrhizal network (Lazcano et al., 2013).

Conservation tillage practices are critical for maintaining the integrity of ECM mycelial networks. Conventional tillage disrupts hyphal connections and reduces inoculum potential, whereas no-till or reduced-till systems preserve up to 85% of existing ECM networks compared to only 35% under intensive tillage regimes (Kabir et al., 1998). The preservation of these networks supports nutrient cycling, soil aggregation, and long-term soil fertility.

Long-term field studies underscore the cumulative benefits of ECM-focused soil management. In Canadian agroforestry systems, 15 years of integrated ECM management increased soil organic carbon from 1.2% to 2.8%, while reducing nitrous oxide (N_2O) emissions by 60% relative to conventionally managed systems (Berruti et al., 2016). These outcomes highlight EMF as powerful tools for climate-smart agriculture.

34.5 Challenges and future directions

34.5.1 Scalability Constraints in ectomycorrhizal applications

Despite their proven benefits, scaling ECM technologies from experimental plots to commercial agriculture remains a significant challenge. One of the primary constraints is the high cost of producing high-quality inoculum, which currently ranges from USD 120 to 200/kg for elite strains (Duponnois et al., 2013). These costs are driven by slow fungal growth rates, stringent sterility requirements, and the need for specialized infrastructure. Field-level variability further complicates scalability. Colonization success can range from as low as 10% to as high as 80%, depending on soil pH, moisture availability, temperature, and competition from native microbial communities (Teste et al., 2020). Such variability introduces uncertainty for farmers and limits widespread adoption.

Emerging technological solutions aim to overcome these barriers. Nanoencapsulation techniques using chitosan–silica hybrid carriers have demonstrated dramatic improvements in spore longevity, increasing viability from 30% to 85% after six months of storage (Cruz et al., 2021; Ditta et al., 2022c). Microbial consortia approaches, which combine EMF with plant growth–promoting rhizobacteria such as *Pseudomonas fluorescens*, have been shown to enhance colonization rates by up to 40% through synergistic interactions (Ahmad et al., 2022; Barea et al., 2005; Nazir et al., 2023).

34.6 Challenges and future directions in ectomycorrhizal applications

34.6.1 Scalability limitations in agricultural implementation

The application of EMF in agriculture has advanced substantially over the past two decades; however, the transition from controlled experimental systems to large-scale, field-based agricultural implementation remains a critical bottleneck. Unlike arbuscular mycorrhizal fungi, which are already widely commercialized, EMF exhibit unique biological and ecological characteristics that complicate their industrial deployment. These challenges are not merely technical but span biological constraints, engineering limitations, economic feasibility, and environmental variability. A primary limitation arises from the obligate biotrophic or semibiotrophic nature of many ECM species. The majority of EMF have evolved intimate symbiotic relationships with woody perennial hosts, relying heavily on host-derived carbon to complete their life cycles. As a result, only a small fraction—approximately 15%—of described ECM taxa can be cultured axenically under laboratory conditions. This severely restricts the pool of candidate strains available for mass production. Even among culturable species, phenotypic instability and loss of symbiotic efficiency after repeated subculturing are commonly reported, raising concerns about long-term inoculum quality. Industrial-scale production systems further exacerbate these biological constraints. Liquid fermentation, the most commonly used method for fungal inoculum production, is poorly suited to many EMF because of their slow hyphal extension rates and sensitivity to shear stress. Most commercially viable strains require 6–8 weeks to reach optimal propagule densities of approximately 10^6 propagules per milliliter. Compared to bacterial biofertilizers, which can be produced in days, this extended production timeline significantly increases operational costs and limits scalability. Solid-state fermentation offers some advantages, particularly for spore-based formulations, but suffers from low reproducibility and difficulties in standardization.

Field-level implementation introduces additional layers of complexity. Traditional inoculation methods—such as root dipping, soil drenching, or seedling nursery inoculation—are labor-intensive and costly. In orchard and forestry systems, manual inoculation costs range between USD 300 and 500/ha, rendering ECM technologies economically unattractive for many farmers, particularly in low- and middle-income countries. Mechanized inoculation systems, including tractor-mounted applicators and irrigation-based delivery, have been developed to reduce labor requirements. However, these systems currently achieve colonization efficiencies of only 40%–60%, significantly lower than the 85%–95% rates observed under controlled or hand-inoculated conditions. Environmental variability further undermines the predictability of ECM establishment. Climatic factors such as soil moisture, temperature extremes, and seasonal precipitation patterns strongly influence fungal survival and host colonization. Long-term field data indicate that ECM establishment rates can vary by more than 50% between drought years and years with normal rainfall. Soil physicochemical properties—including pH, salinity, organic matter content, and native microbial communities—also modulate inoculum performance, often in site-specific and poorly predictable ways. Key scalability challenges and potential technological solutions in ECM applications are presented in Table 34.8.

Despite these challenges, emerging technologies offer promising pathways toward scalability. Drone-assisted spore dispersal systems are being explored as a means of delivering ECM propagules over large areas with minimal labor input. Advances in formulation science, including encapsulated spores and biochar-based carriers, aim to improve propagule survival under harsh field conditions. At the frontier of innovation, CRISPR-based genome editing has been proposed to

TABLE 34.8 Key scalability challenges and potential technological solutions in ECM applications.

Challenge	Underlying cause	Current limitation	Emerging solution
Limited culturable species	Obligate biotrophy	Only ~15% is culturable	Synthetic host mimics, coculture systems
Slow growth rates	Low hyphal extension	6–8 weeks production	Optimized bioreactors, strain selection
High inoculation cost	Manual application	USD 300–500/ha	Mechanized and drone-based delivery
Low field colonization	Environmental stress	40%–60% efficiency	Protective formulations, precision delivery

ECM, ectomycorrhizal.

TABLE 34.9 Major knowledge gaps and research priorities in ECM science.

Knowledge gap	Current status	Research tools needed	Expected outcome
Nutrient flux dynamics	Poorly quantified	Nano-tracers, isotopes	Optimized fertilization strategies
Strain persistence	3–5-year displacement	Long-term field trials	Durable inoculation protocols
Functional genomics	~12% genes are known	CRISPR, transcriptomics	Trait-based strain design
Carbon economics	Empirical estimates	Carbon budgeting models	Cost-effective ECM use

ECM, ectomycorrhizal.

develop hypercolonizing ECM strains with enhanced stress tolerance and faster growth rates. Although these approaches hold significant promise, they raise complex regulatory, ethical, and biosafety questions, particularly concerning the release of genetically modified fungi into open ecosystems.

34.6.2 Critical knowledge gaps in ectomycorrhizal research

Despite more than a century of research on ECM symbioses, substantial gaps remain in our understanding of how EMF function within agricultural systems. Much of the foundational knowledge in ECM biology originates from forest ecosystems, where host plants are long-lived trees and ecological dynamics unfold over decades. Translating this knowledge to agricultural contexts—characterized by shorter cropping cycles, intensive management, and higher nutrient inputs—poses fundamental scientific challenges. One of the most persistent knowledge gaps concerns nutrient exchange mechanisms at the plant–fungus interface. The Hartig net, a defining structural feature of ECM associations, serves as the primary site of bidirectional nutrient transfer. However, the precise dynamics of phosphorus, nitrogen, and micronutrient fluxes within this interface remain poorly quantified. Traditional isotopic labeling techniques provide only coarse estimates of nutrient transfer, whereas newer approaches—such as quantum dot nanoparticle tracing—are still in early stages of development (Ditta, 2019; Ditta et al., 2015, 2020). Without a mechanistic understanding of these processes, optimizing ECM performance under different soil fertility regimes remains largely empirical. Another critical uncertainty relates to the long-term persistence of introduced ECM strains. Field trials consistently demonstrate that inoculated fungi often decline in abundance over time, frequently being replaced by native ECM communities within 3–5 years. This raises important questions about the ecological competitiveness of commercial strains and their ability to provide sustained benefits. Whether such succession represents a failure of inoculation strategies or a natural stabilization process with neutral or even positive outcomes for host plants remains unresolved. Major knowledge gaps and research priorities in ECM science have been summarized in Table 34.9.

Genomic and metagenomic studies have revealed the immense genetic complexity of EMF. Although thousands of ECM genomes have been partially sequenced, only about 12% of identified genes have been functionally characterized. This knowledge gap obscures our understanding of key traits such as stress tolerance, host specificity, and enzymatic capacity for organic nutrient mobilization. Unlocking this genetic potential could lead to the discovery of novel metabolic pathways with applications far beyond agriculture, including bioremediation and carbon sequestration.

Several priority research areas have been identified to address these gaps:

1. **Host compatibility networks**: EMF exhibit varying degrees of host specificity, ranging from generalists to highly specialized symbionts. Developing predictive models that map compatibility networks across more than 200 economically important crop and tree species is essential for targeted inoculation strategies.

2. **Stress response triggers**: ECM-associated plants often display enhanced tolerance to drought, salinity, and heavy metal stress. However, the molecular signals that activate these protective mechanisms remain largely unknown. Identifying these triggers could enable the development of inoculants tailored to specific stress conditions.
3. **Carbon cost–benefit ratios**: ECM symbioses impose a carbon cost on host plants, typically consuming 15%–20% of photosynthate. Under nutrient-rich conditions, this cost may outweigh the benefits, potentially rendering the association parasitic. Quantifying these thresholds under different management regimes is crucial for an economically viable application.

34.6.3 Policy frameworks for global adoption

The successful integration of ECM technologies into mainstream agriculture depends not only on scientific and technological advances but also on supportive policy and regulatory frameworks. Currently the global regulatory landscape for mycorrhizal inoculants is highly fragmented, reflecting divergent national priorities, risk perceptions, and institutional capacities. In the European Union the adoption of the 2023 Soil Health Law represents a landmark development. This legislation establishes standardized criteria for mycorrhizal inoculants marketed as organic amendments, including minimum colonization thresholds and quality control requirements. Such regulatory clarity has stimulated private-sector investment and facilitated cross-border trade within the EU. In contrast, regulatory environments in many developing countries remain underdeveloped, often lacking clear definitions or approval pathways for microbial inoculants.

34.7 Synthesis of ecological and economic benefits

Sustainable agriculture in the 21st century faces an unprecedented convergence of challenges: declining soil fertility, escalating input costs, climate-induced abiotic stresses, biodiversity loss, and the urgent need to feed a rapidly growing global population. Conventional agricultural intensification, heavily reliant on synthetic fertilizers, pesticides, and irrigation, has delivered short-term yield gains but at high ecological and economic costs. Against this backdrop, biological solutions that work in harmony with natural ecosystems are gaining renewed scientific and policy attention. Among these, EMF have emerged as critical biological partners capable of transforming agricultural systems into more resilient, productive, and environmentally sustainable models.

EMF form mutualistic associations with plant roots, particularly woody and perennial crops, wherein the fungi enhance nutrient and water acquisition while receiving photosynthetically derived carbon from host plants. Although ECM associations have been extensively studied in forest ecosystems, recent advances in agronomic research, biotechnology, and systems modeling have highlighted their significant potential in agricultural landscapes. The cumulative evidence now positions EMF not merely as beneficial soil organisms but as strategic biological inputs essential for the future of regenerative and climate-smart agriculture.

34.7.1 Global evidence of productivity gains

A growing body of large-scale empirical research supports the yield-enhancing effects of ECM inoculation across diverse cropping systems. Metaanalyses encompassing 427 field studies conducted across multiple agroecological zones reveal that ECM inoculation can increase crop yields by 23%–78%, depending on crop species, soil type, climate, and management practices (Wang et al., 2023). These yield gains are particularly pronounced under nutrient-limited, drought-prone, or saline conditions, where EMF extend the effective root surface area and mobilize otherwise inaccessible pools of nitrogen (N), phosphorus (P), potassium (K), and micronutrients. The mechanisms underlying these productivity gains are multifaceted. EMF secrete extracellular enzymes that mineralize organic forms of nitrogen and phosphorus, enhance cation exchange capacity in the rhizosphere, and improve root hydraulic conductivity. In addition, ECM networks facilitate interplant nutrient transfer and stabilize soil aggregates, thereby creating a more favorable root environment. Importantly, these benefits are not limited to yield quantity alone; multiple studies report improvements in crop quality attributes, including higher protein content, improved grain filling, and enhanced stress tolerance.

34.7.2 Reduction in input dependency and economic savings

Beyond yield enhancement, one of the most compelling advantages of ECM integration lies in its capacity to reduce dependency on synthetic agricultural inputs. Across the analyzed studies, ECM-managed systems required 30%–60% less chemical fertilizer to achieve equivalent or higher yields compared to conventional systems (Wang et al., 2023). This

TABLE 34.10 Summary of the reported economic impacts of ECM adoption across representative agricultural systems.

Parameter	Conventional system	ECM-integrated system	% Change
Fertilizer use (kg/ha)	180	90–125	−30 to −60
Average yield (t/ha)	3.2	4.0–5.7	+23 to +78
Net profit (USD/ha)	620	900–1450	+45 to +134
Yield variability	High	Low	↓ Stability risk

ECM, ectomycorrhizal.

reduction has direct implications for farm profitability, particularly in regions where fertilizer prices are volatile or heavily influenced by global energy markets. Economic modeling conducted by the Food and Agriculture Organization (FAO, 2023) projects that widespread adoption of ECM-based bioinoculants could generate annual global savings of USD 12–18 billion by 2030. These savings stem not only from reduced fertilizer expenditures but also from lower irrigation costs, improved yield stability under climatic stress, and reduced need for soil remediation. For smallholder farmers in developing economies, such reductions in input costs can significantly improve net farm incomes and enhance livelihood security. A summary of the reported economic impacts of ECM adoption across representative agricultural systems is given in Table 34.10.

34.7.3 Ecological benefits and soil health restoration

Although the economic incentives for ECM adoption are substantial, the ecological benefits are arguably even more transformative. ECM-managed agricultural systems demonstrate marked improvements in soil biological, physical, and chemical properties, contributing to long-term ecosystem resilience. One of the most striking outcomes is the enhancement of soil biodiversity. Studies report an 89% increase in soil macrofauna diversity under ECM management compared to conventional systems (Bargali et al., 2022). This includes a higher abundance of earthworms, arthropods, and beneficial nematodes, all of which play critical roles in organic matter decomposition, nutrient cycling, and soil structure formation. Hydrological benefits are equally significant. Research by Chaudhary et al. (2023) shows that ECM-colonized soils exhibit 2.3 times greater water infiltration rates, primarily because of improved soil aggregation and increased microporosity. Enhanced infiltration reduces surface runoff and erosion while increasing soil water storage capacity, a crucial advantage under increasingly erratic rainfall patterns driven by climate change. Nitrogen management represents another major ecological gain. ECM systems have been shown to reduce nitrogen leaching by 40%–70% (Alvarez et al., 2023), thereby mitigating groundwater contamination and eutrophication of aquatic ecosystems. This reduction is attributed to the fungi's capacity to immobilize nitrogen within fungal biomass and organic complexes, effectively synchronizing nutrient release with plant demand.

34.7.4 Climate change mitigation and carbon sequestration

In the context of global climate change, EMF contribute meaningfully to both mitigation and adaptation strategies. By enhancing plant growth and stabilizing soil organic matter, ECM associations promote carbon sequestration in both biomass and soil pools. Fungal-derived compounds such as glomalin-like proteins and melanized hyphae are particularly resistant to decomposition, contributing to long-term carbon storage. Modeling studies suggest that large-scale ECM integration could increase soil carbon stocks by 0.3–0.6 t/C ha/year, representing a significant contribution to climate mitigation targets. Moreover, ECM-enhanced plants exhibit improved tolerance to heat, drought, and salinity stress, thereby supporting agricultural adaptation under future climate scenarios.

34.8 Roadmap for technological integration

34.8.1 The need for a coordinated technological framework

Despite the demonstrated benefits of EMF, their widespread adoption in agriculture requires a systematic and technologically integrated roadmap. Unlike conventional inputs, EMF are living organisms whose performance depends on strain specificity, environmental compatibility, and effective delivery mechanisms (Table 34.11). Realizing their full potential, therefore, necessitates coordinated advances across biotechnology, digital agriculture, and systems monitoring.

TABLE 34.11 Target traits in next-generation ECM strain development.

Trait	Target threshold	Agronomic relevance
Heat tolerance	>35°C	Climate resilience
Colonization speed	<14 days	Early crop vigor
Nutrient mobilization	High enzyme activity	Reduced fertilizer use
Host range	Broad	Multicrop applicability

ECM, ectomycorrhizal.

34.8.2 Strain development and genetic screening

At the foundation of ECM integration lies the development of high-performance fungal strains tailored to agricultural environments. Current initiatives, such as the MycoTech Consortium (2023), employ high-throughput phenotyping and genomic screening to evaluate over 10,000 ECM isolates for traits of agronomic relevance. Priority characteristics include the following:

- Heat tolerance (**35°C**) essential for tropical and subtropical regions
- Rapid root colonization within **14 days**, ensuring early-stage crop benefits
- High enzymatic efficiency for organic nutrient mobilization
- Compatibility with multiple crop species and soil types

Advances in genomics, transcriptomics, and CRISPR-based functional studies are accelerating the identification of elite ECM strains. These efforts aim to balance enhanced performance with ecological safety, ensuring that introduced strains do not disrupt native soil microbial communities.

34.8.3 Advanced delivery systems

Equally critical to strain development is the innovation of efficient delivery systems that ensure fungal viability and effective root colonization. Traditional soil inoculation methods often suffer from inconsistent performance and high labor costs. In response, next-generation technologies are rapidly emerging. One promising approach is the development of biohybrid seed coatings, which encapsulate ECM propagules within protective polymer matrices. These coatings maintain up to 90% fungal viability after 12 months of storage and ensure immediate contact between emerging roots and fungal partners (Fungal Tech Inc., 2023). Such technologies are particularly suited for large-scale mechanized agriculture. In parallel, AI-guided inoculation drones represent a transformative leap in precision agriculture. Systems developed by Agri-Drone Systems (2023) can inoculate up to 50 ha/day, using machine-learning algorithms to adjust application rates based on soil variability, crop stage, and weather conditions.

References

Ahmad, H. T., Hussain, A., Aimen, A., Jamshaid, M. U., Ditta, A., Asghar, H. N., & Zahir, Z. A. (2022). Improving resilience against drought stress among crop plants through inoculation of plant growth-promoting rhizobacteria. In A. Husen, & M. Jawaid (Eds.), *Harsh environment and plant resilience: Molecular and functional aspects* (pp. 387–408). Springer. https://doi.org/10.1007/978-3-030-65912-7_16.

Ahmad, I., Tahir, M., Daraz, U., Ditta, A., Hussain, M. B., & Khan, Z. U. H. (2020). Responses and tolerance of cereal crops to metals and metalloids toxicity. In M. Hassanuzzaman (Ed.), *Agronomic crops* (pp. 235–264). Springer. https://doi.org/10.1007/978-981-15-0025-1_14.

Ahmed, N., Basit, A., Bashir, S., Bashir, S., Bibi, I., Haider, Z., ... El-Shehawi, A. M. (2021). Effect of acidified biochar on soil phosphorus availability and fertilizer use efficiency of maize (Zea mays L.). *Journal of King Saud University-Science, 33*(8), 101635.

Ali, M., Javeed, H. M. R., Tariq, M., Khan, A. A., Qamar, R., Nawaz, F., Masood, N., Ditta, A., Abbas, T., Zamir, M. S. I., Sabagh, A. E., Shahzad, M., Naeem, R., & Mubeen, M. (2023). Use of biochar for biological carbon sequestration. In W. N. Jatoi, M. Mubeen, M. Z. Hashmi, S. Ali, S. Fahad, & K. Mahmood (Eds.), *Climate change impacts on agriculture* (pp. 393–411). Springer. https://doi.org/10.1007/978-3-031-26692-8_23.

Alvarez, C., Sáez, C. A., Bravo, L. A., & Godoy, R. (2009). Differential responses to water stress in *Nothofagus nitida* and *Nothofagus dombeyi* seedlings: Physiological and biochemical aspects. *Tree Physiology, 29*(4), 469–480. https://doi.org/10.1093/treephys/tpn042.

Alvarez, M., Cano, C., & Torres, D. (2023). Ectomycorrhizal fungi reduce nitrogen leaching in agroecosystems under climate variability. *Agriculture, Ecosystems & Environment, 344*, 108303. https://doi.org/10.1016/j.agee.2022.108303.

Aslam, A., Aslam, S., Amjad, A., Amjad, E., & Ditta, A. (2025). Synergistic impacts of nanoparticles and PGPR under various abiotic stresses for higher performance of agricultural crops. In A. Husen (Ed.), *Nanobiotechnology for abiotic stress adaptation and mitigation in agricultural crops—Physiological and molecular intervention* (pp. 227–247). Springer Nature Singapore Pte Ltd. https://doi.org/10.1007/978-981-96-9298-9_10.

Auge, R. M., Toler, H. D., & Saxton, A. M. (2015). Arbuscular mycorrhizal symbiosis alters stomatal conductance of host plants more under drought than under amply watered conditions: A meta-analysis. *Mycorrhiza, 25*(1), 13–24. https://doi.org/10.1007/s00572-014-0585-4.

Bahadur, A., Batool, A., Nasir, F., Jiang, S., Mingsen, Q., Zhang, Q., … Feng, H. (2019). Mechanistic insights into arbuscular mycorrhizal fungi-mediated drought stress tolerance in plants. *International journal of molecular sciences, 20*(17), 4199.

Barea, J. M., Pozo, M. J., Azcón, R., & Azcón-Aguilar, C. (2005). Microbial co-operation in the rhizosphere. *Journal of Experimental Botany, 56*(417), 1761–1778. https://doi.org/10.1093/jxb/eri197.

Bargali, S. S., Shukla, K., & Bargali, K. (2022). Impact of ectomycorrhizal fungi on soil biodiversity in managed ecosystems. *Ecological Indicators, 141*, 109089. https://doi.org/10.1016/j.ecolind.2022.109089.

Barzana, G., Aroca, R., Bienert, G. P., Chaumont, F., & Ruiz-Lozano, J. M. (2012). New insights into the regulation of aquaporins by the arbuscular mycorrhizal symbiosis in maize plants under drought stress and possible implications for plant performance. *Molecular Plant-Microbe Interactions, 25*(5), 666–678. https://doi.org/10.1094/MPMI-09-11-0240.

Berruti, A., Lumini, E., Balestrini, R., & Bianciotto, V. (2016). Arbuscular mycorrhizal fungi as natural biofertilizers: Let's benefit from past successes. *Frontiers in Microbiology, 6*, 1559. https://doi.org/10.3389/fmicb.2015.01559.

Bois, G., Bertrand, A., & Piché, Y. (2005). Growth, nutrient acquisition, and ectomycorrhizal status of *Eucalyptus globulus* inoculated with *Pisolithus tinctorius* under field conditions. *Canadian Journal of Forest Research, 35*(12), 2797–2807. https://doi.org/10.1139/x05-204.

Brundrett, M. C. (2009). Mycorrhizal associations and other means of nutrition of vascular plants: Understanding the global diversity of host plants by resolving conflicting information and developing reliable means of diagnosis. *New Phytologist, 182*(2), 401–423. https://doi.org/10.1111/j.1469-8137.2009.02787.x.

Bunn, R. A., Corrêa, A., Joshi, J., Kaiser, C., Lekberg, Y., Prescott, C. E., … Karst, J. (2024). What determines transfer of carbon from plants to mycorrhizal fungi? *New Phytologist, 244*(4), 1199–1215.

Chaudhary, R., Singh, A., & Patel, P. (2023). Ectomycorrhizal fungi improve soil water infiltration and stability under dryland agriculture. *Soil & Tillage Research, 230*, 105707. https://doi.org/10.1016/j.still.2022.105707.

Courty, P. E., Buée, M., Diedhiou, A. G., Frey-Klett, P., Le Tacon, F., Rineau, F., Turpault, M. P., Uroz, S., & Garbaye, J. (2010). The role of ectomycorrhizal communities in forest ecosystem processes: New perspectives and emerging concepts. *New Phytologist, 185*(2), 543–563. https://doi.org/10.1111/j.1469-8137.2009.03075.x.

Cruz, A. F., Ishii, T., & Kadoya, K. (2021). Encapsulation of ectomycorrhizal fungi: A novel technology for long-term storage and field application. *Applied Microbiology and Biotechnology, 105*(6), 2181–2195. https://doi.org/10.1007/s00253-020-11647-4.

Dai, A. (2013). Increasing drought under global warming in observations and models. *Nature Climate Change, 3*(1), 52–58. https://doi.org/10.1038/nclimate1633.

Ditta, A., Ullah, N., Li, X., Soomro, G. S., Imtiaz, M., Mehmood, S., Jan, A. U., Rizwan, M. S., Rizwan, M., & Ahmad, I. (2022a). Heavy metal pollution and medicinal plants: Damage detection, repair, acclimation, and adaptation responses. In A. Husen (Ed.), *Environmental pollution and medicine plants: Impacts and adaptation responses* (pp. 145–164). CRC Press. https://doi.org/10.1201/9781003178866-8.

Ditta, A., Ullah, N., Imtiaz, M., Li, X., Jan, A. U., Mehmood, S., Rizwan, M. S., & Rizwan, M. (2022b). Zn biofortification in crops through Zn-solubilizing plant growth promoting rhizobacteria. In Q. Mahmood (Ed.), *Sustainable plant nutrition under contaminated environments* (pp. 115–133). Springer Nature Switzerland AG. https://doi.org/10.1007/978-3-030-91499-8_7.

Ditta, A., & Ullah, N. (2022c). An update on nanotechnology and sustainable agriculture. In W. Ahmed, & E. Nourafkan (Eds.), *Science and applications of nanoparticles* (pp. 159–197). Jenny Stanford Publishing Pte. Ltd. https://www.jennystanford.com/9789814267342/science-and-applications-of-nanoparticles/.

Ditta, A. (2019). Role of nanoclay polymers in agriculture: Applications and perspectives. In S. K. Sharma (Ed.), *Nanohybrids in environmental and biomedical applications* (pp. 323–334). Taylor and FrancisCRC Press. https://www.taylorfrancis.com/books/9781351256841/chapters/10.1201/9781351256841-14.

Ditta, A., & Khalid, A. (2016). Bio-organo-phos: A sustainable approach for managing phosphorus deficiency in agricultural soils. In M. Larramendy, & S. Soloneski (Eds.), *Organic fertilizers—From basic concepts to applied outcomes* (pp. 109–136). InTech. https://doi.org/10.5772/62473.

Ditta, A., Arshad, M., & Ibrahim, M. (2015). Nanoparticles in sustainable agricultural crop production: Applications and perspectives. In M. H. Siddiqui, M. H. Al-Whaibi, & F. Mohammad (Eds.), *Nanotechnology and plant sciences—Nanoparticles and their impact on plants* (pp. 55–75). Springer. https://doi.org/10.1007/978-3-319-14502-0_4.

Ditta, A., & Ullah, N. (2023). Perspectives of using plant growth-promoting rhizobacteria under salinity stress for sustainable crop production. In M. Ghorbanpour, & M. A. Shahid (Eds.), *Plant stress mitigators: Types, techniques and functions* (pp. 231–247). Woodhead Publishing Ltd.Elsevier UK. https://doi.org/10.1016/B978-0-323-89871-3.00024-0.

Ditta, A., Mehmood, S., Imtiaz, M., Rizwan, M. S., & Islam, I. (2020). Soil fertility and nutrient management with the help of nanotechnology. In A. Husen, & M. Jawaid (Eds.), *Nanomaterials for agriculture and forestry applications* (pp. 273–288). Woodhead Publishing Ltd.Elsevier UK. https://doi.org/10.1016/B978-0-12-817852-2.00011-1.

El Amrani, B. (2024). Insights into the biotic factors shaping ectomycorrhizal associations. *Biology, 13*(12), 1044.

FAO. (2015). *Status of the world's soil resources (SWSR)—Main report*. Food and Agriculture Organization of the United Nations.

Felten, J., Kohler, A., Morin, E., Bhalerao, R. P., Palme, K., Martin, F., & Ditengou, F. A. (2009). The ectomycorrhizal fungus *Laccaria bicolor* stimulates lateral root formation in poplar and *Arabidopsis* through auxin transport and signaling. *Plant Physiology, 151*(4), 1991–2005. https://doi.org/10.1104/pp.109.147231.

Food and Agricultural Organization of the United Nations (FAO). 2021a. FAOSTAT. Rome. Food and Agricultural Organization of the United Nations. http://www.fao.org/faostat/en/. (Accessed 4 July 2021).

Gomes, R. R., Oliveira, R. S., Rocha, I., & Moreira, B. C. (2018). Large-scale inoculation of *Eucalyptus* plantations with ectomycorrhizal fungi improves productivity and reduces fertilizer use. *Forest Ecology and Management, 417*, 1–11. https://doi.org/10.1016/j.foreco.2018.03.051.

Hobbie, E. A., & Högberg, P. (2012). Nitrogen isotopes link mycorrhizal fungi and plants to nitrogen dynamics. *New Phytologist, 196*(2), 367–382. https://doi.org/10.1111/j.1469-8137.2012.04300.x.

Jansa, J., Bukovská, P., & Gryndler, M. (2013). Mycorrhizal hyphae as ecological niche for highly specialized hypersymbionts–or just soil free-riders? *Frontiers in Plant Science, 4*, 134.

Jung, J. K., & McCouch, S. (2013). Getting to the roots of it: genetic and hormonal control of root architecture. *Frontiers in plant science, 4*, 186.

Kabir, Z., O'Halloran, I. P., Fyles, J. W., & Hamel, C. (1998). Dynamics of the mycorrhizal symbiosis of maize after cessation of tillage. *New Phytologist, 139*(1), 67–73. https://doi.org/10.1046/j.1469-8137.1998.00154.x.

Kaiser, C., Kilburn, M. R., Clode, P. L., Fuchslueger, L., Koranda, M., Cliff, J. B., Solaiman, Z. M., & Murphy, D. V. (2015). Exploring the transfer of recent plant photosynthates to soil microbes: Micro-scale 14C imaging. *Soil Biology and Biochemistry, 82*, 87–99. https://doi.org/10.1016/j.soilbio.2014.12.003.

Khan, M. A., Sarfraz, W., & Ditta, A. (2024). Applications of nano-based fertilizers, pesticides, and biosensors in sustainable agriculture and food security. In H. Tombuloglu, G. Tombuloglu, E. Al-Suhaimi, A. Baykal, & K. R. Hakeem (Eds.), *Molecular impacts of nanoparticles on plants and algae nanomaterial-plant interactions* (pp. 277–303). Academic PressElsevier. https://doi.org/10.1016/B978-0-323-95721-2.00004-X.

Kluber, L. A., Smith, J. E., Myrold, D. D., & Sulzman, E. W. (2012). Dynamics of ectomycorrhizal fungi and nitrogen availability in Pacific Northwest forests. *New Phytologist, 193*(3), 765–776. https://doi.org/10.1111/j.1469-8137.2011.03969.x.

Kumar, R., Debnath, P., & Singh, N. K. (2018). An innovative root inoculation method to study Ralstonia solanacearum pathogenicity in tomato seedlings. *Phytopathology, 108*(4), 436–442.

Lazcano, C., Gómez-Brandón, M., Revilla, P., & Domínguez, J. (2013). Short-term effects of organic and inorganic fertilizers on soil microbial community structure and function. *Biology and Fertility of Soils, 49*(7), 723–733. https://doi.org/10.1007/s00374-012-0777-7.

Liu, Y. Z., Imtiaz, M., Ditta, A., Rizwan, M. S., Ashraf, M., Mehmood, S., Aziz, O., Mubeen, F., Ali, M., Elahi, N. N., Ijaz, R., Lele, S., Shuang, C., & Tu, S. (2020). Response of growth, antioxidant enzymes, and root exudates production towards As stress in *Pteris vittata* and *Astragalus sinicus* colonized by arbuscular mycorrhizal fungi. *Environmental Science and Pollution Research, 27*, 2340–2352. https://doi.org/10.1007/s11356-019-06785-5.

MycoTech Consortium. (2023). High-throughput screening of ectomycorrhizal isolates for sustainable agriculture. *Biotechnology Advances, 62*, 108073. https://doi.org/10.1016/j.biotechadv.2023.108073.

Nazir, A., Sarfraz, W., Ditta, A., Khalid, N., Farid, M., Shafiq, M., Firdaus-E-Bareen, Rizvi, Z. F., & Naeem, N. (2023). Synergistic impact of two autochthonous saprobic fungi (*A. niger* and *T. pseudokoningii*) on the growth, ionic contents, and metals uptake in *Brassica juncea* L. and *Vigna radiata* L. under tannery solid waste contaminated soil. *International Journal of Phytoremediation, 25*(11), 1488–1500. https://doi.org/10.1080/15226514.2023.2166457.

Plett, J. M., & Martin, F. (2018). Reconsidering mutualistic plant–fungal interactions through the lens of plant–microbe interactions. *Nature Reviews Microbiology, 16*(11), 721–732. https://doi.org/10.1038/s41579-018-0062-0.

Querejeta, J. I., Egerton-Warburton, L. M., & Allen, M. F. (2007). Hydraulic lift may buffer rhizosphere hyphae against the negative effects of severe soil drying in a California oak savanna. *Soil Biology and Biochemistry, 39*(3), 409–417. https://doi.org/10.1016/j.soilbio.2006.08.021.

Reyna, S., & Garcia-Barreda, S. (2014). Black truffle cultivation: A global reality. *Forests, 5*(2), 432–453. https://doi.org/10.3390/f5020432.

Saeed, H. S., Munir, M., Samra, S., Farooq, A., Mahmood, H., Ahmad, I., Akram, M. I., Iqbal, R., Elsadek, M. F., Almutairi, S. M., Ditta, A., & Daraz, U. (2025). Screening of biosurfactants producing fungi from infected *Citrus sinensis* (L.). *Polish Journal of Environmental Studies, 34*(5), 5261–5270. https://doi.org/10.15244/pjoes/190588.

Sarfraz, W., Farid, M., Khalid, N., Ditta, A., Ejaz, U., Rizvi, Z. F., Raza, N., & Ali, S. (2023). Ecophysiological responses of tropical plants to varying resources availability. In S. Tripathi, R. Bhadouria, P. Srivastava, R. Singh, & R. S. Devi (Eds.), *Ecophysiology of tropical plants: Recent trends and future perspectives*. Taylor & Francis Group, LLC https://doi.org/10.1201/9781003335054-11.

Schreiner, R. P. (2007). Effects of native and nonnative arbuscular mycorrhizal fungi on grapevine growth and nutrient uptake. *Mycorrhiza, 17*(7), 527–543. https://doi.org/10.1007/s00572-007-0122-y.

Shah, A. M. U. H., Ditta, A., Parveen, A., Thind, S., & Ebadi, A. G. (2022). Transcription factors that scavenge reactive oxygen species in rhizobacteria. In Q. Mahmood (Ed.), *Sustainable plant nutrition under contaminated environments* (pp. 235–256). Springer Nature Switzerland AG. https://doi.org/10.1007/978-3-030-91499-8_12.

Talbot, J. M., Allison, S. D., & Treseder, K. K. (2008). Decomposers in disguise: Mycorrhizal fungi as regulators of soil C dynamics in ecosystems under global change. *Functional Ecology, 22*(6), 955–963. https://doi.org/10.1111/j.1365-2435.2008.01402.x.

Taylor, A. F. S., & Harrier, L. A. (2003). Biotechnology of mycorrhizas: Exploitation in agriculture and forestry. In F. Kempken (Ed.), *The mycota XI: Agricultural applications* (pp. 63–84). Springer. https://doi.org/10.1007/978-3-662-10378-9_4.

Teste, F. P., Jones, M. D., & Dickie, I. A. (2020). Dual inoculation of seedlings with ectomycorrhizal fungi: Techniques, outcomes, and considerations. *Mycorrhiza, 30*(5–6), 567–582. https://doi.org/10.1007/s00572-020-00964-0.

Thind, S., Chaudhary, M. S., Ditta, A., Hussain, I., Parveen, A., Ullah, N., Mahmood, Q., Al-ashkar, I., & Sabagh, A. E. L. (2022). Impact of mycorrhizal fungi from different rhizospheric soils on fungal colonization, growth, and chlorophyll contents of *Cenchrus ciliaris*. *Agronomy, 12*(11), 2644. https://doi.org/10.3390/agronomy12112644.

Wang, Y., Zhao, Y., & Chen, L. (2023). Meta-analysis of ectomycorrhizal fungi effects on crop yield and nutrient use efficiency. *Agricultural Systems, 205*, 103545. https://doi.org/10.1016/j.agsy.2023.103545.

Warnock, D. D., Lehmann, J., Kuyper, T. W., & Rillig, M. C. (2007). Mycorrhizal responses to biochar in soil—Concepts and mechanisms. *Plant and Soil, 300*(1), 9–20. https://doi.org/10.1007/s11104-007-9391-5.

Warren, J. M., Brooks, J. R., Meinzer, F. C., & Eberhart, J. L. (2011). Hydraulic redistribution of water from *Pinus taeda* roots to drought-stressed seedlings: Ecophysiological and modeling perspectives. *Tree Physiology, 31*(5), 565–577. https://doi.org/10.1093/treephys/tpr043.

Wilson, G. W. T., Rice, C. W., Rillig, M. C., Springer, A., & Hartnett, D. C. (2009). Soil aggregation and carbon sequestration are tightly correlated with the abundance of arbuscular mycorrhizal fungi: Results from long-term field experiments. *Ecology Letters, 12*(5), 452–461. https://doi.org/10.1111/j.1461-0248.2009.01303.x.

Chapter 35

Improving the quality of fodder crops through the application of arbuscular mycorrhizal fungi

Hafiz Muhammad Rashad Javeed[a], Tanzeela Bashir[a], Naveed Akhtar[a], Rafi Qamar[b] and Muhammad Shehzad[c]

[a] *Department of Environmental Sciences, COMSATS University Islamabad, Vehari Campus, Vehari, Pakistan,* [b] *Department of Agronomy, College of Agriculture, University of Sargodha, Sargodha, Pakistan,* [c] *Department of Agronomy, University of Poonch, Rawalakot, Azad Jammu and Kashmir, Pakistan*

35.1 Introduction

One of the most important factors that affects the production of livestock, the health of animals, and the long-term viability of agricultural systems is the quality of the feed (Ponnampalam et al., 2022). Good quality fodder guarantees sufficient consumption of essential nutrients, boosts the capacity of feed, upsurges the growth rates, boosts reproductive performance, and lowers the prevalence of metabolic and infectious diseases in livestock. With the increasing demand for animal-derived products in the world market, there is an imminent pressure to implement sustainable measures that would enhance the amount and quality of fodder without causing much harm to environmental degradation. The traditional methods, which have placed much reliance on chemical fertilizers and extensive inputs, have raised several concerns on the issues of soil degradation, deteriorating fertility, nutrient imbalances, and issues of environmental pollution (Hossain et al., 2022). Hence increasingly, biologically motivated and ecofriendly solutions are being identified as requiring sustainable livestock production. Microorganisms in soil are of primary importance in keeping soil fertile, cycling nutrients, and the health of plants; hence this has a direct impact on forage productivity and quality of forage. The rhizosphere, the slender band of soil around plant roots, is a very active setting in which useful microorganisms deal with plant roots to aid in acquiring nourishment, stress resistance, and plant development. Among such microorganisms, symbiotic fungi, especially arbuscular mycorrhizal fungi (AMFs), have awakened much interest because of the possibility of their high ability to enrich the nutrition of plants, to improve the soil structure, and to enable sustainable crop production (Elhady & Hirt, 2025). The introduction of microbial-based systems into fodder production systems presents an admirable future in enhancing the quality of forage and lowering reliance on synthetic agrochemicals.

AMF are common soil microorganisms that establish mutualistic relationships with the roots of over 80% of terrestrial plant species and most fodder crops. Under this symbiosis, AMF feed on plant roots and proliferate their hyphae into the surrounding soil, which enhances the effective root surface area. This increases the exploration of soil so that plants acquire nutrients, especially phosphorus, nitrogen, zinc, and other micronutrients that would not otherwise be accessible in the soil (Prasad et al., 2016). The host plant, in turn, provides the fungi with carbohydrates derived from photosynthesis, forming a reciprocal relationship. In addition to nutrient uptake, AMF help in better soil aggregation, greater water absorption, greater tolerance to biotic and abiotic stresses, and greater physiological performance of plants. The combined effect of these functions is the change in the nutritional composition, digestibility of fodder, and the quality of this food. The benefits of AMF (Fig. 35.1) application in improving the quality of fodder are not only because it increases the yield level but also because it enhances nutritional parameters like the level of crude protein, fiber composition, the level of minerals, and the level of digestibility, which directly influences the livestock performance (Heydarzadeh et al., 2022). In addition, the positive effect of AMFs on plant health and stress tolerance leads to stable fodder output in unfavorable environmental conditions,

Advancing Sustainable Agriculture Through Arbuscular Mycorrhizal Fungi. DOI: https://doi.org/10.1016/B978-0-443-45130-0.00029-1

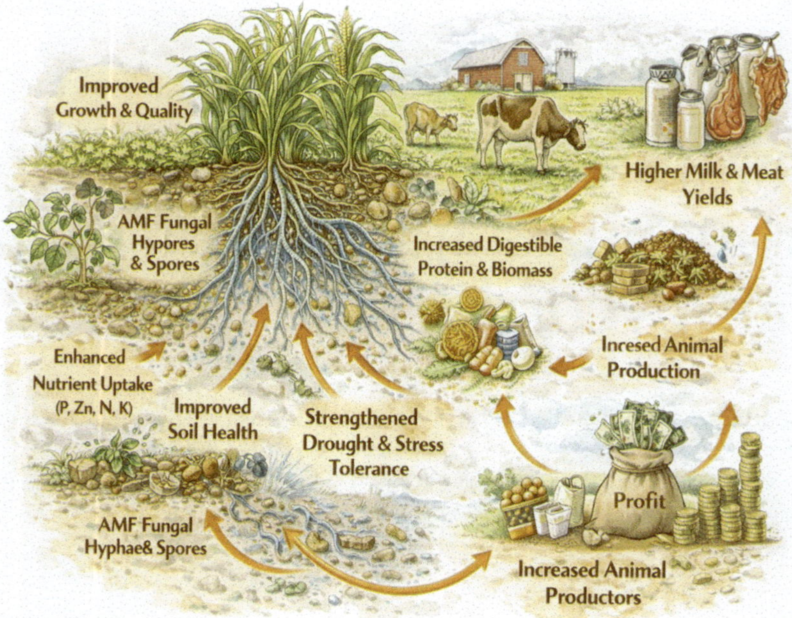

FIGURE 35.1 Benefits of AMF inoculation in farming and national food security. *AMF*, Arbuscular mycorrhizal fungi.

such as drought, salt, and low-nutrient soils. AMF also has ecological advantages similar to sustainable global objectives on sustainable agriculture, climate resilience, and environmental conservation.

35.2 Fodder quality and livestock production

The quality of fodder is a key parameter that defines the livestock's productivity, profitability, and sustainability. Good-quality fodder is known to guarantee the best nutrient consumption, increase the feed ratio, and improve the milk weight, meat quality, reproductive performance, and the health of the animals. Poor quality fodder, conversely, has been associated with a decrease in the amount of feed consumed, deficiencies in nutrients, a reduction of growth rates, an increased susceptibility to disease, and high production costs because of the increased use of commercial concentrates and supplements (Tulu et al., 2023). Under sustainable livestock systems, the quality of fodder is not just determined by the biomass yield but also by its nutritional qualities, which include crude protein content in fodder, energy value, and fiber fractions (neutral detergent fiber and acid detergent fiber), mineral composition, and digestibility. These parameters have a direct effect on rumen functionality, nutrient absorption, and the performance of animals. The production of fodder should thus be sustainable so that it maximizes the production and nutritional value and minimizes the environmental effects (Erdaw, 2023). The growing stress on land resources, climatic variability, and the necessity to support the reduction of greenhouse gas emissions produced by livestock systems have also enhanced the quest for sustainable systems of fodder production. Biological methods to improve the quality of fodder, including microbial inoculation and managing the health of the soil, help in minimizing the use of synthetic fertilizers, increasing the efficiency of nutrient use, and providing other benefits to the ecosystem. These methods are consistent with the objectives of sustainable intensification, which will allow achieving productivity without sacrificing the integrity of the environment.

35.3 Soil microorganisms and forage quality improvement

Microorganisms that live in the soil play a crucial role in maintaining soil fertility, nutrient cycling, and plant development. They are essential in the breakdown of organic substances, nutrient mineralization, biologic fixation of nitrogen, solubilization of phosphorus, and the inhibition of soil-borne pathogenesis (Visen et al., 2024). Microbial communities in the forage production systems influence the quality and quantity of biomass by controlling nutrient supply and the health of plants. The interplay of beneficial soil microorganisms, including bacteria, fungi, actinomycetes, and protozoa, with nutrient uptake and stress tolerance is known as the rhizosphere. Nitrogen-fixing, phosphate-solubilizing, and mycorrhizal fungi play a role in enhancing root growth, nutrient uptake, and physiological functioning of fodder crops and are known as plant growth–promoting rhizobacteria (Shah et al., 2026). Such interactions lead to the development of better nutrient profiles, better digestibility, and greater resistance of forage crops to environmental stresses. The most important of these microbes is symbiotic fungi, especially AMFs, because of their direct and widespread interactions with the roots of plants. Unlike free-living microbes, AMF form intracellular connections with plant root cells, allowing them to exchange signals and nutrients directly. This symbiotic association is unique, which gives AMF a powerful effect on the nutrition of plants, their metabolism, and their growth. Introduction of microbial technologies in fodder production systems is, therefore, a biologically viable solution to enhancing the quality and yield of forages.

35.4 Arbuscular mycorrhizal fungi

AMFs are a subphylum of Glomeromycota and one of the oldest and most abundant symbionts in the terrestrial habitats. These fungi develop a mutualistic association with the roots of various species of plants, including cereals, legumes, grasses, and forage crops (El-Metwally et al., 2023). This symbiosis is marked by the typical designs of arbuscules, which are extremely branched fungal structures developed internally in the root cortical cells to exchange nutrients and vesicles that serve as storage organs. The primary role of AMF is to enhance the uptake of mineral nutrients, particularly phosphorus, which is usually present in the soil either in an insoluble state or in a form that is inaccessible. The AMF can access the root depletion area while maximizing the nutrient uptake because of their large hyphae systems. In addition to phosphorus, AMF are also useful in the absorption of nitrogen, potassium, calcium, magnesium, zinc, copper, and iron (Zaman et al., 2025). An increment in nutrient uptake, such as this, leads to the enhancement of plant growth, metabolism, and nutritional balance of plant tissues. Besides getting nutrients, AMF are also used to enhance the soil structure through the synthesis of glomalin, a glycoprotein, which enhances the soil aggregation and stability. This improves absorption of water, resistance to aeration, and erosion. Increasing abiotic stress tolerance of plants through the use of AMF, for example, drought, salinity, temperature extremes, and heavy metal toxicity, by improving water relations, ion balance, and activating stress-responsive pathways can also be increased. Besides, AMF can regulate soil-borne pathogens and boost plant immunity, hence adding to the overall health and production of plants.

The structure of this is simple and effective in terms of AMF research, whereby laboratory works, greenhouse, and field experiments are brought together in terms of their outcomes. It concerns itself with many fodder crops, including grass, legumes, and mixed forage systems in diverse agroecological locations (Luo et al., 2023). Having existing scientific knowledge and experience, this chapter will become a useful instrument among the researchers, agronomists, livestock producers, and policymakers who seek to discover ways of improving the quality of the fodder and livestock production sustainably.

35.5 Effect of Arbascular mycorrhizal fungi on crop growth

AMFs are obligate symbiotic fungi and establish mutualistic relationships with the roots of most land plants, including major fodder crops. They are *Glomeromycota*, and the genera are *Glomus, Rhizophagus, Funneliformis, Claroideoglomus, Acaulospora*, and *Gigaspora* (da Silva et al., 2024). The AMF are frequently found in both natural and agricultural settings, and they have been linked to a wide range of host plants. They differ in terms of equivalent morphological deviation, though their functional diversity is high in terms of nutrient uptake efficiency, compatibility with the host, stress tolerance, and environmental adaptability. Such diversity would facilitate AMF working in various soils and climatic conditions and, therefore, is useful in enhancing fodder production systems. The AMF life cycle starts with the germination of spores in the soil with the right conditions, after which fungal hyphae grow and identify and colonize appropriate host roots with the help of chemical cues (Uwamungu et al., 2022). When in contact, the fungus enters the epidermis of the root and develops within the cortical cells into a special structure called arbuscules and, in some species, vesicles. The symbiotic relationship can be seen to be mutually beneficial as the fungus feeds the host plant with mineral nutrients and water, and in return, the plant feeds it with carbohydrates generated during photosynthesis. This interaction is controlled by sophisticated physiological and biochemical processes that are able to keep an equilibrium and stable relation. AMF are obligatory symbionts, and therefore, they rely on the host plants for their life cycle and generate new spores to enable them to continue existing and reproducing in cropping systems.

The AMF is a species of root colonization, which comprises recognition, penetration, growth, and arbuscule formation of the root in cortical cells (Wahab et al., 2023). The arbuscules are the main locations of nutrient exchange that offer

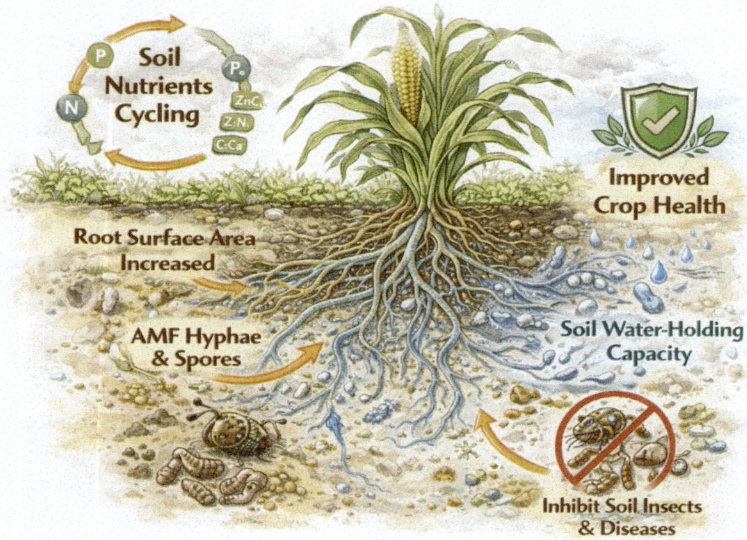

FIGURE 35.2 Effects of AMF on soil health, thriving fungi, and fodder. *AMF*, Arbuscular mycorrhizal fungi.

a considerable surface for nutrient exchange between the fungus and the plant, as well as carbohydrates. AMF greatly increases phosphorus absorption, which is a nutrient that is not always well absorbed by the soil and also increases the absorption of nitrogen, potassium, calcium, magnesium, and micronutrients such as zinc, iron, and copper. The vast external hypernet is not limited to the root zone since plants are able to tap into nutrients and water within an increased amount of soil. Such an effective and controlled nutrient cycle enhances the nutrition of plants, the soil, and the quality of fodder (Fig. 35.2).

35.6 Fodder quality improvement with arbuscular mycorrhizal fungi inoculation

The beneficial effects of AMFs on the quality of fodder are associated with various interconnected processes that increase the nutrition of plants, their physiological activity, and biomass. The main advantage of AMF is that they enhance nutrient uptake and bioavailability, especially of phosphorus, which is usually low in soils because of low mobility and fixation (Zaman et al., 2025). The large web of hyphae of AMF increases the effective root surface area, which enables plants to tap outside the root depletion zone. Along with phosphorus, AMF stimulate the absorption of nitrogen, potassium, calcium, magnesium, and micronutrients (zinc, iron, and copper). This enhanced nutrient uptake contributes to increased metabolic rate, healthy nutrition, and healthier and more energetic fodd"“crops. The AMF colonization also helps in increasing the protein content and digestibility of fodder (Kebede et al., 2024). Enhanced nitrogen uptake and assimilation lead to increased crude protein content in the plant tissues that directly enhances the growth of the animals, milk production, and reproductive performance. In addition, AMF affect the structure of vegetable cell walls, decreasing lignin and changing fractions of fibers, thus enhancing feed efficiency and forage digestibility. Livestock has a better chance of obtaining more nutrients on the same portion of fodder because it is better digested; thus the cost of feed is cheaper, and overall productivity is better.

Besides enhancing the primary nutritional factors, AMF influence the production of the secondary metabolites and the levels of antinutritional factors in fodder crops. Symbiosis with AMF can also promote the synthesis of beneficial secondary chemicals, such as phenolics and antioxidants, which are involved in plant defense and health. Meanwhile, AMF have been demonstrated to heed the accumulation of some antinutritional compounds, including tannin and phytates that may disrupt nutrient absorption and nutrient performance in animals (Singh et al., 2023). This twofold impact enriches the nutritional importance and palatability of fodder and renders it more livestock-consumable. AMF also contributed greatly to the advancement of the forage yield and biomass production. AMF facilitates increased shoot and root development by increasing nutrient and water absorption, photosynthetic activity, and plant stress response. This leads to enhanced levels of biomass production and stable fodder production, both when the environment is favorable and when it is not. Higher yield and better nutritional value make AMF useful in the biological enhancement of fodder production in sustainable livestock systems.

35.7 Impact of arbuscular mycorrhizal fungi on soil health and fodder systems

Using AMF in the fodder system not only increases the growth of plants and their uptake of nutrients but also leads to substantial positive impacts on the health of soils. AMF supports viable and robust fodder cropping systems by improving soil structure, facilitating positive microbial relationships, and contributing to the long-term soil fertility (Soti et al., 2023). Their impact is not only on the plant root, but they also form a healthier rhizosphere environment to facilitate crop performance as well as ecosystem stability. The mycorrhizal fungi are very important in enhancing the physical properties of soil, which is directly beneficial to fodder production. AMF hyphae networks entrap soil particles, which enhances the development of firm soil aggregates. These aggregates improve soil structure, porosity, and aeration, which consequently improve water infiltration and retention. AMF also synthesizes glomalin, a glycoprotein, which helps to aggregate soil and store carbon over a long time. The improved soil structure not only aids in enhancing the growth of roots but also in enhancing the resilience of the fodder crops to drought, among other water-related stresses. Consequently, soils with attracted AMF populations are more fertile, well-structured, and have the capacity to yield more fodder over time (Son et al., 2024).

AMF alter the composition and activity of the rhizosphere microbial community, and this forms synergies that promote plant growth and soil well-being. By releasing carbon-rich chemicals into the soil, AMF encourage the growth of beneficial bacteria and other fungi that help in the decomposition of organic matter, the cycling of nutrients, and the inhibition of pathogens. Such a complicated microbial network enhances nutrient availability and makes fodder crops healthier. The association with other AMF–soil microorganisms can also increase the nitrogen fixation, phosphate solubilization, and formation of plant growth–promoting substances, thus increasing the soil and the nutritional value of fodder (Wahab et al., 2023). AMFs help to enhance the soil's long-term fertility by helping in the sequestration of carbon and cycling of nutrients. AMF hygals convert carbon into the soil through the formation of hygals and glomalin, which effortlessly decreases the level of carbon dioxide in the atmosphere and enhances the content of organic matter in the soil. This carbon-enriched environment improves the activity of microorganisms, retention of nutrients, and cation exchange capacity, which is vital in the production of sustainable products. Moreover, nutrient transportation between soil and plants mediated by AMF will decrease the application of chemical fertilizers, contributing to favorable and economically viable fodder systems. AMF can be important in enhancing sustainable and resilient agricultural practices by enhancing the fertility of soils as well as carbon storage (Niu et al., 2024).

35.8 Arbuscular mycorrhizal fungi application strategies in fodder crops

AMF inoculum can be of different types, such as spores, root pieces that are colonized, and granular or powdered. High-quality inoculum is needed to be applied successfully in fodder crops. Pot culture, trap culture, and in vitro are commonly used techniques of production. Pot culture is the host plants that are grown in sterilized soil or substrates in controlled conditions to form fungal spores and colonized roots. In trap culture, nonsterile soil is employed, and a chosen host plant is used to enhance the multiplication of the fungi in a more natural environment. The production, in vitro, of AMF can be performed in sterile root organ systems or bioreactors, allowing the production of an inoculum of very high purity and free of contaminants (Simsek & Cetin, 2026). The type of inoculum used and the type of production method are determined by the fodder crop of interest, the application area magnitude, storage, and resources at the disposal of the farmers or scientists. The correct choice of AMF species or strain also plays an important role, as not all of the fungi have the same colonization efficiency, nutrient transfer ability, and adaptation to soil and environmental factors.

35.9 Methods of arbuscular mycorrhizal fungi inoculation

Proper inoculation methods are required in the successful colonization of AMFs, and this depends on the type of fodder crops being cultivated under controlled conditions or in the field. Often, inoculation is done at the seed or seedling stage in greenhouse or nursery conditions to ensure early root colonization. AMF spores can be applied to seeds, inoculum can be added to the growth media, or inoculum can be placed in proximity to the emerging root system. In the field, AMF inoculation can be applied directly to planting furrows, mixed with the soil before sowing, or applied around the root zone when transplanting. The aspect of uniform distribution and close contact with roots is important to attain close colonization. The establishment and effectiveness of AMF are highly influenced by the environmental conditions, including soil moisture, soil temperature, soil pH, and soil nutrient status, and these factors need to be well controlled in inoculation (Salomon et al., 2022).

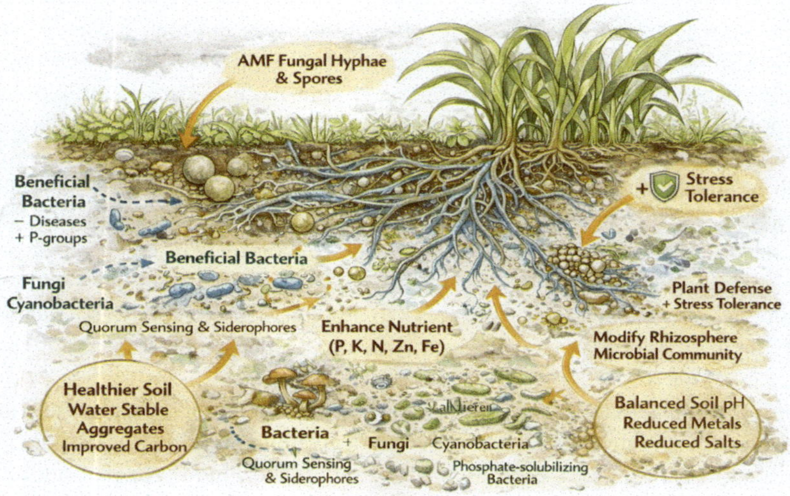

FIGURE 35.3 Effect of AMF on soil microbiomes and soil nutrient availability for fodder crop. *AMF*, Arbuscular mycorrhizal fungi.

35.10 Integration with fertilizer and organic amendments

Combining AMF with fertilizers and organic amendments to the soil boosts the work of fungi and increases the quality of fodder. Although AMF enhance nutrient uptake, phosphorus, which is applied in large amounts, may inhibit the growth of fungi. As such, there should be a balanced management of nutrients to aid the growth of plants as well as fungal activity. The organic amendments (compost, manure, biochar, and crop residues) are other sources of carbon, as well as enhancing soil structure, water-holding capacity, and microbial activities, which support AMF establishment. Integrity of AMF inoculation and prudent use of fertilizers and organic amendments achieve greater nutrient use efficiency, nutritional value of fodder, and sustainability in production without the extensive use of synthetic fertilizers (Hussain & Mehdi, 2026).

35.11 Crop-specific responses in major fodder species

Major fodder species vary greatly in how they react to different crop types, nutritional treatments, and maturation stages. The effect of AMF inoculation on different fodder crops varies according to the root architecture, host compatibility, and nutrient needs. Other legumes, such as beans, have been recorded to greatly increase their fixation of nitrogen and protein content when inoculated with AMF because of synergistic effects with nitrogen-fixing bacteria (Gorgia & Tsikou, 2025). In general, grasses are well placed to gain increased phosphorus levels, improved digestibility, and increased accumulation of biomass. Diverse AMF communities can be taken over by mixed forage systems, which promote complementary nutrient intake and overall improved fodder quality. Knowledge of these crop-specific responses is critical in ensuring that the AMF species, inoculation technique, and integration scheme selected will result in the greatest benefits in terms of yield, nutritional quality, and resistance to changing environmental situations (Sobrinho et al., 2025).

35.12 Arbuscular mycorrhizal fungi and stress tolerance in fodder crops

AMFs are also beneficial as they not only increase the nutrient uptake and fodder quality but also are significant in enabling fodder crops to withstand different environmental stresses (Fig. 35.3). The symbiotic relationship improves the resilience of plants to abiotic and biotic stresses to enable them to grow and be productive in unfavorable conditions. AMF can also help in making the fodder crop more resilient to different environmental stressors by reducing drought, salinity, heavy metal toxicity, and pathogen attacks, in addition to enhancing nutrient uptake and the quality of these fodder crops. The symbiotic relationship they have increases the resilience of the plant to abiotic and biotic stresses, enabling the plant to grow and be productive even in unfavorable conditions. AMFs help to reduce drought, salinity, heavy metal toxicity, and the attacks of pathogens to have more sustainable and resilient fodder production systems (Formenti et al., 2026).

35.13 Drought and salinity stress mitigation

AMFs have been known to make fodder crops highly drought- and salinity-resistant, which are two of the most constraining factors in the forage production in the globe. The large hyphae of AMF increase water uptake because they increase the area of root absorption to enable plants to access water at a greater depth of the soil mass. Moreover, AMF colonization enhances root hydraulic conductivity and stomatal activities to keep plants in water balance in the situation of limited soil moisture. In saline stress, AMF helps in ion homeostasis by selective absorption of the vital nutrients such as potassium and calcium, as well as reducing excessive absorption of sodium. This assists in the osmotic balance and minimizes ionic toxicity, which shields the structure of plant cells. It has been demonstrated that AMF-inoculated fodder crops have a high relative water content, enhanced leaf turgor, and better growth in drought and salinity, leading to continued biomass production and enhanced nutritional content of the forage (El Sabagh et al., 2025).

35.14 Heavy metal and soil contamination tolerance

AMF increases the tolerance of fodder crops to heavy metal-contaminated or other polluted soils. The fungal hyphae are also able to absorb and declare metals such as cadmium, lead, and nickel in the soil or in the buildings of the fungi, decreasing their availability to the roots of the plants. This defense mechanism restricts oxidative stress and damage to cells caused by metal in the host plant. Furthermore, AMF can induce antioxidant enzyme production as well as stress-related metabolites in plants, and this also increases the tolerance of the plant to the toxic soil. AMF help fodder crops to grow in inappropriate soils, enhancing nutrient uptake and physiological balance, increasing cultivation opportunities, and contributing to sustainable forage production in marginal or polluted soils (Ghadge & Trivedi, 2026).

35.15 Disease resistance and plant health promotion

AMF are useful in improving plant health and inhibiting soil-borne pathogens in fodder crops. The AMF colonization triggers systemic defense response in plants, such as the induction of enzymes associated with defense systems and secondary metabolites, which can mitigate pathogenic attack. The existence of AMF also changes the microbial community of the rhizosphere and encourages beneficial microorganisms, which compete or inhibit detrimental pathogens. All these effects lead to a decrease in the occurrence of disease and a general enhancement in the health of plants. The plants that are healthier and associated with AMF not only yield better biomass but also ensure a better quality of nutrition, which produces more digestible and nutritionally rich fodder for the livestock. AMF enhances and leads to more resistant fodder systems because of their increased ability to tolerate stress and resist diseases, which is observed to operate effectively in adverse environmental conditions (Sain et al., 2023).

35.16 Case studies and field evidence

The capability of AMF to improve fodder quality and productivity has been extensively researched, and experimental and field-based studies indicate its effectiveness. It has been shown that AMF inoculation has the potential to enhance nutrient uptake, forage yield, stress tolerance, and general plant health in most fodder crops. Such experiments, both in a controlled and field setting, bring into the limelight of AMF the possibilities of producing forage sustainably. Controlled experiments proved that AMF inoculation can help to increase the growth, nutrient status, and biomass of the forage legumes and grasses significantly. Indicatively, research done on alfalfa, clover, and ryegrass also reported root colonization, higher phosphorus and nitrogen levels, and higher crude protein levels in AMF-inoculated plants as opposed to control groups. On the same note, grasses such as maize, sorghum, and Napier grass had improved root growth, higher biomass buildup, and digestibility when linked to AMF. The results indicate that AMF can respond positively in quantity and quality of fodder when subjected to controlled experimental settings (Wu et al., 2024).

On-farm trials and studies in the field have proven the advantages of AMF. Even with unfavorable soil fertility or soil conditions that are sensitive to drought, inoculated fodder crops have been reported to yield better than noninoculated crops, have greater stress tolerance, and have higher nutritional value. Greater adoption of AMF inoculation, particularly when coupled with organic amendments or fewer chemicals, has resulted in productivity and cost savings. These experiments suggest that AMF can be easily incorporated into the present curdling systems and offer useful advantages to both smallholder and commercial farmers (Soti et al., 2023). Comparison of AMF-inoculated fodder crops and conventional practices shows the benefits of fungi symbiosis. Most of the time, plants treated using AMF record the same or better yields with reduced usage of chemical fertilizers. AMF inoculation has also increased fodder quality with better protein content, digestibility, and mineral composition in addition to better nutrient use efficiency. This shows that as far as AMF application is concerned, it is not only an alternative to traditional methods of fertilization, but also it is an economical approach to enhancing the production systems of fodders in terms of the multitude of productivity and environmental sustainability (Ghorui et al., 2025).

35.17 Environmental and economic benefits

The use of AMF as a fodder crop ingredient has a high environmental and economic benefit in terms of ensuring sustainable farming. Among the main positive changes in the environment, there is a decrease in the usage of chemical fertilizers. AMF increases the absorption efficiency of nutrients, especially phosphorus and nitrogen, which leads to fewer artificial fertilizers. AMF can ensure that nutrients are more available and retained in the soil, thereby reducing nutrient leaching, groundwater pollution, and greenhouse gas emissions caused by excessive use of fertilizers. Not only does this conserve soil and water resources, but it also works toward environmentally friendly and sustainable fodder production systems. From a financial perspective, AMF inoculation may increase the profitability of fodder production. An increase in biomass and reduced input costs by farmers is possible through better utilization of nutrients and enhanced growth in the plants, and it also leads to increased yields of forage and better quality. On-farm experiments have revealed that AMF-treated crops have a tendency to need fewer doses of fertilizers and less irrigation, and, as such, have led to direct cost savings. Moreover, better quality of fodder, such as better protein level and digestibility, enhances livestock performance, which leads to high milk and meat production, which increases farm income. The cost–benefit analysis shows that the economic payoffs of AMF application exceed the initial investment in the inoculum, and it is a feasible and sustainable method of both digital and commercial farm systems (Zaman et al., 2025).

Another aspect that AMF can do to promote climate-resilient and sustainable agriculture is by enhancing soil health, improving carbon sequestration, and making fodder crops more resilient to abiotic stresses like drought and salinity. AMF can enhance the long-term fertility and productivity of agricultural soils by enhancing their stability, increasing the content of organic matter, and maintaining microbial diversity. These ecological and environmental gains, together with the economic gains, point to the dual benefit of AMF use: it improves the quality and yield of fodder, as well as promotes sustainability, low-input, and climate-smart agricultural practices. In general the introduction of AMF in the fodder production systems is an effective and sustainable approach to addressing the increased need for livestock feed and securing the profitability of farms without damaging the natural resources (Le Heiget, 2026).

35.18 Challenges and limitations

Although AMFs have many advantages in enhancing the quality of fodder and ensuring sustainable use of fertile soil, there are several challenges and limitations that should be addressed to successfully employ AMF. Among the main limitations is the variability of AMF performance within various soils, climates, and types of crops. The soil pH, soil texture, soil moisture contents, soil temperature, and nutrient availability are all environmental factors that can greatly impact root colonization, fungal development, and effectiveness in nutrient transfer. As a result the advantages of AMF inoculation may not always be consistent when applied in the field, particularly in harsh or damaged soils. The other difficulty is the production, storage, and quality control of AMF inoculum. Viable spores and root material in colonized form need to be handled carefully, well-stored, and produced in a standard manner to maintain their viability. The low colonization rates and lower effectiveness may be caused by poor-quality inoculum or wrong methods of its application. Moreover, certain AMF strains and host fodder species must be compatible; not all fungi work equally well on all crops, and the selection of the right fungi strain to apply to a particular agroecological environment cannot be done without prior knowledge and research (Sakthivel et al., 2026).

Moreover, large-scale adoption can be restrained by the high start-up expenses of inoculum and the ignorance of farmers. Some of the farmers are not conversant with microbial-based solutions, or they might be unwilling to substitute the traditional fertilizers with bioinoculants. More extension services and practical guidelines are also necessary so that the proper application, combination with fertilization strategies, and monitoring of AMF effectiveness are ensured. Lastly, there are gaps in the research on the long-term effects of AMF on soil microbial communities, nutrient cycling, and fodder system sustainability, which shows that further studies and field experiments are required. To ensure that AMF in sustainable fodder production is fully realized, these challenges will be addressed by enhancing inoculum generation, training the farmers, and crop-differentiated approaches (Kumar, 2018).

35.19 Future perspectives and research directions

The idea of utilizing AMF in fodder production has enormous potential, and the development of technology needs further research, innovation, and association with the use of the latest technological means in agriculture. In further research, attention must be given to the construction of high-quality, crop-specific AMF inoculants that are robust to different environmental factors and can be applied constantly in different kinds of soil. Biotechnological improvements that include the use of molecular screening and selection of high-efficiency AMF strains using genomics can further increase the efficacy of inoculants and enable the creation of multistrain consortia in special fodder crops. Another potential opportunity is the integration of AMF and precision agriculture practices. With the help of soil mapping, nutrient monitoring, and remote sensing, farmers can strategically apply AMF inoculants to ensure they colonize the soil and absorb nutrients, as well as decrease the application of chemical fertilizers. Synergistic interactions between AMF and other useful microorganisms, such as nitrogen-fixing bacteria and phosphate-solubilizing microbes, should also be studied to form optimized microbial consortia to improve the fodder quality and stability under stressful environments (Qiao et al., 2026). Moreover, long-term field research is required to evaluate the effects of AMF on soil health, carbon sequestration, nutrient recycling, and fodder system sustainability. These studies will help understand how AMF will help in climate-smart and environmentally sustainable livestock production. The support of the policy, training of farmers, and awareness programs will also be decisive to encourage the uptake and adaptation of AMF technologies to the traditional farming systems. In general, more research, innovation, and interdisciplinary collaboration of scientists, farmers, and policymakers must be involved to realize the full potential of AMF to enhance the quality of fodder, livestock productivity, and sustainable farming practices in the world (Arif et al., 2025).

35.20 Conclusion

AMFs have been significantly involved in the improvement of the quality of fodder, uptake of nutrients, and sustaining livestock production systems. AMF improve the acquisition of phosphorus and nitrogen, the crude protein content and digestibility of the feeding of a diverse fodder crop, and biomass production by associating mutually beneficially with plant roots. In addition to nutritional benefits, AMF also have a positive effect on soil health by enhancing aggregation, water retention, carbon sequestration, and positive interactions with microbes. They also increase the resistance of plants to drought, salinity, heavy metal toxicity, and soil-borne pathogens, thus increasing the strength of fodder systems to environmental pressures. Although these are the merits, there are still problems such as fluctuating field performance, lack of inoculum quality, host-specific compatibility, and unawareness of the farmers. To maximally use AMF in sustainable agriculture, it is necessary to address these challenges with the help of research, technological innovation, training of farmers, and policy support. The application of AMF in fodder production can be used more productively and have a stronger adoption in the future through crop-specific inoculants, microbial consortia, incorporation of precision agriculture, and long-term field research. In general, AMFs provide a biologically viable and economical approach to enhancing the quality of fodder, livestock productivity, and environmental management and can be regarded as an important instrument for the forage and livestock systems in the future (Ghorui et al., 2025).

References

Arif, M., Shamshad, J., Khalid, F., Kuzyakov, Y., Younas, A., & Li, L. (2025). Biotechnological and sustainable approaches to climate-resilient agriculture. *Journal of Agronomy and Crop Science, 211*(4), e70105.

da Silva, G. A., de Assis, D. M. A., Sieverding, E., & Oehl, F. (2024). Four new families of arbuscular mycorrhizal fungi within the order *Glomerales*. *Taxonomy, 4*(4), 761–779.

El-Metwally, M. M., Mekawey, A. A. I., El-Halmouch, Y., & Naga, N. G. (2023). Symbiotic relationships with fungi: From mutualism to parasitism. In *Plant mycobiome: Diversity, interactions and uses* (pp. 375–413). Springer.

El Sabagh, A., Iqbal, M. A., Wasaya, A., Irfan, M., Rajendran, K., Zahid, R., Yasmin, H., Zahan, T., Younis, U., & Rehman, A. (2025). Strategies to enhance biomass production in forage crops under challenging environments. In *Forage crops in the bioenergy revolution: From fields to fuel* (pp. 157–182). Springer.

Elhady, A., & Hirt, H. (2025). Harnessing plant-soil-microbiome synergy for resilient desert restoration. *Iscience, 28*(11), 113748.

Erdaw, M. M. (2023). Fundamental factors reducing feed quality, efficiency, and carrying-over impacts: A review. *Turkish Journal of Agriculture-Food Science and Technology, 11*(8), 1422–1431.

Formenti, C., Mauromicale, G., Pandino, G., & Lombardo, S. (2026). Arbuscular mycorrhizal fungi mitigate crop multi-stresses under mediterranean climate: A systematic review. *Agronomy, 16*(1), 113.

Ghadge, S. A., & Trivedi, P. (2026). Evaluation on the biological aspect of plant, contaminant types, and application of phytoremediation for environmental and economic sustainability. *Journal of Advances in Biology & Biotechnology, 29*(1), 75–95.

Ghorui, M., Chowdhury, S., & Burla, S. (2025). Recent advances in the commercial formulation of arbuscular mycorrhizal inoculants. *Frontiers in Industrial Microbiology, 3*, 1553472.

Gorgia, P., & Tsikou, D. (2025). Tripartite symbiosis between legumes, arbuscular mycorrhizal fungi and nitrogen fixing rhizobia: Interactions and regulation. *Plant, Cell & Environment*, 1–19. doi:10.1111/pce.15341.

Heydarzadeh, S., Jalilian, J., Pirzad, A., Jamei, R., & Petrussa, E. (2022). Fodder value and physiological aspects of rainfed smooth vetch affected by biofertilizers and supplementary irrigation in an agri-silviculture system. *Agroforestry Systems, 96*(1), 221–232.

Hossain, M. E., Shahrukh, S., & Hossain, S. A. (2022). Chemical fertilizers and pesticides: Impacts on soil degradation, groundwater, and human health in Bangladesh. In *Environmental degradation: Challenges and strategies for mitigation* (pp. 63–92). Springer.

Hussain, M. A. A., & Mehdi, M. M. (2026). Impact of *organic amendments* on *soil structure*, fertility and *growth of* plants. Unique Scientific Publishers.

Kebede, T. G., Birhane, E., Ayimut, K.-M., & Egziabher, Y. G. (2024). Arbuscular mycorrhizal fungi increased biomass, nutritional value, and cochineal resistance of Opuntia ficus-indica plants. *BMC Plant Biology, 24*(1), 706.

Kumar, Pavan (2018). Biofertilizers: Harnessing Microbial Potential for Sustainable Agriculture. *Indian Journal of Life Sciences, 8*(1), 06–10.

Le Heiget, P. (2026). *The effect of nitrogen management on dual-purpose intermediate wheatgrass (Thinopyrum intermedium) stands in western Canada*. University of Manitoba.

Luo, F., Liu, W., Mi, W., Ma, X., Liu, K., Ju, Z., & Li, W. (2023). Legume-grass mixtures increase forage yield by improving soil quality in different ecological regions of the Qinghai-Tibet Plateau. *Frontiers in Plant Science, 14*, 1280771.

Niu, G., He, C., Mao, S., Chen, Z., Ma, Y., & Zhu, Y. (2024). Enhanced soil fertility and carbon sequestration in urban green spaces through the application of Fe-Modified biochar combined with plant growth-promoting bacteria. *Biology, 13*(8), 611.

Ponnampalam, E. N., Kiani, A., Santhiravel, S., Holman, B. W., Lauridsen, C., & Dunshea, F. R. (2022). The importance of dietary antioxidants on oxidative stress, meat and milk production, and their preservative aspects in farm animals: Antioxidant action, animal health, and product quality—Invited review. *Animals, 12*(23), 3279.

Prasad, R., Shivay, Y. S., & Kumar, D. (2016). Interactions of zinc with other nutrients in soils and plants: A Review. *Indian Journal of Fertilisers, 12*(5), 16–26.

Qiao, X., Yan, X., Dong, C., Tao, L., Aili, A., & Waheed, A. (2026). From microbiome collapse to recovery: A roadmap for microbiome-informed grassland restoration under global change. *Frontiers in Microbiology, 17*, 1741287.

Sain, S. K., Dewasi, H., & Singh, A. (2023). Combined application of effective *Trichoderma, Pseudomonas*, and arbuscular mycorrhiza spp. Reduced soil-borne diseases and boosted growth in cotton. *Egyptian Journal of Biological Pest Control, 33*(1), 94.

Sakthivel, A., Balasubramaniam, S., Eswaran, S. P., Ganeshan, S., Jesudass, J. S., Gnanajothi, K., Yin, C., Chiuta, N. E., Chang, S. W., & Balasubramani, R. (2026). Implication of plant growth-promoting microorganisms and nanofertilizer in millets: A review on growth promotion and crop improvement. *Cogent Food & Agriculture, 12*(1), 2617645.

Salomon, M., Demarmels, R., Watts-Williams, S., McLaughlin, M., Kafle, A., Ketelsen, C., Soupir, A., Bücking, H., Cavagnaro, T., & van der Heijden, M. G. (2022). Global evaluation of commercial arbuscular mycorrhizal inoculants under greenhouse and field conditions. *Applied Soil Ecology, 169*, 104225.

Shah, S., Khan, Y., Wang, M., & Zhang, T. (2026). Arbuscular mycorrhizal fungi as mediators of nitrogen, phosphorus, and carbon: Implications for plant growth and development. *Journal of Plant Nutrition, 49*, 687–707.

Simsek, G. K., & Cetin, A. K. (2026). Investigation of growth, protein, and astaxanthin amounts of chlorophyta *Chlorella vulgaris* and *Scenedesmus acutus* species inoculated in different culture media. *Biomass Conversion and Biorefinery, 16*(2), 57.

Singh, P., Pandey, V. K., Sultan, Z., Singh, R., & Dar, A. H. (2023). Classification, benefits, and applications of various anti-nutritional factors present in edible crops. *Journal of Agriculture and Food Research, 14*, 100902.

Sobrinho, R. L., de Oliveira, B. R., Zuffo, A. M., Filho, M. C. M. T., Filho, A. C. M., Zoz, T., Okla, M. K., Alaraidh, I. A., Alwasel, Y. A., & Alhaj Hamoud, Y. (2025). Modeling wheat productivity under elevated CO_2 using fuzzy logic and mycorrhizal inoculation. *BMC Plant Biology, 25*(1), 756.

Son, Y., Martínez, C. E., & Kao-Kniffin, J. (2024). Three important roles and chemical properties of glomalin-related soil protein. *Frontiers in Soil Science, 4*, 1418072.

Soti, P., Kariyat, R., & Racelis, A. (2023). Effective farm management promotes native AMF and benefits organic farming systems. *Agriculture, Ecosystems & Environment, 342*, 108240.

Tulu, D., Gadissa, S., Hundessa, F., & Kebede, E. (2023). Contribution of climate-smart forage and fodder production for sustainable livestock production and environment: Lessons and challenges from Ethiopia. *Advances in Agriculture, 2023*(1), 8067776.

Uwamungu, J. Y., Shi, G., Wang, Y., Paliwal, A., Jadhav, R. R., & Wani, A. W. (2022). Arbuscular mycorrhizal fungi (AMF) for sustainable soil and plant health. In *Microbial and biotechnological interventions in bioremediation and phytoremediation* (pp. 135–152). Springer.

Visen, S. S., Thakur, D., & Atluri, S. (2024). Sustainable management of soil-borne diseases of horticultural crops. In *Advances in postharvest and analytical technology of horticulture crops* (pp. 349–363). Springer.

Wahab, A., Muhammad, M., Munir, A., Abdi, G., Zaman, W., Ayaz, A., Khizar, C., & Reddy, S. P. P. (2023). Role of arbuscular mycorrhizal fungi in regulating growth, enhancing productivity, and potentially influencing ecosystems under abiotic and biotic stresses. *Plants, 12*(17), 3102.

Wu, Y., Chen, C., & Wang, G. (2024). Inoculation with arbuscular mycorrhizal fungi improves plant biomass and nitrogen and phosphorus nutrients: A meta-analysis. *BMC Plant Biology, 24*(1), 960.

Zaman, F., Ali, A., Khattak, W., Khan, H., Anwar, Z., Rahimi, M., & Chen, F.-S. (2025). Diversity of arbuscular mycorrhizal fungi (AMF) and their role in nutrient recycling and availability in the subtropical forest ecosystem of China: A review. *Eurasian Soil Science, 58*(11), 145.

Chapter 36

Mitigating the impact of smog on crops through the application of arbuscular mycorrhizal fungi

Hafiz Muhammad Rashad Javeed[a], Rafi Qamar[b], Muhammad Wakeel[a], Imran Manzoor[a], Misbha Hina[a] and Ayman EL Sabagh[c]

[a] *Department of Environmental Sciences, COMSATS University Islamabad, Vehari Camous, Vehari, Pakistan,* [b] *Department of Agronomy, College of Agriculture, University of Sargodha, Sargodha, Pakistan,* [c] *Department of Field Crops, Faculty of Agriculture, Siirt University, Siirt, Turkey*

36.1 Introduction

Air pollution is no longer a matter of health in the city, but it is also a matter of production in the countryside. Numerous food production areas of the globe are downwind of fast-growing urban-industrial belts where sources of nitrogen oxides (NO_2), volatile organic compounds (VOCs), sulfur dioxide (SO_2), and ammonia (NH_3) develop to foster the development of photochemical oxidants and fine particles. Smog events often coexist with heat and drought stress, which form compound stress conditions where crops must continue being productive and face high oxidative load. One of the most regularly reported phytotoxic agents against crops is ground-level ozone, one of the smog components. System-wide studies of wheat-growing systems at the Indo-Gangetic Plains have measured significant amounts of losses attributed to ozone and found viable adaptations that allow reducing exposure during the sensitive growth periods (Dewan & Lakhani, 2024). On a larger level the flux-based modeling highlights the fact that the risk of ozone is nonuniformly distributed and can also exist in the form of hotspots even when overall risk decreases (Guaita et al., 2025).

Fine particulate matter (PM) contributes a second strain pathway. The deposited particles decrease the light interception, disrupt the stomatal activity, and can transport cocontaminants, which impose biochemical stress reactions (Roy et al., 2024). PM also changes the habitat of the leaf surface (the phyllosphere) that could modify the microbial community structure and interactions with diseases (Xu et al., 2024). Smog, which is often defined by higher ground-level ozone (O_3), fine particulate matter ($PM_{2.5}$), secondary organic aerosol, and other reactive oxidants, is an increasing menace to the productivity of crops, food quality, and stability of agroecosystems (Fig. 36.1). Smog harms crops through inhibition of photosynthesis and advanced senescence, impaired stomatal physiology, and augmentation of oxidative and nitrosative stress. Arbuscular mycorrhizal fungi (AMFs) are common root symbionts of many crop species and may increase stress tolerance by increasing mineral nutrition, water relations, antioxidant potential, and rhizosphere regulation. Moreover, synthesis of the evidence on the topic of smog formation and exposure routes, physiological and yield effects of ozone and particulate matter, and how AMF can mitigate smog damage. This synthesis is translated into practice-based recommendations on how AMF friendly agronomy and well-tested inoculation may be incorporated into the smog-prone production systems and the importance of dose-dependent metrics of ozone.

Concurrently, agriculture is faced with the need to minimize the use of synthetic inputs and enhance climate variability resiliency. AMFs provide a natural pathway to support plant performance through nutrient acquisition, increase water relations, and prepare the defense networks in response to stressors (Sun et al., 2023; Wahab et al., 2023). However, the AMF-based interventions' performance heavily relies on context, such as the genotype of crops, soil fertility, and the quality of inoculant formulations (Ghorui et al., 2025).

Advancing Sustainable Agriculture Through Arbuscular Mycorrhizal Fungi. DOI: https://doi.org/10.1016/B978-0-443-45130-0.00018-7

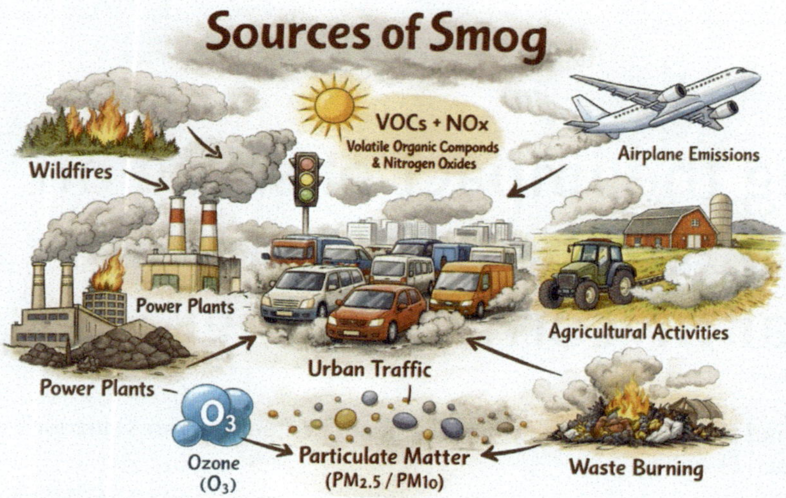

FIGURE 36.1 Sources of smog. *VOCs*, Volatile organic compounds.

36.2 Agroecosystem smog: Formation, composition, and transport

Photochemical smog is caused when the primary emissions combine with sunlight to produce secondary pollutants. The canonical pathway refers to the process of photolysis of nitrogen dioxide (NO_2), which is the formation of atomic oxygen, which is mixed with O_2 to form ozone (O_3). Under conditions of the presence of VOC oxidation chains the peroxyl radicals can oxidize NO to NO_2 in the presence of oxygen and do not consume O_3, maintaining net ozone generation. As these radical cycles are sensitive to the relative abundance of NO_x and VOCs, the formation of ozone can be either VOC-limited or NO_x-limited, which makes the control strategies regionally sensitive (An et al., 2024). Sometimes, secondary aerosol is formed simultaneously: oxidation products of VOCs are a source of secondary organic aerosol, and SO_2 and NO[2] chemistry cause the formation of sulfate and nitrate aerosols. The emissions of ammonia in agriculture may counteract any acidic species, enhancing the formation of ammonium sulfate/nitrate and the formation of the haze. In this fashion, agriculture is a receiver and, in other instances, a source of smog chemistry—especially by way of NH3 emissions and unplanned biomass burning.

Regarding crops the most pertinent smog constituents include those which (1) enter the leaves via stomata (e.g., ozone, NO_2), (2) settle on the leaves (e.g., $PM_{2.5}/PM_{10}$), or (3) alter the radiation environment and the microclimate in the canopy. Ozone is the leading phytotoxic oxidant in most areas, and physical and chemical pressure is brought by fine particles and other cocontaminants (metals and organic compounds) (Roy et al., 2024). Notably, mixtures of smog depend on the season and the region. Photochemical ozone and winter haze episodes with a high content of secondary inorganic aerosol are the primary sources of air-quality issues in some industrial corridors and portions of the atmosphere, respectively. In crops the two trends are important: ozone peaks can occur during flowering and grain filling, and haze can slow the carbon gain daily and delay phenology.

Smog does not exist only in cities. Ozone, which is secondary and relatively long-lived (hours to days), is capable of being transported hundreds of kilometers. Hence low local emissions do not result in low ozone layers on rural agricultural lands. One of the causes of crop losses in the regions that are not seen as contaminated because of traditional indicators of pollution is this spatial decoupling. Seasonality is crucial. The production of the ozone layer is normally robust in high sunlight and in hot weather. The highest levels of smog can be found in the premonsoon season or summer in South Asia and in summer in most temperate areas. Agrimeters should match crop cycles and avoid the times of the day when the ozone is the strongest, hence a realistic adaptation lever in this context (Dewan & Lakhani, 2024).

36.3 Exposure pathways to plants

Exposure to smog is directly related to ozone concentration, which is in turn reacted to by the plant to determine the amount of ozone that really penetrates the leaf (Fig. 36.2). Dose is determined by the stomatal conductance, which is regulated by radiation, temperature, vapor pressure deficit (VPD), and soil moisture, as well as plant phenology. This difference is the reason why the concentration-based indices may be inaccurate at depicting risk under the conditions of part-opening stomata

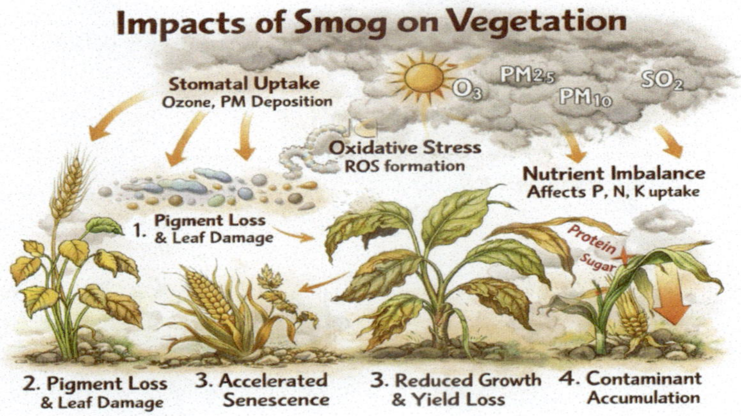

FIGURE 36.2 Impact of smog on vegetation. *ROS*, Reactive oxygen species.

and the reason why stomatal flux and phytotoxic ozone dose (POD) measures are used (Mao et al., 2024; UNECE, 2023). An analysis of methods and techniques demonstrating the superiority of one concentration-based cumulative index over the approaches based on POD reveals that, in the case of stomatal control in question, long-term dynamics and management interpretations may not coincide (Mao et al., 2024). In agronomic mitigation, dose-based framing is especially significant because many interventions (irrigation timing, cultivar properties, and AMF influences) have a physiological mechanism that works via stomata.

Particulate matter usually deposits on crops, unlike the case of ozone. Deposition is influenced by the size of particles, wind speed, the structure of the canopy, and leaf microtraits such as trichomes, wax, grooves, and stomata structure. PM deposits may be resuspended, washed away during rain or irrigation, and are long-lasting. Recently, it has been proposed that not only the surface characteristics of plants control PM retention but also that PM-associated phyllosphere patterns are correlated with those of phyllosphere microbial communities and that PM can reorganize the leaf habitat (Xu et al., 2024). PM exposure is best defined as assessing the leaf load (mass per area), fraction of the particles, and chemical composition, where feasible. Leaf-load measurements allow associating pollution events with vegetation results (such as a slowing down of the photosynthetic rate or chlorophyll level), even in the absence of detailed chemistry (Roy et al., 2024).

Smog usually takes place during stagnant atmospheres and extreme temperatures, which also increases VPD. High VPD can decrease stomatal conductance and hence reduce the dose of ozone, and on the other hand, it also amplifies heat stress. It may decrease carbon assimilation through stomatal restriction. PM-driven haze could cut the radiation and lower the leaf temperature marginally, but it also slows down photosynthesis. The overall effect on a crop is based on the prevailing limitations and their timing of occurrence throughout the season. As AMF could change the water status of plants and their stomatal behaviors, simulating the effects of smog necessitates a combination of atmospheric measurements with the data on plant hydraulic and gas-exchange (Mao et al., 2024; Wahab et al., 2023).

36.4 Smog and crop yield

Ozone penetrates via stomata and quickly reacts in the apoplast and produces reactive oxygen species (ROS). Plants respond by turning on defenses, antioxidant defenses, and pathways, as well as repair mechanisms. Oxidative injury (lipid peroxidation and dysfunctional membrane activity) ensues when the rate of ROS formation exceeds the rate of detoxification. There are consequences such as the decrease of net photosynthesis, carbon allocation, and senescence acceleration (Guaita et al., 2025). On a canopy level the effects of ozone damage can decrease the leaf area period and cut the effective photosynthetic period short. Reduced assimilate supply to growing grains, flowers, or fruit and changed phenology, which breaks the source-sink balance, are common causes of reduced yield. A wheat regional evaluation of the Indo-Gangetic Plains explains the process and measures the effects on yields and suggests ways of mitigating the experience by changing the date of planting (Dewan & Lakhani, 2024).

Ozone layers reduce carbon availability or hasten flag leaf senescence; reproductive tissues are likely to suffer. Moreover, it may be in the form of shortening of grain filling time and decreased thousand-kernel weight in cereals. Stress signaling and carbohydrate limitation Stress signaling and carbohydrate limitation may reduce flower retention or fruit set in legumes

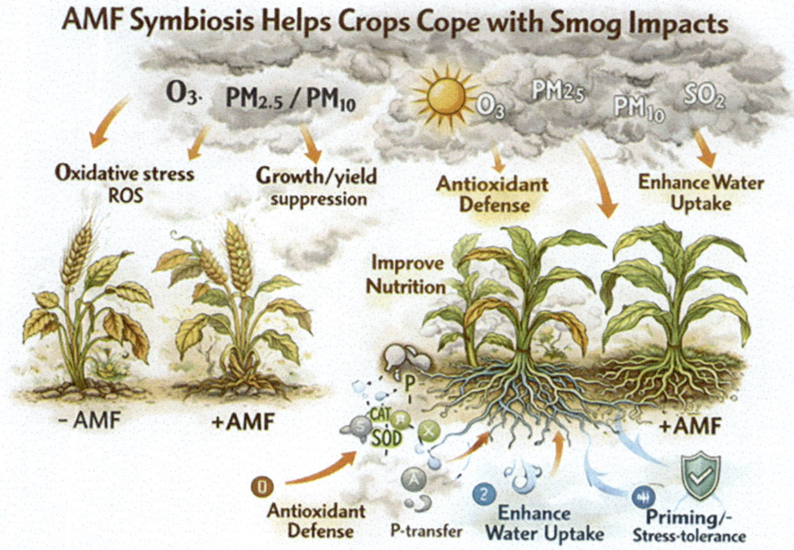

FIGURE 36.3 Arbuscular mycorrhizal fungi symbiosis boosts vegetation resilience against smog. *AMF*, Arbuscular mycorrhizal fungi; *ROS*, reactive oxygen species.

and horticultural crops through the action of ozone. Despite the variation in the intensity of the responses depending on the genotype and the environment the repeated theme is that the presence of ozone changes the plant to defend and repair, as opposed to growth. According to flux-based modeling, risk is not homogeneous. However, the overall risk on a global level decreases in certain conditions; there can still be regional hotspots, and regional interventions can be essential (Guaita et al., 2025).

The effects of PM begin on the surface of leaves. The coated layer of particles has the potential to weaken the light penetration, alter the optical characteristics of leaves, and increase the boundary layer. Particles may also block stomata pores or change the cuticular permeability that might influence transpiration and gaseous exchange. Chemically, PM has the potential to transport metals and organic pollutants that cause oxidative stress reactions. The overall assessment of the effect of PM on plants encompasses pigmentation alterations, enzyme functions, metabolite rates, and transcriptomic reactions (Roy et al., 2024). In addition to physiology, PM may cause plant–microbe interactions on leaves. At the Ulmus the PM load, leaf functional traits, and the microbial community of the Phytotransform show the relationships between pollution and microbial communities, which indicates that pollution can indeed alter microbial communities (Xu et al., 2024). In the case of crops, this puts the potential that PM may indirectly amplify or inhibit disease pressure based on the microbes' preferred and the modulation of host defenses.

Smog may also affect yield and even marketable quality and nutritional profiles. Oxidative stress caused by ozone has the potential to modify the levels of antioxidants, phenolics, and protein structure. In contrast, PM deposition can decrease aesthetic value (e.g., dusting is visible) and increase food safety concerns when particles contain toxic metals or persistent organics (Roy et al., 2024). As edible tissues differ in terms of exposure (e.g., leafy vegetables versus enclosed grains), risk management needs to be crop-specific. Protected cultivation and postharvest washing of leafy vegetables, and yield reduction and quality changes because of ozone layers are the main issues in high-PM environments when it comes to cereals (Dewan & Lakhani, 2024). Climate extremes often go hand in hand with smog. Heat waves may enhance ozone formation and also tend to make crops more stressed, and drought may modify stomatal behavior and also may decrease the dose of ozone and increase vulnerability to other stresses. The management needs to consider, however, compound stresses and not individual pollutants. In addition, it is one of the primary reasons why biological resilience mechanisms, including those that are affected by AMF, are of interest (Wahab et al., 2023).

36.5 Arbuscular mycorrhizal fungi: Functionalities with respect to smog resilience

AMF are compulsory symbionts that proliferate on the roots of plants and develop arbuscules, in which mineral nutrients are exchanged with plant carbon. The external hyphal network expands to reach the soil, which enhances the availability of immobile nutrients (especially phosphorus) and water (Fig. 36.3). Most major crops, such as wheat, maize, rice, legumes, and numerous horticultural species, can be colonized by AM symbioses. However, the level of colonization and benefits depend on the genotype and the soil conditions (Wahab et al., 2023). The AMF relates to smog because it causes injury through three overlaps: (1) smog may result in oxidative stress and metabolic perturbation; (2) smog may impair root activity and nutrient condition; and (3) smog may be accompanied by drought/heat, which AMF can buffer. Therefore AMFs are not directly antismog agents; they are resiliency partners, which can make one less vulnerable to smog-initiated injury pathways.

It has been reviewed that AMF can prime antioxidant defenses and regulate hormone signaling (Absasic Acid (ABA), jasmonates, salicylic acid), which respond faster or stronger to stress when subjected to challenges of plants (Wahab et al., 2023). Nonetheless, there is also the AMF carbon charge: the facility will have to feed fungal partners with photosynthate. The cost of carbon may be less than the net benefit in extreme conditions of ozone or haze that are very depressing to photosynthesis unless AMF can ameliorate nutrient/water conditions sufficiently or decrease oxidative damage. Thus net effects are determined by the difference between the benefit (stress buffering) and cost (carbon allocation). AMF inoculants are increasingly being commercialized; however, recent reviews indicate that there are issues in the development of formulations, shelf life, propagule viability, and performance consistency across soils (Ghorui et al., 2025). Moreover, it is especially applicable in the mitigation of smog since farmers might want to have quick solutions, but the biological inputs need to be checked and adjusted to the management system.

36.6 Smog tolerance mechanisms with arbuscular mycorrhizal fungi

Ozone and PM tend to be combined with oxidative stress. The plants colonized with AMFs tend to exhibit high activities of antioxidant enzymes and redox homeostasis that may suppress lipid peroxidation and preserve photosynthetic activities. AM symbiosis alleviated ozone damage in a tolerant clone of poplar through antioxidant activity of the enzymes, which goes together with the connection that mycorrhiza is connected to redox protection when ozone strikes (Yin et al., 2023). It can be assumed that this mechanism applies to crops since ozone damage is, to a large extent, an oxidative event. The same study, however, also discovered that there was a small benefit of AMF with a sensitive clone of ozone, implying genotype dependence (Yin et al., 2023). In the context of crop breeding and management, it means that AMF might enhance the existing tolerance instead of developing tolerance.

Stomatal conductance is dependent on the dose of ozone. AMF can also affect stomata indirectly through enhancing the water status of plants, as well as by changing hormonal signals. These adaptations can go both ways; the increased hydration can keep stomata open, which might permit more uptake of ozone layers even in the context of supporting growth. In alfalfa, AMF promoted stomatal conductance by enhancing stomatal density, and this may be one of the reasons why alfalfa is more susceptible to ozone; however, ozone damage was reduced when AMF was used in combination with the antiozonant ethylenediurea (EDU) (Yin et al., 2024). The management implication is that dose/flux measures should be used to consider the AMF-based strategies, because dose/flux measures, as opposed to ozone concentration, should be used in systems that are irrigated, where stomata are kept open when ozone is at its maximum (Mao et al., 2024; UNECE, 2023).

AMF enhances the absorption of phosphorus and micronutrients that facilitate chlorophyll production, enzymes of photosynthesis, and antioxidant mechanisms. When it is smoggy, nutrient adequacy may be preserved to guarantee that plants do not experience repair processes and prevent nutrient stress. The research on the application of AMF as microbial biostimulants highlights unanimous growth in nutrient conditions and productivity under different conditions (Sun et al., 2023; Wahab et al., 2023). In situations of PM-induced light attenuation, the nutrient-sufficient vegetation might be more compensated by physiological plasticity, which can be a change in chlorophyll content or a redistribution of nitrogen to photosynthetic proteins. Although direct experiments that can make a connection between AMF and a yield loss caused by PM are few, the overlap in the mechanisms can justify specific studies. Heat and drought are common together with smog episodes. The AMF can enhance efficient root exploration and enhance the water status of plants that stabilize photosynthesis and alleviate secondary stress loads. The canopy temperature can also be lowered through transpiration, which will help relieve heat stress in the case of better hydration. Nevertheless, better water conditions can result in the openness of stomata and higher doses of ozone; therefore the benefits of water relations should be viewed against the backdrop of the ozone flux (Mao et al., 2024; Yin et al., 2024).

36.7 Rhizosphere immobilization and soil deposition

PM deposition causes soil contamination by transporting metals and other pollutants by both dry and wet deposition. Controlling metal stress is possible through AMF by controlling the chemistry of the rhizosphere, binding metals in fungal structures, and supporting the nutritional needs of plants. The systematic review and metaanalysis assess the effectiveness of AMF in the remediation of toxic metals in mine-impacted soils, and it demonstrates that AMF can be used to support plant growth and nutrient uptake under metal stresses (Banerjee et al., 2024). Although my soils are not in a position of cropland deposition, the same processes, such as the metals' immobilization and the enhanced physiological status of plants, might be applicable in the case of smog-derived deposition, increasing the metal load in the soils. AMF friendly management may help in ensuring health and safety in food in the long run in smog-prone areas, where metal deposition is known to occur. Controlled exposure studies give a mechanistic understanding in the most precise way. AMF decreased the destruction of ozone in an ozone-tolerant clone of poplar, and not an ozone-sensitive one, by increasing the activity of antioxidant enzymes (Yin et al., 2023). AMF stimulated stomatal conductance and might have raised the ozone sensitivity, but AMF + EDU decreased the ozone harm, indicating the prospects of interventions (Yin et al., 2024). These papers also demonstrate the importance of including POD when making an assessment instead of assessing concentration alone (Mao et al., 2024; UNECE, 2023).

Physiological damage is converted into production results in the assessments of crop losses, which are regional. In the case of wheat in the Indo-Gangetic Plains, ozone pollution has been considered one of the key limitations, and it has been offered that a realistic measure to mitigate it is the adjustment of planting time (Dewan & Lakhani, 2024). On a global scale an assessment based on flux estimates shows that the risk of ozone to wheat is averaging less in certain future climatic conditions. However, hotspots of risk still exist, supporting the notion of specific adaptation to high-risk areas (Guaita et al., 2025). PM-centered plant science has caught up very fast, with general syntheses of the impact of physical deposition, biochemical responses, and the environmental biomonitoring implications (Roy et al., 2024).

Nevertheless, there is still limited direct evidence of AMF alleviating the damage of crops caused by PM. The existing literature substantiates plausibility: AMF enhances the antioxidant activity and nutrition, which applies to the oxidative stress caused by PM and diminished photosynthesis. However, there is a need to undertake strong field experiments that will relate AMF interventions to yield improvements associated with PM. A new direction is to study the PM impacts on the phyllosphere microbial communities. There was a correlation between PM retention characteristics and leaf microbial communities in Ulmus, and this indicates that pollution can restructure microbial ecology on the leaf (Xu et al., 2024). In the case of corresponding dynamics of crops, AMF may have an indirect effect on results because of systemic signaling and enhanced host vigor.

36.8 Mitigation of smog with arbuscular mycorrhizal fungi

Regional photochemistry cannot be directly controlled by farmers. Still, they can minimize the vulnerability of crops to it by enhancing plant resilience and changing management to avoid the highest exposure. The strategies based on AMF can be discussed as a package of measures that involve such aspects as alignment of the crop calendar, the choice of cultivar, soil-health management, and, in selected cases, precautionary steps in case of the active ozone layer conditions. Where ozone peaks are seasonal and predictable, the time of planting can be changed so that there is no coincidence with the vulnerable stages. In the case of wheat systems in the Indo-Gangetic Plains, it has been suggested that to minimize the losses in yields that are attributed to ozone, it is better to plant at an earlier time (Dewan & Lakhani, 2024). Calibrating to such calendar effects can be an addition to the AMF strategies because they will decrease the intensity of stress that should be buffered by plants. In numerous farms the indigenous AMF communities already exist. The benefits that these communities can deliver can be enhanced by the management through host continuity and by ensuring that practices that disrupt hyphae networks are minimized. Despite situation-specific outcomes the AMF friendly practices are consistent with the general soil-health objectives that are prioritized in recent AMF reviews (Sun et al., 2023; Wahab et al., 2023).

1. Less soil disturbance wherever possible to retain hyphae networks.
2. Diversified rotations with cover crops.
3. Use cover crops to keep AMF alive.
4. Use rotations with living roots.
5. Moderate phosphorus intakes—do not take in excessive amounts, and inhibit symbiosis—maintain adequate micronutrients.
6. Control irrigation to keep an intermediate water level at the highest levels of ozone; analyze benefits with the use of ozone dose scales, where available (Mao et al., 2024).

36.9 Inoculation

Inoculation can be applied in cases where the native AMF has been lost (e.g., because of fumigation, long fallow, or abusive disturbance) or where fast establishment is necessary. Recent commercialization reviews, however, indicate inconsistency

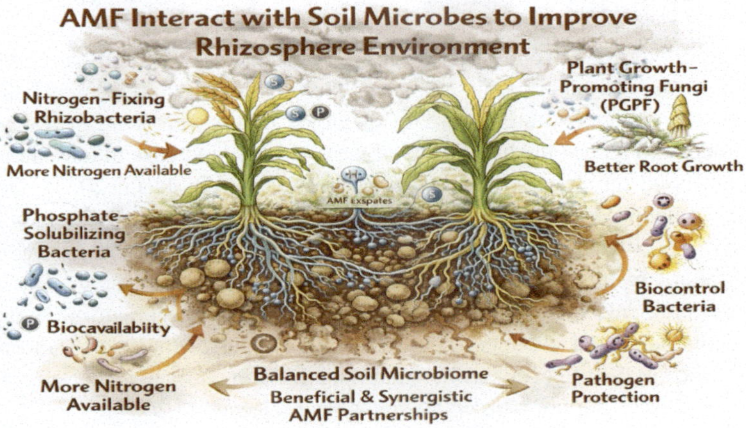

FIGURE 36.4 Improvement in rhizosphere environment with interaction of AMF and soil animals. *AMF*, Arbuscular mycorrhizal fungi.

in quality of formulation, propagule viability, and field performance and call for better standards and quality control (Ghorui et al., 2025).

Determine pretreatment colonization: at the early vegetative stage, sample root to obtain indications of the natural AMF colonization.

Determine the soil fertility: such high phosphorus may inhibit the AMF benefit; ensure that the P management is adjusted first.

Choose tested inoculants: ladies prefer tests that have definite specifics of the species, have been culturally active, and have been working in similar soils (Ghorui et al., 2025).

Inoculate at the right time: inoculate with sowing/transplanting to ensure the greatest contact of inoculated roots; do not apply fungicides that could be harmful to inoculum unless compatibility is known.

Confirm colonization: repeat the sample of roots 4–8 weeks after inoculation. Only with the confirmation of colonization can the yields be interpreted.

As ozone and PM have varying mechanisms of action, joint approaches frequently prove to be rational. Data on alfalfa shows that AMF can be coapplied with an antiozonate (EDU) to mitigate the impact of ozone (Yin et al., 2024). Likewise, we can slow down the calendar effect and select a cultivar, which will decrease the baseline risk (Dewan & Lakhani, 2024). Select genotypes that have been reported as being ozone-tolerant; AMF can give the highest net benefit in ozone-tolerant genotypes (Yin et al., 2023). Implement protection measures at appropriate times, that is, when there is a high ozone in the crop of high value, and at high-value crops, consider the interactions between AMF (Yin et al., 2024). In the case of PM-prone systems, leafy vegetables should be protected by cultivation or physical barriers; the soil and vegetative health should be maintained to counter the effects on physiology (Roy et al., 2024).

Interpretation: In case no increase in colonization is observed in treatment plots, do not assume AMF failed, but rather assume that in such circumstances, the intervention failed in establishing AMF and change the management (Ghorui et al., 2025). The concentration of ozone is not enough to examine biological interventions that affect stomata. The metrics of flux/dose, like POD, clearly consider the conductance of stomata and drivers of the environment. UNECE principles offer standardized methods of assessing the effects of ozone on vegetation by using flux (UNECE, 2023). The review of methodology indicates that concentration-based indices and POD can exhibit divergent long-term behaviors, which supports the argument that the assessment of interventions that can alter gas exchange, such as AMF, requires the measurement of stomatal behavior and microclimate (Mao et al., 2024) (Fig. 36.4).

In the case of PM the measurements of leaf-level deposition are highly informative. In areas where the laboratory capacity is lacking, rudimentary leaf wash-off methods can be used to estimate mass per area deposited. More sophisticated methods are the microscopic counting of particles and chemical profiling. Plant PM reviews are methodologically oriented and contain general pitfalls, for example, confounding because of rainfall wash-off and resuspension (Roy et al., 2024). The percentage of colonization is required, but not enough. Functional indicators could be the phosphorus condition of the plants, water use efficiency proxy, and antioxidant enzyme activity. As the inoculant preparations may differ in their efficacy, it is crucial to verify the colonization and functional consequences (Ghorui et al., 2025).

36.10 Future research thrust

Although mechanistic plausibility is high, to convert AMF into a dependable way of mitigating smog, several gaps need to be addressed. Priority actions include field experimentation in hot spots of smog: focus on cereal belts and peri-urban agriculture where ozone/PM exposure is pronounced and sensitivity of yields is reported (Dewan & Lakhani, 2024; Guaita et al., 2025). Genotype × AMF under ozone: test the hypothesis of the consistent increase in tolerance of AMF in ozone-sensitive and ozone-tolerant cultivars based on clone-specific findings in poplar (Yin et al., 2023). Dose-based testing of effects of AMF on ozone: record stomatal conductance and calculate POD to identify the changes in the efficacy of AMF on the effective dose of ozone; the effect is an increase or decrease (Mao et al., 2024; UNECE, 2023). PM-phyllosphere crop: generalize trait-microbiome results to crops and evaluate PM deposition disease/quality outcomes (Roy et al., 2024; Xu et al., 2024). Inoculant requirements and farmer decision-making instruments: create the protocols of simple verifications (colonization + functional markers) and enhance the consistency of the formulation (Ghorui et al., 2025). Long-term deposition and soil health: measure the effects of soil contaminant burden within each stressor because of smog deposition and the effects of strategies that AMF supports on uptake risks (Banerjee et al., 2024).

36.11 Conclusion

The primary potential impact of smog on crops is through oxidative damage of crops by ozone and deposition of particles, which suppresses light and interferes with the operation of leaves. Recent experiments show that AMF can alleviate ozone injury of certain host genotypes by enhancing antioxidants, and this effect is dependent upon the situation. It can be influenced by stomatal regulation of ozone dose. PM-centered literature records intensive physiological and ecological effects on leaves and points towards the approaches of microtrait and microbiome by which microbiome and AMF-sponsored resilience might cooperate (Roy et al., 2024; Xu et al., 2024). Nevertheless, the direct field evidence of the AMF-mediated damage of the PM injury is scarce.

In practice the AMF strategies must be integrated into the overall management: the crops calendar must be synchronized to prevent the occurrence of high ozone levels when possible, and the soil hands-ons must be respected regarding the native AMF, and inoculation must be thought out with the consideration of the formulation quality and colonization check (Dewan & Lakhani, 2024; Ghorui et al., 2025). Finally, evaluation should follow dose-based ozone metrics and field-relevant PM measures that will allow verifying that the improvement achieved is considered to be a real reduction in smog injuries but not potential confounding growth effects (Mao et al., 2024; UNECE, 2023).

References

An, T., Li, J., Lin, Q., & Li, G. (2024). Ozone formation potential related to the release of volatile organic compounds (VOCs) and nitrogen oxides (NO_x) from a typical industrial park in the Pearl River Delta. *Environmental Science: Atmospheres, 4*(11). https://doi.org/10.1039/D4EA00091A.

Banerjee, S., Mandal, J., Sarkar, D., Datta, R., & Bhattacharyya, P. (2024). A review and meta-analysis of the efficacy of arbuscular mycorrhizal fungi in remediating toxic metals in mine-affected soils. *Frontiers in Environmental Science, 12*, 1532169. https://doi.org/10.3389/fenvs.2024.1532169.

Dewan, S., & Lakhani, A. (2024). Ambient ozone pollution over the Indo-Gangetic Plains: Mechanism and impact on crop yield with mitigation strategies. *Science of the Total Environment, 945*, 173820. https://doi.org/10.1016/j.scitotenv.2024.173820.

Ghorui, M., Chowdhury, S., & Burla, S. (2025). Recent advances in the commercial formulation of arbuscular mycorrhizal inoculants. *Frontiers in Industrial Microbiology, 3*, 1553472. https://doi.org/10.3389/finmi.2025.1553472.

Guaita, P. R., Marzuoli, R., Zhang, L., Turnock, S., Koren, G., Wild, O., Crippa, P., & Gerosa, G. (2025). Global flux-based assessment reveals declining ozone risk for wheat in future climate change scenarios. *Global Change Biology, 31*(12), e70643. https://doi.org/10.1111/gcb.70643.

Mao, H., Felker-Quinn, E., Sive, B., Zhang, L., Ye, Z., & Fang, H. (2024). Examining indicators and methods for quantifying ozone exposure to vegetation. *Atmospheric Environment, 316*, 120195. https://doi.org/10.1016/j.atmosenv.2023.120195.

Roy, A., Mandal, M., Das, S., Popek, R., Rakwal, R., Agrawal, G. K., Awasthi, A., & Sarkar, A. (2024). Particulate matter pollutants in plants: A comprehensive review. *Science of the Total Environment, 914*, 169763. https://doi.org/10.1016/j.scitotenv.2023.169763.

Sun, W., Shahrajabian, M. H., Cheng, Q., & Qi, M. (2023). The application of arbuscular mycorrhizal fungi as microbial biostimulants for agricultural sustainability. *Plants, 12*(17), 3101. https://doi.org/10.3390/plants12173101.

UNECE. (2023). *Guidance on flux-based critical levels for ozone effects on vegetation (UNECE CLRTAP ICP Vegetation; document ECE/EB.AIR/GE.1/2023/INF.10)*. United Nations Economic Commission for Europe.

Wahab, A., Muhammad, M., Munir, A., Abdi, G., Zaman, W., Ayaz, A., Khizar, C., & Reddy, S. P. P. (2023). Role of arbuscular mycorrhizal fungi in regulating growth, enhancing productivity, and potentially influencing ecosystems under abiotic and biotic stresses. *Plants, 12*(17), 3102. https://doi.org/10.3390/plants12173102.

Xu, L., Chen, Y., Li, J., Wang, X., Zhang, J., & Liu, S. (2024). The relationship between atmospheric particulate matter, leaf functional traits, and phyllosphere microbial communities in *Ulmus*. *BMC Plant Biology, 24*, 523. https://doi.org/10.1186/s12870-024-05232-z.

Yin, R., Hao, Z., Yuan, X., Wang, M., Li, S., Zhang, X., & Chen, B. (2023). Arbuscular mycorrhizal symbiosis alleviates ozone injury in the ozone-tolerant poplar clone but not in the ozone-sensitive poplar clone. *Science of the Total Environment, 894*, 165023. https://doi.org/10.1016/j.scitotenv.2023.165023.

Yin, R., Hao, Z., Yuan, X., Zhang, X., Gun, S., Hu, X., Wang, L., & Chen, B. (2024). The combined application of ethylenediurea and arbuscular mycorrhizal fungi alleviates ozone damage to *Medicago sativa* L. *Soil Ecology Letters, 6*(4), 230231. https://doi.org/10.1007/s42832-023-0231-1.

Chapter 37

Arbuscular mycorrhizal fungi and nonessential element uptake by plants

Dong-Xing Guan and Li-Xue Qiu

State Key Laboratory of Soil Pollution Control and Safety, College of Environmental and Resource Sciences, Zhejiang University, Hangzhou, China

37.1 Introduction

37.1.1 Research background

37.1.1.1 Plant–arbuscular mycorrhizal fungi symbiosis: A cornerstone of element acquisition

The symbiotic relationship between plants and arbuscular mycorrhizal fungi (AMF) forms a cornerstone of terrestrial ecosystems, substantially increasing host plant nutrient acquisition, particularly for essential nutrient elements like phosphorus (P) (Shi et al., 2023). AMF increase nutrient acquisition by expanding the root absorptive surface area and secreting biochemical compounds such as glomalin, a glycoprotein produced by AMF hyphae and spores. Soil glomalin facilitates the mobilization and uptake of sparingly soluble nutrients, particularly iron (Fe) and P (Begum et al., 2019; Emran et al., 2017; Miransari, 2010). AMF hyphae extend beyond the root depletion zone, accessing spatially segregated nutrients and delivering them to plants in exchange for photosynthetic carbon, a mutualistic trade that expands the root uptake area by 10%–100% (Etesami et al., 2021). This symbiosis operates through a finely tuned molecular dialogue, where plant-derived strigolactones attract AMF (Kodama et al., 2022), whereas fungal chitin oligosaccharides trigger symbiotic signaling pathways in roots (Volpe et al., 2023), initiating the formation of arbuscules, the primary sites of nutrient exchange (Thind et al., 2022).

Beyond P, AMF symbiosis increases the absorption of other critical nutrient elements. AMF enable plant uptake of nitrogen (N) and potassium (K) through mycorrhizal-specific transport proteins (Hui et al., 2022; Liu et al., 2019). Under AMF inoculation, N is absorbed by the extraradical mycelium, metabolized into arginine, and transported to the intraradical mycelium, where it is released as NH_4^+ (or NH_3) for plant uptake (Hui et al., 2022). Inoculation with the AMF *Rhizophagus irregularis* increases K content in tomato plants, whereas a mycorrhizal symbiosis-induced K transporter is specifically expressed in cells colonized by fungal hyphae and forming arbuscules (Liu et al., 2019). Beyond nutrient elements, AMF colonization reduces heavy metal toxicity (e.g., zinc (Zn), copper (Cu)) in plants by sequestering metals in fungal vesicles or binding them to cell wall components, thereby supporting plant growth in contaminated soils (González-Guerrero et al., 2008; Wu et al., 2018; Zhou et al., 2017).

37.1.1.2 Nonessential elements in plant physiology: Duality, regulation, and ecological significance

In addition to essential nutrients, certain nonessential elements also play regulatory roles in plant growth. Although not classified as essential for plant survival, these elements can influence physiological metabolism, stress resistance, and yield quality. Nonessential elements, including beneficial elements [e.g., silicon (Si), aluminum (Al)] and heavy metals [e.g., arsenic (As), lead (Pb)], exhibit a fascinating duality in plant ecological physiology: acting as both beneficial agents and toxic stressors depending on their concentration, environmental context, and plant adaptation mechanisms.

Advancing Sustainable Agriculture Through Arbuscular Mycorrhizal Fungi. DOI: https://doi.org/10.1016/B978-0-443-45130-0.00009-6

Silicon, though not essential for most plants, plays a critical role in stress mitigation by strengthening mechanical structure, reducing oxidative damage, and increasing water-use efficiency under drought and salinity (Jiang et al., 2022, 2024; Pang et al., 2024). In Gramineae crops like rice, Si deposition in cell walls (phytoliths) reduces pathogen susceptibility and heavy metal toxicity while optimizing carbon allocation for growth (Yang et al., 2024). Conversely, arsenate (AsO_4^{3-}), a toxic metalloid, disrupts cellular metabolism by mimicking phosphate (PO_4^{3-}) in biochemical pathways, leading to oxidative stress and growth inhibition (Sinha et al., 2023). However, certain hyperaccumulating plants can benefit from arsenate exposure, as they detoxify As through vacuolar sequestration or microbial-assisted methylation, reflecting evolutionary adaptation to contaminated environments (Bai et al., 2023).

The ecological implications of these elements extend to agriculture (e.g., Si biofertilizers) and phytoremediation (As removal), emphasizing the need to reconsider their classification as "nonessential" given their functional roles in plant resilience and ecosystem health. Investigating the regulatory mechanisms of these nonessential elements not only aids in optimizing crop cultivation but also provides novel approaches to address soil pollution and climate challenges.

37.1.1.3 Arbuscular mycorrhizal fungi in terrestrial ecosystems: Multifunctional roles and environmental responses

AMF play a pivotal role in terrestrial ecosystems, influencing plant nutrition, soil health, and carbon cycling. Recent studies highlight their multifaceted functions and responses to environmental changes:

1. Nutrient mediation and soil remediation: AMF increase plant P uptake, particularly in P-limited soils, and contribute to heavy metal immobilization (Nazir et al., 2023; Saeed et al., 2025). For instance, AMF combined with biochar markedly reduces cadmium (Cd) bioavailability in contaminated soils, demonstrating potential for ecological restoration (Zhao et al., 2024).
2. Nitrogen deposition and community dynamics: Research in tropical and subtropical forests reveals that low-level N deposition does not alter AMF abundance or diversity, whereas excessive N inputs suppress AMF communities. This challenges earlier assumptions that N deposition consistently disrupts AMF functionality (Li et al., 2025). Another study found that N fertilization, particularly when applied over 11 years, markedly reduced both AMF abundance and plant species richness. Furthermore, N application decreased the complexity and stability of AMF networks, with these effects becoming progressively more pronounced as the duration of N input increased (Zhang et al., 2024).
3. Carbon turnover and sequestration: AMF obtain plant carbon from the root system; decompose litter; and carry out complex processes, such as microbial synthesis and turnover, through biomass, hyphal growth, metabolic products (especially glomalin-related soil proteins), residues, and interactions with adaptive microorganisms (Holátko et al., 2021; Kakouridis et al., 2022; Liang et al., 2019; Rillig et al., 2001). Beyond releasing carbon into the atmosphere, AMF also input decomposed plant residues and microbial-derived carbon into the soil, forming aggregate structures that protect carbon in the soil from decomposition, achieving the inherent mechanism of soil organic carbon sequestration in the AMF pathway (Leifheit et al., 2014).
4. Challenges and future directions: Despite advances, gaps remain in understanding AMF responses under global change scenarios, such as climate warming and land-use shifts. Integrating multiomics approaches and long-term field experiments is crucial to unravel AMF's ecological and agricultural potential.

37.1.2 Research importance

37.1.2.1 Why study arbuscular mycorrhizal fungi–mediated uptake of nonessential elements?

Essential elements refer to plant nutrients that are indispensable for normal growth and development and cannot be replaced by other elements. Nonessential elements include beneficial elements that are required for the growth of certain plants but not all, such as Si, Al, selenium (Se), and cobalt (Co) (Table 37.1). Among these, Si is particularly notable. Although Si has not yet been proven to be an essential plant nutrient, it plays a critical role in increasing nutrient uptake, supporting plant growth, and strengthening resistance to both abiotic and biotic stresses (Coskun et al., 2019). Therefore applying Si to help plants build defense barriers not only increases crop yields and ensures food security but also provides a solution to environmental pollution issues (Pang et al., 2024).

The total Si content in soil is second only to oxygen, but it exists primarily in inorganic forms (mostly insoluble silicates), with a smaller portion in organic forms. Microbial activity, such as that of Si-solubilizing bacteria, can convert insoluble Si into soluble forms (Brucker et al., 2020; Chandrakala et al., 2019). Silicon concentrations in plants vary widely because of differences in their uptake capacity, ranging from 0.1% to 10% (Hodson et al., 2005; Yizhu et al., 2020). Given that AMF

TABLE 37.1 Beneficial elements and their functions in plants.

Beneficial elements	Functions
Silicon	Strengthens cell wall structure to increase plant lodging resistance and pest/disease tolerance; regulates stomatal opening/closing, influencing photosynthesis and transpiration; participates in antioxidant systems to alleviate stress; affects nutrient uptake and transport; and modulates hormone metabolism
Cobalt	A component of rhizobial nitrogenase, supporting nodule formation and nitrogen fixation in leguminous plants; participating in processes related to plant hormone (e.g., ethylene) synthesis; and affecting the activity of certain enzymes, exerting indirect regulatory effects on plant growth and development
Selenium	At low concentrations, it increases plant antioxidant capacity and stress resistance (e.g., drought and heavy metal stress); affects hormonal balance in plants; influences the quality and flavor of certain plants (such as cruciferous species); and participates in the synthesis of some selenium-containing compounds
Aluminum	In acidic soils, at low concentrations, it supports the growth of certain plants (such as tea plants); influence root morphogenesis; participate in regulating the activity of certain enzymes in plants; affect cell wall structure and function, and may be associated with physiological processes such as disease resistance
Sodium	In C4 and crassulacean acid metabolism plants, it participates in photosynthetic carbon assimilation; maintains cellular osmotic pressure; supports water uptake in certain plants (such as halophytes); substitutes for potassium in performing some functions (e.g., maintaining charge balance); and influences enzyme activity and ion transport

TABLE 37.2 Heavy metals/metalloids and their hyperaccumulating plants.

Elements	Hyperaccumulating plants
Cadmium	*Sedum alfredii* and *Sedum plumbizincicola*
Lead	*Arabis paniculata* and *Vetiveria zizanioides*
Copper	*Brassica juncea*
Arsenic	*Pteris vittata*
Chromium	*Leersia hexandra* Swartz
Nickel	*Pycnandra acuminata*
Zinc	*S. alfredii*

increase the absorption of multiple nutrients, if this benefit could also extend to Si, it might help alleviate issues such as low Si availability in soil and poor Si uptake by plants with limited Si transport capacity. This could further assist plants in establishing Si-based defense barriers, thereby strengthening stress resistance across plant species.

In addition to beneficial elements, there is another category of nonessential elements that play a regulatory role in plant growth: heavy metals (and metalloids). Most heavy metals and metalloids are released into the environment through natural processes and/or human activities. As sessile organisms, plants are constantly exposed to these metals in their surroundings. Most plants are highly susceptible to excessive metal concentrations. To cope, they typically employ detoxification strategies such as forming metal-ligand complexes and sequestering metals in root vacuoles.

"Hyperaccumulating plants" refer to plants that absorb high concentrations of heavy metals and grow normally, possessing specialized adaptive mechanisms that allow them to hyperaccumulate or sequester excess heavy metals in their aboveground tissues rather than storing them in the roots (Sytar et al., 2021). For example, *Sedum plumbizincicola* is a Cd-hyperaccumulating plant (Zhang et al., 2023), whereas *Pteris vittata* (Chinese brake fern) is an As-hyperaccumulating plant (Ma et al., 2001) (Table 37.2). The technique of using metal-hyperaccumulating plants to remediate contaminated soil or water is called phytoremediation, which offers advantages such as low cost, high efficiency, ease of operation, and minimal environmental disruption.

If AMF increase heavy metal uptake in hyperaccumulating plants, it may raise their remediation efficiency for soil pollution. In addition, AMF also increase P uptake under As stress by upregulating P transporters, demonstrating a trade-off between nutrient acquisition and toxic element sequestration (Han et al., 2022). Understanding these interactions is vital for optimizing AMF-based strategies to mitigate soil pollution while ensuring crop safety, particularly in regions with cocontamination of heavy metals, where AMF's contrasting effects on element mobility complicate remediation efforts.

Research in this field bridges gaps in plant–microbe ecology, offering insights into how symbiotic relationships evolve under environmental stress and how they can be harnessed for food security and ecosystem health. In general, the study of AMF in regulating the uptake of nonessential elements such as Si and As is critical for advancing sustainable agriculture and environmental remediation.

37.1.2.2 Applications in sustainable agriculture and environmental remediation

AMF offer dual benefits in sustainable agriculture and environmental remediation by increasing Si-mediated stress resilience and mitigating As contamination. In agriculture, AMF synergize with Si to increase P uptake, reduce Cd translocation, and alleviate salinity stress by stabilizing ion homeostasis and reinforcing plant cell walls (Etesami et al., 2022; Khan et al., 2024; Qiu et al., 2025b). For As pollution control, AMF reduce As bioavailability through cell wall and vesicle sequestration (González-Guerrero et al., 2008). Despite promising lab-scale results, field validation and strain-specific optimization remain critical for large-scale implementation.

37.2 Biological foundation of arbuscular mycorrhizal fungi–mediated element uptake

37.2.1 Arbuscular mycorrhizal fungi symbiotic system architecture

AMF form intricate mycelial networks that establish symbiotic relationships with the roots of approximately 70% of terrestrial plant species, creating an exceptionally widespread and ecologically important mutualism. These fungal networks exhibit remarkable structural complexity and sophisticated transport mechanisms that enable efficient nutrient exchange between soil and host plants.

37.2.1.1 Mycelial network structure and function

The structure of AMF is mainly divided into arbuscules, hyphae, vesicles, and spores. The AMF mycelial network consists of two functionally distinct components: the intraradical mycelium within plant root cortical cells and the extraradical mycelium that extends into the surrounding soil. The intraradical structures form characteristic arbuscules, highly branched, tree-like structures that provide an enormous surface area for nutrient exchange with host cells. The extraradical mycelium can extend several centimeters from the root, creating an extensive underground web that effectively increases the plant's root absorption surface area by up to 100% (Etesami et al., 2021). By combining monoxenic cultures with spatially restricted (2D) network growth and automated image analysis, researchers have successfully tracked extraradical mycelium dynamics, uncovering principles of its structural formation (Cargill et al., 2025; Galvez et al., 2025).

AMF express high-affinity phosphate transporters (PTs) in their extraradical hyphae that actively scavenge P from the soil (Rausch et al., 2001). The phosphate is then transported as polyphosphate granules through the hyphal network to the arbuscules, where plant PTs take up the nutrient into root cells. These proteins are exclusively present in mycorrhizal roots and are highly expressed in cortical cells containing arbuscules. The formation of arbuscular mycorrhizae specifically induces the expression of numerous genes, particularly those encoding PTs, which exhibit the closest functional relationship with this symbiosis. This system is so efficient that up to 80% of plant P can be derived from AMF networks (Harrison et al., 2002; Kaur et al., 2022).

The extraradical mycelium of AMF absorbs inorganic (nitrate and ammonium) and organic (amino acids and small peptides) N sources from the soil, rapidly metabolizing them into arginine (Hawkins et al., 2000; Johansen et al., 1993; Tobar et al., 1994). This arginine, along with polyphosphates, is then transported to the intraradical mycelium, where it is broken down into ammonium (NH_4^+/NH_3) and subsequently released into the apoplastic space between the root and the mycorrhizal fungus (Bago et al., 2001; Govindarajulu et al., 2005; Jin et al., 2005). AMF also effectively extend the plant's hydraulic system, facilitating transpiration-driven water transport through an apoplastic pathway along external hyphal surfaces (Kakouridis et al., 2022). This mechanism markedly increases host plant drought resistance.

37.2.1.2 Arbuscular interface: The hub of nutrient exchange

Within plant roots, AMF hyphae invade cortical cells and differentiate into tree-like arbuscules, generating the periarbuscular space, a critical interfacial zone for resource exchange between the symbiotic partners. In the soil environment, extraradical hyphae explore microporous spaces distant from roots, modifying their immediate surroundings into the hyphosphere, a microbially active zone functioning as an ecological hotspot for fungal–bacterial exchanges of carbon and nutrients (Duan et al., 2024).

The arbuscule represents an extensively branched structure formed by intraradical hyphae within specific host cells, constituting a dual-membrane interface between AMF and their symbiotic plants at the cellular level. Internally the arbuscule is bounded by the fungal membrane (arbuscular membrane) while externally enveloped by the plant membrane (periarbuscular membrane), with an intervening apoplastic space serving as the exchange compartment for metabolites (Harrison et al., 2002).

Structurally, this dual-membrane system physically separates the symbiotic partners. The arbuscular membrane demarcates the fungal domain, whereas the periarbuscular membrane delineates the plant territory. Functionally, however, it intimately connects both organisms through numerous transport proteins and ion channels embedded in both membranes, which mediate their metabolic exchanges (Bouwmeester et al., 2007; Harrison et al., 2002; Helber et al., 2011). Although fungal hyphae may also participate in mineral nutrient and monosaccharide transfers between symbionts, their contribution is generally considered negligible.

37.2.1.3 Cellular storage mechanisms

Following arbuscule maturation, AMF initiate the formation of lipid-storage structures called vesicles. As arbuscules begin to senesce, host roots typically accumulate abundant fungal vesicles. Concurrently with arbuscule development, asexual spores are also produced. These spores contain nuclei and store lipids primarily as triacylglycerides, which serve as energy reserves for subsequent spore germination and asymbiotic germ tube growth. Although it remains unclear how the formation of vesicles and spores is regulated, existing evidence suggests that vesicle formation is associated with lipid transfer from the host (Keymer et al., 2017). However, some AMF, such as *Scutellospora calospora*, do not produce vesicles (van Aarle & Olsson, 2003).

37.2.2 Element-specific uptake patterns in arbuscular mycorrhizal fungi symbiosis

AMF generally increase host plant acquisition of essential elements, including P, N, Zn, and Cu through two primary mechanisms: (1) extending the rhizosphere absorption zone via extensive hyphal networks and (2) mobilizing soil nutrients through organic acid secretion. AMF hyphae can penetrate soil micropores inaccessible to roots, facilitating the uptake and translocation of nutrients. All AMF species exchange carbon (supplied by the plant) and mineral nutrients (provided by the fungus) through the specialized arbuscular structure, a process mediated by conserved transporter protein systems.

Mycorrhiza-specific transporters identified in host plants demonstrate fungal-regulated specialization, enabling increased uptake and translocation efficiency for these particular elements. AMF efficiently absorb P primarily as orthophosphate ($H_2PO_4^-/HPO_4^{2-}$) through high-affinity PTs such as *Glomus versiforme* (GvPT), *R. irregularis* (GintPT), and *Glomus mosseae* (GmPT) (Benedetto et al., 2005; Harrison & Van Buuren, 1995; Maldonado-Mendoza et al., 2001), which are strongly upregulated under low-P conditions. AMF colonization is stringently regulated by P availability, representing a distinctive regulatory pattern compared to other elements.

A study identified a gene encoding an AMF-induced ammonium transporter, *ZmAMT3;1*, in maize (*Zea mays*) roots, which is specifically expressed in arbuscule-containing cortical cells, and its encoded protein localizes to the plasma membrane. *ZmAMT3;1* mediates high-affinity ammonium transport (Hui et al., 2022). Liu et al. (2019) identified *SlHAK10*, the sole mycorrhizal symbiosis-induced K transporter gene in tomato (*Solanum lycopersicum*). This gene exhibits specific expression in arbuscule-containing colonized cells. Knockout of *SlHAK10* markedly reduced K uptake via the mycorrhizal pathway and impaired soluble sugar accumulation in roots under low-K stress, compromising symbiotic efficiency. Conversely, overexpression of *SlHAK10* in tomato increased K acquisition under K-deficient conditions, increased root soluble sugar content, and strengthened mycorrhizal colonization efficacy (Liu et al., 2019).

Nevertheless, few other elements have been shown to utilize mycorrhiza-specific transporters, especially for nonessential elements. In addition, when discussing AMF and plant nutrient uptake, the primary focus has been on how AMF increase plant absorption of elements. Conversely, for heavy metals/metalloids, AMF employ vacuolar sequestration or hyphal immobilization to reduce bioavailability, restricting their translocation to plant roots or shoots.

37.3 Silicon acquisition through arbuscular mycorrhizal fungi partnerships

37.3.1 Physiological roles of silicon in crop systems

Silicon absorption by crops, though nonessential, confers considerable physiological benefits through deposition as amorphous silica in plant tissues, where it increases mechanical strength by reinforcing cell walls and preventing lodging while simultaneously strengthening biotic stress resistance through physical barrier formation and defense pathway activation.

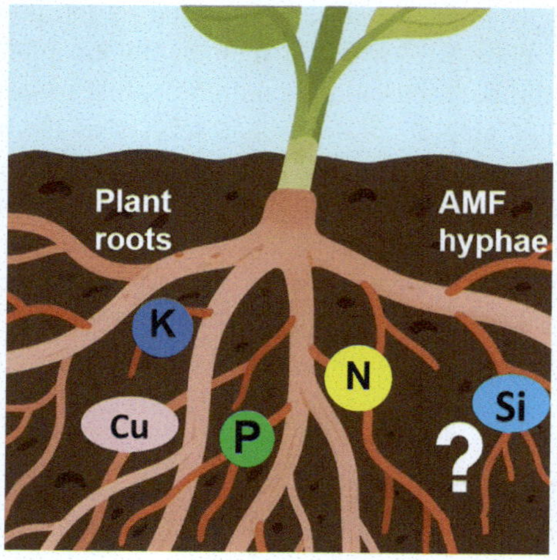

FIGURE 37.1 Schematic representation of AMF-mediated element uptake. AMF hyphae facilitate absorption of multiple elements, including N, P, K, and Cu through established transport mechanisms, whereas Si uptake pathways remain to be elucidated. *AMF*, Arbuscular mycorrhizal fungi.

This beneficial element also mitigates multiple abiotic stresses, including drought (by reducing cuticular transpiration), metal toxicity (via apoplastic sequestration), and salinity (through ion homeostasis regulation), while maintaining photosynthetic efficiency by preserving leaf structure and chlorophyll integrity, making Si fertilization an increasingly important strategy for sustainable crop production and stress resilience in agriculture.

37.3.2 Mechanistic insights into arbuscular mycorrhizal fungi–silicon interactions

Research suggests that Si may increase the symbiotic benefits of AMF through multiple interconnected mechanisms. By increasing photosynthetic rates through elevated chlorophyll content and stomatal conductance (Guntzer et al., 2012; Hajiboland, 2012), Si boosts carbon supply to AMF while strengthening leaf structural stability for optimal light capture. It also stimulates root development in host plants, increasing nutrient uptake and translocation efficiency, thereby supporting arbuscule formation and expanding mycorrhizal colonization (Etesami et al., 2021; Hajiboland et al., 2018; Oye Anda et al., 2016).

Under salinity or drought stress, the synergistic action of Si and AMF stabilizes plant nutrient status, particularly increasing Fe and Zn absorption in low-concentration rhizosphere conditions (Maksimović et al., 2012; Pavlovic et al., 2013). Furthermore, Si enriches soluble sugar pools in roots, likely because of preferential photosynthate allocation or increased CO_2 assimilation, providing energy for AMF colonization while reducing osmotic potential to facilitate water uptake in stressed plants (Moradtalab et al., 2019). In addition, Si optimizes the root microenvironment by suppressing lignin synthesis and modulating phenolic metabolism (e.g., flavonoids), which fosters early AMF establishment (Hajiboland et al., 2018; Mandal et al., 2010; Rodrigues et al., 2004). Collectively, these physiological and biochemical pathways demonstrate how Si strengthens the mutualistic AMF–plant symbiosis.

Silicon nutrition offers benefits to plants, though its efficacy may be limited by the plant's inherent capacity to absorb Si (Etesami et al., 2021; Oye Anda et al., 2016). AMF have been widely demonstrated to increase Si uptake in plants. Multiple AMF species, such as *Glomus etunicatum, Geastrum coronatum*, and *R. irregularis*, have been shown to possess this Si-accumulating ability (Garg & Bhandari, 2016; Hammer et al., 2011; Oye Anda et al., 2016). The mechanism by which Si is transferred from AMF to host plants remains unclear. Although earlier studies hypothesized that AMF hyphae might absorb Si similarly to other elements, this has never been conclusively demonstrated (Fig. 37.1).

Recent studies employing compartmentalized cultivation and diverse Si source treatments have demonstrated three key mechanisms by which AMF increase plant Si uptake: (1) AMF hyphae directly absorb Si through 30-μm nylon mesh, increasing rice root Si content by 113%; (2) AMF secretions raise rhizosphere pH by 0.24 units, solubilizing insoluble silicates (calcium silicate) to boost rice shoot Si by 29%; and (3) AMF upregulate expression of Si transporters *Lsi1* and *Lsi2* in rice roots, increasing transmembrane Si transport, with silicic acid treatment increasing shoot Si by 97% (Qiu et al., 2024). These findings provide a theoretical foundation for leveraging AMF to optimize plant Si nutrition in agriculture.

Furthermore, in the aforementioned study (Qiu et al., 2024), AMF inoculation combined with Si application increased Si content in tomato shoots by over 25%, whereas Si application alone failed to increase Si accumulation. As a typical low-Si accumulator, tomato lacks functional Si efflux transporters (*Lsi2*) in its roots, resulting in limited Si uptake capacity (Sun et al., 2020). Even with exogenous Si supply, Si accumulation in tomato shoots remains markedly lower than in high-accumulating species like rice. This demonstrates that AMF directly absorb Si through their hyphae and transport it to tomato roots, bypassing the plant's inherent transporter limitations. This research reveals that AMF overcome genetic constraints in low-Si plants (e.g., Si transporter deficiency), offering new insights into Si uptake mechanisms. Together, these results suggest that AMF–Si interactions are not limited to Si-accumulating plants, but may also play important roles in low-Si accumulators.

37.3.3 Research gaps and future opportunities

Research on AMF-mediated Si uptake remains considerably understudied compared to other elements. To date, no mycorrhiza-associated Si transporters have been identified, and the proportion of plant Si absorption contributed by AMF hyphae has yet to be quantified. Future studies should prioritize molecular characterization of AMF–Si interactions and develop isotopic tracing methods to quantify hyphal Si contribution. Such advances would bridge critical knowledge gaps in plant–microbe–mineral dynamics, enabling optimized agricultural strategies for Si and P comanagement.

37.4 Arsenic hyperaccumulation: Arbuscular mycorrhizal fungi as facilitators

37.4.1 Arsenic contamination: Environmental and health implications

Arsenic is a metalloid element widely found in nature and is classified as a heavy metal because of its high toxicity. Arsenic in soil can accumulate in the human body through the food chain, leading to diseases such as skin cancer, lung cancer, and liver cancer (Martinez et al., 2011). Currently, many countries and regions around the world are facing severe soil As contamination issues. Soil As pollution has posed a serious threat to food safety and human health (Hou et al., 2025). The most common forms of As in the environment are inorganic arsenate (AsV) and arsenite (AsIII). AsV is the predominant form of As in soil under aerobic conditions, although small amounts of AsIII are also present (Mandal & Suzuki, 2002).

Since As is not an essential element for living organisms, a specific As absorption system has not yet been identified in plants. The absorption of As occurs through pathways that mimic other substrates, utilizing their transport channels, such as phosphate, glucose, and glycerol channels (Garbinski et al., 2019). Studies have shown that arsenate, as a phosphate analog, competes with phosphate absorption. Increasing phosphate concentration reduces the absorption of arsenate by plants, whereas increasing arsenate concentration lowers the phosphate content in the roots (Wang et al., 2002). As(III) is absorbed and transported via Si transporters *Lsi1* and *Lsi2* (Gupta & Khan, 2015). Silicon application effectively reduces As(III) uptake by downregulating *Lsi1* and *Lsi2* expression (Limmer et al., 2018).

37.4.2 Hyperaccumulating plants: Evolutionary adaptations to metal stress

Hyperaccumulating plants represent a remarkable ecological adaptation to metal-rich environments, with *P. vittata* standing out as the most studied As hyperaccumulator (Yan et al., 2019). *P. vittata* is an As hyperaccumulator renowned for its exceptional As tolerance and accumulation capacity. It thrives in soils with As concentrations as high as 1500 mg/kg, with its aboveground tissues accumulating As up to 22,630 mg/kg, far exceeding levels found in nonhyperaccumulating plants (<10 mg/kg) (Zhao et al., 2002). Compared to non-hyperaccumulators, *P. vittata* exhibits highly efficient root-to-shoot As translocation, achieving a remarkable translocation factor of 24. This enables rapid As enrichment in aboveground tissues, substantially increasing total As accumulation (Ma et al., 2001).

The accumulation of As in *P. vittata* involves four key processes: As uptake, reduction, transport, and compartmentalization (Xie et al., 2009; Zhang et al., 2020). In soil, As primarily enters the plant as arsenate through root absorption, where it is subsequently reduced to arsenite (Ma et al., 2001; Wang et al., 2010). Within the roots a small fraction of As is compartmentalized into vacuoles and vesicles, whereas the majority undergoes short-distance transport at the cellular level to reach the vicinity of the xylem (Su et al., 2008; Zhao et al., 2009). From there, As is loaded into the xylem and transported over long distances via xylem vessels to the shoots. Finally, As accumulates in the fronds through vacuolar compartmentalization (Chen et al., 2005; Poynton et al., 2004).

In addition, recent studies have revealed that *P. vittata* can efficiently mobilize insoluble P sources through the secretion of organic acids and phytases, while simultaneously increasing As accumulation (Hu et al., 2025). This discovery provides novel insights for phytoremediation of As-contaminated soils and increasing P utilization efficiency in crops.

37.4.3 Arbuscular mycorrhizal fungi–mediated arsenic acquisition mechanisms

Previous studies have demonstrated that AMF can induce adaptive changes in host plants to mitigate toxic environmental conditions through symbiotic interactions. These fungi increase plant tolerance by (1) increasing water and mineral nutrient uptake, (2) increasing shoot biomass, (3) modifying root architecture, and (4) alleviating heavy metal-induced oxidative stress, collectively optimizing plant resilience while reducing reliance on direct detoxification mechanisms (Riaz et al., 2021). AMF sequester a portion of heavy metals within their fungal structures, reducing heavy metal uptake by host plants. This mechanism mitigates metal toxicity and strengthens host plant tolerance to metal toxicity stress.

Recent studies on *P. vittata* reveal that AMF inoculation increases As accumulation by 2.34-fold in shoots and 2.25-fold in roots, concurrently boosting plant biomass (Pan et al., 2024). Compartmentalized experiments (nylon mesh-separated root/hyphal zones) confirmed AMF hyphae independently absorb and translocate As to the plant, as evidenced by elevated As levels in both tissues when only the hyphal chamber was supplemented. AMF-mediated mechanisms include rhizosphere alkalinization (pH increase from 6.25 to 6.57) and mobilization of As bound to Fe/Al oxides, transforming insoluble As into bioavailable forms. This evidence demonstrates AMF's dual role in augmenting As phytoextraction efficiency while supporting plant growth under contamination stress (Pan et al., 2024).

Another study on AMF inoculation in *P. vittata* similarly found that the AMF-root symbiosis supported plant growth while increasing leaf As content by 64.5% (Sun et al., 2022). Plants absorb and translocate AsV through PTs. In this study, they report the first identification of a mycorrhiza-specific P transporter, *PvPht1;6*. In *P. vittata, PvPht1;6* expression was strongly upregulated in AMF-colonized roots and localized to the plasma membrane of arbusculated cells. This discovery suggests the transporter plays a pivotal role in AsV transport across the periarbuscular membrane of *P. vittata* root cells, elucidating the mechanism underlying increased As accumulation in AMF-associated *P. vittata*. Similarly, Deng et al. (2025) also observed AMF-increased As uptake in *P. vittata*, further corroborating this phenomenon.

A comprehensive study by Qiu et al. (2025a) on rice revealed distinct AMF effects under As-free and As-stressed conditions, demonstrating interactions between nutrient uptake and toxic element dynamics. Without As, AMF increased P content in the aboveground parts of rice by 13%, but with As present, the beneficial effect of AMF was stronger, increasing P content by 38%. This occurs because under As stress, the expression of PTs (such as *PT11* and *PT13*) in rice roots markedly increased, essentially opening more "phosphate absorption channels." The interaction between Si and As transport demonstrates competitive dynamics: When there was no As, AMF and Si together increased the Si content in rice by 44%, but under As conditions, the Si content showed minimal change, as As and Si competed for the same transport proteins (*Lsi1* and *Lsi2*), with As taking priority and blocking Si absorption. Although AMF increased the expression of *Lsi1* under As stress, it also led to more As absorption by rice (38% more in roots and 55% more in aboveground parts), essentially opening As entry channels. AMF demonstrated protective mechanisms by potentially storing absorbed As in their hyphae, preventing excessive As transfer to rice tissues, so roots contained more As, but aboveground parts were relatively unaffected. Simultaneously, AMF helped reduce oxidative damage in rice roots, evidenced by a 28% decrease in malondialdehyde content. This study suggests that AMF absorb As through their hyphae and regulate transporters to allow rice to absorb more As while mitigating As toxicity by storing As and increasing antioxidant enzyme activity and that selecting specific AMF strains may help reduce As transfer while increasing P transfer, thereby raising rice yield and safety in contaminated soils (Qiu et al., 2025a).

This indicates that AMF's influence on As uptake may vary across plant species, with translocation to shoots occurring only when plants require As (Fig. 37.2). This also presents a potential risk: under AMF mediation, increased As could be transported to edible plant parts. Future studies should quantitatively analyze As dynamics in AMF–plant associations, particularly by determining the precise ratio of As allocated to plant tissues versus stored in fungal hyphae, while elucidating the distinct mechanisms governing Si and As uptake. Such investigations will provide insights for optimizing nutrient absorption and developing strategies to mitigate heavy metal stress in plant–fungal symbiotic systems.

37.4.4 Translating research to field applications

Previous studies have mostly selected a specific AMF strain; future research should focus on identifying AMF strains that maximize As immobilization while minimizing plant transfer, understanding molecular mechanisms underlying AMF-mediated As speciation changes, and developing AMF-based strategies tailored to different crops and soil types. The dual

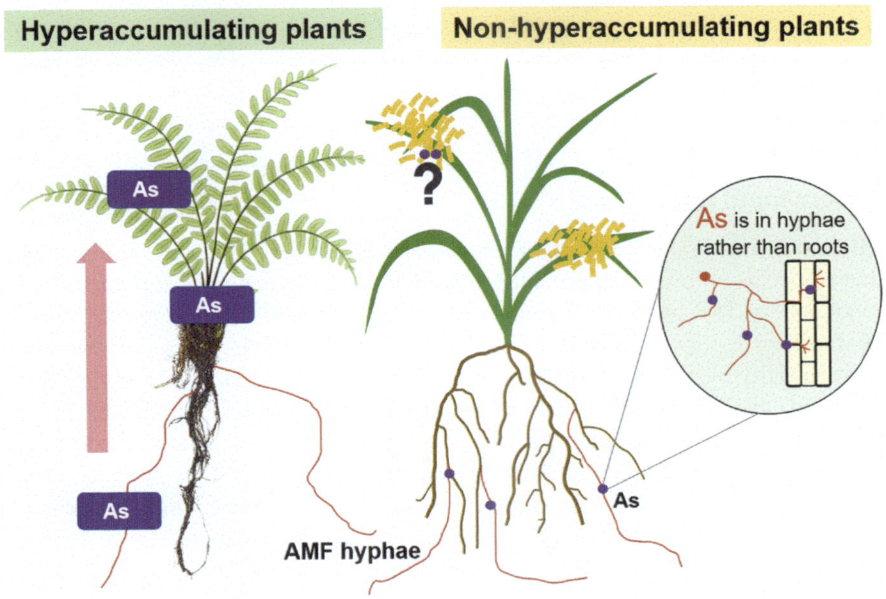

FIGURE 37.2 AMF-mediated As dynamics in different plant systems. The diagram illustrates how AMF regulate As uptake, translocation, and distribution patterns in hyperaccumulating versus nonhyperaccumulating plant species. *AMF*, Arbuscular mycorrhizal fungi.

nature of AMF–As interactions highlights the need for context-specific applications in agriculture and remediation, where AMF can either serve as a solution or a potential risk depending on environmental conditions and management practices.

37.5 Practical implications and applications

37.5.1 Environmental remediation potential

37.5.1.1 Increased phytoextraction efficiency through arbuscular mycorrhizal fungi–plant partnerships

The hyphal network of AMF extends into soil regions beyond root reach, directly absorbing As and transporting it to plant tissues through symbiotic structures. In the As hyperaccumulator *P. vittata*, AMF hyphae contribute approximately 32% of total As accumulation, increasing shoot As content by 234% (Pan et al., 2024). This additional uptake pathway expands the plant's As acquisition range and is particularly beneficial for remediating deep-layer contaminated soils. AMF increase the dissolution of insoluble As in soil through mechanisms such as secreting organic acids and altering soil pH. Inoculation with AMF increased the rhizospheric soil pH of *P. vittata* by 0.39 and elevated bioavailable As content by 116%, facilitating the transformation of As from Fe/Al oxide-bound forms into plant-absorbable species (Pan et al., 2024). In rice, although AMF suppress Si uptake because of As competing for transporters like *Lsi1*, they indirectly strengthen As tolerance by upregulating P transporters (*PT11* and *PT13*), thereby maintaining plant vitality during remediation efforts.

37.5.1.2 Integrated nutrient-toxicity management for sustainable remediation

Under As stress, AMF mitigate oxidative damage in plants by increasing antioxidant enzyme activity and reducing lipid peroxidation products (Qiu et al., 2025a). Rice inoculated with AMF exhibited a 28% decrease in malondialdehyde content in roots while simultaneously increasing P uptake (from 13% to 38% under As stress), thereby maintaining plant biomass and ensuring remediation sustainability. AMF upregulate P transporters (e.g., *PT2* and *PT6*) to increase P assimilation. Because of the chemical similarity between P and As [particularly As(V)] (Sinha et al., 2023), they compete for absorption pathways, indirectly reducing As uptake. This balanced regulation of "nutrition-toxicity" ensures crop yield while facilitating remediation.

Although Si deposition in plant cell walls helps combat various environmental stresses (Exley, 2015), not all plants effectively utilize Si. Most dicotyledonous plants contain minimal Si, likely because of deficiencies in transporter proteins or specific cell wall components (Burton & Fincher, 2009). Even with exogenous Si application, Si concentrations in these plants rarely increase markedly. Thus considerable potential remains for increasing Si use efficiency to strengthen plant stress resistance (Ma & Yamaji, 2015). The increased Si content observed in low-Si plants following AMF inoculation offers a novel solution to this challenge. This approach provides a new pathway for establishing Si-based defense barriers in low-Si-accumulating species.

37.5.2 Advancing low-input agricultural systems

The integration of AMF into agricultural systems offers a biological approach to reducing external inputs while maintaining productivity, addressing key goals of low-input agriculture. AMF hyphae can penetrate soil micropores and directly absorb Si that plant roots cannot access. This mechanism enables plants to acquire Si from deeper soil layers through the hyphal network even when Si fertilizer input is reduced, decreasing reliance on external Si supplementation. AMF increase Si transmembrane transport capacity by upregulating the expression of Si transporters (*Lsi1* and *Lsi2*) in rice roots. Even with reduced Si fertilization, plants can maintain Si accumulation through the efficient expression of these transporters.

The regulation of Si uptake by AMF operates via a composite mechanism: expanded absorption pathways, increased transport efficiency, synergistic P utilization, and strengthened stress resistance. This provides a biological solution for low-input agriculture systems. Its core value lies in optimizing nutrient use efficiency through microbe–plant interactions, maintaining crop yield and environmental adaptability while minimizing chemical fertilizer reliance. These insights offer both theoretical support and practical pathways for achieving fertilizer reduction and efficiency enhancement in low-input agricultural systems.

37.6 Current limitations and research frontiers

37.6.1 Bridging laboratory-field knowledge gaps

37.6.1.1 Environmental complexity versus controlled conditions

Laboratory studies often employ sterilized sand or artificial soil substrates, neglecting the influence of clay minerals, organic matter, and native microbial communities present in field soils. For instance, the binding forms of As with Fe/Al oxides in field soils are more complex, and the ability of AMF to dissolve Si and As by altering pH may be reduced because of differences in soil buffering capacity. In addition, laboratory experiments typically apply a single concentration of As stress, whereas As contamination in field conditions often exhibits patchy distribution and fluctuates with redox potential [e.g., the proportion of As(III) increases under flooded paddy conditions]. This variability may lead to inconsistent regulatory effects of AMF on Si/As transport.

37.6.1.2 Microbial community interactions versus single-strain studies

Laboratory studies often employ single-species inoculations (e.g., *R. irregularis*), whereas field conditions naturally host diverse AMF species and other microorganisms (e.g., bacteria and saprophytic fungi), which may interfere with AMF functionality through competition for carbon sources or alterations in soil enzyme activity. Synergistic interactions between AMF and phosphate-solubilizing bacteria in the field may indirectly influence plant Si uptake, a mechanism not yet incorporated into laboratory investigations, despite both microbes affecting Si acquisition pathways (Lee et al., 2019; Qiu et al., 2024). Different genotypes of the same AMF species (e.g., various strains of *R. irregularis*) may exhibit distinct regulatory effects on plant Si/As uptake (Yu et al., 2010). One genotype may preferentially increase the expression of the Si transporter *Lsi1*, whereas another might focus on reducing As translocation to shoots through hyphal barriers. Current studies have not yet explored functional differentiation at the AMF genotype level.

37.6.1.3 Temporal scales: Short-term effects versus long-term dynamics

Current research cycles are often relatively short, whereas the colonization rate of AMF and the expansion of hyphal networks in field conditions require several months to years. Under long-term conditions the sustained regulatory capacity of AMF on Si transporters and the dynamics of As cycling between plants and soil remain unclear. For instance, the short-term effect of AMF increasing Si uptake by 44% in laboratory settings may vary in field conditions because of plant growth stages, as demand for Si deposition increases during maturity.

37.6.2 Emerging research priorities

37.6.2.1 Methodological innovations for field validation

Field-based *in situ* experiments should be conducted by establishing AMF inoculation comparative trials in diverse soil types (e.g., red soil and paddy soil), combined with isotope tracing techniques to quantify Si transport pathways. Multispecies AMF communities should be reconstructed in artificial microcosms to investigate the influence of community structure on Si/As regulation and screen functionally complementary microbial combinations.

37.6.2.2 Expanding the elemental scope beyond silicon and arsenic

In addition to As and Si, investigating the relationship between AMF and plant uptake of heavy metals (Cd, Pb, and mercury) and beneficial elements (Al and Se) will contribute to a more comprehensive understanding of AMF symbiotic mechanisms. Lead contamination is widespread in soils and can inhibit plant growth and disrupt metabolic processes (Ur Rahman et al., 2024). Studies have shown that AMF reduce Pb concentration in maize leaves while increasing Zn concentration (Hovsepyan & Greipsson, 2004), suggesting that AMF can simultaneously increase plant micronutrient uptake and reduce nonessential element absorption, thereby strengthening heavy metal resistance in plants. Further research is needed to clarify these patterns and develop more effective biological strategies for remediating Pb-contaminated soils.

In acidic soils, increased Al solubility causes toxicity to plants, inhibiting root growth. AMF can alleviate Al toxicity by secreting organic acids, among other strategies, supporting plant growth (Alotaibi et al., 2021). Further exploration of their interactions will help harness AMF to increase plant adaptability and productivity in acidic soils.

37.6.2.3 Spatial visualization of arbuscular mycorrhizal fungi–mediated element dynamics

Recent studies have employed soil zymography and diffusive gradients in thin-film imaging to elucidate the spatial mobilization dynamics of P in maize-AMF systems, deepening our understanding of the spatial distribution of rhizospheric processes (Liu et al., 2025). When investigating AMF's role in the uptake of Si, As, and other nonessential elements in the future, such high-resolution spatial techniques could unravel the localized interactions between fungal hyphae and elemental bioavailability. Furthermore, integrating these methods with isotopic labeling may quantify the contributions of AMF to element translocation under varying soil conditions.

References

Alotaibi, M. O., Saleh, A. M., Sobrinho, R. L., Sheteiwy, M. S., El-Sawah, A. M., Mohammed, A. E., & Abdelgawad, H. (2021). Arbuscular mycorrhizae mitigate aluminum toxicity and regulate proline metabolism in plants grown in acidic soil. *Journal of Fungi, 7*(7), 531. doi:10.3390/jof7070531.

Bago, B., Pfeffer, P., & Shachar-Hill, Y. (2001). Could the urea cycle be translocating nitrogen in the arbuscular mycorrhizal symbiosis? *New Phytologist, 149*(1), 4–8. doi:10.1046/j.1469-8137.2001.00016.x.

Bai, Y., Wan, X., Lei, M., Wang, L., & Chen, T. (2023). Research advances in mechanisms of arsenic hyperaccumulation of *Pteris vittata*: Perspectives from plant physiology, molecular biology, and phylogeny. *Journal of Hazardous Materials, 460*, 132463. doi:10.1016/j.jhazmat.2023.132463.

Begum, N., Qin, C., Ahanger, M. A., Raza, S., Khan, M. I., Ashraf, M., … Zhang, L. (2019). Role of arbuscular mycorrhizal fungi in plant growth regulation: Implications in abiotic stress tolerance. *Frontiers in Plant Science, 10*, 1068. doi:10.3389/fpls.2019.01068.

Benedetto, A., Magurno, F., Bonfante, P., & Lanfranco, L. (2005). Expression profiles of a phosphate transporter gene (*GmosPT*) from the endomycorrhizal fungus *Glomus mosseae*. *Mycorrhiza, 15*(8), 620–627. doi:10.1007/s00572-005-0006-9.

Bouwmeester, H. J., Roux, C., Lopez-Raez, J. A., & Bécard, G. (2007). Rhizosphere communication of plants, parasitic plants and AM fungi. *Trends in Plant Science, 12*(5), 224–230. doi:10.1016/j.tplants.2007.03.009.

Brucker, E., Kernchen, S., & Spohn, M. (2020). Release of phosphorus and silicon from minerals by soil microorganisms depends on the availability of organic carbon. *Soil Biology and Biochemistry, 143*, 107737. doi:10.1016/j.soilbio.2020.107737.

Burton, R. A., & Fincher, G. B. (2009). (1,3;1,4)-β-D-glucans in cell walls of the Poaceae, lower plants, and fungi: A tale of two linkages. *Molecular Plant, 2*(5), 873–882. doi:10.1093/mp/ssp063.

Cargill, R. I. M., Shimizu, T. S., Kiers, E. T., & Kokkoris, V. (2025). Cellular anatomy of arbuscular mycorrhizal fungi. *Current Biology, 35*(11), R545–R562. doi:10.1016/j.cub.2025.03.053.

Chandrakala, C., Voleti, S. R., Bandeppa, S., Sunil Kumar, N., & Latha, P. C. (2019). Silicate solubilization and plant growth promoting potential of *Rhizobium* sp. Isolated from rice rhizosphere. *Silicon, 11*(6), 2895–2906. doi:10.1007/s12633-019-0079-2.

Chen, T., Yan, X., Liao, X., Xiao, X., Huang, Z., Xie, H., & Zhai, L. (2005). Subcellular distribution and compartmentalization of arsenic in *Pteris vittata* L. *Chinese Science Bulletin, 50*(24), 2843–2849. doi:10.1360/982005-943.

Coskun, D., Deshmukh, R., Sonah, H., Menzies, J. G., Reynolds, O., Ma, J. F., … Bélanger, R. R. (2019). The controversies of silicon's role in plant biology. *New Phytologist, 221*(1), 67–85. doi:10.1111/nph.15343.

Deng, S., Yang, Y., Hu, C. Y., Xiao, S., Kuzyakov, Y., Liu, C., & Ma, L. Q. (2025). Arsenic uptake and metabolism in mycorrhizal As-hyperaccumulator *Pteris vittata*: Symbiotic P transporters and As reductases. *Environmental Science & Technology, 59*(11), 5556–5567. doi:10.1021/acs.est.4c11639.

Duan, S., Feng, G., Limpens, E., Bonfante, P., Xie, X., & Zhang, L. (2024). Cross-kingdom nutrient exchange in the plant–arbuscular mycorrhizal fungus–bacterium continuum. *Nature Reviews Microbiology, 22*(12), 773–790. doi:10.1038/s41579-024-01073-7.

Emran, M., Rashad, M., Gispert, M., & Pardini, G. (2017). Increasing soil nutrients availability and sustainability by glomalin in alkaline soils. *International Journal of Agricultural and Biosystems Engineering, 2*, 74–84.

Etesami, H., Jeong, B. R., & Glick, B. R. (2021). Contribution of arbuscular mycorrhizal fungi, phosphate–solubilizing bacteria, and silicon to P uptake by plant. *Frontiers in Plant Science, 12*, 699618. doi:10.3389/fpls.2021.699618.

Etesami, H., Li, Z., Maathuis, F. J. M., & Cooke, J. (2022). The combined use of silicon and arbuscular mycorrhizas to mitigate salinity and drought stress in rice. *Environmental and Experimental Botany, 201*, 104955. doi:10.1016/j.envexpbot.2022.104955.

Exley, C. (2015). A possible mechanism of biological silicification in plants. *Frontiers in Plant Science, 6*, 853. doi:10.3389/fpls.2015.00853.

Galvez, L. O., Bisot, C., Bourrianne, P., Cargill, R., Klein, M., van Son, M., … Shimizu, T. S. (2025). A travelling-wave strategy for plant–fungal trade. *Nature, 639*(8053), 172–180. doi:10.1038/s41586-025-08614-x.

Garbinski, L. D., Rosen, B. P., & Chen, J. (2019). Pathways of arsenic uptake and efflux. *Environment International, 126*, 585–597. doi:10.1016/j.envint.2019.02.058.

Garg, N., & Bhandari, P. (2016). Silicon nutrition and mycorrhizal inoculations improve growth, nutrient status, K^+/Na^+ ratio and yield of *Cicer arietinum* L. genotypes under salinity stress. *Plant Growth Regulation, 78*(3), 371–387. doi:10.1007/s10725-015-0099-x.

González-Guerrero, M., Melville, L. H., Ferrol, N., Lott, J. N. A., Azcón-Aguilar, C., & Peterson, R. L. (2008). Ultrastructural localization of heavy metals in the extraradical mycelium and spores of the arbuscular mycorrhizal fungus *Glomus intraradices*. *Canadian Journal of Microbiology, 54*(2), 103–110. doi:10.1139/W07-119.

Govindarajulu, M., Pfeffer, P. E., Jin, H., Abubaker, J., Douds, D. D., Allen, J. W., Bücking, H., Lammers, P. J., & Shachar-Hill, Y. (2005). Nitrogen transfer in the arbuscular mycorrhizal symbiosis. *Nature, 435*(7043), 819–823. doi:10.1038/nature03610.

Guntzer, F., Keller, C., & Meunier, J. D. (2012). Benefits of plant silicon for crops: A review. *Agronomy for Sustainable Development, 32*(1), 201–213. doi:10.1007/s13593-011-0039-8.

Gupta, M., & Khan, E. (2015). Mechanism of arsenic toxicity and tolerance in plants: Role of silicon and signalling molecules. In *Stress responses in plants: Mechanisms of toxicity and tolerance* (pp. 143–157). Springer International Publishing. doi:10.1007/978-3-319-13368-3_6.

Hajiboland, R. (2012). Effect of micronutrient deficiencies on plants stress responses. In *Abiotic stress responses in plants: Metabolism, productivity and sustainability* (pp. 283–329). Springer New York. doi:10.1007/9781461406341_16.

Hajiboland, R., Moradtalab, N., Aliasgharzad, N., Eshaghi, Z., & Feizy, J. (2018). Silicon influences growth and mycorrhizal responsiveness in strawberry plants. *Physiology and Molecular Biology of Plants, 24*(6), 1103–1115. doi:10.1007/s12298-018-0533-4.

Hammer, E. C., Nasr, H., Pallon, J., Olsson, P. A., & Wallander, H. (2011). Elemental composition of arbuscular mycorrhizal fungi at high salinity. *Mycorrhiza, 21*(2), 117–129. doi:10.1007/s00572-010-0316-4.

Han, R., Chen, J., He, S., Dai, Z., Liu, X., Cao, Y., & Ma, L. Q. (2022). Arsenic-induced up-regulation of P transporters PvPht1;3–1;4 enhances both As and P uptake in As-hyperaccumulator *Pteris vittata*. *Journal of Hazardous Materials, 438*, 129430. doi:10.1016/j.jhazmat.2022.129430.

Harrison, M. J., Dewbre, G. R., & Liu, J. (2002). A phosphate transporter from *Medicago truncatula* involved in the acquisition of phosphate released by arbuscular mycorrhizal fungi. *The Plant Cell, 14*(10), 2413–2429. doi:10.1105/tpc.004861.

Harrison, M. J., & Van Buuren, M. L. (1995). A phosphate transporter from the mycorrhizal fungus *Glomus versiforme*. *Nature, 378*(6557), 626–629. doi:10.1038/378626a0.

Hawkins, H. J., Johansen, A., & George, E. (2000). Uptake and transport of organic and inorganic nitrogen by arbuscular mycorrhizal fungi. *Plant and Soil, 226*(2), 275–285. doi:10.1023/A:1026500810385.

Helber, N., Wippel, K., Sauer, N., Schaarschmidt, S., Hause, B., & Requena, N. (2011). A versatile monosaccharide transporter that operates in the arbuscular mycorrhizal fungus *Glomus* sp is crucial for the symbiotic relationship with plants. *The Plant Cell, 23*(10), 3812–3823. doi:10.1105/tpc.111.089813.

Hodson, M. J., White, P. J., Mead, A., & Broadley, M. R. (2005). Phylogenetic variation in the silicon composition of plants. *Annals of Botany, 96*(6), 1027–1046. doi:10.1093/aob/mci255.

Holátko, J., Brtnický, M., Kučerík, J., Kotianová, M., Elbl, J., Kintl, A., Kynický, J., Benada, O., Datta, R., & Jansa, J. (2021). Glomalin—Truths, myths, and the future of this elusive soil glycoprotein. *Soil Biology and Biochemistry, 153*, 108116.

Hou, D., Jia, X., Wang, L., McGrath, S. P., Zhu, Y. G., Hu, Q., … Nriagu, J. (2025). Global soil pollution by toxic metals threatens agriculture and human health. *Science, 388*(6744), 316–321. doi:10.1126/science.adr5214.

Hovsepyan, A., & Greipsson, S. (2004). Effect of arbuscular mycorrhizal fungi on phytoextraction by corn (*Zea mays*) of lead-contaminated soil. *International Journal of Phytoremediation, 6*(4), 305–321. doi:10.1080/16226510490888820.

Hu, C.-Y., Xiao, S., Menezes-Blackburn, D., Turner, B. L., Cao, Y., Liu, C., & Ma, L. Q. (2025). Phosphorus limitation enhances plant growth and arsenic accumulation in As-hyperaccumulator *Pteris vittata*: Insights from insoluble calcium phytate and phosphate rock. *Soil & Environmental Health, 3*(3), 100158. doi:10.1016/j.seh.2025.100158.

Hui, J., An, X., Li, Z., Neuhäuser, B., Ludewig, U., Wu, X., … Yuan, L. (2022). The mycorrhiza-specific ammonium transporter ZmAMT3;1 mediates mycorrhiza-dependent nitrogen uptake in maize roots. *The Plant Cell, 34*(10), 4066–4087. doi:10.1093/plcell/koac225.

Jiang, H., Song, Z., Su, Q. W., Wei, Z. H., Li, W. C., Jiang, Z. X., … Wu, Z. H. (2022). Transcriptomic and metabolomic reveals silicon enhances adaptation of rice under dry cultivation by improving flavonoid biosynthesis, osmoregulation, and photosynthesis. *Frontiers in Plant Science, 13*, 967537. doi:10.3389/fpls.2022.967537.

Jiang, H., Xu, X., Sun, A., Bai, C., Li, Y., Nuo, M., Shen, X., Li, W., Wang, D., Tian, P., Wei, X., Wang, G., Yang, M., & Wu, Z. (2024). Silicon nutrition improves the quality and yield of rice under dry cultivation. *Journal of the Science of Food and Agriculture, 104*(4), 1897–1908. doi:10.1002/jsfa.13098.

Jin, H., Pfeffer, P. E., Douds, D. D., Piotrowski, E., Lammers, P. J., & Shachar-Hill, Y. (2005). The uptake, metabolism, transport and transfer of nitrogen in an arbuscular mycorrhizal symbiosis. *New Phytologist, 168*(3), 687–696. doi:10.1111/j.1469-8137.2005.01536.x.

Johansen, A., Jakobsen, I., & Jensen, E. S. (1993). Hyphal transport by a vesicular-arbuscular mycorrhizal fungus of N applied to the soil as ammonium or nitrate. *Biology and Fertility of Soils, 16*(1), 66–70. doi:10.1007/bf00336518.

Kakouridis, A., Hagen, J. A., Kan, M. P., Mambelli, S., Feldman, L. J., Herman, D. J., … Firestone, M. K. (2022). Routes to roots: Direct evidence of water transport by arbuscular mycorrhizal fungi to host plants. *New Phytologist, 236*(1), 210–221. doi:10.1111/nph.18281.

Kaur, S., Campbell, B. J., & Suseela, V. (2022). Root metabolome of plant–arbuscular mycorrhizal symbiosis mirrors the mutualistic or parasitic mycorrhizal phenotype. *New Phytologist, 234*(2), 672–687. doi:10.1111/nph.17994.

Keymer, A., Pimprikar, P., Wewer, V., Huber, C., Brands, M., Bucerius, S. L., … Gutjahr, C. (2017). Lipid transfer from plants to arbuscular mycorrhiza fungi. *eLife, 6*, e29107. doi:10.7554/eLife.29107.

Khan, S. R., Ahmad, Z., Khan, Z., Khan, U., Asad, M., & Shah, T. (2024). Synergistic effect of silicon and arbuscular mycorrhizal fungi reduces cadmium accumulation by regulating hormonal transduction and lignin accumulation in maize. *Chemosphere, 346*, 140507. doi:10.1016/j.chemosphere.2023.140507.

Kodama, K., Rich, M. K., Yoda, A., Shimazaki, S., Xie, X., Akiyama, K., … Kyozuka, J. (2022). An ancestral function of strigolactones as symbiotic rhizosphere signals. *Nature Communications, 13*(1), 3974. doi:10.1038/s41467-022-31708-3.

Lee, K. E., Adhikari, A., Kang, S. M., You, Y. H., Joo, G. J., Kim, J. H., … Lee, I. J. (2019). Isolation and characterization of the high silicate and phosphate solubilizing novel strain *Enterobacter ludwigii* GAK2 that promotes growth in rice plants. *Agronomy, 9*(3), 144. doi:10.3390/agronomy9030144.

Leifheit, E. F., Veresoglou, S. D., Lehmann, A., Morris, E. K., & Rillig, M. C. (2014). Multiple factors influence the role of arbuscular mycorrhizal fungi in soil aggregation-a meta-analysis. *Plant and Soil, 374*(1–2), 523–537. doi:10.1007/s11104-013-1899-2.

Li, A., Meidl, P., Wang, S., Tang, B., Rillig, M. C., Yu, G., Chen, J., Liu, R., Lie, Z., Wu, A., Rong, L., Peng, C., Liu, Z., Zhang, W., Lu, X., Liu, J., Ye, Q., Mo, J., & Zheng, M. (2025). Atmospheric nitrogen deposition has minor impacts on the abundance and diversity of arbuscular mycorrhizal fungi and their contribution to soil carbon stock in tropical forests. *Soil Biology and Biochemistry, 204*, 109746. doi:10.1016/j.soilbio.2025.109746.

Liang, C., Amelung, W., Lehmann, J., & Kästner, M. (2019). Quantitative assessment of microbial necromass contribution to soil organic matter. *Global Change Biology, 25*(11), 3578–3590. doi:10.1111/gcb.14781.

Limmer, M. A., Wise, P., Dykes, G. E., & Seyfferth, A. L. (2018). Silicon decreases dimethylarsinic acid concentration in rice grain and mitigates straighthead disorder. *Environmental Science & Technology, 52*(8), 4809–4816. doi:10.1021/acs.est.8b00300.

Liu, J., Liu, J., Liu, J., Cui, M., Huang, Y., Tian, Y., Chen, A., & Xu, G. (2019). The potassium transporter SlHAK10 is involved in mycorrhizal potassium uptake. *Plant Physiology, 180*(1), 465–479. doi:10.1104/pp.18.01533.

Liu, Y. W., Guan, D. X., Qiu, L. X., Luo, Y., Liu, F., Teng, H. H., … Ma, L. Q. (2025). Spatial dynamics of phosphorus mobilization by mycorrhiza. *Soil Biology and Biochemistry, 206*, 109797. doi:10.1016/j.soilbio.2025.109797.

Ma, J. F., & Yamaji, N. (2015). A cooperative system of silicon transport in plants. *Trends in Plant Science, 20*(7), 435–442. doi:10.1016/j.tplants.2015.04.007.

Ma, L. Q., Komar, K. M., Tu, C., Zhang, W., Cai, Y., & Kennelley, E. D. (2001). A fern that hyperaccumulates arsenic. *Nature, 409*(6820), 579. doi:10.1038/35054664.

Maksimović, J. D., Mojović, M., Maksimović, V., Römheld, V., & Nikolic, M. (2012). Silicon ameliorates manganese toxicity in cucumber by decreasing hydroxyl radical accumulation in the leaf apoplast. *Journal of Experimental Botany, 63*(7), 2411–2420. doi:10.1093/jxb/err359.

Maldonado-Mendoza, I. E., Dewbre, G. R., & Harrison, M. J. (2001). A phosphate transporter gene from the extra-radical mycelium of an arbuscular mycorrhizal fungus glomus intraradices is regulated in response to phosphate in the environment. *Molecular Plant-Microbe Interactions, 14*(10), 1140–1148. doi:10.1094/MPMI.2001.14.10.1140.

Mandal, B. K., & Suzuki, K. T. (2002). Arsenic round the world: A review. *Talanta, 58*(1), 201–235. doi:10.1016/S0039-9140(02)00268-0.

Mandal, S. M., Chakraborty, D., & Dey, S. (2010). Phenolic acids act as signaling molecules in plant-microbe symbioses. *Plant Signaling and Behavior, 5*(4), 359–368. doi:10.4161/psb.5.4.10871.

Martinez, V. D., Vucic, E. A., Becker-Santos, D. D., Gil, L., & Lam, W. L. (2011). Arsenic exposure and the induction of human cancers. *Journal of Toxicology, 2011*, 431287. doi:10.1155/2011/431287.

Miransari, M. (2010). Contribution of arbuscular mycorrhizal symbiosis to plant growth under different types of soil stress. *Plant Biology, 12*(4), 563–569. doi:10.1111/j.1438-8677.2009.00308.x.

Moradtalab, N., Hajiboland, R., Aliasgharzad, N., Hartmann, T. E., & Neumann, G. (2019). Silicon and the association with an arbuscular-mycorrhizal fungus (*Rhizophagus clarus*) mitigate the adverse effects of drought stress on strawberry. *Agronomy, 9*(1), 41. doi:10.3390/agronomy9010041.

Nazir, A., Sarfraz, W., Allah, D., Khalid, N., Farid, M., Shafiq, M., Firdaus-E-Bareen, Rizvi, Z. F., & Naeem, N. (2023). Synergistic impact of two autochthonous saprobic fungi (*A. niger* and *T. pseudokoningii*) on the growth, ionic contents, and metals uptake in *Brassica juncea* L. and *Vigna radiata* L. under tannery solid waste contaminated soil. *International Journal of Phytoremediation, 25*(11), 1488–1500. doi:10.1080/15226514.2023.2166457.

Oye Anda, C. C., Opfergelt, S., & Declerck, S. (2016). Silicon acquisition by bananas (c.V. Grande Naine) is increased in presence of the arbuscular mycorrhizal fungus *Rhizophagus irregularis* MUCL 41833. *Plant and Soil, 409*(1–2), 77–85. doi:10.1007/s11104-016-2954-6.

Pan, G., Li, W., Huang, L., Mo, G., & Wang, X. (2024). Arbuscular mycorrhizal fungi promote arsenic accumulation in *Pteris vittata* L. through arsenic solubilization in rhizosphere soil and arsenic uptake by hyphae. *Journal of Hazardous Materials, 466*, 133579. doi:10.1016/j.jhazmat.2024.133579.

Pang, Z., Peng, H., Lin, S., & Liang, Y. (2024). Theory and application of a Si-based defense barrier for plants: Implications for soil-plant-atmosphere system health. *Critical Reviews in Environmental Science and Technology, 54*(9), 722–746. doi:10.1080/10643389.2023.2267939.

Pavlovic, J., Samardzic, J., Maksimović, V., Timotijevic, G., Stevic, N., Laursen, K. H., Hansen, T. H., Husted, S., Schjoerring, J. K., Liang, Y., & Nikolic, M. (2013). Silicon alleviates iron deficiency in cucumber by promoting mobilization of iron in the root apoplast. *New Phytologist, 198*(4), 1096–1107. doi:10.1111/nph.12213.

Poynton, C. Y., Huang, J. W., Blaylock, M. J., Kochian, L. V., & Elless, M. P. (2004). Mechanisms of arsenic hyperaccumulation in *Pteris* species: Root As influx and translocation. *Planta, 219*(6), 1080–1088. doi:10.1007/s00425-004-1304-8.

Qiu, L.-X., Guan, D.-X., Liu, Y.-W., Luo, Y., Teng, H. H., Kuzyakov, Y., & Ma, L. Q. (2025b). Interactions of silicon and arbuscular mycorrhizal fungi on phosphorus uptake during rice vegetative growth. *Geoderma, 454*, 117184.

Qiu, L. X., Guan, D. X., Liu, Y. W., Teng, H. H., Li, Z. B., Lux, A., … Ma, L. Q. (2024). Mechanisms of arbuscular mycorrhizal fungi increasing silicon uptake by rice. *Journal of Agricultural and Food Chemistry, 72*(30), 16603–16613. doi:10.1021/acs.jafc.4c01487.

Qiu, L. X., Xu, K. X., Guan, D. X., Liu, Y. W., Luo, Y., Zhu, X. Y., Teng, H. H., Kuzyakov, Y., & Ma, L. Q. (2025a). Contrasting effects of arsenic on mycorrhizal-mediated silicon and phosphorus uptake by rice. *Journal of Environmental Management, 373*. doi:10.1016/j.jenvman.2024.124005.

Rausch, C., Daram, P., Brunner, S., Jansa, J., Laloi, M., Leggewie, G., Amrhein, N., & Bucher, M. (2001). A phosphate transporter expressed in arbuscule-containing cells in potato. *Nature, 414*(6862), 462–465. doi:10.1038/35106601.

Riaz, M., Kamran, M., Fang, Y., Wang, Q., Cao, H., Yang, G., Deng, L., Wang, Y., Zhou, Y., Anastopoulos, I., & Wang, X. (2021). Arbuscular mycorrhizal fungi-induced mitigation of heavy metal phytotoxicity in metal contaminated soils: A critical review. *Journal of Hazardous Materials, 402*, 123919. doi:10.1016/j.jhazmat.2020.123919.

Rillig, M. C., Wright, S. F., Nichols, K. A., Schmidt, W. F., & Torn, M. S. (2001). Large contribution of arbuscular mycorrhizal fungi to soil carbon pools in tropical forest soils. *Plant and Soil, 233*(2), 167–177. doi:10.1023/A:1010364221169.

Rodrigues, F. A., McNally, D. J., Datnoff, L. E., Jones, J. B., Labbé, C., Benhamou, N., … Bélanger, R. R. (2004). Silicon enhances the accumulation of diterpenoid phytoalexins in rice: A potential mechanism for blast resistance. *Phytopathology, 94*(2), 177–183. doi:10.1094/PHYTO.2004.94.2.177.

Saeed, H. S., Munir, M., Samra, S., Farooq, A., Mahmood, H., Ahmad, I., … Daraz, U. (2025). Screening of biosurfactants producing fungi from infected *Citrus sinensis* (L.). *Polish Journal of Environmental Studies, 34*(5), 5261–5270. doi:10.15244/pjoes/190588.

Shi, J., Wang, X., & Wang, E. (2023). Mycorrhizal symbiosis in plant growth and stress adaptation: From genes to ecosystems. *Annual Review of Plant Biology, 74*(1), 569–607. doi:10.1146/annurev-arplant-061722-090342.

Sinha, D., Datta, S., Mishra, R., Agarwal, P., Kumari, T., Adeyemi, S. B., … Chen, J. T. (2023). Negative impacts of arsenic on plants and mitigation strategies. *Plants, 12*(9), 1815. doi:10.3390/plants12091815.

Su, Y. H., McGrath, S. P., Zhu, Y. G., & Zhao, F. J. (2008). Highly efficient xylem transport of arsenite in the arsenic hyperaccumulator *Pteris vittata*. *New Phytologist, 180*(2), 434–441. doi:10.1111/j.1469-8137.2008.02584.x.

Sun, D., Zhang, X., Liao, D., Yan, S., Feng, H., Tang, Y., … Ma, L. Q. (2022). Novel mycorrhiza-specific P transporter PvPht1;6 contributes to As accumulation at the symbiotic interface of As-hyperaccumulator *Pteris vittata*. *Environmental Science & Technology, 56*(19), 14178–14187. doi:10.1021/acs.est.2c04367.

Sun, H., Duan, Y., Mitani-Ueno, N., Che, J., Jia, J., Liu, J., Guo, J., Ma, J. F., & Gong, H. (2020). Tomato roots have a functional silicon influx transporter but not a functional silicon efflux transporter. *Plant, Cell & Environment, 43*(3), 732–744. doi:10.1111/pce.13679.

Sytar, O., Ghosh, S., Malinska, H., Zivcak, M., & Brestic, M. (2021). Physiological and molecular mechanisms of metal accumulation in hyperaccumulator plants. *Physiologia Plantarum, 173*(1), 148–166. doi:10.1111/ppl.13285.

Thind, S., Chaudhary, M. S., Ditta, A., Hussain, I., Parveen, A., Ullah, N., Mahmood, Q., Al-Ashkar, I., & El-Sabagh, A. (2022). Impact of mycorrhizal fungi from different rhizospheric soils on fungal colonization, growth, and chlorophyll contents of *Cenchrus ciliaris*. *Agronomy, 12*(11), 2644. doi:10.3390/agronomy12112644.

Tobar, R., Azcón, R., & Barea, J. M. (1994). Improved nitrogen uptake and transport from ^{15}N-labelled nitrate by external hyphae of arbuscular mycorrhiza under water-stressed conditions. *New Phytologist, 126*(1), 119–122. doi:10.1111/j.1469-8137.1994.tb07536.x.

Ur Rahman, S., Qin, A., Zain, M., Mushtaq, Z., Mehmood, F., Riaz, L., Naveed, S., Javed Ansari, M., Saeed, M., Ahmad, I., & Shehzad, M. (2024). Pb uptake, accumulation, and translocation in plants: Plant physiological, biochemical, and molecular response: A review. *Heliyon, 10*(6), e27724. doi:10.1016/j.heliyon.2024.e27724.

van Aarle, I. M., & Olsson, P. A. (2003). Fungal lipid accumulation and development of mycelial structures by two arbuscular mycorrhizal fungi. *Applied and Environmental Microbiology, 69*(11), 6762–6767. doi:10.1128/AEM.69.11.6762-6767.2003.

Volpe, V., Chialva, M., Mazzarella, T., Crosino, A., Capitanio, S., Costamagna, L., Kohlen, W., & Genre, A. (2023). Long-lasting impact of chitooligosaccharide application on strigolactone biosynthesis and fungal accommodation promotes arbuscular mycorrhiza in *Medicago truncatula*. *New Phytologist, 237*(6), 2316–2331. doi:10.1111/nph.18697.

Wang, J., Zhao, F. J., Meharg, A. A., Raab, A., Feldmann, J., & McGrath, S. P. (2002). Mechanisms of arsenic hyperaccumulation in *Pteris vittata*. Uptake kinetics, interactions with phosphate, and arsenic speciation. *Plant Physiology, 130*(3), 1552–1561. doi:10.1104/pp.008185.

Wang, X., Ma, L. Q., Rathinasabapathi, B., Liu, Y., & Zeng, G. (2010). Uptake and translocation of arsenite and arsenate by *Pteris vittata* L.: Effects of silicon, boron and mercury. *Environmental and Experimental Botany, 68*(2), 222–229. doi:10.1016/j.envexpbot.2009.11.006.

Wu, S., Vosátka, M., Vogel-Mikus, K., Kavčič, A., Kelemen, M., Šepec, L., … Komárek, M. (2018). Nano zero-valent iron mediated metal(loid) uptake and translocation by arbuscular mycorrhizal symbioses. *Environmental Science & Technology, 52*(14), 7640–7651. doi:10.1021/acs.est.7b05516.

Xie, Q. E., Yan, X. L., Liao, X. Y., & Li, X. (2009). The arsenic hyperaccumulator fern *Pteris vittata* L. *Environmental Science & Technology, 43*(22), 8488–8495. doi:10.1021/es9014647.

Yan, H., Gao, Y., Wu, L., Wang, L., Zhang, T., Dai, C., … He, Z. (2019). Potential use of the *Pteris vittata* arsenic hyperaccumulation-regulation network for phytoremediation. *Journal of Hazardous Materials, 368*, 386–396. doi:10.1016/j.jhazmat.2019.01.072.

Yang, X., Ni, Y., Li, Z., Yue, K., Wang, J., Li, Z., Yang, X., & Song, Z. (2024). Silicon in paddy fields: Benefits for rice production and the potential of rice phytoliths for biogeochemical carbon sequestration. *Science of the Total Environment, 929*, 172497. doi:10.1016/j.scitotenv.2024.172497.

Yizhu, L., Imtiaz, M., Ditta, A., Rizwan, M. S., Ashraf, M., Mehmood, S., … Tu, S. (2020). Response of growth, antioxidant enzymes and root exudates production towards As stress in *Pteris vittata* and in *Astragalus sinicus* colonized by arbuscular mycorrhizal fungi. *Environmental Science and Pollution Research, 27*(2), 2340–2352. doi:10.1007/s11356-019-06785-5.

Yu, Y., Zhang, S., Huang, H., & Wu, N. (2010). Uptake of arsenic by maize inoculated with three different arbuscular mycorrhizal fungi. *Communications in Soil Science and Plant Analysis, 41*(6), 735–743. doi:10.1080/00103620903563964.

Zhang, C., Xiang, X., Yang, T., Liu, X., Ma, Y., Zhang, K., Liu, X., & Chu, H. (2024). Nitrogen fertilization reduces plant diversity by changing the diversity and stability of arbuscular mycorrhizal fungal community in a temperate steppe. *Science of the Total Environment, 918*, 170775. doi:10.1016/j.scitotenv.2024.170775.

Zhang, T., Yan, H., & He, Z. (2020). Advances in molecular mechanisms of arsenic hyperaccumulation of *Pteris vittata* L. *Shengwu Gongcheng Xuebao/Chinese Journal of Biotechnology, 36*(3), 397–406. doi:10.13345/j.cjb.190374.

Zhang, Y., Wu, X., Mo, Y., Han, L., Peng, X., Sun, Z., & Xu, W. (2023). Suppression of highly efficient cadmium translocation by exogenous gibberellin provides insights into the molecular regulation mechanism of hyperaccumulation in *Sedum plumbizincicola*. *Environmental and Experimental Botany, 209*, 105296. doi:10.1016/j.envexpbot.2023.105296.

Zhao, F. J., Dunham, S. J., & McGrath, S. P. (2002). Arsenic hyperaccumulation by different fern species. *New Phytologist, 156*(1), 27–31. doi:10.1046/j.1469-8137.2002.00493.x.

Zhao, F. J., Ma, J. F., Meharg, A. A., & McGrath, S. P. (2009). Arsenic uptake and metabolism in plants. *New Phytologist, 181*(4), 777–794. doi:10.1111/j.1469-8137.2008.02716.x.

Zhao, T., Wang, L., & Yang, J. (2024). Synergistic effects of combined application of biochar and arbuscular mycorrhizal fungi on the safe production of rice in cadmium contaminated soil. *Science of the Total Environment, 951*, 175499. doi:10.1016/j.scitotenv.2024.175499.

Zhou, X., Fu, L., Xia, Y., Zheng, L., Chen, C., Shen, Z., & Chen, Y. (2017). Arbuscular mycorrhizal fungi enhance the copper tolerance of *Tagetes patula* through the sorption and barrier mechanisms of intraradical hyphae. *Metallomics, 9*(7), 936–948. doi:10.1039/c7mt00072c.

Chapter 38

Role of biostimulants and biocontrol agents for enhanced mycorrhizal symbiosis

Sana Ashraf[a], Meshal Saeed[a], Muhammad Kashif Irshad[b,c], Sajid Rashid Ahmad[a], Eshal Zainab[a], Sobia Ashraf[b] and Umer Farooq[d]

[a] *College of Earth and Environmental Sciences, University of the Punjab, Quaid-e-Azam Campus, Lahore, Pakistan,* [b] *Department of Environmental Sciences, Government College University Faisalabad, Faisalabad, Pakistan,* [c] *Department of Environmental and Energy Engineering, Yonsei University, Wonju, South Korea,* [d] *Department of Food Engineering, University of Agriculture Faisalabad, Faisalabad, Pakistan*

38.1 Introduction

The most prevalent, most abundant, and most important beneficial fungal group in mycorrhiza is arbuscular mycorrhizal fungi (AMFs), casting a specific inhibitory or antagonistic effect on soil-borne diseases (du Jardin et al., 2015). Many edaphic fungi, which Frank (1885) referred to as "mycorrhizal," establish a symbiotic connection with plants at the root level of the majority of terrestrial plants. Von Nägeli was the first to describe them in 1842. According to Spatafora et al. (2016), most AMF species are members of the phylum Mucoromycota's subphylum Glomeromycotina (Redecker et al., 2013). As obligatory biotrophs, they need lipids and plant photosynthetic products to complete their life cycle (Jiang et al., 2017).

Since their appearance in the Ordovician period, 480 million years ago (Wang et al., 2022), AMFs have coevolved with various plant species, allowing them to show better growth performance even in stressed environments through various complex mechanisms between host species and AMF, resulting in enhanced photosynthetic activity (Birhane et al., 2012) and increased relative water content. Ahanger et al. (2014) reported that AMF can form stable and complex relationships with more than 90% of the known plant species.

AMF's inadequate exoenzymatic apparatus prevents them from mineralizing soil organic materials. They must therefore form a mutualistic symbiotic relationship with host plants that is considered compatible for the completion of their development cycle, making AMF an obligatory biotroph (Allsup et al., 2021).

When AMF creates a hyphal association with plant root nodules (Razak & Gange, 2021), it enhances the plant's nutritional content and improves soil health by increasing the accessibility to different nutrients present in soil or through the modification of soil texture (Aseel et al., 2019; Li et al., 2021; Nazir et al., 2023; Saeed et al., 2025; Thind et al., 2022; Yizhu et al., 2020). The breakdown of organic oil materials can be accelerated by fungal hyphae (Li et al., 2021). In addition, it also influences the CO_2 fixation process in root nodules and controls the transfer of photoassimilates in different parts of the host body.

AMF can greatly enhance nutrient acquisition and resilience to various environmental stressors (Eke et al., 2020). In addition to these benefits, AMF also protects the plants from several fungal diseases. Consequently, AMFs are essential endosymbionts that contribute significantly to plant productivity, ecological function, and sustainable crop development (Guzman et al., 2021).

The market for agricultural products is currently focusing on a significant increase in the use of biostimulants, which include microbial species or products that enhance the host's ability to respond efficiently to stress factors, leading to improved nutritional quality. Among these, beneficial microorganisms are being utilized more as biostimulating agents, particularly in horticulture, forestry, and agricultural fields (Bubici et al., 2019).

Advancing Sustainable Agriculture Through Arbuscular Mycorrhizal Fungi. DOI: https://doi.org/10.1016/B978-0-443-45130-0.00010-2

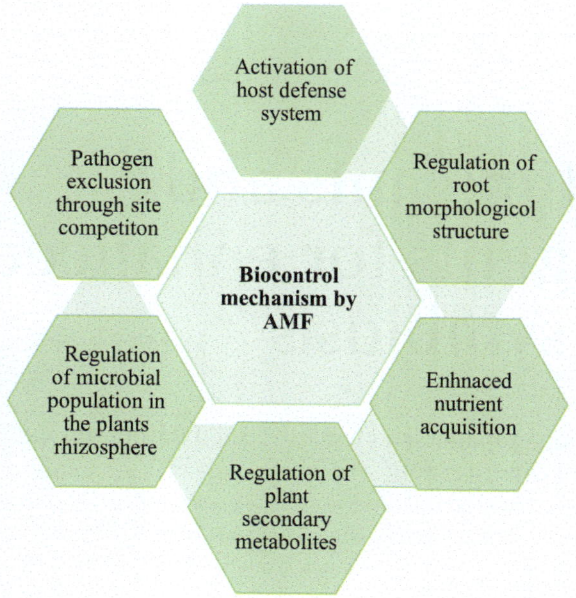

FIGURE 38.1 Mechanism of arbuscular mycorrhizal fungi (AMFs) as a biocontrol agent.

The role of AMF in the environment as a biostimulating agent is often widely discussed, as it plays a diversified role against several agricultural issues, including biotic and abiotic stressors (Sudhasha et al., 2020). It is reported to have been effective against various biotic stressors alongside abiotic stressors, as it develops a defensive mechanism against the attack of biotic stress-causing agents, including pests (Kryzanark et al., 2021).

For many years, chemical pesticides and plant breeding have been the primary methods employed in conventional agriculture to control pests and crop diseases. As an alternative, biocontrol solutions integrate a variety of helpful bacteria, including plant growth–promoting rhizobacterium (PGPR) and plant growth–promoting fungi (Cai et al., 2021), which limit the growth of pests and disease by employing direct approaches (Rashad et al., 2020). AMF may be employed as biocontrol agents because of their capacity to activate the oxidative defense mechanism of plants.

In farming ecosystems, various naturally occurring soil microbial populations have declined, resulting in higher pest populations, as a variety of these populations have developed resistance to pesticides because of their extensive application (Chen et al., 2021). They can persist in plants and animals, make their way into the living organisms, and alter growth patterns and develop health risks (Tatsumi et al., 2021). Food safety issues brought on by pesticides and fertilizers have drawn a lot of attention as people's living standards continue to rise (Li et al., 2021). Nowadays, researchers are focusing on developing pest management strategies that lead to a reduction in plant diseases (Ma et al., 2021).

By utilizing natural mechanisms, biocontrol products and agents provide sustainable methods for controlling pests and diseases. It comprises natural materials derived from plants, animals, and minerals, chemical mediators, and microorganisms. Microorganisms, such as bacteria, oomycotes, viruses, and fungi, are also permitted for sale on the market as biocontrol products (Liang et al., 2021).

The application of microorganisms for biological control has advanced significantly. Several plant species can all be protected from various disease-causing agents by using AMF, thereby reducing the need for pesticides (Kuila & Ghosh, 2022). Plant soil-borne diseases can be effectively controlled by AMF species. Studying the role of AMF is considered valuable both theoretically and practically for better plant growth (Cai et al., 2021).

38.2 Mechanism of biocontrol by arbuscular mycorrhizal fungus

Various studies regarding the role of AMF as a biocontrol agent in controlling plant diseases have been conducted. Examining its associated mechanism primarily entails enhancing plant nutrition, altering the morphological structure of plant roots, controlling the production of secondary metabolites, enhancing the microenvironment of the plant rhizosphere, and promoting the defense mechanism in plants (Li et al., 2021) as depicted in Fig. 38.1.

38.3 Regulation of plant morphological structure

AMF regulates plant morphological structure in various ways to protect plants from disease-causing agents. Pathogens that infect plant roots enter the plant cells through the cell wall (Sudhasha et al., 2020). AMF symbiosis enhances cell wall lignification and causes the host plant's root system to grow, thicken, and branch. It can also alter the root structure, which effectively slows down the process of pathogenic infection (Cai et al., 2021).

38.4 Enhanced nutrition content

AMF is responsible for enhancing the nutrients and moisture absorption in host plants. Begum et al. (2019) reported a mechanism for improving nutrition absorption, the AMF extraroot mycelia, which are extensively dispersed in the medium and form a large mycelium network, increasing the root system's capacity to absorb water and nutrients, particularly phosphate and nitrate, as well as for several plants at once. To a certain degree, plants have efficient water and nutrient pathways for the redistribution of water and nutrients (Israel et al., 2022). Pathogens and AMF fight for photosynthetic products from the host plant's root system; once AMF uses photosynthetic products first, the pathogen acquisition chance is diminished, which restricts pathogen development and reproduction (Chen et al., 2018). By boosting the plants' ability to absorb nutrients and water, mycorrhizae support host plants to make up for the loss of root biomass and function brought on by the pathogenic infection. This indirectly lessens the harm that pathogenic microorganisms inflict and increases the host plants' resistance to illness (Hasami et al., 2015).

38.5 Regulation of the synthesis of plant secondary metabolites

Secondary metabolites also play a crucial role in controlling plant diseases by altering the physicochemical characteristics of plant bodies (McNear et al., 2024). AMF produces various chemicals, including phytochemicals, callouses, alkaloids, and phenols in host plants. These are suitable for host species and help to withstand the harmful impacts brought on by infections (Kuila & Ghosh, 2022).

The production rate and accumulation of phytoprotectins, a family of resistant chemicals generated when pathogenic bacteria infect plants, are correlated with the plants' ability to withstand disease (Israel et al., 2022). Phytophanins build up around diseased cells and act as a barrier for the prevention of further spreading of viruses. *Glomus mosseae* can increase plant disease resistance by inducing a phytotoxin stress response in the roots of plants. Cucumbers can be shielded from *Colletotrichum orbiculare* poisoning after being inoculated with *Glomus intraradices* (Hashem et al., 2016). Following AMF inoculation, leaves had a much higher vinblastine concentration, which enhances the plant's ability to withstand biotic stress. Many chemicals, including flavonoids and phenolic carboxylic acids, belong to the phenolic substance family and are associated with plant defense systems and signaling molecules (Cai et al., 2021).

38.6 Microbial growth management

AMF plays the role of biocontrol agent in host plants through the regulation of microbial entities found in their rhizosphere (Eke et al., 2020). Its growth pattern changes the roots' chemical composition, permeability to cell membranes, the quantity and composition of root exudates, the physical and chemical properties of the rhizosphere, and eventually the quantity and population structure of rhizosphere microorganisms once AMF and plants establish a symbiotic relationship (Li et al., 2021; Rashad et al., 2020). AMF can have a synergistic effect by fostering a partnership with beneficial soil microbes (Kathikeyan et al., 2016). By lowering the quantity of pathogens and the likelihood that they would infect plants, these advantageous bacteria also indirectly increase plant resistance to infections (Allsup et al., 2021).

38.7 Competition with pathogenic microorganisms for invasion sites

AMFs usually reside along with several symbiotic bacteria in a similar soil rhizosphere (Aseel et al., 2019). AMF and pathogens must therefore have a spatially competitive interaction in natural habitat settings, and their primary biocontrol function is to prevent roots from pathogenic infection (Razak & Gange, 2021).

38.8 Activation of the host defense system

Plant hormones, which are tiny signaling molecules, are crucial for controlling how plants grow, develop, and react to their surroundings (Colla et al., 2015). AMF leads to hormonal production during the growth and development of symbiotic relationships with host plants (Rouphael et al., 2017). These hormones are responsible for developing a better defensive mechanism in host species via their association with AMF.

38.9 Practical applications and case studies of biocontrol

AMF have the greatest impact on soil-borne diseases because they reside in the soil and infect plant roots (Li et al., 2021). To manage a variety of phytopathogenic fungi, AMFs have been utilized extensively as a biological management technique (Li et al., 2021). For instance, in comparison to the plants with sterile roots, mycorrhizal cotton plants were more resilient to infection by the pathogen *Thielaviopsis basicola*. According to further research the degree of mycorrhizal infection was inversely correlated with *T. basicola*'s chlamydospore yield.

AMF colonization on apple seedlings' roots decreased actinomycetes' infection of the roots, which, in turn, decreased apple seedling illness (Aseel et al., 2019). AMF has been shown in numerous studies to reduce worm illnesses. *G.* mosseae preinoculation of tobacco can improve plant disease resistance to pathogenic nematodes and drastically decrease the quantity of *T. basicola* and *Meloidogyne incognita* propagules (De Sá et al., 2020). According to Silva et al. (2021), AMF can parasitize soybean cyst nematode cysts and a variety of nematode illnesses that affect various edible crops.

38.10 Influence of arbuscular mycorrhizal fungus and soil microbes in combating plant disease

The microbial population residing in the soil, such as AMFs and other beneficial microbes, is important in fighting plant diseases, as it helps to enhance plant defense systems and increases soil health (Ghadbani et al., 2021). Moreover, microorganisms of soil, like PGPR, inhibit disease-causing organisms by competition, antibiosis, and induction of systemic resistance in plants. AMF and soil microbes also work together to provide a protective rhizosphere environment to reduce infection by the pathogen and enhance sustainable plant health (Jia-Dong et al., 2019). The combination of AMF and microbial inoculation significantly reduces the incidence of infection and harshness in host plants, outperforming the single inoculation. AMF prevents plants from pathogen colonies by competing with pathogens over available root colonization sites and resources, which inhibits the development of pathogens (Jia-Dong et al., 2019). They also induce systemic resistance through the stimulation of plant defense-related genes and increased synthesis of phenolics, phytoalexins, and defense enzymes (Ghadbani et al., 2021). Moreover, AMF enhances the root structure and cell wall density, which forms a physical and biochemical barrier against the invasion of pathogens. Caser et al. (2019) reported infectious reduction in tomato plants by inoculating them with *Fusarium oxysporum* (70%), and it was only 10% after inoculation with one AMF and *Trichoderma viride*. When combined, two AMF inoculations and *Trichoderma* can completely stop the infectious growth. However, there were some differences in how specific AMF and *Trichoderma* combinations managed plant diseases.

38.11 Role of arbuscular mycorrhizal fungus as biostimulants in plant growth

By developing mutualistic relationships with plant roots, AMF can actively interact with their surroundings. These fungi improve the uptake of nutrients and water, change the levels of phytohormones, and strengthen plants' resistance to various abiotic stresses (Jia-Dong et al., 2019). In addition, especially in aromatic and medicinal plants, AMF after mutualistic association with host plants can enhance several characteristics, including their growth, production of secondary metabolites, and morphological attributes. By strengthening plant resistance and physiological functioning under harsh stress conditions the AMFs are crucial for improving sustainable crop production and quality (Zamljen et al., 2023).

Being phytoremediative and nutrient-enriched in nature, AMFs are considered to play an important role as growth-stimulating biofactors by making plants more tolerant to various abiotic conditions, including drought, salinity, and heavy metal toxicity (Rouphael et al., 2015). As AMFs are present in over 90% of plant species, integrating them with plant species for improved crop performance is advised (Ghadbani et al., 2021). However, several conventional agricultural techniques, such as tillage and fertilization, can alter the amount of AMF in the soil, with ameliorating functions for plant species (Nacoon et al., 2020). Currently, there are some biostimulation products available based on mycorrhizal fungi, primarily *Glomus* sp. variations, as well as other mycorrhizal inoculants such as the spores of *Rhizophagus irregularis, Funneliformis mosseae*, and *Claroideoglomus etunicatum* (Zamljen et al., 2023).

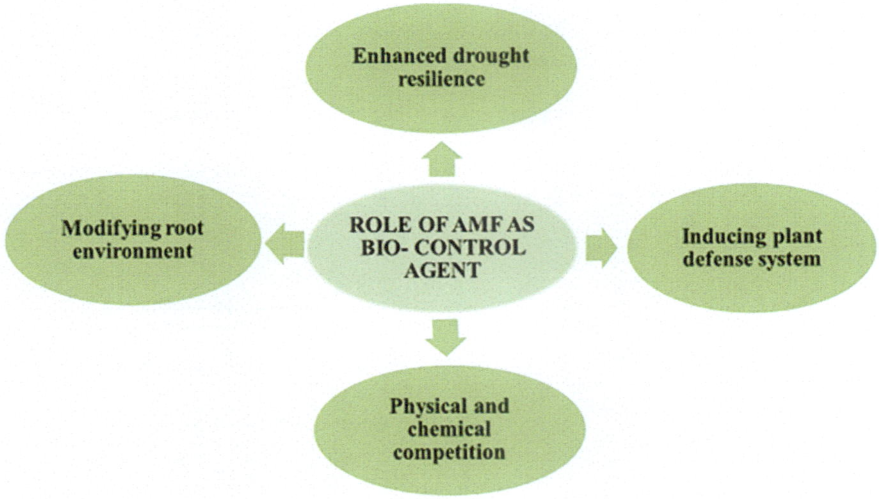

FIGURE 38.2 Mechanisms of AMF as a biocontrol agent. *AMF*, Arbuscular mycorrhizal fungus.

The phylum Glomeromycota contains AMF, which are considered obligatory symbionts and are responsible for boosting growth, nutrient absorption, and enhancing biomass production under stressful circumstances (Nacoon et al., 2020). In addition, it has been demonstrated that inoculation of AMF in a variety of horticultural plants is a profitable technique (Rouphael et al., 2015). The bacterial activity of the plant rhizosphere has also been positively and directly impacted by various fungal species, making it suitable for application in organic farming (Jia-Dong et al., 2019). As a result, using a combination of PGPRs and AMFs is a viable strategy for boosting plant growth. Study revealed that coinoculation of AMF and plant growth–promoting bacteria improved horticulture crop yield and quality in comparison to the single inoculations (Ghadbani et al., 2021).

The application of microbial stimulants like AMFs has recently been investigated in vegetable production and horticulture plants because they can enhance plant nutrient uptake, leading to enhanced productivity and better quality while suppressing the influence of phytopathogens (Mondal et al., 2020). AMF boosts plants' access to nutrients and water, enhances their production and quality, and strengthens their resilience to diseases and environmental stress. By influencing root chemical characteristics, it indirectly influences the characteristics of macropores. In addition, AMF is crucial for nitrogen cycling in the soil as it reduces soil nitrous oxide emissions and boosts the nitrogen level of plants and enhances their biomass (Jain et al., 2021). Studies regarding the utilization of AMF in controlling plant diseases are expanding continuously (Chen et al., 2021). The role of AMF as a biostimulant in plant growth and disease reduction is presented in Fig. 38.2.

Because AMF increases the anabolic breakdown of molecular compounds that play a crucial role in maintaining oxidative stress and leading to enhanced product quality and better yield of fruits and vegetables (Thiub et al., 2019). Inoculation of host species with various strains of AMF resulted in enhancing their water content, photosynthetic activity, and nutrient content, as stated by Chen et al. (2018).

There are two distinct ways in which microbial biostimulants function (Fig. 38.3). The primary mechanisms that microorganisms target, according to different biostimulants, include root targets such as root zone water availability, hormone enhancement, membrane stability, and osmoprotection, as well as shoot targets such as stomatal regulation and xylem hydraulic conductance (Comby et al., 2017). Biostimulants can affect host plants at different growth levels, bringing about a positive change, including increased plant height and biomass and better morphological attributes (Qiu et al., 2022).

38.12 Advantages of arbuscular mycorrhizal fungus as biocontrol and biostimulating agent

AMFs have dual roles of being a highly effective biocontrol and biostimulant agent and have a significant role in improving plant growth and resilience. These defense mechanisms increase the actions of protective enzymes, stimulate the synthesis of phytoalexins, and lignify the cell wall, making the disease less severe (Krishna et al., 2020). AMFs also assist in inhibiting malicious nematodes by modifying their pathways of penetration as well as raising root resistance (Baghaie et al., 2019). AMF also plays an important role as a biostimulant in enhancing the nutrition of plants, by promoting the uptake of phosphorus, nitrogen, potassium, and other beneficial nutrients through its extensive hyphal system. This improves the

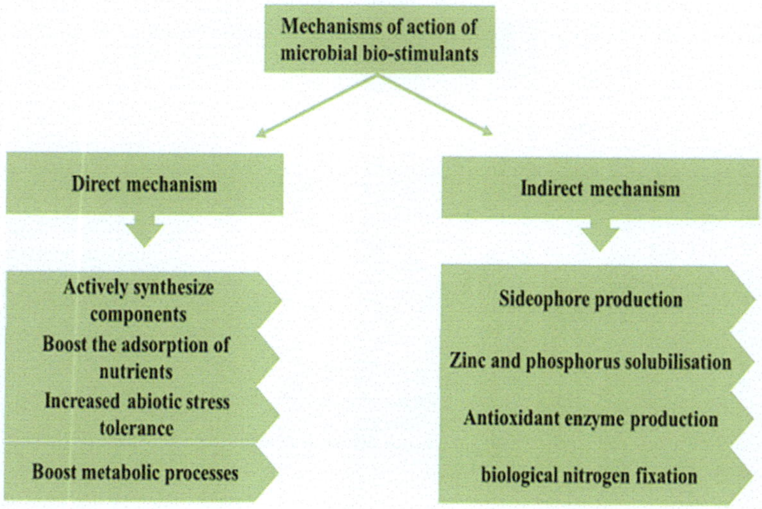

FIGURE 38.3 Modes of action of AMF as a biosimulant in plant growth. *AMF*, Arbuscular mycorrhizal fungus.

nutrient uptake, leading to an increase in root system development, root branching, and high shoot and root biomass (Pérez-De-Luque et al., 2017). AMF-colonized plants have better photosynthetic performance, which is enhanced by the increased chlorophyll content and enhanced stomatal regulation that leads to enhanced growth. Moreover, AMF can increase the drought resistance of plants by increasing water uptake, osmotic balance, and osmoprotectants (Khalloufi et al., 2017). The production of glomalin aids in stabilizing soil structure to enhance soil aggregation and water-holding capacity. In general, synergistic effects of AMF as a biocontrol and biostimulation agent reinforce plant resistance, improve growth and yield, and contribute to the sustainability of agricultural production (Nacoon et al., 2020).

38.13 Case studies on the role of arbuscular mycorrhizal fungus as biostimulants

Table 38.1 presents the various case studies where various strains of AMF are inoculated alone or in combination with other microbial biostimulants in plant species, including horticulture plants, vegetables, edible crops, and ornamental plants, thereby enhancing their growth.

38.14 Case studies related to biocontrol

Adopting sustainable agricultural practices could maximize the benefits of AMF application because it supports plant nutrition by moving and absorbing mineral nutrients outside of the plant rhizosphere's depletion zones and may alter secondary metabolism, increasing the content of nutraceutical compounds (Rouphael et al., 2015; Tarantino et al., 2018). Table 38.2 presents a list of case studies conducted by various researchers to demonstrate the influence of AMF as biocontrol agents in plant growth.

38.15 Conclusion

AMFs, being biostimulants and biocontrol agents, are environmentally friendly and sustainable products that may boost enzymatic and metabolic processes in plants for enhancing yield and productivity of crops, as well as boost resistance to several abiotic stresses. They can naturally enhance plant growth because of their unique formulations or different microbial populations. In addition to being significant substitutes for mineral fertilizers, they are considered a unique and significant approach in sustainable agriculture systems for sustainable crop production management. Among the most notable outcomes of AMF application in sustainable agriculture systems are enhanced morphological attributes of host plants, reduction of fertilizer application, and development of better defensive mechanisms. Their associations with other useful microorganisms such as PGPR and *Trichoderma* make them more effective as biostimulants and biocontrol agents. The case studies reported in this chapter have been a great testimony that AMF inoculation enhances plant growth characteristics, plant yields, and protection against diseases in a large diversity of crop species. To sum up, AMF-derived biostimulants and biocontrol products are environmentally safe and sustainable substitutes for synthetic agrochemicals. They can be used in modern

TABLE 38.1 Case studies indicating the role of AMF as biostimulants in plant growth.

Host species	Fungal species	Observations	References
Solanum lycopersicum	*Rhizoglomus irregulare, Funnerliformis mosseae*	Enhanced biosynthesis, nutrient uptake of important molecules involved in cellular pH and oxidative stress, increased growth, productivity, and nutritional quality	Khalloufi et al. (2017)
Chloris gayana	*Glomus mosseae, Glomus manihotis,* Gigaspora gigantea	Improved nutritional status, relative water content, photosynthetic rate, and increased nutritional content	Krishna et al. (2020)
Cupressus arizonica	*Rhizophagus irregularis, Pseudomonas fluorescens, Funneliformis mosseae*	Increases ascorbate peroxidase and glutathione peroxidase enzyme activities, improves plant growth parameters, reduces water deficit damage, and lowers hydrogen peroxidation and lipid peroxidation	Aalipour et al. (2020)
Pisum sativum	*Glomus mosseae, Rhizobium leguminosarum, Arthrobacter protophormiae*	Reduced proline content and lipid peroxidation, increased pigment content, improved plant weight, improved nutrient uptake, and reduced salt stress	Alam et al. (2020)
Schizolobium parahyba	*Acaulospora* sp., *Claroideoglomus etunicatum, Rhizobium* sp., *Burkholderia* sp.	Improves the absorption of chemical fertilizers and increases the wood yield	Mohanta et al. (2017)
Triticum aestivum	*Azoarcus* sp.	Increased plant nitrogen nutrition and root growth and reduced nutrient deficiency	Baghaie et al. (2019)
T. aestivum	*Azorhizobium* sp.	boosted plant nitrogen nutrition and root growth and alleviated nutrient deficiency	Tofighi et al. (2017)
Arachis hypogaea	*Rhizobium meliloti*	Improved plant growth, nitrogen use efficiency, and the quality of the pods of peanuts	Mondal et al. (2020)
Glycine max	*R. leguminosarum*	Boosted growth and yield performance under drought stress	Abdel-Fttah et al. (2011)
Lonicera japonica	*R. irregulare, Aoufous consortium*	More absorption of N and P content, increased protein and sugar content, alleviated soil pH, and increased the electrical conductivity, organic matter, and total organic carbon	Jiang et al. (2016)
Phoenix dactylifera	*Glomus* spp.	Enhanced morphological attributes, including root and shoot fresh and dry weight, plant height, and germination index	Ghadbani et al. (2021)
Cymbopogon martinii	AMF	Enhanced plant growth	Pankaj et al. (2021)

AMF, arbuscular mycorrhizal fungus.

TABLE 38.2 Case studies indicating the role of AMF as biocontrol in plant growth.

Strains	Host species	Advantages	References
Gigaspora gigantea, Glomus mosseae	*Solanum melongena*	Improved plant growth, reduced the influence of chemical pesticides	Sharma et al. (2021)
Bacillus subtilis	*Zinnia elegans*	Enhanced growth attributes	Lin et al. (2021)
PGPR and humic acids	*Solanum lycopersicum*	Phytohormonal production, form stable plant root colonization	Ziane et al. (2021)
Glomus intraradices, Trichoderma atroviride	*Olea europaea*	Enhanced seedling establishment, Enhanced crop productivity	Tekaya et al. (2022)
AMF	*Tamarindus indica*	Enhanced chlorophyll and carotenoid content	Ndiate et al. (2022)
Myccorihazae	*Vitis vinifera L.*	Enhanced nutritional status, relative water content, and physiological status	Krishna et al. (2005)
F. mosseae, Rhizophagus intraradices	*Thymus daenensis, T. vulgaris L.*	Reduction in essential oil biosynthesis alleviated drought stress in plant bodies	Arpanahi et al. (2020)
AMF	*Crocus sativus L.*	Reduces the incidence of fungal diseases	Caser et al. (2020)
Rhizophagus intraradices	*Ocimum tenuiflorum*	Enhanced productivity and promoted product quality	Thokchom et al. (2020)
AMF	*Piper nigrum*	Reduction in cell membrane leakage under salinity stress	Kaya, Ashraf et al. (no dat)
AMF	*S. lycopersicum*	Promoted yield, reduction in salinity stress on plant bodies	Al-Karaki (2006)
Rhizophagus irregularis	*Punica granatum* L.	Adjusted hormonal balance under salinity stress, improved growth production	Singh et al. (2012)
AMF	*Ocimum basilicum*	Increased chlorophyll content and water use efficiency under salinity stress	Elhindi et al. (2017)
AMF and dopamine	*Malus domestica*	Higher nutrient absorption from the soil	Gao et al. (2020)
AMF	*Abelmoschus esculentus*	Increased plant growth, root morphological characteristics, and chlorophyll content	Jabborova et al. (2021)

AMF, arbuscular mycorrhizal fungus; *PGPA*, plant growth-promoting rhizobacteria.

agricultural systems with a significant effect of decreasing the reliance on chemical fertilizers and pesticides and also promote the growth of crops and improve soil health. Thus the use of AMF in sustainable agriculture is highly promising in improving food security and environmental sustainability.

List of abbreviations

AMF	Arbuscular mycorrhizal fungus
AMFs	Arbuscular mycorrhizal fungi
PGPR	Plant growth–promoting rhizobacterium
PGPRs	Plant growth–promoting rhizobacteria
PGPF	Plant growth–promoting fungi
PGPB	Plant growth–promoting bacteria
GRSP	Glomalin-related soil protein
ROS	Reactive oxygen species
CO_2	Carbon dioxide
IAA	Indole-3-acetic acid
SOD	Superoxide dismutase
CAT	Catalase
APX	Ascorbate peroxidase
Cu	Copper
Ca	Calcium

Fe	Iron
Zn	Zinc
N	Nitrogen
P	Phosphorus
K	Potassium
Eh	Redox potential
sp.	Single species (unspecified)
spp.	Multiple species
F.	*Fusarium* (genus abbreviation)
T.	*Trichoderma*
G.	*Glomus*
R.	*Rhizophagus*

References

Aalipour, H., Nikbakht, A., Etemadi, N., Rejali, F., & Soleimani, M. (2020). Biochemical response and interactions between arbuscular mycorrhizal fungi and plant growth promoting rhizobacteria during establishment and stimulating growth of Arizona cypress (*Cupressus arizonica* G.) under drought stress. *Scientia Horticulturae, 261*, 108923.

Abdel-Fattah, G. M., El-Haddad, S. A., Hafez, E. E., & Rashad, Y. M. (2011). Induction of defense responses in common bean by arbuscular mycorrhizal fungi. *Microbiological Research, 166*, 268–281.

Ahanger, M. A., Alyemeni, M. N., Wijaya, L., Alamri, S. A., Alam, P., Ashraf, M., & Ahmad, P. (2018). Potential of exogenously sourced kinetin in protecting *Solanum lycopersicum* from NaCl-induced oxidative stress through up-regulation of the antioxidant system, ascorbate-glutathione cycle and glyoxalase system. *PLoS One, 13*(9), e0202175.

Alam, M. Z., Hoque, M. A., Ahammed, G. J., & Carpenter-Boggs, L. (2020). Effects of arbuscular mycorrhizal fungi, biochar, selenium, silica gel, and sulfur on arsenic uptake and biomass growth in *Pisum sativum* L. *Emerging Contaminants, 6*, 312–322. www.keaipublishing.com/en/journals/emerging-contaminants/. doi:10.1016/j.emcon.2020.08.001.

Al-Karaki, G. N. (2006). Nursery inoculation of tomato with arbuscular mycorrhizal fungi and subsequent performance under irrigation with saline water. *Scientia Horticulturae, 109*, 1–7.

Allsup, C. M., Lankau, R. A., & Paige, K. N. (2021). Herbivory and soil water availability induce changes in arbuscular mycorrhizal fungal abundance and composition. *Microbial Ecology, 84*, 141–152.

Arpanahi, A. A., Feizian, M., Mehdipourian, G., & Khojasteh, D. N. (2020). Arbuscular mycorrhizal fungi inoculation improve essential oil and physiological parameters and nutritional values of *Thymus daenensis* Celak and *Thymus vulgaris* L. under normal and drought stress conditions. *European Journal of Soil Biology, 100*. https://www.journals.elsevier.com/european-journal-of-soil-biology. doi:10.1016/j.ejsobi.2020.103217.

Aseel, D. G., Rashad, Y. M., & Hammad, S. M. (2019). AM fungi trigger transcriptional expression of flavonoid and chlorogenic pathways in tomato. *Scientific Reports, 9*, 9692.

Baghaie, A. H., Aghili, F., & Jafarinia, R. (2019). Soil-indigenous arbuscular mycorrhizal fungi and zeolite addition synergistically increase grain yield and reduce cadmium uptake in bread wheat. *Environmental Science and Pollution Research, 26*, 30794–30807.

Begum, N., Qin, C., Ahanger, M. A., Raza, S., Khan, M. I., Ashraf, M., Ahmed, N., & Zhang, L. (2019). Role of arbuscular mycorrhizal fungi in plant growth regulation: Implications in abiotic stress tolerance. *Frontiers in Plant Science, 10*, 1068.

Birhane, E., Sterck, F., Fetene, M., Bongers, F., & Kuyper, T. (2012). Arbuscular mycorrhizal fungi enhance photosynthesis, water use efficiency, and growth of frankincense seedlings under pulsed water availability conditions. *Oecologia, 169*, 895–904.

Bubici, G., Kaushal, M., Prigigallo, M. I., Cabanás, C.-L., & Mercado-Blanco, J. (2019). Biological control agents against *Fusarium* wilt of banana. *Frontiers in Microbiology, 10*. doi:10.3389/fmicb.2019.00616.

Cai, X., Zhao, H., Liang, C., Li, M., & Liu, R. (2021). Effects and mechanisms of symbiotic microbial combination agents to control tomato *Fusarium* crown and root rot disease. *Frontiers in Microbiology, 12*, 629793.

Caser, M., Demasi, S., Victorino, I. M. M., Berruti, A., Donno, D., Lumini, E., … Scariot, V. (2019). Arbuscular mycorrhizal fungi modulate the crop performance and metabolic profile of tomato subjected to Fusarium oxysporum. *Scientia Horticulturae, 256*, 108574.

Caser, M., Victorino, I. M. M., Demasi, S., Berruti, A., Lumini, E., Bianciotto, V., & Scariot, V. (2020). Arbuscular mycorrhizal fungi association promotes corm multiplication in potted saffron (*Crocus sativus* L.) plants. *Acta Horticulturae, 1287*, 441–446. https://www.actahort.org/books/1287/1287_56.htm. doi:10.17660/ActaHortic.2020.1287.56.

Chen, M., Arato, M., Borghi, L., Nouri, E., & Reinhardt, D. (2018). Beneficial services of arbuscular mycorrhizal fungi – from ecology to application. *Frontiers in Plant Science, 9*, 1270.

Chen, Q., Wu, W. W., Qi, S. S., Cheng, H., Li, Q., Ran, Q., Dai, Z. C., Du, D. L., Egan, S., & Thomas T. (2021). Arbuscular mycorrhizal fungi improve the growth and disease resistance of the invasive plant *Wedelia trilobata*. *Journal of Applied Microbiology, 130*, 582–591.

Colla, G., Rouphael, Y., Di Mattia, E., El-Nakhel, C., & Cardarelli, M. (2015). Co-inoculation of *Glomus intraradices* and *Trichoderma atroviride* acts as a biostimulant to promote growth, yield and nutrient uptake of vegetable crops. *Journal of the Science of Food and Agriculture, 95*, 1706–1715.

Comby, M., Mustafa, G., Magnin-Robert, M., Randoux, B., Fontaine, J., Reignault, P., & Lounès-Hadj Sahraoui, A. (2017). Arbuscular mycorrhizal fungi as potential bioprotectants against aerial phytopathogens and pests. *Arbuscular mycorrhizas and stress tolerance of plants* (pp. 195–223). Springer. doi:10.1007/978-981-10-4115-0_9.

De Sá, C. S. B., & Campos, M. A. S. (2020). AM fungi decrease *Meloidogyne enterolobii* infection in guava seedlings. *Journal of Helminthology, 94*, e183.

du Jardin, P. (2015). Plant biostimulants: Definition, concept, categories, and regulation. *Scientia Horticulturae, 196*, 3–14.

Eke, P., Adamou, S., Fokom, R., Nya, V. D., Tsouh Fokou, P. V., Nana Wakam, L., Nwaga, D., & Fekam Boyom, F. (2020). Arbuscular mycorrhizal fungi alter antifungal potential of lemongrass essential oil against *Fusarium solani*, causing root rot in common bean (*Phaseolus vulgaris* L.). *Heliyon, 6*, e05737.

Elhindi, K. M., El-Din, A. S., & Elgorban, A. M. (2017). Impact of AM fungi in mitigating salt-induced stress in sweet basil. *Saudi Journal of Biological Sciences, 24*, 170–179.

Frank, A. B. (1885). Über die auf Wurzelsymbiose beruhende Ernährung gewisser Bäume durch unterirdische Pilze. *Berichte der Deutschen Botanischen Gesellschaft, 3*, 128–145.

Gao, S., Wang, G., Liu, N., Chen, M., Zhang, J., & Wang, Z. (2020). Dopamine alleviates salt stress and improves nutrient uptake in apple plants inoculated with arbuscular mycorrhizal fungi. *Scientia Horticulturae, 264*, 109175. doi:10.1016/j.scienta.2020.109175.

Ghadbane, M., Medjekal, S., Benderradji, L., Belhadj, H., & Daoud, H. (2021). Assessment of arbuscular mycorrhizal fungi status and rhizobium on date palm (*Phoenix dactylifera* L.) in Pb-contaminated soil. *Recent advances in environmental science* (pp. 703–707). Springer.

Guzman, A., Montes, M., Hutchins, L., De La Cerda, G., Yang, P., Kakouridis, A., Dahlquist-Willard, R. M., Firestone, M. K., Bowles, T., & Kremen, C. (2021). Crop diversity enriches arbuscular mycorrhizal fungal communities in an intensive agricultural landscape. *New Phytologist, 231*(1), 447–459. http://onlinelibrary.wiley.com/journal/10.1111/(ISSN)1469-8137. doi:10.1111/nph.17306.

Hasani, H., & Aminpanah, H. (2015). Effect of *Pseudomonas fluorescens* inoculation on yield and yield components of rice (*Oryza sativa* L.) under different levels of phosphorus fertilizer. *Journal of Agricultural Science, 48*, 157–163.

Hashem, A., Abd-Allah, E. F., Alqarawi, A. A., Al-Huqail, A. A., & Shah, M. A. (2016). Induction of osmoregulation and modulation of salt stress in *Acacia gerrardii* Benth. by arbuscular mycorrhizal fungi and *Bacillus subtilis* (BERA 71). *BioMed Research International, 2016*, 6294098.

Israel, A., Langrand, J., Fontaine, J., & Lounès-Hadj Sahraoui, A. (2022). Significance of arbuscular mycorrhizal fungi in mitigating abiotic environmental stress in medicinal and aromatic plants: A review. *Foods, 11*, 2591.

Jabborova, D., Annapurna, K., Al-Sadi, A. M., Alharbi, S. A., Datta, R., & Zuan, A. T. K. (2021). Biochar and arbuscular mycorrhizal fungi mitigate drought stress and enhance growth of okra (Abelmoschus esculentus). *Saudi Journal of Biological Sciences, 28*(10), 5490–5499. doi:10.1016/j.sjbs.2021.05.039.

Jain, D., Sharma, J., Kaur, G., Bhojiya, A. A., Chauhan, S., Sharma, V., … Maharjan, E. (2021). Phenetic and molecular diversity of nitrogen fixating plant growth promoting Azotobacter isolated from semiarid regions of India. *BioMed Research International, 2021*(1), 6686283.

Jia-Dong, H., Tao, D., Hui-Hui, W., Zou, Y.-N., Wu, Q.-S., & Kamil, K. (2019). Mycorrhizas induce diverse responses of root *TIP* aquaporin gene expression to drought stress in trifoliate orange. *Scientia Horticulturae, 243*, 64–69.

Jiang, Y., Wang, W., Xie, Q., Liu, N., Liu, L., Wang, D., … Wang, E. (2016). Arbuscular mycorrhizal fungi enhance nutrient acquisition and improve soil properties in Lonicera japonica. *Applied Soil Ecology, 108*, 148–155.

Jiang, Y., Wang, W., Xie, Q., Liu, N., Liu, L., Wang, D., … Wang, E. (2017). Plants transfer lipids to sustain colonization by mutualistic mycorrhizal and parasitic fungi. *Science, 356*(6343), 1172–1175. doi:10.1126/science.aam9970.

Karthikeyan, B., Abitha, B., Henry, A. J., Sa, T., & Joe, M. M. (2016). Interaction of rhizobacteria with AMF in stress reduction. *Fungal biology* (pp. 117–142). Springer.

Kaya, C., Ashraf, M., Sonmez, O., Aydemir, S., Tuna, A. L., & Cullu, M. A. The influence of arbuscular mycorrhizal colonisation on key growth parameters and fruit yield of pepper plants grown at high salinity. Scientia Horticulturae. ***121***, 1–6.

Khalloufi, M., Martinez-Andujar, C., Lachaal, M., Karray-Bouraoui, N., Perez-Alfocea, F., & Albacete, A. (2017). The interaction between foliar GA3 application and arbuscular mycorrhizal fungi inoculation improves growth in salinized tomato (*Solanum lycopersicum* L.) plants by modifying the hormonal balance. *Journal of Plant Physiology, 214*, 134–144.

Krishna, H., Singh, S. K., Sharma, R. R., Khawale, R. N., Grover, M., & Patel, V. B. (2005). Biochemical changes in micropropagated grape (*Vitis vinifera* L.) plantlets due to arbuscular-mycorrhizal fungi (AMF) inoculation during ex vitro acclimatization. *Scientia Horticulturae, 106*(4), 554–567. doi:10.1016/j.scienta.2005.05.009.

Krishna, H., Singh, S. K., Sharma, R. R., Patel, V. B., & Grover, M. (2020). *Arbuscular mycorrhizal fungi enhance nutrient uptake, photosynthesis, and stress tolerance in plants Journal of Plant Nutrition, 43*, 729–746.

Krzyzaniak, Y., Magnin-Robert, M., Randoux, B., Fotaine, J., & Sahraoui, A. L. (2021). Beneficial bacteria and AMF for biocontrol of plant diseases. *Symbiotic soil microorganisms* (pp. 429–468). Springer.

Kuila, D., & Ghosh, S. (2022). AM fungi as biofertilizers in sustainable agriculture. *Current Research in Microbial Sciences, 3*, 100107.

Li, Y., Duan, T., Nan, Z., & Li, Y. (2021). Arbuscular mycorrhizal fungus alleviates alfalfa leaf spots caused by Phoma medicaginis revealed by RNA-seq analysis. *Journal of Applied Microbiology, 130*, 547–560.

Liang, M., Shi, L., Burslem, D. F. R. P., Johnson, D., Fang, M., Zhang, X., & Yu, S. (2021). Soil fungal networks moderate density-dependent survival and growth of seedlings. *New Phytologist, 230*, 2061–2071.

Lin, P., Zhang, M., Wang, M., Li, Y., Liu, J., & Chen, Y. (2021). AM fungi modulate defense-related genes in banana. *Plant Signaling & Behavior, 16*, 1884782.

Ma, X., Li, X., & Ludewig, U. (2021). AM colonization outcompetes root hairs in maize under low P. *Annals of Botany, 127*, 155–166.

McNear, D. H., Jr. (2013). The rhizosphere—roots, soil, and everything in between. *Natural Sciences Education, 4*, 1.

Mohanta, T. K., Bashir, T., Hashem, A., & Abd Allah, E. F. (2017). Systems biology approach in plant abiotic stresses. *Plant Physiology and Biochemistry, 121*, 58–73. http://www.journals.elsevier.com/plant-physiology-and-biochemistry/. doi:10.1016/j.plaphy.2017.10.019.

Mondal, M., Skalicky, M., Garai, S., Hossain, A., Sarkar, S., Banerjee, H., Kundu, R., Brestic, M., Barutcular, C., & Erman, M. (2020). Nitrogen supplementation with rhizobium improves peanut production. *Agronomy, 10*, 1513.

Nacoon, S., Jogloy, S., Riddech, N., Mongkolthanaruk, W., Kuyper, T. W., & Boonlue, S. (2020). Interaction between phosphate solubilizing bacteria and arbuscular mycorrhizal fungi on growth promotion and tuber inulin content of *Helianthus tuberosus* L. *Scientific Reports, 10*, 4916.

Nazir, A., Sarfraz, W., Allah, D., Khalid, N., Farid, M., Shafiq, M., Firdaus-E-Bareen, Rizvi, Z. F., & Naeem, N. (2023). Synergistic impact of two autochthonous saprobic fungi (*A. niger* and *T. pseudokoningii*) on the growth, ionic contents, and metals uptake in *Brassica juncea* L. and *Vigna radiata* L. under tannery solid waste contaminated soil. *International Journal of Phytoremediation, 25*(11), 1488–1500. doi:10.1080/15226514.2023.2166457.

Ndiate, N. I., Qun, C. L., & Nkoh, J. N. (2022). Soil amendments with biochar and/or AM fungi mitigate aluminum toxicity in tamarind. *Heliyon, 8*, e09009.

Pankaj, U., Kurmi, A., Lothe, N. B., & Verma, R. K. (2021). Influence of the seedlings emergence and initial growth of palmarosa (*Cymbopogon martinii* (Roxb.) Wats. var. Motia Burk) by arbuscular mycorrhizal fungi in soil salinity conditions. *Journal of Applied Research on Medicinal and Aromatic Plants, 24*, 100317.

Pérez-De-Luque, A., Tille, S., Johnson, I., Pascual-Pardo, D., Ton, J., & Cameron, D. D. (2017). The interactive effects of arbuscular mycorrhiza and plant growth-promoting rhizobacteria synergistically enhance host plant defences against pathogens. *Scientific Reports, 7*, 16409.

Qiu, Y. J., Zhang, N. L., Zhang, L. L., Zhang, X. L., Wu, A. P., Huang, J. Y., … Wang, Y. H. (2022). Biostimulants induce positive changes in plant growth, biomass, and morpho-physiological attributes. *Frontiers in Plant Science, 13*, 950393. doi:10.3389/fpls.2022.950393.

Rashad, Y. M., Abbas, M. A., Soliman, H. M., Abdel-Fattah, G. G., & Abdel-Fattah, G. M. (2020). Synergy between endophytic *Bacillus amyloliquefaciens* GGA and arbuscular mycorrhizal fungi induces plant defense responses against white rot of garlic and improves host plant growth. *Phytopathologia Mediterranea, 59*, 169–186.

Razak, N. A., & Gange, A. C. (2021). Multitrophic interactions between arbuscular mycorrhizal fungi, foliar endophytic fungi and aphids. *Microbial ecology, 85*(1), 146–156.

Redecker, D., Kodner, R., & Graham, L. E. (2000). Glomalean fungi from the Ordovician. *Science, 289*, 1920–1921.

Sharma, S., Prasad, R., Varma, A., & Sharma, A. K. (2021). Arbuscular mycorrhizal fungi and biocontrol agents improve growth and reduce pesticide dependency in brinjal (Solanum melongena L.). *Biological Control, 152*, 104451. doi:10.1016/j.biocontrol.2020.104451.

Silva, M. T. R. E., Calandrelli, A., Miamoto, A., Rinaldi, L. K., Moreno, B. P., da Silva, C., & Dias-Arieira, C. R. (2021). Pre-inoculation with arbuscular mycorrhizal fungi affects essential oil quality and the reproduction of root lesion nematode in *Cymbopogon citratus*. *Mycorrhiza, 31*, 613–623.

Rouphael, Y., Cardarelli, M., Bonini, P., & Colla, G. (2017). Synergistic action of a microbial-based biostimulant and a plant derived-protein hydrolysate enhances lettuce tolerance to alkalinity and salinity. *Frontiers in Plant Science, 8*, 131.

Rouphael, Y., Franken, P., Schneider, C., Schwarz, D., Giovannetti, M., Agnolucci, M., Pascale, S. D., Bonini, P., & Colla, G. (2015). Arbuscular mycorrhizal fungi act as biostimulants in horticultural crops. *Scientia Horticulturae, 196*, 91–108. Www.elsevier.com/inca/publications/store/5/0/3/3/1/6. doi:10.1016/j.scienta.2015.09.002.

Saeed, H. S., Munir, M., Samra, S., Farooq, A., Mahmood, H., Ahmad, I., Akram, M. I., Iqbal, R., Elsadek, M. F., Almutairi, S. M., Ditta, A., & Daraz, U. (2025). Screening of biosurfactants producing fungi from infected *Citrus sinensis* (L.). *Polish Journal of Environmental Studies, 34*(5), 5261–5270. https://www.pjoes.com/pdf-190588-115894?filename=Screening%20of.pdf. doi:10.15244/pjoes/190588.

Singh, N. V., Singh, S. K., Singh, A. K., Meshram, D. T., Suroshe, S. S., & Mishra, D. C. (2012). Arbuscular mycorrhizal fungi (AMF) induced hardening of micropropagated pomegranate (*Punica granatum* L.) plantlets. *Scientia Horticulturae, 136*, 122–127. doi:10.1016/j.scienta.2012.01.005.

Spatafora, J. W., Chang, Y., Benny, G. L., Lazarus, K., Smith, M. E., Berbee, M. L., … Stajich, J. E. (2016). *A phylum-level phylogenetic classification of zygomycete fungi based on genome-scale data Mycologia, 108*, 1028–1046. doi:10.3852/16-042.

Sudhasha, S. (2020). Constructiveness of biocontrol agents on *Fusarium* wilt of tomato. *Current research and innovations in plant pathology*. AkiNik Publications.

Tarantino, A., Lops, F., Disciglio, G., & Lopriore, G. (2018). Effects of plant biostimulants on fruit set, growth, yield and fruit quality attributes of 'Orange rubis®' apricot (*Prunus armeniaca* L.) cultivar in two consecutive years. *Scientia Horticulturae, 239*, 26–34.

Tatsumi, C., Hyodo, F., Taniguchi, T., Shi, W., Koba, K., Fukushima, K., Du, S., Yamanaka, N., Templer, P., & Tateno, R. (2021). Arbuscular mycorrhizal community in roots and nitrogen uptake patterns of understory trees beneath ectomycorrhizal and non-ectomycorrhizal overstory trees. *Frontiers in Plant Science, 11*, 583585.

Tekaya, M., Dabbaghi, O., Guesmi, A., Attia, F., Chehab, H., Khezami, L., Algathami, F. K., Ben Hamadi, N., Hammami, M., Prinsen, E., & Mechri, B. (2022). Arbuscular mycorrhizas modulate carbohydrate, phenolic compounds and hormonal metabolism to enhance water deficit tolerance of olive trees (*Olea europaea*). *Agricultural Water Management, 274*. http://www.journals.elsevier.com/agricultural-water-management/. doi:10.1016/j.agwat.2022.107947.

Thind, S., Chaudhary, M. S., Ditta, A., Hussain, I., Parveen, A., Ullah, N., Mahmood, Q., Al-Ashkar, I., & El-Sabagh, A. (2022). Impact of mycorrhizal fungi from different rhizospheric soils on fungal colonization, growth, and chlorophyll contents of *Cenchrus ciliaris*. *Agronomy, 12*(11), 2644. doi:10.3390/agronomy12112644.

Thiub, M., Ewusi-Mensah, N., Sarkodie-Addo, J., & Adjei-Gyapong, T. (2019). AMF increases P-use efficiency and soybean productivity. *Soil & Tillage Research, 192*, 174–186.

Thokchom, S. D., Gupta, S., & Kapoor, R. (2020). Arbuscular mycorrhiza augments essential oil composition and antioxidant properties of *Ocimum tenuiflorum*. *Industrial Crops and Products, 153*, 112418.

Tofighi, C., Khavari-Nejad, R. A., Najafi, F., Razavi, K., & Rejali, F. (2017). Brassinosteroid (BR) and arbuscular mycorrhizal (AM) fungi alleviate salinity in wheat. *Journal of Plant Nutrition, 40*(8), 1091–1098. Www.tandf.co.uk/journals/titles/01904167.asp. doi:10.1080/01904167.2016.1263332.

Wang, Y., He, X., & Yu, F. (2022). Non-host plants and their role in mycorrhizal networks. *Plant Diversity, 44*, 127.

Yizhu, L., Imtiaz, M., Ditta, A., Rizwan, M. S., Ashraf, M., Mehmood, S., Aziz, O., Mubeen, F., Ali, M., Elahi, N. N., Ijaz, R., Lele, S., Shuang, C., & Tu, S. (2020). Response of growth, antioxidant enzymes and root exudates production towards as stress in *Pteris vittata* and in *Astragalus sinicus* colonized by arbuscular mycorrhizal fungi. *Environmental Science and Pollution Research, 27*(2), 2340–2352. https://link.springer.com/journal/11356. doi:10.1007/s11356-019-06785-5.

Zamljen, T., Medic, A., Veberic, R., Hudina, M., Grohar, M. C., & Slatnar, A. (2023). Influence of hydrolyzed animal protein-based biostimulant on primary, soluble and volatile secondary metabolism of Genovese and Greek-type basil grown under salt stress. *Scientia Horticulturae, 319*, 112178.

Ziane, H., Hamza, N., & Meddad-Hamza, A. (2021). AM fungi and fertilization optimize tomato growth and yield in Mediterranean agroecosystems. *Journal of the Saudi Society of Agricultural Sciences, 20*, 454–458.

Chapter 39

Challenges, opportunities, and future directions for arbuscular mycorrhizal fungi research and development

Salma Noureen[a], Tabassum Yaseen[a,*], Allah Ditta[b,c], Javed Iqbal[a], Banzeer Ahsan Abbasi[d], Zakir Ullah[e], Shumaila Ijaz[e,f], Kamran Akbar[a], Farishta Zarshan[a], Naila Inayat[g] and Ghazala Zamani[h]

[a] *Department of Botany, Bacha Khan University Charsadda, Charsadda, Pakistan,* [b] *Department of Environmental Sciences, Shaheed Benazir Bhutto University Sheringal, Sheringal, Pakistan,* [c] *School of Biological Sciences, The University of Western Australia, Perth, WA, Australia,* [d] *Department of Botany, Rawalpindi Women University, Rawalpindi, Pakistan,* [e] *Department of Plant Sciences, Faculty of Biological Sciences, Quaid-i-Azam University Islamabad, Islamabad, Pakistan,* [f] *School of Biomedical Engineering, Shenzhen University Medical School, Shenzhen University, Shenzhen, China,* [g] *Department of Botany, Shaheed Benazir Bhutto Woman University Peshawar, Pakistan,* [h] *Department of Biotechnology, Bacha Khan University Charsadda, Pakistan*

* *Corresponding author. e-mail address: tabassumyaseen@bkuc.edu.pk.*

39.1 Introduction

Plants growing in the soil rely on a complex system of nutrient uptake to support their growth and improvement. Their roots absorb vital nutrients from the soil, which are then transported to the leaves and other portions of the plant. Baron Justus von Liebig was the first to discover plant nutrients in the 19th century. There are 16 elements in total, out of which 9 are macro elements and the remaining are micronutrients. Macronutrients are classified into groups (hydrogen, carbon, and oxygen); primary macronutrients are N (nitrogen), P (phosphorus), and K (potassium); and secondary macronutrients are calcium (Ca), sulfur (S), and magnesium (Mg) (Reddy et al., 2025). They are necessary in comparatively large quantities and play an important function in meeting most of a plant's nutritional needs, despite their percentage in the soil being much lower than 90%. For example, nitrogen is essential for protein synthesis, the most limiting nutrient in soils. Similarly, phosphorus supports root and flower development, whereas potassium regulates various plant functions, such as water uptake and enzyme activity (Yahaya et al., 2023). On the other hand, micronutrients are important and required equally in small amounts, such as Zn (zinc), Mn (manganese), Mo (molybdenum), Cu (copper), B (boron), and Fe (iron). However, their percentage is less than 0.1% in the soil, but they are essential for a variety of processes such as photosynthesis, enzymatic activity, and maintaining overall plant health (Sharma et al., 2025).

Plant physiology and agricultural results are significantly affected through the compound ecological connections among plants and microorganisms, particularly rhizospheric microbes (Yizhu et al., 2020). Soil microbes, including fungi and bacteria, interact closely with plants to enhance nutrient availability. Legumes' roots develop symbiotic relationships through nitrogen-fixing bacteria, such as *Rhizobium*, which convert atmospheric nitrogen into forms that plants can utilize (Nag & Das, 2025). Through enhancing nutrient uptake, stress acceptance, and general development, these microbes can improve plant performance and help create more productive and sustainable agricultural systems. As the main decomposers in terrestrial ecosystems, soil biota are rich in species diversity and functions and are the most active part of soil structure. Soil biota not only promote energy flow and nutrient cycling, but they also maintain structural and functional stability of soil ecosystems and store nutrients for aboveground vegetation by establishing symbiotic relationships

Advancing Sustainable Agriculture Through Arbuscular Mycorrhizal Fungi. DOI: https://doi.org/10.1016/B978-0-443-45130-0.00037-0

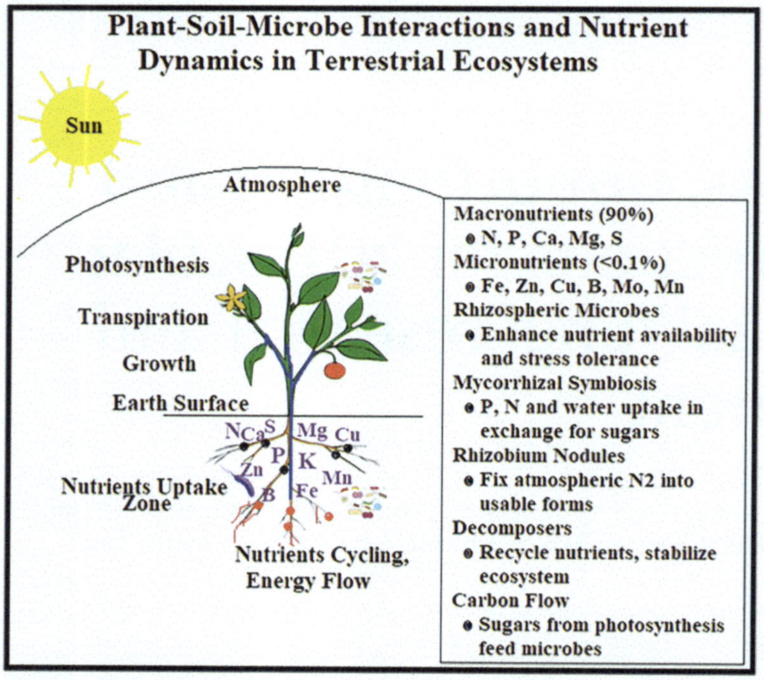

FIGURE 39.1 Plant–microbe interactions driving soil nutrient cycling in terrestrial ecosystems.

(Nazir et al., 2023; Rathna et al., 2025). This interaction is essential for nutrient cycling, maintaining soil fertility, promoting a healthy environment, and providing a balanced environment for plant growth (Fig. 39.1). Plant–soil microbe interactions are widespread in nature and greatly influence the survival of symbionts (Thind et al., 2022). For example, by forming symbiotic associations through arbuscular mycorrhizal (AM) fungi, plants can accept nutrients, such as phosphorus, while providing sugars produced via photosynthesis to the fungi. Mycorrhizal fungi also play a key role, forming mutualistic relationships with plant roots to enhance phosphorus, nitrogen, and water uptake; however, the crops provide carbohydrates to the fungi in return (Ahmed et al., 2025).

39.2 Mycorhizae

The term “mycorrhiza” refers to fungal roots. More precisely, mycorrhizae are fungi that live in symbiotic associations with many plant roots. In the soil, fungi that frequently develop mycorrhizal associations with plants are constantly present. As many mycorrhizal fungi are obligately symbiotic, they cannot survive in the wild for long periods without their host. The fungus and plant have a symbiotic connection, meaning that both parties gain something from the other. The association has no negative consequences for either the mildew or the host crop (Mazumder et al., 2025). The fungus is unable to fix its own carbon because it does not engage in photosynthetic activity. And so the host plant provides it with all the carbohydrates it needs. As a plant’s roots typically make up merely 0.5% of the topsoil capacity and even fewer of the subsoil, the mycorrhiza reciprocates by absorbing nutrients from the earth (soil) that are then moved to the houseplant (Beechey-Gradwell et al., 2025). The mycorrhizal fungus may encounter more soil per volume because its hyphae are thinner than the plant’s roots. The mycorrhizal fungus has a system of mycelium that spreads into the soil outside of the tree roots and resembles a root. After absorbing nutrients, this mycelium returns them to the host plant. Consequently the absorption surface area of the roots increases (Zhang et al., 2025a). Certain plant nutrients may appear to be inaccessible to the plant because they flow slowly through the soil. This slow movement causes the plant to exhibit symptoms of a deficit, which is especially significant when it comes to phosphorus. Most of the P (phosphorus) in the soil is insoluble, and plant roots without mycorrhizal relationships cannot access insoluble phosphorus. In addition, these same mycorrhizal roots associate with organisms that break down trash, allowing them to access nutrients that would not otherwise be available (Singh et al., 2025). Mycorrhizal deficit can happen in soils that have undergone fumigation, in places where a lot of topsoil has been removed, or in places where trees have never been planted before. The latter is especially true for more host-specific ectomycorrhiza and evergreens. Nitrogen, phosphorus, or full fertilizer additions will decrease mycorrhizal activity and presence. Depending on the fungal strain, this impact varies. Despite their hidden nature, their effect is remarkable, particularly in their absence (Putnam et al., 2025).

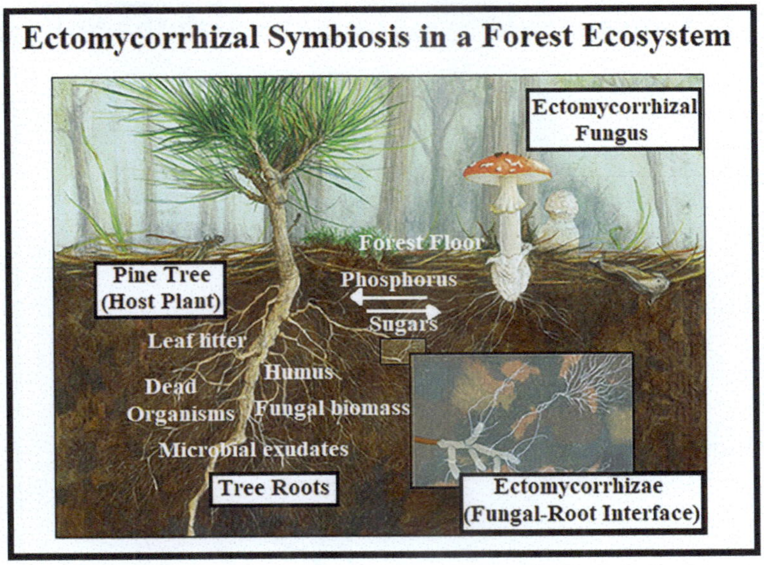

FIGURE 39.2 Ectomycorrhizal symbiosis shows nutrient exchange between tree roots and fungi in a forest ecosystem.

Mycorrhizae are an essential symbiotic partnership between fungus and plant roots that completely alters soil structure, nutrient uptake, and ecosystem resilience. Essentially, this symbiosis splits into two opposing but complementary forms: ectomycorrhizae and endomycorrhizae. Through the penetration of root cortical cells, endomycorrhizae—typified by AM fungi—form complex arbuscules that function as close-knit exchange sites for carbon and mineral nutrients, particularly phosphorus, allowing plants to flourish in nutrient-limited situations (Liu & Chen, 2024). Ectomycorrhizal (ECM), on the other hand, creates an external fungal mantle that envelops the tips of the roots and weaves the Hartig net across the gaps between the cells of the roots. They mostly associate woody plants to maximize the intake of nitrogen and other minerals while controlling the dynamics of the root–soil interface. Though their evolutionary histories, host specificity, and ecological functions are quite different, both types are important drivers of feedback loops in biogeochemical cycles (Eagar et al., 2024). By disturbing the makeup of plant communities, stabilizing soil carbon, and enhancing their ability to respond to environmental disturbances, the interaction of these mycorrhizal species highlights the intricacy of belowground networks and binds terrestrial ecosystems into a web of fungal-plant interdependence (Nizamani et al., 2024).

Another name for the ectomycorrhiza is sheathing mycorrhizae, which coexist with several evergreen shrubs and trees (Fig. 39.2). The genera *Betula, Fagus, Salix, Quercus, Populus, Castanea*, and *Tilia* are among the deciduous plants that have also invaded the area. Young roots are covered by the fungus, which only enters the cortex through the cell wall and stops there. The class of fungi known as Basidiomycetes includes ECM fungi. As their fruiting structures, fungi known as basidiomycetes frequently create mushrooms (Hou et al., 2025). This explains why mushrooms grow in the root zone under a tree's dripline. Ectomycorrhizae play a vital part in the absorption of zinc and ammonium, in addition to absorbing phosphate from the soil. Relatively host-specific fungi coexist with plants in symbiotic relationships. Moreover, flying spores or the spread of diseased plant material are the two ways that ECM fungi spread. ECM linkages are identified by the existence of a Hartig net, which is made up of complex hyphae linking root compartments (Florence et al., 2025). The idea that the Hartig net serves as the main region of nutrient transmission in these interactions is supported by correlations found between host development reactions to strains of injected fungus and the thickness of the Hartig net. Through thorough microscopic examination, ECM relationships may be differentiated from saprotrophic mildews rising on the outward of roots or incidental connections among nonhost species and ECM fungi (Shi et al., 2025). It is not possible to identify ECM relationships only by observing the fruiting of suspected fungal partners close to a possible host plant. These names might be inaccurate if fungi are thought to be ECM companions or if they fruit far from their host tree. These issues are demonstrated by citations to the literature that mistakenly identify vesicular AM (VAM) trees such as *Ulmus, Fraxinus*, and *Acer* as ECM masses. Bead-trimmed VAM roots are present in numerous plants that are incorrectly identified as ECM hosts because, without close microscopic inspection, they can be confused with ECM short roots (Chaudhury et al., 2024).

ECM can be divided into two basic morphological categories: (1) gymnosperm associations, like those of the Pinaceae, wherever the Hartig net dwells in numerous layers of chambers in the cortex and (2) angiosperm associations, like those of *Betula, Eucalyptus, Fagus, Shorea*, and *Populus*, where the Hartig net is restricted to epidermal chambers (Fig. 39.3). The

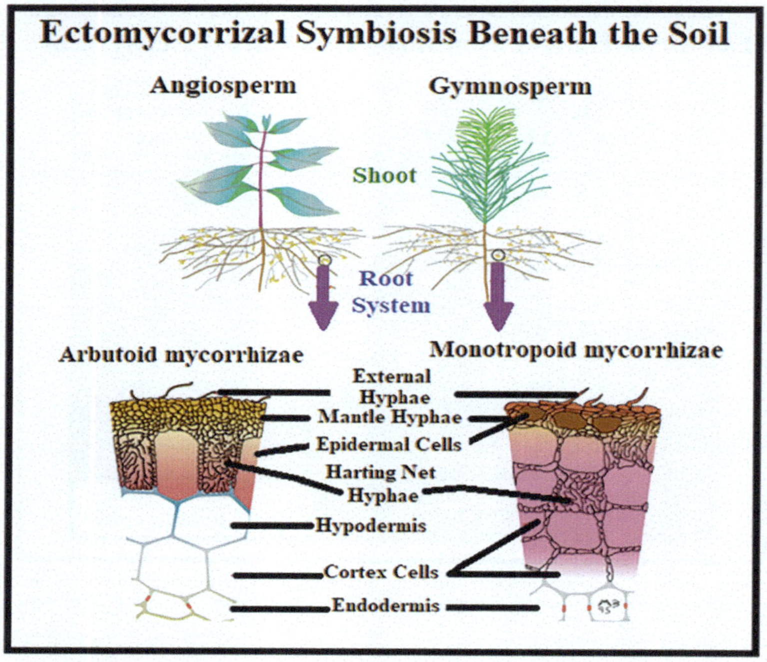

FIGURE 39.3 Ectomycorrhizal network beneath forest soils and arbutoid and monotropoid mycorrhizae symbiosis.

angiosperm *Dryas integrifolia* is one of the rare exceptions to this rule, as it possesses a cortical Hartig net (Antunes et al., 2025). These classifications are based on the host root's anatomical characteristics, and the same fungus can develop both kinds with distinct hosts. Heterorhizic (dimorphic) root organizations, in which small roots contain few apical growths and many diverging densities, are the consequence of the convergent development of plants through ECM. ECM produced by one host interacting with various fungi has significant differences in both structure and function (Sugiyama et al., 2025). From fungi with a small host assortment that interact with a particular host class to mildews with a vast host range that associate with several plant families, plant–fungus specificity in ECM relationships varies significantly. Complementary host–fungus pairings are those that establish functional relationships. It is known that several ECM morphotypes with increasingly thicker mantles and Hartig nets indicate different levels of host–fungus compatibility (Leyva et al., 2025).

Even while arbutoid and monotropoid mycorrhizae show many similarities, they have historically been thought to be distinct from ECM connections. The host plant must regulate the special characteristics of these partnerships as the fungi that establish them are usually also ECM companions of other hosts (Sugiyama et al., 2025). Arbutoid mycorrhizas are inconsistent, according to a review of plants in the Arecaceae family. It suggested that arbutoid and monotropoid mycorrhizae are categorized as subdivisions of epidermal ectomycorrhizae in the hierarchical categorization of mycorrhizal types. The present definition of ecto- and endomycorrhizae is based on fungi rather than plants, and it does not differentiate them morphologically from arbutoid mycorrhizas (Sharma et al., 2025). In addition, there could be instances in which the hyphae that penetrate host cells are not from the same fungus as the one that forms the Hartig net or in which cell dissemination only takes place in senescent relationships. In the categorization system, ectendomycorrhizae are at the same level as superficial and tuberculate mycorrhizae and do not belong to a distinct plant lineage (Naturally, 2025).

39.3 Endomycorrhizae (vesicular arbuscular mycorrhiza)

In contrast, endomycorrhizae enter the plant through its roots and grow inside it (Fig. 39.4). Deciduous trees, annual agronomic crops, and other herbaceous plants are frequently home to vesicular arbuscular endomycorrhizae (V-A type). The mycelium of the endomycorrhizae pierces the cortical chambers of the root, in contrast to the ectomycorrhiza. Within the root the hyphae develop both intracellularly and between cells. Only the short roots of woody ornamentals are impacted (Majhi et al., 2025). Water molds, or Phycomycetes, are a type of fungus that includes vesicular arbuscular endomycorrhizae. *Phytophthora*, which causes crown rot and root rot, and *Pythium*, which causes seedlings to wilt, are two prevalent fungi in this class. Zinc and phosphate are absorbed by the fungus through the formation of specialized absorptive organs known as haustoria. The fungus's spores spread through infected plant matter in the soil after germinating in the rhizosphere, the

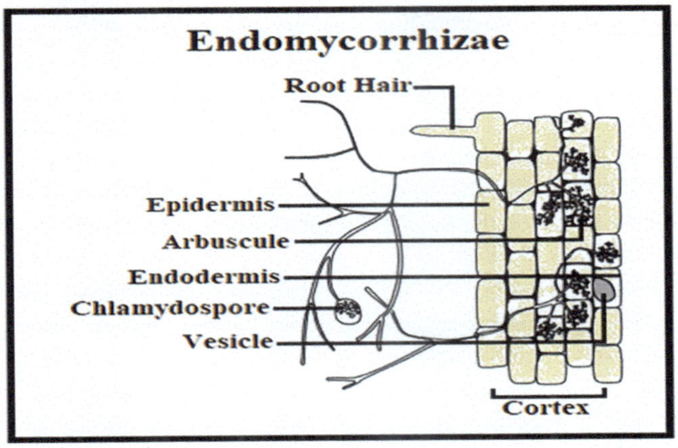

FIGURE 39.4 Endomycorrhizal symbiosis.

region of topsoil immediately next to the root's outward growth (Pan et al., 2025). They will infect any appropriate plant species because they are often agnostic about the host they need. Arbuscular vesicular mycorrhizas' arbuscules are typically employed to identify VAM relationships; their quantity is typically connected with the extent to which VAM fungi have colonized young roots, and they may be measured using standard microscopic techniques. The presence of distinctive hyphal branching patterns in host plants can reliably identify old VAM relationships without arbuscules, provided the observer has sufficient expertise. Glomeromycota fungi's colonization of nonhost plants is frequently referred to as vesicular mycorrhizal colonization; however, it is better described as endophytic activity (Mikiciuk et al., 2024). The following are morphological and functional kinds of relationships that Glomeromycota fungi create in plant organs and that field surveys may come across: (1) characteristic stable VAM in new plant structures with arbuscules and hyphae with characteristic growth patterns colonizing roots widely and uniformly. (2) Mature, composed VAM in roots lacking entire arbuscules but exhibiting typical hyphal colonization patterns. (3) Endophytic relationships with widespread hyphae development in nonhost backgrounds and additional nonmycorrhizal (NM) plant tissues, which lead to sparse, irregular colonization devoid of arbuscules. (4) Manipulative VAM in plants that are incompletely or completely myco-heterotrophic, where specialized hyphae intensely colonize reduced roots or stems to generate unique patterns that occasionally lack arbuscules (uncommon and confined to particular plant families) (Hosamani, 2025). In each of these groups, vesicles may or may not exist. The specific phylogenetic, structural, or functional method utilized to categorize the components of VAM relationships determines the best nomenclature to employ. Certain plants that contain coiling VAM fungus have "inner cortex" connections, in which arbuscules only develop in one inner-cortex cell layer, although hyphae are found throughout the cortex. A distinct subgroup of coiling VAM is plants with "beaded roots," which contain several small segments divided by constrictions (Garcia Massini et al., 2025). Another name for these roots is nodular roots; however, this confuses them with nitrogen-fixing connections. Only plants with twining VAM are known to have internal cortex, beaded, and exploitative relations; further subgroups of winding VAM are seen in plants through manipulative links or with extremely irregular coils (Bikang et al., 2025). Some fungal morphotypes propagate in roots with linear VAM via mutually linear and coiled hyphae, it is particularly uncertain if these mildews produce tiny groups. Approximately, researchers have proposed that floras having roots, including linear and coiled hyphae, must be designated a transitional type (Mikiciuk et al., 2024). However, these contradictions appear to be caused by a specific fungus and hence do not represent valid categories of VAM as described. Mycorrhizal processes can help plants withstand root infections in the following ways: (1) the existence of a layer of hyphae (mantle) that serves as a physical barrier to pathogen penetration. (2) Mycorrhizal fungi absorb more carbohydrates from the roots before excreting them in the form of root exudates, preventing infections from developing. (3) The production of antibiotics is secreted to limit the growth of infections. (4) Encourage the growth of saprophytic microorganisms surrounding the roots, and (5) plant roots infected with pathogenic fungi that exhibit competitiveness (Priyadarshani et al., 2025).

39.4 AMF occurrence and diversity

The presence of AM fungi (AMF) depends on the houseplant type, soil conditions, and ecosystem type: The most common form, found in about 80% of vascular plants (tracheophytes), comprises most crops, grasses, and tropical plants (Moora et al., 2025). Ectomycorrhizae are mainly found in warm and boreal ecosystems, notably with trees such as pines, oaks, birches, and eucalyptus. Instead of entering cells, ECM fungi develop a cover around the root tips and a system linking cells. They are often from fungal groups such as Basidiomycota and Ascomycota. Ericoid mycorrhizae are specialized associations found in plants of the Ericaceae family, common in acidic and nutrient-poor soils such as heath lands and peat bogs (Tedersoo et al., 2024). Orchid mycorrhizae are critical for orchids, whose tiny seeds lack sufficient nutrients and must be associated with fungi (often Basidiomycetes) to germinate and grow. There are roughly 42 AMF species connected with Cerrado biome plants, with the leading genera being *Acaulospora, Glomus*, and *Gigaspora*. AMF incidence studies have been conducted at the regional and local levels. Though coincidence arrangements among AMFs and plant-dependent nematodes need further definition in terms of the ecological factors that determine those (Akbar et al., 2025; Ma et al., 2025).

Mycorrhizae are multidimensional connections that fall into several morphological, functional, and evolutionary groups. Some mycorrhizas are parallel and portion plant ancestries, but others have unique structural characteristics and evolutionary antiquities. VAM links, also termed arbuscular mycorrhizas or Glomeromycotean mycorrhizas, are the maximum prevalent and frequent root–fungus relationships (Drijber & McPherson, 2021). ECM associations are moreover essential in numerous ecosystems, although they are limited to specific plant groups. Other mycorrhizas are limited to Ericales or Orchidaceae, but certain angiosperm groups often have NM roots. AM and ECM plants have different nitrogen (N) acquisition techniques and cost-benefit balances (Zhang et al., 2025b). These changes have a significant impact on plant production, soil organic carbon storage, and the connection between AM and ECM-dominated ecosystems. The variety of AM classes, such as *Funneliformis mosseae, Claroideoglomus etunicatum,* and *Rhizophagus intraradices*, affects plant stress acceptance and nutrient achievement. Nevertheless, their efficiency differs depending on the ecology. For instance, *R. intraradices* improves lead acceptance in soybeans, but *R. intraradices* and *C. etunicatum* increase nutrient and biomass absorption in further plants. Yet, some types behave inversely in diverse eco-friendly situations (Wang et al., 2025a, 2025b). *R. intraradices* tolerates molybdenum better than *C. etunicatum*, whereas *Claroideoglomus claroideum* and *Rhizophagus irregularis* improve nutrient absorption and minimize manure dependency. However, these advantages are not worldwide, because similar kinds may behave differently across habitats or crop kinds because of varying environmental circumstances and host plant partialities. AMFs play a significant role in weathering stressors in desert habitats by enhancing the moisture and nutrient availability to desert plants (Yang & Wang, 2024). The potential significance of intraspecific diversity in ecosystems outlined four scenarios in which intraspecific diversity might have significant implications for ecosystem functioning. (1) When a few keystone species dominate the community. (2) When one species' genetic diversity directly influences the abundance, distribution, or function of a keystone species, (3) when phenotypic diversity reflects genetic diversity, and (4) in changing and varied settings (Farah & Fisher, 2025).

39.5 AMF establishment and signaling

Complicated molecular networks that organize fungal invasion and nutrient interchange establish AM symbiosis. The relationship starts with the production of strigolactones by plant roots, which promotes fungal spore development and hyphal division. The collective symbiosis-signaling pathway, which comprises receptor-like kinases like lysin motif receptor-like kinases (LysM RLKs), is responsible for identifying these signals. Key controlling genes perform crucial parts in this procedure (Naz et al., 2025). This activation sets off calcium signaling cascades that are mediated by Cyclops proteins and calcium and calmodulin-dependent protein kinase (CCaMK) (a DNA-obligated transcriptional activator). These forces are essential for transcriptional rearrangement and the expansion of association. For instance, RAM1 (reduced arbuscular mycorrhiza 1) controls the genes intricate in nutrient conveyance and arbuscule construction, helping as a principal controller of mycorrhizal development (Pimprikar & Gutjahr, 2018).

The PHT1 family of crop phosphate carriers, which is expressed chiefly through settlement, encourages phosphate absorption, typical of AM symbiosis. Arbuscule manufacture is troubled by these carriers' dysfunction, highlighting their vigorous consequence in nutrient conservation (Gao et al., 2025). Through an arrangement of metabolic procedures that comprise equally straight absorption and exchange, AM helps in the carrying of nitrogen. Over its extraradical mycelium (ERM), AM initially takes up nitrogen from the soil as NH_4^+ (ammonium) or NO_3^- (nitrate). The external hyphae take nitrate straight and use nitrate reductase enzymes to comportment a decrease method that converts it into ammonium. Preceding additional decrease to NH_4^+ (ammonium), nitrate is primarily compact to NO_2^- (nitrite) (Wang et al., 2025a, 2025b). As AM fungi are further modified to obtain nitrogen, this exchange was supplementary there than in nonmycorrhizal fungi. Subsequently, absorption, ammonium is transformed into amino acids, with arginine aiding as the major nitrogen carrier. In the ERM, arginine is made, assisted by the enzymes carbamoyl-phosphate synthetase and argininosuccinate lyase. Next, arginine is conveyed to the intraradical mycelium, where it is influenced and changed at the symbiotic boundary into free ammonium and other amino acids (Ren et al., 2025). Definite carriers, including ammonium transporters and NPFs (nitrate Transporter 1/peptide Transporter Family), are improved in response to AM establishment and discharge these nitrogen molecules for the crop to absorb. To encourage plant improvement and domain symbiotic stability, this harshly measured technique ensures that nitrogen is available to the crop in an actual and organized way (Ma et al., 2025).

Plant cellular signaling, which comprises dynamic and regular rises in calcium absorption inside the perinuclear cytoplasm and nucleus, is influenced analytically by calcium stress. The calcium assets near the nuclear cover, most probably the endoplasmic reticulum, are the cause of these calcium alternations. Calcium networks for discharge; counterion channels such as potassium or anion channels to stabilize responsibilities throughout release; and calcium ATPases that pump calcium reverse into the accumulation for replacement are thought to be the vital molecular technology obligatory to assist calcium points (Diaz et al., 2025). Between these crucial basics, it has been resolved that the POLLUX and CASTOR networks, occasionally referred to as MtDMI1, which control calcium signaling inside the nuclear envelope, are counterion channels. In symbiotic associations among floras and microorganisms such as AMF and rhizobia, calcium signaling is important. One of the primary conspicuous responses that plants have when they classify these symbiotic associates is the nucleus's calcium fluctuations. This procedure shocks for rhizobia when the plant observes Nod features, which sources calcium spikes in a problem of action. The complexity of this signaling way is painted by the contribution of receptor developments related to the plasma membrane, as well as additional fundamentals associated with the nuclear envelope (Zheng & Jiang, 2022). Similarly, even before direct contact with fungal hyphae, calcium spiking is seen in plant roots reacting to AMF. This implies that the calcium response is triggered by diffusible fungal signals, including Myc factors. These signals include lipochitooligosaccharides and chitooligosaccharides (LCOs and COs), which are important molecules. It has been demonstrated that sulfated and nonsulfated LCOs generated from fungi, or too short-chain COs, cause calcium oscillations in the epidermal compartments of houseplant roots. Furthermore, during symbiotic establishment, patterns of calcium spikes change. For instance, preceding poisoning, the external cortical cells display little frequency oscillations, whereas symbionts pass in host cells and show greater frequency spiking (Luan & Wang, 2021). This improvement advises that calcium signaling is severely measured and adjusts to the numerous periods of symbiotic development, organizing the host plant's cellular rearrangement to support the symbiotic method. The consequence of calcium signaling on plants' ability to differentiate among many symbiotic bacteria in the primary stages of interaction is emphasized by these responses. Future investigation presented that these responses are identical and analogous, particularly when formed by LCOs, notwithstanding the early theory that AMF and rhizobia might yield diverse calcium spiking marks (Sun et al., 2015). In addition, the beginning outlines of short-chain LCOs and COs contrast by cell kind, with atrichoblasts and trichoblasts reacting in a different way based on their groupings and absorptions. The combinatorial nature of these indications is revealed by the information that, in rice, calcium spiking in trichoblasts is single seen when mutually LCOs and COs exist (Diaz et al., 2025).

39.6 Phytohormones and arbuscular mycorrhizal symbiosis

Through monitoring the development and maintenance of AM relationships, phytohormones have an important effect on establishment, stress tolerance, and arbuscule creation. Auxins, for example, play a key role in the primary stages of establishment by monitoring the production of strigolactone, which stimulates fungal hyphal division and supports the growth of a symbiotic assembly (Naz et al., 2025). Via directing auxin carriers, strigolactones themselves are utilized as significant signaling molecules during the presymbiotic age, inducing auxin movement and fungal metabolism. Moreover, AM influences comprise cytokinin (CK), and an improved AM foundation is regularly connected with difficult CK stages in the host plant. To possess an affirmative response loop that cheers occupation, AM fungus can produce CKs, increase host CK biosynthesis, or avoid CK deprivation. On the other hand, AM colonization produces unfair gibberellic acid (GA), with mycorrhizal roots commonly displaying larger quantities (Ahmed et al., 2025). Even though earlier they were supposed to stop AM symbiosis, GAs have a complex character, and their properties change, allowing for additional hormones and circumstances. Moreover, intricate in AM connections is jasmonic acid (JA), which may have too encouraging and inhibitory effects on absorption and ecological features. To switch plant responses and exploit symbiosis in stress, JA includes hormonal signals approximating CKs and GA (Samanta & Roychoudhury, 2025). It has been revealed that salicylic acid improves AM symbiosis by stimulating plant development and generating metabolic pathways to facilitate multiplicative processes, particularly under demanding conditions resembling salt stress. Abiotic stressors, such as dehydration and heavy metals exterminating, further obscure the connections concerning these hormones. For illustration, abscisic acid (ABA) ranks increase in response to drought stress, associated with AM establishment and utility. Similarly, AM can increase plant acceptance to heavy metal stress by persuading the production of ABA, GA, and Indole-3-acetic acid (IAA) (Khan et al., 2020). As it comprises gene transcription, protein synthesis, hormonal equilibrium, and hormonal crosstalk, it is very dynamic, with numerous ways cooperating at dissimilar stages. On the molecular stage, AM symbiosis helps the appearance of vital genes linked to decontamination and stress retorts, containing heavy metal ATPase genes, which reduce

Cu poisoning. This involved hormonal regulator assurances that AM can tolerate plant fitness in severe ecological conditions, signifying the energetic interrelationship among hormones and AM that maintains productive symbiosis (Yusuf et al., 2025).

39.7 sRNAs' role in AMF symbiosis

A controlling instrument that is continuous in eukaryotes, RNA interference (RNAi) is critical for stress retorts and improvement. Arrangement complementarity regulates whether sRNAs, which facilitate this appliance, direct gene silencing transcriptionally (TGS) or posttranscriptionally (PTGS). In connections among biotics and plants, containing symbiosis and pathogen resistance, plant sRNAs are indispensable. They stop viral infections, defend the reliability of the genome in contrast to transposable elements (TEs), and regulate endogenous genes (Diaz et al., 2025). Plant sRNAs, such as miR166 and miR159 in cotton, can translocate into pathogens and affect their virulence. In addition to viruses, plant sRNAs react to fungal and bacterial elicitors such as chitin and flagellin, stimulating antimicrobial resistance. Viral dsRNAs are converted into viral sRNAs through host RNAi technology, which aims viral genomes to prevent contaminations or, in approximately cases, modifies host transcriptions for viral determination (Jia et al., 2025). AMF invasion dramatically alters plant sRNA expression, a trait shared by all species. MiRNAs influenced by AM colonization in *Medicago truncatula* target transcription factors that control immune suppression and AM development. Likewise, Argonaute protein (AGO7) controls AM-induced sRNAs in *Nicotiana attenuata*, affecting colonization via phytohormone signaling and nutrition absorption pathways. AGO7 loss promotes mycorrhizal colonization, which is consistent with findings in *M. truncatula* mutants, wherever nodulation is also elevated, indicating that symbiotic relationships share processes (Silvestri et al., 2025). Important components of RNAi machinery, including Argonaute (AGO) proteins, RNA-dependent RNA polymerases, and Dicer-like (DCL) proteins, are found in almost all fungal lineages, including AM fungi. Multiple genes per genome frequently encode these components. AMF similarly holds the sRNA methyl transferase HEN1, which alleviates sRNAs, a gene misplaced in some other fungal lines (Felix et al., 2024). The vigorous appearance and direction of RNAi-related genes in AMF throughout their lifetime were confirmed by proteome and transcriptome countenance inquiries. AMF, between other belongings, produced sRNAs with characteristic distances of 24–26 nucleotides (nt), which are similar to the action of a single DCL enzyme, according to sRNA sequencing (Mierziak & Wojtasik, 2024). Confident sRNAs have a 5′ uracil, and HEN1 homologs facilitate the 2-*O*-methylation at the 3′ termination. Two key classes of sRNA-generating loci have remained originate over genomic studies: those linked to TEs, which create sRNAs with a span of 22–26 nt, and those related to protein-coding genes, which yield sRNAs with changing measurements (Silvestri et al., 2019).

39.8 Techniques used for mass production of AMF

In recent decades a variety of methods for producing AM fungi in large quantities have been developed. The obligatory nature of this biotrophic fungus and the inability to identify AM fungal species at the species level during the initial period of development are the primary barriers to mass production approaches (Marysol et al., 2025). For the large production of AM fungus, in vitro cultivating techniques, including root tissue culture and hydroponic systems, are currently commonly utilized. In addition to preserving the quality of AM fungal propagules, these techniques have the potential to become economically viable means of mass multiplication of AM fungi. Three main, well-known methods are frequently used in AM fungal mass manufacturing. These three categories of manufacturing organizations are in vitro, substrate-based, and substrate-free (Jo et al., 2025).

39.8.1 Substrate-based production system

The traditional technique for manufacturing AM fungus is another name for this process. First a soil- or sand-based substrate is used to grow the plants and the symbionts that are connected to them. These molds grow for frame multiplications operating a single class or a group of recognized AM fungal types in soil or artificial jars, or they may be grown up to average-sized masses and dishes and massively raised or beached beds after the first creation of AM fungal inoculums (Liu et al., 2025). To manage the temperature and humidity the whole arrangement configuration grows in a glasshouse or plant development cavities in semicontrolled or controlled conditions. A distinct group of spores and diseased root sections often make up the beginning inoculum. Damp filtering and pouring procedures are utilized to extract the spores, whereas the root segments are dried and finely diced to produce the mixed inoculums for the subsequent inoculum preparation. When producing AM fungal species that can create intraradical spores and vesicles, mixed inocula are frequently utilized (Sellappan & Thangavel, 2025).

39.8.2 Substrate free production system

Abundant nutrient stream procedures or substrate-free farming arrangements are working. These approaches may differ in how the nutrient solution functions and how the drying passes out. To stop oxygen deficiency in the roots, a ventilation force exposes the nourishing solution in the stationary care of the scheme. To decrease the movement of nutritive solutions and air bubbles, which can damage the development of the delicate further radial hyphae, the pumps essentially revolve on and off on a steady basis (Drewnowski et al., 2019). Another approach is the current nutrient method, where the roots of the AM fungus and plant develop in a thin layer of nutritive solution that shelters the roots and enlarges the comparative range for gas exchange. A kind of hydroponic organization called aeroponics includes host plant roots and immersing AM fungus propagules in a nutrient solution. The fluid layer that surrounds the roots enables gas exchange, and the use of microdroplets advances the method's aeration (Reddy et al., 2025). When constructing AM fungi in great amounts the ultrasonic vaporizer process is assumed to be enhanced way than atomizing disks. The root remains that comprise several infective propagules influence working as an inoculum accurately. The production of algal pollutants is quite likely to occur in liquid nutrient solutions. Lack of a transporter substrate may also have an impact on the amount of spore generation (Lalsare, 2024).

39.8.3 In vitro production system

To propagate AM fungus, several researchers have employed a variety of intricate in vitro systems. One such system is one in which the young branch is constantly external to the Petri plate. In the meantime the AM fungus and roots are housed in Petri dishes that have been filled with an appropriate gel medium. In addition, in a closed system containing nutrient solution, the preinoculated plants generated separately delivered AM fungal inoculum in a sterile growing tube (Subhash et al., 2025). This process is more suited for producing high-quality AM fungal inoculum in large quantities because of the absence of undesirable bacteria. It is constantly necessary to monitor and control the cultures in this system. Laboratory equipment and competent staff are also needed to make this method cost-effective. However, the requirement for frequent additions of growth media in in vitro plant cultures raises the possibility of cross-contamination (Indu, Hemamalini, Nagamani, & Anuradha, 2025).

39.9 Isolation of vesicular arbuscular mycorrhiza

39.9.1 Isolation from plant roots

The following procedures are used to isolate VAM species from root samples: The root samples must (1) be cleaned under running water, (2) be cut into 1-cm fragments, and (3) be preserved with a 10% KOH solution for 30 minutes at 40°C. (4) Rinse the roots well in a beaker with at least three full changes of tap water, or until the rinse water is completely free of brown. (5) Trypan blue (0.01% trypan blue) is used to stain root pieces for 24 hours at room temperature, following a thorough washing. Under a microscope the dyed root fragments were mounted in lactoglycerol and checked for AMF colonization (Dudhane et al., 2024).

39.9.2 Isolation of vesicular arbuscular mycorrhizal spores from rhizosphere soil

The "damp filtering and pouring technique," created by J.W. Gerdemann and T.H. Nicolson in 1963, is frequently used to separate spores. The following steps are involved:

39.10 Extinct and cryptic species of (AMF)

With the advent of environmental DNA sequencing and metagenomics, researchers have uncovered hundreds, potentially thousands, of genetically distinct AMF lineages that do not yet have formal taxonomic names (Fig. 39.5). Many of these cryptic species exhibit significant genetic divergence, ecological specificity, or geographic isolation, suggesting they are biologically distinct despite their morphological similarities. This unseen variety reveals the earliest evolutionary antiquity of AMF, which dates back over 400 million years, and highlights their adaptive characters in different environments (Cahill et al., 2024). As taxonomic implements develop further advances, it is estimated that a variety of these unidentified, then genetically different objects would finally be defined as a novel type, inspiring our appreciation of the difficulty and environmental significance of this symbiotic fungal assembly. The extermination of AM fungi (AMF) classes is a matter that residues mostly hypothetical, as continuous indication of vanished AMF kinds is particularly restricted because of their microscopic size and the insufficiency of well-conserved fungal remains (Wadas & Domingues, 2025). The absence of

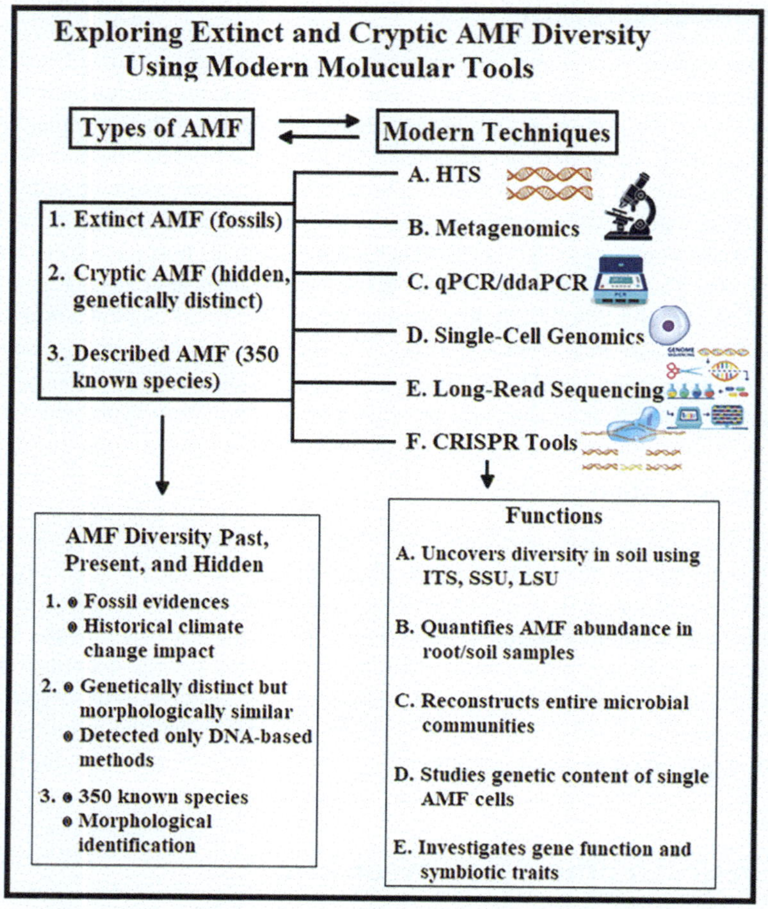

FIGURE 39.5 AMF Diversity and use of molecular tools. ddPCR, Digital droplet PCR; *HTS*, High throughput sequencing; *ITS*, internal transcribed spacer; *LSU*, large subunit; *qPCR*, quantitative PCR; *SSU*, small subunit.

explanations for indefinite taxa can be recognized as three dynamics. (1) They could be strains that are hard to cultivate. (2) In many earth-terrestrial locations, there may not have been enough AMF sampling to detect them. (3) The molecular and morphological classification of this collection of fungi is now handled by a limited number of specialists (Zhang et al., 2021).

However, scientists believed that a significant number of AMF species might have gone extinct over the course of their evolutionary history, which spans over 450 million years. Fossil evidence, such as that found in the Rhynie chert of Scotland, dating back to the Devonian period, suggests that early AMF-like symbioses were previously present during the initial colonization of land by floras (Antoine et al., 2021). Given this immense timescale, it is likely that numerous species evolved and later vanished because of major shifts in climate, mass extinction events, geological changes, or the transformation of plant communities. One key driver of AMF extinction could have been the **evolution of nonmycorrhizal plant lineages** or those that form alternative types of mycorrhizae (e.g., ectomycorrhizae), which would have reduced the ecological niches available to AMF (Liu & Chen, 2024). In addition, **changes in soil chemistry,** such as the rise in oxygen levels, nutrient availability, and pH fluctuations over geological time, could have altered the suitability of certain habitats for specific AMF lineages. The **coextinction** of host plants is another crucial factor because, as obligatory symbionts, AMFs depend on plant roots for carbon; the destruction of a plant class might easily lead to the loss of its associated AMF species (Xiao et al., 2024). Moreover, in modern times, human activities such as intensive agriculture, pollution, deforestation, and habitat fragmentation are likely causing local or even global extinctions of AMF taxa, though these losses often go undocumented because of the cryptic nature of these fungi. Therefore, although it is difficult to put an exact number on extinct AMF species, it is highly probable that a substantial and underrecognized portion of their historical diversity has been lost over time, both in deep evolutionary history and because of ongoing anthropogenic pressures (Izar et al., 2025).

To date, scientists have identified approximately **300–350 validly described species** of AMF, belonging primarily to the subphylum "Glomeromycotina." These species have been classified through a combination of morphological traits such as hyphal adhesion, wall structure, and spore size, as well as molecular markers, particularly rDNA sequences. Traditionally, AMFs are identified based on the physical characteristics of their spores and other structures (such as hyphae, vesicles, and arbuscules) (Zhang et al., 2025c). Important morphological features include size, color, and form of the spores; wall construction and number of layers; attachment structures (such as subtending hyphae); and surface ornamentations. Spores are usually extracted from soil using the damp filtering and pouring technique methods and surveyed by sucrose centrifugation to separate spores for microscopic observation (Marysol et al., 2025). Because morphology can sometimes be ambiguous (e.g., similar-looking spores from different species), molecular methods are now prevalent (Noreen et al., 2023). Molecular characterization consuming DNA arrangements presented in the 1990s 14 to categorize and calculate AMF in the roots of floras. An often-used portion for determining thoroughly connected species includes the whole internal transcribed spacer (ITS) rDNA region, the small subunit (SSU) rRNA gene, comprising the partial large subunit (LSU) rRNA gene, and 5.8 S rRNA genes, denoted as SSU–ITS–LSU (Corsaro, 2024).

39.10.1 High-throughput sequencing

This includes next-generation sequencing platforms, approximating Ion Torrent, Illumina, and PacBio, which allow massive parallel sequencing of DNA. Specific regions such as ITS, SSU rRNA, or LSU rRNA are amplified and sequenced to characterize AMF communities. Sequencing entire genomes of AMF species is challenging because of their obligate biotrophy and complex genomes (Lanfranco & Bonfante, 2023).

39.10.2 Metagenomics and metatranscriptomics

DNA sequencing is extracted from soil and root samples without PCR bias and helps to understand not just "who is there" but also potential functions such as nutrient cycling genes. RNA sequencing is used to study active AMF and their gene expression during symbiosis. Both provide insights into the construction and function of AMF in natural surroundings (Jiang et al., 2024).

39.10.3 Quantitative PCR and digital droplet PCR

Quantitative PCR, or polymerase chain reaction, is utilized to measure the profusion of specific AMF taxa in a tester. Digital droplet PCR is a newer, more sensitive method that partitions a PCR reaction into many droplets, enabling precise measurement without a standard curve. AMF populations are highly sensitive and precise measurements when they are at low abundance (Richter et al., 2022).

39.10.4 Single-cell genomics

Arbuscular mycorrhizae fungi are multinucleate and heterogeneous; single-nucleus or single-spore sequencing is being explored. Individual nuclei or spores are isolated and sequenced to study genetic variation within a single AMF individual. Techniques like whole genome amplification are used to amplify tiny amounts of DNA. Demonstrate that intraorganismal genetic diversity is very important for AMF, which do not follow simple diploid genetics (Manyara et al., 2024).

39.10.5 Long-read sequencing technologies

Platforms such as Oxford Nanopore and PacBio provided sequencing across entire rRNA gene regions, such as complete 18S-ITS-28S, in one read. Better resolution in identifying closely related spores of AMF species and de novo genome assemblies, overcoming the problems of short-read fragmentation and better handling highly repetitive AMF genomes (Mordret et al., 2024).

39.10.6 Clustered regularly interspaced short palindromic repeats Cas-based techniques

The techniques identified as clustered regularly interspaced short palindromic repeats were established to edit AMF genomes. To study gene function directly by knocking out genes that are involved in symbiosis, including nutrient transporters and signaling proteins. Functional genomics and synthetic symbiosis engineering are enabling (Garg et al., 2025).

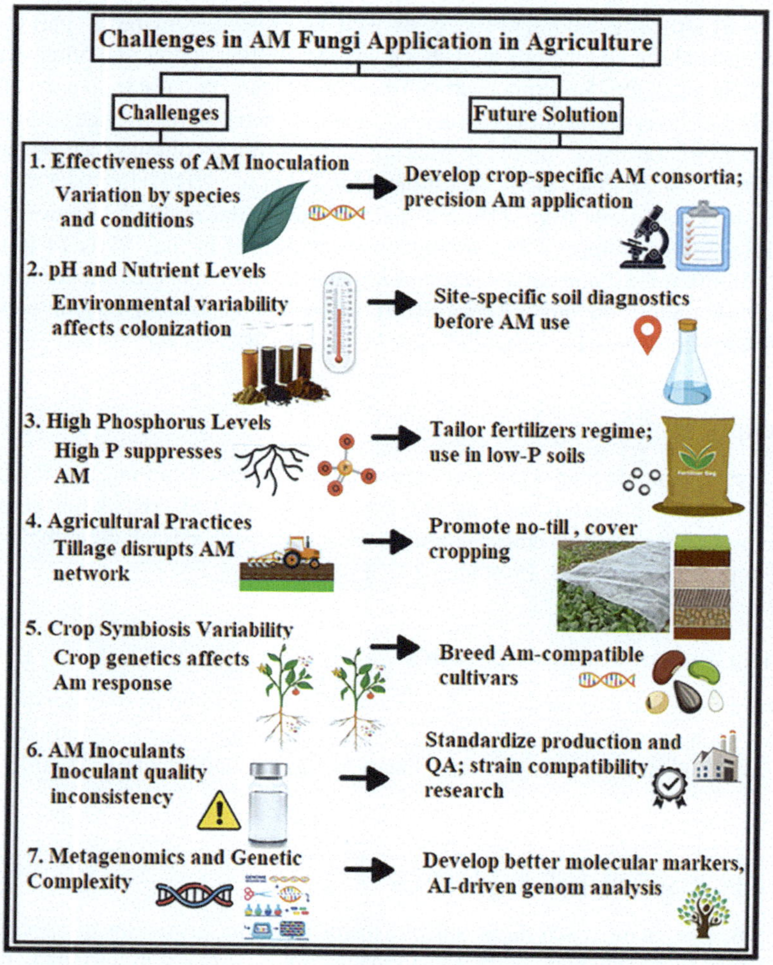

FIGURE 39.6 AMF fungal challenges and potential solutions. *AM*, Arbuscular mycorrhizal.

39.11 Challenges

The following issues plague the use of AM in agriculture, even though it has several advantages in terms of improving soil structure, nutrient absorption, and stress tolerance (Ahmed et al., 2025) (Fig. 39.6).

39.11.1 Fungal coinoculation and culture medium effects

Coimmunization of AMF and *Penicillium* caused in advanced root colonization than either treatment alone, suggesting a definite reciprocal fungal stimulation during the interaction between AMF and the endophyte. However, when the consortium's *Penicillium* was replaced with the same fungus but cultivated in a different culture medium, coinfection stress led to slower development, leaf drooping, and death. This suggests that when using AMF and assistance organisms, the culture medium of the inoculum is a vital feature to take into consideration (Basiru & Hijri, 2022).

39.11.2 Mycorrhizal networks and biotic resistance

In addition, it has been shown that soil planted with coffee seedlings derived from a mix of grafting and mycorrhizal symbiotic procedures had a reduced Plant Parasitic Nematodes (PPN) density. A common mycorrhizal network (CMN) that connects many plants through their roots facilitates the redistribution of carbon and nutrients as well as the transfer of disease resistance and defense signals across plants. Monoculture, tillage, and genetically engineered crops can all have an impact on AMF colonization. It is necessary to assess the AM association's performance in light of the shifting climate (Marrassini et al., 2025).

39.11.3 Elevated CO_2, AMF interaction, and herbivory

Frew and Price (2019) have anticipated certain results in plant–herbivore interactions under high CO_2 settings. They expect that both C3 and C4 plants will produce more photosynthates in response to increased CO_2. Although C4 plants are better known for sharing photosynthates with AMF symbionts, C3 plants are known to use extra photosynthates for their own growth. As a result, C4 plants will have stronger AM colonization than C3 plants, which will improve their ability to absorb nutrients (Zhao, 2024). However, compared to C3 plants, the insect herbivores that consume those C4 plants would also receive superior nutrition. Furthermore the diversity of AMF species, renowned for superior nutrient uptake, is predicted to decline because of increased CO_2. To satisfy their nutritional needs, insect herbivores will consume more if that occurs, which would have a detrimental effect. This suggests that the efficacy of AM will be significantly influenced by shifting agroclimatic conditions, particularly increased CO_2 concentration and altered soil physicochemical characteristics (Purohit et al., 2024).

39.11.4 Factors influencing AMF effectiveness against pathogens

The mycorrhizal species may not be the sole element influencing how well AMF protects a particular crop against a disease; additional considerations include rhizosphere chemistry, related microbes, and agroclimatic conditions. An efficient AMF for giving biotic stress resistance to a comprehensive variety of plants that are affected by various diseases and herbivores is the genus *Glomus* (Sellappan & Thangavel, 2025). On the other hand, it has been observed that many AMF species can effectively manage a single plant disease. In these situations, it is possible to evaluate the efficacy of several AMFs in managing a specific illness under shifting circumstances. Lastly the most effective AMFs for a certain ailment may be found and used for greater effectiveness (Fang et al., 2024).

39.11.5 Effectiveness of arbuscular mycorrhizal inoculation

Contingent on the yield species, soil category, and ecological variables, AM immunization has an inclusive series of properties. Supplementary yields, such as legumes and rice, may perhaps respond to AM partially or asymmetrically, whereas crops resembling wheat and maize always advantage from it. Besides, AM usefulness might fluctuate based on eco-friendly aspects, even in the same yield species. *R. intraradices*, for example, advances drought acceptance in certain plants, nevertheless not in others (Liu et al., 2024).

39.11.6 PH and nutrient level

AM establishment is likewise influenced by soil features approximating nitrogen levels, pH, and temperature; later, it is difficult to transform AM application systems to an ensemble of definite ecological conditions (Ghorui et al., 2025).

39.11.7 High phosphorus level

Definitely, developed phosphorus stages can stop AM from settling, but inferior points of microscopic biomass and soil gradual carbon can increase AM's significance for plant improvement (Ahmed et al., 2025).

39.11.8 Agricultural practices

Tillage is a unique farming practice that can interrupt AM systems, decreasing their efficiency. The supplementary methods, such as cover yields and fewer tillage, can support AM establishment (Chaudhary et al., 2025).

39.11.9 Crop species form effective arbuscular mycorrhizal symbiosis

Moreover, the ability of diverse yield species to create resourceful AM symbioses fluctuates. Contrasting with wheat, potatoes react to AM frequently, and present cultivars may develop less accessible uncertainty; they are selectively farmed for crop-connected characters (Ahmed et al., 2025).

39.11.10 Arbuscular mycorrhizal inoculants

AM inoculants that are retailed commercially can also diverge in superiority; some of these crops comprise nonfeasible propagules, which makes their expenditure level extra interesting. Important problems include increasing the production of AM inoculants, enhancing their compatibility with a variety of crops, and resolving the diversity in field performance (Ghorui et al., 2025).

39.11.11 Metagenomic approaches

The depth-of-coverage problem can be addressed with higher throughput sequencing, but this can result in additional issues connected to the great computational difficulties for huge datasets. Nevertheless, metagenomics approaches remain difficult for accumulating complete fungal genomes from topsoil models because they often denote a very small amount of soil DNA (Ahmed et al., 2025). The difficulty in assessing the degree of intraspecific physiological variation is especially evident in AM fungi because of our incomplete and even contradictory understanding of their distinct reproduction strategies. They are multigenomic, which resources that nuclei may separate and activate recombination in newly made spores. In addition, recombination may further outcome through the combination of heritarily diverse spores into a solitary mycelial system by a direct involvement method (Hou et al., 2025). According to current information, some characters from the identical atmosphere have knowledgeable recombination in wildlife, but others have fallen away from clonal ancestries. There is a slight demand that there be recurrently important genetic variety in a recognized AM fungal type; however, it may be stimulating to relate predictable thoughts of individuals and species to AMF. This influence suggests that AM fungi are extra vital for "intraspecific" multiplicity as it is nowadays defined than last mycorrhizal forms and that genomic matters may change over time (Aguirre-Carvajal et al., 2025).

The asset of molecular indications active for interspecific and intraspecific AMF empathy and quantification is individual of the principal problems in AMF study. Evaluating the variety of AMF in molecular field examinations and conveying a single gene order to a fungal strain through other problematic aspects by the important intraisolate distinction of nuclear ribosomal genes. The massive task of manufacturing the nuclear genome of *R. irregularis* has delayed the search for unique indicator genes (Aparicio Chacon et al., 2024).

39.12 Future perspectives

To evaluate the AM fungi procedure in applied environments and generate superior practices, widespread field exploration is necessary. AMF multiplicity and useful possibilities may be planned with the usage of cutting-edge approaches, such as high-throughput sequencing and metagenomics (Kumar et al., 2025) (Fig. 39.7).

39.12.1 Quality control and environmental influences on AMF inoculant effectiveness

It is fundamental to set up superiority regulator processes for the inoculants to resolve these difficulties. Ecological foundations that inspire AMF's efficiency embrace water and temperature accessibility. Greater temperatures have the probability to reduce establishment in certain states, although cumulative mycorrhizal wealth in others (Moora et al., 2025).

39.12.2 Enhancing AMF application through sustainable practices and targeted inoculation strategies

In addition, the conditions for exterior inoculants may be diminished by associated limited AM inhabitants through sustainable approaches, such as yield divergence and organic agriculture. To assurance its efficiency, AM inoculation must be modified for specific yields, soils, and microbial inhabitants. In command of determination the residual uncertainties near AM applications, advanced study, and field testing are essential. This research contributes to improving systems and their effectiveness in a diversity of farming surroundings (Mikiciuk et al., 2024).

39.12.3 Biotechnological advances and future of AMF in agriculture

Novel enlargements in biotechnological skills, such as genetic modifications and bioformulations, exist, inspiring trials for successful AM's efficiency. In addition, in a diversity of agro-environmental settings, exact cultivation expertise can increase yield flexibility, reserve productivity, and AM usage. AM has the prospective to mark a considerable influence on nutrition

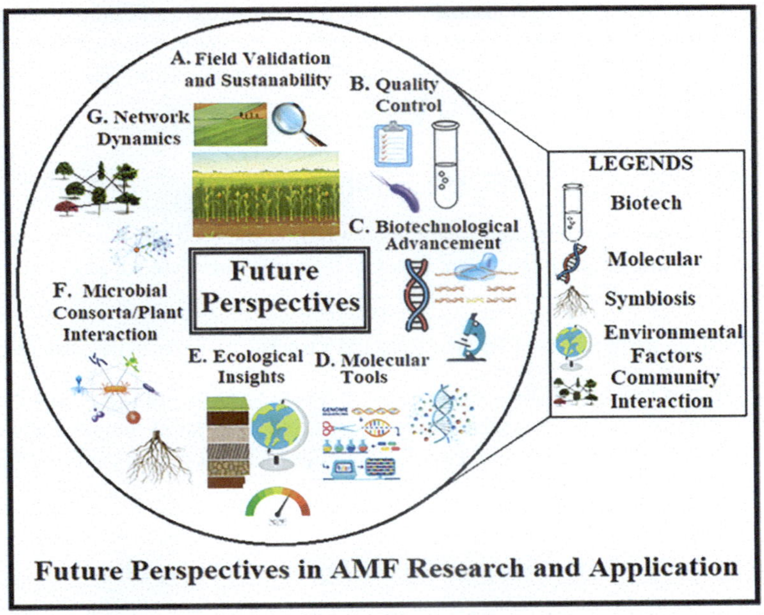

FIGURE 39.7 AMF investigation predictions for the prospect.

safety, sustainable cultivation, and weather alteration qualification with more learning, enhanced inoculation systems, and broader application. To integrate AM into global farming organizations, interdisciplinary collaboration, governmental care, and agriculturalist learning are essential (Athuman, 2023).

39.12.4 Future direction in AMF signaling research

Forthcoming education should emphasize several significant topics, such as an improved understanding of communication among the CO/LCO and SL signaling ways and the purpose of other root exudates. Specifically the instruments supervise the carrying of calcium spiking apparatus constituents, and the purposeful insinuations of unalike spiking subtleties amid rhizobial and AM fungal exchanges need extra study into the downstream gesturing falls activated by these particles, containing transcriptional instruction and calcium signaling (Diaz et al., 2025).

39.12.5 Exploring GRAS protein

The purposes of different Gibberellic acid insensitive Repressor of Gibberellic acid Amidosynthetase (GRAS) subfamilies in varied plant classes, as well as posttranslational and posttranscriptional alterations, should be the emphasis of upcoming education on GRAS-domain proteins. Lastly, it is critical to functionally characterize various sRNA inhabitants and RNAi mechanisms in AM fungi and plants. This includes validating in silico sRNA–mRNA relations, investigating mycovirus-mediated variation, and focusing on the tools of sRNA conversion (cross-kingdom RNAi) via extracellular vesicles. Considering climate change, these studies will offer a more thorough knowledge of AM symbiosis as well as a foundation for creating plans to improve plant–AMF exchanges for sustainable farming (Firdaus et al., 2025).

39.12.6 Harnessing beneficial microbes for improved legume cultivation

Future studies are expected to contribute to the segregation of naturally occurring beneficial microbes and test the application of enhanced mixtures of designated strains that transmit remarkable and harmonious characters in legume yields (Compant et al., 2025).

39.12.7 Translating omics insights into practical applications for AMF

A good acceptance of the hereditary, physiological, and ecological instruments fundamental to these mutualistic relationships may be obtained by studying mycorrhizal symbioses using genomics and other -omics techniques. The information ought to be converted into useful applications that support ecosystem preservation and functioning, aid in the general slowing down of climate change, and guarantee sustainable agricultural land. Utilizing genomes and other omics methods to enhance our comprehension of mycorrhizal relationships, comprising their development, progression, utility, and environmentalism, still presents several long-term problems (Antunes et al., 2025).

39.12.8 Genomic exploration of mycorrhizal diversity

Certain gene sets linked to adaptation to understudied habitats, including the cold, alpine, tropical biomes, and desert, may be discovered by examining the genomes of mycorrhizal types in these regions (Martin et al., 2024).

39.12.9 Advance molecular approaches to unravel mycorrhizal function and evolution

It is possible to link genes to known biological processes and possibly forecast evolutionary associations and environmental adaptation by employing techniques such as multiplexed single-cell and 3D spatial gene expression examination of mycorrhizal roots; proteomics; established transcriptomic expression study; and protein–protein interaction studies (Xia et al., 2025).

39.12.10 Uncovering the role of effector proteins in mycorrhizae symbiosis

Protein buildings generated by symbiosis-controlled genes with unidentified roles, including effector-comparable secreted proteins, can be identified with the use of methods employing high-throughput protein-folding calculations. It is essential to comprehend the genesis of these effector proteins, although they originated from saprotrophic lineages or developed independently to promote establishment in plant tissues (Snelders et al., 2022).

39.12.11 Exploring signaling pathways and master genes in mycorrhizal symbiosis

The complex chemical interactions between mycorrhizal partners and the signaling pathways and particular molecules involved in symbiotic improvement must be addressed. The master genes that initiate ECM relationships in plant partners and fungi (if any) are still indefinite (Martin et al., 2024).

39.12.12 Investigating host range variability in mycorrhizal fungi

In addition, it is critical to look at the causes of the various host ranges that mycorrhizal species have. The question of why some mycorrhizal fungus species can colonize a variety of hosts, whereas others have preferences that are more specific, is still fascinating and needs more research (Akbar et al., 2025; Xia et al., 2025).

39.12.13 RNA profiling to explore symbiotic coexistence in mycorrhizal fungi

On the other hand, we now determine if several symbiotic classes can concur on (or under) the identical host tree by using the similar properties and carrying out the same environmental occupations (i.e., showing high purposeful resemblance) or by using dissimilar capitals and providing different functions because of RNA sketching of soil hyphae and mycorrhizal root tips (Moora et al., 2025).

39.12.14 Metatranscriptomics for understanding mycorrhizal networks and fungal colonization

Mycorrhizal networks, such as the CMN, can be better understood by using the metatranscriptomics of soil fungal communities. This is because these studies can determine whether dissimilar tree characters indoors a plot are occupied by the same fungus and if or not different tree individuals express the same sets of genes (Martin & Van, 2024).

39.12.15 Managing mycorrhizal fungal diversity for sustainable agriculture and restoration

The variety of fungal taxa existing and the variables influencing their variety and dissemination at the district and continental stages are critical for the effective management of mycorrhizal fungal communities and the advancement of environmentally sustainable farming or restoration projects (Gufwan et al., 2025).

39.12.16 Optimizing mycorrhizal fungi and microbial consortia for plant growth

Metagenomics, metatranscriptomics, and unchanging isotope analytical methods might offer useful visions and categorize the role of mycorrhizal assistant bacteria and other microbial associations that contribute to plant development. Mycorrhizal fungi can enhance the development and establishment of plants in damaged environments (Jagadesh et al., 2024). It is imperative to examine the precise conditions that render the presentation of mycorrhizal fungi advantageous, the well-matched host plants (or genotypes), and if taxa of mycorrhizal fungi can be utilized independently or in microbial groupings, alongside other microorganisms, including protists and mycorrhiza collaborator bacteria, or using a compound pitch inoculum (Marrassini et al., 2025).

There might not be any more advantages to merely broadening AMF variety. Nonetheless the chosen AMF needs to incorporate distant relatives, because this is likely to enhance the likelihood of discovering complementary characteristics. More additive effects are produced by consortia of distant AMF families than by those of closely related ones (Wang et al., 2025a, 2025b). This means that the grouping of more detached species, such as *Gigaspora* sp. and *Rhizophagus fasciculatus*, had a higher impact on pea crop (50%) than did *F. mosseae* and *R. intraradices* (40%). Upcoming research should thus concentrate on screening and finding suitable phylogenetically distant AMF strains to create commercial AMF inoculants that are successful in both wild and greenhouse settings (Salomon et al., 2022).

39.13 Conclusions and recommendations

Numerous commercially available mycorrhizal products are either inefficient or nonviable, and it is frequently forgotten to check if the fungi included in these inoculants have made it to the field. The inoculum must, therefore, be DNA barcoded; the formation and existence of the mycorrhizal fungal separate or isolates in the ground must be tracked; and the ecological and economic viability of their use must be assessed.

Future research on networks in nature will benefit from the use of fluorescent dyes, microsatellite methods, morphological data, and isotopic tracers. Although research in this field is still in its infancy, mycorrhizal networks may have an impact on the designs of seedling formation, opposition, interplant plant variety, and plant community dynamics. Simultaneous investigational application of fungal system controls, isotopic tagging, straight reflection of interplant links, and long-period field thought would be beneficial for future plant community inquiries. Both intrinsic (such as inoculant excellence and capability) and extrinsic (such as soil and weather situations) factors affect the determination, long-term performance, and effects of familiarized AMF on native species. These factors must be taken into consideration when choosing AMF strains for commercial production. By implementing a regulatory framework that may assist in preventing the spread of subpar products and gain the trust of consumers, the intrinsic factors—that is, inoculant capability—can be perfectly handled. Monitoring the effectiveness of the inoculant in the field under various climatic and edaphic conditions, however, requires constant work. This research will help to increase the openness of product recommendations and standards, as well as regulations on the management of macro- and microbiodiversity.

Furthermore, physical characteristics of the soil, such as precipitation, nitrogen, and soil organic matter, can provide hints about the mycorrhizal potential of the soil and frequently correlate with the productivity and abundance of natural AMF populations. Predicting immunization achievement and delivering inoculation impacts on the resident microbiota can be done by modeling such parameters. Furthermore, AMFs are distributed and abundant differently in the rhizosphere and root endophytic zone. Although advanced endosphere inhabitants may suggest stronger creation and durability, this does not always imply greater benefits for plants.

Key point: Damp filtering and pouring technique

First, the course materials, that is, straw, debris, and rocks, must be removed with a sieve (2 mm),

In a glass container, combine approximately 100 g of air-dried root rhizosphere soil with 1000 mL of tap water,

Using a glass rod, aggressively stir the root–soil combination for approximately 30 seconds,

Allow the organic matter and heavier particles to settle, then slowly pour through a set of sieves (with pores of different diameters),

The extracts are washed, and spores are collected from the sieves into Petri dishes,

By using a microscope, AMF spores are isolated with the help of a dropper and needle, and

To keep the cover slip from drying out and allowing air bubbles to enter, place a drop of polyvinyl lactoglycerol in the middle of a dry, clean slide. Then, seal the edge of the cover slip with clear nail polish.

Lastly, examine the spores under a microscope and take photographs (Ahmed et al., 2025).

References

Aguirre-Carvajal, K., Cardenas, S., Munteanu, C. R., & Armijos-Jaramillo, V. (2025). Rampant interkingdom horizontal gene transfer in Pezizomycotina? An updated inspection of anomalous phylogenies. *International Journal of Molecular Sciences, 26*(5), 1795.

Ahmed, N., Li, J., Li, Y., Deng, L., Deng, L., Chachar, M., & Tu, P. (2025). Symbiotic synergy: How arbuscular mycorrhizal fungi enhance nutrient uptake, stress tolerance, and soil health through molecular mechanisms and hormonal regulation. *IMA Fungus, 16*, e144989.

Akbar, K., Yaseen, T., Abbasi, B. A., Iqbal, J., Ullah, Z., Ijaz, S., Mustafa, A. E.-Z. M. A., Elshikh, M. S., Ercisli, S., Harsonowati, W., & Iqbal, R. (2025a). Role and genetic diversity of arbuscular mycorrhizal fungi in improving productivity and enhancing nutrient absorption in chickpea (*Cicer arietinum* L.) genotypes. *Genetic Resources and Crop Evolution, 72*, 2459–2473. https://doi.org/10.1007/s10722-024-02132.

Akbar, K., Yaseen, T., Attia, K. A., Abbasi, B. A., Iqbal, J., Mohammed, A. A., Fiaz, S., Noreen, S., Ijaz, S., & Aziz, E. (2025b). Identification of indigenous arbuscular mycorrhizae fungi in chickpea rhizosphere and their role in nutrient uptake. *Functional Plant Biology, 52*, FP24326.

Antoine, S., Heriche, M., Boussageon, R., Noceto, P. A., van Tuinen, D., Wipf, D., & Courty, P. E. (2021). A historical perspective on mycorrhizal mutualism emphasizing arbuscular mycorrhizas and their emerging challenges. *Mycorrhiza, 31*, 637–653.

Antunes, P. M., Sturmer, S. L., Bever, J. D., Chagnon, P. L., Chaudhary, V. B., Deveautour, C., & Zhang, H. (2025). Enhancing consistency in arbuscular mycorrhizal trait-based research to improve predictions of function. *Mycorrhiza, 35*(2), 1–25.

Aparicio Chacon, M. V., Hernandez Luelmo, S., Devlieghere, V., Robichez, L., Leroy, T., Stuer, N., & Van Dingenen, J. (2024). Exploring the potential role of four *Rhizophagus irregularis* nuclear effectors: Opportunities and technical limitations. *Frontiers in Plant Science, 15*, 1384496.

Athuman, J. J. (2023). Fostering sustainable agriculture through integrated agricultural science education: General overview and lessons from studies. In V. M. Khale (Ed.), *Research and reviews in agriculture science: 1* (p. 1). Bhumi Publishing.

Basiru, S., & Hijri, M. (2022). The potential applications of commercial arbuscular mycorrhizal fungal inoculants and their ecological consequences. *Microorganisms, 10*(10), 1897.

Beechey-Gradwell, Z., Mackay, A., Condron, L., Bowatte, S., Agrelo, F. D. L., Brock, S., & Newton, P. (2025). Loss of P fertilizer effectiveness in raising soil P availability in a grazed grassland enriched with CO_2 for 24 years. *Global Change Biology, 31*(4), e70150.

Bikang, A. N. N., Tekam, L. M., & Sontsa-Donhoung, A. M. (2025). Enhancing coating inoculation for improved nitrogen-fixing efficiency of rhizobia in soybean (*Glycine max* L.). *American Journal of Plant Sciences, 16*(2), 245–262.

Cahill, A. E., Meglecz, E., & Chenuil, A. (2024). Scientific history, biogeography, and biological traits predict presence of cryptic or overlooked species. *Biological Reviews, 99*(2), 546–561.

Chaudhary, A., Poudyal, S., & Kaundal, A. (2025). Role of arbuscular mycorrhizal fungi in maintaining sustainable agroecosystems. *Applied Microbiology, 5*(6), 1.

Chaudhury, R., Chakraborty, A., Rahaman, F., Sarkar, T., Dey, S., & Das, M. (2024). Mycorrhization in trees: Ecology, physiology, emerging technologies and beyond. *Plant Biology, 26*(2), 145–156.

Compant, S., Cassan, F., Kostic, T., Johnson, L., Brader, G., Trognitz, F., & Sessitsch, A. (2025). Harnessing the plant microbiome for sustainable crop production. *Nature Reviews Microbiology, 23*(1), 9–23.

Corsaro, D. (2024). Learning from the rDNA operon: A reanalysis of the *Acanthamoeba palestinensis* group. *Microorganisms, 12*(10), 2105.

Diaz, V., Villalobos, M., Arriaza, K., Flores, K., Hernandez-Saravia, L. P., & Velasquez, A. (2025). Decoding the dialog between plants and arbuscular mycorrhizal fungi: A molecular genetic perspective. *Genes, 16*(2), 143.

Drewnowski, J., Remiszewska-Skwarek, A., Duda, S., & Łagod, G. (2019). Aeration process in bioreactors as the main energy consumer in a wastewater treatment plant. Review of solutions and methods of process optimization. *Processes, 7*(5), 311.

Dudhane, M., Borde, M., & Thomas, S. (2024). Advances in AMF research: Isolation, histochemical staining, enumeration, morphological and molecular techniques. In *Arbuscular mycorrhizal fungi in sustainable agriculture: Inoculum production and application* (pp. 37–55). Singapore.

Eagar, A. C., Abu, P. H., Brown, M. A., Moledor, S. M., Smemo, K. A., Phillips, R. P., Case, A. L., & Blackwood, C. B. (2024). Setting the stage for plant–soil feedback: Mycorrhizal influences over conspecific recruitment, plant and fungal communities, and coevolution. *Journal of Ecology, 113*(6), 1327–1344.

Enriquez-Felix, E. E., Perez-Salazar, C., Rico-Ruiz, J. G., Calheiros de Carvalho, A., Cruz-Morales, P., Villalobos-Escobedo, J. M., & Herrera-Estrella, A. (2024). Argonaute and Dicer are essential for communication between *Trichoderma atroviride* and fungal hosts during mycoparasitism. *Microbiology Spectrum, 12*(4), e0316523.

Escudero-Leyva, E., Belle, M., Dadkhahtehrani, A., Culver, J. N., Araya-Salas, M., Kutza, J. P., & Chaverri, P. (2025). Genomic insights reveal community structure and phylogenetic associations of endohyphal bacteria and viruses in fungal endophytes. *bioRxiv*, 2025–2032. doi:10.1101/2025.02.07.637084.

Fang, X., Lee, X., Twagirayezu, G., Cheng, H., Lu, H., Huang, S., & Ji, B. (2024). A critical review of the effectiveness of biochar coupled with arbuscular mycorrhizal fungi in soil cadmium immobilization. *Journal of Fungi, 10*(3), 182.

Farah, U. A., & Fisher, K. M. (2025). Environmental gradients shape genetic variation in the desert moss, *Syntrichia caninervis* Mitt. (Pottiaceae). *Scientific Reports, 15*(1), 2064.

Firdaus, N., Rachmawati, D., Khastini, R. O., Fitri, S. G. S., Amelia, E., Fitriana, D. E. N., & Wahyuni, I. (2025). Exploring the educational potential of mutualistic interaction between *Cynometra cauliflora* and arbuscular mycorrhiza in PT Chandra Asri's Biodiversity Park: A content analysis for high school biology learning. *International Journal of Biology Education Towards Sustainable Development, 5*(1), 34–53.

Florence, L., Rose, T. J., Rose, M. T., & Truong, C. (2025). Dual mycorrhizal associations in tea tree (*Melaleuca alternifolia*) differ between Australian temperate shrublands and subtropical forests. *Plant and Soil, 512*, 1067–1082.

Frew, A., & Price, J. N. (2019). Mycorrhizal-mediated plant–herbivore interactions in a high CO2 world. *Functional Ecology, 33*(8), 1376–1385.

Gao, Y., Long, X., Liao, Y., Lin, Y., He, Z., Kong, Q., & He, X. (2025). Influence of arbuscular mycorrhizal fungi on nitrogen dynamics during *Cinnamomum camphora* litter decomposition. *Microorganisms, 13*(1), 151.

Garcia Massini, J. L., Nunes, G. C., Yanez, A., Escapa, I. H., & Guido, D. (2025). Jurassic osmundaceous landscapes in Patagonia: Exploring the concept of ecological stasis in the Deseado Massif, Argentina. *Plants, 14*(2), 165.

Garg, S., Kim, M., & Romero-Suarez, D. (2025). Current advancements in fungal engineering technologies for Sustainable Development Goals. Trends in. *Microbiology, 33*(3), 285–301.

Ghorui, M., Chowdhury, S., & Burla, S. (2025). Recent advances in the commercial formulation of arbuscular mycorrhizal inoculants. *Frontiers in Industrial Microbiology, 3*, 1553472.

Gufwan, L. A., Peng, L., Gufwan, N. M., Lan, S., & Wu, L. (2025). Enhancing soil health through biocrusts: A microbial ecosystem approach for degradation control and restoration. *Microbial Ecology, 88*(1), 1–26.

Hosamani, P. (2025). Arbuscular mycorrhizal fungi (AMF) a promising biofertilizer for sustainable plant growth. *Life Sciences Leaflets, 177*, 8–25.

Hou, H., Yan, H., Bai, X., Zhang, Y., Guo, Y., Zhou, J., & Gao, S. (2025a). A synthetic review of feedbacks and drivers of shrub–grass interaction in the process of grassland shrub encroachment. *Plants, 14*(4), 605.

Hou, Z., Merenyi, Z., Yang, Y., Zhang, Y., Csernetics, A., Balint, B., & Nagy, L. G. (2025b). An evolutionarily ancient transcription factor drives spore morphogenesis in mushroom-forming fungi. *Current Biology, 35*(7), 1470–1483.

Indu, B.K., Hemamalini, Nagamani, J.E., & Anuradha, M. (2025). In vitro strategies for enhanced secondary metabolite synthesis. In Anuradha, M., & Balasubramanya, S. (Eds.), In vitro Production of Plant Secondary Metabolites: Theory and Practice (pp. 179–198). Singapore.

Izar, P., van de Waal, E., & Robbins, M. M. (2025). Integrating culture into primate conservation. *Philosophical Transactions B, 380*(1925), 20240135.

Jagadesh, M., Dash, M., Kumari, A., Singh, S. K., Verma, K. K., Kumar, P., Bhatt, R., & Sharma, S. K. (2024). Revealing the hidden world of soil microbes: Metagenomic insights into plant, bacteria, and fungi interactions for sustainable agriculture and ecosystem restoration. *Microbiological Research, 285*, 127764.

Jia, H., Li, P., Li, M., Liu, N., Dong, J., Qu, Q., & Cao, Z. (2025). Trans-kingdom RNA dialogues: MiRNA and milRNA networks as biotechnological tools for sustainable crop defense and pathogen control. *Plants, 14*(8), 1250.

Jiang, X. R., Pan, J. F., Zhao, M., Guo, X. Y., Wang, Q., Zhang, L., & Liu, W. (2024). Transcriptome analysis reveals the impact of arbuscular mycorrhizal symbiosis on *Toona ciliata* var. *pubescens* seedlings. *Forests, 15*(4), 673.

Jo, H., Gu, B., Jeon, M., & Lim, S. I. (2025). In vitro enzymatic production of 3-hydroxypropionic acid from glycerol via CoA-independent pathway. *Process Biochemistry, 149*, 288–294.

Khan, N., Bano, A., Ali, S., & Babar, M. A. (2020). Crosstalk amongst phytohormones from planta and PGPR under biotic and abiotic stresses. *Plant Growth Regulation, 90*, 189–203.

Kumar, A., Kumar, R., Singh, P., Kalaichelvan, S., Santos-Villalobos, S. D. L., Kumar, N., … Babalola, O. O. (2025). Emerging role of arbuscular mycorrhizal fungi in sustainable agriculture: from biology to field application. *MicrobiologyOpen, 14*(5), e70082.

Lalsare, S. S. (2024). Comparative evaluation of nebulizer devices present in market and factors influencing nebulizer performance. *Research Journal of Pharmacy and Technology, 17*(11), 5605–5615.

Lanfranco, L., & Bonfante, P. (2023). Lessons from arbuscular mycorrhizal fungal genomes. *Current Opinion in Microbiology, 75*, 102357.

Liu, C. Y., Hao, Y., Wu, X. L., Dai, F. J., Abd-Allah, E. F., Wu, Q. S., & Liu, S. R. (2024a). Arbuscular mycorrhizal fungi improve drought tolerance of tea plants via modulating root architecture and hormones. *Plant Growth Regulation, 102*(1), 13–22.

Liu, R., & Chen, Y. (2024). Mycorrhizal symbiosis: Evolution, opportunities, challenges, and prospects. In Parihar, M., Rakshit, A., Adholeya, A., & Chen, Y. (Eds.), Arbuscular mycorrhizal fungi in sustainable agriculture: Inoculum production and application (pp. 1–35). Springer.

Liu, R., Li, M., Guo, S., & Chen, Y. (2024b). The role of AMF community composition, diversity, and distribution in sustainable agroecosystems. In *Arbuscular mycorrhizal fungi in sustainable agriculture: Inoculum production and application* (pp. 281–317). Springer.

Luan, S., & Wang, C. (2021). Calcium signaling mechanisms across kingdoms. *Annual Review of Cell and Developmental Biology, 37*(1), 311–340.

Ma, Y., Xiao, C., Liu, J., & Ren, G. (2025). Nutrient-dependent regulation of symbiotic nitrogen fixation in legumes. *Horticulture Research, 12*(3), uhae321.

Majhi, P., Pradhan, U., Toppo, A., & Shukla, A. K. (2025). Fungal endophytes: An insight into diversity, stress tolerance, biocontrol and plant growth-promoting potentials. *Current Microbiology, 82*(6), 1–20.

Manyara, D., Sanchez-Garcia, M., Montoliu-Nerin, M., & Rosling, A. (2024). Detection of rare variants among nuclei populating the arbuscular mycorrhizal fungal model species *Rhizophagus irregularis* DAOM197198. *G3: Genes, Genomes, Genetics, 14*(6), jkae074.

Marrassini, V., Ercoli, L., Paredes, A. V. A., & Pellegrino, E. (2025). Positive response to inoculation with indigenous arbuscular mycorrhizal fungi as modulated by barley genotype. *Agronomy for Sustainable Development, 45*(2), 1–20.

Martin, F. M., & van Der Heijden, M. G. (2024). The mycorrhizal symbiosis: Research frontiers in genomics, ecology, and agricultural application. *New Phytologist, 242*(4), 1486–1506.

Martin-Diáz, F., Baeza-Aranzaez, J., Giraldo, J. D., Schoebitz, M., Urrutia, H., & Lopez-Belchí, M. D. (2025). Innovative JetCutter technology to scale the production of a solid *Bacillus pumilus* biofertilizer to transit to sustainable agriculture. *ACS Agricultural Science & Technology, 5*(2), 188–200.

Marysol, R. C., Antonio, A. T., Evelyn, A. S., & Laura Yesenia, S. R. (2025). Establishing monoxenic culture of arbuscular mycorrhizal fungus *Glomus* sp. through in vitro root organ culture and *Swietenia macrophylla* king in vitro cultures. *Agriculture, 15*(7), 673.

Mazumder, S., Bhattacharya, D., Lahiri, D., & Nag, M. (2025). Rhizobacteria and arbuscular mycorrhizal fungi (AMF) community in growth management and mitigating stress in millets: A plant-soil microbe symbiotic relationship. *Current Microbiology, 82*(6), 242.

Mierziak, J., & Wojtasik, W. (2024). Epigenetic weapons of plants against fungal pathogens. *BMC Plant Biology, 24*(1), 175.

Mikiciuk, G., Miller, T., Kisiel, A., Cembrowska-Lech, D., Mikiciuk, M., Łobodzinska, A., & Bokszczanin, K. (2024a). Harnessing beneficial microbes for drought tolerance: A review of ecological and agricultural innovations. *Agriculture, 14*(12), 2228.

Mikiciuk, S., Kefale, H., & Gelaye, Y. (2024b). Application of precision agriculture technologies for sustainable crop production and environmental sustainability: A systematic review. *The Scientific World Journal, 2024*(1), 2126734.

Moora, M., Davison, J., Kohout, P., & Zobel, M. (2025). The importance of the plant mycorrhizal collaboration niche across scales. *Nature Reviews Biodiversity, 1*, 262–273.

Mordret, S., MacKinnon, J., Behnke, J., O'Leary, S. J., & Chenard, C. (2024). Identification of phytoplankton isolates from the eastern Canadian waters using long-read sequencing. *Journal of Plankton Research, 46*(6), 527–541.

Nazir, A., Sarfraz, W., Allah, D., Khalid, N., Farid, M., Shafiq, M., Firdaus-E-Bareen, Rizvi, Z. F., & Naeem, N. (2023). Synergistic impact of two autochthonous saprobic fungi (*A. niger* and *T. pseudokoningii*) on the growth, ionic contents, and metals uptake in *Brassica juncea* L. and *Vigna radiata* L. under tannery solid waste contaminated soil. *International Journal of Phytoremediation, 25*(11), 1488–1500. https://doi.org/10.1080/15226514.2023.2166457.

Nag, P., & Das, S. (2025). Nitrogen-fixing bacteria: The friends in need. In Bhar, A. (Ed.), Plant-microbe interactions: A comprehensive review (pp. 174–185). Bentham Science Publishers.

Naturally, C. Y. (2025). Fungal biostimulants. In *Mycological inventions for sustainable agriculture and food production* (p. 97). IGI Global.

Naz, M., Afzal, M. R., Raza, M. A., Tariq, M., Yan, M., Dai, Z., Qi, S., & Du, D. (2025). The significant effects of Strigolactones on plant growth and microbe interactions: A review. *Plant Growth Regulation, 105*, 937–957.

Nizamani, M. M., Hughes, A. C., Qureshi, S., Zhang, Q., Tarafder, E., Das, D., Acharya, K., Wang, Y., & Zhang, Z. G. (2024). Microbial biodiversity and plant functional trait interactions in multifunctional ecosystems. *Applied Soil Ecology, 201*, 105515.

Noreen, S., Yaseen, T., Iqbal, J., Abbasi, B. A., Farouk Elsadek, M., Eldin, S. M., Ijaz, S., & Ali, I. (2023). Morphological and molecular characterizations of arbuscular mycorrhizal fungi and their influence on soil physicochemical properties and plant nutrition. *ACS Omega, 8*(36), 32468–32482.

Pan, X., Yue, Y., Zhao, F., Song, T., Xu, B., Li, Z., & Liu, Y. (2025). Rhizosphere microbes facilitate the break of chlamydospore dormancy and root colonization of rice false smut fungi. *Cell Host & Microbe, 33*(5), 731–744.

Pimprikar, P., & Gutjahr, C. (2018). Transcriptional regulation of arbuscular mycorrhiza development. *Plant and Cell Physiology, 59*(4), 678–695.

Priyadarshani, T.D.C., Weerasinghe, P.A., Law, S., & Zakeel, M.C.M. (2025). Hidden allies: Exploring the cutting-edge insights into the role of mycorrhizal fungi in plant health. In Abd-Elsalam, K.A., & Hashem, A.H. (Eds.), Fungal endophytes volume II: Applications in agroecosystems and plant protection (pp. 551–585). Springer Nature.

Purohit, H. J., Pandit, P., Pal, R., Warke, R., & Warke, G. M. (2024). Soil microbiome: An intrinsic driver for climate smart agriculture. *Journal of Agriculture and Food Research, 18*, 101433.

Putnam, E. E., May, R., Freeman, N., Arrigan, D., Boylan, A., Childs, L. H., & Wolfe, B. E. (2025). Fungi shape genome evolution of bacteria even in the absence of major growth phenotypes. *The ISME Journal, 19*, wraf081.

Rathna, V., Prasanna, S.L., Anandakumar, J., Swathi, B., Patil, S.S., & Kujur, R.S. (2025). Forest alchemists: Microbe-mediated nutrient recycling. In Kalia, R.K., & Pathak, R. (Eds.), Tree biology and biotechnology (pp. 67–86). Springer Nature.

Reddy, M.P., Sen, A., Swamy, C.T., Chelladurai, P., Sivadurga, K., Pandey, A., & Purewal, S.S. (2025). Chemistry of macro-and micronutrients. In Purewal, S.S, & Singh, R.S. (Eds.), Colored cereals: Properties, processing, health benefits, and industrial uses (p. 96). CRC Press.

Ren, H., Pu, Q., Yang, X., Kashyap, S., & Liu, S. (2025). Regulatory mechanisms of nitrogen homeostasis in insect growth and development. *Insect Science, 32*(4), 485–499.

Richter, F., Bindschedler, S., Calonne-Salmon, M., Declerck, S., Junier, P., & Stanley, C. E. (2022). Fungi-on-a-chip: Microfluidic platforms for single-cell studies on fungi. *FEMS Microbiology Reviews, 46*(6), fuac039.

Salomon, M. J., Demarmels, R., Watts-Williams, S. J., McLaughlin, M. J., Kafle, A., Ketelsen, C., & van der Heijden, M. G. (2022). Global evaluation of commercial arbuscular mycorrhizal inoculants under greenhouse and field conditions. *Applied Soil Ecology, 169*, 104225.

Samanta, S., & Roychoudhury, A. (2025). Molecular crosstalk of jasmonate with major phytohormones and plant growth regulators during diverse stress responses. *Journal of Plant Growth Regulation, 44*(1), 62–88.

Sellappan, R., & Thangavel, K. (2025). Role of arbuscular mycorrhizal fungi (AMF) in organic vegetables production. In Rana, M.K. (Ed.), Organic production of vegetable crops (pp. 245–266). Apple Academic Press.

Shi, W., Chen, H., Xu, Z., Liu, C., Yeerken, N., & Ma, R. (2025). Prediction of the potential geographic distribution of the Wilsonomyces carpophilus under multiple climate change scenarios. *Scientific Reports, 15*(1), 32199.

Silvestri, A., Fiorilli, V., Miozzi, L., Accotto, G. P., Turina, M., & Lanfranco, L. (2019). In silico analysis of fungal small RNA accumulation reveals putative plant mRNA targets in the symbiosis between an arbuscular mycorrhizal fungus and its host plant. *BMC Genomics, 20*, 1–18.

Silvestri, A., Ledford, W. C., Fiorilli, V., Votta, C., Scerna, A., Tucconi, J., & Lanfranco, L. (2025). A fungal sRNA silences a host plant transcription factor to promote arbuscular mycorrhizal symbiosis. *New Phytologist, 246*(3), 924–935.

Singh, V., Gamage, H., Jones, A., Wood, H. V., Bruning, B., James, A., & Dalal, R. C. (2025). Fungal endophytes influence soil organic carbon and nitrogen fractions promoting carbon sequestration and improving grain yield in soybean. *Scientific Reports, 15*(1), 11402.

Snelders, N. C., Rovenich, H., & Thomma, B. P. (2022). Microbiota manipulation through the secretion of effector proteins is fundamental to the wealth of lifestyles in the fungal kingdom. *FEMS Microbiology Reviews, 46*(5), fuac022.

Subhash, A. P., Veena, S. S., Makeshkumar, T., & Anith, K. N. (2025). *Piriformospora indica* and arbuscular mycorrhizal fungus suppress fungal root rot and mosaic diseases of cassava. *Symbiosis, 95*, 241–254.

Sugiyama, Y., Matsuoka, S., Ishizuka, W., & Sugai, T. (2025). Reduction of the α and β diversity of ectomycorrhizal fungal community under snowmelt: Highlights from a common garden trial using *Abies sachalinensis* with differing host origins and light condition. *Mycorrhiza, 35*(2), 1–13.

Sun, J., Miller, J. B., Granqvist, E., Wiley-Kalil, A., Gobbato, E., Maillet, F., & Oldroyd, G. E. (2015). Activation of symbiosis signaling by arbuscular mycorrhizal fungi in legumes and rice. *The Plant Cell, 27*(3), 823–838.

Thind, S., Chaudhary, M. S., Ditta, A., Hussain, I., Parveen, A., Ullah, N., Mahmood, Q., Al-ashkar, I., & El-Sabagh, A. (2022). Impact of mycorrhizal fungi from different rhizospheric soils on fungal colonization, growth, and chlorophyll contents of *Cenchrus ciliaris*. *Agronomy, 12*(11). https://doi.org/10.3390/agronomy12112644.

Tedersoo, L., Drenkhan, R., Abarenkov, K., Anslan, S., Bahram, M., Bitenieks, K., & Mikryukov, V. (2024). The influence of tree genus, phylogeny, and richness on the specificity, rarity, and diversity of ectomycorrhizal fungi. *Environmental Microbiology Reports, 16*(2), e13253.

Wadas, I., & Domingues, I. (2025). Systematic review of phylogenetic analysis techniques for RNA viruses using bioinformatics. *International Journal of Molecular Sciences, 26*(5), 2180.

Wang, C., Zhao, S., & Qiao, S. (2025a). Achieving high-level nitrogen removal performance for unstable partial nitrification effluent through endogenous dissimilatory nitrate reduction to ammonium function of anammox bacteria. *ACS ES&T Water, 5*(3), 1270–1280.

Wang, J., Shen, Y., Chen, Y., Gao, S., Xue, W., Chen, X., & Li, J. (2025b). Arbuscular mycorrhizal fungi regulate the diversity–invasion resistance relationship by influencing the role of complementarity and selection effects. *New Phytologist, 246*(1), 317–330.

Xia, S., Chen, J., Arsala, D., Emerson, J. J., & Long, M. (2025). Functional innovation through new genes as a general evolutionary process. *Nature Genetics, 57*, 295–309.

Xiao, D., Tang, Y., Zhang, W., Hu, P., & Wang, K. (2024). Lithology and niche habitat have significant effect on arbuscular mycorrhizal fungal abundance and their interspecific interactions. *Science of the Total Environment, 919*, 170774.

Yahaya, S. M., Mahmud, A. A., Abdullahi, M., & Haruna, A. (2023). Recent advances in the chemistry of nitrogen, phosphorus and potassium as fertilizers in soil: A review. *Pedosphere, 33*(3), 385–406.

Yang, D., & Wang, L. (2024). Molybdenum-mediated nitrogen accumulation and assimilation in legumes stepwise boosted by the coexistence of arbuscular mycorrhizal fungi and earthworms. *Science of the Total Environment, 927*, 171840.

Yizhu, L., Imtiaz, M., Ditta, A., Shahid Rizwan, M., Ashraf, M., Mehmood, S., Aziz, O., Mubeen, F., Ali, M., Noor Elahi, N., Ijaz, R., Lele, S., Shuang, C., & Tu, S. (2020). Response of growth, antioxidant enzymes and root exudates production towards as stress in *Pteris vittata* and in *Astragalus sinicus* colonized by arbuscular mycorrhizal fungi. *Environmental Science and Pollution Research, 27*(2), 2340–2352. https://doi.org/10.1007/s11356-019-06785-5.

Yusuf, A., Li, M., Zhang, S. Y., Odedishemi-Ajibade, F., Luo, R. F., Wu, Y. X., & Duan, S. (2025). Harnessing plant–microbe interactions: Strategies for enhancing resilience and nutrient acquisition for sustainable agriculture. *Frontiers in Plant Science, 16*, 1503730.

Zhang, M., Sayer, E. J., Ye, J., Yuan, Z., Lin, F., Hao, Z., & Wang, X. (2025a). Tree mycorrhizal associations regulate relationships between plant and microbial communities and soil organic carbon stocks at local scales in a temperate forest. *Functional Ecology, 39*(6), 1550–1566.

Zhang, M., Shi, Z., Gao, J., Yan, J., Xu, S., & Wang, S. (2025b). Pulling nutrients from Mo-polluted soil by arbuscular mycorrhizal fungi extra radical mycelia is quenching thirsty with poison. *Plant Physiology and Biochemistry, 220*, 109488.

Zhang, S. X., Babady, N. E., Hanson, K. E., Harrington, A. T., Larkin, P. M., Leal, S. M., Jr, & Lockhart, S. R. (2021). Recognition of diagnostic gaps for laboratory diagnosis of fungal diseases: Expert opinion from the fungal diagnostics laboratories consortium (FDLC). *Journal of Clinical Microbiology, 59*(7), 10–1128.

Zhang, Y., Liang, Z., Zheng, L., Wang, X., Chen, H., Xu, T., & Tang, M. (2025c). Impact of long-term loquat cultivation on rhizosphere soil characteristics and AMF community structure: Implications for fertilizer management. *Frontiers in Plant Science, 16*, 1549384.

Zhao, G. (2024). Plant compensatory growth: Its mechanisms and implications to agricultural sustainability under global environmental changes. *International Journal of Sustainable Development & World Ecology, 31*(3), 348–360.

Zheng, H. C., & Jiang, H. M. (2022). Shuttling of cellular proteins between the plasma membrane and nucleus. *Molecular Medicine Reports, 25*(1), 1–12.

Sharma, N., Chaudhary, A., Sharma, N., Shukla, S., Joshi, N.C., Vishwakarma, K., & Mishra, A. (2025). Plant nutrient requirements and nutrient homeostasis in plants. In Joshi, N.C., Leustek, T., & Singh, P.K. (Eds.), Soil Health and Nutrition Management (pp. 1–20). CABI.

Drijber, R.A., & McPherson, M.R. (2021). Mycorrhizal symbioses. In Gentry, T.J., Fuhrmann, J.J., & Zuberer, D.A. (Eds.), Principles and applications of soil microbiology (3rd ed., pp. 303-325). Elsevier.

Index

Page numbers followed by "*f*" and "*t*" indicate, figures and tables respectively.

B

C

D

E

F

G

H

T

U

V

W

Printed and bound by CPI Group (UK) Ltd, Croydon, CR0 4YY
22/09/2026
02220469-0013